Table of Contents

Contributors to Volume 395   ix

Preface   xv

Volumes in Series   xvii

Section I. Comparing Macromolecules: Exploring Biological Diversity

Subsection A. Accessing the Templates

1. Isolation of DNA from Plants with Large Amounts of Secondary Metabolites — Elizabeth A. Friar — 3

2. Nucleic Acid Isolation from Environmental Aqueous Samples — Klaus Valentin, Uwe John, and Linda Medlin — 15

3. Nucleic Acid Isolation from Ecological Samples—Vertebrate Gut Flora — Lise Nordgård, Terje Traavik, and Kaare M. Nielsen — 38

4. Nucleic Acid Isolation from Ecological Samples—Fungal Associations, Lichens — Martin Grube — 48

5. Nucleic Acid Isolation from Ecological Samples—Fungal Associations, Mycorrhizae — Roger T. Koide — 58

6. Nucleic Acid Isolation from Ecological Samples—Animal Scat and Other Associated Materials — Lori S. Eggert, Jesús E. Maldonado, and Robert C. Fleischer — 73

7. Isolation and Analysis of DNA from Archeological, Clinical, and Natural History Specimens — Connie J. Mulligan — 87

Subsection B. The Markers

8. Animal Phylogenomics: Multiple Interspecific Genome Comparisons — Rob DeSalle — 104

9. ISSR Techniques for Evolutionary Biology — Andrea D. Wolfe — 134

10. Use of Amplified Fragment Length Polymerphism (AFLP) Markers in Surveys of Vertebrate Diversity — AURÉLIE BONIN, FRANÇOIS POMPANON, AND PIERRE TABERLET — 145

11. Use of AFLP Markers in Surveys of Arthropod Diversity — TAMRA C. MENDELSON AND KERRY L. SHAW — 161

12. Use of AFLP Markers in Surveys of Plant Diversity — CHIKELU MBA AND JOE TOHME — 177

13. Isolating Microsatellite DNA Loci — TRAVIS C. GLENN AND NANCY A. SCHABLE — 202

14. Use of Microsatellites for Parentage and Kinship Analyses in Animals — MICHAEL S. WEBSTER AND LETITIA REICHART — 222

15. Use of Capillary Array Electrophoresis Single-Strand Conformational Polymorphism Analysis to Estimate Genetic Diversity of Candidate Genes in Germplasm Collections — DAVID N. KUHN AND RAYMOND J. SCHNELL — 238

16. Ribosomal RNA Probes and Microarrays: Their Potential Use in Assessing Microbial Biodiversity — KATJA METFIES AND LINDA MEDLIN — 258

17. The Role of Geographic Analysis in Locating, Understanding, and Using Plant Genetic Diversity — ANDY JARVIS, SAM YEAMAN, LUIGI GUARINO, AND JOE TOHME — 279

18. In Situ Hybridization of Phytoplankton Using Fluorescently Labeled rRNA Probes — RENÉ GROBEN AND LINDA MEDLIN — 299

Subsection C. The Genomes

19. Sequencing and Comparing Whole Mitochondrial Genomes of Animals — JEFFREY L. BOORE, J. ROBERT MACEY, AND MÓNICA MEDINA — 311

20. Methods for Obtaining and Analyzing Whole Chloroplast Genome Sequences — ROBERT K. JANSEN, LINDA A. RAUBESON, JEFFREY L. BOORE, CLAUDE W. dePAMPHILIS, TIMOTHY W. CHUMLEY, ROSEMARIE C. HABERLE, STACIA K. WYMAN, ANDREW J. ALVERSON, RHIANNON PEERY, SALLIE J. HERMAN, H. MATTHEW FOURCADE, JENNIFER V. KUEHL, JOEL R. McNEAL, JAMES LEEBENS-MACK, AND LIYING CUI — 348

21. Construction of Bacterial Artificial Chromosome Libraries for Use in Phylogenetic Studies   ANDREW G. McCUBBIN AND ERIC H. ROALSON   384

22. Comparative EST Analyses in Plant Systems   QUNFENG DONG, LORI KROISS, FREDRICK D. OAKLEY, BING-BING WANG, AND VOLKER BRENDEL   400

23. Isolation of Genes from Plant Y Chromosomes   DMITRY A. FILATOV   418

24. Preparation of Samples for Comparative Studies of Plant Chromosomes Using *In Situ* Hybridization Methods   JASON G. WALLING, J. CHRIS PIRES, AND SCOTT A. JACKSON   443

25. Preparation of Samples for Comparative Studies of Arthropod Chromosomes: Visualization, *In Situ* Hybridization, and Genome Size Estimation   ROB DeSALLE, T. RYAN GREGORY, AND J. SPENCER JOHNSTON   460

Section II. Comparing Macromolecules: Functional Analyses

26. Experimental Methods for Assaying Natural Transformation and Inferring Horizontal Gene Transfer   JESSICA L. RAY AND KAARE M. NIELSEN   491

27. Use of Confocal Microscopy in Comparative Studies of Vertebrate Morphology   ANDRES COLLAZO, OLIVIER BRICAUD, AND KALPANA DESAI   521

28. PrimerSelect: A Transcriptome-Wide Oligonucleotide Primer Pair Design Program for Kinetic RT-PCR–Based Transcript Profiling   KENNETH J. GRAHAM AND MICHAEL J. HOLLAND   544

29. Detecting Differential Expression of Parental or Progenitor Alleles in Genetic Hybrids and Allopolyploids   CRAIG S. PIKAARD, SASHA PREUSS, KEITH EARLEY, RICHARD J. LAWRENCE, MICHELLE S. LEWIS, AND Z. JEFFREY CHEN   554

30. Methods for Genome-Wide Analysis of Gene Expression Changes in Polyploids   JIANLIN WANG, JINSUK J. LEE, LU TIAN, HYEON-SE LEE, MENG CHEN, SHEETAL RAO, EDWARD N. WEI, R. W. DOERGE, LUCA COMAI, AND Z. JEFFREY CHEN   570

31. Designing Experiments Using Spotted   JEFFREY P. TOWNSEND AND
    Microarrays to Detect Gene   JOHN W. TAYLOR   597
    Regulation Differences Within
    and Among Species

32. Methods for Studying the Evolution   ELENA M. KRAMER   617
    of Plant Reproductive Structures:
    Comparative Gene Expression
    Techniques

33. Developing Antibodies to Synthetic   ROGER H. SAWYER,
    Peptides Based on Comparative DNA   TRAVIS C. GLENN,
    Sequencing of Multigene Families   JEFFREY O. FRENCH, AND
                                                  LOREN W. KNAPP   636

34. Applications of Ancestral Protein   BELINDA S. W. CHANG,
    Reconstruction in Understanding   JUAN A. UGALDE, AND
    Protein Function: GFP-Like Proteins   MIKHAIL V. MATZ   652

Section III. Comparing Macromolecules: Phylogenetic Analysis

35. Advances in Phylogeny Reconstruction   BERNARD M. E. MORET
    from Gene Order and Content Data   AND TANDY WARNOW   673

36. Analytical Methods for Detecting Paralogy   JAMES A. COTTON   700
    in Molecular Datasets

37. Analytical Methods for Studying the   SARAH MATHEWS   724
    Evolution of Paralogs Using Duplicate
    Gene Datasets

38. Supertree Construction in the Genomic Age   OLAF R. P. BININDA-EMONDS 745

39. Maximum-Likelihood Methods for   JACK SULLIVAN   757
    Phylogeny Estimation

40. Context Dependence and Coevolution   ZHENGYUAN O. WANG
    Among Amino Acid Residues in Proteins   AND DAVID D. POLLOCK   779

AUTHOR INDEX   791

SUBJECT INDEX   839

Contributors to Volume 395

Article numbers are in parentheses and following the names of contributors.
Affiliations listed are current.

ANDREW J. ALVERSON (20), *Section of Integrative Biology, The University of Texas at Austin, Austin, Texas 78712-0253*

OLAF BININDA-EVANS (38), *Lehrstuhl für Tierzucht, Technical University of Munich, 85354 Freising-Weihenstephan, Germany*

AURÉLIE BONIN (10), *Laboratoire d'Ecologie Alpine (LECA), Université Joseph Fourier, F-38041 Grenoble Cedex 9, France*

JEFFREY L. BOORE (19, 20), *Evolutionary Genomics Department, Department of Energy Joint Genome Institute & Lawrence, Berkeley National Lab, Walnut Creek, California 94598*

VOLKER BRENDEL (22), *Department of Genetics, Iowa State University, Development and Cell Biology, Ames, Iowa 50011-3260*

OLIVIER BRICAUD (27), *Department of Cell and Molecular Biology, House Ear Institute, Los Angeles, California 90057*

BELINDA S. W. CHANG (34), *Department of Zoology, University of Toronto, Toronto, Ontario M5S 3G5, Canada*

Z. JEFFREY CHEN (29, 30), *Molecular Genetics/MS2474, Department of Soil and Crop Sciences, Texas A&M University, College Station, Texas 77843-2474*

MENG CHEN (30), *Department of Soil and Crop Sciences, Texas A&M University, College Station, Texas 77843-2474*

TIMOTHY W. CHUMLEY (20), *Section of Integrative Biology, The University of Texas at Austin, Austin, Texas 78712-0253*

ANDRES COLLAZO (27), *Department of Cell and Molecular Biology, House Ear Institute, Los Angeles, California 90057*

LUCA COMAI (30), *Department of Biology, University of Washington, Seattle, Washington 98195*

JAMES A. COTTON (36), *Department of Zoology, The Natural History Museum, London SW7 5BD, United Kingdom*

LIYING CUI (20), *Department of Biology, Huck Institutes of Life Science, The Pennsylvania State University, University Park, Pennsylvania 16802*

CLAUDE W. DEPAMPHILIS (20), *Department of Biology, Huck Institutes of Life Science, The Pennsylvania State University, University Park, Pennsylvania 16802*

ROB DESALLE (8, 25), *Department of Invertebrate Zoology, American Museum of Natural History, New York, New York 10024*

KALPANA DESAI (27), *Department of Cell and Molecular Biology, House Ear Institute, Los Angeles, California 90057*

R. W. DOERGE (30), *Department of Statistics, Purdue University, West Lafayette, Indiana 47907*

QUNFENG DONG (22), *Department of Genetics, Development and Cell Biology, Iowa State University, Ames, Iowa 50011-3260*

KEITH EARLEY (29), *Biology Department, Washington University, Saint Louis, Missouri 63130*

LORI S. EGGERT (6), *Genetics Program, Department of Systematic Biology, National Museum of Natural History, Smithsonian Institution, Washington, DC 20008-0551*

DMITRY A. FILATOV (23), *School of Biosciences, University of Birmingham, Birmingham B15 2TT, United Kingdom*

ROBERT C. FLEISCHER (6), *Genetics Program, Department of Systematic Biology, National Museum of Natural History, Smithsonian Institution, Washington, DC 20008-0551*

H. MATTHEW FOURCADE (20), *Evolutionary Genomics Department, Department of Energy Joint Genome Institute & Lawrence, Walnut Creek, California 94598*

JEFFREY O. FRENCH (33), *Department of Biological Sciences, Coker Life Sciences, University of South Carolina, Columbia, South Carolina 29208*

ELIZABETH A. FRIAR (1), *Rancho Santa Ana Botanic Garden, Claremont, California 91711*

TRAVIS C. GLENN (13, 33), *Savannah River Ecology Laboratory, Savannah River Site, Aiken, South Carolina 29803; Department of Biological Sciences, Coker Life Sciences, University of South Carolina, Columbia, South Carolina 29208*

KENNETH J. GRAHAM (28), *Department of Biochemistry and Molecular Medicine, University of California, School of Medicine, Davis, California 95616*

T. RYAN GREGORY (25), *Department of Entomology, The Natural History Museum, London SW7 5BD, United Kingdom*

RENÉ GROBEN (18), *Alfred Wegener Institute, D-27570 Bremerhaven, Germany*

MARTIN GRUBE (4), *Institute of Plant Sciences, Karl-Franzens-University Graz, 8010 Graz, Austria*

LUIGI GUARINO (17), *Secretariat of the Pacific Community (SPC), Suva, Fiji*

ROSEMARIE C. HABERLE (20), *Section of Integrative Biology, The University of Texas at Austin, Austin, Texas 78712-0253*

SALLIE J. HERMAN (20), *Department of Biological Sciences, Central Washington University, Ellensburg, Washington 98923*

MICHAEL J. HOLLAND (28), *Department of Biochemistry and Molecular Medicine, University of California, School of Medicine, Davis, California 95616*

SCOTT A. JACKSON (24), *Department of Agronomy, Purdue University, West Lafayette, Indiana 47907*

ROBERT K. JANSEN (20), *Section of Integrative Biology, The University of Texas at Austin and Institute of Cellular and Molecular Biology, Austin, Texas 78712-0253*

ANDY JARVIS (17), *International Center for Tropical Agriculture (CIAT); International Plant Genetic Resources Institute (IPGRI), AA6713 Cali, Colombia*

UWE JOHN (2), *Alfred Wegener Institute, D-27570 Bremerhaven, Germany*

L. SPENCER JOHNSTON (25), *Department of Entomology, Texas A&M University, College Station, Texas 77843-2474*

LOREN W. KNAPP (33), *Department of Biological Sciences, Coker Life Sciences, University of South Carolina, Columbia, South Carolina 29208*

ROGER T. KOIDE (5), *Department of Horticulture, The Pennsylvania State University, University Park, Pennsylvania 16802*

ELENA M. KRAMER (32), *Department of Organismic and Evolutionary, Biology, Harvard University, Cambridge, Massachusetts 02138*

LORI KROISS (22), *Department of Plant Pathology, Buckhout Laboratory, University Park, Pennsylvania 16802*

JENNIFER V. KUEHL (20), *Evolutionary Genomics Department, Department of Energy Joint Genome Institute & Lawrence, Walnut Creek, California 94598*

DAVID N. KUHN (15), *Department of Biological Sciences, Florida International University, Miami, Florida 33199*

RICHARD J. LAWRENCE (29), *Biology Department, Washington University, Saint Louis, Missouri 63130*

HYEON-SE LEE (30), *Department of Soil and Crop Sciences, Texas A&M University, College Station, Texas 77843-2474*

JINSUK J. LEE (30), *Department of Soil and Crop Sciences, Texas A&M University, College Station, Texas 77843-2474*

JAMES LEEBENS-MACK (20), *Department of Biology, Huck Institutes of Life Science, The Pennsylvania State University, University Park, Pennsylvania 16802*

MICHELLE S. LEWIS (29), *Biology Department, Washington University, Saint Louis, Missouri 63130*

J. ROBERT MACEY (19), *Evolutionary Genomics Department, Department of Energy Joint Genome Institute & Lawrence, Berkeley National Lab, Walnut Creek, California 94598*

JESÚS E. MALDONADO (6), *Genetics Program, Department of Systematic Biology, National Museum of Natural History, Smithsonian Institution, Washington, DC 20008-0551*

SARH MATTHEWS (37), *Arnold Arboretum of Harvard University, Cambridge, Massachusetts 02138*

MIKHAIL V. MATZ (34), *Whitney Laboratory, University of Florida, St. Augustine, Florida 32080*

CHIKELU MBA (12), *Plant Breeding Unit, Joint FAO/IAEA, Agriculture and Biotechnology Laboratory, International Atomic Energy Agency, Laboratories, A-2444 Seibersdorf, Austria*

ANDREW G. MCCUBBIN (21), *School of Biological Sciences, Washington State University, Pullman, Washington 99164-4236*

JOEL R. MCNEAL (20), *Department of Biology, Huck Institutes of Life Science, The Pennsylvania State University, University Park, Pennsylvania 16802*

MÓNICA MEDINA (19), *Evolutionary Genomics Department, Department of Energy Joint Genome Institute & Lawrence, Berkeley National Lab, Walnut Creek, California 94598*

LINDA MEDLIN (2, 16, 18), *Alfred Wegener Institute, D-27570 Bremerhaven, Germany*

TAMRA C. MENDELSON (11), *Department of Biology, University of Maryland, College Park, Maryland 20742*

KATJA METFIES (16), *Alfred Wegener Institute, D-27570 Bremerhaven, Germany*

BERNARD M. E. MORET (35), *Department of Computer Sciences, University of New Mexico, Albuquerque, New Mexico 87131*

CONNIE J. MULLIGAN (7), *Department of Anthropology, University of Florida, Gainesville, Florida 32611*

KAARE M. NIELSEN (3, 26), *Department of Pharmacy, Faculty of Medicine, University of Tromso; Norwegian Institute of Gene Ecology, N9037 Tromso, Norway*

LISE NORDGÅRD (3), *Norwegian Institute of Gene Ecology, 9294 Tromso, Norway*

FREDERICK D. OAKLEY (22), *Department of Genetics, Development and Cell Biology, Iowa State University, Ames, Iowa 50011-3260*

RHIANNON PEERY (20), *Department of Biological Sciences, Central Washington University, Ellensburg, Washington 98923*

CRAIG S. PIKAARD (29), *Biology Department, Washington University, Saint Louis, Missouri 63130*

J. CHRIS PIRES (24), *Department of Agronomy, University of Wisconsin Madison, Madison, Wisconsin 53706*

DAVID D. POLLOCK (40), *Department of Biological Sciences, Biological Computation and Visualization Center, Louisiana State University, Baton Rouge, Louisiana 70803*

FRANÇOIS POMPANON (10), *Laboratoire d'Ecologie Alpine (LECA), Université Joseph Fourier, F-38041 Grenoble, Cedex 9, France*

SASHA PREUSS (29), *Biology Department, Washington University, Saint Louis, Missouri 63130*

SHEETAL RAO (30), *Department of Soil and Crop Sciences, Texas A&M University, College Station, Texas 77843-2474*

LINDA A. RAUBESON (20), *Department of Biological Sciences, Central Washington University, Ellensburg, Washington 98923*

JESSICA L. RAY (26), *Department of Pharmacy, Faculty of Medicine, University of Tromso, N9037 Tromso, Norway*

LETITIA REICHART (14), *School of Biological Sciences, Washington, State University, Pullman, Washington 99164-4236*

ERIC H. ROALSON (21), *School of Biological Sciences, Washington State University, Pullman, Washington 99164-4236*

ROGER H. SAWYER (33), *Department of Biological Sciences, Coker Life Sciences, University of South Carolina, Columbia, South Carolina 29208*

NANCY A. SCHABLE (13), *Savannah River Ecology Laboratory, Savannah River Site, Aiken, South Carolina 29803*

RAYMOND J. SCHNELL (15), *US Department of Agriculture, Agriculture Research Service, Subtropical Horticulture Research Station, Miami, Florida 33158*

KERRY L. SHAW (11), *Department of Biology, University of Maryland, College Park, Maryland 20742*

JACK SULLIVAN (39), *Department of Biological Sciences; The Initiative for Bioinformatics and Evolutionary Studies, University of Idaho, Moscow, Idaho 83844-3051*

PIERRE TABERLET (10), *Laboratoire d'Ecologie Alpine (LECA), Université Joseph Fourier, F-38041 Grenoble Cedex 9, France*

JOHN W. TAYLOR (31), *Plant and Microbial Biology Department, University of California, Berkeley, Berkeley, California 94720*

Lu Tian (30), *Department of Soil and Crop Sciences, Texas A&M University, College Station, Texas 77843-2474*

Joe Tohme (12, 17), *Agrobiodiversity and Biotechnology Project, International Center for Tropical Agriculture (CIAT), AA6713 Cali, Columbia*

Jeffrey P. Townsend (31), *Plant and Microbial Biology Department, University of California, Berkeley, Berkeley, California 94720*

Terje Traavik (3), *Department of Medical Biology, Faculty of Medicine, University of Tromso; Norwegian Institute of Gene Ecology, N9037 Tromso, Norway*

Juan A. Ugalde (34), *Whitney Laboratory, University of Florida, St. Augustine, Florida 32080*

Klaus Valentin (2), *Alfred Wegener Institute, D-27570 Bremerhaven, Germany*

Jason G. Walling (24), *Department of Agronomy, Purdue University, West Lafayette, Indiana 47907*

Bing-Bing Wang (22), *Iowa State University, Department of Genetics, Development and Cell Biology, Ames, Iowa 50011-3260*

Jianlin Wang (30), *Department of Soil and Crop Sciences, Texas A&M University, College Station, Texas 77843-2474*

Zhengyuan O. Wang (40), *Department of Biological Sciences, Biological Computation and Visualization Center, Louisiana State University, Baton Rouge, Louisiana 70803*

Tandy Warnow (35), *The Department of Computer Sciences, The University of Texas at Austin, Austin, Texas 78712*

Michael S. Webster (14), *School of Biological Sciences, Washington State University, Pullman, Washington 99164-4236*

Edward N. Wei (30), *Department of Soil and Crop Sciences, Texas A&M University, College Station, Texas 77843-2474*

Andrea D. Wolfe (9), *Department of Evolution, Ecology, and Organismal Biology, The Ohio State University, Columbus, Ohio 43210-1293*

Stacia K. Wyman (20), *Department of Computer Sciences, The University of Texas at Austin, Austin, Texas 78712-0253*

Sam Yeaman (17), *International Center for Tropical Agriculture (CIAT); International Plant Genetics Resources Institute (IPGRI), AA6713 Cali, Colombia*

Preface

Volume 395 of Methods in Enzymology, "Molecular Evolution: Producing the Biochemical Data," Part B, reflects the significant increase in the kinds of data available to researchers in the biological sciences as well as new and refined methods for analyzing these and traditional data types since the publication of Part A, Volume 224 in 1993. These advances range from a refined ability to extract DNA samples from a wide variety of environmental sources to extraction of complete genomes. Additionally, our toolbox for studying macromolecules using functional and phylogenetic analyses has expanded significantly, to include transcript and microarray profiling, application of confocal microscopy to understanding morphology, and more refined detection and analyses of gene family paralogs.

In this book we have brought together a wide range of methodologies for studying molecular evolution. The first section addresses three different aspects of exploring biological diversity, namely, how we gather macromolecular templates (Chapters 1–8), how different kinds of molecular markers are developed (Chapters 9–19), and how we explore genomes (Chapters 20–26). Sections II and III each address aspects of analysis of macromolecular data, partitioned into analyses of function (Chapters 27–35), such as gene expression studies, and phylogenetic analyses (Chapters 36–41), exploring various methods for understanding molecular evolution in a phylogenetic context.

By no means do we expect that the coverage of this volume is exhaustive. The breadth of application of new markers and analyses is beyond any one volume. We have not been able to address some techniques here that have recently come into common usage, particularly the application of Bayesian inference methodologies to phylogenetic analyses. Recent reviews of the application of this technique are available in the literature, particularly, Huelsenbeck, J. P., Larget, B., Miller, R. E., and Ronquist, F., 2002, Potential applications and pitfalls of Bayesian inference of phylogeny, *Systematic Biology* **51**, 673–688.

We wish to thank all of the authors for their valuable contributions to this volume. We also wish to thank Cindy Minor and Noelle Gracy at Elsevier/ Academic Press for all of their help in preparing this volume and Mel Simon and Tom White for their helpful comments on the organization of the volume.

ELIZABETH A. ZIMMER
ERIC H. ROALSON

METHODS IN ENZYMOLOGY

VOLUME I. Preparation and Assay of Enzymes
Edited by SIDNEY P. COLOWICK AND NATHAN O. KAPLAN

VOLUME II. Preparation and Assay of Enzymes
Edited by SIDNEY P. COLOWICK AND NATHAN O. KAPLAN

VOLUME III. Preparation and Assay of Substrates
Edited by SIDNEY P. COLOWICK AND NATHAN O. KAPLAN

VOLUME IV. Special Techniques for the Enzymologist
Edited by SIDNEY P. COLOWICK AND NATHAN O. KAPLAN

VOLUME V. Preparation and Assay of Enzymes
Edited by SIDNEY P. COLOWICK AND NATHAN O. KAPLAN

VOLUME VI. Preparation and Assay of Enzymes *(Continued)*
Preparation and Assay of Substrates
Special Techniques
Edited by SIDNEY P. COLOWICK AND NATHAN O. KAPLAN

VOLUME VII. Cumulative Subject Index
Edited by SIDNEY P. COLOWICK AND NATHAN O. KAPLAN

VOLUME VIII. Complex Carbohydrates
Edited by ELIZABETH F. NEUFELD AND VICTOR GINSBURG

VOLUME IX. Carbohydrate Metabolism
Edited by WILLIS A. WOOD

VOLUME X. Oxidation and Phosphorylation
Edited by RONALD W. ESTABROOK AND MAYNARD E. PULLMAN

VOLUME XI. Enzyme Structure
Edited by C. H. W. HIRS

VOLUME XII. Nucleic Acids (Parts A and B)
Edited by LAWRENCE GROSSMAN AND KIVIE MOLDAVE

VOLUME XIII. Citric Acid Cycle
Edited by J. M. LOWENSTEIN

VOLUME XIV. Lipids
Edited by J. M. LOWENSTEIN

VOLUME XV. Steroids and Terpenoids
Edited by RAYMOND B. CLAYTON

VOLUME XVI. Fast Reactions
Edited by KENNETH KUSTIN

VOLUME XVII. Metabolism of Amino Acids and Amines (Parts A and B)
Edited by HERBERT TABOR AND CELIA WHITE TABOR

VOLUME XVIII. Vitamins and Coenzymes (Parts A, B, and C)
Edited by DONALD B. McCORMICK AND LEMUEL D. WRIGHT

VOLUME XIX. Proteolytic Enzymes
Edited by GERTRUDE E. PERLMANN AND LASZLO LORAND

VOLUME XX. Nucleic Acids and Protein Synthesis (Part C)
Edited by KIVIE MOLDAVE AND LAWRENCE GROSSMAN

VOLUME XXI. Nucleic Acids (Part D)
Edited by LAWRENCE GROSSMAN AND KIVIE MOLDAVE

VOLUME XXII. Enzyme Purification and Related Techniques
Edited by WILLIAM B. JAKOBY

VOLUME XXIII. Photosynthesis (Part A)
Edited by ANTHONY SAN PIETRO

VOLUME XXIV. Photosynthesis and Nitrogen Fixation (Part B)
Edited by ANTHONY SAN PIETRO

VOLUME XXV. Enzyme Structure (Part B)
Edited by C. H. W. HIRS AND SERGE N. TIMASHEFF

VOLUME XXVI. Enzyme Structure (Part C)
Edited by C. H. W. HIRS AND SERGE N. TIMASHEFF

VOLUME XXVII. Enzyme Structure (Part D)
Edited by C. H. W. HIRS AND SERGE N. TIMASHEFF

VOLUME XXVIII. Complex Carbohydrates (Part B)
Edited by VICTOR GINSBURG

VOLUME XXIX. Nucleic Acids and Protein Synthesis (Part E)
Edited by LAWRENCE GROSSMAN AND KIVIE MOLDAVE

VOLUME XXX. Nucleic Acids and Protein Synthesis (Part F)
Edited by KIVIE MOLDAVE AND LAWRENCE GROSSMAN

VOLUME XXXI. Biomembranes (Part A)
Edited by SIDNEY FLEISCHER AND LESTER PACKER

VOLUME XXXII. Biomembranes (Part B)
Edited by SIDNEY FLEISCHER AND LESTER PACKER

VOLUME XXXIII. Cumulative Subject Index Volumes I-XXX
Edited by MARTHA G. DENNIS AND EDWARD A. DENNIS

VOLUME XXXIV. Affinity Techniques (Enzyme Purification: Part B)
Edited by WILLIAM B. JAKOBY AND MEIR WILCHEK

VOLUME XXXV. Lipids (Part B)
Edited by JOHN M. LOWENSTEIN

VOLUME XXXVI. Hormone Action (Part A: Steroid Hormones)
Edited by BERT W. O'MALLEY AND JOEL G. HARDMAN

VOLUME XXXVII. Hormone Action (Part B: Peptide Hormones)
Edited by BERT W. O'MALLEY AND JOEL G. HARDMAN

VOLUME XXXVIII. Hormone Action (Part C: Cyclic Nucleotides)
Edited by JOEL G. HARDMAN AND BERT W. O'MALLEY

VOLUME XXXIX. Hormone Action (Part D: Isolated Cells, Tissues,
and Organ Systems)
Edited by JOEL G. HARDMAN AND BERT W. O'MALLEY

VOLUME XL. Hormone Action (Part E: Nuclear Structure and Function)
Edited by BERT W. O'MALLEY AND JOEL G. HARDMAN

VOLUME XLI. Carbohydrate Metabolism (Part B)
Edited by W. A. WOOD

VOLUME XLII. Carbohydrate Metabolism (Part C)
Edited by W. A. WOOD

VOLUME XLIII. Antibiotics
Edited by JOHN H. HASH

VOLUME XLIV. Immobilized Enzymes
Edited by KLAUS MOSBACH

VOLUME XLV. Proteolytic Enzymes (Part B)
Edited by LASZLO LORAND

VOLUME XLVI. Affinity Labeling
Edited by WILLIAM B. JAKOBY AND MEIR WILCHEK

VOLUME XLVII. Enzyme Structure (Part E)
Edited by C. H. W. HIRS AND SERGE N. TIMASHEFF

VOLUME XLVIII. Enzyme Structure (Part F)
Edited by C. H. W. HIRS AND SERGE N. TIMASHEFF

VOLUME XLIX. Enzyme Structure (Part G)
Edited by C. H. W. HIRS AND SERGE N. TIMASHEFF

VOLUME L. Complex Carbohydrates (Part C)
Edited by VICTOR GINSBURG

VOLUME LI. Purine and Pyrimidine Nucleotide Metabolism
Edited by PATRICIA A. HOFFEE AND MARY ELLEN JONES

VOLUME LII. Biomembranes (Part C: Biological Oxidations)
Edited by SIDNEY FLEISCHER AND LESTER PACKER

VOLUME LIII. Biomembranes (Part D: Biological Oxidations)
Edited by SIDNEY FLEISCHER AND LESTER PACKER

VOLUME LIV. Biomembranes (Part E: Biological Oxidations)
Edited by SIDNEY FLEISCHER AND LESTER PACKER

VOLUME LV. Biomembranes (Part F: Bioenergetics)
Edited by SIDNEY FLEISCHER AND LESTER PACKER

VOLUME LVI. Biomembranes (Part G: Bioenergetics)
Edited by SIDNEY FLEISCHER AND LESTER PACKER

VOLUME LVII. Bioluminescence and Chemiluminescence
Edited by MARLENE A. DeLUCA

VOLUME LVIII. Cell Culture
Edited by WILLIAM B. JAKOBY AND IRA PASTAN

VOLUME LIX. Nucleic Acids and Protein Synthesis (Part G)
Edited by KIVIE MOLDAVE AND LAWRENCE GROSSMAN

VOLUME LX. Nucleic Acids and Protein Synthesis (Part H)
Edited by KIVIE MOLDAVE AND LAWRENCE GROSSMAN

VOLUME 61. Enzyme Structure (Part H)
Edited by C. H. W. HIRS AND SERGE N. TIMASHEFF

VOLUME 62. Vitamins and Coenzymes (Part D)
Edited by DONALD B. McCORMICK AND LEMUEL D. WRIGHT

VOLUME 63. Enzyme Kinetics and Mechanism (Part A: Initial Rate and
Inhibitor Methods)
Edited by DANIEL L. PURICH

VOLUME 64. Enzyme Kinetics and Mechanism (Part B: Isotopic Probes and
Complex Enzyme Systems)
Edited by DANIEL L. PURICH

VOLUME 65. Nucleic Acids (Part I)
Edited by LAWRENCE GROSSMAN AND KIVIE MOLDAVE

VOLUME 66. Vitamins and Coenzymes (Part E)
Edited by DONALD B. McCORMICK AND LEMUEL D. WRIGHT

VOLUME 67. Vitamins and Coenzymes (Part F)
Edited by DONALD B. McCORMICK AND LEMUEL D. WRIGHT

VOLUME 68. Recombinant DNA
Edited by RAY WU

VOLUME 69. Photosynthesis and Nitrogen Fixation (Part C)
Edited by ANTHONY SAN PIETRO

VOLUME 70. Immunochemical Techniques (Part A)
Edited by HELEN VAN VUNAKIS AND JOHN J. LANGONE

VOLUME 71. Lipids (Part C)
Edited by JOHN M. LOWENSTEIN

VOLUME 72. Lipids (Part D)
Edited by JOHN M. LOWENSTEIN

VOLUME 73. Immunochemical Techniques (Part B)
Edited by JOHN J. LANGONE AND HELEN VAN VUNAKIS

VOLUME 74. Immunochemical Techniques (Part C)
Edited by JOHN J. LANGONE AND HELEN VAN VUNAKIS

VOLUME 75. Cumulative Subject Index Volumes XXXI, XXXII, XXXIV–LX
Edited by EDWARD A. DENNIS AND MARTHA G. DENNIS

VOLUME 76. Hemoglobins
Edited by ERALDO ANTONINI, LUIGI ROSSI-BERNARDI, AND EMILIA CHIANCONE

VOLUME 77. Detoxication and Drug Metabolism
Edited by WILLIAM B. JAKOBY

VOLUME 78. Interferons (Part A)
Edited by SIDNEY PESTKA

VOLUME 79. Interferons (Part B)
Edited by SIDNEY PESTKA

VOLUME 80. Proteolytic Enzymes (Part C)
Edited by LASZLO LORAND

VOLUME 81. Biomembranes (Part H: Visual Pigments and Purple Membranes, I)
Edited by LESTER PACKER

VOLUME 82. Structural and Contractile Proteins (Part A: Extracellular Matrix)
Edited by LEON W. CUNNINGHAM AND DIXIE W. FREDERIKSEN

VOLUME 83. Complex Carbohydrates (Part D)
Edited by VICTOR GINSBURG

VOLUME 84. Immunochemical Techniques (Part D: Selected Immunoassays)
Edited by JOHN J. LANGONE AND HELEN VAN VUNAKIS

VOLUME 85. Structural and Contractile Proteins (Part B: The Contractile Apparatus and the Cytoskeleton)
Edited by DIXIE W. FREDERIKSEN AND LEON W. CUNNINGHAM

VOLUME 86. Prostaglandins and Arachidonate Metabolites
Edited by WILLIAM E. M. LANDS AND WILLIAM L. SMITH

VOLUME 87. Enzyme Kinetics and Mechanism (Part C: Intermediates, Stereo-chemistry, and Rate Studies)
Edited by DANIEL L. PURICH

VOLUME 88. Biomembranes (Part I: Visual Pigments and Purple Membranes, II)
Edited by LESTER PACKER

VOLUME 89. Carbohydrate Metabolism (Part D)
Edited by WILLIS A. WOOD

VOLUME 90. Carbohydrate Metabolism (Part E)
Edited by WILLIS A. WOOD

VOLUME 91. Enzyme Structure (Part I)
Edited by C. H. W. HIRS AND SERGE N. TIMASHEFF

VOLUME 92. Immunochemical Techniques (Part E: Monoclonal Antibodies and General Immunoassay Methods)
Edited by JOHN J. LANGONE AND HELEN VAN VUNAKIS

VOLUME 93. Immunochemical Techniques (Part F: Conventional Antibodies, Fc Receptors, and Cytotoxicity)
Edited by JOHN J. LANGONE AND HELEN VAN VUNAKIS

VOLUME 94. Polyamines
Edited by HERBERT TABOR AND CELIA WHITE TABOR

VOLUME 95. Cumulative Subject Index Volumes 61–74, 76–80
Edited by EDWARD A. DENNIS AND MARTHA G. DENNIS

VOLUME 96. Biomembranes [Part J: Membrane Biogenesis: Assembly and Targeting (General Methods; Eukaryotes)]
Edited by SIDNEY FLEISCHER AND BECCA FLEISCHER

VOLUME 97. Biomembranes [Part K: Membrane Biogenesis: Assembly and Targeting (Prokaryotes, Mitochondria, and Chloroplasts)]
Edited by SIDNEY FLEISCHER AND BECCA FLEISCHER

VOLUME 98. Biomembranes (Part L: Membrane Biogenesis: Processing and Recycling)
Edited by SIDNEY FLEISCHER AND BECCA FLEISCHER

VOLUME 99. Hormone Action (Part F: Protein Kinases)
Edited by JACKIE D. CORBIN AND JOEL G. HARDMAN

VOLUME 100. Recombinant DNA (Part B)
Edited by RAY WU, LAWRENCE GROSSMAN, AND KIVIE MOLDAVE

VOLUME 101. Recombinant DNA (Part C)
Edited by RAY WU, LAWRENCE GROSSMAN, AND KIVIE MOLDAVE

VOLUME 102. Hormone Action (Part G: Calmodulin and Calcium-Binding Proteins)
Edited by ANTHONY R. MEANS AND BERT W. O'MALLEY

VOLUME 103. Hormone Action (Part H: Neuroendocrine Peptides)
Edited by P. MICHAEL CONN

VOLUME 104. Enzyme Purification and Related Techniques (Part C)
Edited by WILLIAM B. JAKOBY

VOLUME 105. Oxygen Radicals in Biological Systems
Edited by LESTER PACKER

VOLUME 106. Posttranslational Modifications (Part A)
Edited by FINN WOLD AND KIVIE MOLDAVE

VOLUME 107. Posttranslational Modifications (Part B)
Edited by FINN WOLD AND KIVIE MOLDAVE

VOLUME 108. Immunochemical Techniques (Part G: Separation and Characterization of Lymphoid Cells)
Edited by GIOVANNI DI SABATO, JOHN J. LANGONE, AND HELEN VAN VUNAKIS

VOLUME 109. Hormone Action (Part I: Peptide Hormones)
Edited by LUTZ BIRNBAUMER AND BERT W. O'MALLEY

VOLUME 110. Steroids and Isoprenoids (Part A)
Edited by JOHN H. LAW AND HANS C. RILLING

VOLUME 111. Steroids and Isoprenoids (Part B)
Edited by JOHN H. LAW AND HANS C. RILLING

VOLUME 112. Drug and Enzyme Targeting (Part A)
Edited by KENNETH J. WIDDER AND RALPH GREEN

VOLUME 113. Glutamate, Glutamine, Glutathione, and Related Compounds
Edited by ALTON MEISTER

VOLUME 114. Diffraction Methods for Biological Macromolecules (Part A)
Edited by HAROLD W. WYCKOFF, C. H. W. HIRS, AND SERGE N. TIMASHEFF

VOLUME 115. Diffraction Methods for Biological Macromolecules (Part B)
Edited by HAROLD W. WYCKOFF, C. H. W. HIRS, AND SERGE N. TIMASHEFF

VOLUME 116. Immunochemical Techniques (Part H: Effectors and Mediators of Lymphoid Cell Functions)
Edited by GIOVANNI DI SABATO, JOHN J. LANGONE, AND HELEN VAN VUNAKIS

VOLUME 117. Enzyme Structure (Part J)
Edited by C. H. W. HIRS AND SERGE N. TIMASHEFF

VOLUME 118. Plant Molecular Biology
Edited by ARTHUR WEISSBACH AND HERBERT WEISSBACH

VOLUME 119. Interferons (Part C)
Edited by SIDNEY PESTKA

VOLUME 120. Cumulative Subject Index Volumes 81–94, 96–101

VOLUME 121. Immunochemical Techniques (Part I: Hybridoma Technology and Monoclonal Antibodies)
Edited by JOHN J. LANGONE AND HELEN VAN VUNAKIS

VOLUME 122. Vitamins and Coenzymes (Part G)
Edited by FRANK CHYTIL AND DONALD B. MCCORMICK

VOLUME 123. Vitamins and Coenzymes (Part H)
Edited by FRANK CHYTIL AND DONALD B. MCCORMICK

VOLUME 124. Hormone Action (Part J: Neuroendocrine Peptides)
Edited by P. MICHAEL CONN

VOLUME 125. Biomembranes (Part M: Transport in Bacteria, Mitochondria, and Chloroplasts: General Approaches and Transport Systems)
Edited by SIDNEY FLEISCHER AND BECCA FLEISCHER

VOLUME 126. Biomembranes (Part N: Transport in Bacteria, Mitochondria, and Chloroplasts: Protonmotive Force)
Edited by SIDNEY FLEISCHER AND BECCA FLEISCHER

VOLUME 127. Biomembranes (Part O: Protons and Water: Structure and Translocation)
Edited by LESTER PACKER

VOLUME 128. Plasma Lipoproteins (Part A: Preparation, Structure, and Molecular Biology)
Edited by JERE P. SEGREST AND JOHN J. ALBERS

VOLUME 129. Plasma Lipoproteins (Part B: Characterization, Cell Biology, and Metabolism)
Edited by JOHN J. ALBERS AND JERE P. SEGREST

VOLUME 130. Enzyme Structure (Part K)
Edited by C. H. W. HIRS AND SERGE N. TIMASHEFF

VOLUME 131. Enzyme Structure (Part L)
Edited by C. H. W. HIRS AND SERGE N. TIMASHEFF

VOLUME 132. Immunochemical Techniques (Part J: Phagocytosis and Cell-Mediated Cytotoxicity)
Edited by GIOVANNI DI SABATO AND JOHANNES EVERSE

VOLUME 133. Bioluminescence and Chemiluminescence (Part B)
Edited by MARLENE DELUCA AND WILLIAM D. MCELROY

VOLUME 134. Structural and Contractile Proteins (Part C: The Contractile Apparatus and the Cytoskeleton)
Edited by RICHARD B. VALLEE

VOLUME 135. Immobilized Enzymes and Cells (Part B)
Edited by KLAUS MOSBACH

VOLUME 136. Immobilized Enzymes and Cells (Part C)
Edited by KLAUS MOSBACH

VOLUME 137. Immobilized Enzymes and Cells (Part D)
Edited by KLAUS MOSBACH

VOLUME 138. Complex Carbohydrates (Part E)
Edited by VICTOR GINSBURG

VOLUME 139. Cellular Regulators (Part A: Calcium- and Calmodulin-Binding Proteins)
Edited by ANTHONY R. MEANS AND P. MICHAEL CONN

VOLUME 140. Cumulative Subject Index Volumes 102–119, 121–134

VOLUME 141. Cellular Regulators (Part B: Calcium and Lipids)
Edited by P. MICHAEL CONN AND ANTHONY R. MEANS

VOLUME 142. Metabolism of Aromatic Amino Acids and Amines
Edited by SEYMOUR KAUFMAN

VOLUME 143. Sulfur and Sulfur Amino Acids
Edited by WILLIAM B. JAKOBY AND OWEN GRIFFITH

VOLUME 144. Structural and Contractile Proteins (Part D: Extracellular Matrix)
Edited by LEON W. CUNNINGHAM

VOLUME 145. Structural and Contractile Proteins (Part E: Extracellular Matrix)
Edited by LEON W. CUNNINGHAM

VOLUME 146. Peptide Growth Factors (Part A)
Edited by DAVID BARNES AND DAVID A. SIRBASKU

VOLUME 147. Peptide Growth Factors (Part B)
Edited by DAVID BARNES AND DAVID A. SIRBASKU

VOLUME 148. Plant Cell Membranes
Edited by LESTER PACKER AND ROLAND DOUCE

VOLUME 149. Drug and Enzyme Targeting (Part B)
Edited by RALPH GREEN AND KENNETH J. WIDDER

VOLUME 150. Immunochemical Techniques (Part K: *In Vitro* Models of B and T Cell Functions and Lymphoid Cell Receptors)
Edited by GIOVANNI DI SABATO

VOLUME 151. Molecular Genetics of Mammalian Cells
Edited by MICHAEL M. GOTTESMAN

VOLUME 152. Guide to Molecular Cloning Techniques
Edited by SHELBY L. BERGER AND ALAN R. KIMMEL

VOLUME 153. Recombinant DNA (Part D)
Edited by RAY WU AND LAWRENCE GROSSMAN

VOLUME 154. Recombinant DNA (Part E)
Edited by RAY WU AND LAWRENCE GROSSMAN

VOLUME 155. Recombinant DNA (Part F)
Edited by RAY WU

VOLUME 156. Biomembranes (Part P: ATP-Driven Pumps and Related Transport: The Na, K-Pump)
Edited by SIDNEY FLEISCHER AND BECCA FLEISCHER

VOLUME 157. Biomembranes (Part Q: ATP-Driven Pumps and Related Transport: Calcium, Proton, and Potassium Pumps)
Edited by SIDNEY FLEISCHER AND BECCA FLEISCHER

VOLUME 158. Metalloproteins (Part A)
Edited by JAMES F. RIORDAN AND BERT L. VALLEE

VOLUME 159. Initiation and Termination of Cyclic Nucleotide Action
Edited by JACKIE D. CORBIN AND ROGER A. JOHNSON

VOLUME 160. Biomass (Part A: Cellulose and Hemicellulose)
Edited by WILLIS A. WOOD AND SCOTT T. KELLOGG

VOLUME 161. Biomass (Part B: Lignin, Pectin, and Chitin)
Edited by WILLIS A. WOOD AND SCOTT T. KELLOGG

VOLUME 162. Immunochemical Techniques (Part L: Chemotaxis and Inflammation)
Edited by GIOVANNI DI SABATO

VOLUME 163. Immunochemical Techniques (Part M: Chemotaxis and Inflammation)
Edited by GIOVANNI DI SABATO

VOLUME 164. Ribosomes
Edited by HARRY F. NOLLER, JR., AND KIVIE MOLDAVE

VOLUME 165. Microbial Toxins: Tools for Enzymology
Edited by SIDNEY HARSHMAN

VOLUME 166. Branched-Chain Amino Acids
Edited by ROBERT HARRIS AND JOHN R. SOKATCH

VOLUME 167. Cyanobacteria
Edited by LESTER PACKER AND ALEXANDER N. GLAZER

VOLUME 168. Hormone Action (Part K: Neuroendocrine Peptides)
Edited by P. MICHAEL CONN

VOLUME 169. Platelets: Receptors, Adhesion, Secretion (Part A)
Edited by JACEK HAWIGER

VOLUME 170. Nucleosomes
Edited by PAUL M. WASSARMAN AND ROGER D. KORNBERG

VOLUME 171. Biomembranes (Part R: Transport Theory: Cells and Model Membranes)
Edited by SIDNEY FLEISCHER AND BECCA FLEISCHER

VOLUME 172. Biomembranes (Part S: Transport: Membrane Isolation and Characterization)
Edited by SIDNEY FLEISCHER AND BECCA FLEISCHER

VOLUME 173. Biomembranes [Part T: Cellular and Subcellular Transport: Eukaryotic (Nonepithelial) Cells]
Edited by SIDNEY FLEISCHER AND BECCA FLEISCHER

VOLUME 174. Biomembranes [Part U: Cellular and Subcellular Transport: Eukaryotic (Nonepithelial) Cells]
Edited by SIDNEY FLEISCHER AND BECCA FLEISCHER

VOLUME 175. Cumulative Subject Index Volumes 135–139, 141–167

VOLUME 176. Nuclear Magnetic Resonance (Part A: Spectral Techniques and Dynamics)
Edited by NORMAN J. OPPENHEIMER AND THOMAS L. JAMES

VOLUME 177. Nuclear Magnetic Resonance (Part B: Structure and Mechanism)
Edited by NORMAN J. OPPENHEIMER AND THOMAS L. JAMES

VOLUME 178. Antibodies, Antigens, and Molecular Mimicry
Edited by JOHN J. LANGONE

VOLUME 179. Complex Carbohydrates (Part F)
Edited by VICTOR GINSBURG

VOLUME 180. RNA Processing (Part A: General Methods)
Edited by JAMES E. DAHLBERG AND JOHN N. ABELSON

VOLUME 181. RNA Processing (Part B: Specific Methods)
Edited by JAMES E. DAHLBERG AND JOHN N. ABELSON

VOLUME 182. Guide to Protein Purification
Edited by MURRAY P. DEUTSCHER

VOLUME 183. Molecular Evolution: Computer Analysis of Protein and Nucleic Acid Sequences
Edited by RUSSELL F. DOOLITTLE

VOLUME 184. Avidin-Biotin Technology
Edited by MEIR WILCHEK AND EDWARD A. BAYER

VOLUME 185. Gene Expression Technology
Edited by DAVID V. GOEDDEL

VOLUME 186. Oxygen Radicals in Biological Systems (Part B: Oxygen Radicals and Antioxidants)
Edited by LESTER PACKER AND ALEXANDER N. GLAZER

VOLUME 187. Arachidonate Related Lipid Mediators
Edited by ROBERT C. MURPHY AND FRANK A. FITZPATRICK

VOLUME 188. Hydrocarbons and Methylotrophy
Edited by MARY E. LIDSTROM

VOLUME 189. Retinoids (Part A: Molecular and Metabolic Aspects)
Edited by LESTER PACKER

VOLUME 190. Retinoids (Part B: Cell Differentiation and Clinical Applications)
Edited by LESTER PACKER

VOLUME 191. Biomembranes (Part V: Cellular and Subcellular Transport: Epithelial Cells)
Edited by SIDNEY FLEISCHER AND BECCA FLEISCHER

VOLUME 192. Biomembranes (Part W: Cellular and Subcellular Transport: Epithelial Cells)
Edited by SIDNEY FLEISCHER AND BECCA FLEISCHER

VOLUME 193. Mass Spectrometry
Edited by JAMES A. MCCLOSKEY

VOLUME 194. Guide to Yeast Genetics and Molecular Biology
Edited by CHRISTINE GUTHRIE AND GERALD R. FINK

VOLUME 195. Adenylyl Cyclase, G Proteins, and Guanylyl Cyclase
Edited by ROGER A. JOHNSON AND JACKIE D. CORBIN

VOLUME 196. Molecular Motors and the Cytoskeleton
Edited by RICHARD B. VALLEE

VOLUME 197. Phospholipases
Edited by EDWARD A. DENNIS

VOLUME 198. Peptide Growth Factors (Part C)
Edited by DAVID BARNES, J. P. MATHER, AND GORDON H. SATO

VOLUME 199. Cumulative Subject Index Volumes 168–174, 176–194

VOLUME 200. Protein Phosphorylation (Part A: Protein Kinases: Assays,
Purification, Antibodies, Functional Analysis, Cloning, and Expression)
Edited by TONY HUNTER AND BARTHOLOMEW M. SEFTON

VOLUME 201. Protein Phosphorylation (Part B: Analysis of Protein
Phosphorylation, Protein Kinase Inhibitors, and Protein Phosphatases)
Edited by TONY HUNTER AND BARTHOLOMEW M. SEFTON

VOLUME 202. Molecular Design and Modeling: Concepts and Applications
(Part A: Proteins, Peptides, and Enzymes)
Edited by JOHN J. LANGONE

VOLUME 203. Molecular Design and Modeling: Concepts and Applications
(Part B: Antibodies and Antigens, Nucleic Acids, Polysaccharides,
and Drugs)
Edited by JOHN J. LANGONE

VOLUME 204. Bacterial Genetic Systems
Edited by JEFFREY H. MILLER

VOLUME 205. Metallobiochemistry (Part B: Metallothionein and
Related Molecules)
Edited by JAMES F. RIORDAN AND BERT L. VALLEE

VOLUME 206. Cytochrome P450
Edited by MICHAEL R. WATERMAN AND ERIC F. JOHNSON

VOLUME 207. Ion Channels
Edited by BERNARDO RUDY AND LINDA E. IVERSON

VOLUME 208. Protein–DNA Interactions
Edited by ROBERT T. SAUER

VOLUME 209. Phospholipid Biosynthesis
Edited by EDWARD A. DENNIS AND DENNIS E. VANCE

VOLUME 210. Numerical Computer Methods
Edited by LUDWIG BRAND AND MICHAEL L. JOHNSON

VOLUME 211. DNA Structures (Part A: Synthesis and Physical Analysis of DNA)
Edited by DAVID M. J. LILLEY AND JAMES E. DAHLBERG

VOLUME 212. DNA Structures (Part B: Chemical and Electrophoretic Analysis of DNA)
Edited by DAVID M. J. LILLEY AND JAMES E. DAHLBERG

VOLUME 213. Carotenoids (Part A: Chemistry, Separation, Quantitation, and Antioxidation)
Edited by LESTER PACKER

VOLUME 214. Carotenoids (Part B: Metabolism, Genetics, and Biosynthesis)
Edited by LESTER PACKER

VOLUME 215. Platelets: Receptors, Adhesion, Secretion (Part B)
Edited by JACEK J. HAWIGER

VOLUME 216. Recombinant DNA (Part G)
Edited by RAY WU

VOLUME 217. Recombinant DNA (Part H)
Edited by RAY WU

VOLUME 218. Recombinant DNA (Part I)
Edited by RAY WU

VOLUME 219. Reconstitution of Intracellular Transport
Edited by JAMES E. ROTHMAN

VOLUME 220. Membrane Fusion Techniques (Part A)
Edited by NEJAT DÜZGÜNEŞ

VOLUME 221. Membrane Fusion Techniques (Part B)
Edited by NEJAT DÜZGÜNEŞ

VOLUME 222. Proteolytic Enzymes in Coagulation, Fibrinolysis, and Complement Activation (Part A: Mammalian Blood Coagulation Factors and Inhibitors)
Edited by LASZLO LORAND AND KENNETH G. MANN

VOLUME 223. Proteolytic Enzymes in Coagulation, Fibrinolysis, and Complement Activation (Part B: Complement Activation, Fibrinolysis, and Nonmammalian Blood Coagulation Factors)
Edited by LASZLO LORAND AND KENNETH G. MANN

VOLUME 224. Molecular Evolution: Producing the Biochemical Data
Edited by ELIZABETH ANNE ZIMMER, THOMAS J. WHITE, REBECCA L. CANN, AND ALLAN C. WILSON

VOLUME 225. Guide to Techniques in Mouse Development
Edited by PAUL M. WASSARMAN AND MELVIN L. DePAMPHILIS

VOLUME 226. Metallobiochemistry (Part C: Spectroscopic and Physical Methods for Probing Metal Ion Environments in Metalloenzymes and Metalloproteins)
Edited by JAMES F. RIORDAN AND BERT L. VALLEE

VOLUME 227. Metallobiochemistry (Part D: Physical and Spectroscopic Methods for Probing Metal Ion Environments in Metalloproteins)
Edited by JAMES F. RIORDAN AND BERT L. VALLEE

VOLUME 228. Aqueous Two-Phase Systems
Edited by HARRY WALTER AND GÖTE JOHANSSON

VOLUME 229. Cumulative Subject Index Volumes 195–198, 200–227

VOLUME 230. Guide to Techniques in Glycobiology
Edited by WILLIAM J. LENNARZ AND GERALD W. HART

VOLUME 231. Hemoglobins (Part B: Biochemical and Analytical Methods)
Edited by JOHANNES EVERSE, KIM D. VANDEGRIFF, AND ROBERT M. WINSLOW

VOLUME 232. Hemoglobins (Part C: Biophysical Methods)
Edited by JOHANNES EVERSE, KIM D. VANDEGRIFF, AND ROBERT M. WINSLOW

VOLUME 233. Oxygen Radicals in Biological Systems (Part C)
Edited by LESTER PACKER

VOLUME 234. Oxygen Radicals in Biological Systems (Part D)
Edited by LESTER PACKER

VOLUME 235. Bacterial Pathogenesis (Part A: Identification and Regulation of Virulence Factors)
Edited by VIRGINIA L. CLARK AND PATRIK M. BAVOIL

VOLUME 236. Bacterial Pathogenesis (Part B: Integration of Pathogenic Bacteria with Host Cells)
Edited by VIRGINIA L. CLARK AND PATRIK M. BAVOIL

VOLUME 237. Heterotrimeric G Proteins
Edited by RAVI IYENGAR

VOLUME 238. Heterotrimeric G-Protein Effectors
Edited by RAVI IYENGAR

VOLUME 239. Nuclear Magnetic Resonance (Part C)
Edited by THOMAS L. JAMES AND NORMAN J. OPPENHEIMER

VOLUME 240. Numerical Computer Methods (Part B)
Edited by MICHAEL L. JOHNSON AND LUDWIG BRAND

VOLUME 241. Retroviral Proteases
Edited by LAWRENCE C. KUO AND JULES A. SHAFER

VOLUME 242. Neoglycoconjugates (Part A)
Edited by Y. C. LEE AND REIKO T. LEE

VOLUME 243. Inorganic Microbial Sulfur Metabolism
Edited by HARRY D. PECK, JR., AND JEAN LEGALL

VOLUME 244. Proteolytic Enzymes: Serine and Cysteine Peptidases
Edited by ALAN J. BARRETT

VOLUME 245. Extracellular Matrix Components
Edited by E. RUOSLAHTI AND E. ENGVALL

VOLUME 246. Biochemical Spectroscopy
Edited by KENNETH SAUER

VOLUME 247. Neoglycoconjugates (Part B: Biomedical Applications)
Edited by Y. C. LEE AND REIKO T. LEE

VOLUME 248. Proteolytic Enzymes: Aspartic and Metallo Peptidases
Edited by ALAN J. BARRETT

VOLUME 249. Enzyme Kinetics and Mechanism (Part D: Developments in Enzyme Dynamics)
Edited by DANIEL L. PURICH

VOLUME 250. Lipid Modifications of Proteins
Edited by PATRICK J. CASEY AND JANICE E. BUSS

VOLUME 251. Biothiols (Part A: Monothiols and Dithiols, Protein Thiols, and Thiyl Radicals)
Edited by LESTER PACKER

VOLUME 252. Biothiols (Part B: Glutathione and Thioredoxin; Thiols in Signal Transduction and Gene Regulation)
Edited by LESTER PACKER

VOLUME 253. Adhesion of Microbial Pathogens
Edited by RON J. DOYLE AND ITZHAK OFEK

VOLUME 254. Oncogene Techniques
Edited by PETER K. VOGT AND INDER M. VERMA

VOLUME 255. Small GTPases and Their Regulators (Part A: Ras Family)
Edited by W. E. BALCH, CHANNING J. DER, AND ALAN HALL

VOLUME 256. Small GTPases and Their Regulators (Part B: Rho Family)
Edited by W. E. BALCH, CHANNING J. DER, AND ALAN HALL

VOLUME 257. Small GTPases and Their Regulators (Part C: Proteins Involved in Transport)
Edited by W. E. BALCH, CHANNING J. DER, AND ALAN HALL

VOLUME 258. Redox-Active Amino Acids in Biology
Edited by JUDITH P. KLINMAN

VOLUME 259. Energetics of Biological Macromolecules
Edited by MICHAEL L. JOHNSON AND GARY K. ACKERS

VOLUME 260. Mitochondrial Biogenesis and Genetics (Part A)
Edited by GIUSEPPE M. ATTARDI AND ANNE CHOMYN

VOLUME 261. Nuclear Magnetic Resonance and Nucleic Acids
Edited by THOMAS L. JAMES

VOLUME 262. DNA Replication
Edited by JUDITH L. CAMPBELL

VOLUME 263. Plasma Lipoproteins (Part C: Quantitation)
Edited by WILLIAM A. BRADLEY, SANDRA H. GIANTURCO, AND JERE P. SEGREST

VOLUME 264. Mitochondrial Biogenesis and Genetics (Part B)
Edited by GIUSEPPE M. ATTARDI AND ANNE CHOMYN

VOLUME 265. Cumulative Subject Index Volumes 228, 230–262

VOLUME 266. Computer Methods for Macromolecular Sequence Analysis
Edited by RUSSELL F. DOOLITTLE

VOLUME 267. Combinatorial Chemistry
Edited by JOHN N. ABELSON

VOLUME 268. Nitric Oxide (Part A: Sources and Detection of NO; NO Synthase)
Edited by LESTER PACKER

VOLUME 269. Nitric Oxide (Part B: Physiological and Pathological Processes)
Edited by LESTER PACKER

VOLUME 270. High Resolution Separation and Analysis of Biological Macromolecules (Part A: Fundamentals)
Edited by BARRY L. KARGER AND WILLIAM S. HANCOCK

VOLUME 271. High Resolution Separation and Analysis of Biological Macromolecules (Part B: Applications)
Edited by BARRY L. KARGER AND WILLIAM S. HANCOCK

VOLUME 272. Cytochrome P450 (Part B)
Edited by ERIC F. JOHNSON AND MICHAEL R. WATERMAN

VOLUME 273. RNA Polymerase and Associated Factors (Part A)
Edited by SANKAR ADHYA

VOLUME 274. RNA Polymerase and Associated Factors (Part B)
Edited by SANKAR ADHYA

VOLUME 275. Viral Polymerases and Related Proteins
Edited by LAWRENCE C. KUO, DAVID B. OLSEN, AND STEVEN S. CARROLL

VOLUME 276. Macromolecular Crystallography (Part A)
Edited by CHARLES W. CARTER, JR., AND ROBERT M. SWEET

VOLUME 277. Macromolecular Crystallography (Part B)
Edited by CHARLES W. CARTER, JR., AND ROBERT M. SWEET

VOLUME 278. Fluorescence Spectroscopy
Edited by LUDWIG BRAND AND MICHAEL L. JOHNSON

VOLUME 279. Vitamins and Coenzymes (Part I)
Edited by DONALD B. MCCORMICK, JOHN W. SUTTIE, AND CONRAD WAGNER

VOLUME 280. Vitamins and Coenzymes (Part J)
Edited by DONALD B. MCCORMICK, JOHN W. SUTTIE, AND CONRAD WAGNER

VOLUME 281. Vitamins and Coenzymes (Part K)
Edited by DONALD B. MCCORMICK, JOHN W. SUTTIE, AND CONRAD WAGNER

VOLUME 282. Vitamins and Coenzymes (Part L)
Edited by DONALD B. MCCORMICK, JOHN W. SUTTIE, AND CONRAD WAGNER

VOLUME 283. Cell Cycle Control
Edited by WILLIAM G. DUNPHY

VOLUME 284. Lipases (Part A: Biotechnology)
Edited by BYRON RUBIN AND EDWARD A. DENNIS

VOLUME 285. Cumulative Subject Index Volumes 263, 264, 266–284, 286–289

VOLUME 286. Lipases (Part B: Enzyme Characterization and Utilization)
Edited by BYRON RUBIN AND EDWARD A. DENNIS

VOLUME 287. Chemokines
Edited by RICHARD HORUK

VOLUME 288. Chemokine Receptors
Edited by RICHARD HORUK

VOLUME 289. Solid Phase Peptide Synthesis
Edited by GREGG B. FIELDS

VOLUME 290. Molecular Chaperones
Edited by GEORGE H. LORIMER AND THOMAS BALDWIN

VOLUME 291. Caged Compounds
Edited by GERARD MARRIOTT

VOLUME 292. ABC Transporters: Biochemical, Cellular, and Molecular Aspects
Edited by SURESH V. AMBUDKAR AND MICHAEL M. GOTTESMAN

VOLUME 293. Ion Channels (Part B)
Edited by P. MICHAEL CONN

VOLUME 294. Ion Channels (Part C)
Edited by P. MICHAEL CONN

VOLUME 295. Energetics of Biological Macromolecules (Part B)
Edited by GARY K. ACKERS AND MICHAEL L. JOHNSON

VOLUME 296. Neurotransmitter Transporters
Edited by SUSAN G. AMARA

VOLUME 297. Photosynthesis: Molecular Biology of Energy Capture
Edited by LEE MCINTOSH

VOLUME 298. Molecular Motors and the Cytoskeleton (Part B)
Edited by RICHARD B. VALLEE

VOLUME 299. Oxidants and Antioxidants (Part A)
Edited by LESTER PACKER

VOLUME 300. Oxidants and Antioxidants (Part B)
Edited by LESTER PACKER

VOLUME 301. Nitric Oxide: Biological and Antioxidant Activities (Part C)
Edited by LESTER PACKER

VOLUME 302. Green Fluorescent Protein
Edited by P. MICHAEL CONN

VOLUME 303. cDNA Preparation and Display
Edited by SHERMAN M. WEISSMAN

VOLUME 304. Chromatin
Edited by PAUL M. WASSARMAN AND ALAN P. WOLFFE

VOLUME 305. Bioluminescence and Chemiluminescence (Part C)
Edited by THOMAS O. BALDWIN AND MIRIAM M. ZIEGLER

VOLUME 306. Expression of Recombinant Genes in Eukaryotic Systems
Edited by JOSEPH C. GLORIOSO AND MARTIN C. SCHMIDT

VOLUME 307. Confocal Microscopy
Edited by P. MICHAEL CONN

VOLUME 308. Enzyme Kinetics and Mechanism (Part E: Energetics of
Enzyme Catalysis)
Edited by DANIEL L. PURICH AND VERN L. SCHRAMM

VOLUME 309. Amyloid, Prions, and Other Protein Aggregates
Edited by RONALD WETZEL

VOLUME 310. Biofilms
Edited by RON J. DOYLE

VOLUME 311. Sphingolipid Metabolism and Cell Signaling (Part A)
Edited by ALFRED H. MERRILL, JR., AND YUSUF A. HANNUN

VOLUME 312. Sphingolipid Metabolism and Cell Signaling (Part B)
Edited by ALFRED H. MERRILL, JR., AND YUSUF A. HANNUN

VOLUME 313. Antisense Technology (Part A: General Methods, Methods of
Delivery, and RNA Studies)
Edited by M. IAN PHILLIPS

VOLUME 314. Antisense Technology (Part B: Applications)
Edited by M. IAN PHILLIPS

VOLUME 315. Vertebrate Phototransduction and the Visual Cycle (Part A)
Edited by KRZYSZTOF PALCZEWSKI

VOLUME 316. Vertebrate Phototransduction and the Visual Cycle (Part B)
Edited by KRZYSZTOF PALCZEWSKI

VOLUME 317. RNA–Ligand Interactions (Part A: Structural Biology Methods)
Edited by DANIEL W. CELANDER AND JOHN N. ABELSON

VOLUME 318. RNA–Ligand Interactions (Part B: Molecular Biology Methods)
Edited by DANIEL W. CELANDER AND JOHN N. ABELSON

VOLUME 319. Singlet Oxygen, UV-A, and Ozone
Edited by LESTER PACKER AND HELMUT SIES

VOLUME 320. Cumulative Subject Index Volumes 290–319

VOLUME 321. Numerical Computer Methods (Part C)
Edited by MICHAEL L. JOHNSON AND LUDWIG BRAND

VOLUME 322. Apoptosis
Edited by JOHN C. REED

VOLUME 323. Energetics of Biological Macromolecules (Part C)
Edited by MICHAEL L. JOHNSON AND GARY K. ACKERS

VOLUME 324. Branched-Chain Amino Acids (Part B)
Edited by ROBERT A. HARRIS AND JOHN R. SOKATCH

VOLUME 325. Regulators and Effectors of Small GTPases (Part D: Rho Family)
Edited by W. E. BALCH, CHANNING J. DER, AND ALAN HALL

VOLUME 326. Applications of Chimeric Genes and Hybrid Proteins (Part A:
Gene Expression and Protein Purification)
Edited by JEREMY THORNER, SCOTT D. EMR, AND JOHN N. ABELSON

VOLUME 327. Applications of Chimeric Genes and Hybrid Proteins (Part B:
Cell Biology and Physiology)
Edited by JEREMY THORNER, SCOTT D. EMR, AND JOHN N. ABELSON

VOLUME 328. Applications of Chimeric Genes and Hybrid Proteins (Part C:
Protein–Protein Interactions and Genomics)
Edited by JEREMY THORNER, SCOTT D. EMR, AND JOHN N. ABELSON

VOLUME 329. Regulators and Effectors of Small GTPases (Part E: GTPases
Involved in Vesicular Traffic)
Edited by W. E. BALCH, CHANNING J. DER, AND ALAN HALL

VOLUME 330. Hyperthermophilic Enzymes (Part A)
Edited by MICHAEL W. W. ADAMS AND ROBERT M. KELLY

VOLUME 331. Hyperthermophilic Enzymes (Part B)
Edited by MICHAEL W. W. ADAMS AND ROBERT M. KELLY

VOLUME 332. Regulators and Effectors of Small GTPases (Part F: Ras Family I)
Edited by W. E. BALCH, CHANNING J. DER, AND ALAN HALL

VOLUME 333. Regulators and Effectors of Small GTPases (Part G: Ras Family II)
Edited by W. E. BALCH, CHANNING J. DER, AND ALAN HALL

VOLUME 334. Hyperthermophilic Enzymes (Part C)
Edited by MICHAEL W. W. ADAMS AND ROBERT M. KELLY

VOLUME 335. Flavonoids and Other Polyphenols
Edited by LESTER PACKER

VOLUME 336. Microbial Growth in Biofilms (Part A: Developmental and Molecular Biological Aspects)
Edited by RON J. DOYLE

VOLUME 337. Microbial Growth in Biofilms (Part B: Special Environments and Physicochemical Aspects)
Edited by RON J. DOYLE

VOLUME 338. Nuclear Magnetic Resonance of Biological Macromolecules (Part A)
Edited by THOMAS L. JAMES, VOLKER DÖTSCH, AND ULI SCHMITZ

VOLUME 339. Nuclear Magnetic Resonance of Biological Macromolecules (Part B)
Edited by THOMAS L. JAMES, VOLKER DÖTSCH, AND ULI SCHMITZ

VOLUME 340. Drug–Nucleic Acid Interactions
Edited by JONATHAN B. CHAIRES AND MICHAEL J. WARING

VOLUME 341. Ribonucleases (Part A)
Edited by ALLEN W. NICHOLSON

VOLUME 342. Ribonucleases (Part B)
Edited by ALLEN W. NICHOLSON

VOLUME 343. G Protein Pathways (Part A: Receptors)
Edited by RAVI IYENGAR AND JOHN D. HILDEBRANDT

VOLUME 344. G Protein Pathways (Part B: G Proteins and Their Regulators)
Edited by RAVI IYENGAR AND JOHN D. HILDEBRANDT

VOLUME 345. G Protein Pathways (Part C: Effector Mechanisms)
Edited by RAVI IYENGAR AND JOHN D. HILDEBRANDT

VOLUME 346. Gene Therapy Methods
Edited by M. IAN PHILLIPS

VOLUME 347. Protein Sensors and Reactive Oxygen Species (Part A: Selenoproteins and Thioredoxin)
Edited by HELMUT SIES AND LESTER PACKER

VOLUME 348. Protein Sensors and Reactive Oxygen Species (Part B: Thiol Enzymes and Proteins)
Edited by HELMUT SIES AND LESTER PACKER

VOLUME 349. Superoxide Dismutase
Edited by LESTER PACKER

VOLUME 350. Guide to Yeast Genetics and Molecular and Cell Biology (Part B)
Edited by CHRISTINE GUTHRIE AND GERALD R. FINK

VOLUME 351. Guide to Yeast Genetics and Molecular and Cell Biology (Part C)
Edited by CHRISTINE GUTHRIE AND GERALD R. FINK

VOLUME 352. Redox Cell Biology and Genetics (Part A)
Edited by CHANDAN K. SEN AND LESTER PACKER

VOLUME 353. Redox Cell Biology and Genetics (Part B)
Edited by CHANDAN K. SEN AND LESTER PACKER

VOLUME 354. Enzyme Kinetics and Mechanisms (Part F: Detection and Characterization of Enzyme Reaction Intermediates)
Edited by DANIEL L. PURICH

VOLUME 355. Cumulative Subject Index Volumes 321–354

VOLUME 356. Laser Capture Microscopy and Microdissection
Edited by P. MICHAEL CONN

VOLUME 357. Cytochrome P450, Part C
Edited by ERIC F. JOHNSON AND MICHAEL R. WATERMAN

VOLUME 358. Bacterial Pathogenesis (Part C: Identification, Regulation, and Function of Virulence Factors)
Edited by VIRGINIA L. CLARK AND PATRIK M. BAVOIL

VOLUME 359. Nitric Oxide (Part D)
Edited by ENRIQUE CADENAS AND LESTER PACKER

VOLUME 360. Biophotonics (Part A)
Edited by GERARD MARRIOTT AND IAN PARKER

VOLUME 361. Biophotonics (Part B)
Edited by GERARD MARRIOTT AND IAN PARKER

VOLUME 362. Recognition of Carbohydrates in Biological Systems (Part A)
Edited by YUAN C. LEE AND REIKO T. LEE

VOLUME 363. Recognition of Carbohydrates in Biological Systems (Part B)
Edited by YUAN C. LEE AND REIKO T. LEE

VOLUME 364. Nuclear Receptors
Edited by DAVID W. RUSSELL AND DAVID J. MANGELSDORF

VOLUME 365. Differentiation of Embryonic Stem Cells
Edited by PAUL M. WASSAUMAN AND GORDON M. KELLER

VOLUME 366. Protein Phosphatases
Edited by SUSANNE KLUMPP AND JOSEF KRIEGLSTEIN

VOLUME 367. Liposomes (Part A)
Edited by NEJAT DÜZGÜNEŞ

VOLUME 368. Macromolecular Crystallography (Part C)
Edited by CHARLES W. CARTER, JR., AND ROBERT M. SWEET

VOLUME 369. Combinational Chemistry (Part B)
Edited by GUILLERMO A. MORALES AND BARRY A. BUNIN

VOLUME 370. RNA Polymerases and Associated Factors (Part C)
Edited by SANKAR L. ADHYA AND SUSAN GARGES

VOLUME 371. RNA Polymerases and Associated Factors (Part D)
Edited by SANKAR L. ADHYA AND SUSAN GARGES

VOLUME 372. Liposomes (Part B)
Edited by NEJAT DÜZGÜNEŞ

VOLUME 373. Liposomes (Part C)
Edited by NEJAT DÜZGÜNEŞ

VOLUME 374. Macromolecular Crystallography (Part D)
Edited by CHARLES W. CARTER, JR., AND ROBERT W. SWEET

VOLUME 375. Chromatin and Chromatin Remodeling Enzymes (Part A)
Edited by C. DAVID ALLIS AND CARL WU

VOLUME 376. Chromatin and Chromatin Remodeling Enzymes (Part B)
Edited by C. DAVID ALLIS AND CARL WU

VOLUME 377. Chromatin and Chromatin Remodeling Enzymes (Part C)
Edited by C. DAVID ALLIS AND CARL WU

VOLUME 378. Quinones and Quinone Enzymes (Part A)
Edited by HELMUT SIES AND LESTER PACKER

VOLUME 379. Energetics of Biological Macromolecules (Part D)
Edited by JO M. HOLT, MICHAEL L. JOHNSON, AND GARY K. ACKERS

VOLUME 380. Energetics of Biological Macromolecules (Part E)
Edited by JO M. HOLT, MICHAEL L. JOHNSON, AND GARY K. ACKERS

VOLUME 381. Oxygen Sensing
Edited by CHANDAN K. SEN AND GREGG L. SEMENZA

VOLUME 382. Quinones and Quinone Enzymes (Part B)
Edited by HELMUT SIES AND LESTER PACKER

VOLUME 383. Numerical Computer Methods (Part D)
Edited by LUDWIG BRAND AND MICHAEL L. JOHNSON

VOLUME 384. Numerical Computer Methods (Part E)
Edited by LUDWIG BRAND AND MICHAEL L. JOHNSON

VOLUME 385. Imaging in Biological Research (Part A)
Edited by P. MICHAEL CONN

VOLUME 386. Imaging in Biological Research (Part B)
Edited by P. MICHAEL CONN

VOLUME 387. Liposomes (Part D)
Edited by NEJAT DÜZGÜNEŞ

VOLUME 388. Protein Engineering
Edited by DAN E. ROBERTSON AND JOSEPH P. NOEL

VOLUME 389. Regulators of G-Protein Signaling (Part A)
Edited by DAVID P. SIDEROVSKI

VOLUME 390. Regulators of G-protein Sgnalling (Part B)
Edited by DAVID P. SIDEROVSKI

VOLUME 391. Liposomes (Part E)
Edited by NEGAT DÜZGÜNES

VOLUME 392. RNA Interference
Edited by ENGELKE ROSSI

VOLUME 393. Circadian Rhythms
Edited by MICHAEL W. YOUNG

VOLUME 394. Nuclear Magnetic Resonance of Biological Macromolecules (Part C)
Edited by THOMAS L. JAMES

VOLUME 395. Producing the Biochemical Data (Part B)
Edited by ELIZABETH A. ZIMMER AND ERIC H. ROALSON

VOLUME 396. Nitric Oxide (Part E) (in preparation)
Edited by LESTER PACKER AND ENRIQUE CADENAS

VOLUME 397. (In Preparation)

VOLUME 398. Ubliquitin and Protein Degradation (Part A) (in preparation)
Edited by RAYMOND J. DESHAIES

Section I

Comparing Macromolecules:
Exploring Biological Diversity

Subsection A
Accessing the Templates

[1]   Isolation of DNA from Plants with Large Amounts
of Secondary Metabolites

By ELIZABETH A. FRIAR

Abstract

Many plant species have high contents of polysaccharides, polyphenols, or other secondary metabolites that can interfere with DNA extraction and purification. These contaminating compounds can lead to poor DNA yield and prevent access by modifying enzymes, such as restriction endonucleases and *Taq* polymerase. A number of factors, including choice of plant tissue, tissue preparation, and modifications of the extraction buffer, can help in DNA extraction for difficult plant species. This chapter presents some of the DNA extraction protocols developed for various plants.

Introduction

The tissues of many plant species contain large amounts of secondary compounds. These compounds may be derived from various biosynthetic pathways and play many roles in the biology of the organism. Secondary metabolites may play important roles in plant defense, as chemotoxins to potential fungal, insect, or vertebrate predators. Other plant secondary compounds, particularly complex polysaccharides, play an important role in plant osmoregulation and protection from desiccation. Consequently, complex polysaccharides can make up a high percentage of tissue wet weight in cacti and other plants adapted to very dry environments.

The presence of these plant secondary compounds can make DNA extraction problematic. Some of them will coprecipitate with DNA during extraction and inhibit further enzymatic modification of the DNA, including restriction endonuclease digestion and polymerase chain reaction (PCR) (Guillemaut and Marechal-Drouard, 1992). Large amounts of complex polysaccharides can make extraction of usable DNA impossible, rendering the aqueous portion of the extraction too viscous to allow for efficient separation of DNA from the contaminating polysaccharides. These polysaccharides can also tightly adhere to the DNA, preventing access by modifying enzymes (de la Cruz *et al.*, 1997).

A number of methods have been developed to dilute, selectively precipitate, or inactivate the contaminating substances. For removal of polysaccharides, higher concentrations of cetyltrimethylammonium bromide (CTAB), either in the initial extraction step or in a stepwise fashion over several steps, help to selectively precipitate DNA. Higher concentrations of CTAB and the addition of polyvinylpyrrolidone (PVP) or polyvinylpolypyrrolidone (PVPP) can help to remove polyphenols. Additionally, some protocols suggest the use of ascorbic acid, diethyldithiocarbamic acid (DIECA), and 2-mercaptoethanol to reduce oxidation and prevent DNA degradation.

The relative success of different protocols appears to be largely taxon dependent; what may work for one taxon may not work for another. Here, I present a number of protocols optimized for several taxonomic groups. It may be necessary to try several approaches to achieve high yields of clean DNA from a given species.

Tissue Choice

The choice of tissue for DNA extraction can be critical. Typically, the younger the tissue, the lower the amounts of secondary compounds. Thus, very young leaf tissue or newly germinated seedlings may be good choices for DNA extraction. Additionally, flower buds, petals, epidermal layer, or other soft tissues may be superior, particularly for plants such as cacti in which stem tissue is highly mucilaginous. However, at least one protocol has been developed to extract DNA from bark or young wood of woody species, allowing for identification of cultivars or rootstocks in winter condition (Cheng *et al.*, 1997; see below for protocol).

Tissue Preparation

Typically, the more finely ground the tissue before DNA extraction, the better the extraction of DNA from the contaminating compounds. This can be accomplished by fine grinding in a mortar and pestle of tissues frozen in liquid nitrogen. Highly mucilaginous tissues, however, may become too hard to grind by hand. In those cases, it is possible to coarse chop frozen tissues in a blender or food processor, followed by hand grinding. Lyophilization of plant tissues can also aid in DNA extraction and minimize the amount of contaminating substances. In this case, the plant tissue can be lyophilized and stored, then ground to a fine powder in a bead mill, and extracted using a standard CTAB extraction protocol.

DNA Yield

Yield of DNA from several of these extraction methods can be quite low. Barnwell *et al.* (1998) found that using the same protocol, pea shoots yielded 300 µg of DNA per gram of fresh tissue, but *Sedum telephium* yielded only 20 µg/g of fresh tissue. Some of this difference can be attributed to the high water content of many succulent plants. In addition, some DNA will be inevitably lost because it has been complexed with polysaccharides or polyphenols and lost during the extraction procedure. The more purification steps requiring differential precipitation or extraction, the more DNA will be inevitably lost. Thus, protocols that can successfully remove contaminants from genomic DNA in as few steps as possible will provide the best yields.

It should also be noted that samples yielding low amounts of DNA should be redissolved in distilled water rather than TE buffer, if PCR applications will be used downstream. If yields of DNA are low, the normal practice of dissolving the pellet in TE for a stable stock solution and diluting an aliquot in water for PCR may lead to levels of ethylenediaminetetraacetic acid (EDTA) in the PCR stock sufficient to inhibit *Taq* polymerase activity.

RNA Removal

All of the protocols presented here will coprecipitate DNA and RNA. Retention of contaminating RNA in DNA extractions can inflate measures of DNA concentration, because of the overlapping fluorescence peaks. RNA may also complicate downstream PCR protocols by selectively amplifying shorter complementary DNA (cDNA) copies rather than the desired genomic copies of genes. RNA contamination may be particularly problematic in random amplified polymorphic DNA (RAPD) protocols.

RNA can be removed by incubation with RNase A, either after the nucleic acids have been extracted or by inclusion of RNase A in the initial extraction buffer. RNA may also be removed by differential precipitation using lithium chloride. This protocol will remove larger RNA fragments, and smaller RNA species (e.g., tRNAs) may remain in the solution with the genomic DNA. (See detailed protocols in the following sections.)

Methods: Isolation of DNA

Method 1: Isolation of DNA from Mucilaginous Tissues (Large Preparation)

1. Preheat 2× CTAB buffer (2% CTAB, 1.4 M NaCl, 0.2% 2-mercaptoethanol, 20 mM EDTA, 0.1 M Tris–HCl, pH 8.0) to 60°. Aliquot 25 ml of preheated buffer into a 50-ml conical tube for each DNA extraction.

2. Grind about 5 g of fresh tissue to a fine powder using one of the methods described earlier in the section "Tissue Preparation."

3. Immediately brush ground powder into the preheated isolation buffer and place in a 60° water bath.

4. Incubate at 60° for 30 min.

5. Add an equal volume of chloroform:isoamyl alcohol (24:1) to each tube and agitate for 10 min. Adequate mixing of both layers is vital at this stage.

6. Centrifuge tubes at 6000 rpm for 5 min.

7. Transfer supernatant to a new 50-ml tube, being careful not to disturb the white layer between phases.

8. Add 2/3 volume of ice-cold isopropanol to supernatant and gently rock tube to precipitate DNA.

9. Collect DNA by pelleting or spooling. Pellet DNA by centrifugation for 2 min at 4000g. Spool by collecting long strands of DNA using a glass rod or pipette with a bent tip (a DNA hook can be made by gently heating the narrow end of a disposable glass pipette and bending it into a hook shape). Spooling selectively extracts high-molecular-weight DNA and reduces the amount of coprecipitating contaminants.

10. Transfer pellet to one or two 1.5-ml Eppendorf tubes filled with 0.75 ml of 60° 2× CTAB buffer. Use of two tubes allows for better agitation and increases yield. Mix well.

11. Incubate tubes at 60° for 30 min, mixing occasionally.

12. Add an equal volume of chloroform:isoamyl alcohol (24:1) and agitate well for 10 min.

13. Centrifuge in a Microfuge at maximum speed for 5 min.

14. Transfer supernatant to a new 1.5-ml tube.

15. Add 2/3 volume of ice-cold isopropanol and gently rock tubes to precipitate DNA.

16. Collect DNA by centrifugation in a microcentrifuge at maximum speed for 3 min.

17. Carefully pour off isopropanol, rinse with 70% ethanol, and centrifuge at maximum speed for 2 min.

18. Pour off ethanol and dry pellet thoroughly in a vacuum centrifuge or vacuum oven.

19. Resuspend pellet in 200 μl of TE or diH$_2$O.

Method 2: Isolation of DNA Using Starch Digestion

1. Mix 2× CTAB (2% CTAB, 1.4 M NaCl, 20 mM EDTA, 0.1 M Tris–HCl, pH 8.0) buffer with 1% w/v caylase (Cayla, Inc.). Just before using, add 1% 2-mercaptoethanol.

2. Cut approximately 1 g of leaf tissue into small pieces and then grind in liquid nitrogen. Add 7 ml of the CTAB mixture to the powder and mix to form a slurry. Pour the slurry into a 15-ml polypropylene tube.

3. Cap the tube loosely and incubate at 65° for 30–60 min without shaking.

4. Add 5 ml of chloroform:isoamyl alcohol (24:1) to each tube, cap tightly, and mix well.

5. Centrifuge tubes at 7000g for 10 min.

6. Using a wide-bore pipette, transfer the supernatant to a clean, labeled 30-ml Corex tube. Add 5 ml of ice-cold isopropanol. Cap the tube and swirl gently to precipitate the DNA. Place at −20° for at least 1 h.

7. Centrifuge the tubes at 7000g at 4° for 30 min to pellet the DNA. Pour off the supernatant and dry pellets in a vacuum oven for 10 min.

8. Dissolve the pellet in 1.5 ml of distilled water. Place in a 37° water bath for 10–30 min, mixing the tube occasionally to dissolve the pellet. The pellet will probably not dissolve completely. Transfer pellet and solution to a 15-ml polypropylene tube.

9. Add 7 ml of CTAB to each tube (without caylase or 2-mercaptoethanol). Cap tubes loosely and incubate at 65° for 30 min.

10. Repeat steps 4, 5, and 6.

11. Centrifuge tubes at 7000g at 4° for 30 min. Add 5 ml of ice-cold EtOH-NH$_4$OAc (10 mM NH$_4$OAc in 76% EtOH) to each tube to pellet DNA. Swirl gently, then let stand at room temperature for 10 min. Pour off supernatant and dry the pellet in a vacuum oven for 10 min. Be careful not to pour off the pellet, because it may be loose.

12. Resuspend pellet in 0.2 ml of diH$_2$O or TE at 37° and transfer solution to a 1.5-ml Microfuge tube.

Method 3: DNA Isolation Using Nucleon PhytoPure Resin (Micropreparation)

1. Grind 0.1 g of frozen plant tissue in liquid nitrogen to form a fine powder.

2. Transfer powder to a 1.5-ml Microfuge tube.

3. Add 500 μl of CTAB extraction buffer 1 to each tube (500 μl 2× CTAB [0.1 M Tris–HCl, pH 8.0, 1.4 M NaCl, 20 mM EDTA, pH 8.0, 2% w/v CTAB, 1% w/v PVP-40, 1% w/v sodium bisulfite, 0.2% 2-mercaptoethanol], 2 μl of 20 mg/ml proteinase K, 2 μl 2-mercaptoethanol). Mix well by vortexing or by inversion.

4. Incubate tubes at 50° for 20–30 min, mixing every 5 min.

5. Transfer tubes to 65° and incubate another 15 min.

6. Remove tubes from water bath and incubate on ice for 2–3 min.

7. Add 500 μl of ice-cold 100% chloroform to each tube, then add 100 μl nucleon resin, using a wide-bore pipette tip. Invert tubes to mix, degas once, and then close lids tightly.

8. Gently shake or rock tubes for 15 min at room temperature.

9. Centrifuge tubes in a microcentrifuge at maximum speed for 10 min.

10. Transfer supernatant to a 2-ml tube. (Do not use a 1.5-ml tube, as the additional volume is necessary for optimal mixing.)

11. Add 1 ml of ice-cold 95% ethanol to each tube to precipitate DNA. Incubate tubes at −20° for at least 1 h or overnight.

12. Centrifuge tubes in a microcentrifuge at maximum speed for 10 min to pellet DNA.

13. Pour off supernatant.

14. Dry pellets in a vacuum centrifuge for 10–20 min.

15. Resuspend pellets in 150 μl of diH$_2$O and incubate at 37° for 10–20 min, mixing every 5 min.

16. Add 500 μl of CTAB extraction buffer 2 to each tube (500 μl 2× CTAB, 1% w/v caylase) and invert to mix. Incubate tubes at 65° for 30 min.

17. Remove tubes from water bath and place on ice for 3–5 min.

18. Add 500 μl of ice-cold 100% chloroform to each tube. Invert tubes to mix, degas once, and then close lids tightly.

19. Centrifuge tubes in a microcentrifuge at maximum speed for 10 min.

20. Transfer supernatant to new 2-ml tube.

21. Add 1 ml of ice-cold 95% ethanol to each tube, invert gently to precipitate DNA. Incubate tubes at −20° for at least 1 h. An overnight incubation is preferable.

22. Centrifuge tubes in a microcentrifuge at maximum speed for 10 min. Discard supernatant.

23. Wash pellet by adding 500 μl of 75% ethanol to each tube and rocking gently for 10 min.

24. Centrifuge tubes in a microcentrifuge at maximum speed for 5 min. Discard supernatant.

25. Dry pellet in a vacuum centrifuge for 10–20 min.

26. Resuspend pellets in 100 μl of TE or diH$_2$O and incubate at 37° for 10–20 min, mixing every 5 min.

Method 4: Fragaria DNA Extraction (Porebski et al., 1997)

1. Grind 0.5 g of frozen leaf tissue in liquid nitrogen until finely ground. Transfer powder to 15-ml polypropylene tubes.

2. Add 5 ml of 60° extraction buffer (100 mM Tris–HCl, pH 8.0, 1.4 M NaCl, 20 mM EDTA, pH 8.0, 2% w/v CTAB, 0.3% 2-mercaptoethanol) and 50 mg of PVP-40 to each tube. Mix by inversion and incubate at 60° with shaking for 25–60 min.

3. Remove tubes from oven and allow to cool to room temperature.

4. Add 6 ml of chloroform:octanol (24:1) to each tube and mix well.

5. Centrifuge tubes at 3000 rpm for 20 min.

6. Transfer aqueous phase to a new polypropylene tube and repeat chloroform:octanol extraction once more to remove cloudiness in the aqueous phase caused by the PVP. Repeat steps 4, 5, and 6 until the aqueous phase is no longer cloudy.

7. Add 1/2 volume of 5 M NaCl to final aqueous solution. Mix well. Add 2 volumes of ice-cold 95% ethanol to precipitate DNA. If required, incubate tubes at −20° for 20 min to overnight to aid precipitation.

8. Centrifuge tubes at 3000 rpm for 6 min to pellet DNA.

9. Pour off supernatant and wash pellet with ice-cold 70% ethanol. Dry pellet in 37° oven or vacuum oven.

10. Dissolve pellets in 300 μl of TE overnight at 4° to 6°. Transfer DNA to 1.5-ml Eppendorf tubes. (Note: DNA is usually not amplifiable at this stage.)

11. Add 3 μl of RNase A (10 mg/ml) to each tube and incubate at 37° for approximately 1 h. Add 3 μl proteinase K (1 mg/ml), and incubate for an additional 15–30 min.

12. Add 300 μl of phenol:chloroform (1:1) to each tube. Vortex well to mix.

13. Centrifuge tubes in microcentrifuge at maximum speed for 10–20 min.

14. Remove top layer to a new 1.5-ml tube. Add 50 μl of TE to the phenol phase and recentrifuge as above. Collect top layer and add to the previously collected top layer.

15. Add 1/10 volume 2 M NaOAc and 2 volumes absolute ethanol to sample and mix well. Incubate tubes overnight at −20°.

16. Centrifuge tubes in a microcentrifuge at maximum speed for 10–20 min. Pour off supernatant and wash pellet with 70% ethanol. Dry pellet in 37° oven or vacuum oven.

17. Add 100–200 μl of TE or diH$_2$O to each tube. Incubate overnight at 4° for complete resuspension.

Method 5: Cactus DNA Extraction (de la Cruz et al., 1997)

1. Grind 3 g of tissue to a fine powder in liquid nitrogen.

2. Add 4 ml of CTAB extraction buffer (100 mM Tris–HCl, pH 8.0, 20 mM EDTA, pH 8.0, 4% CTAB, 1.5 M NaCl, 4% PVP-40, 500 g ascorbic acid, 500 g DIECA, 10 mM 2-mercaptoethanol) with further grinding to produce a slurry.

3. Add 15 ml of STE extraction buffer (100 mM Tris–HCl, pH 8.0, 50 mM EDTA, pH 8.0, 100 mM NaCl, 10 mM 2-mercaptoethanol) and transfer the solution to a 50-ml tube.

4. Add 1 ml of 20% sodium dodecyl sulfate (SDS) and shake vigorously for 7 min.

5. Incubate at 65° for 10 min.

6. Add 5 ml of cold 5 M potassium acetate and incubate at 0° for 40 min.

7. Centrifuge tubes at 20,000 rpm for 20 min to remove debris.

8. Filter the aqueous phase through a Miracloth filter into a clean 50-ml tube.

9. Add 7/10 volume of ice-cold isopropanol, mix gently, and incubate at −20° for 10 min to precipitate genomic DNA.

10. Centrifuge tubes at 20,000 rpm for 15 min and discard the supernatant.

11. Air-dry the pellet and resuspend in 1 ml of TE.

12. Transfer the solution to a 1.5-ml tube and centrifuge in a microcentrifuge at full speed for 10 min.

13. Transfer the supernatant to a new 1.5-ml tube and add 65 μl of 3 M sodium acetate and 600 μl of ice-cold isopropanol and gently mix.

14. Incubate at −20° for 10 min.

15. Centrifuge tube in a microcentrifuge at full speed for 30 s.

16. Wash the pellet carefully with 76% ethanol.

17. Resuspend pellet in 1 ml of TE or diH$_2$O.

Method 6: Cactus DNA Extraction 2, from Roots (Tel-Zur et al., 1999)

1. Rinse 0.5–1.0 g of root material with distilled water.

2. Grind the tissue to a fine powder in a mortar and pestle under liquid nitrogen. Transfer the powder to a 50-ml tube.

3. Add 20 ml of extraction buffer to each tube (100 mM Tris–HCl, pH 8.0, 0.35 M sorbitol, 5 mM EDTA, pH 8.0, 1% 2-mercaptoethanol), keeping tubes on ice.

4. Centrifuge tubes at 10,000g for 10 min at 4°.

5. Pour off the supernatant carefully. Redissolve the pellet with 20 ml of extraction buffer and recentrifuge. Repeat once more.

6. Pour off the supernatant again and redissolve the pellet in 5 ml of extraction buffer. Add 3.5 ml of high-salt CTAB buffer (50 mM Tris–HCl, pH 8.0, 4 M NaCl, 1.8% CTAB, 25 mM EDTA, pH 8.0) and 0.3 ml of Sarkosyl 30% to each tube. Incubate mixture at 55° for 60–90 min.

7. Add equal volume of chloroform:isoamyl alcohol and mix well. Centrifuge tubes at 10,000g for 10 min. Transfer supernatant to a Corex tube.

8. Add 2/3 volume ice-cold absolute isopropanol and then 1/10 volume 3 M sodium acetate, pH 5.2.

9. Pour off supernatant and wash the pellet with ice-cold 75% ethanol.

10. Air-dry the pellets and dissolve in 200 μl of TE or diH$_2$O.

11. Perform RNase digestion as below.

Method 7: Sedum DNA Extraction (Barnwell et al., *1998)*

1. Grind leaves to a fine powder in a mortar and pestle under liquid nitrogen. Transfer powder to a 50-ml tube.

2. Add 5 volumes of extraction buffer (100 mM Tris–HCl, pH 8.0, 20 mM Na$_2$EDTA, 2% w/v CTAB, 1.4 M NaCl, 1% w/v PVP-40) and mix to form a paste.

3. Incubate mixture at 65° with occasional shaking for 10 min.

4. Centrifuge tubes at 3000g for 5 min.

5. Transfer the supernatant to a new tube and add 1.25 times the original tissue volume of 10% w/v CTAB (in 0.7 M NaCl). Vortex the mixture well.

6. Centrifuge tubes at 3000g for 5 min.

7. Transfer supernatant to a new tube and add 3 volumes of precipitation buffer (50 mM Tris–HCl, pH 8.0, 1 mM Na$_2$EDTA, 1.0 M NaCl) and mix well.

8. Incubate tubes at room temperature (22°) for 30 min.

9. Centrifuge tubes at 5000g for 15 min. Discard the supernatant.

10. Dissolve the pellet in high-salt TE buffer (10 mM Tris–HCl, pH 8.0, 1 mM Na$_2$EDTA, 1 M NaCl). Transfer to a microcentrifuge tube.

11. Add 2 volumes of ice-cold 95% ethanol to precipitate nucleic acids. Incubate at −20° for 1 h.

12. Centrifuge tubes in a microcentrifuge at full speed for 5 min to pellet nucleic acids.

13. Wash the pellet twice with 70% ethanol.

14. Dissolve pellet in 100 μl of diH$_2$O.

Method 8: DNA Extraction from Woody Material (Bark, Dormant Buds, Young Stems; Cheng et al., 1997)

1. Grind 0.05–0.1 g of material to a fine powder with liquid nitrogen.
2. Add 1 ml of extraction buffer (100 mM Tris–HCl, pH 8.0, 20 mM EDTA, pH 8.0, 1.5 M NaCl, 2% w/v CTAB, 2% w/v PVP-40T) to the powder and transfer solution to a 1.5-ml tube.
3. Incubate tubes at 65° for 30 min.
4. Add enough chloroform:isoamyl alcohol (24:1) to almost fill the tube. Shake vigorously to form an emulsion.
5. Centrifuge tubes at 8000g for 5 min.
6. Transfer aqueous phase to a new 1.5-ml tube and add 1 ml of ice-cold 95% ethanol.
7. Incubate tubes at −20° for 5 min to precipitate DNA.
8. Centrifuge tubes at 8000g for 5 min. Pour off supernatant.
9. Add 600 μl 1 M NaCl and incubate at 65° for 10 min to dissolve DNA.
10. Add 300 μl of phenol and mix gently.
11. Centrifuge tubes at 8000g for 3 min.
12. Transfer supernatant to a new 1.5-ml tube and add 500 μl of chloroform to each tube.
13. Centrifuge tubes at 8000g for 5 min.
14. Transfer supernatant to a new 1.5-ml tube and add 1 ml of ice-cold 95% ethanol.
15. Incubate at −20° for 30 min.
16. Centrifuge tubes at 8000g for 10 min. Pour off supernatant.
17. Dissolve pellet in 600 μl of diH$_2$O and add 6 μl of 0.1 M spermine. Incubate on ice for 20 min.
18. Centrifuge tubes at 8000g for 10 min. Pour off supernatant.
19. Add 500 μl 1× spermine pellet extraction buffer (10×: 3 M sodium acetate, 0.1 M magnesium acetate), diluted in 75% ethanol. Incubate tubes on ice for 1 h.
20. Drain off liquid and wash pellet with 500 μl of 75% ethanol.
21. Dry pellet either by air-drying or in a vacuum centrifuge.
22. Redissolve pellet in TE or diH$_2$O.

Method 9: Isolation and Purification of DNA Using Sephacryl (Li et al., 1994)

1. Grind 0.2 g of fresh tissue to a fine powder under liquid nitrogen.
2. Transfer powder to a 13-ml Röhrer tube containing 6 ml of extraction buffer (100 mM Tris–HCl, pH 8.0, 20 mM EDTA, 500 mM NaCl) on ice.

3. Add 10 μl of 2-mercaptoethanol and 400 μl of 20% SDS. Mix gently.

4. Incubate tubes at 65° for 10 min.

5. Add 2 ml of 3 M potassium acetate buffer, pH 5.2. Mix gently and incubate on ice for 10 min.

6. Centrifuge tubes at 10,000g for 10 min.

7. Pour supernatant through 2 layers of Miracloth into a clean Röhrer tube. Add 4 ml of isopropanol and mix gently. Incubate at −20° for 30 min.

8. Centrifuge tubes at 10,000g for 10 min to pellet DNA. Pour off the supernatant and dry by inverting the tube over paper towels.

9. Dissolve pellet in 200 μl buffer and transfer solution to a 1.5-ml tube.

10. Add 1 μl RNase A (10 mg/ml) and incubate tubes at 65° for 10 min.

11. Put a small amount of glasswool in the bottom of a 1-ml plastic syringe.

12. Fill the syringe to 0.7 ml with Sephacryl S-1000.

13. Place the syringe in a 13-ml Röhrer tube and centrifuge at 150–300g for 1 min.

14. Run approximately 300 μl of STE buffer (100 mM NaCl, 10 mM Tris–HCl, pH 8.0, 1 mM EDTA, 20% w/v PEG 8000 in 1.2 M NaCl) through the syringe twice, centrifuging at 150–300g for 2 min each time.

15. Place the prepared spin column into a 13-ml Röhrer tube, making sure the tip of the syringe is in the Eppendorf tube.

16. Load the DNA solution into the column and centrifuge at 150–300g for 2 min.

17. Add 200 μl of STE to the column and recentrifuge as above.

18. Add 200 ml of 20% PEG 8000 in 1.2 M NaCl to the DNA solution and incubate on ice for 20 min.

19. Centrifuge tubes at 10,000g for 15 min at 4°.

20. Wash the pellets once using 70% ethanol.

21. Dissolve pellet in 100–500 μl of TE or diH$_2$O.

Methods: Removal of RNAs from Genomic DNA

Method 1: Differential Precipitation with Lithium Chloride

1. Add 0.2 volumes of ice-cold 12 M lithium chloride (LiCl) to resuspended DNA.
2. Incubate at 4° for 30 min.
3. Centrifuge tubes for 10 min at 4°.
4. Discard pellet and retain supernatant.

Method 2: Digestion with RNase A (Tel-Zur et al., 1999)

1. Add 10 μl of 5 mg/ml RNase A to 200 μl of resuspended DNA in TE.
2. Incubate tubes for 40 min at 37°.
3. Transfer solution to a 1.5-ml Eppendorf tube and add an equal volume of phenol:chloroform (1:1 v/v).
4. Centrifuge tubes at maximum speed in a microcentrifuge for 10 min.
5. Transfer aqueous phase to a new 1.5-ml tube and add an equal volume of cold chloroform.
6. Centrifuge tubes at maximum speed in a microcentrifuge for 10 min.
7. Remove the aqueous phase and precipitate the DNA with 2 volumes of ice-cold absolute ethanol mixed with 1/10 volume of 3 M sodium acetate, pH 5.2.
8. Incubate tubes at −20° for 30 min.
9. Centrifuge tubes at maximum speed in a microcentrifuge for 15 min.
10. Rinse the pellet with ice-cold 75% ethanol and allow pellets to air-dry.
11. Dissolve pellet in 50–100 μl of TE or diH$_2$O.

References

Barnwell, P., Blanchard, A. N., Bryant, J. A., Smirnoff, N., and Weir, A. F. (1998). Isolation of DNA from the highly mucilaginous succulent plant *Sedum telephium. Plant. Mol. Biol. Reptr.* **16,** 133–138.

Cheng, F. S., Brown, S. K., and Weeden, N. F. (1997). A DNA extraction protocol from various tissues in woody species. *HortScience* **32,** 921–922.

de la Cruz, M., Ramirez, F., and Hernandez, H. (1997). DNA isolation and amplification from cacti. *Plant Mol. Biol. Reptr.* **15,** 319–325.

Guillemaut, P., and Marechal-Drouard, L. (1992). Isolation of plant DNA: A fast, inexpensive and reliable method. *Plant Mol. Biol. Reptr.* **10,** 60–65.

Li, Q.-B., Cai, Q., and Guy, L. (1994). A DNA extraction method for RAPD analysis from plants rich in soluble polysaccharides. *Plant Mol. Biol. Reptr.* **12,** 215–220.

Porebski, S., Bailey, L. G., and Baum, B. R. (1997). Modification of a CTAB DNA extraction protocol for plants containing high polysaccharide and polyphenol components. *Plant Mol. Biol. Reptr.* **15,** 8–15.

Tel-Zur, N., Abbo, S., Myslabodski, D., and Mizrahi, Y. (1999). Modified CTAB procedure for DNA isolation from epiphytic cacti of the genera *Hylocereus* and *Selenicereus* (Cactaceae). *Plant Mol. Biol. Reptr.* **17,** 249–254.

[2] Nucleic Acid Isolation from Environmental Aqueous Samples

By Klaus Valentin, Uwe John, and Linda Medlin

Abstract

The application of molecular techniques has revolutionized freshwater and marine ecology, especially for plankton research. Methods, such as denatured gradient gel electrophoresis (DGGE), temperature gradient gel electrophoresis (TGGE), and single-strand conformation polymorphism (SSCP), together with environmental clone libraries, have unraveled an unexpected biodiversity of organisms in the water column. Molecular probes are just entering the field of commercialization for monitoring toxic algal blooms. Genomics and metagenomics were recently introduced into marine biology. At the basis of all molecular approaches is the isolation of nucleic acids from cultures, tissue, or environmental samples. Here, we summarize methods, quality controls, and hints for sample treatment to reliably isolate nucleic acids, both DNA and RNA, from environmental aqueous samples. This chapter not only is directed to researchers inexperienced with such methods but also is an aid to those already working in the field. It may be used as a step-by-step guide for nucleic acid isolation from field samples, and we make suggestions for subsequent use of the DNA/RNA.

Introduction

During the last 2–3 decades, molecular techniques have become more and more integrated into all fields of biology, including traditional areas, such as in systematics and ecology, but especially in biodiversity studies. Understanding and preserving biodiversity has been one of the most important global challenges for the past 20 years and will continue to be an important scientific issue. The global environment has been experiencing rapid and accelerating changes, largely originating from human activity, whether they originate from local causes or from the more diffuse effects of global climate change. Widespread realization that biodiversity can be strongly modified by such changes has generated plans on a global scale to conserve and protect biodiversity in many parts of the world that were previously subjected to rampant salvaging for natural resources. Adequate ecosystem functioning, and therefore, the continued use of the goods and services that ecosystems provide to humans depend on how important biodiversity is perceived and preserved. Thus, it follows that knowing and recognizing biodiversity at all levels is an essential strategy for preserving biodiversity.

Sequence comparisons and sequence-based phylogenies have revolutionized our understanding of biodiversity and how that biodiversity has evolved on earth. Molecular techniques have even crept into ecological studies because now genomewide gene expression profiles can explain adaptation of organisms to ecological niches. Biodiversity is no longer only explained as the organisms that can be found by biologists in a natural sample but is now widened by the term *hidden biodiversity*, referring to the fact that in most ecosystems only a small percentage of species, sometime less than 10%, are known. The remaining 90% or so may only be found using molecular methods, such as environmental clone library construction, sequencing of many global isolates of a single species, or fingerprinting techniques. Consequently, biological sampling on field trips today frequently has a molecular aspect in that more and more samples will later be analyzed by nucleic acid methods in addition to the determination of physiochemical parameters and the classic counting of organisms present in the sample.

Molecular tools in general offer the possibility to estimate biodiversity at all levels, for example, kingdom/class/family/species level, in a comparatively small environmental sample. In some cases even a few milliliters of water or a few milligrams of tissue may be enough. Moreover, some of the techniques are very sensitive, that is, offer the possibility to detect single cells in a sample. Depending on the question(s) being asked, the molecular tools to answer them differ greatly. One may wish to detect as many species as possible in a given sample. In this case the establishment of a rRNA clone library with subsequent sequencing of as many clones as possible can uncover the biodiversity in the sample in great detail. General assessment of comparative biodiversity in a larger number of samples can be achieved with fingerprinting methods based on amplified fragment length polymorphisms (AFLPs) (Vos *et al.*, 1995), denaturing gradient gel electrophoresis (DGGE), temperature gradient gel electrophoresis (TGGE) (Muyzer *et al.*, 1993), or single-strand conformation polymorphisms (SSCPs) (Schwieger and Tebbe, 1998). Presence or absence of a known species can be monitored with species-specific probes using chemiluminescent detection with dot-blot techniques or with the more sophisticated fluorescence *in situ* hybridization (FISH) (Groben and Medlin, 2005). Distinction of individuals at the species or even the strain level can be done using highly variable molecular markers, such as inter-transcribed spacer (ITS) sequences or microsatellites.

The starting point of any molecular analysis is the isolation and investigation of nucleic acids, their sequences and abundance, from a sample. Whereas such isolation is rather simple with cell lines or *Escherichia coli*, more care is necessary for environmental samples. Here, we give an overview of how one can isolate DNA or RNA from such samples and address common difficulties step by step. We outline a basic routine that would enable a rather inexperienced biologist to take a natural sample, store it for later use, isolate DNA or RNA from it, check the quality of the preparation, remove possible contaminants, and do some basic experiments with it. It is not aimed at highly sophisticated analyses, although some of the hints presented here will also be of help for experienced workers. The structure of our chapter is logical in that the steps required for nucleic acid isolation are addressed as they appear practically; we start with the size of the sample and continue through the treatment of the sample, and then some thoughts are given on the quality requirements for the nucleic acid, the lysis of the organisms in the sample, the treatment of typical contaminants in the sample, and the actual nucleic acid preparation. We describe how the nucleic acid is to be checked and how certain contaminants may be removed. A short description of gel electrophoresis and storage of nucleic acids follows, and we provide an extra section on subsequent experiments.

In some cases, it may be of interest to decide whether the species was physiologically active when collected. Here, one would isolate RNA rather than DNA, back-translate the RNA into complementary DNA (cDNA) via a reverse-transcriptase (RT) step and then amplify by polymerase chain reaction (PCR) the target gene from the cDNA. The entire process is called *RT-PCR* and allows estimation, semiquantitatively or quantitatively, of the amount of specific messenger RNA (mRNA) present in the cell. The underlying assumption is that only in active cells are genes transcribed into mRNA, which can then produce a PCR product. Therefore, we also include a section on RNA isolation from natural samples and how to perform RT-PCR from it. The extraction of RNA requires additional precautions because it is easily degraded by the ever abundant ribonucleases (RNases).

How to Isolate DNA/RNA from a Natural Sample

Sampling

Before taking a sample, one has to consider what amount of DNA or RNA is needed for subsequent analyses (see the Section "Subsequent Steps (PCR, Restriction Enzyme Digestion, Hybridization," later in this chapter). For any single PCR analysis, only a few nanograms is necessary, whereas for hybridization experiments, several micrograms may be required. In most cases, one will be interested in isolating nucleic acids from organisms rather than from viruses. For "large" organisms (e.g., fish, copepods, and macroalgae), the question then is how many of these organisms or how much tissue is necessary to isolate enough DNA/RNA. This differs greatly depending on the organisms to be studied. Generally, animals contain more nucleic acid per gram of fresh weight than algae because the bodies of the latter predominantly consist of cell walls and vacuoles, both of which contain no DNA. An exponentially growing organism has a higher RNA:DNA ratio and would give more mRNA (polyA+ in the case of eukaryotes) compared to the extracted total RNA. In general, organisms contain 10 times as much RNA than DNA. Of this RNA, more than 90% is ribosomal RNAs (rRNAs). Most subsequent steps for RNA analysis require about 10 times more total RNA than DNA. For example, 1–2 μg of DNA is enough for a single Southern blot analysis, whereas 10–20 μg of total RNA is needed for a single Northern blot. For example, 100 mg of animal tissue, or a single copepod, will yield enough DNA for perhaps 10–20 PCRs, whereas gram amounts will be needed from a macroalga for the same purpose. The amount of sample needed for RT-PCR from RNA will be perhaps twofold that amount. For the isolation of planktonic microorganisms, a water sample typically is filtered through one filter or a series of filters. The amount that can be filtered onto one filter then strongly depends on the population density in the water (oligotrophic vs. eutrophic waters), the pore size (typically between 0.2 and 50 μm), the filter type (polycarbonate, GF/F, etc.), filter size, and the amount of particles (sand, mud, debris) in the water body. To give extremes, for eutrophic waters that are filtered onto a 0.2-μm/2.5-cm diameter nucleopore filter, a few milliliters of water can be enough to clog the filter, whereas many liters of oligotrophic waters can be filtered onto a single 3-μm GF/F filter. As rule of thumb, a filter that is visibly stained with microorganisms will produce enough DNA for 10–100 PCRs, assuming that the staining was primarily not caused by particles like sand, mud, or debris. We suggest filtering the type of water to be analyzed onto a filter until it clogs and then isolating DNA from it. This will give an estimate of the amount needed. One should sample at least triple the amount of what is estimated in this manner. When isolating RNA, one should sample at least fivefold this amount.

Treatment of Samples

When taking samples in the field, there is usually a significant amount of time between sampling and sample processing, that is, more than minutes or hours, and in the case of cruises up to several months. Because most organisms will start to lyse and to degrade their nucleic acids immediately after sampling, the cells or tissue must be preserved immediately. This is especially true for RNA. Ideally, this is achieved by freezing the sample in liquid nitrogen directly after sampling and subsequent storage at $-20°$ for months or at $-80°$ for years (RNA: $-80°$). However, liquid nitrogen is not always available in the field, and in these cases, crushed dry ice can also be used to freeze the sample as quickly as possible (dry ice can be stored longer, safer, and easier than liquid nitrogen) before storing at $-20°$ or $-80°$. In case neither liquid nitrogen nor dry ice is available, we do not suggest just to put the samples into a $-20°$ freezer or on ice because this frequently leads to significant degradation of nucleic acids, especially from organisms adapted to low temperatures and for RNA. In such a situation, probably typical for many field trips, the sample can be preserved chemically and by heat. The basis for this is the following assumptions: (1) Most DNA degrading enzymes require Mg^{2+} ions, which can be chelated by treatment with ethylenediaminetetraacetic acid (EDTA). However, an important enzyme for lysis of cells, proteinase K (see later discussion), is also partially inhibited by EDTA; (2) most proteins are denatured by detergents, such as sodium dodecyl sulfate (SDS), which also lyse membranes, and (3) most DNA degrading enzymes are not stable at temperatures above $50°$, unless the organisms are thermophilic (e.g., *Taq* polymerase). This leads to a simple protocol for sample treatment when no liquid nitrogen or dry ice is available: An equal volume of autoclaved 20 mM Tris–Cl pH 8.5, 100 mM EDTA, and 2% SDS (note: SDS precipitates below $8°$, so do not store in a fridge) is added to the sample and heated to $67°$ for 30 min or $100°$ for 10 min. The sample can then be stored on ice for days or at $-20°$ for weeks. Keep in mind that SDS will precipitate under these conditions and will dissolve only after reheating to $50°$. Preservation with 80% ethanol is also acceptable if heating and cooling facilities are not available. However, ethanol treatment precipitates DNA, RNA, and polysaccharides and may result in problems dissolving high-molecular-weight DNA. Preservation of RNA requires extra precaution because of the robust RNA-specific nucleases that are difficult to inhibit. The addition of a few milliliters of "RNA Later" (Ambion) will preserve the DNA and the RNA, and this preservation can be easily shipped, whereas those processed in ethanol cannot. Macroalgae can also be stored in a plastic bag with a desiccant (silica gel, also used to dry vascular plant tissue that can be purchased in many craft stores in the United States) until processing in the laboratory. Simple drying of specimens may also be acceptable. Dried radiolarians produced better quality DNA than specimens preserved in ethanol or formaldehyde (Holzmann and Pawlowski, 1996). DNA extraction from herbarium specimens works in many cases

(Taylor and Swann, 1994). DNA can also be isolated from small pieces of tissue, such as fish scales (Yue and Orban, 2001). It is generally difficult to obtain good DNA from fixed samples, that is, samples treated with Lugol's solution, paraformaldehyde, or glutaraldehyde, but see Yue and Orban (2001). However, environmental samples preserved with Lugol's solution or paraformaldehyde are usable for FISH hybridizations (Toebe *et al.*, 2001).

General Advice for Consumables While Working with Nucleic Acids

In general, normal laboratory safety is required. Always wear a lab coat, disposable gloves, and protective goggles. These will not only safeguard your health but also protect your nucleic acids from degrading enzymes, which are abundant on the skin and, in the case of RNases, are very stable and active. There is no need to autoclave plastic consumables; they come clean and nuclease free from the company and should be aliquoted using gloves. Autoclaving does not destroy DNA and RNA and the steam in the autoclave may distribute contaminating DNA on all pipette tips and reaction vials. In case you work with RNA, glassware can be baked for 4 h at $180°$, and plastic ware such as gel electrophoresis chambers can be incubated in 0.1 N NaOH solution for 1 h and then washed carefully with RNase-free water. RNase-free water is commercially available, or it can be prepared by adding 0.1% diethyl pyrocarbonate (DEPEC) (note: DEPEC is toxic) to deionized water, allow it to sit overnight, and autoclave before use.

Subsequent Use of Nucleic Acids and Implications for Its Quality

Today, most analyses of environmental DNA will be done with PCR approaches, that is, by amplifying certain genes, especially the rRNA genes (Medlin *et al.*, 1988). For this purpose, the quality of DNA is not too critical. However, other methods, such as fingerprinting techniques (random amplified polymorphic DNA [RAPD] and amplified fragment length polymorphism [AFLP]), need high-quality DNA for the standardization of their banding patterns (Fig. 1), which would otherwise be inconsistent if the DNA were fragmented. If high-quality DNA is not required, then it may be enough simply to boil the sample for 5–15 min, centrifuge or filter it to remove debris, and use the supernatant directly for a PCR. In our hands, this works for at least 50% of all samples, and it will lyse most organisms and degrade RNA. Shortcomings are that this method does not remove any contaminants and that the DNA cannot be quantified. In addition, this DNA is less stable when stored at $-20°$ or in the refrigerator. Sometimes varying amounts of substances are present that inhibit the PCR. Then

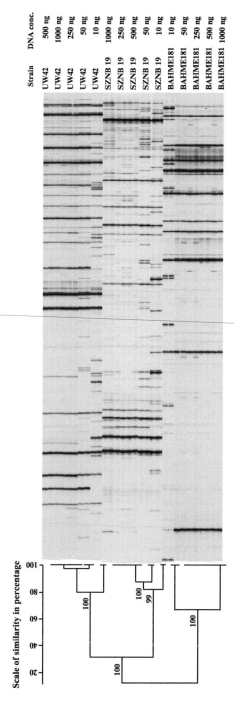

Fig. 1. Amplified fragment length polymorphism (AFLP) analysis of five DNA concentrations from three strains of the *Alexandrium tamarense* species complex using specific primers *Eco*RI + AAG and *Mse*I + CTT. The dendrogram was constructed using unweighted pair group method with arithmetic means (UPGMA) and the scale indicates percentages of similarity as determined with the Dice similarity coefficient (BioNumerics cluster analysis). Reproduced, with permission, from John *et al.* (2004); courtesy Gustav Fischer.

smaller amounts of starting material will amplify better than large amounts. This can be because of the dilution of the contaminants. Compounds, such as acetamide (10%), can be added to PCRs to block contaminants that prevent the amplification of some groups (Lange *et al.*, 2002), although in the original publication of this method, it was used to reduce the melting temperature of templates with high G/content that could not be amplified (Reysenbach *et al.*, 1992). T4-G32 Protein can also be added to relax the secondary structure to improve PCR amplification (Schwieger and Tebbe, 1997). The quick boiling method described is not very reliable or reproducible, and the DNA extraction cannot be stored long term. However, because it is simple, fast, and cheap, it can be used for large sample numbers. We have successfully applied it to mass screening of cultures for crude identification by sequencing a portion of their rRNA. Of course, results are more reliable and reproducible when using a "normal" DNA preparation (see "Standard DNA Extractions," later in this chapter) that systematically removes proteins, small molecules, and RNA before the PCR. For expression analysis using real-time PCR approaches, the integrity and purity of the RNA is crucial, and the quick boiling method cannot be applied here.

Lysis of Organisms

Not all organisms will readily lyse just by freezing and thawing. Therefore, some thought should be given to a lysis procedure. Generally, two approaches can be used or combined: chemical procedures or mechanical forces. Chemical methods are more gentle and will yield DNA of higher molecular weight but will not work with all organisms and do not work for RNA. A chemical lysis can also be better standardized to give reproducible results for large sample numbers. Mechanical forces, on the other hand, when used rigorously will lyse all organisms. They are more difficult to standardize, and mechanical forces will shear DNA, in the worst case to molecules that can no longer be amplified. Another drawback of rigorously treated samples is that the membrane-bound polysaccharides go into solution. They are very difficult to remove from the sample because once in solution they co-purify with nucleic acids and can be the reason for failure of subsequent enzymatic treatments of the nucleic acids.

A standard chemical lysis for DNA extraction combines a detergent (cetyltrimethylammonium bromide [CTAB] [see later discussion; Doyle and Doyle, 1990] and/or SDS [1%]), proteinase K (20 ng/μl) and heat (65°) for 1–2 h. EDTA can optionally be added (up to 50 mM), but this will partially inhibit proteinase K. We suggest trying this first and checking lysis microscopically. A good indication for cell lysis of photosynthetic organisms is a change of color to a more "greenish" appearance, which is produced when the plastids are lysed and the light-harvesting complexes are disrupted. This chemical lysis may be improved by a freeze–thaw cycle ideally using liquid nitrogen. The proteinase K concentration may be increased, and the treatment extended for up to 12 h. We suggest including RNase in the lysis buffer to remove RNA before the DNA preparation. For chemical lysis, entire filters with microorganisms on them can be used.

The standard mechanical lysis is done by grinding with a mortar and pestle under liquid nitrogen and a small amount of quartz sand for a few minutes up to an hour. This method is applicable to both RNA and DNA extractions. The progress of lysis should be checked periodically by eye. This procedure is tedious but works with most organisms. In addition, the DNA produced from it is of reasonable integrity for standard PCR applications. One should take care that the cell powder produced by this method does not thaw but is added instantly to a chemical lysis buffer (see previous discussion), to liquid phenol, to CTAB extraction buffer (EB), or to a binding buffer for commercial purification columns. Mechanical disruption can also be done by sonication or bead beating with glass beads. With both of these latter methods, disruption should be done in short bursts and with visual inspection between each replicate treatment because they both tend to shear the DNA significantly into small fragments. These methods are only suitable for RNA extraction if done on ice.

In case DNA for subsequent steps must be of very high molecular weight (e.g., for pulse field electrophoresis of intact chromosomes), lysis has to be done chemically without mechanical treatment. This implies that the organisms have to be embedded into agarose and treated with proteinase K therein. However, this will rarely be necessary with field samples.

Typical Contaminants: The two major contaminants that may cause problems with extractions are polysaccharides and phenols. Polysaccharides can make the lysate too viscous for proper handling and may behave like DNA in ethanol precipitation. Once a DNA is co-precipitated with polysaccharides, one will observe a large white flocculate pellet after centrifugation, which will be difficult to dry, and which will not resolve any more. A good way to eliminate polysaccharides is to use CTAB as an extraction protocol, but most commercial DNA purification columns also work fine (see later discussion). Phenols, even small amounts, may inhibit subsequent enzymatic steps, such as PCR amplification or restriction enzyme digestion (Fig. 2). Phenols are common in many plants or macroalgae and often serve as a protection against herbivores. After lysis and exposure to oxygen, they quickly oxidize, resulting in a brownish color. An indication for phenol contamination, therefore, can be a brownish color of the DNA. Most commercial DNA purification columns will be able to remove phenols,

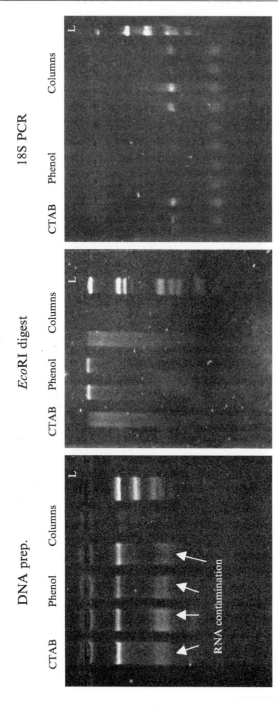

Fig. 2. Influence of different DNA preparations on DNA quality. We compared cetyltrimethylammonium bromide (CTAB) extraction, phenol extraction, and two types of commercial columns. (Left) Shown are the size, integrity, and yield of DNA from equal field sample (∼50 ml of filtered seawater from Helgoland) volumes; the preparations were not treated with ribonuclease (RNase). Four of five preparations produced undegraded DNA of high molecular weight with perhaps the highest yield with the CTAB extraction. (Middle) DNA from CTAB and column preparations was readily digestible with EcoRI, whereas the phenol preparations was not. Similarly, CTAB and column DNA easily produced full-length 18S rDNA products, whereas phenol preparations did not. L, the DNA size ladder. CTAB extraction produced best results and is cheap but is less suitable for standardization (see text).

but one can also add a small amount (milligrams per gram of fresh weight) of water-insoluble polyvinylpyrrolidone (PVP) suspended in water before the lysis procedure (*http://www.mrcgene.com/tb7-plant.htm*). After lysis, the PVP can be centrifuged away. A problem that might occur with PVP is a decrease in yield because it also weakly binds DNA. Humic substances also can be difficult to remove from samples and will in most cases inhibit PCRs. One can try to use acetamide (10%) in the PCR to block the inhibition by humic acids or M4 G32 protein (Tebbe and Vahjen, 1993).

Standard DNA Extractions

Basically there are three types of DNA extractions commonly used: (1) phenol extraction, (2) CTAB extraction, and (3) commercially available purification columns or DNA binding resins for batch use. (For small DNA molecules, there are also columns available that are based on filtering, but these are of no use for high-molecular-weight DNA.) Whereas methods 1 and 2 invoke a precipitation step, method 3 keeps the DNA in solution throughout the preparation. This implies that method 3 usually is much faster and has a higher yield, but unfortunately it is also much more expensive.

1. Phenol extraction (Maniatis) is the longest used method, and it is reliable and cheap. There is also no danger of DNA (or even RNA, see later discussion) degradation during the preparation. It is based on the complete removal of proteins including RNase by protein degradation with phenol and subsequent centrifugation. Medlin *et al.* (1991) found, however, that for the diatoms, the phenol extraction must be done on ice. They presumed that because they often saw degradation of their DNA after phenol extraction, there must be nucleases present in the diatom that escaped removal by phenol extraction. Certain diatom species, e.g., *Aulacoseira* spp., must have their DNA aliquoted for use in PCRs. A single aliquot of DNA (100 ng/ml) can be used perhaps two to three times before it repeatedly fails in a PCR. In general, we suggest paying special attention to the following points when extracting high-molecular-weight DNA: (1) In case the lysis was done mechanically by grinding in the presence of liquid nitrogen, the lysate should still be frozen when added to the phenol. (2) To avoid shearing of the DNA, all mixing should be done very gently (not necessary with RNA, see later discussion) (e.g., by turning the tube about 10–20 times/min for about 10–15 min until a milky suspension is observed). (3) All centrifugation steps for phase separation, especially the first ones, should be extended to at least 1 h and great care should be taken not to remove any of the interphase between the phenol and the aqueous phase. Centrifugation at 10,000 rpm will compact the interphase so that you can

avoid contamination. (4) It is crucial to remove completely the phenol before DNA precipitation by repeated extraction with chloroform:isoamyl alcohol. (See below for the ethanol precipitation step.) The major disadvantages of phenol extraction are the possible contamination of the DNA preparation with phenol, the time needed, and the need to work in a fume hood (phenol is toxic and most people do not like its smell). The major advantages are its reliability—when used properly you will always get pure DNA—and its low price. Be sure to store the phenol under buffer (50 mM Tris pH 8.5, 1 mM EDTA, 100 mM NaCl) and check the pH of the phenol before each use. If it falls below 7, then remove the buffer and add new buffer, mix vigorously with the phenol and then allow the phases to separate, and recheck the pH of the phenol. If not above 7.5, then repeat until this pH is reached. If the phenol is acidic, then the nucleic acids will remain in the organic phase rather than in the aqueous phase and will be lost.

2. CTAB extraction (Doyle and Doyle, 1990) is excellent for removing polysaccharides, which are the major components of the cell wall and thus very common in the algae. It is also faster than phenol extraction. The underlying principle is the complexing of the DNA with CTAB, removal of contaminants, and subsequent precipitation of the DNA (see below). In our hands, the DNA will be slightly less stable than DNA from phenolization, but it is nevertheless the method of choice for most applications. It can also be used in case a DNA preparation is contaminated with polysaccharides (see the Section "Nucleic Acid Quality Control and Determination of Concentration," later in this chapter). The yield from a CTAB extraction will exceed that obtained by a phenol extraction.

3. Commercially available DNA purification columns or binding resins for batch methods are based on binding of the DNA to a matrix, typically under high salt conditions. Then, contaminants are washed away from the binding matrix by buffers containing ethanol. The ethanol keeps the DNA from resolving in the washing buffer. Then the DNA is eluted from the binding matrix under low salt conditions and sometimes elevated temperatures. The major advantages of this method are its speed (it takes <1 h) and its potential for standardization. The DNA from these columns is less pure and less stable than from phenolization or CTAB extraction, and the method is much more expensive than the former (i.e., in the range of $2–10 per preparation. We recommend this method for its speed and in cases in which one wants to isolate DNA in a standardized way from many samples (e.g, for time series or when different labs cooperate and want to use exactly the same DNA isolation protocol). A critical step in the preparation is the complete removal of the ethanol-containing washing buffer (e.g., by extended [5–10 min] or repeated [two to three times]

centrifugation steps). In case the ethanol is not completely removed, the yield will decrease, the DNA will slip out of the pocket in agarose gel electrophoresis (see later discussion), and subsequent enzymatic steps can be inhibited.

We do not recommend a certain company for DNA purification columns here because different brands have worked fine in our hands. We suggest contacting a few companies, asking for free samples of columns, and testing them with your samples.

DNA Precipitation

There are a number of reasons why one might want to precipitate DNA (see last section), including concentration, removal of small molecules, such as nucleotides or ions, or long-term storage. When doing a precipitation step, one should be aware that this would cause some loss of DNA ranging up to 30%. Although precipitation of DNA with ethanol or other alcohols is quite trivial and has been used in laboratories for decades now, a number of mistakes are possible, which still can be found in laboratory manuals. The most common mistakes concern the temperature, the time necessary to precipitate DNA, and the procedure of redissolving a DNA pellet. Many protocols suggest cooling the sample, maybe even to $-20°$, and allowing precipitation for prolonged periods or overnight. In our experience, this is even counterproductive because after addition of ethanol–sodium acetate, DNA will instantly precipitate at room temperature and can be collected by centrifugation directly. Any longer incubation will only cause precipitation of substances other than DNA (e.g., salts, polysaccharides, or other contaminants) and will produce a less pure DNA. Another point is the g-force and time necessary for centrifugation. We suggest centrifuging the DNA directly after precipitation for at least 30 min at the highest possible g-force (e.g., $10,000g$) or maximum speed in a small benchtop Eppendorf centrifuge to achieve high yields of DNA. Protocols sometimes ask for only 5–10 min, which certainly is too short. Especially when small amounts of DNA are to be collected, one may centrifuge for 1–2 h. Once the pellet is washed with 70% ethanol, it can be stored in 70% ethanol for extended time periods or may be dried and redissolved. For the latter, it is critical to let the pellet dry until the ethanol is completely evaporated, but not until the pellet has completely dried. Ethanol in the DNA sample inhibits subsequent enzymatic steps and causes a DNA sample to "slip out" of the pocket of an agarose gel (see later discussion). A completely dried DNA sample, especially if it is of high molecular weight or is contaminated with polysaccharides, might need days or even a week to completely redissolve. The best way to dry a DNA pellet is to leave it open under a clean bench and check for ethanol smell. Once the sample no longer smells of ethanol and still looks wet, it is ready for redissolving. Precipitation of the DNA with 1/2 volume of 7 M ammonium acetate instead of 1/10 volume 2 M sodium acetate will bring down smaller pieces of DNA. Thus, this method is an inexpensive way of purifying and concentrating PCR products and is the method of choice if you need to precipitate sequencing reactions. If you have only a small amount of DNA, we recommend the concentration of this DNA using columns with molecular weight cutoffs, such as Microcentricon columns from Qiagen.

Removal of RNA Contamination

Removal of RNA contamination is a critical and often underestimated step in DNA preparation. Because RNA is present in all living cells at much higher concentrations than DNA, and because it behaves similarly to DNA, in many DNA isolation steps, virtually all DNA preparations are contaminated with RNA. This is true despite that RNA, especially mRNA, is much less stable than DNA and that many DNA preparation kits aim at the specific isolation of DNA rather than RNA. The problems RNA contamination in DNA preparations cause are numerous:

1. Separation of DNA in agarose gels can be hampered by RNA resulting in smeary bands.

2. When quantification of DNA is done photometrically (optical density [OD] at 260 nm), RNA interferes in two ways: It leads to an overestimation of DNA content because it absorbs at similar wavelengths. For most subsequent steps, one should use a certain amount of DNA or, for example, similar amounts of DNA from various samples. In case the initial determination of DNA content was wrong, subsequent steps will also produce wrong results. In addition, RNA produces higher OD 260:280 ratios than DNA (typically 2.0–2.2 for good RNA vs. 1.6–1.8 for reasonable DNA) and this ratio is a measure for the quality of the preparation. Also, the OD 260:280 ratio influences the determination of DNA concentration. As a result, an RNA contamination will increase the OD 260:280 ratio, implying a better DNA quality than actually exists, and again will lead to an overestimation of DNA content.

3. Many subsequent steps, such as PCR or restriction enzyme digestion, are severely hampered or even inhibited by RNA contamination (Fig. 2).

The best way to remove RNA is by adding RNase (40 μl of a 10-mg/ml solution) before the DNA preparation, ideally even before the lysis step.

RNase is a very active enzyme that is barely influenced by heat, SDS, or EDTA, and thus, it has enough time to completely degrade the RNA. Essentially all DNA isolation protocols remove proteins so the RNase will also be purified away, as will the nucleotides and small RNA molecules. RNA can also be removed after the DNA preparation by adding RNase (40 μl of a 10-mg/ml solution) and incubation at 50° for 10 min or overnight at 4°. Nucleotides and small RNA fragments then should be removed by precipitation (see the section "DNA Precipitation," earlier in this chapter) or with a commercial column.

Standard RNA Preparation

The amount and the quality of extracted RNA is always strongly dependent on the lysis/disruption of the cells (for methods see earlier discussion). Again, we do have three principle extraction methods. First, there are the phenol-based methods (such as TRIzol), which work well, but because of the toxicity of phenol, we try to avoid them. Second are the silica gel–based membrane and the glass fiber filter-based methods such as RNeasy and RNAqueous from Qiagen and Ambion, respectively. These kits are good for RNA extractions from aqueous environments and especially from plants, because they do not have an alcohol precipitation step, which in many cases leads to co-precipitations of contaminants. Both companies offer PVP-based buffers to bind contaminants, such as polyphenols and polysaccharides. If these kits do not work, for example, with red and brown algae, we recommend as a third option the CTAB method modified according to Chang *et al.* (1993).

TRIzol RNA Isolation Method. For the TRIzol RNA isolation method, you can either buy the TRIzol solution (Gibco BRL, catalogue no. 15596-026) or make it yourself (in 100:38 ml buffered phenol [38%], 12 g of guanidine thiocyanate [0.8 M], 7.6 g of ammonium thiocyanate [0.4 M], sodium acetate pH 5 3.3 ml of 3 M stock [0.1 M], 5 ml glycerol, and fill up to 100 ml with H_2O). Grind 1–2 g of fresh tissue in liquid nitrogen in a mortar. Put the tissue powder into a 50-ml Falcon tube containing 15 ml of TRIzol reagent and incubate samples for 5 min at 30–60°. In some cases, it might be helpful to homogenize the tissue 2–15 s with a homogenizer. Centrifuge the samples above 8000g for 30 min at 4°. Put the supernatant into a new tube and discard the pellet. Add 3 ml of chloroform:isoamyl alcohol to each tube and mix vigorously but do not vortex. Incubate the tubes for 5 min at room temperature and centrifuge them for at least 30 min above 8000g at 4°. Transfer carefully the colorless upper aqueous phase into a new centrifuge tube. Add 0.5 volume of isopropanol and 0.8 M sodium citrate/1.2 M NaCl and mix gently by inversion. Incubate for

5–10 min at room temperature and centrifuge for 30 min at 8000g at room temperature. After discarding the supernatant, wash the RNA pellet two times with 75% ethanol, vortex briefly, and centrifuge for 10 min at room temperature. Discard the supernatant and let the RNA pellet dry no longer than 10 min to remove residual ethanol, but do not completely dry the pellet because this will make resuspension more difficult. Resuspend the pellet in 100–250 μl of RNase-free H_2O by pipetting up and down a few times. Heat the RNA solution for 10 min at 55–60° (note: you can add 1 μl of an RNase inhibitor before the incubation) to fully resuspend the pellet. Transfer samples into a 1.5-ml reaction tube (Eppendorf) and centrifuge for 5–10 min at 8000g to pellet the non-resuspended material. Put the supernatant into a new tube and measure an aliquot with a spectrophotometer (see earlier discussion).

CTAB Method (Chang et al., *1993).* The CTAB method is a simple and effective method for isolation of RNA from samples containing high amounts polyphenolic and polysaccharide contaminants and is, therefore, recommended for problematic organisms, such as dinoflagellates and red and brown algae, when high-quality RNA is needed (cDNA library construction, etc.). Warm 15 ml of EB (2% CTAB, 2% PVP K30, 100 mM Tris–HCl [pH 8]), 25 mM EDTA, 2 M NaCl, 0.5 g/liter spermidine [mix and autoclave] just before use, add 2% β-mercaptoethanol). One or two grams of fresh-weight sample should be ground with a mortar and pestle to powder under liquid nitrogen and put into 50-ml Falcon tubes, quickly supplied with 15 ml of EB, and mixed by inverting and vortexing. Extract two times with an equal volume of chloroform:isoamyl alcohol and separate the phases by centrifuging at 8000g for 30 min at room temperature each time. Carefully combine the colorless upper aqueous phase into a new centrifuge tube and add a quarter volume of 10 M lithium chloride (LiCl) (~6 ml, final concentration of 2 M) to the supernatant and mix. The RNA will precipitate overnight at 4° and can be harvested by centrifugation at 8000g, 4° for 30 min. Discard the supernatant carefully and resuspend the pellet with 500 μl SSTE buffer (1 M NaCl, 0.5% SDS, 10 mM Tris–HCl, pH 8, 1 mM EDTA, pH 8) and pipette the sample into a 2-ml Eppendorf reaction vial. Extract the sample with 500 μl of chloroform:isoamyl alcohol mix by inverting and carefully vortexing and spin the samples by 10,000g for 15 min. Again, carefully transfer the upper aqueous phase into a new Eppendorf vial and add 2 volumes (~1 ml) of 96–100% ethanol for alcohol precipitation. After mixing the sample, incubate at −20° for at least 2 h and spin the sample 30 min in a microcentrifuge at room temperature to pellet the RNA. Wash pellet twice with 75% ethanol. Dry the pellet and resuspend in 100–200 μl of buffered RNase-free H_2O (see earlier discussion).

RNeasy (Silica-Gel Membrane). Kits such as RNeasy (Qiagen) and RNAqueous (Ambion) are particularly practical if many samples have to be processed. In most cases, theses kits work well, and they offer standardization of processing even in a 96-well format. In general, for RNeasy, the binding capacity of the minicolumns is advertised to be 100 μg of RNA, but in the case of our type of samples, 20–50 μg seems more reasonable. We generally follow the introductions of the manufacturer's protocol, but we included some slight modifications. It is critical not to overload the buffer capacity of the lysis buffer and the columns, because this will lead to a lower yield and less pure RNA. Therefore, we strongly recommend not using more than 30 mg of sample material for an extraction. β-Mercaptoethanol has to be added to the lysis buffer (RLT buffer) 10 μl/ml prior to use. Dissolve the sample pellet with 500 μl of RLT buffer and add 1/3 volume glass beads into the vial. Bead beat the sample in a bead beater until the sample material is completely disrupted. Check the lysis carefully because too vigorous beating can disrupt the cell walls, releasing polysaccharides that will contaminate the extracted RNA. Transfer the sample solution onto a QIAshredder, wash the glass beads with 250 μl RLT buffer, and add that to the material in the QIAshredder column. Spin at maximum speed for 15 min. This step will remove cell debris, homogenize the lysate, and shear the genomic DNA. Carefully transfer the filtrate to a new 2-ml Eppendorf vial without disturbing the pellet. Add 0.5 volume ($\sim$300 μl) 96–100% ethanol to the sample and mix immediately by pipetting. Transfer the samples to an RNeasy minicolumn and centrifuge twice. Wash the column-bound RNA once with 700 μl of RW1 buffer. We recommend using the RNase Free DNase set (Qiagen) for DNA removal and following the manufacturer's instructions. Wash a second time with 700 μl of RW1 buffer and then follow the protocol until you elute the RNA with 40 μl of RNase-free H_2O.

Poly A+ RNA Isolation Method. We prefer a two-step protocol to a direct poly A+ extraction, because in our hands, we obtain higher yields and better quality. We routinely use two kits (Oligotex mRNA kit [Qiagen] and the Poly(A)Purist mRNA purification kit [Ambion]), which work well with the manufacturer's protocol.

Removal of DNA Contamination

DNA contamination can be removed from RNA preparations with RNase-free DNase or an additional LiCl precipitation step (add 1 volume of RNase-free 4 *M* LiCl, incubate over night at 4°, etc.; see earlier discussion). The enzymatic removal is more expensive, and one has to make sure that the enzyme truly is RNase free (e.g., by testing with a less valuable

RNA). In case your RNA preparation contains a small RNase contamination, the procedure may degrade your RNA even in the absence of RNase in the DNase. You should test this with a small aliquot of your RNA. The LiCl purification will cause some loss of RNA (10–30%).

Nucleic Acid Quality Control and Determination of Concentration

It is quite common to just measure the optical density of a nucleic acid preparation at 230/260/280 nm and to calculate the amount and purity from these values. Whereas nucleic acids absorb maximally at about 260 nm. the values for 230 and 280 nm indicate the contamination with polysaccharides and proteins, respectively. A measure for the purity of nucleic acids will be the ratio between 260:280 nm, which should be between 1.6 and 1.8 for DNA and 1.8 and 2.1 for RNA, respectively. The ratio between 260 nm (nucleic acids) and 230 (polysaccharides) should be at least more than 2. We strongly suggest not to rely on these numbers in cases in which DNA/RNA concentration and quality really matter for subsequent steps (e.g., in case large fragments have to be amplified or equal amounts of DNA/RNA from different samples have to be used). For example, if a DNA preparation is contaminated with RNA (see earlier discussion), the absorption at 260 nm and the absorption ratio 260/280 nm, respectively, will increase and the actual concentration and purity will be overestimated. Another problem is that small DNA or RNA fragments will produce the same kind of optical density as large molecules so a degraded preparation will look fine as judged by its optical density. However, most genes, especially long amplicons, cannot be amplified from degraded DNA. A degraded RNA will be even worse for subsequent use. So, in case the concentration and quality of a DNA or RNA preparation is important, we suggest to always check it on an agarose gel (see later discussion). A good DNA should produce a sharp band of equal size or larger than the 23-kb band of commonly used lambda-size markers and no smear should be visible. If RNA bands are visible, typically more diffuse than the DNA band, and in the range of 1–3 kb, there is probably heavy RNA contamination. That is because DNA is stained much more efficiently than RNA because of its double-stranded nature. Staining of the pockets of the gel indicates contamination with polysaccharides. Both contamination should be removed (see "Standard DNA Extractions" and "Removal of RNA Contamination" sections, respectively). A smear usually indicates a contamination with DNA degrading enzymes. Such DNA will likely be difficult to amplify by PCR, and the amount of DNA should be increased because many small fragments are present. Any protein contamination can be removed by phenolization. Because of the sensitivity of RNA to degradation because of the denaturation-resistant RNases, it is strongly recommended to check the quality and integrity of RNA as described for DNA. RNA is single stranded and frequently has secondary structure; so a denaturing formaldehyde gel is the best way to check the RNA for its integrity. A typical RNA preparation will produce a smear between about 7 and 0.1 kb with about two to six prominent reasonably sharp bands representing the ribosomal RNAs. Sometimes additional weaker bands stemming from abundant mRNAs may occur. An additional diffuse massive band will be present at the bottom of the gel caused by tRNAs (Fig. 3). In general, an RNA gel looks more smeary than a DNA gel, so the ladder is less sharp than a DNA ladder. Degradation of RNA is indicated by a heavy smear with barely visible rRNA bands and no smear above 3 kb. Such an RNA preparation should be discarded. If a sharp high-molecular-weight band (above the 9-kb band of commercial RNA ladders) is visible on the RNA gel, the preparation is likely contaminated with DNA. Such contamination has to be removed for subsequent steps, such as RT-PCR or quantitative RT-PCR (QRT-PCR), because the DNA molecules will serve as a template in the PCR and will lead to an overestimation of the expression level (see earlier discussion for removal of DNA contaminations).

Gel Electrophoresis. Gel electrophoresis is a very basic method to analyze nucleic acid preparations (i.e., the separation of nucleic acid molecules of different sizes by an electric field in a gel). Two gel types are commonly used: agarose and polyacrylamide gels. Polyacrylamide gels can

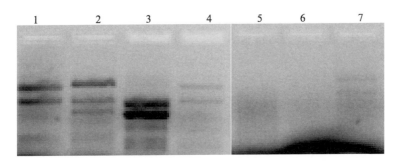

Fig. 3. Typical RNA preparations. Lanes 1–4 show RNA preparations (250–500 ng total RNA ea.) from various species: column 1, *Alexandrium ostenfeldii*; column 2, *Prymnesium parvum*; column 3, *Chrysochromulina polylepis*; and column 4, *Alexandrium tamarense*. This demonstrates that there is no "typical" band pattern, but that every new species may differ from others in that its rRNAs migrate differently and different numbers of rRNA bands may be visible. To give a negative example, lanes 5–7 are progressively degraded RNA preparations, showing smeary bands from *Prorocentrum lima*, which should not be used for experiments.

separate DNA molecules between about 1 and 1000 base pairs (bp) and agarose gels nucleic acids between 20 and 1000 bp (nusieve agarose), between 100 and 50 kbp (standard agarose), or up to millions of bp when using pulsed fields and different angles for the electric field, respectively. For the purpose of this chapter, we describe agarose gel electrophoresis not using pulsed fields because this is by far the easiest method to apply. Polyacrylamide gels are limited to small molecules, separation takes much longer, staining is more tedious, and the monomer (acrylamide) is highly toxic. Pulsed field electrophoresis, though powerful, is a highly sophisticated method and requires expensive equipment.

A typical agarose gel can be poured and will be ready for use within minutes (see textbooks for details). Apparatuses are available in many sizes from microgels of only a few centimeters wide with a few slots to large ones that can separate up to 100 samples at a time. To analyze a DNA preparation from an environmental sample, we suggest using the smallest possible gel (e.g., 4 by 6 cm to 6 to 10 cm) with small slots (e.g., slot volume 2–10 μl). The gel should be run only for a few minutes (e.g., until the bromophenol blue band has moved 1–2 cm). The stain, for example, ethidium bromide, should be poured directly into the gel (e.g., 0.1 μg/ml of gel). By doing so, one will be able detect about 2 ng of DNA as a faint band and the run will take only a few minutes at 10 V/cm gel size, leading to the lowest possible loss of DNA. Quantification can be done by comparing the band intensities to those of the DNA ladder, which have a known concentration. Most gel documentation systems come with a software for this. One can also load different amounts of their DNA and do a comparison by eye. As a rule of thumb, the amount of DNA that produces a clear band in an average size agarose gel ($\sim$6 by 10 cm, slot volume $\sim$10 μl) is in the range of 50 ng; at about 500 ng, the band begins to smear.

The following protocol for formaldehyde–agarose (FA) gel electrophoresis is routinely used in our group and is based on the Sambrook protocol (Sambrook et al., 1989). For 120 ml of a 1% FA gel, add 1.2 g of agarose into 81.5 ml of RNase-free water (see earlier discussion) and heat the mixture to melt the agarose in a microwave oven. Cool the solution to approximately 50–60° and add 22.1 ml of 37% (12.3 M) formaldehyde and 12 ml 10X MOPS (3-[N-morpholino]propanesulfonic acid). Mix thoroughly and pour onto gel support. Before running the gel, equilibrate in 1X MOPS gel running buffer for at least 15 min. The RNA sample preparation for gel electrophoresis differs from the one for DNA because RNA will be loaded onto the gel denatured. For an integrity test, 100–500 ng should be enough, but do not use more than 10 μg because of overloading problems. The RNA/H_2O mix should be 10 μl, and added to the RNA is 1.8 μl of 37% formaldehyde, 0.75 μl of 10X MOPS, 0.5 μl of ethidium bromide

(40 μg/ml), and 1.5 μl of RNA stop buffer (saturated aqueous bromophenol blue solution, formamide, and glycerol). Incubate for 15 min at 56°, chill on ice, and load the samples onto the equilibrated gel (3–4 V/cm^2 gel).

Storage of Nucleic Acids

Storage of DNA is not very difficult because it is a rather stable molecule, but some care should be taken for long-term storage, particularly if RNA is to be stored. A pure DNA preparation will be stable at room temperature for hours or days, in the refrigerator for days or weeks, at −20° for months, at −80° for years, and precipitated in ethanol at room temperature for decades. DNA can degrade upon frequent freeze–thaw cycles (i.e., >10–20). Therefore, we suggest freezing large amounts of DNA in small batches and only let it thaw on ice. The best way to ship DNA is precipitated in ethanol or frozen on dry ice. RNA should be stored at −80° in RNase-free H$_2$O, which should be buffered with TE (0.1 mM Tris, 0.1 mM EDTA, pH 7–8), and it should be stored undiluted, if possible, in aliquots. EDTA should be added in small amounts because MgCl$_2$ and some other metal ions can catalyze nonspecific breaks in the nucleic acids.

Subsequent Steps (PCR, Restriction Enzyme Digestion, Hybridization)

By far the most common use of DNA from environmental samples will be for PCR. Fortunately, for this application, the quality requirements are low as compared with other ones (see later discussion), and sometimes even boiled samples, as mentioned earlier, will produce good PCR templates. Numerous textbooks are available on PCR, so we will give only a few hints on how to perform subsequent PCR steps. A few important issues to consider when running a PCR with environmental DNA include (1) the size of the amplification product, (2) the possible presence of secondary structures, (3) the copy number of the gene of interest in the target organisms, (4) the amount of DNA available, and (5) the possibility of generating PCR chimeras of two different organisms. The first and second factors concern the length of the extension and denaturing cycles, respectively. As a rule of thumb, one may use an extension cycle of 1 min/kb of the amplicon, but not shorter than 30 s. For very long amplicons (i.e., >3 kb), the extension time should be doubled. The same is true for amplicons with a complex secondary structure, such as those coding for rRNAs or tRNAs. The denaturing cycle can be 10–30 s for normal templates and 30–60 s for ribosomal or tRNAs. The third factor (copy number) influences the amount of DNA necessary for successful amplification within 30–35 cycles. For eukaryotic DNA and single copy genes, this is in the range of 100 ng, but for multicopy genes such as the 18S rDNA, as low as 1–10 ng may be sufficient (values are for a standard 50 μl PCR, but this volume can be decreased to 20 μl). If only small amounts of DNA are available (factor 4), one may increase the number of cycles up to 45–50. With such a high number of PCR cycles, there is, however, an increasing danger of amplifying false positives, which will generate an amplification product in the negative control. Factor 5 can be checked at the ribosomal database project *http://rdp.cme.msu.edu/html/* with their option of "check chimera" program, and sequences that are chimeras of two organisms may then be eliminated from further analysis.

For hybridization experiments of specific labeled probes against environmental DNA spotted onto a slide or a membrane, one should use amounts between 100 and 1000 ng per spot, depending on whether single or multicopy genes are used as a target. For this application, DNA needs to be purified and should not contain too much protein or RNA. Just boiling the sample will not produce good templates.

Even higher amounts of DNA are required for restriction enzyme digestion and subsequent Southern blotting (e.g., RFLP analyses). Here, at least 500 ng is necessary for single digests, and the digests should be done with a large amount of enzyme (e.g., 10–50 units) overnight. A test should be performed by incubating a small amount of the DNA (50 ng) in restriction enzyme buffer at 37° over night without an enzyme. This should not produce a smear on an agarose gel. This prevents the danger of DNA degradation caused by DNase contamination. One should not be confused by the fact that the restriction enzyme digestion will likely also produce a smear on the gel. This is because a high-molecular-weight DNA will produce thousands of different bands with common restriction enzymes, which will appear as a smear on the gel. Obviously for this application, DNA needs to be of very high purity, because even low contaminations with DNase will lead to complete degradation, small contaminations with RNA, phenols, or salt can inhibit restriction enzyme digestion (Fig. 2), and contamination with polysaccharides can inhibit proper separation in an agarose gel. The amounts of DNA necessary for a certain application discussed in this chapter are important for the amount of sample initially taken (see "Sampling" Section).

References

Chang, S., Puryear, J., and Crainey, J. (1993). A simple an efficient method for isolating RNA from pine trees. *Plant Mol. Biol. Rep.* **11**(2), 113–116.
Doyle, J. J., and Doyle, J. L. (1990). Isolation of plant DNA from fresh tissue. *Focus* **12**, 13–15.
Groben, R., and Medlin, L. K. (2005). Molecular tools and approaches in eukaryotic microbial ecology: *In situ* hybridisation of phytoplankton using fluorescently-labelled rRNA probes.

In "Molecular Microbial Ecology" (A. M. Osborn, ed.). Bios Scientific, Oxford, United Kingdom.

Holzmann, M., and Pawlowski, J. (1996). Preservation of Foraminifera for DNA extraction and PCR amplification. *J. Forman. Res.* **26**, 264–267.

John, U., Groben, R., Beszteri, B., and Medlin, L. K. (2004). Utility of amplified fragment length polymorphisms (AFLP) to analyse genetic structures within the *Alexandrium tamarense* species complex. *Protist* **155**(2), 169–179.

Lange, M., Chen, Y.-Q., and Medlin, L. K. (2002). Molecular genetic delineation of *Phaeocystis* species (Prymnesiophyceae) using coding and non-coding regions of nuclear and plastid genomes. *Eur. J. Phycol.* **37**, 77–92.

Medlin, L., Elwood, H. J., Stickel, S., and Sogin, M. L. (1988). The characterization of enzymatically amplified eukaryotic 16S-like rRNA coding regions. *Gene* **71**, 491–499.

Medlin, L. K., Elwood, H. J., Stickel, S., and Sogin, M. L. (1991). Genetic and morphological variation within the diatom *Skeletonema costatum* (Bacillariophyta): Evidence for a new species, *Skeletonema pseudocostatum*. *J. Phycol.* **27**, 514–524.

Muyzer, G., De Waal, E. C., and Uittrtlinden, A. G. (1993). Profiling of complex microbial populations by denaturing gradient gel electrophoresis analysis of polymerase chain reaction–amplified genes coding of the 16S rRNA. *Appl. Environ. Microbiol.* **59**, 695–700.

Reysenbach, A. L., Giver, L. J., Wickham, G. S., and Pace, N. R. (1992). Differential amplification of rRNA genes by polymerase chain reaction. *Appl. Environ. Microbiol.* **58**, 3417–3418.

Sambrook, J., Fritsch, E. F., and Maniatis, T., eds. (1989). "Molecular Cloning: A Laboratory Manual," 4th Ed. Cold Spring Harbor Laboratory Press, Cold Spring Harbor, NY.

Schwieger, F., and Tebbe, C. C. (1997). Efficient and accurate PCR amplification and detection of a recombinant gene in DNA directly extracted from soil using the Exand High Fidelity PCR system and T4 gene 32 protein. In A. Boehringer Manheim (Hrsg.) Biochemica Information. Sonderheft: PCR Bibliographie. (ISSN 0942-556X). *Kapitel* **3**, (PCR-Highlights), S73–S75.

Schwieger, F., and Tebbe, C. C. (1998). A new approach to utilize PCR–single-strand-conformation polymorphism for 16S rRNA gene-based microbial community analysis. *Appl. Environ. Microbiol.* **64**, 4870–4876.

Tebbe, C. C., and Vahjen, D. (1993). Interference of humic acids and DNA extracted directly from soil in detection and transformation of recombinant DNA from bacteria and a yeast. *Appl. Environ. Microbiol.* **59**, 2657–2665.

Taylor, J. W., and Swann, E. (1994). DNA from herbarium specimens. *In* "Ancient DNA" (B. Herrmann and S. Hummel, eds.), pp. 166–181. Springer-Verlag, New York.

Töbe, K., Ferguson, C., Kelly, M., Gallacher, S., and Medlin, L. K. (2001). Seasonal occurrence at a Scottish PSP monitoring site of purportedly toxic bacteria originally isolated from the toxic dinoflagellate genus *Alexandrium*. *Eur. J. Phycol.* **36**, 243–256.

Vos, P., Hogers, R., Bleeker, M., Reijans, M., Vandelee, T., Hornes, M., Frijters, A., Pot, J., Peleman, J., Kuiper, M., and Zabeau, M. (1995). AFLP—a new technique for DNA fingerprinting. *Nucl. Acids Res.* **23**, 4407–4414.

Yue, G. H., and Orban, L. (2001). Rapid isolation of DNA from fresh and preserved fish scales for polymerase chain reactions. *Mar. Biotech.* **3**, 199–204.

[3] Nucleic Acid Isolation from Ecological Samples—Vertebrate Gut Flora

By LISE NORDGÅRD, TERJE TRAAVIK, and KAARE M. NIELSEN

Abstract

The utility of DNA molecules in identifying and characterizing intestinal microorganisms depends on methods that facilitate access to DNA of sufficient purity, quantity, and integrity. An efficient and unbiased extraction of DNA is thus critical to the validity of the subsequent analysis of the prevalence and diversity of the DNA sources in the sample. The highly heterogeneous composition of the diet of vertebrates makes DNA isolation challenging for this environment. Here, we consider the key steps involved in DNA isolation from vertebrate gut microflora including sample homogenization, lysis of bacterial cells, and extraction and precipitation of DNA. A detailed protocol for DNA isolation of the microbial contents of intestine and feces is also provided. In addition, we refer to commercially available methods for DNA extraction from the vertebrate gut flora.

Introduction

The gastrointestinal (GI) tract (gut) of vertebrates harbors complex microbial communities that provide essential functions for the host, including food digestion, breakdown of toxic substances, production of essential vitamins, and prevention of gut colonization by microbial pathogens (Mackie *et al.*, 1999; Wang *et al.*, 1996, 2002).

Traditional methods for identifying and characterizing bacteria in the gut flora are based on various cultivation techniques (e.g., selective media), biochemical tests (e.g., enzymatic assays, Biolog plates), morphological examination (e.g., microscopy), and analysis of fatty acid production (e.g., fatty acid methyl ester [FAME] and phospholipid fatty acid [PFLA]) (Tannock, 1999; Wang *et al.*, 1996; Zoetendal *et al.*, 2004). More recently, DNA-based methods for identifying and characterizing intestinal microorganisms have been adopted. These methods include 16S ribosomal DNA (rDNA)–targeted polymerase chain reaction (PCR) amplification and sequencing, and PCR-based community profiling techniques such as denaturing gradient gel electrophoresis (DGGE), temperature gradient gel electrophoresis (TGGE), and terminal restriction fragment length polymorphism. (T-RFLP) (Gong *et al.*, 2002; Mackie *et al.*, 1999; Matsuki *et al.*,

METHODS IN ENZYMOLOGY, VOL. 395   0076-6879/05 $35.00

2002; Simpson *et al.*, 1999; Withford *et al.*, 1998; Zhu *et al.*, 2002). DNA-based methods require that DNA can be extracted from the GI tract in sufficient quantity, free from excess fragmentation and contaminants, and in a way that is representative of the true species distribution and abundance in the specific environment. Incomplete bacterial cell lysis, DNA sorption to particulate material, enzymatic or physical degradation, and chemical modifications are some common problems that can negatively affect the efficiency of DNA extraction (Monterio *et al.*, 1997; Santini *et al.*, 2001; Zoetendal *et al.*, 2001). The variable composition of the diet of vertebrates makes nucleic acids isolation challenging for this heterogeneous environment and excludes the adoption of a single protocol (Blaut *et al.*, 2002; Mackie *et al.*, 1999).

Several methods for the extraction of DNA from the vertebrate gut have been described (Anderson and Lebepe-Mazur, 2003; Avgustin *et al.*, 1994; Bruce *et al.*, 1992; Li *et al.*, 2003; Pryde *et al.*, 1999; Santini *et al.*, 2001; Simpson *et al.*, 1999; Stahl *et al.*, 1988; Tsai and Olsen, 1991; Wilson *et al.*, 1996; Wood *et al.*, 1998; Zhu *et al.*, 2002). In addition, a number of commercial kits are available (Table I).

Here, we summarize the various steps involved in DNA isolation from vertebrate gut microflora and provide a detailed protocol for DNA isolation of intestinal contents and feces. The extraction and purification procedures described in this chapter have, with modifications, been shown to be suitable for identification of microorganisms in the gut flora of different animals, like chickens, pigs, and cattle (Anderson and Lebepe-Mazur, 2003; Gong *et al.*, 2002; Li *et al.*, 2003; Withford *et al.*, 1998).

TABLE I

SOME COMMERCIAL KITS USED FOR NUCLEIC ACID ISOLATION OF VERTEBRATE GUT SAMPLES

Kit	Supplier	References
FastDNA Kit	Bio101 (Carlsbad, CA)	McOrist *et al.*, 2002
IsoQuick	Orca Research (Bothell, WA)	Holland *et al.*, 2000
Nucleospin + T Kit	Macherey-Nagel (Germany)	McOrist *et al.*, 2002
NucliSens Isolation Kit	Organon Teknika (The Netherlands)	Holland *et al.*, 2000; Mahony *et al.*, 2001
QIAamp DNA stool minikit	Qiagen (Germany)	Heritage *et al.*, 2001; Higgins *et al.*, 2001; Li *et al.* 2003; Vandenberg *et al.*, 2002
The Wizard SV Genomic DNA Purification System	Promega (Madison, WI)	Liang *et al.*, 2003
UltraClean Soil DNA extraction kit	Mo Bio Laboratories (Solona Beach, CA)	Clement and Kitts, 2000; Higgins *et al.*, 2001

Steps in DNA Isolation

The basic steps in the extraction of nucleic acids from vertebrate gut flora for the purpose of molecular analysis are summarized in Fig. 1. DNA isolation protocols found in the literature provide alternative approaches and combinations of these steps. Below, the steps are considered in more detail to allow system-specific optimization of the DNA isolation procedure.

Preparation of Sample before DNA Extraction

Samples from different compartments of the gut vary in composition, consistency, amount of material present, and bacterial density and diversity in most animals (Apajalahti *et al.*, 1998; Blaut *et al.*, 2002; Mackie *et al.*, 1999). In general, the stomach and proximal small intestine contain few microorganisms (0–10^4 g^{-1}), and those that are present are aerobes and facultative anaerobes. The content of the colon is characterized by large populations (10^{10}–10^{12} g^{-1}) and consists predominantly of anaerobes (Mackie *et al.*, 1999). The composition of the starting material will determine the yield and quality of the isolated DNA (Anderson and Lebepe-Mazur, 2003; Holland *et al.*, 2000; Liang and Redlinger, 2003). Optimally, freshly sampled material should be used because even short periods of storage or air exposure may alter the microbial composition of the sample. Alternatively, the sample can be frozen in liquid nitrogen and stored at $-70°$ (Withford *et al.*, 1998). Repeated thawing cycles should be avoided to prevent microbial activity in the stored sample. Depending on the consistency of the sample, it may be recommended to remove large particles to enhance the subsequent extraction of DNA. This can be achieved by homogenizing and suspending the sample material in a liquid solution and then removing the larger food particles by low-speed centrifugation (Apajalahti *et al.*, 1998; Li *et al.*, 2003).

Lysis of Bacterial Cells

The next step in the nucleic acid isolation procedure is lysis of the bacterial cells. An efficient lysis is essential because both the yield of the DNA and the representativeness of the DNA sample depend on the composition of the cell lysate. Inefficient or species-dependent lysis will introduce methodological bias when the species composition is examined through total DNA analysis.

Lysis may be achieved by rapid freeze–thaw cycles, chemical, mechanical, or enzymatic procedures. Some protocols use a combination of the different methods (Li *et al.*, 2003; Simpson *et al.*, 1999; Tsai and Olsen, 1991;

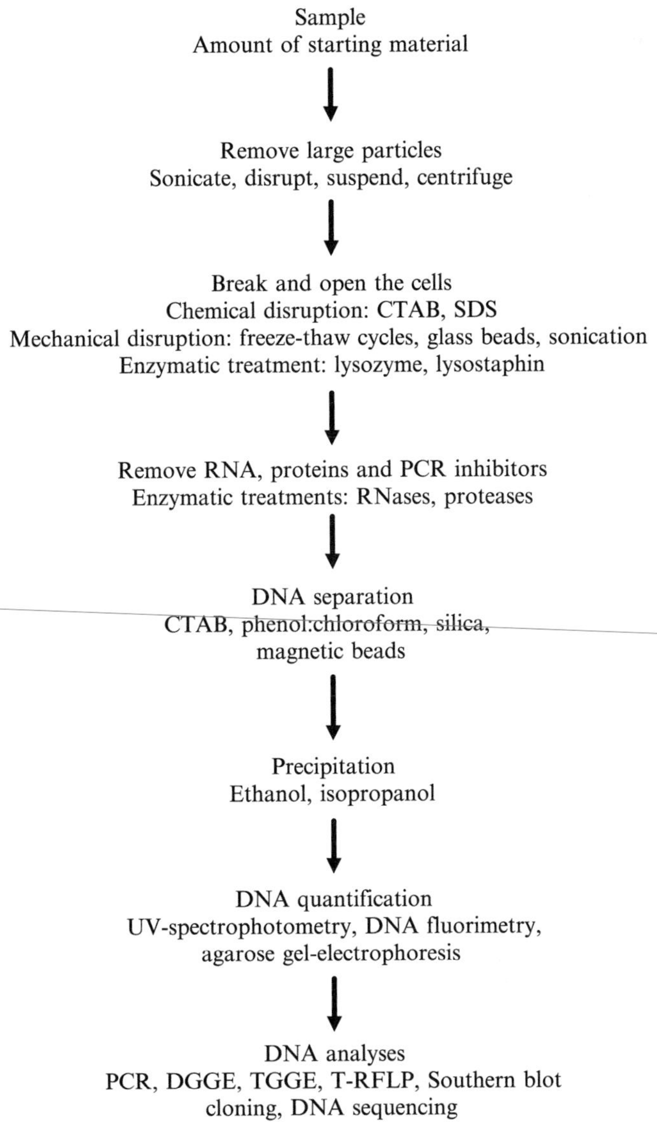

Sample
Amount of starting material

Remove large particles
Sonicate, disrupt, suspend, centrifuge

Break and open the cells
Chemical disruption: CTAB, SDS
Mechanical disruption: freeze-thaw cycles, glass beads, sonication
Enzymatic treatment: lysozyme, lysostaphin

Remove RNA, proteins and PCR inhibitors
Enzymatic treatments: RNases, proteases

DNA separation
CTAB, phenol:chloroform, silica,
magnetic beads

Precipitation
Ethanol, isopropanol

DNA quantification
UV-spectrophotometry, DNA fluorimetry,
agarose gel-electrophoresis

DNA analyses
PCR, DGGE, TGGE, T-RFLP, Southern blot
cloning, DNA sequencing

Fig. 1. Different steps involved in extraction and analysis of microbial DNA from vertebrate gut flora.

Withford *et al.*, 1998). A common approach is the use of a lysis buffer that contains a detergent (e.g., sodium dodecyl sulfate [SDS]) that disrupts cellular membranes and a protease (e.g., proteinase K) for digestion of cellular and intracellular proteins. Other enzymes such as lysozyme is also used for the breakdown of the cell wall, for example, from gram-positive bacteria (Apajalahti *et al.*, 1998; Wang *et al.*, 2002; Zhu *et al.*, 2002). Another common method uses physical disruption through bead beating, in which the sample material is disrupted at high speed in the presence of glass beads (Stahl *et al.*, 1988). The freeze–thaw method is also commonly used to lyse bacterial cells. The technique involves freezing a cell suspension in a dry ice–ethanol bath or freezer and then thawing the material at a higher temperature. This method of lysis causes cells to swell and ultimately break as ice crystals form during the freezing process and then contract during thawing. Multiple cycles are necessary for efficient lysis (Li *et al.*, 2003).

Nucleic Acid Extraction

After cell lysis, it is necessary to remove RNA, cell debris, various proteins including DNases, and other compounds that may inhibit subsequent molecular analysis. The enzyme ribonuclease (RNase) is added to break down messenger RNA (mRNA) and ribosomal RNA (rRNA). The standard way to remove proteins from nucleic acid solutions is by extraction with phenol and/or chloroform, because phenol denatures proteins efficiently. Cell lysates are extracted by the addition of 1 volume of phenol (pH 8.0) to the sample, followed by gentle mixing to form an emulsion. After centrifugation, the aqueous (top) phase containing the nucleic acids is collected with a wide-bore pipette tip and transferred to a new tube. Residual phenol is removed from the nucleic acid preparation by extraction with chloroform (Stahl *et al.*, 1988; Wood *et al.*, 1998).

Alternative DNA extraction methods based on co-precipitants such as cetyltrimethylammonium bromide (CTAB) and polyvinylpolypyrrolidone (PVPP) may be desirable to avoid phenol exposure. As with phenol extraction, denatured proteins and polysaccharides complexed with CTAB and/or PVPP are removed from the DNA-containing liquid phase by precipitation and centrifugation (Griffiths *et al.*, 2000; Picard *et al.*, 1992; Steffan *et al.*, 1988).

Other methods developed involve binding of released DNA in cell lysates onto a solid matrix made of silica. Proteins that are not adsorbed to the matrix are removed by washing. Commercially available products such as the Nucleospin + T kit (Macherey-Nagel, Germany) and the QIAamp DNA stool minikit (Qiagen, Germany) use silica matrices for nucleic acid isolation.

Nucleic Acid Precipitation

Nucleic acids are recovered from aqueous solution by precipitation using either 2 volumes of 100% ice-cold ethanol or 1 volume of isopropanol. The solution should be incubated on ice (ethanol) or at room temperature (isopropanol) to allow the DNA to precipitate. After an incubation period of 10 min to 2 h the nucleic acids are recovered by centrifugation. The pellet containing the nucleic acids is washed with 70% ethanol to remove salts and residual chemicals such as SDS, ethylenediaminetetraacetic acid [EDTA], chloroform, and phenol. The DNA pellet is briefly air-dried to allow all traces of alcohol to evaporate before suspension of the DNA in an appropriate buffer such as TE buffer (10 mM Tris–Cl, 5 mM EDTA, pH 8) or sterile water (Sambrook and Russell, 2001). DNA samples should be stored buffered at $-20°$ to prevent self-degradation.

DNA Yield Determination

The DNA concentration and purity in an aqueous solution is routinely measured with a ultraviolet (UV) spectrophotometer. The concentration of DNA is determined by measuring the absorbance at 260 nm in a 1-cm light-path length quartz cuvette. Usually, 100 μl of a 1/20 dilution of the DNA solution is measured. An abs$_{260\,nm}$ of 1.0 corresponds to about 50 μg of DNA/ml for double-stranded DNA. Absorbance readings should fall between 0.1 and 1.0 to be accurate. Sample dilution should be adjusted accordingly. The ratio of the absorbance at 260 nm and 280 nm (A_{260}:A_{280}) is often used as a measure for contamination of the DNA solution with RNA and protein. Pure DNA has an A_{260}:A_{280} ratio between 1.8 and 2.0 (Sambrook and Russell, 2001). Ratios lower than 1.8 are usually indicative of protein contamination. Additional phenol extraction steps or kit applications may be necessary to remove contaminating proteins.

Protocol for Bacterial DNA Extraction from Vertebrate Gut Flora[1]

Preparation of Sample before DNA Extraction

1. Suspend 1–10 g of sample (depending on bacterial density) in 30 or 70 ml of sterile saline (0.85% NaCl) containing 0.1% Tween 80.

2. Centrifuge at low speed (200g) for 5 min at 4° to remove larger particulate matter. Collect the upper phase, transfer the supernatant to a clean sterile tube, and keep on ice. Save the pellet.

[1] Li *et al.*, 2003; Stahl *et al.*, 1988; Wilson *et al.*, 1996; Withford *et al.*, 1998; Wood *et al.*, 1998; Zhu *et al.*, 2002.

3. Resuspend the pellet in saline containing 0.1% Tween 80 and repeat centrifugation.

4. Repeat step 3 three times and pool the supernatants. Pellets may now be discarded.

5. Centrifuge the pooled supernatants at 9000g for 15 min at 4° to pellet the bacterial cells. Discard supernatant.

6. Store the pellet at −70° until further processing.

Disruption of Cells by Freeze–Thaw Cycles and Bead Beating

7. Add 1.2 ml of TE buffer (10 mM Tris–Cl and 5 mM EDTA, pH 8) containing β-mercaptoethanol[2] (5 μl/ml) to the bacterial cells and resuspend the pellet thoroughly.

8. Freeze and thaw the cells, alternating between 5-min immersion in liquid nitrogen and 5-min immersion in a 65° water bath, for five cycles.

9. Centrifuge the sample at 12,000g for 10 min at 4° to pellet unbroken cells, and then collect the cell lysate.

10. Suspend the unbroken cells in 0.5 ml of lysis buffer (0.2 M NaOH, 2 mg/ml lysozyme and 1% SDS[3]), and transfer the solution into a 2-ml screw-capped microcentrifuge tube containing 0.6 ml of phenol, 0.1% SDS, and 0.5 g each of 0.1- and 0.5-mm zirconium beads (BioSpec Products, Bartlesville, OK). Perform bead beating twice for 2 min using the Mini-Beadbeater-8 Cell Disrupter (BioSpec Products, Bartsville, OK). Keep the sample at room temperature for 5 min between each bead beating.

11. Combine the cell lysate from the freeze–thaw cycles and the bead-beating treatment in a centrifuge tube.

Nucleic Acid Isolation by Phenol: Chloroform:Isoamyl Alcohol Extraction

12. Add 1 volume of phenol:chloroform:isoamyl alcohol[4] (25:24:1), gently mix by inverting the tubes, and centrifuge for 6 min at 12,000g.

13. Transfer the upper aqueous layer to a new centrifuge tube and repeat step 12 three times. Avoid collecting the white protein-rich precipitate that is present at the interphase.

14. Transfer the upper aqueous layer to a new tube.

---

[2] β-mercaptoethanol is toxic. Wear appropriate gloves and safety glasses and work in a fume hood.

[3] SDS is toxic. Wear gloves and do not breathe dust.

[4] Phenol is toxic and highly corrosive and can cause severe burns. Chloroform is carcinogenic. Isoamyl alcohol may be harmful by inhalation or skin absorption and presents a risk of serious damage to eyes. Wear gloves and safety goggles and work in a fume hood.

Precipitation of Nucleic Acids by Ethanol

15. Add 2 volumes of ice-cold 96% ethanol, invert the tubes, and incubate for 1 h at −70°, or overnight at −20°.

16. Centrifuge the samples at 12,000g for 30 min at 4°.

17. Discard the supernatant and wash the DNA pellet with 1 ml of ice-cold 70% ethanol.

18. Centrifuge the samples for 15 min at 12,000g, remove the supernatant, and air-dry the pellet for 10–15 min.

19. Resuspend the pellet in 50–100 μl of TE buffer (pH 8.0) and store at −20° or −70°.

Further Considerations

The DNA yield and quality obtained using several of the published methods have been compared (Anderson and Lebepe-Mazur, 2003; Holland *et al.*, 2000; Li *et al.*, 2003; McOrist *et al.*, 2002; Santini *et al.*, 2001). Anderson and Lebepe-Mazur (2003) compared 19 noncommercial methods that are used for nucleic acid extraction of gut microflora and concluded that four of the methods were preferable with regards to DNA yield and purity. One of these methods, with modifications, has been described in this chapter (Withford *et al.*, 1998). Some studies have also compared the efficiency of the different commercially available DNA extraction kits and the efficiency of commercial versus noncommercial approaches (Holland *et al.*, 2000; Li *et al.*, 2003; McOrist *et al.*, 2002; Vandenberg *et al.*, 2002). These studies demonstrate that different extraction methods result in different quality DNA. Among the commercial methods, QIAamp DNA Stool Kit gave the highest yield and purity of DNA (Holland *et al.*, 2000; McOrist *et al.*, 2002). The choice of a DNA isolation method depends on many factors including the DNA size, yield, and purity obtained, the repeatability of the technique, the ease of performance, and the number of samples that can be processed per time unit. Commercially available extraction kits (Table I) circumvent some of the drawbacks of noncommercial methods, including the risk of transmission of nucleic acids from sample-to-sample preparation, lack of quality-controlled reagents, and elimination of toxic agents such as phenol. They may also reduce the time required to obtain DNA isolates of sufficient purity (Holland *et al.*, 2000; Li *et al.*, 2003; McOrist *et al.*, 2002). Some of the kits have been specifically developed for fecal analysis, because feces contains contaminants that may inhibit subsequent molecular analysis (Monterio *et al.*, 1997, 2001; Vandenberg *et al.*, 2002).

Acknowledgments

The authors acknowledge financial support from the Research Council of Norway (140890/720 and 140870/130). We thank Jessica L. Ray for comments on the manuscript.

References

Anderson, K. L., and Lebepe-Mazur, S. (2003). Comparison of rapid methods for the extraction of bacterial DNA from colonic and faecal lumen contents of the pigs. *J. Appl. Microbiol.* **94**, 988–993.

Apajalahti, J. H. A., Särkilathi, L. K., Mäki, B. R. E., Heikkinen, P., Nurminen, P. H., and Holben, W. E. (1998). Effective recovery of bacterial DNA and percent-guanine-plus-cytocine-based analysis of community structure in the gastrointestinal tract of broiler chickens. *Appl. Environ. Microbiol.* **64**, 4084–4088.

Avgustin, G., Wright, F., and Flint, H. J. (1994). Genetic diversity and phylogenetic relationship among strains of *Prevotella (Bacteroides) ruminicola* from the rumen. *Int. J. Syst. Bacteriol.* **44**, 246–255.

Blaut, M., Collins, M. D., Welling, G. W., Dorè, J., van Loo, J., and de Vos, W. (2002). Molecular biological methods for studying the gut microbiota: the EU human gut flora project. *Br. J. Nutr.* **87**, S203–S211.

Bruce, K. D., Hiorns, W. D., Hobman, J. L., Osborn, A. M., Strike, P., and Ritchie, D. A. (1992). Amplification of DNA from native populations of soil bacteria by using the polymerase chain reaction. *Appl. Environ. Microbiol.* **58**, 3413–3415.

Clement, B. G., and Kitts, C. L. (2000). Isolating PCR-quality DNA from human feces with a soil DNA kit. *BioTechniques.* **28**, 640–646.

Gong, J., Forster, R. J., Yu, H., Chambers, J. R., Wheatcroft, R., Sabour, P. M., and Chen, S. (2002). Molecular analysis of bacterial populations in the ileum of broiler chickens and comparison with bacteria in the cecum. *FEMS Microbiol. Ecol.* **41**, 171–179.

Griffiths, R. I., Whiteley, A. S., O'Donnell, G., and Bailey, M. J. (2000). Rapid method for coextraction of DNA and RNA from natural environments for analysis of ribosomal DNA- and rRNA-based microbial community composition. *Appl. Environ. Mircrobiol.* **66**, 5488–5491.

Heritage, J., Ransome, N., Chambers, P. A., and Wilcox, M. H. (2001). A comparison of culture and PCR to determine the prevalence of ampicillin-resistant bacteria in the faecal flora of general practice patients. *J. Antimicrob. Chemother.* **48**, 287–289.

Higgins, J. A., Fayer, R., Trout, J. M., Xiao, L., Lal, A. A., Kerby, S., and Jenkins, M. C. (2001). Real-time PCR for the detection of *Cryptosporidium parvum*. *J. Microbiol. Meth.* **47**, 323–337.

Holland, J. L., Louie, L, Simor, A. E., and Louie, M. (2000). PCR Detection of *E. coli* O157:H7 directly from stools: Evaluation of commercial extraction methods for purifying fecal DNA. *J. Clin. Microbiol.* **28**, 4108–4113.

Li, M., Gong, J., Cottrill, M., Yu, H., deLange, C., Burton, J., and Topp, E. (2003). Evaluation of QIAamp DNA stool mini kit for ecological studies of gut microbiota. *J. Microbiol. Meth.* **54**, 13–20.

Liang, S., and Redlinger, T. (2003). A protocol for isolating putative *Helicobacter pylori* from fecal specimens and genotyping using *vacA* alleles. *Helicobacter.* **5**, 561–567.

Mackie, R. I., Minov, R. I., Gaskins, H. R., and White, B. A. (1999). Molecular microbial ecology in gut ecosystems. *In* "Microbial Biosystems: New Frontiers Proceedings of the 8th International Symposium on Microbial Ecology," (C. R. Bell, M. Brylinska, and P. Johnson-Green, eds), pp. 1–9. Atlantic Canada Society for Historical Ecology. Halifax, Canada.

Mahony, J. B., Song, X., Faught, M., Salonga, T., and Kapala, J. (2001). Evaluation of the nuclisens basic kit for detection of *Chlamydia trachomatis* and *Neisseria gonorrhoeae* in genital tract specimens using nucleic acid sequence-based amplification of 16S rRNA. *J. Clin. Microbiol.* **39**, 1429–1435.

Matsuki, T., Watanabe, K., Fujimoto, J., Miyamoto, Takada, T., Matsumoto, K., Oyaizu, H., and Tanaka, R. (2002). Development of 16S rRNA-gene-target group-specific primers for the detection and identification of predominant bacteria in human feces. *Appl. Environ. Microbiol.* **68**, 5445–5451.

McOrist, A. L., Jackson, M., and Bird, A. R. (2002). A comparison of five methods for extraction of bacterial DNA from human faecal samples. *J. Microbiol. Meth.* **50**, 131–139.

Monterio, L., Bonnemaison, D., Vekris, A., Petry, K. G., Bonnet, J., Vidal, R., Cabrita, J., and Mègraud, F. (1997). Complex polysaccharides as PCR inhibitors in feces: *Helicobacter pylori* model. *J. Clin. Microbiol.* **35**, 995–998.

Monterio, L., Gras, N., Vidal, R., Cabrita, J., and Mègraud, F. (2001). Detection of *Helicobacter pylori* DNA in human feces by PCR: DNA stability and removal of inhibitors. *J. Microbiol. Meth.* **45**, 89–94.

Picard, C., Ponsonnet, C., Paget, E., Nesme, X., and Simonet, P. (1992). Detection and enumeration of bacteria in soil by direct DNA extraction and polymerase chain reaction. *Appl. Environ. Microbiol.* **58**, 2717–2722.

Pryde, S. E., Richardson, A. J., Stewart, C. S., and Flint, H. J. (1999). Molecular analysis of the microbial diversity present in the colonic wall, colonic lumen, and cecal lumen of a pig. *Appl. Environ. Microbiol.* **65**, 5372–5377.

Sambrook, J., and Russell, D. W. (2001). *In* "Molecular Cloning: A Laboratory Manual," 3rd Ed., Vol. 3. Cold Spring Harbor Laboratory Press, Cold Spring Harbor, NY.

Santini, M. P., Renz, D., and Doerfler, W. (2001). A comparison of methods to extract pure DNA from mammalian intestinal contents and from feces. *Gene Funct. Dis.* **1**, 51–57.

Simpson, J. M., McCracken, V. J., White, B. A., Gaskins, H. R., and Mackie, R. I. (1999). Application of denaturant gradient gel electrophoresis for the analysis of the porcine gastrointestinal microbiota. *J. Microbiol. Meth.* **36**, 167–179.

Stahl, D. A., Flesher, B., Mansfield, H. R., and Montgomery, L. (1988). Use of phylogenetically based hybridization probes for studies of ruminal microbial ecology. *Appl. Environ. Microbiol.* **54**, 1079–1084.

Steffan, R. J., Goksøyr, J., Bej, A. K., and Atlas, R. M. (1988). Recovery of DNA from soils and sediments. *Appl. Environ. Microbiol.* **54**, 2908–2915.

Tannock, G. W. (1999). Analysis of the intestinal microflora: A renaissance. *Antonie van Leewenhoek* **76**, 265–278.

Tsai, Y. L., and Olson, B. H. (1991). Rapid method for direct extraction of DNA from soil and sediments. *Appl. Environ. Microbiol.* **57**, 1070–1074.

Vandenberg, N., and Oorschot, R. A. H. (2002). Extraction of human nuclear DNA from feces samples using the QIAamp DNA stool mini kit. *J. Forens. Sci.* **47**, 1–3.

Wang, R.-F., Beggs, M. L., Robertson, L. H., and Cerniglia, C. R. (2002). Design and evaluation of oligonucleotide-microarray method for the detection of human intestinal bacteria in fecal samples. *FEMS Microbiol. Lett.* **213**, 175–182.

Wang, R.-F., Cao, W.-W., and Cerniglia, C. E. (1996). PCR detection and quantitation of predominant anaerobic bacteria in human and animal fecal samples. *Appl. Environ. Microbiol.* **62**, 1242–1247.

Withford, M. F., Forster, R. J., Beard, C. E., Gong, J., and Teather, R. M. (1998). Phylogenetically Analysis of rumen bacteria by comparative sequence analysis of cloned 16S rRNA genes. *Anaerobe* **4**, 153–163.

Wood, J., Scott, K. P., Avgustin, G., Newbold, C. J., and Flint, H. J. (1998). Estimation of the relative abundance of different *Bacteroides* and *Prevotella* ribotypes in gut samples by restriction enzyme profiling of PCR-Amplified 16S rRNA gene sequences. *Appl. Environ. Microbiol.* **64,** 3683–3689.

Wilson, K. H., and Blitchington, R. B. (1996). Human colonic biota studied by ribosomal DNA sequence analysis. *Appl. Environ. Microbiol.* **62,** 2273–2278.

Zhu, X. Y., Zhong, T., Pandya, Y., and Joerger, R. D. (2002). 16S rRNA-bases analysis of microbiota from the cecum of broiler chickens. *Appl. Environ. Microbiol.* **68,** 124–137.

Zoetendal, E. G., Ben-Amor, K., Akkermans, A. D. L., Abee, T., and deVos, W. M. (2001). DNA isolation protocols affect the detection limit of PCR approaches of bacteria in samples from the human gastrointestinal tract. *Syst. Appl. Microbiol.* **24,** 405–410.

Zoetendal, E. G., Collier, C. T., Koike, S., Mackie, R. I., and Gaskins, H. R. (2004). Molecular ecological analysis of the gastrointestinal microbiota: A review. *J. Nutr.* **134,** 465–472.

[4] Nucleic Acid Isolation from Ecological Samples—Fungal Associations, Lichens

By Martin Grube

Abstract

Ecological samples of fungal associations pose particular challenges for nucleic acid extraction due to the presence of several genomes. Thorough examination of the samples prior to extraction is important to assess the risks of contamination. If manual separation of symbionts or their axenic cultivation is not feasible, symbiont-specific primers can be applied in PCR experiments. A basic protocol is suggested here which can be optimized for specific applications.

Introduction

Ecological samples, as we may delimit this term in this chapter, are materials taken directly from nature. They usually include their immediate biological context and are therefore sometimes very complex in composition. This is especially true for most fungi, because they live in intricate association with their immediate environment by formation of vegetative bodies that grow in optimized ways to exploit the nutritional resources, by either yeastlike or hyphal growth habits. On the other hand, fungi have adapted to a great diversity of ecological niches. Some grow on largely abiotic anorganic substrates such as rocks or soils (a few specialists adapted to high salt tolerance to live in salterns). Others are saprobes on dead organisms or on processed materials such as food. A substantial fraction of fungi, however, live in close associations with other life forms, thus forming symbioses in the widest sense. This includes mycorrhizal fungi, lichens, and fungi that are pathogenic. Because of their general ecological importance, symbiotic fungi are increasingly focused on by molecular approaches. I concentrate in the following parts on lichens, but most considerations discussed here also apply to other fungal associations.

Complexity of Fungal Associations, with a Focus on Lichens

Special attention is needed when ecological samples may contain a mixture of several organisms. Lichen thalli often not only include a heterotrophic fungal and an autotrophic algal partner (either eukaryotic, prokaryotic, or both) but may also include a range of other co-occurring organisms. Many of these will be found growing on external parts of lichen thalli, whereas others are hidden inside. This encompasses particularly other fungi, which are on one hand host-specific lichenicolous fungi, while symptomless or cryptic endogeneous fungi may also occur. In addition, a number of heterotrophic bacteria may be attached to lichen thalli. For these reasons, it is practically impossible to obtain absolutely pure material of lichen mycobionts in ecological samples. Rather, the extracts will contain a mixture of genomes, of which some occur in predominant concentrations.

Strategies of Material Preparation

The strategy of material preparation depends largely on the subsequent procedures. For a PCR using symbiont- or gene-specific primers, it will be sufficient to remove attached organisms, either by washing with water and subsequent drying or by scraping off visible contaminants, such as mosses, substrate particles, superficial fungi, and so on, from the surface layers. For amplifications using nonspecific primers (e.g., random amplified polymorphic DNA [RAPD] primers), it will be important to use appropriate material of a particular symbiont only.

One strategy to obtain sufficiently clean mycobiont material in nucleic acid extractions of lichen samples is either to use thallus parts that lack the algal partner or to mechanically remove algal layers. The latter is easily possible only in a few cases, such as in *Usnea* or certain *Cladonia* species, in which the cortical and algal layers can be scraped off from the central medulla. Other lichens develop conspicuous stipitate structures devoid of algae, such as the genera *Baeomyces* and *Icmadophila*. Fungal structures that connect foliose thalli to the substrate (e.g., rhizines), can also be used, although they may contain substantial amounts of attached organisms.

Smaller amounts of fungal-only material are clearly present in most ascomatal structures (except, for example, in *Endocarpon*, which also contain co-distributed algae in the hymenia). Although there were spectacular reports on DNA amplification from single spores (Lee and Taylor, 1990), my experience is that this procedure has failed to produce consistent results in lichens. A larger number of ascospores is usually required, which can easily be done with in species that form mazaedia. In these cases, the ascal walls evanesce and all spores are deposited freely on top of the hymenia. A special protocol has previously been developed for DNA isolation from entire lichen fruit bodies (Grube *et al.*, 1995), which relies on the binding of DNA to glass beads. This approach also circumvents substantial problems caused by the great amount of polysaccharides in ascomatal structures. The ascomata used for DNA extraction can be prepared in various ways. Rigid apothecial ascomata are usually dissected using a razor blade (sometimes facilitated by moistening the fruit bodies before), while the hymenial contents of large pyrenocarpous fruit bodies can be poked out using a needle after moistening them.

If the lichen symbionts cannot be separated and the isolate contains DNA of different sources, discriminative primers are used during PCR to obtain products from the organisms of interest. Until now, most phylogenetic investigations of ribosomal DNA from lichen mycobionts apply primers that do not amplify the respective genes from the algal symbionts (Gardes and Bruns, 1993; Gargas and DePriest, 1996; Gargas and Taylor, 1992). Only recently, primers for the selective amplification of rDNA from trebouxioid photobionts became available (Kroken and Taylor, 2000), and similar is the case for 16S rDNA of cyanobacterial photobionts (Lohtander *et al.*, 2003). With the ever increasing amount of genomic data on fungi, it is possible to design symbiont-specific primers also for a range of protein-coding genes. In any case, if an accessory closely related fungal species occurs in a lichen thallus, the design of specific primers may not be sufficient. Contaminations can be recognized either by the presence of multiple bands (different sizes of the amplicons), by unresolved sequence electropherograms, or by unexpected outcomes in phylogenetic studies. The latter can be the case for example with lichen thalli that contain considerable amounts of endogeneously occuring heterobasidiomycetes.

More complicated techniques need to be pursued in such cases, which include the cloning of PCR products or the excision of specific bands from agarose gels. However, the most elegant solution is to cultivate the relevant organisms of an ecological sample. The reason this has not become the routine approach in studies of fungal symbioses is that the symbionts usually grow very slowly, if at all, under axenic conditions. Also, there is a high contamination risk of axenic cultures of lichen thallus fragments by other fast-growing fungi or bacteria, especially when surface sterilization is not adequately performed. Even with surface-sterilized material of lichen thalli, a number of internally co-occurring fungi or bacteria may start to grow in the cultures. Despite their growth in cultures, these co-occurring organisms are not necessarily involved in symbiotic interactions, but they may cause misinterpretations of subsequent molecular results. Taking into account the slow growth rate of mycobionts, it will nevertheless be optimal to use single-spore isolates for absolutely pure material of lichen mycobionts.

The failure of PCR using extracts of entire ecological samples such as lichen thalli may lead to underestimation of the diversity of involved symbionts. A good example is when PCR fails to amplify a symbiont that is present in rather low amounts as indicated by microscopic examination of the material. For example, with the available specific primers for trebouxioid photobionts, it is sometimes difficult to amplify algal ITS sequences directly from small thallus parts of certain foliose lichens where the algae are present in a very thin layer and in a small number. Likewise, photobiont diversity may remain cryptic in thalli that contain one predominant and other rare photobiont species (of the same taxonomic group). In the latter cases, it will be necessary to cultivate the photobionts axenically and to sample individual thalli appropriately.

Quality of Starting Material

The age of the starting material to be used for DNA isolation is an important parameter of ecological samples that needs to be considered carefully, partially because symbiotic organisms may be of different age stages within a sample. For most phylogenetic studies of lichens, herbarium specimens are the main source of material. Fresh material would be optimal for any kind of nucleic acid isolation and is of course particularly important for isolation of RNA. If fresh material of lichens is available, only a too high hydration of the thalli may be a problem for isolation, because it may be difficult to grind such materials to powder with liquid nitrogen in the initial step (this applies, for example, to hydrated central cords of fresh *Usnea*). Table I presents a brief overview of maximum average age of the lichen samples that have been used for DNA isolation. It is generally recommended to store lichen material needed for DNA isolation in a freezer, because this can significantly delay DNA degradation. The oldest material from which I extracted DNA successfully was a 36-year-old herbarium specimen of *Multiclavula mucida* (Gargas *et al.*, 1995). Under which conditions DNA is best preserved in fungal samples is not clear, but the conditions may vary from species to species and among mycelial contexts.

TABLE I

MAXIMUM AGE OF STARTING MATERIAL IN SOME LICHENS

Taxon	Maximum age of successfully isolated DNA using the unmodified standard protocol
Arthonia spp.	Several months to 2 yr (depending on the taxonomic group)
Buellia spp.	10 yr
Lecanora spp.	10 yr
Multiclavula mucida	36 yr
Porina (foliicolous species) spp.	Several months
Rinodina spp.	10–15 yr
Usnea filipendula	4 yr
Trebouxioid photobionts	~10–15 yr

The overall quality of DNA can also be evaluated with histochemical staining of the nuclei. With 4'6-diamidino-2-phenylindole-2HCl (DAPI) staining, the nuclei of suitable material are usually compact and well delimited, whereas DNA appears to be severely degraded when the stained nuclei are diffuse or fragmented. Such material may contain considerable amounts of fragmented DNA strands, which will contribute to undesired and nonspecific priming in subsequent PCRs, and which results in typical smears on the control gels.

It might also be noted that decaying material may become the substrate of other fungi. This was the case with lichens that were collected from the front of retreating glaciers (DePriest *et al.*, 2000).

A General Protocol

In the following protocol, a general method of DNA isolation is described. It is comparatively cheap when compared to commercial kits that use extraction columns. This method (Cubero and Crespo, 2002; Cubero *et al.*, 1999) has been used for many years with a wide range of ecological samples of fungal associations. The key step is a nonalcoholic precipitation step with cetyltrimethylammonium bromide (CTAB) to remove most of the polysaccharides that may be particularly abundant in lichen associations. The protocol may also serve as a basis for trying various modifications. For example, additives may be included to account for the presence of polyphenols (see later discussion), and in cases in which polysaccharides do not represent a problem, the CTAB precipitation step may be omitted and may be followed directly by an ethanol (EtOH) or isopropanol precipitation of genomic DNA.

Nucleic Acids Isolation

Remove any rings, bracelets, and watches before working with liquid nitrogen, and use gloves at all stages to prevent contamination of reagents.

1. Turn on water bath at 65°.
2. Prepare labeled 1.5-ml Eppendorf tubes.
3. Place uncontaminated material from lichens into an Eppendorf tube. Without significant effect on the DNA quality, dried material of lichens can also be rinsed with acetone to extract lichen compounds before DNA isolation.
4. Grind material to powder in presence of liquid nitrogen using precooled minipestils that fit exactly into the Eppendorf tube. For example, pour nitrogen over a tray and place tube into a slot filled with liquid nitrogen for 1–2 min before grinding the frozen material inside the tube. Take care that powderized material will not contaminate gloves during rigorous grinding, to avoid cross-contamination of samples.
5. Add 500 µl of lysis buffer and incubate for 1 h at 65° (mix from time to time). The incubation may be prolonged to increase the yield of DNA (especially for old herbarium samples).
6. Add 500 µl of (chloroform:isoamyl alcohol, 24:1) in a fume hood, vortex briefly, and centrifuge for 5 min at 12,000 rpm (optionally repeat this step).
7. Take upper phase (with DNA) carefully with a pipette and transfer it into a new tube.
8. Add 1 ml of precipitation buffer, mix gently, and let rest for 1 h at room temperature. This step can be prolonged to increase yield of nucleic acids (always keep at room temperature, because polysaccharides will precipitate in the cold).
9. Centrifuge for 15 min at 12,000 rpm (at room temperature).
10. Discard supernatant, resuspend pellet in 350 µl of 1.2 M NaCl.
11. Add 500 µl of chloroform:isoamyl alcohol, vortex and centrifuge for 5 min at 12,000 rpm.
12. Carefully transfer upper phase into a new tube.
13. Add 210 µl of cold isopropanol and mix carefully. Incubate for 15 min to overnight at −20°.
14. Centrifuge for 20 min at 4° and 12,000 rpm.
15. Discard supernatant (the pellet is usually not visible).
16. Add 200 µl of 80% EtOH (optionally let rest for 5 min at 4°) and centrifuge for 2 min at 12,000 rpm.
17. Discard supernatant.

18. Dry pellet in vacuum centrifuge (for 10 min) or in an oven (for ~20 min) at 45° (avoid excessive drying because this may cause problems for redissolving DNA).

19. Resuspend dried pellet in 25 μl of aqua dest or TE buffer (the latter may require Mg^{2+} ion adjustment in subsequent PCR, as ethylenediaminetetraacetic acid [EDTA] chelates magnesium).

20. Prepare a dilution of the DNA stock solution for subsequent use in PCR.

Solutions

    Lysis buffer:
        CTAB 1.4%
        NaCl 1 *M*
        Tris 7 m*M*
        EDTA 30 m*M*
    Precipitation buffer:
        CTAB 0.5%
        NaCl 40 m*M*
Chloroform:isoamyl alcohol: 24:1

Comments

Polyvinylpolypyrrolidone (PVPP) powder can be added directly to the initial extraction mixture up to an amount of 1% w/v. This efficiently eliminates polyphenolic compounds (Pitch and Schubert, 1993). However, in most cases, good isolation results were achieved without the use of PVPP.

The results of the nucleic acids extractions can be evaluated on an agarose gel (0.8–1.2%). In the ideal case, a clear band of heavy-weight genomic DNA will be recognized in addition to coisolated ribosomal RNA (rRNA), which appears as two distinct bands of small and large rRNA, as well as a smear of messenger RNA (mRNA) in the lower molecular weight part of the gel. Pure RNA can thus be obtained after enzymatic digestion of DNA after the ultimate isolation step. However, special care is needed for RNA isolation to avoid the degradation by ribonucleases (RNases), which includes the use of RNase-free chemicals or RNase inhibitors.

Lack of the genomic DNA bands on the control gel does not necessarily predict that a subsequent PCR will fail. PCR may work well, because genomic DNA may often be present at amounts that are too low for detection in agarose gels, especially when the starting material was scant. However, if the isolation fails despite the use of a sufficient amount of starting material, the most likely cause is that the DNA pellet is lost during the cleaning step after the alcoholic precipitation. This is more likely—according to my experience—when impurities such as polysaccharides are still present at this stage.

If poor yield is suspected because of the age of the samples, it may help to substantially prolong the initial incubation step in the extraction buffer.

An extreme reduction of the protocol is to perform the disruption of cells in liquid solutions and to use part of the resulting solute directly for PCR. Disruption of cells in liquids is possible by mechanical treatment (e.g., grinding in the presence of glass beads) or by ultrasonic irradiation, if not extended to periods that would cause significant shearing damage to DNA. Alternatively, thermal cycling may lead DNA to diffuse from disrupted cells in the extraction buffer, which can optionally be improved by prior partial enzymatic digestion of fungal cell walls. It may be added here that enzyme mixtures for cell wall degradation differ considerably in their effectiveness in different fungal groups, because of variations in the composition of hyphal cell walls.

A simplified version of the heat-cycling treatment is possible with unicellular trebouxioid photobionts in lichens. In this case, it is sufficient to take a small amount of the algal layer of the lichen and to heat the sample to 95° in 50–100 μl of water for some minutes. A few microliters of this crude extract is usually sufficient for a 30–50 μl PCR. It is apparent that this and similar simplifications do not result in DNA isolates suitable for long-term storage.

Simultaneous Isolation and Amplification of DNA

Good results are in several cases obtained when the extraction of nucleic acid is performed simultaneously with PCR, a technique usually called *direct PCR*. This approach is especially useful if only very minute amounts of material are available. Direct PCR can be performed on both fragments or sections taken from ascomata or from minute amounts of thallus material (De los Rios *et al.*, 2000; Wolinski *et al.*, 1999). The sections can be prepared by hand, using a thin razor blade (e.g., Gillette) and should be as thin as about 10 μm to adhere properly on glass. Microtome sections are preferred and can be produced from resin-embedded plectenchyma, but kryo-sections are also suitable. Small fragments of thalli can be detached from the sample using a cleaned forceps with sufficiently fine tips.

In a direct PCR approach, it is essential to avoid any parts that contain PCR inhibitors, such as anthraquinones and melanin-like polymeric pigments. The latter often occur in the margins of fungal fruit bodies. These mycelial parts can be removed by microdissection if sections are fixed on glass slides. Not all lichen compounds, however, are detrimental to PCR. DNA of medullar parts of *Usnea* can readily be amplified, although they contain diverse phenolic compounds. Direct PCR approaches are also useful when samples are taken from species that grow in very dense and small-scaled communities, as, for example, lichens on leaves.

Previously, direct PCR has been carried out with standard microscopic slides. Handling and space problems when a large number of samples are processed lead to the alternative use of so-called *microslides* (Hindkjaer *et al.*, 1996). These are about 2 by 10 mm large and can be cut from standard coverslips using a diamond cutter. They fit well in a 0.5-ml Eppendorf tube, and if the sectioned material is mounted close to the edge of the microslides, it is readily covered by the PCR mix. In my laboratory, we have also used the same slide twice for amplification. After removal from the finished PCR and a brief rinse in distilled water, we successfully amplified DNA from a different genomic locus.

Before mounting the sections, we need to coat the glass slides. For this purpose, the glass slides are washed in sterile water, 2–5 min in 100% EtOH, air-dried, and immersed in a gelatin/chromalum solution (0.25% gelatin, 0.025% chromium III potassium sulfate) for 5 s under sterile conditions. Afterwards, the coated microslides are dried on aluminum foil, before rehydrated sections can be placed on them. After mounting the sections, we may place the slides for about 1 min at 40–50° on a heating plate, which increases the fixation of the sections to the slides.

Microslide mounts can also be used for various purposes, for example, to obtain PCR products from a lichenicolous fungus (Wolinski *et al.*, 1999). We have also successfully amplified from minute amounts of cultured mycelium of a *Ceratobasidium* (*Rhizoctonia* morph) species involved in the symbiosis with orchids (*Nigritella*). To achieve this, we place uncoated microslides in the vicinity of the growing culture of the mycobiont. After a few days, the mycelium starts to overgrow the slides, which are then removed from the cultures (De los Rios *et al.*, 2000). A prerequisite for this approach is that the hyphae of the fungus grow on the agar surface. Material sampled this way was also used for *in situ* hybridization of rRNA of these fungi.

Concluding Remarks

Commercial kits that are specifically adapted to the extraction of nucleic acids from samples that contain high amounts of polysaccharides and polyphenols perform equally as well as the method outlined in this chapter. However, if costs are a limiting factor, the latter or its modifications may be preferred. In addition, other protocols published elsewhere will result in more or less pure DNA isolates of fungal associations. However, the selective isolation of nucleic acids from organelles, such as mitochondria, plastids in the case of lichen photobionts, plasmids, or fungal viruses using gradient centrifugation will require particular considerations beyond the scope of this chapter and usually more amounts of preferably cultured material. Amplification with specific primers (e.g., for gene loci of mitochondrial DNA) is the more efficient solution in most cases.

The investigation of the transcriptome by analysis of the mRNA pool is another problem when working with fungal association. RNA isolations will contain nucleic acids from various organisms, which makes the interpretation of results difficult, unless symbiont- and gene-specific primers are applied. A study of separately cultured symbionts, on the other hand, will not adequately reflect the transcriptome of the organisms in their natural symbiotic states. The efficient separation of symbionts before the extractions will, therefore, be a challenge for future functional studies of fungal associations.

References

Cubero, O. F., and Crespo, A. (2002). Isolation of nucleic acids from lichens. *In* "Protocols in Lichenology. Culturing, Biochemistry, Ecophysiology and Use in Biomonitoring" (I. Kranner, R. P. Beckett, and A. K. Varma, eds.), pp. 381–392. Springer-Verlag, Berlin, Heidelberg.

Cubero, O. F., Crespo, A., Fatehi, J., and Bridge, P. D. (1999). DNA extraction and PCR amplification method suitable for fresh, herbarium-stored, lichenized, and other fungi. *Plant Systematics Evol.* **216,** 243–249.

De los Rios, A., Deutsch, G., and Grube, M. (2000). Efficient genetic analysis of fungal samples. *Prep. Biochem. Biotechnol.* **30,** 145–153.

DePriest, P. T., Ivanova, N. V., Fahselt, D., Alstrup, V., and Gargas, A. (2000). Sequences of psychrophilic fungi amplified from glacier-preserved ascolichens. *Can. J. Botany* **78,** 1450–1459.

Gardes, M., and Bruns, T. D. (1993). ITS primers with enhanced specificity for Basidiomycetes. Application to the identification of mycorrhizae and rusts. *Mol. Ecol.* **2,** 113–118.

Gargas, A., and Taylor, J. (1992). Polymerase chain reaction (PCR) primers for amplifying and sequencing nuclear 18S rDNA from lichenized fungi. *Mycologia* **84,** 589–592.

Gargas, A., and DePriest, P. T. (1996). A nomenclature for fungal PCR primers with examples from intron-containing SSU rDNA. *Mycologia* **88,** 745–748.

Gargas, A., DePriest, P. T., Grube, M., and Tehler, A. (1995). Multiple origins of lichen symbioses in fungi suggested by SSU rDNA phylogeny. *Science* **268,** 1492–1495.

Grube, M., Gargas, A., DePriest, P. T., and Hafellner, J. (1995). DNA isolation from lichen ascomata. *Mycol. Res.* **99,** 1321–1324.

Hindkjaer, J., Terkelsen, C., Kølvraa, S., Koch, J., and Bolund, L. (1996). Detection of nucleic acids (DNA and RNA) *in situ* by single and cyclic primed *in situ* labelling (PRINS): Two alternatives to traditional *in situ* hybridization methods. *In* "*In situ* Hybridization" (M. Clark, ed.). pp. 45–66. Chapman and Hall, London.

Kroken, S., and Taylor, J. W. (2000). Phylogenetic species, reproductive mode, and specificity of the green alga *Trebouxia* forming lichens with the fungal genus *Letharia*. *Bryologist* **103,** 645–660.

Lee, S. B., and Taylor, J. W. (1990). Isolation of DNA from fungal mycelia and single spores. *In* "PCR Protocols: A Guide to Methods and Applications" (M. A. Innis, D. H. Gelfand, J. J. Sninsky, and T. J. White, eds.), pp. 282–287. Academic press, San Diego.

Lohtander, K., Oksanen, I., and Rikkinen, J. (2003). Genetic diversity of green algal and cyanobacterial photobionts in *Nephroma* (Peltigerales). *Lichenologist* **35,** 325–339.

Pitch, U., and Schubert, I. (1993). Midiprep method for isolation of DNA from plant with a high content of polyphenolics. *NAR* **21,** 14.

Wolinski, H., Grube, M., and Blanz, P. (1999). Direct PCR of symbiotic fungi using microslides. *Biotechniques* **26,** 454–455.

[5] Nucleic Acid Isolation from Ecological Samples—Fungal Associations, Mycorrhizae

By ROGER T. KOIDE

Abstract

Mycorrhizal fungi are among the most common symbioses found in terrestrial ecosystems, both natural and managed. They are important for many reasons, but most notably because of their positive effects on plant growth, which are mediated by their uptake of nutrients from the soil and transport of these to the roots. Moreover, many edible fungi are mycorrhizal. The study of mycorrhizal fungi has been hampered by the inability to identify species and individuals in the soil. This has been greatly aided by DNA-based methods, which first require the extraction of DNA. Herein, I discuss some general concerns that must be considered when extracting and purifying DNA from ecological samples and offer specific methods for soil, mycorrhizal roots, and fruiting bodies. These methods are rapid, safe, effective, relatively inexpensive, and convenient because they are based on commercially available kits.

Introduction

Mycorrhizal fungi are terrestrial fungi that colonize plant roots, forming ecologically and agronomically important symbioses called mycorrhizae, which are among the most common symbioses in nature. The several mycorrhizal types include arbuscular, ectomycorrhizal, ectendomycorrhizal, arbutoid, monotropoid, ericoid, and orchid (Smith and Read, 1997). The symbioses may be mutually beneficial, and this has been particularly well documented for the arbuscular mycorrhizae, ectomycorrhizae, and ericoid mycorrhizae. The nutrient status of the host plant may be improved as the fungus absorbs nutrients from the soil and transfers them to the root. The fungus may also derive a significant amount of carbohydrate from the photosynthetic plant. From an agricultural standpoint, the arbuscular mycorrhizae and ectomycorrhizae are probably the most important mycorrhizae because these involve the majority of important food, fiber, and timber plant species, as well as many of the important edible fungi. These symbioses are also very important ecologically because most of the earth's land surface is dominated by vegetation that is largely arbuscular mycorrhizal or ectomycorrhizal.

Arbuscular mycorrhizae are formed when roots are colonized by members of the fungal phylum Glomeromycota (Schüssler et al., 2001), of which there are currently approximately 150 described species. The plants involved include many mosses, ferns, gymnosperms, and angiosperms. The arbuscular mycorrhiza is named for the arbuscule, a highly branched fungal organ usually produced within cortical cells of colonized plant roots. This is the organ across which the fungus absorbs carbohydrate from the plant and from which the plant absorbs various nutrients from the fungus including copper and zinc, possibly nitrogen (N), and especially phosphate. Some arbuscular mycorrhizal fungi may produce other structures within the roots of colonized plants including vesicles, which are capable of storing high concentrations of lipid. Thus, some of the arbuscular mycorrhizal fungi are vesicular–arbuscular (VA). In some cases the arbuscular mycorrhizal fungi may be far more important than the roots themselves as organs of phosphate uptake (Smith et al., 2003). In phosphate-deficient soils, arbuscular mycorrhizal colonization can significantly improve plant growth and yield (Koide, 1991). The main reason for this is that phosphate usually occurs in low concentrations in the soil and diffuses slowly. The hyphae of the mycorrhizal fungi extending from colonized roots compensate for this by exploring a greater volume of soil than the roots can themselves and by presenting a greater surface area for phosphate uptake.

In the ectomycorrhizal symbiosis, the ability to associate with roots to form ectomycorrhizae has developed independently in more than one fungal lineage. The several thousands of fungal species that form ectomycorrhizae include members of the Basidiomycota, Ascomycota, and Zygomycota (Bruns, 1995). The plants forming ectomycorrhizae are mostly angiosperm and gymnosperm shrubs and trees, including members of the economically important and ecologically dominant Pinaceae, Fagaceae, and Myrtaceae. No fungal structures are produced within root cells, but the fungi grow between the outer cortical and epidermal cells and may form a dense mantle outside of the epidermis. This often results in characteristic short ectomycorrhizal roots. Just as in the arbuscular mycorrhiza, the additional surface area provided by hyphae extending from the surface of the mantle into the soil can significantly increase the phosphate status of the plant. N absorption may also be enhanced (Smith and Read, 1997). Some fungal species are also capable of transporting significant amounts of water to the plant (Brownlee et al., 1983).

Because of the large agricultural and ecological significance of mycorrhizal fungi, there is a growing need to study them in both agricultural and natural settings. For example, there is an interest in the study of the persistence of fungi that are introduced into soils (Armstrong et al., 1989; Gardes et al., 1991; Henrion et al., 1992). In addition, the composition of the mycorrhizal fungal community in the soil may influence plant species composition in natural ecosystems and plant productivity (Bever *et al.*, 2002; Hart and Klironomos, 2002; van der Heijden, 1999), as well as other ecosystem properties such as soil aggregation and carbon and nitrogen storage (Miller and Jastrow, 2000).

The study of mycorrhizal fungi has been hampered by the inability to identify them when they are growing in soil. Irrespective of species, nearly all hyphae look similar. Moreover, identification cannot be based on growth on selective media, because many species cannot be cultured. Serological methods (Aldwell and Hall, 1987) of identification have proven either too complex or unreliable. Moreover, "signatures" or "fingerprints" based on fatty acids (Bentivenga and Morton, 1994; Vestal and White, 1989) or carbohydrates (Koide *et al.*, 2000) may vary somewhat among fungal species, but they also vary depending on environmental conditions (Dart, 1976; Koide *et al.*, 2000) and are generally insufficient to use as reliable markers for purposes of identifying species or strains. DNA-based methods, however, have proven valuable in agronomic and ecological studies of mycorrhizal fungi (Horton and Bruns, 2001). DNA-based methods are now quite rapid, and even for extensive ecological or agronomic studies, their costs are tolerable.

I report on methods for extraction and purification of mycorrhizal fungal DNA from several kinds of field samples including fruiting bodies of ectomycorrhizal fungi, hyphae found in soil, and mycorrhizal roots. For polymerase chain reaction (PCR), the DNA from spores of arbuscular mycorrhizal fungi does not have to be extracted (see later discussion). Comparisons of some methods and general discussions of DNA extraction from environmental samples are also given by Leung *et al.* (1995), Rochelle *et al.* (1995), Saano *et al.* (1995), Cullen and Hirsch (1998), and Miller *et al.* (1999).

General Considerations in the Extraction of Mycorrhizal Fungal DNA from Ecological Samples

General Remarks

The extraction and purification of DNA from fungal tissues such as fruiting bodies of ectomycorrhizal fungi is relatively easy, and for the purposes of PCR, extraction of DNA from chlamydospores of arbuscular mycorrhizal fungi is not even necessary (see later discussion). Other ecological samples are more difficult to handle because the fungi grow within a living (roots) or a nonliving (soil) matrix. Knowing the species composition of the mycorrhizal fungal community is important because of its effects on plant productivity and ecosystem function (see earlier discussion). Moreover, the hyphae growing in soil are responsible for nutrient and water uptake. Their physical location in the soil may determine in large measure their effects on plant growth (Bending and Read, 1996; Smith and Read, 1997).

Most published methods for the extraction of DNA from soil were developed for bacteria. It is interesting to note that use of the misleading word *microbial* in the literature generally signifies *bacterial*. The general considerations are the same for fungi, of course, but special consideration must be given to the thick cell wall of the fungi, which differs chemically from bacterial walls. Historically two approaches have been taken to extract DNA from microbes in soil. The so-called *indirect* method involves the removal of microbes from the soil before extraction of their DNA (Torsvik, 1980). One significant challenge with this method is to remove all the fungi so as not to bias the result. Moreover, this method obviously involves an extra time-consuming step compared to the so-called *direct* method (Bruce *et al.*, 1992; Ogram *et al.*, 1987) of extracting DNA from the fungi while still within the soil. The direct method (i.e., that of extracting DNA directly from soil) is usually more rapid than the indirect method but may result in the extraction of extracellular DNA adsorbed onto soil surfaces (Frostegård *et al.*, 1999; Paget *et al.*, 1992; Romanowski *et al.*, 1993). Moreover, the direct method may result in the coextraction of problematic phenolic compounds that may inhibit hybridization, PCR, or restriction digestion. The direct method has always been applied to mycorrhizal fungi colonizing roots. I am not aware of any research in which the fungi were first separated from root tissues before DNA extraction, although this is possible, in principle, using plant cell-wall–hydrolyzing enzymes. Like soils, roots may contain high concentrations of phenolic compounds (Koide *et al.*, 1998).

The Coextraction of Inhibitors

Phenolic compounds, including tannins and humic substances, are major components of many plant organs such as leaves and roots and thus are contained within litter and more decomposed forms of soil organic matter. When they occur in ecological samples, extraction of DNA inevitably results in the extraction of the phenolics, which may inhibit the PCR and restriction digestion (Cullen and Hirsch, 1998; Krause, 2001; Tebbe and Vahjen, 1993; Wilson, 1997). The exact mode of action is not entirely understood (Wilson, 1997) but may involve binding to the DNA or to enzymes (Young *et al.*, 1993). Moreover, some phenolic compounds absorb light at 260 nm, making the standard spectrophotometric quantification of

DNA impossible without their removal (Cullen and Hirsch, 1998; Rochelle *et al.*, 1995). Arbuscular mycorrhizae are often found predominantly in mineral soils (Read, 1993), in which the concentration of phenolic compounds may be rather low. However, ectomycorrhizae and, especially, ericoid mycorrhizae are often found in soils of high organic matter concentration (Read, 1993), in which the concentration of phenolics compounds can be high. Therefore, consideration of phenolic inhibitors that coextract with DNA from soil is a high priority.

If DNA concentration in the extract is high enough, one approach to dealing with inhibitors when PCR is the next step is simply to dilute the extract many fold to reduce inhibitor concentrations to a level that is tolerable (Miler *et al.*, 1999; Wintzingerode *et al.*, 1997). Another approach is to add soluble polyvinylpyrrolidone (PVP) (Xin *et al.*, 2003) or protein such as bovine serum albumin (Kreader, 1996; Xin *et al.*, 2003) to the PCR mixture to inactivate the inhibitors by adsorption or other means. However, because dilution does not always solve the problem, and because inhibitor concentrations may vary from sample to sample and from season to season, a more reliable approach is to purify the DNA sample before subsequent use. Several methods have been devised and tested for their efficiency in separating phenolic inhibitors from DNA.

DNA Purification

Some of the postextraction methods of purification include hydroxyapatite chromatography (Ogram *et al.*, 1987; Steffan *et al.*, 1988), ion exchange chromatography (Tebbe and Vahjen, 1993), cesium chloride–ethidium bromide density centrifugation (Ogram *et al.*, 1987; Porteous and Armstrong, 1991), DNA concentration by selective filtration followed by washing (Bruns *et al.*, 1999; Zhou *et al.*, 1996), precipitation of DNA by ethanol or isopropanol followed by washing (Steffan *et al.*, 1988; Zhou *et al.*, 1996), adsorption of inhibitors by polyvinylpolypyrrolidone (PVPP) (Berthelet *et al.*, 1996), gel electrophoresis of extract followed by cutting of bands and elution of DNA (Bruns *et al.*, 1999; Knaebel and Crawford, 1995; Miller *et al.*, 1999), electrophoresis through a gel containing soluble (PVP) (Young *et al.*, 1993), size exclusion chromatography with Sephadex resin (Cullen and Hirsch, 1998; Leung *et al.*, 1995; Tsai and Olson, 1992), adsorption of DNA onto silica followed by washing and desorption (Porteous *et al.*, 1991; Rochelle *et al.*, 1995; Saano *et al.*, 1995; Smalla *et al.*, 1993; Zhou *et al.*, 1996), or extraction with organic solvents (Henrion *et al.*, 1992; Leung *et al.*, 1995; Ogram *et al.*, 1987). The more rapid among these tend to be less effective (Leung *et al.*, 1995), whereas the more effective of these methods tend to be expensive or time consuming. All of them may significantly reduce DNA yield (Bruns *et al.*, 1999; Lloyd-Jones and Hunter, 2001).

To save time, a popular approach to purification has been to attempt to selectively precipitate or adsorb inhibitors during the extraction process itself. Some precipitation of inhibitors may be accomplished using cetyltrimethylammonium bromide (CTAB) (Armstrong *et al.*, 1989; Gardes and Bruns, 1993; Gardes *et al.*, 1992; Murray and Thompson, 1980; Zhou *et al.*, 1996), ammonium acetate, or polyethylene glycol (PEG) (Krause *et al.*, 2001). Adsorption during extraction is often accomplished with PVPP (Cullen and Hirsch, 1998; Leung *et al.*, 1995; Porteous and Armstrong, 1991; Rochelle *et al.*, 1995; Steffan *et al.*, 1988; Zhou *et al.*, 1995). The selective flocculation of phenolic inhibitors by $AlNH_4(SO_4)_2$ has been shown to be effective under a wide range of circumstances (Braid *et al.*, 2003).

Cell Lysis

For DNA to be extracted, the fungal cells must first be lysed. Cell lysis has been accomplished by both physical and chemical methods. Vigorous shaking of the sample with mineral beads (bead beating) has been frequently used to disrupt cells and usually results in high yields of DNA (Cullen and Hirsch, 1998; Lloyd-Jones and Hunter, 2001; Moré *et al.*, 1994; Ogram *et al.*, 1987; Smalla *et al.*, 1993). Ultrasonication (Bakken and Olsen, 1989) has been used for cell lysis. Freeze–thaw cycles (Gardes and Bruns, 1993; Moré *et al.*, 1994; Tsai and Olson, 1992; Zhou *et al.*, 1996) and grinding, typically with liquid N_2 (Armstrong *et al.*, 1989; Frostegård *et al.*, 1999; Gardes *et al.*, 1991; Henrion *et al.*, 1992; Zhou *et al.*, 1996) may be effective. Heating or microwaving can also be used to lyse fungal cells (Cullen and Hirsch, 1998; Gardes *et al.*, 1991; Porteous and Armstrong, 1991), but heating may release large concentrations of phenolics from soils or roots. Bead beating is thought to be more effective than chemical methods (see later discussion), particularly with thick-walled fungi, and it may result in less extraction of phenolic compounds than heating (Cullen and Hirsch, 1998), although it may shear DNA more than chemical methods (Lloyd-Jones and Hunter, 2001; Miller *et al.*, 1999) and more than grinding (Zhou *et al.*, 1996). Ultrasonication may also shear DNA (Bakken and Olsen, 1989). Lysis has also been accomplished with chemicals including detergents such as sodium dodecyl sulfate (SDS) (Gardes *et al.*, 1991; Henrion *et al.*, 1992; Moré *et al.*, 1994; Ogram *et al.*, 1987; Porteous and Armstrong, 1991; Zhou *et al.*, 1996), with high salt concentrations (Edwards *et al.*, 1997; Zhou *et al.*, 1996), or with enzymes such as with Novozyme 234 (Claassen *et al.*, 1996; Leung *et al.*, 1995; Porteous and Armstrong, 1991).

Another commonly used enzyme, lysozyme, is of doubtful effectiveness for fungi. One disadvantage of enzymes is that they require relatively long incubations. Chemical methods may result in some degree of species-selective lysis (Frostegård *et al.*, 1999; Rochelle *et al.*, 1995). Thus, it is very common for chemical and physical methods of lysis to be used together. Typically chelators of metals such as Chelex-100 resin (Kjøller and Rosendahl, 2000) or ethylenediaminetetraacetic acid (EDTA) are used in the extraction mixture to minimize the activity of nucleases. Ironically, DNA yields actually can be reduced with the use of Chelex-100 (Miller *et al.*, 1999).

Utility of Commercially Available Kits

The published methods for DNA extraction and purification from ecological samples that involve fungal cell lysis and DNA purification vary tremendously in cost, effectiveness, and rapidity. Most published methods possess two of the three traits (inexpensive, effective, rapid), but few appear to possess all three. Porteous and Armstrong (1991) reported that it took 48 h to process eight samples. Methods tested by Zhou *et al.* (1996) required 2 h to many hours. Other methods required a few hours to 2 days to process a few samples (Saano *et al.*, 1995). The availability of relatively inexpensive DNA extraction kits has simplified and reduced the time needed for DNA extraction and purification from ecological samples. As with other methods, one must appreciate that any kit may not work for all species of fungi in all soils. Some modifications may prove necessary. Nevertheless, the methods for extraction and purification of mycorrhizal fungal DNA from soils, mycorrhizal roots, and fruiting bodies, detailed in the following sections, which are based on kits from Mo Bio Laboratories (Solana Beach, CA), have worked well for us on various fresh, previously frozen, or dried samples (Dickie *et al.*, 2002). These rapid, relatively inexpensive, and effective methods employ a combination of physical (bead beating) and chemical (detergent) lysis, a combination of chemical flocculation of phenolic inhibitors, ammonium acetate precipitation, and DNA washing on a silica column to remove impurities and require only a benchtop microcentrifuge and a common laboratory vortex type of shaker. Because freeze–thaw, manual grinding, chromatographic, and lengthy incubation steps have been eliminated, the methods are rapid. Using a single vortex shaker, 24 samples can be processed in less than 2 h. Because organic solvents are not used, the methods are also quite safe. Because all chemicals and supplies necessary for extraction and purification are supplied in the kit, the methods are also extremely convenient. The FastDNA SPIN Kit for Soil (Qbiogene, Carlsbad, CA) is another kit for extraction of DNA from soil samples. It was tested by Borneman *et al.* (1996) and Lloyd-Jones and Hunter (2001) and was used by Chelius and Triplett (1999) on arbuscular mycorrhizal roots and Guidot *et al.* (2003) on forest soil containing ectomycorrhizal fungi. Another kit designed for soil samples is the SoilMaster DNA Extraction Kit (Epicentre, Madison, WI).

Extraction and Purification of Mycorrhizal Fungal DNA from Soil Samples

This method is based on the Mo Bio Laboratories Soil DNA Isolation Kit. Because we recommend a few modifications to the published procedure from the manufacturer, detailed steps and explanations are given here.

1. Add soil samples to extraction tubes.[1] For mineral soil, use approximately 0.25 g/tube. For organic substrates (e.g., litter and forest floor layers), use approximately 0.06 g.[2]

2. Gently shake (by hand) or vortex the extraction tubes for a few seconds to moisten the samples with the buffer in the tubes.

3. Add 60 μl detergent lysis solution (SD1) to each extraction tube.

4. Add 200 μl inhibitor removal solution (IRS) to each extraction tube.[3]

5. Attach extraction tubes to vortex adapter[4] and shake on the highest setting for 15 min.

6. Centrifuge the extraction tubes at 10,000g for 30 s.

7. Transfer supernatant containing DNA (approximately 400 μl) to a clean Microfuge tube. This may still contain some particulate matter.

---

[1] Soil samples may be either fresh, previously frozen, or dried by air or with the aid of silica gel. We have found that for ectomycorrhizal fungi fresh and frozen soil samples (not more than 2 mo old) work well. For arbuscular mycorrhizal fungi, fresh or dry soil samples work well. Extraction tubes (included in the kit) are 2-ml Microfuge tubes with screw caps and contain extraction buffer and small garnet particles.

[2] This should be equivalent to approximately 0.25 ml. If more volume is added, there may not be sufficient free volume in the extraction tube to produce the large forces for bead beating.

[3] IRS is a chemical flocculent that effectively removes phenolic inhibitors. Koide and Dickie (2002) demonstrated its efficacy.

[4] The vortex adapter is a flat plate that attaches to the head of the vortex shaker, allowing one to secure 12 extraction tubes. Mo Bio Laboratories manufactures adapters to fit either a Genie 2 (Scientific Industries, Bohemia, NY) or a LabNet (LabNet International, Woodbridge, NJ) vortex mixer. We have used both and prefer to use the adapter on the LabNet because the head is brass, allowing the adapter to be attached by screws. With the Genie 2, the adapter attaches via a plastic head, which becomes loose with wear.

8. Add 250 μl SD2 (ammonium acetate) to each tube, vortex 5 s, and incubate at low temperature ($-20°$ to $5°$ for 10 min) to precipitate proteins and other impurities.

9. Centrifuge Microfuge tubes at 10,000g for 1 min.

10. Transfer supernatant ($\sim$600 μl) to clean Microfuge tubes.

11. Add 1300 μl SD3[5] to supernatant in each Microfuge tube and vortex 5 s.

12. Load approximately 650 μl of the solution into a spin filter placed inside a clean Microfuge tube. Centrifuge at 10,000g for 1 min. Discard the flow-through. Repeat for three batches until all the solution has been processed.[6]

13. Wash the adsorbed DNA by adding 300 μl SD4[7] (ethanol solution) to the spin filter and centrifuging at 10,000g for 30 s. Repeat for three washes.

14. Centrifuge once again at 10,000g for 1 min to remove all traces of ethanol.

15. Carefully place the spin filter into a clean Microfuge tube.

16. Add 50 μl SD5 (Tris) to the center of the spin filter, wait 1 min, then centrifuge at 10,000g for 30 s.

17. Discard the spin filter and retain the filtrate, which contains the extracted and purified DNA.

18. Store DNA at $-20°$ or colder.

Extraction and Purification of Mycorrhizal Fungal DNA from Mycorrhizal Roots

Gardes and Bruns (1993) developed a popular method for extraction of mycorrhizal fungal DNA from ectomycorrhizal roots. It consists of extraction of tissues in a CTAB-based solution by freeze–thaw cycling, crushing with a micropestle, and incubating at an elevated temperature, followed by organic solvent extraction and isopropanol pelleting and washing of DNA. Henrion et al. (1992) described an SDS extraction of ground tissue followed by organic solvent extraction and isopropanol pelleting and washing of DNA. These methods have been applied successfully to arbuscular mycorrhizal roots (Lanfranco et al., 1999; Yamato 2001). Both methods were developed before kits were readily available, but they are still frequently used, because they are inexpensive, effective, and moderately rapid. Edwards et al. (1997) tested a number of methods for extracting fungal

---

[5] SD3 is a guanidine salt and allows the DNA in the sample to be adsorbed to the spin filter.

[6] The DNA is now adsorbed onto the spin filter, ready for washing in the next step.

[7] The kit comes with enough ethanol wash to perform this step only once. Repeated washings have proven beneficial. We use 70% ethanol for this.

DNA from arbuscular mycorrhizal roots including the method of Gardes and Bruns (1993). The most effective method to extract homogenized roots used potassium ethyl xanthate and a high concentration of NaCl and incubation for 1 h at an elevated temperature, which was followed by organic solvent extraction and isopropanol pelleting and washing of DNA. Koide and Dickie (2002) published a method for extraction of fungal DNA from ectomycorrhizal roots that is more rapid than those published methods. It was based on the Mo Bio Microbial DNA kit. That method requires a freeze–thaw step because the combination of bead beating and detergent lysis does not reliably release ectomycorrhizal fungal DNA. Moreover, that method is probably not very effective in extracting DNA from arbuscular mycorrhizal fungi from roots as the fungal tissues occur primarily within the root. The following modified method, which also uses the Mo Bio Laboratories Microbial DNA kit, has proven to work well with both ectomycorrhizal and arbuscular mycorrhizal roots and does not require the freeze–thaw step.

1. Place two or three 2.4-mm Zirconia beads[8] into each extraction tube.

2. Place root sample[9] into the extraction tube.

3. Attach extraction tubes to vortex adapter and vortex on high for 10 min to pulverize the root sample.

4. Add 300 μl bead solution, 50 μl IRS[10] solution, and 50 μl MD1 (detergent lysis solution). Vortex again for 15 min on high.

5. Centrifuge the extraction tubes at 10,000g for 1 min.

6. Transfer supernatant containing DNA ($\sim$300 μl) to a clean Microfuge tube.

7. Add 100 μl MD2 (ammonium acetate) to each Microfuge tube, vortex 5 s, and incubate at low temperature ($-20°$ to $5°$ for 10 min) to precipitate proteins and other impurities.

---

[8] Extraction tubes already contain garnet beads supplied by Mo Bio. These can either be removed before use or retained and used in conjunction with the zirconia beads. Zirconia beads are supplied by Biospec Products, Bartlesville, OK. We thank Dr. Zhihua Zhou for sharing with us her idea to use zirconia beads for this purpose.

[9] The root sample may be as small as a single ectomycorrhizal root tip or several millimeters of arbuscular mycorrhizal root, depending on the need. As for the extraction of soil samples, there must be sufficient room to allow the beads to do their work. Fresh roots may be used. We usually dry ectomycorrhizal roots with silica gel and then store them in the freezer ($-20°$). The same procedure can be followed for arbuscular mycorrhizal roots, but we have simply kept them at room temperature in a dried state.

[10] IRS is a chemical flocculent that effectively removes phenolic inhibitors. Koide and Dickie (2002) demonstrated its efficacy. It is necessary for many roots as they are known to contain high concentrations of tannins and other phenolic compounds (Koide et al., 1998).

8. Centrifuge the Microfuge tubes at 10,000g for 1 min.

9. Transfer supernatant ($\sim$450 μl) to clean Microfuge tubes.

10. Add 900 μl MD3[11] to supernatant in each Microfuge tube and vortex 5 s.

11. Load approximately 700 μl of the solution into a spin filter placed inside a clean Microfuge tube. Centrifuge at 10,000g for 30 s. Discard the flow-through. Repeat until all the solution has been processed.[12]

12. Wash the adsorbed DNA by adding 300 μl MD4[13] (ethanol solution) to the spin filter and centrifuging at 10,000g for 30 s. Repeat for two washes.

13. Centrifuge once again at 10,000g for 1 min to remove all traces of ethanol.

14. Carefully place the spin filter into a clean Microfuge tube.

15. Add 50 μl MD5 (Tris) to the center of the spin filter, wait 1 min, then centrifuge at 10,000g for 30 s.

16. Discard the spin filter and retain the filtrate, which contains the extracted and purified DNA.

17. Store DNA at $-20°$ or colder.

Extraction and Purification of Mycorrhizal Fungal DNA from Sporocarp Tissue

DNA can be extracted from fresh or frozen or dried sporocarps. In principle, this is accomplished as for the roots, except that it is necessary to add only a single zirconium bead to the garnet beads in each extraction tube.

Extraction of DNA from Arbuscular Mycorrhizal Fungal Chlamydospores

Some authors have used simple procedures for extraction of DNA from chlamydospores, usually without purification (Lanfranco et al., 1999; van Tuinen et al., 1998). For the purpose of PCR, however, special extraction steps are not necessary. The reaction can be accomplished by placing crushed spores directly into the reaction mixture (Kjøller and Rosendahl, 2000; Renker et al., 2003). We have used this method successfully on a wide range of species.

---

[11] See footnote 5.
[12] See footnote 6.
[13] See footnote 7.

Concluding Remarks

We are the beneficiaries of pioneers who developed methods for the extraction and purification of DNA from ecological samples, which in many cases were soil or underwater sediments. Although much of the previous method was for bacteria, slight modifications of those methods also made it possible to extract DNA effectively from fungal tissues despite their thick cell walls. Relatively inexpensive, rapid, and effective commercially available kits have now made it possible for the nonspecialist to do this research. This has opened important areas of research in agriculture and ecology that only recently were impossible to address.

Acknowledgments

I thank the A. W. Mellon Foundation and the U.S. Department of Agriculture for research funding. I also thank Ylva Besmer, Ian Dickie, Jori Sharda, and Bing Xu for help in optimizing these methods.

References

Aldwell, F. E. B., and Hall, I. R. (1987). A review of serological techniques for the identification of mycorrhizal fungi. *In* "Mycorrhizae in the Next Decade. Practical Applications and Research Priorities. Proceedings of the 7th North American Conference on Mycorrhizae" (D. M. Sylvia, L. L. Hung, and J. H. Graham, eds.), pp. 305–307. Institute of Food and Agricultural Sciences, University of Florida, Gainesville, FL.

Armstrong, J. L., Fowles, N. L., and Rygiewicz, P. T. (1989). Restriction fragment length polymorphisms distinguish ectomycorrhizal fungi. *Plant Soil* **116**, 1–7.

Bakken, L. R., and Olsen, R. A. (1989). DNA content of soil bacteria of different cell size. *Soil Biol. Biochem.* **21**, 789–793.

Bending, G. D., and Read, D. J. (1996). Nitrogen mobilization from protein-polyphenol complex by ericoid and ectomycorrhizal fungi. *Soil Biol. Biochem.* **28**, 1603–1612.

Bentivenga, S. P., and Morton, J. B. (1994). Stability and heritability of fatty acid methyl ester profiles of Glomalean endomycorrhizal fungi. *Mycol. Res.* **98**, 1419–1426.

Berthelet, M., Whyte, L. G., and Greer, C. W. (1996). Rapid, direct extraction of DNA from soils for PCR analysis using polyvinylpolypyrrolidone spin columns. *FEMS Microbiol. Lett.* **138**, 17–22.

Bever, J. D., Pringle, A., and Schultz, P. A. (2002). Dynamics within the plant-arbuscular mycorrhizal fungal mutualism: Testing the nature of community feedback. *In* "Mycorrhizal Ecology" (M. G. A. van der Heijden and I. R. Sanders, eds.), pp. 267–292. Springer, Berlin.

Borneman, J., Skroch, P. W., O'Sullivan, K. M., Palus, J. A., Rumjanek, N. G., Jansen, J. L., Nienhuis, J., and Triplett, E. W. (1996). Molecular microbial diversity of an agricultural soil in Wisconsin. *Applied Environ. Microbiol.* **62**, 1935–1943.

Braid, M. D., Daniels, L. M., and Kitts, C. L. (2003). Removal of PCR inhibitors from soil DNA by chemical flocculation. *J. Microbiol. Methods* **52**, 389–393.

Brownlee, C., Duddridge, J. A., Malibari, A., and Reed, D. J. (1983). The structure and function of mycelial systems of ectomycorrhizal roots with special reference to their role in assimilate and water transport. *Plant Soil* **71**, 433–443.

Bruce, K. D., Hirons, W. D., Hobman, J. L., Osborn, A. M., Strike, P., and Ritchie, D. A. (1992). Amplification of DNA from native populations of soil bacteria by using the polymerase chain reaction. *Applied Environ. Microbiol.* **58,** 3413–3416.

Bruns, M. A., Stephen, J. R., Kowalchuk, G. A., Prosser, J. I., and Paul, E. A. (1999). Comparative diversity of ammonia oxidizer 16S rRNA gene sequences in native, tilled, and successional soils. *Applied Environ. Microbiol.* **65,** 2994–3000.

Bruns, T. D. (1995). Thoughts on the processes that maintain local species diversity of ectomycorrhizal fungi. *Plant Soil* **170,** 63–73.

Chelius, M. K., and Triplett, E. W. (1999). Rapid detection of arbuscular mycorrhizae in roots and oil of an intensively managed turfgrass system by PCR amplification of small subunit rDNA. *Mycorrhiza* **9,** 61–64.

Claassen, V. P., Zasoski, R. J., and Tyler, B. M. (1996). A method for direct soil extraction and PCR amplification of endomycorrhizal fungal DNA. *Mycorrhiza* **6,** 447–450.

Cullen, D. W., and Hirsch, P. R. (1998). Simple and rapid method for direct extraction of microbial DNA from soil for PCR. *Soil Biol. Biochem.* **30,** 983–993.

Dart, R. T. (1976). Effect of temperature on the fatty-acid composition of *Sporotrichum thermophile. Trans. Br. Mycol. Soc.* **66,** 532–533.

Dickie, I. A., Xu, B., and Koide, R. T. (2002). Vertical niche differentiation of ectomy-corrhizal hyphae in soil as shown by T-RFLP analysis. *New Phytol.* **156,** 527–535.

Edwards, S. G., Fitter, A. H., and Young, J. P. W. (1997). Quantification of an arbuscular mycorrhizal fungus, *Glomus mosseae,* within plant roots by competitive polymerase chain reaction. *Mycol. Res.* **101,** 1440–1444.

Frostegård, Å., Courtois, S., Ramisse, V., Clerc, S., Bernillon, D., Le Gall, F., Jeannin, P., Nesme, X., and Simonet, P. (1999). Quantification of bias related to the extraction of DNA directly from soils. *Applied Environ. Microbiol.* **65,** 5409–5420.

Gardes, M., and Bruns, T. D. (1993). ITS primers with enhanced specificity for basidiomycetes – application to the identification of mycorrhizae and rusts. *Mol. Ecol.* **2,** 113–118.

Gardes, M., White, T. J., Fortin, J. A., Bruns, T. D., and Taylor, J. W. (1991). Identification of indigenous and introduced symbiotic fungi in ectomycorrhizae by amplification of nuclear and mitochondrial ribosomal DNA. *Can. J. Botany* **69,** 180–190.

Guidot, A., Debaud, J.-C., Effosse, A., and Marmeisse, R. (2003). Below-ground distribution and persistence of an ectomycorrhizal fungus. *New Phytol.* **161,** 539–547.

Hart, M., and Klironomos, J. N. (2002). Diversity of arbuscular mycorrhizal fungi and ecosystem functioning. *In* "Mycorrhizal Ecology" (M. G. A. van der Heijden and I. R. Sanders, eds.), pp. 225–242. Springer, Berlin.

Henrion, B., Le Tacon, F., and Martin, F. (1992). Rapid identification of genetic variation of ectomycorrhizal fungi by amplification of ribosomal RNA genes. *New Phytol.* **122,** 289–298.

Horton, T. R., and Bruns, T. D. (2001). The molecular revolution in ectomycorrhizal ecology: Peeking into the black box. *Mol. Ecol.* **10,** 1855–1871.

Kjøller, R., and Rosendahl, S. (2000). Detection of arbuscular mycorrhizal fungi (Glomales) in roots by nested PCR and SSCP (single stranded conformation polymorphism). *Plant Soil* **226,** 1889–1896.

Knaebel, D. B., and Crawford, R. L. (1995). Extraction and purification of microbial DNA from petroleum-contaminated soils and detection of low numbers of toluene, octane and pesticide degraders by multiplex polymerase chain reaction and Southern analysis. *Mol. Ecol.* **4,** 579–591.

Koide, R. T. (1991). Nutrient supply, nutrient demand and plant response to mycorrhizal infection. *New Phytol.* **117,** 365–386.

Koide, R. T. and Dickie, I. A. (2002). Kit-based, low-toxicity method for extracting and purifying fungal DNA from ectomycorrhizal roots. *BioTechniques* **32,** 52–56.

Koide, R. T., Shumway, D. L., and Stevens, C.M. (2000). Soluble carbohydrates of red pine (*Pinus resinosa*) mycorrhizas and mycorrhizal fungi. *Mycol. Res.* **104,** 834–840.

Koide, R. T., Suomi, L., and Berghage, R. (1998). Tree-fungus interactions in ectomycorrhiza symbiosis. *In* "Phytochemical Signals and Plant-Microbe Interactions" (J. T. Romeo, K. R. Downum, and R. Verpoorte, eds.), pp. 57–70. Plenum Press, New York.

Krause, D. O., Smith, W. J., and McSweeney, C. S. (2001). Extraction of microbial DNA from rumen contents containing plant tannins. *BioTechniques* **31,** 294–299.

Kreader, C. A. (1996). Relief of amplification inhibition in PCR with bovine serum albumin or T4 gene 32 protein. *Applied Environ. Microbiol.* **62,** 1102–1106.

Lanfranco, L., Delpero, M., and Bonfante, P. (1999). Intrasporal variability of ribosomal sequences in the endomycorrhizal fungus *Gigaspora margarita*. *Mol. Ecol.* **8,** 37–45.

Leung, K., Trevors, J. T., and van Elsas, J. D. (1995). Extraction and amplification of DNA from the rhizosphere and rhizoplane of plants. *In* "Nucleic Acids in the Environment" (J. T. Trevors and J. D. van Elsas, eds.), pp. 69–87. Springer, Berlin.

Lloyd-Jones, G., and Hunter, D. W. F. (2001). Comparison of rapid DNA extraction methods applied to contrasting New Zealand soils. *Soil Biol. Biochem.* **33,** 2053–2059.

Miller, D. N., Bryant, J. E., Madsen, E. L., and Ghiorse, W. C. (1999). Evaluation and optimization of DNA extraction and purification procedures for soil and sediment samples. *Applied Environ. Microbiol.* **65,** 4715–4724.

Miller, R. M., and Jastrow, J. D. (2000). Mycorrhizal fungi influence soil structure. *In* "Arbuscular Mycorrhizas: Physiology and Function" (Y. Kapulnik and D. D. Douds, Jr., eds.), pp. 3–18. Kluwer, Dordrecht.

Moré, M. I., Herrick, J. B., Silva, M. C., Ghiorse, W. C., and Madsen, E. L. (1994). Quantitative cell lysis of indigenous microorganisms and rapid extraction of microbial DNA from sediment. *Applied Environ. Microbiol.* **60,** 1572–1580.

Murray, H. G., and Thompson, W. F. (1980). Rapid isolation of high molecular weight DNA. *Nucleic Acids Res.* **8,** 4321–4325.

Ogram, A., Sayler, G. S., and Barkay, T. (1987). The extraction and purification of microbial DNA from sediments. *J. Microbiol. Methods* **7,** 57–66.

Paget, E., Monrozier, L. J., and Simonet, P. (1992). Adsorption of DNA on clay minerals: Protection against DNaseI and influence on gene transfer. *FEMS Microbiol. Lett.* **97,** 31–39.

Porteous, L. A., and Armstrong, J. L. (1991). Recovery of bulk DNA from soil by a rapid, small-scale extraction method. *Current Microbiol.* **22,** 345–358.

Read, D. J. (1993). Mycorrhiza in plant communities. *In* "Mycorrhiza Synthesis" (I. Tommerup, ed.), pp. 1–32. Academic Press, San Diego.

Renker, C., Heinrichs, J., Kaldorf, M., and Buscot, F. (2003). Combining nested PCR and restriction digest of the internal transcribed spacer region to characterize arbuscular mycorrhizal fungi on roots from the field. *Mycorrhiza* **13,** 191–198.

Rochelle, P. A., Will, J. A. K., Fry, J. C., Jenkins, G. J. S., Parkes, R. J., Turley, M., and Weightman, A. J. (1995). Extraction and amplification of 16S rRNA genes from deep marine sediments and seawater to assess bacterial community diversity. *In* "Nucleic Acids in the Environment" (J. T. Trevors and J. D. van Elsas, eds.), pp. 219–239. Springer, Berlin.

Romanowski, G., Lorenz, M. G., and Wackernagle, W. (1993). Use of polymerase chain reaction and electroporation of *Escherichia coli* to monitor the persistence of extracellular plasmid DNA introduced into natural soils. *Applied Environ. Microbiol.* **59,** 3438–3446.

Saano, A., Tas, E., Pippola, S., and Lindström, K, and van Elsas, J. D. (1995). Extraction and analysis of microbial DNA from soil. *In* "Nucleic Acids in the Environment" (J. T. Trevors and J. D. van Elsas, eds.), pp. 49–67. Springer, Berlin.

Schüssler, A., Schwarzott, D., and Walker (2001). A new fungal phylum, the *Glomeromycota*: Phylogeny and evolution. *Mycol. Res.* **105**, 1413–1421.

Smalla, K., Cresswell, N., Mendonca-Hagler, L. C., Wolters, A., and van Elsas, J. D. (1993). Rapid DNA extraction protocol from soil for polymerase chain reaction-mediated amplification. *J. Applied Bacteriol.* **74**, 78–85.

Smith, S. E., and Read, D. J. (1997). "Mycorrhizal Symbiosis," 2nd Ed. Academic Press, San Diego.

Smith, S. E., Smith, F. A., and Jakobsen, I. (2003). Mycorrhizal fungi can dominate phosphate supply to plants irrespective of growth responses. *Plant Physiol.* **133**, 16–20.

Steffan, R. J., Goksoyr, J., Bej, A. K., and Atlas, R. M. (1988). Recovery of DNA from soils and sediments. *Applied Environ. Microbiol.* **54**, 2908–2915.

Tebbe, C.C., and Vahjen, W. (1993). Interference of humic acids and DNA extracted directly from soil in detection and transformation of recombinant DNA from bacteria and a yeast. *Applied Environ. Microbiol.* **59**, 2657–2665.

Torsvik, V. L. (1980). Isolation of bacterial DNA from soil. *Soil Biol. Biochem.* **12**, 15–21.

Tsai, Y. L., and Olson, B. H. (1992). Detection of low numbers of bacterial cells in soils and sediments by polymerase chain reaction. *Applied Environ. Microbiol.* **58**, 754–757.

v. Wintzingerode, F., Göbel, U. B., and Stackebrandt, E. (1997). Determination of microbial diversity in environmental samples: Pitfalls of PCR-based rRNA analysis. *FEMS Microbiol. Rev.* **21**, 213–229.

van der Heijden, M. G. A. (1999). Arbuscular mycorrhizal fungi as a determinant of plant diversity: In search for underlying mechanism and general principles. *In* "Mycorrhizal ecology" (M. G. A. van der Heijden and I. R. Sanders, eds.), pp. 243–266. Springer, Berlin.

van Tuinen, D., Jacquot, E., Zhao, B., Gollotte, A., and Gianinazzi-Pearson, V. (1998). Characterization of root colonization profiles by a microcosm community of arbuscular mycorrhizal fungi using 25S rDNA-targeted nested PCR. *Mol. Ecol.* **7**, 879–887.

Vestal, J. R., and White, D. C. (1989). Lipid analysis in microbial ecology. *BioScience* **39**, 535–541.

Wilson, I. G. (1997). Inhibition and facilitation of nucleic acid amplification. *Applied Environ. Microbiol.* **63**, 3741–3751.

Xin, Z., Velten, J. P., Oliver, M. J., and Burke, J. J. (2003). High-throughput DNA extraction method suitable for PCR. *BioTechniques* **34**, 820–825.

Yamato, M. (2001). Identification of a mycorrhizal fungus in the roots of achlorophyllous *Sciaphila tosaensis* Makino (Triuridaceae). *Mycorrhiza* **11**, 83–88.

Young, Burghoff, R. L., Keim, L. G., Minak-Bernero, V., Lute, J. R., and Hinton, S. M. (1993). Polyvinylpyrrolidone-agarose gel electrophoresis purification of polymerase chain reaction-amplifiable DNA from soils. *Applied Environ. Microbiol.* **59**, 1972–1974.

Zhou, J., Bruns, M. A., and Tiedje, J. M. (1996). DNA recovery from soils of diverse composition. *Applied Environ. Microbiol.* **62**, 316–322.

[6] Nucleic Acid Isolation from Ecological Samples—Animal Scat and Other Associated Materials

By LORI S. EGGERT, JESÚS E. MALDONADO, and ROBERT C. FLEISCHER

Abstract

Noninvasive sampling is very attractive to field biologists and has tremendous potential for studying secretive species and being a cost-effective method of increasing sample sizes in studies of large, dangerous animals. Extracting DNA from noninvasively collected samples can be challenging, and the methods have been developed mainly through modification of previously developed protocols for other sample types. We present the most commonly used methods along with modifications used by some researchers to deal with the problem of coextraction of polymerase chain reaction (PCR) inhibitors. Although it is difficult to generalize about which methods should be used on particular sample types, we discuss the success of the methods in studies to date. We close with general suggestions for dealing with potential problems associated with the analysis of DNA obtained from noninvasively collected samples.

Introduction

The field of molecular ecology has grown exponentially. New classes of polymorphic genetic markers, such as microsatellites, can be used to genotype individuals from free-ranging populations, providing insights into questions that would have required years of field studies (Kohn *et al.*, 1999). Previously, these genetic studies required blood or fresh tissue, making it particularly challenging to study animals with low population densities or those that are elusive or dangerous. The development of polymerase chain reaction (PCR) ushered in a new era in which it became possible to conduct studies using the DNA obtained from noninvasively collected samples.

In the past, the term *noninvasive* has been used somewhat ambiguously to describe any situation in which the animal is not destroyed to obtain a sample for analysis. Here, we follow the definition of Taberlet *et al.* (1999), considering noninvasively collected samples as those that do not require either capture or handling of the animal. For instance, plucking hairs from a captured small mammal for DNA analysis is termed *nondestructive* sampling, whereas collecting hairs from tapes or snags placed around its

METHODS IN ENZYMOLOGY, VOL. 395

burrow is *noninvasive*. Studies have presented data obtained from DNA that was isolated from noninvasively collected samples such as shed hairs, shed feathers, shed snake skins, sloughed whale skin, buccal cells from partially eaten fruits, feces, urine, scent markings, and eggshells.

Noninvasive sampling is very attractive to field biologists as it is easier and safer for both the animal and the researcher. It has tremendous potential for studying secretive species and for being a cost-effective method of increasing sample sizes in studies of large, dangerous mammals. There is, however, a trade-off between ease of sample collection and the quality of the DNA obtained. Noninvasively collected samples often contain small amounts of degraded DNA from the target species that is mixed with bacterial DNA and substances that can inhibit the PCR. The risk of contamination during DNA extraction and amplification is similar to that of ancient DNA (see Chapter 7). Because the DNA is likely to be degraded, only short fragments can be amplified, and the risk of obtaining incorrect genotypes is increased. Another obvious downside to noninvasive sampling is that no museum voucher is kept for reference. However, some portion of the sample could potentially be retained in collections for future verification of the identification.

Many studies have reported the amplification of mitochondrial DNA (mtDNA) fragments from noninvasively collected samples. Because sequences are analyzed for these fragments, there is no question that they represent the target species. When the goal is to genotype nuclear fragments that will not be sequenced, such as microsatellites, incorrect genotypes can result (Taberlet *et al.*, 1996, 1999). The problems of *allelic dropout* and *null alleles* occur when one or another of the alleles at a heterozygous locus fails to amplify in the PCR. Allelic dropout is a common problem that usually results from the degraded and dilute nature of DNA extracts from noninvasively collected samples. Null alleles are more often the result of primer mismatch, particularly when primers from one species are used in related species. The less common but equally serious problem of *spurious alleles* occurs when DNA other than that of the species of interest is amplified. When these alleles are coamplified in homozygous individuals, they appear to be genuine and may bias the results of a study if they are not detected. In a population study, however, they may be identified when they are seen as a third allele in a true heterozygote.

Despite the potential problems, the benefits of noninvasive sampling for answering important questions in ecology and conservation biology are such that its use has continued to grow. Several DNA extraction methods have been used by those in the field, and they are outlined with reference to the type of sample for which they have been found to be appropriate. These methods are followed by a discussion of general issues associated with the analysis of DNA obtained from noninvasively collected samples.

Extraction Methods

Modified Phenol–Chloroform Extraction

The modified phenol–chloroform extraction method is only slightly modified from standard phenol–chloroform extraction methods (Sambrook *et al.*, 1989) and has been used successfully to extract total genomic DNA from hair roots or feathers (Table I). Although it has been used to extract DNA from feces, in most cases other methods provided superior results (Reed *et al.*, 1997).

Prior to DNA extraction, specimens should be kept dry and away from ultraviolet (UV) light to retard further degradation of cells. Brown collecting envelopes are ideal for this purpose. To remove any contaminating debris, rinse the feather tip or hair root gently with sterile water and allow to air-dry.

- *Feather tip:* With a sterile razor blade, cut the tip away from the feather. Using sterile tweezers, secure the tip while cutting it lengthwise into two pieces, exposing the entire inside of the tip.
- *Hair:* Use a sterile razor blade to cut close to the root. Discard the hair shaft, which contains pigments that could inhibit the PCR. If you are confident that you have multiple hairs from the same animal, they may be extracted together.

Place both halves of the feather tip or the hair root(s) in a 1.5-ml microcentrifuge tube containing 500 μl of lysis buffer (0.2 M NaCl, 0.05 M Na$_2$EDTA, pH 8.0). Add sodium dodecyl sulfate (SDS) to 0.5% of the final volume and proteinase K (PK) at 100–200 μg/ml. Mix and incubate overnight at 55°. (*Optional*: Add ribonuclease A at 100–200 μg/ml of solution, mix, and incubate for 4 h at 55°.)

Extract one time each with equal volumes of phenol and chloroform: isoamyl alcohol (24:1). To precipitate DNA, add 1/10 volume of 3 M NaOAc and 2.5 volumes of cold absolute ethanol to the aqueous layer from the extractions. Incubate overnight in a −20° freezer, then centrifuge for 10 min at 13,000 rpm to pellet DNA. Wash once with 70% ethanol, air-dry the pellet, and resuspend in either sterile water or suspension buffer (0.05 M Tris, pH 7.5, 0.1 M NaCl, 0.025 M Na$_2$EDTA).

In our experience, amplification success is enhanced by purification of the extracted DNA with Microcon Centrifugal Filter Units (Millipore, Billerica, MA).

TABLE I

STUDIES USING ECOLOGICAL SAMPLES AS SOURCES OF DNA HAVE TARGETED A
NUMBER OF SPECIES[a]

Extraction method	DNA source	Species	Target loci	Citation
Modified Phenol Chloroform	Hair	Asian and African elephants	Nuclear introns	Greenwood and Pääbo, 1999
	Hair	Northern hairy nosed wombat	Microsatellites	Alpers et al., 2003
Chelex	Hair	Black and brown bears	mtDNA, Y, microsatellites	Woods et al., 1999
	Hair	Brown bear	Y, microsatellites	Taberlet et al., 1993, 1997
	Hair	Brown bear	Microsatellites	Mowat and Strobeck, 2000
	Feces	Canids	mtDNA	Paxinos et al., 1997
	Hair	Chimpanzee	Microsatellites	Goossens et al., 2002
	Hair	Chimpanzee	Microsatellites	Morin and Woodruff, 1992
	Feathers	Spectacled eider	Microsatellites	Pearce et al., 1997
	Hair	Chimpanzee	Microsatellites	Constable et al., 2001
	Scent mark	Giant panda	mtDNA	Ding et al., 1998
	Hair	Northern hairy nosed wombat	Y, microsatelllites	Sloane et al., 2000
Qiagen Standard DNA Kits	Feces	Common wombat	Microsatellites	Banks et al., 2002
	Hair	Marten	Microsatellites	Mowat and Paetkau, 2002
	Feces	Black and sun bears	mtDNA, X/Y, microsatellites	Wasser et al., 1997
	Feces	Asian elephant	mtDNA, microsatellites	Fernando, et al., 2003
	Feces	Brown bear	mtDNA, X/Y, microsatellites	Murphy et al., 2003
	Feces	Lynx	mtDNA	Pires and Fernandes 2003
	Feces	San Joaquin kit fox	mtDNA, X/Y	Ortega et al., 2004; Smith et al., 2003
	Hair, Feces	Chimpanzee	Microsatellites	Constable et al., 2001
	Feces	Arctic fox, red fox, wolverine	mtDNA	Dalén et al., 2004
	Feces	Carnivores	mtDNA	Farrell et al., 2000

(continued)

TABLE I   (*continued*)

Extraction method	DNA source	Species	Target loci	Citation
	Feces	Reindeer, sheep	mtDNA, microsatellites	Flagstad *et al.*, 1999
	Hair	Lynx	mtDNA	Mills *et al.*, 2000
	Feces	Mustelids	mtDNA	Murakami, 2002
	Hair	Mustelids	mtDNA	Riddle *et al.*, 2003
DNA QIAamp DNA Stool Kit	Feces	Wolf	Microsatellites	Creel *et al.*, 2003
	Feces	Baboon	Microsatellites	Bayes *et al.*, 2000
	Feces	Chimpanzee, gorilla, gibbon	X/Y, microsatellites	Bradley *et al.*, 2000, 2001
	Feces	Mandrill	mtDNA	Telfer *et al.*, 2003
	Feces	Red wolf and coyote	mtDNA	Adams *et al.*, 2003
	Feces	Black rhinoceros	Microsatellites	Garnier *et al.*, 2001
GuSCN/Silica	Feces	Great bustard	mtDNA, CHD	Idaghdour *et al.*, 2003
	Feces	Brown bear	Microsatellites	Taberlet *et al.*, 1996, 1997
	Feces	Dolphins	mtDNA, microsatellites	Parsons *et al.*, 1999; Parsons, 2001
	Feces	Eurasian otter	Y, microsatellites	Dallas *et al.*, 2000, 2003
	Feces	African elephant	mtDNA, Y, microsatellites	Eggert *et al.*, 2002, 2003
	Feces	Brown bear	mtDNA, Y	Kohn *et al.*, 1995
	Feces	Seals	mtDNA, Y, microsatellites	Reed *et al.*, 1997
	Feces	Baboon	mtDNA, nuclear introns	Frantzen *et al.*, 1998
	Feces	Eurasian badger	Microsatellites	Frantz *et al.*, 2003; Wilson *et al.*, 2003
	Feces	Red deer	Y, microsatellites	Huber *et al.*, 2002
	Feces	Bonobo	Microsatellites	Gerloff *et al.*, 1995
	Feces	Wolf	Microsatellites X/Y, mtDNA	Lucchini *et al.*, 2002
Other Methods				
IsoQuick Kit	Feces	Coyote	mtDNA, Y, microsatellites	Kohn *et al.*, 1999
IsoQuick Kit	Feces	Lynx	mtDNA	Pires and Fernandes, 2003

[a] These studies differ by DNA source and target loci, factors that should be considered when selecting the extraction method.

Chelex-100 (InstaGene)

The Chelex-100 method is a slightly modified version of the protocol of Walsh *et al.* (1991) and uses a commercial resin preparation known as InstaGene (formerly Chelex-100; BioRad, Hercules, CA). It has been used for extracting DNA from hair, feathers, and scent markings (Table I) and modified somewhat for use in feces. For preparation of feather tips or hair before DNA extraction, see the previous section on the modified phenol–chloroform extraction method.

Place both halves of the feather, or all hair roots, into a 1.5-ml micro-centrifuge tube with 250 μl of 5% Chelex-100 or InstaGene. Vortex 10–15 s, make sure all of the DNA sources are in the liquid. Incubate overnight at 56°. Vortex again for 10–15 s, making sure all DNA sources are in the liquid. Boil 15 min at 100°, and centrifuge 3 min at 10,000 rpm. Some researchers have removed the DNA extract from the InstaGene beads at this point, placing it in a separate tube for storage, whereas others leave the extract with the beads.

Use 1–10 μl in 25 μl PCR. The extract should be stored at −20° between uses. Before each use, vortex 5 s and centrifuge at 10,000 rpm for 3 min. The DNA may eventually be degraded by vortexing, so it should be done as seldom as possible.

To adapt this method for use with feces, Paxinos *et al.* (1997) suspended 0.1–0.3 μg of fecal material in 500 μl 5% Chelex-100. Samples were boiled for 7 min, vortexed at full speed, boiled another 7 min, and centrifuged at full speed for 5 min. The supernatant was removed from the beads and aliquoted into sterile tubes. Four μl were used in 100 μl volume reactions.

Commercially Available Kits

In recent years, the use of commercially available kits for isolation of DNA from animal feces has become more and more widespread (Table I). Several of these kits were originally designed for isolation of genomic DNA, mtDNA, and viral DNA from various sample sources including fresh or frozen animal tissues and cells, yeasts, or bacteria, whole blood, plasma, serum, buffy coat, bone marrow, other body fluids, lymphocytes, cultured cells, tissue, and forensic specimens. However, because these kits provide fast and easy methods for purification of total DNA for PCR, molecular ecologists have incorporated minor modifications to the protocols for isolating fecal DNA from various mammal species (Table I). The purified DNA resulting from these kits is suited for PCR, random amplified polymorphic DNA (RAPD), amplified fragment length poly-morphism (AFLP), and restriction fraction length polymorphism (RFLP) applications.

Qiagen Extraction Kits. The DNeasy Tissue kit and QIAamp DNA Mini kit are two of the most widely used commercial kits for fecal DNA isolation and can recover DNA fragments as small as 100 bp. Both have buffer systems that have been optimized to allow direct cell lysis followed by selective binding of DNA to a silica gel membrane. Simple centrifugation processing removes contaminants and enzyme inhibitors and allows simultaneous processing of multiple samples. In addition, these procedures are suitable for a wide range of sample sizes. Both kits are designed to minimize the possibility of sample-to-sample cross-contamination and to allow safe handling of potentially infectious samples.

A small amount of the external portion of the feces is placed in a 1.5-ml microcentrifuge tube, to which lysis buffer and PK are added (mechanical homogenization is not necessary). The buffer is used to provide optimal DNA binding conditions; after incubation, the lysate is loaded onto a spin column. During a brief centrifugation, DNA is selectively bound to the silica gel membrane as contaminants pass through. Remaining contaminants and enzyme inhibitors are removed during two wash steps in which conditions are optimized to remove residual contaminants without affecting DNA binding. DNA is then eluted in water or buffer and is ready for direct addition to PCR or other enzymatic reactions.

We have found that certain modifications enhance the effectiveness of extractions from fecal samples using these Qiagen kits. DNA yield is increased if the spin column is incubated with the elution buffer at room temperature for 5 min before centrifugation. Although the DNA may be eluted in water, elution with the kit buffer (AE) reduces the chances of degradation by hydrolysis. Storage of the samples at $-20°$ will further reduce the chance of degradation.

Qiagen has marketed a kit specifically designed to extract DNA from fecal samples. The QIAamp DNA Stool Mini Kit was designed for purification of total DNA from fresh or frozen human stool samples and is aimed at clinical work. The protocol has been adopted by several molecular ecologists, primarily those using fecal DNA to study primates, but has also been used successfully in fecal DNA studies of carnivores and herbivores (Table I).

The main difference between the stool kit and the two mentioned earlier lies in the first steps of the protocol. First, stool samples are placed in a lysis buffer and bacterial cells and those of other pathogens in the stool are lysed by incubating the homogenate at $70°$ (if necessary, this temperature can be increased to $95°$). After lysis, DNA-damaging substances and PCR inhibitors are adsorbed to a unique reagent provided in a tablet form called *InhibitEX*. The reagent is then pelleted by centrifugation and the DNA in the supernatant is purified on spin columns. This is very similar to the Qiagen procedures described earlier and involves digestion of proteins, binding DNA to a silica gel membrane, washing away impurities, and elution of purified DNA from the spin column.

The QIAamp DNA Stool Mini Kit is optimized for use with up to 220 mg of fresh or frozen feces but can also be used with larger or smaller amounts. Starting with larger amounts of feces is recommended when the target DNA is not distributed homogeneously throughout the sample or is at a low concentration; a larger amount of starting material will increase the likelihood of purifying DNA from low-titer sources in fecal samples. The QIAamp protocols can also be used for samples of less than 180 mg, which is often the case in ecological studies and forensic samples. In such cases the amounts of buffers and other reagents must be reduced proportionally. DNA of up to 20 kb can be purified, and the yield is typically 15–60 μg (depending on the sample and the way it was stored, this may range from 5 to 100 μg) at a concentration of 75–300 ng/μl.

Some researchers have made minor modifications to these protocols, including (but not limited to) preextraction steps to remove inhibitors (Banks *et al.*, 2002) and postextraction purification with Microcon Centrifugal Filter Units (Millipore, Billerica, MA).

Other Extraction Kits. The IsoQuick Kit (ORCA Research, Inc., Botell, WA) has also been used for extracting DNA from feces. This kit uses the chaotropic salt guanidine-thiocyanate (GuSCN), which lyses cells and inhibits nuclease activities, along with a nuclease binding matrix. During extraction, the aqueous phase, containing nucleic acids, is separated from the organic phase, which contains proteins and other cellular contaminants. The nucleic acids are then precipitated with alcohol and resuspended in water or buffer. At least one study followed this method with a purification step using Microcon Centrifugal Filter Units.

Guanidine Thiocyanate/Silica

This approach has been successfully used to extract DNA from the feces of a number of mammals, including both carnivores and herbivores. It is based on the protocols and reagents of Boom *et al.* (1990), in which nucleic acids are bound to silica in the presence of GuSCN. The silica and bound nucleic acids are washed with buffers containing GuSCN, after which purified DNA is eluted in either water or buffer.

If feces are in preservation buffer, mix completely and add 1.5 ml of the slurry to a 1.5-ml centrifuge tube. Centrifuge at full speed for 15 min and discard the supernatant. If feces are dry or frozen, scrape a small amount of the outside layer into a 1.5-ml microcentrifuge tube. Add extraction buffer (lysis buffer L6) until volume is approximately 1.5 ml and all feces are in suspension. Incubate overnight at 60°. Centrifuge at 5000 rpm just long enough to pellet the debris (2–5 min), and pipette approximately 750 μl of the supernatant into a sterile tube containing 250 μl of fresh L6 buffer. Add 50 μl of the silica suspension, mix well, and incubate at least 1 h at room temperature with moderate shaking. Centrifuge 3 min at highest speed and discard the supernatant.

To wash the silica pellet, add 1 ml of wash buffer and mix well by stirring with a pipette tip (vortexing may cause DNA shearing). Centrifuge 3 min, discard the liquid, and repeat. Add 1 ml of 70% ethanol, mix well, centrifuge 3 min, and discard the liquid. Dry the pellet at 56° to remove any residue of ethanol.

To elute the DNA, resuspend the pellet in 200 μl of sterile water, heat to 56°, centrifuge 1 min at full speed, carefully remove the supernatant, making sure not to disturb the silica, and place the extracted DNA in a labeled sterile tube. This step can be repeated with 100 μl of water if the feces were very fresh and DNA yield is likely to be good.

Aliquot a small amount of the extract into a tube for your immediate needs. All extracts should be stored at −20°, and only the aliquot should be thawed for testing because each freeze–thaw cycle may further degrade the DNA.

The preparation of these reagents and buffers is described in detail in Boom *et al.* (1990). One study reported that using diatomaceous earth (DE) in place of silica gives better results (Wasser *et al.*, 1997), and others have added a preextraction incubation step with a 2% cetyltrimethyl-ammonium bromide (CTAB) solution (Huber *et al.*, 2002; Parsons *et al.*, 2001) to remove inhibitors that may be coextracted from plant material.

Extracting DNA from Ancient Feces (Molecular Coproscopy)

Studies have shown that DNA extracted from trace fossils, such as Ice Age coprolites (fossilized feces) of extinct animals and archaic humans, can be used to identify both the defecator and their diets, complementing pollen, cuticle, and macrofossil analyses (Hofreiter *et al.*, 2000; Poinar *et al.*, 1998, 2001). Among the richest archives of ancient plant and animal DNA are the fossil rodent feces preserved in middens in arid regions worldwide, and studies of pooled fecal pellets have been used to conduct studies of phylogeography and population biology in the fossil record (Kuch *et al.*, 2002).

Poinar *et al.* (1998) demonstrated that coprolites contain DNA that can be released upon the addition of a chemical compound *N*-phenacylthiazo-lium bromide (PTB) during the extraction procedure. This compound breaks crosslinks between reducing sugars and primary amines formed by the Maillard reaction (Ledl and Scleischer, 1992). PTB treatment has been demonstrated to allow DNA sequences from the animal that deposited the coprolite and the plants it had ingested to be determined (Hofreiter *et al.*, 2000).

Kuch et al. *(2002) Method.* Pooled fecal pellets are ground to a fine powder in liquid nitrogen in a Spex freezer grinding mill. Approximately 0.1 g of pellet powder is incubated in a standard PK buffer (100 μl PK [10 mg/ml] and 100 μl of 0.1 *M* PTB). Samples then receive 800 μl of 0.5% CTAB/2% PVP solution and are incubated overnight in a 37° oven on a rotary wheel. The tubes are then centrifuged at 16,000*g* for 2 min and the supernatant is extracted with phenol, phenol:chloroform, and chloroform. The supernatant is concentrated with Centricon 30s (Millipore) to approximately 120 μl.

Samples are further extracted via a silica purification step to remove inhibitors from the extraction.

Modified Poinar et al. *(1998) Method.* Coprolite pieces are ground to a fine powder under liquid nitrogen with a pestle and mortar. To 0.2 g of powder, 1.4 ml of extraction buffer (0.1 *M* Tris–HCl, pH 8.0, 2 m*M* EDTA, 0.7 *M* NaCl, 1% SDS, 50 m*M* DTT, 0.2 mg/ml PK) is added. Samples are incubated for 24 h at 37° on a rotary shaker and 100 μl of a 100-m*M* PTB solution are added. The incubation is continued for another 72 h. The samples are extracted twice with chloroform, and DNA is recovered by binding the supernatant to silica and washing (Höss and Pääbo, 1993; Poinar *et al.*, 1998).

General Recommendations

Sample preservation methods strongly affect the results of fecal DNA extraction. Although those methods are outside the scope of this chapter, they have been discussed in several reviews including (but not limited to) those of Frantzen *et al.* (1998), Murphy *et al.* (2002), Wasser *et al.* (1997), Whittier *et al.* (1999).

For all noninvasively collected samples, the risk of contamination during DNA extraction and amplification is similar to that of ancient DNA. In the laboratory, a separate space (preferably a separate room) and set of reagents should be devoted to extraction of DNA from noninvasively collected samples. This area must be free of contamination from PCR products and from DNA-rich samples (blood or other tissues). All surfaces should be cleaned with a 10% bleach solution before extracting DNA and all laboratory equipment and supplies should be sterilized, either by cleaning with bleach or by autoclaving.

With every group of extractions, include a sample of each of the reagents and perform all extraction steps on these controls. These extractions should be included in the PCR for detection of contaminants during the extraction process.

In general, for optimum results, use the minimum amount of DNA eluate possible in the PCR. The volume of eluate used as template should not exceed 10% of the final volume of the PCR mixture. It is highly recommended that 0.1 $\mu g/\mu l$ bovine serum albumin (BSA) is added to PCR assays to maximize robustness. To maximize PCR specificity, we recommend using Hot Start PCR. It is best to amplify only short fragments (i.e., 100–300 bp.). The amplification of longer mtDNA fragments has been reported, but it is more likely the result of the higher copy number of mtDNA in mammalian cells rather than of improved methods of DNA preservation and extraction.

Some noninvasively collected samples will contain DNA from species other than the target species. For instance, when extracting DNA from feces, it is impossible to avoid extracting the DNA of enteric bacteria and some foods, thus making it difficult to determine the amount of DNA available from the species of interest. Morin *et al.* (2001) have proposed an interesting method, quantitative PCR analysis, for estimating the amount of target DNA. This procedure uses a standard PCR with the addition of a double-labeled oligonucleotide probe and reports the fluorescence levels at each cycle of the PCR. When compared with the fluorescence curves for a set of samples of known DNA amounts, the amount of DNA template for the target species can be estimated. This allows researchers to detect samples that have such low quantities of DNA from the species of interest that genotypes would likely be unreliable.

Whether or not this test is used, it is important to confirm genotypes that are obtained from noninvasively collected samples. The most commonly used approach is the "multiple-tubes" method of Taberlet *et al.* (1996), which involves the scoring of each extract at each locus seven times. Although this should reduce the problem of genotyping error to very low levels, it requires a large investment in time and reagents and may limit the number of loci that can be used as it quickly exhausts the small amount of DNA than can be extracted from noninvasively collected samples.

Acknowledgments

NSF grant no. DEB0083944, Alternatives Research and Development Foundation (ARDF), and the Smithsonian Institution supported this research.

References

Adams, J. R., Kelly, B. T., and Waits, L. P. (2003). Using faecal DNA sampling and GIS to monitor hybridization between red wolves (*Canis rufus*) and coyotes (*Canis latrans*). *Mol. Ecol.* **12,** 2175–2186.
Alpers, D. L., Taylor, A. C., Sunnucks, P., Bellman, S. A., and Sherwin, W. B. (2003). Pooling hair samples to increase DNA yield for PCR. *Cons. Gen.* **4,** 779–788.

Banks, S. C., Piggott, M. P., Hansen, B. D., Robinson, N. A., and Taylor, A. C. (2002). Wombat coprogenetics: Enumerating a common wombat population by microsatellite analysis of faecal DNA. *Aust. J. Zool.* **50**, 193–204.

Bayes, M. K., Smith, K. L., Alberts, S. C., Altmann, J., and Bruford, M. W. (2000). Testing the reliability of microsatellite typing from faecal DNA in the Savannah baboon. *Cons. Gen.* **1**, 173–176.

Boom, R., Sol, C. J. A., Salimans, M. M. M., Jansen, C. L., van Dillen Werthien, P. M. E., and van der Noordaa, J. (1990). Rapid and simple method for purification of nucleic acids. *J. Clin. Microbiol.* **28**, 495–503.

Bradley, B. J., Boesch, C., and Vigilant, L. (2000). Identification and redesign of human microsatellite markers for genotyping wild chimpanzee (*Pan troglodytes verus*) and gorilla (*Gorilla gorilla gorilla*) DNA from faeces. *Cons. Gen.* **1**, 289–292.

Bradley, B. J., Chambers, K. E., and Vigilant, L. (2001). Accurate DNA-based sex identification of apes using non-invasive samples. *Cons. Gen.* **2**, 179–181.

Constable, J. L., Ashley, M. V., Goodall, J., and Pusey, A. E. (2001). Noninvasive paternity assignment in Gombe chimpanzees. *Mol. Ecol.* **10**, 1279–1300.

Creel, S., Spong, G., Sands, J. L., Rotella, J., Zeigle, J., Joe, L., Murphy, K. M., and Smith, D. (2003). Population size estimation in Yellowstone wolves with error-prone noninvasive microsatellite genotypes. *Mol. Ecol.* **12**, 2003–2009.

Dalén, L., Götherström, A., and Angerbjörn, A. (2004). Identifying species from pieces of faeces. *Cons. Gen.* **5**, 1–3.

Dallas, J. F., Carss, D. N., Marshall, F., Koepfli, K.-P., Kruuk, H., Piertney, S. B., and Bacon, P. J. (2000). Sex identification of the Eurasian otter *Lutra lutra* by PCR typing of spraints. *Cons. Gen.* **1**, 181–183.

Dallas, J. F., Coxon, K. E., Sykes, T., Chanin, P. R. F., Marshall, F. A., Carss, D. N., Bacon, P. J., Piertney, S. B., and Racey, P. A. (2003). Similar estimates of population genetic composition and sex ratio derived from carcasses and faeces of Eurasian otter *Lutra lutra*. *Mol. Ecol.* **12**, 275–282.

Ding, B., Zhang, Y.-P., and Ryder, O. A. (1998). Extraction, PCR amplification, and sequencing of mitochondrial DNA from scent mark and feces in the giant panda. *Zoo Biol.* **17**, 499–504.

Eggert, L. S., Rasner, C. A., and Woodruff, D. S. (2002). The evolution and phylogeography of the African elephant (*Loxodonta africana*), inferred from mitochondrial DNA sequence and nuclear microsatellite markers. *Proc. R. Soc. Lond. B.* **269**, 1993–2006.

Eggert, L. S., Eggert, J. A., and Woodruff, D. S. (2003). Estimating population sizes for elusive animals: The forest elephants of Kakum National Park, Ghana. *Mol. Ecol.* **12**, 1389–1402.

Farrell, L. E., Roman, J., and Sunquist, M. E. (2000). Dietary separation of sympatric carnivores identified by molecular analysis of scats. *Mol. Ecol.* **9**, 1583–1590.

Fernando, P., Vidya, T. N. C., Rajapakse, C., Dangolla, A., and Melnick, D. J. (2003). Reliable noninvasive genotyping: Fantasy or reality? *J. Heredity* **94**, 115–123.

Flagstad, Ø., Røed, K., Stacy, J. E., and Jakobsen, K. S. (1999). Reliable noninvasive genotyping based on excremental PCR of nuclear DNA purified with a magnetic bead protocol. *Mol. Ecol.* **8**, 879–883.

Frantz, A. C., Pope, L. C., Carpenter, P. J., Roper, T. J., Wilson, G J., Delahay, R. J., and Burke, T. (2003). Reliable microsatellite genotyping of the Eurasian badger (*Meles meles*) using faecal DNA. *Mol. Ecol.* **12**, 1649–1661.

Frantzen, M. A. J., Silk, J. B., Ferguson, J. W. H., Wayne, R. K., and Kohn, M. H. (1998). Empirical evaluation of preservation methods for faecal DNA. *Mol. Ecol.* **7**, 1423–1428.

Garnier, J. N., Bruford, M. W., and Goossens, B. (2001). Mating system and reproductive skew in the black rhinoceros. *Mol. Ecol.* **10,** 2031–2041.

Gerloff, U., Schlötterer, C., Rassmann, K., Rambold, I., Hohmann, F., Fruth, B., and Tautz, D. (1995). Amplification of hypervariable simple sequence repeats (microsatellites) from excremental DNA of wild living bonobos (*Pan paniscus*). *Mol. Ecol.* **4,** 515–518.

Goossens, B., Funk, S. M., Vidal, C., Latour, S., Jamart, A., Ancrenaz, M., Wickings, E. J., Tutin, C. E. G., and Bruford, M. W. (2002). Measuring genetic diversity in translocation programmes: Principles and application to a chimpanzee release project. *Anim. Conserv.* **5,** 225–236.

Greenwood, A. D., and Pääbo, S. (1999). Nuclear insertion sequences of mitochondrial DNA predominate in hair but not in blood of elephants. *Mol. Ecol.* **8,** 133–137.

Hofreiter, M., Poinar, H. N., Spaulding, W. G., Bauer, K., Martin, P. S., Possnert, G., and Pääbo, S. (2000). A molecular analysis of ground sloth diet through the last glaciation. *Mol. Ecol.* **9,** 1975–1984.

Höss, M., and Pääbo, S. (1993). DNA extraction from Pleistocene bones by a silica-based purification method. *Nucleic Acids Res.* **21,** 3913–3914.

Huber, S., Bruns, U., and Arnold, W. (2002). Sex determination of red deer using polymerase chain reaction of DNA from feces. *Wildlife Soc. Bull.* **30,** 208–212.

Idaghdour, Y., Broderick, D., and Korrida, A. (2003). Faeces as a source of DNA for molecular studies in a threatened population of great butards. *Cons. Gen.* **4,** 789–792.

Kohn, M., Knauer, F., Stoffella, A., Schröder, W., and Pääbo, S. (1995). Conservation genetics of the European brown bear—a study using excremental PCR of nuclear and mitochondrial sequences. *Mol. Ecol.* **4,** 95–103.

Kohn, M. H., York, E. C., Kamradt, D. A., Haught, G., Sauvajot, R. M., and Wayne, R. K. (1999). Estimating population size by genotyping faeces. *Proc. Roy. Soc. Lond. B.* **266,** 657–663.

Kuch, M., Rohland, N., Betancourt, J. L., Latorre, C., Steppan, S., and Poinar, H. N. (2002). Molecular analysis of a 11,700-year-old rodent midden from the Atacama Desert, Chile. *Mol. Ecol.* **11,** 913–924.

Ledl, F., and Schleicher, E. (1992). Die Maillard Reaktion in Lebensmitteln und im menschlichen Körper—neue Ergebnisse zur Chemie, Biochemie und Medizin. *Angewandte Chem.* **102,** 597–626.

Lucchini, V., Fabbri, E., Marucco, F., Ricci, S., Boitani, L., and Randi, E. (2002). Noninvasive molecular tracking of colonizing wolf (*Canis lupus*) packs in the western Italian Alps. *Mol. Ecol.* **11,** 857–868.

Mills, L. S., Pilgrim, K. L., Schwartz, M. K., and McKelvey, K. (2000). Identifying lynx and other North American felids based on mtDNA analysis. *Cons. Gen.* **1,** 285–288.

Morin, P. A., Chambers, K. E., and Boesch, C. and Vigilant, L. (2001). Quantitative polymerase chain reaction analysis of DNA from noninvasive samples for accurate microsatellite genotyping of wild chimpanzees (*Pan troglodytes verus*). *Mol. Ecol.* **10,** 1835–1844.

Morin, P. A., and Woodruff, D. S. (1992). Paternity exclusion using multiple hypervariable microsatellite loci amplified from nuclear DNA of hair cells. *In* "Paternity in Primates: Genetic Tests and Theories" (R. D. Martin, A. F. Dixson, and E. J. Wickings, eds.), pp. 63–81. Karger, Basel.

Mowat, G., and Paetkau, D. (2002). Estimating marten *Martes americana* population size using hair capture and genetic tagging. *Wildlife Biol.* **8,** 201–209.

Mowat, G., and Strobeck, C. (2000). Estimating population size of grizzly bears using hair capture, DNA profiling, and mark-recapture analysis. *J. Wildlife Manag.* **64,** 183–193.

Murakami, T. (2002). Species identification of mustelids by comparing partial sequences on mitochondrial DNA from fecal samples. *J. Vet. Med. Sci.* **64,** 321–323.

Murphy, M. A., Waits, L. P., Kendall, K. C., Wasser, S. K., Higbee, J. A., and Bogden, R. (2002). An evaluation of long-term preservation methods for brown bear (*Ursus arctos*) faecal DNA samples. *Cons. Gen.* **3**, 435–440.

Murphy, M. A., Waits, L. P., and Kendall, K. C. (2003). The influence of diet on faecal DNA amplification and sex identification in brown bears (*Ursus arctos*). *Mol. Ecol.* **12**, 2261–2265.

Ortega, J., Franco, M. d. R., Adams, B. A., Ralls, K., Maldonado, J. E. (2005). A reliable, noninvasive method for sex determination in the endangered San Joaquin kit fox. *Cons. Gen.* **5**, 715–718.

Parsons, K. (2001). Reliable microsatellite genotyping of dolphin DNA from faeces. *Mol. Ecol.* **1**, 341–344.

Parsons, K. M., Dallas, J. F., Claridge, D. E., Durban, J. W., Balcomb, K. C., III, Thompson, P. M., and Noble, L. R. (1999). Amplifying dolphin mitochondrial DNA from faecal plumes. *Mol. Ecol.* **8**, 1753–1768.

Paxinos, E., McIntosh, C., Ralls, K., and Fleischer, R. (1997). A noninvasive method for distinguishing among canid species: Amplification and enzyme restriction of DNA from dung. *Mol. Ecol.* **6**, 483–486.

Pearce, J. M., Fields, R. L., and Scribner, K. T. (1997). Nest materials as a source of genetic data for avian ecological studies. *J. Field Ornith.* **68**, 471–481.

Pires, A. E., and Fernandes, M. L. (2003). Last lynxes in Portugal? Molecular approaches in a pre-extinction scenario. *Cons. Gen.* **4**, 525–532.

Poinar, H. N., Hofreiter, M., Spaulding, W. G., Martin, P. S., Stankiewicz, B. A., Bland, H., Evershed, R. P., Possnert, G., and Pääbo, S. (1998). Molecular coproscopy: Dung and diet of the extinct ground sloth *Nothrotheriops shastensis*. *Science* **281**, 402–406.

Poinar, H. N., Kuch, M., Sobolik, K., Barnes, I., Stankiewicz, B. A., Kuder, T., Spaulding, W. G., Bryant, V. M., Cooper, A., and Pääbo, S. (2001). A molecular analysis of dietary diversity for three archaic Native Americans. *Proc. Natl. Acad. Sci. USA* **98**, 4317–4322.

Reed, J. Z., Tollit, D. J., Thompson, P. M., and Amos, W. (1997). Molecular scatology: The use of molecular analysis to assign species, sex and individual identity to seal faeces. *Mol. Ecol.* **6**, 225–234.

Riddle, A. E., Pilgrim, K. L., Mills, L. S., McKelvey, K .S., and Ruggiero, L. F. (2003). Identification of mustelids using mitochondrial DNA and non-invasive sampling. *Cons. Gen.* **4**, 241–243.

Sambrook, J., Fritsch, E. F., and Maniatis, T. (1989). "Molecular Cloning: A Laboratory Manual." Cold Spring Harbor Laboratory Press, New York.

Sloane, M. A., Sunnucks, P., Alpers, D., Beheregaray, L. B., and Taylor, A. C. (2000). Highly reliable genetic identification of individual northern hairy nosed wombats from single remotely collected hairs: A feasible censusing method. *Mol. Ecol.* **9**, 1233–1240.

Smith, D. A., Ralls, K., Hurt, A., Adams, B., Parker, M., Davenport, B., Smith, M. C., and Maldonado, J. E. (2003). Detection and accuracy rates of dogs trained to find scats of San Joaquin kit foxes (*Vulpes macrotis mutica*). *Anim. Cons.* **6**, 339–346.

Taberlet, P., Matlock, H., Dubois-Paganon, C., and Bouvet, J. (1993). Sexing free-ranging brown bears *Ursus arctos* using hairs found in the field. *Mol. Ecol.* **2**, 399–403.

Taberlet, P., Griffin, S., Goossens, B., Questiau, S., Manceau, V., Escaravage, N., Waits, L. P., and Bouvet, J. (1996). Reliable genotyping of samples with very low DNA quantities using PCR. *Nucleic Acids Res.* **24**, 3189–3194.

Taberlet, P., Camarra, J.-J., Griffin, S., Uhrès, E., Hanotte, O., Waits, L. P., Dubois-Paganon, C., Burke, T., and Bouvet, J. (1997). Noninvasive genetic tracking of the endangered Pyrenean brown bear population. *Mol. Ecol.* **6**, 869–876.

Taberlet, P., Waits, L. P., and Luikart, G. (1999). Noninvasive genetic sampling: Look before you leap. *Trends Ecol. Evol.* **14**, 323–327.

Telfer, P. T., Souquière, S., Clifford, S. L., Abernethy, K. A., Bruford, M. W., Disotell, T. R., Sterner, K. N., Roques, P., Marx, P. A., and Wickings, E. J. (2003). Molecular evidence for deep phylogenetic divergence in *Mandrillus sphinx. Mol. Ecol.* **12,** 2019–2024.

Walsh, P. S., Metzger, D. A., and Higuchi, R. (1991). Chelex 100 as a medium for simple extraction of DNA for PCR-based typing from forensic material. *BioTechniques* **10,** 506–513.

Wasser, S. K., Houston, C. S., Koehler, G. M., Cadd, G. G., and Fain, S. R. (1997). Techniques for application of faecal DNA methods to field studies of Ursids. *Mol. Ecol.* **6,** 1091–1097.

Whittier, C. A., Dhar, A. K., Stem, C., Goodall, J., and Alcivar-Warren, A. (1999). Comparison of DNA extraction methods for PCR amplification of mitochondrial cytochrome c oxidase subunit II (COII) DNA from primate fecal samples. *Biotechnol. Tech.* **13,** 771–779.

Wilson, G. J., Frantz, A. C., Pope, L. C., Roper, T. J., Burke, T. A., Cheeseman, C. L., and Delahay, R. J. (2003). Estimation of badger abundance using faecal DNA typing. *J. Applied Ecol.* **40,** 658–666.

Woods, J. G., Paetkau, D., Lewis, D., McLellan, B. N., Proctor, M., and Strobeck, C. (1999). Genetic tagging of free-ranging black and brown bears. *Wildlife Soc. Bull.* **27,** 616–627.

[7] Isolation and Analysis of DNA from Archaeological, Clinical, and Natural History Specimens

By CONNIE J. MULLIGAN

Abstract

The use of ancient DNA (aDNA) in the reconstruction of population origins and evolution is becoming increasingly common. Novel methods exist for the isolation, purification, and analysis of aDNA because these DNA templates are likely to be damaged, fragmented and/or associated with non-nucleic acid material. However, contamination of ancient specimens and DNA extracts with modern DNA is more widespread than is generally acknowledged and remains a significant problem in aDNA analysis. Studies of human aDNA are uniquely sensitive to contamination due to the continual presence of potential contamination sources. Meticulous authentication of results and careful selection of polymorphic markers capable of distinguishing between aDNA and probable DNA contaminants are critical to a successful aDNA study.

Introduction

Ancient DNA (aDNA) technology has emerged within the past decade as one of the breakthroughs borne of the revolution in molecular biology that began with the polymerase chain reaction (PCR). The enormous capacity of PCR has stimulated the study of large numbers of individuals that is fundamental to population genetics and has permitted the investigation of DNA samples that are too degraded or damaged for analysis by traditional cloning methods. "Ancient" samples are generally those that were not collected for the purpose of immediate DNA analysis and include archaeological, clinical, and natural history specimens. Because these specimens were not originally collected or preserved for nucleic acid analysis, endogenous DNA is often damaged to an extent that enzymatic amplification can be quite difficult, if not impossible, to achieve. The types of DNA damage that are primarily encountered include modifications of pyrimidine and sugar residues, baseless sites, intermolecular crosslinks, and fragmented DNA (Paabo, 1989).

Damage of aDNA increases the potential for another characteristic of aDNA PCR: that of contamination. Because PCR analysis involves the exponential generation of new, synthetic DNA products from a few molecules, contamination with exogenous DNA in one of the initial PCR cycles can result in exclusive amplification of the contaminating DNA. This possibility is increased in aDNA analysis in which the contaminant is likely to be undamaged DNA that will be amplified preferentially over the damaged endogenous DNA. The growing number of published aDNA studies and number of assayed samples and polymorphic sites may give the impression that all technological hurdles associated with aDNA technology have been overcome. However, identification of contamination and authentication of results remain the most critical issues in aDNA methodology. Early spectacular claims of successful DNA extraction from extremely old specimens, such as 17–20-million-year-old *Magnolia* leaf fossils (Golenberg *et al.*, 1990), 25–135-million-year-old specimens preserved in amber (Cano *et al.*, 1993; DeSalle *et al.*, 1992), and 80-million-year-old dinosaur bones (Woodward *et al.*, 1994), have generally been disproved or cast into serious doubt (Austin *et al.*, 1997; DeSalle *et al.*, 1993; Hedges and Schweitzer 1995; Sidow *et al.*, 1991). Table I lists the controls essential for an aDNA study and suggests additional experiments in the case of questionable or controversial results.

Human specimens are uniquely sensitive to contamination simply because every person involved in the study represents a potential source of contaminating DNA. For instance, analysis of DNA extracted from the Neanderthal type specimen (Krings *et al.*, 1997) revealed two distinct sets of mitochondrial DNA (mtDNA) sequences: one significantly different from modern humans and proposed to be Neanderthal in origin and one identical to the human reference sequence (Anderson *et al.*, 1981) and presumed to reflect modern human contamination. Regardless of the difficulties associated with aDNA analysis, in the past year we have seen a

TABLE I
CONTAMINATION CONTROLS FOR ANCIENT DNA STUDIES

Essential controls
  No studies of modern specimens/physically isolated ancient DNA lab
  Disposable labware wherever possible
  Disposable lab coats, gloves, breathing masks, and head coverings
  Dedicated pipetmen
  Filtered pipet tips
  Purchase of molecular biology–grade water and solutions
  Frequent treatment of bench and equipment with 20% bleach solution (autoclaving will
    destroy living organisms but will not destroy DNA; UV irradiation of solid objects
    is maximally effective only at a perpendicular angle)
  Multiple DNA extractions for each specimen (from the same and multiple samples,
    if possible)
  No positive PCR controls
  UV irradiation (254 nm for 20 min) of PCRs before DNA + enzyme addition
  Multiple blank extract and negative PCR controls
  Multiple PCRs of each locus for each specimen
  Appropriate molecular behavior of aDNA, i.e., PCR success is inversely related
    to amplification fragment size
  Cloning and DNA sequence analysis
  Appropriate phylogenetic sense of results
Additional controls when results are questionable or controversial
  Replication of results in an independent lab
  Analysis of associated remains to demonstrate feasibility of DNA isolation
  DNA quantitation or amino acid analysis/racemization analysis of extract to demonstrate
    organic preservation consistent with DNA survival

Note: aDNA, ancient DNA; PCR, polymerase chain reaction; UV, ultraviolet.

range of aDNA success stories, including (1) a population study of mitochondrial and nuclear markers (single-copy nuclear markers are more difficult to amplify than multicopy mtDNA) in which partial genealogies were reconstructed in a human population about 2000 years old (Keyser-Tracqui *et al.*, 2003); (2) sequence analysis of two Cro-Magnon humans that confirmed the independence of modern human and Neanderthal genetic lineages (Caramelli *et al.*, 2003); and (3) identification of extreme sexual dimorphism in a single species of extinct New Zealand moa that was previously thought to be three distinct species based on overall size differences (Bunce *et al.*, 2003).

Methods: Isolation of DNA

DNA can be isolated from any organism. The types of tissue typically available for DNA extraction can be classified as hard (e.g., bone or teeth) or soft (e.g., muscle or skin). In general, hard tissues are preferred over soft tissues for DNA isolation. Hard tissues typically have not been subjected to any preservation methods. In contrast, soft tissues have often been air-dried (museum skins, mummies, insects in amber) or fixed in formalin (insects, fish, organs) and these preservation methods often increase post-mortem DNA damage (De Giorgi *et al.*, 1994). Both types of tissue are subject to introduction of contaminants whether from bacterial growth on dry remains or groundwater permeation of bones or human handling of museum specimens. Different DNA isolation protocols have been developed for each type of tissue. Hard tissues must be decalcified through use of ethylenediaminetetraacetic acid (EDTA) to release the bound nucleic acid. Preserved soft tissues generally require greatly increased concentrations of and prolonged incubation times with proteinase K (PK) to digest the tissue and release the nucleic acid. A special purification protocol has been developed to facilitate extraction of DNA from formalin-fixed soft tissues (C. Mulligan and N. Tuross, unpublished results). Methods for isolation of DNA from hard and soft tissues are presented in this chapter, with possible modifications listed. In general, a minimalist approach to DNA extraction and purification is recommended to reduce the number of steps in which contamination may be introduced.

Method 1: Isolation of Nucleic Acid from Hard Tissues (Bone or Teeth)

1. *Sample preparation.* Brush or wipe obvious dirt from chosen specimens. Break up bone into pieces sufficiently small to fit into a 15- or 50-ml conical tube. If bone can be broken by hand, specimen preservation is likely to be poor. A hammer can be used to fragment well-preserved bone (after placing specimen in plastic bags of sufficient thickness to withstand hammering). Alternatively, bone can be cut using a bone saw or Dremel tool or material can be powdered using a mortar and pestle or Spex mill (Spex Industries, Los Angeles, CA), being careful to control for scatter of dust, which could serve as a contamination source.

2. *Decontamination.* Place fragments or powder in a 15- or 50-ml conical tube so the tissue accounts for only 1/5 of the volume in the tube. Add 20% (v/v) Clorox bleach solution (1.2% sodium hypochlorite) until all material is covered. Gently rock tube in your hand. Friable or powdered material should be exposed to the bleach solution for 45–60 s, whereas well-preserved dense fragments can be exposed for 2–10 min. Pour off bleach solution and fill tube with purified water. Shake tube intermittently for a total wash time of 2–4 min. Repeat water wash two more times. Kemp and Smith (2003) have found that treatment with 33% bleach for

10–15 min or use of "DNA Away" (E&K Scientific, Campbell, CA) is also effective at reducing contamination.

3. *Decalcification.* Place fragments/powder in a fresh 15-ml conical tube. Add sufficient 0.5 M EDTA to cover all material. Ensure that tubes are securely capped and will not leak. Place tubes on a rocking platform (e.g., Nutator [Becton Dickinson, Sparks, MD], and secure tubes with tape so they will not move. Leave tubes at room temperature for 2–4 days (well-preserved material requires more time for complete decalcification). If material has been excavated and stored at cold temperatures, such as excavated from permafrost sediments, DNA isolation should be performed at 4°.

4. N-*phenacylthiazolium bromide (PTB)/PK digestion.* After 2–4 days, add PTB (Trace Genetics, Davis, CA) to a final concentration of 20 mM and add PK) (Sigma, St. Louis, MO) to a final concentration of 1 μg/ml. Place tubes on a rocking platform at 65° for an overnight incubation. Wearing gloves, check bone/tooth material for status of decalcification. If there are hard pieces remaining in the tissue and complete decalcification is desired, continue DNA isolation at 65° until all hard calcified bits are gone.

5. *Final step for well-preserved tissue.* If tissue was not well preserved, skip to step 6. If tissue was well preserved, further purification may be unnecessary. Because minimal purification is optimal, the DNA extract may be ready to test for DNA polymerase inhibition at this point. All EDTA must be removed because it will inhibit amplification reactions. First, centrifuge tubes to pellet solid material, and then transfer liquid to dialysis tubing and place tubing in a container with 4 liters of water. The number of water changes required to dilute the EDTA to negligible levels can be calculated based on the volume of the DNA extract(s) and the volume of water used in the dialysis. Typically, two to three changes in 4 liters of water will be sufficient for a 2–4 ml sample. Dialysis against water will increase the volume of the sample by as much as an order of magnitude (e.g., a 3-ml sample will increase to 30 ml, so use sufficient dialysis tubing to accommodate the increase in sample volume). After dialysis, samples must be concentrated to reduce volumes. Samples can be concentrated by partial lyophilization or by filtration through Centricon filters (Millipore, Billerica, MA). Lyophilization is preferable if the necessary equipment is available because the opportunity for introduction of contaminants is reduced and because large volumes are dealt with more easily. The DNA extract is now complete and ready to be tested for the presence of DNA polymerase inhibitors (see below).

6. *Further purification of poorly preserved specimens: Organic extractions.* Add an equal volume of 25:24:1 phenol:chloroform:isoamyl alcohol

(PCI) to the EDTA/PTB/PK solution (be sure to use polypropylene tubes because phenol will destroy polystyrene). Rock tubes for 5–15 min, ensuring that an emulsion forms. Centrifuge the mixture for 15 min at 3000 rpm to separate the aqueous and organic layers (higher speed spins are possible if smaller tubes are used). If the two layers are not well separated, centrifuge again for a longer time. Use a pipette to transfer the top aqueous layer to a fresh tube. Leave some of the aqueous layer behind to ensure that the interface between the two layers, where the denatured proteins collect, is not disturbed. Repeat PCI extraction until no protein is visible in the interface (typically two PCI extractions is sufficient). Add an equal volume of 24:1 chloroform:isoamyl alcohol to the aqueous layer (a final extraction with chloroform removes any lingering traces of phenol from the DNA extract). Rock for 5–20 min, ensuring that an emulsion forms. Centrifuge the mixture for 15 min at 3000 rpm. Use a pipette to transfer the top aqueous layer to a fresh tube.

7. *Ethanol precipitation and washes.* If total sample volume is greater than 500 μl, divide the sample into separate Microfuge tubes with equal volumes of 500 μl or less. Add 1/10 volume 3 M sodium acetate and mix solution well. Add 2 volumes of ice-cold 100% ethanol (do not use <100% ethanol) and mix solution well. Store the solution at −20° overnight to allow a precipitate of DNA to form. Recover the DNA by centrifugation at maximum speed in a Microfuge tube for 10 min at 4°. Remove the ethanol supernatant taking care not to disturb the DNA pellet (which may be invisible if you are lucky and have very clean DNA). Add 300 μl of 70% ethanol and recentrifuge at maximum speed in a Microfuge tube for 2 min at 4°. Carefully remove the supernatant taking care not to disturb the DNA pellet. Repeat ethanol wash. Store the open tube on the bench at room temperature until all traces of fluid have evaporated. Dissolve the DNA pellet in the desired volume (usually 50–100 μl) of low TE buffer (10 mM Tris base, 0.1 mM EDTA). The DNA extract is now ready to be tested for the presence of DNA polymerase inhibitors (see below).

Additional Notes on DNA Isolation and Purification from Hard Tissues.

a. *Choice of bone or tooth specimens.* Bone or tooth specimens should be chosen that are as complete and as well preserved as possible. In general, bones are preferred over teeth because of the larger quantity of material available in bones. Teeth are often assumed to be sealed against the surrounding environment and, therefore, resistant to contamination, but this is not true as groundwater and associated organisms can permeate the external surface of teeth. Large robust bones, such as femur or tibia, are preferable over small light bones, such as clavicle or rib, because of the increased density and increased amounts of material in larger bones. When possible, specimens should be chosen that are light colored, have intact epiphyses or edges, have good "heft" to them (assuming the specimens are not fossilized, in which case they will be heavy but lacking in organic material), and have a general appearance of good preservation. If a specimen is small, discolored, or friable, the chances of obtaining amplifiable DNA are lower, and larger amounts of material should be used in the extraction procedure to increase the probability of successful DNA extraction.

   b. *Modification of isolation protocol.* Many DNA isolation protocols using skeletal or dental material have been reported in the literature (see Table II for examples of different protocols). The protocol presented here has been used successfully with bovid and human bones and teeth for extraction of DNA representing a total range from 300 to 2500 years before present (YBP) (Kolman, 1999; Kolman and Tuross, 2000; Kolman *et al.*, 1999; M. Ascunce and C. Mulligan, unpublished results). Many modifications to the extraction protocol are possible and some are listed here.

   i. If the material to be extracted is in good condition and endogenous DNA is presumed to be present, smaller amounts of bone or teeth may be used and the extraction protocol can be performed in Microfuge tubes for added convenience.

   ii. It is possible that different amounts of DNA and inhibitors extract at different times during EDTA decalcification (Kolman and Tuross, 2000; Kolman *et al.*, 1999). During the decalcification step, the EDTA can be changed after a few days, replaced with fresh EDTA, and both EDTA extracts processed as separate DNA isolates.

   iii. PTB is a relatively new chemical in the aDNA literature and was first used in the extraction of DNA from late Pleistocene coprolites (Poinar *et al.*, 1998) and from the third published Neanderthal specimen (Krings *et al.*, 2000). PTB is thought to improve DNA yields by cleaving sugar-derived protein cross-links that may entrap DNA. Regardless of the exact mechanism of PTB, PTB does not appear to harm DNA isolations and can be easily incorporated into the extraction protocol at the PK step. Initially, PTB concentrations of 1 mM were tested, although concentrations as high as 200 mM have been used successfully with teeth (Gilbert *et al.*, 2003). It has not been reported in the literature, but PTB may improve DNA amplification if added directly to the PCR.

   iv. Silica-GuSCN is also frequently used in the initial steps of DNA isolation in combination with low levels of EDTA (instead

TABLE II
SUMMARY OF DNA ISOLATION METHODS USED IN SUCCESSFUL ANCIENT DNA STUDIES

Material extracted	EDTA extraction buffer	Complete decalcification	Silica-GuSCN buffer	Proteinase K digestion	N-phenacyl thiazolium bromide	Organic extraction	Silica purification	Centricon purification/ concentration	EtOH precipitation	References
Hard tissue										
Human bone (300–1000 YBP)	X	X		X						Kolman, 1999; Kolman and Tuross, 2000; Kolman et al., 1999
Human bone (300–9000 YBP)	X			X		X		X		Kaestle and Smith, 2001
Human bone (~700 YBP)			X	X		X		X		Stone and Stoneking, 1993, 1998, 1999
Late Pleistocene animal remains, Neanderthal (~40,000 YBP)	X			X		X	X	X	X	Greenwood et al., 1999; Hoss and Paabo, 1993; Krings et al., 1997
Bear bone (14,000–42,000 YBP)	X			X		X		X		Leonard et al., 2000
Soft tissue										
Coprolites from extinct ground sloth (~20,000 YBP)					X	X	X	X		Poinar et al., 1998
Frozen sediment cores (600–400,000 YBP)				X		X	X			Willerslev et al., 2003

relying solely on high concentrations of EDTA) (Hoss and Paabo, 1993). Many DNA purification kits are also based on a silica-GuSCN protocol, such as the Wizard Plus Megaprep DNA Purification System (Promega, Madison, WI). In these protocols, GuSCN facilitates the binding of DNA to the silica so unbound impurities can be washed away. Silica-based methods should be used with caution, however, because aDNA may be bound to protein and may not interact properly with DNA binding resins, resulting in a loss of DNA (Kolman and Tuross, 2000).

v. Complete decalcification of well-preserved bones may require several weeks if DNA isolation is performed at 4° (Kolman and Tuross, 2000). More typically, complete decalcification requires 3–4 days when performed at room temperature. The level of decalcification can be monitored by feeling the material in EDTA solution (while wearing gloves) and checking for hard decalcified pieces among the soft decalcified material. Complete decalcification of skeletal material may be necessary only for older specimens, as Fisher *et al.* (1993) reported that no decalcification was necessary for bones up to 125 years old.

c. *Purification of DNA extracts.* Minimal purification of DNA extracts will yield maximal recovery of DNA because all purification methods result in the loss of some nucleic acid. Therefore, the number and type of purifications to be performed after initial decalcification should be considered carefully. Although extensive purification seems to be the norm in aDNA studies, I have found that the DNA extracts often can be used immediately after decalcification (and dialysis and sample concentration) (Kolman and Tuross, 2000; Kolman *et al.*, 1999; M. Ascunce and C. Mulligan, unpublished results). If additional purification is deemed necessary, organic extractions most often yield amplifiable DNA extracts and are included in the purification method in the aforementioned protocol. However, phenol extractions can result in losses of up to 50% of total DNA, so organic extractions should be performed only until the organic–aqueous interface is clear of proteinaceous material. Filtration through Centricon filters should not result in appreciable loss of nucleic acid, but filters should be batch-checked for human DNA contamination if human aDNA is being analyzed. Ethanol precipitation of DNA can result in significant loss of low-molecular-weight DNA, such as typical aDNA. DNA binding resin-based methods of purification, such as Glass Milk/GeneClean (Qbiogene, Inc., Carlsbad, CA) and Wizard Megaprep DNA Purification System (Promega, Madison, WI), have been cited extensively in the aDNA literature (Hoss and Paabo, 1993) but should be used cautiously. Kolman and Tuross (2000) reported that silica resin purification resulted in exclusive recovery of contaminating DNA from extracts containing both endogenous and contaminating DNA. The high probability that aDNA is crosslinked to proteins and may not interact as expected with DNA binding resins provides a cautionary note for purification protocols based on these resins.

Method 2: Isolation of Nucleic Acid from Soft Tissues (Formalin-Fixed Tissues, Museum Skins, Mummies)

For the initial steps of DNA isolation from soft tissues, a standard sodium dodecyl sulfate (SDS)/PK digestion is used with two modifications: 1000 times the typical quantity of PK and extended PK incubations are used. Standard phenol:chloroform:isoamyl alcohol (PCI) extraction typically fails to remove all PCR inhibitors from DNA extracts based on formalin-fixed tissues. Therefore, a novel PCI extraction, described by White and Densmore (1992), has been further modified and results in a significant reduction of PCR inhibition (C. Mulligan and N. Tuross, unpublished results).

1. *Sample preparation.* Using a fresh scalpel, dissect a portion of tissue from the chosen specimen that can fit easily into a 2.0-ml tube with a screw cap, about 200–400 mg.

2. *Cell lysis.* Add sufficient lysis buffer (100 mM Tris, pH 8.0, 10 mM EDTA, 100 mM NaCl, 0.1% (w/v) SDS, 50 mM DTT, 0.5 mg/ml PK) to completely cover tissue, typically about 500 μl. Incubate at 50° until tissue is completely digested. End-over-end rotation of tubes during incubation will facilitate digestion. If more than 1 day is required for complete digestion, add an aliquot of PK for each extra day.

3. *Frozen PCI extraction.* After digestion, centrifuge tubes to precipitate particulate matter. Transfer supernatant to a clean 15-ml Corex tube (tubes should be muffled at 400° for 15 min or treated with a 20% bleach solution to destroy contaminating DNA). Add 3.5 volumes of SDS:urea buffer (1%/0% [w/v] SDS, 10 M urea, 240 mM Na$_2$HPO$_4$, pH 6.8, 1 mM EDTA) to the supernatant. Incubate tubes at room temperature for 15–30 min with occasional vortex mixing. Add an equal volume of PCI and incubate at room temperature for 5–10 min with gentle occasional vortex mixing. Spin tubes in a precooled centrifuge at 15,000 rpm for 15 min at −13°. A solid crystallized urea interface must form to obtain maximum extraction of proteins. Tubes should be removed immediately after centrifugation so the organic and aqueous layers do not freeze. If the urea does not crystallize and form a solid interface, a higher concentration of urea can be used or the tubes can be precooled immediately before centrifugation. Remove the aqueous layer and dialyze against two to three

changes of low TE buffer (10 mM Tris, pH 7.5, 0.1 mM EDTA) at 4°. DNA extracts are now ready to test for the presence of DNA polymerase inhibitors (see later discussion).

Coextraction of DNA Polymerase Inhibitors during DNA Isolation. During DNA isolation, compounds may co-purify that inhibit the DNA polymerase used in amplification reactions. These compounds may derive from the soil in buried bones or from formaldehyde in fixed soft tissues. Generally, researchers do not attempt to identify the exact nature and source of an inhibitor but focus on eliminating the inhibitor(s). One strategy is to dilute the DNA extracts until the inhibitor is no longer present at inhibitory concentrations. However, excessive dilution may also dilute the DNA to nonamplifiable concentrations. A second strategy is to purify the DNA extracts in an attempt to eliminate the inhibitor without also eliminating the DNA. In practice, both methods are typically used.

Ideally, one should test for presence of inhibition throughout the DNA isolation and purification procedure so the minimum number of purification steps is performed to maximize DNA yield. Presence of inhibition is assayed by amplifying an unrelated DNA template in the presence of varying amounts of the aDNA extract. Dilutions of the aDNA extract are tested to monitor the inhibition (typical dilutions are 1:10, 1:100, and 1:1000). By determining which dilutions are required to eliminate inhibition, one can determine the most effective purification methods and determine when the DNA extract is free of inhibiting compounds. In theory, DNA from any organism that is not studied in the laboratory can serve as a template in the inhibition assay (to avoid introducing contaminating DNA). In the past, manufacturers included control DNA template and primers in commercial PCR kits that could be used as an inhibition assay, but this is no longer true. Currently, my lab uses an inhibition assay with Sendai virus DNA and primers for the P protein (SM464mod = 5'-TGCAGCTGAGAGCAGTCCCC-3' and SM324 = 5'-GATGCCT-CACCCGGGATCTAGTTG-3'; PCR profile consists of 10 in at 94°, followed by 28 cycles of 30 s at 94°, 30 s at 55°, 30 s at 72°, and a final 5 min extension step at 72°; S. Moyer and S. Smallwood, personal communication). Magnitude of inhibition is determined by visual comparison on an agarose gel of a control PCR, lacking aDNA extract to a series of PCRs with control DNA plus varying amounts of aDNA extract. To achieve successful amplification of the locus of interest, additional dilution may be required beyond the level indicated in the inhibition assay, for example, in a study of African bovid bone samples approximately 1000 years old, the inhibition assay indicated absence of inhibitors at a 1:10 dilution, but successful amplification was achieved only with a 1:100 dilution (M. Ascunce and C. Mulligan, unpublished results).

Methods to Improve PCR Success. Certain chemicals have been found to increase the likelihood of successful DNA amplification. These reagents may be used by themselves or in combination. In some cases, the exact concentration of reagent must be determined empirically. Changes to the PCR profile may also enhance amplification. If initial PCRs do not result in successful amplification of the target DNA, it is recommended to try several or all of the following PCR modifications. Modifications marked by an asterisk are those that appear most frequently in the aDNA literature, suggesting that they most often result in successful DNA amplification.

a. *↑ *Taq* polymerase, ↑ # of amplification cycles: Increased amounts of Taq polymerase and increased number of amplification cycles are used when amplifying aDNA. Typical amounts are 1–2 units of polymerase and 40–50 cycles. Reamplification of a PCR aliquot or a purified amplified DNA band is not advisable because of the increased likelihood of introducing DNA contamination.

b. *Hot-start PCR: Hot-start PCR (withholding an essential component of the PCR until the reaction has reached a temperature that inhibits nonspecific priming and primer oligomerization) is used to optimize the yield of desired PCR product and suppress nonspecific amplification. Hot-start PCR can be achieved by either using polymerase–antibody complexes in which the enzyme is inactive until it has been heated at 95° for 5–10 min (e.g., Amplitaq Gold; Applied Biosystems, Foster City, CA) or by using wax beads in which the embedded Mg^{2+} is released only upon prolonged heating (e.g., HotWax Beads; Invitrogen, Carlsbad, CA).

c. *Spermidine: Low levels of spermidine may facilitate amplification of certain DNA samples (Wan and Wilkins, 1993). However, excess spermidine can inhibit amplification, so samples should be tested first for amplification in the absence of spermidine. In two studies, optimal concentration of spermidine was determined to be $400–800\mu M$ (Kolman and Tuross, 2000; Kolman *et al.*, 1999).

d. *Bovine serum albumin (BSA), glycerol, dimethylsulfoxide (DMSO), formamide: BSA and glycerol are used to stabilize protein–DNA interactions, and reagents such as DMSO and formamide facilitate DNA strand separation by disrupting base pairing. However, the mechanism of these interactions is not known and the best additive must be determined by trial and error. Typical additive concentrations are $10–100\mu g/ml$ BSA, 10% glycerol, 5–10% DMSO, and 2.5–10.0% formamide (Fisher *et al.*, 1993). BSA and DMSO are the most widely used additives, most likely because they most often result in successful amplification.

e. *Nested PCR: Nested PCR is the amplification of a single PCR product through the use of two primer pairs in sequential amplification reactions so the second PCR target is completely contained within the first PCR product. This strategy allows a weak (or invisible) PCR product to be reamplified but increases the specificity of the reaction through the use of two primer pairs.

f. Betaine: It has been suggested that betaine may help stabilize proteins against thermal denaturation and facilitate DNA strand separation by isostabilization of the DNA. Betaine at concentrations of 0.5–2.2 M, in the presence or absence of DMSO, has been found to increase DNA amplification and improve consistency of amplification (Baskaran et al., 1996; Pergams et al., 2003). Betaine has been found to inhibit some DNA polymerases, so addition of betaine may require further modification of the PCR protocol (Baskaran et al., 1996).

g. PTB: The use of PTB to improve aDNA isolation and amplification is still in its infancy. The rationale of using PTB to cleave protein crosslinks and free entrapped DNA may also be reasonable in the context of an amplification reaction.

h. ↑ Mg^{2+}: If residual EDTA in the DNA extract is left over from the decalcification step, the EDTA may reduce the Mg^{2+} concentration in the PCR to suboptimal levels.

i. Touchdown PCR: Touchdown PCR (reducing the annealing temperature by 5–10° during the first 10 PCR cycles) is a method used when the annealing temperature of the PCR primers with the DNA template is not known with certainty. Although there should be no uncertainty in this regard with aDNA, this strategy may facilitate the amplification of a complex DNA extract characterized by fragmented or complexed DNA and presence of unknown, possibly inhibitory, chemicals.

Postmortem Damage to DNA

Since the advent of aDNA analysis, it has been known that DNA isolated from ancient remains frequently shows evidence of molecular damage (Paabo et al., 1988). Postmortem damage to DNA can take the form of modification of pyrimidine or sugar residues, baseless sites, intermolecular crosslinks, and an average low molecular weight of aDNA due to strand breakage (Hofreiter et al., 2001b). The effects of postmortem damage include (1) fragmented DNA that limits the possible length of PCR product, (2) low to undetectable levels of aDNA that allow preferential amplification of contaminating DNA, (3) random insertion of a nucleotide by the DNA polymerase at a damaged baseless site, and (4) PCR "jumping" in which discontiguous DNA fragments are joined through the amplification reaction. The average lifetime for DNA under physiological conditions and a temperature of 15° has been estimated to be approximately 100,000 years

(Hofreiter *et al.*, 2001b), although low temperatures and low humidity may improve preservation and extend this time limit. Research into postmortem damage has focused on the specific types of lesions that are most prevalent and the distribution patterns of postmortem damage. The most common chemical damage to DNA is deamination of cytosine to uracil and deamination of adenine to hypoxanthine, meaning that transitions ($\leftrightarrow$ T and G $\leftrightarrow$ A) are much more common than transversions (Gilbert *et al.*, 2003; Hofreiter *et al.*, 2001a). Comparison of DNA sequences derived directly from PCRs compared to sequences derived from cloned PCR products should identify DNA damage sites (and will detect multiple templates in the DNA extract or PCR indicating contamination). Gilbert *et al.* (2003) identified specific sites in the human mitochondrial control region I (HVRI) that are particularly prone to postmortem damage, and the authors summarize probable misidentification of human haplogroups based on their reported distribution of postmortem damage.

Contamination Control and Authentication of aDNA Results

Despite more than a decade of aDNA research, contamination by modern DNA remains a problem because the many sources and modes of contamination are still not known or understood and, therefore, cannot be completely controlled or eliminated. Identification of contamination and authentication of results in aDNA studies rely on the ability to discriminate between aDNA and contaminating DNA. Experiments must be designed with a goal of identifying all DNA contaminants to distinguish convincingly between endogenous and contaminating DNA. Careful selection of polymorphic markers capable of discriminating between aDNA and probable DNA contaminants is critical. Table I lists the controls essential for an aDNA study and presents additional experiments in the case of questionable or controversial results. Sequence analysis of cloned PCR products has emerged as a particularly useful control because it allows one to directly evaluate individual amplification products for evidence of multiple templates in the reaction (indicating contamination) or DNA damage sites (characteristic of aDNA).

Studies of human populations are uniquely prone to contamination because every researcher, archaeologist, or curator represents a potential source of contamination. Richards *et al.* (1995) reported that approximately 50% of animal bones from a Holocene site in England exhibited contamination with human sequences. Human bones should be assumed similarly contaminated. Analysis of human populations in which genetic markers exist that distinguish between the ancient population and probable contaminants should focus on those discriminatory markers to ensure that contamination is identified quickly. aDNA studies wherein the prehistoric population and likely sources of contamination are genetically similar will be much more difficult to conduct.

Only in nonhuman studies is there control over the presence or absence of contamination sources. However, this control exists only if the studies are performed in a lab in which modern specimens of the organism of interest have never been studied. This restriction is a bit paradoxical because the researchers studying modern populations of an organism are those most likely to be interested in ancient populations of that organism. Typically, this requirement is met by the physical separation of ancient and modern laboratories, although ideally completely different labs would work on modern and ancient specimens. Furthermore, positive PCR controls must never be used because use of modern undamaged DNA represents the purposeful introduction of a potential DNA contaminant. Addition of control DNA to PCRs or storage of control DNA in a separate lab is unlikely to eliminate contamination because it is typically the same researcher handling both modern and aDNA samples. All experiments involving modern DNA, including primer design and PCR optimization, should be conducted in an independent laboratory if possible. These conditions are most easily met by a collaborative project in which a researcher experienced in the analysis of modern populations of the organism teams up with a researcher experienced in aDNA analysis.

Concluding Remarks

The incredible amount of time and resources necessary for aDNA studies mandate that only questions that cannot be addressed with modern samples should be undertaken. Each aDNA research project must be custom designed with the study populations in mind. Polymorphisms that differentiate between DNA from ancient specimens and any potential sources of contamination must be identified and analyzed. The impressive comparative database of DNA data available for a wide variety of organisms assures the merit of a project assaying those markers in ancient specimens, but only if extensive precautions have been taken to ensure the accuracy and reliability of the data.

References

Anderson, S., Bankier, A. T., Barrell, B. G., de Bruijn, M. H., Coulson, A. R., Drouin, J., Eperon, I. C., Nierlich, D. P., Roe, B. A., Sanger, F., Schreier, P. H., Smith, A. J., Staden, R., and Young, I. G. (1981). Sequence and organization of the human mitochondrial genome. *Nature* **290,** 457–465.

Austin, J. J., Ross, A. J., Smith, A. B., Fortey, R. A., and Thomas, R. H. (1997). Problems of reproducibility—does geologically ancient DNA survive in amber-preserved insects? *Proc. R. Soc. Lond. B Biol. Sci.* **264,** 467–474.

Baskaran, N., Kandpal, R. P., Bhargava, A. K., Glynn, M. W., Bale, A., and Weissman, S. M. (1996). Uniform amplification of a mixture of deoxyribonucleic acids with varying GC content. *Genome Res.* **6,** 633–638.

Bunce, M., Worthy, T. H., Ford, T., Hoppitt, W., Willerslev, E., Drummond, A., and Cooper, A. (2003). Extreme reversed sexual size dimorphism in the extinct New Zealand moa Dinornis. *Nature* **425,** 172–175.

Cano, R. J., Poinar, H. N., Pieniazek, N. J., Acra, A., and Poinar, G. O., Jr. (1993). Amplification and sequencing of DNA from a 120–135-million-year-old weevil. *Nature* **363,** 536–538.

Caramelli, D., Lalueza-Fox, C., Vernesi, C., Lari, M., Casoli, A., Mallegni, F., Chiarelli, B., Dupanloup, I., Bertranpetit, J., Barbujani, G., and Bertorelle, G. (2003). Evidence for a genetic discontinuity between Neandertals and 24,000-year-old anatomically modern Europeans. *Proc. Natl. Acad. Sci. USA* **100,** 6593–6597.

De Giorgi, C., Sialer, M. F., and Lamberti, F. (1994). Formalin-induced infidelity in PCR-amplified DNA fragments. *Mol. Cell Probes* **8,** 459–462.

DeSalle, R., Barcia, M., and Wray, C. (1993). PCR jumping in clones of 30-million-year-old DNA fragments from amber preserved termites (Mastotermes electrodominicus). *Experientia* **49,** 906–909.

DeSalle, R., Gatesy, J., Wheeler, W., and Grimaldi, D. (1992). DNA sequences from a fossil termite in Oligo-Miocene amber and their phylogenetic implications. *Science* **257,** 1933–1936.

Fisher, D. L., Holland, M. M., Mitchell, L., Sledzik, P. S., Wilcox, A. W., Wadhams, M., and Weedn, V. W. (1993). Extraction, evaluation, and amplification of DNA from decalcified and undecalcified United States Civil War bone. *J. Forensic Sci.* **38,** 60–68.

Gilbert, M. T., Hansen, A. J., Willerslev, E., Rudbeck, L., Barnes, I., Lynnerup, N., and Cooper, A. (2003). Characterization of genetic miscoding lesions caused by postmortem damage. *Am. J. Hum. Genet.* **72,** 48–61.

Golenberg, E. M., Giannasi, D. E., Clegg, M. T., Smiley, C. J., Durbin, M., Henderson, D., and Zurawski, G. (1990). Chloroplast DNA sequence from a miocene Magnolia species. *Nature* **344,** 656–658.

Greenwood, A. D., Capelli, C., Possnert, G., and Paabo, S. (1999). Nuclear DNA sequences from late Pleistocene megafauna. *Mol. Biol. Evol.* **16,** 1466–1473.

Hedges, S. B., and Schweitzer, M. H. (1995). Detecting dinosaur DNA. *Science* **268,** 1191–1192, 1194.

Hofreiter, M., Jaenicke, V., Serre, D., Haeseler Av, A., and Paabo, S. (2001a). DNA sequences from multiple amplifications reveal artifacts induced by cytosine deamination in ancient DNA. *Nucleic Acids Res.* **29,** 4793–4799.

Hofreiter, M., Serre, D., Poinar, H. N., Kuch, M., and Paabo, S. (2001b). Ancient DNA. *Nat. Rev. Genet.* **2,** 353–359.

Hoss, M., and Paabo, S. (1993). DNA extraction from Pleistocene bones by a silica-based purification method. *Nucleic Acids Res.* **21,** 3913–3914.

Kaestle, F. A., and Smith, D. G. (2001). Ancient mitochondrial DNA evidence for prehistoric population movement: The Numic expansion. *Am. J. Phys. Anthropol.* **115,** 1–12.

Kemp, B. M., and Smith, D. G. (2003). Tackling (some of) the vagaries of ancient DNA work. *Am. J. Phys. Anthropol. Supp.* **36,** 127.

Keyser-Tracqui, C., Crubezy, E., and Ludes, B. (2003). Nuclear and mitochondrial DNA analysis of a 2,000-year-old necropolis in the Egyin Gol valley of Mongolia. *Am. J. Hum. Genet.* **73**, 247–260.

Krings, M., Capelli, C., Tschentscher, F., Geisert, H., Meyer, S., von Haeseler, A., Grosschmidt, K., Possnert, G., Paunovic, M., and Paabo, S. (2000). A view of Neandertal genetic diversity. *Nat Genet.* **26**, 144–146.

Kolman, C. J. (1999). Molecular Anthropology: Progress and Perspectives on Ancient DNA Technology. *In* "Genomic Diversity: Applications in Human Populations Genetics" (S. S. Papiha, R. Deka, and R. Chakraborty, eds.), pp. 183–200. Kluwer Academic/Plenum Publishers, New York.

Kolman, C. J., Centurion-Lara, A., Lukehart, S. A., Owsley, D. W., and Tuross, N. (1999). Identification of Treponema pallidum subspecies pallidum in a 200-year-old skeletal specimen. *J. Infect. Dis.* **180**, 2060–2063.

Kolman, C. J., and Tuross, N. (2000). Ancient DNA analysis of human populations. *Am. J. Phys. Anthropol.* **111**, 5–23.

Krings, M., Stone, A., Schmitz, R. W., Krainitzki, H., Stoneking, M., and Paabo, S. (1997). Neandertal DNA sequences and the origin of modern humans. *Cell* **90**, 19–30.

Leonard, J. A., Wayne, R. K., and Cooper, A. (2000). Population genetics of ice age brown bears. *Proc. Natl. Acad. Sci. USA* **97**, 1651–1654.

Paabo, S. (1989). Ancient DNA: Extraction, characterization, molecular cloning, and enzymatic amplification. *Proc. Natl. Acad. Sci. USA* **86**, 1939–1943.

Paabo, S., Gifford, J. A., and Wilson, A. C. (1988). Mitochondrial DNA sequences from a 7000-year old brain. *Nucleic Acids Res.* **16**, 9775–9787.

Pergams, O. R., Barnes, W. M., and Nyberg, D. (2003). Mammalian microevolution: Rapid change in mouse mitochondrial DNA. *Nature* **423**, 397.

Poinar, H. N., Hofreiter, M., Spaulding, W. G., Martin, P. S., Stankiewicz, B. A., Bland, H., Evershed, R. P., Possnert, G., and Paabo, S. (1998). Molecular coproscopy: Dung and diet of the extinct ground sloth Nothrotheriops shastensis. *Science* **281**, 402–406.

Richards, M. B., Sykes, B. C., and Hedges, R. E. M. (1995). Authenticating DNA extracted from ancient skeletal remains. *J. Arch. Sci.* **22**, 291–299.

Sidow, A., Wilson, A. C., and Paabo, S. (1991). Bacterial DNA in Clarkia fossils. *Philos. Trans. R. Soc. Lond. B Biol. Sci.* **333**, 429–433.

Stone, A. C., and Stoneking, M. (1993). Ancient DNA from a pre–Columbian Amerindian population. *Am. J. Phys. Anthropol.* **92**, 463–471.

Stone, A. C., and Stoneking, M. (1998). mtDNA analysis of a prehistoric Oneota population: Implications for the peopling of the New World. *Am. J. Hum. Genet.* **62**, 1153–1170.

Stone, A. C., and Stoneking, M. (1999). Analysis of ancient DNA from a prehistoric Amerindian cemetery. *Philos. Trans. R. Soc. Lond. B Biol. Sci.* **354**, 153–159.

Wan, CC. Y., and Wilkins, T. A. (1993). Spermidine facilitates PCR amplification of target DNA. *PCR Methods Appl.* **3**, 208–210.

White, P. S., and Densmore, L. D. (1992). Mitochondrial DNA isolation. *In* "Molecular Genetic Analysis of Poplulations, A Practical Approach" (A. R. Hoelzel, ed.), pp. 29–58. Oxford University Press, New York.

Willerslev, E., Hansen, A. J., Binladen, J., Brand, T. B., Gilbert, M. T., Shapiro, B., Bunce, M., Wiuf, CC., Gilichinsky, D. A., and Cooper, A. (2003). Diverse plant and animal genetic records from Holocene and Pleistocene sediments. *Science* **300**, 791–795.

Woodward, S. R., Weyand, N. J., and Bunnell, M. (1994). DNA sequence from Cretaceous period bone fragments. *Science* **266**, 1229–1232.

Subsection B

The Markers

[8] Animal Phylogenomics: Multiple Interspecific Genome Comparisons

By ROB DESALLE

Abstract

The utility of DNA sequence information for phylogenetics and phylo-geography is now well known. Rather than attempt to summarize studies addressing this well-demonstrated utility, this chapter focuses on funda-mental approaches and techniques that implement the collection of DNA sequence data for comparative phylogenetic purposes in a genomic context (phylogenomics). Whole genome sequencing approaches have changed the way we think about phylogenetics and have opened the way for new perspectives on "old" phylogenetics concerns. Some of these concerns are which gene regions to use and how much sequence information is needed for robust phylogenetic inference. Whole genome sequences of a few animal model organisms have gone a long way to implement ap-proaches to better understand these important phylogenetic concerns. This chapter also addresses how genomics has made it more important for a clear understanding of orthology of gene regions in comparative biology. Finally, genome-enabled technologies that are affecting comparative biology are also discussed.

Background

The moral of the story of the early historical phases of evolutionary gene and genome comparison (using such techniques as immunoprecipita-tion, chromosome comparisons, allozymes, and a range of other ap-proaches) is that the most popular methods used by comparative biologists will be those with the greatest technical accessibility to the broader community of researchers in the area. The Polymerase Chain Reaction (PCR) (Mullis *et al.*, 1986; Saiki *et al.*, 1988) and automated dye sequencing (Hunkapiller *et al.*, 1991) have produced a degree of accessibil-ity to automated DNA sequencing that has allowed for an unprecedented expansion of sequencing into comparative biology. At the same time that the data-gathering techniques have advanced, new ways of quantifying gene comparisons have also been developed. Phylogenetic methodology has been the subject of several books and manuals (Desalle *et al.*, 2002a,b; Felsenstein, 2004; Hillis *et al.*, 1996; Ronquist and Huelsenbeck, 2003) and are not reviewed here (for a superb compilation of the various sites on the world wide web dedicated to phylogenetic analysis programs, see *http:// evolution.genetics.washington.edu/phylip/software.html*). However, certain advances in phylogenetic techniques are relevant to our discussion of multiple gene and genomic comparisons. Because the first step in comparing gene and protein sequences is to establish what part of the gene is the "same" from one species to the next, the molecular sequence alignment problem was expanded. Over the past 20 years, alignment problems have become a major preoccupation of bioinformaticists and several reviews addressing alignment procedures have appeared (Baxevanis and Ouellette, 2001; Mount, 2001; Phillips *et al.*, 2000; Thompson *et al.*, 1999). These should be consulted for comparisons of different approaches both in the models underpinning the approaches and for their speed and efficacy. In addition, the following web sites offer excellent summaries of many of the alignment tools available on the web: *http://ca.expasy.org/tools/*, *http:// www.techfak.uni-bielefeld.de/bcd/Curric/MulAli/welcome.html*, and *http:// pbil.univ-lyon1.fr/alignment.html*.

Some of the more important developments in phylogenetic approaches concern indicators of the robustness of phylogenetic hypotheses made from both molecular and morphological data. In addition, methods for the quantification of congruence of partitioned data have also been developed. These methods include bootstrap (Felsenstein, 1985), jackknife (Felsenstein *et al.*, 1996), Bremer support (Bremer, 1988), parametric bootstrap (Huelsenbeck *et al.*, 1996), incongruence length difference (Farris *et al.*, 1994), partitioned Bremer support (Baker and DeSalle, 1997; Baker *et al.*, 1998), and many others (Gatesy *et al.*, 1999; Lee and Hugall, 2003).

Most of these methods have led to personal computer program packages that allow accessibility to a broad range of researchers. As cluster computing becomes more and more important in the area of phylogenetics and phylogeography, many of these programs are being adapted to parallelization (Janies and Wheeler, 2001, 2002). Again, once the accessibility issue is solved for computing, it will be possible for a wide range of comparative questions to be approached by a broad group of researchers.

Gene + omics = Genomics

The sequencing of the first whole genome of an organism (Fleischmann *et al.*, 1995) and the ensuing onslaught of published completed genomes of nearly 200 bacterial and archaeal genomes and 25 or so eukaryotic genomes began a new era of comparative biology. Even more exciting is that in 2004 there are another 1000 whole genome sequences in the pipeline (see Appendix 1 and *http://www.genomesonline.org/*). High-throughput sequencing technology, though not immediately accessible to a broad range of researchers, has allowed for the transfer of many approaches and techniques to comparative biologists, speeding up data collection. Large-scale PCR sequencing studies (Murphy *et al.*, 2001a,b; Whiting, 2002; Zilversmit *et al.*, 1998) and expressed sequence tag (EST) sequencing studies (Brenner *et al.*, 2003; Bult *et al.*, 1997; Rudd, 2003; Theodorides *et al.*, 2002) in comparative biology owe their genesis to the development of high-throughput methods originally developed for whole genome sequencing.

A similarly important outcome of whole genome sequencing has been the availability of the database to the general scientific community for two reasons. First, whole genome sequences are a great source to "mine" for sequences that can inform interesting comparative or evolutionary questions. An example of this data-mining approach to a phylogenetic question is the Rokas *et al.* (2003) study in which eight yeast species with whole genomes sequenced originally for developmental and genetic purposes were data mined for phylogenetic purposes. Other examples of the utility of whole genomes and the large database now available include those in which functionality of proteins is examined. Second, whole genome sequences can inform researchers about gene choice for PCR studies and other kinds of "shallow genomic" (Phillips, 2002) approaches.

Phylo + genomics = Phylogenomics

The term *phylogenomics* arose from the whole genome sequencing frenzy of the last decade. It emanated from The Institute for Genomic Research (TIGR) and the writings of Eisen and colleagues (Eisen, 1998; Eisen and Fraser, 2003; Eisen *et al.*, 1995, 1997). Eisen (1998, p. 164), first defined phylogenomics as the process whereby "functions of uncharacterized genes are predicted by their phylogenetic position relative to characterized genes." In a clearly articulated definition, Eisen (1998) describes the approach of phylogenomics as consisting mostly of the tools of phylogenetics and accentuated by the existence of reams and reams of whole genome sequences. The approach of phylogenomics in this context is important in establishing homology (orthology) relationships among the various genes and gene regions discovered in the whole genome approach. Orthologous genes or protein domains (see later discussion) could then be used to examine function of such groups of molecules. In its original formulation, phylogenomics used phylogenetics to accomplish these gene nomenclatural and gene functional problems.

Eisen and Fraser (2003) broadened the definition of the term *phylogenomics* to embrace more of a feedback or reciprocal illumination relationship of genomics and phylogenetics. In addition, Eisen and Fraser (2003) include any problem—biological, genomic, or bioinformatic—that results from the burgeoning database of whole genomic sequences and that can be approached using phylogenetics as part of phylogenomics. They cite genome annotation, lateral gene transfer, genomewide expression studies, and gene discovery in unexplored organisms as some of the major roles of phylogenomics in modern biology. Regulatory element discovery via "phylogenetic shadowing" (Boffelli *et al.*, 2003) should also be added to this list.

Phylogenomic Space

Determining the number and kinds of gene partitions to use in gene comparisons for phylogenetic purposes has been a major preoccupation of phylogeneticists for more than a decade. This question, though of tantamount importance in phylogenetics, is being answered by shear brute force. Because high-throughput techniques make DNA sequencing approaches more accessible, it is now possible to compare the impact of many gene regions on robustness and precision of the phylogenetic inferences we make and to move systematics and comparative biology into new frontiers. The availability of whole genome sequences for a number of model organisms has allowed for the expansion of phylogenetics into other areas of what can be called the "phylogenome space" (Box 1). Several domains in the phylogenomic space are relevant to multiple gene and genome comparisons.

DNA barcoding: The availability of whole genome sequences of organisms begs the question of how much sequence is needed to complete a meaningful comparative study. Box 1 addresses this problem and introduces some of the newer areas of comparative molecular studies. At the leftmost end of the graph is a comparative approach that has recently gained momentum in the study of biology—barcoding (Box 2). As originally defined, barcoding would used a single gene (COI in animals) to develop unique identifiers for as many organisms as possible (Box 2). Given that there are 1.7 million named species, this does not seem an impossible task mostly because the number of potential combinations of unique DNA motifs that could exist in this stretch of DNA is very large (Hebert *et al.*, 2003a,b; Stoeckle, 2003).

BOX 1

PHYLOGENOMIC ANALYSIS SPACE

The y-axis in Fig. A indicates the number of species involved in a study. The number of species and number of base pairs are read as a power of 10 on the graph. The graph ends at 10 million (10^7) species because the number of named species is 1.7 (10^6). The x-axis indicates the number of base pairs in the particular kind of study indicated on the graph. For instance, the mitogenomic oval sits on the graph at around 10^4 because the typical animal mitochondrial genome is about 15,000 base pairs in length. The arrows in the graph indicate the direction of "growth" of the particular approach. Mitogenomics, microbial (meaning *bacterial* and *archaeal*) genomics and eukaryotic genomics can grow in the number of species to be analyzed. Phylogenetics can grow in both directions. DNA barcoding is indicated on the left of the graph. Because proponents of this approach suggest that only one or a few genes are needed to accomplish its goals, this technique will most likely only grow in the number of taxa analyzed. In Fig. B the fate of phylogenomics is denoted in the phylogenomics analysis space. Axes are the same as in A. By using the techniques of genomics, the number of eukaryotic taxa in phylogenetic studies will grow, but only partial aspects (shallow genomics) of genomes will be used in broad-scale phylogenomic studies. Microbial phylogenomics will most likely be a possibility. Loosely adapted from Phillips (2002).

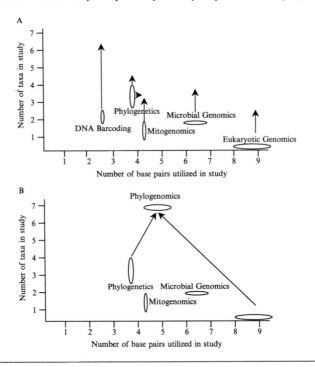

BOX 2

DNA BARCODING

The dynamics of DNA sequence change have led some researchers to suggest that short DNA sequences can be used as a source of information for obtaining unique identifiers (barcodes) in organisms. DNA barcodes are developed by sequencing specific genes such as cytochrome oxidase I for animals, or ITS2 for plants, for a reasonable number of individuals from a species. There are several ways that barcodes can then be "read" once these sequences have been obtained (multivariate statistical, tree building, or with diagnostic single nucleotide polymorphisms [SNPs]).

Excitement for and criticism of DNA barcoding have arisen in the scientific community. Several meetings have been conducted to address the prospects and problems of DNA barcoding, two of which were held at the Banbury Center at Cold Spring Harbor Laboratories, sponsored by the Alfred P. Sloan Foundation in 2003, and one held at the Zoologische Staatssammlung in Munich, Germany, sponsored by Bundesministeriu fur Bildung und Forschung. A DNA Barcoding Secretariat has been established after a barcoding meeting in May of 2004 (*http://www.barcoding.si.edu/*). A web site dedicated to barcoding developments has been constructed (*http://phe.rockefeller.edu/Barcode Conference/*).

Criticism of the initiative has come mostly from classic taxonomists with lively debate ensuing from both perspectives. Debate also exists concerning the source of DNA barcodes. One approach supported by Hebert *et al.* (2003a,b) is to use a single gene region such as mitochondrial COI. Supporters of this approach suggest that there will be enough DNA sequence variability to accomplish barcoding across all animals. Other approaches would employ a larger number and different sources of gene regions for sequencing. We suggest that any source of character information that can be shown to be a unique identifier of a species should be considered a potential barcode. Such molecular tools as restriction fragment length polymorphism [RFLP], amplified fragment length polymorphism [AFLP], random amplified polymorphic DNA [RAPD], microsatellites, gene order, gene presence or absence, or SNP would be appropriate methods to discover unique identifiers for DNA barcoding. DNA barcoding can also be complemented and enriched with character information from other sources such as morphological or allozyme approaches. Several initiatives toward barcoding specific groups have already been launched. Systems for cetaceans (*http://www.dna-surveillance.auckland.ac.nz/html/mainIndex.jsp*) and bacteria (*http://www.dsmz.de/bactnom/bactname.htm*) are prominent taxonomic groups being examined using barcoding methods.

Mitogenomics: To the right of "barcoding" is "mitogenomics," which is the study of whole mitochondrial genomes. As of June 2004, there were more than 1600 complete animal mitochondrial genomes that have been sequenced (*http://www.mitomap.org/mitomap/euk_mitos.html*). High-throughput methods using multiple primer systems or "long PCR" are the favored approaches to whole animal mitochondrial genome sequences. The multiple-primer approach requires that sets of primers be designed for different groups of organisms to implement amplification across the whole genome. One major problem with this approach is that there are no truly universal primer sets that can cover the entire animal mitochondrial genome and so whole mitochondrial genome long PCR has been developed and used to obtain DNA sequences for whole mitochondrial genomes from groups with no closely related complete genomes.

Transforming molecular systematics: Areas of the graph in Box 1 to the right of "mitogenomics" can be contrasted with the area of the graph representing modern eukaryotic genomics, where the amount of sequence information is sacrificed at the expense of the number of terminal taxa involved in a comparison. Box 1 also separates bacterial and archaeal genomics from eukaryotic genomics, because the sequencing of microbial genomes can be implemented rather rapidly, and over the past 10 years about 1000 microbial sequencing projects have been finished or initiated. The microbial genomics point on the graph will continue to move upward at a fairly rapid pace. The eukaryotic genomics point has moved upward fairly slowly, but there are nearly 100 eukaryotic genome projects in the pipeline (Appendix 1) and so this point will also move upward but not as quickly as for bacteria and archaea.

Until whole genomes were sequenced, phylogenetics in eukaryotes consisted of using single targeted "neutrally evolving" genes such as nuclear rDNA gene regions and mitochondrial cytochrome oxidase *b* in animals and *rbcL* (chloroplast), *atpA* (mitochrondrial), and ITS (nuclear) gene regions in plants (Savolainen and Chase, 2003). These gene regions were used primarily because of their ease of isolation and sequencing, relative sequence conservation across species that facilitated design of "universal" PCR primers, ease of assessing orthology, and overall information content. A quick perusal of the literature in 1995 reveals that most of the molecular phylogenetic articles focused on single genes or at most two genes to accomplish the studies (Box 1). One rule of thumb from these various studies is that although several genes (e.g., 18S rDNA, ITS, COI, and *rbcL*) have been used heavily in phylogenetic analysis, there is no panacea gene or gene region for robustly resolving all phylogenetic questions. Box 1 shows where in the genomic space will phylogenomic studies most likely expand. This area of the phylogenomic space has been termed "shallow genomics" (Phillips, 2002).

Apples and Oranges: Orthology

All of comparative biology relies on establishing homology among the entities and the parts of the entities being compared. In terms of genes and gene families between species, this amounts to the classic molecular biology orthology–paralogy problem (Fitch, 2000). Establishing orthology of genes in comparative studies is extremely critical, and those studies in which the orthology has been carefully determined using whole genomes and synteny are perhaps the best (Rokas *et al.*, 2003).

Determining gene orthology: In general, procedures for detection of membership in a gene family and for subsequent orthology detection are to use some similarity score cutoff. The most widely used score is the BLAST score, which most sequence databases incorporate into their search engines. In these comparisons, a query sequence is compared to a database of sequences by alignment and a score is given to how similar the query sequence is to the database sequences based on length of the similar region and the overall similarity of the region. The score is called an "e value" and is really a probability that two sequences show similarity because of randomness, hence very small e values are preferred when doing BLAST searches. An arbitrary cutoff is usually chosen to establish membership in a gene family and the comparison with the smallest e value is usually chosen to represent the ortholog of the query sequence. Because the best BLAST score will not always accurately pinpoint the nearest neighbor or more concisely the true ortholog (Koski and Golding, 2001), other criteria such as synteny of two suspected orthologs are used to determine orthology. Phylogenetic tree building methods have been used to determine orthologs (Sarkar *et al.*, 2002; *http://orthologid.amnh.org/*). In this approach, the position of a member in a gene family in a phylogenetic tree indicates its orthology relationships. These authors used homeobox genes from the HOX 9-13 paralog group as a test case and the results of their approach indicate it is accurate and operational.

From a technical standpoint, the likelihood of accurately targeting orthologs in shallow genomics studies for phylogenetic or phylogeographic purposes can be enhanced by careful design of primers (to be specific for orthologs across a wide range of organisms). This care to primer design is assisted by referring to PCR-based sequencing studies and nesting primers in areas of high conservation for orthologs to avoid primers that will amplify paralogs. Comparison of sequences in EST studies to sequences from model organisms with whole sequenced genomes and assessing their nearest neighbor in a phylogeny is an approach to assessing orthology.

Nu-mtDNA Problems. Several examples of problems with orthology determination do exist. The most prominent of these are the problematic nu-mt sequences that are acquired in mtDNA studies that actually emanate from mitochondrial fragments that have been transferred to the nucleus. Due to the horizontal transfer, orthology of such gene sequences becomes suspect, and less than optimal phylogenetic and phylogeographic inferences are usually obtained when these sequences are inadvertently included in analyses.

To Combine or Not to Combine, That is the Question: Total Evidence =
Simultaneous Analysis = Concatenation

Recent approaches to assessing character congruence (total evidence, simultaneous analysis, or concatenated analysis) versus taxonomic congruence (consensus among trees produced from single partition analysis) and phylogenetic analysis have shown that concatenated analyses are best for extracting phylogenetic inferences (Gatesy *et al.*, 2002, 2003; Rokas *et al.*, 2003).

Several studies have increased the number of DNA sequence regions examined in phylogenetic analyses. The most prominent of these is a study by Rokas *et al.* (2003) in which eight species of yeast with whole genome sequences were placed into a phylogenetic context. Using a strict orthology definition based on synteny, Rokas *et al.* (2003) were able to use more than 106 DNA sequence regions (in this case, whole genes because they had whole genome sequences at their disposal) and this resulted in the availability of 127,026 nucleotides or 42,342 amino acids as characters for phylogenetic analysis. Other studies on the relationships of the three superkingdoms in the tree of life (Korbel *et al.*, 2002; Rivera and Lake 2004; Snel *et al.*, 1999, 2002), bacterial phylogeny (Planet *et al.*, 2003; Wolf *et al.*, 2002), and several studies using more than 25 gene regions in mammals (Gatesy *et al.*, 2002; Murphy *et al.*, 2001b) as sources for phylogenetic analysis do exist.

One of the major questions approached in most of these studies is whether it is most appropriate or informative to analyze the characters as single partitions (a *partition* here defined as a single gene region) or to combine all information into a single phylogenetic analysis. Again the most prominent study to address this issue is the Rokas *et al.* (2003) study in which they showed rampant incongruence among the various single gene phylogenies. Previous phylogenetic studies have addressed this problem of incongruence among single gene regions and why it might be more appropriate to combine information from all partitions in phylogenetic analysis from both a philosophical and empirical standpoint.

Phylogenetic Hide and Go Seek. The concept of hidden support was first characterized several years ago (Gatesy *et al.*, 1999), and methods for measuring this emergent support exist. Box 3 lists the various partitioned support measures and the general applicability of the measure to understanding gene region contribution to phylogenetic hypotheses. In addition to using these partitioned support methods in a parsimony framework, likelihood methods have also been developed (Lee and Hugall, 2003). Early studies using partitioning methods have shown clearly that even though certain partitions can show a high degree of incongruence with

---

BOX 3

CHARACTER/NODE SUPPORT

Character-based approaches to measuring data set, node, and character stability in phylogenetic analysis. For detailed explanations of the measures and their sources, as well as methods of calculation, see Gatesy *et al.* (1999).

Branch Support (BS)	Measures robustness of inference of node
Partitioned Branch Support (PBS)	Measures the relative contribution of a data partition at a node in a phylogenetic tree
Character Support (CS)	Measures individual character support on a tree; also used to test the significance of a particular node in a phylogenetic tree
Hidden Branch Support (HBS)	An indicator of the increased support at a node due to combining all character information
Partitioned Hidden Branch Support (PHBS)	Measures the distribution of HBS in difference partitions
Clade Stability Index (CSI)	Measures the stability of a clade as a function of removing characters from the data matrix
Data Set Removal Index (DRI)	Measures the stability of a clade as a function of removing entire data partitions
Data Set Influence (DI)	Measures the influence of a data partition on an analysis

other partitions, there might also exist a large amount of hidden support for the combined analysis tree. Box 4 gives an example of how the hidden support measures are used to interpret incongruence in the Rokas *et al.* (2003) study.

Don't Force It; Use a Bigger Hammer. In addition to understanding the nature of hidden support, the technical tools and approaches to obtaining DNA sequence information for multiple gene and genome comparisons also are contingent upon the comparative questions. If the entities being compared are recently diverged, then completely different tools than the ones used for deep phylogenetic questions need to be found. Matching a phylogenetic question with the proper or appropriate genetic probe is a matter of quantifying the degree of divergence among sequences in the database. In general, when we attempt to determine whether a gene might be appropriate for phylogenetic purposes, we sequence two closely related individuals for the gene and several phylogenetically distant species. The distant species are chosen to span the phylogenetic range of the specific question being approached. For instance, if the systematics of *Diptera* is the

BOX 4

HIDDEN SUPPORT AND YEAST PHYLOGENOMICS

Partitioned support approaches explain why conflict among data partitions is prevalent and why, as Rokas *et al.* (Rokas *et al.*, 2003) concluded from an empirical analysis of 106 genes from yeast, that a single tree based on all of the available data is the best approach to phylogenetic analysis. A parsimony-based partitioned analysis of the Rokas *et al.* (2003) data set indicates that there is no single gene partition that shows a net negative partitioned branch support value, suggesting that no single gene shows overall conflict in the combined or concatenated analysis. The reason for this observation is found in the analysis of hidden branch support, in which the overall hidden support on the tree is 18% (1709/9480 in branch support units), meaning that nearly one in every five character state changes that supports the tree in the "concatenated" analysis does not support the tree in separate analyses of individual genes. More importantly, there is positive hidden branch support at each of the five nodes in the tree and in 84 of the 106 genes; more than 55% of the character support at node 3 is emergent and can be explained by hidden support. The analysis shown here and in other studies of hidden support (Gatesy *et al.*, 1999) in phylogenetic analysis suggests that there is a great amount of latent support for concatenated phylogenetic hypotheses in gene partitions that do not result in the same topology when analyzed singly. Such hidden support can contribute to the overall total evidence hypothesis and strengthens the argument for combined or concatenated analyses in modern phylogenetics. It is recommended that before one rejects a gene region for phylogenetic purposes that partitioned analysis be performed to justify rejection.

The tree in this box is a simultaneous or concatenated analysis tree from Rokas *et al.* (2003) generated by parsimony analysis using PAUP* v. 4.0b (Swofford, 2002). Branch support measures were calculated using TreeRot v.2 (Sorenson, 1999) and PAUP* v. 4.0b (Swofford, 2002). The numbers in the boxes on the tree refer to various partitioned branch support measures (Gatesy *et al.*, 1999; Box 3) that are relevant to understanding the role of hidden support in phylogenetic analysis. The values are from left to right-Leftmost refers to the node numbers as assigned in Rokas *et al.* (2003). Second from left refers to the total branch support at each node; the summed branch support over the whole tree is 9480. Third from left refers to the hidden branch support for each node (total of 1709). Fourth from left is the percentage of branch support that is hidden in the individual genes. Fifth from left refers to the percent of gene partitions (out of 106) that show negative partitioned branch support in the combined analysis. The rightmost box refers to the percentage of gene partitions with positive hidden support at that node.

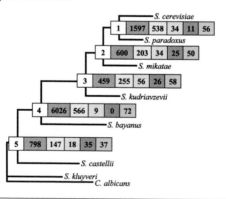

goal of the gene design, then two relatively closely related *Drosophila* are sequenced for the gene followed by sequencing of a non-drosophilid Brachycera (the major suborder that *Drosophila* belongs to) and some Nematoceran (the other suborder [paraphyletic] of *Diptera*). The sequences obtained from these organisms are then aligned and degree of divergence among the various species included in the preliminary analysis can be compared to determine whether there will be enough variation in the gene to impart some information to the nodes in the tree.

Current and Future Approaches

Currently, the approaches to gene and genome comparison in animals are varied. The most obvious technical approach is to generate whole genomes for subjects of comparative analysis. While sequencing large animal genomes is feasible, the bang for the buck in the context of comparative evolutionary studies is often lost. Box 1 again shows where broadscale phylogenomic approaches should strive with respect to taxon number and amount of sequence information. This intermediate or shallow genomics can be accomplished in many ways.

Whole Genomes

There are many ways to approach nuclear gene comparisons in organisms. The most obvious approach is to obtain whole genome sequences of the organisms of interest. Although this might be possible for whole genomes of microbial species and plastid genomes (mitochondrial and chloroplast) because of their small genome sizes, the feasibility of this approach for nuclear genomes of all life on the planet is low and perhaps even unwarranted.

Aside from studies in which whole genome sequences have been used as the primary source of phylogenetic information as best exemplified by Rokas *et al.* (2003), two other major kinds of approaches are used in comparing the genomes of the so far sequenced organisms in the "tree of life." The first approach concerns the comparison of gene presence/absence and gene order in genomes. For the tree of life, the presence/absence of genes turns out to be a relatively good source of characters for detailing the relationships of organisms. The only exceptions to this positive statement about whole genome phylogenetics are the reduced genome species mentioned earlier and the possibility that the tree of life is actually a ring due to horizontal transfer. Plastid genomes have also been treated in this way to infer the relationships of the green, red, and glaucophyta lineages of eukaryotes (Nozaki *et al.*, 2003), and gene presence, absence, and order.

A second genomic approach that is somewhat similar to assessing the presence and absence of gene regions in whole genomes is the approach of phylogenetic shadowing (Boffelli *et al.*, 2003). This approach is important for determining the position of conserved regulatory elements in non-transcribed or processed regions (such as 5′ regions and introns) of genes. The approach simply aligns these noncoding regions and looks for stretches of conserved nucleotide sequences. In some cases, regions of conservation in noncoding gene regions are inferred to be potential regulatory sequences. Although any noncoding region that is conserved across all species in a phylogenetic shadowing analysis is most likely an indication of a regulatory region, other patterns of conservation will be indicative of other interesting regulatory patterns (Iafrate *et al.*, 2004; Li *et al.*, 2004; Sebat *et al.*, 2004).

Gene Mining from Whole Genomes

In addition to the study by Rokas *et al.* (2003), several other authors have settled on sets of genes from genomes that are in the database (Zeigler, 2003). This approach has its highest visibility in bacterial phylogenetics, but the principles being established for these studies should be carried over into eukaryotic studies such as the Rokas *et al.* (2003) study, when enough genomes are finished. Brown and Doolittle (1997) have used this approach to establish a phylogenetic framework for a broad range of archaea, bacteria, and eukarya. This approach has also been used within the kingdoms to establish phylogenetic relationships. Baldauf *et al.* (2000) used α-tubulin, β-tubulin, actin, and elongation factor-α sequences from more than 60 eukaryotes to construct a eukaryotic phylogeny. Within the archaea, Matte-Tailliez *et al.* (2002) used 45 non-transferred ribosomal protein genes to infer relationships among the 14 archaeal species sequenced at the time. Brochier *et al.* (2002) used proteins involved in the translational apparatus (57 genes) in more than 40 species to examine relationships of the major groups of bacteria. Finally, this approach has also been used to examine horizontal gene transfer (Eisen, 2000; Stanhope *et al.*, 2001). Within the bacteria, Planet *et al.* (2003) used more than 40 orthologous genes mined from the database over a wide range of bacteria to examine horizontal transfer dynamics.

Primer Mining from Whole Genome Comparisons

Another database-mining technique is to take advantage of whole genome sequences of phylogenetically disparate model organisms (Bonacum *et al.*, 2001). The procedure is simple and is shown in Protocol 1. An excellent example and resource for animal phylogenetics is the web-based ortholog search for an "Arthropod phylogenetics program" (Cunningham *et al.*, *http://rush.genetics.duke.edu/orthologs/*).

PROTOCOL 1

Designing Primers Using Genome Databases; Primer Design for
"Shallow Genomics"

The following "protocol" describes how a "universal" primer pair
would be designed for a broad array of animal taxa.

1. Take the human, fly, and worm genomes and blast them sequentially
   against each other.
2. As in Box 4, an arbitrarily chosen cutoff score is set.
3. These "blast" searches will produce a list of genes and gene regions
   that are conserved among the three taxa and hence most likely
   conserved in the many taxa that are within the phylogenetic range of
   the three starting genomes.
4. Once the gene regions in which primer design has been designated as
   desirable have been localized, any number of oligonucleotide primer
   algorithms can be used to design degenerate primers for a specific
   gene region (see below for a list of some of the more widely used
   web-based primer design systems and Zeigler [2003] for a review of
   degenerate primer design).
5. Examples of the approach in animals have been very effective at
   mining the genome databases for potential primers (Aitken *et al.*,
   2004; Lyons *et al.*, 1997; Murphy *et al.*, 2001a,b; Zilversmit *et al.*, 1998).
6. Once primer sequences are designated and synthesized, gradient
   polymerase chain reaction (PCR) is performed with a broad range of
   species to determine whether the primers are "universal."
7. Although this approach widens the range of gene regions that can be
   used in phylogenetics, the process is often hit or miss, and many primer
   pairs for different gene regions need to be tested for efficacy.

List of primer design web sites (from *http://www.hgmp.mrc.ac.uk/
GenomeWeb/nuc-primer.html*):
   The PCR Jump Station
   GeneFisher
   GeneWalker
   Web Primer
   Primer3
   CODEHOP
   NetPrimer
   rawprimer
   ExonPrimer
   PrimerX
   RevTrans

Autoprime
SNPbox
Primclad

The target sequences for amplification can be for protein coding sequences, in which case design of the primers must take into consideration the location of intron regions to avoid these highly variable areas (Aitken *et al.*, 2004; Lyons *et al.*, 1997; Murphy *et al.*, 2001a,b). If phylogeographic or population genetic approaches are the focus of a study, then the rapidly evolving intronic regions are a desired target for amplification. In this case the conserved exonic regions can be used for primer design for fragments that will span introns (the so-called exon priming intron crossing [EPIC] primers; Palumbi, 1995; Palumbi and Baker, 1994). More recently, 5′ noncoding regions have been shown to have highly conserved regions (Dermitzakis *et al.*, 2002, 2003, 2004), called conserved nongenic sequences. These regions could also be used as regions for primer design for both phylogenetic and phylogeographic studies.

The approach to more targeted groups is similar. For instance, if one wants to examine phylogeny or phylogeography in a group of closely related organisms such as the family Drosophilidae, whole genome sequences from flies closely related to *Drosophila melanogaster* can be mined for genes and gene regions for primer sites. In the *Drosophila* case (Zilversmit *et al.*, 1998), the *D. melanogaster* and *Anopheles gambiae* genomes were compared to the human genome for regions of conservation and primers designed to highly conserved regions of those genomes.

As more and more genomes become available, the design of primers for this category of gene comparison will greatly facilitate the design of primers for specific groups and enhance the design of primers for higher phylogenetic analysis. Appendix 1 lists all of the finished and officially in progress genome projects for animals as of July 2004. Again, the Drosophilidae are a good example of the increase in accuracy of primer design for comparative studies, because the probability of developing a working primer system increases as the number of species from a group is increased. As Appendix 1 demonstrates, flies in the family Drosophilidae are the animal genomes with the most concentration, with fungal genomes also showing a strong concentrated sequencing focus. The intensity of sequencing in these two groups allows for greater accessibility to shallow genome-level sequencing projects for large numbers of taxa.

High-Throughput "Targeted Genome" Sequencing

The high-throughput targeted genome sequencing approach takes advantage of the fact that while at the gross level there is limited synteny, at local levels there is a high degree of conservation of gene order (Iafrate *et al.*, 2004; Li *et al.*, 2004; Sebat *et al.*, 2004). Protocol 2 describes the protocol by Thomas and Touchman (2002) for the isolation of targeted genomic regions using the human genome as a starting point. This approach or modifications of it have been used to examine vertebrate genome evolution (Thomas *et al.*, 2003) and drosophilid evolution (Bergman *et al.*, 2002). In the mammalian study 13 vertebrate species were used to examine the dynamics of genome sequence change.

PROTOCOL 2

Targeted Genome Sequencing

1. Human genomic sequences are first used to find orthologous sequences in existing mammalian or vertebrate sequence databases such as ESTs, BAC ends, or WGS (Whole Genome Sequencing).

2. An arbitrary *e*-value cutoff is used. Depending on the phylogenetic relatedness of the species under examination, some *e*-value cutoffs will be more appropriate than others. For instance, if closely related mammals are the species of interest, a relatively small *e*-value cutoff ($e = 10-20$) should be used. Broader phylogenetic comparisons will warrant more flexible *e*-value scores (i.e., larger *e* values).

3. Construct and screen a BAC library with orthologous "overgo" probes (overgo probes are defined as "being produced using two oligos that are complementary and anneal to form a double-stranded region" [Zheng, 2003]). A BAC library screened with several "overgo" probes will yield information for step 4 below. An overgo oligonucleotide design website can be found at *http://www.mouse-genome.bcm.tmc.edu/webovergo/Overgo Description.asp*.

4. Construct an orthologous BAC tiling path for sequencing orthologous genomic sequences from other organisms.

5. Shotgun sequence the BAC tiling path.

Thomas *et al.* (2003) reported results on more than 12 megabases of sequence from these vertebrate species. All sequences were derived from the orthologous region of about 1.8 megabase stretch on human chromosome 7 that contains 10 genes. These comparisons revealed that there were substantial stretches of noncoding regions that are conserved across all vertebrate species in the study (human, mouse, rat, baboon, chimpanzee, cat, dog, pig, cow, chicken, Fugu, Tetraodon, and zebrafish) and demonstrated the utility of the approach to obtaining large amounts of sequence information for evolutionary studies from a broad range of species. Bergman *et al.* (2002) examined five drosophilid species (*Drosophila melanogaster, Drosophila erecta, Drosophila pseudoobscura, Drosophila willistoni*, and *Drosophila littoralis*) and *Anopheles gambiae* for 10 gene regions, totaling about 500,000 base pairs for each species. Their results were similar to the vertebrate results in that there was a high degree of conservation in some noncoding regions. The phylogenetic utility of this approach in both studies was not assessed in detail, but they both demonstrate the feasibility of generating large regions of syntenic (orthologous) regions of the genomes of related organisms for phylogenetic analysis without resorting to WGS shotgun sequencing.

Expressed Sequence Tag Studies

EST generation involves the isolation of mRNA from extracts of target tissues. The isolated mRNA is then either normalized (transcripts with high mRNA concentration are reduced to the same concentration of the more rare transcripts) or used directly to produce copy DNA (cDNA). The cDNA is then cloned and a large number of individual clones are picked and sequenced from the large pool of clones. This cloning step and the choice of vector for cloning are critical to the kind of transcript that can be obtained. In some studies full-length cDNA sequences are not necessary and some shortcuts can be taken. Protocol 3 shows the general outline for how an EST library is constructed. When full-length cDNA libraries are needed, the cDNA synthesis step can be altered to ensure that full-length transcripts are isolated. There are several ways to obtain full-length cDNA (Carninci *et al.*, 1997, 1998; Chenchik *et al.*, 1996; Maruyama and Sugano, 1994; Ying, 2003), and these methods can be inserted into Protocol 3 in step 5. The final steps include processing and annotating the sequences obtained (Ewing *et al.*, 1998; Gordan *et al.*, 1998; Huan and Madan, 1999; Liang *et al.*, 2000; Pearson and Lipman, 1988). For a clear description of this approach, see Protocol 4, which is based on information in *http://www.genome.clemson.edu/projects/peach/est/est_process.html*).

PROTOCOL 3

General Outline for cDNA Library Preparation (not full length; from *http://www.brunel.ac.uk/depts/bl/project/genome/moltec/library/exprot.htm*)

1. Extract total RNA from target tissue.
2. Purify messenger RNA (mRNA) away from structural RNA by oligo-dT cellulose chromatography.
3. Complementary DNA (cDNA) synthesis proceeds by initial first strand synthesis followed by a reverse-transcriptase step and then a second strand synthesis.
   a. A free oligo dT primer or an oligo dT-tailed vector is annealed to the mRNA or the mRNA is annealed to oligo-dT–loaded

magnetic beads, which will greatly facilitate the subsequent recovery of the cDNA.

b. Reverse-transcriptase treatment is then performed, which gives rise to a DNA/RNA hybrid.

4. The DNA/RNA hybrid is tailed at the 3′ end of the DNA strand with oligo-dG (or dC), to prevent self-annealing during subsequent steps.

5. At this point in the procedure one of many protocols for maximizing full-length cDNA synthesis can be used (Carninci et al., 1997, 1998; Chenchik et al., 1996; Maruyama and Sugano, 1994; Ying, 2003). These approaches will minimize the risk of loosing the 5′ end of the gene in subsequent steps.

6. Second strand synthesis then requires the removal of the RNA strand by either alkaline treatment or by ribonuclease H (RNase H) treatment followed by reverse transcriptase or *Escherichia coli* DNA polymerase treatment.

7. At this point the cloning step is implemented. If the first strand synthesis has been done free in solution as described earlier, then the ends of the DNA have to be manipulated before insertion into a cloning vector is possible; if the DNA is already linked to the vector, then the molecule has to be circularized via T4-DNA ligase treatment.

8. A battery of cloning vectors can be used to randomly clone the cDNA fragments and the clones are picked and arrayed in microtiter plates.

9. Clones are then colony PCR amplified and sized on an agarose gel. Clones that show no product, more than a single insert, or inserts less than 150 base pairs are usually rejected and the rest are sequenced.

10. After sequencing, the analysis phase includes the following: the removal of any sequence less than 100 bases; the removal of sequences with microsatellite repeats only; the trimming away of long poly(A) sequences (>100 bases) to less than 50 bases; the removal of sequences less than 100 bases after removal of poly(A) tail; the removal of all sequences with more than 40 bases of poly(T)s at the beginning of the read; the removal of all sequences if the read has more than 35% of total base call below phred score 2; and the removal of rRNA sequence contamination.

PROTOCOL 4

Expressed Sequence Tag (EST) Processing Protocol (from *http:// www.genome.clemson.edu/projects/peach/est/est_process.html*)

Processing occurs in three stages:
Stage I: Trace File Processing

1. Sequence trace files are converted into fasta files and a quality score files using the phred (Enard et al., 2002) base-calling program.

2. Vector and host contamination are identified and masked using the sequence comparison program cross_match (Huan and Madan, 1999).

3. Vector trimming excises the longest nonmasked sequence and further trimming removes low-quality bases (phred score <20) at both ends of a read.

4. Sequences are discarded if they have more than 5% ambiguous bases, more than 40 poly(A) or poly(T) bases, or less than 100 high-quality bases (minimum phred score of 20).

5. The Fasta library is further filtered to remove reads having significant similarity with the species-specific mitochondrial, rRNA, tRNA, or snoRNA sequences downloaded from the Genbank nucleotide database.

Stage II: Assembly of High-Quality Sequences

1. In stage II processing, the filtered library file is assembled using the contig assembly program CAP3 (Huan and Madan, 1999).

2. More stringent parameters (- p 95. -d 60) are typically used to prevent overassembly and help identify potential paralogs.

Stage III: Annotation

1. Annotation consists of pairwise comparison of both the filtered library and the contig consensus library file against the Genbank nr protein database using the fastx3.4 algorithm (Pearson and Lipman, 1988).

2. The most significant matches (EXP < 1e-9) for each contig and individual clones in the library are recorded.

3. The unigene data set is derived by selecting the clone that best represents the contig and singletons that have either unique protein matches (EXP < 1e-9) or no known significant matches.

4. Contig assembly is stored in a database that can allow efficient access to the EST database and the genome sequence databases for BLAST or other database queries.

   Although the EST process is used widely in genomics, it has also found its way to evolutionary studies. Theodorides *et al.* (2002) used seven beetle species in an EST survey. They concluded that the EST approach is a viable method for generating target genes for phylogenetic analysis. Examples of other evolutionary studies using the EST approach at broadly different taxonomic levels indicate a likely great utility of the approach in plant and animal phylogenetics and evolutionary biology and at higher levels (Brenner *et al.*, 2003; Campbell, 2004; Kortschak *et al.*, 2003; Krylov *et al.*, 2003; Poustka *et al.*, 2003).

A comprehensive up-to-date list of organisms in which ESTs exist is given at *http://www.ncbi.nlm.nih.gov/dbEST/dbEST_summary.html*. Some researchers have also used this expanding database to devise methods for detecting single nucleotide polymorphisms (SNPs) and to use this information to design primers to accomplish comparative analysis of specific organisms (He *et al.*, 2003).

SNP Studies and the Expansion of Microarray Applications

Once a sequence database has been constructed and data mining allowed, focus on rapid detection of sites in the sequences that might be useful at both higher and lower phylogenetic or phylogeographic levels is possible. A wide range of approaches can be used to develop rapid detection methods. These approaches in general take advantage of the inability of PCR to proceed if there is a mismatch in the 3′-most base of an oligonucleotide primer (Birstein *et al.*, 1998) and the design of primers that either have a match (to diagnose the presence of the SNP) or do not match on the 3′ end (to demonstrate the absence of the SNP). Several modifications of this approach exist such as the TAQman assay (Livak *et al.*, 1995; Tyagi *et al.*, 1998) and molecular beacons (Marras *et al.*, 1999; Tyagi *et al.*, 1996, 1998).

These rapid assessment approaches are used mainly in phylogeographic or population genetic studies, and DNA sequencing approaches are used to detect SNPs at higher levels (Aitken *et al.*, 2004; Lyons *et al.*, 1997; Murphy *et al.*, 2001a,b; Zilversmit *et al.*, 1998). For phylogeographic or population-level studies, very few markers are usually needed to make clear inferences and so design and development of TAQman and Beacon systems is feasible, whereas for phylogenetic studies in which large numbers of phylogenetically informative SNPs are desired, the design of large batteries of SNP detecting systems is not feasible. An exciting new technology that promises to expand animal comparative biology is the microarray. Several studies exist that use microarrays to detect expression patterns that can be used in a phylogenetic context (Enard *et al.*, 2002; Rifkin *et al.*, 2003). However, it is the ability of microarrays to detect SNPs (Hacia, 1999; Mir and Southern, 2000; Southern, 1996) that has led some authors to suggest the relevance of this genome-enabled approach to phylogenetics. Several studies on bacteria (Hinchliffe *et al.*, 2003; Read *et al.*, 2003) and viruses (Boonham *et al.*, 2003; Wang *et al.*, 2002) have shown that the microarray sequencing approach is viable. A human MitoChip has been developed (Maitra *et al.*, 2004) that allows for the detection of mitochondrial DNA SNPs in the control region.

The combining of microarray technology with representational difference analysis (RDA) (Lisitsyn *et al.*, 1993; Lucito *et al.*, 2003) has also provided for a new method to detect large-scale genomic differences among animal genomes. The gene discovery RDA method allows researchers to "clone the differences" between any two different genomes. Used mostly to identify oncogenes in humans, RDA has been adapted to microarrays via representational oligonucleotide microarray analysis (ROMA) (Lucito *et al.*, 1998). Because it takes advantage of the whole human genome sequence, this technique has the unprecedented ability to detect gene copy number variation on a genomewide scale. These technologies relevant to human disease such as the human MitoChip and ROMA are being transferred to accommodate phylogenetic and DNA barcoding (Box 2) questions (Locke *et al.*, 2003; Pfunder *et al.*, 2004).

Conclusion

The sequencing of whole genomes of a wide variety of organisms has revolutionized how new data are being collected and interpreted for comparative biologists. The burgeoning database of sequences and technologies springing from this revolution are providing exciting new insight into phylogenetics and phylogeography and guiding the choice of genome-enabled strategies for these two areas. Because orders of magnitude more gene regions can now be tapped by researchers, newly developed analytical approaches are being developed for phylogenetic analysis, and these new methods give new insight into genome-enabled strategies for comparative biology. Adaptation of high-throughput approaches to human genetics and disease to comparative approaches such as phylogenetics and phylogeography (rapid SNP detection methods, microarray SNP detection, EST approaches, and targeted genomics) has also allowed for the promise of the influx of orders of magnitude more data into comparative biology.

Appendix 1. Animal (and Fungal) Genome Projects: Finished, First Drafts, and Proposed

Compiled from *http://www.genome.gov/10002154*, *http://www.genome.gov/10001843*, *http://www.genome.gov/10002154*, *http://mosquito.colostate.edu/tikiwiki/tiki-page.php?page Name=Mosquito_genomics*, and *http://www.ncbi.nlm.nih.gov/genomes/FUNGI/funtab.html*.

Finished and First Drafts
Bos taurus
Caenorhabditis briggsae
Caenorhabditis elegans

Canis familiaris
Ciona savignyi
Drosophila rerio
Drosophila melanogaster
Gallus gallus
Homo sapiens
Macaca mulatta
Mus musculus
Pan troglodytes
Rattus norvegicus
Saccharomyces cerevisiae
Schizosaccharomyces pombe
Tetraodon nigroviridis
Tribolium castaneum
Fungal Genome Projects
Aspergillus fumigatus
Aspergillus nidulans FGSC A4
Aspergillus parasiticus
Aspergillus terreus
Candida albicans
Coccidioides posadasii C735
Coprinopsis cinerea okayama7#130
Cryptococcus neoformans var. grubii H99
Encephalitozoon cuniculi
Eremothecium gossypii
Fusarium sporotrichioides
Gibberella zeae PH-1
Kluyveromyces waltii NCYC 2644
Magnaporthe grisea 70-15
Naumovia castellii NRRL Y-12630
Neurospora crassa strain OR74A
Phanerochaete chrysosporium
Saccharomyces bayanus 623-6C
Saccharomyces bayanus MCYC 623
Saccharomyces cerevisiae
Saccharomyces kluyveri NRRL Y-12651
Saccharomyces kudriavzevii IFO 1802
Saccharomyces mikatae IFO 1815
Saccharomyces paradoxus NRRL Y-17217
Schizosaccharomyces pombe
Ustilago maydis 521

Proposed High Priority
Drosophilidae
  Drosophila ananassae
  Drosophila erecta
  Drosophila grimshawi
  Drosophila mojavensis
  Drosophila persimilis
  Drosophila sechellia
  Drosophila simulans
  Drosophila virilis
  Drosophila willistoni
  Drosophila yakuba
Others
  Apis mellifera
  Caenorhabditis remanei
  Caenorhabditis japonica
  Caenorhabditis n. sp. CB5161
  Monodelphis domestica
  Oxytricha trifallax
  Schmidtea mediterranea
  Ornithorhynchus anatinus
  Strongylocentrotus purpuratus
  Gasterosteus aculeatus
  Tetrahymena thermophila
  Macaca mulatta
  Saccoglossus kowalesvskii
Fungal Genome Initiative I Proposal
  Aspergillus nidulans
  Coccidioides immitis
  Coprinus cinereus
  Cryptococcus neoformans, Serotype A
  Pneumocystis carinii
  Rhizopus oryza
  Ustilago maydis
Fungal Genome Initiative II
  Candida albicans
  Candida tropicalis
  Lodderomyces elongisporus
  Saccharomyces cerevisiae RM11-1A
Fungal Genome Initiative III
  Candida lusitaniae
  Candida guilliermondii

Chaetomium globosum
Cryptococcus neoformans serotype B
Histoplasma capsulatum
Podospora anserina
Unicinocarpus reesii
Moderate Priority Proposals
Aedes aegypti
Aedes albopictus
Branchiostoma floridae
Felis catus
Culex pipiens
Macropus eugenii
Ochlerotatus
Sus scrofa
Trichoplax adhaerens

References

Aitken, N., Smith, S., Schwarz, C., and Morin, P. A. (2004). Single nucleotide polymorphism (SNP) discovery and genotyping in mammals: A targeted-gene approach. *Mol. Ecol.* **13**, 1423–1431.

Baker, R. H., Yu, X. B., and DeSalle, R. (1998). Assessing the relative contribution of molecular and morphological characters in simultaneous analysis trees. *Mol. Phylogenet. Evol.* **9**, 427–436.

Baker, R. H., and DeSalle, R. (1997). Multiple sources of character information and the phylogeny of Hawaiian drosophilids. *Syst. Biol.* **46**, 654–673.

Baldauf, S. L., Roger, A. J., Wenk-Siefert, I., and Doolittle, W. F. (2000). A Kingdom-level phylogeny of eukaryotes based on combined protein data. *Science* **290**, 972–977.

Baxevanis, A. D., and Ouellette, B. F. F. (2001). "Bioinformatics: A Practical Guide to the Analysis of Genes and Proteins," 2nd Ed. Wiley Interscience, New York.

Bergman, C. M., Pfeiffer, B. D., Rincón-Limas, D. E., Hoskins, R. A., Gnirke, A., Mungall, C. J., Wang, A. M., Kronmiller, B., Pacleb, J., Park, S., Stapleton, M., Wan, K., George, R. A., de Jong, P. J., Botas, J., Rubin, G. M., and Celniker, S. E. (2002). Assessing the impact of comparative genomic sequence data on the functional annotation of the *Drosophila* genome. *Genome Biol.* **3**, 1–20.

Birstein, V. J., Doukakis, P., Sorkin, B., and DeSalle, R. (1998). Population aggregation analysis of three caviar-producing species of stugeons and implications for the species identification of black caviar. *Conserv. Biol* **12**, 766–775.

Boffelli, D., McAuliffe, J., Ovcharenko, D., Lewis, K. D., Ovcharenko, I., Pachter, L., and Rubin, E. M. (2003). Phylogenetic shadowing of primate sequences to find functional regions of the human genome. *Science* **299**, 1391–1394.

Bonacum, J., Stark, J., and Bonwich, E. (2001). PCR methods and approaches. *In* "Techniques in Molecular Systematics and Evolution," pp. 302–328. Birkhauser Verlag.

Boonham, N., Walsh, K., Smith, P., Madagan, K., Graham, I., and Baker, I. (2003). Detection of potato viruses using microarray technology: Towards a generic method for plant viral disease diagnosis. *J. Virol. Methods* **108**, 181–187.

Bremer, K. (1988). The limits of amino-acid sequence data in angiosperm phylogenetic reconstruction. *Evolution* **42**, 795–803.

Brenner, E. D., Stevenson, D. W., McCombie, W. R., Katari, M. S., Rudd, S. A., Mayer, K. F. X., Pelenchar, P. M., Runko, S. J., Twigg, R. W., Dai, G., Martienssen, R. A., Benfey, P. N., and Coruzzi, G. M. (2003). EST analysis in *Cycas*, the oldest living seed plant. *Genome Biol.* **4**, R78.

Brochier, C., Bapteste, E., Moreira, D., and Philippe, H. (2002). Eubacterial phylogeny based on translational apparatus proteins. *Trends Genet.* **18**, 1–5.

Brown, J. R., and Doolittle, W. F. (1997). Archaea and the prokaryote-to-eukaryote transition. *Microbiol. Mol. Biol. Rev.* **61**, 456–502.

Bult, C., Blake, J. A., Adams, M. D., White, O., Sutton, G., Kerlavage, R., Fields, C., and Venter, J. C. (1997). The impact of rapid gene discovery technology on studies of evolution and biodiversity. *In* Biodiversity II: Understanding and Protecting Our Biological Resources, pp. 289–300. John Henry Press, Washington, DC.

Campbell, N. (2004). Comparative genomics: This year's model? *Nature Reviews in Genetics* **5**, 82–83.

Carninci, P., Westover, A., Nishiyama, Y., Ohsumi, T., Itoh, M., Nagaoka, S., Sasaki, N., Okazaki, Y., Muramatsu, M., Schneider, C., and Hayashizaki, Y. (1997). High efficiency selection of full-length cDNA by improved biotinylated cap trapper. *DNA Res.* **4**, 61–66.

Carninci, P., Nishiyama, Y., Westover, A., Itoh, M., Nagaoka, S., Sasaki, N., Okazaki, Y., Muramatsu, M., and Hayashizaki, Y. (1998). Thermostabilization and thermoactivation of thermolabile enzymes by trehalose and its application for the synthesis of full length cDNA. *Proc. Natl. Acad. Sci. USA* **95**, 520–524.

Chenchik, A., Diachenko, L., Moqadam, F., Tarabykin, V., Lukyanov, S., and Siebert, P. D. (1996). Full-length cDNA cloning and determination of mRNA 5′ and 3′ ends by amplification of adaptor-ligated cDNA. *Biotechniques* **21**, 526–534.

Dermitzakis, E. T., Reymond, A., Scamuffa, N., Ucla, C., Kirkness, E., Rossier, C., and Antonarakis, S. E. (2003). Evolutionary discrimination of mammalian conserved non-genic sequences. *Science* **302**, 1033–1035.

Dermitzakis, E. T., Reymond, A., Lyle, R., Scamuffa, N., Ucla, C., Deutsch, S., Stevenson, B. J., Flegel, V., Bucher, P., Jongeneel, C. V., and Antonarakis, S. E. (2002). Numerous potentially functional but non-genic conserved sequences on human chromosome 21. *Nature* **420**, 578–582.

Dermitzakis, E. T., Kirkness, E., Schwarz, S., Birney, E., Reymond, A., and Antonarakis, S. E. (2004). Comparison of human chromosome 21 conserved non-genic sequences (CNGs) with the mouse and dog genomes shows that their selective constraint is independent of their genic environment. *Genome Res.* **14**, 852–859.

Desalle, R., Giribet, G., and Wheeler, W. (2002a). Molecular Systematics and Evolution: Theory and Practice, p. 309. Birkhäuser Verlag, Basel Switzerland.

Desalle, R., Giribet, G., and Wheeler, W. (2002b). Techniques in Molecular Systematics and Evolution, p. 450. Birkhäuser Verlag, Basel Switzerland.

Eisen, J. A. (1998). Phylogenomics: Improving functional predictions for uncharacterized genes by evolutionary analysis. *Genome Res.* **8**, 163–167.

Eisen, J. A. (2000). Horizontal gene transfer among microbial genomes: New insights from complete genome analysis. *Curr. Opin. Genet. Dev.* **10**, 606–611.

Eisen, J. A., and Fraser, C. M. (2003). Phylogenomics: Intersection of evolution and genomics. *Science* **300**, 1706–1707.

Eisen, J. A., Kaiser, D., and Myers, R. M. (1997). Gastrogenomic delights: A movable feast. *Nature Med.* **3**, 1076–1078.

Eisen, J. A., Sweder, K. S., and Hanawalt, P. C. (1995). Evolution of the SNF2 family of proteins: Subfamilies with distinct sequences and functions. *Nucleic Acids Res.* **23**, 2715–2723.

Enard, W., Khaitovich, P., Klose, J., Zöllner, S., Heissig, F., Giavalisco, P., Nieselt-Struwe, K., Muchmore, E., Varki, A., Ravid, R., Doxiadis, G. M., Bontrop, R. E., and Pääbo, S. (2002). Intra- and interspecific variation in primate gene expression patterns. *Science* **296**, 340–343.

Ewing, B., Hiller, L., Wendl, M., and Green, P. (1998). Basecalling of automated sequence traces using phred. I. Accuracy assessment. *Genome Res.* **8**, 175–185.

Farris, J. S., Källersjö, M., Kluge, A. G., and Bult, C. (1994). Testing significance of incongruence. *Cladistics* **10**, 315–319.

Farris, J. S., Albert, V. A., Kallersjo, M., Lipscomb, D., and Kluge, A. G. (1996). Parsimony jackknifing outperforms neighbour-joining. *Cladistics* **12**, 99–124.

Felsenstein, J. (1985). Confidence limits on phylogenies: An approach using the bootstrap. *Evolution* **39**, 783–791.

Felsenstein, J. (2004). Inferring Phylogenies. Sinauer Associates, Sunderland, MA.

Fitch, W. M. (2000). Homology: A personal view of some of the problems. *Trends Genet.* **16**, 227–234.

Fleischmann, R. D., Adams, M. D., White, O., Clayton, R. A., Kirkness, E. F., Kerlavage, A. R., Bult, C. J., Tomb, J. -F., Dougherty, B. A., Merrick, J. M., McKenney, K., Sutton, G., FitzHugh, W., Fields, C., Gocayne, J. D., Scott, J., Shirley, R., Liu, L.-I., Glodek, A., Kelley, J. M., Weidman, J. F., Phillips, C. A., Spriggs, T., Hedblom, E., Cotton, M. D., Utterback, T. R., Hanna, M. C., Nguyen, D. T., Saudek, D. M., Brandon, R. C., Fine, L. D., Fritchman, J. L., N. Geohagen, S. M., Gnehm, C. L., McDonald, L. A., Small, K. V., Fraser, C. M., Smith, H. O., and Venter, J. C. (1995). Whole-genome random sequencing and assembly of *Haemophilus influenzae* Rd. *Science* **269**, 496–512.

Gatesy, J., Matthee, C., DeSalle, R., and Hayashi, C. (2002). Resolution of a supertree/supermatrix paradox. *Syst. Biol.* **51**, 652–664.

Gatesy, J., Amato, G., Norell, M., DeSalle, R., and Hayashi, C. (2003). Combined support for wholesale taxic atavism in gavialine crocodylians. *Syst. Biol.* **52**, 403–422.

Gatesy, J. E., O'Grady, P. M., and Baker, R. (1999). Corroboration among data sets in simultaneous analysis: Hidden support for phylogenetic relationships among higher level Artiodactyl taxa. *Cladistics* **15**, 271–313.

Gordon, D., Abanjian, C., and Green, P. (1998). Consed: A graphical tool for sequence finishing. *Genome Res.* **8**, 195–202.

Hacia, J. C. (1999). Resequencing and mutational analysis using oligonucleotide microarrays. *Nat. Genet.* **21**(Suppl.), 42–47.

He, C., Chen, L., Simmons, M., Li, P., Kim, S., and Liu, Z. J. (2003). SNP discovery in interspecific hybrids of catfish by comparative EST analysis. *Anim. Genet.* **34**, 445–448.

Hebert, P. D. N., Cywinska, A., Ball, S., and DeWaard, J. (2003a). Biological identifications through DNA barcodes. *Proc. R. Soc. Lond. B.* **270**, 313–322.

Hebert, P. D. N., Ratnasingham, S., and DeWaard, J. R. (2003b). Barcoding animal life: Cytochrome *c* oxidase subunit 1 divergences among closely related species. *Proc. R. Soc. Lond. B Suppl.* **270**, S96–S99.

Hillis, D. M., Moritz, C., and Mable, B. K. (1996). "Molecular Systematics," 2nd Ed. Sinauer Associates, Inc., Sunderland, MA.

Hinchliffe, S. J., Isherwood, K. E., Stabler, R. A., Prentice, M. B., Rakin, A., Nichols, R. A., Oyston, P. C., Hinds, J., Titball, R. W., and Wren, B. W. (2003). Application of DNA microarrays to study the evolutionary genomics of *Yersinia pestis* and *Yersinia pseudotuberculosis*. *Genome Res.* **13**, 2018–2029.

Huan, X., and Madan, A. (1999). CAP3: A DNA sequence assembly program. *Genome Res.* **9,** 868–877.

Huelsenbeck, J. P., Hillis, D. M., and Jones, R. (1996). Parametric bootstrapping in molecular phylogenetics: Applications and performance. *In* "Molecular Zoology: Advances, Strategies, and Protocols" (J. D. Ferraris and S. R. Palumbi, eds.), pp. 19–46. Wiley & Liss, New York.

Hunkapiller, T., Kaiser, R. J., Koop, B. F., and Hood, L. (1991). Large-scale and automated DNA sequence determination. *Science* **254,** 59–67.

Iafrate, A. J., Feuk, L., Rivera, M. N., Listewnik, M. L., Donahoe, P. K., Qi, Y., Scherer, S. W., and Lee, C. (2004). Detection of large-scale variation in the human genome. *Nat. Genet.* **36,** 949–952.

Janies, D. A., and Wheeler, W. C. (2002). Theory and practice of parallel direct optimization. *In* "Techniques in Molecular Systematics and Evolution" (R. DeSalle, G. Giribet, and W. Wheeler, eds.). Birkhauser Verlag, Basel.

Janies, D., and Wheeler, W. (2001). Efficiency of parallel direct optimization. *Cladistics* **17,** S71–S82.

Korbel, J. O., Snel, B., Huynen, M. A., and Bork, P. (2002). SHOT: A webserver for the construction of genome phylogenies. *Trends Genet.* **18,** 158–162.

Kortschak, R. D., Samuel, G., Saint, R., and Miller, D. J. (2003). EST analysis of the cnidarian *Acropora millepora* reveals extensive gene loss and rapid sequence divergence in the model invertebrates. *Curr. Biol.* **13,** 2190–2195.

Koski, J., and Golding, B. (2001). The closest BLAST hit is often not the nearest neighbor. *J. Mol. Evol.* **52,** 540–542.

Krylov, D. M., Wolf, Y. I., Rogozin, I. B., and Koonin, E. V. (2003). Gene loss, protein sequence divergence, gene dispensability, expression level, and interactivity are correlated in eukaryotic evolution. *Genome Res.* **13,** 2229–2235.

Lee, M. S. Y., and Hugall, A. F. (2003). Partitioned likelihood support and the evaluation of data set conflict. *Syst. Biol.* **52,** 11–22.

Li, J., Jiang, T., Mao, J.-H., Balmain, A., Peterson, L., Harris, C., Rao, P. H., Havlak, P., Gibbs, R., and Cai, W.-W. (2004). Genomic segmental polymorphisms in inbred mouse strains. *Nat. Genet.* **36,** 952–955.

Liang, F., Holt, I., Pertea, G., Karamycheva, S., Salzberg, S. L., and Quackenbush, J. (2000). Gene index analysis of the human genome estimates approximately 120,000 genes. *Nat. Genet.* **25,** 239–240.

Lisitsyn, N., Lisitsyn, N., and Wigler, M. (1993). Cloning the differences between two complex genomes. *Science* **258,** 946–951.

Livak, K. J., Flood, S. A. J., Marmaro, J., Giusti, W., and Deetz, K. (1995). Oligonucleotides with fluorescent dyes at opposite ends provide a quenched probe system useful for detecting PCR product and nucleic acid hybridization. *PCR Methods and Applications* **4,** 357–362.

Locke, D. P., Segraves, R., Carbone, L., Archidiacono, N., Albertson, D. G., Pinkel, D., and Eichler, E. E. (2003). Large-scale variation among human and great ape genomes determined by array comparative genomic hybridization. *Genome Res.* **13,** 347–357.

Lucito, R., Healy, J., Alexander, J., Reiner, A., Esposito, D., Chi, M., Rodgers, L., Brady, A., Sebat, J., Troge, J., West, J. A., Rostan, S., Nguyen, K. C., Powers, S., Ye, K. Q., Olshen, A., Venkatraman, E., Norton, L., and Wigler, M. (2003). Representational oligonucleotide microarray analysis: A high-resolution method to detect genome copy number variation. *Genome Res.* **13,** 2291–2305.

Lucito, R., Nakimura, M., West, J. A., Han, Y., Chin, K., Jensen, K., McCombie, R., Gray, J. W., and Wigler, M. (1998). Genetic analysis using genomic representations. *Proc. Natl. Acad. Sci. USA* **95,** 4487–4492.

Lyons, L. A., Laughlin, T. F., Copeland, N. G., Jenkins, N. A., Womack, J. E., and O'Brien, S. J. (1997). Comparative anchor tagged sequences (CATS) for integrative mapping of mammalian genomes. *Nat. Genet.* **15,** 47–56.

Maitra, A., Cohen, Y., Gillespie, S. E., Mambo, E., Fukushima, N., Hoque, M. O., Shah, N., Goggins, M., Califano, J., Sidransky, D., and Chakravarti, A. (2004). The Human MitoChip: A high-throughput sequencing microarray for mitochondrial mutation detection. *Genome Res.* **14,** 812–819.

Matte-Tailliez, O., Brochier, C., Forterre, P., and Philippe, H. (2002). Archaeal phylogeny based on ribosomal proteins. *Mol. Biol. Evol.* **19,** 631–639.

Marras, S. A. E., Kramer, F. R., and Tyagi, S. (1999). Multiplex detection of single-nucleotide variations using molecular beacons. *Genet. Anal.* **14,** 151–156.

Maruyama, K., and Sugano, S. (1994). Oligo-capping: A simple method to replace the cap structure of eukaryotic mRNAs with oligoribonucleotides. *Gene* **138,** 171–174.

Mir, K. U., and Southern, E. M. (2000). Sequence variation in genes and genomic DNA: Methods for large-scale analysis. *Annu. Rev. Genomics Human Genet.* **1,** 329–360.

Mount, D. W. (2001). "Bioinformatics: Sequence and Genome Analysis." Cold Spring Harbor Press, Cold Spring Harbor.

Mullis, K., Faloona, F., Scharf, S., Saki, R., Horn, G., and Erlich, H. (1986). Specific enzymatic amplification of DNA *in vitro*: The polymerase chain reaction. *Cold Spring Harbor Symposia on Quantitative Biology* **51,** 263.

Murphy, W. J., Eizirik, E., O'Brien, S. J., Madsen, O., Scally, M., Douady, C. J., Teeling, E., Ryder, O. A., Stanhope, M. J., de Jong, W. W., and Springer, M. S. (2001a). Resolution of the early placental mammal radiation using Bayesian phylogenetics. *Science* **294,** 2348–2351.

Murphy, W. J., Eizirik, E., Johnson, W. J., Zhang, Y. P., Ryder, O. A., and O'Brien, S. J. (2001b). Molecular phylogenetics and the origins of placental mammals. *Nature* **409,** 614–618.

Nozaki, H., Ohta, N., Matsuzaki, M., Misumi, O., and Kuroiwa, T. (2003). Phylogeny of plastids based on cladistic analysis of gene loss inferred from complete plastid genome sequences. *J. Mol. Evol.* **57,** 377–382.

Palumbi, S. R. (1995). Nucleic acids II: The polymerase chain reaction. *In* "Molecular Systematics" (D. Hillis and C. Moritz, eds.), 2nd Ed., pp. 205–247. Sinauer, Sunderland, MA.

Palumbi, S. R., and Baker, C. S. (1994). Contrasting population structure from nuclear intron sequences and mtDNA of humpback whales. *Mol. Biol. Evol.* **11,** 426–435.

Pearson, W. R., and Lipman, D. J. (1988). Improved tools for biological sequence comparison. *Proc. Natl. Acad. Sci. USA* **4,** 2444–2448.

Pfunder, M., Holzgang, O., and Frey, J. E. (2004). Development of microarray-based diagnostics of voles and shrews for use in biodiversity monitoring studies, and evaluation of mitochondrial cytochrome oxidase I vs. cytochrome b as genetic markers. *Mol. Ecol.* **13,** 1277–1285.

Phillips, A., Janies, D., and Wheeler, W. (2000). Multiple sequence alignment in phylogenetic analysis. *Mol. Phylogenet. Evol.* **16,** 317–330.

Phillips, A. (2002). Sub-genomic sequencing projects for phylogenetic analysis. *In* "Techniques in Molecular Systematics and Evolution" (R. DeSalle, G. Giribet, and W. Wheeler, eds.), pp. 132–146. Birkhauser.

Planet, P. J., Kachlany, S. C., Fine, D. H., DeSalle, R., and Figurski, D. H. (2003). The widespread colonization island of *Actinobacillus actinomycetemcomitans. Nat. Genet.* **34,** 193–198.

Poustka, A. J., Groth, D., Hennig, S., Thamm, S., Cameron, A., Beck, A., Reinhardt, R., Herwig, R., Panopoulou, G., and Lehrach, H. (2003). Generation, annotation, evolutionary analysis, and database integration of 20,000 unique sea urchin EST clusters. *Genome Res.* **13**, 2736–2746.

Read, T. D., Peterson, S. N., Tourasse, N., Baillie, L. W., Paulsen, I. T., Nelson, K. E., Tettelin, H., Fouts, D. E., Eisen, J. A., Gill, S. R., Holtzapple, E. K., Okstad, O. A., Helgason, E., Rilstone, J., Wu, M., Kolonay, J. F., Beanan, M. J., Dodson, R. J., Brinkac, L. M., Gwinn, M., DeBoy, R. T., Madpu, R., Daugherty, S. C., Durkin, A. S., Haft, D. H., Nelson, W. C., Peterson, J. D., Pop, M., Khouri, H. M., Radune, D., Benton, J. L., Mahamoud, Y., Jiang, L., Hance, I. R., Weidman, J. F., Berry, K. J., Plaut White, O., Salzberg, S. L., Thomason, B., Friedlander, A. M., Koehler, T. M., Hanna, P. C., Kolsto, A. B., and Fraser, C. M. (2003). The genome sequence of *Bacillus anthracis* Ames and comparison to closely related bacteria. *Nature* **423**, 81–86.

Rifkin, S. A., Kim, J., and White, K. P. (2003). Evolution of gene expression in the *Drosophila melanogaster* subgroup. *Nat. Genet.* **33**, 138–144.

Rivera, M. C., and Lake, J. A. (2004). The ring of life provides evidence for a genome fusion origin of eukaryotes. *Nature* **432**, 152–155.

Rokas, A., Williams, B. L., King, N., and Carroll, S. B. (2003). Genome-scale approaches to resolving incongruence in molecular phylogenies. *Nature* **425**, 798–804.

Ronquist, F., and Huelsenbeck, J. P. (2003). MrBayes 3: Bayesian phylogenetic inference under mixed models. *Bioinformatics* **19**, 1572–1574.

Rudd, S. (2003). Expressed sequence tags: Alternative or complement to whole genome sequences? *Trends Plant Sci.* **8**, 321–328.

Saiki, R. K., Gelfand, D. H., Stoffel, S., Scharf, S. J., Higuchi, R., Horn, G. T., Mullis, K. B., and Erlich, H. A. (1988). Primer-directed enzymatic amplification of DNA with a thermostable DNA polymerase. *Science* **239**, 487.

Sarkar, I. N., Thornton, J. W., Planet, P. J., Figurski, D. H., Schierwater, B., and DeSalle, R. (2002). An automated phylogenetic key for classifying homeoboxes. *Mol. Phylogenet. Evol.* **24**, 388–399.

Savolainen, V., and Chase, M. W. (2003). A decade of progress in plant molecular phylogenetics. *Trends Genet.* **19**, 717–724.

Sebat, J., Lakshmi, B., Troge, J., Alexander, J., Young, J., Lundin, P., Månér, S., Massa, H., Walker, M., Chi, M., Navin, N., Lucito, R., Healy, J., Hicks, J., Ye, K., Reiner, A., Gilliam, T. C., Trask, B., Patterson, N., Zetterberg, A., and Wigler, M. (2004). Large-scale copy number polymorphism in the human genome. *Science* **305**, 525–528.

Snel, B., Bork, P., and Huynen, M. (1999). Genome phylogeny based on gene content. *Nat. Genet.* **21**, 108–110.

Snel, B., Bork, P., and Huynen, M. (2002). Genomes in flux: The evolution of archaeal and protobacterial gene content. *Genome Res.* **12**, 17–25.

Sorenson, M. D. (1999). TreeRot, version 2. Boston University, Boston, MA.

Southern, E. M. (1996). DNA chips: Analyzing sequence by hybridization to oligonucleotides on a large scale. *Trends Genet.* **12**, 110–115.

Stanhope, M. J., Lupas, A., Italia, M. J., Koretke, K. K., Volker, C., and Brown, J. R. (2001). Phylogenetic analyses do not support horizontal gene transfers from bacteria to vertebrates. *Nature* **411**, 940–944.

Stoeckle, M. (2003). Taxonomy, DNA, and the bar code of life. *BioScience* **53**, 2–3.

Swofford, D. L. (2002). PAUP* version 4.0b. Sinauer, Sunderland, MA.

Theodorides, K., De Riva, A., Gómez-Zurita, J., Foster, P. G., and Vogler, A. P. (2002). Comparison of EST libraries from seven beetle species: Towards a framework for phylogenomics of the Coleoptera. *Insect Mol. Biol.* **11**, 467–475.

Thomas, J. W., Touchman, J. W., Blakesley, R. W., Bouffard, G. G., Beckstrom-Sternberg, S. M., Margulies, E. H., Blanchette, M., Siepel, A. C., Thomas, P. J., McDowell, J. C.,

Maskeri, B., Hansen, N. F., Schwartz, M. S., Weber, R. J., Kent, W. J., Karolchik, D., Bruen, T. C., Bevan, R., Cutler, D. J., Schwartz, S., Elnitski, L., Idol, J. R., Prasad, A. B., Lee-Lin, S.-Q., Maduro, V. V. B., Summers, T. J., Portnoy, M. E., Dietrich, N. L., Akhter, N., Ayele, K., Benjamin, B., Cariaga, K., Brinkley, C. P., Brooks, S. Y., Granite, S., Guan, X., Gupta, J., Haghighi, P., Ho, S.-L., Huang, M. C., Karlins, E., Laric, P. L., Legaspi, R., Lim, M. J., Maduro, Q. L., Masiello, C. A., Mastrian, S. D., McCloskey, J. C., Pearson, R., Stantripop, S., Tiongson, E. E., Tran, J. T., Tsurgeon, C., Vogt, J. L., Walker, M. A., Wetherby, K. D., Wiggins, L. S., Young, A. C., Zhang, L.-H., Osoegawa, K., Zhu, B., Zhao, B., Shu, C. L., de Jong, P. J., Lawrence, C. E., Smit, A. F., Chakravarti, A., Haussler, C., Green, P., Miller, M., and Green, E. D. (2003). Comparative analyses of multi-species sequences from targeted genomic regions. *Nature* **424,** 788–793.

Thomas, J. W., and Touchman, J. W. (2002). Vertebrate genome sequencing: Building a backbone for comparative genomics. *Trends Genet.* **18,** 104–108.

Thompson, J. D., Plewniak, F., and Poch, O. (1999). A comprehensive comparison of multiple sequence alignment programs. *Nucleic Acids Res.* **27,** 2682–2690.

Tyagi, S., and Kramer, F. R. (1996). Molecular beacons: Probes that fluoresce upon hybridization. *Nat. Biotechnol.* **14,** 303–308.

Tyagi, S., Bratu, D. P., and Kramer, F. R. (1998). Multicolor molecular beacons for allele discrimination. *Nat. Biotechnol.* **16,** 49–53.

Wang, D., Coscoy, L., Zylberberg, M., Avila, P. C., Boushey, H. A., Ganem, D., and DeRisi, J. L. (2002). Microarray-based detection and genotyping of viral pathogens. *Proc Natl. Acad. Sci. USA* **99,** 15687–15692.

Whiting, M. F. (2002). High throughput DNA sequencing for systematic applications. *In* "Techniques in Molecular Systematics and Evolution" (R. DeSalle, W. C. Wheeler, and G. Giribet, eds.). Birkhauser Press.

Wolf, Y. I., Rogozin, I. B., Grishin, N. V., and Koonin, E. V. (2002). Genome trees and the tree of life. *Trends Genet.* **18,** 472–479.

Ying, S.-Y. (2003). "Generation of cDNA Libraries: Methods and Protocols." Humana Press, Los Angeles.

Zeigler, R. (2003). Gene sequences useful for predicting relatedness of whole genomes in bacteria. *Int. J. Syst. Evol. Microbiol.* **53,** 1893–1900.

Zheng, J. (2003). "Efficient Selection of Unique and Popular Oligos for Large EST Databases. Proceedings of Symposium on Combinatorial Pattern Matching (CPM'03)," pp. 273–283. LNCS 2676, Morelia, Mexico.

Zilversmit, M., O'Grady, P. M., and DeSalle, R. (1998). Shallow genomics, phylogenetics, and evolution in the family Drosophilidae. *In* "Pacific Symposium on Biocomputing" (R. Altman, A. K. Dunker, L. Hunter, K. Lauderdale, and T. Klein, eds.), pp. 512–523. University of Hawaii Press, Honolulu.

[9] ISSR Techniques for Evolutionary Biology

By ANDREA D. WOLFE

Abstract

Inter-simple sequence repeat (ISSR) markers were originally devised for differentiating among closely related plant cultivars but have become extremely useful for studies of natural populations of plants, fungi, insects, and vertebrates. The markers are easily generated using minimal equipment and are hypervariable, yielding a large amount of data for a reasonable cost to the researcher. The methods for Miniprep DNA extraction and cleanup, polymerase chain reaction (PCR) amplification, optimization, data gathering and scoring, and data analyses are outlined.

Introduction

Inter-simple sequence repeat (ISSR) markers are generated from single-primer polymerase chain reaction (PCR) amplifications in which primers are based on dinucleotide or trinucleotide repeat motifs. The number of dinucleotide or trinucleotide repeats varies but is generally sufficiently long to make a primer sequence of at least 14 nucleotides. The microsatellite primer sequence may be anchored with one or two nucleotides on either the 5′ or the 3′ end of the oligonucleotide. An anchoring sequence on the 3′ end of the primer will eliminate the detection of fragment length differences resulting from simple sequence repeat variation, but this would be detectable only with the finest sieving gels and is not an issue for standard agarose gel techniques.

ISSR markers were introduced in 1994 (Gupta *et al.*, 1994; Zietkiewicz *et al.*, 1994) for studies of cultivated plants but have been used for studies of hybridization and hybrid speciation (Archibald *et al.*, 2004; Wolfe *et al.*, 1998a,b), population and conservation genetics (Culley and Wolfe, 2001; Esselman *et al.*, 1999), and systematic investigations in natural populations (Crawford *et al.*, 2001; Mort *et al.*, 2003; Wolfe and Randle, 2001). ISSRs are also being used in population studies of fungi (Kerrigan *et al.*, 2003; Sawyer *et al.*, 2003) and animals (Chatterjee and Mohandas, 2003; Haig *et al.*, 2003; Nagy *et al.*, 2002). The hypervariable nature of ISSRs combined with minimal equipment requirements and ease of use has made them extremely useful and cost-effective molecular markers for many ecological and systematic investigations (Wolfe *et al.*, 1998b; Yang *et al.*, 1996).

The amplification and data-scoring protocols used for ISSR markers are similar to those used for random amplified polymorphic DNA (RAPD) markers with the exception that the annealing temperature for ISSR amplification is generally higher, resulting in a higher degree of stringency for amplified fragments (Wolfe and Liston, 1998). The number of amplification fragments (bands) generated per reaction varies with the primer–template combination but generally is easily scored using agarose gel technology. Finer sieving gels and radioactive labeling of primers will result in the separation of additional bands that are undetectable with standard agarose media (Godwin *et al.*, 1997).

Visible bands are defined as dominant markers and are assigned to genetic loci with two alleles: 1 = present and 0 = absent. This should not be confused with dominance as it relates to mendelian inheritance patterns. The dominant allele of dominant markers can result from a heterozygous or homozygous genotype for a particular priming location. Inheritance patterns of ISSR markers can be established through genetic studies (Tsumura *et al.*, 1996). In plants it is unlikely that ISSR markers result from amplification of plastid DNA because the microsatellites found in this genome are predominantly mononucleotide repeats (Powell *et al.*, 1995).

Protocols for Plant ISSR Markers

The protocols presented here were refined for plant DNAs but are easily modified for other organisms. Because the common microsatellite motifs differ among phyla (Wolfe and Liston 1998), one should use ISSR primers based on the common dinucleotide or trinucleotide motifs for a particular group of organisms.

Plant DNA Miniprep Extraction Protocols

Sample Preparation and Grinding

- Weigh or estimate a small amount of tissue (e.g., 10–50 mg of dried leaves).
- Warm 2X cetyltrimethylammonium bromide (CTAB) extracting buffer (1 ml × number of samples to be extracted) to 60°; add about 30 μl 2-mercaptoethanol to 15 ml extracting buffer just before grinding.
- Measure a small volume of sterilized sand and approximately an equal amount of polyvinylpyrrolidone (PVP) into a mortar.
- Add tissue and grind thoroughly in mortar with pestle. Add 500 μl of extracting buffer to mortar and grind again.
- Add 500 μl of extracting buffer (for a final volume of 1000 μl) and grind again. Carefully pour the ground material into a 1.5-ml Microfuge tube.

Extraction and Isolation of DNA

- Incubate for 1 h at 65–70°.
- Add an equal volume (~700 μl) of chloroform:isoamyl alcohol (24:1; this will fill a 1.5-ml Microfuge tube); shake the tube to produce a homogeneous suspension.
- Spin for 20 min in a high-speed microcentrifuge until the aqueous layer of the supernatant is clear.
- Transfer the top aqueous phase to a new Microfuge tube.
- Optional: Repeat the chloroform:isoamyl extraction if the separation seems poor.

DNA Precipitation

- Add 540 μl of cold isopropanol; mix and store overnight at −20° or 2–3 h at −80°.
- Spin in a microcentrifuge for 1–3 min at maximum speed.
- For samples with clear pellets, decant the supernatant and drain using a Kim wipe. For samples with gelatinous material above the pellet, remove all but a thin (~1 mm) band of isopropanol with a micropipette, leaving the gelatinous portion and pellet intact.
- Wash the pellet with 500 μl of 75% ethanol (EtOH); invert the tube several times (gelatinous material should dissolve or precipitate).
- Spin for 3 min; remove the supernatant; dry with a Kim wipe or on a paper towel, leaving the pellet intact.
- Dry with a speed vac or in a vacuum oven (20 min to 1h) at room temperature.

Final Cleaning and Resuspension of DNA

- Use wide-mouth 200 μl pipette tips (cut the tip off ~2 mm from the opening).
- Resuspend the pellet in 80 μl of 1X TE buffer; gently pumping the pipetter to break up the pellet.
- Add 8 μl of 5 M 7.5 M ammonium acetate and 180 μl of 100% EtOH.
- Mix and place the tube in a −20° freezer overnight or at −80° for 2 h.
- Spin 2–3 min and decant the supernatant (optional: add final wash of 75% EtOH and spin again).
- Vacuum dry the DNA pellet, and resuspend in 100 μl of TE. Incubate at 37–40° for 15–30 min to ensure DNA goes into solution.
- Verify presence of DNA using a test gel. Aliquot and store DNA at −80° (long term), −20° (temporary), or 4° (active use).

Note: Commercially available kits can also be used for DNA extraction, but we've found this protocol to be effective and inexpensive.

Elu-Quik DNA Cleanup Protocols

- Add 60 μl of binding buffer to 25 μl of DNA sample to be purified.
- After resuspending glass using a plastic transfer pipette, add 15 μl of glass concentrate. Tap the tube briefly to mix the contents.
- Let the solution stand at room temperature for 10 min, gently tapping the tube at 1-min intervals.
- Centrifuge at 7000g for 30 s in a microcentrifuge. Discard as much supernatant as possible using a transfer pipette.
- Add 300–500 μl of wash buffer (300 works well). Using a nonshear pipette tip (supplied with kit), tease the glass pellet from the wall of the microcentrifuge tube. Gently pipette the glass pellet up and down until pellet becomes flocculent. Invert the tube once. (*Note:* Achieving a flocculent pellet is important for buffer exchange.)
- Centrifuge at 7000g for 30 s. Discard the supernatant.
- Repeat the previous two steps.
- Add 300–500 μl (300 works well) of salt-reduction buffer. Gently resuspend the glass pellet until flocculent with the nonshear pipette tip as described above. Invert the tube once, and centrifuge at 7000g for 2 min. Discard the supernatant.
- Centrifuge the glass pellet at 7000g for 30 s to condense the pellet. Using a 20-μl pipette tip, carefully withdraw as much residual supernatant as possible and discard.
- Add 70 μl of 1X TE or distilled deionized water to the glass pellet and resuspend by flicking the tube repeatedly. Incubate the mixture at 50° for 5 min, occasionally flicking the tube. Centrifuge at 7000g for 30 s in a microcentrifuge and withdraw the supernatant containing the isolated DNA to a fresh tube.
- Quantify the DNA concentration before use.

The Elu-Quik kit is available from Schliecher & Schuell (Keene, NH).

PCR Amplification Protocols

A reaction volume of 25 μl works well for ISSR primers. The entire reaction volume is loaded into a well in an agarose gel for a data run. The protocol listed below works for hot bonnet thermocyclers and thermocyclers requiring a mineral oil overlay. All reagents used should be prepared for PCR by ultraviolet (UV) irradiation of ddH$_2$O and stock chemical solutions before adding dNTPs, enzyme, or DNA. Standardization of DNA concentrations should also be done before PCR amplification. This can be accomplished by comparing the ethidium bromide (EtBr) staining intensity of the DNA extracts in the test gel and adjusting the samples to have similar band intensities under standardized conditions (1–2 μl aliquots in the test gel). Exact DNA concentration determination through use of a fluorometer is unnecessary but would yield a more exact standardization of template concentrations. Some DNA extractions may contain secondary compounds that interfere with PCR amplification using ISSR primers (e.g., parasitic plant DNAs). When amplification is problematic, the DNA stocks can be diluted with ddH$_2$O (1–100-fold) or cleaned using the Elu-Quik kit.

A number of primers (Table I) work on a wide range of angiosperm groups, including Acanthaceae, Asteraceae (Mort et al., 2003), Ericaceae, Lactoridaceae (Crawford et al., 2001), Orchidaceae, Orobanchaceae (Wolfe and Randle, 2001), Poaceae (Esselman et al., 1999), Scrophulariaceae (Archibald et al., 2004; Wolfe et al., 1998a,b), and Violaceae (Culley and Wolfe, 2001). Frequently, the initial amplification protocol used for testing a primer is not the most optimal for the taxon–primer combination and the reactions will need to be optimized before scorable data can be obtained. Furthermore, once the optimization for a particular primer has been done, it is important to run the reactions and duplicate reactions for all taxa in a timely fashion. ISSR primers will degrade in the freezer over long periods; aliquoting the stock solution in small volumes and storing it in an ultracold freezer will reduce the degradation of primer. Working solutions of primers can be stored at −20° for several weeks.

TABLE I

ISSR Primers, Annealing Temperatures, and MgCl$_2$ Routinely Used in My Lab for Studies of Plant Populations

Primer sequence (5' to 3')	Annealing temperature range (°C)	MgCl$_2$ range (mM)
(CA)$_6$-RY	44–46	1.4–3.0
(CA)$_6$-RG	44–47	1.4–3.0
(AG)$_7$-YC	44–47	1.4–3.0
(GT)$_6$-YR	44–45	3.0
(GT)$_6$-AY	44–46	1.5–3.0
(CT)$_8$-RG	45–47	2.0–3.0
(CTC)$_4$-RC	44–47	1.4–3.0
(GT)$_6$-RG	45–47	2.0
(GT)$_6$-RC	45–47	1.5–3.0

The primer optimization process can be initiated using the following thermocycler program and reaction conditions:

Thermocycler program:

- 1.5 min at 94°
- 35 cycles of 40 s at 94°, 45 s at 44° or 45°, 1.5 min at 72°
- 45 s at 94°, 5 min at 72°
- Soak at 6°

Reaction conditions:

- 0.5 μl DNA
- 0.5 μl primer (20–50 μM)
- 2.5 μl Taq polymerase buffer (10X)
- 0.1 μl Taq polymerase (5 U/μl)
- 4.0 μl dNTPs (1.25 mM)
- 1.5 μl MgCl$_2$ (50 mM)
- 15.9 μl ddH$_2$0

After the DNA amplification is complete, the entire volume of each reaction tube is mixed with loading buffer and loaded into separate wells of 1.2–1.5% agarose gels. ISSR bands are separated using low voltage until the loading buffer runs 10 cm from the origin (e.g., 4 V/cm). The gel is stained for 20 min in EtBr (70 μl of 1% EtBr added to 1 L of H$_2$O), and then rinsed in H$_2$O for 2 × 20 min. The destained gel is placed on a UV transilluminator and the image is captured digitally or photographically for data scoring.

Optimization Protocols

Most of the ISSR primers listed in Table I have optimal conditions with an annealing temperature range of 44–47° and 0.7–1.5 mM MgCl$_2$ for plant DNA amplifications. However, each taxon–primer combination will need to be optimized. A temperature gradient thermocycler with 1° gradations is useful for establishing the optimal annealing temperature range (42–48° works well). It is important to optimize for one reaction condition at a time to establish the best protocol for a particular taxon–primer combination. An effective sequence of optimization reactions starting from the amplification protocol listed above is (1) establish an annealing temperature that produces bands or smears of bands, (2) use that temperature and systematically increase or decrease the concentration of MgCl$_2$ in the reaction mix, and (3) take the best result from step 2 and systematically increase or decrease the volume of Taq polymerase. One additional step may be beneficial where the best reaction from step 3 is used, but the sample

DNA concentration is increased or decreased. The goal is to produce bands that are of similar intensity across the gel.

Data Gathering and Scoring

The standard methodology for gathering and scoring ISSR data includes PCR and replication of ISSR primer amplifications, running the PCRs on agarose gel, staining and destaining of the DNA using EtBr, visualization and image capture of stained DNA bands on a UV transilluminator, and assignment of ISSR bands to genetic loci. The latter can be done manually or semiautomatically using image analysis software, with the latter being more accurate. Each amplification experiment should be replicated to verify the amplification products. If a band for a particular DNA template–primer combination is produced in one replicate, but not the second, the band should not be scored as a data point.

There are different approaches for data gathering and scoring, depending on the type of study to be conducted. For example, in studies in which one is examining patterns of gene flow (e.g., hybridization and parentage analysis) among populations or species, the type of data needed are qualitative as opposed to quantitative. For qualitative analyses, one needs to identify ISSR markers typical of individuals, populations, or species that can be used to assess the amount and directionality of gene flow (Wolfe et al., 1998a,b). To identify these markers, numerous ISSR primers are screened in an exemplar sampling across the taxa. For population-specific markers, one would screen several individuals for each population included in the study to determine whether there are ISSR markers that differ among the populations; for species-specific markers, one would screen individuals from multiple populations of each species to determine whether there are ISSR markers that differ among the species. If population- or species-specific markers are identified, the relevant primers would then be used to amplify DNA from all individuals included in the study, and only the relevant ISSR bands would be scored for analysis.

Studies of genetic diversity (e.g., population and conservation biology) require a quantitative approach in which the goal is to obtain as many ISSR bands as possible in as many individuals and populations as possible. We have found that 25 individuals per population will yield a sufficient amount of data to assay patterns of genetic diversity robustly. Systematic investigations of species boundaries will focus on intensive sampling across the geographic range of populations, but with fewer individuals (e.g., five) per population included in the study. A typical ISSR study will include 3–10 primers and several hundred individuals. Candidate ISSR primers are screened by testing exemplars from each taxon, and the primers producing identifiable bands on the initial amplification are optimized for further use. Because of the hypervariability in ISSR markers, it is usually possible to genotype all individuals in a study with one to three ISSR primers.

Both qualitative and quantitative approaches require the assignment of ISSR bands to genetic loci. In practice, each ISSR band visualized on the gel is assigned as a locus identified by its molecular weight. This is calculated by comparing the migration of the band against DNA fragments of known molecular weight (i.e., a standardized DNA ladder). The software provided in most image analysis packages will calculate molecular weights of each band identified on a gel. To provide reference points among different gels, it is helpful to have DNA templates from several individuals (reference accessions) constant across all amplification experiments conducted for each ISSR primer. The loci assigned to the reference taxa are used to calibrate the molecular weights calculated for bands from the other individuals included in the study. A typical gel setup in my lab has three DNA ladders and three reference accessions (Fig. 1). After all molecular weights have been calculated and loci assigned, the data are converted to a matrix of 1s and 0s where 1 = band present and 0 = band absent for each locus–individual combination.

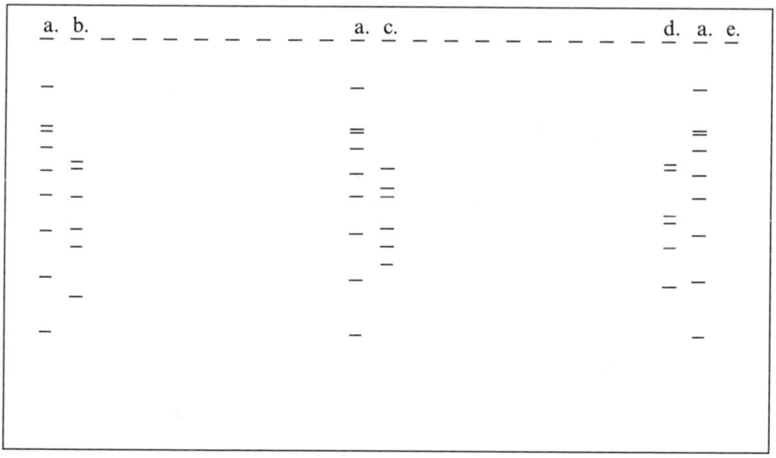

FIG. 1. Typical inter-simple sequence repeat (ISSR) gel layout. a, DNA ladder; b, reference taxon 1; c, reference taxon 2; d, reference taxon 3; e, negative control. The other lanes are used for the reactions from the taxa to be assayed. The same reference taxa are used for each gel in the study.

Data Analysis

Qualitative Approaches. When ISSRs are used to study patterns of gene flow, it is important to assess the relative frequency of a particular ISSR band in each taxon to make a decision about whether it can be used as a marker for a particular taxon (Wolfe *et al.*, 1998a,b). Patterns of marker additivity or distribution are assessed in gene flow studies. For example, if there is a hypothesis of hybrid speciation in which taxon C is the purported derivative species resulting from hybridization between taxon A and taxon B, one would expect to see species-typical markers from each of the purported parental species in the hybrid-derivative species. If the hybrid-derivative species is relatively young, one would expect to see few novel alleles in the derivative species. However, with the hypervariable nature of ISSRs, one would expect to see more novel alleles in a hybrid-derivative species than one would see in an allozyme study.

Asymmetrical patterns of gene flow are easily detected using ISSR markers (Datwyler, 2001; Wolfe *et al.*, 1998b). Low-frequency markers are sometimes more useful in detecting these patterns than species-typical markers (Archibald *et al.*, 2004). After the identification of particular marker alleles, the frequency of occurrence of the markers in the taxon of interest is measured, and the hypothesis of gene flow is evaluated based on the patterns detected.

Quantitative Approaches. ISSR data can be analyzed using descriptive or comparative statistics to measure levels of diversity and can be used for describing population structure (Camacho and Liston, 2001; Culley and Wolfe, 2001; Esselman *et al.*, 1999; Holsinger *et al.*, 2002; Lutz, 2001; Mort *et al.*, 2003; Wolfe *et al.*, 1998b). There are assumptions made in some methods of dominant marker analyses that require the estimation of allele frequencies, because it is not possible to establish the genotype from the ISSR data alone. The bands are scored as diallelic loci in which band presence (dominant allele) could represent two alleles or just one. Band absence implies homozygosity for a null allele, but this is very problematic because the absence of a band may result from a primer mismatch at either end of the SSR motif or an insertion–deletion event that changes the molecular weight of the amplified fragment. Allele frequencies are calculated from assessing the frequency of the null allele:

$$q = (\text{\# individuals lacking the band at a particular locus} \div \text{total \# individuals assayed})^{1/2}$$

$$p = 1 - q,$$

where q is the frequency of the null allele and p is the frequency of the dominant allele.

The descriptive statistics often reported include the total number of genetic loci scored, the percentage of polymorphic loci, average number of bands per primer, the percentage of polymorphic bands per population, the number of species- or population-typical markers, and the number of shared and unique genotypes. The comparative statistics used in studies of natural populations include genetic similarities calculated by a band-matching algorithm (e.g., Jaccard or Dice). Calculated distances are then used in cluster analyses for examining within- and among-population or -taxa relationships. The use of a band-matching algorithm is important because of the difficulty in determining null allele homology. Genetic structure among populations can be estimated using AMOVA in Arlequin 2.00 (Schneider *et al.*, 2000) or F-statistics using HICKORY (Holsinger *et al.*, 2002).

Acknowledgments

The DNA Miniprep protocol was derived from protocols established by J. Doyle, R. Jansen, J. Francisco-Ortega, and P. Soltis as modified in my lab by J. Xiang, S. Kephart, S. Datwyler, and N. Arguedas. I especially thank N. Arguedas, C. Randle, and J. Morawetz for helpful discussion of this manuscript during its preparation, and L. Wallace for helpful discussion on dominant marker analyses. ISSR studies conducted in my lab were funded by NSF grants DEB 9632675, DEB 9708332, and DEB 0089640.

References

Archibald, J., Wolfe, A. D., and Johnson, S. (2004). Hybridization and gene flow between a day- and night-flowering species of *Zaluzianskya* (*Scrophulariaceae s.s., tribe Manuleeae*). *Am. J. Botany* **91,** 1333–1344.

Camacho, F. J., and Liston, A. (2001). Population structure and genetic diversity of *Botrychium pumicola* (*Ophioglassaceae*) based on inter-simple sequence repeats (ISSR). *Am. J. Botany* **88,** 1065–1070.

Chatterjee, S. N., and Mohandas, T. P. (2003). Identification of ISSR markers associated with productivity traits in silkworm, *Bombyx mori* L. *Genome* **46,** 438–447.

Crawford, D., Tago-Nakazawa, M., Stuessy, T. F., Anderson, G. J., Bernardello, G., Ruiz, E., Jensen, R. J., Baeza, C. M., Wolfe, A. D., and Silva, O. M. (2001). Intersimple sequence repeat (ISSR) variation in *Lactoris fernandeziana* (*Lactoridaceae*), a rare endemic of the Juan Fernandez Archipelago, Chile. *Plant Species Biol.* **16,** 185–192.

Culley, T. M., and Wolfe, A. D. (2001). Population genetic structure of the cleistogamous plant species *Viola pubescens* Aiton (*Violaceae*), as indicated by allozyme and ISSR markers. *Heredity* **86,** 545–556.

Datwyler, S. L. (2001). "Evolution and dynamics of hybridization in *Penstemon* subgenus *Dasanthera* (*Schrophulariaceae s.l.*)," Ph.D. Thesis, Ohio State University.

Esselman, E. J., Jianqiangt, L., Crawford, D. J., Windus, J. L., and Wolfe, A. D. (1999). Clonal diversity in the rare *Calamagrostis proteri* ssp. *insperata* (*Poaceae*): Comparative results for allozymes and RAPD and ISSR markers. *Mol. Ecol.* **8,** 443–451.

Godwin, I. D., Aitken, E. A. B., and Smith, L. W. (1997). Application of inter-simple sequence repeat (ISSR) markers to plant genetics. *Electrophoresis* **18,** 1524–1528.

Gupta, M., Chyi, Y.-S., Romero-Severson, J., and Owen, J. L. (1994). Amplification of DNA markers from evolutionarily diverse genomes using single primers of simple-sequence repeats. *Theoret. Applied Genet.* **89,** 998–1006.

Haig, S. M., Mace, T. R., and Mullins, T. D. (2003). Parentage and relatedness in polyandrous comb-crested jacanas using ISSRs. *J. Heredity* **94,** 302–309.

Holsinger, K. E., Lewis, P. O., and Dey, D. K. (2002). A Bayesian approach to inferring population structure from dominant markers. *Mol. Ecol.* **11,** 1157–1164.

Kerrigan, J., Smith, M. T., Rogers, J. D., Poot, G. A., and Douhan, G. W. (2003). *Ascobotryozyma cognata* sp. nov., a new ascomycetous yeast associated with nematodes from wood-boring beetle galleries. *Mycol. Res.* **107,** 1110–1120.

Lutz, A. W. (2001). "Patterns of genetic diversity in *Penstemon caryi* Pennel. (*Scrophulariaceae s.l.*), an endemic to limestone substrates," Ms Thesis, Ohio State University.

Mort, M. E., Crawford, D. J., Santos-Guerra, A., Francisco-Ortega, J., Esselman, E. J., and Wolfe, A. D. (2003). Relationships among the Macaronesian members of *Tolpis* (*Asteraceae: Lactuceae*) based upon analyses of inter-simple sequence repeat (ISSR) markers. *Taxon* **52,** 511–518.

Nagy, Z. T., Joger, U., Guicking, D., and Wink, M. (2002). Phylogeography of the European whip snake *Coluber* (*Hierophis*) *viridflavus* as inferred from nucleotide sequences of the mitochondrial cytochrome b gene and ISSR genomic fingerprinting. *Biota (Race)* **3,** 109–118.

Powell, W., Morgante, M., Andre, C., McNicol, J. W., Machray, G. C., Doyle, J. J., Tingey, S. V., and Rafalski, J. A. (1995). Hypervariable microsatellites provide a general source of polymorphic DNA markers for the chloroplast genome. *Curr. Biol.* **5,** 1023–1029.

Sawyer, N. A., Chambers, S. M., and Cairney, J. W. G. (2003). Distribution of *Amanitai* spp. genotypes under eastern Australian sclerophyll vegetation. *Mycol. Res.* **107,** 1157–1162.

Schneider, S., Roessli, D., and Excoffier, L. (2000). Arlequin: A software for population genetics data analysis, version 2.000. Genetics and Biometry Lab, Department of Anthropology, University of Geneva.

Tsumura, Y., Ohba, K., and Strauss, S. H. (1996). Diversity and inheritance of inter-simple sequence repeat polymorphisms in Douglas-fir (*Pseudotsuga menziesii*) and sugi (*Cryptomeria japonica*). *Theoret. Applied Genet.* **92,** 40–45.

Wolfe, A. D., and Liston, A. (1998). Contributions of PCR-based methods to plant systematics and evolutionary biology. *In* "Plant Molecular Systematics II" (D. E. Soltis, P. S. Soltis, and J. J. Doyle, eds.), pp. 43–86. Kluwer, Boston.

Wolfe, A. D., and Randle, C. P. (2001). Relationships within and among species of the holoparasitic genus *Hyobanche* (*Orobanchaceae*) inferred from ISSR banding patterns and nucleotide sequences. *Syst. Botany* **26,** 120–130.

Wolfe, A. D., Xiang, Q.-Y., and Kephart, S. R. (1998a). Diploid hybrid speciation in *Penstemon* (*Scrophulariaceae*). *Proc. Natl. Acad. Sci. USA* **95,** 5112–5115.

Wolfe, A. D., Xiang, Q.-Y., and Kephart, S. R. (1998b). Assessing hybridization in natural populations of *Penstemon* (*Scrophulariaceae*) using hypervariable inter simple sequence repeat markers. *Mol. Ecol.* **7,** 1107–1125.

Yang, W., de Oliveira, A. C., Godwin, I., Schertz, K., and Bennetzen, J. L. (1996). Comparison of DNA marker technologies in characterizing plant genome diversity: Variability in Chinese sorghums. *Crop Sci.* **36,** 1669–1676.

Zietkiewicz, E., Rafalski, A., and Labuda, D. (1994). Genome fingerprinting by simple sequence repeat (SSR)–anchored polymerase chain reaction amplification. *Genomics* **20,** 176–183.

[10] Use of Amplified Fragment Length Polymorphism (AFLP) Markers in Surveys of Vertebrate Diversity

By AURÉLIE BONIN, FRANÇOIS POMPANON, and PIERRE TABERLET

Abstract

The amplified fragment length polymorphism (AFLP) technique is one of the most informative and cost-effective fingerprinting methods. It produces polymerase chain reaction (PCR)–based multi-locus genotypes helpful in many areas of population genetics. This chapter focuses on technical laboratory information to successfully develop the AFLP technique for vertebrates. Several AFLP protocols are described, as well as recommendations about important factors of the procedure such as the choice of enzyme and primer combinations, the choice and scoring of markers, the influence of the genome size on the AFLP procedure, and the control and estimation of genotyping errors. Finally, this chapter proposes a troubleshooting guide to help resolve the main technical difficulties encountered during the AFLP procedure.

Introduction

Recently developed by Vos *et al.* (1995), the amplified fragment length polymorphism (AFLP) technique has become one of the most reliable and promising DNA fingerprinting methods, producing hundreds of informative polymerase chain reaction (PCR)–based genetic markers to provide a wide multi-locus screening of any genome. The AFLP analysis has been largely documented in the literature (Blears *et al.*, 1998; Jones *et al.*, 1997; Mueller *et al.*, 1999; Savelkoul *et al.*, 1999); here, we emphasize one of its more overlooked aspects—technical information. We discuss the important factors of the procedure (enzyme, primer, and marker choice; influence of genome size; genotyping errors) and give several recommendations and protocols to successfully develop AFLP markers for vertebrates.

AFLP Features and Applications

The essence of the AFLP procedure lies in the combined use of two basic tools in molecular biology: restriction, which reduces the total genomic DNA into a pool of fragments, and PCR, which amplifies a subset of these restriction fragments thanks to primers with arbitrary selective extensions (Mueller *et al.*, 1999; Savelkoul *et al.*, 1999). Three kinds of AFLP polymorphisms can then be observed: a mutation in the restriction site, a mutation in the sequence adjacent to the restriction site and complementary to the primer extensions, or a deletion/insertion within the amplified fragment (Ajmone-Marsan *et al.*, 2001; Matthes *et al.*, 1998). Polymorphisms are revealed by the presence of a fragment of a given size in some AFLP profiles versus its absence from other profiles.

AFLP fingerprinting has been of great interest in population genetics because of several advantageous characteristics. First, it is the method of choice for studies of non-model organisms (Blears *et al.*, 1998; Vos *et al.*, 1995). Theoretically, it can be performed on any genome, regardless of its complexity and structure and without any prior sequence knowledge, in contrast to other kinds of molecular markers like microsatellites that require taxon-specific primers (Dogson *et al.*, 1997). Practically, commercial AFLP primer sets are available that work on most organisms. Second, large numbers (up to several hundreds) of AFLP markers can be typed quickly and at low cost, offering fine-scale genome coverage (Blears *et al.*, 1998; Mueller *et al.*, 1999), although several studies have reported AFLP clustering in centromeric regions (Lindner *et al.*, 2000; Young *et al.*, 1998). AFLP markers are also largely independent, because 90% of them reflect point mutations in enzyme restriction site (Buntjer *et al.*, 2002) that remove the fragments from the AFLP profile rather than change its size (Albertson *et al.*, 1999). Third, AFLP markers usually reveal a greater amount of diversity compared to simple sequence repeats (SSRs) and random amplified polymorphic DNAs (RAPDs) (Archak *et al.*, 2003; Barker *et al.*, 1999) and provide valuable fingerprints of organisms like birds in which microsatellite markers are difficult to obtain (Dogson *et al.*, 1997; Knorr *et al.*, 1999). Fourth, thanks to stringent hybridization conditions and relative insensitivity to template DNA concentration, the AFLP fingerprint is highly reproducible and reliable (Ajmone-Marsan *et al.*, 1997; Bagley *et al.*, 2001; Jones *et al.*, 1997). As a result, it can be standardized, reproduced easily between different technicians and laboratories, and computer-scored for subsequent comparisons (Hong and Chuah, 2003). This makes it particularly well-adapted for large-scale studies involving several research centers (Jones *et al.*, 1997). Fifth, only small amounts of genomic DNA are necessary to generate several informative AFLP profiles with different primer combinations (Blears *et al.*, 1998; Savelkoul *et al.*, 1999; Vos *et al.*, 1995). Finally, AFLP markers have been shown to follow mendelian inheritance in plants (Blears *et al.*, 1998; Savelkoul *et al.*, 1999), as well as in animals (Ajmone-Marsan *et al.*, 1997; Otsen *et al.*, 1996).

Despite its attractiveness, the AFLP method has some detrimental aspects. First, AFLP markers should be considered as dominant biallelic markers: fragment presence versus absence, with the fragment presence allele dominant over the absence allele (Ajmone-Marsan *et al.*, 2001; Mueller *et al.*, 1999). It is indeed difficult to distinguish between heterozygous individuals and individuals homozygous for the presence allele because of differential efficiencies between distinct PCR amplifications, unless exact genotypes can be inferred by means of pedigree studies (Van Haeringen *et al.*, 2002). AFLP data are thus of poor information contents in analyses requiring precise estimations of heterozygosity. Nonetheless, several studies have managed to score up to 65% of the markers in a codominant way by rigorous standardization of profile intensities (Ajmone-Marsan *et al.*, 1997), and new protocols have been developed to investigate AFLP-like codominant markers (Bradeen and Simon, 1998; Hakki and Akkaya, 2000). Second, fragments originating from distinct loci may have the same length by chance (homoplasy of size) (O'Hanlon and Peakall, 2000; Vekemans *et al.*, 2002). Such fragments display exactly the same electrophoretic mobility and thus overlap on the AFLP profile, introducing an undesirable source of artifacts. However, comigration of distinct fragments has proven to be a rare event (Mechanda *et al.*, 2003; Rosendahl and Taylor, 1997). Third, the AFLP procedure is particularly sensitive to contamination by exogenous DNA; even low and unobtrusive levels of bacterial or fungal contaminants, for example, may alter the AFLP profiles (Dyer and Leonard, 2000; Savelkoul *et al.*, 1999). When working with organisms prone to such kinds of contaminations, one should take special precautions to ensure the reliability of the results.

Originally worked out for plants and microorganisms, the AFLP analysis now finds more and more applications within the animal kingdom, especially in vertebrate species. Because their resolution power extends from the individual to the species level, AFLP markers have proven to be valuable tools in individual identification (Ovilo *et al.*, 2000), sex determination (Griffiths and Orr, 1999; Questiau *et al.*, 2000), parentage analysis (Questiau *et al.*, 1999), genetic diversity assessment (Ajmone-Marsan *et al.*, 2001, 2002; Mickett *et al.*, 2003; Mock *et al.*, 2002), population assignments (Campbell *et al.*, 2003), investigations of population structure and estimations of gene flow (Dearborn *et al.*, 2003; Jorde *et al.*, 1999), hybridization studies (Bensch *et al.*, 2002b; Nijman *et al.*, 2003), and taxonomic and phylogenetic inferences (Albertson *et al.*, 1999; Buntjer *et al.*, 2002; Giannasi *et al.*, 2001; Ogden and Thorpe, 2002). For higher taxonomic levels (e.g., infrageneric), the multi-locus fingerprint becomes too variable, increasing the risk of size homoplasy for the fragments generated

(Vekemans *et al.*, 2002) and rendering the analysis of AFLP profiles too complex and largely meaningless.

In addition, AFLP markers have encountered considerable success in production of high-resolution genetic and quantitative trait loci (QTL) maps, in fish (Lindner *et al.*, 2000; Liu *et al.*, 2003; Naruse *et al.*, 2000; Ransom and Zon, 1999; Young *et al.*, 1998), amphibians (Kochan *et al.*, 2003; Voss *et al.*, 2001), birds (Groenen *et al.*, 2000; Herbergs *et al.*, 1999; Knorr *et al.*, 1999), and mammals (Otsen *et al.*, 1996; Van Haeringen *et al.*, 2002). The AFLP technique has found a new and productive application in the search for informative single nucleotide polymorphisms (SNPs) in non-model vertebrates (Bensch *et al.*, 2002a; Meksem *et al.*, 2001; Nicod and Largiader, 2003).

AFLP Basic Steps for a Complex Genome

Genomic DNA Extraction and Preparation

Any DNA extraction method is suitable to isolate total genomic DNA for subsequent AFLP analysis as long as it provides good quality DNA (no or limited degradation). For example, the DNeasy Tissue Kit (Qiagen) usually gives good results. After the extraction, an extra purification step might be necessary for DNA extracts containing restriction or PCR inhibitors.

Genomic DNA Digestion

The objective of restriction digestion is to reduce the big genomic DNA molecules into a mixture of fragments enabling posterior amplification and electrophoretic detection. The DNA amount needed for the AFLP procedure depends mainly on the genome size and structure, as well as on the DNA quality (see below for further details). The DNA concentration should be standardized among samples to yield comparable and homogeneous fingerprints.

Restriction fragments of genomic DNA are generated using two restriction enzymes, a rare cutter like *Eco*RI (6-bp restriction site) and a frequent cutter like *Mse*I (4-bp restriction site). The enzyme choice is discussed further in this chapter. After digestion, three categories of fragments exist in the mixture: fragments with *Eco*RI cuts at both ends (longer ones on average), fragments with *Mse*I cuts at both ends (smaller ones on average), and fragments with an *Eco*RI cut at one end and a *Mse*I cut at the other end. The AFLP protocol is designed to amplify and preferentially detect this last kind of fragment.

Ligation of Oligonucleotide Adapters

Using a T4 ligase enzyme, restriction fragments are ligated to double-stranded adapters specific to one particular restriction enzyme. The adapter structure is composed of a core sequence followed by an enzyme-specific sequence. These adapters are conceived so that the restriction site is not recreated after ligation, which eventually allows simultaneous digestion and ligation if restriction enzymes and T4 ligase are active at the same temperature. If ligation is performed after digestion, restriction enzymes should not be denatured after digestion, to prevent the formation of adapter concatenates. Once the ligation reaction is achieved, only one strand of the adapters is ligated to the restriction fragments, as the T4 ligase enzyme lacks a 3' to 5' activity.

Preselective Amplification

Preselective amplification aims to decrease the complexity of the initial fragment mixture by amplifying only a subset of fragments. It is conducted with a set of primers whose structure consists of a core sequence, an enzyme-specific sequence, and a selective single-base extension at the 3' end. As a result, the adapter sequences offer primer binding sites and the selective base will recognize the fragments having the matching nucleotide after the restriction site. Fragment amplification can occur only if two primers can bind perfectly at both ends of the fragment, so statistically, 1 fragment out of 16 (4 × 4) originally present in the mixture will be amplified.

Before the preselective amplification, the preselective mix undergoes an initial incubation at 72°, taking advantage of the 3' to 5' ligase activity of the DNA polymerase to complete the ligation of the adapters to the restriction fragments before first denaturation.

Selective Amplification

The selective amplification is based on the same principle as the preselective one. Using primers identical to the preselective primers, plus one or two extra selective bases at the 3' end, a second complexity reduction is performed on the pool of preselective fragments. Selective primers thus have two or three selective bases, depending on the genome complexity. Finally, after two rounds of amplification, the mixture complexity is divided by a 256 (4^4) factor (selective primers both carrying 2 selective bases) or by a 4096 (4^6) factor (selective primers both carrying 3 selective bases). During this step, hybridization conditions are very stringent: a "touchdown" PCR (i.e., with a particularly high annealing temperature decreasing progressively on the first cycles) ensures highly specific amplification and thus good reproducibility of the technique.

*Eco*RI selective primers are specially designed to have a higher annealing temperature than *Mse*I selective primers. As a result, the "touchdown" PCR allows a preferential amplification of *Eco*RI/*Mse*I versus *Mse*I/*Mse*I fragments. In addition, because only *Eco*RI primers are fluorescently or radio-labeled, the AFLP patterns display *Eco*RI/*Mse*I fragments exclusively in a 50–500 bp detection range, with *Eco*RI/*Eco*RI fragments being statistically longer. Finally, after the complexity reduction and the preferential amplification and detection of *Eco*RI/*Mse*I fragments, 20–150 discrete bands can be visualized on a typical AFLP profile.

Electrophoresis and Analysis of the AFLP Profiles

After selective amplification, the restriction fragments are denatured, separated by electrophoresis, and visualized by either fluorescence or radioactivity (autoradiography) detection. Autoradiography may enable easier characterization of codominant markers, given that the detected radioactive signal is known to increase linearly with the number of labeled fragments (Hawkins *et al.*, 1992).

Automated sequencers are now more exploited for the separation and detection of fluorescent AFLP fragments. Multi-locus fingerprints can then be visualized using software packages like GeneScan Analysis 3.1 (Perkin-Elmer). This program determines the size of amplified restriction fragments with the help of an internal size standard (Rox 500, Perkin-Elmer), then it classifies them according to their size with single-base resolution. The fragments (bands) can thus be scored as present/absent for a given size, producing a binary matrix. Some software packages can construct this binary matrix semiautomatically, on which subsequent analyses can be performed (Genographer, available at *http://hordeum.oscs.montana.edu/genographer/*) (Papa *et al.*, 2005). Capillary electrophoresis with an automated sequencer has been shown to increase the AFLP data throughput and reliability (Papa *et al.*, 2005). Certain parameters should then be optimized according to the manufacturer's instructions for high-quality fingerprints. For example, longer injection times improve the peak intensities for longer fragments, which tend to be loaded with more difficulty during the electrokinetic process.

Important Factors in the AFLP Procedure

Choice of Enzyme and Primer Combinations

Restriction enzymes and primer pairs are key parameters in the AFLP procedure, influencing the number of amplified fragments, the level of polymorphism detected, and the possibility of comparing AFLP profiles from different studies. In theory, any restriction enzyme can be used in an

AFLP protocol. However, some of them cut too often, generating comigrating non-homologous fragments, or too rarely, reducing the probability of polymorphism detection. Thus, the choice of enzyme combinations for the AFLP protocol consists of a compromise between adequate levels of polymorphism and readable AFLP profiles. Restriction patterns for several enzyme combinations can be prescreened on agarose gel, and the range of fragment sizes (ideally between 50 and 500 bp) provides an insight into the complexity of the mixture produced. The *Eco*RI/*Mse*I enzyme combination originally published by Vos *et al.* (1995) is traditionally used for AFLP analysis, although several studies of vertebrate diversity are based on other enzyme combinations, for example, *Not*I/*Hpa*II (Voss *et al.*, 2001), *Sse*I/*Mse*I (Van Haeringen *et al.*, 2002), or *Eco*RI/*Msp*I (Knorr *et al.*, 1999). It appears that the *Eco*RI/*Taq*I combination is particularly efficient in generating high-quality AFLP profiles for avian and mammalian genomes (Ajmone-Marsan *et al.*, 1997; Knorr *et al.*, 1999). This may be because the *Taq*I restriction site contains a CpG dinucleotide, which is a hot spot for mutations in the genome of hot-blooded animals (Ajmone-Marsan *et al.*, 1997; Gardiner-Garden and Frommer, 1987). When selecting enzymes for an AFLP use, one should exclude enzymes sensitive to DNA methylation, which is a phenomenon known to be tissue- and age-specific in plants and in animals (Bird, 2002), unless AFLPs are being used to investigate DNA methylation patterns (Cervera *et al.*, 2002).

The length and bases composition of primer pairs appropriate for the AFLP procedure are also important, because the fine-tuning of polymorphisms detected relies mainly on the addition or suppression of 3' selective bases. As a result, multiple combinations of different selective primers give access to hundreds of polymorphic markers. Vos *et al.* (1995) have established that selectivity is good for selective primers with elongations constituted of one or two extra bases. With three or more selective bases, nonspecific annealing may occur at the first selective base after enzyme-specific sequence, leading to artifactual AFLP bands. As for the choice of appropriate primer pairs to use in organisms never analyzed with AFLP markers, there are unfortunately no general rules except that in most cases, an extensive screening of different primer combinations is necessary. Trying all pairwise combinations is a reasonable and effective approach, but it can rapidly become expensive and time-consuming. If nucleotide frequencies are available for the organism under consideration, an alternative strategy is to test in priority primer pairs with similar base frequencies. In practice, the choice of primer pairs should be performed on four to eight individuals representing the widest possible range of genetic diversity. Testing two or three dozen different primer pairs then is generally sufficient to select three to five clear primer combinations at the end.

Influence of Genome Size and Complexity

The size and complexity of the studied genome can have a tremendous influence on the AFLP analysis in three ways. First, they will condition the total quantity of DNA required for the initial restriction. Indeed, a general rule in the AFLP procedure is that the larger the genome is, the larger the required amount of DNA is. It is indeed essential to ensure a sufficient number of AFLP loci copies, even if this number may vary according to the proportion of repeated sequences in the genome. Although the AFLP protocol is relatively insensitive to the template concentration (Jones *et al.*, 1997; Vos *et al.*, 1995), stochastic amplifications can occur if DNA is limiting, so we recommend using DNA slightly in excess, with appropriate concentrations of reactants and enzymes. For example, 400 ng of DNA is necessary to produce clear, intense AFLP patterns for the common frog (*Rana temporaria*, genome size: 4.3×10^9 bp) with the protocol mentioned below. Genome size for many vertebrate species can be checked in the Animal Genome Size Database (*http://www.genomesize.com*).

Second, the genome size and complexity will determine the number of selective bases used for the preselective and selective amplifications. Traditionally, preselective primers count one selective base, whereas the selective primers count three selective bases. However, more complex genomes may require the use of additional selective bases, that is, two selective bases in the first amplification step and four in the second one. For some high-complexity genomes, it is impossible to achieve clear AFLP profiles only by varying the enzyme choice or the number of selective bases. In this case, a variant of the AFLP procedure can be considered: the three endonuclease (TE)-AFLP technique (Van der Wurff *et al.*, 2000). Compared to traditional AFLP, this technique, based on the use of three restriction enzymes, has an extra reduction step due to a selective ligation. Finally, the ploidy level is an important factor to consider when developing AFLP markers for vertebrates, because differences in ploidy levels can occur within natural populations and species in fish, amphibians, and reptiles, although this phenomenon is less common than in the plant kingdom (Otto and Whitton, 2000). It is, therefore, a potential source of bias for AFLP pattern comparisons and allele frequencies estimates and should remain a major concern when working with individuals of unknown karyotypes.

Choice and Scoring of Markers

The choice and scoring of markers are perhaps the most critical points of the AFLP procedure—or at least the most exacting steps in terms of experience and rigor. They largely determine the reproducibility and reliability of the technique. A good marker has to fulfill several requirements. First, it has to be polymorphic enough to be informative. A peak appearing for only one individual must raise suspicion, because contamination and technical artifacts can never be totally ruled out. Second, it must be clearly distinct (at least 1 bp) from other peaks on the profiles, so that it can be scored properly. Overlapping peaks should be discarded as early as the pilot study for primer choice, because dealing with more individuals usually adds more confusion to the profiles. Third, as many individuals as possible should be scored unambiguously for the chosen marker. AFLP profiles usually display many potential markers, but most of them may not be helpful for subsequent analysis, when their scoring is doubtful.

When it comes to the scoring of the markers, several protocols can be adopted. One can choose to score very high and very low peaks the same way. Nevertheless, this method is uncertain and not conservative enough, as a considerable number of low peaks can be attributed to background noise. An absolute or relative intensity threshold can also be set up, under which a peak is considered to be absent. To determine a scoring threshold, we personally scrutinize a drop in intensity among the peaks corresponding to the marker under consideration. A clear discontinuity indicates the frontier between non-selective (i.e. background) and selective amplifications and it usually shows up around 10% of the highest peak's intensity. Whatever the scoring protocol established, it appears essential to follow it strictly, and because of the amount of subjectivity entering in this process, the same person should be charged to score all the data, to ensure consistency of the results. Normalization of the profiles obtained from different runs and double reading of the data by two experimenters are also efficient measures to take to facilitate and confirm the scorings.

AFLP and Genotyping Errors

Even if several studies have reported high levels of reproducibility of the AFLP technique (Ajmone-Marsan et al., 1997; Bagley et al., 2001; Jones et al., 1997), the problem of genotyping errors should not be overlooked, because every AFLP data set includes typing errors that might greatly bias the final results. Methods to track and monitor such errors are not the purpose of this chapter, so we will mention only major recommendations to fulfill this objective. First, when developing AFLP markers for a new organism, a systematic pilot study should be carried out before any extensive investigation, providing the opportunity to acquire experience with the AFLP technique and to achieve reproducibility. Second, including blind samples throughout the procedure provides a reliable way to (1) estimate the error rate, (2) detect contamination, tube mixings, or biochemical anomalies, and (3) eliminate unreliable markers (i.e., makers that are unstable or difficult to score). Third, automation, capillary electrophoresis, and semiautomatic scoring have been proven to limit the overall experimental error (Papa *et al.*, 2005). Like every other genotyping method, the AFLP procedure is subject to common errors (chance, human factors, technical artifacts), but the vast majority of errors are specific to the AFLP technique. Indeed, differences in peak intensities and comigration of non-homologous fragments cause most of them, whereas contamination, for example, counts for a marginal number (Bonin *et al.*, 2004). In the literature, reproducibility of AFLP data in vertebrates is usually higher than 95% (Ajmone-Marsan *et al.*, 1997, 2001; Bagley *et al.*, 2001; Ovilo *et al.*, 2000).

Methods: AFLP Protocols for Vertebrates

Method 1. AFLP Procedure with Taq*I/Eco*R1 *Enzymes Combination (Birds, Mammals)*

Volumes are indicated for one sample.

1. Digestion of total DNA. Digest genomic DNA (50–500 ng) in a 25-μl reaction containing DNA, 2.5 μl of 10× *Taq*I buffer (New England Biolabs [NEB]), 5 units of restriction endonuclease *Taq*I (NEB) and q.s.p ultrahigh-quality (UHQ) water. Incubate up to 2 h at 65°, then add 1.5 μl of 10× *Eco*RI buffer (NEB), 5 units of restriction endonuclease *Eco*RI (NEB), and make up to 40 μl with UHQ water. Incubate up to 2 h at 37°.

2. Preparation of 10 μM double-stranded adapters. Prepare 10 μM double-stranded adapters by mixing equal volumes of 10 μM individual synthetic oligonucleotides. Denature by heating 5 min at 65° in a hot block and cool slowly down to room temperature. Store adapters at −20°. When being aliquoted for subsequent ligation, adaptors should be kept in a refrigerated rack to avoid denaturation.

3. Ligation of adapters to restriction fragments. Ligate adaptors to 40 μl of the digested genomic DNA by adding 1 μl of 10 μM *Eco*RI adapter, 5 μl of 10 μM *Taq*I adapter, 1 μl of 10 mM ATP, 0.5 μl of 1 mg/μl bovine serum albumin (BSA), 1 μl of 10× T4 ligase buffer (NEB), 100 units of T4 DNA ligase (NEB) and UHQ water to 50 μl. Incubate another 3 h at 37°.

4. Preparation of template DNA for preselective amplification. Dilute the ligation reaction mixture 5–10 times with UHQ water. Store diluted DNA at −20°.

5. Preselective amplification. Prepare the preselective mix with the following components: 3 μl of diluted template DNA, 2.5 of μl 10× Amplitaq buffer (Applied Biosystems), 1.5 μl of 25 mM MgCl$_2$, 2 μl of

10 mM dNTPs, 0.5 μl of 10 μM *Eco*RI preselective primer, 0.5 μl of 10 μM *Taq*I preselective primer, 1 unit of Amplitaq DNA polymerase, and UHQ water to 25 μl. Preamplify using the following program: initial incubation 2 min at 72°; 25–30 cycles of 30 s at 94°, 30 s at 56°, and 2 min at 72°; final extension 10 min at 72°; storage at 4°. For extended periods (few days to several months), preselective product should be stored at −20°. After amplification, the preselective PCR product can be monitored on a 2% agarose gel and usually gives a faint smear in the 100–1000 bp range.

6. Preparation of template DNA for selective amplification. Dilute the preselective product 20 times with UHQ water. Store diluted DNA at −20°.

7. Selective amplification. Prepare the selective mix with following components: 5 μl of diluted preselective product, 2.5 μl of 10× Amplitaq buffer, 2.5 μl of 25 mM MgCl₂, 2 μl of 10 mM dNTPs, 0.5 μl of 10 μM labeled *Eco*RI selective primer, 0.5 μl of 10 μM *Taq*I selective primer, 1 unit of Amplitaq Gold DNA polymerase, UHQ water to a final volume of 25 μl. Amplify using the following program: initial incubation 10 min at 95°; 13 cycles of 30 s at 94°, 1 min at 65° (first cycle, then decrease of 0.7° for the 12 last cycles) and 1 min at 72°; 23 cycles of 30 s at 94°, 1 min at 56°, and 1 min at 72°; final extension 10 min at 72°; storage at 4°. Selective product can be stored several days at −20°. On a 2% agarose gel, the selective PCR product gives a clearly distinguishable smear in the 100–500 bp range.

8. Electrophoresis of AFLP products. AFLP products can be separated, detected, and sized with any automated DNA sequencer. Refer to the manufacturer for further instructions.

Method 2. AFLP Procedure with Mse*I/*EcoR1 *Enzymes Combination (All Vertebrates except Mammals)*

1. Preparation of 10 μM double-stranded adapters. See Method 1, step 2.
2. Simultaneous digestion and ligation of total DNA. Digest and ligate total genomic DNA (50–500 ng) in an 11-μl reaction containing DNA, 1.1 μl of 10× T4 ligase buffer (NEB), 1.1 μl of 0.5 M NaCl, 0.55 μl of 1 mg/μl⁻¹ BSA, 1 μl of 10 μM *Mse*I adapter, 1 μl of 10 μM *Eco*RI adapter, 5 units of restriction endonuclease *Mse*I (NEB), 5 units of restriction endonuclease *Eco*RI (NEB), 5 units of T4 DNA ligase (NEB), and UHQ water to 11 μl. Incubate 2 h at 37°.
3. For the followings steps, see Method 1, steps 4–8.

TABLE I

TECHNICAL PROBLEMS ENCOUNTERED DURING AFLP PROCEDURE, THEIR POSSIBLE
CAUSES, AND SOLUTIONS

Problem	Possible causes	Solutions
• No DNA restriction and/or no amplification	• DNA extract contains restriction or PCR inhibitor • Restriction enzymes or T4 DNA ligase or adapters are limited • Adapters are denatured	• Dilute the DNA extract • Purify the DNA extract before the AFLP procedure • Check that restriction enzymes, T4 DNA ligase and adapters are in excess • Heat the adapters 5 min at 65° and let cool down to room temperature; store in the freezer before use
• No amplification after the preselective PCR	• Ligation of the adapters to the restriction fragments is not completely achieved before first denaturation	• Check that the preselective PCR program includes an initial 2-min incubation at 72°
• No amplification after the selective PCR	• Amplitaq Gold DNA polymerase has not been activated before amplification	• Check that the selective PCR program includes an initial 10-min activation at 95°
• No or few polymorphism	• The chosen enzymes do not detect polymorphic sites	• Test other enzyme combinations
• Profiles with not enough peaks	• Only a few peaks can be amplified	• Discard some of the selective bases
• Profiles with not enough peaks for a large genome	• Enzymes cut within repeated sequences	• Test other enzyme combinations
• Profiles with too many peaks	• Too many peaks can be amplified	• Add some selective bases
• Profiles with too many peaks, even with three selective bases	• Genome too large	• Test preselective primers with two selective bases and selective primers with four selective bases • Test TE-AFLP

(continued)

TABLE I (*continued*)

Problem	Possible causes	Solutions
• No reproducibility	• Too much/too few DNA leading to stochastic amplification • Degraded DNA • "Star activity" of the restriction enzymes, which cut nonspecifically	• Use less/more DNA for the whole procedure • Check for DNA quality • Shorten the digestion time or modify the enzyme buffer composition
• Parasite peaks	• Contamination during one or several steps of the procedure	• Use disposable tips and tubes throughout the procedure • Monitor contamination using negative controls at each step of the protocol and sample references
• Weak profiles	• Too little DNA • Low fluorescence/ radioactivity levels	• Try the AFLP procedure with more DNA • Load more AFLP products for the electrophoresis • Check the quality of primer labeling
• Presence of peak doublets	• Incomplete addition by the *Taq* polymerase of additional adenine residues at the 3′ end of amplified fragments	• Add a final 10-min extension at 72° after amplification
• Difficulties when comparing different profiles	• Profiles come from different runs and display different levels of intensities	• Normalize the intensities in the Genographer software • Use some samples as internal standards

Troubleshooting Guide

Table I lists the solutions to the principal technical problems encountered when developing AFLP markers for vertebrates.

References

Ajmone-Marsan, P., Valentini, A., Cassandro, M., Vecchiotti-Antaldi, G., Bertoni, G., and Kuiper, M. (1997). AFLP markers for DNA fingerprinting in cattle. *Anim. Genet.* **28,** 418–426.

Ajmone-Marsan, P., Negrini, R., Crepaldi, P., Milanesi, E., Gorni, C., Valentini, A., and Cicogna, M. (2001). Assessing genetic diversity in Italian goat populations using AFLP markers. *Anim. Genet.* **32,** 281–288.

Ajmone-Marsan, P., Negrini, R., Milanesi, E., Bozzi, R., Nijman, I. J., Buntjer, J. B., Valentini, A., and Lenstra, J. A. (2002). Genetic distances within and across cattle breeds as indicated by biallelic AFLP markers. *Anim. Genet.* **33,** 280–286.

Albertson, R. C., Markert, J. A., Danley, P. D., and Kocher, T. D. (1999). Phylogeny of a rapidly evolving clade: The cichlid fishes of Lake Malawi, East Africa. *Proc. Natl. Acad. Sci. USA* **96,** 5107–5110.

Archak, S., Gaikwad, A. B., Gautam, D., Rao, E. V., Swamy, K. R., and Karihaloo, J. L. (2003). Comparative assessment of DNA fingerprinting techniques (RAPD, ISSR and AFLP) for genetic analysis of cashew (*Anacardium occidentale* L.) accessions of India. *Genome* **46,** 362–369.

Bagley, M. J., Anderson, S. L., and May, B. (2001). Choice of methodology for assessing genetic impacts of environmental stressors: Polymorphism and reproducibility of RAPD and AFLP fingerprints. *Ecotoxicology* **10,** 239–244.

Barker, J. H., Matthes, M., Arnold, G. M., Edwards, K. J., Ahman, I., Larsson, S., and Karp, A. (1999). Characterization of genetic diversity in potential biomass willows (*Salix* spp.) by RAPD and AFLP analyses. *Genome* **42,** 173–183.

Bensch, S., Akesson, S., and Irwin, D. E. (2002a). The use of AFLP to find an informative SNP: Genetic differences across a migratory divide in willow warblers. *Mol. Ecol.* **11,** 2359–2366.

Bensch, S., Helbig, A. J., Salomon, M., and Seibold, I. (2002b). Amplified fragment length polymorphism analysis identifies hybrids between two subspecies of warblers. *Mol. Ecol.* **11,** 473–481.

Bird, A. (2002). DNA methylation patterns and epigenetic memory. *Genes Dev.* **16,** 6–21.

Blears, M. J., de Grandis, S. A., Lee, H., and Trevors, J. T. (1998). Amplified fragment length polymorphism (AFLP): A review of the procedure and its applications. *J. Ind. Microbiol. Biotech.* **21,** 99–114.

Bonin, A., Bellemain, E., Bronken Eidesen, P., Pompanon, F., Brochmann, C., and Taberlet, P. (2004). How to track and assess genotyping errors in population genetics studies. *Mol. Ecol.* **13,** 3261–3273.

Bradeen, J. M., and Simon, P. W. (1998). Conversion of an AFLP fragment linked to the carrot Y2 locus to a simple, codominant, PCR-based marker form. *Theor. Appl. Genet.* **97,** 960–967.

Buntjer, J. B., Otsen, M., Nijman, I. J., Kuiper, M. T., and Lenstra, J. A. (2002). Phylogeny of bovine species based on AFLP fingerprinting. *Heredity* **88,** 46–51.

Campbell, D., Duchesne, P., and Bernatchez, L. (2003). AFLP utility for population assignment studies: Analytical investigation and empirical comparison with microsatellites. *Mol. Ecol.* **12,** 1979–1991.

Cervera, M. T., Ruiz-Garcia, L., and Martinez-Zapater, J. M. (2002). Analysis of DNA methylation in *Arabidopsis thaliana* based on methylation-sensitive AFLP markers. *Mol. Genet. Genomics* **268,** 543–552.

Dearborn, D. C., Anders, A. D., Schreiber, E. A., Adams, R. M., and Mueller, U. G. (2003). Inter-island movements and population differentiation in a pelagic seabird. *Mol. Ecol.* **12,** 2835–2843.

Dogson, J. B., Cheng, H. H., and Okimoto, R. (1997). DNA marker technology: A revolution in animal genetics. *Poultry Sci.* **76,** 1108–1114.

Dyer, A. T., and Leonard, K. J. (2000). Contamination, error, and nonspecific molecular tools. *Phytopathology* **90,** 565–567.

Gardiner-Garden, M., and Frommer, M. (1987). CpG islands in vertebrate genomes. *J. Mol. Biol.* **196,** 261–282.

Giannasi, N., Thorpe, R. S., and Malhotra, A. (2001). The use of amplified fragment length polymorphism in determining species trees at fine taxonomic levels: Analysis of a medically important snake, *Trimeresurus albolabris. Mol. Ecol.* **10,** 419–426.

Griffiths, R., and Orr, K. (1999). The use of amplified fragment length polymorphism (AFLP) in the isolation of sex-specific markers. *Mol. Ecol.* **8,** 671–674.

Groenen, M. A., Cheng, H. H., Bumstead, N., Benkel, B. F., Briles, W. E., Burke, T., Burt, D. W., Crittenden, L. B., Dodgson, J., Hillel, J., Lamont, S., de Leon, A. P., Soller, M., Takahashi, H., and Vignal, A. (2000). A consensus linkage map of the chicken genome. *Genome Res.* **10,** 137–147.

Hakki, E. E., and Akkaya, M. S. (2000). Microsatellite isolation using amplified fragment length polymorphism markers: No cloning, no screening. *Mol. Ecol.* **9,** 2152–2154.

Hawkins, T. L., Du, Z., Halloran, N. D., and Wilson, R. K. (1992). Fluorescence chemistries for automated primer-directed DNA sequencing. *Electrophoresis* **13,** 552–559.

Herbergs, J., Siwek, M., Crooijmans, R. P., Van der Poel, J. J., and Groenen, M. A. (1999). Multicolour fluorescent detection and mapping of AFLP markers in chicken (*Gallus domesticus*). *Anim. Genet.* **30,** 274–285.

Hong, Y., and Chuah, A. (2003). A format for databasing and comparison of AFLP fingerprint profiles. *Bioinformatics* **4,** 7.

Jones, C. J., Edwards, K. J., Castaglione, S., Winfield, M. O., Sala, F., Van de Wiel, C., Bredemeijer, G., Vosman, B., Matthes, M., Daly, A., Brettschneider, R., Bettini, P., Buiatti, M., Maestri, E., Malcevschi, A., Marmiroli, N., Aert, R., Volckaert, G., Rueda, J., Linacero, R., Vazquez, A., and Karp, A. (1997). Reproducibility testing of RAPD, AFLP and SSR markers in plants by a network of European laboratories. *Mol. Breeding* **3,** 381–390.

Jorde, P. E., Palm, S., and Ryman, N. (1999). Estimating genetic drift and effective population size from temporal shifts in dominant gene marker frequencies. *Mol. Ecol.* **8,** 1171–1178.

Knorr, C., Cheng, H. H., and Dodgson, J. B. (1999). Application of AFLP markers to genome mapping in poultry. *Anim. Genet.* **30,** 28–35.

Kochan, K. J., Wright, D. A., Schroeder, L. J., Shen, J., and Morizot, D. C. (2003). Genetic linkage maps of the West African clawed frog *Xenopus tropicalis. Dev. Dynam.* **227,** 155–156.

Lindner, K. R., Seeb, J. E., Habicht, C., Knudsen, K. L., Kretschmer, E., Reedy, D. J., Spruell, P., and Allendorf, F. W. (2000). Gene-centromere mapping of 312 loci in pink salmon by half-tetrad analysis. *Genome* **43,** 538–549.

Liu, Z., Karsi, A., Li, P., Cao, D., and Dunham, R. (2003). An AFLP-based genetic linkage map of channel catfish (*Ictalurus punctatus*) constructed by using an interspecific hybrid resource family. *Genetics* **165,** 687–694.

Matthes, M. C., Daly, A., and Edwards, K. J. (1998). Amplified length polymorphism (AFLP). *In* "Molecular Tools for Screening Biodiversity: Plants and Animals" (A. Karp, P. G. Isaac, and D. S. Ingram, eds.). Chapman and Hall, London.

Mechanda, S. M., Baum, B. M., Johnson, D. A., and Arnason, J. T. (2003). Sequence assessment of comigrating AFLP bands in *Echinacea*—implications for comparative biological studies. *Genome* **47,** 15–25.

Meksem, K., Ruben, E., Hyten, D., Triwitayakorn, K., and Lightfoot, D. A. (2001). Conversion of AFLP bands into high-throughput DNA markers. *Mol. Genet. Genomics* **265,** 207–214.

Mickett, K., Morton, C., Feng, J., Li, P., Simmons, M., Cao, D., Dunham, R. A., and Liu, Z. (2003). Assessing genetic diversity of domestic populations of channel catfish (*Ictalurus punctatus*) in Alabama using AFLP markers. *Aquaculture* **228,** 91–105.

Mock, K. E., Theimer, T. C., Rhodes, O. E., Jr., Greenberg, D. L., and Keim, P. (2002). Genetic variation across the historical range of the wild turkey (*Meleagris gallopavo*). *Mol. Ecol.* **11,** 643–657.

Mueller, U. G., and Wolfenbarger, L. L. (1999). AFLP genotyping and fingerprinting. *Trends Ecol. Evol.* **14,** 389–394.

Naruse, K., Fukamachi, S., Mitani, H., Kondo, M., Matsuoka, T., Kondo, S., Hanamura, N., Morita, Y., Hasegawa, K., Nishigaki, R., Shimada, A., Wada, H., Kusakabe, T., Suzuki, N., Kinoshita, M., Kanamori, A., Terado, T., Kimura, H., Nonaka, M., and Shima, A. (2000). A detailed linkage map of medaka, *Oryzias latipes*: Comparative genomics and genome evolution. *Genetics* **154,** 1773–1784.

Nicod, J. C., and Largiader, C. R. (2003). SNPs by AFLP (SBA): A rapid SNP isolation strategy for non-model organisms. *Nucleic Acids Res.* **31,** e19.

Nijman, I. J., Otsen, M., Verkaar, E. L., de Ruijter, C., Hanekamp, E., Ochieng, J. W., Shamshad, S., Rege, J. E., Hanotte, O., Barwegen, M. W., Sulawati, T., and Lenstra, J. A. (2003). Hybridization of banteng (*Bos javanicus*) and zebu (*Bos indicus*) revealed by mitochondrial DNA, satellite DNA, AFLP and microsatellites. *Heredity* **90,** 10–16.

Ogden, R., and Thorpe, R. S. (2002). The usefulness of amplified fragment length polymorphism markers for taxon discrimination across graduated fine evolutionary levels in Caribbean Anolis lizards. *Mol. Ecol.* **11,** 437–445.

O'Hanlon, P. C., and Peakall, R. (2000). A simple method for the detection of size homoplasy among amplified fragment length polymorphism fragments. *Mol. Ecol.* **9,** 815–816.

Otsen, M., den Bieman, M., Kuiper, M. T., Pravenec, M., Kren, V., Kurtz, T. W., Jacob, H. J., Lankhorst, A., and Van Zutphen, B. F. (1996). Use of AFLP markers for gene mapping and QTL detection in the rat. *Genomics* **37,** 289–294.

Otto, S. P., and Whitton, J. (2000). Polyploid incidence and evolution. *Annu. Rev. Genet.* **34,** 401–437.

Ovilo, C., Cervera, M. T., Castellanos, C., and Martinez-Zapater, J. M. (2000). Characterization of Iberian pig genotypes using AFLP markers. *Anim. Genet.* **31,** 117–122.

Papa, R., Troggio, M., Ajmone-Marsan, P., and Nonnis Marzano, F. (2005). An improved protocol for the production of AFLP markers in complex genomes by means of capillary electrophoresis. *J. Anim. Breed. Genet.* **122,** 62–68.

Questiau, S., Escaravage, N., Eybert, M. C., and Taberlet, P. (2000). Nestling sex ratios in a population of Bluethroats *Luscinia svecica* inferred from AFLP analysis. *J. Avian Biol.* **31,** 8–14.

Questiau, S., Eybert, M. C., and Taberlet, P. (1999). Amplified fragment length polymorphism (AFLP) markers reveal extra-pair parentage in a bird species: The bluethroat (*Luscinia svecica*). *Mol. Ecol.* **8,** 1331–1339.

Ransom, D. G., and Zon, L. I. (1999). Mapping zebrafish mutations by AFLP. *Method Cell Biol.* **60,** 195–211.

Rosendahl, S., and Taylor, J. W. (1997). Development of multiple genetic markers for studies of genetic variation in arbuscular mycorrhizal fungi using AFLP. *Mol. Ecol.* **6,** 821–829.

Savelkoul, P. H., Aarts, H. J., de Haas, J., Dijkshoorn, L., Duim, B., Otsen, M., Rademaker, J. L., Schouls, L., and Lenstra, J. A. (1999). Amplified-fragment length polymorphism analysis: The state of an art. *J. Clin. Microbiol.* **37,** 3083–3091.

Van der Wurff, A. W., Chan, Y. L., Van Straalen, N. M., and Schouten, J. (2000). TE-AFLP: Combining rapidity and robustness in DNA fingerprinting. *Nucleic Acids Res.* **28**, e105.

Van Haeringen, W. A., Den Bieman, M. G., Lankhorst, A. E., Van Lith, H. A., and Van Zutphen, L. F. (2002). Application of AFLP markers for QTL mapping in the rabbit. *Genome* **45**, 914–921.

Vekemans, X., Beauwens, T., Lemaire, M., and Roldan-Ruiz, I. (2002). Data from amplified fragment length polymorphism (AFLP) markers show indication of size homoplasy and of a relationship between degree of homoplasy and fragment size. *Mol. Ecol.* **11**, 139–151.

Vos, P., Hogers, R., Bleeker, M., Reijans, M., Van de Lee, T., Hornes, M., Frijters, A., Pot, J., Peleman, J., Kuiper, M., and Zabeau, M. (1995). AFLP: A new technique for DNA fingerprinting. *Nucleic Acids Res.* **23**, 4407–4414.

Voss, S. R., Smith, J. J., Gardiner, D. M., and Parichy, D. M. (2001). Conserved vertebrate chromosome segments in the large salamander genome. *Genetics* **158**, 735–746.

Young, W. P., Wheeler, P. A., Coryell, V. H., Keim, P., and Thorgaard, G. H. (1998). A detailed linkage map of rainbow trout produced using doubled haploids. *Genetics* **148**, 839–850.

[11] Use of AFLP Markers in Surveys of Arthropod Diversity

By TAMRA C. MENDELSON and KERRY L. SHAW

Abstract

Arthropods comprise the most diverse group of animals on earth and as such have been the subject of considerable evolutionary research. For example, much of our understanding of the genetic basis of evolutionary change is derived from the insect genus *Drosophila,* one of the most well-studied organisms in biology. Arthropods are also of tremendous economic importance as both providers and chief destroyers of food for human consumption. Thus, the genetic diversity of arthropods is of interest from both a pure research perspective and for practical economic reasons. The amplified fragment length polymorphism (AFLP) method of genetic analysis, developed in the early and mid-1990s (Vos *et al.*, 1995; Zabeau, 1992; Zabeau and Vos, 1993), offers a relatively new method for assessing genetic diversity and has been increasingly applied in studies of arthropods. Originally coined selective restriction fragment amplification (SRFA) (Zabeau and Vos, 1993), the method was renamed (Vos *et al.*, 1995) presumably to reflect its similarity to restriction fragment length polymorphism (RFLP). Since then, AFLPs have become a popular tool in both population genetics to estimate population parameters such as heterozygosity, F-statistics, migration rates, and genetic distances, as well as phylogenetics, to infer relationships among closely related taxa. In arthropods, AFLPs have been used to assess genetic variation both within and between species in various taxa including crustaceans, chelicerates, and insects, often yielding novel insights. In this chapter, we briefly describe the AFLP method and its strengths and limitations. We then discuss the use of AFLPs in surveys of arthropod diversity, highlighting the specific questions addressed using AFLPs. Finally, a section on experimental design and methods, based on research in our laboratory, is provided.

Introduction

The amplified fragment length polymorphism (AFLP) method has become a common tool for studies of genetic diversity. The method is best described in two stages, where in the first stage, genomic DNA is subject to digestion by a pair of restriction enzymes, generating many thousands of DNA fragments. The availability of restriction sites in the genome will dictate whether the DNA is cleaved; thus, genetic differences among individuals at restriction sites will yield different sizes and numbers of fragments. In the second stage, subsets of the total population of fragments are visualized through polymerase chain reaction (PCR) amplification using selective primers that bind with ligated adaptors, the restriction site, and 1–3 base pairs of genomic DNA flanking the fragment. The size and number of visualized fragments, or "bands," generated in a selective PCR comprises an individual's profile (Mueller and Wolfenbarger, 1999; Vos et al., 1995).

Each visualized fragment in a profile is referred to as a "locus" or "marker." A locus is characterized by a unique size (i.e., number of base pairs) and is represented in an individual in either of two states (alleles): present or absent (null allele). Present alleles of the same size are assumed to represent the same physical location in the genome and to be identical by descent, that is, same-sized bands are assumed to be homologous. Individuals who share an allele must, therefore, share the same nucleotide sequence at the two restriction sites responsible for that allele. Each locus, thus, has the potential to offer an independent estimate of genetic similarity (depending on degree of linkage) and its converse genetic diversity.

The AFLP method yields a vast number of genetic markers with high repeatability at relatively low cost. It is similar to the restriction fragment length polymorphism (RFLP) method of analysis but has the advantage of sampling variation throughout the entire genome, rather than restricting analysis to the mitochondrial genome or to a particular locus. AFLPs are also similar to the random amplified polymorphic DNA (RAPD) method but have the advantage of being far more repeatable (Mueller and Wolfenbarger, 1999). AFLPs may be particularly useful for establishing relationships among closely related taxa. By sampling such a large number of markers throughout the genome, AFLPs are more likely to reveal the rare genetic divergences that characterize isolated populations. Given this suite of advantages, the AFLP method holds great promise for assessing genetic diversity and may become a standard tool in studies of population genetics and phylogenetics.

Of course, no method of genetic analysis has proven ideal for all questions, so as for any technique, one must be aware of the limitations. For example, the assumption of homology of same-sized bands is critical for all parameter estimates derived from AFLPs. This is probably a realistic assumption, although the probability of homology has been shown theoretically to vary with the size of the band, with smaller-sized fragments more likely to be homoplasious (Vekemans et al., 2002). The assumption of homology may be tested, however, by sequencing a subsample of bands (see the methodological details, later in this chapter).

The assumption of homology of the null, or absent, allele presents a greater challenge, because a band may be absent for various reasons. A nucleotide substitution, an insertion, or a deletion at any position in the flanking restriction sites or a large PCR prohibitive insertion between the sites will fail to produce a band. In addition, deletions or small insertions within the fragment can generate various different-sized bands; therefore, the genetic similarity of two individuals that lack a particular band may not be a robust assumption. This problem is mitigated in analyses of genetic distance by using the equation of Nei and Li (1979) or any equation developed for restriction site data that derives estimates of similarity only from shared present and not shared absent alleles.

Another caveat to its usefulness is that not all AFLP loci necessarily are neutral. Some may contain genes under strong selection and thus affect estimates of neutral genetic divergence. Such "outlier" loci, characterized by unusually high (or low) F_{ST} values, will bias estimates of population demography and phylogenetic relationships but should not present a serious problem if they are identified and removed from the analysis. Moreover, identifying these outliers may help in the search for candidate genes involved in adaptive phenotypes (Luikart et al., 2003).

Perhaps the most vexing limitation of the AFLP method is that loci primarily are "dominant." As with RAPD data, it is difficult, if not impossible, to determine reliably whether an individual with a band present has one or two copies of the allele (i.e., whether it is homozygous or heterozygous). Thus, for dominant markers, only the frequency of null homozygotes can be estimated with reliability. Dominance is a problem for estimating population parameters that require knowledge of allele frequencies (including heterozygosity) because using the square root of the frequency of null homozygotes as an estimate of q (frequency of the null allele) requires the assumption that loci are in Hardy-Weinberg equilibrium (HWE), which may not be valid in all cases. Various methods have been developed to address this limitation and have been used to generate what appear to be reliable estimates of allele frequencies in many cases (Campbell *et al.*, 2003; Hardy, 2003; Krauss, 2000; Lynch and Milligan, 1994; Zhivotovsky, 1999). In addition, distance measures based on shared present bands avoid the dominance problem because they are measures of phenetic similarity and do not require the assumption of HWE.

Although the AFLP method has limitations, it has important strengths, which have led to its increased use in surveys of biodiversity. Here, we summarize several studies that have used AFLPs to assess genetic diversity in arthropods.

Applications in Surveys of Arthropod Diversity

Estimating Heterozygosity within Populations

Average heterozygosity ($\overline{H}$) is one of the most commonly used estimators of within-population genetic variation (Hartl and Clark, 1989). Despite some controversy in estimating heterozygosity based on allele frequencies and assuming HWE (i.e., average expected heterozygosity, $\overline{H}_E$; see earlier discussion), a few studies in arthropods have reported such estimates using AFLPs. Salvato *et al.* (2002) estimated average expected heterozygosity for several populations of the winter pine processionary moth (*Thaumetopoea pityocampa—wilkinsoni* complex), and Gomez-Uchida *et al.* (2003) estimated $\overline{H}_E$ for three populations of the hairy edible crab *Cancer setosus* from the coast of Chile. In the latter study, estimates based on AFLP data were found to be significantly higher than those based on allozyme data. However, as only two allozymes were analyzed, as compared with 109 AFLP loci (of which 99 were polymorphic), the AFLP results may be more accurate.

A study of *Pieris* butterflies reported estimates of expected heterozygosity for multiple populations of two species, *P. rapae* and *P. melete*, in Japan and Korea (Takami *et al.*, 2004). These estimates were used to test the prediction that urban populations, due to a reduction in population size, exhibit reduced genetic diversity compared to their rural relatives. Allele frequencies were estimated using the Bayesian method proposed by Zhivotovsky (1999) for RAPD and AFLP data. Contrary to expectation, results indicated no significant difference in $\overline{H}_E$ between rural and urban populations in either of the two species.

Yan *et al.* (1999), however, suggest caution in the interpretation of heterozygosity estimates based on AFLP data. These authors estimated $\overline{H}_E$ for several populations of the yellow fever mosquito (*Aedes aegypti*), using both AFLP and RFLP data. The latter method yields codominant markers, which means the number of heterozygotes can be directly observed and the assumption of HWE can be tested. Of eight random cDNA loci sampled across three populations in the RFLP analysis, one-third deviated significantly from HWE. This calls into question the chief assumption of most methods used to derive $\overline{H_E}$ from AFLP data. More-over, estimates of $\overline{H_E}$ derived from AFLPs were lower, in all three popula-tions, than those derived from RFLP data. It is unclear, however, to what extent these data are comparable. RFLP data necessarily represent a small sample of the genome and, in the study of Yan *et al.*, were derived from only a few individuals per locus. On the other hand, AFLP markers were derived from 116 individuals and more than 15 times the number of loci. It should be possible to directly test assumptions of HWE by converting dominant AFLP markers into codominant markers to further probe this issue.

Estimating Intraspecific Genetic Differentiation

A more common application of AFLPs in surveys of arthropod diversi-ty is the assessment of genetic structure among populations within species, especially in species of economic importance. As discussed, the AFLP method is well suited for this level of analysis as large sample sizes increase the likelihood of sampling rare genetic divergence that characterizes re-cently isolated populations. AFLPs have been used to assess intraspecific genetic differentiation in several arthropod species. Differentiation among populations largely has been examined using Wright's F-statistics or their derived analogs, including θ-statistics (Weir and Cockerham, 1984) and ϕ-statistics (Excoffier *et al.*, 1992). The latter is estimated using analyses of molecular variance (AMOVA) based on similarity coefficients (e.g., Dice's similarity coefficient) (Sneath and Sokal, 1973) or genetic distance, typically calculated using the distance equation of Nei and Li (1979). These estimators indicate the extent to which variation among subpopulations accounts for overall genetic diversity, with greater values indicating a greater degree of genetic differentiation (Gomez-Uchida *et al.*, 2003; Miller *et al.*, 2002; Salvato *et al.*, 2002; Takami *et al.*, 2004; Van der Wurff *et al.*, 2003). Similarity and distance estimates also are used to construct UPGMA or neighbor-joining dendrograms, which can indicate close and/or hierarchical relationships among individuals or populations (Cervera *et al.*, 2000; Katiyar *et al.*, 2000; Lu *et al.*, 2000; McMichael and Prowell, 1999; Mendelson and Shaw, 2002; Parsons and Shaw, 2001; Reineke *et al.*, 1999; Schneider *et al.*, 2002; Takami *et al.*, 2004; Triantaphyllidis *et al.*, 1997).

Several studies have found evidence of intraspecific genetic differen-tiation in arthropod taxa. Genetic structure in these studies typically is correlated with geographic origin, but genetic differentiation also is observed among ecological forms or biotypes. Early studies using AFLPs include that of Reineke *et al.* (1999), which examined population genetic structure of the hardwood forest pest *Lymantria dispar* (gypsy moth). Sampling 50 sites throughout Europe, Asia, and North America, the authors found that individuals and populations clustered according to geographic location. Results indicated well-supported clades corresponding to each of the three continents and to smaller geographic areas within each continent. Their results confirmed hypotheses based on morphological and behavioral differences, as well as those from early studies of allozymes, mitochondrial DNA (mtDNA), and RAPDs. In comparison to these other markers, AFLPs indicated greater differentiation among populations within Europe and a variation between North American and European populations that had not been previously detected.

Another early study (Katiyar *et al.*, 2000) examined genetic structuring among 15 populations representing several biotypes of the Asian gall midge cereal pest, *Orseolia oryzae*. This study revealed two genetically distinct groups corresponding to separate geographic locations within Asia and discovered that a new Indian biotype, initially thought to have originated within India, appears to be derived from a Chinese form.

Strong evidence of geographic structuring also was observed in the winter pine processionary moth, *Thaumetopoea pityocampa*, in Europe (Salvato *et al.*, 2002). In this study, which included eight *T. pityocampa* populations and one population of its putative sister species *Thaumetopoea wilkinsoni*, AFLP variation was compared to mtDNA variation. Although mtDNA and AFLPs both indicated strong interspecific differentiation (*T. pityocampa–T. wilkinsoni*), only the AFLP data showed geographic structuring among all sampled populations of *T. pityocampa*, with eight populations throughout Europe forming differentiated largely monophyletic clades. Estimates of gene flow derived from F-statistics and measured as the number of migrants (*Nm*) also differed between the two types of data, with AFLP data yielding an estimate of *Nm* roughly twice that of mtDNA. Based on these data, the authors hypothesize that males, who, unlike females, do not pass mitochondria to the next generation, disperse more than females.

Cervera *et al.* (2000) examined genetic differentiation among biotypes of the white fly *Bemisia tabaci* using both AFLPs and RAPDs. In this study, analysis of AFLP data revealed three main genetic groups within the species, each of which contain a variety of biotypes but cluster more or less according to geographic origin. Importantly, results based on AFLP and RAPD data were compared using a Mantel test and proved indistinguishable. AFLPs also have been used to study population differentiation within species on a microgeographic (microallopatric) scale. Van der Wurff *et al.* (2003) found some evidence for local differentiation in the flightless Collembolan soil dweller *Orchesella cincta,* although microsatellite analyses indicated a greater degree of differentiation within and between forest patches than AFLPs.

Differentiation also has been investigated in sympatric biotypes to determine whether closely related coexisting ecological forms represent distinct gene pools. McMichael and Prowell (1999) found evidence of genetic differentiation between two biotypes of the fall armyworm *Spodoptera frugiperda* (rice and corn specialized). Individuals did not always cluster according to biotype, however (i.e., some corn-specialized individuals were more similar to rice-specialized individuals and vice versa). It is difficult to draw firm conclusions from this study as only 10 loci were used in the analysis. It has been demonstrated that increasing the number of sampled loci will increase the resolution of relationships among taxa (Albertson *et al.,* 1999), and although the number of loci needed for reliable resolution may vary from case to case, AFLP studies typically report results from no fewer than 100 loci.

Finally, a few studies have reported that intraspecific AFLP structuring correlated with gender (Gomez-Uchida *et al.,* 2003; Katiyar *et al.,* 2000; Reineke *et al.,* 1999). In one study of hairy edible crabs *Cancer setosus* (Gomez-Uchida *et al.,* 2003), a particular locus was identified that statistically distinguished males from females, being present in 9% of males and 96% of females. Interestingly, this locus appeared to function as an outlier, biasing estimates of geographic structuring. Because geographically isolated populations in this species varied in sex ratio, genetic differentiation associated with geographic location was detected only when this locus was included in the analysis and disappeared when this outlier was removed.

Estimating Interspecific Genetic Differentiation

Another use of AFLPs in arthropods has been to examine interspecific variation, using much the same methods as in studies described earlier, typically to test taxonomic hypotheses (i.e., species designations) among closely related species. One of the first AFLP studies in arthropods was that of Triantaphyllidis *et al.* (1997), in which the species status of several populations of brine shrimp (genus *Artemia*) was investigated. Debate centered on the status of populations found throughout the Mediterranean basin, with some authors recognizing one, and others two, distinct species. Results of AFLP analyses indicate strong similarity among all Mediterranean populations as compared to closely related species from other parts of Europe, Asia, and North America. Bolstered by the fact that

Mediterranean *Artemia* share a key morphological feature (the absence of a spiny growth on the base of the penes), the authors supported the conspecificity of Mediterranean populations.

Other interspecific analyses using AFLPs include those of Lu *et al.* (2000), in which (sub-)species of the crustacean genera *Eriocheir* and *Machrobrachium* (Decapoda) were examined. Both RAPD and AFLP data were used but yielded different results. RAPD data indicated greater intraspecific versus interspecific similarity among three species of *Erocheir*. Conversely, AFLP data indicated no significant differences among the species. For the genus *Machrobrachium,* only AFLP data were used, and significant differences among species were identified.

Testing Evolutionary and Ecological Hypotheses

Although primarily used in arthropods to investigate genetic structure as a problem in itself, AFLPs also have been used to address specific ecological and evolutionary hypotheses. The method is particularly well suited for questions that require resolution of closely related populations or species and has been applied successfully to an increasing number of interesting ecological and evolutionary questions. For example, Miller *et al.* (2002) predicted that adult dispersal distance in aquatic insects should correlate (inversely) with genetic differentiation of populations. They examined four species of aquatic insect that varied in dispersal distance and compared estimates of dispersal with the strength of genetic differentiation. Results generally supported their hypothesis, with one widely dispersing species exhibiting low F_{ST} values and insignificant isolation by distance and a species with shorter dispersal distances exhibiting relatively higher local differentiation and significant isolation by distance. The two other species, however, exhibited minimal dispersal distances but weak F_{ST} and insignificant isolation by distance. In these latter cases, factors other than dispersal may play a role in the lack of genetic differentiation.

Schneider *et al.* (2002) investigated the origin of asexual phenotypes in a species of parasitoid wasp (*Venturia canescens*) distributed along the southeast coast of France. In this study, most sampled populations contained both sexual and asexual individuals. The authors, therefore, tested two alternative hypotheses, one being that asexuality has arisen independently in multiple populations and the other that asexuality arose once within a single source population. Results suggested that many of the asexual individuals, sampled from across different populations, form a clade, albeit weakly supported. A few asexual individuals appeared more closely related to sexual types, but in general the data suggest fewer rather than many evolutionary origins of the asexual trait. This study undoubtedly would have benefited, however, from using a greater number of loci (n = 83 polymorphic bands) and perhaps by testing for outlier loci. Loci responsible for the asexual phenotype, if sampled, could have biased estimates of evolutionary relatedness.

The study of *Pieris* butterflies used AFLPs to test an evolutionary prediction, namely that urban populations exhibit reduced genetic diversity compared to their rural relatives (Takami *et al.*, 2004). Although this study revealed no significant difference in diversity between urban and rural populations, the authors did discover significant intraspecific genetic variation. However, this variation was detected between urban populations sampled in different seasons. Urban populations sampled from different seasons exhibited significant F_{ST} values, whereas rural populations did not, appearing instead to remain genetically similar over time. These results were interpreted to indicate a difference in the dispersal patterns of urban and rural populations, with urban populations being derived from different source populations each season.

Parsons and Shaw (2001) used AFLP data to investigate cryptic species boundaries among four sympatric and allopatric species of the Hawaiian cricket genus *Laupala*. Previous investigations of species relationships in this genus yielded conflicting results, with mtDNA data suggesting very different relationships than those suggested by a nuclear marker (Shaw, 2002). Results of their study were consistent with species relationships indicated by nuclear DNA (nDNA) and with hypotheses based on behavioral and morphological data. Mendelson and Shaw (2002) more thoroughly investigated differentiation between two sympatric *Laupala* species and found a high degree of differentiation based on AFLPs despite similarities in mtDNA. Thus, in this system, AFLPs have been instrumental in interpreting the conflicting results of different genetic markers and have provided the highest resolution for very close species relationships.

Phylogeny Reconstruction

Although investigations of interspecific relationships in arthropods have focused on relatively few species, AFLPs hold promise for resolving species relationships on a broader scale, especially for closely related groups and recent species radiations. Efforts to reconstruct evolutionary relationships in such groups have been hindered both by a paucity of variable nuclear markers and by hybridization, which appears in many cases to facilitate interspecific introgression of mtDNA and/or chloroplast DNA (Semerikov *et al.*, 2003; Shaw, 2002). However, several studies in plants have shown that AFLPs are able to uncover sufficient genetic variation to indicate species relationships, providing robust phylogenetic hypotheses consistent with expectations (Beardsley *et al.*, 2003; Despres *et al.*, 2003; Pelser *et al.*, 2003).

Research in our laboratory has yielded one of the first broad-scale phylogenetic studies in arthropods. The cricket genus *Laupala* is a Hawaiian endemic consisting of 38 morphologically cryptic species. Using AFLPs, we reconstructed a strongly supported phylogeny estimate including 25 of these species plus a Hawaiian outgroup (Mendelson and Shaw, 2005). Species relationships estimated with AFLPs are largely consistent with, but more finely resolved than, those based on an anonymous nuclear region of 1049 base pairs (Shaw, 2002) and suggest that speciation in this genus has proceeded in a manner consistent with the geology of the region. The most basal species are found on the oldest island, Kauai, and the most derived species are found on the youngest island, the Big Island of Hawaii. Importantly, AFLP data suggest significantly different species relationships than those estimated based on mtDNA, a pattern likely due to hybridization.

As the AFLP method becomes more broadly applied, we expect to see an increasing number of phylogenetic studies that investigate relationships among very closely related species, particularly for taxa in which little variation in nuclear or cytoplasmic markers has been found.

Experimental Design and Methods

How Many Bands?

The number of AFLP markers required to successfully address a given question will vary from case to case. Here, we address the question of phylogeny reconstruction, or genetic relatedness. The number of markers required to estimate a well-supported phylogeny (e.g., with most nodes receiving >50% bootstrap support) will depend on the relatedness of species or populations, with more recently diverged taxa requiring more markers primarily because of the inheritance of ancestral polymorphism. Three studies of *Laupala* that vary in the number of taxa sampled, and for which sampled taxa span a range of genetic relatedness, provide points of reference.

The first is a study of the relatedness of *Laupala cerasina* and *Laupala kohalensis,* two sympatric species on Hawaii for which mtDNA and nDNA sequence data yielded contradictory results (Mendelson and Shaw, 2002; Shaw, 2002). Based on their respective positions in an mtDNA phylogeny, the two species are each other's closest relatives. Conversely, an nDNA tree suggests that these species are comparatively distant relatives. Nineteen individuals of *L. cerasina* and seventeen individuals of *L. kohalensis* were sampled for an AFLP analysis, and 77 bands, of which 59 were polymorphic, were scored. Only 30 bands were necessary for the two species to form differentiated genetic groups with 100% bootstrap support, likely because of their comparatively distant relationship as evidenced by the nDNA data.

A second study examined phylogenetic relationships among eight allopatric populations, representing four species of *Laupala*, to examine geographic pathways of speciation (Mendelson *et al.*, 2004). A total of 81 individuals were sampled, and 631 bands were scored, 494 of which were polymorphic. All nodes but one were resolved with more than 50% bootstrap support using 432 bands, and resolution did not increase with additional data (Fig. 1A). Whether increased sampling effort would resolve the final weakly supported node or whether current interpopulation migration is preventing genetic differentiation is unclear.

Finally, we conducted a phylogenetic study of 209 individuals, representing 25 species of *Laupala* plus an additional outgroup (Mendelson and Shaw, 2005). Most of the diversification among these lineages has occurred within the last 5.1 million years (Clague and Dalrymple, 1987). A total of 1173 bands were scored, of which 1148 were polymorphic. Of a possible 49 nodes, 42 were resolved with more than 50% bootstrap support; this resolution was achieved with 1048 bands and resolution did not increase with additional data (Fig. 1B).

Our studies have shown that strong phylogenetic signal can and does exist despite observing a small number of "fixed clade–specific" bands (i.e., bands for which a particular character state [present or absent] is found in all individuals of a given clade and no others). Mendelson *et al.* (2004) found that the average number of fixed clade–specific bands was three; thus, on average, only three loci unambiguously identified an individual to a particular clade. Thus, large allele frequency differences among clades contributed to a substantial degree to the strong phylogenetic signal observed in that study.

The number of primer pairs required to generate the desired number of bands will depend on the size of the restriction site, or the number of nucleotides in the recognition site of the restriction enzyme used. For example, using two 6-base ("rare") cutters will generate fewer fragments than using two 4-base cutters. The number of fragments per primer pair will in turn affect the probability that comigrating fragments are non-homologous, as the number of bands in each size class will vary proportionately with the overall number of bands. The number of same-sized fragments will in turn affect the accuracy of homology assessment, so a balance must be struck between the number of scorable bands per primer pair and the reliability of homology assessment.

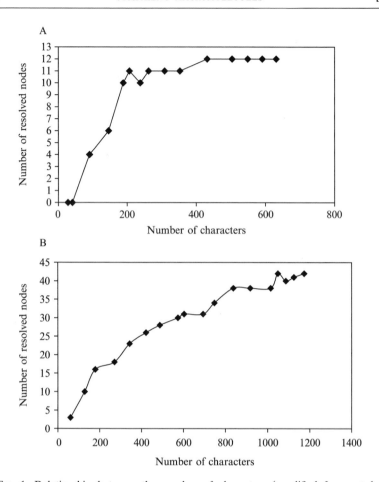

FIG. 1. Relationship between the number of characters (amplified fragment length polymorphic [AFLP] bands) and the number of nodes with >50% bootstrap support for two phylogenetic studies of *Laupala*. (A) A study of eight populations resulted in 12 nodes resolved within 500 AFLP bands (reprinted, with permission, from Mendelson *et al.*, 2004). (B) A study of 26 species (209 individuals) resulted in 42 nodes resolved within approximately 1000 bands (from Mendelson and Shaw, 2005).

Sequencing AFLP Bands to Test Homology

As stated, a critical assumption of AFLP studies is that same-sized bands represent homologous regions of the genome. The most straightforward way to test the assumption of homology is to directly sequence the fragments and determine whether the bands share similar nucleotide sequences. The expected nucleotide similarity will depend on the relatedness of sampled individuals, but the difference between homologous and non-homologous fragments should be clear. AFLPs are primarily used to determine structure among closely related populations or species, so homologous fragments should not be highly differentiated. In *Laupala,* we found that same-sized bands differed by 0.4–0.9% of nucleotide sites between species separated by 3.7 million years. Homologous fragments from populations that diverged less than 0.43 million years ago differed at 0.2–0.5% of nucleotide sites. Fragments identified as non-homologous could not be aligned for comparison.

At least two methods are used in our laboratory for obtaining sequences from AFLP fragments. One is fast and easy, and the other is labor intensive but more reliable. For both methods, at least 20 μl total selective PCR product is loaded per individual in a 5% polyacrylamide gel (this may require up to eight separate lanes). The gel is run as long as necessary to achieve adequate separation of fragments, and then silver stained using standard protocols (Silver Sequence staining reagents, Promega, Inc.). After drying overnight, the gel is resoaked in water for a minute or less, and a 200 μl pipette tip is used to scrape the bands from the gel.

For the first method, bands are placed into a 0.5-ml Eppendorf tube with 20–50 μl water. Tubes are placed in a hot water bath at 95° for at least 1 h, in which heat allows the DNA to elute from the acrylamide. The eluate is then used as PCR template with the same selective primers that produced the fragment. The resulting product is purified using *ExoSAP-IT* protocol (usb Corp.) and then used as template in a standard sequencing reaction with one of the selective primers.

The more labor-intensive method, modified from that of Sambrook *et al.* (1989), has yielded greater success. For this method, excised bands are placed in a 0.5-ml Eppendorf tube with 100 μl elution buffer (0.5 M ammonium acetate, 10 mM magnesium acetate, 1 mM ethylenediaminetetraacetic acid [EDTA] [pH 8.0], 0.1% sodium dodecyl sulfate [SDS]). Bands are soaked overnight at 37° with shaking, tubes are spun for 1 min at 14,000 rpm, and the supernate is transferred to new tubes. Remaining acrylamide is rinsed once with 50 μl fresh elution buffer, spun, and the supernates are combined. To the supernate, 300 μl of 100% chilled (4°) ethanol is added and incubated 4° for 1 h, then spun at 4° for 15 min (13,000 rpm). The supernate is discarded, the precipitate is rinsed once with 300 μl of 70% ethanol (4°), and spun for 1 min. The supernate is again discarded and the remaining precipitate is left to air-dry for 15 min and then resuspended in 20 μl double distilled water or TE buffer. The product is used as a template in a PCR or sequencing reaction as stated earlier.

In a pilot study, we tested the homology of eight sets of same-sized bands, representing eight unique size classes ranging from 170 to 500 bp in length. Bands were derived from three primer pairs and homology was tested by sequencing the bands from several individuals, sampled from at least two species. Each size class was identified *a priori* as "unambiguously" homologous based on a visual assessment through silver staining; for each size class, the bands were precisely aligned on the gel with no detectable variation in size. Of these eight sets of bands, seven were confirmed through sequencing as homologous regions of the genome. Parsons and Shaw (2001) reported similar results through sequencing 10 different size classes of AFLP bands, also sampled from at least two species of *Laupala*. Sequence length ranged from 260 to 509 nucleotides per band and it was concluded that all 10 sets of same-sized bands were sampled from homologous genomic regions. These studies, therefore, support the assumption that same-sized bands represent homologous genomic regions. The increased use of automated fragment analysis will likely reduce ambiguity in band sizing and thus the probability of misidentifying homologous bands. Sequencing of bands identified through automated fragment analysis remains an objective of research.

The methods described earlier to assess homology by band sequencing do not allow the homology of the restriction site itself to be evaluated. However, sequencing the internal portion of a band can nonetheless indicate nucleotide identity of same-sized fragments. An additional benefit of sequencing AFLP fragments is that it may reveal sequence variation (e.g., single nucleotide polymorphisms) among closely related species for which genetic variation can be rare.

Concluding Remarks

The AFLP method of genetic analysis has become an increasingly popular tool for assessing genetic diversity. Although early studies focused primarily on plants, the number of studies using AFLPs to assess genetic diversity in arthropods is steadily increasing, and we are certain to miss several important studies published between this writing and publication. Although issues surrounding the dominant nature of AFLPs and associated problems with heterozygosity estimates remain to be fully resolved, it appears that AFLPs can be successfully used to estimate relationships among closely related species. In some cases, results of such analyses have confirmed those obtained from other molecular markers, but in many cases AFLPs indicate greater differentiation and, therefore, finer resolution than previously observed. We predict that AFLPs will be used increasingly for testing of ecological and evolutionary hypotheses and for a greater number of phylogeny estimates, particularly of recent species radiations.

References

Albertson, R. C., Markert, J. A., Danley, P. D., and Kocher, T. D. (1999). Phylogeny of a rapidly evolving clade: The cichlid fishes of Lake Malawi, East Africa. *Proc. Natl. Acad. Sci. USA* **96**, 5107–5110.

Beardsley, P. M., Yen, A., and Olmstead, R. G. (2003). AFLP phylogeny of *Mimulus* section Erythranthe and the evolution of hummingbird pollination. *Evolution* **57**, 1397–1410.

Campbell, D., Duchesne, P., and Bernatchez, L. (2003). AFLP utility for population assignment studies: Analytical investigation and empirical comparison with microsatellites. *Mol. Ecol.* **12**, 1979–1991.

Cervera, M. T., Cabezas, J. A., Simon, B., Martinez-Zapater, M. M., Beitia, F., and Cenis, J. L. (2000). Genetic relationships among biotypes of *Bemisia tabaci* (*Hemiptera: Aleyrodidae*) based on AFLP analysis. *Bull. Ent. Res.* **90**, 391–396.

Clague, D. A., and Dalrymple, G. B. (1987). The Hawaiian Emperor volcanic chain. *In* "Volcanism in Hawaii" (R. W. Decker, T. L. Wright, and P. H. Stauffer, eds.), pp. 5–54. U. S. Government Printing Office, Washington, D. C.

Despres, L., Gielly, L., Redoutet, W., and Taberlet, P. (2003). Using AFLP to resolve phylogenetic relationships in a morphologically diversified plant species complex when nuclear and chloroplast sequences fail to reveal variability. *Mol. Phyl. Evol.* **27**, 185–196.

Excoffier, L., Smouse, P. E., and Quattro, J. M. (1992). Analysis of molecular variance inferred from metric distances among DNA haplotypes: Application to human mitochondrial DNA restriction data. *Genetics* **131**, 479–491.

Gomez-Uchida, D., Weetman, D., Hauser, L., Galleguillos, R., and Retamal, M. (2003). Allozyme and AFLP analyses of genetic population structure in the Hairy Edible Crab *Cancer setosus* from the Chilean Coast. *J. Crust. Biol.* **23**, 486–494.

Hardy, O. J. (2003). Estimation of pairwise relatedness between individuals and characterization of isolation-by-distance processes using dominant genetic markers. *Mol. Ecol.* **12**, 1577–1588.

Hartl, D. L., and Clark, A. G. (1989). "Principles of Population Genetics," 2nd Ed. Sinauer Associates Inc., Sunderland, MA.

Katiyar, S. K., Chandel, G., Tan, Y., Zhang, Y., Huang, B., Nugaliyadde, L., Fernando, K., Bentur, J. S., Inthavong, S., Constantino, S., and Bennett, J. (2000). Biodiversity of Asian rice gall midge (*Orseolia oryzae* Wood Mason) from five countries examined by AFLP analysis. *Genome* **43**, 322–332.

Krauss, S. L. (2000). Accurate gene diversity estimates from amplified fragment length polymorphism (AFLP) markers. *Mol. Ecol.* **9**, 1241–1245.

Lu, R., Qiu, T., Xiang, C., Xie, H., and Zhang, J. (2000). RAPD and AFLP techniques for the analysis of genetic relationships in two genera of Decapoda. *Crustaceana* **73**, 1027–1036.

Luikart, G., England, P. R., Tallmon, D., Jordan, S., and Taberlet, P. (2003). The power and promise of population genomics: From genotyping to genome typing. *Nat. Rev. Gen.* **4**, 981–994.

Lynch, M., and Milligan, B. G. (1994). Analysis of population genetic structure with RAPD markers. *Mol. Ecol.* **3**, 91–99.

McMichael, M., and Prowell, D. P. (1999). Differences in amplified fragment-length polymorphisms in Fall Armyworm (*Lepidoptera: Noctuidae*) host strains. *Ent. Soc. Am.* **92**, 175–181.

Mendelson, T. C., and Shaw, K. L. (2002). Genetic and behavioral components of the cryptic species boundary between *Laupala cerasina* and *L. kohalensis* (*Orthoptera: Gryllidae*). *Genetica* **116**, 301–310.

Mendelson, T. C., and Shaw, K. L. (2005). Sexual behaviour: Rapid speciation in an arthropod. *Nature* **433**, 375–376.

Mendelson, T. C., Siegel, A. M., and Shaw, K. L. (2004). Testing geographic pathways of speciation in a recent island radiation. *Mol. Ecol.* **13**, 3787–3796.

Miller, M. P., Blinn, D. W., and Keim, P. (2002). Correlations between observed dispersal capabilities and patterns of genetic differentiation in populations of aquatic insect species from the Arizona White Mountains, USA. *Freshwater Biol.* **47**, 1660–1673.

Mueller, U. G., and Wolfenbarger, L. L. (1999). AFLP genotyping and fingerprinting. *Trends Ecol. Evol.* **14**, 389–394.

Nei, M., and Li, W. -H. (1979). Mathematical model for studying genetic variation in terms of restriction endonucleases. *Proc. Natl. Acad. Sci. USA* **76**, 5269–5273.

Parsons, Y. M., and Shaw, K. L. (2001). Species boundaries and genetic diversity among Hawaiian crickets of the genus *Laupala* identified using amplified fragment length polymorphism. *Mol. Ecol.* **10**, 1765–1772.

Pelser, P. B., Gravendeel, B., and van der Meijden, R. (2003). Phylogeny reconstruction in the gap between too little and too much divergence: The closest relatives of *Senecio jacobaea* (*Asteraceae*) according to DNA sequences and AFLPs. *Mol. Phyl. Evol.* **29**, 613–628.

Reineke, A., Karlovsky, P., and Zebitz, C. P. W. (1999). Amplified fragment length polymorphism analysis of different geographic populations of the gypsy moth, *Lymantria dispar* (*Lepidoptera: Lymantriidae*). *Bull. Ent. Res.* **89**, 79–88.

Salvato, P., Battisti, A., Concato, S., Masutti, L., Patarnello, T., and Zane, L. (2002). Genetic differentiation in the winter pine processionary moth (*Thaumetopoea pityocampa–wilkinsoni* complex), inferred by AFLP and mitochondrial DNA markers. *Mol. Ecol.* **11**, 2435–2444.

Sambrook, J., Fritsch, E. F., and Maniatis, T. (1989). "Molecular Cloning: A Laboratory Manual," 2nd Ed. Cold Spring Harbor Laboratory Press.

Semerikov, V. L., Zhang, H. Q., Sun, M., and Lascoux, M. (2003). Conflicting phylogenies of *Larix* (*Pinaceae*) based on cytoplasmic and nuclear DNA. *Mol. Phyl. Evol.* **27**, 173–184.

Schneider, M. V., Beukeboom, L. W., Driessen, G., Lapchin, L., Bernstein, C., and Van Alphen, J. J. M. (2002). Geographical distribution and genetic relatedness of sympatrical thelytokous and arrhenotokous populations of the parasitoid *Venturia canescens* (*Hymenoptera*). *J. Evol. Biol.* **15**, 191–200.

Shaw, K. L. (2002). Conflict between nuclear and mitochondrial DNA phylogenies of a recent species radiation: What mtDNA reveals and conceals about modes of speciation in Hawaiian crickets. *Proc. Natl. Acad. Sci.* **99**, 16122–16127.

Sneath, S. K., and Sokal, R. R. (1973). "Numerical Taxonomy." Freeman and Company, San Francisco, California.

Takami, Y., Koshio, C., Ishii, M., Fujii, H., Hidaka, T., and Shimizu, I. (2004). Genetic diversity and structure of urban populations of Pieris butterflies assessed using amplified fragment length polymorphism. *Mol. Ecol.* **13**, 245–258.

Triantaphyllidis, G. V., Criel, G. R. L., Abatzopoulos, T. J., Thomas, K. M., Peleman, J., Beardmore, J. A., and Sorgeloos, P. (1997). International study on *Artemia*. LVII. Morphological and molecular characters suggest conspecificity of all bisexual European and North African *Artemia* populations. *Mar. Biol.* **129**, 477–487.

van der Wurff, A. W. G., Isaaks, J. A., Ernsting, G., and van Straalen, N. M. (2003). Population substructures in the soil invertebrate *Orchesella cincta*, as revealed by microsatellite and TE-AFLP markers. *Mol. Ecol.* **12**, 1349–1359.

Vekemans, X., Beauwens, T., Lemaire, M., and Roldan-Ruiz, I. (2002). Data from amplified fragment length polymorphism (AFLP) markers show indication of size homoplasy and of a relationship between degree of homoplasy and fragment size. *Mol. Ecol.* **11,** 139–151.

Vos, P., Hogers, R., Bleeker, M., Reijans, M., van de Lee, T., Hornes, M., Frijters, A., Pot, J., Peleman, J., Kuiper, M., and Zabeau, M. (1995). AFLP: A new technique for DNA fingerprinting. *Nucleic Acids Res.* **23,** 4407–4414.

Weir, B. S., and Cockerham, C. C. (1984). Estimating *F*-statistics for the analysis of population structure. *Evolution* **38,** 1358–1370.

Yan, G., Romero-Severson, J., Walton, M., Chadee, D. D., and Severson, D. W. (1999). Population genetics of the yellow fever mosquito in Trinidad: Comparisons of amplified fragment length polymorphism (AFLP) and restriction fragment length polymorphism (RFLP) markers. *Mol. Ecol.* **8,** 951–963.

Zabeau, M. (1992). Selective restriction fragment amplification: A general method for DNA fingerprinting. European patent application no. 0534858 A1.

Zabeau, M., and Vos, P. (1993). Selective restriction fragment amplification: A general method for DNA fingerprinting. European patent application no. 92402629.7, publication no. 0534858 A1.

Zhivotovsky, L. A. (1999). Estimating population structure in diploids with multilocus dominant DNA markers. *Mol. Ecol.* **8,** 907–913.

[12] Use of AFLP Markers in Surveys of Plant Diversity

By CHIKELU MBA and JOE TOHME

Abstract

The collection of the available range of genetic variation in a gene pool, usually made of the cultivated species and their undomesticated relatives, is referred to as a *germplasm* collection. Increasingly, discriminator data generated using molecular genetic markers are either complementing or completely replacing those from morphological characters (known as *descriptors*) in surveys of genetic diversity. In addition to highlighting the state of knowledge on the specific applications of amplified fragment length polymorphisms (AFLPs) in surveys of plant genetic diversity, an attempt at a brief description and comparison of the different marker systems in use has also been made in this chapter. We have also attempted a description of the AFLP marker technology, its strengths and weaknesses, methodologies for generating reliable AFLP data, available resources (hardware, software, consumables); the kinds of questions for which AFLP data provide valid answers; and data management options. This chapter also highlights salient considerations that would guide decisions on the adoption of molecular marker assays.

Introduction

Genetic Diversity

The existence of definable and in turn exploitable variation within a gene pool is the basis for developing strategies for domestication, genetic improvement, conservation, and sustainable use of plant species. "Genetic resources" is a term used to describe this variation and encompasses the spectrum of genetic diversity among cultivated species and their undomesticated relatives. Wild relatives would mainly include species that occur within or outside the centers of genetic diversity of cultivated plants; weeds existing in crop-weed complexes in such centres of diversity; land races or local varieties that have not arisen due to deliberate improvement strategies; and modern breeding lines and genetic stocks and obsolete cultivars (Ford-Lloyd, 2002).

A collection of this range of variation is referred to as a germplasm collection and ideally is conserved in a manner that makes access to its component members easy and meaningful. The aim for conserving germplasm is to have a repository of useful genes that could be employed in addressing immediate or future constraints to the species. A trove of genetic variation for many plant species is now available in many parts of the world. However, to be able to fully harness the potential that is locked up within these invaluable resources, an accurate assessment of genetic variation within and between populations is required. A germplasm collection in its formative stages, usually the product of several collection trips to different places and entries by different individuals or immediately after the collections and establishment, would be expected to include an appreciable number of redundant accessions. Such redundancy makes the conservation of the genetic resources more expensive and confounds the selection of sources of genes. It is customary, therefore, to characterize any germplasm collection. The process of this characterization is referred to as *genotyping* or *fingerprinting*.

In general, plants are genotyped for the purpose of establishing identities in commerce, for example, for the protection of the plant breeder's rights and patents or for quality control in the identification of genetically modified organisms or mutants. In plant improvement programs, plants are genotyped in marker-aided selection, in selection of parents for use in crosses, and for the survey of progenies. In basic research, data arising from fingerprints are used in phylogenetic studies and other areas of evolutionary and population genetics. The scope of this chapter is, however, restricted to the application of fingerprints in the studies of genetic variation.

Molecular Characterization of Germplasm Collections

Before the discovery of molecular marker technologies, germplasm collections were characterized using a checklist of morphological characters or descriptor lists, normally made up of quantitative measures such as height, weight, and quality parameters such as color. This, however, had the inherent risk of being insensitive as most morphological characters are influenced to varying extents by the environment. This handicap becomes the more acute considering that a germplasm collection is usually made up of accessions from different environmental regions. There is a need, therefore, to have a valid germplasm characterization that efficiently assesses diversity, identifies representative samples, mislabeling, and quite importantly eliminates duplicates.

Molecular markers, the road signs in the genome, directly assay the hereditary material and thereby circumvent the confounding effect that genotype-by-environment interaction has on the phenotype. Also, molecular markers can produce very reliable results even in the absence of measurable phenotypic characters. Increasingly, molecular markers are being used to complement morphological or phenotypic data in assessing the range of variation within the germplasm. Also, molecular data can be a replacement for phenotypic data (Marshall, 1997). Indeed, most public and private plant breeding stations and companies as well as academic laboratories currently rely on molecular markers for producing the fingerprints of their materials (Edwards and Mogg, 2002).

These markers are able to detect variations to a degree that is not achievable using other methods including biochemical assays such as isozymes. They are robust rendering them immune to environmental effects. Finally, some of them are able to exploit the seemingly infinite possibilities of the polymerase chain reaction (PCR) to speedily assay even very minute samples (Karp, 2002). Karp (2002) also summarized the types of information usually derived from molecular marker assays of genetic diversity to include the following:

- *A numerical measure of the genetic diversity present in the population being studied.* This estimates the diversity at both at the interspecific and intraspecific levels.
- *The structure of the genetic diversity.* This looks at how the diversity is distributed among and within populations in temporal and spatial contexts.
- *Evolutionary history.* This examines population history, genetic lineages, and gene genealogies and discriminates between ongoing and past events.

- *Genetic distance.* This estimates the extent of difference between genes, sequences, populations, or species.
- *Relatedness.* This is a measure of kinship.
- *Identity.* This answers the query of whether individuals are genetically identical or distinct.
- *Gene flow.* This is an estimate of the exchange of genes between individuals.
- *Linkage disequilibria and the interlocus correlation of allelic variation.* This estimates departures from randomness in the population resulting from factors such as natural selection, inbreeding, mutations, genetic drift, or differential migration.

Molecular marker-based genetic diversity assessments therefore offer a promising complementary or even alternative approach for designing more efficient strategies for the use of plant genetic diversity.

Some Commonly Used Molecular Genetic Markers

Several types of marker systems are available for use in molecular characterization of plants and can be categorized into three broad groups based on whether they are PCR-based and, if they are, on whether the PCR is driven by arbitrary or target-specific primer pairs. The non–PCR-based molecular marker types include isozyme and restriction fragment length polymorphism (RFLP) methods (Feinberg and Vogelstein, 1983). The most commonly used PCR-based marker types that use arbitrary primers include random amplified polymorphic DNA (RAPD) (Liu and Furnier, 1993; Williams *et al.*, 1990) and amplified fragment length polymorphism (AFLP) (Vos *et al.*, 1995; Zabeau and Vos, 1993). Simple sequence repeats (SSRs), or microsatellites, first discovered in 1984 (Tautz and Rentz, 1984) and used as markers in the 1990s, are the most commonly used locus-specific marker system. Single nucleotide polymorphisms (SNPs), first used in 1998 (Schafer and Hawkins, 1998), and DNA sequencing are also molecular fingerprinting methods but require significant prior optimization.

The utility of all these techniques is based on their ability to identify variation in DNA sequence information (polymorphisms) between the individuals being assayed. Such polymorphisms, the analyses of which form the basis for discriminating between individuals, usually result from mutation events in the form of base substitutions, insertions, deletions, or even inversions. Although a detailed treatment of each individual technique is beyond the scope of this chapter, it is still necessary to attempt a brief comparison of the more commonly used types (Table I).

TABLE I
COMPARISON OF MARKER TYPES

	AFLP	RFLP	RAPD	SSR	SNPs
Principle	PCR of a subset of restriction fragments from extended adapter primers	Restriction endonuclease digestion and Southern blotting	DNA amplification with random primers	PCR simple sequence regions	Identification of single nucleotide variations through PCR, sequencing, hybridization, etc.
Polymorphism	Single base changes (insertions and deletions)	Single base changes (insertions and deletions)	Single base changes (insertions and deletions)	Repeat length changes	Single base changes (insertions, deletions or inversions)
Abundance	High	High	Very high	Medium	Very high
Level of polymorphism	Medium	Medium	Medium	High	Very high
Dominance	Dominant/codominant[b]	Codominant	Dominant	Codominant	Dominant/codominant
Multiplex	30–100	1–2	5–20	1–5[a]	$1-10^3$
DNA quantity	Medium	High	Low	Low	Low
Sequence information required?	No	No	No	Yes	Yes
Development costs	Medium	Medium	Low	High	High
Operational costs	Medium	High	Low	Low to medium	High
Technical demands	Medium	High	Low	Low to medium	High
Automation	Yes	No	Yes	Yes	Yes

[a] The number of primer pairs to be multiplexed depends on the visualization method. A fluorescence-based system could handle assays of up to five primer pairs depending on the differences among their product sizes.

[b] AFLP markers could produce codominant alleles if it is possible to determine that the alleles have the same origin.

From this listing of the features of the most commonly used molecular markers, it is clear that no one marker type reliably can produce all the possible types of information required for a thorough study of genetic diversity, necessitating at times an assessment strategy that uses a combination of marker systems. For instance, some markers can detect only dominant alleles, that is, their recessive alleles cannot be visualized and scored. On the other hand, others are codominant, meaning that the different variants of an allele can be seen. Some markers are capable of assaying several loci at the same time, while others can examine only one locus at a time. It is, therefore, obvious that these individual marker types have their merits and demerits and have particular situations for which they are best suited. In the end, the marker choice will depend on the precise issue to be addressed by the information to be generated. For instance, when large numbers of samples are being assayed, the researcher's choice of technique will, in addition to budgetary considerations and other logistics, be guided by considerations for a system that is amenable to high-throughput assays (including multiplexing of markers and handling of multiple samples); may be automated; does not necessarily rely on a gel-based system for visualization; and whose data can be exported easily to computer-based data management platforms.

In general, for most diversity studies, one wants to use a marker system that is robust, accurate, reproducible, highly informative, and cost effective and produces data that can be easily input into database and suitable for automation (Karp, 2002; Sergovia-Lerma et al., 2003). Therefore, the AFLP method is fast becoming the preferred compromise marker type in meeting all or most of these requirements (Hill et al., 1996; Karp et al., 1996, 1997; Lu et al., 1996; Mueller and Wolfenbarger, 1999; Paul et al., 1997; Qamaruz-Zaman et al., 1997; Sharma et al., 1996; Travis et al., 1996).

AFLP Marker Technology

The AFLP method (a trademark of Keygene, Wageningen, The Netherlands) is one of the DNA fingerprinting technologies that is PCR based and relies on its ability to amplify a limited subset of DNA restriction fragments from a specific DNA sample (Blears et al., 1998; Vos et al., 1995). AFLPs are DNA fragments, usually in the range of 80–500 bp that are obtained from endonuclease restriction, followed by ligation of oligonucleotide adapters to the fragments and selective amplification by PCR. In simple terms, it is a combination of the ubiquity of endonuclease restriction digestion sites throughout the genome and the power of the PCR and made up of the following sequential five steps (Fig. 1):

AFLP procedure

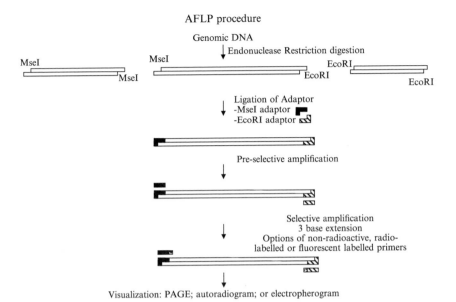

FIG. 1. Amplified fragment length polymorphism procedure (courtesy B. P. Forster).

- *DNA restriction.* The sample genomic DNA or cDNA is digested using two restriction enzymes: a frequent cutter (tetra-cutter, e.g., the four-base restriction enzyme *Mse*I) and a rare cutter (hexa-cutter, e.g., the six-base restriction enzyme *Eco*RI). This produces three types of restriction fragments, those with *Eco*RI cuts at both ends; those cut at one end by *Eco*RI and at the other end by *Mse*I; and the rest are the ones that have been cut at both ends by *Mse*I.
- *Ligation of adaptors.* Double-stranded adaptors, specific to *Eco*RI and *Mse*I sequences, are added with the help of a ligase (joining enzyme) to the cut ends of the restriction fragments. These adaptors are made up of a core sequence and the restriction enzyme–specific sequence. The function of the adaptors is to provide known sequences for the annealing of the PCR primers given that the overhangs from the restriction are not enough for PCR primer design. In practice, the restriction and ligation take place in a single reaction tube so the ligation of the adaptor to the cut end of the DNA fragment alters the topography so a second round of restriction cannot occur after ligation. As expected, there are far too many fragments to have their amplification products resolvable on a single gel at this stage.

- *Preselective PCR amplification.* It is customary to reduce the complexity (i.e., number of fragments that are amplifiable) in a two-step procedure, the first being referred to as *preselective* amplification to be contrasted with the *selective* amplification of the second step. To reduce the complexity of the ligated restriction fragments, a PCR primer that incorporates the known adaptor sequence, an enzyme-specific sequence, and an additional "preselective" single-base extension at the 3′ end are used to amplify the fragments. Such primers function because they are able to anneal to the fragments carrying the complementary nucleotides of the adaptors and endonuclease restriction sites. However, the complexity reduction is achieved because this primer will recognize and anneal only to the fragments carrying this extra preselective nucleotide in the appropriate location relative to the core adaptor and restriction site sequences. The majority of the amplicons from this preselective PCR, therefore, are those fragments bearing one *Mse*I cut and one *Eco*RI cut, and that also contain the core adaptor sequence. These then account for a 16-fold reduction of the complexity of the fragment patterns produced from the DNA digestion step, but even this is still irresolvable on a gel, hence the smeary appearance of the products on a denaturing polyacrylamide gel. A further complexity reduction step is, therefore, necessary.
- *Selective PCR amplification.* This step is identical to the preselective PCR amplification step, except that the PCR primers including an identical sequence to that at the preselective stage now have two additional selective nucleotides at the 3′ end, making a total of three selective nucleotides. These two additional nucleotides can be any of the 16 possible combinations derivable from four nucleotides. The complexity reduction at this step, estimated to be in the range of a 256-fold reduction, is achieved by the selective annealing of the PCR primer to only that subset of restriction fragments that carry these three nucleotides in the corresponding positions relative to the adaptor and restriction site sequences. The reasoning is that there is an inverse relationship between the number of fragments that are amplified and the number of nucleotides that are added to the primer. However, any addition of nucleotides beyond the maximum of 3 yields unsatisfactory results. Depending on the method for visualizing the products, one of the primers may be radiolabeled or fluorescent labeled for autoradiography and automated sequencing procedures, respectively. Nonlabeled primers are also available. In general, the PCR produces 50–100 fragments that are then separated on a denaturing polyacrylamide gel by electrophoresis. Alternatively,

non–gel-based methods of separation are now available using automated capillary sequencers.

• *Visualization.* The electrophoretically separated amplicons are visualized after exposure to X-ray film or photographic paper when radiolabeled primers have been used. For nonlabeled primers, the products are visualized through silver staining procedures. The technology has been further developed with fluorescent-labeled primers, however. In this case, the separated fragments are visualized as a reconstructed gel image, electropherograms, or tabular data on an automated sequencer such as the Perkin-Elmer/Applied Biosystems, Inc., range of sequencers. High throughput is achieved through the possibility of loading in the same gel lane a set of multiple samples that have been PCR amplified with separate primer sets and labeled with different fluorescent dyes along with an internal DNA size standard. This is referred to as *multiplexing*, a strategy that significantly reduces the operational costs.

cDNA AFLP

In an adaptation of the AFLP technique, whole-genomic DNA is replaced with reverse transcriptase PCR (RT-PCR)–derived cDNA preparations and this is termed *cDNA AFLP*. This technique is also fragment based and assays genomewide gene expression. Theoretically, it is capable of analyzing all genes involved in a particular biological process or expressed under certain conditions. Unique transcript tags of expressed genes are PCR amplified and visualized on polyacrylamide gels. This is used principally for transcript profiling and permits the display and quantification of transcripts based on AFLP fingerprinting of double-stranded cDNA. This is applied in the identification of differentially expressed messenger RNA (mRNA). Its amenability to automation also makes it a high-throughput technique. However, detailed explanation is beyond the scope of this chapter, because this is more a gene expression analytical technique rather than a genetic diversity marker type.

Data Collection

A critical part of any assay is data management. This is true also for AFLPs, particularly because an appreciable number of data points are generated on a typical gel. Since the development of this technique, several analytical methods have arisen and this chapter attempts a survey of the most commonly used ones.

For fluorescent-based AFLPs, the outputs from GeneScan are exportable to the analytical software, Genotyper, which is capable of identifying and measuring bands within the size range of 50–500 bp. The GeneScan software itself has the capacity for analyzing four fluorescent labels that have the pseudo-colors of blue, green, yellow, and red. In the case of autoradiograms or silver stained displays, the bands could be manually read off the X-ray film, photographic paper, the gel, or a scanned image of the gel. The AFLP-Quantar and the update AFLP-Quantar-Pro from Keygene, developed specifically for AFLP image analyses, are the most widely used AFLP data management software. In addition to scoring for presence/absence of bands, this software is also capable of using the band intensities to score for heterozygosity or homozygosity of the bands. Such other software as Gene ImagIR and SAGA GT (both from LI-COR, Inc., Lincoln, NE) and RFLP Scan (Scanalytics, Inc., Fairfax, VA) also may be used for collecting and scoring AFLP data. In general, a binary matrix is constructed based on scoring the migrating bands as present or absent.

Summary of the Bench Procedures of the AFLP Technique

The analytical procedures involved in AFLP assays involve the following sequential steps:

- DNA extraction.
- Endonuclease restriction digestion of the DNA sample.
- Ligation of adaptors to the DNA fragments.
- Preselective amplification of a subset of the DNA fragments (achieving a 16-fold complexity reduction).
- Selective amplification of a subset of the DNA fragments (achieving a further 256-fold complexity reduction).
- Separation of the fragments by denaturing polyacrylamide gel electrophoresis.
- Visualization of the polymorphisms by autoradiography, silver staining, or fluorescence.

Minimum Requirements for AFLP Assays

At the minimum, the laboratory should be set up for the following:

- *PCR.* Thermocyclers are needed for the amplification of the DNA fragments and are readily available under different name brands. The most commonly used thermocyclers are manufactured by such companies as Applied Biosystems, Biometra, Cold Fire, Eppendorf, Idaho Tech, MJ Research, Inc., Protedyne, Stratagene, Thermo Hybaid, and Thistle.

- *Fragment separation.* It is customary to visualize the products of the preselective amplification step on an agarose gel to confirm that this step worked before proceeding to the next and more stringent selective amplification stage. It is, therefore, necessary to have the setup for agarose gel electrophoresis (electrophoresis chamber, power source, gel trays, ultraviolet [UV] light source, etc.). The main electrophoresis step is after the selective PCR amplification and the requirements are a bit more sophisticated. In general, however, an automated sequencer facility would be required for the separation of fluorescence-labeled fragments. Automated sequencers are readily available in both the slab gel and capillary formats (at some significant cost). The most commonly used ones come from such manufacturers as Applied Systems (ABI series), Amersham Bioscience (MegaBACE, ALFexpress series), MJ Research (Basestation series), LI-COR (Global IR and other series), Beckman Coulter (CEQ series), and Biotage AB (PSQ series). For laboratories without a capillary sequencer, the preferred means would be electrophoresis on denaturing polyacrylamide gels. In this case, the requirements include a power source and gel rigs. Several brands are on the market by manufacturers like BIORAD, Amersham Pharmacia Biotech, and Cole Parmer Instrument Co.
- *Visualization.* If radiolabeled primers are being used in the laboratory, there must exist explicit procedures for safe handling and disposal of the commonly used radionuclides (^{32}P, ^{33}P, and ^{35}S). These would normally conform to the local laws. Requirements include a dedicated lab space and equipment for handling these hazardous materials. Facilities have to be in place for exposure of the gels to X-ray films and the subsequent development. For a nonradioactive visualization method, usually the simplest strategy, the capacity to carry out silver nitrate staining of the gels, is necessary. This requires trays (vertical or horizontal) for holding the different reagents used at this step. It is also advisable to have these trays coupled to a shaking device.

Advantages

One of the greatest advantages of the AFLP technique is its capacity to generate many polymorphic bands per assay. Robinson and Harris (1999) cited many studies showing that in general, the AFLP technique produced up to four times more polymorphic loci per primer combination than the comparable techniques of RAPDs, RFLP, and SSRs. In addition, where the genome of the organism being assayed has not been characterized sufficiently so as to have target sequence information, AFLP will remain the molecular marker of choice. It shares this advantage with RAPDs and RFLP, but the high reproducibility of the polymorphisms (as compared with RAPDs) and the ease of use, particularly its being PCR-based (as compared with RFLP), make this technique quite superior to the other ones. Other desirable attributes deriving from the PCR-based technique include its amenability to automation. This high-throughput feature implies that a lot of data points can be generated within a relatively short period. In addition, because AFLP markers are distributed throughout the genome, each assay is in itself a multiplex reaction. Several regions of the genome are being assayed at the same time in the same reaction mixture (one primer combination). With SSR, on the other hand, only one locus is assayed per reaction.

Disadvantages

Compared to other marker systems, one of the major drawbacks for the use of the AFLP fingerprinting technique is the requirement for pure high-molecular-weight DNA, usually in the range of 0.2–0.25 μg. This becomes a hindrance when DNA extraction or sample availability is limiting. Also, except where gel scan facilities are available and optimized, it is usually difficult to interpret banding profiles in terms of loci and alleles. This is because co-migrating bands with similar sizes are not necessarily of the same origin. Also, the bands are generally considered dominant alleles.

Robinson and Harris (1999) surveyed the use of the AFLP marker technique and categorized the problems associated with it as being broadly divided into practical considerations, the nature of the data, and analysis challenges. One of the practical problems that merits consideration is the running costs. Although recognizing that this is a rather subjective matter as it depends on the availability of resources for the individual or group of researchers, several steps in the AFLP technique could still be considered expensive. Using radiolabeled or fluorescent-labeled primers would normally require a substantial investment. Also, primers could be considered expensive depending on whether the researcher is using custom primers or off-the-shelf primers. Robinson and Harris (1999) also inferred that the choice of restriction enzymes and primers (which are at times dependent on nothing else other than available resources) could greatly affect both the quality and the quantity of data generated. However, with the availability of AFLP kits, most researchers use only the enzymes and primers in such kits and this is usually sufficient for their needs. From the surveyed studies, it is also the opinion of Robinson and Harris (1999) that although AFLP banding patterns are generally considered reproducible (e.g., vis-à-vis RAPDs), several concerns remain unresolved, with the primary one being that on reruns, different banding profiles have been encountered. Also, the scoring of banding profiles not being explicit is open to subjectivity: Different individuals may score the same bands differently. Also, such factors as partial endonuclease DNA digestion due to either poor DNA quality or insufficient (or faulty) restriction enzymes could ultimately affect the banding profiles produced.

One of the major disadvantages of the AFLP technique (just as with RAPDs) is that the bands are considered dominant (as compared to the codominant markers produced with isozyme or RFLP markers), implying that polymorphism is scored only in terms of presence or absence. This implies, therefore, that 2–10 times more individuals would be required for this type of marker than for the co-dominant ones for typical population genetics studies (Karp *et al.*, 1997; Lynch and Milligan, 1994; Robinson and Harris, 1999). This problem is probably mitigated by the number of polymorphisms that are generated per assay, as well as emerging technologies that are able to differentiate heterozygotes from homozygotes based on the intensity of the bands. Also, according to Robinson and Harris (1999), the identification of homology may be the greatest criticism for AFLP as co-migrating bands are not always necessarily from the same genome regions and similar-sized missing bands in different individuals may be results of different mutation events. This problem may, however, not be very grave considering the stringent complexity reduction steps that lead to these bands. Another negative factor ironically is the high level of polymorphisms that are generated by AFLP because this tends to generate so much variation that current analytical tools may wrongly predict relationships. Robinson and Harris (1999) also inferred that without proper knowledge of the mutation events leading to the variations, which is usually impossible to determine *a priori,* the scoring of the bands is usually distorted, leading to "overscoring." A mutation event, say, a deletion or an insertion, would be seen as two different-sized bands in two individuals, so with the presence or absence scoring method, this singular event would be scored twice. This would confound the analysis as the nonindependence assumption is violated. Ploidy levels of the individuals also may confound the scoring system, especially when populations of individuals with different ploidy levels are included in the same analysis (Robinson and Harris, 1999).

Summary of Advantages. There is a consensus that the following are some of the advantages AFLP has over other comparable techniques.

- No prior information on the sequence of target loci is required.
- There is high genomic abundance of this marker type.
- Markers are randomly distributed throughout the genome.

- Each assay yields many polymorphic bands.
- The assay results are highly reproducible.
- AFLP lends itself to a very wide range of applications in molecular assays.
- The technique is amenable to automation.

Summary of Disadvantages. The drawbacks to the technique include the following:

- It is difficult to interpret banding profiles in terms of loci and alleles.
- The alleles are generally considered dominant.
- Homology cannot be inferred on the basis of co-migration of fragments.

Data Analysis

The different types of population genetics information that can be derived from the analysis of molecular data are presented in an earlier section of this chapter. These parameters estimate genetic relatedness and diversity and partition them into their corresponding component parts. The starting point in the analysis is the creation of the data matrix that sets out the character state of each marker for each sample being assayed.

In the case of AFLP, this is a binary matrix in which for each marker, the character state is either presence or absence. The sample-by-marker matrix of character states is then used to construct a sample-by-sample matrix of pairwise genetic distances or similarities (Karp *et al.*, 1997) through the generation of several estimates of distances between the samples. Once the data matrix has been generated, and on the assumption of independence for the bands, the analysis of the AFLP data could proceed in either of three ways, similarity, frequency, and character measures (Robinson and Harris, 1999). Although the similarity or frequency measures convert the binary data matrix into a series of distance measures between taxa, the third and less frequently used character-based method usually is more suitable for deriving phylogenetic hypotheses. Making a choice between the use of similarity or frequency measures is usually a function of the number of samples and the questions the study aims at answering. For few samples of less than 50 and in which the emphasis is on the variation between individuals, similarity measures are usually adopted. This is contrasted with the use of frequency measures for larger sample sizes and to study variations between populations.

1. *Similarity estimates.* AFLP data are subjected to similarity measures using the simple matching coefficient (SMC); Jaccard's coefficient; or the Nei and Li (NL) coefficient. Although the first two parameters estimate

the proportion of shared present or absent bands, the NL coefficient measures the probability that any two bands observed in two samples are the same on account of the common ancestry of the two samples. (Harris, 1999; Karp et al., 1997; Robinson and Harris, 1999). The incidence of false positives and false negatives and the problems of nonhomology for co-migrating bands are some of the factors that could compromise the validity of these estimates.

2. *Frequency estimates.* Like the similarity estimates, the levels and structure of genetic diversity are calculated using the AFLP band frequencies. Analysis of molecular variance (AMOVA) and Shannon's measure are common ways for estimating genetic diversity (Robinson and Harris, 1999). In addition to the sources of bias for the similarity measures, the frequency measures are further biased on account of the dominance of the AFLP bands. This is because the variance at any single loci is increased with dominance.

Karp et al. (1997) listed the two main ways for analyzing distance or similarity matrices and displaying the results. The first group, based on the estimation of the overall estimate of distance or similarity, and which also corresponds to classification of Robinson and Harris (1999) into similarity and frequency measures, are called the "phenetic" methods. These are principal coordinate analysis (PCO) or the generation of a so-called den-drogram (or tree diagram). The PCO produces a two- or three-dimensional scatter plot of the samples. This is interpreted by viewing the geometric distances between the samples as being representative of their genetic distances. Samples that fall within the same or nearby clusters are deemed genetically similar. This is similar to the dendrogram approach in which clusters are linked to each other at progressively decreasing genetic simi-larity levels (or the inverse increasing genetic distance) until all the samples are aggregated into one global cluster. Some of the more commonly used algorithms for performing cluster analysis include the unweighted pair-group arithmetic averages (UPGMA), the neighbor-joining method, and Ward's method.

3. *Character estimates.* This is the "cladistics" method for the analysis of genetic relationships and, like the phenetic method, begins with the production of a sample-by-marker character-state matrix, ultimately resulting in dendrograms. Such trees often are, however, referred to as 'cladograms' to distinguish them from the phenograms of cluster analysis. The major difference between phenograms and cladograms is that in the former clustering is on the basis of overall similarity for all markers, and in the latter, two samples are assigned to the same clade or cluster on the basis of sharing the same character state for a given marker (or markers). Because AFLPs are easily scored as a discrete binary data matrix, the

characters being the AFLP bands and their states as being either present or absent, it would then seem ideal to analyze such data using character measures. However, for character measures to be valid, the assumptions of variability, independence, and homology of the character states must be met. These are as yet unresolved for AFLP, so further discussions on the use of parsimony or maximum likelihood methods to analyze AFLP data and generate dendrograms/cladograms would at best be academic and therefore are not included in this chapter.

AFLP fingerprinting techniques generate numerous data points in one assay and have the advantage of multiplexing. There is abundant information in the literature of their use in varied genetic studies. However, like RAPDs, although AFLP assays several regions of the genome in one reaction, the locations of the loci being assayed are unknown. Karp *et al.* (1997) listed four considerations that always must be kept in mind while generating and using AFLP data:

1. These are dominant markers.
2. Where pedigree information is lacking (as is almost the norm), the identity of the individual bands should be considered unknown and the assigning of bands to loci are also uncertain.
3. The presence of co-migrating bands (of same molecular weight) in two individuals does not connote homology of the fragments.
4. Single bands can sometimes consist of two or more co-migrating amplicons because it is difficult to distinguish between heterozygotes and homozygotes.

Relative to codominant markers such as RFLPs and SSRs, it is plausible to conclude that these shortcomings do reduce the efficiency of AFLP and indeed any arbitrary primer-based molecular marker technique in genetic diversity and other population genetic analyses. This is because they negate the earlier stated assumptions, especially that of the independence of markers. As has been pointed out earlier, increasing the number of individuals assayed per locus 2- to 10-fold can enhance their utility.

Applications of AFLP in Plant Diversity Surveys

Going by the large volume of published data on the use of the AFLP technique, we can infer that it has caught on as the marker of choice, especially where the sequence information for the genome is lacking. The literature is replete with these accounts of the application of the AFLP technique in plant diversity studies. However, for reasons of brevity, we shall not attempt the rather daunting task of documenting anything close to an exhaustive survey of these reports. We have, therefore, only attempted to look at some of these works with the aim of highlighting the most common methodologies being adopted by researchers in generating and analyzing AFLP data for plant genetic diversity studies.

AFLP Data Generation

The reasons for carrying out molecular fingerprinting of plant samples are many and varied and have been mentioned in the earlier parts of this chapter. In some of the studies reviewed in this chapter, the main aim was the partitioning of breeding materials into possible heterotic groups to develop efficient breeding strategies. This was the case with the study of genetic variability in New Guinea *Impatiens* by Carr *et al.* (2003), in which the AFLP information generated was compared to that of the pedigree method. This was akin to the comparison of AFLP- and pedigree-based methods for genetic diversity studies using wheat cultivars by Barrett *et al.* (1998). Although the latter showed that the use of AFLP data led to reliable inferences, the former study was inconclusive because the pedigree information was limited. AFLP analysis of phenetic organization and genetic diversity has also been used to study domestication trends in crops such as cowpea (Coulibaly *et al.*, 2002). On the other hand, Tohme *et al.* (1996) had applied this technique in analyzing the gene pools of a wild bean collection. Information on genetic diversity usually based on ecotypic differences was also generated using AFLP assays for wild barley (Turpeinen *et al.*, 2003), alfalfa (Sergovia-Lerma *et al.*, 2003), *Brassica juncea* (Srivastava *et al.*, 2001), *Carica papaya* (Kim *et al.*, 2002), *Curcubita pepo* (Ferriol *et al.*, 2003), aromatic grapevines (Fanizza *et al.*, 2003), *Arabidopsis thaliana* (Erschadi *et al.*, 2000), plantain (Ude *et al.*, 2003), wheat (Barrett and Kidwell, 1998), faba beans (Zeid *et al.*, 2003), chicory (van Custem *et al.*, 2003), rhodesgrass (Ubi *et al.*, 2003), rapeseed (Lombard *et al.*, 2000), *Caladium bicolor* (Loh *et al.*, 1999), and sugar beet (de Riek *et al.*, 2001).

Based on these studies, it seems plausible that the methodologies for generating AFLP data are becoming standardized. Some researchers use custom reagents for the different steps of the AFLP assays (Barrett and Kidwell, 1998; Coulibaly *et al.*, 2002; Erschadi *et al.*, 2000; Fanizza *et al.*, 2003; Kim *et al.*, 2002; Turpeinen *et al.*, 2003). It would appear that those researchers that used custom reagents also used radiolabeled primers (Coulibaly *et al.*, 2002; Erschadi *et al.*, 2000; Fanizza *et al.*, 2003; Kim *et al.*, 2002). However, Turpeinen *et al.* (2003) used primers whose 5' ends had been fluorescent labeled (with 700- or 800-nm infrared dyes). Still, another group (Barrett and Kidwell, 1998) did not use labeled primers but opted for the silver staining protocol. There is no evidence that any of these techniques has any advantages in terms of the quality of the products visualized. For the majority (based on this survey) that opted for commercially available kits, Sergovia-Lerma *et al.* (2003), van Custem *et al.* (2003), and de Riek *et al.* (2001) used the kit from Perkin Elmer–Applied Biosystems (Foster City, CA) for fluorescent fragment reaction. On the other hand, Ferriol *et al.* (2003) used the AFLP Core Reagent Kit (Invitrogen). Still, in most of the surveyed reports, the AFLP kit from GIBCO–Life Technologies, Inc. (Rockville, MD; Carlsbad, CA; Gaithersburg, MD), had been used (Carr *et al.*, 2003; Loh *et al.*, 1999; Lombard *et al.*, 2000; Srivastava *et al.*, 2001; Ude *et al.*, 2003; Zeid *et al.*, 2003). Some of these kits contained already dye-labeled or radiolabeled primers, whereas some still were not labeled. The AFLP fingerprint profiles from the last group were visualized on silver stained polyacrylamide gels, whereas those for the radiolabeled primers were visualized after exposure to X-rays. The bands with high intensity and usually of more than 100 bp in size were then scored manually by eye and entered into a spreadsheet sample—by marker matrix. To reduce bias, some researchers had two persons independently confirm these scores (Erschadi *et al.*, 2000). Most of the researchers who adopted the fluorescent-labeled method separated the amplicons on the ABI series of automated DNA Sequencer from Perkin Elmer Applied Biosystems, whereas others still opted for the LI-COR series automated DNA Sequencer (Lincoln, NE). For the former, gel images were visualized using the GeneScan software of Perkin Elmer–Applied Biosystems. This software obtains the detection time, signal peak height, and size of the amplicons. Scoring of these bands was then subsequently performed using the Genotyper software of Perkin Elmer–Applied Biosystems. This creates a table showing the fragment sizes for each of the samples. For the LI-COR system, gel images were visualized by the program Base ImagIR (LI-COR, Inc., Lincoln, NE). Some lesser known band information extraction software also were used. Turpeinen *et al.* (2003) reported using the Cross Checker software, whereas another group (Ferriol *et al.*, 2003) reported introducing the GeneScan trace files to the Genographer software for scoring the bands. The AFLP-Quantar (KeyGene Product, Wageningen, The Netherlands) also seemed to be regularly used as a band scoring software. In some instances, it was necessary to visually confirm the scoring profile generated by the software (Kim *et al.*, 2002). It is, thus, plausible to conclude that just as there now exists quite some flexibility with the choice of bench analytical procedures, the researcher also has a wide array of options ranging from the fully automated to the manual to choose from in deciding how to manage AFLP data. In general, the data, irrespective of how they are generated, are usually readied for further statistical analyses by being entered into a spreadsheet in a binary matrix form. Bands are entered as either 1 (present) or 0 (absent) in a genotype-by-marker matrix.

AFLP Data Management

The strategies for gathering the desired information from a body of AFLP fingerprinting data are also becoming quite routine. The aim usually is to generate estimates of heterozygosity of the markers, levels of genetic diversity among the samples, structure of the genetic diversity, and the significance of the genetic diversity structure. Based on this information, valid inferences on the relationships between the samples are made. These are the basis for deciding the management practices for the germplasm. In breeding, this would guide the selection of parents for use in crossing programs.

Polymorphism Information Content

Once the data have been transformed into a binary matrix, the analyses usually commence with ascertaining the discriminating power of each marker through a determination of the polymorphism information content (PIC). Lombard *et al.* (2000), Ude *et al.* (2003), and de Riek *et al.* (2001) made such an assessment. PIC estimates the degree of polymorphism of a marker, which essentially is the proportion of individuals that are heterozygous for the marker. PIC is a good measure of heterozygosity in that it is an index of how many alleles a certain marker has and how these alleles divide. For instance, a marker that has few alleles or that has many alleles but with only one of them being frequent among the samples would have a low PIC and would, therefore, be of minimal value in assaying diversity. Such markers would then be excluded from further statistical analysis. In essence, the PIC value of a marker indicates its ability to distinguish between genotypes and is analogous to the estimate of gene diversity, a measure of heterozygosity best suited for inbred lines where there are not many heterozygotes (Weir, 1996).

Assessment of the Levels of Genetic Diversity

The genetic relationships among samples are elucidated through ascertaining the levels of genetic diversity in the population. The sample-by-marker matrix of character states generated from scoring the AFLP banding profiles is at this point converted into a sample-by-sample matrix of pairwise genetic distances (or similarities). There are several ways for estimating the genetic distance (or the converse, similarity) between any two samples on the basis of the differences (or similarities) between them in the marker states. Many workers use a combination of these in a given study. The most commonly used indices of genetic diversity include the Jaccard (1908) coefficient (J), Sokal and Michener's (1958) simple matching (SM) coefficient, and a modified Sokal and Michener's (MSM) coefficient, Nei's (1972) similarity index, Nei's (1973) genetic distance (D), and the Nei and Li (1979) (NL) coefficient of dissimilarity. Euclidean distances are also calculated as estimates of genetic diversity (Mielke, 1986).

The software used for these analyses were mainly NTSYS-PC (Rohlf, 1993, 2000); SAS (1999); PHYLIP (Felsenstein, 1989); R-Package (Legendre and Vaudor, 1991); and POPGENE (Yeh et al., 1997). These indices were used either singly or in combinations in AFLP surveys of plant genetic diversity (Barrett and Kidwell, 1998; Barrett et al., 1998; Carr et al., 2003; Coulibaly et al., 2002; de Riek et al., 2001; Erschadi et al., 2000; Fanizza et al., 2003; Ferriol et al., 2003; Kim et al., 2002; Loh et al., 1999; Lombard et al., 2000; Sergovia-Lerma et al., 2003; Srivastava et al., 2001; Turpeinen et al., 2003; Ubi et al., 2003; Ude et al., 2003; van Custem et al., 2003; Zeid et al., 2003). Also, van Custem et al. (2003) used the AFLP-SURV (Vekemans et al., 2002) software for estimating pairwise relatedness. Bootstrapping, usually with 1000 permuted data sets, is being used frequently to estimate the sampling variance of each pairwise distance (Efron and Tibshirani, 1991, 1993; Felsenstein, 1985). The accuracy of the estimated distance also may be further estimated by a computation of the mean coefficient of variation for each distance estimator (Lombard et al., 2000).

Description of the Genetic Structure

The structure of the relationships between the samples in the populations studied is usually elucidated through clustering methods. Dendrograms and principal component analysis were the methods used for assigning samples into clusters based on levels of sample-by-sample similarity estimates. These were commonly adopted in all the reports surveyed (see earlier sections). Dendrograms are normally constructed on the basis of the three distance matrices: the Ward (1963) method and the UPGMA cluster analysis method (Sokal and Michener, 1958). The robustness of the nodes in the clusters was also tested by a bootstrap method (1000 permuted runs, usually).

Testing the Significance of the Genetic Structure

To test the significance of the structure of genetic diversity, the AMOVA was carried out. This is a population genetics statistical tool (Excoffier, 2001; Excoffier *et al.*, 1992) that is based on the analysis of variance principle. In molecular assays, it is usually applied to the pairwise distance matrix to partition the sources of the observed variation and thereby ascertain the proportion of contribution to the total variation by the component parts. Virtually all the reports surveyed (see earlier sections) had adopted the AMOVA strategy for testing this significance.

Concluding Remarks

Information on the use of AFLP as a DNA fingerprinting technique under various circumstances is available in the literature. Given the trend of adoption and usage of this flexible and versatile technique, it will continue to be popular. It combines the ubiquity of endonuclease restriction sites throughout the genome (as in RFLPs) with the power and ease of use of the PCR (as in RAPDs). Also, so long as the genomes of many species of plants are not characterized to the extent that the development of locus-specific markers at a sufficient density level becomes feasible, researchers will continue to rely on AFLP profiling. This undeniably would remain the case for the so-called *orphan crops* such as cassava that constitute the major staples of many countries in Africa, South and Central America, and Asia. Sequence information on these crops will take a while in coming, and until such information is available to facilitate the development of locus-specific molecular markers, AFLP will remain the marker of choice.

The fact that the range of adaptation of the AFLP technology is literally limitless, varying from very basic facilities to highly sophisticated ones, ensures that virtually any lab that has PCR capacity can apply the technique. For instance, although more well-equipped labs could adapt the technique to more sophisticated processes (e.g., using radionuclide and fluorescent labeling and high throughput), there is no evidence that AFLP assays carried out under more rustic conditions (e.g., using silver staining) are any less reliable. Amplification products can be electrophoretically separated on automated DNA sequencers (capillary or gel slab systems) and on manual sequencing gel rigs. Finally, bands can be scored manually by eye or by the use of scanners and imaging software that are hooked up to analytical platforms and the end results would be comparable. Having said all this, the best guide to the choice of AFLP as a fingerprinting technique should derive from how this technique fits into the framework of the five key criteria of Karp *et al.* (1997). These are the nature of the questions being asked in the diversity study; the anticipated level of polymorphisms in the population being studied; the availability of probes or primers; issues relating to time constraints; and financial and other resource considerations.

Acknowledgment

We gratefully acknowledge the perspectives provided by Dr. Brian P. Fortser of the Scottish Crops Research Institute, Dundee, Scotland.

References

Barrett, B. A., and Kidwell, K. K. (1998). AFLP-based genetic diversity assessment among wheat cultivars from the Pacific Northwest. *Crop Sci.* **38**, 1261–1271.

Barrett, B. A., Kidwell, K. K., and Fox, P. N. (1998). Comparison of AFLP and pedigree-based genetic diversity assessment methods using wheat cultivars from the Pacific Northwest. *Crop Sci.* **38**, 1271–1278.

Blears, M. J., De Grandis, S. A., Lee, H., and Trevors, J. T. (1998). Amplified fragment length polymorphism (AFLP): A review of the procedure and its applications. *J. Industrial Microbiol. Biotech.* **21**, 99–114.

Carr, J., Xu, M., Dudley, J. W., and Korban, S. S. (2003). AFLP analysis of genetic variability in New Guinea impatiens. *Theor. Appl. Genet.* **106**, 1509–1516.

Coulibaly, S., Pasquet, R. S., Papa, R., and Gepts, P. (2002). AFLP analysis of the phenetic organization and genetic diversity of *Vigna* L. Walp. Reveals extensive gene flow between wild and domesticated types. *Theor. Appl. Genet.* **104**, 358–366.

de Riek, J., Calsyn, E., Everaert, I., van Bockstaele, E., and de Loose, M. (2001). AFLP based alternatives for the assessment of distinctness, uniformity and stability of sugar beet varieties. *Theor. Appl. Genet.* **103**, 1254–1256.

Edwards, K. J., and Mogg, R. (2002). Plant genotyping by analysis of single nucleotide polymorphisms. *In* "Plant Genotyping: The DNA Fingerprinting of Plants" (R. G. Henry, ed.), pp. 1–13. CABI Publishing, Oxon, UK; New York.

Efron, B., and Tibshirani, R. (1991). Statistical analysis with computer age. *Science* **253**, 390–395.

Efron, B., and Tibshirani, R. J. (1993). "An Introduction to the Bootstrap." Chapman and Hall, New York.

Erschadi, S., Haberer, G., Schöniger, M., and Torres-Ruiz, R. A. (2000). Estimating genetic diversity of *Arabidopsis thaliana* ecotypes with amplified fragment length polymorphisms (AFLP). *Theor. Appl. Genet.* **100**, 633–640.

Excoffier, L. (2001). Analysis of population subdivision. *In* "Handbook of Statistical Genetics" (D. J. Balding, M. Bishop, and C. Cannings, eds.), pp. 271–307. John Wiley & Sons, Chichester, UK.

Excoffier, L., Smouse, P. E., and Quattro, J. M. (1992). Analysis of molecular variance inferred from metric distances among DNA haplotypes: Application to human mitochondrial DNA restriction data. *Genetics* **131**, 479–491.

Fanizza, G., Chaabane, R., Lamaj, F., and Ricciardi, L. (2003). AFLP analysis of genetic relationships among aromatic grapevines (*Vitis vinifera*). *Theor. Appl. Genet.* **107**, 1043–1047.

Felsenstein, J. (1985). Confidence limits on phylogenies: An approach using the bootstrap. *Evolution* **39**, 783–791.

Felsenstein, J. (1989). PHYLIP—Phylogeny Inference Package (version 3.2). *Cladistics* **5**, 164–166.

Feinberg, A. P., and Vogelstein, B. (1983). A technique for radiolabeling DNA restriction endonuclease fragments to high specific activity. *Anal. Biochem.* **132**(1), 6–13.

Ferriol, M., Pico, B., and Nuez, F. (2003). Genetic diversity of a germplasm collection of *Curcubita pepo* using SRAP and AFLP markers. *Theor. Appl. Genet.* **107**, 271–282.

Ford-Lloyd, B. V. (2002). Genotyping in plant genetic resources. *In* "Plant Genotyping: The DNA Fingerprinting of Plants" (R. J. Henry, ed.), pp. 59–81. CABI Publishing, Oxon, UK; New York.

Harris, S. A. (1999). RAPDs in systematics—a useful methodology? *In* "Molecular Systematics, Plant and Evolution" (P. M. Hollingsworth, R. M. Bateman, and R. J. Gornall, eds.), pp. 221–228. Taylor and Francis, London.

Hill, M., Witsenboer, H., Zabeau, M., Vos, P., Kesseli, R., and Michelmore, R. (1996). PCR-based fingerprinting using AFLPs as a tool for studying genetic relationships in *Lactuca* spp. *Theor. Appl. Genet.* **93**, 1202–1210.

Jaccard, P. (1908). Nouvelles recherches sur la distribution florale. *Bull. Soc. Vaud. Sci. Nat.* **44**, 223–270.

Karp, A. (2002). The new genetic era: will it help us in managing genetic diversity? *In* "Managing Plant Genetic Diversity" (J. M. M. Engels, V. Ramanatha Rao, A. H. D. Brown, and M. T. Jackson, eds.), pp. 43–56. IPGRI, Rome, Italy.

Karp, A., Kresovich, S., Bhat, K. V., Ayad, W. G., and Hodgkin, T. (1997). Molecular Tools in Plant Genetic Resources Conservation: A Guide to the Technologies. IPGRI Technical Bulletin no. 2, p. 47. IPGRI, Rome, Italy.

Karp, A., Seberg, O., and Buiatti, M. (1996). Molecular techniques in the assessment of botanical diversity. *Ann. Bot.* **78**, 143–149.

Kim, M. S., Moore, P. H., Zee, F., Fitch, M. M. M., Steiger, D. L., Manshardt, R. M., Paull, R. E., Drew, R. A., Sekioka, T., and Ming, R. (2002). Genetic analysis of *Carica papaya* as revealed by AFLP markers. *Genome* **45**, 503–512.

Legendre, P., and Vaudor, A. (1991). "The R Package: Multidimensional Analysis, Spatial Analysis." Department de Sciences Biologiques, Universite de Montreal, Montreal, Canada.

Loh, J. P., Kiew, R., Kee, L., Gan, L. H., and Gan, Y. (1999). Amplified fragment polymorphism (AFLP) provides molecular markers for the identification of *Caladium bicolour* varieties. *Ann. Bot.* **84**, 155–161.

Lombard, V., Baril, C. P., Dubreuil, P., Blouet, F., and Zhang, D. (2000). Genetic relationships and fingerprinting of rapeseed cultivars by AFLP: Consequences for varietal registration. *Crop Sci.* **40**, 1417–1425.

Liu, Z., and Furnier, G. R. (1993). Comparison of allozyme, RFLP, and RAPD markers for revealing genetic variation within and between trembling aspen and bigtooth aspen. *Theor. Appl. Genet.* **87**, 97–105.

Lu, J., Knox, M. R., Ambrose, M. J., Brown, J. K. M., and Ellis, T. H. N. (1996). Comparative analysis of genetic diversity in peas assessed by RFLP and PCR-based methods. *Theor. Appl. Genet.* **93**, 1103–1111.

Lynch, M., and Milligan, B. G. (1994). Analysis of population genetic structure with RAPD markers. *Mol. Ecol.* **3**, 91–99.

Marshall, D. F. (1997). Meeting training needs in developing countries. *In* "Molecular Genetic Techniques for Plant Genetic Resources" (W. G. Ayad, T. Hodgkin, A. Jaradat, and V. R. Rao, eds.), pp. 128–132. IPGRI, Rome, Italy.

Mielke, P. W. (1986). Non-metric statistical analyses: Some metric alternatives. *J. Stat. Planning Inf.* **13**, 377–387.

Mueller, U. G., and Wolfenbarger, L. L. (1999). AFLP genotyping and fingerprinting. *Tree* **14**(10), 389–394.

Nei, M. (1972). Genetic distances between populations. *Am. Nat.* **106**, 283–292.

Nei, M. (1973). Analysis of gene diversity in subdivided populations. *Proc. Natl. Acad. Sci. USA* **70**, 3321–3323.

Nei, M., and Li, W. H. (1979). Mathematical models for studying genetic variation in terms of restriction endonucleases. *Proc. Natl. Acad. Sci. USA* **76**, 5269–5273.

Paul, S., Wachira, F. N., Powell, W., and Waught, R. (1997). Diversity and genetic differentiation among populations of Indian and Kenyan tea (*Camellia sinensis* [L.] O. Kuntze) revealed by AFLP markers. *Theor. Appl. Genet.* **94**, 255–263.

Qamaruz-Zaman, F., Fay, M. F., Parker, J. S., and Chase, M. W. (1997). The Use of AFLP Fingerprinting in Conservation Genetics: A Case Study of Orchis simia (Orchidaceae). Preliminary report. Royal Botanical Gardens, Kew.

Robinson, J. P., and Harris, S. A. (1999). Amplified fragment length polymorphisms and microsatellites: A phylogenetic perspective. *In* "Which DNA Marker for Which Purpose? Final Compendium of the Research Project Development, Optimisation and Validation of Molecular Tools for Assessment of Biodiversity in Forest Trees in the European Union DGXII Biotechnology FW IV Research Programme Molecular Tools for Biodiversity" (E. M. Gillet, ed.). http://webdoc.sub.gwdg.de/ebook/y/1999/whichmarker/index.htm.

Rohlf, F. J. (1993). NTSYS-pc numerical taxonomy and multivariate analysis system. Exeter Software, Setauket, NY.

Rohlf, F. J. (2000). NTSYS-pc: Numerical taxonomy and multivariate analysis system, version 2.1. Exeter Software: State University of New York, Setauket, NY.

SAS Institute (1999). The SAS System for Windows, version 7.0. SAS Institute, Inc., NC.

Schafer, A. J., and Hawkins, J. R. (1998). DNA variation and the future of human genetics. *Nat. Biotechnol.* **16**, 33–39.

Sergovia-Lerma, A., Cantrell, R. G., Conway, J. M., and Ray, I. M. (2003). AFLP-based assessment of genetic diversity among nine alfalfa germplasms using bulk DNA templates. *Genome* **46**, 51–58.

Sharma, S. K., Knox, M. R., and Ellis, T. H. N. (1996). AFLP analysis of the diversity and phylogeny of *Lens* and its comparison with RAPD analysis. *Theor. Appl. Genet.* **93**, 751–758.

Sokal, R. R., and Michener, C. D. (1958). A statistical method for evaluating systematic relationships. *Univ. Kansas Sci. Bull.* **38**, 1409–1438.

Srivastava, A., Gupta, V., Pental, D., and Pradhan, A. K. (2001). AFLP-based egentic diversity assessment amongst agronomically important natural and some newly synthesized lines of *Brassica juncea*. *Theor. Appl. Genet.* **102**, 193–199.

Tautz, D., and Rentz, M. (1984). Simple sequences are ubiquitous repetitive elements of eukaryote genomes. *Nucleic Acids Res.* **12**, 4127–4138.

Tohme, J., Gonzalez, D. O., Beebe, S., and Duque, M. C. (1996). AFLP analysis of gene pools of a wild bean core collection. *Crop Sci.* **36**, 1375–1384.

Travis, S. E., Maschinski, J., and Keim, P. (1996). An analysis of genetic variation in *Astragalus cremnophylax* var. *cremnophylax*, a critically endangered plant, using AFLP markers. *Mol. Ecol.* **5**, 735–745.

Turpeinen, T., Vanhala, T., Nevo, E., and Nissila, E. (2003). AFLP genetic polymorphism in wild barley (*Hordeum spontaneum*) population in Israel. *Theor. Appl. Genet.* **106**, 1333–1339.

Ubi, B. E., Kolliker, R., Fujimori, M., and Komatsu, T. (2003). Genetic diversity in diploid cultivars of Rhodes grass determined on the basis of amplified fragment length polymorphism markers. *Crop Sci.* **43**(4), 1516–1522.

Ude, G., Pillay, M., Ogundiwin, E., and Tenkouano, A. (2003). Genetic diversity in an African core collection using AFLP and RAPD markers. *Theor. Appl. Genet.* **107**, 248–255.

van Custem, P., du Jardin, P., Boutte, C., Beauwens, T., Jacqmin, S., and Vekemans, X. (2003). Distinction between cultivated and wild chicory gene pools using AFLP markers. *Theor. Appl. Genet.* **107**, 713–718.

Vekemans, X., Beauwens, T., Lemaire, M., and Roldan-Ruiz, I. (2002). Data from amplified fragment length polymorphism (AFLP) markers show indication of size homoplasy and of a relationship between degree of homoplasy and fragment size. *Mol. Ecol.* **11**, 139–151.

Vos, P., Hogers, R., Bleeker, M., Rijans, M., Van de Lee, T., Hornes, M., Frijters, A., Pot, J., Kuiper, M., and Zabeau, M. (1995). AFLP: A new technique for DNA fingerprinting. *Nucleic Acids Res.* **23**, 4407–4414.

Ward, J. H. (1963). Hierarchical grouping to optimize an objective function. *Am. Static. Assoc. J.* **56**, 236–244.

Weir, B. (1996). Genetic Data Analysis II. Methods for Discrete Population Genetic Data, p. 445. Sinauer Associates, Inc., Sundernland, MA.

Williams, J. G. K., Kubelik, A. R., Livak, K. J., Rafalski, J. A., and Tingey, S. V. (1990). DNA polymorphisms amplified by arbitrary primers are useful as genetic markers. *Nucleic Acids Res.* **18**, 6531–6535.

Yeh, F. C., Yang, R.-C., Boyle, T. J. B., Ye, Z.-H., and Mao, J. X. (1997). "POPGENE, The User-Friendly Shareware for Population Genetic Analysis." Molecular Biology and Biotechnology Centre, University of Alberta, Edmonton, Canada.

Zabeau, M., and Vos, P. (1993). Selective restriction fragment amplification: A general method for DNA fingerprinting. European patent publication 92402629 (publication no. EP0534858A1).

Zeid, M., Schön, C., and Link, W. (2003). Genetic diversity in recent elite faba bean lines using AFLP markers. *Theor. Appl. Genet.* **107**, 1304–1314.

[13] Isolating Microsatellite DNA Loci

By Travis C. Glenn and Nancy A. Schable

Abstract

A series of techniques are presented to construct genomic DNA libraries highly enriched for microsatellite DNA loci. The individual techniques used here derive from several published protocols but have been optimized and tested in our research laboratories as well as in classroom settings at the University of South Carolina and University of Georgia, with students achieving nearly 100% success. Reducing the number of manipulations involved has been a key to success, decreasing both the failure rate and the time necessary to isolate loci of interest. In our lab during the past 3 years alone, these protocols have been successfully used to isolate microsatellite DNA loci from at least 55 species representing three kingdoms. These protocols have made it possible to reduce the time to identify candidate loci for primer development from most eukaryotic species to as little as 1 week.

Introduction

Microsatellite DNA loci have become important sources of genetic information for a variety of purposes (Goldstein and Schlotterer, 1999 see Chapter 14). To amplify microsatellite loci by polymerase chain reaction (PCR), primers must be developed from the DNA that flanks specific microsatellite repeats. These regions of DNA are among the most variable in the genome, thus primer-binding sites are not well conserved among distantly related species (Moore *et al.*, 1991; Pepin *et al.*, 1995; Primmer *et al.*, 1996; Zhu *et al.*, 2000). Although microsatellite loci have now been developed for hundreds of species (indeed the journal *Molecular Ecology Notes* is largely devoted to their description), these loci have not been isolated from many additional species of interest and remain to be developed.

Many strategies for obtaining microsatellite DNA loci have been described. The simplest approach, cloning small genomic fragments and using radiolabeled oligonucleotide probes of microsatellite repeats to identify clones with microsatellites, was the first described and works well in organisms with abundant microsatellite loci (Tautz, 1989; Weber and May, 1989; Weissenbach *et al.*, 1992). Unfortunately, this approach does not work well when microsatellite repeats are less abundant. Thus, two classes of enrichment strategies have been developed: (1) uracil-DNA selection

(Ostrander *et al.*, 1992) and (2) hybridization capture (Armour *et al.*, 1994; Kandpal *et al.*, 1994; Kijas *et al.*, 1994). Hybridization capture is the predominant strategy in use because it allows selection before cloning and, thus, is faster and easier to do with multiple samples than uracil-DNA selection, which requires passage of each library through two bacterial strains.

We have refined the hybridization capture approach described by Hamilton *et al.* (1999), which derives from Armour *et al.* (1994), Fisher and Bachmann (1998), Kandpal *et al.* (1994), Kijas *et al.* (1994), and others (Zane *et al.*, 2002). In theory, this protocol will work for any eukaryotic organism (i.e., anything with an appreciable number of microsatellite loci) or any other piece of DNA that may be captured using an oligonucleotide. The biggest differences among DNA samples (i.e., species from which the DNA derives) are (1) how the initial DNA sample is isolated and (2) which microsatellite repeats occur most frequently in a particular organism and, thus, are targeted for enrichment and isolation. In practice, we and our collaborators have used this protocol to construct libraries and determine flanking sequences of microsatellite DNA loci in amphibians, birds, fish, mammals, reptiles, insects, nematodes, and various other invertebrates, fungi, plants, and coral. The protocol has been outlined in several publications (Hauswaldt and Glenn, 2003; Korfanta *et al.*, 2002; Prince *et al.*, 2002; Schable *et al.*, 2002) but has many fine points that are not likely to be obvious from those publications.

The most unique feature in this protocol is the incorporation of a GTTT "pig tail" on the SuperSNX linker and modification of the originally described SNX sequence. The SuperSNX linker has PCR characteristics even better than the SNX linker in the study by Hamilton *et al.* (1999) and ensures efficient A-tailing of each PCR product yielding good results from TA cloning. Interestingly and most importantly, amplification of DNA using the SuperSNX linker/primer is biased against producing small PCR products. Thus, PCR products obtained after enrichment can be cloned directly without obtaining a large proportion of small DNA fragments. Most of the other details in this protocol have been reported previously or have been generously provided by colleagues; we have simply compiled the best specific approaches from many protocols to reduce the time and steps required to isolate microsatellite DNA loci.

Step I: Extracting DNA

Goal: To Isolate about 10 μg of high-molecular-weight DNA (ideally 50^+ μl of 100^+ ng/μl). About 2–3 μg of good DNA will suffice and considerably less can be used, but it is a good idea to have much more than minimal amounts of DNA available. For most organisms, it is best to perform a PCI extraction, followed by an ethanol precipitation

(Sambrook *et al.*, 1989). Many people also have very good success with silica-based protocols such as DNeasy kits (Qiagen, Valencia, CA), Wizard Preps (Promega Corp., Madison, WI), or homemade equivalents (http:// www.uga.edu/srel/DNA_Lab/MUD_DNA'00.rtf_.rtf). The only caution offered regarding the use of such protocols is that the recovered concentration of DNA is usually 50–100 ng/μl and is often less than that. Thus, you may need to do an ethanol precipitation of the DNA recovered from a Qiagen kit (or comparable kits or method used) to increase the concentration of DNA. It is also best to destroy the RNA by performing the "optional" RNase treatment during the DNA extraction.

Detailed steps: It is imperative to check the concentration and quality of the DNA before proceeding.

1. Quantify the DNA concentration and examine its quality by diluting 2 μl of DNA with 3 μL of TLE (10 m*M* Tris, pH 8.0, 0.2 m*M* EDTA), 2 μl of loading buffer (Sambrook and Russell, 2001; Sambrook *et al.*, 1989), and loading onto a 1% agarose gel containing ethidium bromide. Use 50 and 200 ng of uncut lambda DNA as standards. DNA quantity must be at least as bright as the 50-ng band and ideally as bright as the 200-ng band of lambda DNA. DNA quality is assessed by the absence (high quality) or presence (lower quality) of a smear down the gel when compared to lambda DNA. Any remaining RNA will also appear as a smear much smaller than the lambda DNA.

Choosing DNA for Marker Development

It is best to use DNA of the highest quality that can be reasonably obtained. In practice, most projects start with fewer than 10 DNA samples, and the best 2 among those are used. A small to modest amount of DNA smearing down below 5000 base pairs (bp) is generally fine. If a substantial proportion of the DNA is less than 5000 bp (especially if <2000 bp), then it will be worthwhile to do additional DNA extractions to obtain higher quality DNA.

There are some advantages and some disadvantages to mixing DNA from multiple individuals. In general, we recommend using DNA from one individual (heterogametic sex if there is an interest in possibly obtaining a sex-specific marker) of one species. The use of markers from any one individual may create an ascertainment bias (especially when the markers developed are applied to other species), but at least the researcher will be aware of the potential bias. If a mixture of DNA from different individuals is used, then it may never be known from which individual any particular clone (locus) came. Subsequently, it may be difficult or impossible to resolve the source of problems that may be encountered further in the development process as a result of using multiple DNAs. If one would like to use the markers for multiple species for species that are closely related, then development of one library will likely be sufficient (all else being equal, it may be best to choose the basal species). Another approach is to develop loci from two species—ideally the two least related (i.e., most distant phylogenetically).

Step II: Restriction Enzyme Digest

Goal: To fragment the DNA into approximately 500 bp fragments. After several steps, these fragments will be inserted into a plasmid and then bacteria. Fragments of this size are small enough to sequence easily while retaining a high probability of having enough DNA flanking the microsatellites that primers can be designed. Restriction enzymes are an easy way to fragment the DNA. Restriction enzymes recognize specific sequences and will cut the DNA at this site, leaving a known end that will prove helpful later in this protocol. The following restriction enzymes have been used: *Rsa*I and *Bst*UI. These restriction enzymes can be purchased through New England BioLabs (NEB). To learn more about these or other enzymes, NEB has an informative web site (http://www.neb.com). Any frequent cutting restriction enzyme that leaves a blunt end could be used, although it is best if they are heat labile and work in NEB buffer no. 2.

It is best to begin by setting up a digest on two DNA samples using *Rsa*I. If the resulting smear is not continuous with most of the DNA ranging from approximately 300 to 1000 bp, then it is wise to attempt another digest on uncut DNA using *Bst*UI.

Recipe

*Rsa*I or *Bst*UI[1]

2.50 μl NEB 10× ligase buffer (*note:* heat to 50° or 65° to get all components in solution)

---

[1] *Rsa*I recognizes GT^AC and *Bst*UI recognizes CG^CG; so one may work better than another in any particular organism. *Hae*III (GG^CC) could also be used, but it has a recognition site in SuperSNX. Linker ligations are still generally successful on DNAs digested using *Hae*III because only a small proportion of the linker is digested in the subsequent linker ligation reaction. Using *Hae*III is less than optimal, so its use is not recommended unless other restriction enzymes have failed to yield DNAs cut to an average of about 500 bp in length. If these enzymes result in fragments that are too small, 6-base cutting alternatives include the following: *Eco*R V (GAT^ATC; only 75% efficient in this buffer but still adequate; (*Ssp*I) (AAT^ATT), *Stu*I (AGG^CCT), and *Sfo*I (GGC^GCC).

0.25 µl 100× bovine serum albumin (BSA) (BSA supplied with enzymes from NEB)

0.25 µl 5 M NaCl (50 mM final)

1.00 µl RsaI (NEB catalog no. R0167S) or BstUI (NEB catalog no. R0518S)

1.00 µl XmnI (NEB catalog no. R0194S; *note:* XmnI can be added at step III.2, later in this chapter instead)

20.0 µl[2] genomic DNA (100 ng/ µL).

If plenty of DNA is available and saving time is important, then one may perform digests with each enzyme (in separate tubes) simultaneously. Because some enzymes may give biased results, it is potentially helpful to combine ligations from multiple enzymes. Note that it is unwise to cut the DNA with multiple 4-base cutting enzymes at once or to combine the DNA until after the linker ligation. If digests are combined before the linkers are ligated, then one will not be able to determine whether multiple unrelated DNA fragments have been joined (i.e., ligated into chimeras), potentially resulting in unamplifiable loci.

Detailed steps:

1. To make master mixes for *Rsa* I and *Bst*U I, multiply the volume of each of the components in the preceding list by the number of DNA samples to be digested, plus half a sample to account for pipetting error and add to a 1.5-ml tube.

2. Prepare the restriction enzyme digest for *Rsa* I and *Bst*U I by adding 5 µl of master mix into a new tube (0.2 or 0.5 ml depending on thermal cycler available) and use a thermal cycler for all incubations. Add 20 µl of DNA to each tube. Pipette up and down to mix the solution.

3. Incubate all samples (*Rsa* I and *Bst*U I) at 37° for 30–60 min.

4. While the restriction digest is incubating, pour a 1% agarose gel, including ethidium bromide (Sambrook *et al.*, 1989).

5. Set aside a small aliquot (4 µl) of the digested DNA.

6. Immediately proceed to step III.

Note: You will run the aliquot of the restriction enzyme digest on a 1% agarose gel to verify that the restriction enzyme digest was successful at step III.4 below.

---

[2] Assumes a DNA concentration of about 100 ng/µl (i.e., ~2 µg of DNA). Adjust accordingly if the DNA is significantly more concentrated (i.e., if >200 ng/µl, use less and make up the volume in water). This recipe may still be used if less than 20 µl of 100 ng/µl of DNA is available, but it may be necessary to amplify the DNA with the SuperSNX24 primer before enrichment (especially if the amount of DNA available is <100 ng). It is important to note that amplifying the linker ligated DNA before enrichment may bias the enrichment results.

Step III: Ligating Linkers to DNA Fragments

Goal: To ligate a double-stranded linker onto both ends of each DNA fragment. The linkers will provide the primer-binding site for subsequent PCR steps. They also provide sites to ease cloning of the fragments into the vectors that will subsequently be used. The linkers are, therefore, compatible with the restriction sites in the vector's multiple cloning site. The SuperSNX also incorporates a GTTT "pigtail" to facilitate nontemplate A addition by *Taq* DNA polymerase during PCR, which can be used for TA cloning.

Note: This protocol is written with the assumption that there is only a need for two enrichments per linker-ligated DNA. There is enough linker ligation for three enrichments using the recipes below. If one plans to perform more than three enrichments, it is important to scale up reaction volumes, set up additional linker ligations, decrease the amount of linker-ligated DNA used in step IV or use PCR-amplified linker-ligated DNA (from step III.6).

SuperSNX24 Forward: 5'GTTTAAGGCCTAGCTAGCAGAATC

SuperSNX24 + 4P Reverse: 5'pGATTCTGCTAGCTAGGCCTTAAA CAAAA

```
The most stable 3'-dimer: 24 bp, -43.6 kcal/mol

   5' GTTTAAGGCCTAGCTAGCAGAATC 3'
      ||||||||||||||||||||||||
3' AAAACAAATTCCGGATCGATCGTCTTAG 5'
```

Note: The phosphate (p) on SuperSNX24 + 4P Reverse allows ligation of the linkers to each other or the digested DNA, but it is not shown in the Fig.

Detailed steps:

1. Preparation of double-stranded (ds) SuperSNX linkers:
   Mix equal volumes of equal molar amounts of SuperSNX24 and SuperSNX24 + 4p primers (e.g., 100 μl of 10 μM each). Add salt to a final concentration of 100 mM (i.e., 4 μl of 5 M NaCl for 200 μl of primers). Heat this mixture to 95°, and let it cool slowly to room temperature to form the ds SuperSNX linkers.

2. Linker-ligation recipe: (*note:* add Xmn I if it was not added above]:
   7.0 μl ds SuperSNX linkers
   1.0 μl 10× ligase buffer (ensure components are in solution—warm if necessary)
   2.0 μl DNA ligase (NEB #M0202S; 400 units/μl)
   _____
   10.0 μl total

If multiple DNAs are being ligated, make a master mix of the components listed above the line and add them (10 μl of mix) to the cut DNAs.

3. Incubate at room temperature for 2 or more h or ideally at 16° overnight.

4. While the ligation is proceeding, run the small aliquots of restriction enzyme–digested DNA (from step II.5) on the 1% minigel (from step II.4) to ensure the DNA samples were successfully digested. A successful reaction should yield a smear of fragments centered at approximately 500 bp, but at least with most fragments at or below 1000 bp.

5. To ensure ligation was successful, perform a PCR on the linker ligation using the following recipe for a 25-μl reaction (a 50-μl reactions volume should be used when enrichment will be performed using the PCR products):

2.5 μl 10× PCR buffer (optimal buffer for *Taq* used below)

2.5 μl BSA (250 μg/ml → 25 μg/ml final)

1.3 μl SuperSNX24 (10 μM → 0.5 μM final)

1.5 μl deoxyribonucleic triphosphates (dNTPs) (2.5 mM each → 150 μM final)

2.0 μl MgCl$_2$ (25 mM → 2.0 mM final)

13.0 μl dH$_2$O

0.2 μl *Taq* DNA polymerase (5 units/μl)

2.0 μl linker-ligated DNA fragments

If multiple DNAs are being tested, make a master mix of the components above the solid line and add them (23 μl of mix) to the linker-ligated DNAs (2 μl).

Note: Only one primer is used, see note V. 1.

Cycling: 95° for 2 min; then 20 cycles at 95° for 20 s, 60° for 20 s, 72° for 1.5 min. Hold at 15°.

Note: The same program as the enrichment recovery (step V) may be used, but it takes longer.

6. Run 4 μl of PCR product on a 1.0% minigel to see if the linker ligation was successful using a 100-bp ladder as a size standard. A successful reaction should yield a smear of fragments centered at approximately 500 bp. This PCR product can be used for enrichment if insufficient amounts of original linker-ligated DNA are available.

Notes:

- All restriction enzymes must be kept on ice until use and immediately placed back in the −20° freezer after use.
- Three to five times the amount of linkers relative to each fragment increases the odds that the linkers will ligate to a DNA fragment instead of the latter to each other.

- *Conversion factors:* 1 μg of 1-kb fragments = 3.3 pmol fragment ends. Also, 1 μl of X μM linkers = X pmol of linker ends.
- The *Xmn I* prevents the dimerization (self-ligation) of linkers, *so it is vital for success.*

Step IV: Dynabead Enrichment for Microsatellite-Containing DNA Fragments

Goal: To capture DNA fragments with microsatellite sequences complementary to the microsatellite oligos (probes) and wash away all other DNA fragments.

Note: This protocol is written with the assumption that there is a need for only two enrichments per linker-ligated DNA (see note III.1 above). If one plans to perform serial (double) enrichments, it is a good idea to set up replicate enrichments.

Materials and Solutions

Washed Dynabeads (see step 7—wash twice in TE [10 mM Tris pH 8.0, 2 mM EDTA] and twice in 1× Hyb solution) *note:* each 50 μl of Dynabeads (Dynal, Oslo, Norway) can capture 100 pmol of biotinylated oligo. It is critical to have an excess of bead capacity relative to the amount of biotin/oligo added. If beads from other manufacturers are used, the amount of beads should be adjusted to account for variation in biotin-binding capacity.

2× Hyb Solution: 12× SSC, 0.2% SDS (warmed; stock solution 20× SSC: 3.0 M NaCl, 0.3 M sodium citrate, pH 7.0).

1× Hyb Solution: 6× SSC, 0.1% SDS (warmed to get everything into solution).

Washing Solutions: 2× SSC, 0.1% SDS (warmed to get everything into solution), 1× SSC, 0.1% SDS (warmed to get everything into solution).

Biotinylated oligos: Mixtures of 3' biotinylated oligos are used with this protocol (see http://www.uga.edu/srel/Msat_Devmt/Probe_List.htm) (note there are underscores [not spaces] in that web address). 3' labeling is used because it has the highest efficiency of labeling (each oligo synthesis starts with a biotin). A large number of oligos may be used in a mix together when their lengths are varied to achieve similar melting temperatures (T_ms). We use oligos purified by standard desalting methods (i.e., no additional purification by high-performance liquid chromatography [HPLC], gels, etc.), because we order large numbers of oligos and the additional purification would be quite expensive. The critical factor to keep in mind when using biotinylated oligos purified by standard desalting methods is that the solution will contain many "free" biotins, so it is critical to ensure the amount of biotin (estimated from the oligo concentration) added is *far* less than the bead-binding capacity.

NaOAc EDTA Solution: To a 50-ml conical, make 20 ml of 3 *M* NaOAc from the dry chemical stock. Do not adjust the pH. Add 20 ml of 500 m*M* EDTA, pH 8.0. This makes a solution that is 1.5 *M* NaOAc and 250 m*M* EDTA. Aliquot into 1.5-ml microcentrifuge tubes and/or 0.2-ml strip tubes and freeze.

Detailed Steps

1. In a 0.2-ml PCR tube, add:

25.0 μl 2× Hyb solution (warmed to get everything into solution)
10.0 μl biotinylated microsatellite probe (mix of oligos at 1 μM each)
10.0 μl linker-ligated DNA from step III (or PCR product if <2 μg DNA initially used)
5.0 μl dH$_2$O

50.0 μl Total

2. Use thermal cycler program *OligoHyb*. This program denatures the DNA–probe mixture at 95° for 5 min. It then quickly ramps to 70° and steps down 0.2° every 5 s for 99 cycles (i.e., 70° for 5 s, 69.8° for 5 s, 69.6° for 5 s, ... down to 50.2°), and stays at 50° for 10 min. It then ramps down 0.5° every 5 s for 20 cycles (i.e., 50° for 5 s, 49.5° for 5 s, 49° for 5 s, ... down to 40°), and finally quickly ramps down to 15°. The idea is to denature everything, quickly go to a temperature slightly above the annealing temperatures of the oligos in the mixes used, and then slowly decrease, allowing the oligos the opportunity to hybridize with DNA fragments that they most closely match (i.e., hopefully, long perfect repeats) when the solution is at or near the oligo's T$_m$.

3. While the DNA–probe mixture is in the thermal cycler, wash 50 μl of Dynabeads (Dynal, Oslo, Norway). Resuspend the beads in their original tube, and transfer to a 1.5-ml tube. Add 250 μl of TE. Shake. Capture beads using the Magnetic Particle Concentrator (MPC) (Dynal, Oslo, Norway). Repeat with TE, and twice with 1× Hyb solution. Resuspend the final beads in 150 μl of 1× Hyb solution.

4. Pulse-spin your DNA–probe mix and add all of it to the 150 μl of washed, resuspended Dynabeads (i.e., to the 1.5-ml tube).

5. Incubate on rotator or sideways in orbital shaker on slow speed at room temperature for 30 or more min.

6. Capture beads using the MPC. Remove the supernatant by pipetting with a P200 pipetter (*Optional:* Save supernatant for troubleshooting purposes).

7. Wash the Dynabeads two times with 400 μl 2× SSC, 0.1% SDS each time using the MPC to collect the beads and removing the supernatant by pipetting with a P200 pipetter (which can be saved for troubleshooting purposes). Resuspend beads well (i.e., flick or gently vortex) in next wash each time.

8. Wash two additional times using 400 μl 1× SSC, 0.1% SDS.

9. Wash two final times using 400 μl 1× SSC, 0.1% SDS, and heating the solution to within 5–10° of the T_m for the oligo mix used (usually 45 or 50°). *Note:* More stringent washes may increase the relative proportion of long microsatellite sequences in cloned colonies, but they may also cause the loss of microsatellite sequences via undesired elution during washing.

10. Add 200 μl TLE, vortex, and incubate at 95° for 5 min. Label a new tube while incubating. Capture beads using the MPC. Quickly remove the supernatant by pipetting to the new tube. This supernatant contains the enriched fragments (i.e., "the gold").

Note: It is important to remove the supernatant from the beads reasonably quickly after removing from the 95° heat block. It is not unusual for the supernatant to have slight discoloration from the beads (appears that the magnet is not working well). A *very small* amount of discoloration (leading to a colored pellet following precipitation) does not seem to be harmful.

11. Add 22 μl of NaOAc/EDTA solution (see "Recipe" above). Mix by pipetting up and down.

12. Add 444 μl of 95% EtOH. Mix by inverting the tube and place on ice for 15 min or more (or store in the −20° freezer for as long as desired).

13. Centrifuge at full speed for 10 min.

14. Discard supernatant and add approximately 0.5 ml of 70% EtOH. Centrifuge for 1 min.

15. Carefully pipette off *all* the supernatant and air-dry the sample. If there is any visible trace of EtOH, pulse-spin the tube and use a pipette to remove any residual EtOH. Dry until there is no trace (smell) of EtOH.

16. Resuspend the pellet in 25 μl of TLE. This is the "pure gold." Let the pellet hydrate while setting up PCRs in step V (at least 20 min). It may be best to allow for overnight rehydration to be sure that the DNA is in solution. Inadequate rehydration is the most common reason for failure of the next step.

Step V: PCR Recovery of Enriched DNA

Goal: To increase the amount of "pure gold" DNA. To do serial (double) enrichments, use the resulting PCR products (step V.2) for the second enrichment.

Detailed steps:

1. Perform PCR on supernatant (step IV.16) to recover the enriched DNA fragments:

2.5 μl 10× PCR buffer (optimal buffer for *Taq* used below)
2.5 μl BSA (250 μg/ml → 25 μg/ml final)
1.5 μl dNTPs (2.5 mM each → 150 μM final)
1.3 μl SuperSNX-24 (10μM → 0.5 μM final)
2.0 μl MgCl$_2$ (25 mM → 2.0 mM final)
13.0 μl dH$_2$O
0.2 μl *Taq* DNA polymerase (5 units/μl)

2.0 μl eluted DNA fragments ("pure gold") (*note:* Ensure gold pellet has hydrated for 20 min or longer—longer is better; lots of mixing ensures the pellet becomes rehydrated and the PCR will be successful)

If multiple DNAs are being tested, make a master mix of the components above the solid line and add them (23 μl of mix) to the eluted DNAs (2 μl). It is often wise to perform a second PCR, using half as much eluted DNA (i.e., 1.0 μl of eluted DNA + 1.0 μl of dH$_2$O).

Cycling: 95° for 2 min; then 25 cycles of 95° for 20 s, 60° for 20 s, 72° for 1.5 min; then 72° for 30 min; then hold at 15°.

Note: It is correct that only one primer is used (the SuperSNX24 Forward primer). If both forward and reverse primers are used, then the reaction will fail. If one draws an example of the linker-ligated DNA in double-stranded form, it is easier to visualize why using only one primer works.

5′ForwardPrimer::DNAofInterest::ReversePrimer3′
3′ReversePrimer::DNAofInterest::ForwardPrimer5′

2. Run 4 μl of PCR product on a 1.0% minigel next to a 100-bp ladder as a standard to verify whether DNA recovery was successful. *Note:* The smear of fragments should be visible, centered at approximately 500 bp. If bands are visible, it is likely few microsatellite loci are present; setting up multiple PCRs may increase the chances that one of the reactions will not have defined bands. If setting up multiple PCRs fails to yield bandless smears of product, enrich using a different mix of oligos.

Optional: For troubleshooting and to verify enrichment success, use 2 μl of PCR product for a dot-blot analysis (see step XII of *Msat_Easy_Isolation _2000.rtf* available at http://www.uga.edu/srel/DNA_Lab/protocols.htm)

to ensure that microsatellite containing fragments have been recovered. Use genomic DNA and PCR product from the linker-ligation check as rough controls (the enriched PCR product resulting dot should be much darker than either the genomic DNA or the linker-ligation check PCR dots). For the best comparison, one should add an equal number of nanograms of DNA from the enriched PCR and the linker-ligation check PCR. Diluting each PCR product a few times (e.g., five, ten, or fifty times) will most likely enhance the ability to see differences among dots. The leftover linker-ligated DNA includes fragments that are too long to be recovered by PCR (thus, biasing the comparison somewhat). If the linker-ligation check PCR dot is not very dark and the linker-ligated DNA is very dark, it may be wise to use a different mix of microsatellite oligos or to use a different restriction enzyme for the linker ligation.

3. Make S-Gal Amp or LB Amp plates in preparation for transformation (transformation step 2 below). Four or more S-Gal Amp or LB Amp (50–100 μg/ml) bacterial plates will be needed for each successfully enriched PCR. Follow the protocol from Sigma for S-Gal, in the Invitrogen Topo-TA manual, or Sambrook *et al.* (1989).

Step VI: Ligating-Enriched DNA into Plasmids

Goal: To incorporate (ligate) the enriched/ recovered DNA (amplified pure gold) into a cloning vector. The idea is to place one fragment of the DNA into one vector and to do this for as many fragments as possible. Once ligated into the vector, the DNA is known as an *insert*.

This protocol assumes that it is best to use the fastest and most reliable method available. In our experience and the experience of our colleagues, the TA cloning kits from Invitrogen (Carlsbad, CA) are the quickest and most robust. It is important to point out, however, that there is nothing wrong with using TA cloning kits from other vendors (e.g., Promega, Madison, WI) or homemade preps (Holton and Graham, 1991; Marchuk *et al.*, 1991). Both the TOPO TA Cloning Kit containing pCR 2.1-TOPO with TOP 10 cells (catalog K4500-40) as well as the kit with TOP 10F' cells (catalog K4550-40) have successfully been used with this protocol. In general, the former option is recommended because it does not require the use of IPTG. If only a small number of ligations and transformations will be performed, one can purchase the 20 reaction kit (K4500-01), rather than the 40 reaction kit.

Detailed steps:

1. Follow the directions supplied with Invitrogen's TOPO-TA cloning kit *exactly*! If another TA cloning protocol is used, follow the appropriate directions for ligation and transformation.

2. Warm S-Gal/LB Agar (Sigma) Amp plates before starting transformation (LB Amp plates spread with X-Gal and IPTG may be substituted).

Note: It is important to know that the restriction sites in the SuperSNX linker can be used for sticky-end cloning, exactly as described by Hamilton *et al.* (1999). If many enrichments will be performed, it may be wise to invest the time to use that approach, which is a superior method in many ways. The major disadvantage is that the enriched DNA must be cleaned after digestion of the linker end, which requires additional time and more steps (increasing the likelihood of the product getting lost or having other "handling tragedies," especially in the hands of inexperienced workers). In addition, some DNA fragments will contain *Nhe*I restriction sites, and thus, the number of inserts with no microsatellites and those with little flanking DNA will be increased (this can be countered to some extent by adding *Nhe*I to the original restriction enzyme digest—step II).

Step VII: Transforming Plasmid DNA

Goal: To incorporate your enriched/recovered DNA (amplified pure gold; or insert) + cloning vector into a bacterial host. The idea is to place one vector (which, ideally, has one fragment of amplified pure gold [insert]) into one bacterial host, and do this for as many vectors–inserts as possible. Usually ampicillin (*amp*)-sensitive bacteria and a vector that carries a gene conferring *amp* resistance are used. When a bacterium incorporates the vector, the vector transforms the phenotype of the bacterium from *amp* sensitive to *amp* resistant. Thus, when a mixture of bacteria is plated on media containing *amp*, only bacteria with *amp* resistance (i.e., those that have incorporated the vector) can grow and form colonies. *Note:* This step continues with the assumption that the TA cloning kits from Invitrogen are used. Other standard transformation protocols are available (Sambrook and Russell, 2001; Sambrook *et al.*, 1989).
Detailed steps:

1. Follow Invitrogen's TOPO-TA cloning kit *exactly*!
2. Following the 1-h incubation in SOC, plate out 25 μl of transformed bacteria onto two plates and 50 μl of transformed bacteria onto two other plates. This will ensure that plenty of colonies will be present, but that they are not growing on top of each other. This will probably yield enough colonies, but it is reasonable to plate out the entire amount, or one may save the remaining transformed bacteria in broth (at 4°) to be plated the next day (waiting to plate is particularly valuable if one is unsure of the quality of the plates or is trying to minimize the number of plates used;

the number of colonies obtained per microliter will, however, be reduced by plating at a later time).

3. Grow colonies overnight at 37°.

Step VIII: PCR and Storing Positive Colonies

Goal: To determine the number and proportion of colonies with vector and vector–insert, to amplify inserts from the bacteria/vectors, and to archive bacteria from each colony of interest (i.e., those with inserts).

Detailed steps:

Day 1

1. Count the number of blue (or black, if S-Gal was used) colonies and the number of positive (white) colonies on each plate. If more than a few hundred are present, simply note that fact rather than trying to count each colony. The proportion of colonies with inserts (i.e., vector ligation efficiency) can be determined from the number of white colonies divided by the total number of colonies.

2. Prepare 50 ml of LB broth with ampicillin by adding 50 μl of ampicillin (50 mg/ml stock) to a 50-ml conical tube full of broth. Add 300 μl of LB broth + ampicillin to each well of a sterilized 0.65 ml deep-well plate.

3. Lift isolated white colonies from the LB plate using the end of a sterile toothpick and transfer each colony to one well of the sterilized deep-well plate (spin the toothpick in your fingers while the end with the colony is immersed in the LB broth).

4. Cover the 96 deep-well plate loosely with Saran Wrap or a loose-fitting 96-well mat. Incubate overnight at 37° with semivigorous shaking. It is often beneficial to incubate an additional 24 h ($\sim$40 h total) to achieve high-density cell growth.

Day 2

5. Set up the following 25-μl PCRs:
For one 96-well tray, add the following to a clean V-bottom trough:

275.00 μl	250 μg/ml BSA
275.00 μl	10× PCR buffer
110.00 μl	10 μM M13 forward primer
110.00 μl	10 μM M13 reverse primer
220.00 μl	25 mM MgCl$_2$
165.00 μl	2.5mM dNTPs (2.5 mM each)
1408.00 μl	dH$_2$O
22.00 μl	*Taq* DNA polymerase (2.5 units/μl)

Using a multichannel pipetter, dispense 23.5 μl to each well of 96-well thermal plate, and then add 1.5 μl DNA template from bacteria colony grown up in LB broth. If setting up fewer than 96 reactions, the 25-μl reaction recipe (per reaction) is as follows:

2.50 μl	250 μg/ml BSA
2.50 μl	10x PCR buffer
1.00 μl	10 μM M13 forward primer
1.00 μl	10 μM M13 reverse primer
2.00 μl	25 mM MgCl$_2$
1.50 μl	2.5 mM dNTPs (2.5 mM each)
12.80 μl	dH$_2$O
0.20 μl	*Taq* DNA polymerase (2.5 units/μl)
1.50 μl	DNA template from bacteria colony grown up in LB broth

Cover the reactions using a mat or caps and place the PCRs in the thermal cycler. Store bacteria colonies in LB broth at 4° until PCR product has been observed.

Cycling: 95° for 3 min; then 35 cycles of 95° for 20 s, 50° for 20 s, 72° for 1.5 min. Hold at 15°.

After the PCR is finished, the product will need to be examined for the presence of inserts in each plasmid.

6. Pour a 1% agarose gel on the centipede rig (Owl Scientific, Portsmouth, NH).

7. Run 2 μl of the M13/bacterial PCR product on the agarose gel along with a 100-bp ladder and 2 μl of several lambda concentration standards (λ10 ng/μl, λ25 ng/μl, λ50 ng/μl, and λ100 ng/μl). A 10-μl multichannel pipetter may be used to save time. It loads every other lane, so it is important to keep notes on where each sample is located.

8. Run at 80 V for approximately 30–40 min.

9. Examine PCR results using a visual imaging system and save the results. Ensure that bands are clearly visible, but that they are not saturated/overintegrated (red). If DNA concentration varies a lot (i.e., the brightest samples are red when you are exposing enough to see the dimmest samples, then save multiple exposures).

10. The desired insert range is from 300 to 1000 bp. Because the pCR2.1 vector contains about 200 bp of DNA between the M13 forward and reverse priming sites, the total fragment size of desired PCR products is 500–1200 bp. Proceed to purification step using only samples that are the target size.

After the bacteria have grown overnight and the PCRs have been examined, perform the following:

11. To a 50-ml conical tube, add 15 ml of glycerol, fill to 50 ml with LB broth, and add 50 μl of ampicillin (50 mg/ml stock). Mix thoroughly by shaking.
12. Remove the bacterial cultures from the refrigerator or incubator.
13. Using the multichannel pipette, add 300 μl of prepared broth to each culture (being careful not to contaminate samples), tightly seal the cap mat on the tray, mix gently by inverting several times, label well, and store at -70°.

Step IX: Prepare PCR samples for Sequencing using ExoSAP

Goal: To determine PCR product concentration and size and to purify the PCR product for subsequent sequencing. There are many ways to prepare PCR products for sequencing. If the PCR products are good and strong, dilution (i.e., using no more than 0.5 μl of PCR product) is efficient and usually works well. However, cleaning the PCR using exonuclease I and SAP is a preferred option, which improves consistency among experiments and researchers in our lab.

Materials and Solutions

  Premixed ExoSAP: catalog no. 78201, U.S. Biochemical Corp., Cleveland, OH, or Homemade ExoSAP: combine 5 μl of 20 units/μl exonuclease I (NEB catalog no. M0293L) with 15 μl of 1 unit/μl SAP (U.S. Biochemical Corp., catalog no. 70092Z).

1. Quantify PCR product concentration and size. For a single sequencing reaction, the desired amount of template is 10 ng of PCR product per 100 bp of length (i.e., for a 500-bp product, 50 ng is needed). Generally, we purify enough PCR product for two sequencing reactions. Use the visual imaging results saved from step VIII.9 above.
2. Add 1 μl ExoSAP mixture to 6–10 μl of PCR product. If PCR products are of varying concentration, the concentrations can be standardized by adding H$_2$O (*note:* Volumes should be adjusted to account for differences in lengths of the PCR products so that the molar concentrations are approximately equal).
3. Incubate the samples at 37° for 15 min, 80° for 15 min, and then hold at 15°. The samples are now purified and ready for use as sequencing reaction template.

Step X: DNA Cycle Sequencing Reactions

Goal: To complete sequencing reactions that can be used to determine the DNA sequence of the fragments that contain microsatellite repeats.

This protocol is optimized for 0.2-ml strip tubes and a titer plate centrifuge but can be used with 96-well plates with only minor modifications.

Alternative 1[3]: One-Fourth Reaction Recipe

2.0 μl BigDye Terminator version 3.1 mix[3]
1.0 μl 5× Sequencing Dilution Buffer[4]
1.0 μl primer[5] (3.3 μM)
Note: Make a master mix of the above for the number of reactions being set up. If the template concentration is constant, water may also be added into the master mix.

2.0 μl DNA template (10 ng/ 100 bp of product length; adjust volume as appropriate)[6]
4.0 μl H_2O (adjust volume as appropriate to make a total of 10.0 μl)

Alternative 2[3]: One-Eighth Reaction Recipe

1.0 μl BigDye Terminator version 3.1 mix[3]
1.5 μl 5× Sequencing Dilution Buffer[4]
1.0 μl Primer[5] (3.3 μM)
Note: Make a master mix of the above for the number of reactions being set up. If the template concentration is constant, water may also be added into the master mix

2.0 μl DNA template (10 ng/100 bp of product length; adjust volume as appropriate)[6]
4.5 μl H_2O (adjust volume as appropriate to make a total of 10.0 μl)
Cycling: 50 cycles at 96° for 10 s, 50° for 5 s, 60° for 4 min. Hold at 15°.
Note: No initial denaturation is necessary.

---

[3] Depending on level of experience, instruments available, and success, one may want to use standard, half, quarter, one-eighth, one-twelfth, or one-sixteenth reactions. Adjust BigDye and dilution buffer appropriately (*note:* the BigDye is at 2.5× and dilution buffer is at 5×).

[4] A homemade version of Sequencing Dilution Buffer is 400 mM Tris–HCl, pH 9.0; 10 mM $MgCl_2$ (see http://www.genome.ou.edu/proto.html for details), which we have used successfully as a 5× or 2.5× buffer.

[5] M13 forward or reverse, or other primers closer to the insertion site, as appropriate.

[6] If templates are consistent in size and concentration, water may be combined in the master mix and a constant volume of template may be used.

Place the tubes in a 96-well holder and store the reactions at −20° in a *non*–frost-free freezer until they can be precipitated. The reactions are stable for days at this stage, but it is best to keep them cold or frozen and away from light.

Step XI: Precipitation of Sequencing Reactions

Goal: To remove the unincorporated fluorescent ddNTPs and stabilize the labeled DNA until it can be run on an automated DNA sequencer. There are several options for cleaning sequencing reactions. Column puri-fication (Sephadex G50/ Centri-Sep columns from Princeton Separations, Adelphia, NJ) is often superior in that bases close to the primer are more likely to be recovered and the remaining salts are reduced, which is best when using capillary sequencers. The ABI BigDye version 3.1 manual includes a protocol very similar to the one below, except that NaOAc is not used. We use the following protocol with good success on ABI 377, 3700, and 3730 sequencers. The trick to it seems to be to use the recipe given in step IV (above) for the 1.5 M NaOAc 250-mM EDTA solution.

This protocol is optimized for 0.2-ml strip tubes and a titer plate centri-fuge but can be used with 96-well plates with only minor modifications.

1. If evaporation has occurred in any of the tubes, add dH$_2$O until it matches the others. Total volume should be about 10 μl.

2. Add 1 μl of 1.5 M NaOAc, 250 mM of EDTA (pH 8.0, from step IV), using the 0.5–10.0 μl multichannel pipetter and mix by pipetting up and down (i.e., sklooshing). (*Note:* 1.5 M NaOAc pH should *not* be adjusted to pH 5.2).

3. Add 40 μl of 95% ethanol using the 5–50 μl multichannel pipetter (dripping down the sides of the tubes, tips do not need to be changed between samples).

4. Recap the tubes, invert several times, and incubate for 15 min at −20°.

5. Centrifuge at 1500g for 45 min.

6. Remove caps, setting them aside on a clean Kim wipe.

7. Ensure the 96 deep-well block (*note:* Use the 96 deep-well blocks with *square* holes; these came from a Qiagen DNA prep kit) is dry by whipping out any liquids.

8. Carefully place a dry 96 deep-well block (*note:* Use the 96 deep-well blocks with *square* holes; these came from a Qiagen DNA prep kit) over the top of the tubes and flip (i.e., invert), leaving the tube holder in place over the tubes.

9. Centrifuge at 300g for 1 min, balancing with an empty deep-well block and tube holder.

10. Carefully pull the tubes straight up, off the 96 deep-well block. If any of the tubes "stick," put them back into the holder in the correct orientation.

11. Recap the tubes and store them at $-20°$ (*non*–frost-free freezer) until ready to sequence.

12. The reactions are stable for many weeks at this point. It is best to store them frozen and away from light.

Concluding Remarks

At this point, the DNA can be sequenced on several commercially available DNA sequencers (e.g., ABI, Amersham Biosciences, or Spectrumedix). Following DNA sequencing, vector and linker sequences should be removed. We screen sequences for microsatellite repeats using a simple program, Ephemeris 1.0, written in Perl by N. Dean Pentcheff (download from http://www.uga.edu/srel/DNA_Lab/programs.htm). Sequences containing microsatellites identified on at least one strand are processed further. Both strands are then contiged and edited to ensure accuracy of the sequence. After editing, primers for PCR are designed from the sequences flanking DNA using standard methodology or a three-primer system (Boutin-Ganache *et al.*, 2001); see *5'PrimerTags3.doc* at http://www.uga.edu/srel/DNA_Lab/protocols.htm for details).

All protocols used in our lab are available by following links from http://www.uga.edu/srel/ or http://gator.biol.sc.edu/. Updates to this protocol will be posted on the SREL DNA lab web site (http://www.uga.edu/srel/DNA_Lab/protocols.htm). Additional background information, steps in obtaining genotypes from microsatellite loci, and data analysis are available in *MsatMan2000.rtf* (download from http://www.uga.edu/srel/DNA_Lab/protocols.htm) and the microsatellite list-serve and associated web pages (http://www.uga.edu/srel/Microsat/Microsat-L.htm).

Acknowledgments

We have been working on this refinement for several years. Too many people and funding agencies have contributed to the protocol presented here to acknowledge everyone. We would, however, especially like to thank and acknowledge Ryan Thum, Kari Schilling, Pam Svete, Susanne Hauswaldt, Lisa Davis, Cris Hagen, and the many visiting students to our labs at SREL and USC, as well as the students enrolled in Biology 656 at USC for their contributions. Financial assistance was primarily provided by award DE-FC09-96SR18546 from the Environmental Remediation Sciences Division of the Office of Biological and Environmental Research, U.S. Department of Energy to the University of Georgia Research Foundation, with additional support from projects funded by DOE, NIH, NOAA, NSF, USDA, and various state agencies.

References

Armour, J. A. L., Neumann, R., Gobert, S., and Jefferys, A. J. (1994). Isolation of human simple repeat loci by hybridization selection. *Human Mol. Gen.* **3**, 599–605.

Boutin-Ganache, I., Raposo, M., Raymond, M., and Deschepper, C. F. (2001). M13-tailed primers improve the readability and usability of microsatellite analyses performed with two different allele sizing methods. *BioTechniques* **31**, 24–28.

Fisher, D., and Bachmann, K. (1998). Microsatellite enrichment in organisms with large genomes (*Allium cepa L.*). *BioTechniques* **24**, 796–802.

Goldstein, D. B., and Schlotterer, C. (1999). "Microsatellites: Evolution and Applications." Oxford Press, Oxford, UK.

Hamilton, M. B., Pincus, E. L., DiFiore, A., and Fleischer, R. C. (1999). Universal linker and ligation procedures for construction of genomic DNA libraries enriched for micro-satellites. *BioTechniques* **27**, 500–507.

Hauswaldt, J. S., and Glenn, T. C. (2003). Microsatellite DNA loci from the diamondback terrapin (*Malaclemys terrapin*). *Mol. Ecol. Notes* **3**, 174–176.

Holton, T. A., and Graham, M. W. (1991). A simple and efficient method for direct cloning of PCR product using ddT-tailed vectors. *Nucleic Acids Res.* **19**, 1156.

Kandpal, R. P., Kandpal, G., and Weissman, S. M. (1994). Construction of libraries enriched for sequence repeats and jumping clones, and hybridization selection for region-specific markers. *Proc. Natl. Acad. Sci. USA* **91**, 88–92.

Kijas, J. M. H., Fowler, J. C. S., Garbett, C. A., and Thomas, M. R. (1994). Enrichment of microsatellites from the citrus genome using biotinylated oligonucleotide sequences bound to streptavidin-coated magnetic particles. *BioTechniques* **16**, 656–662.

Korfanta, N. M., Schable, N. A., and Glenn, T. C. (2002). Isolation and characterization of microsatellite DNA primers in burrowing owl (*Athene cunicularia*). *Mol. Ecol. Notes* **2**, 584–585.

Marchuk, D., Drumm, M., Saulino, A., and Collins, F. S. (1991). Construction of T-vectors, a rapid and general system for direct cloning of unmodified PCR products. *Nucleic Acids Res.* **19**, 1154.

Moore, S. S., Sargeant, L. L., King, T. J., Mattick, J. S., Georges, M., and Hetzel, J. S. (1991). The conservation of dinucleotide microsatellites among mammalian genomes allows the use of heterologous PCR primer pairs in closely related species. *Genomics* **10**, 654–660.

Ostrander, E. A., Jong, P. M., Rine, J., and Duyk, G. (1992). Construction of small-insert genomic DNA libraries highly enriched for microsatellite repeat sequences. *Proc. Natl. Acad. Sci. USA* **89**, 3419–3423.

Pepin, L., Amigues, Y., Lepingle, A., Berthier, J., Bensaid, A., and Vaiman, D. (1995). Sequence conservation of microsatellites between Bos Taurus (cattle), Capra hircus (goat) and related species. Examples of use in parentage testing and phylogeny analysis. *Heredity* **74**, 53–61.

Primmer, C. R., Moller, A. P., and Ellengren, H. (1996). A wide-range survey of cross species microsatellite amplifications in birds. *Mol. Ecol.* **5**, 365–378.

Prince, K. L., Glenn, T. C., and Dewey, M. J. (2002). Cross-species amplification among peromyscines of new microsatellite DNA loci from the oldfield mouse (*Peromyscus polionotus subgriseus*). *Mol. Ecol. Notes* **2**, 133–136.

Sambrook, J., Fritsch, E. F., and Maniatis, T. (1989). "Molecular Cloning: A Laboratory Manual." Cold Spring Harbor Laboratory Press, Cold Spring Harbor, NY.

Sambrook, J., and Russell, D. W. (2001)). "Molecular Cloning: A Laboratory Manual." Cold Spring Harbor Laboratory Press, Cold Spring Harbor, NY.

Schable, N. A., Fischer, R. U., and Glenn, T. C. (2002). Tetranucleotide microsatellite DNA loci from the dollar sunfish (*Lepomis marginatus*). *Mol. Ecol. Notes* **2,** 509–511.

Tautz, D. (1989). Hypervariability of simple sequences as a general source for polymorphic DNA markers. *Nucleic Acids Res.* **17,** 6463–6471.

Weber, J. L., and May, P. E. (1989). Abundant class of human DNA polymorphisms which can be typed using the polymerase chain reaction. *Am. J. Human Genet.* **44,** 388–396.

Weissenbach, J., Gyapay, G., Dib, C., Vignal, A., Morissette, J., Millasseau, P., Vaysseix, G., and Lathrop, M. (1992). A second-generation linkage map of the human genome. *Nature* **359,** 794–801.

Zane, L., Bargelloni, L., and Patarnello, T. (2002). Strategies for microsatellite isolation: A review. *Mol. Ecol.* **11,** 1–16.

Zhu, Y., Queller, D. C., and Strassmann, J. E. (2000). A phylogenetic perspective on sequence evolution in microsatellite loci. *J. Mol. Evol.* **50,** 324–338.

[14] Use of Microsatellites for Parentage and Kinship Analyses in Animals

By Michael S. Webster and Letitia Reichart

Abstract

Microsatellite markers are quickly becoming the molecular marker of choice for studies of parentage and kinship in animals. In this chapter, we review methods and give protocols for screening potential microsatellite markers, as well as protocols for genotyping individuals with useful markers once they have been identified. In addition, we explain how microsatellites can be used to assess parentage and kinship, give basic analytical methods, and briefly review more sophisticated approaches that can be used to circumvent many of the problems that arise in any real empirical study.

Introduction

The application of molecular genetic methods to the study of natural populations has allowed researchers to directly examine kinship and parent–offspring relationships and thereby ushered in a revolution in our understanding of mating systems and social behavior. During the early phase of this "molecular revolution," most researchers used protein allozymes or multilocus DNA fingerprinting (Burke, 1989). Microsatellites have become the marker of choice for studies of kinship and parentage.

Microsatellites are tandem repeats of short (~2–6 bp) genetic elements, in which differences between alleles are primarily in the number of repeats (Goldstein and Schlötterer, 1999; Jarne and Lagoda, 1996). These markers provide a powerful approach to analyses of parentage and kinship, with many advantages over other approaches (Webster and Westneat, 1998). The principal advantage of microsatellites over multilocus approaches is that microsatellites are codominant markers, so heterozygotes can be distinguished from homozygotes. This allows for exact genotyping and more precise genetic comparisons among individuals. In addition, microsatellite loci have high mutation rates, and as a consequence, a large number of alleles typically exist at a single locus. This allows for highly powerful analyses of kinship because unrelated individuals will be unlikely to share alleles. Finally, because microsatellite analyses are polymerase chain reaction (PCR) based, only small amounts of DNA are needed, and highly degraded DNA can be used. This allows for DNA to be used from nontraditional sources, including feces and hair (Constable *et al.*, 2001; Morin *et al.*, 2001), feathers (Pearce *et al.*, 1997), and museum skins (Bouzat *et al.*, 1998).

Despite these advantages, microsatellites also carry a number of significant disadvantages that must be weighed against the benefits. First, the primers used to amplify microsatellites tend to be fairly species specific (i.e., the primers that work for species X may not work for species Y). As a consequence, primers often need to be isolated for each study species, which can be very laborious, particularly for taxa with a relatively low frequency of microsatellites (Primmer *et al.*, 1997). This is beginning to change, as a growing number of species now have microsatellites available (see any issue of *Molecular Ecology Notes*), and techniques are now available to facilitate the process of microsatellite isolation (Zane *et al.*, 2002). Second, because microsatellite analyses are PCR based, mutations in primer regions can lead to nonamplifying "null alleles" (Ishibashi *et al.*, 1996; Paetkau and Strobeck, 1995). These can pose problems for parentage assignments (Pemberton *et al.*, 1995) because a parent and offspring who share a null allele will appear as a mismatch. Techniques for estimating the frequency of null alleles are straightforward (Brookfield, 1996), but not necessarily very sensitive.

The analysis of parentage and kinship via microsatellites typically involves PCR amplification of a locus from a number of individuals followed by gel electrophoresis to distinguish alleles of different size. Depending on how PCR products are labeled, electrophoresis can be done on an automated sequencer, which greatly facilitates scoring and comparison of individual genotypes. In this chapter, we provide general methodologies for microsatellite analysis, including protocols of the methods employed in our laboratory. In addition, we provide a brief outline of analytical approaches to assess parentage and other types of kinship relationship from microsatellite genotypes.

Testing and Optimizing Loci

There are two alternative sources for potentially useful microsatellite loci. First, one may test microsatellite primers isolated from other species. The likelihood that a particular primer pair will work for the target species appears rather stochastic, although some microsatellites show highly conserved flanking/primer regions (Rico et al., 1996; Slate et al., 1998; Zardoya et al., 1996), and detailed comparisons show that the probability that primers from one species work for another increases if they are closely related (Primmer et al., 1996). Second, one may isolate microsatellite markers from the genome of the study organism itself. Methods for isolating useful microsatellites are beyond the scope of this chapter but are reviewed elsewhere (Zane et al., 2002). In our lab, we have had good success using the method of Hammond et al. (1998) to develop and screen genomic libraries enriched for simple sequence repeats (SSRs).

Once a set of potentially useful loci have been identified, it is necessary to optimize PCR conditions for the study organism, because conditions often vary from one organism to the next. After optimization, useful primers should show one (homozygotes) or two (heterozygotes) bands per individual on agarose gels, although in reality two bands will often blur together to form a single fuzzy band. Useful loci will show variation across individuals in the population, and this variation can often be detected on an agarose test gel (Fig. 1).

The general approach for optimizing PCR conditions is to run multiple reactions under varying conditions and then visualize the resulting products on agarose gels stained with ethidium bromide (EtBr). Typical PCR recipes include a pair of microsatellite primers to be tested/optimized, 10X PCR buffer, salt ($MgCl_2$ or KCl), Taq polymerase, and deoxyribonucleic triphosphates (dNTPs). No recipe will optimize all microsatellite primers, and PCR conditions will likely vary for each pair. The most useful strategies for optimizing loci involve testing different annealing temperatures (increasing or decreasing 1° or 2° at a time) and varying salt concentrations (usually in the range 0.5–3.0 mM). Varying concentrations of dNTPs, primers, and/or Taq polymerase can affect the quality of the PCR products obtained. Manipulating these components typically allows complete optimization of loci. Below, we give the standard protocol used in our lab for optimization of PCR conditions (Protocol 1).

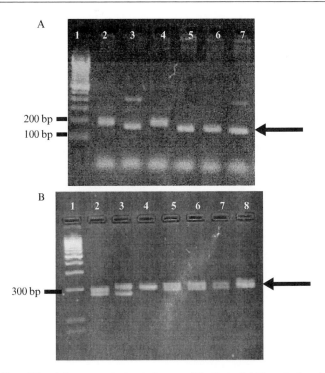

FIG. 1. Two 3% mini–agarose test gels for amplification of (A) a single microsatellite optimized for Bicknell's thrush (*Catharus bicknelli*), and (B) a single microsatellite optimized for milkweed (*Ascelpias* sp.). Lane one in each gel contains a 100-bp DNA ladder used to estimate product sizes present in the remaining lanes. Large arrows (on right) indicate location of variable microsatellite bands. Both gels depict polymerase chain reaction (PCR) products for highly polymorphic loci ready for visualization on a 4.5% polyacrylamide gel to determine individual genotypes.

PROTOCOL 1

OPTIMIZING PCR CONDITIONS FOR MICROSATELLITES USING COLD PCR

Cold PCR is a general PCR procedure in which the reaction begins at room temperature and then the temperature is raised to 94–96° for DNA denaturation. The initial denaturation period is followed by a specific number of cycles of denaturation, annealing, and extension. After finishing a specified number of cycles, the program finishes with a final extension (72°) period. Annealing temperatures can range from 45° to 65° and will vary for each primer set. The following list is a cold PCR protocol used in our laboratory.

1. For each individual to be run, pipette 1 μl of genomic DNA into a well-labeled (individual, primer pair, date) PCR tube.

2. Make up the following master mix:

Item	Initial concentration	Per sample	Final concentration
Sterile water	—	11.8 μl	
PCR buffer II (no MgCl$_2$)	10×	2.5 μl	1×
dNTP mix (2.5 mM)	2.5 mM	1.5 μl	0.15 mM
Forward primer	5 μM	2.5 μl	0.50 μM
Reverse primer	5 μM	2.5 μl	0.50 μM
MgCl$_2$	25 mM	3.0 μl	3.00 mM
Taq polymerase	5 U/μl	0.2 μl	1.00 Units

3. Pipette 24.0 μl of the master mix into each of the tubes with the 1.0-μl DNA template, for a total volume of 25.0 μl.

4. Cycle the PCRs as follows:

1 cycle:	94° for 3 min
30 cycles:	94° for 60 s
	$X°$ for 60 s (annealing temperature to be varied)
	72° for 45 s
1 cycle:	72° for 5 min

The annealing temperature is to be varied on different runs. Our lab usually does reactions at 50°, 55°, and 60° for each primer pair. Some primer pairs might require other annealing temperatures.

5. Run the products out on a 2% or 3% agarose/1x TBE buffer (1× TBE buffer diluted from 10× TBE stock, 1 M Tris, 0.9 M boric acid, 0.02 M EDTA) minigel and stain with ethidium bromide (EtBr) to test for useful amplification.

6. If PCR products are not satisfactory (multiple bands per lane, little or no product, etc.), repeat procedure using different annealing temperatures. Improved PCR results might also result from altering concentration of salt (MgCl$_2$) and/or polymerase.

An alternative PCR procedure for primer testing and optimization is "touchdown PCR" (TD-PCR). Under normal PCR protocols, primers can produce spurious bands caused by nonspecific binding of the primers. These spurious bands can increase scoring difficulty and make a locus less useful for genotyping. Don et $al.$ (1991) developed TD-PCR to help eliminate spurious bands and increase the quantity of target DNA produced. For TD-PCR, cycles begin with a very high annealing temperature, well above the expected annealing temperature. The TD-PCR program is designed to decrease the annealing temperature in small increments (e.g., 1°) every second cycle to the expected annealing temperature (the "touchdown" temperature). Once the reaction reaches the touchdown temperature, 10 cycles are run at this annealing temperature before final extension. This method reduces the number of spurious bands because less nonspecific annealing occurs at higher temperatures, so only the target region should amplify during early cycles, exponentially increasing the amount of target DNA available in later cycles. TD-PCR can be used with a "cold start" for initial denaturation (i.e., *Taq* polymerase added to reactions at room temperature before cycling). However, Roux (2003) recommends that TD-PCR be used with a "hot-start" reaction in which *Taq* polymerase is not added to the reaction until samples are near denaturation temperature (at least 85°), thereby avoiding most low-temperature priming altogether.

TD-PCR has been used for optimizing primers in few parentage analyses (but see Hughes *et al.* [2003]) and has been used mostly for primer pairs that are difficult to amplify. Roux (2003) suggests that TD-PCR can be used as a faster method for primer optimization because it could potentially reduce the number of test PCRs needed to optimize PCR conditions for a primer pair. TD-PCR may be most useful when testing primer combinations across species, because the likelihood of nonspecific binding increases with slight differences in homologous regions being amplified.

Once PCR products have been obtained, they should be visualized on a mini–agarose gel (test gel), which is a quick way to assess whether PCR conditions produce clear bands and whether a microsatellite is polymorphic (i.e., useful for genotyping). Mini–agarose gels (2% or 3% agarose in 1× TBE buffer, weight by volume) are useful for visualizing test PCR products (see Protocol 1). Test gels should include amplification products from 10 to 20 individuals and should reveal one or two clear bands per individual (from homozygotes and heterozygotes, respectively). Highly polymorphic loci are most useful for kinship analyses because the precision of relatedness estimates increases as the level of polymorphism increases.

Visualizing PCR Products and Determining Individual Genotypes

After microsatellite primers have been optimized, several methods can be used to visualize PCR products and determine individual genotypes at each locus. The three most common methods use radioactivity, silver staining, or fluorescent markers. Radioactivity and silver staining have been used most often, and protocols for each of these methods are described elsewhere

(Strassmann *et al.*, 1996; Tegelström, 1992). The most popular method for visualization and scoring is to fluorescently end-label one PCR primer of a pair and quantify the size of PCR products on an automated sequencer.

Products amplified using labeled primers should be run on agarose test gels, as the dye-labeled primers may require slight alterations to PCR conditions (e.g., annealing temperature), and because PCRs sometimes (and inexplicably) fail. Several fluorescent dye labels are available for end-labeling primers, and final PCR products are visualized using automated sequencers (e.g., Applied Biosystems, Inc. [ABI], Prism models) and gel scanners. For visualizing PCR products on an ABI 377, our lab typically orders end-labeled primers with FAM (blue), HEX (yellow), and TET (green) fluorescent markers, and we also run an internal size standard in each lane (TAMRA, red). By varying locus color, multiple loci can be run on a single gel and products for each locus are easily distinguished. Our standard protocol for running PCR products on an automated sequencer is given in Protocol 2.

PROTOCOL 2

RUNNING FLUORESCENTLY LABELED PCR PRODUCTS ON THE ABI

1. Run optimized PCR protocol with fluorescently end-labeled forward primer. Run 5 μl of PCR product on mini–agarose test gel to double-check product amplification. The fluorescent dyes are light sensitive, so PCR products (and labeled size standards) should be kept out of light at all times (wrap PCR tubes in aluminum foil).

2. Pour (4.5%) polyacrylamide gel. Mix 45 ml automatrix polyacrylamide (National Diagnostics SequaGel Automatrix 4.5) and 5 ml 10× TBE. Just before pouring gel, add 250 μl 10% ammonium persulfate (APS) and 30 μl TEMED. Add entire solution (45 ml) to syringe and pour gel.

3. Sample preparation for 48 lanes: Sample preparation is done on ice. Loading buffer (90 μl) is mixed by adding 50 μl formamide, 24 μl blue dye, and 16 μl TAMRA. Add 1.5 μl of loading buffer mix to each sample tube (one tube per sample). Then 0.8–1.0 μl of PCR product is added into each sample tube. (*Note:* The amount of PCR product can be decreased or diluted if fluorescence is too strong for scoring purposes.) After PCR product has been added to the loading buffer, heat shock the samples at 96° for 5 min to denature DNA. After heat shocking, samples are immediately placed on ice.

4. Loading samples: We load 1.0–1.2 μl of sample in each lane. Samples can be loaded one at time into a shark-tooth comb or can be

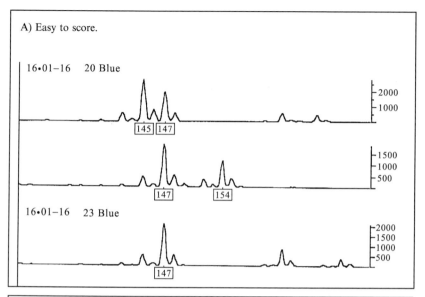

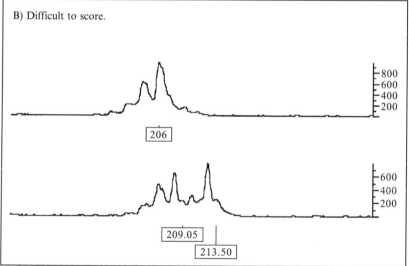

FIG. 2. Genotyper chromatogram profiles of microsatellite loci illustrating varying degrees of scoring difficulty. The top panel shows locus T2 amplified from three individual Bicknell's thrush (two heterozygotes and one homozygote [bottom profile] are shown). This locus is relatively easy to score because of the single large "peak" characterizing each allele. The bottom panel shows locus Dca24 amplified from two individual black-throated blue warblers (*Dendroica caerulescens*; one homozygote [upper profile] and one heterozygote [lower profile] are shown). This locus is more difficult to score because of the complex profiles and lower amplification.

loaded by using a membrane comb protocol developed by the Gel Company. Our lab uses the membrane comb technique because samples can be loaded quickly and have a higher probability of running straight (compared to when using a traditional shark-tooth comb).

5. Microsatellite gels are called GeneScan runs in the ABI system and are run for 2.5–3.0 h. Data are automatically recorded by the system and saved in a gel file that will be used for analysis.

GeneScan and GenoTyper (ABI) are used to analyze microsatellite data from an ABI automated sequencer (i.e., assign base-pair sizes to individual alleles at each locus). GeneScan is used to align lanes, assign size standards for each lane, and then extract lanes for use in GenoTyper, which is then used to assign sizes to bands (PCR products) for each individual at each locus. The ease of scoring allele sizes from sequencer profiles can vary dramatically across loci (Fig. 2). We offer suggestions on how to score microsatellite alleles using these software packages (Protocol 3).

PROTOCOL 3

Tips for Scoring Gels with ABI Software

After gel data are collected from a run, GeneScan is used to view the gel file. Adjust gel contrast for the size standard (red) and for the color of the labeled microsatellite products run on the gel. Then track lanes to ensure the lane assignment matches the correct lane. Lanes should be tracked with 70% confidence or higher. After tracking lanes and checking alignment, lanes can be extracted. After extracting lanes, an analysis window will list each lane to be analyzed with a size standard. To analyze lanes, first choose the color of the labeled microsatellite, then choose one to two lanes as size standard lanes for comparison of all microsatellite allele sizes in each lane. To assign size standards, ABI will provide a chromatogram of base-pair sizes for bands produced by TAMRA. You will view the bands (as peaks in a chromatogram) and assign base-pair sizes for size standard peaks on your gel by comparing the chromatogram provided by ABI with your gel chromatogram. Assigning size standards is important because the size standard determines how well allele sizes can be compared between individuals at a locus. After size standards have been selected for all lanes, analyze lanes.

GeneScan data must now be imported into GenoTyper for further analysis. Import all lanes from the gel file you want to view. Select the color of the fluorescently labeled microsatellite and you will see a window with a chromatogram of all the lanes. Zoom in on the section of the gel where you see the highest concentration of peaks (this should be the microsatellite region and will correspond to the base-pair size estimated using test gels). After zooming in, select the lighting bolt symbol on the right-hand side of the screen. This will show individual chromatograms for each lane (Fig. 2). Now, create a category with the color of the labeled microsatellite, and label the two highest peaks on your chromatogram.

After the program labels the peaks, the researcher checks the labeled peaks and designs scoring criteria for identifying homozygotes and heterozygotes. Typically, the highest peaks (most fluorescence) represent an allele, because the microsatellite region should have produced the most product during PCR analysis and should produce the strongest signal. The key to assigning allele sizes is to look for patterns between individuals and to identify patterns produced by each allele. Allelic patterns will be consistent. Also, the type of repeat region (i.e., dinucleotide or trinucleotide) produced for each microsatellite locus is known and the distance between allele sizes can be estimated. By using this criteria, some false alleles can be avoided. Each individual will have one or two peaks (homozygote or heterozygote) per locus.

Some problems can arise during scoring if a "false allele" is produced. A false allele is a shadow peak (i.e., a peak that looks as if it could be an allele). This peak can be identified after viewing peaks for lots of individuals and will show up as a third peak for heterozygotes who clearly have two distinct alleles. The most difficult part of scoring is creating criteria to assign allele sizes. Once allele size criteria are assigned, at least 10 individuals should be run on every gel to ensure that allele sizes are assigned correctly. Consistently rerunning these individuals will allow you to compare individuals across gels and detect any problem gels.

Analysis of Microsatellite Genotype Data

As microsatellites have become more popular for parentage and population studies, the number and sophistication of analytical approaches has increased exponentially. Here, we briefly review some of the most basic analyses necessary and give references for more sophisticated approaches. This chapter focuses on analyses of parentage or kinship and does not include the large number of analytical approaches that have been developed to estimate population structure and other population parameters from microsatellite data (Sunnucks, 2000).

Before conducting an analysis of kinship or parentage, one should characterize the genetic variation (polymorphism) seen at each locus

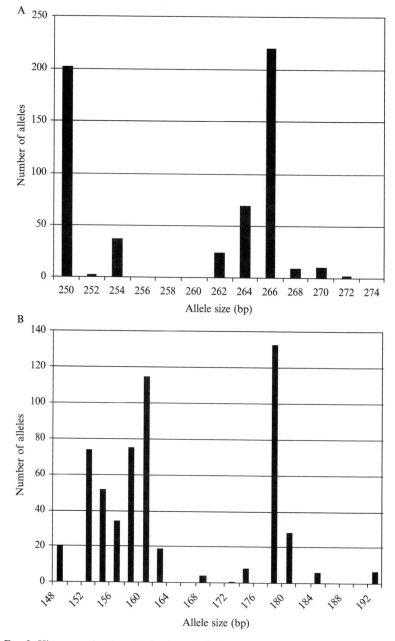

FIG. 3. Histogram showing distribution of allele sizes for a two microsatellite loci, (A) *Msp*6 and (B) *Msp*10, assayed in a population of splendid fairy wrens (*Malurus splendens*) sampled in 1998 (n = 290 adults).

(Fig. 3). The frequency of each allele (x_i) can be calculated from the total population of adults genotyped ($\sum x_i = 1.00$), and the expected frequency of heterozygotes (h_e) can be calculated as (Nei, 1987):

$$h_e = 1 - \sum (x_i)^2$$

This expected frequency can be compared to the observed frequency of heterozygotes (h_o) using a standard goodness-of-fit test and a continuity correction suggested for Hardy-Weinberg tests (Lessios, 1992), and more sophisticated tests of significance are available (e.g., see approach employed in GENEPOP at *http://wbiomed.curtin.edu.au/genepop*). A significant difference between h_e and h_o suggests the presence of a null (i.e., nonamplifying) allele, and in these cases the frequency of the null allele (r) can be estimated. Under the assumption of no null homozygotes, the frequency of the null allele is given by (Brookfield, 1996; Krauss, 2000; Summers and Amos, 1997):

$$r = (h_e - h_o)/(1 + h_e)$$

For parentage studies, the most basic analysis is based on a simple mendelian principle: At any given locus, an offspring should possess one allele inherited from its mother and another allele inherited from its father. Therefore, it should be possible to identify the biological parents of an offspring by comparing its genotype to that of potential parents. Adults who do not match the offspring (i.e., do not possess an allele found in the offspring) can be excluded as biological parents, and the combination of two parents (male and female) should explain all of the alleles found in the offspring (although sometimes a single mismatch may be explained by mutation).

To properly analyze parentage using microsatellites (or any other codominant marker), one must understand the probability that an offspring would match an adult by chance. This probability can be high for loci with relatively few alleles and for common alleles at more polymorphic loci. Several methods for estimating this probability have been developed (Dodds *et al.*, 1996; Jamieson, 1994; Selvin, 1980; Usha *et al.*, 1995), and some of these address various problems that may arise in any real study. For example, the probability of a random match can be corrected for systems in which population substructure leads to some potential parents being genetically related to each other (Double *et al.*, 1997; Waits *et al.*, 2001).

In many research projects, one parent will be known (often the mother, e.g., in mammals where maternity is not ambiguous), but the other (the father) will not, so the biological father will be a male who possesses all of the offspring's nonmaternal alleles. In these cases, the average probability of paternal exclusion (P_{ej}) for each polymorphic locus can be calculated as follows (Jamieson, 1994):

$$P_{ej} = 1 - 2\sum(x_i)^2 + \sum(x_i)^3 + 2\sum(x_i)^4 - 3\sum(x_i)^5$$
$$+ 2\left(\sum(x_i)^2\right)^2 + 3\sum(x_i)^2\sum(x_i)^3$$

This is the probability, averaged over all i alleles at the jth locus, that a randomly chosen non-sire male will *not* possess the paternal allele found in an offspring (i.e., will not match), given that the mother of the offspring is known with certainty. It is also possible to calculate the total probability of exclusion (P_{et}), which is the probability that a randomly chosen male will not possess the paternal allele of an offspring at one or more of the loci surveyed (i.e., it is equal to one minus the probability that the male will match at all loci surveyed):

$$P_{et} = 1 - \Pi(1 - P_{ej})$$

Consideration of these probabilities makes clear the difference between a parentage *exclusion* analysis and a parentage *assignment* analysis. In the former, one is interested in determining whether a particular adult is or is not a biological parent of a particular offspring. Such comparisons are meaningful only if the probability of sharing alleles by chance is low. Thanks to the high polymorphism at most microsatellites, in most cases a small number of loci will give a very low probability of sharing by chance. For an assignment analysis, on the other hand, one is interested in determining which of the many possible adults in the population are the biological parents of an offspring. In this case, the offspring is being compared to multiple adults and some of those adults will likely match by chance. Consider, for example, the situation in which the mother of an offspring is known but the father can be any male in the population, and the offspring is compared to multiple males at several microsatellites. If the probability that the offspring will falsely match a nonparent male at all loci is 0.01, the probability that at least one of 100 males will match the offspring by chance is 0.634. Thus, false matches are a serious concern in parentage assignment studies, and consequently, several microsatellite markers should be used.

Though straightforward, parentage analyses can be complicated by a number of factors, including null alleles, genotype scoring errors, mutations, and incomplete sampling. Many of these problems can be addressed in the analysis. For example, in exclusion analyses, null alleles can create apparent mismatches between a potential sire and an offspring, but these can be recognized (because the adult male and the offspring will appear to be homozygous for different alleles) and ignored. Similarly, incomplete sampling of candidate parents can create analysis problems, but Neff *et al.* (2000a,b) discuss ways to deal with this problem, including guidelines on the number of loci to use and ways of estimating the number of parents contributing to a brood. It is also possible to reconstruct the genotype of unsampled parents, particularly if a large number of offspring have been sampled from a brood/litter (Jones, 2001).

Marshall *et al.* (1998) give a very lucid account of the difficulties of inferring parentage from genotype data and describe a likelihood approach for assigning parentage that circumvents many such problems. These authors have developed a software package (*CERVUS*) for conducting such analyses, and this software is quickly becoming widely used in parentage studies. However, although an extremely useful advance over simple exclusion analyses, in our experience the likelihood approach will often assign parentage to an adult who, for other (nongenetic) reasons, is very unlikely to be a true parent (Prodöhl *et al.*, 1998). This is because the likelihood approach does not take nongenetic data into account, and because males who are homozygous for alleles possessed by an offspring will have an increased probability of being assigned as the sire. Therefore, we find it useful to closely scrutinize the results of a likelihood analysis (i.e. the output from *CERVUS*) to ensure that each paternity assignment makes biological sense (Prodöhl *et al.*, 1998; Webster *et al.*, 2004). Neff *et al.* (2001) developed a Bayesian approach that can objectively do this by including nongenetic data in the calculation of previous and posterior probabilities. Though not widely used, this method holds great promise for the analysis of parentage because it may reduce the chances of assigning parentage to a biologically implausible adult.

Finally, in many studies researchers will be interested in kinship relationships outside of the parent–offspring relationship. Determining such relationships from genetic data can prove challenging because any given category of relationship can show considerable variation in the degree of genetic similarity between any two individuals, and so different categories of relationship may overlap broadly in genetic similarity scores (Blouin *et al.*, 1996; Queller and Goodnight, 1989; Webster and Westneat, 1998). Nevertheless, estimates of kinship can be robust if a number of loci are used in the analysis, and several methods for inferring relatedness/kinship have been developed (Blouin, 2003). The most widely used approach is to use a software package (*RELATEDNESS*) that employs a likelihood approach to calculate the most likely category of relationship for two individuals based on their genotypes and the distribution of genotypes in the population (Goodnight and Queller, 1999).

References

Blouin, M. S. (2003). DNA-based methods for pedigree reconstruction and kinship analysis in natural populations. *Trends Ecol. Evol.* **18,** 503–511.

Blouin, M. S., Parsons, M., Lacailee, V., and Lotz, S. (1996). Use of microsatellite loci to classify individuals by relatedness. *Mol. Ecol.* **5,** 393–401.

Bouzat, J. L., Lewin, H. A., and Paige, K. N. (1998). The ghost of genetic diversity past: Historical DNA analysis of the greater prairie chicken. *Am. Naturalist* **152,** 1–6.

Brookfield, J. F. Y. (1996). A simple new method for estimating null allele frequency from heterozygote deficiency. *Mol. Ecol.* **5,** 453–455.

Burke, T. (1989). DNA fingerprinting and other methods for the study of mating success. *Trends Ecol. Evol.* **4,** 139–144.

Constable, J. L., Ashley, M. V., Goodall, J., and Pusey, A. E. (2001). Noninvasive paternity assignment in Gombe chimpanzees. *Mol. Ecol.* **10,** 1279–1300.

Dodds, K. G., Tate, M. L., McEwan, J. C., and Crawford, A. M. (1996). Exclusion probabilities for pedigree testing in animals. *Theor. Applied Genet.* **92,** 966–975.

Don, R. H., Cox, P. T., Wainwright, B. J., Baker, K., and Mattick, J. S. (1991). 'Touchdown' PCR to circumvent spurious priming during gene amplification. *Nucleic Acids Res.* **19,** 4008.

Double, M. C., Cockburn, A., Barry, S. C., and Smouse, P. E. (1997). Exclusion probabilities for single-locus paternity analysis when related males compete for matings. *Mol. Ecol.* **6,** 1155–1166.

Goldstein, D. B., and Schlötterer, C. (1999). "Microsatellites: Evolution and Applications." Oxford University Press, New York.

Goodnight, K. F., and Queller, D. C. (1999). Computer software for performing likelihood tests of pedigree relationship using genetic markers. *Mol. Ecol.* **8,** 1231–1234.

Hammond, R. L., Saccheri, I. J., Ciofi, C., Coote, T., Funk, S. M., McMillan, W. O., Bayes, M. K., Taylor, E., and Bruford, M. W. (1998). Isolation of microsatellite markers in animals. *In* "Molecular Tools for Screening Biodiversity: Plants and Animals" (A. Karp, P. G. Isaac, and D. S. Ingram, eds.), pp. 279–285. Chapman and Hall, London.

Hughes, J. M., Mather, P. B., Toon, A., Ma, J., Rowley, I., and Russell, E. (2003). High levels of extra-group paternity in a population of Australian magpies *Gymnorhina tibicen:* Evidence from microsatellite analysis. *Mol. Ecol.* **12,** 3441–3450.

Ishibashi, Y., Saitoh, T., Abe, S., and Yoshida, M. C. (1996). Null microsatellite alleles due to nucleotide sequence variation in the grey-sided vole *Clethrionomys rufocanus. Mol. Ecol.* **5,** 589–590.

Jamieson, A. (1994). The effectiveness of using co-dominant polymorphic allelic series for (1) checking pedigrees and (2) distinguishing full-sib pair members. *Anim. Genet.* **25**(suppl. 1), 37–44.

Jarne, P., and Lagoda, J. L. (1996). Microsatellites, from molecules to populations and back. *Trends Ecol. Evol.* **11,** 424–429.

Jones, A. G. (2001). GERUD 1.0: A computer program for the reconstruction of parental genotypes from progeny arrays using multilocus DNA data. *Mol. Ecol. Notes* **1,** 215–218.

Krauss, S. L. (2000). Accurate gene diversity estimates from amplified fragment length polymorphism (AFLP) markers. *Mol. Ecol.* **9,** 1241–1245.

Lessios, H. A. (1992). Testing electrophoretic data for agreement with Hardy-Weinberg expectations. *Marine Biol.* **112,** 517–523.

Marshall, T. C., Slate, J., Kruuk, L. E. B., and Pemberton, J. M. (1998). Statistical confidence for likelihood-based paternity inference in natural populations. *Mol. Ecol.* **7,** 639–655.

Morin, P. A., Chambers, K. E., Boesch, C., and Vigilant, L. (2001). Quantitative polymerase chain reaction analysis of DNA from noninvasive samples for accurate microsatellite genotyping of wild chimpanzees (*Pan troglodytes verus*). *Mol. Ecol.* **10**, 1835–1844.

Neff, B. D., Repka, J., and Gross, M. R. (2000a). Parentage analysis with incomplete sampling of candidate parents and offspring. *Mol. Ecol.* **9**, 515–528.

Neff, B. D., Repka, J., and Gross, M. R. (2000b). Statistical confidence in parentage analysis with incomplete sampling: How many loci and offspring are needed? *Mol. Ecol.* **9**, 529–539.

Neff, B. D., Repka, J., and Gross, M. R. (2001). A Bayesian framework for parentage analysis: The value of genetic and other biological data. *Theor. Popul. Biol.* **59**, 315–331.

Nei, M. (1987). "Molecular Evolutionary Genetics." Columbia University Press, New York.

Paetkau, D., and Strobeck, C. (1995). The molecular basis and evolutionary history of a microsatellite null allele in bears. *Mol. Ecol.* **4**, 519–520.

Pearce, J. M., Fields, R. L., and Scribner, K. T. (1997). Nest materials as a source of genetic data for avian ecological studies. *J. Field Ornithol.* **68**, 471–481.

Pemberton, J. M., Slate, J., Bancroft, D. R., and Barrett, J. A. (1995). Nonamplifying alleles at microsatellite loci: A caution for parentage and population studies. *Mol. Ecol.* **4**, 249–252.

Primmer, C. R., Møller, A. P., and Ellegren, H. (1996). A wide-range survey of cross-species microsatellite amplification in birds. *Mol. Ecol.* **5**, 365–378.

Primmer, C. R., Raudsepp, T., Chowdhary, B. P., Møller, A. R., and Ellegren, H. (1997). Low frequency of microsatellites in the avian genome. *Genome Res.* **7**, 471–482.

Prodöhl, P. A., Loughry, W. J., McDonough, C. M., Nelson, W. S., Thompson, E. A., and Avise, J. C. (1998). Genetic maternity and paternity in a local population of armadillos assessed by microsatellite DNA markers and field data. *Am. Naturalist* **151**, 7–19.

Queller, D. C., and Goodnight, K. F. (1989). Estimating relatedness using genetic markers. *Evolution* **43**, 258–275.

Rico, C., Rico, I., and Hewitt, G. (1996). 470 million years of conservation of microsatellite loci among fish species. *Proc. R. Soc. Lon. B.* **263**, 549–557.

Roux, K. H. (2003). Optimization and troubleshooting in PCR. In "PCR Primer: A Laboratory Manual" (C. W. Dieffenbach and G. S. Dveksler, eds.), 2nd Ed., pp. 35–41. Cold Spring Harbor Laboratory Press, Cold Spring Harbor, New York.

Selvin, S. (1980). Probability of nonpaternity determined by multiple allele codominant systems. *Am. J. Hum. Genet.* **32**, 276–278.

Slate, J., Coltman, D. W., Goodman, S. J., MacLean, I., Pemberton, J. M., and Williams, J. L. (1998). Bovine microsatellite loci are highly conserved in red deer (*Cervus elaphus*), sika deer (*Cervus nippon*) and Soay sheep (*Ovis aries*). *Anim. Genet.* **29**, 307–315.

Strassmann, J. E., Solís, C. R., Peters, J. M., and Queller, D. C. (1996). Strategies for finding and using highly polymorphic DNA microsatellite loci for studies of genetic relatedness and pedigrees. In "Molecular Zoology: Advances, Strategies and Protocols" (J. D. Ferraris and S. R. Palumbi, eds.), pp. 163–180. Wiley, New York.

Summers, K., and Amos, W. (1997). Behavioral, ecological, and molecular genetic analyses of reproductive strategies in the Amazonian dart-poison frog, *Dendrobates ventrimaculatus*. *Behav. Ecol.* **8**, 260–267.

Sunnucks, P. (2000). Efficient genetic markers for population biology. *Trends Ecol. Evol.* **15**, 199–203.

Tegelström, H. (1992). Detection of mitochondrial DNA fragments. In "Molecular Genetic Analysis of Populations" (A. R. Hoelzel, ed.). Oxford University Press, Oxford.

Usha, A. P., Simpson, S. P., and Williams, J. L. (1995). Probability of random sire exclusion using microsatellite markers for parentage verification. *Anim. Genet.* **26**, 155–161.

Waits, L. P., Luikart, G., and Taberlet, P. (2001). Estimating the probability of identity among genotypes in natural populations: Cautions and guidelines. *Mol. Ecol.* **10**, 249–256.

Webster, M. S., Tarvin, K. A., Tuttle, E. M., and Pruett-Jones, S. (2004). Reproductive promiscuity in the splendid fairy-wren: Effects of group size and reproduction by auxiliaries. *Behav. Ecol.* **15,** 907–915.

Webster, M. S., and Westneat, D. F. (1998). The use of molecular markers to study kinship in birds: Techniques and questions. *In* "Molecular Approaches to Ecology and Evolution" (R. DeSalle and B. Schierwater, eds.), pp. 7–35. Birkhäuser, Boston.

Zane, L., Bargelloni, L., and Patarnello, T. (2002). Strategies for microsatellite isolation: A review. *Mol. Ecol.* **11,** 1–16.

Zardoya, R., Vollmer, D., Craddock, C., Streelman, J. T., Karl, S., and Meyer, A. (1996). Evolutionary conservation of microsatellite flanking regions and their use in resolving the phylogeny of cichlid fishes (Pisces: Perciformes). *Proc. R. Soc. Lond. B.* **263,** 1589–1598.

[15] Use of Capillary Array Electrophoresis Single-Strand Conformational Polymorphism Analysis to Estimate Genetic Diversity of Candidate Genes in Germplasm Collections

By David N. Kuhn and Raymond J. Schnell

Abstract

Capillary array electrophoresis single-strand conformation polymorphism (CAE-SSCP) analysis provides a reliable high-throughput method to genotype plant germplasm collections. Primers designed for highly conserved regions of candidate genes can be used to amplify DNA from plants in the collection. These amplified DNA fragments of identical length are turned into useful markers by assaying sequence differences by CAE-SSCP analysis. Sequence differences affect the electrophoretic mobility of single-stranded DNA under non-denaturing conditions. By collecting the mobility data for both strands assayed at two temperatures, alleles can be defined by mobility alone. For a germplasm collection with an unknown number of alleles at a locus, such mobility data of homozygotes can be used to determine the number of unique alleles without the necessity of cloning and sequencing each allele.

Introduction

The assessment of genetic diversity of germplasm collections and the genotyping of individuals for identification are important goals in the maintenance of such collections. As collections have grown, identification of parents for future breeding programs or for expanding current research populations has become more difficult, requiring sufficient polymorphic markers to resolve questions of identity and/or relatedness. Traditional markers such as restriction fragment length polymorphism (RFLP) are better suited for laboratories that commonly use radioactivity. In laboratories in which most sequencing and fragment analysis are dependent on high-throughput capillary array electrophoresis (CAE), a polymerase chain reaction (PCR)–based method such as microsatellite analysis or short sequence repeats provides high-resolution genotyping due to numerous alleles at each locus. However, the cost of developing microsatellite markers for small crops may be prohibitive; not all microsatellite loci found will be polymorphic in the target population; the markers may not be useful to assess genetic diversity in closely related species and usually are not in identifiable genes, so synteny or positional information from other plants cannot be used.

An alternative approach is to use candidate genes that share significant homology across taxonomic families, design degenerate primers to these candidate genes that allow amplification of portions of these genes from a wide range of plant genera, and convert the amplified regions into markers using single-strand conformation polymorphism (SSCP) analysis. In this way, the same loci can be studied in populations of a wide range of plant species and greater coverage of the genome can be accomplished. In addition, candidate genes can be chosen that reflect specific breeding interests, such as disease resistance, flower morphology, flowering time, secondary compound biosynthesis, or any other trait of interest.

Single nucleotide polymorphisms (SNPs), both as single base substitutions and as single base pair insertion/deletions (indels), are the most common sequence differences found between alleles. In plants, SNPs in candidate genes have been found at a frequency of one every 139 bp (Schneider et al., 2001). The frequency, stability, distribution, and presence within coding regions make SNPs attractive as markers for detecting intraspecific sequence diversity. Potential applications of SNP-based markers include developing saturated genetic maps, mapping ESTs, detecting the genetic associations of phenotypes controlled by multiple loci, studying genetic diversity, and screening for disease susceptibility (Baba et al., 2003).

Methods have been developed for high-throughput detection of SNPs, but these methods require a priori knowledge of the SNP being assayed or sequence information surrounding the SNP (Kirk et al., 2002; Landegren et al., 1998). Typically, SNP discovery requires either an extensive investment in generating sequence information from genetically unrelated individuals or data-mining sequence information available from genomic and/or EST sequencing projects. This is an effective strategy for organisms with well-characterized genomes (i.e., Oryza and Arabidopsis) or when large EST libraries have been created from genetically distinct individuals.

Our research involves tropical plants that are minor crops in the United States. Typically, their genomes are not well characterized, the genetic diversity of collections is based predominantly on phenotypic data, and few or no large well-defined populations, families, or inbred lines exist. An efficient method for SNP discovery and characterization in such organisms would be one that (1) reliably and reproducibly detects polymorphisms based solely on mobility, (2) can be accomplished on a high-throughput platform, (3) can be performed at different temperatures to capture most of the polymorphisms, and (4) can be automated for the identification of novel alleles from mobility data.

A method readily applied to detect novel polymorphisms without *a priori* knowledge is SSCP analysis, which is a sensitive economical procedure that indirectly detects sequence differences in amplified DNA fragments of the same length (Orita *et al.*, 1989). Polymorphisms are detected as alterations of mobility induced by nucleotide differences that cause stable changes in conformation of the single-stranded DNA (ssDNA). Although the exact identity of the polymorphism cannot be determined, SSCP can be employed to detect and map unknown polymorphisms as they are codominant and can be as robust as other sequence-specific markers (Borrone *et al.*, 2004; Kuhn *et al.*, 2003). In addition, because SSCP analyzes sequence differences in amplified DNA, it requires little template DNA to do the analysis. We typically use DNA isolated from leaves that have been harvested and sent in paper envelopes from tropical countries. Herbarium samples can also be used as the source for DNA.

Initially developed for polyacrylamide gel electrophoresis (Orita *et al.*, 1989), SSCP has been adapted to capillary electrophoresis (CE) (Inazuka *et al.*, 1997; Ren and Ueland, 1999) and now to capillary array electrophoresis (CAE) (Kukita *et al.*, 2002; Larsen *et al.*, 1999, 2000), allowing the high-throughput analysis of hundreds of samples per day. The application of SSCP to CAE has been relatively straightforward with the exceptions of (1) establishing a convenient internal standard to control for capillary-to-capillary and run-to-run variations in allele mobility at different temperatures and (2) overcoming the limitations of readily available software to allow automated analysis from mobility data alone.

To make CAE-SSCP of candidate genes as useful as other marker systems, it must be reliable, sensitive, capable of automation for high-throughput analysis of large germplasm collections, and most importantly affordable. The method is described in three parts: primer design and testing, CAE-SSCP, and data analysis. An excellent introduction to high-throughput fluorescent SSCP analysis and a detailed protocol are available at the Applied Biosystems, Inc. (ABI), web site (http://docs.appliedbiosystems.com, User

Bulletin 4339844A High-Throughput Fluorescent SSCP Analysis: ABI PRISM 3100, and 3100-Avant Genetic Analyzers: User Bulletin: Rev A).

Primer Design and Testing

Candidate Genes from Other Plants

Designing primers for candidate genes can be done from GenBank sequences for the candidate gene from other plants. Amino acid sequences for the gene of interest are aligned, regions of identity defined, and nucleotide sequences generated from the regions of greatest conservation by reverse translation. Degenerate primers are usually designed and a useful resource for their design is DePiCt1.0 (http://www.cs.fiu.edu/~giri/bioinf/ DePiCt1.0/WebVersion/depict.htm). The DePiCt algorithm allows the design of primer sets that are not highly degenerate by breaking the original alignment into subgroups and designing primers for conserved regions in each subgroup. Primers can then be tested against the collection and fragments of the appropriate size sequenced to identify the successful primers.

When designing primers for candidate genes, single-copy genes should be considered first, because they are easier to convert into markers. Multigene families such as resistance gene homologues (NBS/LRR genes) or WRKY transcription factors can also be used. Because of the frequent duplication and clustering of resistance gene homologues, they pose greater difficulties in analyzing alleles in germplasm collections, unless primers that are specific to single loci can be designed. The WRKY transcription factors are a group of single-copy loci that share the WRKY motif. Amplification with degenerate primers can often generate a number of independent loci that can be converted into markers. At least 72 independent single-locus WRKY genes are in the plant genome.

Once sequences of the candidate genes of interest from the target plant have been obtained, specific primers should be designed to include known SNPs, if any. Products should be in the range of 150 to 250 bp, if possible, to improve the chances of identifying all possible sequence variants in the product. This conclusion comes from our investigation of the effect of fragment length, sequence difference, and temperature on the resolving ability of CAE (Kuhn et al., 2005).

Designing Primers to EST Sequences

If EST sequences are available, they are an excellent source for candidate gene sequences to turn into SSCP markers. Primers should be designed to include known SNPs and products should be in the range of 150 to 250 bp to allow accurate identification of all SSCP variants. Primers should be designed to have annealing temperatures in the same range for each locus so that amplification of multiple primer sets can be performed under the same amplification conditions. This allows greater flexibility in multiplexing.

Testing Primers (Unlabeled)

Fluorescent labels add $75–150 to the cost of each primer. In developing SSCP markers, it may be of interest to produce primers to hundreds of loci from EST sequences. To avoid excessive expense and labeling of primers that are not polymorphic in the target population, primers should be initially produced unlabeled and tested for polymorphism. Members of the germplasm collection that are believed to be the most genetically diverse should be used to test the primer pairs.

The initial test is to amplify the test cultivars with the primers and determine whether a single amplicon of the correct size is produced. These amplicons can be denatured and run on acrylamide gels under nondenaturing conditions to determine whether the amplicons contain sequence differences. Electrophoresis is usually on acrylamide gels and ssDNA conformers are determined by silver staining, if radioactivity is not commonly used in the laboratory (Kong et al., 2003).

We have attempted testing of the unlabeled primers by amplification in the presence of fluorescently labeled nucleotides. This method is effective when testing microsatellite primers but does not work for SSCP primers for the following reasons. Incorporation of a fluorescently labeled nucleotide labels both strands with the same fluorescent dye, so strands cannot be distinguished. In addition, incorporation of different numbers of labeled nucleotides into the strand significantly alters the mobility, producing broad humps with numerous closely spaced peaks. Thus, the results are so variable that polymorphism cannot be determined.

Methods to label the strands after amplification have been proposed that could be used to inexpensively determine polymorphism (Inazuka et al., 1997). However, we are often faced with the tradeoff between incurring greater expense by labeling the primers before testing them and the longer times needed to test the cold primers before ordering labeled primers. In general, if hundreds of individuals are to be screened, it is probably more efficient to simply order the labeled primers and screen them against the population.

Testing the Fluorescently Labeled Primers

Primers that amplify fragments that are SSC polymorphic are labeled with different fluorescent compounds for both the forward and the reverse strands. If amplicons differ in size by 50 bp, they can be multiplexed.

If multiplexing of amplicons of the same size is necessary, primers can be labeled with up to four fluorescent compounds. However, for less commonly used labels, it will be more difficult to produce the matrix required to analyze the fragments on the ABI3100 (see Calibration of the ABI 3100 for SSCP).

The fluorescent-labeled primer testing should involve running individuals from a population that are known to be different (ideally by some other marker). Analysis at three temperatures is required to determine two temperatures that give the optimal separation of peaks of both strands. We commonly use 20°, 25°, and 30°. These temperatures will have to be empirically determined for each locus. The products from this initial test of the primers can be used as the plate controls when the larger population is tested as described below.

CAE-SSCP

Amplification

Amplifications should be set up in 96-well plates with the same set of cultivars in the same order in each plate. Because of plate-to-plate variation, we recommend leaving two wells open in the plates that will be for a mixture of the products that will serve as an allele control for each locus. The empty wells should be one in the first 16 wells (A1..H2) and one in the last 16 (A11..H12) to observe run-to-run variation across the plate. The empty wells should not be in the same position in the two sections so that capillary-to-capillary differences can also be observed.

Although we have used the following protocol, any typical PCR protocol is suitable. PCR was performed in 10-μl volumes containing 1× PCR buffer with 1.5 mM MgCl$_2$, 200 μM each deoxyribonucleic triphosphate (dNTP), 200 nM forward and reverse primers, 1 mg/ml bovine serum albumin (BSA), 0.4 units Amplitaq DNA polymerase (Perkin Elmer), and 2.5-ng template DNA. After an initial denaturation at 94° for 2 min, 32 cycles of PCR at 94° for 30 s, 49° for 40 s, and 72° for 1 min were performed with a final 72° extension step of 5 min.

Sample Preparation

The approved method for sample preparation for SSCP on either a single capillary (ABI310) or capillary array (ABI3100) is as follows. Before electrophoresis, 1 μl of PCR product was added to 10.5 μl Hi Di Formamide (ABI), 0.5 μl GeneScan ROX 1000 internal size standard (ABI), and 0.5 μl of freshly prepared 0.3 N NaOH. Samples were denatured in a heat block at 95° for 5 min and thereafter were immediately placed into an ice-water bath for at least 2 min.

This method gives good results, but we have modified our standard analysis procedure so that we no longer use formamide or NaOH. We use the following protocol for sample preparation. After amplification, products are diluted 1:10 by addition of water to the 96-well plate. One microliter of diluted products is transferred to the half-skirt 96-well plate that is used for sample loading on the ABI3100. Nine microliters of water is added to each well, the plate is briefly vortexed, centrifuged for 20 s, and heat denatured in a thermocycler at 95° for 5 min. The plate is removed and snap cooled in an ice/water slurry. Ten microliters of a 1:200 dilution of non-denatured GeneScan ROX 2500 molecular-weight standards is added to each well, the plate is vortexed briefly, centrifuged for 20 s, and placed in the ABI3100.

We compared results with both sample preparation methods and noted no differences in peak appearance, peak migration, or presence of artefactual peaks. The second method has become our standard method because it is easier and much less expensive with regard to amount of molecular-weight standards used per sample, and it allows the use of non-denatured molecular-weight standards.

Molecular-Weight Standards

Molecular-weight standards are included in each sample to reduce the capillary-to-capillary variation and to allow comparison of samples run in different capillaries at different times. Three commercially available molecular-weight standards are GeneScan ROX 500, GeneScan ROX 1000, and GeneScan ROX 2500 (ABI). ROX 500 is labeled only on one strand, whereas ROX 1000 and 2500 are labeled on both strands.

Molecular-weight standards can be run denatured as ssDNA or non-denatured as double-stranded DNA (dsDNA). The ABI-approved method is to mix the standards with the samples before denaturing and thus run the standards as ssDNA. Mobilities of the denatured molecular-weight standards run under SSCP conditions are variable from temperature to temperature and are so much so that individual peaks cannot be identified from the mobility pattern. To avoid this variability, to be able to compare mobilities of products across temperatures, and to permit calling of alleles by mobility, we have used non-denatured molecular-weight standards. The non-denatured dsDNA standards have an identifiable pattern at each temperature, which allows most peaks to be identified and, thus, alignment of electropherograms across temperatures. In addition, the actual molecular weight (length in nucleotides) of the fragments can be assigned, which allows us to assign a "molecular weight" to the mobility of the sample ssDNA fragments. By using dsDNA standards, we have been able to analyze data from large germplasm collections, identify novel alleles, and calculate allele frequency without having to visually assess each allele or compare mobilities to known allele standards.

There are several technical problems with non-denatured standards. The ROX 2500 internal standard consists of 28 fragments ranging from 55 to 14,097 bp (ABI, product information). From 55 to 1740 bp, the electrophoretic profile of the non-denatured molecular-weight standards at each temperature was identical to the profile provided by ABI. In this range, the ROX 2500 non-denatured standards have some peaks that are oddly shaped (544 nt) or run anomalously (508 nt). If product peaks fall precisely in this range, it may be better to use denatured standards, either ROX 1000 or ROX 2500. Thus, in the testing of the fluorescent-labeled primers (see earlier discussion), both non-denatured standards and denatured standards should be tested to determine which are most suitable for the range of the product peaks for a particular locus.

For the higher molecular-weight standards (1740–14,097 bp range), the assignment of molecular-weight size for the 2026–2499 bp range was equivocal at all temperatures. In addition, the electrophoretic profiles of the higher molecular-weight standards were not identical with one another and were not identical with the profile presented in the product literature. Thus, dsDNA standards were not useful for this range, which is approximately the mobility of ssDNA of 250 nt or greater.

For both denatured and non-denatured molecular-weight standards, there can be two sources of error that require control. First, the relative mobility of the DNA fragments is affected as the polymer ages, and so there may be drift in the mobility values from early runs to late runs. This drift may not be noticeable in the first 100 runs, which is the suggested number for any capillary array. However, capillary arrays are expensive, and when they are working well, with no lost capillaries or other problems, we tend to continue to use them, sometimes for 500 or more runs. Drift can be reduced by making up the polymer in small batches (usually 5 ml, enough to fill the syringe) and using it without long periods of inactivity.

However, we have also observed a plate-to-plate variation, which can cause mobility shifts in either the larger or the smaller direction. These can occur in an overnight run of two consecutive plates when polymer age or buffer exchange does not play a significant role. Because of these two types of variation and their effect on the mobilities, we suggest including a plate standard on each plate. The plate standard should be a mixture of the products from the initial testing of the primers on different individuals in the population. The products from as many different well-separated alleles as possible should be mixed and run in two positions in the plate. The two positions should be one in the first run and one in the last run but not in wells that are injected into the same capillary. This will allow control of drift from plate to plate and capillary array to capillary array. Plate standards should be denatured as other samples and run with internal molecular weight standards.

Electrophoresis Parameters

In previous experiments using a single capillary machine, we have altered the polymer concentration and adjusted the capillary length, in addition to adjusting the electroinjection parameters. However, with a capillary array, this has not been possible. We use the 36-cm array, as ssDNA under SSCP conditions have relatively long retention times. For example, single strands of a 437-nt fragment had mobilities slower than a 14,097-nt dsDNA fragment. Thus, for large ssDNA fragments, there is no advantage to longer capillaries. For smaller fragments in which allele mobilities may be close, a 50-cm capillary array may provide better resolution. However, it is probably less expensive to redesign the fragment length to allow it to be separated on the 36-cm capillary array than to use two arrays to assay loci in a germplasm collection. Usually small fragments are very well separated, and if not, altering the temperature can usually improve resolution. Thus, the 36-cm capillary arrays seem to be the most broadly useful for SSCP. Lowering the polymer percentage below 5% may also be advantageous for resolving larger fragments, but we had difficulty with the ABI3100 when we tried this because of frequent run aborts due to a spurious "syringe leak detection" error. Users who are comfortable altering program files in the machine's software using a text editor can overcome this problem by adjusting the pumping rate for capillary refill. In the following sections, we discuss electrophoresis parameters that the average user can alter through menus in the software: run temperature and electroinjection time and voltage.

Run Temperature

A key to success in resolving ssDNA by SSCP is finding a temperature at which the different alleles have the greatest difference in mobility. Unfortunately, mobility of ssDNA conformers is unpredictable with regard to temperature, so there is no single temperature that is optimal for resolving all possible alleles in a population. Thus, we suggest running all samples at two temperatures at least 5° apart in the range of 20–30°. Such runs can be set up consecutively on the ABI3100, and the lower temperature should always be run first. We have used 18° (the lowest value the ABI3100 can achieve), but if the previous run was 30° or higher, the run may abort because the oven cannot be cooled in time. We have also used temperatures above 30°, but capillary-to-capillary variation increases, causing error in identification of individual alleles (see the section "Data Analysis," later in this chapter). By running a subset of samples at several temperatures, two suitable temperatures can be empirically chosen. Once these temperatures are chosen, it will be convenient to use them for all individuals and all marker loci for a particular population. Collecting the mobility data for alleles from both temperatures should increase the sensitivity of detection of genetic variability in the population.

Platform Choice

A number of multicapillary machines can perform SSCP. We prefer to use the ABI3100 because it allows us to alter the temperature of the run below the ambient room temperature. As we have developed a method that collects data for both strands from two temperatures, this flexibility is important in discovering the maximum number of alleles at a single locus. We use GeneScan and GenoTyper software to analyze our electropherograms from the ABI3100. However, software has been developed to handle data from any multicapillary machine (QUISCA) (Higasa *et al.*, 2002).

Calibration of the ABI3100 for SSCP

Automated sequencers that use capillary electrophoresis usually separate ssDNA under denaturing conditions based on length. The ssDNA is usually labeled fluorescently by modification of a nucleotide in the 5' end of the primer or by termination at the 3' end with a fluorescent dideoxynucleotide. A spectral calibration of the dyes used for SSCP must be done for each polymer and buffer combination used. This can often lead to difficulty when using the same conditions under which SSCP of the samples will be performed, because under some conditions (usually temperatures <30°), the spectral calibration will fail. Using the spectral calibration performed for another polymer–buffer combination (e.g., one used for sequencing or fragment analysis) will cause shifting of the peaks of the products, because the spectral calibration basically makes a software filter to identify the individual fluorescent dyes. Running the spectral calibration standards non-denatured does not solve this problem. If the spectral calibration fails repeatedly at 30°, run the spectral standards, denatured, at 60°, because the exact spectral characteristics of the fluorescent dyes in the standards under the polymer buffer combination used should not change with regard to temperature. Fortunately, spectral calibration need only be done for a particular polymer–buffer combination once. However, if the percentage of polymer is changed or the buffer is changed with the addition of glycerol, it would be best to rerun the spectral calibration.

Polymer and Buffer Choice

Whichever polymer is used, it cannot contain a denaturant that will abolish any secondary structure of the ssDNA. All polymers used for sequence analysis or for fragment analysis contain a denaturant and are not suitable for SSCP. Numerous groups have created their own polymers and prepared them without denaturants for SSCP (Albarghouthi and Barron, 2000). We strive here to keep the method within what is easily commercially available. For the ABI3100, the original GeneScan polymer is still available and we have used it throughout our development of the method. In theory, any neutral polymer available for capillary electrophoresis can be used.

A historic use of glycerol in the buffer for running SSCP from the use of acrylamide gels has been carried over to capillary electrophoresis. We experimented with other buffer systems, with and without glycerol, as have others (Ren and Ueland, 1999). We settled on 5% GeneScan polymer with 10% glycerol and 1× TBE for reasons of convenience. Using lower percentages of GeneScan polymer or GeneScan polymer without glycerol decreased the length of time of the runs. However, as described earlier, the capillary refilling on the ABI3100 is sensitive to the viscosity of the polymer being pumped and requires a modification of a program file to correctly fill the capillaries without giving a syringe leak error. Because shortening run times by a few minutes has not been important, we have done all our analysis with 5% GeneScan with 10% glycerol. This was viscous enough to refill the capillaries correctly and gave us acceptable resolution of fragments up to 250 nt in length. In addition, because changing buffers and polymers requires performing a new spectral calibration, choosing one buffer–polymer combination that gives acceptable performance is important to standardizing results.

Polymer should be made up in small batches and not used if more than 2 wk old. To prepare 5 ml of 5% GeneScan polymer with 10% glycerol, combine 3.57 g of 7% GeneScan polymer with 0.5 g of 100% glycerol and 0.5 ml of 10× TBE and add water to a total volume of 5 ml. Filter the diluted polymer through a 0.2-micron filter.

Older polymer will slow migration of peaks, and peaks get rounded and oddly shaped. If 5 ml of polymer is made and used within 2 wk, this reduces the variability in peak mobilities from polymer batch to batch. Old polymer will often give run characteristics that might cause concern about the age of the capillary array. Because capillary arrays are expensive, change the polymer first before changing arrays. We frequently get more than 500 runs per capillary array.

Electroinjection Parameters

A good place to start with electroinjection parameters is 1.5 kV for 22 s. If the amount of fluorescence is too great, the time can be shortened. If the amount of fluorescence is too low, the voltage can be raised. Fluorescence should be in the 300–4000 rfu range for both the sample peaks and the molecular-weight standards.

Injecting identical samples into all 16 capillaries will demonstrate that fluorescence is greater in capillaries 1 and 16 than in 7 and 8. This can sometimes be as much as a fivefold difference. Thus, in any run, some samples may be too high or too low, whereas the majority are within the range. Because fluorescence is not uniform from sample to sample in real population data, the bias in capillaries may not be apparent. However, if the same population is run with many primers, do not be misled into thinking that certain DNA samples (i.e., those that are always run in either capillary 1 or capillary 16) actually amplify better, and so on. The capillary bias may be due to a software filter that attempts to balance the fluorescence from the two lasers by taking into account distance from the laser.

Data Analysis

GeneScan 3.1 (ABI) software was used to analyze all runs with the Large Fragment Analysis option enabled. In June of 2004, ABI discontinued GeneScan and GenoTyper software but will support it for 5 years. The software that has replaced GeneScan and GenoTyper, GeneMapper, does not yet have the Large Fragment Analysis capability, so it is not capable of analyzing ssDNA fragments that have apparent mobilities greater than 500 nt.

The Local Southern method was used for all alignments. The ROX 2500 molecular-weight standard peaks were assigned either their actual molecular weight or a pseudo–molecular weight to align the data. The actual molecular weight is the size, in base pairs, of each standard peak, based on a comparison with the profile of the ROX 2500 standards, run non-denatured at 30° on 3% GeneScan polymer, supplied by the manufacturer (ABI, product information). The pseudo–molecular weight was generated by the standard method used to analyze capillary electrophoresis SSCP (ABI, GeneScan product information). During the analysis of the mobility data, it is best to use a short range of the available molecular-weight standard peaks flanking the product peaks, ensuring that at least two molecular-weight standard peaks are on each flank to satisfy the requirements of the sizing algorithm.

A single sample file (data from one capillary) from each temperature was chosen and a value assigned to each molecular-weight standard peak based on the scan number of that peak for that run. These assignments were used as a "size standard" to analyze all the sample files run at the same temperature. Enabling the Large Fragment Analysis Update (ABI) allowed the actual scan number to be applied to each peak. When analyzing run data to assign molecular weights, individual runs may fail to analyze correctly for various reasons. The range for the molecular weights may have drifted so that peaks are not included. This occurs frequently when shorter specific ranges are chosen. The advantage of choosing smaller ranges is that the interpolation of assigned values is linear and much more accurate, usually to 0.1–0.2% of the mobility value. Runs that have failed to analyze can be reanalyzed with expanded ranges without reanalyzing all the runs. This allows a recursive method to slightly alter the analysis parameters and to get all runs analyzed after several passes.

The other most common reason for failure to analyze is low peak height of the molecular-weight standards. Again, the parameters can be altered to lower the minimum threshold height and only the failed runs reanalyzed. The simplest method to determine whether all runs have correctly analyzed is to view the molecular-weight standards for all runs in GenoTyper, which allows the display of up to 250 runs simultaneously. If all molecular-weight standards have been applied correctly for any one temperature, all molecular-weight standard peaks from all runs should align. Any misaligned peaks are easily detected and these can be reanalyzed in GeneScan.

Examples of Data Analysis

Once the molecular-weight standards have been correctly applied, there are two ways to analyze the electrophoretic data: visually and by tables of mobility data. The user will need to do an initial visual analysis of the data to determine how many different alleles may be present, but when the project is scaled up to hundreds of individuals, an automated approach will be necessary.

Figure 1 has an output plot of electropherograms collected from several experiments that demonstrate some of the problems of calling novel alleles in a germplasm collection. The fragments generated are 437 nt long and are amplified from cultivars of *Theobroma cacao* (Apa4, ICS1, MXC76, P7, Sca6, Spec54). Samples were analyzed at 30°. Denatured molecular-weight standards (ROX 1000) were assigned pseudo–molecular weight values based on their retention time in scan units (the scale listed at the top of the figure). A single run file was used to define the molecular-weight standards.

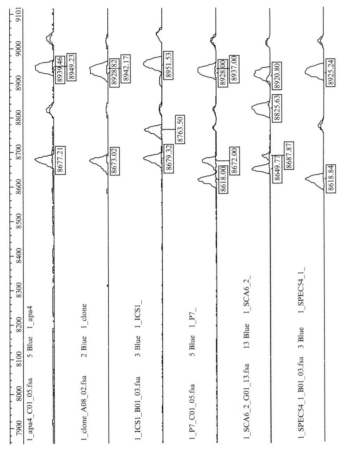

FIG. 1. Screen capture of output from GenoTyper, showing both forward and reverse strand peaks of RGH1 locus on the same electropherogram. Horizontal scale at the top is pseudo–molecular weight or scan number. Peaks are labeled beneath with assigned pseudo–molecular weight. Samples were run at 30°. Cultivars are as follows: row 1, *Theobroma cacao* Apa4; row 2, *T. cacao* cloned allele of RGH1; row 3, *T. cacao* ICS1; row 4, *T. cacao* P7; row 5, *T. cacao* Sca6; row 6, *T. cacao* Spec54.

A cloned single allele sequence from *T. cacao* is also included as a control for the shape and number of peaks that are attributable to a single sequence (line 2 of Fig. 1). In this example, there is a single peak at 8673 (forward strand) for the clone, but two identifiable peaks at 8929 and 8942 (reverse strand, green dye) that are not separated by baseline. Thus, individual sequences may have several stable conformers that may be confused as separate alleles. When cloned individual sequences are not available, homozygotes should provide information on the shape and number of conformers for any sequence. There are also two minor peaks whose mobilities are not labeled. Because peak patterns can be quite complicated, it is usually best to ignore minor peaks and to not be too concerned about relative peak heights of blue and green peaks.

Apa4 is a homozygote of the same allele as the clone (line 1 of Fig. 1). Note that the 8677 peak (forward) and the two reverse peaks are not exactly coincident with the mobilities of the clone peaks. They are within 14 "scan units" of one another, which represents about a 2-s difference in retention times of greater than 20 min. ICS1 is a heterozygote of the Apa4 allele and another allele (8763, forward) but could not be reliably identified as a heterozygote by the reverse peak (8951) alone. It may seem that the reverse strand is not informative for this locus, but in the case of Sca6, the reverse peaks add confidence to the assignment of two novel alleles (8650, 8688, forward) to the locus. In a heterozygote, it is not always possible to link a forward strand peak with a reverse strand peak. Strand mobility for alleles can be quite variable, and it is frequently the case that the forward strand with the greatest mobility has a complementary strand with the slowest mobility in a heterozygote. Thus, if no homozygotes of these alleles are present, positive assignment of a forward and reverse strand peak to an allele is not possible.

Alleles of this locus in different species should be different in sequence, and this appears to be the case for *Theobroma angustifolium* and *Theobroma grandiflorum*, where distinct mobility shifts are seen for both forward and reverse peaks (Fig. 2). The forward peaks for *T. angustifolium* (8594, 8604) and *T. grandiflorum* (8541) have mobilities distinct from the fastest *T. cacao* allele (8618). The reverse peak mobilities are distinct from all but the unique Sca6 allele. Primers specific for *T. cacao* also amplified DNA from the closely related genus *Herrania*. For *Herrania nitida* and *Herrania nycterodendron*, the forward strand peaks have identical mobilities (8773), but the reverse strand peaks are clearly separated (8893 and 8993). Thus, it is important to collect the mobility data for both strands to be able to identify novel alleles.

To increase the amount of information about alleles, mobility data for both strands can be collected over more than one temperature. Because the

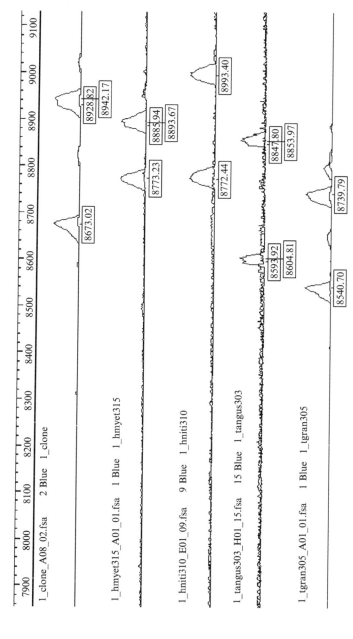

FIG. 2. Screen capture of output from GenoTyper, showing both forward and reverse strand peaks of RGH1 locus on the same electropherogram. Horizontal scale at the top is pseudo–molecular weight or scan number. Peaks are labeled beneath with assigned pseudo–molecular weight. Samples were run at 30°. Row 1, *Theobroma cacao* cloned allele of RGH1; row 2, *Herrania nycterodendron*; row 3, *Herrania nitida*; row 4, *Theobroma angustifolium*; row 5, *Theobroma grandiflorum*.

mobilities of individual ssDNA fragments are strongly and unpredictably affected by temperature, this increases the possibility of identifying hetero-zygotes, as can be seen in Fig. 3. Figure 3 shows an analysis of the WRKY8 locus in TSH516, an important cultivar because it is the parent of the only F_2 population of cacao trees. The internal molecular-weight standards are non-denatured ROX 2500, and they have been assigned their dsDNA molecular-weight values, shown on the scale at the top of Fig. 2. At 24°, TSH516 appears to be homozygous both for the blue (forward) strand and the green (reverse) strand, but at 30°, the reverse strand is clearly hetero-zygous. Thus, from a single amplification, much more information can be collected about the potential alleles by combining the mobility information from both strands at two temperatures.

Allele Calling Based on Mobility for Germplasm Collections

After the application of molecular-weight standards to the GeneScan files, the files can be brought into GenoTyper, a program that allows the automated assignment of alleles based on mobility. GenoTyper works best for microsatellite data in which fragments fall into well-defined categories based on length. In SSCP data, all the fragments are the same length and the mobility of either strand of one allele or the mobilities of the same strand of different alleles cannot be predicted. Thus, it is not usually possible to set up categories or bins that will allow the automated calling of alleles.

Table I shows a subset of data from the WRKY8 locus assayed in 92 individuals from the cacao germplasm collection. The WRKY8 amplicon is 283 nt and at least 16 alleles have been identified. The table contains data exported from GenoTyper into Microsoft Excel. The columns are the pseu-do–molecular weight for the first blue peak at 24°, the first green peak at 24°, the first blue peak at 30°, the first green peak at 30°, and so on. The mobility values are duplicated for the second peak if the individual is a homozygote.

The values for the first peaks of Silecia1 and Tap6 are essentially identical. However, Silecia1 is heterozygous and Tap6 is homozygous. The second allele for Silecia1 is easily detected when present in a hetero-zygote but would probably go undetected or be considered identical to the Tap6 allele as a homozygote. A similar example is seen when comparing the first peaks from EQX3360, Tip4, and UF273. Are they three separate alleles? Thus, even though collecting data over two temperatures increases the resolution of the method, in this case, the fragment is simply too large, and the mobility differences caused by sequence differences are too small to reliably identify alleles in homozygotes. The amplicon size must be reduced to less than 250 nt to allow accurate assessment of the genetic diversity at this locus.

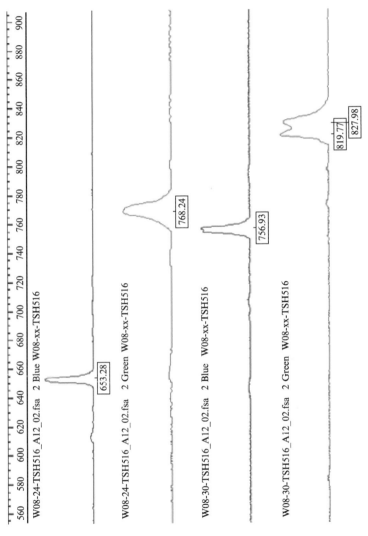

Fig. 3. Screen capture of output from GenoTyper, showing forward and reverse strands peaks of WRKY08 locus of *T. cacao* TSH516 on separate electropherograms. Horizontal scale at the top is molecular weight from un-denatured ROX 2500 standards. Peaks are labeled beneath with assigned molecular weight. Row 1, forward strand at 24°; row 2, reverse strand at 24°; row 3, forward strand at 30°; row 4, reverse strand at 30°.

TABLE I
Identification of Alleles by Mobility from Individuals in a Germplasm Collection

Accession name	24° forward peak 1	24° reverse peak 1	30° forward peak 1	30° reverse peak 1	24° forward peak 2	24° reverse peak 2	30° forward peak 2	30° reverse peak 2
JMC47	5978	6185	5301	5398	5995	6198	5318	5409
Silecia1	5994	6197	5316	5406	6001	6209	5325	5416
Tap6	5998	6199	5317	5407	5998	6199	5317	5407
3C10 cloned allele	6003	6200	5333	5418	6003	6200	5333	5418
IMC60	6003	6201	5336	5421	6012	6213	5336	5428
Nap21bi	6012	6205	5325	5413	6012	6205	5325	5413
P12	5980	6185	5314	5408	5996	6197	5331	5417
Ama9	5987	6189	5309	5402	6003	6202	5324	5413
UNAP2	6007	6202	5321	5410	6022	6202	5339	5410
EQX3360-3	6011	6198	5349	5426	6011	6198	5349	5426
Tip4	6014	6210	5326	5416	6014	6210	5326	5416
UF273	6020	6213	5345	5427	6052	6240	5380	5455
IFC5	6042	6237	5380	5460	6042	6237	5380	5460

Future Developments

We are developing a method that collects mobility data for both strands at two temperatures for cultivars in a collection. The data for homozygotes are then used to define mobilities for both forward and reverse strands at two temperatures, basically giving a four-number name to each allele (forward temp 1, reverse temp 1, forward temp 2, reverse temp 2). Using a program that is still in the testing phase, the data for each homozygote is compared to the data for all the other homozygotes and the alleles are identified. The comparison of allele mobility data for homozygotes is based on an empirically defined threshold that takes into account mobility drift. Once the number of unique alleles has been defined, this process is then used to define the alleles in heterozygotes in the population. The method can use mobility data from either non-denatured or denatured molecular-weight standards. In addition, the assignment of a mobility based on a dsDNA molecular weight reduces the amount of variability in allele assignment when a second population is analyzed at a different time.

References

Albarghouthi, M. N., and Barron, A. E. (2000). Polymeric matrices for DNA sequencing by capillary electrophoresis. *Electrophoresis* **21,** 4096–4111.

Baba, S., Kukita, Y, Higasa, K., Tahira, T., and Hayashi, K. (2003). Single-stranded conformational polymorphism analysis using automated capillary array electrophoresis apparatuses. *Biotechniques* **34,** 746–750.

Borrone, J. W., Kuhn, D. N., and Schnell, R. J. (2004). Isolation, characterization, and development of WRKY genes as useful genetic markers in *Theobroma cacao. Theor. Applied Genet.* **109,** 495–507.

Higasa, K., Kukita, Y., Baba, S., and Hayashi, K. (2002). Software for machine-independent quantitative interpretation of SSCP in capillary array electrophoresis (QUISCA). *Biotechniques* **33,** 1342–1348.

Inazuka, M., Wenz, H. M., Sakabe, M., Tahira, T., and Hayashi, K. (1997). A streamlined mutation detection system: Multicolor post-PCR fluorescence labeling and single-strand conformational polymorphism analysis by capillary electrophoresis. *Genome Res.* **7,** 1094–1103.

Kirk, B. W., Feinsod, M., Favis, R., Kliman, R. M., and Barany, F. (2002). Single nucleotide polymorphism seeking long term association with complex disease. *Nucleic Acids Res.* **30,** 3295–3311.

Kong, P., Hong, C., Richardson, P. A., and Gallegly, M. E. (2003). Single-strand-conformation polymorphism of ribosomal DNA for rapid species differentiation in genus Phytophthora. *Fungal Genet. Biol.* **39,** 238–249.

Kuhn, D. N., Heath, M., Wisser, R. J., Meerow, A., Brown, J. S., Lopes, U., and Schnell, R. J. (2003). Resistance gene homologues in *Theobroma cacao* as useful genetic markers. *Theor. Applied Genet.* **107,** 191–202.

Kuhn, D. N., Borrone, J., Meerow, A. W., Motamayor, J. C., Brown, J. S., and Schnell, R. J. (2005). Single-strand conformation polymorphism analysis of candidate genes for reliable identification of alleles by capillary array electrophoresis. *Electrophoresis* **26,** 112–125.

Kukita, Y., Higasa, K., Baba, S., Nakamura, M., Manago, S., Suzuki, A., Tahira, T., and Hayashi, K. (2002). A single-strand conformation polymorphism method for the large-scale analysis of mutations/polymorphisms using capillary array electrophoresis. *Electrophoresis* **23**, 2259–2266.

Landegren, U., Nilsson, M., and Kwok, P. Y. (1998). Reading bits of genetic information: Methods for single-nucleotide polymorphism analysis. *Genome Res.* **8**, 769–776.

Larsen, L. A., Christiansen, M., Vuust, J., and Andersen, P. S. (1999). High-throughput single-strand conformation polymorphism analysis by automated capillary electrophoresis: Robust multiplex analysis and pattern-based identification of allelic variants. *Human Mutat.* **13**, 318–327.

Larsen, L. A., Christiansen, M., Vuust, J., and Andersen, P. S. (2000). High throughput mutation screening by automated capillary electrophoresis. *Comb. Chem. High Throughput Screen* **3**, 393–409.

Orita, M., Suzuki, Y., Sekiya, T., and Hayashi, K. (1989). Rapid and sensitive detection of point mutations and DNA polymorphisms using the polymerase chain reaction. *Genomics* **5**, 874–879.

Ren, J. C., and Ueland, P. M. (1999). Temperature and pH effects on single-strand conformation polymorphism analysis by capillary electrophoresis. *Human Mutat.* **13**, 458–463.

Schneider, K., Weisshaar, B., Borchardt, D. C., and Salamini, F. (2001). SNP frequency and allelic haplotype structure of beta vulgaris expressed genes. *Mol. Breeding* **8**, 63–74.

[16] Ribosomal RNA Probes and Microarrays: Their Potential Use in Assessing Microbial Biodiversity

By Katja Metfies and Linda Medlin

Abstract

The awareness that global biological diversity is affected by numerous, mostly human-made threats has made biodiversity assessment an important scientific issue for decades. Biodiversity includes different levels of complexity, such as community diversity, habitat diversity, genetic diversity, and species diversity. The application of molecular methods to answer ecological questions permits issues of biodiversity to be addressed at all levels. Microorganisms dominate global biological diversity in terms of their species numbers. However, their small size and limited morphological features make it challenging to obtain a comprehensive view of their biodiversity. The application of ribosomal RNA (rRNA) probes contributes significantly to the assessment of biodiversity at the molecular level. DNA microarrays offer a great potential to facilitate the application of molecular probes and other DNA analytical methods to answer ecological and biodiversity questions. We provide an introduction into the application of rRNA probes and DNA microarrays for the assessment of microbial biodiversity, as well as protocols for the implementation of DNA microarrays.

METHODS IN ENZYMOLOGY, VOL. 395

Introduction

Biodiversity is a popular and widely discussed topic. It is generally accepted that the global biological diversity is under threat. Despite or because of the broad public attention biodiversity has received, the term is often used too simply. Most people associate it with total numbers of different species, but biodiversity is more than that: It also comprises habitat diversity, genetic diversity, and community diversity (Harper and Hawksworth, 1994; Purvis and Hector, 2000). Although biodiversity has been the focus of popular and scientific interest for the past two decades, scientists can only guess at global biodiversity if just species numbers are considered. These estimates range from 10 million to 30 million species. In general, there is a consensus in total numbers at about 13–14 million species on earth (Cruz, 1996; Eldrege, 1998; Mann and Plummer, 1996; Myers, 1998). Even though people want to explore other planets and outer space, our knowledge about the species on earth is far from being complete (Fig. 1). The number of scientifically described species is as few as 13%, or 1.7 million species (Myers, 1998). Among the described species, insects are the largest group, but it is estimated that the microbes are numerically the most dominant organisms on the planet. Nevertheless, most microbes are still unknown. These microscopic aerobic and anaerobic cell factories are the essential catalysts for all chemical reactions within biogeochemical cycles. Macroscopic life and planetary habitability completely depend on the transformations mediated by complex microbial communities. Considering the vital role of microorganismal ecology, it is believed that the microbial diversity is an important prerequisite for ecosystem stabilization. This is of particular interest for the marine environment, where protists and other microorganisms dominate the biodiversity and represent the basis of the food chain. However, it appears to be very difficult to study the

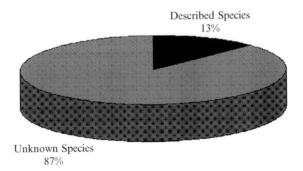

FIG. 1. Current status of the knowledge of global biological diversity in terms of species numbers (Myers, 1998).

biodiversity of microbial organisms by conventional methods. Knowing what "kinds" of organisms exist within a microbial population and how the community structure changes in response to environmental shifts challenges even the most advanced genetic technology and evolutionary theory. One of the fundamental tenets of biodiversity assessment regarding species composition is a reliable taxonomy that provides a hierarchical system for the identification and classification of diverse biological species. However, because of their small size and the presence of only limited or in some cases uninformative morphological features, microbial organisms are taxonomically very challenging. During the first half of the past century, microbial taxonomy based on morphological and physiological characteristics was controversial (Palleroni, 2003). With the introduction of molecular methods into microbial biology, methods became available that address questions related to microbial biodiversity with respect to compositional, structural, or functional aspects.

Assessment of Microbial Species Diversity with Ribosomal RNA

It was first demonstrated in the 1960s that ribosomal genes (ribosomal DNA [rDNA]) and their gene products (ribosomal RNA [rRNA]) could be used for a taxonomic classification of microbial species (Doi and Igarashi, 1965; Dubnau et al., 1965; Pace and Campbell, 1971a,b). This was a big step forward in the study of microbial species diversity. Since then, the comparative analysis of rRNA sequences has become an indispensable method to gain new insights into the phylogeny and diversity of microbial organisms (Díez et al., 2001; John et al., 2003; Moon-van der Staay et al., 2001). Homologous gene sequences possess far more phylogenetic information than phenotypic features. In comparison to other genes, the genes coding for the rRNA are particularly well suited for phylogenetic analysis, because they are universal—found in all cellular organisms; they are of relatively large size; and they contain both highly conserved and variable regions with no evidence for lateral gene transfer (Woese, 1987). The evolutionary information of the rRNA can be uncovered by comparative sequence analysis. Numerous publications have shown the power of the analysis of ribosomal sequences to identify both prokaryotic and eukaryotic microorganisms (Amann et al., 1990; Groben et al., 2004; Simon et al., 2000). Direct cloning and sequencing of the small subunit rDNA (18S rDNA) from natural samples has, for example, permitted a broader view of the structure and composition of picoplankton communities (Giovannoni et al., 1990; Giuliano et al., 1999; López-Garcia et al., 2001). Because of their small size, picoplankton samples are particularly difficult to analyze by conventional methods (Díez et al., 2001; Moon-van der Staay *et al.*, 2001). The analysis of the 18S rDNA circumvents the selective step of laboratory cultivation (Giovannoni *et al.*, 1990).

Ribosomal RNA–Targeted Molecular Probes

The continually growing number of available algal 18S rDNA sequences (e.g., in the Ribosomal Database Project [RDP]) (Maidak *et al.*, 2001) makes it possible to design hierarchical sets of probes that specifically target the 18S rDNA from higher taxonomic levels down to the species level (Groben *et al.*, 2004; Guillou *et al.*, 1999; Lange *et al.*, 1996). This approach makes it possible to quantify representatives from very broad to circumscribed phylogenetic groups in a systematic fashion. In general, the principle of molecular probes identifies an organism by targeting any specific sequence in the genome of the organism. However, the binding sites of phylogenetically derived probes are mainly located within the polymorphic regions of the rRNA gene.

These characters make it possible to detect organisms specifically by the binding to a homologous site within the ribosomal gene sequence of the target species. The design of molecular probes involves the identification of unique sites within the ribosomal sequence of a taxon of interest. Molecular probes usually have a length of approximately 15–25 bases and at least one mismatch to the same region in all other known sequences. A useful tool for the design of molecular probes is the ARB software package (Ludwig *et al.*, 2004). It comprises tools for the handling of sequence information and any other type of additional data linked to the respective sequence. The database that is produced with the ARB software package can be structured according to phylogeny, thus providing the possibility to develop a hierarchical set of specific molecular probes at different taxonomic levels with the help of a probe design function in the program. The ARB software package is available as freeware from the Department of Microbiology, Technical University, Munich (http://www.arb-home.de). The program initially ran on a UNIX platform, but Linux versions are now available.

Applications of Molecular Probes

Molecular probes can be applied for the quantitative analysis of microbial communities with detection by flow cytometry, fluorescence *in situ* hybridization (FISH) (Lim *et al.*, 1999; Miller and Scholin, 1998) or other methods that take advantage of the hybridization principle (Fig. 2). With respect to the application of molecular probes for FISH, it is advantageous if the probes are specific for rRNA because rRNA is a highly abundant molecule in proliferating cells. The efficiency and practicality of molecular probe techniques have been successfully proven in numerous examples from

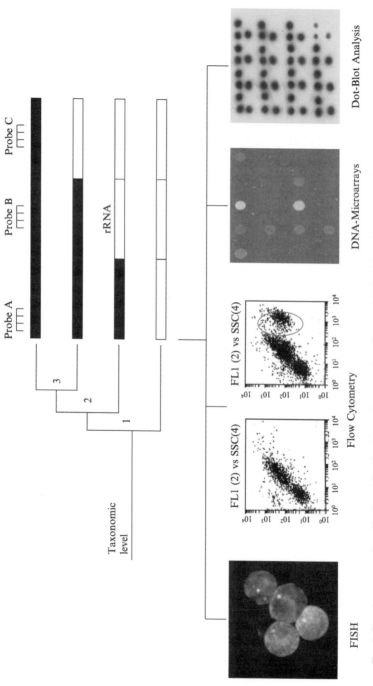

Fig. 2. Development and applications of molecular probes targeting ribosomal RNA (rRNA) sequences. Sequence databases are screened for short stretches of sequences within the rRNA that are shared among the organisms belonging to the same taxon. Hierarchical sets of probes target the rRNA at different taxonomic levels. The probes can be detected by various means.

both prokaryotic and eukaryotic target organisms (Biegala *et al.*, 2003; Sekar *et al.*, 2003). Nevertheless, these well-established approaches have the major disadvantage that they can be used to identify only one or a few organisms at a time and are not yet automated (DeLong *et al.*, 1989). This makes it very time consuming to get a broader view of a microbial sample.

DNA Microarrays

The introduction of the concept of DNA microarrays about 10 years ago suggests a solution for the limitations of molecular probe applications (Lockhart *et al.* 1996). They present an opportunity for high-throughput analysis of hybridization-based methods by miniaturization, and they possess the potential for highly multiplexed assays. DNA microarrays allow the simultaneous analysis of almost infinite numbers of probes simultaneously on just one DNA microchip (Brown and Botstein, 1999; Lockhart *et al.*, 1996). The DNA microchip is at the heart of the technology. It contains a high-density array of oligonucleotides, polymerase chain reaction (PCR) products or cDNAs spotted in an ordered way onto the surface of the DNA microchip (Lockhart and Winzeler, 2000). There are two fundamentally different ways for the manufacturing of gene chips. In the first, oligonucleotides are synthesized directly on the chip surface (Singh-Gasson *et al.*, 1999), which enables the production of gene chips with a high spot density. Currently gene chips exist with numbers as high as 100,000/cm^2 spots. Alternatively, it is possible to print or spot nucleic acids onto the surface of a glass slide (Okamato *et al.*, 2000) via a spotting device that features a high-speed robotic arm fitted with a number of pins or a piezo-electric pipette. Several companies (e.g., Qiagen Operon [Germany] or Affymetrix [USA]) produce customized DNA microarrays. In either case, the most broadly used support of rDNA microchips is a glass slide in the format of a commonly known microscope slide. Glass is an appropriate material for the production of DNA microarrays because it has low background fluorescence and is, therefore, compatible with fluorescence labeling. This is of particular importance because target nucleic acids like cDNA, RNA, or PCR products are usually detected via fluorescence. The fluorescence can be either incorporated in the nucleic acid directly via a fluorescent dye or indirectly by some other moiety such as biotin, which permits detection with a secondary fluorescent label (Cheung *et al.*, 1999; Southern *et al.*, 1999). Subsequent to the hybridization of the target nucleic acid to the immobilized probes on the DNA microchip, the analysis of the fluorescence pattern is performed with a microarray scanner. A microarray scanner is a device that bears a laser or a polychromatic light source for the specific excitation of the fluorophore (DeRisi *et al.*, 1997).

Assessment of Microbial Species Composition with DNA Microarrays

Microarray technology provides a promising tool based on molecular probes to identify species in samples from complex environments quickly without a cultivation step (Figs. 3 and 4). Consequently, it has great potential as an application for the assessment of microbial biodiversity in terms of species composition. This is of special interest for the identification of prokaryotic and eukaryotic cells with very small sizes and few distinct morphological features (Figs. 3 and 4). In addition, because the number of taxonomists in phycology has been decreasing steadily, DNA microarrays also might be of special value for phycological studies because they represent a tool that does not require a broad taxonomic knowledge to identify cells. Consequently, a growing number of publications report the use of microarrays bearing molecular probes that target the rRNA for the identification of microbial species. They have been used successfully in combination with an amplification of the rRNA gene for the identification of phytoplankton, nitrifying bacteria, bacterial fish pathogens, and sulfate-reducing prokaryotes (Call *et al.*, 2003; Guschin *et al.*, 1997; Loy *et al.*, 2002; Metfies and Medlin, 2004).

Even though DNA microarrays appear to have great potential to facilitate the analysis of the species composition of microbial samples, it has to be considered that amplification of nucleic acids from complex environmental samples can be very susceptible to biases leading to under-representation of targets or problems with the amplification of nucleic acids from certain species. One possibility to circumvent these possible

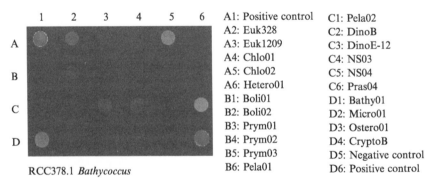

	A1: Positive control	C1: Pela02
	A2: Euk328	C2: DinoB
	A3: Euk1209	C3: DinoE-12
	A4: Chlo01	C4: NS03
	A5: Chlo02	C5: NS04
	A6: Hetero01	C6: Pras04
	B1: Boli01	D1: Bathy01
	B2: Boli02	D2: Micro01
	B3: Prym01	D3: Ostero01
	B4: Prym02	D4: CryptoB
	B5: Prym03	D5: Negative control
RCC378.1 *Bathycoccus*	B6: Pela01	D6: Positive control

Fig. 3. Image of a scanned DNA microarray. The DNA chip contained the indicated probes. A polymerase chain reaction (PCR) fragment of the 18S ribosomal DNA (rDNA) of a species belonging to the picoplanktonic genus *Bathycoccus* was identified with a hierarchical set of probes that targeted the 18S rDNA at the level of kingdom (Euk1209), phylum (Chlo02), class (Pras04), and genus (Bathy01).

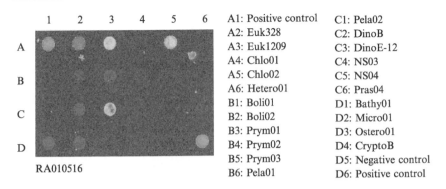

A1: Positive control	C1: Pela02
A2: Euk328	C2: DinoB
A3: Euk1209	C3: DinoE-12
A4: Chlo01	C4: NS03
A5: Chlo02	C5: NS04
A6: Hetero01	C6: Pras04
B1: Boli01	D1: Bathy01
B2: Boli02	D2: Micro01
B3: Prym01	D3: Ostero01
B4: Prym02	D4: CryptoB
B5: Prym03	D5: Negative control
B6: Pela01	D6: Positive control

FIG. 4. Analysis of a picoplankton sample without previous cultivation. The 18S ribosomal DNA (rDNA) was amplified by polymerase chain reaction (PCR) and hybridized to a DNA chip that contained the indicated probes. The sample included Chlorophytes (Prasinophytes), Prymnesiophytes, Dinoflagellates, and Bolidophytes.

biases introduced by PCR amplification is to directly detect the rRNA from the species in the environmental sample. The use of rRNA as the target molecule for the organisms of choice could circumvent the PCR amplification because the molecule is present in high numbers in a cell. In prokaryotes, up to 80% of the RNA in a cell is rRNA (Woese, 1987). As another potential bias, it also must be considered that the specificity of the 18S rDNA–targeted probes on the DNA chip may have changed since the probes were initially designed. Probe development must always be subject to refinement. Because the known rRNA sequence database is only a small fraction of the total diversity, it is important to reevaluate the specificity of rRNA-targeted probes on a regular basis with respect to the continually increasing number of 18S rRNA sequences in public databases.

Gene Expression Profiling with DNA Microarrays

The application of microarrays for the identification of organisms is a younger application of the technology. However, the assessment of gene expression profiles was one of the first applications of microarray technology and is probably the most widely distributed application (Brown and Botstein, 1999). Gene expression profiling with DNA microarrays has numerous potential applications to address questions in ecology and evolution. The success of a species in any given environment is determined by its ability to recognize and respond to environmental changes. This involves the interaction of an organism with the environment, as well as the interaction of an organism with other organisms in a habitat. A response to an environmental change usually produces significant change in the gene expression and protein composition of an organism. The estimation of the messenger RNA (mRNA) composition of an organism is one of the keys to understanding the cellular response to environmental forces. In this respect, the widely distributed methods for gene expression profiling in biomedicine have been adapted to address ecologically related questions. Gene expression profiling generates information about gene expression on a genomewide scale for different organisms at certain phases of development and under different environmental conditions. The response of the yeast, *S. cerevisiae,* to environmental changes was one of the first applications of expression profiling and has been intensively exploited (DeRisi *et al.,* 1997; Holstege *et al.,* 1998). In *S. cerevisiae,* it could be shown that the transcriptional activators, Msn2/Msn4, are involved in the activation of almost all of the genes in the cell that were induced by environmental stresses (Causton *et al.,* 2001). With the continuously growing number of completed genomes, it is to be expected that the number of this kind of study is going to increase, and it is conceivable that it could be possible to identify specific stress-induced genes or genes related to particular biochemical processes for a variety of organisms (Fig. 5).

In terms of biochemical microbial processes, the productivity of the ocean is an important issue. The photosynthetic carbon dioxide fixation by oceanic phytoplankton is one of the most important global carbon sinks (Chisholm, 2000). The measurement of the expression of carbon fixation genes in phytoplankton can serve as an indicator for the community photosynthesis. In this respect, the species-specific determination of the gene expression of RuBisCo (ribulose-1, 5-bisphosphate carboxylase/oxygenase)

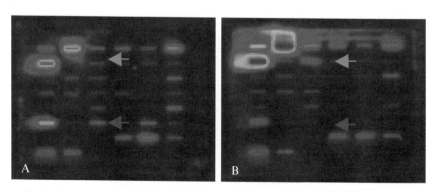

Fig. 5. Gene expression profiles of a subset of genes of the Antarctic diatom *Fragilariopsis cylindrus* at different temperatures (from Mock [2003]). (A) The diatom was grown at 2°. (B) The diatom was grown at 5°. Light arrows indicate a gene that is upregulated in panel B in comparison to panel A and dark arrows indicate a gene that is downregulated in panel B in comparison to panel A.

provided an insight into which phytoplankton groups were photosynthetically active at an oligotrophic site of the Gulf of Mexico (Wawrik *et al.*, 2002). However, understanding the response of an organism to its environment involves the elucidation of gene functions. Understanding gene functions may provide important insights to how organisms function and interact in the environment. Expression profiling could contribute to the understanding of other gene functions by the comparison of transcription profiles of unknown genes to transcription profiles of known genes. As probes for the elucidation of gene function and genes expressed specifically under certain environmental conditions or in the course of certain biochemical processes are under development, the possibility exists that a DNA microarray could be used to characterize an ecosystem in terms of the abundance and biodiversity of organisms that perform specific processes or respond to the stresses that are impacting populations in the field.

Applications of DNA Microarrays for Genotyping

Besides the determination of species numbers, biodiversity assessment may involve the estimation of genetic diversity and the investigation of population structure at the subspecies level. In the premolecular era, the study of microbial genetic diversity and population structures was often hampered by the same limitations as the identification of species, for example, the small size of the cells, a lack of considerable morphological features, and the inability to bring cells into culture (Medlin and Simon, 1998). The introduction of molecular methods was not only a quantum leap in regard to microbial phylogenetics, but it also made it possible to increase the resolution of the analysis of microbial population structure. The application of molecular marker techniques represents a means to study population structures at subspecies level unlike that available with conventional methods. Molecular markers can be locus specific and polymorphic in a studied population; therefore, they are very well suited to differentiate organisms in a given population. Molecular markers revealed, for example, the general trend that multiple isolates of a single species are often geographically related (Bakker *et al.*, 1995), but sometimes the geographic groupings uncover polyphyletic or paraphyletic taxa (Bakker *et al.*, 1995). In the past 2 decades, a number of genotyping methods based on molecular markers have been developed, and in theory most of them could be adapted to an analysis in combination with DNA microarrays. Fingerprinting methods such as restriction fragment length polymorphism (RFLP), random amplified polymorphic DNA (RAPD), and amplified fragment length polymorphism (AFLP) are widely used and have been applied in the identification of genetic variation of different microorganism species

(Gomez and Gonzalez, 2001; Jackson *et al.*, 1999; Johannsson *et al.*, 1995). However, these methodologies based on the size separation of DNA fragments with gel electrophoresis are labor intensive and time consuming and are difficult to analyze statistically (Kingsley *et al.*, 2002). Moreover, they possess difficulties if bands on the gel have to be precisely correlated with genetic variants, because in some cases two bands on a gel that have the same size do not necessarily also have the same sequence.

The limitations of gel-based methods for genotyping can be overcome by the application of DNA microarray–based methods for genomic finger-printing. DNA microarrays make it feasible to analyze target organisms in a sequence-specific and quantitative way, allowing molecular assessment of genetic variation among organisms in a population. With respect to genetic variation, the analysis of single nucleotide polymorphisms (SNPs) has become a focus of interest. SNPs are polymorphic positions in the genome of organisms at which nucleotides differ because of single nucleotide substitutions or single nucleotide insertions/deletions (Wang *et al.*, 1998). The analysis of SNPs with DNA microarrays could serve as a high-throughput method for the analysis of genetic variation (Lindblad-Toh *et al.*, 2000; Raitio *et al.*, 2001). However, a prerequisite of the analysis of SNPs by DNA microarrays is detailed sequence information of a number of variants of organisms from one species. Because of missing sequence information, the application of SNPs for genotyping of organisms is not very widely distributed if the variety of species is considered. Currently, genotyping by SNPs is mostly used to address questions related to the human genome. In particular, the amount of money that is required for the development of SNPs exceeds, in many cases, the budget of laboratories that are not involved in medical biotechnology or intensive genomics studies in other model organisms. With regard to microorganisms, a collection of 186 SNPs has been developed for the green alga *Chlamydomonas reinhardtii* (Vysotskaia *et al.*, 2001).

The application of DNA microarrays for genotyping is not restricted to the analysis of SNPs. In fact, there are examples in the literature where DNA microarrays facilitate the assessment of genetic variation. For example, DNA microarrays have been applied for the analysis of microsatellites (Radtkey *et al.*, 2000). Microsatellites are specific DNA sequences that consist of tandemly repeated monomers of 1–5 bp and are highly variable among individuals in a population. Hence, they possess the possibility to make inferences about a number of questions related to the assessment of biodiversity, including population structure, the level of genetic drift, sudden changes in populations size, effects of population fragmentation and the interaction with different populations, and the identification of new and incipient populations. Microsatellites have been identified for a wide variety of eukaryotic genomes (Toth *et al.*, 2000). With respect to the assessment of microbial diversity, microsatellites have been identified for phytoplankton (Kang and Fawley, 1997). As with other fingerprinting techniques, the analysis of microsatellites is usually discriminated on gels, which are time consuming to prepare and pose problems for a reliable quantitative analysis. The application of DNA microarrays to the analysis of microsatellites would facilitate the assay and improve the accuracy of the analysis. For the analysis of microsatellites, a DNA chip would contain probes that target sequences of microsatellite loci that have been identified previously. Thus, the development of microsatellites is also a cost-intensive issue but may be worth the time investment for large-scale screenings.

To circumvent the need for sequence information before genotyping, DNA microarrays have been developed that reproduce a broadly applicable fingerprinting technique. They contain nanomers with random sequences. Such a DNA microarray has been used to display differences between closely related bacterial organisms belonging to the genus *Xanthomonas* and different strains of *Escherichia coli* (Kingsley *et al.*, 2002). DNA chips that contain nanomers can be used to generate a fingerprint of any microorganism, because on average nanomers occur only once every 131 kbp in double stranded DNA (Kingsley *et al.*, 2002). DNA microarrays with random oligonucleotides possess strong potential for fingerprinting microorganisms in different ways. Thousands of probes could be immobilized on one DNA chip, making it possible to generate fingerprints of microorganisms by using all possible nanomers on that single DNA chip. This would greatly increase the resolution of the fingerprinting method for the study of genetic variation in populations. In comparison to gel-based fingerprinting, all DNA chip–based fingerprinting methods have the major advantage that the DNA chip offers the possibility for reliable quantitative analysis of the different fingerprinting patterns. Genetic differences among isolates and closely related species can also be identified with microarrays that contained probes targeting the whole genome of organisms. In this case, polymorphisms and insertions can be detected as a reduction or elevation of a hybridization signal if two different genomic DNA samples are used for hybridization (Gibson, 2002). The comparison of whole genomes of different strains of various microbes indicates that polymorphism for gene content is not uncommon (Riley and Serres, 2000), suggesting genomic adaptations to certain habitats.

Protocol DNA Microarray Analysis

If it comes to the application of microarrays in practice, any DNA microarray analysis involves simplified four steps until, for example, target species in a sample or differentially displayed genes in an expression analysis can be identified. Subsequent to the experimental design, the first step of a microarray analysis is the preparation of sufficient target nucleic acid. This step differs slightly between the different approaches and is discussed in detail separately for species identification or gene expression profiling. However, the principles of the following steps are shared among all potential applications of DNA microarray analysis in biodiversity assessment. The steps are labeling of the nucleic acids, hybridization, and analysis of the hybridization pattern on the DNA microarray.

Preparation of Target Nucleic Acid for Species Identification

In terms of species composition, environmental samples can be analyzed with a DNA microarray containing probes that specifically target the small subunit of the rRNA of a diverse set of microbial taxa at different taxonomic levels (Fig. 2). In the case of prokaryotes, such a chip would contain probes that specifically target the 16S rDNA (Loy *et al.*, 2002). However, a chip dedicated to the identification of eukaryotes would contain specific probes that bind to specific motives in the 18S rDNA (Metfies and Medlin, 2004). The analysis could be either carried out with rRNA isolated from the samples or with DNA fragments that were amplified from genomic DNA of the sample. The isolation of high-quality genomic DNA or rRNA from environmental samples is a prerequisite for a successful microarray analysis. This issue is discussed in detail in Chapter 2 of this volume (Valentin *et al.*).

Preparation of Target Nucleic Acid for Gene Expression Profiling

The focus of gene expression profiling is the identification of differentially expressed genes of the same type of cells grown under different environmental conditions on a genomewide level. The target nucleic acid of gene expression profiling is mRNA, because the response of an organism to environmental changes goes along with broad changes of the gene expression, which is reflected by the mRNA composition. It was mentioned before that a successful microarray analysis requires high-quality target nucleic acids. This is of particular importance for gene expression profiling. mRNA is very susceptible to degradation. Partial degradation of the RNA may result in a wrong estimation of the relative proportions of mRNAs hybridized to DNA microarrays.

Labeling of Nucleic Acids

Subsequent to the isolation of RNA or DNA, the next step of the microarray analysis is the generation of labeled target nucleic acid. A number of different strategies can be used to label nucleic acids. First, nucleic acids can be directly labeled randomly without an amplification step. For DNA, Nick translation or the application of the Klenow fragment for random-sequence oligonucleotide-primed synthesis is the classic way to directly label nucleic acids (Ausubel et al., 1998). Alternatively, there are easy-to-handle labeling kits like the BrightStar Psoralen-Biotin labeling kit (Ambion, Austin, TX) or the BiotinULS labeling kit (Fermentas, Germany), which make it possible to chemically incorporate fluorophores or a biotin moiety into both DNA and RNA. The principle of these direct labeling kits is that the fluorophore or the biotin is coupled to a compound that intercalates or binds covalently to nucleotides within the target nucleic acid. These labeling methods are quick and simple and make it possible to incorporate more than one label to the nucleic acid. However, a drawback to these direct labeling methods is that they do not select for degraded RNAs. Partially degraded RNAs are shorter than their cognate display elements. Shorter RNAs lead to a reduction in signal intensity. This displays, consequently, differences in fluorescent signal intensities between two different RNA samples that reflect differences in the quality of the two RNA preparations rather than differences in numbers of molecules. The application of oligo (dT)-primer for the synthesis of labeled cDNA can help to circumvent this problem in gene expression profiling, where mRNA is the target nucleic acid. However, if rRNA is the target nucleic acid, synthesizing fluorescently labeled PCR fragments of the ribosomal gene can minimize this problem. This could be done by the incorporation of nucleotides that carry a fluorophore or a biotin moiety in the course of the PCR. Alternatively, labeled primers could be used for the amplification. As an example, biotinylated target DNA for the analysis of eukaryotic microbial species diversity can be generated with a set of primers that amplify about an 1800 bp PCR fragment of the 18S rDNA. The primer set encounters the biotinylated primer 1F-biotin (5′-AAC CTG GTT GAT CCT GCC AGT-3′) and the unlabeled primer 1528R (5′-TGA TCC TTC TGC AGG TTC ACC TAC-3′) (Medlin et al., 1988).

Hybridization

One of the goals of a successful hybridization is to minimize cross-hybridization between arrayed elements on the chip and nontarget nucleic acids. It is of significant importance to achieve high signal-to-noise ratios and to ensure that the signal intensity on the chip is proportional to the amount of nucleic acid bound to the chip. These goals can be achieved by optimizing the hybridization conditions in respect to a number of

different parameters like ionic and buffering conditions, hybridization volume, or time of prehybridization and hybridization. A hybridization
buffer typically contains blocking reagents such as bovine serum albumin
(BSA) or a detergent like sodium dodecyl sulfate (SDS) to minimize
background noise. However, the important step to minimize background
noise is the prehybridization. Typically, this involves an incubation of the
DNA chips for a minimum of 30 min before the actual hybridization at
hybridization temperature in hybridization buffer lacking target nucleic
acid. During hybridization of the nucleic acid, the annealing to the arrayed
elements on the DNA chip is promoted by a high ionic strength of the
hybridization buffer. The concentration of the nucleic acid in the hybridization mixture is important for the outcome of the microarray experiment.
If the concentration is too low, annealing will not proceed quick enough
and it will be difficult to detect a sufficient signal for analysis. As a rule of
thumb, the hybridization mixture should consist of the nucleic acid at a
concentration of about 30 ng/μl in hybridization buffer (1 M NaCl/10 mM
Tris; pH 8/0.005% Triton X-100/1 mg/ml BSA/0.1 μg/μl Herring Sperm
DNA). If these amounts are not available, it is recommended to amplify
the nucleic acid (Bowtell and Sambrook, 2003). The volume of the hybridization mixture should not exceed about 35–50 μl. During hybridization,
the reaction is protected from evaporation by a coverslip, which is floating
on the DNA chip and dislocated from the hybridization area, if the hybridization volume is too high. Before the application to the microarray, the
hybridization mixture has to be denatured for 5 min at 94°. This step is
important to dissolve secondary structures that might have been formed by
the target nucleic acid. The application of the hybridization mixture is a
crucial step in a microarray experiment. It is recommended to perform this
step in a clean, dust-free environment, because dust particles on the chip
can interfere with the hybridization signals. The hybridization mixture
should be applied evenly onto the microarray and a coverslip should be
deposited over the solution right after the application of the hybridization
solution to prevent it from evaporating. An incubation of the DNA chip in
a humid environment during hybridization additionally counteracts evaporation that occurs at the edges of the coverslip. Humid environments can be
generated, for example, by filling the lower part of an empty pipette box
with 5–10 ml of water. In this arrangement, the DNA chip rests on the
perforated surface of the pipette box. Hybridization should be carried out
for 12–16 h at hybridization temperature. The hybridization temperature is
another important factor that strongly affects the hybridization efficiency
in a microarray experiment. However, it depends on the nature of the
hybridization buffer and the melting temperatures of the arrayed elements

on the chip. If molecular probes are spotted on the chip, the hybridization temperature should be selected and optimized in respect to the average melting temperature of the probes. In respect to hybridization buffers, optimal hybridization temperatures that contain 50% formamide are on average about 20° lower than optimal hybridization temperatures of aqueous hybridization buffers. If aqueous hybridization buffers are used, it is recommended to optimize the hybridization temperature in a range between 55° and 65°. Subsequent to the hybridization, cross-hybridized nucleic acids are washed off the chip by the application of washing buffers that have lower ionic strength than the hybridization buffer. Typically, washing buffers contain between 2× SSC and 0.01× SSC. Best results are achieved if a sequence of different washings with gradually decreasing SSC concentrations is performed. The first buffer after the hybridization can contain additionally SDS as a detergent at concentrations between 0.01% and 0.05%. However, the buffer used for the last washing step should not contain SDS. The presence of SDS in the buffer leads to strong background intensities and consequently a decrease of the signal-to-noise ratio.

Acquisition of Hybridization Signals. Before the measurement of signal intensities, hybridized biotinylated target nucleic acids that are labeled with biotin have to be visualized by staining the DNA microchips with streptavidin–fluorophore conjugates (e.g., streptavidin-Cy5 or streptavidin-Cy3 [Amersham Biosciences, Germany]). Dissolved in hybridization buffer, the streptavidin–fluorophore complex binds covalently to the biotin label on the nucleic acids. The staining works well if the DNA chips are incubated at room temperature for about 30 min. The concentration of the dye should be optimized in a range of 50–400 ng/ml. To prevent evaporation during the staining procedure, a coverslip is deposited onto the microarray. For the same reasons discussed earlier for the volume of the hybridization solution, the volume of the staining solution should not exceed 35–50 μl. Fluorescence images of the hybridized microarray are taken with a microarray reader, such as the Genepix 4000B-Scanner (Axon Instruments, Inc., USA). This device is a laser-based system that uses two lasers as excitation sources for the fluorescent dyes. Alternatively, some microarray readers use a charged-coupled device (CCD) camera. These systems use a beam of white light that is directed through optical filters to gain the adequate excitation wavelength. In contrast to the laser-based systems that are restricted to certain excitation wavelengths, the CCD-based systems are more flexible in this respect. The optical filters in the systems are usually relatively easy to replace and allow choosing between wider ranges of excitation wavelengths.

Image Analysis

The fluorescent signal intensities on the DNA chip are quantified using image analysis software packages, such as the GenePix 4.0 software (Axon Instruments, Inc., Union City, CA). The first step to precisely quantify each single spot is the superimposition of a grid of individual circles onto the microarray image. In this step, the location of a spot is defined. This step is very important, because spots might be dislocated from the expected position or have different sizes due to variations during the printing process. It follows the segmentation of the image, which is the classification of pixels on the image in foreground or background. This allows fluorescence intensities to be calculated and displayed as pixel values. Each pixel value represents the level of hybridization at a certain location on the slide. The amount of hybridized nucleic acid at a certain spot of the microarray is proportional to the fluorescent signal at the spot. For later calculations, the mean signal intensity is determined for each spot. The measurement of spot intensities includes signals originating from nonspecific hybridization or fluorescence emitted by other chemicals on the DNA chip. Therefore, it is finally important to assess the background signal. If segmentation has been performed, it is possible to determine local background intensities for each spot. The values of the fluorescent intensities are typically stored as 16-bit images and can be used for data analysis.

Concluding Remarks

Microarray technology has various potential applications for biodiversity assessment at all levels. The biodiversity of natural complex microbial samples can be addressed in terms of species numbers and composition without a cultivation step by the application of microarrays that contain hierarchical sets of molecular probes like rRNA probes, which allow the quantification of representatives from very broad to circumscribed phylogenetic groups in a systematic fashion. Gene expression profiles should provide insights into how organisms interact with their environment and how they respond to environmental changes. Biodiversity in terms of genetic diversity at the subspecies level can be addressed with microarrays that target molecular markers, such as RAPDs, AFLPs, or microsatellites. A microarray that contains random oligonucleotides could be used to generate fingerprints of populations that can be easily interpreted. A polymorphism in genomic DNA content of organisms of one species can be visualized by a microarray that targets whole genomes of organisms. Usually, however, this kind of microarray is most widely used to generate expression profiles of organisms. A protocol for the application of DNA microarrays in biodiversity assessment was described in detail in this chapter.

References

Amann, R. I., Binder, B. J., Olson, R. J., Chisholm, S. W., Devereux, R., and Stahl, D. A. (1990). Combination of 16S rRNA–targeted oligonucleotide probes with flow cytometry for analyzing mixed microbial populations. *Appl. Environ. Microbiol.* **56,** 1919–1925.

Ausubel, F. M., Brent, R., Kingston, R. E., Moore, D. D., Seidman, J. G., Smith, J. A., and Struhl, K. (1998). "Current Protocols in Molecular Biology." John Wiley & Sons, New York.

Bakker, F. T., Olsen, J. L., and Stam, W. T. (1995). Global phylogeography in the cosmopolitan species *Cladophora vagabunda* (Chlorophyta) based on nuclear DNA ITS sequences. *Eur. J. Phycol.* **30,** 187–210.

Biegala, I. C., Not, F., Vaulot, D., and Simon, N. (2003). Quantitative assessment of picoeukaryotes in the natural environment by using taxon-specific oligonucleotide probes in association with tyramide signal amplification-fluorescence *in situ* hybridization and flow cytometry. *Appl. Environ. Microbiol.* **69**(9), 5519–5529.

Bowtell, D., and Sambrook, J. (2003). "DNA Microarrays." Cold Spring Harbor Laboratory Press, Cold Spring Habor, NY.

Brown, P. O., and Botstein, D. (1999). Exploring the new world of the genome with DNA microarrays. *Nat. Genet.* **21,** 33–37.

Call, D. R., Borucki, M. K., and Loge, F. J. (2003). Detection of bacterial pathogens in environmental samples using DNA microarrays. *J. Microbiol. Methods* **53**(2), 235–243.

Causton, H. C., Ren, B., Koh, S. S., Harbison, C. T., Kanin, E., Jennings, E. G., Lee, T. I., True, H. L., Lander, E. S., and Young, R. A. (2001). Remodeling of yeast genome expression in response to environmental changes. *Mol. Biol. Cell* **12**(2), 323–337.

Cheung, V. G., Morley, M., Aguilar, F., Massimi, A., Kucherlapati, R., and Childs, G. (1999). Making and reading microarrays. *Nat. Genet.* **21**(Suppl. 1), 15–19.

Chisholm, S. W. (2000). Stirring times in the southern ocean. *Nature* **407,** 685–689.

Cruz, M. (1996). Management options for biodiversity protection and population. *In* "Human Population, Biodiversity and Protected Areas: Science and Policy Issues" (V. Dompka, ed.), pp. 71–89. American Association for the Advancement of Science, Washington, DC.

DeLong, E. F., Wickham, G. S., and Pace, N. R. (1989). Phylogenetic stains: Ribosomal RNA–based probes for the identification of single cells. *Science* **243,** 1360–1363.

DeRisi, J. L., Iyer, V. R., and Brown, P. O. (1997). Exploring the metabolic and genetic control of gene expression on a genomic scale. *Science* **278**(5338), 680–686.

Díez, B., Pedrós-Alió, C., and Massana, R. (2001). Study of genetic Diversity of eukaryotic picoplankton in different oceanic regions by small-subunit rRNA Gene Cloning and Sequencing. *Appl. Environ. Microbiol.* **67,** 2942–2951.

Doi, R. H., and Igarashi, R. T. (1965). Conservation of ribosomal and messenger ribonucleic acid cistrons in *Bacillus* species. *J. Bacteriol.* **90,** 384–390.

Dubnau, D., Smith, I., Morell, P., and Marmur, J. (1965). Gene conservation in Bacillus species. I. Conserved genetic and nucleic acid base sequence homologies. *Proc. Natl. Acad. Sci. USA* **54**(2), 491–498.

Eldrege, N. (1998). "Life in the Balance: Humanity and the Biodiversity Crisis." Princeton University Press, Princeton, NJ.

Gibson, G. (2002). Microarrays in ecology and evolution: A preview. *Mol. Ecol.* **11,** 17–24.

Giovannoni, S. J., Britschgi, T. B., Moyer, C. L., and Field, K. G. (1990). Genetic diversity in Sargasso Sea bacterioplankton. *Nature* **345,** 60–63.

Giuliano, L., De Domenico, M., De Domenico, E., Hofle, M. G., and Yakimov, M. M. (1999). Identification of culturable oligotrophic bacteria within naturally occurring bacterioplankton communities of the Ligurian Sea by 16S rRNA sequencing and probing. *Microb. Ecol.* **37**(2), 77–85.

Gomez, P. I., and Gonzalez, M. A. (2001). Genetic polymorphism in eight Chilean strains of the carotenogenic microalga *Dunaliella salina* Teodoresco (Chlorophyta). *Biol. Res.* **34**(1), 23–30.

Groben, R., John, U., Eller, G., Lange, M., and Medlin, L. K. (2004). Using fluorescently-labelled rRNA probes for hierarchical estimation of phytoplankton diversity. *Nova Hedwigia* **79,** 313–320.

Guillou, L., Moon-van-der-Staay, S. Y., Claustre, H., Partensky, F., and Vaulot, D. (1999). Diversity and abundance of Bolidophyceae (Heterokonta) in two oceanic regions. *Appl. Environ. Microbiol.* **65,** 4528–4536.

Guschin, D. Y., Mobarry, B. K., Proudnikov, D., Stahl, D. A., Rittmann, B. E., and Mirzabekov, A. D. (1997). Oligonucleotide microchips as genosensors for deter-minative and environmental studies in microbiology. *Appl. Environ. Microbiol.* **63,** 2397–2402.

Harper, J. L., and Hawksworth, D. L. (1994). Biodiversity: Measurement and estimation. *Philos Trans. R. Soc. Lond. B Biol. Sci.* **345**(1311), 5–12.

Holstege, F. C., Jennings, E. G., Wyrick, J. J., Lee, T. I., Hengartner, C. J., Green, M. R., Golub, T. R., Lander, E. S., and Young, R. A. (1998). Dissecting the regulatory circuitry of a eukaryotic genome. *Cell* **95**(5), 717–728.

Jackson, P. J., Hill, K. K., Laker, M. T., Ticknor, L. O., and Keim, P. (1999). Genetic comparison of *Bacillus anthracis* and its close relatives using amplified fragment length polymorphism and polymerase chain reaction analysis. *J. Appl. Microbiol.* **87**(2), 263–269.

Johannsson, M. L., Molin, G., Pettersson, B., Hlen, M., and Ahrne, S. (1995). Characterization and species recognition of *Lactobacillus plantarum* strains by restriction fragment length polymorphism (RFLP) of the 16S rRNA gene. *J. Appl. Bacteriol.* **79,** 536–541.

John, U., Fensome, R. A., and Medlin, L. K. (2003). The application of a molecular clock based on molecular sequences and the fossil record to explain biogeographic distributions within the *Alexandrium tamarense* "species complex" (Dinophyceae). *Mol. Biol. Evol.* **20,** 1015–1027.

Kang, T. J., and Fawley, M. W. (1997). Variable (CA/GT)n simple sequence repeat DNA in the alga *Chlamydomonas*. *Plant Mol. Biol.* **35**(6), 943–948.

Kingsley, M. T., Straub, T. M., Call, D. R., Daly, D. S., Wunschel, S. C., and Chandler, D. P. (2002). Fingerprinting closely related *Xanthomonas* pathovars with random nonamer oligonucleotide microarrays. *Appl. Environ. Microbiol.* **68**(12), 6361–6370.

Lange, M., Guillou, L., Vaulot, D., Simon, N., Amann, R. I., Ludwig, W., and Medlin, L. K. (1996). Identification of the class Prymnesiophyceae and the genus *Phaeocystis* with ribosomal RNA–target nucleic acid probes detected by flow cytometry. *J. Phycol.* **32,** 858–868.

Lim, E. L., Dennett, M. R., and Caron, D. A. (1999). The ecology of *Paraphysomonas* imperforata based on studies employing oligonucleotide probe identification on coastal water samples and enrichment cultures. *Limnol. Oceanogr.* **44,** 37–51.

Lindblad-Toh, K., Winchester, E., Daly, M. J., Wang, D. G., Hirschhorn, J. N., Laviolette, J. P., Ardlie, K., Reich, D. E., Robinson, E., Sklar, P., Shah, N., Thomas, D., Fan, J. B.,

Gingeras, T., Warrington, J., Patil, N., Hudson, T. J., and Lander, E. S. (2000). Large-scale discovery and genotyping of single-nucleotide polymorphisms in the mouse. *Nat. Genet.* **24**(4), 381–386.

Lockhart, D. J., Dong, H., Byrne, M. C., Follettie, M. T., Gallo, M. V., Chee, M. S., Mittmann, M., Wang, C., Kobayashi, M., Horton, H., and Brown, E. L. (1996). Expression monitoring by hybridization to high-density oligonucleotide arrays. *Nat. Biotechnol.* **14**, 1675–1680.

Lockhart, D. J., and Winzeler, E. A. (2000). Genomics, gene expression and DNA arrays. *Nature* **405**, 827–836.

López-Garcia, P., Rodriguez-Valera, F., Pedros-Alio, C., and Moreira, D. (2001). Unexpected diversity of small eukaryotes in deep-sea Antarctic plankton. *Nature* **409**, 603–607.

Loy, A., Lehner, A., Lee, N., Adamczyk, J., Meier, H., Ernst, J., Schleifer, K. H., and Wagner, M. (2002). Oligonucleotide microarray for 16S rRNA gene-based detection of all recognized lineages of sulfate-reducing prokaryotes in the environment. *Appl. Environ. Microbiol.* **68**, 5064–5081.

Ludwig, W., Strunk, O., Westram, R., Richter, L., Meier, H., Yadhukumar, Buchner, A., Lai, T., Steppi, S., Jobb, G., Forster, W., Brettske, I., Gerber, S., Ginhart, A. W., Gross, O., Grumann, S., Hermann, S., Jost, R., Konig, A., Liss, T., Lussmann, R., May, M., Nonhoff, B., Reichel, B., Strehlow, R., Stamatakis, A., Stuckmann, N., Vilbig, A., Lenke, M., Ludwig, T., Bode, A., and Schleifer, K. H. (2004). ARB: A software environment for sequence data. *Nucleic Acids Res.* **32**(4), 1363–1371.

Maidak, B. L., Cole, J. R., Lilburn, T. G., Parker, C. T. J., Saxman, P. R., Farris, R. J., Garrity, G. M., Olson, G. J., Schmidt, T. M., and Tiedje, J. M. (2001). The RDP-II (Ribosomal Database Project). *Nucleic Acids Res.* **29**, 173–174.

Mann, C. C., and Plummer, M. L. (1996). "Noah's Choice: The Future of Endangered Species." Alfred Knopf, New York.

Medlin, L., Elwood, H. J., Stickel, S., and Sogin, M. L. (1988). The characterization of enzymatically amplified eukaryotic 16S-like rRNA-coding regions. *Gene* **71**, 491–499.

Medlin, L., and Simon, N. (1998). Phylogenetic analysis of marine phytoplankton. *In* "Molecular Approaches to the Study of the Ocean" (K. E. Cooksey, ed.), pp. 161–179. Chapman and Hall, London.

Metfies, K., and Medlin, L. (2004). DNA microchips for phytoplankton: The fluorescent wave of the future. *Nova Hedwigia* **79**, 321–327.

Miller, P. E., and Scholin, C. A. (1998). Identification and enumeration of cultured and wild *Pseudo-Nitzschia* (Bacillariophyceae) using species-specific LSU rRNA-targeted fluorescent probes and filter based whole cell hybridization. *J. Phycol.* **34**, 371–382.

Mock, T. (2003). "Photosynthesis in Antarctic sea-ice diatoms." Dissertation, University Bremen, Germany.

Moon-van der Staay, S. Y., De Wachter, R., and Vaulot, D. (2001). Oceanic 18S rDNA sequences from picoplankton reveal unsuspected eukaryotic diversity. *Nature* **409**, 607–610.

Myers, N. (1998). Securing the fabric of life. *People Planet* **7**(4), 6–9.

Okamato, T., Suzuki, T., and Yamamoto, N. (2000). Microarray fabrication with covalent attachment of DNA using bubble jet technology. *Nat. Biotechnol.* **18**, 438–441.

Pace, B., and Campbell, L. L. (1971a). Homology of ribosomal ribonucleic acid of *Desulfovibrio* species with *Desulfovibrio vulgaris*. *J. Bacteriol.* **106**(3), 717–719.

Pace, B., and Campbell, L. L. (1971b). Homology of ribosomal ribonucleic acid diverse bacterial species with *Escherichia coli* and *Bacillus stearothermophilus*. *J. Bacteriol.* **107**(2), 543–547.

Palleroni, N. J. (2003). Prokaryote taxonomy of the 20th century and the impact of studies on the genus *Pseudomonas*: A personal view. *Microbiology* **Part 1**, 1–7.

Purvis, A., and Hector, A. (2000). Getting the measure of biodiversity. *Nature* **405**(6783), 212–219.

Raitio, M., Lindroos, K., Laukkanen, M., Pastinen, T., Sistonen, P., Sajantila, A., and Syvanen, A. C. (2001). Y-chromosomal SNPs in Finno-Ugric-speaking populations analyzed by minisequencing on microarrays. *Genome Res.* **11**(3), 471–482.

Radtkey, R., Feng, L., Muralhidar, M., Duhon, M., Canter, D., DiPierro, D., Fallon, S., Tu, E., McElfresh, K., Nerenberg, M., and Sosnowski, R. (2000). Rapid, high fidelity analysis of simple sequence repeats on an electronically active DNA microchip. *Nucleic Acids Res.* **28**(7), E17.

Riley, M., and Serres, M. H. (2000). Interim report on genomics of *Escherichia coli*. *Annu. Rev. Microbiol.* **54**, 341–411.

Sekar, R., Pernthaler, A., Pernthaler, J., Warnecke, F., Posch, T., and Amann, R. (2003). An improved protocol for quantification of freshwater *Actinobacteria* by fluorescence *in situ* hybridization. *Appl. Environ. Microbiol.* **69**(5), 2928–2935.

Simon, N., Campbell, L., Ornolfsdottir, E., Groben, R., Guillou, L., Lange, M., and Medlin, L. K. (2000). Oligonucleotide probes for the identification of three algal groups by dot blot and fluorescent whole-cell hybridization. *J. Eukaryot. Microbiol.* **47**, 76–84.

Singh-Gasson, S., Green, R. D., Yue, Y., Nelson, C., Blattner, F., Sussman, M. R., and Cerrina, F. (1999). Maskless fabrication of light-directed oligonucleotide microarrays using a digital micromirror array. *Nat. Biotechnol.* **17**, 974–978.

Southern, E., Mir, K., and Shchepinov, M. (1999). Molecular interactions on microarrays. *Nature Genet.* **21**(Suppl. 1), 5–9.

Toth, G., Gaspari, Z., and Jurka, J. (2000). Microsatellites in different eukaryotic genomes: Survey and analysis. *Genome Res.* **10**(7), 967–981.

Wang, D. G., Fan, J. B., Siao, C. J., Berno, A., Young, P., Sapolsky, R., Ghandour, G., Perkins, N., Winchester, E., Spencer, J., Kruglyak, L., Stein, L., Hsie, L., Topaloglou, T., Hubbell, E., Robinson, E., Mittmann, M., Morris, M. S., Shen, N., Kilburn, D., Rioux, J., Nusbaum, C., Rozen, S., Hudson, T. J., Lipshutz, R., Chee, M., and Lander, E. S. (1998). Large-scale identification, mapping, and genotyping of single-nucleotide polymorphisms in the human genome. *Science* **280**(5366), 1077–1082.

Vysotskaia, V. S., Curtis, D. E., Voinov, A. V., Kathir, P., Silflow, C. D., and Lefebvre, P. A. (2001). Development and characterization of genome-wide single nucleotide polymorphism markers in the green alga *Chlamydomonas reinhardtii*. *Plant Physiol.* **127**(2), 386–389.

Wawrik, B., Paul, J. H., and Tabita, F. R. (2002). Real-time PCR quantification of rbcL (ribulose-1,5-bisphosphate carboxylase/oxygenase) mRNA in diatoms and pelagophytes. *Appl. Environ. Microbiol.* **68**(8), 3771–3779.

Woese, C. R. (1987). Bacterial evolution. *Microbiol. Rev.* **51**, 221–271.

[17]   The Role of Geographic Analysis in Locating, Understanding, and Using Plant Genetic Diversity

By Andy Jarvis, Sam Yeaman, Luigi Guarino, and Joe Tohme

Abstract

The genetic structure of an organism is shaped by various factors, many of which vary significantly over space. In this chapter, we provide insight on how studying geographic patterns may contribute to an improved understanding of variability in genetic structure. We first review the theoretical background on how differences in genetic structure may be generated through processes that are inherently variable over space. We then present novices with some basics on how geographic information systems (GIS) may be adopted to study this variation, including advice on software, data, and the type of research questions that might be addressed. The chapter finishes with a brief review of how spatial analysis has contributed to the conservation and use of plant genetic resources, through an understanding of spatial patterns in species distribution and genetic structure. We conclude that spatial variation is a factor often overlooked in genetic studies and one that merits greater consideration. With the advent of functional genomics and improved quantification of adaptive traits, spatial analysis may be key in understanding variation in genetic structure through careful analysis of genotype–environment interactions.

Introduction

A genome arises through the processes of mutation, selection, gene flow, and genetic drift. Differences in how these processes unfold from one place to another result in the development and maintenance of distinct genotypes, population differentiation, and ultimately, in some cases, the emergence of new species with distinct adaptations and distributions. Spatial considerations are, therefore, key to understanding evolution. This chapter outlines a potential role for spatial analysis using geographic information systems (GIS) in understanding spatial variation in biological data. By visualizing and analyzing spatial patterns in genetic and ecological data, GIS can provide a tool for investigating the processes that shape genomes and for conserving and using genetic diversity as effectively and efficiently as possible. Although applying spatial analysis to the conservation and use of agricultural plant genetic diversity is our focus, the principles can be applied to other organisms and different objectives.

The chapter is aimed at the GIS novice, outlining the tools and potential data sources available, and providing some examples of their use. We start by introducing some theory of how genetic structure may vary over space and how some researchers have analyzed it. We provide some practical information, describing the types of georeferenced biological and nonbiological data available to plant genetic resources (PGR) workers. Next, we outline some of the PGR questions that can be investigated through spatial analysis, providing the reader with an overview of the kinds of insights GIS can provide. We outline the GIS tools available and conclude with a set of examples of GIS-based geographic analysis applied to PGR research.

Understanding Variation in Genetic Structure

Genetic structure is shaped by the interaction between the evolutionary forces of selection, gene flow, mutation, and drift. Variation in the effects of these forces in different parts of a species' range results in characteristic spatial patterns in the genetic structure of populations. Understanding these can be of practical utility in prioritizing where and how to conserve and use genetic resources.

The simplest of genetic structure tends to occur when there is no spatial variation in any of the evolutionary processes. In this case, genetic drift causes the gradual accumulation of genetic differences between different areas of a species range, and all areas appear equally related. Departures from this simple pattern are caused by spatial variations in the evolutionary processes of drift, selection, and gene flow, and can provide information relevant to conservation planning.

While drift can be influenced by spatial variation in population density, selection and gene flow often are explicitly shaped by environmental and geographic variables. Selection, in particular, may be related directly to landscape features acting on traits that confer adaptive responses to environmental stresses. These can include climatic factors such as cold or drought stress (Bekessy et al., 2003), edaphic factors such as soil texture, mineral availability, or toxicity (Wu and Antonovics, 1976) and biological factors such as vegetation cover (Brouat et al., 2003). Hedrick et al. (1976) and Linhart and Grant (1996) have reviewed links between genetics and environment extensively. Selection also can result in indirect correlation between genotype and environment in cases in which neutral genes or gene complexes are linked to selected alleles. The results of numerous studies by Nevo (2001) and colleagues suggest extensive linkage and indirect selection on many nonadaptive regions of the genome in a wide array of species and environments.

Gene flow results in the redistribution of alleles between populations or areas of a species' range, modifying patterns caused by drift and selection.

This can take place either through the physical movement and reproduction of individuals or through the dispersal of seeds or pollen. Although the degree of displacement will be affected primarily by geographical distance, other factors can modify the manner in which gene flow redistributes alleles, including barriers or restrictions to migration (Arnaud, 2003; Pfenninger, 2002), regional differences in phenology and pollination time (Galen et al., 1997), and seed/pollen dispersal mechanisms (Ennos, 1994). Since genetic patterns resulting from drift and/or selection can be greatly modified by gene flow, it is important to account for both.

Because evolutionary and conservation biologists are interested in understanding these processes, many methods have been developed to model their effect on spatial genetic patterns. For the most part, these rely on a hypothesis about how a given spatial process or feature should affect genetic structure and then compare landscape features to observed patterns in genetic structure.

An often examined hypothesis in the study of geographic effects on genetic structure is that of isolation by distance, suggesting that more geographically distant populations will also tend to be more genetically distinct (Wright, 1943). Areas of a species' range that are fragmented and not connected by gene flow tend to evolve distinct genetic patterns by the gradual accumulation of genetic differences by drift and mutation. Mutual exchange of alleles by gene flow tends to counteract such differentiation. A range of analytical techniques have been developed to analyze this effect of gene flow by comparing genetic structure and relatedness among several populations with measures of geographic or biological distance between them.

One of the simplest ways to analyze patterns of isolation by distance has been through comparing measures of relatedness or genetic differentiation between populations with the geographic distance separating them. Both Mantel tests and spatial autocorrelation methods use this approach to analyze whether populations have genetic structure expected from the isolation by distance model (Escudero et al., 2003; Heywood, 1991). These methods have been applied to many natural populations and have been used to identify minimum sampling distances for collection of neutral diversity in wild soybeans (*Glycine soja* Siebold and Zucc.) (Jin et al., 2003) and to study gene linkage in Norway spruce (*Picea abies* [L.] H. Karst.) (Bucci and Menozzi, 2002). Often, however, migration is not linear or equal in all directions, and euclidean distances are not appropriate. To account for the effects of barriers to migration, studies have used connectivity networks (Arnaud, 2003; Pfenninger, 2002). Although this approach has not been applied widely to plant populations, geographical barriers such as mountains and phonological barriers such as differences in climate or altitude could be incorporated. Other methods such as wombling and the Monmonier algorithm also can be used to analyze observed genetic structure for evidence of barriers to gene flow (Manel *et al.*, 2003). It is important to note, however, that all of these methods rely on the use of neutral molecular markers. Because selection can alter spatial patterns in adaptive traits, isolation by distance typically is studied only for neutral traits or for adaptive traits in which selection pressure is homogenous across a species' range (Lande, 1991; Nagylaki, 1994).

Another very extensively investigated hypothesis is that of the link between adaptive diversity and environmental heterogeneity. Although there is no standard methodology for such investigations, identifying such interactions typically has required the identification of significant differences between population trait means and the detection of correlations between trait means and selection pressures (but see Volis *et al.*, 2004). Reciprocal transplant experiments can also be used to assess local adaptation. Varying approaches to this have been reviewed extensively (Hedrick, 1986; Hedrick *et al.*, 1976; Linhart and Grant, 1996; Nevo, 2001).

The methods described here are only a small sample of a wide variety of approaches used for interpreting ecological effects on genetic structure. They have proven useful for understanding why genetic structure tends to look the way it does and have provided the basis for a secondary type of analysis, namely reversing the approach and using ecological and geographical information to predict patterns in spatial genetic structure. This approach has obvious uses in conservation, where the collection of genetic data is often prohibitively time-consuming and expensive. Although rules of thumb are derived easily from the results of the aforementioned investigations (i.e., genetic differences will be high between populations that are geographically isolated), an increasing number of methods are being developed to quantitatively predict and assess these patterns. Whereas spatial autocorrelation and similar methods rely on explicit sampling and analytical design to account for geographical patterns, mathematical approaches based on ecological mapping are a means of implicitly incorporating geography into analysis. Some of these advances are outlined in this chapter after some practical information on how to analyze biological data in a GIS.

Georeferenced Biological Data

Before discussing tools and analyses, it is important to describe the types of data generated and used in PGR conservation and use. PGR collections are typically sets of samples (or accessions) of seeds, live plants, or tissue cultures of cultivated plants or of wild crop relatives, sometimes with associated herbarium specimens. The information normally associated with these collections includes so-called "passport data" (i.e., species name and perhaps local land race name, plus data about the plant collector, the date of collection, and descriptive information about the collecting site, including its geographic coordinates). Many PGR collection databases also include characterization data, which may refer to the phenotype (morphology, phenology) and/or genotype (molecular markers, isoenzymes, quantitative trait loci [QTLs], etc.) of the accessions. If geographic coordinate data are available or can be obtained, they may be said to be georeferenced in that the passport, characterization, and other associated data may all be linked to a particular location on the earth's surface.

Nonbiological Georeferenced Data

Both environmental factors such as climate, topography, and soils and anthropogenic factors such as habitat destruction and artificial selection help shape genetic patterns and structure in crops and related wild species.

Over the past decade, many global georeferenced datasets of such environmental and socioeconomic variables have been produced. For example, WORLDCLIM (http://biogeo.berkeley.edu/worldclim/worldclim. htm) is a global database of monthly climate variables (maximum temperature, minimum temperature, and rainfall) in the form of grid surfaces with a spatial resolution (cell size) of 1 km. These surfaces have been produced through the interpolation of up to 46,000 meteorological stations distributed across the globe. Topography is a fundamental environmental factor that affects soil characteristics, hydrology, and climate, among others. Global datasets of topography are available at very high resolution from the Shuttle Radar Topography Mission (SRTM) (http://srtm.usgs.gov/). This is a global dataset with a cell size of 3-arc s (~100 m in the tropics) and is available from a number of sources (U.S. Geological Survey (USGS) FTP server [ftp://edcsgs9.cr.usgs.gov/pub/data/srtm/], USGS National Map Seamless Distribution System [http://seamless.usgs.gov/], or the Land Use Project of the International Center for Tropical Agriculture [CIAT] [http://srtm.csi. cgiar.org]). Global datasets of soil exist but are considered fairly crude and inexact. The most popularly used is the Food and Agriculture Organization (FAO) Digital Soil Map of the World (http://www.fao.org/ag/agl/agll/ dsmw.htm), which includes variables such as soil classification unit, pH, organic carbon content, C/N ratio, clay mineralogy, and soil depth. However, when working at the regional scale, this dataset does not contain sufficient detail, and national or regional scale soil surveys should be consulted.

Land cover also is important when attempting to locate habitats or understand the degree of fragmentation to which a habitat might have been subjected. Land-cover datasets are available at a variety of scales.

Global datasets exist with 1-km cell sizes that classify the land surface into land-cover classes, including forest (evergreen/deciduous, needleleaf/broadleaf), savannas, grasslands, water bodies, croplands, and others. Some examples of these are the USGS 1-km Land Cover Characterization dataset (http://edcdaac.usgs.gov/glcc/glcc.asp) and the European Union Joint Research Council Global Land Cover 2000 Project (http://www.gvm.sai.jrc.it/glc2000/defaultGLC2000.htm). Other products are available on a more regional scale, often derived from Landsat or SPOT imagery. The Global Land Cover Facility (http://glcf.umiacs.umd.edu/index.shtml) provides an array of land-cover measurements and free satellite images to download. The National Aeronautics and Space Administration (NASA) Mr. SID Image Server offers Landsat data for the entire globe (https://zulu.ssc.nasa.gov/mrsid/).

Humans also play an important role in shaping genetic diversity and species distributions of both wild plants and cultivated species. Basic socioeconomic datasets on the distribution of roads, administrative boundaries and towns and cities are available form the Digital Chart of the World (http://www.maproom.psu.edu/dcw/). Human influence can play a diversifying role for cultivated species but can be a cause of genetic erosion. The human footprint dataset (http://wcs.org/humanfootprint) is an integrated map with global coverage, at 1-km cell resolution, rating the degree to which human activities have influenced the land surface. Population surfaces exist for most parts of the world (the Consortium for International Earth Science Information Network's [CIESIN's] Global Gridded Population, http://sedac.ciesin.columbia.edu/plue/gpw/index.html; 1-km gridded population for Latin America, http://gisweb.ciat.cgiar.org/population/). Livestock grazing is a documented cause of genetic erosion (Williams, 2001), and 1-km datasets of cattle density exist for Asia and Africa (http://ergodd.zoo.ox.ac.uk/livatl2/).

This is only a brief review of the most fundamental spatial datasets of environmental and socioeconomic variables on a global scale. The Internet is a huge resource for locating spatial datasets and should be consulted to find the most up-to-date and detailed datasets for the study. Web portals, such as http://www-sul.stanford.edu/depts/gis/bookmark.htm and http://unr.edu.homepage/daved/gislinks.html are useful starting points for novice users.

Some Spatial Questions

- Where might I find a given species?
- Where might I find the greatest intraspecific and/or interspecific diversity?
- Where might I find germplasm with a specific genetic adaptation?

• Where should I collect samples of a species to accurately reflect its intraspecific diversity?

Plant genetic resource workers always are asking themselves these kinds of "where?" questions. Given the resource constraints under which the PGR community operates, it clearly is important to be able to target interventions as accurately as possible in space, to prioritize areas for germplasm collecting or target a particular region for the introduction and testing of a new improved cultivar. These spatial questions are important because answers to them—and others like them—are necessary to make the conservation and subsequent use of genetic resources as effective and efficient as possible. Geographic information systems may be used to analyze georeferenced data from genetic resource collections, either on their own or in conjunction with the other georeferenced data described earlier.

Geographic Information Systems and Spatial Analysis
   Tools for Biologists

A GIS may be defined as a database management system that can simultaneously handle digital spatial data (e.g., a map of the countries of the world) and logically attached, nonspatial, attribute data (e.g., the names and populations of the countries) (Guarino et al., 2002). In our application, the digital spatial data would be the locations where genetic resource accessions were collected and the attribute data would include the species name, collector, and characterization information associated with each accession. In the past, the adoption of GIS technology required significant investments in hardware, software, and human resources. Nowadays, GIS is within the reach of most interested biologists, given a computer and some good ideas. The tools available include generic GIS software (which will be used in diverse fields ranging from surveying to land planning to mineral exploration), Internet mapping technologies for publishing maps on the web, and specialist software tailored to the spatial analysis of biological phenomena.

Generic GIS software include the Environmental Systems Research Institute's (ESRI's) range of products (such as ArcGIS, ArcInfo, and ArcView 3.2), IDRISI, MapMaker, and the open-source program GRASS. These GIS tools include basic visualization of spatial data, in the form of points (e.g., towns), lines (e.g., roads), polygons (shapes such as country boundaries), and grids (continuous surfaces based on an array of cells, e.g., topography). Once visualized on screen, these software packages provide means of locating specific conditions, analyzing spatial patterns, and combining spatial datasets. Given that the biological researcher has clear spatially related questions in mind, these generic tools often provide the means to analyze the relevant data and provide a useful answer.

However, there also are various tools for GIS beginners, tailored specifically to biologists, which incorporate established methodologies for the spatial analysis of biological data and facilitate their application. Some tools relevant to PGR conservation and use are presented in the following sections, but many more are available for diverse applications.

Data Checking

In many cases, the locality data are missing or erroneous, especially for older collections, making it important to complete and check the coordinates in the plant collection database before performing any analysis (Hijmans *et al.*, 1999). Collecting localities often are distributed nonrandomly in space, showing distinct geographical biases. Hijmans *et al.* (2000) analyzed gene-bank collections of wild potato for bias in their geographic representativeness and detected strong overcollecting along roads and within areas previously identified as hotspots for the gene pool. Herbarium collections focus on diversity at the species level, with a strong taxonomic bias reflecting the specialization of botanists. These biases must be acknowledged in any analysis of PGR data.

Diversity Analysis Tools

Species-level and genetic diversity are not distributed randomly over the surface of the earth, and knowing where they are greatest obviously is a key consideration in targeting conservation efforts. However, diversity is a difficult parameter to map and analyze. Diversity studies usually begin by dividing the target area into a number of smaller zones, for each of which a measure of diversity is then calculated (Jarvis *et al.*, 2003; Müller *et al.*, 2003). Different geometric, political, or socioeconomic spatial units have been used (Csuti *et al.*, 1997), although ideally, areas of equal shape and size (to reduce the area effect on diversity measures), such as square grid cells, are best. For each grid cell, either richness (number of different categories) or an array of diversity indices (combining richness with evenness in different ways) can be calculated, resulting in a diversity surface.

Grid-based mapping of diversity from point data is not a trivial analysis and can be done using various methodologies, all with associated assumptions and caveats. For example, moving the origin of the grid or changing the size of the grid cells can change the final result significantly. Nelson (2004) examined in detail the issue of scale in diversity mapping, identifying a method for selecting the most appropriate scale of analysis using Monte Carlo simulations and statistical analyses of confidence.

Two GIS tools that can "map" diversity using grids in this way are freely available. DIVA-GIS (http://www.diva-gis.org) has a user-friendly interface that permits integrated analysis of PGR data, from mapping of diversity (employing different methods and offering various diversity indices) to understanding environmental adaptations and predicting species distribution (using the DOMAIN and BioClim methods described later in this chapter). DIVA-GIS contains global datasets of climatic variables for both the present and the projected future climate of 2055, as well as land-cover data, topography, and population. DIVA-GIS includes other useful functions for spatial analysis of biological data, many of which are discussed in Hijmans *et al.* (2002). WORLDMAP (http://www.nhm.ac.uk/science/projects/worldmap/) also maps diversity using the grid-based approach, including among other functions a means of mapping diversity weighted for the distinctness of taxonomic units, calculated from a phylogeny based on cladistic analysis (Vane-Wright *et al.*, 1991).

Predictive Species Distribution Modeling

Identifying the precise geographic range of a species is often a fundamental step in locating, conserving, and using PGR. Specialist plant collectors use vegetation maps and previous experience to define the geographic range of a species, but this is both subjective and reliant on the availability and quality of these maps. For many species, knowledge is just insufficient to accurately map the geographic distribution. Anderson *et al.* (2002) state that shaded outline maps ranging between and beyond known localities are likely to overestimate species distribution, whereas dot maps of known localities portray species distribution conservatively. Geographic bias in collecting efforts [e.g., along roads (Hijmans *et al.*, 2000)] creates further uncertainty in defining species range.

Much effort has gone into the development of methods for predicting the geographic distribution of species and now many of these have been incorporated into user-friendly tools. Typically, these methods use the conditions at points where the species has been found in order to construct a statistical model of the adaptation range of the species, based on a set of user-defined environmental variables. The statistical model then is applied over a wide region to locate other areas where the environmental conditions are potentially suitable for the species in question. These methods have been found to minimize the risk of overestimation and underestimation of geographic range (Franklin, 1995). Although they have been applied only at the species level, they can be adapted to work at the genetic level if there is good reason to believe that the trait being studied is likely to be distributed nonrandomly with reference to given environmental variables.

Many of these range estimation methods assume that climatic variables are the principal drivers of geographic distribution (Franklin, 1995; Guisan and Zimmerman, 2000; Walker and Cocks, 1991), although other factors also have been used, including soils (Anderson et al., 2002), topography (Draper et al., 2003), and specific habitat conditions (Reutter et al., 2003).

Guisan and Zimmerman (2000) discuss some of the applications of species distribution modeling and the various algorithms that have been applied to the problem. Perhaps the most widely recognized method uses generalized linear models (GLMs), specifically logistic regression, to predict species distribution (Cumming, 2000; Draper et al., 2003; Guisan et al., 2002; Osborne and Suárez-Seoane, 2002; Pearce and Ferrier, 2000). This method requires not only input points detailing where a species has been found but also points of reported absences. In many cases, especially with PGR databases, these absence data are not available and are difficult to generate. Confirming an absence is also difficult and can often lead to false negatives (Jarvis et al., in press). No specific tool exists for performing species distribution modeling with the logistic regression method, but this analysis can be made easily with IDRISI in conjunction with a standard statistical software package (Draper et al., 2003).

Another algorithm for predictive species distribution modeling uses principal components analysis (PCA) (Jones et al., 1997; Robertson et al., 2001). This method involves performing a PCA on the environmental data at the points where a species has been collected and then uses the PC loadings to compute a probability distribution for all other environments in the study area. The result is a map of probabilities of finding the species. This method has been incorporated in the FloraMap software (Jones and Gladkov, 1999) (http://www.floramap-ciat.org/), which has been used in the study of wild crop relatives (Jarvis et al., 2003, in press; Segura et al., 2003). Further information about FloraMap is available in Jones et al. (2002).

Factor analysis also has been adopted for species distribution modeling (Hirzel et al., 2002) and is incorporated in BioMapper (http://www.unil.ch/biomapper/), which uses the Ecological Niche Factor Analysis (ENFA) algorithm. Other species distribution modeling tools worthy of mention are BIOM, which combines habitat suitability methods with distance-based calculations (Henning Sommer et al., 2003); DOMAIN (http://www.cifor.cgiar.org/scripts/default.asp?ref=research_tools/domain/index.htm), which uses the Gower metric to calculate similarity and distance from the conditions at known points of presence; GARP (http://www.lifemapper.org/

desktopgarp), which uses a neural network, specifically a genetic algorithm, to calculate the fitness of each area based on the calibration dataset (Anderson *et al.*, 2002); and BIOCLIM, which uses a bounding box technique to define the environmental envelope that the species inhabits.

An important issue with species distribution modeling is validation of the results. Evaluations of species distribution models typically use presence/absence data to test how well the prediction fits with reality (Fielding and Bell, 1997). However, sampling issues complicate this, because absence is difficult to confirm, especially if the study covers a large area (Jarvis *et al.*, in press). Both the kappa statistic and the area under the curve (AUC) (derived from the threshold Receiver Operating Characteristic [ROC]) have been used in the literature to validate presence/absence distributions (Cumming, 2000; Osborne and Suárez-Seoane, 2002; Pearce and Ferrier, 2000; Robertson *et al.*, 2001). Manel *et al.* (2001) conclude that Cohen's kappa provides an appropriate statistical evaluation, benefiting from its simplicity to calculate and interpret.

As can be seen, many methodologies and tools are available for predicting species distributions. Manel *et al.* (1999) compare different methodologies, concluding that model performance differs only marginally and that the choice of method should depend more on the research questions being asked and the type of data that are available. Some of the criteria that might be used to decide among methods include the following:

- Whether the user needs to provide presence and absence data or presence only
- Whether the environmental variables can be categorical and continuous
- The degree of explanation that the method provides in terms of the environmental adaptation of the species
- Ease of use of the software and the inclusion of built-in datasets

Table I provides a brief review of four of the most common methods and tools, with some critical analyses of their relative advantages and disadvantages.

Genetic Diversity Models

A large body of work also endeavors to understand genetic structure of populations through simulation modeling of ecological processes at the genetic level. These models often include a spatial component that takes into account the effect of distance and population distribution on ecological and genetic processes. Typically, they require large calibration datasets to run simulations on real-life biological populations.

TABLE I

ADVANTAGES AND DISADVANTAGES OF FOUR COMMON METHODS AND TOOLS FOR PREDICTING SPECIES DISTRIBUTION

Method	Tool	Ease of use	Type of input biological data	Associated environmental data	Validation	Explanatory power
Logistic regression	No tool performs logistic regression; requires generic geographic information systems (GIS) and statistical software	Use of statistical package and generic GIS tool requires some basic knowledge	Presence and absence	Continuous data only, not necessarily independent variables; user must select, find, and provide associated variables for this analysis through use of a generic GIS tool	Split-sample validation can be made using the presence/absence data within a statistical package	Detailed analysis of statistical coefficients permits the user to understand the environmental factors most important in defining the species distribution
Principal components analysis (PCA)	FloraMap	User-friendly tool with example dataset and manual	Presence only	Continuous data only, not necessarily independent variables; FloraMap uses built-in datasets of 36 climate variables (monthly rainfall, temperature, and diurnal range in temperature), near-global coverage; users cannot use other datasets	No validation method incorporated in FloraMap; validation is difficult in FloraMap because of the difficulty in confirming an absence given the large cell size of the climate data (1–18 km)	Principal component scores and loadings allow the user to assess which climatic factors are most important in defining the species distribution

Method	Software	User-friendliness	Data type	Data requirements	Validation	Results/Notes
Ecological niche factor analysis (ENFA)	BioMapper	User-friendly tool with example dataset and manual; importing data can be difficult without some GIS experience	Presence only	Continuous data only, not necessarily independent variables; BioMapper does not include any data, so data must be imported, requiring some basic GIS skills	BioMapper contains a detailed set of validation techniques, including Cohen's kappa and area-under-the-curve (AUC) plots	Results of the factor analysis permit the user to examine which environmental variables are most important in defining the distribution
Genetic algorithm	Desktop GARP	User-friendly tool	Presence only	Desktop GARP contains built-in global datasets of climate and topography but also offers users the ability to import their own data (requires some GIS experience)	Desktop GARP offers automated split-sample validation techniques with statistical reports	Result is presence/absence (0/1), and provides no evidence for goodness of fit; model is black box and provides little insight into the contributing variables, although multiple runs can be made using jackknifing to define the most contributing variables

An example of this is the simulation model, ECO-GENE (http:// www.ecogene21.org/), developed to study temporal and spatial dynamics of the genetic structure of tree populations (Degen *et al.*, 1996). It is a distance-dependent model that combines elements of population genetics, demographical dynamics, forest growth, and management models. Overlapping or separate generations can be created, and different processes such as gene flow, mating systems, flowering phenology, selection, random drift, and competition can be simulated. It has been applied to study the impact of different silvicultural practices and the effect of air pollution on the genetic structure of tree populations (Degen *et al.*, 1997, 2002; Takahashi *et al.*, 2000). In its current form, ECO-GENE only deals with neutral traits, but the results provide a means of testing the possible effects of different management options.

Examples of GIS Use in Genetic Resources Conservation and Use

As has been shown, genetic diversity is in some part shaped by the environment, and in many cases, adaptations to local environments are of most interest to gene banks and germplasm users. Several concrete studies have shown how spatial analysis might prioritize conservation intervention, optimizing genetic conservation.

Jones *et al.* (1997) used the FloraMap program to predict the geographic distribution of wild bean (*Phaseolus vulgaris* L.) based on the distribution of germplasm and herbarium specimens. The results correctly predicted areas where wild bean had not been collected but was reported to occur in the literature. Given the success of this research, Segura *et al.* (2003) also used FloraMap to predict the distribution of five species of the genus *Passiflora* in the northern Andes. The results fitted closely with areas of known distribution of the species and identified two separate climatic adaptations within one species. Isoenzyme studies identified different zymotypes that were closely related to the two climatic clusters within the species *Passiflora tripartita* var. *mollissima* (Kunth) Holm-Niels. and P. Jorg. The study also identified collection gaps where *ex situ* germplasm collection should be focused.

Jarvis *et al.* (in press) also used FloraMap, this time in combination with land-cover maps, to locate potential collection sites for the rare wild pepper, *Capsicum flexuosum* Sendtn. In a controlled experiment, plant collectors visited 10 predefined sites where the species was predicted to be present and 10 sites where its absence was predicted. This methodology aimed to allow a formal validation of the method. Six new populations were found, representing a significant improvement over two previous collecting missions for the species in the same region, undertaken without the use of GIS targeting. *C. flexuosum* was found at five of seven points predicted to harbor the species and not found at four of five points predicted not to harbor the species. Genetic analyses of this species are now being used to examine genetic diversity of collections from different climatic regions.

Draper *et al.* (2003) used cluster analysis of a suite of sites to define ecogeographic units to stratify germplasm collections of various species within a region. This method was applied to ensure that the germplasm collection covered the full environmental gradient, with the aim of conserving the greatest genetic diversity. This type of analysis could lead to the generation of sampling strategies to conserve the greatest intraspecific diversity in the least number of accessions.

Also at the species level, Draper *et al.* (2001) used logistic regression to select translocation sites for the rare species *Narcissus cavanillesii* Barra and G. López in Portugal. The natural habitat of the species was under threat from the construction of a dam, so spatial analysis was used to identify a region where the species might survive translocation. Climatic and ecological variables were used to find the optimum sites, and the survivorship of the species in their new habitat is being monitored.

A number of ecogeographic studies of crop wild relatives have been made using GIS-based approaches, with the aim of describing the biogeography of the gene pool and prioritizing potential conservation programs. Hijmans and Spooner (2001) constructed a database of more than 6000 collections of wild potatoes (*Solanaceae sect.* Petota) and analyzed the distribution of each species, locating the continental hotspots of species diversity. They used the grid-based diversity mapping method in DIVA-GIS to locate the areas with the greatest number of species, as well as Rebelo's (1994) complementarity algorithms to select the least number of grid cells to capture all species. Analyses of species distribution were also made, quantifying the spatial area that each species occupied and defining the maximum distance between collection points. Careful examination of these distribution characteristics permitted the identification of species that were undercollected relative to their geographic range.

Maxted *et al.* (2004) used this work as a model to analyze the biogeography of wild *Vigna* species in Africa. In this case, conservation priorities were assessed through comparing the actual species richness of germplasm and herbarium collection with the potential species richness calculated through predictive species distribution modeling. The grid-based diversity mapping method was used to identify the currently known hotspots of *Vigna* diversity, based on the existing collections. Then FloraMap was used to map the potential distribution of each of the 70 species. If the probability of finding the species was more than 0.5, presence was assumed and the results of each species were combined to create a map of potential species richness. Comparing the "potential" with the "actual" species richness permitted the authors to identify new areas for germplasm collection or areas already visited but that were identified as potentially containing more species.

Jarvis *et al.* (2002, 2003) made a similar analysis of the biogeography of wild peanuts in Latin America, also prioritizing areas for *ex situ* and *in situ* conservation. Using the same dataset of wild peanuts, Ferguson *et al.* (2005) analyzed the climatic adaptation of each species through the extraction of climate data for each collection point in the database. Multivariate statistics permitted the authors to identify clustering of species adaptations and provided insights into the potential evolution of the cultivated peanut (*Arachis hypogaea* L.) from its wild relatives. This supported molecular evidence indicating that the species *Arachis duranensis* Krapov and W. C. Gregory and *A. ipaënsis* Krapov and W. C. Gregory are the wild progenitors of the cultivated peanut. Analysis of climatic adaptations like this provides key information to improve the use of genetic resources and can feed into crop improvement programs.

A limited amount of literature of diversity mapping is available at the genetic level. Hoffmann *et al.* (2003) used molecular data of the number of variable positions in the alignments and the distribution of recombinant sequence blocks to map genetic level diversity of *Arabidopsis thaliana* (L.) Heynh. In this case, the method of Kriging (a form of spatial interpolation) was applied to accessions with 13 sequenced loci to identify areas of greater diversity. The Atlantic Coast in Europe, from the western Iberian Peninsula to southern Great Britain, was found to have the greatest genetic variability. Although this type of analysis provides no insight into the processes creating the pattern of genetic diversity, it did detect spatial patterns in diversity that had not been identified in the data before geographic analysis.

Concluding Remarks

This chapter has shown how georeferenced biological information, analyzed with georeferenced environmental and socioeconomic data, can be used to understand the processes that generate genetic diversity. Such knowledge is necessary to answer the "where" questions that PGR researchers and users must address to be able to target their interventions most effectively and efficiently. Most of the practical examples of spatial analysis in PGR research have been made at the species or gene-pool level, locating areas of high species diversity, or using species distribution modeling. These are important contributions to PGR work, but more examples at the genetic level are needed now.

Despite an established body of theory about how adaptive genetic structure may vary over space and along environmental gradients, experiments have tended to focus on simple and controlled examples to test the theories. As such, few examples translate this into general methods for prediction and analysis in complex environments more typical of most plant species. In part, this can be explained by the limitations of analytical models in simultaneously accounting for various spatially varying evolutionary forces. GIS-based analysis is one means of overcoming this limitation, because maps can inherently represent spatial processes, greatly simplifying the models required for analysis. Likely as a result of the difficulty of simultaneously accounting for varying selection pressure and gene flow, most genetic analyses in PGR have until now used neutral markers. Though useful for understanding gene flow and population demographics, neutral markers are of little use in assessing the adaptive traits that are of importance in conservation and PGR. Coupled with the improvement of methods for identifying adaptive traits (e.g., QTLs and SNPs), GIS-based spatial analysis will enable the rapid assessment of genetic diversity without costly field-based sampling and laboratory-based genetic analyses. As knowledge of adaptive genetics increases, tools such as FloraMap and DIVA-GIS will be invaluable in revealing spatial patterns, as has been done at the species level, thus guiding efforts to conserve and use this diversity. In order for genetic analyses to benefit from spatial analysis, it is important that germplasm collections are accurately georeferenced and that the geneticist considers spatial variation from the point of defining the sampling strategy through to analysis and interpretation of results.

References

Anderson, R., Gomez-Laverde, M., and Peterson, A. (2002). Geographical distributions of spiny pocket mice in South America: Insights from predictive models. *Global Ecol. Biogeogr.* **11,** 131–141.

Arnaud, J.-F. (2003). Metapopulation genetic structure and migration pathways in the land snail *Helix aspera*: Influence of landscape heterogeneity. *Landscape Ecol.* **18,** 333–346.

Bekessy, S. A., Ennos, R. A., Burgman, M. A., Newton, A. C., and Ades, P. K. (2003). Neutral DNA markers fail to detect genetic divergence in an ecologically important trait. *Biol. Conserv.* **110,** 267–275.

Brouat, C., Sennedot, F., Audiot, P., Leblois, R., and Rasplus, J. (2003). Fine-scale genetic structure of two *carabid* species with contrasted levels of habitat specialization. *Mol. Ecol.* **12,** 1731–1745.

Bucci, G., and Menozzi, P. (2002). Spatial autocorrelation and linkage of mendelian RAPD markers in a population of *Picea abies* Karst. *Mol. Ecol.* **11,** 305–315.

Csuti, B., Polasky, S., Williams, P. H., Pressey, R. L., Camm, J. D., Kershaw, M., Kiester, A. R., Downs, B., Hamilton, R., Huso, M., and Sahr, K. (1997). A comparison of reserve selection algorithms using data on terrestrial vertebrates in Oregon. *Biol. Conserv.* **80,** 83–97.

Cumming, G. (2000). Using between-model comparisons to fine-tune linear models of species ranges. *J. Biogeogr.* **27,** 441–455.

Degen, B., Gregorius, H.-R., and Scholz, F. (1996). ECO-GENE, a model for simulation studies on the spatial and temporal dynamics of genetic structures of tree populations. *Silvae Genet.* **45,** 323–329.

Degen, B., Roubik, D. W., and Loveless, M. D. (2002). Impact of selective logging and forest fragmentation on the seed cohorts of an insect-pollinated tree: A simulation study. *In* "Modelling and Experimental Research on Genetic Processes in Tropical and Temperate Forests" (B. Degen, M. D. Loveless, and A. Kremer, eds.), pp. 108–119. Empresa Brasileira de Pesquisa Agropecuária (EMBRAPA), Amazonia Oriental, Belém, Brazil.

Degen, B., Streiff, R., Scholz, F., and Kremer, A. (1997). Analyzing the effects of regeneration regime on genetic diversity and inbreeding in oak populations by use of the simulation model ECO-GENE. *In* "Diversity and Adaption in Oak Species" (K. C. Steiner, ed.), pp. 9–21. Pennstate College of Agricultural Sciences, Pennsylvania.

Draper, D., Rossello-Graell, A., and Iriondo, J. M. (2001). A translocation action in Portugal: Selecting a new location for *Narcissus cavanillesii*. A. Barra and G. López. Poster presented at the third Planta Europa Conference, 23–28 June 2001, Pruhonice, Czech Republic. http://www.plantaeuropa.org/html/conference_2001/conference_poster_pres.htm.

Draper, D., Rossello-Graell, A., Garcia, C., Gomes, C., and Sergia, C. (2003). Application of GIS in plant conservation programmes in Portugal. *Biol. Conserv.* **113,** 337–349.

Ennos, R. A. (1994). Estimating the relative rates of pollen and seed migration among plant populations. *Heredity* **72,** 250–259.

Escudero, A., Iriondo, J. M., and Torres, M. E. (2003). Spatial analysis of genetic diversity as a tool for plant conservation. *Biol. Conserv.* **113,** 351–365.

Ferguson, M. E., Jarvis, A., Stalker, H. T., Valls, J. F. M., Pittman, R. N., Simpson, C. E., Bramel, P., Williams, D., Guarino, L. (2005). Biogeography of wild *Arachis*: Distribution and environmental characterization. *Biodivers. Conserv.*

Fielding, A., and Bell, J. (1997). A review of methods for the assessment of prediction errors in conservation presence/absence models. *Environ. Conserv.* **24,** 38–49.

Franklin, J. (1995). Predictive vegetation mapping: Geographic modelling of biospatial patterns in relation to environmental gradients. *Prog. Phys. Geogr.* **19,** 474–499.

Galen, C., Stanton, M. L., Shore, J. S., and Sherry, R. A. (1997). Source-sink dynamics and the effect of an environmental gradient on gene flow and genetic substructure of the alpine buttercup *Ranunculus adoneus*. *Opera Bot.* **132,** 179–188.

Guarino, L., Jarvis, A., Hijmans, R. J., and Maxted, N. (2002). Geographic information systems (GIS) and the conservation and use of plant genetic resources. *In* "Managing Plant Genetic Diversity" (J. E. A. Engels, ed.), pp. 387–404. CAB International, Wallingford.

Guisan, A., and Zimmermann, N. (2000). Predictive habitat distribution models in ecology. *Ecol. Model* **135,** 147–186.

Guisan, A., Edwards, T. C., Jr., and Hastie, T. (2002). Generalized linear and generalized additive models in studies of species distribution: Setting the scene. *Ecol. Model* **157,** 89–100.

Hedrick, P. W. (1986). Genetic polymorphism in heterogeneous environments: A decade later. *Annu. Rev. Ecol. Syst.* **17,** 535–566.

Hedrick, P. W., Ginevan, M. E., and Ewing, E. P. (1976). Genetic polymorphism in heterogeneous environments. *Annu. Rev. Ecol. Syst.* **7**, 1–32.

Henning Sommer, J., Nowicki, C., Rios, L., Barthlott, W., and Ibisch, P. L. (2003). Extrapolating species range and biodiversity in data-poor countries: The computerized model BIOM. *Rev. Soc. Boliv. Bot.* **4**, 171–190.

Heywood, J. S. (1991). Spatial analysis of genetic variation in plant populations. *Annu. Rev. Ecol. Syst.* **22**, 335–355.

Hijmans, R. J., and Spooner, D. (2001). Geographic distribution of wild potato species. *Am. J. Bot.* **88**, 2101–2112.

Hijmans, R. J., Schreuder, M., De la Cruz, J., and Guarino, L. (1999). Using GIS to check co-ordinates of gene bank accessions. *Genet. Resour. Crop Evol.* **46**, 291–296.

Hijmans, R. J., Garrett, K., Huaman, Z., Zhang, D., Schreuder, M., and Bonierbale, M. (2000). Assessing the geographic representativeness of gene bank collections: The case of Bolivian wild potatoes. *Conserv. Biol.* **14**, 1755–1765.

Hijmans, R. J., Guarino, L., Cruz, M., and Rojas, E. (2002). Computer tools for spatial analysis of plant genetic resources data: 1. DIVA-GIS. *Plant Genet. Res. Newsl.* **127**, 15–19.

Hirzel, A., Hausser, J., Hessel, D. C., and Perrin, N. (2002). Ecological-niche factor analysis: How to compute habitat-suitability maps without absence data? *Ecology* **83**, 2027–2036.

Hoffman, M. H., Glas, A. S., Tomiuk, J., Schmuths, H., Fritsch, R. M., and Bachmann, K. (2003). Analysis of molecular data of *Arabidopsis thaliana* (L.) Heynh. (Brassicaceae) with geographical information systems (GIS). *Mol. Ecol.* **12**, 1007–1019.

Jarvis, A., Guarino, L., Williams, D., Williams, K., and Hyman, G. (2002). Spatial analysis of wild peanut distributions and the implications for plant genetic resource conservation. *Plant Genet. Res. Newsl.* **131**, 29–35.

Jarvis, A., Ferguson, M., Williams, D., Guarino, L., Jones, P., Stalker, H., Valls, J., Pittman, R., Simpson, C., and Bramel, P. (2003). Biogeography of wild *Arachis*: Assessing conservation status and setting future priorities. *Crop Sci.* **43**, 1100–1108.

Jarvis, A., Williams, K., Williams, D., Guarino, L., Caballero, P., Mottram, G. (In press). Use of GIS for optimizing a collecting mission for a rare wild pepper (*Capsicum flexuosum* Sentn.) in Paraguay. *Genet. Resour. Crop Evol.*

Jin, Y. M., He, T., and Lu, B. R. (2003). Fine scale genetic structure in a wild soybean (*Glycine soja*) population and the implications for conservation. *New Phytol.* **159**, 513–519.

Jones, P., and Gladkov, A. (1999). "FloraMap: A Computer Tool for the Distribution of Plants and Other Organisms in the Wild." Centro Internacional de Agricultura Tropical (CIAT), Cali, Colombia.

Jones, P., Beebe, S., Tohme, J., and Galwey, N. (1997). The use of geographical information systems in biodiversity exploration and conservation. *Biodivers. Conserv.* **6**, 947–958.

Jones, P., Guarino, L., and Jarvis, A. (2002). Computer tools for spatial analysis of plant genetic resources data: 2. FloraMap. *Plant Genet. Res. Newsl.* **130**, 1–6.

Lande, R. (1991). Isolation by distance in a quantitative trait. *Genetics* **128**, 443–452.

Linhart, Y. B., and Grant, M. C. (1996). Evolutionary significance of local genetic differentiation in plants. *Annu. Rev. Ecol. Syst.* **27**, 237–277.

Manel, S., Dias, J. M., Buckton, S. T., and Ormerod, S. J. (1999). Alternative methods for predicting species distribution: An illustration with Himalayan river birds. *J. Appl. Ecol.* **36**, 734–747.

Manel, S., Williams, H., and Ormerod, S. (2001). Evaluating presence-absence models in ecology: The need to account for prevalence. *J. Appl. Ecol.* **38**, 921–931.

Manel, S., Schwartz, M. K., Luikart, G., and Taberlet, P. (2003). Landscape genetics: Combining landscape ecology and population genetics. *Trends Ecol. Evol.* **18**, 189–197.

Maxted, N., Mabuza-Dlamini, P., Moss, H., Padulosi, S., Jarvis, A. Guarino, L. (2004). African *Vigna*: An ecogeographic study. International Plant Genetic Resources Institute (IPGRI), Italy.

Müller, R. T., Nowicki, C., Barthlott, W., and Ibisch, P. L. (2003). Biodiversity and endemism mapping as a tool for regional conservation planning—case study of the Pleurothallidinae (Orchidaceae) of the Andean rain forests in Bolivia. *Biodivers. Conserv.* **12**, 2005–2024.

Nagylaki, T. (1994). Geographical variation in a quantitative character. *Genetics* **136**, 361–381.

Nelson, A. (2004). "The Spatial Analysis of Socio-Economic and Agricultural Data Across Geographic Scales: Examples and Applications in Honduras and Elsewhere." Ph.D. Thesis, School of Geography, University of Leeds, Leeds, UK.

Nevo, E. (2001). The evolution of genome–phenome diversity under environmental stress. *Proc. Natl. Acad. Sci.* **98**, 6233–6240.

Osborne, P., and Suárez-Seoane, S. (2002). Should data be partitioned spatially before building large-scale distribution models? *Ecol. Model.* **157**, 249–259.

Pearce, J., and Ferrier, S. (2000). Evaluating the predictive performance of habitat models developed using logistic regression. *Ecol. Model.* **133**, 225–245.

Pfenninger, M. (2002). Relationship between microspatial population genetic structure and habitat heterogeneity in *Pomatias elegans* (O. F. Müller, 1774) (Caenogastropoda, Pomatiasidae). *Biol. J. Linn. Soc.* **76**, 565–575.

Rebelo, A. G. (1994). Iterative selection procedures: Centres of endemism and optimal placement of reserves. *Strelitzia* **1**, 231–257.

Reutter, B. A., Helfer, V., Hirzel, A. H., and Vogel, P. (2003). Modelling habitat-suitability using museum collections: An example with three sympatric *Apodemus* species from the Alps. *J. Biogeogr.* **30**, 581–590.

Robertson, M., Caithness, N., and Villet, M. (2001). A PCA-based modelling technique for predicting environmental suitability for organisms from presence records. *Divers Distrib.* **7**, 15–27.

Segura, S., Coppens d'Eeckenbrugge, G., Lopez, L., Grum, M., and Guarino, L. (2003). Mapping the potential distribution on five species of *Passiflora* in Andean countries. *Genet. Res. Crop Evol.* **50**, 555–566.

Takahashi, M., Mukouda, M., and Koono, K. (2000). Differences in genetic structure between two Japanese beech (*Fagus crenata* Blume) stands. *Heredity* **84**, 103–115.

Vane-Wright, R. I., Humphries, C. J., and Williams, P. H. (1991). What to protect? Systematics and the agony of choice. *Biol. Conserv.* **55**, 235–254.

Volis, S. (2004). The influence of space in genetic-environmental relationships when environmental heterogeneity and seed dispersal occur at similar scale. *Am. Nat.* **163**, 312–327.

Walker, P., and Cocks, K. (1991). Habitat: A procedure for modelling a disjoint environmental envelope for a plant or animal species. *Global. Ecol. Biogeogr. Lett* **1**, 108–118.

Williams, D. (2001). New directions for collecting and conserving peanut genetic diversity. *Peanut Sci.* **28**, 135–140.

Wright, S. (1943). Isolation by distance. *Genetics* **28**, 114–138.

Wu, L., and Antonovics, J. (1976). Experimental ecological genetics in *Plantago*. II. Lead tolerance in *Plantago lanceolata* and *Cynodon dactylon* from a roadside. *Ecology* **57**, 205–208.

[18]   *In Situ* Hybridization of Phytoplankton Using Fluorescently Labeled rRNA Probes

By RENÉ GROBEN and LINDA MEDLIN

Abstract

Phytoplankton are one of the major components of ecosystem processes and play an important role in many biogeochemical cycles in the marine and freshwater environment. Despite their importance, many microalgae are poorly described and little is known of broad spatial and temporal scale trends in their abundance and distribution. Reasons for this are that microalgae are often small, lack distinct morphological features, and are unculturable, which make analyses difficult. It is now possible by using molecular biological techniques to advance our knowledge of aquatic biodiversity and to understand how biodiversity supports ecosystem structure, dynamics, and resilience. We present in this chapter a brief review of the progress that has been made in analyzing microalgae from populations to the species level. The described methods range from DNA fingerprinting techniques, such as random amplified polymorphic DNA (RAPD), amplified fragment length polymorphisms (AFLPs), and simple sequence repeats (SSRs), to microsatellites, which are used in population studies, to sequence analysis, which help to reconstruct the evolutionary history of organisms and to examine relationships at various taxonomic levels. Special emphasis is given to the application of molecular probes for the identification and characterization of microalgal taxa. The fast and secure identification of phytoplankton, especially of toxic species, is important from an ecological and economical point of view and whole-cell hybridization with specific fluorochrome-labeled probes followed by fluorescence microscopy or flow cytometry offers a fast method for this purpose. In this context, we present a detailed protocol for fluorescence *in situ* hybridization (FISH) of ribosomal RNA (rRNA) probes that can be applied to many algal cell types and discuss practical considerations of its use.

Introduction

Molecular biological techniques have greatly enhanced our ability to analyze all types of organisms, including the microalgae. This represents a major step forward in marine oceanography, because many microalgae are small, lack distinct morphological markers, and are unculturable, which makes it difficult to estimate their biodiversity. The lack of knowledge of their breeding systems makes genetic or demographic studies difficult, and long-term seasonal studies in aquatic environments are problematic for logistic reasons. This has hindered our understanding of microalgal diversity and their population structure. Despite this, physiological/biochemical measurements have been used to infer the existence of significant genetic diversity within and between microalgal populations (Brand 1989; Partensky *et al.*, 1993; Waterbury *et al.*, 1979). With these data, researchers have speculated on hidden biodiversity and temporal and spatial structuring of genetic diversity or gene flow. Now molecular techniques can present a quantitative framework through which the diversity, structure, and evolution of microalgal populations can be analyzed, predictive models of the dynamics of aquatic ecosystems formulated, and the idea of functional groups in the plankton proven.

Nevertheless, molecular analysis of microalgal population structure is behind other groups and has been usually inferred from physiological data determined from relatively few clones. This unfortunately is a very naive approach because nearly every physiological measurement has shown that no single clone of any microalgal species can be considered truly representative of that species (Wood and Leatham, 1992).

The interaction of a species with environmental parameters is influenced by the genetic diversity at the population level of a species. Spatial and temporal partitioning of genetic diversity will occur because these interactions structure the ecosystem. Such structuring has seldom been measured in the microalgal community and very few studies of genetic diversity existent in pelagic ecosystems. All evidence of geographically isolated populations would be erased if we continue to assume that microalgae with high dispersal capacities are genetically homogeneous over their entire range. Support for this assumption has come mainly from phenotypic comparisons based initially on net phytoplankton biogeographic studies and later on isozyme studies. It is clear that the same morphotype/ species may be endemic or cosmopolitan (Kristiansen, 2001), but it is more likely that cosmopolitan species will exhibit regional differentiation when examined with molecular techniques (Medlin *et al.*, 2000a).

Karp *et al.* (1998) provide an excellent introduction into the various molecular techniques available for use in studying biodiversity at all taxonomic levels, whereas other useful reviews deal with the biodiversity in the marine environment (Ormond *et al.*, 1998) and in the marine phytoplankton (Medlin *et al.*, 2000b). We present a brief review of the progress in analyzing microalgal populations, beginning at the population level and building to higher taxonomic levels with the use of rRNA probes. A detail protocol of fluorescence *in situ* hybridization (FISH) methods applicable for most algal cell types so far tested is presented at the end of the chapter.

DNA Fingerprinting

Our limited knowledge of microalgal genetic diversity is a direct consequence of the difficulties of finding polymorphic markers for ecological genetic studies. Isozymes, the molecular markers used in early studies, evolve so slowly that closely related populations appear identical. The early viewpoints suggesting the absence of genetic diversity in microalgae have undoubtedly been propagated from these studies. The use of high-resolution DNA fingerprinting techniques *sensu lato* circumvents these problems and has, thus, opened areas previously considered unreachable for the microalgae.

DNA fingerprinting is a generic term for different molecular techniques that can produce multilocus banding patterns and can be used to analyze populations down to individuals. The most commonly used techniques are random amplified polymorphic DNA (RAPD), amplified fragment length polymorphisms (AFLP), variable number of tandem repeats (VNTRs, or minisatellites), and simple sequence repeats (SSRs, or microsatellites) (Karp *et al.*, 1998). Although these methods are used for population studies in many higher eukaryotic organisms, their employment for analyzing microalgae is still limited. The few examples of studies of biodiversity and population genetics in phytoplankton include those of the prymnesiophyte *Emiliania huxleyi* (Barker *et al.*, 1994) and the dinoflagellate *Symbiodinium* (Baillie *et al.*, 2000), both analyzed by RAPDs, as well as the dinoflagellate species *Alexandrium tamarense* (John *et al.*, 2004), which was investigated using AFLPs. The latter example showed that these types of fingerprints can ironically provide too much variation in the case of population studies because of the high variability at each locus and the large number of loci. Banding patterns can quickly become so complex that they cannot be analyzed in terms of allele frequencies (the data of population genetic measures). Obviously, more effort is necessary to establish these techniques, and more knowledge about the genetic structure of most phytoplankton species must be gained before DNA fingerprinting methods can be routinely used to investigate phytoplankton biodiversity and population structures. Thus far, DNA fingerprinting with VNTRs has not been used in microalgal analysis to our knowledge, but single-locus microsatellite markers were applied in analyzing the diatom *Ditylum brightwellii* (Rynearson and Armbrust, 2000) and *E. huxleyi* (Iglesias-Rodriguez *et al.*, 2002). In both cases, unique individuals were found and population structure was revealed.

Sequence Data

Sequence data for both coding and noncoding regions of the genome can be used to reconstruct the evolutionary history of organisms and to examine relationships at all taxonomic levels. The rRNA genes are commonly used for phylogenetic analyses, although many genes are potentially available. The rRNA genes have special attributes that make them ideally suited as molecular markers. They are of a relatively large size, contain both variable and highly conserved regions, which can be used to address both close and distant evolutionary relationships, respectively, and are of an universally conserved function with no evidence to suggest that they are laterally transferred (Woese, 1987).

Noncoding regions separating genes are termed *spacer regions*. In some operons, such as in the ribosomal operon, they function in the final processing of the mature rRNA molecule, but in most other genes, their function is not well understood. They can evolve at a faster rate because they are not subjected to the same evolutionary constraints as coding regions. To resolve closely related species or population-level genetic structure, these faster-evolving noncoding regions are best used, but even they can be conserved at the genus level or higher in some algae (Medlin *et al.*, 2000a).

Analysis can be performed on the sequences obtained from mixed natural samples/communities. Whole DNA extraction of a water sample, followed by cloning of polymerase chain reaction (PCR)–amplified rDNA sequences and screening of random clones from this library can provide insights into the genetic diversity, which are unobtainable using more traditional means of community analyses. This is especially true for unculturable groups. In every such study novel groups have been found such that the biodiversity of the picoeukaryotic fraction in oceanic samples is considerably underestimated, as has found to be the case for the prokaryotic fraction.

Biases in the phylogenetic results obtained will vary with choice of gene used, the geological age of the taxa investigated, the rate of evolution in the gene of choice, the number of nucleotides used in the analysis, and the number of outgroups, the evolutionary model and the taxa selected for analysis. The Modeltest program (Posada and Crandall, 1998; http://bioag.byu.edu/zoology/crandall_lab/modeltesr.htm) tests 56 types of models of evolution to determine the model that best fits the data so that this model and all of its parameters can be imported into PAUP* (Swofford, 2002) for analysis, but nevertheless phylogenies have been generated for most microalgal groups and they are under frequent refinement. In the last 10 years, three new microalgal classes have been either defined or recognized from sequence analysis (Anderson *et al.*, 1993; Guillou *et al.*, 1999a; Kawachi *et al.*, 2002).

Oligonucleotide Probes for the Detection of Phytoplankton

The fast and secure identification of phytoplankton, especially of toxic species, is important from an ecological and economical point of view, but the aforementioned problems for most picoplankton and nanoplankton species makes this difficult. Identification usually necessitates other time and cost-intensive techniques, such as electron microscopy, pigment analysis with high-performance liquid chromatography (HPLC), or sequencing of conserved genes before a definitive identification can be made of particularly difficult taxa. Phytoplankton species identification by whole-cell hybridization with specific fluorochrome-labeled probes followed by fluorescence microscopy or flow cytometry offers a faster alternative for species identification. Based on conserved and variable regions of the RNA of the ribosomal small and large subunit (SSU, LSU rRNA), signature sequences of varying specificity can be found, which has been used to develop probes for the identification of phytoplankton at various taxonomic levels from classes down to species or strains. The vast amount of rapidly accumulating sequence data for all kinds of organisms makes it possible to develop these probes for a broad spectrum of taxa. Although these techniques have been largely used for bacteria (Amann, 1995; Stahl and Amann, 1991), there is already a growing number of probes for eukaryotic picophytoplankton and nanophytoplankton. To date, they have, for example, been developed for classes including Chlorophyceae (Simon *et al.*, 1995, 2000), Prymnesiophyceae (Lange *et al.*, 1996; Simon *et al.*, 2000), Pelagophyceae (Simon *et al.*, 2000), Dinophyceae (John *et al.*, 2003), and Bolidophyceae (Guillou *et al.*, 1999b), taxonomic clades like those for toxic and nontoxic *Chrysochromulina/Prymnesium* species (Simon *et al.*, 1997) and for ecologically and economically important genera and species like *Chrysochromulina polylepis* (Simon *et al.*, 1997), *Alexandrium tamarense* (Miller and Scholin, 1998), *A. ostenfeldii* (John *et al.*, 2003), *Phaeocystis* (Lange *et al.*, 1996), *Emiliania huxleyi* (Moon-van der Staay *et al.*, 2000), and various *Pseudonitzschia* species (Miller and Scholin, 1998). As mentioned before, even more probes are available for numerous prokaryotic groups and strains, including many marine and limnic bacteria. An overview of the analysis of these groups by rRNA probes is given by Amann *et al.* (2001). Successful application of molecular probes to field samples have already demonstrated their use in characterization of phytoplankton abundance, for example, for *Pseudo-nitzschia* species in coastal waters from Louisiana (Parsons *et al.*, 1999), Bolidophyceae in the Mediterranean Sea, and the Pacific Ocean (Gulliou *et al.*, 1999b) or groups of Prymnesiophytes in the Pacific Ocean (Moon-van der Staay *et al.*, 2000).

Experimental Considerations

The broad diversity we face in the phytoplankton makes it difficult to develop an *in situ* protocol capable of analyzing all kinds of algal cells. For example, different types of cell walls and membranes may require different conditions for probe penetration. Also, cell autofluorescence, especially from chlorophyll, can become a problem when it is very strong and, therefore, masks the probe signal. Taking these problems into account, we adapted an existing protocol (Scholin *et al.*, 1996, 1997) for *in situ* hybridization with specific, fluorescent-labeled probes to suit a broader range of phytoplankton species.

For the fixation and hybridization of phytoplankton using fluorescent-labeled probes, various protocols are in use (Scholin *et al.*, 1996; Simon *et al.*, 2000) that are mainly derived from those developed for bacteria and often use paraformaldehyde (PFA) as the fixative (Amann, 1995). Comparing different protocols, we found that the approach of Scholin *et al.* (1996, 1997) gave the best results with most species tested. This method eliminates PFA and replaces it with a saline–ethanol fixative. Nevertheless, this method was originally developed for probes for two genera only and including at least two mismatches between target and nontarget sequence. We found the conditions in our own experiments not to be stringent enough for a broader range of species and probes. Thus, we modified the hybridization conditions to increase stringency by addition of formamide (FA) to the hybridization buffer and reducing the salt concentration in the last washing step. FA concentrations must be established empirically for every probe and normally range between 0% and 50%.

During the testing of different fixation/hybridization protocols, we found that one of the most important components was the type of detergent used. Sodium dodecyl sulfate (SDS), for example, which is often used in hybridization buffers (Simon *et al.*, 2000), destroys the more fragile cells like unarmored dinoflagellates, whereas IGEPAL-CA630 (or the chemically identical NONIDET-P40) maintains cell stability while enabling efficient probe penetration into the cell. The use of the latter detergent in the hybridization buffer enables the investigation of some of the most delicate dinoflagellates, such as *Karenia mikimotoi* (Fig. 1A and B).

In addition to the fixation, the saline ethanol in the Scholin *et al.* protocol (1996, 1997) extracts the chlorophyll from the cells and bleaches them, thus enabling good visualization of probe signals. This is an advantage for *in situ* hybridization experiments with phytoplankton in which autofluorescence can be problematic. Nevertheless, sometimes the autofluorescence of some species is so strong and persistent that a sole ethanol treatment, even for a prolonged time, is not sufficient for probe detection

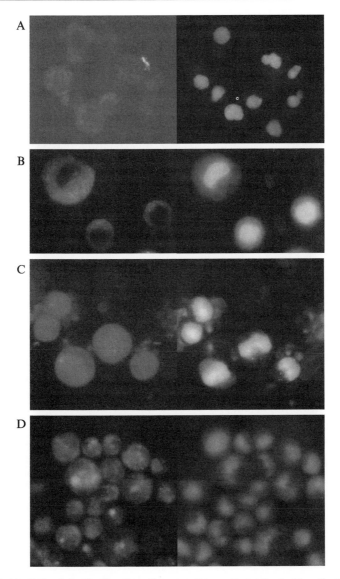

Fig. 1. (A) Cells of the dinoflagellate *Karenia mikimotoi* hybridized with a dinoflagellate-specific fluorescein-labeled probe (left) and the same field seen with DAPI staining (right). Cells were preserved with paraformaldehyde (PFA), and sodium dodecyl sulfate (SDS) detergent was used in the hybridization buffer. The cells were ruptured after the hybridization and no signals could be detected. (B) Cells of the dinoflagellate *Karenia mikimotoi* hybridized with a dinoflagellate-specific fluorescein-labeled probe (left) and the same field seen with DAPI staining (right). Cells were preserved with saline–ethanol fixative, and IGEPAL-CA630

using fluorescence microscopy (Fig. 1C). In these cases, an additional treatment with 50% dimethylformamide (DMF) can help, because it bleaches the chlorophyll from the cells far better than ethanol alone (Fig. 1D). Because DMF is toxic, this procedure should be added to the protocol only if autofluorescence of cells is expected to be a problem.

Using this modified protocol, it is possible to analyze a very broad range of laboratory cultures and field samples with *in situ* hybridization using probes ranging from single- to multiple-base mismatches. This provides researchers with a powerful tool for investigating the occurrence and biodiversity of phytoplankton.

In Situ Hybridization Protocol

Materials

Reagents
1. 25X SET buffer: 3.75 M NaCl, 25 mM EDTA, 0.5 M Tris–HCl (pH 7.8). Filter-sterilize through 0.2-μm pore-size filters and store at room temperature.
2. Saline EtOH fixative: 100% ethanol: distilled water: 25X SET (25:2:3) (v/v). Prepare fresh for every experiment.
3. Hybridization buffer: 5X SET, 0.1% (v/v) IGEPAL-CA630 (or Nonidet-P40), X% (v/v) FA. Filter-sterilize through 0.2-μm pore-size filters, add 30 μg/ml^{-1} poly(A), and store at room temperature. The percentage of FA (X) in the buffer depends on the probe requirements.
4. 50% DMF: not mandatory. Dilute DMF with deionized water, and use glassware only because concentrated DMF can dissolve plastic.
5. Citifluor–DAPI mix: Citifluor (Citifluor Ltd, Cambridge, UK): distilled water (2:1) (v/v) containing DAPI at a final concentration of 1 μg/ml^{-1}.
6. Fluorescein-labeled oligonucleotide probes. Store at $-20°$ in the dark.
7. Nail varnish.

detergent was used in the hybridization buffer. The cells were intact and gave clear hybridization signals. (C) Cells of the dinoflagellate *Alexandrium tamarense* hybridized with a dinoflagellate-specific fluorescein-labeled probe (left) and the same field seen with DAPI staining (right). Without DMF treatment. Cells show strong autofluorescence that masks the probe signal and makes them impossible to analyze. (D) Cells of the dinoflagellate *A. tamarense* hybridized with a dinoflagellate-specific fluorescein-labeled probe (left) and the same field seen with DAPI staining (right). After DMF treatment, autofluorescence is largely reduced in the cells and probe signals are clearly visible.

Equipment
1. Filtration unit with vacuum pump.
2. Polycarbonate filters (0.2 or 3.0 μm pore size).
3. Moisture chamber (i.e., plastic box with its inside covered by 3MM Whatman paper that is soaked with hybridization buffer).
4. Hybridization oven.
5. Microscopic slides and coverslips.
6. Fluorescence microscope equipped with ultraviolet (UV) lamp and appropriate filter sets for probe label and DAPI.

Methods

Fixation of samples
1. Set up filtration unit. Use polycarbonate filters with pore sizes matching the expected cell sizes (0.2 μm pore size for picoplankton and bacteria, 3 μm pore size for larger phytoplankton). The cells should be filtered onto the shiny side of the filter.
2. Pour sample into filtration unit using enough liquid to evenly cover the whole filter.
3. Filter sample through by using the lowest amount of vacuum possible while maintaining a constant flow to prevent breakage of delicate cells.
4. Close connecting piece to pump with parafilm so that no liquid is dripping through the filter during incubation and the filter does not run dry.
5. Add 5–15 ml of freshly prepared saline–EtOH fixative so that the whole filter is covered and incubate at room temperature for 1 h. The solution will normally show strong precipitation, but this does not influence the fixation. For samples that contain species with strong autofluorescence, incubate for 2 h. Afterwards, remove the solution by vacuum filtration.
6. Add 5–15 ml of hybridization buffer (without FA), incubate at room temperature for 5 min, and filter buffer through.
7. If strong autofluorescence is expected, add 50% DMF to cover the filter and incubate at room temperature for 1 h. Repeat step 6 afterwards.
8. Briefly air-dry the filter on Whatman paper. Wrapped filters can be stored at 4° for a couple of months or processed immediately.

Hybridization
1. Cut filter into pieces. A filter of 25-mm diameter can easily be cut into six pieces if necessary, one of 47-mm diameter into at least 12 pieces.

2. Put a filter piece onto a microscope slide, and apply 54 μl of hybridization buffer with 6 μl of probe (50 ng/μl^{-1} stock) directly onto the filter piece. Ensure that the whole filter piece is covered with liquid, and if necessary, use more buffer–probe mixture. The fluorescent-labeled probe is light sensitive, so keep the filters in the dark for the rest of the procedure (e.g., cover them during incubation times) and minimize exposure to light when handling them.

3. Hybridize slides at hybridization temperature (normally 50° in a moisture chamber in the dark for 1 h). FA concentration for new probes needs to be determined empirically. Hybridize a filter piece each with 0%, 10%, 20%, 30%, 40%, and 50% of FA in the buffer. Choose the highest FA concentration for which target cells from defined lab cultures give clear positive signals while nontarget species show no signal. A moisture chamber can be made using hybridization buffer without FA.

4. Wash filter briefly in a small tray in 1X SET buffer prewarmed to hybridization temperature and dry the filter on Whatman paper. Put the filter back onto a clean slide.

5. Apply 100 μl of prewarmed 1X SET buffer onto the filter and incubate again in the moisture chamber at the hybridization temperature for 5 min.

6. Briefly dry the filter on Whatman paper, then put it back onto a clean slide.

7. Apply 20–30 μl of Citifluor–DAPI mixture directly onto the filter, mount coverslip carefully so that the liquid covers the filter piece without any air bubbles, and seal the sides of the coverslip with nail varnish.

8. View slides by fluorescence microscopy. Look at the specimens using the filter for the DAPI stain and then switch to the one for the fluorochrome; only target cells should give positive emerald green signals, whereas nontarget cells show a brown-yellowish color.

Acknowledgments

We thank G. Kirst (University of Bremen) for his helpful ideas about the DMF treatment. This work was funded in part by EU PICODIV EVK3-CT-1999-00021.

References

Amann, R. I. (1995). *In situ* identification of micro-organisms by whole cell hybridization with rRNA-targeted nucleic acid probes. *In* "Molecular Microbial Ecology Manual 3.3.6" (A. D. L. Akkermans, J. D. van Elsas, and F. J. de Bruijn, eds.), pp. 1–15. Kluwer Academic Publishers, Dordrecht, NL.

Amann, R., Fuchs, B. M., and Behrens, S. (2001). The identification of microorganisms by fluorescence *in situ* hybridization. *Curr. Opin. Biotechnol.* **12,** 231–236.

Anderson, R. A., Saunders, G. W., Paskind, M. P., and Sexton, J. P. (1993). The ultrastructure and 18S rRNA gene sequence for *Pelagomonas calceolata* gen. & sp. nov., and the description of a new algal class, the Pelagophyceae classis nov. *J. Phycol.* **29,** 701–715.

Baillie, B. K., Belda-Baille, C. A., Silvestre, V., Sison, M., Gomez, A. V., and Monje, V. (2000). Genetic variation in *Symbiodinium* isolates from giant clams based on random-amplified-polymorphic DNA (RAPD) patterns. *Mar. Biol.* **136,** 829–836.

Barker, G. L. A., Green, J. C., Hayes, P. K., and Medlin, L. K. (1994). Preliminary results using the RAPD analysis to screen bloom populations of *Emiliania huxleyi* (Haptophyta). *Sarsia* **79,** 301–306.

Brand, L. E. (1989). Review of genetic variation in marine phytoplankton species and the ecological implications. *Biol. Oceanogr.* **6,** 397–409.

Guillou, L., Chretiennot-Dinet, M.-J., Medlin, L. K., Claustre, H., Goeer, S. L.-D., and Vaulot, D. (1999a). *Bolidomonas*: A new genus with two species belonging to a new algal class, the Bolidophyceae (Heterokonta). *J. Phycol.* **35,** 368–381.

Guillou, L., Moon-van-der-Staay, S. Y., Claustre, H., Partensky, F., and Vaulot, D. (1999b). Diversity and abundance of Bolidophyceae (Heterokonta) in two oceanic regions. *Appl. Environ. Microbiol.* **65,** 4528–4536.

Iglesias-Rodriguez, M. D., Garcia Sáez, A., Groben, R., Edwards, K. J., Batley, J., Medlin, L. K., and Hayes, P. K. (2002). Polymorphic microsatellite loci in global populations of the marine coccolithophorid. *Emiliania huxleyi. Mol. Ecol. Notes* **2,** 495–497.

John, U., Cembella, A., Hummert, C., Elbrächter, M., Groben, R., and Medlin, L. K. (2003). Discrimination of the toxigenic dinoflagellate species *Alexandrium tamarense* and *Alexandrium ostenfeldii* in co-occurring natural populations from Scottish coastal waters. *Eur. J. Phycol.* **38,** 25–40.

John, U., Groben, R., Beszteri, B., and Medlin, L. K. (2004). Utility of amplified fragment length polymorphisms (AFLP) to analyse genetic structures within the *Alexandrium tamarense* species complex. *Protist* **155,** 169–179.

Karp, A., Isaac, P. G., and Ingram, D. S. (1998). "Molecular Tools for Screening Biodiversity." Chapman & Hall, London, UK.

Kawachi, M., Inouye, I., Honda, D., O'Kelly, C. J., Bailey, J. C., Bidigare, R. R., and Andersen, R. A. (2002). The Pinguiophyceae classis nova, a new class of photosynthetic strameophiles whose member produce large amounts of omega-3 fatty acid. *Phycol. Res.* **50,** 31–47.

Kristiansen, J. (2001). Biogeography of silica-scaled chrysophytes. *Nova Hedwigia Beih.* **132,** 23–39.

Lange, M., Guillou, L., Vaulot, D., Simon, N., Amann, R. I., Ludwig, W., and Medlin, L. K. (1996). Identification of the class Prymnesiophyceae and the genus *Phaeocystis* with ribosomal RNA-targeted nucleic acid probes detected by flow cytometry. *J. Phycol.* **32,** 858–868.

Medlin, L. K., Lange, M., Edvardsen, B., and Larsen, A. (2000a). Cosmopolitan flagellates and their genetic links. *In* "The Flagellate Algae" (J. C. Green and B. S. C. Leadbeater, eds.), pp. 288–308. Francis and Taylor, London.

Medlin, L. K., Lange, M., and Noethig, E. V. (2000b). Genetic diversity of marine phytoplankton: A review and a look to Antarctic phytoplankton. *Antarctic. Sci.* **12,** 325–331.

Miller, P. E., and Scholin, C. A. (1998). Identification and enumeration of cultures and wild *Pseudo-Nitzschia* (Bacillariophyceae) using species-specific LSU rRNA-targeted fluorescent probes and filter-based whole cell hybridization. *J. Phycol.* **34,** 371–382.

Moon-van der Staay, S. Y., van der Staay, G. W. M., Guillou, L., Claustre, H., and Vaulot, D. (2000). Abundance and diversity of Prymnesiophyceae in picoplankton communities from the equatorial Pacific Ocean inferred from 18S rDNA sequences. *Limnol. Oceanogr.* **45,** 98–109.

Ormond, R. F., Gage, J. D., and Angel, M. V. (1998). "Marine Biodiversity: Patterns and Processes." Cambridge University Press, Cambridge.

Parsons, M. L., Scholin, C. A., Miller, P. E., Doucette, G. J., Powell, C. L., Fryxell, G. A., Dortch, Q., and Soniat, T. M. (1999). *Pseudo-nitzschia* species (Bacillariophyceae) in Louisiana coastal waters: Molecular probe field trials, genetic variability, and domoic acid analyses. *J. Phycol.* **35**(Suppl), 1368–1378.

Partensky, F., Hoepffner, N., Li, W. K. W., Ulloa, O., and Vaulot, D. (1993). Photoacclimation of *Prochlorococcus* sp. (Prochlorophyta) strains isolated from the North Atlantic and the Mediterranean Sea. *Plant Physiol.* **101,** 285–296.

Posada, D., and Crandall, K. A. (1998). Modeltest: Testing the model of DNA substitution. *Bioinformatics* **14,** 817–818.

Rynearson, T. A., and Armbrust, E. V. (2000). DNA fingerprinting reveals extensive genetic diversity in a field population of the centric diatom. *Ditylum brightwellii. Limnol. Oceanogr.* **45,** 1329–1340.

Scholin, C. A., Buck, K. R., Britschgi, T., Cangelosi, G., and Chavez, F. P. (1996). Identification of *Pseudo-nitzschia australis* (Bacillariophyceae) using rRNA-targeted probes in whole cell and sandwich hybridization formats. *Phycologia* **35,** 190–197.

Scholin, C., Miller, P., Buck, K., Chavez, F., Harris, P., Haydock, P., Howard, J., and Cangelosi, G. (1997). Detection and quantification of *Pseudo-nitzschia australis* in cultured and natural populations using LSU rRNA-targeted probes. *Limnol. Oceanogr.* **42,** 1265–1272.

Simon, N., Brenner, J., Edvardsen, B., and Medlin, L. K. (1997). The identification of *Chrysochromulina* and *Prymnesium* species (Haptophyta, Prymnesiophyceae) using fluorescent or chemiluminescent oligonucleotide probes: A means for improving studies on toxic algae. *Eur. J. Phycol.* **32,** 393–401.

Simon, N., Campbell, L., Ornolfsdottir, E., Groben, R., Guillou, L., Lange, M., and Medlin, L. K. (2000). Oligonucleotide probes for the identification of three algal groups by dot blot and fluorescent whole-cell hybridization. *J. Euk. Microbiol.* **47,** 76–84.

Simon, N., LeBot, N., Marie, D., Partensky, F., and Vaulot, D. (1995). Fluorescent *in situ* hybridization with rRNA-targeted oligonucleotide probes to identify small phytoplankton by flow cytometry. *Appl. Environ. Microbiol.* **61,** 2506–2513.

Stahl, D. A., and Amann, R. (1991). Development and application of nucleic acid probes. *In* "Nucleic Acids Techniques in Bacterial Systematics" (E. Stackebrandt and M. Goodfellow, eds.), pp. 205–248. John Wiley & Sons, Chichester, UK.

Swofford, D. L. (2002). *PAUP*, Phylogenetic Analysis Using Parsimony.* Version 4.0 Beta version 8, program and documentation. Illinois Natural History Survey, University of Illinois, Champaign, IL.

Waterbury, J. B., Watson, S. W., Guillard, R. R. L., and Brand, L. E. (1979). Widespread occurrence of a unicellular, marine planktonic cyanobacterium. *Nature* **277,** 293–294.

Woese, C. R. (1987). Bacterial evolution. *Microbiol. Rev.* **51,** 221–271.

Wood, A. M., and Leatham, T. (1992). The species concept in phytoplankton ecology. *J. Phycol.* **28,** 723–729.

Subsection C

The Genomes

[19] Sequencing and Comparing Whole Mitochondrial
Genomes of Animals

By Jeffrey L. Boore, J. Robert Macey, and Mónica Medina

Abstract

Comparing complete animal mitochondrial genome sequences is becoming increasingly common for phylogenetic reconstruction and as a model for genome evolution. Not only are they much more informative than shorter sequences of individual genes for inferring evolutionary relatedness, but these data also provide sets of genome-level characters, such as the relative arrangements of genes, which can be especially powerful. We describe here the protocols commonly used for physically isolating mitochondrial DNA (mtDNA), for amplifying these by polymerase chain reaction (PCR) or rolling circle amplification (RCA), for cloning, sequencing, assembly, validation, and gene annotation, and for comparing both sequences and gene arrangements. On several topics, we offer general observations based on our experiences with determining and comparing complete mitochondrial DNA sequences.

Introduction

Mitochondria are subcellular organelles of nearly all eukaryotes, descended from α-Proteobacteria that took up residence inside an early member of the eukaryotic lineage (Lang *et al.*, 1999). They still contain their own much diminished genomes and have systems for transcription, message processing, and translation that are separate from those of the cytoplasm. For animals, these mitochondrial genomes are almost always circular (for exceptions in the Cnidaria, see Bridge *et al.*, 1992; in a crustacean, see Raimond *et al.*, 1999) and usually contain the same set of 37 genes, encoding 13 proteins, 2 rRNAs, and 22 tRNAs (see Boore, 1999; for some exceptions, see Armstrong *et al.*, 2000; Beagley *et al.*, 1995, 1998; Beaton *et al.*, 1998; Helfenbein *et al.*, 2004; Hoffmann *et al.*, 1992; Keddie *et al.*, 1998; Le *et al.*, 2000; Nickisch-Rosenegk *et al.*, 2001; Okimoto *et al.*, 1991; Wolstenholme *et al.*, 1987; Yokobori *et al.*, 1999, 2003). Typically, they are about 16 kb in size, and so are very gene dense and without introns except in cnidarians (Beagley *et al.*, 1995, 1998; Beaton *et al.*, 1998), which also contain one extra gene, a homologue to bacterial *mutS*. Some mtDNAs have all genes on one strand; for others they are distributed between both. In the few cases where it has been studied, transcription produces a single large transcript for each DNA strand, which is then enzymatically cut into (mostly) gene specific RNAs (Clayton, 1992).

There are several merits to comparing these diminutive genomes. Their small size and compact arrangements facilitate broad comparisons for many animals. Comparisons can include the homologous genes found in the mtDNAs of plants, protists, and fungi, and in the genomes of prokaryotes (Adams and Palmer, 2003; Gray, 1999; Gray *et al.*, 1998; Lang *et al.*, 1997). Their (usually) circular structure enables physical isolation from nuclear DNA. Their biochemistry is relatively well understood and they are known to play important roles in cellular metabolism (Nieminen, 2003), development (Krakauer and Mira, 1999; Yost *et al.*, 1995), aging (Nagley and Wei, 1998), and human disease (Wallace, 1999). Their products interact with those from hundreds of nuclear genes, inviting studies of coevolution with these interacting factors (Wu *et al.*, 2000). Many of the processes of genome evolution can be studied in these relatively simple systems, including genome rearrangements (Boore, 2000), tRNA editing (Lavrov *et al.*, 2000), tRNA gene "identity theft" (Rawlings *et al.*, 2003), and the causes of mutational biases and their effects on amino acid substitution patterns (Helfenbein *et al.*, 2001).

Mitochondrial genome comparisons have successfully addressed a broad range of phylogenetic questions. Rapidly evolving portions of noncoding DNA are used for forensic identifications (Budowle *et al.*, 2003) and addressing population structure (Nyakaana *et al.*, 2002). Although mtDNA sequences generally evolve more rapidly than those of nuclear genes (Brown *et al.*, 1979; Gissi *et al.*, 2000), leading some to question their resolving power, they have produced robust phylogenies even at very deep levels (Helfenbein *et al.*, 2004). Whole mtDNA sequence comparisons are much more powerful for phylogenetic reconstruction than single gene comparisons (Boore *et al.*, 2004; Ingman *et al.*, 2001; Macey *et al.*, 2004; Parsons and Coble, 2001). Complete mitochondrial genome sequences also provide a set of "genome-level characters" (Nikaido *et al.*, 1999; Schmitz *et al.*, 2001), such as RNA secondary structures (Macey *et al.*, 1997c; Macey *et al.*, 2000b), modes of control of replication and transcription (Clayton, 1992), mtDNA physical structures (Bridge *et al.*, 1992), and especially the relative arrangements of genes, which can be a reliable indicator of common ancestry (see later discussion).

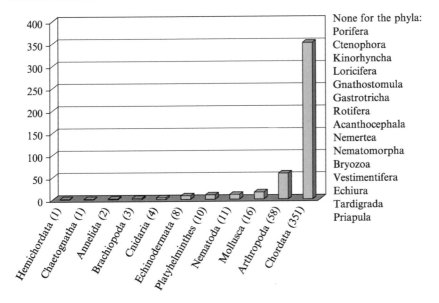

FIG. 1. Phylogenetic distribution, by phylum, of the complete mitochondrial DNA sequences in GenBank as of June 2004, with a list of those phyla remaining unrepresented.

For these reasons, and because high-throughput genome sequencing is becoming a mature technology, comparing complete mtDNA sequences is becoming increasingly common. In no other part of the genome could one so easily obtain the sequences of 37 unambiguously orthologous genes so densely packed. The rate of production of complete mtDNA sequences is increasing exponentially and, as of June 2004, GenBank holds 464 complete animal mtDNA sequences. Although taxonomic sampling is highly biased (Fig. 1), we are moving toward a thorough sampling across life of this small genome for better understanding of the evolution of genomes and organisms. (Table I lists some of the resources on the web for learning more about mitochondrial systems.) It is imperative that we streamline the processes of acquiring, analyzing, and comparing these data.

Making the Templates

Differences in body structures among various animal groups make it impossible to comprehensively generalize the best tissues for isolating mtDNA. Eggs, gonads, muscle, liver, and brain are commonly used, generally in this order of preference, although other tissues are often

TABLE I
WEB RESOURCES FOR FURTHER INFORMATION ON mtDNA AND MITOCHONDRIAL SYSTEMS

URL	Description
http://www.jgi.doe.gov/programs/comparative/ top_level/organelles.html	Organelle Genomics at DOE Joint Genome Institute
http://evogen.jgi.doe.gov/dogma/	Tools for gene annotation of mtDNAs and cpDNAs
http://megasun.bch.umontreal.ca/ogmpproj.html	Organelle Genome Megasequencing Project
http://www.ncbi.nlm.nih.gov/genomes/ organelles/organelles.html	Organelle genome resources at GenBank
http://www.mitomap.org/	Mitomap: Human mitochondria and disease
http://www-lecb.ncifcrf.gov/mitoDat/	Nuclear genes with mitochondrial products database

acceptable. All methods are most reliable when using fresh tissue, although tissues frozen (quickly, if possible) at $-80°$ are commonly used. Working with tissues stored in ethanol is variably successful, but those in formalin have never yielded useful DNA in our hands [but see Kearney and Stuart (2004) for a case with obtaining small fragments from formalin-fixed samples].

There are numerous methods, including the use of several commercial products, for isolating total DNA from tissue that are described elsewhere. For study of any particular region of the genome, the next step must be to physically isolate that portion of the DNA. For the case of mtDNA, there are several methods in common usage: (1) physically isolating mtDNA from nuclear DNA by exploiting the differential masses of the nucleus and organelle and/or the differing properties of linear versus circular DNA; (2) creating and identifying clones that contain copies of the mtDNA; or (3) generating identical copies of the mtDNA in high molarity by long polymerase chain reaction (PCR) or rolling circle amplification (RCA).

Physical Isolation

It is possible to generate about (nominally, and with high variance) 1 μg of mtDNA per gram of tissue using standard cell fractionation techniques (Protocol 1). Especially if this DNA is ever to be used for PCR, it is critical to minimize the potential for contamination between samples or otherwise. Briefly, the tissue is disrupted using either a dounce homogenizer or a Tissuemizer (Tekmar, Inc.), trying to maximize the breaking of fibrous matrix and cell membranes while minimizing the breaking of nuclear and mitochondrial membranes. This step may need optimizing, because too vigorous treatment will break more nuclei, reducing the amount of nuclear DNA removed by centrifugation, and too gentle treatment will incompletely lyse cells, which will pellet in this centrifugation, so removing mitochondria and reducing yield. It is critical to minimize enzymatic damage to DNA by working quickly, maintaining dilute solutions, and keeping reagents and materials cold. Nuclei are pelleted by low-speed centrifugation, but nuclear DNA from any that have been inadvertently lysed will contaminate the supernatant (an amount typically larger than the isolated mtDNA itself after ultracentrifugation). It is optional at this point to layer mitochondria at the interface of a 1.5–1.0 M sucrose step gradient; this is most useful if a large amount of tissue was used. Mitochondria are pelleted and lysed, then the mtDNA is separated from any contaminating nuclear DNA by centrifugation in the presence of CsCl (to produce the primary gradient) and propidium iodide (PI) in an ultracentrifuge. The PI intercalates between base pairs, causing unwinding of the DNA helix, which is physically resisted by circular DNA, but not linear DNA, because the latter can unwind without limit. This renders circular DNA more dense than linear, which seems counterintuitive, but the dye binding lessens the mass because it introduces more "space" than weight. Detailed explanation and a description of the critical parameters for the ultracentrifugation can be found under sections for separating plasmids from genomic DNA in Sambrook and Russell (2001) and Ausubel *et al.* (2004). The mtDNA fraction will be a few millimeters below the nuclear band. After collecting this with a needle, the PI is removed by extraction with butanol and the CsCl removed by dialysis or ultrafiltration.

PROTOCOL 1

Physical isolation of mtDNA

Solutions (and Storage Conditions)
  Homogenization buffer (4°):
    210 mM Mannitol (adds viscosity for more gentle cell lysis)
    70 mM Sucrose (enhances mitochondrial integrity)
    50 mM Tris–HCl, pH 7.5 (buffers pH changes)
    3 mM CaCl$_2$ (protects nucleoprotein complexes for nuclear
      membrane integrity)
  (This cannot be autoclaved but can be sterilized by passage through a 0.22-micron filter.)

500 mM EDTA (4°)

TE: 10 mM Tris, 1 mM EDTA, pH 8.0 (room temperature)

2% bleach solution (room temperature)

For a sucrose step gradient: 1 M sucrose in TE and 1.5 M sucrose in TE (4°)

10% sodium dodecyl sulfate (SDS): 10 g SDS, water to 100 ml (room temperature)

Saturated CsCl is 7 M in water

Propidium iodide (PI) solution: 2 mg/ml in water

CsCl/PI solution, density 1.57 g/ml (makes 100 ml, room temperature):

74.2 g	CsCl
10.0 ml	10 × TE
54.9 ml	water
16.6 ml	PI solution (2 mg/ml)

For a "velocitization," make this solution at two other densities also: For 1.4 g/ml, modify to 53.3 g of CsCl and 71.7 ml of water. For 1.7 g/ml, modify to 93.3 g of CsCl and 61.7 ml of water.

Water-saturated butanol: Mix equal portions, shake, and let the layers form. (Butanol is on top.)

Procedure

This is for small amounts of tissue, less than about 0.3 grams. Scale up for larger amounts.

1. Homogenize tissue in 3 ml of cold homogenization buffer using three 5-s strokes of the Tissuemizer (Tekmar). A dounce homogenizer can be used instead. This may need optimizing.

2. Add 600 μl of 500 mM EDTA. Keep on ice. (Clean the Tissuemizer between samples by running for 15 s in cold 2% bleach solution, then rinsing with three iterations of cold, high-purity water.)

3. Pellet nuclei at 1200g for 5 min at 4°. Remove the supernatant into a fresh sterile tube with a pipette. If working with a large sample, this step can be repeated.

4. Pellet mitochondria at 23,000g for 25 min at 4°. Pour off and discard the supernatant. Allow the tube to drain for a few minutes, then put on ice.

Some would add a sucrose step gradient at this point to separate mitochondria from other cellular constituents, but this is not generally necessary unless the amount of tissue was very large.

A. Prepare step gradients by placing 10 ml of 1 M sucrose in TE in swinging bucket centrifuge tube. Using a pipette, underlayer this with 8 ml of 1.5 M sucrose in TE.

B. Resuspend pellet in 20 ml of homogenization buffer and carefully layer on top of the gradient.

C. Centrifuge at 33,000g for 60 min at 4° in a swinging bucket rotor. (Microsomes collect at the top of the 1 M sucrose, mitochondria at the interface of 1.5 M and 1 M sucrose, and nuclei at the bottom.)

D. Aspirate off the excess at the top, then collect the interface layer between the 1.5 M and 1 M of sucrose with a short pipette, trying for minimum volume (<5 ml).

E. Add about 4 volumes of cold TE to this collected fraction to reduce the sucrose concentration.

F. Pellet mitochondria at 23,000g for 25 min at 4°, then pour off and discard the supernatant. Allow the tube to drain for a few minutes, then put on ice.

5. Resuspend pellet in 1.6 ml of TE with vigorous vortexing at room temperature.

6. Lyse the mitochondria by adding 0.4 ml of 10% SDS and mixing gently, then let stand for 10 min at room temperature.

7. Add 0.33 ml of saturated CsCl, mix, and place on ice for at least 15 min. (Can remain at 4° overnight.)

8. Pellet the mitochondrial membrane debris at 17,000g for 10 min at 4°, then collect the supernatant into a sterile culture tube.

9. Add 0.5 ml of PI solution. Measure the volume and add solid CsCl according to the following schedule, interpolating as necessary: 2.2 ml/2.04 g; 2.3 ml/2.13 g; 2.4 ml/2.22 g; 2.5 ml/2.32 g.

10. Invert several times to dissolve the CsCl. Check the density by weighing 1 ml to ensure it is 1.57 ± 0.01 g. Adjust using solid CsCl or water to 1.57 g/ml. Top off the tube with the CsCl/PI stock solution of density 1.57 g/ml. Mix. Balance the tubes in pairs to within 0.02 g.

11. Ultracentrifuge to separate linear and circular DNAs. The vertical rotor spins harder than the swinging bucket rotor so the gradient sets up faster; however, the shorter gradient produces a less pure product. The possibilities are, in order of speed versus purity: (1) single spin in the vertical rotor; (2) single spin in the swinging bucket rotor; (3) spin in the vertical rotor followed by the swinging bucket rotor; and (4) spin in the vertical rotor, a "velocitization" removing small DNA fragments, then spin in the swinging bucket rotor. Parameters for spinning in the vertical rotor: 55,000 rpm for 24 h at 20°. Parameters for spinning in the swinging bucket rotor: 36,000 rpm for 36–48 h at 20°. Consult the machine instructions for details. For the swinging bucket rotor, layer light mineral oil on top to within 3 mm of the top, balancing them using the mineral oil.

A velocitization is sometimes used between two density-gradient ultra-centrifugations to remove small DNA fragments: (A) Measure the volume of the sample (V) and calculate $X = 3.8 - (2V + 0.7)$. Add X ml of a 1.4 g/ml CsCl/PI solution to a tube for the swinging bucket rotor. (B) Underlayer 0.7 ml of a 1.7 g/ml CsCl/PI solution by placing a Pasteur pipette tip against the side of the tube, gently releasing a drop, then sliding the pipette down to the bottom of the tube. Slowly withdraw the pipette as the last bit of liquid is delivered, keeping it against the side of the tube. (C) Add V ml of TE to the sample and layer this on top of the gradient. Top off with light mineral oil and balance. (D) Swinging bucket rotor run parameters: 45,000 rpm for 3 h and 30 min at 20°. Brake should be on while the machine is accelerating, but *must* be turned off before the spin comes down. (E) Collect by puncturing tube and collecting the bottom 1.4 ml *without* using ultraviolet (UL) light. The intact mtDNA will have been at the interface of the two CsCl density solutions.

12. After each ultracentrifugation (except a velocitization), collect sample using a needle or a tube dripping apparatus and a UV light. There will be as many as four fluorescent bands. Against the bottom of the tube is RNA. At the top are carbohydrates. Nuclear DNA is about the middle of the tube. About 5 mm below the nuclear will be the mtDNA. Even if not visible, collect this region of the gradient. See details in Sambrook and Russell (2001). If another ultracentrifugation is desired and the mtDNA band is not visible, consider collecting a small amount of the nuclear band to use as a trace.

13. Add about 500 μl of water saturated butanol to the sample, shake, then let the layers form. The butanol will absorb much of the propidium iodide. Remove this with a Pasteur pipette and discard, then repeat about five times, until the sample appears clear, then one more time. Check for fluorescence with UV light. If volume reduction is desired, extract with straight butanol, which will adsorb water.

14. Remove the CsCl either by using ultrafiltration (e.g., using ultrafree spin columns; Millipore) or dialysis. If dialyzing, use a large volume of dialysate (TE or water), at least 200 ml, with four buffer changes, over about 24 h, mixing gently on a stir plate.

This process requires relatively large amounts of tissue, a great effort, and often extensive optimization. However, it can be very reliable, especially when it can be optimized for a single tissue type. Figure 2 (lanes 2, 3, and 7) shows the results of a physical purification from the very mtDNA-rich eggs of *Xenopus*.

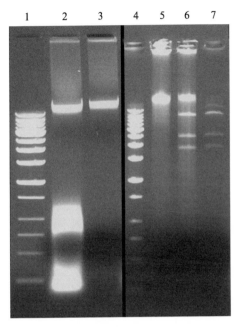

Fig. 2. Products of physical purification and of a rolling circle amplification reaction. Lanes are as follows: (1) KB ladder standard; (2) mitochondrial DNA purified from *Xenopus* eggs using the technique in Protocol 1; (3) This same preparation treated with RNAse; (4) KB ladder standard; (5) RCA product using the technique in Protocol 3 from 1/100 of the material in lane 2; (6) this RCA product digested with *Spe*I; (7) the purified mtDNA shown in lane 2 digested with *Spe*I.

Cloning

The optimal insert size for plasmids is too small, and for fosmids is too large to accommodate most mtDNAs. The vector of choice is a replacement type of phage vector optimal for a mtDNA-sized insert such as EMBL3 or EMBL4 (available from Stratagene and others). The greatest problem with cloning an entire mtDNA into a phage is that one must begin with a single break, ordinarily generated by cutting with a restriction enzyme recognizing the mtDNA at only one site and compatible with (or modifiable to be compatible with) a site in the polylinker of the vector. This can be determined by constructing a restriction enzyme map of isolated mtDNA, if it is very pure, by radiolabeling fragments (Protocol 2) or by doing a Southern blot (Southern, 1975) of total DNA using (normally heterologous) mtDNA probes. Detailed protocols for this and other molecular techniques not individually referenced here can be found in Sambrook and Russell (2001) and Ausubel *et al.* (2004). Briefly, cloning into phage requires ligation of the vector and insert, packaging into infectious particles, transfecting bacteria, and plating, where the clones form plaques on a lawn of growing bacteria. If the library is made from isolated mtDNA, then it is normally sufficient to randomly pick a few clones and test each for a mtDNA insert. If the library is made from total DNA instead, this would be inefficient, and one would ordinarily probe plaque lifts to identify the correct clone(s). The best probe DNA for this would be short PCR fragments amplified from the mtDNA itself, so that the hybridization can use very stringent conditions, although a heterologous probe from other mtDNAs can work well. In either case, the candidate clone should be verified as containing the entire mtDNA by checking that the restriction enzyme map made from DNA produced from the clone, ordinarily visualized on an agarose gel stained with ethidium bromide (i.e., rather than by radiolabeling), matches that determined for the native mtDNA.

PROTOCOL 2

Mapping mtDNA by Radiolabeling Fragments

Perform restriction-enzyme digests of purified mtDNA as desired, each in 10 μl total volume.

Inactivate the enzyme by heating. In most cases, 10 min at 70° is sufficient. Spin to bottom of tube.

For calculations below, n = number of digests + 2 for standards + 1 for pipetting error.

This is acceptable for up to 20 reactions; scale up for more.

In addition to filling in at 5' overhangs, the polymerase has an exonuclease activity that erodes 3' overhanging or blunt ends, which are filled back in, and will also translate from any nicks in the DNA.

The acrylamide gel allows very accurate sizing of fragments from about 20 to 1500 base pairs, while the agarose gel best sizes larger fragments. Choose size standards with this in mind.

Carefully follow good procedures for safely handling radioactive materials.

Reaction Mix

1.5n μl reaction buffer
2 μl equal mixture of alpha-labeled radionucleotides, 300 μCi/ml
1 μl Klenow polymerase
Add water to bring volume to 5n μlAdd 5 μl of reaction mix to each restriction enzyme digest, including 2 for size standards.

Incubate for 30 min, 5' overhang digests on ice, 3' overhang or blunt end digests at 37°.

Divide this evenly into two fractions, add 2 μl of gel loading buffer ("bluecrose"), and load onto each of an acrylamide (3.5 to 6%) and agarose (0.8 to 1.2%) gel, using TBE, normally in vertical gel rigs. Load bluecrose into empty lanes.

After gel is run, remove to 3-MM paper, cover with plastic wrap, and dry in a vacuum gel dryer.

Expose to X-ray film, nominally for 24 h, although this can be repeated for shorter or longer times.

The result looks like a negative of an ethidium bromide–stained gel.

10× reaction buffer: 60 mM KCl, 100 mM Tris (pH 7.2), 100 mM MgCl$_2$, 70 mM beta-mercaptoethanol.

PCR Amplification

Because of the difficulties in purifying and/or cloning mtDNAs, many have adopted PCR amplification from whole genomic DNA extractions. Until a few years ago, PCR could amplify only fragments of a few kilobases, but now techniques are available for amplifying much longer regions (Cheng et al., 1994). Although we have been able to amplify an entire mtDNA from a single set of primers (Boore et al., 2004), it is a more efficient strategy with more common success to amplify each in two or three overlapping portions (Fig. 3). In some cases, we use primers designed to match conserved regions for long PCR. However, it is generally more effective to amplify and determine a short sequence and then make primers of perfect match to the mtDNA and facing "out" from the fragment, because the success rate for long PCR is lower than that for shorter fragments, at least in our hands, and this eliminates the variable of the extent of primer matching. In either case, where mtDNAs have large-scale gene rearrangements, it is often necessary to try many combinations because it is impossible to tell a priori which primers are opposed on the mtDNA and of acceptable separation. In cases of difficulty, we sometimes amplify and sequence one portion, nominally half, and then make specific primers to amplify the remainder.

We routinely use either Takara LA (Takara Bio, Inc.) or rTth-XL polymerase (Perkin-Elmer) essentially according to suppliers' instructions. (Other enzymes are available, including Elongase [Invitrogen] and Herculase [Stratagene], with which we have less experience.) Optimization of reaction conditions is often required, especially the magnesium concentration and primer annealing temperature. It is important that sufficient time be allowed for the extension step, especially in later cycles, so we

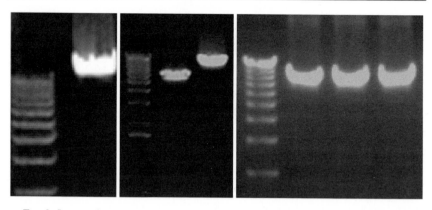

Fɪɢ. 3. Long polymerase chain reaction (PCR) amplification products on agarose gels stained with ethidium bromide. Multibanded standards in each case are KB ladder. The first amplification is the nearly complete mitochondrial DNA (14,465 nts) of the scaphopod mollusk *Graptacme eborea* (Boore *et al.*, 2004). Although this was useful, we much more commonly amplify mtDNAs in two or three generously overlapping fragments, as shown in the remaining panels.

typically start with 1 min/kb to amplify for the first 15 cycles and then use the "autoextend" feature of the PCR machine to lengthen this step by an additional 15 s/cycle, usually to a total of 37 cycles.

Although more complex methods are available, we find that it is sufficient to estimate primer annealing temperature by this simple formula: $(2\times$ the number of As or Ts) plus $(4\times$ the number of Gs or Cs) minus 5. The two primers should be as close as possible to the same estimated annealing temperature. Primer sequences should avoid long runs of homopolymers, be 40–60% G + C, and not have lengthy sequences at their 3′ ends that are in reverse complement to one another. When designing primers to conserved regions, start by aligning amino acid sequences of other animals, being sure to include those distant enough to see saturation of substitutions for nonessential portions of the genes. Choose portions about eight amino acids long that are well conserved and about the desired distance apart. Favor regions that include amino acids that are of minimum possible codon variation, such as tryptophan and methionine, and avoid those that can be coded by many variations, such as serine and leucine. Primers can be made that are degenerate for all possible codon possibilities or that simply use the codons expected to be most common for the amino acids; each strategy seems about equally likely to be successful. It is most important that there are no mismatches at the 3′ end, so we recommend ending a primer on a second codon position of a universally conserved amino acid. Although these are useful guidelines, in practice, some primers work well that do not fully conform, and others fail even when conforming to all of these rules, for reasons that are not obvious. Table II lists a set of primers with which we have been routinely successful at amplifying short fragments from many animal mtDNAs spanning the diversity of the Metazoa.

One of the common problems we have experienced is amplifying regions that include the large noncoding region, sometimes called the *control region*, which generally contains signals for origin and termination of replication. We speculate that in some cases, the polymerase used for the PCR may respond to the replication termination signal for the mtDNA, but there is no obvious similarity among these difficult regions either for sequence, base composition, or potential secondary structures. We have sometimes overcome difficulties by switching to a different polymerase or by reducing the size of the region to amplify by determining flanking sequences. Another effective strategy can be to use the greatly diluted product of an unsuccessful (as judged by having either multiple bands or nothing at the level of detection on the gel) PCR as template for a subsequent amplification with primers that are internally nested. Not uncommonly, using less of any template DNA results in a higher success rate of amplification, because this also dilutes any impurities that might

TABLE II

PRIMERS IN COMMON USE IN OUR LABORATORY THAT ARE DESIGNED TO MATCH CONSERVED REGIONS AND HAVE BEEN BROADLY SUCCESSFUL ACROSS THE METAZOA

Gene	Primer	Sequence	Reference
rrnS	12SaL	AAACTGGGATTAGATACC CCACTAT	Palumbi *et al.*, 1991
	12SaiL	AAACTAGGATTAGATACCCTATTAT	
	12SbH	GAGGGTGACGGGCGGTGTGT	
rrnL	16SarL	CGCCTGTTTATCAAAAACAT	
	16SbL	ACGTGATCTGAGTTCAGACCGG	
	16SaH	ATGTTTTTGATAAACAGGCG	
	16SbrH	CCGGTCTGAACTCAGATCACGT	
	16S1148H	ATTAYGCTACCTTWGCACRGTCARRRT ACYGCGG	This publication
	16S1148L	CCGCRGTAYYYTGACYGTGCWAAGGTA GCRTAAT	
cox1	LCO1490	GGTCAACAAATCATAAAGATATTGG	Folmer *et al.*, 1994
	HCO2198	TAAACTTCAGGGTGACCAAAAAATCA	
cob	cobF424	GGWTAYGTWYTWCCWTGRGGWCARAT	Boore and Brown, 2000
	cobR876	GCRTAWGCRAAWARRAARTAYCAYTC WGG	
cox3	cox3F	TGGTGGCGAGATGTKKTNCGNGA	
	cox3R	ACWACGTCKACGAAGTGTCARTATCA	

be inhibitory; although amount of template DNA added can, at best, boost amplification product proportionally, inhibitory elements affect the reaction product exponentially.

Rolling Circle Amplification

RCA is a technique for producing *in vitro* long, double-stranded, multiple tandem copies of circular DNA molecules (Fire and Xu, 1995; Liu *et al.*, 1996; Lizardi *et al.*, 1998). The technique is dependent on use of a DNA polymerase with high processivity and strong strand-displacement activity, such as that from Phi29 bacteriophage (Lizardi *et al.*, 1998). This enzyme allows extension of more than 70,000 bases from a single priming event (Blanco *et al.*, 1989) and has high fidelity, yielding only 1 error in 10^7–10^8 bases (Estaban *et al.*, 1993). After heat denaturation, multiple primers (typically random hexamers) are annealed and then extended simultaneously at constant temperature by the polymerase. For a circular DNA molecule—like a mitochondrial genome—as each growing strand reaches its origin, it displaces itself, spooling off long single-stranded tandem copies. These are converted during the RCA reaction into double-stranded form by further primer annealing and extension of the complementary strand. RCA has been used to augment the signal from oligonucleotide probes (Baner *et al.*, 1998), to amplify plasmid templates for high-throughput DNA sequencing (Dean *et al.*, 2001; Detter *et al.*, 2002), and even to amplify large genomes (Dean *et al.*, 2002; Detter *et al.*, 2002).

It is important to note that linear DNA undergoes strand displacement amplification in this same reaction, and that this is an isothermal reaction at low temperature, so there is limited specificity of amplification. Nonetheless, we have had success using six of the oligonucleotides to *rrnS, rrnL*, and *coxI* (Table II) on crude DNA preparations. To prevent exonucleolytic activity on the oligonucleotides, the two most 3′ bonds must be phosphorothiol bonds rather than phosphodiesters. We have also successfully amplified mtDNAs that have been purified using ultracentrifugation as above by RCA using random hexamers. Although these mtDNAs were presumably pure, they were of such small quantity that nothing was visible below the nuclear DNA band with ultraviolet irradiation of the CsCl gradient. Figure 2 (lanes 5 and 6) shows one result using purified *Xenopus* mtDNA. For RCA using random hexamer primers, we use the protocols available with the kit (Amersham, Molecular Staging, or Epicentre); Protocol 3 describes the method we use for RCA with specific primers. The best measure of RCA success at amplifying an organelle genome is the ratio of DNA appearing in bands versus smearing on a gel after digestion; this correlates well with the proportion of organelle DNA found in random sequencing reads. We have successfully produced several complete mtDNA sequences from random shotgun plasmid libraries (below) produced from these RCA products.

PROTOCOL 3

RCA with Specific Primers

Denaturing reaction:

4 μl	DNA template
4 μl	10 μM primer mix (at least one for each strand)
2 μl	5× denaturing buffer (from TempliPhi kit; Amersham)

10 μl   Final volume
Denature at 95° for 1 min. Cool to 34°.

Mix:

2 μl	10× reaction buffer
5 μl	High-quality water
2 μl	2 mM dNTPs
1 μl	Phi29 enzyme

10 μl   Final volume
Add this mix to the denatured sample at 34° for 15 h, then heat to 65°, 15 min. Store at 4°.

10× reaction buffer: 75 mM NaCl, 60 mM MgCl$_2$, 1 mM DTT, 0.01% Tween 20, 25 mM Tris pH 8.1 (pH adjusted with boric acid rather than HCl).

Although both approaches have been successful, each has shortcomings and advantages. With specific primers, we can use total DNA preparations but may be limited by endogenous enzyme inhibitors or by the ability to accurately design conserved primers. Further, several commercially available kits do not provide pure Phi29 enzyme, but a reaction mix that contains both enzyme and random hexamers. (Another problem with these kits is that there is nearly always amplification in an attempted negative control reaction, presumably caused by contamination of the enzyme mix with DNA from the Phi29 plasmid clone.) Alternatively, although randomly primed RCA requires highly purified mtDNA, it must be available only in miniscule amounts. Protocols for purifying mtDNA have emphasized quantitative recovery, as has been necessary for direct cloning or restriction enzyme analyses, but RCA enables the use of alternatives that would generate only minute amounts of pure mtDNA.

Work in progress is attempting to increase the specificity of amplification even at the low RCA temperatures by modification of reaction conditions. Other experiments are testing the potential for transposon-mediated cloning of mtDNAs from total DNA preparations, purification of mtDNAs using biotinylated primers matching mtDNA sequences coupled to streptavidin-coated magnetic beads, and isolating intact mitochondria using a fluorescence-activated cell sorter (FACS).

Sequencing the Templates

No techniques are available for reading more than about 1 kb of DNA sequence in a single reaction, so whether the template is purified mtDNA, an mtDNA-containing phage clone, a long PCR product, or the result of an RCA reaction, the task must be broken into smaller components by one of two methods: (1) The DNA to be sequenced is physically broken into smaller pieces that are cloned for "shotgun" sequencing (below), then the sequence that is determined for each is assembled based on overlap; or (2) primers are designed to incrementally walk through a longer fragment. These methods and their variations will be described separately.

DNA is broken into smaller fragments by one of three methods. Restriction enzymes can cut the DNA into precise and reproducible fragments, but these vary greatly in size and can cut only at specific points, which may not be conveniently distributed. To minimize these problems, it is common to use a four base–recognizing enzyme, expected to cut about every 256 nucleotides, but under conditions that result in only partial digestion. This is difficult to control and requires optimization of conditions. Secondly, the DNA can be sonicated, that is, broken by intense sound waves using a specially constructed device. Breaks are random, but the distribution of fragment sizes is wide, so a very large amount of starting material is required. The best method is generally to break the DNA by driving it repeatedly through a narrow aperture, like the one presented in a HydroShear device (Gene Machines; protocol at http://www.jgi.doe.gov/prod/SCLIB.html). The DNA stretches as it passes through under high pressure and breaks if it is longer than a size specified by the pressure, typically to a size between 1 and 1.5 kb for mtDNAs. Figure 4 shows the mechanism of action and a typical result. (A common misconception is that we intend to sequence completely through these clones, leading to the comment that they are too large; actually, the sequencing reads from each end of a clone do not overlap, but we rely on reads from many clones for the contig assembly [see later discussion]).

In our process at JGI, after enzymatic end repair and electrophoretic size selection, these fragments are ligated into pUC18 and transformed into

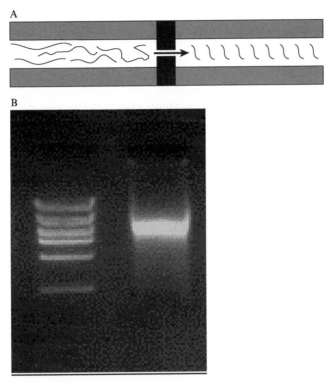

FIG. 4. Breaking DNA into random fragments with a HydroShear device (GeneMachines). (A) Long strands of DNA are loaded into the device, then driven under pressure repeatedly through a narrow aperture, shearing them to a size specified by the pressure applied. (B) Agarose gel stained with ethidium bromide showing a typical result. The standard in lane 1 has bands at 1, 2, 3, 4, 6, and 10 kb. Note the narrow distribution of product size.

Escherichia coli DH10b to create plasmid libraries. These are plated and grown overnight, then an automated colony picker (Genetix) is used to select colonies into 384-well plates of LB with 10% glycerol. These are incubated overnight in a static incubator, without shaking or enhanced aeration, then a small aliquot is processed robotically through plasmid amplification using RCA, sequencing reactions, reaction cleanup using SPRI (Elkin *et al.*, 2002) and processing on an automated capillary DNA sequencer. Detailed protocols are available at http://www.jgi.doe.gov/sequencing/protocols. Because these sequenced fragments are from random starting points, we determine at least 10 times the amount of sequence in the target template and then assemble these sequencing reads together for a complete mtDNA sequence (see later discussion).

Another option is to primer walk through long PCR fragments or clones. Although many alternatives are possible, we load PCRs onto ultrafiltration spin columns (Ultrafree 30,000 NMWL, Millipore), fill with water to 400 μl, and spin according to supplier's instructions through three repetitions. This removes primers, buffer, and unincorporated nucleotides and provides template directly for sequencing reactions. After each sequencing reaction, a primer is selected for the next round until a high-quality, complete sequence is determined from each strand. Primers are typically 18 nucleotides in length, although can be made longer to increase annealing temperature if necessary. In some cases, when sequencing multiple related mtDNAs, researchers have made a series of primers to conserved regions to streamline the process (Inoue *et al.*, 2001; Sorenson *et al.*, 1999).

Assembly of the Sequencing Reads

The raw sequencing reads are first processed with PHRED (Green, 1996), which generates chromatograms, base calls, and quality scores. The latter are expressed as "Q values" that correspond to the probability of error according to the scale that Q10 = 1/10, Q20 = 1/100, Q30 = 1/1000, and so on. Each quality file is linked to the corresponding base call file and these can be moved between platforms. Although there is other commercial software for generating chromatograms and base calling, PHRED's quality scores are an outstanding feature.

If the method of choice is primer walking, then assembly becomes trivial and can be done manually, because each subsequent sequencing read begins immediately beyond the primer annealing site in the previous read. The more complex issues are for the assembly of a large collection of sequencing reads from random clones. The most commonly used computer programs for this are Sequencher (Gene Codes) or PHRAP (Green, 1996). In most cases, the first step is to trim, to remove low-quality sequence at the end of each read, and to remove the small amount of vector sequences (part of the multiple cloning site) at the beginning of each read. This can be done automatically in either system. Details for the use of these particular programs can be found at http://www.genecodes.com/ and http://www.phrap.org/.

Common difficulties are gaps remaining either from having an insufficient number of sequencing reads, from cloning biases, or from misassemblies due to repeated sequences. Gaps can be closed by directed approaches, either by PCR amplifying and sequencing the missing portions or, if a plasmid clone can be seen to span the gap, by primer walking with this clone as a template. Misassemblies caused by repeats often appear as unusually deep coverage of a region by sequencing reads and a pattern where reads from opposing ends of clones are closer together than the typical clone size. This can be definitively resolved by using Southern hybridizations (Ausubel *et al.*, 2004; Sambrook and Russell, 2001; Southern, 1975) or PCR amplifications from unique flanking sequences.

All assemblies and sequence quality should be verified by eye using either Sequencher or CONSED (Green, 1996). The former is available for the Macintosh and is somewhat simpler to use; CONSED (and PHRED and PHRAP) is based in Unix but has more features, especially for resolving misassemblies (and all three are free). Even if the assembly was generated in PHRAP, the entire set of files, including quality scores, can be imported into Sequencher for verification and viewing of this assembly. Throughout the assembly, there should be significant overlap of reads and multiple reads of high quality. One common problem is in gap handling for the consensus sequence, especially if multiple sequences of low quality are included.

If any portion of the sequence was generated from PCR fragments, it is critical to remove the sequences of the primers from the ends before assembly, because these may not exactly match the mtDNA sequence. If the mtDNA is circular, any assembly will show it arbitrarily linearized, with some sequence repeated on each end of the assembly. It is critical to identify this and to trim one end back to unique sequence; the use of a dot-matrix sequence identity plot, such as is available in MacVector (Accelrys) or other packages, can help.

An iterative approach to verifying sequence quality and assembly is sometimes needed, where one pays particular attention to deficiencies discovered during gene annotation (see later discussion), such as a frame shift or stop codon within a coding region or mismatches in paired nucleotides in tRNA genes, to ensure that no error was generated. To the best of our knowledge, no software is effective in allowing visualization of many complex features while viewing assembled sequences, so this is a highly manual and iterative process.

Annotation of the Genes

Identifying tRNA genes is the most challenging aspect of annotating animal mtDNAs, because there is little sequence similarity except among closely related animals, many features common to cytoplasmic tRNAs are absent, and many are of aberrant secondary structure. Figure 5 shows a labeled schematic diagram of a tRNA. Universally present is a seven-member anticodon loop, a 5-bp anticodon stem, and a seven-member acceptor stem. Either stem may contain some mismatched nucleotides and, rarely, there is potential for a longer anticodon stem. Nearly always found are two nucleotides between the acceptor and D arm, one nucleotide

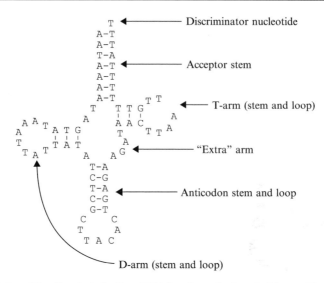

FIG. 5. An arbitrarily selected valine tRNA in schematic cloverleaf form with each section labeled.

between the D arm and anticodon arm, three to five nucleotides in the extra arm, and 2–6 bp, plus 3–12 loop nucleotides, in the D and T arms. Commonly the nucleotides before and after the anticodon are T and A, respectively, the two nucleotides between the acceptor and D arm are TA, and the two most proximal nucleotides of the D loop are As. Some tRNA genes lack the potential for base pairing in the D or T arms and one or both of the tRNAs for serine, in particular, almost always lack paired D arms. There is occasionally apparent overlap of tRNA genes, without obvious explanation of how these might be processed from the presumed polycis-tron. This is so commonly the case for the discriminator nucleotide that it is unclear how often it is encoded by the tRNA gene. There have been particular difficulties with annotating tRNAs that are heavily edited post-transcriptionally, such as for centipede (Lavrov *et al.*, 2000) and presum-ably for jumping spider (Masta and Boore, 2004) mtDNAs. With only a few exceptions (Beagley *et al.*, 1995, 1998; Beaton *et al.*, 1998; Helfenbein *et al.*, 2004), a complete set of 22 tRNA genes is found, one for each amino acid plus an additional one for each of serine and leucine because these can be encoded by two different codon families. An extra *trnM*, perhaps to sepa-rate the roles of this tRNA as initiator and elongator, has been found in the mtDNAs for each of *Mytilus* (Hoffmann *et al.*, 1992), some platyhelminths (although this was not pointed out in the manuscript; Le *et al.*, 2000), and

Ciona (Yokobori *et al.*, 2003). *Ciona* and *Halocynthia* (Yokobori *et al.*, 1999) mitochondria also have an additional tRNA for glycine that recognizes AGR codons, causing a modification of the genetic code. In practice, the search centers on finding the best possibility for each of the 22 expected tRNAs, followed by an effort to find any others in any remaining unassigned sequences.

Protein and rRNA encoding genes are found easily by similarity to other animal mitochondrial sequences using BLAST or a similar program. It is critical to consider that there are several genetic code variations for mtDNAs (Wolstenholme, 1992). There can be some ambiguity in assigning the precise ends of these genes without information from their transcripts. This is particularly acute for the rRNA genes, and one must generally estimate the ends within a few nucleotides based on similarity to other animals' rRNA sequences or assume that they extend to the boundaries of adjacent genes.

For protein-encoding genes, there can be ambiguity in either end. For genes with complete stop codons not overlapping the downstream gene that would produce a protein of typical length and well matched, there can be little doubt of the 3′ end assignment. However, some end at abbreviated stop codons (i.e., a T at a first codon position or a TA in the first and second codon positions), after which the transcript is enzymatically cleaved, with the stop codon completed to TAA by polyadenylation (Ojala *et al.*, 1980). The commonality of such a T exactly juxtaposed to the first nucleotide of the downstream gene, in cases where the reading frame remains open for a great many more codons, argues strongly that this is a common mechanism. For mtDNAs of vertebrates, AGA is a stop codon, so here abbreviated stop codons can be AG in the first two codon positions that would be completed by the same mechanism.

Mitochondrial proteins initiate not only as in the "universal" code (ATG), but in some cases with alternative ATN (ATA, ATT, and ATC) and NTG (GTG, TTG, and CTG) codons. Particular animal groups may use only some of these variations, and this is codified into the translation tables available at GenBank. All match at least two of the nucleotides in the CAT anticodon of the methionyl tRNA. This normally singular tRNA must deliver formyl-methionine to the initiator position during mitochondrial protein translation and methionine to internal positions, so whatever mechanism allows this discrimination presumably also enables the looser codon matching during initiation. Further, *cox1* uniquely has been found rarely to use bizarre initiation codons, including the four-member ATAA for *Drosophila* (Clary and Wolstenholme, 1985). These variations complicate the assignment of the beginnings of genes. In practice, we assume that the correct initiation site is the first eligible in frame start codon as close as possible to the extent of similarity matching and without overlapping the upstream gene.

Of course, it is necessary to do a thorough search of both strands for these genes and to report it correctly to the sequence databases. GenBank has a very large number of errors where genes are not correctly annotated as being on the reverse complement strand.

A significant aid in annotating mitochondrial (and chloroplast, see Chapter 20) genomes is called DOGMA (Wyman *et al.*, 2004; http:// evogen.jgi.doe.gov/dogma/; Fig. 6). This accepts user sequences into a password-protected file, searches using COVE methods (Eddy and Durbin, 1994; Wyman and Boore, 2003) for tRNAs and presents potential secondary structures, along with the alternatives found for each, and presents aligned protein sequences. There are many user-interactive features, including the ability to add or remove genes from the automated annotation and to designate start and stop codons by clicking on the highlighted alternatives while viewing all annotations on a graphical display. DOGMA can extract genes or intergenic regions for subsequent analysis and can save all annotations in a table that can be read into Sequin, GenBank's submission program. Annotated sequences can also be saved in a format that can be read directly into commercial software such as MacVector (Accelrys) for subsequent analysis.

Software for Analysis and Comparisons

Once an accurate sequence is assembled and verified and the genes annotated, the analysis begins. For aspects of molecular evolution, this usually starts with a set of measurements, such as amino acid composition and hydrophilicity (and so on) of inferred proteins, nucleotide composition, codon usage patterns, and strand skew. The latter is the bias between the strands for G versus C and T versus A and commonly reported as G-skew $(G - C)/(G + C)$ and T-skew $(T - A)/T + A)$ (Perna and Kocher, 1995), such that zero indicates perfect balance and 1 or -1 indicates total skew. (A misconception often heard is that the heavy and light strands, caused by strand skew, are somehow homologous; actually, there is no homology of heavy strands [or light strands] across animals.) Noncoding regions are often searched for potential secondary structures, which may mediate replication or transcription control, and for repeats, both direct and inverted, typically using a dot-matrix plot. These tasks are easily accomplished with MacVector (Accelrys) or other commercially available software packages (e.g., VectorNTI, GCG).

The comparison of many of these molecular features has become very difficult with so many mitochondrial sequences available. We are working

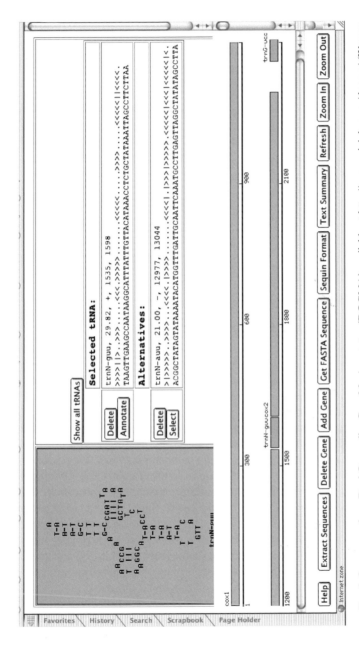

FIG. 6. One screen from Dual Organellar GenoMe Annotator (DOGMA) available at http://evogen.jgi.doe.gov/dogma/ (Wyman et al., 2004) showing the folding of an identified tRNA gene along with alternative possibilities for this same gene elsewhere in the mtDNA. Shown also is part of the graphical overview of the gene annotation produced by the software.

to facilitate this by building searchable databases that will enable broad comparisons of all mtDNA sequences, complete or by individual gene, for specified sets of taxa, for features such as codon usage, nucleotide, and amino acid content, gene arrangement, anticodon identities, and tRNA structures. Look for updates by following the Evolutionary Genomics/Organelles links at http://www.jgi.doe.gov/.

For phylogenetic analysis, the relevant sequences must be aligned. Multiple sequence alignment and phylogenetic analysis of sequences is beyond the scope of this chapter, but we offer a few general comments. Because of both real variability and uncertainty of annotation, the ends of protein and rRNA encoding genes are often ambiguously aligned and are best trimmed back to a region of confidence. Alignment of genes for tRNAs and rRNAs can be guided by their potential secondary structures (Hickson et al., 1996; Kumazawa and Nishida, 1993; Macey and Verma, 1997). Regions with many gaps can contain significant alignment errors and should be eliminated or used only with caution.

Gene-Order Comparisons

There are many strengths to using mitochondrial gene rearrangements for phylogenetic reconstruction of animals. Nearly all animal mtDNAs contain an identical set of genes and these can potentially be reordered into an enormous number of states, so it is unlikely that reversion or convergence would occur. In many lineages, these are slowly rearranging, enhancing the possibility that there will be signal at deep levels, but the finding of radically rearranged mtDNAs shows that these are not under strong selection. (Rather, it seems likely that the infrequency of rearrangement is due to the lack of recombination and the paucity of intergenic nucleotides, such that rearrangements commonly interrupt genes.) The finding of lineages that have rapidly rearranging mtDNAs does not undermine the utility of comparing gene arrangements for phylogenetic reconstruction—they are not rearranging into convergent states—but determines the taxonomic level at which the signal of relatedness would most likely be found. This common misunderstanding is well exemplified by the "cautionary tale" of Le et al. (2000), where a surprising number of rearrangements are described for some platyhelminths, but without demonstrating evidence that this would compromise phylogenetic reconstruction. It is specifically this lack of clocklike behavior of mitochondrial gene rearrangements that makes them most useful for addressing the most difficult situation for phylogenetic reconstruction, having a short internode and a long subsequent time of divergence; there is a brief time to have accumulated signal of relatedness and a long time to erase it, where a perfectly regular rate of change is most likely to fail to reveal the signal.

Although DNA sequence comparisons continue to revolutionize our understanding of organismal relationships, they have also created an expectation that the signal of relatedness must be teased by ever more complex methods from a large body of homoplasious noise. They have revealed much of the pattern of evolution, but many branches of life that were ambiguous in early studies have remained recalcitrant, partly because of limitations such as having only four (or 20, for amino acids) character states, alignment ambiguities, compositional bias, convergent selection, extreme rate variation, uncertainty over weighting changes of nucleotides that are paired versus unpaired in RNA secondary structures, and especially difficulties with short internodes. Although gene rearrangements will define only a small number of evolutionary groups, because there may have been no rearrangements during the period of shared history, or because subsequent rearrangements may have erased the signal, these can be such strong synapomorphies as to be singularly convincing. This is similar to the situation for many groups that are well accepted based on sharing very strong morphological synapomorphies, such as Tetrapoda or Mammalia.

Reconstructing phylogenetic relationships from gene arrangements can be likened to a card-shuffling exercise. Imagine a stack of cards for each genome so that each card represents a single gene, which can be face up or down, analogous to transcriptional orientation. Transformation processes among these decks can be removal and replacement of one or a block of genes, with or without turning them over. (If the genome is circular, the deck must be continuous, i.e., have the bottom card also above the top card.) One can imagine constructing a network connecting these decks with branch lengths proportional to the (perhaps weighted) number of transformations. The shortest possible network would constitute a phylogenetic reconstruction.

Unfortunately, despite considerable effort (Bader et al., 2001; Blanchette et al., 1997, 1999a,b; Cosner et al., 2000; El-Mabrouk and Sankoff, 1999; Moret et al., 2001a,b; Sankoff and Blanchette, 1998a,b, 1999; Sankoff et al., 1990, 1992; Tang and Moret, 2003; Wang et al., 2002), no available method is completely satisfactory. Those based on distance matrices are highly subject to artifactual clustering of taxa into biologically unrealistic groupings and those that retain character information are computationally unfeasible. Further, some methods use models such as allowing only gene inversions (Bader et al., 2001; El-Mabrouk and Sankoff, 1999) that are not biologically realistic for animal mtDNAs. Other work in this volume will detail these efforts, so here we will limit ourselves to the method we most commonly use, a gene adjacency matrix [equivalent to "maximum parsimony

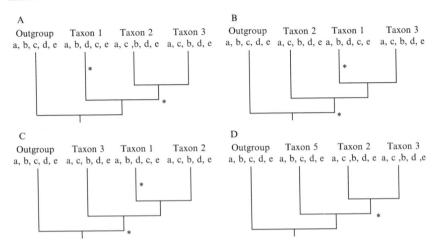

Fig. 7. Having identical gene arrangements does not necessarily indicate close relationship. Lowercase letters label genes in this hypothetical example. The gene arrangement for taxa 2 and 3 are identical, but trees A, B, and C are equally parsimonious. Two changes (*) are required for each tree. (This also shows the primary shortcoming of the gene adjacency method, and the reason single-gene boundaries are weak phylogenetic characters; independent movements of gene "c" creates gene boundary "b, d" in parallel.) A valid phylogenetic reconstruction requires the situation shown in tree D, where (at least) two taxa are in one arrangement, with (at least) two in an alternative arrangement. This tree requires only one change, whereas any other relationship among these taxa would be less parsimonious.

of multiple encodings" of Wang *et al.* (2002)], and to some general observations about gene rearrangements for phylogenetic inference.

Simply sharing even identical gene arrangements does not necessarily indicate close evolutionary relationship. As shown in Fig. 7, gene arrangements can be shared as ancestral states (i.e., sympleisiomorphies) retained from an ancestor, whereas other lineages have changed independently. It is important that taxa are united only when their shared arrangement can be shown to be evolutionarily derived (i.e., a synapomorphy) by comparison with that of outgroup taxa (Fig. 7).

In an attempt to do this, we have developed a method (Boore *et al.*, 1995) based on gene adjacencies as phylogenetic characters. There are 72 characters scored, each being "upstream of" or "downstream of" each of the 37 genes. Character states then are the 3′ or 5′ end of the adjacent gene, and so a matrix is filled by these, considering each animal. This matrix is then subjected to parsimony analysis (Swofford, 2001). An obvious shortcoming is that single gene boundaries are not very reliable phylogenetic characters; longer shared gene blocks are much stronger (Macey *et al.*, 1997a). This matrix and analysis, however, forms an excellent bookkeeping system that can be used to search for these longer shared arrangements by eye. In practice, the phylogenetic questions we have addressed have included gene arrangements at two extremes, either well conserved with so few gene rearrangements that synapomorphies are apparent by eye or so rearranged that no convincing signal is likely to be recovered using any method. Table III outlines several phylogenetic relationships, some controversial and others not, well supported at this point by mitochondrial gene rearrangements.

What generalities can we infer from comparing animal mitochondrial gene rearrangements? First, they do not follow to any extent a molecular clock. For some lineages, rearrangements are few over hundreds of millions of years. For example, the gene arrangements of sharks and humans are identical, and those of *Drosophila* and horseshoe crab differ only by the location of one tRNA gene. For others, rearrangements are much more rapid. For example, there are nearly no gene boundaries in common between a polyplacophoran (Boore and Brown, 1994) and bivalve (Hoffmann *et al.*, 1992) mollusk mtDNAs. Gene rearrangements are found even within a genus of amphisbaenian reptiles (Macey *et al.*, 2004) and among closely related gastropod mollusks (Rawlings *et al.*, 2001).

Second, the most common mode of change can be modeled by "duplication-random loss" (Boore, 2000; Lavrov *et al.*, 2002) (Fig. 8). In this model, a duplication is first generated, perhaps by slipped strand mispairing, illegitimate recombination, or errors during replication (so the termination point overruns its initiation), RNA processing, or topoisomerase activity (Boore, 2000). The supernumerary genes can then be lost without functional consequence, which may restore the original order or lead to rearrangement. In contrast to the mode in chloroplast genomes, inversions seem to have been rare. To the best of our knowledge, none of the algorithms in development for reconstructing phylogeny incorporates gene duplication intermediates in their modeling.

Although it seems unlikely that identical duplications followed by an identical pattern of losses would occur separately in different lineages, it is possible that homoplasious rearrangements would occur if a duplication persisted through several lineage splits, so each of several descending lineages would inherit the identically duplicated genes. Then one could imagine that identical losses might occur in less related groups. However, the infrequency with which we have observed duplicated genes in animal mtDNAs suggests that this state is usually short lived.

Third, for vertebrates, the loss of the stem-loop structure between *trnN* and *trnC* that defines the light-strand origin of replication (Clayton, 1992) is correlated with aberrant tRNA structures and rearrangement of tRNA genes in several lineages (Desjardins and Morais, 1990; Kumazawa and Nishida, 1995; Lee and Kocher, 1995; Macey *et al.*, 1997a,b, 2000b; Rest *et al.*, 2003).

TABLE III

SOME OF THE PHYLOGENETIC RELATIONSHIPS THAT ARE WELL SUPPORTED BY SHARED-DERIVED
REARRANGEMENTS OF MITOCHONDRIAL GENES (EVEN THOUGH MONOPHYLY OF SOME GROUPS—
CROCODILIANS, BIRDS, VERTEBRATES, AND ECHINODERMS—WERE NOT CONTROVERSIAL) OR WELL
SUPPORTED BY COMPARISONS OF COMPLETE OR NEARLY COMPLETE MTDNA SEQUENCES FOR CASES
IN WHICH OTHER STUDIES HAD BEEN EQUIVOCAL OR CONTRADICTORY[a]

Relationship supported	Reference
Based on gene rearrangements:	
Cestode platyhelminths are within the Eutrochozoa	Nickisch-Rosenegk *et al.* (2001)
Opisthobranchia unites with Pulmonata within Gastropoda	Kurabayashi and Ueshima (2000)
Annelida is more closely related to Mollusca than to Arthropoda	Boore and Brown (2000)
Onychophora does not group with Hexapoda to form the "Uniramia" and Arthropoda and Pancrustacea are each monophyletic	Boore *et al.* (1995); Lavrov *et al.* (2004)
Myriapoda is outside Pancrustacea, not the sister group to insects	Boore *et al.* (1998)
Pentasomida is not a phylum, but a type of crustacean	Lavrov *et al.* (2004)
Phoronida is part of the Eutrochozoa	Helfenbein and Boore (2004)
Monophyly of Crocodylidae	Kumazawa and Nishida (1995)
Monophyly of Aves	Mindell *et al.* (1998)
Monophyly of Vertebrata to the exclusion of Cephalochordata	Boore *et al.* (1999)
Monophyly of Echinodermata	Scouras and Smith (2001)
Hydrozoa, Scyphozoa, and Cubozoa group within Cnidaria	Bridge *et al.* (1992) (shared mtDNA structure)
Monophyly of acrodont lizards	Macey *et al.* (1997a,b, 1998, 2000a,c); Melville *et al.* (2001); Schulte *et al.* (2002, 2003, 2004); Townsend and Larson (2002)
Monophyly of the lizard family Iguanidae	Macey *et al.* (1997b); Schulte *et al.* (1998)
Bipedid amphisbaenians are derived relative to limbless groups	Macey *et al.* (2004)
Based on mtDNA sequence comparisons:	
Chaetognatha is basal to protostomes	Helfenbein *et al.* (2004)
Pogonophora is not a phylum, but a member of Annelida	Boore and Brown (2000)
Sipuncula is more closely related to Annelida than to Mollusca	Boore and Staton (2002)
Collembolida is basal to the Pancrustacea	Nardi *et al.* (2003)
Brachiopoda is part of the Eutrochozoa	Helfenbein and Boore (2004)
Salamanders group with frogs to the exclusion of caecilians	Zardoya and Meyer (2001)

[a] In a few of these references, the phylogenetic conclusions were not emphasized, but the data presented can easily be interpreted to support these conclusions. See Boore and Brown (1998) and Boore (1999) for a discussion of some of these earlier results.

A

Eutherian arrangement	nad2,	W,	A,	N,	C,	Y,	cox1
Hypothetical duplication	nad2,	W,	A,	N,	C,	W,	A,	N,	C,	Y,	cox1
To restore original order	nad2,	W,	A,	-,	-,	-,	-,	N,	C,	Y,	cox1
For marsupial order	nad2,	-,	A,	-,	C,	W,	-,	N,	-,	Y,	cox1
Marsupial arrangement	nad2,	A,	C,	W,	N,	Y,	cox1

B

I, Q, I, Q ⟶ I, Q, -, - ⟶ I, Q   Restores original order

⟶ -, -, I, Q ⟶ I, Q   Restores original order

I, -, -, Q ⟶ I, Q   Restores original order

Duplication,
followed by losses   -, Q, I, - ⟶ Q, I   Rearrangement

C

Ancestral echinoderm	rrnS,	E,	T,	120,	P,	Q,	N,	L,	A,	W,	C,	V,	M,	D,	Y,	G,	L,	nad1
Cucumaria region 1	rrnS,	64,	T,	459,		Q,	98,		A,	30,		C,	60,	M,	D,	Y,	G,	L,	nad1
Cucumaria region 2	1030,	E,		410,	P,	20,	N,	L,	63,	W,	58,	V,	79,	nad4L

D

Hypothetical ancestor	nad5,	cob,	T,	NC,	P,	nad6,	E,	NC,	F,	rrnS
Bird order B	nad5,	cob,	T,	--,	P,	nad6,	E,	NC,	F,	rrnS
Bird order C	nad5,	cob,	T,	NC,	P,	nad6,	E,	NC,	F,	rrnS

FIG. 8. Illustration of the duplication-random loss model of gene rearrangements (see Boore, 2000). Genes for tRNAs are abbreviated by the one-letter code for the corresponding amino acid. No annotation is made here for transcriptional orientation. (A) Eutherian and marsupial mammals differ in the arrangement of several genes (see Boore, 1999). A hypothetical intermediate includes the duplication of four of the genes. Losses of the supernumerary genes, then, would be expected to have little functional consequence. Some patterns of loss could potentially have restored the original order, but others could have led to the rearrangement in marsupials. (B) A model of tandem duplication of a two gene block, followed by random losses of supernumerary genes (Macey *et al.*, 1997a, 1998), could have led to the rearrangement found in acrodont lizards (references in Table III). Even after duplication, three of the four possible patterns of loss would have led to restoration of the original order. A similar case has been found for *trnP* and *trnT* in the amphisbaenian genus *Bipes* (Macey *et al.*, 2004). (C) An intermediate in this type of process seems to have been found in the sea cucumber *Cucumaria* (Arndt and Smith, 1998). Here, one region of the mtDNA matches partially the arrangement easily inferred to be ancestral for this echinoderm class, with blocks of unassignable nucleotides (indicated by numerals) in place of genes missing from the array. At another location in the mtDNA are the "missing" genes, again separated by blocks of unassignable nucleotides. One might infer that this stems from an ancestral duplication and movement of the tRNA gene cluster, followed by random gene degradation leading to this arrangement, with the unassigned blocks being unrecognizable vestiges of tRNA genes. (D) The rearrangement at the base of birds was a switch in order of the blocks *cob, trnT, trnP* and *nad6, trnE* to generate the order shown for the "hypothetical intermediate", except here we speculate that there was an additional "NC" (noncoding region) between *trnT* and *trnP*. Random losses of these two noncoding regions could generate the conditions found for modern birds. Although Mindell, Sorenson, and Dimcheff (1998) interpret this as a gene rearrangement (and do not view the NC between *trnE* and *trnF* in order C as homologous to the corresponding region in order B, even though there is significant sequence similarity), it is really the position of these noncoding regions that vary.

The hypothesis has been advanced (Macey *et al.*, 1997a,b, 2004) that the loss of this stem-loop enables the use of multiple regions, perhaps structures that are a compromise between encoding a tRNA and serving this function, for replication origins. These weaker origins then may lead to local duplications because of errors in aligning termination with the initiation point, which then leads to gene rearrangements through subsequent random losses.

Fourth, convergence of animal mitochondrial gene arrangements is seldom found. In a very large study, 540 partial mtDNA sequences have been determined for iguanian reptiles, with 199 sharing a derived switch in the order of *trnI* and *trnQ* (Macey *et al.*, 1997a,b, 1998, 2000a,c; Melville *et al.*, 2001; Schulte *et al.*, 2002, 2003, 2004; Townsend and Larson, 2002), from which we can infer the monophyly of the Acrodonta, and 341 sharing the ancestral condition, without homoplasy or additional changes observed. Even in cases where many of the genes are "scrambled" relative to other taxa, as is the case, for example, for a bivalve mollusk (Hoffmann *et al.*, 2000), a phthirapteran insect (Shao *et al.*, 2001), nematodes (Keddie *et al.*, 1998; Okimoto *et al.*, 1991, 1992), and platyhelminths (Le *et al.*, 2000; Nickisch-Rosenegk *et al.*, 2001), there is no case of these separately arriving at an identical arrangement for any of the genes. Although widely cited as an example of convergent rearrangement (Mindell *et al.*, 1998), birds do not constitute such an example; in fact, all birds studied to date share an identical arrangement of all 37 mitochondrial genes (Fig. 8) (see explanation in Boore and Brown, 1998). One case has been found of convergence in protein–gene rearrangement (Macey *et al.*, 2004), a switch in gene rearrangement of the block *cob, trnT, trnP* with the block *nad6, trnE* in rhineurid reptiles that is identical to the rearrangement shared by birds. These genes are immediately downstream of the origin of heavy strand replication. There has been a case of convergent rearrangement of two nearest neighbor tRNA genes (*trnK* and *trnD*) in orthopteran insects (Flook *et al.*, 1995) and of the same tRNA pair in hymenopteran insects (Dowton and Austin, 1999). As was pointed out by Boore and Brown (1998), these are the two types of rearrangements (i.e., rearrangements immediately downstream of an origin of replication and exchange of position of nearest neighbor tRNA genes) that should be given less weight for phylogenetic inference, because they are the most subject to duplications that might lead to rearrangement in the duplication-random loss mode.

Acknowledgments

We are grateful to lab members and guests who have worked to refine many of these bench techniques, including Ronald Bonett, David Engle, Jonathan Fong, H. Matthew Fourcade, Matthew Fujita, Kevin Helfenbein, Jennifer Kuehl, Kirsten Lindstrom, Susan

Masta, Jenna Morgan, Rachel Mueller, Dan Mulcahy, James Parham, Gabriela Parra, Marco Passamonti, Marcos Perez-Losada, Ernesto Recuero, Inaki Ruiz-Trillo, Wes Savage, Renfu Shao, Brian Simison, Matthias Stoeck, Tori Takaoka, and Yvonne Vallès. Thanks to Douda Bensasson, Jeff Froula, Allen Haim, and Stacia Wyman for work on databases and software for determining, cataloging, and comparing gene arrangements. Thanks to the National Science Foundation for support (DEB-0089624, EAR-0120646, DEB-9807100, DEB-9726064). This is LBNL-55278; part of this work was performed under the auspices of the U.S. Department of Energy, Office of Biological and Environmental Research, by the University of California, Lawrence Berkeley National Laboratory, under contract no. DE-AC03-76SF00098.

References

Adams, K. L., and Palmer, J. D. (2003). Evolution of mitochondrial gene content: Gene loss and transfer to the nucleus. *Mol. Phylogenet. Evol.* **29**, 380–395.

Armstrong, M. R., Blok, V. C., and Phillips, M. S. (2000). A multipartite mitochondrial genome in the potato cyst nematode *Globodera pallida*. *Genetics* **154**, 181–192.

Arndt, A., and Smith, M. J. (1998). Mitochondrial gene rearrangement in the sea cucumber genus *Cucumaria*. *Mol. Biol. Evol.* **15**(8), 1009–1016.

Ausubel, F. M., Brent, R., Kingston, R. E., Moore, D. D., Seidman, J. G., Smith, J. A., and Struhl, K. (eds.) (2004). *In* "Current Protocols in Molecular Biology." John Wiley & Sons, New York.

Bader, D. A., Moret, B. M., and Yan, M. (2001). A linear-time algorithm for computing inversion distance between signed permutations with an experimental study. *J. Comput. Biol.* **8**(5), 483–491.

Baner, J., Nilsson, M., Mendel-Hartvig, M., and Landegren, U. (1998). Signal amplification of padlock probes by rolling circle replication. *Nucleic Acids Res.* **26**, 5073–5078.

Beagley, C. T., MacFarlane, J. L., Pont-Kingdon, G. A., Okimoto, R., Okada, N., and Wolstenholme, D. R. (1995). Mitochondrial genomes of Anthozoa (Cnidaria). *In* "Progress in Cell Research" (F. Palmieri, ed.), Vol. 5, pp. 149–153. Elsevier, New york.

Beagley, C. T., Okimoto, R., and Wolstenholme, D. R. (1998). The mitochondrial genome of the sea anemone *Metridium senile* (Cnidaria): Introns, a paucity of tRNA genes, and a near-standard genetic code. *Genetics* **148**, 1091–1108.

Beaton, M. J., Roger, A. J., and Cavalier-Smith, T. (1998). Sequence analysis of the mitochondrial genome of *Sarcophyton glaucum*: Conserved gene order among octocorals. *J. Mol. Evol.* **47**, 697–708.

Blanco, L., Bernad, A., Lazaro, J. M., Martin, G., Garmendia, C., and Salas, M. (1989). Highly efficient DNA synthesis by the phage phi29 DNA polymerase: Symmetrical mode of DNA replication. *J. Biol. Chem.* **264**, 8935–8940.

Boore, J. L. (1999). Animal mitochondrial genomes. *Nucleic Acids Res.* **27**, 1767–1780.

Boore, J. L. (2000). The duplication/random loss model for gene rearrangement exemplified by mitochondrial genomes of deuterosome animals. *In* "Comparative Genomics" (D. Sankoff and J. Nadeau, eds.), Vol. 1, pp. 133–147. Kluwer Academic Publishers, Dordrecht, Netherlands.

Boore, J. L., and Brown, W. M. (1994). The complete DNA sequence of the mitochondrial genome of the black Chiton *Katharina tunicata*. *Genetics* **138**, 423–443.

Boore, J. L., and Brown, W. M. (1998). Big trees from little genomes: Mitochondrial gene order as a phylogenetic tool. *Curr. Opin. Genet. Dev.* **8**, 668–674.

Boore, J. L., and Brown, W. M. (2000). Mitochondrial genomes of *Galathealinum, Helobdella*, and *Platynereis*: Sequence and gene arrangement comparisons indicate that Pogonophora is not a phylum and Annelida and Arthropoda are not sister taxa. *Mol. Biol. Evol.* **17**(1), 87–106.

Boore, J. L., Daehler, L. L., and Brown, W. M. (1999). Complete sequence, gene arrangement and genetic code of mitochondrial DNA of the cephalochordate *Branchiostoma floridae* ("Amphioxus"). *Mol. Biol. Evol.* **16**(3), 410–418.

Boore, J. L., and Staton, J. (2002). The mitochondrial genome of the sipunculid *Phascolopsis gouldii* supports its association with Annelida rather than Mollusca. *Mol. Biol. Evol.* **19**(2), 127–137.

Boore, J. L., Collins, T. M., Stanton, D., Daehler, L. L., and Brown, W. M. (1995). Deducing arthropod phylogeny from mitochondrial DNA rearrangements. *Nature* **376**, 163–165.

Boore, J. L., Lavrov, D. V., and Brown, W. M. (1998). Gene translocation links insects and crustaceans. *Nature* **392**, 667–668.

Boore, J. L., Medina, M., and Rosenberg, L. A. (2004). Complete sequences of two highly rearranged molluscan mitochondrial genomes, those of the scaphopod *Graptacme eborea* and of the bivalve *Mytilus edulis*. *Mol. Biol. Evol.* **21**(8), 1492–1503.

Blanchette, M., Bourque, G., and Sankoff, D. (1997). Breakpoint phylogenies. *Genome Inform. Ser. Workshop Genome Inform.* **8**, 25–34.

Blanchette, M., Kunisawa, T., and Sankoff, D. (1999a). Gene order breakpoint evidence in animal mitochondrial phylogeny. *J. Mol. Evol.* **49**(2), 193–203.

Blanchette, M., Kunisawa, T., and Sankoff, D. (1999b). Parametric genome rearrangement. *Gene* **172**(1), GC11–17.

Bridge, D., Cunningham, C. W., Schierwater, B., DeSalle, R., and Buss, L. W. (1992). Class-level relationships in the phylum Cnidaria: Evidence from mitochondrial genome structure. *Proc. Natl. Acad. Sci. USA* **89**, 8750–8753.

Brown, W. M., George, M. Jr., and Wilson, A. C. (1979). Rapid evolution of animal mitochondrial DNA. *Proc. Natl. Acad. Sci. USA* **76**(4), 1967–1971.

Budowle, B., Allard, M. W., Wilson, M. R., and Chakraborty, R. (2003). Forensics and mitochondrial DNA: Applications, debates, and foundations. *Annu. Rev. Genomics Hum. Genet.* **4**, 119–141.

Cheng, S., Fockler, C., Barnes, W. M., and Higuchi, R. (1994). Effective amplification of long targets from cloned inserts and human genomic DNA. *Proc. Natl. Acad. Sci. USA* **91**, 5695–5699.

Clary, D. O., and Wolstenholme, D. R. (1985). The mitochondrial DNA molecule of *Drosophila yakuba:* Nucleotide sequence, gene organization, and genetic code. *J. Mol. Evol.* **22**, 252–271.

Clayton, D. A. (1992). Transcription and replication of animal mitochondrial DNAs. *Intl. Rev. Cytol.* **141**, 217–232.

Cosner, M. E., Jansen, R. K., Moret, B. M., Raubeson, L. A., Wang, L. S., Warnow, T., and Wyman, S. (2000). A new fast heuristic for computing the breakpoint phylogeny and experimental phylogenetic analyses of real and synthetic data. *Proc. Int. Conf. Intell. Syst. Mol. Biol.* **8**, 104–115.

Dean, F. B., Nelson, J. R., Giesler, T. L., and Lasken, R. S. (2001). Rapid amplification of plasmid and phage DNA using phi29 DNA polymerase and multiply-primed rolling circle amplification. *Genome Res.* **11**, 1095–1099.

Dean, F. B., Hosono, S., Fang, L., Wu, X., Faruqi, A. F., Bray-Ward, P., Sun, Z., Zong, Q., Du, Y., Du, J., Driscoll, M., Song, W., Kingsmore, S. F., Egholm, M., and Lasken, R. S. (2002). Comprehensive human genome amplification using multiple displacement amplification. *Proc. Natl. Acad. Sci. USA* **99**, 5261–5266.

Desjardins, P., and Morais, R. (1990). Sequence and gene organization of the chicken mitochondrial genome. *J. Mol. Biol.* **212**, 599–634.

Detter, J. C., Jett, J. M., Lucas, S. M., Dalin, E., Arellano, A. R., Wang, M., Nelson, J. R., Chapman, J., Lou, Y., Rokhsar, D., Hawkins, T. L., and Richardson, P. M. (2002). Isothermal strand-displacement amplification: Applications for high-throughput genomics. *Genomics* **80,** 691–698.

Dowton, M., and Austin, A. D. (1999). Evolutionary dynamics of a mitochondrial rearrangement "hot spot" in the Hymenoptera. *Mol. Biol. Evol.* **16**(2), 298–309.

Eddy, S. R, and Durbin, R. (1994). RNA sequence analysis using covariance models. *Nucleic Acids Res.* **22,** 2079–2088.

Elkin, C., Kapur, H., Smith, T., Humphreis, D., Pollard, M., Hammon, N., and Hawkins, T. L. (2002). Magnetic bead purification of labeled DNA fragments for high-throughput capillary electrophoresis sequencing. *Biotechniques* **32,** 1296–1302.

El-Mabrouk, N., and Sankoff, D. (1999). On the reconstruction of ancient doubled circular genomes using minimum reversals. *Genome Inform. Ser. Workshop Genome. Inform.* **10,** 83–93.

Estaban, J. A., Salas, M., and Blanco, L. (1993). Fidelity of phi29 DNA polymerase. Comparison between protein-primed initiation and DNA polymerization. *J. Biol. Chem.* **268,** 2719–2726.

Flook, P., Rowell, H., and Gellissen, G. (1995). Homoplastic rearrangements of insect mitochondrial tRNA genes. *Naturwissenschaften* **82,** 336–337.

Fire, A., and Xu, S. Q. (1995). Rolling replication of short DNA circles. *Proc. Natl. Acad. Sci. USA* **92,** 4641–4645.

Folmer, O., Black, M., Hoeh, W., Lutz, R., and Vrijenhoek, R. (1994). DNA primers for amplification of mitochondrial cytochrome c oxidase subunit I from diverse metazoan invertebrates. *Mol. Marine Biol. Biotechnol.* **3,** 294–299.

Gray, M. W. (1999). Evolution of organellar genomes. *Curr. Opin. Genet. Dev.* **9,** 678–687.

Gray, M. W., Lang, B. F., Cedergren, R., Golding, G. B., Lemieux, C., Sankoff, D., Turmel, M., Brossard, N., Delage, E., Littlejohn, T. G., Plante, I., Rioux, P., Saint-Louis, D., Zhu, Y., and Burger, G. (1998). Genome structure and gene content in protist mitochondrial DNAs. *Nucleic Acids Res.* **26**(4), 865–878.

Gissi, C., Reyes, A., Pesole, G., and Saccone, C. (2000). Lineage-specific evolutionary rate in mammalian mtDNA. *Mol. Biol. Evol.* **17**(7), 1022–1031.

Green, P. (1996). http://bozeman.mbt.washington.edu/phrap.docs/phrap.html.

Helfenbein, K. G., Brown, W. M., and Boore, J. L. (2001). The complete mitochondrial genome of a lophophorate, the brachiopod *Terebratalia transversa. Mol. Biol. Evol.* **18**(9), 1734–1744.

Helfenbein, K. G., and Boore, J. L. (2004). The mitochondrial genome of *Phoronis architecta*—Comparisons demonstrate that phoronids are lophotrochozoan protostomes. *Mol. Biol. Evol.* **21**(1), 153–157.

Helfenbein, K. G., Fourcade, H. M., Vanjani, R. G., and Boore, J. L. (2004). The mitochondrial genome of *Paraspadella gotoi* is highly reduced and reveals that chaetognaths are a sister-group to protostomes. *Proc. Natl. Acad. Sci. USA* **101**(29), 10639–10643.

Hickson, R. E., Simon, C., Cooper, A., Spicer, G. S., Sullivan, J., and Penny, D. (1996). Conserved sequence motifs, alignment, and secondary structure for the third domain of animal 12S rRNA. *Mol. Biol. Evol.* **13,** 150–169.

Hoffmann, R. J., Boore, J. L., and Brown, W. M. (1992). A novel mitochondrial genome organization for the blue mussel, *Mytilus edulis. Genetics* **131,** 397–412.

Inoue, J. G., Miya, M., Tsukomoto, K., and Nishida, M. (2001). Mitogenomic perspective on the basal teleostean phylogeny: Resolving higher-level relationships with longer DNA sequences. *Mol. Phylogenet. Evol.* **20,** 275–285.

Ingman, M., Kaessmann, H., Pääbo, S., and Gyllensten, U. (2001). Mitochondrial genome variation and the origin of modern humans. *Nature* **408**, 708–713.

Kearney, M., and Stuart, B. (2004). Repeated evolution of limblessness and digging heads in worm lizards revealed by DNA from old bones. *Proc. R. Soc. Lond* **271**, 1677–1683.

Keddie, E. M., Higazi, T., and Unnasch, T. R. (1998). The mitochondrial genome of *Onchocerca volvulus*: Sequence, structure and phylogenetic analysis. *Mol. Biochem. Parasitol.* **95**(1), 111–127.

Krakauer, D. C., and Mira, A. (1999). Mitochondria and germ-cell death. *Nature* **400**, 125–126.

Kumazawa, Y., and Nishida, M. (1993). Sequence evolution of mitochondrial tRNA genes and deep-branch animal phylogenetics. *J. Mol. Evol.* **37**, 380–398.

Kumazawa, Y., and Nishida, M. (1995). Variations in mitochondrial tRNA gene organization of reptiles as phylogenetic markers. *Mol. Biol. Evol.* **12**, 759–772.

Kurabayashi, A., and Ueshima, R. (2000). Complete sequence of the mitochondrial DNA of the primitive opisthobranch gastropod *Pupa strigosa*: Systematic implication of the genome organization. *Mol. Biol. Evol.* **17**, 266–277.

Lang, B. F., Burger, G., O'Kelly, C. J., Cedergren, R., Golding, G. B., Lemieux, C., Sankoff, D., Turmel, M., and Gray, M. W. (1997). An ancestral mitochondrial DNA resembling a eubacterial genome in miniature. *Nature* **387**, 493–497.

Lang, B. F., Gray, M. W., and Burger, G. (1999). Mitochondrial genome evolution and the origin of eukaryotes. *Annu. Rev. Genet.* **33**, 351–397.

Lavrov, D. V., Brown, W. M., and Boore, J. L. (2000). A novel type of RNA editing occurs in the mitochondrial tRNAs of the centipede *Lithobius forticatus*. *Proc. Natl. Acad. Sci. USA* **97**, 13738–13742.

Lavrov, D. V., Boore, J. L., and Brown, W. M. (2002). Complete mtDNA sequences of two millipedes suggest a new model for mitochondrial gene rearrangements: Duplication and non-random loss. *Mol. Biol. Evol.* **19**(2), 163–169.

Lavrov, D., Brown, W. M., and Boore, J. L. (2004). Phylogenetic position of the Pentastomida and (pan)crustacean relationships. *Proc. R. Soc. Lond. B Biol. Sci.* **271**, 537–544.

Le, T. H., Blair, D., Agatsuma, T., Humair, P. F., Campbell, N. J., Iwagami, M., Littlewood, D. T., Peacock, B., Johnston, D. A., Bartley, J., Rollinson, D., Herniou, E. A., Zarlenga, D. S., and McManus, D. P. (2000). Phylogenies inferred from mitochondrial gene orders—a cautionary tale from the parasitic flatworms. *Mol. Biol. Evol.* **17**(7), 1123–1125.

Lee, W.-J., and Kocher, T. D. (1995). Complete sequence of a Sea Lamprey (*Petromyzon marinus*) mitochondrial genome: Early establishment of the vertebrate genome organization. *Genetics* **139**, 873–887.

Liu, D., Dubendiek, S. L., Zillman, M. A., Ryan, K., and Kool, E. T. (1996). Rolling circle DNA synthesis: Small circular oligonucleotides as efficient templates for DNA polymerases. *J. Am. Chem. Soc.* **118**, 1587–1594.

Lizardi, P. M., Huang, X., Zhu, Z., Bray-Ward, P., Thomas, D. C., and Ward, D. C. (1998). Mutation detection and single-molecule counting using isothermal rolling-circle amplification. *Nat. Genet.* **19**, 225–232.

Macey, J. R., Larson, A., Ananjeva, N. B., Fang, Z., and Papenfuss, T. J. (1997a). Two novel gene orders and the role of light-strand replication in rearrangement of the vertebrate mitochondrial genome. *Mol. Biol. Evol.* **14**, 91–104.

Macey, J. R., Larson, A., Ananjeva, N. B., and Papenfuss, T. J. (1997b). Evolutionary shifts in three major structural features of the mitochondrial genome among Iguanian lizards. *J. Mol. Evol.* **44**, 660–674.

Macey, J. R., Larson, A., Ananjeva, N. B., and Papenfuss, T. J. (1997c). Replication slippage may cause parallel evolution in the secondary structures of mitochondrial transfer RNAs. *Mol. Biol. Evol.* **14**, 30–39.

Macey, J. R., Schulte, J. A. II, Ananjeva, N. B., Larson, A., Rastegar-Pouyani, N., Shammakov, S. M., and Papenfuss, T. J. (1998). Phylogenetic relationships among agamid lizards of the *Laudakia caucasia* species group: Testing hypotheses of biogeographic fragmentation and an area cladogram for the Iranian Plateau. *Mol. Phylogenet. Evol.* **10**, 118–131.

Macey, J. R., Schulte, J. A. II, Kami, H. G., Ananjeva, N. B., Larson, A., and Papenfuss, T. J. (2000a). Testing alternative hypotheses of vicariance in the agamid lizard *Laudakia caucasia* in mountain ranges on the northern Iranian Plateau. *Mol. Phylogenet. Evol.* **14**, 479–483.

Macey, J. R., Schulte, J. A. II, and Larson, A. (2000b). Evolution and phylogenetic information content of mitochondrial genomic structural features illustrated with acrodont lizards. *Syst. Biol.* **49**, 257–277.

Macey, J. R., Schulte, J. A. II, Larson, A., Ananjeva, N. B., Wang, Y., Pethiyagoda, R., Rastegar-Pouyani, N., and Papenfuss, T. J. (2000c). Evaluating trans-Tethys migration: An example using acrodont lizard phylogenetics. *Syst. Biol.* **49**, 233–256.

Macey, J. R., Papenfuss, T. J., Kuehl, J. V., Fourcade, H. M., and Boore, J. L. (2004). Phylogenetic relationships among amphisbaenian reptiles based on complete mitochondrial genome sequences. *Mol. Phylogenet. Evol.* **33**(1), 22–31.

Macey, J. R., and Verma, A. (1997). Homology in phylogenetic analysis: Alignment of transfer RNA genes and the phylogenetic position of snakes. *Mol. Phylogenet. Evol.* **4**, 272–279.

Masta, S. E., and Boore, J. L. (2004). The complete mitochondrial genome sequence of the spider *Habronattus oregonensis* reveals rearranged and extremely truncated tRNAs. *Mol. Biol. Evol.* **21**, 893–902.

Melville, J., Schulte, J. A. II, and Larson, A. (2001). A molecular phylogenetic study of ecological diversification in the Australian lizard genus *Ctenophorus*. *J. Exp. Zool.* **291**, 339–353.

Mindell, D., Sorenson, M. D., and Dimcheff, D. E. (1998). Multiple independent origins of mitochondrial gene order in birds. *Proc. Natl. Acad. Sci. USA* **95**, 10693–10697.

Moret, B. M., Wang, L. S., Warnow, T., and Wyman, S. K. (2001a). New approaches for reconstructing phylogenies from gene order data. *Bioinformatics* **17**(Suppl 1), S165–S173.

Moret, B. M., Wyman, S., Bader, D. A., Warnow, T., and Yan, M. (2001b). A new implementation and detailed study of breakpoint analysis. *Pac. Symp. Biocomput.* 583–594.

Nagley, P., and Wei, Y.–H. (1998). Ageing and mammalian mitochondrial genetics. *Trends Genet.* **14**, 513–517.

Nardi, F., Spinsanti, G., Boore, J. L., Carapelli, A., Dallai, R., and Frati, F. (2003). Hexapod origins, monophyletic or paraphyletic? *Science* **299**, 1887–1889.

Nickisch-Rosenegk, M. von, Brown, W. M., and Boore, J. L. (2001). Sequence and structure of the mitochondrial genome of the tapeworm *Hymenolepis diminuta*: Gene arrangement indicates that platyhelminths are derived eutrochozoans. *Mol. Biol. Evol.* **18**(5), 721–730.

Nieminen, A. L. (2003). Apoptosis and necrosis in health and disease: Role of mitochondria. *Intl. Rev. Cytol.* **224**, 29–55.

Nikaido, M., Rooney, A. P., and Okada, N. (1999). Phylogenetic relationships among cetartiodactyls based on insertions of short and long interspersed elements: Hippopotamuses are the closest extant relatives of whales. *Proc. Natl. Acad. Sci. USA* **96**, 10261–10266.

Nyakaana, S., Arctander, P., and Siegismund, H. (2002). Population structure of the African savannah elephant inferred from mitochondrial control region sequences and nuclear microsatellite loci. *Heredity* **89**(2), 90–98.

Ojala, D., Merkel, C., Gelfand, R., and Attardi, G. (1980). The tRNA genes punctuate the reading of genetic information in human mitochondrial DNA. *Cell* **22**, 393–403.

Okimoto, R., Chamberlin, H. M., MacFarlane, J. L., and Wolstenholme, D. R. (1991). Repeated sequence sets in mitochondrial DNA molecules of root knot nematodes (*Meloidogyne*): Nucleotide sequences, genome location and potential for host race identification. *Nucleic Acids Res.* **19**, 1619–1626.

Okimoto, R., MacFarlane, J. L., Clary, D. O., and Wolstenholme, D. R. (1992). The mitochondrial genomes of two nematodes, *Caenorhabditis elegans* and *Ascaris suum*. *Genetics* **130**(3), 471–498.

Palumbi, S., Martin, A., Romano, S., McMillan, W. O., Stice, L., and Grabowski, G. (1991). "The Simple Fool's Guide to PCR," Version 2.0, University of Hawaii, Department of Zoology and Kewalo Marine Laboratory.

Parsons, T. J., and Coble, M. D. (2001). Increasing the forensic discrimination of mitochondrial DNA testing through the analysis of the entire mitochondrial DNA genome. *Croatian Med. J.* **42**, 304–309.

Perna, N. T., and Kocher, T. D. (1995). Patterns of nucleotide composition at fourfold degenerate sites of animal mitochondrial genomes. *J. Mol. Evol.* **41**, 353–358.

Raimond, R., Marcade, I., Bouchon, D., Rigaud, T., Borry, J.-P., and Souty-Grosset, C. (1999). Organization of the large mitochondrial genome in the isopod *Armadillidium vulgare*. *Genetics* **151**, 203–210.

Rawlings, T., Collins, T., and Bieler, R. (2001). A major mitochondrial gene rearrangement among closely related species. *Mol. Biol. Evol.* **18**(8), 1604–1609.

Rawlings, T., Collins, T., and Bieler, R. (2003). Changing identities: tRNA duplication and remolding within animal mitochondrial genomes. *Proc. Natl. Acad. Sci. USA* **100**, 15700–15705.

Rest, J. S., Ast, J. C., Austin, C. C., Waddell, P. J., Tibbetts, E. A., Hay, J. M., and Mindell, D. P. (2003). Molecular systematics of primary reptilian lineages and the tuatara mitochondrial genome. *Mol. Phylogenet. Evol.* **29**, 289–297.

Sambrook, J., and Russell, D. W. (2001). "Molecular Cloning: A Laboratory Manual." Cold Spring Harbor Laboratory Press, Cold Spring Harbor, NY.

Sankoff, D., and Blanchette, M. (1998a). Multiple genome rearrangement and breakpoint phylogeny. *J. Comput. Biol.* **5**(3), 555–570.

Sankoff, D., and Blanchette, M. (1998b). Phylogenetic invariants for metazoan mitochondrial genome evolution. *Genome Inform. Ser. Workshop Genome Inform.* **9**, 22–31.

Sankoff, D., and Blanchette, M. (1999). Phylogenetic invariants for genome rearrangements. *J. Comput. Biol.* **6**, 431–445.

Sankoff, D., Cedergren, R., and Abel, Y. (1990). Genomic divergence through gene rearrangement. *Methods Enzymol.* **183**, 428–438.

Sankoff, D., Leduc, G., Antoine, N., Paquin, B., Lang, B. F., and Cedergren, R. (1992). Gene order comparisons for phylogenetic inference: Evolution of the mitochondrial genome. *Proc. Natl. Acad. Sci. USA* **89**(14), 6575–6579.

Schmitz, J., Ohme, M., and Zischler, H. (2001). SINE insertions in cladistic analyses and the phylogenetic affiliations of *Tarsius bancanus* to other primates. *Genetics* **157**, 777–784.

Schulte J. A., II, Macey, J. R., Larson, A., and Papenfuss, T. J. (1998). Molecular tests of phylogenetic taxonomies: A general procedure and example using four subfamilies of the lizard family Iguanidae. *Mol. Phylogenet. Evol.* **10**, 367–376.

Schulte, J. A., II, Macey, J. R., Pethiyagoda, R., and Larson, A. (2002). Rostral horn evolution among agamid lizards of the genus *Ceratophora* endemic to Sri Lanka. *Mol. Phylogenet. Evol.* **22**, 111–117.

Schulte, J. A., II, Melville, J., and Larson, A. (2003). Molecular phylogenetic evidence for ancient divergence of lizard taxa on either side of Wallace's Line. *Proc. R. Soc. Lond. B Biol. Sci.* **270**, 597–603.

Schulte, J. A. II, Vindum, J. V., Win, H., Thin, T., Lwin, K. S., Shein, A. K., and Tun, H. (2004). Phylogenetic relationships of the genus *Ptyctolaemus* (Squamata: Agamidae), with a description of a new species from the Chin Hills of Western Myanmar. *Proc. Calif. Acad. Sci.* **55**, 222–247.

Scouras, A., and Smith, M. J. (2001). A novel mitochondrial gene order in the crinoid echinoderm *Florometra serratissima*. *Mol. Biol. Evol.* **18**(1), 61–73.

Shao, R., Campbell, N. J. H., and Barker, S. C. (2001). Numerous gene rearrangements in the mitochondrial genome of the wallaby louse, *Heterodoxus macropus* (Phthiraptera). *Mol. Biol. Evol.* **18**(5), 858–865.

Sorenson, M. D., Ast, J. C., Dimcheff, D. E., Yuri, T., and Mindell, D. P. (1999). Primers for a PCR-based approach to mitochondrial genome sequencing in birds and other vertebrates. *Mol. Phylogenet. Evol.* **12**(2), 105–114.

Southern, E. M. (1975). Detection of specific sequences among DNA fragments separated by gel electrophoresis. *J. Mol. Biol.* **98**(3), 503–517.

Swofford, D. L. (2001). "PAUP*, Phylogenetic Analysis Using Parsimony (*and Other Methods)," Beta Version 4.0b8. Sinauer, Sunderland, MA.

Tang, J., and Moret, B. M. (2003). Scaling up accurate phylogenetic reconstruction from gene-order data. *Bioinformatics* **19**(Suppl. 1), i305–i312.

Townsend, T., and Larson, A. (2002). Molecular phylogenetics and mitochondrial genomic evolution in the Chamaeleonidae (Reptilia, Squamata). *Mol. Phylogenet. Evol.* **23**, 22–36.

Wallace, D. C. (1999). Mitochondrial diseases in man and mouse. *Science* **283**, 1482–1488.

Wang, L. S., Jansen, R. K., Moret, B. M., Raubeson, L. A., and Warnow, T. (2002). Fast phylogenetic methods for the analysis of genome rearrangement data: An empirical study. *Pac. Symp. Biocomput.* 524–535.

Wolstenholme, D. R. (1992). Animal mitochondrial DNA: Structure and evolution. *In* "Mitochondrial Genomes" (K. W. Jeon and D. R. Wolstenholme, eds.), pp. 173–216. Academic Press, New York.

Wolstenholme, D. R., MacFarlane, J. L., Okimoto, R., Clary, D. O., and Wahleithner, J. A. (1987). Bizarre tRNAs inferred from DNA sequences of mitochondrial genomes of nematode worms. *Proc. Natl. Acad. Sci. USA* **84**, 1324–1328.

Wu, W., Schmidt, T. R., Goodman, M., and Grossman, L. I. (2000). Molecular evolution of cytochrome c oxidase subunit I in primates: Is there coevolution between mitochondrial and nuclear genomes? *Mol. Phylogenet. Evol.* **17**(2), 294–304.

Wyman, S. K., and Boore, J. L. (2003). Annotating animal mitochondrial tRNAs: An experimental evaluation of four methods. *In* "Proceedings of the European Conference on Computational Biology (ECCB)" pp. 44–46.

Wyman, S. K., Jansen, R. K., and Boore, J. L. (2004). Automatic annotation of organellar genomes with DOGMA. *Bioinformatics* **20**(17), 3252–3255.

Yokobori, S.-I., Rakuya, U., Feldmaier-Fuchs, G., Pääbo, S., Ueshima, R., Kondow, A., Nishikawa, K., and Watanabe, K. (1999). Complete DNA sequence of the mitochondrial genome of the ascidian *Halocynthia roretzi* (Chordata, Urochordata). *Genetics* **153**, 1851–1862.

Yokobori, S., Watanabe, Y., and Oshima, T. (2003). Mitochondrial genome of *Ciona savignyi* (Urochordata, Ascidiacea, Enterogona): Comparison of gene arrangement and tRNA genes with *Halocynthia roretzi* mitochondrial genome. *J. Mol. Evol.* **57**, 574–587.

Yost, H. J., Phillips, C. R., Boore, J. L., Bertman, J., Whalen, B., and Danilchik, M. V. (1995). Relocation of mitochondrial RNA to the prospective dorsal midline during *Xenopus* embryogenesis. *Dev. Biol.* **170**, 83–90.

Zardoya, R., and Meyer, A. (2001). On the origin of and phylogenetic relationships among living amphibians. *Proc. Natl. Acad. Sci. USA* **98**(13), 7380–7383.

[20] Methods for Obtaining and Analyzing Whole Chloroplast Genome Sequences

By Robert K. Jansen, Linda A. Raubeson, Jeffrey L. Boore, Claude W. dePamphilis, Timothy W. Chumley, Rosemarie C. Haberle, Stacia K. Wyman, Andrew J. Alverson, Rhiannon Peery, Sallie J. Herman, H. Matthew Fourcade, Jennifer V. Kuehl, Joel R. McNeal, James Leebens-Mack, and Liying Cui

Abstract

During the past decade, there has been a rapid increase in our understanding of plastid genome organization and evolution due to the availability of many new completely sequenced genomes. There are 45 complete genomes published and ongoing projects are likely to increase this sampling to nearly 200 genomes during the next 5 years. Several groups of researchers including ours have been developing new techniques for gathering and analyzing entire plastid genome sequences and details of these developments are summarized in this chapter. The most important developments that enhance our ability to generate whole chloroplast genome sequences involve the generation of pure fractions of chloroplast genomes by whole genome amplification using rolling circle amplification, cloning genomes into Fosmid or bacterial artificial chromosome (BAC) vectors, and the development of an organellar annotation program (Dual Organellar GenoMe Annotator [DOGMA]). In addition to providing details of these methods, we provide an overview of methods for analyzing complete plastid genome sequences for repeats and gene content, as well as approaches for using gene order and sequence data for phylogeny reconstruction. This explosive increase in the number of sequenced plastid genomes and improved computational tools will provide many insights into the evolution of these genomes and much new data for assessing relationships at deep nodes in plants and other photosynthetic organisms.

Introduction

Historical Overview of Chloroplast Genomics

The study of chloroplast genomes dates back to the 1950s when plant biologists first discovered that chloroplasts contain their own DNA (see Sugiura (2003) for a review). Early work used electron microscopy, cloning, comparative restriction site mapping, and gene mapping to characterize genome structure–gene order and organization (Palmer, 1991; Sugiura, 1992). Such comparisons yielded numerous phylogenetic studies based on restriction site polymorphisms and gene order changes (Downie and Palmer, 1992; Jansen et al., 1998; Olmstead and Palmer, 1994). The publication of complete plastid sequences for *Nicotiana* (Shinozaki et al., 1986) and *Marchantia* (Ohyama et al., 1986) provided the first opportunity for nucleotide-level whole genome comparisons (Morton, 1994; Wolfe et al., 1987). Currently the list of completely sequenced plastid genomes has increased to 45 and includes a wide diversity of taxonomic groups. The number of sequenced chloroplast genomes is growing rapidly: 19 of these 45 genomes (Table I) have appeared in the last two years. In spite of the availability of so many complete genome sequences, our understanding of chloroplast genome evolution is still limited because this remains a very small sampling of plastid-containing species and because previous sequencing efforts were not designed to address phylogenetic or molecular evolutionary issues. A number of groups (e.g., algae and various lineages of land plants, including bryophytes, ferns and fern allies, gymnosperms, and certain angiosperm groups, especially monocots other than the cereal grasses) remain poorly sampled. However, several groups of scientists are now focusing their sequencing efforts at filling these gaps, and the number of completely sequenced chloroplast genomes will continue to increase dramatically in the next few years (for details of three such projects see http://megasun.bch.umontreal.ca/ogmp/projects/sumprog.html, http://www.jgi.doe.gov/programs/comparative/second_levels/chloroplasts/jansen_project_home/chlorosite.html, and http://ucjeps.berkeley.edu/TreeofLife/).

Brief Overview of Chloroplast Genome Structure and Evolution

Plastid genomes vary in size from 35 to 217 kilobases (kb), but the vast majority from photosynthetic organisms are between 115 and 165 kb (Table I). The 45 completely sequenced genomes (Table I) encode from 63 (*Toxiplasma*) to 209 (*Porphyra*) genes with most containing 110–130 genes. Most of these genes code for proteins, mostly involved in photosynthesis or gene expression, with the remainder being transfer RNA or ribosomal

TABLE I
ALPHABETICAL LIST OF 45 COMPLETE PLASTID GENOME SEQUENCES AS OF FEBRUARY 17, 2005[a]

Species	NCBI classification	Accession number	Year completed	Genome size (bp)
Adiantum capillus-veneris	Embryophyta	AY178864	2003	150,568
Amborella trichopoda	Embryophyta	AJ506156	2003	162,686
Anthoceros formosae	Embryophyta	AB086179	2003	161,162
Arabidopsis thaliana	Embryophyta	AP000423	1999	154,478
Atropa belladonna	Embryophyta	AJ316582	2003	156,687
Calycanthus fertilis var. ferax	Embryophyta	AJ428413	2003	153,337
Chaetosphaeridium globosum	Streptophyta	AF494278	2002	131,183
Chlamydomonas reinhardtii	Chlorophyta	BK000554	2004	203,828
Chlorella vulgaris	Chlorophyta	AB001684	1997	150,613
Cyanidioschyzon merolae	Rhodophyta	AB002583/	2003/	149,987
		AY286123	2004	149,705
Cyanidium caldarium	Rhodophyta	AF022186	1999	164,921
Cyanophora paradoxa	Glaucocystophyceae	U30821	1995	135,599
Eimeria tenella[b]	Alveolata	AY217738	2003	34,750
Epifagus virginiana[c]	Embryophyta	M81884	1993	70,028
Euglena gracilis	Euglenozoa	X70810	1993	143,171
Euglena longa	Euglenozoa	AJ294725	2001	73,345
Gracilaria tenuistipitata	Rhodophyta	AY673996	2004	183,883
Guillardia theta	Cryptophyta	AF041468	1998	121,524
Huperzia lucidula	Embryophyta	AY660566	2005	154,373
Lotus corniculatus	Embryophyta	AP002983	2001	150,519
Marchantia polymorpha	Embryophyta	X04465	1986	121,024
Medicago truncatula	Embryophyta	AC093544	2001	124,033
Mesostigma viride	Chlorophyta	AF166114	2000	118,360
Nephroselmis olivacea	Chlorophyta	AF137379	1999	200,799
Nicotiana tabacum	Embryophyta	Z00044	1986	155,939
Nymphaea alba	Embryophyta	AJ627251	2004	159,930
Odontella sinensis	Stramenopiles	Z67753	1996	119,704
Oenothera elata	Embryophyta	AJ271079	2000	163,935
Oryza nivara	Embryophyta	AP006728	2004	134,494
Oryza sativa	Embryophyta	X15901/	1989/	134,525/
		AY522329/	2004/	134,496/
		AY522331	2004	134,551
Ponax schinsing	Embryophyta	AY582139	2004	156,318
Physcomitrella patens	Embryophyta	AP005672	2003	122,890
Pinus koraiensis	Embryophyta	AY228468	2003	116,866
Pinus thunbergii	Embryophyta	D17510	1996	119,707
Porphyra purpurea	Rhodophyta	U38804	1996	191,028
Psilotum nudum	Embryophyta	AP004638	2002	138,829
Saccharum hybrid	Embryophyta	AE009947	2004	141,182
Saccharum officinarum	Embryophyta	AP006714	2004	141,182
Spinacia oleracea	Embryophyta	AJ400848	2000	150,725
Toxoplasma gondii[b]	Alveolata	U87145	1999	34,996
Triticum aestivum	Embryophyta	AB042240	2001	134,545
Zea mays	Embryophyta	X86563	1995	140,384

RNA genes. Although the number of genes may be similar between even distantly related lineages, the exact gene complement may be quite different. Although gene content is largely consistent within land plants, Martin *et al.* (2002) found only 44 protein-coding genes to be common among 15 chloroplast genomes representing all major lineages of photosynthetic organisms. A few genes have evidently been gained during plastid genome evolution, but the vast majority of gene content changes represent gene losses, some of which have been lost independently in different lineages (Martin *et al.*, 2002; Maul *et al.*, 2002). In all plastid genomes, most genes are part of polycistronic transcription units, suggestive of bacterial operons (Fig. 1) (Mullet *et al.*, 1992; Palmer, 1991). Plastid operons often have multiple promoters that enable a subset of genes to be transcribed within the operon (Kuroda and Maliga, 2002; Miyagi *et al.*, 1998). Both group I and group II types of self-splicing introns are found in cpDNAs; the majority are group II (Palmer, 1991). A unique intron type (known as a "twintron") that contains an intron within an intron is found in *Euglena* (Copertino and Hallick, 1991) and possibly other organisms (Maier *et al.*, 1995). Although intron content is quite variable among algal genomes, it is highly conserved among land plant cpDNAs.

Most land plant (and some algal) genomes have a quadripartite organization (Fig. 1), composed of two copies of a large inverted repeat (IR) and two sections of unique DNA, which are referred to as the "large" and "small single copy regions" (LSC and SSC, respectively). The gene content and organization of the chloroplast genome change by several mechanisms. Transposition has been suggested as a mechanism of genomic change in chloroplasts (e.g., in *Trachelium* in the Campanulaceae (Cosner *et al.*, 1997) and in *Trifolium* in the Fabaceae (Milligan *et al.*, 1989), but few definitive examples have been documented. Only one clear case of transpositional gain has been documented in *Chlamydomonas* (Fan *et al.*, 1995), where a transposable element that is no longer active has been characterized. The frequency of the other types of rearrangements, including gene and intron gains and losses, expansion, and contraction of the IR, and inversions, varies from group to group. Most genomes have very few gene order changes, at least in comparison to close relatives. However, several lineages have cpDNAs that are highly rearranged. The most notable examples are in the algae (e.g., *Chlamydomonas*) (Maul *et al.*, 2002), conifers

---

[a] See http://megasun.bch.umontreal.ca/ogmp/projects/other/cp_list.html, http://www.ncbi. nlm.nih.gov:80/genomes/static/euk_o.html, and http://www.rs.noda.tus.ac.jp/~kunisawa/ order/front.html for access to these genomic sequences. All listed genomes are chloroplasts except as noted.

[b] Plastid genome remnant, nonphotosynthetic protist.

[c] Plastid genome, nonphotosynthetic flowering plant.

FIG. 1. Gene map of tobacco chloroplast genome (from Raubeson and Jansen [2005]). The inner circle shows the four major regions of the genome: the two copies of the inverted repeat (IRA and IRB) and the large and small single-copy regions (LSC and SSC). The outer circle represents the tobacco genome with the transcribed regions shown as boxes proportional to gene size. Genes inside the circle are transcribed in a clockwise direction, and genes outside of the circle are transcribed counterclockwise. The IR extent is shown by the increased width of the circle representing the tobacco genome. Genes with introns are marked with asterisks (*). Arrows between the gene boxes and gene names show those operons known to occur in tobacco cpDNA. Genes coding for products that function in protein synthesis are dark gray; genes coding for products that function in photosynthesis are stippled; and genes coding for products with various other functions are lighter gray.

(e.g., *Pinus*) (Wakasugi *et al.*, 1994), and several angiosperm lineages (e.g., Campanulaceae [Cosner *et al.*, 1997], Fabaceae [Milligan *et al.*, 1989], Geraniaceae [Palmer *et al.*, 1987], and Lobeliaceae [Knox and Palmer,

1998]). Two reviews summarize the types of genomic rearrangements in cpDNAs of algae (Simpson and Stern, 2002)and land plants (Raubeson and Jansen, 2005). Gene order changes in plastid genomes have proven useful for resolving phylogenetic relationships within a number of plant groups (Raubeson and Jansen, 2005).

Overview

This chapter focuses on the methods used to gather and analyze plastid genomic sequences. This includes methods for (1) isolating chloroplasts and purified cpDNA, (2) amplifying, cloning, and sequencing cpDNA, (3) assembling drafts and finishing genomes, (4) annotating chloroplast genomes, and (5) analyzing genome sequence and structure. Most of the steps are equally applicable to the plastid genomes of nonphotosynthetic plants, except for the initial isolation steps, which typically involve generation of a large insert genomic library. In our treatment of genomic analysis, we focus on evolutionary issues, and even then we will not be able to be comprehensive. In addition to reviewing methods that others have used, this chapter provides some more detailed protocols used by our group in an ongoing project for which we are sequencing 60 plastid genomes from seed plants.

Whole Chloroplast Genome Sequencing

Chloroplast genomes had been sequenced by cloning cpDNA into plasmid vectors, selecting cpDNA-containing clones, and then sequencing the clones using both plasmid and chloroplast-specific primers. This process is very labor intensive and involves isolation of highly purified cpDNA, which can be quite difficult for many taxa. Now, faster and more cost-effective approaches have been developed. There are four basic approaches to sequencing entire chloroplast genomes: (1) isolation of pure cpDNA, followed by random shearing, shotgun cloning, and sequencing; (2) amplification using long polymerase chain reaction (PCR) of large segments of the genome, followed by cloning, and then sequencing of the products using chloroplast-specific primers; (3) amplification of the entire genome using rolling circle amplification (RCA) followed by shearing of the RCA product and shotgun cloning and sequencing of the fragments; and (4) construction of bacterial artificial chromosome (BAC) or Fosmid libraries from total DNA preparations, preferably ones that are enriched for cpDNA, followed by shearing, cloning, and sequencing. We first outline our general genomic sequencing methods and then go on to describe the unique parts of each of the four aforementioned approaches, with an emphasis on those used by our group.

Draft sequences of chloroplast genomes from our group are being produced at the DOE Joint Genome Institute (JGI) in Walnut Creek, California. This facility is a very high-throughput operation that relies on robotics for many of the steps in the process. Details of JGI protocols can be found at http://www.jgi.doe.gov/Internal/protocols/prots_production. html, but a general description is given here. Our approach is to shear the DNA, select approximately 3-kb fragments, and clone these fragments into plasmid vectors. *Escherichia coli* are then transformed with the recombinant plasmids and spread onto large plates from which colonies are robotically picked and placed into 384-well plates containing the appropriate growth medium. Picking of colonies from the library is random, so the percentage of wells in the plates that contain cpDNA clones will be proportional to the percentage of cpDNA (as opposed to nuclear or mitochondrial "contaminant") in the DNA sample used to create the library. The inserts are sequenced from the 384-well plates using forward and reverse plasmid primers, yielding about 500–800 bp of sequence from each end of the insert. Sequencing proceeds until the depth of coverage, from many overlapping sequence reads, enables the assembly of the reads into one contiguous genomic sequence. In this approach, most steps are performed robotically minimizing human effort compared to earlier methods. The tradeoff is that unlike directed approaches such as chromosome walking with custom primers, the genome must be sequenced to a depth of 6–10× coverage to ensure accurate characterization of the entire genome.

Isolation of Chloroplast DNA

If pure cpDNA can be obtained in sufficient quantity, it can serve as the template for the sequencing approach just described. Many methods have been developed for isolating purified cpDNA from plants (Palmer, 1986). Most of these methods involve three basic steps: separation of plastids from other organelles, lysis of the chloroplasts, and purification of DNA. The most commonly applied methods use sucrose or Percoll gradients (Palmer, 1986), DNAse I treatment (Kolodner and Tewari, 1979), or high salt buffers (Bookjans *et al.*, 1984) to isolate purified cpDNA (or more realistically, a total DNA preparation enriched for cpDNA). The use of sucrose gradients is most generally applicable at least in land plants and a detailed protocol is provided in Table II. Basically, sucrose step gradients are used to obtain chloroplasts that are then lysed and the DNA is recovered from the lysate. We include several modifications of the basic method that have been used by our group to improve the quality and quantity of cpDNA. Consistent problems are encountered with two aspects of cpDNA isolations, using this method or any other: (1) collecting a sufficient quantity

TABLE II
ISOLATION OF CHLOROPLASTS OR cpDNA BY SUCROSE STEP-GRADIENT CENTRIFUGATION (SEE PALMER (1986) AND SANDBRINK *ET AL.* (1989))

1. Before extraction, place plants in the dark for 1–2 days to reduce chloroplast starch levels. Approximately ≥100 g of leaf tissue is required to get sufficient quantities of cpDNA. If the chloroplast isolation is being prepared for rolling circle amplification (RCA), at least 10 g of leaf tissue is generally necessary. The quality of the plant tissue is probably the most important criterion for a successful isolation. Leaves that are fresher and younger are far superior to older senescing leaves.

2. Wash healthy green leaves in tap water if visibly dirty and cut into small pieces ($\sim$2–10 cm^2 in surface area).

3. Place 10–100 g of cut leaves in 400 ml of ice-cold isolation buffer. Steps 3–5 are done in a cold room at 4° or on ice. We have found that the isolation buffer in Sandbrink *et al.* (1989) often yields a much purer chloroplast pellet (see recipes at end of protocol). This buffer contains higher concentrations of salts and 2-mercaptoethanol.

4. Homogenize in a prechilled blender for five 5-s bursts at high speed.

5. Filter through four layers of cheesecloth and squeeze remaining liquid through the cloth. Then filter through one layer of Miracloth (Calbiochem, catalog no. 475855) without squeezing.

6. Divide filtrate into multiple centrifuge bottles and centrifuge at 1000g for 15 min at 4°. Pour off supernatant.

7. Resuspend pellet 7 ml of ice-cold wash buffer using a soft paintbrush and by vigorous swirling.

8. Gently load the resuspended pellet onto a step gradient consisting of 18 ml of 52% sucrose, overlayered with 7 ml of 30% sucrose. The overlay should be added with sufficient mixing to create a diffuse interface. It is best to pour the sucrose gradients 1–2 days before the extraction and allow them to sit at 4° to allow for mixing of the interface. To enhance the purity of your cpDNA isolation, it is best to use more sucrose gradients, each with material from a smaller amount of tissue so the nuclei can better penetrate the chloroplast band. At least six sucrose gradients are recommended for up to 200 g of starting material. When preparing chloroplasts (rather than cpDNA), we will use three gradients for just 20 g of tissue. We also have experimented with modifying the percentage of sucrose in the step gradients. We have found that the optimal percentage varies from one taxon to the next. For example, 52/30% gradients work well for most angiosperms, *Ginkgo,* and conifers, but we found that a 44–48% sucrose in the bottom layer yielded DNA with a much higher proportion of cpDNA for cycads.

9. Centrifuge the step gradients at 25,000 rpm for 30–60 min at 4° in an SW-27 (Beckman) or AH-627 (Sorvall) swinging bucket rotor.

10. Remove the chloroplast band from the 30–52% interface using a wide-bore pipette, dilute with 3–10 volumes of wash buffer, and centrifuge at 1500g for 15 min at 4°. We have found that the use of the Sandbrink wash buffer often improves the purity of the cpDNA. Multiple cycles of washing, pelleting, and resuspending of the chloroplasts often renders much purer cpDNA.

11. Resuspend the chloroplast pellet in wash buffer to a final volume of 2 ml. Depending on the size of the final pellet, it may be necessary to resuspend the pellet in a larger volume and then divide resuspended pellet into separate tubes with no more than 3 ml per tube. If you are planning to use the chloroplasts for RCA, this is the point at which you proceed to the RCA protocol in Table III.

(continued)

TABLE II (continued)

12. Add one-tenth volume of a 10 mg/ ml solution of self-digested (2 h at 37°) Pronase (Calbiochem, catalog no. 537088) and incubate for 2 min at room temperature.

13. Gently add one-fifth volume of 1× lysis buffer and mix in by slowly inverting the tube several times over a period of 10–15 min at room temperature. We experimented with higher concentrations of lysis buffer (a 5× lysis buffer vs. the normal 1× buffer) and with doing the lysis at higher temperatures for longer periods (37° for 15–60 min). In general, we found that the 5× lysis buffer incubated at 37° gave much higher yields of cpDNA. We also tried several alternative lysis buffers that used cetyltrimethylammonium bromide (CTAB) (Milligan et al., 1989) or sodium dodecyl sulfate (SDS) (Triboush et al., 1998), but in general we did not have much success with these buffers.

14. Centrifuge for 10 min at room temperature in a clinical centrifuge to remove residual starch and cell-wall debris from the chloroplast lysate. Transfer lysate to a new tube. This step is optional.

15. Add 1.0 g of technical-grade cesium chloride (CsCl) per 1 ml of lysate and add ethidium bromide (EtBr) to a final concentration of 200 mg/ml. Fill remaining volume of ultracentrifuge tubes with a premixed solution of 1 g CsCl per 1 ml of TE buffer.

16. Centrifuge the small CsCl/EtBr gradients (5 ml) in a vertical rotor for 5–8 h at 65,000 rpm at 20°.

17. Remove the band from gradient, and if necessary, reband in a second gradient or move on to step 18. High-molecular-weight chloroplast DNA will be very viscous and easily removed "en masse" from near the center of the gradient.

18. Remove EtBr by at least three extractions with isopropanol saturated with NaCl and H_2O and dialyze against at least three changes of 2 liters of dialysis buffer over a period of 1–2 days.

19. Check purity of cpDNA by doing restriction digests and agarose gel electrophoresis.

20. Store the chloroplast DNA at 4° for short-term and at −20° for long-term use. Digests of cpDNA produce well-defined bands, whereas nuclear DNA produces so many bands that it appears as a smear on the gel.

Standard isolation buffer	Sandbrink isolation buffer
0.35 M sorbitol	1.25 M NaCl
50 mM tris-HCl, pH 8.0	50 mM Tris–HCl, pH 8.0
5 mM EDTA	5 mM EDTA
0.1% BSA (w/v, Sigma A-4503)	1% BSA (w/v, Sigma A-4503)
1.5 mM 2-mercaptoethanol	10 mM 2-mercaptoethanol
	5% poly pyrrolidone (PVP-40)
Standard wash buffer	Sandbrink wash buffer
0.35 M sorbitol	10 mM Tris–HCl, pH 8.0
50 mM Tris–HCl, pH 8.0	5 mM EDTA
25 mM EDTA	10 mM 2-mercaptoethanol
	100 μg/ml proteinase K
52% sucrose solution	30% sucrose solution
52% Sucrose (w/v)	30% Sucrose (w/v)
50 mM Tris pH8.0	50 mM Tris pH 8.0
25 mM EDTA	25 mM EDTA

(continued)

TABLE II (*continued*)

Standard isolation buffer	Sandbrink isolation buffer
1× lysis buffer	5× lysis buffer
5% sodium sarcosinate (w/v)	20% sodium sarcosinate (w/v)
50 mM Tris pH 8.0	50 mM Tris pH 8.0
25 mM EDTA	25 mM EDTA
Dialysis buffer	
10 mM Tris, pH 8.0	
10 mM NaCl	
0.1 mM EDTA	

of chloroplasts while eliminating nuclear contamination and (2) lysing the chloroplasts and releasing the membrane-bound cpDNA. Nuclear DNA tends to adhere to the outer chloroplast membrane, leading to the first challenge. Regarding the second challenge, chloroplasts can be surprisingly difficult to lyse. If harsh enough detergents are used to lyse the chloroplasts abruptly, then the DNA is degraded. Because the DNA is bound to the thylakoid membranes, the membranes must be solubilized to release the DNA, but if the chloroplast is lysed too gently, the DNA remains bound to the membrane and is lost. Our modifications to the basic procedure help reduce these problems but do not totally overcome them.

Two other approaches to cpDNA isolation are the DNAse I (Kolodner and Tewari, 1979), which is used as a modification of the sucrose gradient technique, and the high salt (Bookjans *et al.*, 1984) methods (see http://www.jgi.doe.gov/programs/comparative/second_levels/chloroplasts/jansen_project_home/cpDNA_protocols.html for protocols). In the DNAse I method, the chloroplast pellet in step 7 (Table I) is treated with DNAse I to destroy nuclear DNA. This treatment also will destroy any cpDNA that is not protected within intact plastids. Thus, although the purity of cpDNA is very high, the yield is much lower and much more leaf material is needed to obtain sufficient cpDNA. In our experience, this method yields very pure cpDNA when it works, but it has only worked for two species of the many that we have attempted (*Lactuca sativa* [Fig. 2] and *Ginkgo biloba*). Even in those cases, sufficient quantities of cpDNA for shearing and shotgun cloning were not always recovered. The second alternative method employs a high NaCl (1.25 M) concentration in the isolation and wash buffers, and it does not involve any step-gradient centrifugation. The high salt concentration is supposed to significantly reduce nuclear contamination. According to Bookjans *et al.* (1984), the undissociated

Lactuca   Ranuculus

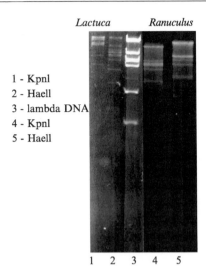

1 - Kpnl
2 - Haell
3 - lambda DNA
4 - Kpnl
5 - Haell

1   2   3   4   5

Fɪɢ. 2. Gel photo showing chloroplast DNA isolations for *Lactuca* (Asteraceae) using DNAse I method and *Ranunuculus* using the NaCl method (see the section "Isolation of Chloroplast DNA"). Lanes 1 and 2 and 4 and 5 were digested with KpnI and HaeII, respectively; lane 3 is a lambda DNA digest used as a size marker.

chromatin or nuclear DNA tends to stick to chloroplast membranes because of electrostatic interactions. The high salt concentration diminishes these electrostatic interactions, yielding a DNA prep that is enriched in cpDNA. We have had only limited success with this approach; one isolation by this method yielded cpDNA of sufficient purity and quantity to proceed to genomic sequencing (*Ranunculus macranthus*, Fig. 2). However, the use of high-salt wash buffers in combination with the sucrose gradient technique has proven quite valuable for decreasing nuclear DNA contamination in chloroplast preps.

The methods just described can also be used (stopping prior to lysis) to collect chloroplasts for use in whole genome amplifications (described later in this chapter). Other workers are experimenting with the use of a fluorescence-activated cell sorter (FACS) to separate chloroplasts from mitochondria and nuclei (D. Mandoli, personal communication, 2004). This method may be particularly valuable when limited tissue is available. Once purified chloroplasts have been obtained from the FACS, they can be further processed using one of the methods described below. Another advantage of the FACS approach is that it may also provide purified fractions of both mitochondria and nuclei in addition to chloroplasts.

Whole Genome Amplification

If purified chloroplasts can be obtained, they can serve as a template from which to produce abundant cpDNA via RCA, a powerful approach for performing whole genome amplification. This process involves an iso-thermal amplification using bacteriophage Phi29 polymerase, which is capable of performing strand-displacement DNA synthesis for more than 70 kb without disassociating from the template (Dean *et al.*, 2002). This feature, combined with the stability of this polymerase and its low error rate, makes this enzyme a powerful tool for template preparation. RCA involves the use of random hexamer primers that are exonuclease resistant, necessary because the DNA polymerase has a 3′–5′ exonuclease proofread-ing activity. Most applications of RCA have been directed toward per-forming human genome amplification and a kit for this purpose (Repli-G) is available from Qiagen. Our group has been using this kit routinely for amplifying entire chloroplast genomes, and we have modified the Repli-G protocol to improve cpDNA amplification (see Table III for protocol). We have had considerable success with the RCA approach for a wide diversity of seed plants. Figure 3 shows restriction digests of RCA products for two taxa that had sufficient quality and quantity of cpDNA to proceed with genome sequencing. One possible further modification of this protocol would be to develop genome-specific primers for chloroplast or mitochon-drial genomes, which would enable the amplification of the chloroplast and mitochondrial genomes from total DNA isolations. Although the low tem-perature of the RCA reaction limits the specificity of annealing for these primers, experiments are in progress, focusing on buffer modifications that show promise for increasing the specificity of the amplification.

Long PCR and Sequencing

A third approach for obtaining DNA template from which to generate whole chloroplast genome sequences involves PCR amplifying of large fragments of the genome using conserved chloroplast primers. This ap-proach has been employed to sequence three basal angiosperm genomes (Goremykin *et al.*, 2003a,b, 2004). Goremykin *et al.* developed conserved primers by aligning sequences from seven seed-plant genomes (*Arabidopsis, Nicotiana, Oenothera, Oryza, Pinus, Spinacia*, and *Zea*). These primers then were used to amplify long fragments ranging in size from 4 to 20 kb and covering the entire chloroplast genome. The long PCR products were then sheared into smaller pieces, shotgun cloned, and sequenced. Although this approach worked well for Goremykin's group, it does have several disad-vantages: (1) The primer combinations may not work for seed-plant gen-omes that have experienced gene order changes or substantial sequence

TABLE III
WHOLE CHLOROPLAST GENOME AMPLIFICATION USING RCA

A. Setting Up the RCA Reaction

1. Thaw RCA kit (Repli-g, Qiagen, Inc.) reaction components ($1\times$ PBS, $4\times$ mix, solution B, polymerase) on ice. Prepare the alkaline lysis solution (solution A) if necessary.

2. Activate solution A by adding DTT (must be made fresh before using): For each reaction, 31.5 μl of solution A and 3.5 μl 1 M of DTT is needed. This can be done while waiting for lysis in the next step or while components are thawing in the previous step.

3. Add 3 μl of the $5\times$ lysis buffer to 15 μl of isolated chloroplasts (from step 11 in Table II) and incubate for 15 min at 37°. We have attempted to quantify the amount of chloroplasts in this step but it turns out that this is futile. The success of subsequent steps is more dependent on the quality and purity of the chloroplasts rather than on the number of chloroplasts that are added to the lysis reaction. We have found that the amount of the chloroplast prep added needs to be optimized for each taxon.

4. Add 50 μl of $1\times$ PBS to the lysate.

5. Add 35 μl of the resulting solution to 35 μl of activated solution A and incubate on ice for 10 min.

6. While alkaline lysis is proceeding, prepare the reaction cocktail (35 μl H_2O + 12.5 μl $4\times$ mix + 0.5 μl polymerase) and aliquot it to the reaction tubes. This is based on 2 μl of lysate being added to each reaction—adjust volume of water accordingly if using more or less of the lysate.

7. Stop the alkaline lysis by adding 35 μl of neutralization solution B to the lysate.

8. Take 2 μl of lysate and add to each reaction.

9. Incubate at 30° for 16 h; terminate with 3 min at 65°. Generally the solution looks cloudy if the reaction has worked. Store in refrigerator or freezer until proceeding to B.

B. Checking for RCA product

1. Run 2 μl of product on minigel to determine whether the RCA was successful.

2. If there is product on the minigel, proceed with restriction digests.

3. Do restriction enzyme digests of 2 μl of RCA product using *BstBl* and *EcoRl* following the manufacturers recommendations in 20-μl reactions. Some enzymes do not digest RCA product very well. We have tested a number of enzymes and found that *BstBl* and *EcoRl* work best.

    2 μl RCA product

    2 μl of appropriate enzyme buffer

    Sufficient H_2O to end up with a volume of 20 μl

    10–20 units of enzyme

4. Load entire digest into 1% agarose gel and run dye marker to 10 cm

5. Stain, visualize, and photograph gel to assess the quality of the RCA product (see Fig. 2 for an example).

    Stocks: 5 M KOH (28 g KOH pellets + H_2O to 100 ml; exothermic!)

    0.5 M EDTA (18.6 g EDTA + 80 ml H_2O, pH to 8.0; raise volume to 100 ml)

    Lysis solution (solution A): 0.4 ml 5 M KOH + 0.1 ml 0.5 M EDTA + 4.5 ml H_2O

    $5\times$ lysis buffer: 20% sarcosyl, 50 mM Tris pH 8, 25 mM EDTA.

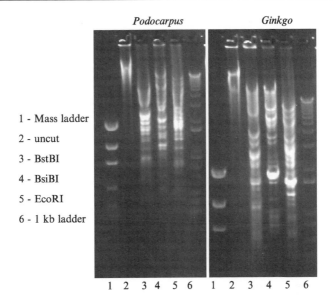

FIG. 3. Gel photo showing results of whole chloroplast genome amplification using rolling circle amplification (RCA) of isolated chloroplasts of *Ginkgo* and *Podocarpus*. Lane 2 shows uncut RCA product, and lanes 3–5 show 2 μl of RCA product cut with restriction enzymes. Lanes 1 and 6 are two different-size markers. Quality of RCA product can be assessed by performing digests and running gels such as those shown here. Nuclear contamination would appear as a smear while the cpDNA forms discrete bands. The relative proportion of smear to bands is assessed visually from the gel photo. Upon sequencing, this *Podocarpus* RCA product was found to be >80% cpDNA and the *Ginkgo* product >60% cpDNA.

divergence at priming sites; (2) the method relies on PCR, which can sometimes be problematic for some DNAs or segments of the genome; (3) it would be difficult to extend this approach to algae or spore-bearing plants because little or no published chloroplast genome sequence information is available to direct primer design in these groups; and (4) numerous PCR and cloning reactions are required, consuming more time than some of the other available methods.

Cloning Chloroplast Genomes for Sequencing

Finally, a more labor-intensive but highly useful approach for obtaining sequencing template involves cloning the genome into either BAC or Fosmid vectors. This approach is superior to plasmid cloning because the insert is much larger, 40–150 kb. The larger insert reduces the amount of screening involved and allows the clones to be sequenced via the JGI method described earlier. A number of BAC and Fosmid cloning kits are available commercially; we have been using the Epicentre CopyControl Kit (catalog no. CCF0S110). Our group has used the Fosmid cloning approach to sequence plastid genomes from parasitic plants and normal photosynthetic plants. The details of our Fosmid protocol can be found in McNeal *et al.* (submitted), but here we provide a general outline for this procedure. DNA is isolated using a modified cetyltrimethylammonium bromide (CTAB) method (Doyle and Doyle, 1987) with 1% PEG 8000 in the extraction buffer. The DNA must then be end-repaired for cloning into vectors that require blunt-ended, 5′ phosphorylated ends. Pulse field gel electrophoresis (PFGE) is used to separate fragments in the 40–50 kb range for Fosmid cloning and in the 100–150 kb range for BAC cloning. DNA of the correct size is excised and recovered from the gel, and its concentration is measured, preferably by fluorometry, to ensure the proper ratio of template to vector for efficient ligation. Clones are plated and then transferred to 384-well plates for easy referencing and gridding onto nylon filters. We use robotics to pick, transfer, and grid clones quickly and efficiently. Plants with larger nuclear genome sizes have a proportionally higher ratio of nuclear-to-plastid clones and, thus, require a greater number of clones to be arrayed for screening to ensure enough plastid clones will be found to cover the entire plastid genome. When the DNA used for Fosmid or BAC cloning is enriched for cpDNA, fewer clones need to be screened. Macroarrays are screened using hybridization probes generated by PCR amplification of genes scattered throughout the plastid genome. Once positively hybridizing plastid clones have been identified, a minimal set of Fosmid clones are selected that cover the entire plastid genome (usually 2–5). End-sequencing and PCR assays of each clone aid in the selection of minimally overlapping clones, which together cover the genome completely. One caveat of this method is that the macroarray hybridizations may also detect recent mitochondrial or nuclear plastid gene transfers. However, single- or low-copy nuclear transfers are much less likely to be found than true plastid genome fragments, which occur in many more copies per cell. End-sequencing and PCR assays of each clone should eliminate all but the largest and most recent mitochondrial transfers from passing as plastid clones. For BAC libraries, only one or two clones are needed to get complete coverage of the genome, depending on genome size. The clones are then sheared, shotgun cloned, and sequenced as described for other methods. One 384-well plate is sequenced for each Fosmid clone (with both plasmid primers to yield 768 reads) or two to three plates for each BAC clone to obtain 6–10× coverage of the insert. Additional sequencing may be required to close gaps or verify regions with low coverage.

Our group has used the Fosmid cloning method to successfully create libraries for a number of photosynthetic (*Ipomoea, Lindenbergia,* and *Yucca*) and non-photosynthetic parasitic and mycotrophic plants (*Corynaea, Cuscuta, Cytinus, Monotropa, Orobanche,* and *Prosopanche*). Researchers preparing BAC libraries typically screen for "contaminant" clones containing chloroplast genome fragments. In collaboration with Pietro Piffanelli (CIRAD-AMIS, Montpellier, France), we have obtained plastid genome sequences from cpDNA-containing BACs identified from his *Musa* and *Elaeis* libraries. While Fosmid or BAC library construction is certainly more technically demanding and time consuming than cpDNA isolation or RCA amplification of plastid genomes, the libraries will have a broader utility, and we have found, generally, less finishing of draft genome sequences is required when the shotgun sequencing libraries are made from well-chosen Fosmid or BAC templates.

Assembling, Finishing, and Annotating Genomes

Assembling Draft Genome Sequences

When preparing a draft genomic sequence from cpDNA or RCA product, we first generate one 384-well plate of sequences using both forward and reverse primers (768 reads). Vector and quality trimming of the resulting sequences is performed using PHRED (Ewing and Green, 1998). Using BLASTN (Altschul *et al.*, 1997), trimmed reads are then used to query a nucleotide sequence database of previously sequenced chloroplast genomes. If the BLASTN search indicates that 60% or more of the reads are chloroplast sequences, we then proceed to sequence four more plates for a total of 3840 reads, although additional plates are sometimes required. If less than 60% of the library is cpDNA, we do not proceed with additional sequencing but work to obtain purer cpDNA preps from which to construct a new library. When sequencing from Fosmid/BAC clones, we prepare a separate library for each clone. One plate per Fosmid clone library or two plates per BAC clone library usually provide sufficient coverage.

Individual reads generated from the plates are assembled into contiguous sequences ("contigs") using PHRAP (Ewing and Green, 1998) and the resulting contigs are analyzed in CONSED, a powerful software package used for sequence finishing (http://www.phrap.org/consed/consed.html) (Gordon *et al.*, 1998). CONSED has numerous useful features (Fig. 4), including an overview of the assembly, numerous editing options, a method for tearing contigs into pieces and performing mini-reassembly, an option for designing finishing primers, and options for adding new reads. The Assembly View option (see Fig. 4 for an example) provides a wealth of information to evaluate the draft genome sequence, including the depth of coverage, the possible arrangement of the contigs, and crossmatches of

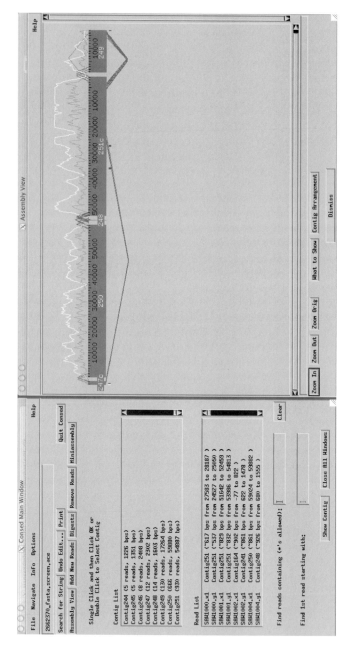

FIG. 4. Screen shot of two Consed (Gordon *et al.*, 1998) windows. The left panel shows the main window with the list of contigs, individual reads, and several other features. The right panel shows the assembly view of four contigs of *Nuphar*, illustrating contig order, read depth, and inconsistent forward-reverse subclone pairs.

sequences between contigs. For chloroplast genomes that are not highly rearranged, one generally does not encounter many problems with the assembly, but highly rearranged genomes often require considerable work interactively reassembling the sequences because of the high frequency of repeated sequences. This will require examination of each of the contigs to identify possible misassemblies and the removal and/or relocation of misplaced reads. Examination of the assembly will also reveal regions of the draft where there are few high-quality reads and more sequencing is needed. Effective integrated use of PHRED, PHRAP, and CONSED takes considerable time to master. PHRED and PHRAP, however, are necessary for sequence assembly, and CONSED is extremely valuable for assessment of draft assemblies and identifying regions where directed sequencing is necessary to finish the genome sequence. The most finishing will likely be required when purified cpDNA or RCA product is sheared, shotgun cloned, and sequenced "randomly," whereas the least finishing is required when Fosmid or BAC clones are used as the template.

Finishing Genomic Sequences

Finishing draft chloroplast genomic sequences involves four basic steps: (1) make a preliminary identification of genes occurring in each contig using the chloroplast genome annotation program Dual Organellar GenoMe Annotator (DOGMA) (Wyman *et al.*, 2004); (2) examine depth of coverage within each contig to identify regions of low sequence coverage; (3) design primers that flank gaps and regions of low coverage and perform PCR and sequencing to fill in necessary regions; and (4) determine the extent of the IR and if necessary confirm using PCR and sequencing across the IR–single-copy (SC) junctions. These four finishing steps are described in more detail in the following sections. All of these steps may not be necessary. For example, drafts generated by sequencing BAC or Fosmid clones often do not require finishing if the screening and selection of clones was done correctly. However, when purified cpDNA or RCA product is used, some finishing will be necessary. The amount of finishing will depend on the purity of the cpDNA or RCA product. High-purity cpDNA could yield the entire chloroplast genome in one contig with no areas of low coverage, although in our experience this rarely happens unless the purity of the cpDNA or RCA product is exceptionally high or more than five plates of sequences are done. Even in these cases, it is still necessary to confirm the boundaries of the IR because both copies of the IR will assemble together in CONSED.

1. Identify genes in contigs with DOGMA: DOGMA is a web-based program developed by our group that makes this step in the finishing

process very easy (see the section "Phylogenetic and Molecular Evolutionary Analysis of Genomic Sequences," later in this chapter, for more details about the program) (Wyman *et al.*, 2004). DOGMA identifies those genes that are likely to occur in each contig. Knowledge of the gene content assists in determining the arrangement of the contigs so that primer pairs that span gaps can be developed. For genomes where previous gene mapping data are available, one simply compares the gene content of the contigs to the gene map to arrange the contigs. When no gene map is available, comparison of gene orders in the contigs to already sequenced chloroplast genomes often can provide valuable information for deciding how the contigs likely are arranged. It is often possible to use the already sequenced genomes to estimate the location and sizes of gaps and to develop more universal primers to amplify through the gaps.

2. Examine depth of coverage in contigs: Generally our methods generate contigs that have 6–10× coverage, but this certainly will depend on which genome sequencing method we have employed and the quality of the sequencing template. Our group has decided that for each nucleotide, a minimum of two reads, each with a PHRED/PHRAP quality score (q value) exceeding 20, is necessary for satisfactory genome coverage. In general, coverage is much higher except in those regions where we fill in gaps. However, if areas with sparse coverage occur within contigs, primers are designed and additional sequence data are gathered.

3. Design primers to fill in gaps by PCR and sequencing: Once all gaps and areas of low coverage have been identified, primers are designed that flank these regions. We generally design 18–20 bp primers in coding regions that are adjacent to the gaps or regions of low coverage. We attempt to make the primers as universal as possible by comparing the primer sequences with previously sequenced chloroplast genomes so that the primers could be used in the finishing of other chloroplast genomes. In some cases, we need to design primers that are not in coding regions. This is more difficult because primers in nongenic regions may have multiple priming sites. We usually can avoid this problem by searching the genome for the primer sequence using the CONSED. For larger gaps, additional primers must be made to sequence through the gap. In many cases, the size of the gap is unknown, so it may be a matter of trial and error to determine what extension time to use in the PCR.

4. Confirm extent of IRs: Chloroplast genomes that are sequenced using long PCR or BAC/Fosmid clones may include each copy of the IR in separate contigs, in which case defining the extent of the IR is straightforward. However, in many cases, all the individual IR sequencing reads generated by shotgun cloning of purified cpDNA or RCA product will be assembled together, making it difficult to determine the precise IR–SC

boundaries. Several tricks can be used to get a general idea of these boundaries, especially if parts of the IR are present in different contigs. In general, the assembly view in CONSED shows a higher depth of reads in the IR (Fig. 4). Also, another very useful feature of CONSED shows subclone pairing. This provides information about the positions of forward and reverse reads from the same clone. If ends of the same clone match in distant regions or in different contigs, this may be due to a sequence being part of one of the IR–SC junctions. In most cases, these methods for identifying possible IR boundaries are not definitive, and it is necessary to design primer pairs (two for each of the four IR–SC boundaries) that span the IR–SC junctions. Amplification and sequencing of these regions is needed to confirm the boundaries. Once this has been confirmed, the IR sequence must be copied, inverted, and inserted into the appropriate location to complete the chloroplast genome sequence.

Annotation Using DOGMA

Annotation of chloroplast genomes traditionally has been a very tedious and error-prone task. The annotations in GenBank are not consistent in terms of gene names, and they are not usually updated when the identities and functions of hypothetical chloroplast reading frames (ycfs) or open reading frames (ORFs) are clarified. In the past, most chloroplast genome sequences were annotated by performing BLASTN and BLASTX searches on GenBank. Many of these problems were alleviated upon completion of DOGMA, a web-based program designed by our group to assist in the annotation of chloroplast and animal mitochondrial genomes (see http://evogen.jgi-psf.org/dogma/) (Wyman et al., 2004). This program takes a FASTA-formatted input file of the complete (or partial) genomic sequences and identifies putative protein-coding genes by performing BLASTX searches against a custom database of 16 published chloroplast genomes of green plants (Fig. 5A). Errors in the GenBank entries have been corrected in the database, and names of genes and their products have been standardized following Martin et al. (2002). Sequence identity is highly conserved for both tRNAs and rRNAs in chloroplast genomes, so these genes are identified by BLASTN searches against a database of the same 16 chloroplast genomes. DOGMA also uses a custom program to infer the stem loop structure of tRNAs and draw candidate secondary structure diagrams.

DOGMA has many other features to aid in annotation of chloroplast genomes (Fig. 5B). One DOGMA panel displays all of the putative genes color-coded by gene type. Selection of a gene in this lower panel generates an upper panel that shows the five or more most similar

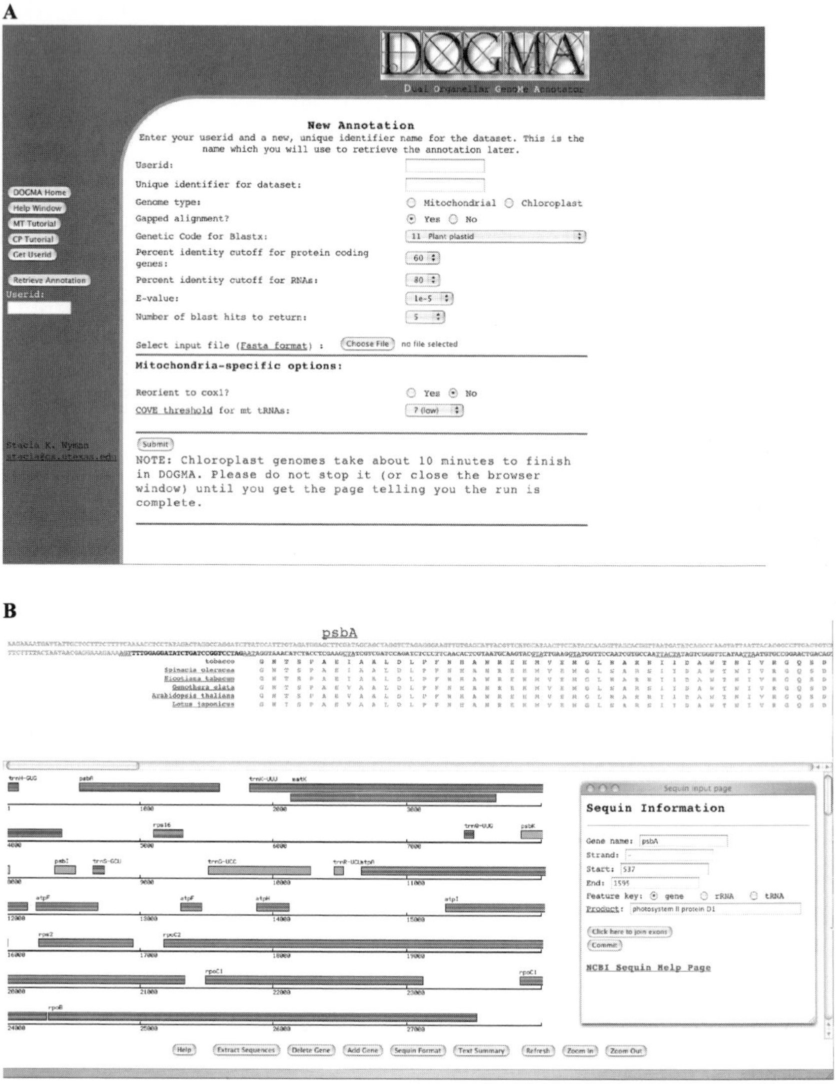

A

B

FIG. 5. Two web browser windows from DOGMA (Wyman *et al.*, 2004). (A) The main window for submitting FASTA-formatted input files of complete genome sequences or contigs of portions of the genome. A number of optional settings are available for the genetic code for BLASTX, percent identity for protein coding genes and RNAs, e value, and the number of BLAST hits to return. (B) A view of the annotation window with three panels: Lower panel has several option buttons for extracting sequences, deleting/adding genes, and generating a Sequin-formatted file or text file; middle panel shows tentative gene identifications, clicking on a gene will display that gene, its BLAST hits, and putative start

sequences from the database compared to the sequence under analysis, along with potential start and stop codons (Fig. 5B). The user then must select the most likely start and stop codons to identify each putative gene. For genes with introns, DOGMA will identify putative exon boundaries by BLAST; the user must verify these boundaries and use DOGMA to connect the exons. Another window appears that records the annotation information that can be used to generate a Sequin file for submitting the annotation to GenBank. Selection of the gene name in the top panel also generates a window with the actual BLAST results. In the lower panel of the annotation window, there are additional buttons (Fig. 5B). The "extract sequences" button enables the user to extract certain sets of sequences from the annotation, including protein-coding genes (either nucleotide or amino acid sequences), intergenic regions, introns, tRNAs, or rRNAs (Fig. 5B). This feature is particularly useful for extracting sequences to add to a data matrix for phylogenetic analyses. The text summary button generates a tabular form of the annotation with coordinates for the genes and other information about each gene. More details about the features of this program can be found by downloading the cp tutorial at http://evogen.jgi-psf.org/dogma/.

In the future, DOGMA will be modified in several ways: (1) The chloroplast database will be expanded to include more sequences, especially from underrepresented groups such as algae, (2) mitochondrial genomes from plants, fungi, and protists will be added, (3) an option will be included to allow users to develop a custom database, (4) an ORF finder will be added to search for putative new genes, and (5) methods will be developed to deal with RNA editing of start and stop codons, a phenomenon that is common to plant mitochondrial genomes and chloroplast genomes of some plants (Bock, 2000).

Analysis of Genome Sequences

The analysis of whole genome sequences is an immense scientific field for which numerous databases and computational tools exist, some relevant to the study of chloroplast genomes (Table IV). Some of these are simply a listing of available genome sequences with accession numbers to access the sequences on GenBank, whereas others provide additional information about chloroplast gene names, details of the characteristics

---

and stop codons in the upper panel; upper panel shows the BLAST hits for the *psbA* gene and some putative stop codons. The sequin information window is also shown here. This is the window used to commit to the start and stop codon and it generates an entry compatible with Sequin.

TABLE IV
CHLOROPLAST GENOME ONLINE DATABASES AND SOFTWARE

Database/ software	Web address	Features
Organelle Genome Megasequencing Program (OGMP)	http://megasun.bch.umontreal.ca/ ogmp/projects/other/ cp_list.html	Lists all sequenced chloroplast genomes with NCBI classification, accession numbers, and links to GenBank
ExPASy	http://us.expasy.org/txt/plastid.txt	Lists names of chloroplast and cyanelle proteins with abbreviations; also gives list of completely sequenced plastid genomes
NCBI: Organelle Genomes	http://www.ncbi.nlm.nih.gov:80/ genomes/static/euk_o.html	Lists all completely sequenced organelle genomes with accession numbers, genome size, and date of submission
Dual Organellar GenoMe Annotator (DOGMA)	http://evogen.jgi-psf.org /dogma/	A program for annotation of chloroplast and animal mitochondrial genomes
Plastid Gene Order Database	http://www.rs.noda.tus.ac.jp/ ~kunisawa/order/front.html	Provides corrected annotations for chloroplast genomes with tools to view gene orders and extracting sequences
Genomemine	http://www.genomics.ceh.ac.uk/ cgi-bin/gmine/gminemenu. cgi?action=listorganelles &sort=genome	Provides list of all sequenced genomes with details of accession number, size, numbers of ORFs, percent coding, and base frequency
DOE Joint Genome Institute (JGI) Organelle Genomics	http://www.jgi.doe.gov/programs/ comparative/top_level/ organelles.html	Provides access to several ongoing projects in organelle genomics and access to various tools for annotating and analyzing chloroplast and mitochondrial genomes
A database of PCR primers for the study of the chloroplast genome in plants	http://fbva.forvie.ac.at/200/ 1859.html	Contains information about universal primers for chloroplast genomes
BPAnalysis	http://www.cs.washington.edu/ homes/blanchem/software.html	A program that computes minimal breakpoint trees from gene order data
Derange2	http://www.cs.washington.edu/ homes/blanchem/software.html	A program that computes an approximation of minimal edit distances between pairs of gene orders

(continued)

TABLE IV (*continued*)

Database/ software	Web address	Features
Genome Rearrangements In Man and Mouse (GRIMM)	http://www-cse.ucsd.edu/groups/ bioinformatics/GRIMM/ index.html	Rearrangement algorithms for genomes, which compute the minimum possible number of rearrangement steps, and determine a possible evolutionary scenario using this number of steps
Genome Rearrangements Analysis under Parsimony and other Phylogenetic Algorithms (GRAPPA)	http://www.cs.unm.edu/~moret/ GRAPPA/	A program for constructing phylogenies using gene order data
Multiple Genome Rearrangements (MGR)	http://www-cse.ucsd.edu/groups/ bioinformatics/MGR/index.html	A tool for constructing phylogenies based on gene order data
PipMaker and MultiPipmaker	http://pipmaker.bx.psu.edu/ pipmaker/	Used to align two (PipMaker) or multiple (MultiPipmaker) genomes and provide dot-matrix and percent identity plot (PIP) diagrams of whole genomes
REPuter	http://www.genomes.de/	A program for identifying repeated sequences in genomes and provides an excellent visualization of the location and sequence of various types of repeats
FootPrinter	http://bio.cs.washington.edu/ software.html	A program to identify putative regulatory elements in DNA sequences that requires a phylogeny
RepeatFinder	http://www.tigr.org/software/	Organizes repeats into classes
RepeatMasker	http://www.repeatmasker.org/	A program that screens DNA sequences for interspersed repeats
Chloroplast Genome Database	http://cbio.psu.edu/chloroplast/ index.html	Contains all plastid genomes; allows searches for genes, downloading genes, proteins, or whole genomes; and performs BLAST searches

of the genomes, databases of corrected annotations, gene orders, universal primer sequences, and searchable databases. All of these are valuable resources for anyone working on comparative chloroplast genomics. In the following sections, we discuss chloroplast genome analysis in terms of phylogenetic comparisons of gene content and gene order, detection of repeats, and use of coding sequences for phylogenetic studies.

Whole Genome Comparisons and Repeat Analysis

A number of computational tools exist for whole genome comparisons, although most of these were not designed specifically for chloroplast genomes. We have used several of these tools to compare gene content, examine genome wide sequence similarity, look for repeated sequences, and identify putative regulatory motifs with the primary goal of improving our understanding of genome evolution. Our primary goal in using these programs has been to improve our understanding of both the patterns and the mechanisms of chloroplast genome evolution. We briefly review a few of these tools and discuss how we have applied them to comparisons of chloroplast genomes. Table IV includes information about accessing these programs.

MultiPipmaker (Table IV) (Schwartz et al., 2003) allows the user to compare multiple chloroplast genomes. The program generates alignments of whole genomes in comparison to a reference genome. The output from MultiPipmaker includes a stacked set of percent identity plots (Fig. 6), referred to as a "MultiPip," which illustrate sequence similarity among the genomes in coding and noncoding regions. This output is helpful in identifying potential genes and regulatory elements. Visual inspection of the Multi-Pip also is useful for identifying putative gene losses or gene duplications, for identifying unannotated genes or conserved nongenic regions, and for assessing overall sequence similarity among genomes (see Maul et al. [2002] for a chloroplast genome comparison). PipMaker (Elnitski et al., 2002), the pairwise version of this tool, also has been used to identify repeated sequences by aligning a genome against itself (see Pombert et al. [2004] for an example of this application using a plant mitochondrial genome).

With the exception of the large IR present in most taxa, chloroplast genomes generally are considered to have very few repeated sequences (Palmer, 1991). However, repeated sequences have been identified in a number of genomes, including *Chlamydomonas* (Maul et al., 2002), *Pseudotsuga* (Hipkins et al., 1995), *Trachelium* (Cosner et al., 1997), *Trifolium* (Milligan et al., 1989), wheat (Bowman and Dyer, 1986; Howe, 1985), and *Oenothera* (Hupfer et al., 2000; Sears et al., 1996; Vomstein and Hachtel, 1988). The most striking example is the *Chlamydomonas* chloroplast genome, of which more than 20% is composed of short dispersed repeats. In

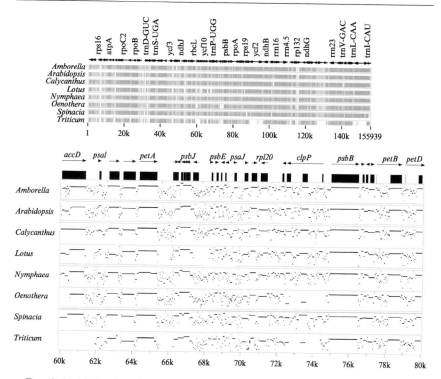

FIG. 6. MultiPipmaker (Schwartz *et al.*, 2003) output of various published chloroplast genome sequences. The reference genome *Nicotiana* (Z00044) was analyzed against eight other genome sequences, including *Amborella* (AJ506156), *Arabidopis* (AP000423), *Calycanthus* (AJ428413), *Lotus* (AP002983), *Nymphaea* (AJ627251), *Oenothera* (AJ271079), *Spinacia* (AJ400848), and *Triticum* (AB042240). (A) MultiPip view showing sequencing identity (50–100%) among genomes with identity increasing with darker shades. Positions of genes and selected gene names are shown at top, names of taxa are on left. (B) Selected region of the MultiPip showing sequence identity between 50% and 100%. Arrows on top of map indicate position of selected genes and numbers above gene indicate the exons for genes with introns. Note that this diagram shows that *accD* is absent from *Triticum*.

most of these cases, repeats appeared to be associated with rearranged blocks of genes. Thus, characterization of repeat structure in chloroplast genomes could provide insights into mechanisms of gene order changes.

Several programs were designed to identify repeats and group them into classes. The two programs that we have found most useful are RE-Puter and RepeatFinder (Table IV). REPuter (Kurtz and Schleiermacher, 1999; Kurtz *et al.*, 2001) includes a search algorithm that finds various types of repeats, including direct repeats and IRs (Fig. 7). The user specifies the desired repeat type, minimum repeat length, and the percent identity (Hamming distance) and the program locates all repeats that meet these

A

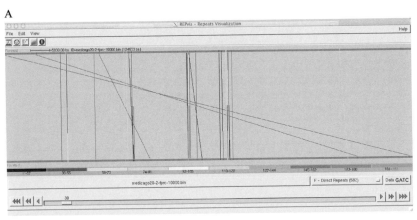

B

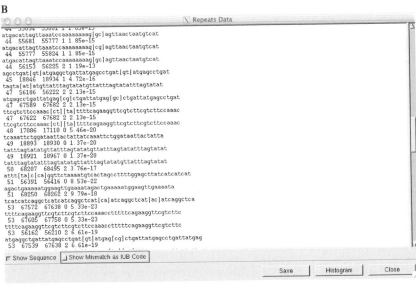

FIG. 7. REPuter (Kurtz and Schleiermacher, 1999) output views of an analysis of the *Medicago* chloroplast genome (AC093544). The search examined forward and inverted repeats >20 bp in length with 90% sequence identity. (A) The visualization window is shown for forward repeats >30 bp in length. (B) A portion of the display of repeats found with the size of repeat, the coordinates in the genome, the hamming distance, e value, and the DNA sequence of the repeat given.

criteria. The program also provides a graphic visualization of the location of the repeats in the genome (Fig. 7A). REPuter can be accessed and run directly using a web browser (http://www.genomes.de/), although this platform does not allow the user to modify the default options. We recommend that users download the standalone version, which is available for Unix platforms for free. RepeatFinder (Volfvovsky *et al.*, 2001) is a software tool for clustering repeats into classes. It takes as input repeats that have been identified by another program such as REPuter. This program must be downloaded and set up on a Unix platform. Both REPuter and Repeat-Finder have been used together to examine repeat structure in plant mitochondrial genomes (Bartoszewski *et al.*, 2004; Pombert *et al.*, 2004).

Gene Content and Order for Phylogeny Reconstruction

Chloroplast genomes in many groups are highly conserved in gene content, although there are significant differences in these features in comparisons between algal and land-plant genomes (Raubeson and Jansen, 2005; Simpson and Stern, 2002). Martin *et al.* (2002) estimated that only 44 of the 274 plastid-encoded genes are retained in all plastid genomes and approximately half (117) of the ones that are missing have been lost or transferred to the nucleus. Among green plants, there is considerable conservation of both gene content and gene order. For example, the gene organization of the earliest diverged green alga sequenced so far, *Mesostigma*, is very similar in structure to land-plant cpDNAs, with 81% of its genes being found in the same clusters as in land plants (Lemieux *et al.*, 2000). Other comparisons with the green alga *Chlamydomonas* also revealed a high incidence of gene loss among algal chloroplast genomes but a much higher level of similarity among green plants (Maul *et al.*, 2002; Simpson and Stern, 2002). The large number of gene losses among plastid genomes, often occurring in parallel in different lineages (Martin *et al.*, 2002; Maul *et al.*, 2002), suggests that the use of gene content for phylogeny reconstruction may be of limited value, and in most cases, the utility of these types of characters may be restricted to selected groups.

Gene order of the chloroplast genome is generally highly conserved, especially among land plants. Previous studies have demonstrated the phylogenetic utility of gene rearrangements for resolving relationships at deep nodes, although in most cases, only one or a few characters were available. Some notable examples include a 30-kb inversion that identified the lycopsids as the basal lineage of vascular plants (Raubeson and Jansen, 1992), three inversions that supported monophyly of the Poaceae and indicated its relationship to Joinvilleaceae and Restionaceae (Doyle *et al.*, 1992), and a 22-kb inversion that identified the basal clade in the Asteraceae (Jansen and Palmer, 1987). These types of changes make powerful phylogenetic markers, and subsequent phylogenetic studies using DNA sequence data corroborated these relationships first identified by gene order changes. The best example of the utility of gene order data for phylogeny reconstruction is in the angiosperm family Campanulaceae (Cosner *et al.*, 1994, 2000, 2004). Gene mapping studies of 18 genera in this family identified numerous changes in gene order, which were caused by inversion, expansion, and contraction of the IR, and possibly transposition. The situation in the Campanulaceae is so complicated that it is not possible to define clearly the evolutionary events responsible for these rearrangements. However, phylogenetic analyses of the gene order data have generated a well-resolved phylogeny for 18 taxa (Fig. 8), and the dataset exhibits lower levels of homoplasy than phylogenies inferred from *rbcL* or ITS sequences for the same taxa (Cosner *et al.*, 2004).

A number of groups have been developing computational methods for using gene order data for phylogeny reconstruction (Table IV) (Bourque and Pevzner, 2002; Cosner *et al.*, 2000, 2004; Larget *et al.*, 2002; Moret *et al.*, 2001; Wang *et al.*, 2002). The approaches are designed to analyze highly rearranged genomes using several phylogenetic approaches, including distance, parsimony, and Bayesian methods. Most of these algorithms are designed for genomes that have a single chromosome with equal gene content, although other studies have begun to implement methods for multiple chromosomes (Bourque and Pevzner, 2002) and unequal gene content (Tang and Moret, 2003). The utility of most of these algorithms has been tested using simulation studies; however, the Campanulaceae chloroplast genomes have been used as a benchmark empirical dataset for assessing speed and accuracy of these methods (Bourque and Pevzner, 2002; Moret *et al.*, 2001). A more detailed review of algorithms for phylogenetic analysis of gene order data can be found in Chapter 35. The availability of many new completely sequenced chloroplast genomes should provide a much expanded empirical dataset for the development of new algorithms that use gene order data for phylogeny reconstruction.

Phylogenetic and Molecular Evolutionary Analysis of Genomic Sequences

Completely sequenced chloroplast genomes provide a rich source of nucleotide and amino acid sequence data that can be used to address phylogenetic and molecular evolutionary questions. Several studies have attempted to use entire suites of sequences (e.g., all shared protein-coding genes) from completely sequenced genomes to resolve a number of phylogenetic issues, including relationships among grasses (Matsuoka *et al.*, 2002), identification of the basal lineage of flowering plants (Goremykin *et al.*, 2003a,b, 2004; Leebens-Mack *et al.*, submitted) and land plants (Kugita *et al.*, 2003), and relationships among land plants and green algae (Lemieux *et al.*, 2000; Turmel *et al.*, 1999). Phylogenies based on all or

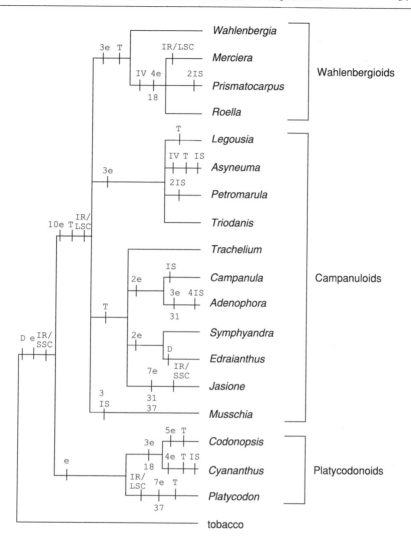

Fig. 8. Campanulaceae phylogeny based on a maximum parsimony analysis of gene order changes (modified from Cosner *et al.* [2004]). Number and type of each genomic change are indicated as e = endpoint of IR, IV = inversion, IS = insertion >5 kb, T = transposition, and D = deletion/divergence. Only three endpoint characters are homoplasious, changing twice on the tree. Brackets indicate the major clades of Campanulaceae.

at least many of the shared genes among completely sequenced chloroplast genomes also have been used to address questions about the origins of plastids and patterns of gene loss or transfer (Chu *et al.*, 2004; Martin *et al.*, 2002; Maul *et al.*, 2002). The latter studies have supported several phylogenetic conclusions: (1) There has been a single primary endosymbiotic origin of plastids; (2) extensive gene loss and/or transfer to the nucleus has occurred; and (3) multiple, independent secondary endosymbiotic events have occurred.

Use of many or all of the genes from the chloroplast genome provides many more characters for phylogeny reconstruction in comparison with previous studies that have relied on only a few genes to address the same questions. However, one current problem with the whole genome approach is that taxon sampling is quite limited and can result in misleading estimates of relationship. A recent example of this problem is the study by Goremykin et al. (2003b) that suggests that *Amborella* may not be the basal angiosperm, a result that contradicts many phylogenies based on sequences of a few genes (Barkman et al., 2000; Graham and Olmstead, 2000; Mathews and Donoghue, 1999; Parkinson et al., 1999; Qiu et al., 1999; Soltis et al., 1999). Phylogenetic analyses of expanded taxon sets have demonstrated that inadequate taxon sampling caused Goremykin's anomalous result (Leebens-Mack et al., submitted; Soltis and Soltis, 2004). In the future, increased availability of more completely sequenced chloroplast genomes will facilitate phylogenetic inference. Much denser taxon sampling is necessary before many of the advantages of whole genome sequencing can be fully realized and investigators must seriously consider the effects of long branch attractions (Felsenstein, 1978). Three other problematic issues have been identified and must be considered, especially in broad phylogenetic comparisons. Compositional bias among plastids from divergent lineages can generate incorrect tree topologies (Lockhart et al., 1999); alignment of coding regions can be very difficult, especially when addressing phylogenetic issues at deep nodes (Chu et al., 2004); and tree topologies are very sensitive to the model of evolution being used (Leebens-Mack et al., submitted; Soltis and Soltis, 2004). A number of studies have attempted to address these issues by developing more realistic models of amino acid substitutions for chloroplast-encoded genes (Adachi et al., 2000; Morton and So, 2000), by examining lineage and locus-specific rate heterogeneity among chloroplast genomes (Muse and Gaut, 1997), and by developing alternative methods for using sequences from whole chloroplast genomes (Chu et al., 2004; Lockhart et al., 1999; Rivas et al., 2002).

Summary and Future Directions

It is a very exciting time for the field of comparative chloroplast genomics. The first chloroplast genome sequences were published 18 years ago, and now there are 45 genomes available, almost two-thirds of which have been completed during the past 4 y (Table I). In this chapter, we have described many developments, which by improving methods for gathering and analyzing chloroplast genome sequences, are providing the necessary framework for greatly expanding the number of sequenced genomes. The most significant advancements include RCA for amplification of entire genomes and DOGMA software (Wyman *et al.*, 2004) for annotation. Several ongoing projects on seed plants, land plants, and algae are likely to result in the availability of nearly 200 completely sequenced genomes during the next 5 y (see http://megasun.bch.umontreal.ca/ogmp/projects/ sumprog.html, http://www.jgi.doe.gov/programs/comparative/second_levels/ chloroplasts/jansen_project_home/chlorosite.html, and http://ucjeps.berke ley.edu/TreeofLife/ for more detailed information about ongoing projects). This increased taxon sampling to include more representatives of all of the major lineages of plants ultimately will provide unprecedented opportunities for addressing phylogenetic questions at deep nodes. These data also will provide important new insights into both patterns and mechanisms of chloroplast genome evolution. Another outcome of these efforts will be the development of new algorithms, new models of chloroplast sequence and genome evolution, and improved computational tools for using both gene order and sequence data for phylogeny reconstruction. Finally, the chloroplast genomic data and the computational methods will be of great value to plant molecular biologists interested in the functional attributes of chloroplast genes and their interaction with other plant organelles.

Acknowledgments

Our research on chloroplast genome sequencing and gene order phylogeny is supported by NSF grants (Biocomplexity grant DEB 0120709 to RKJ, LAR, JB, and CWD; IIS 0113654 to RKJ, T. Warnow, and B. Moret; and RUI/DEB 0075700 to LAR). SKW acknowledges support from an NSF IGERT fellowship (DGE 0114387) and RCH acknowledges Research Internship support from the Graduate School at UT-Austin. We thank Ms. Sheila Plock for technical assistance.

References

Adachi, J., Waddell, P. J., Martin, W., and Hasegawa, M. (2000). Plastid genome phylogeny and a model of amino acid substitution for proteins encoded by chloroplast DNA. *J. Mol. Evol.* **50,** 348–358.

Altschul, S. F., Madden, T. L., Schaffer, A. A., Zhang, J. H., Zhang, Z., Miller, W., and Lipman, D. J. (1997). Gapped BLAST and PSI-BLAST: A new generation of protein database search programs. *Nucleic Acids Res.* **25,** 3389–3402.

Barkman, T. J., Chenery, G., McNeal, J. R., Lyons-Weiler, J., Ellisens, W. J., Moore, G., Wolfe, A. D., and dePamphilis, C. W. (2000). Independent and combined analyses of sequences from all three genomic compartments converge on the root of flowering plant phylogeny. *Proc. Natl. Acad. Sci. USA* **97,** 13166–13171.

Bartoszewski, G., Katzir, N., and Havey, M. J. (2004). Organization of repetitive DNAs and the genomic regions carrying ribosomal RNA, *cob*, and *atp9* genes in the cucurbit mitochondrial genomes. *Theor. Applied Genet.* **108**, 982–992.

Bock, R. (2000). Sense from nonsense: How the genetic information of chloroplasts is altered by RNA editing. *Biochimie* **82**, 549–557.

Bookjans, G., Stummann, B. M., and Henningsen, K. W. (1984). Preparation of chloroplast DNA from pea plastids isolated in a medium of high ionic-strength. *Anal. Biochem.* **141**, 244–247.

Bourque, G., and Pevzner, P. A. (2002). Genome-scale evolution: Reconstructing gene orders in the ancestral species. *Genome Res.* **12**, 26–36.

Bowman, C. M., and Dyer, T. (1986). The location and possible evolutionary significance of small dispersed repeats in wheat ctDNA. *Curr. Genet.* **10**, 931–941.

Chu, K. H., Qi, J., Yu, Z. G., and Anh, V. (2004). Origin and phylogeny of chloroplasts revealed by a simple correlation analysis of complete genomes. *Mol. Biol. Evol.* **21**, 200–206.

Copertino, D. W., and Hallick, R. B. (1991). Group-II Twintron—an intron within an intron in a chloroplast cytochrome-B-559 gene. *EMBO J.* **10**, 433–442.

Cosner, M. E., Jansen, R. K., and Lammers, T. G. (1994). Phylogenetic relationships in the Campanulales based on *rbcL* sequences. *Plant Syst. Evol.* **190**, 79–95.

Cosner, M. E., Jansen, R. K., Moret, B. M. E., Raubeson, L. A., Wang, L.-S., Warnow, T., and Wyman, W. S. (2000). A new fast heuristic for computing the breakpoint phylogeny and experimental analyses of real and synthetic data. 8th International Conference on Intelligent Systems for Molecular Biology, La Jolla, CA.

Cosner, M. E., Jansen, R. K., Palmer, J. D., and Downie, S. R. (1997). The highly rearranged chloroplast genome of *Trachelium caeruleum* (Campanulaceae): Multiple inversions, inverted repeat expansion and contraction, transposition, insertions/ deletions, and several repeat families. *Curr. Genet.* **31**, 419–429.

Cosner, M. E., Raubeson, L. A., Jansen, R. K. (2004). Chloroplast DNA rearrangements in Campanulaceae: Phylogenetic utility of highly rearranged genomes. *BMC Evol. Biol.* **4**, 27.

Dean, F. B., Hosono, S., Fang, L. H., Wu, X. H., Faruqi, A. F., Bray-Ward, P., Sun, Z. Y., Zong, Q. L., Du, Y. F., Du, J., Driscoll, M., Song, W. M., Kingsmore, S. F., Egholm, M., and Lasken, R. S. (2002). Comprehensive human genome amplification using multiple displacement amplification. *Proc. Natl. Acad. Sci. USA* **99**, 5261–5266.

Downie, S. R., and Palmer, J. D. (1992). Use of chloroplast DNA rearrangements in reconstructing plant phylogeny. *In* "Molecular Systematics of Plants" (P. S. Soltis, D. E. Soltis, and J. J. Doyle, eds.), pp. 14–35. Chapman and Hall, NY.

Doyle, J. J., Davis, J. I., Soreng, R. J., Garvin, D., and Anderson, M. J. (1992). Chloroplast DNA inversions and the origin of the grass family (Poaceae). *Proc. Natl. Acad. Sci. USA* **89**, 7722–7726.

Elnitski, L., Riemer, C., Petrykowska, H., Florea, L., Schwartz, S., Miller, W., and Hardison, R. (2002). PipTools: A computational toolkit to annotate and analyze pairwise comparisons of genomic sequences. *Genomics* **80**, 681–690.

Ewing, B., and Green, P. (1998). Base-calling of automated sequencer traces using phred. II. Error probabilities. *Genome Res.* **8**, 186–194.

Fan, W. H., Woelfle, M. A., and Mosig, G. (1995). 2 copies of a DNA element, Wendy, in the chloroplast chromosome of *Chlamydomonas reinhardtii* between rearranged gene clusters. *Plant Mol. Biol.* **29**, 63–80.

Felsenstein, J. (1978). Cases in which parsimony or compatibility methods will be positively misleading. *Syst. Zool.* **27**, 401–410.

Gordon, D., Abajian, C., and Green, P. (1998). Consed: A graphical tool for sequence finishing. *Genome Res.* **8**, 195–202.

Goremykin, V., Hirsch-Ernst, K. I., Wolfl, S., and Hellwig, F. H. (2003a). Analysis of the *Amborella trichopoda* chloroplast genome sequence suggests that *Amborella* is not a basal angiosperm. *Mol. Biol. Evol.* **20**, 1499–1505.

Goremykin, V., Hirsch-Ernst, K. I., Wolfl, S., and Hellwig, F. H. (2003b). The chloroplast genome of the "basal" angiosperm *Calycanthus fertilis*—structural and phylogenetic analyses. *Plant Syst. Evol.* **242**, 119–135.

Goremykin, V., Hirsch-Ernst, K. I., Wolfl, S., and Hellwig, F. H. (2004). The chloroplast genome of *Nymphaea alba*: Whole-genome analyses and the problem of identifying the most basal angiosperm. *Mol. Biol. Evol.* **21**, 1445–1454.

Graham, S. W., and Olmstead, R. G. (2000). Utility of 17 chloroplast genes for inferring the phylogeny of the basal angiosperms. *Am. J. Botany* **87**, 1712–1730.

Hipkins, V. D., Marshall, K. A., Neale, D. B., Rottmann, W. H., and Strauss, S. H. (1995). A mutation hotspot in the chloroplast genome of a conifer (Douglas-Fir, *Pseudotsuga*) is caused by variability in the number of direct repeats derived from a partially duplicated transfer-RNA gene. *Curr. Genet.* **27**, 572–579.

Howe, C. J. (1985). The endpoints of an inversion in wheat chloroplast DNA are associated with short repeated sequences containing homology to att-lambda. *Curr. Genet.* **10**, 139–145.

Hupfer, H., Swaitek, M., Hornung, S., Herrmann, R. G., Maier, R. M., Chiu, W.-L., and Sears, B. (2000). Complete nucleotide sequence of the *Oenothera elata* plastid chromosome, representing plastome 1 of the five distinguishable *Euoenthera* plastomes. *Mol. Gen. Genet.* **263**, 581–585.

Jansen, R. K., and Palmer, J. D. (1987). A chloroplast DNA inversion marks an ancient evolutionary split in the sunflower family (Asteraceae). *Proc. Natl. Acad. Sci. USA* **84**, 5818–5822.

Jansen, R. K., Wee, J. L., and Millie, D. (1998). Comparative utility of restriction site and DNA sequence data for phylogenetic studies in plants. *In* "Molecular Systematics of Plants II: DNA Sequencing" (D. E. Soltis, P. S. Soltis, and J. J. Doyle, eds.), pp. 87–100. Chapman and Hall, NY.

Knox, E. B., and Palmer, J. D. (1998). Chloroplast DNA evidence on the origin and radiation of the giant lobelias in eastern Africa. *Syst. Botany* **23**, 109–149.

Kolodner, R., and Tewari, K. K. (1979). Inverted repeats in chloroplast DNA from higher plants. *Proc. Natl. Acad. Sci. USA* **76**, 41–45.

Kugita, M., Kaneko, A., Yamamoto, Y., Takeya, Y., Matsumoto, T., and Yoshinaga, K. (2003). The complete nucleotide sequence of the hornwort (*Anthoceros formosae*) chloroplast genome: Insight into the earliest land plants. *Nucleic Acids Res.* **31**, 716–721.

Kuroda, H., and Maliga, P. (2002). Overexpression of the *clpP* 5′-untranslated region in a chimeric context causes a mutant phenotype, suggesting competition for a *clpP*-specific RNA maturation factor in tobacco chloroplasts. *Plant Physiol.* **129**, 1600–1606.

Kurtz, S., Choudhuri, J. V., Ohlebusch, E., Schleiermacher, C., Stoye, J., and Giegerich, R. (2001). REPuter: The manifold applications of repeat analysis on a genomic scale. *Nucleic Acids Res.* **29**, 4633–4642.

Kurtz, S., and Schleiermacher, C. (1999). REPuter: Fast computation of maximal repeats in complete genomes. *Bioinformatics* **15**, 426–427.

Larget, B., Simon, D. L., and Kadane, J. B. (2002). Bayesian phylogenetic inference from animal mitochondrial genome arrangements. *J. R. Stat. Soc. Series B Stat. Method* **64**, 681–693.

Leebens-Mack, J. H., Raubeson, L. A., Cui, L., Kuehl, J. V., Fourcade, M., Chumley, T. W., Boore, J. L., Jansen, R. K., dePamphilis, C. W. (submitted). Identifying the basal angiosperm lineages in chloroplast genome phylogenics: Sampling one's way out of the Felsenstein zone. *Mol. Biol. Evol.*

Lemieux, C., Otis, C., and Turmel, M. (2000). Ancestral chloroplast genome in *Mesostigma viride* reveals an early branch of green plant evolution. *Nature* **403,** 649–652.

Lockhart, P. J., Howe, C. J., Barbrook, A. C., Larkum, A. W. D., and Penny, D. (1999). Spectral analysis, systematic bias, and the evolution of chloroplasts. *Mol. Biol. Evol.* **16,** 573–576.

Maier, U. G., Rensing, S. A., Igloi, G. L., and Maerz, M. (1995). Twintrons are not unique to the *Euglena* chloroplast genome—structure and evolution of a plastome *cpn60* Gene from a cryptomonad. *Mol. Gen. Genet.* **246,** 128–131.

Martin, W., Rujan, T., Richly, E., Hansen, A., Cornelsen, S., Lins, T., Leister, D., Stoebe, B., Hasegawa, M., and Penny, D. (2002). Evolutionary analysis of *Arabidopsis*, cyanobacterial, and chloroplast genomes reveals plastid phylogeny and thousands of cyanobacterial genes in the nucleus. *Proc. Natl. Acad. Sci. USA* **99,** 12246–12251.

Martin, W., Stoebe, B., Goremykin, V., Hansmann, S., Hasegawa, M., and Kowallik, K. V. (1998). Gene transfer to the nucleus and the evolution of chloroplasts. *Nature* **393,** 162–165.

Mathews, S., and Donoghue, M. J. (1999). The root of angiosperm phylogeny inferred from duplicate phytochrome genes. *Science* **286,** 947–950.

Matsuoka, Y., Yamazaki, Y., Ogihara, Y., and Tsunewaki, K. (2002). Whole chloroplast genome comparison of rice, maize, and wheat: Implications for chloroplast gene diversification and phylogeny of cereals. *Mol. Biol. Evol.* **19,** 2084–2091.

Maul, J. E., Lilly, J. W., Cui, L., dePamphilis, C. W., Miller, W., Harris, E. H., and Stern, D. B. (2002). The *Chlamydomonas reinhardtii* plastid chromosome: Islands of genes in a sea of repeats. *Plant Cell* **14,** 1–22.

McNeal, J. R., Leebens-Mack, J. H., dePamphilis, C. W. (submitted). Utilization of fosmid genomic libraries for sequencing complete organellar genomes. *Biotechniques.*

Milligan, B. G., Hampton, J. N., and Palmer, J. D. (1989). Dispersed repeats and structural reorganization in subclover chloroplast DNA. *Mol. Biol. Evol.* **6,** 355–368.

Miyagi, T., Kapoor, S., Sugita, M., and Sugiura, M. (1998). Transcript analysis of the tobacco plastid operon *rps2/ atpI/H/F/A* reveals the existence of a non-consensus type II (NCII) promoter upstream of the *atpI* coding sequence. *Mol. Gen. Genet.* **257,** 299–307.

Moret, B. M. E., Wang, L. S., Warnow, T., and Wyman, S. K. (2001). New approaches for reconstructing phylogenies from gene order data. 9th International Conference on Intelligent Systems in Molecular Biology, Copenhagen, Denmark.

Morton, B. R. (1994). Codon use and the rate of divergence of land plant chloroplast genes. *Mol. Biol. Evol.* **11,** 231–238.

Morton, B. R., and So, B. G. (2000). Codon usage in plastid genes is correlated with context, position within the gene, and amino acid content. *J. Mol. Evol.* **50,** 184–193.

Mullet, J. E., Christopher, D., Rapp, J., Meng, B. Y., Kim, M. Y., Kim, J. M., and Laflamme, D. (1992). Dynamic regulation of plastid gene expression during chloroplast biogenesis. *Photosynth. Res.* **34,** 94–94.

Muse, S. V., and Gaut, B. S. (1997). Comparing patterns of nucleotide substitution rates among chloroplast loci using the relative ratio test. *Genetics* **146,** 393–399.

Ohyama, K., Fukuzawa, H., Kohchi, T., Shirai, H., Sano, T., Sano, S., Umesono, K., Shiki, Y., Takeuchi, M., Chang, Z., Aota, S., Inokuchi, H., and Ozeki, H. (1986). Chloroplast gene organization deduced from complete sequence of Liverwort *Marchantia-Polymorpha* chloroplast DNA. *Nature* **322,** 572–574.

Olmstead, R. G., and Palmer, J. D. (1994). Chloroplast DNA systematics—a review of methods and data analysis. *Am. J. Botany* **81,** 1205–1224.

Palmer, J. D. (1986). Isolation and structural analysis of chloroplast DNA. *Meth. Enzymol.* **118,** 167–186.

Palmer, J. D. (1991). Plastid chromosomes: Structure and evolution. *In* "The Molecular Biology of Plastids. Cell Culture and Somatic Cell Genetics of Plants" (R. G. Hermann, ed.), 7A, pp. 5–53. Springer–Verlag, Vienna.

Palmer, J. D., Nugent, J. M., and Herbon, L. A. (1987). Unusual structure of Geranium chloroplast DNA—a triple-sized inverted repeat, extensive gene duplications, multiple inversions, and 2 repeat families. *Proc. Natl. Acad. Sci. USA* **84,** 769–773.

Parkinson, C. L., Adams, K. L., and Palmer, J. D. (1999). Multigene analyses identify the three earliest lineages of extant flowering plants. *Curr. Biol.* **9,** 1485–1488.

Pombert, J. F., Otis, C., Lemieux, C., and Turmel, M. (2004). The complete mitochondrial DNA sequence of the green alga *Pseudendoclonium akinetum* (Ulvoplayceae) highlights distinctive evolutionary trends in the chlorophyta and suggests a sister–group relationship between the Ulvophyceae and Chlorophyceae. *Mol. Biol. Evol.* **21,** 922–935.

Qiu, Y. L., Lee, J. H., Bernasconi-Quadroni, F., Soltis, D. E., Soltis, P. S., Zanis, M., Zimmer, E. A., Chen, Z. D., Savolainen, V., and Chase, M. W. (1999). The earliest angiosperms: Evidence from mitochondrial, plastid and nuclear genomes. *Nature* **402,** 404–407.

Raubeson, L. A., and Jansen, R. K. (1992). Chloroplast DNA evidence on the ancient evolutionary split in vascular land plants. *Science* **255,** 1697–1699.

Raubeson, L. A., and Jansen, R. K. (2005). Chloroplast genomes of plants. *In* "Plant Diversity and Evolution: Genotypic and Phenotypic Variation in Higher Plants" (R. J. Henry, ed.), pp.45–68. CAB International, Wallingford, UK.

Rivas, R. D., Lozano, J. J., and Ortiz, A. R. (2002). Comparative analysis of chloroplast genomes: Functional annotation, genome-based phylogeny, and deduced evolutionary patters. *Genome Res.* **12,** 567–583.

Sandbrink, J. M., Vellekoop, P., Vanham, R., and Vanbrederode, J. (1989). A method for evolutionary studies on RFLP of chloroplast DNA, applicable to a range of plant species. *Biochem. Syst. Ecol.* **17,** 45–49.

Schwartz, S., Elnitski, L., Li, M., Weirauch, M., Riemer, C., Smit, A., Program, N. C. S., Green, E. D., Hardison, R. C., and Miller, W. (2003). MultiPipMaker and supporting tools: Alignments and analysis of multiple genomic DNA sequences. *Nucleic Acids Res.* **31,** 3518–3524.

Sears, B. B., Stoike, L. L., and Chiu, W. L. (1996). Proliferation of direct repeats near the *Oenothera* chloroplast DNA origin of replication. *Mol. Biol. Evol.* **13,** 850–863.

Shinozaki, K., Ohme, M., Tanaka, M., Wakasugi, T., Hayashida, N., Matsubayashi, T., Zaita, N., Chunwongse, J., Obokata, J., Yamaguchishinozaki, K., Ohto, C., Torazawa, K., Meng, B. Y., Sugita, M., Deno, H., Kamogashira, T., Yamada, K., Kusuda, J., Takaiwa, F., Kato, A., Tohdoh, N., Shimada, H., and Sugiura, M. (1986). The complete nucleotide sequence of the tobacco chloroplast genome–its gene organization and expression. *EMBO J.* **5,** 2043–2049.

Simpson, C. L., and Stern, D. B. (2002). The treasure trove of algal chloroplast genomes. Surprises in architecture and gene content, and their functional implications. *Plant Physiol.* **129,** 957–966.

Soltis, D. E., and Soltis, P. S. (2004). *Amborella* not a "basal angiosperm"? Not so fast. *Am. J. Botany* **91,** 997–1001.

Soltis, P. S., Soltis, D. E., and Chase, M. W. (1999). Angiosperm phylogeny inferred from multiple genes as a tool for comparative biology. *Nature* **402,** 402–404.

Sugiura, M. (1992). The chloroplast genome. *Plant Mol. Biol.* **19,** 149–168.

Sugiura, M. (2003). History of chloroplast genomics. *Photosynth. Res.* **76,** 371–377.

Tang, J. J., and Moret, B. M. E. (2003). Phylogenetic reconstruction from gene-rearrangement data with unequal gene content. *In* "Proceedings of the eighth workshop on Algorithms and Data Structures, Proceedings" 2748, pp. 37–46.

Triboush, S. O., Danilenko, N. G., and Davydenko, O. G. (1998). A method for isolation of chloroplast DNA and mitochondrial DNA from sunflower. *Plant Mol. Biol. Reporter* **16,** 183–189.

Turmel, M., Otis, C., and Lemieux, C. (1999). The complete chloroplast DNA sequence of the green alga Nephroselmis olivacea: Insights into the architecture of ancestral chloroplast genomes. *Proc. Natl. Acad. Sci. USA* **96,** 10248–10253.

Volfvovsky, N., Hass, B. J., and Salzberg, S. L. (2001). A clustering method for repeat analysis in DNA sequences. *Genome Biol.* **2,** 1–11.

Vomstein, J., and Hachtel, W. (1988). Deletions, insertions, short inverted repeats, sequences resembling att-lambda, and frame shift mutated open reading frames are involved in chloroplast DNA differences in the genus *Oenothera* subsection *Munzia. Mol. Gen. Genet.* **213,** 513–518.

Wakasugi, T., Tsudzuki, J., Ito, T., Nakashima, K., Tsudzuki, T., and S. M. (1994). Loss of all *ndh* genes as determined by sequencing the entire chloroplast genome of the black pine *Pinus thunbergii. Proc. Natl. Acad. Sci. USA* **91,** 9794–9798.

Wang, L.-S., Jansen, R. K., Moret, B. M. E., Raubeson, L. A., and Warnow, T. (2002). Fast phylogenetic methods for analysis of genome rearrangement data: An empirical study. Proceedings of the 7th Pacific Symposium Biocomputing PSB 2002. Lihue, Hawaii.

Wolfe, K. H., Li, W. H., and Sharp, P. M. (1987). Rates of nucleotide substitution vary greatly among plant mitochondrial, chloroplast, and nuclear DNAs. *Proc. Natl. Acad. Sci. USA* **84,** 9054–9058.

Wyman, S. K., Jansen, R. K., Boore, J. L. (2004). Automatic annotation of organellar genomes with DOGMA. *Bioinformatics* **20,** 3252–3255.

[21] Construction of Bacterial Artificial Chromosome Libraries for Use in Phylogenetic Studies

By Andrew G. McCubbin and Eric H. Roalson

Abstract

Bacterial artificial chromosome (BAC) libraries are emerging as valuable tools for investigating phylogenetic relationships at the level of genome structure. To date, BAC library construction has been restricted to a fairly small number of laboratories and species that represent a not insignificant, but a fairly small, fraction of diversity in the plant kingdom. This chapter is intended to contribute to rectifying this situation by providing protocols that facilitate BAC library construction in laboratories possessing basic molecular biology skills.

Introduction

The development and deployment of methodologies for genomic research is providing data that are greatly expanding our insight into the nature of changes associated with the origin and diversification of taxa. Studies at levels ranging from single genes to entire genomes have provided information on interspecific differences in gene number and location (Bernacchi and Tanksley, 1997; Bradshaw *et al.*, 1998; Westerbergh and Doebley, 2002), the number and interactions of genes involved in reproductive isolation (Reiseberg *et al.*, 1996; Wu and Hollocher, 1998), and synteny between diverse genera, which reveals a surprising degree of structural conservation (Gale and Devos, 1998; Ku *et al.*, 2000; Paterson *et al.*, 1996; Tanksley *et al.*, 1988, 1992). Extending this research to changes in chromosomal structure is an emerging facet of genomics (Nadeau and Snakoff, 1998; Paterson *et al.*, 2000), which promises to provide an avenue to test models for reproductive isolation arising from genetic factors associated with karyotypic alterations (Noor *et al.*, 2001; Reiseberg, 2001). Data generated from such studies will improve our understanding of the mechanisms by which diversification and speciation are initiated at the genomic level.

Comprehensive, stable large-insert (>100-kb) DNA libraries are essential for the physical mapping and molecular analysis of complex eukaryotic genomes. Initial efforts focused on yeast artificial chromosomes (YACs) (Burke *et al.*, 1987), but for several reasons, bacterial artificial chromosomes (BACs) (Shizuya *et al.*, 1992) have become the method of choice. The ease of construction and manipulation of BAC relative to YAC libraries, combined with the fact that BACs do not suffer from the problems of deletions or chimeras associated with YACs, has contributed to their widespread adoption in the scientific community. BACs are not actually artificial chromosomes but are based on modified bacterial F-factors. Potentially, these vectors are capable of carrying inserts approaching 500 kb, but for largely technical reasons, BAC libraries generally have insert sizes of 80–200 kb. Libraries have now been constructed for a considerable number of crop species and their wild relatives (e.g., see http://hbz.tamu.edu) and have become an integral part of genome mapping, sequencing, and gene identification.

Although the number of BAC libraries available is sizable, these resources represent only a small fraction of taxonomic diversity in the plant kingdom and it is frequently desirable to develop new BAC library resources. In most cases, published protocols are not sufficiently detailed to enable research groups without previous experience in BAC library construction to easily adopt these techniques, with one recent notable exception (Peterson *et al.*, 2000). This chapter is intended to contribute to rectifying this situation, particularly with regard to protocol modifications to address additional challenges faced when using wild species for which inbred lines are not available and plant material may be limited.

General Considerations

Vector Choice

A considerable number of vectors are now available for constructing BAC libraries, most possessing standard plasmid selection features such as antibiotic resistance and a polycloning site within a reporter gene (allowing insertional inactivation) (for a review of BAC vectors, see Choi and Wing, 1999). Probably the most widely used vector has been pBeloBAC 11 (Shizuya *et al.*, 1992; available from New England Biolabs), which allows blue/white selection, although color development is not always distinct even after extended incubation. Several vectors with enhanced blue/white selection have been developed, such as TrueblueBacII (Genomics One) and pIndigoBac-5 (Epicenter), which significantly improve selection of insert-containing clones and in addition are commercially available predigested and dephosphorylated. A recent development in vector technology—that of conditionally amplifiable BACs—provides additional advantages. Through the incorporation of a conditional and tightly controlled *oriV*/TrfA amplification system, Wild *et al.* (2002) have developed a vector that is maintained as single copy but can be conditionally induced by L-arabinose to a yield of about 100 copies/cell in *Escherichia coli*. This feature provides stability in a single-copy state for library maintenance but has the considerable benefit of allowing the induction of high copy number, leading to increased yield and quality in downstream applications. This technology has been combined with enhanced blue/white selection and is available commercially under the name CopyControl pCC1BAC (Epicentre).

The preparation of high-quality digested, dephosphorylated vector DNA is a critical step in BAC library construction. For laboratories experienced in molecular protocols, this is certainly attainable and we provide a protocol later in this chapter. However, commercial sources of cloning-ready vectors are available and are worth considering. We tested products from several sources and found cloning-ready pIndigoBAC-5 and Copy-Control pCC1BAC vectors (Epicentre) to be consistently effective and high quality (McCubbin, unpublished data, 2004).

Plant Material

The quantity of leaf material required to construct a BAC library depends largely on genome size, a rough guide being about 30 g for species with genomes of more than 2000 Mb to about 100 g for species with genomes less than 200 Mb (the smaller the genome, the more nuclei required to generate sufficient DNA for cloning). The ideal starting material for BAC library construction is a genetically uniform population of young preflowering plants, which have been dark treated for 3 days to reduce starch content. This is easily attainable if inbred lines are available, but for wild species, this would require multiple generations of inbreeding before library construction. A more rapid alternative applicable to many species is to bulk up particular lines by clonal propagation and to harvest leaves from mature plants. This renders purification of high-quality large DNA more challenging because of the increased levels of secondary compounds, particularly polyphenolics, but is achievable using some of the recently developed protocols, which incorporate polyvinylpyrrolidone (PVP) and antioxidants during nuclei isolation and processing (Peterson *et al.*, 2000).

Library Size and Enzyme Choice

The number of clones (N) required for a particular library depends on the desired genome coverage (C), the genome size (G) of the species in question, and the average insert size of the library (I), the relationship between them being represented by $C = (N \times I)/G$. Clearly, it is desirable to have comprehensive genome coverage. Theoretically, the probability (P) of any particular piece of the genome being represented in the library is determined by the formula $N = \ln(1 - P)/\ln(1 - [I/G])$, three times genome coverage providing approximately 95% probability of representation and five times coverage 99%. In reality, restriction sites are not evenly distributed throughout genomes and partial digestion is not completely random, so it is advisable to generate libraries that are as large as is technically feasible, with 7.5–10 times genome coverage generally being adequate to provide a high probability of finding any sequence of interest.

The restriction enzyme used to construct the library should ideally cut the genomic DNA randomly at 40–60 kb intervals. The optimum enzyme varies according to species, but those most frequently employed are *Bam*H I, *Eco*R I, and *Hin*d III. The protocols in this chapter are described using *Hin*d III but can readily be adapted for other enzymes simply by substituting the appropriate restriction enzyme buffers.

Specialist Equipment Required

> *CHEF gel apparatus*: For pulse field gel (PFG) separation of high-molecular-weight DNA, the most commonly used units are the CHEF-DRII, CHEF-DRIII, or CHEF-MAPPER-XA (Bio-Rad), attached to a chiller and with associated gel casting stands, combs, and so on.
>
> *Electroporation apparatus*: Several systems are commercially available. The protocols in this text are designed for use with the Gene Pulser II (Bio-Rad).
>
> *Multichannel pipette*: An 8-, 12-, or 16-channel pipette for dispensing media into 384-well plates.
>
> *384-well handheld replicator*: Stainless-steel (reusable) versions are most economical and are available from several companies including V&P Scientific, Nunc, and Genetix.

Methods

Vector Isolation and Preparation

1. Prepare vector DNA from two 500-ml *E. coli* cultures grown with the appropriate antibiotic selection, using a NucleoBond BAC Maxi Kit (BD Biosciences) according to the manufacturer's instructions. Check that the vector is free from contamination by *E. coli* genomic DNA by electrophoresis on a 0.7% agarose gel before proceeding.

2. Digest 10 μg of plasmid with 30 units of *Hin*d III (New England Biolabs) in NEB buffer B with the addition of spermidine to 2 mM (final concentration) for 2 h at 37°, followed by addition of a further 10 units and incubation for an additional hour.

3. Treat digested plasmid DNA with calf intestinal alkaline phosphatase (Promega) (0.1 U/mole of vector DNA), for 30 min at 37°. Stop the reaction by adding 2 μl of 0.5 M ethylenediaminetetraacetic acid (EDTA), 5 μl of 20% sodium dodecyl sulfate (SDS) and 20 μl of 1 mg/ml proteinase K (Fisher BioReagents) and incubate at 56° for 30 min.

4. Extract the reaction mix with an equal volume of phenol (preequilibrated with Tris–HCl pH 8.0). Partition by centrifugation at 14,000g for 10 min and remove the aqueous phase to a fresh tube. Add 1/10 volume of 3 M sodium acetate (pH 5.2) and 2 volumes of 95% ethanol, mix thoroughly, and incubate at −20° for at least 2 h. Pellet the precipitated vector DNA by centrifuging at 14,000g for 10 min and remove the supernatant. Carefully wash the pellet with 70% ethanol and air-dry. Resuspend vector DNA in molecular biology grade (MBG) water at a final concentration of ∼25 ng/μl.

Assay of Vector Quality

As dephosphorylation reactions are somewhat difficult to control, it is important to determine whether the vector DNA prepared is of sufficient quality for library construction. This is generally achieved through a test ligation with λ-phage restriction fragments. It is not a good idea to perform this test ligation with DNA from which you plan to construct the library, because if the ligation fails, it will be unclear whether the fault lies with the vector or the insert DNA.

1. Set up a ligation reaction as described below:
   Ligation reaction:
       20 ng *Hin*d III cut dephosphorylated vector DNA
       10 μl 10X T4 DNA ligase buffer
       200 ng *Hin*d III digested λ DNA
       MBG water to give a final reaction volume of 98 μl
       2 μl T4 DNA ligase (6 units)
2. Gently tap each reaction tube to mix the tube's contents.
3. Incubate the ligation reactions at 4° overnight.

Note: Ligated DNA is stable at 4° for 1–2 wk.

E. coli Transformation: Electroporation

1. Before electroporation, desalt each ligation mix by drop dialysis using a 0.025-μm VSWP filter (Millipore, Bedford, MA). Float the filter, shiny surface upwards, in a Petri plate containing 30 ml of sterile MBG water, apply the ligation mixture to the center of the filter using a wide-bore pipette tip and incubate for 45 min, after which carefully retrieve the sample droplet from the filter surface and place on ice (desalted ligation mixes can be stored at 4°, *do not freeze*).

2. Electroporation is carried out using electrocompetent GeneHog cells (Invitrogen) or equivalent and a Gene Pulser II (Bio-Rad). Each electroporation consists of 2.2 μl of ligation mix and 20 μl of cells, with the conditions of 1.8 kV, 200 Ω, 25 μF (using a 0.1-cm gap electroporation cuvette).

3. After electroporation, add 600 μl SOC media (Invitrogen) directly to the cuvette and mix gently by pipetting up and down. Withdraw the cell suspension and place in a sterile culture tube.

4. Incubate at 37° and 250 rpm for 45 min to 1 h (but no longer) to allow the cells to recover. After incubation, place 50 ml onto prewarmed Luria Broth (LB) agar plates supplemented with the appropriate antibiotic, X-gal (40 mg/L) and IPTG (0.4 m*M*). Use a glass plating rod to spread the culture, moving the rod across the agar until all the liquid has been absorbed.

5. Incubate plates for about 20 h at 37°. Colonies should appear within 15 h, and reach a diameter of 1–2 mm by 20 h.

6. Determine the number of colonies per plate and the percentage of clones that are blue. Determine the number of CM-resistant bacteria (colony forming units = cfu) per microliter of liquid culture.

Interpreting the Results. A good vector preparation should yield more than 2 cfu/μl with less than 10% blue colonies. A low yield of colonies may be indicative of excessive dephosphorylation of the vector (causing damage to the overhanging ends), and a high percentage of blue colonies is indicative of either insufficient digestion or dephosphorylation of vector.

Preparation of DNA Inserts

Nuclei Isolation. Nuclei isolation is based on that of Liu and Whittier (1994) incorporating modifications made by Peterson *et al.* (2000), in particular the addition of PVP-40 and antioxidants that function to reduce the risks of oxidative cross-linking by polyphenolics during the isolation. We have found this relatively simple procedure to be effective for both monocot and dicot species (*Carex, Petunia,* and *Primula*), although it should be noted that Peterson *et al.* (2000) suggest that an alternative, rather more complex, procedure may be more suitable for some dicot species but requires considerably more leaf material.

1. Grind the tissue to a fine powder in liquid nitrogen using a mortar and pestle, then leave at room temperature for 10 min. Add the powder to sucrose extraction buffer (SEB: 10 mM Tris–HCl [pH 9.5], 100 mM KCl, 10 mM EDTA, 0.5 M sucrose, 4 mM spermidine, 1 mM spermine, 0.1 mM ascorbic acid, 0.13 mM sodium diethyldithiocarbamate, 0.1% β-mercaptoethanol), using 100 ml of SEB for every 10 g of starting leaf material, and stir gently with a plastic spatula. Incubate on ice for 15 min, stirring gently every 2 min.

2. Filter the homogenate sequentially through a single layer of Miracloth (previously soaked in 0.5 M EDTA and autoclaved) and a double layer of pretreated Miracloth.

3. Slowly add 1/20 volume of 10% Triton X-100 in SEB while gently swirling (to promote chloroplast lysis).

4. Transfer to 50-ml polypropylene tubes and pellet nuclei by centrifugation at 650*g* for 15 min at 4°.

5. Gently resuspend nuclei pellets in 10 ml of SEB, combine into a single tube, and bring the volume to about 50 ml.

6. Filter suspension through 20 μm nylon mesh (this step is optional but significantly reduces the amount of cell wall debris in the final

preparation). This is most easily achieved by using 20-μm nylon mesh disk (Millipore) and a screened cap for 50-ml polypropylene tubes (Bio-Rad). Cut off the bottom of a 50-ml polypropylene tube, place the nylon mesh on top of the screened cap, and screw the tube tightly into the screened cap, creating a supported mesh filter apparatus with the cut-off tube acting as a reservoir. Pour the nuclei suspension through the filter apparatus into a fresh 50-ml tube (gently tap the filter apparatus or replace the mesh filter if it becomes clogged). Pellet nuclei by centrifugation at 650g for 15 min at 4°.

7. Remove supernatant and resuspend pellets in SEB to a final volume of about 1.2 ml.

8. Equilibrate nuclei suspension to 45° in a waterbath (5 min) and then add an equal volume of 1.5% low melting point agarose (Incert agarose, FMC) (prepared in SEB and equilibrated to 42°). Immediately gently transfer the mixture into 200-μl reusable plug molds (Bio-Rad) using a wide-bore 1-ml pipette tip, wrap the molds in Saran Wrap and chill on ice for 30 min.

9. Transfer the solidified agarose plugs to a 50-ml polypropylene tube and add 50 ml of Lysis buffer (LyB) (1% sarcosyl, 0.25 M EDTA pH 8.0, 2% PVP-40, 0.1% ascorbic acid, 0.13% sodium diethyldithiocarbamate, 0.1 mg/ml proteinase K). Incubate at 50° for 48 h with one change of lb (after 24 h).

10. Decant LyB and replace with 50 ml of wash buffer A (0.1% w/v ascorbic acid, 2.0% w/v PVP-40, 0.13% w/v sodium diethyldithiocarbamate dissolved in 0.5 M EDTA [pH 9.1]). Incubate at 50° for 1 h.

11. Decant the supernatant and replace with 50 ml of wash buffer B (0.1% w/v ascorbic acid, 2.0% w/v PVP-40, 0.13% w/v sodium diethyldithio-carbamate dissolved in 50 mM EDTA [pH 8.0]). Incubate at 50° for 1 h.

12. Repeat the process with 50 ml of wash buffer C (WBC: 0.1% w/v ascorbic acid, 2.0% w/v PVP-40, and 0.13% w/v sodium diethyldithiocar-bamate, 10 mM EDTA, 10 mM Tris–HCl [pH 8.0]).

13. Remove this buffer and wash twice with 50 ml of fresh WBC supplemented with 50 μl of 0.1 M phenylmethylsulfonyl fluoride (PMSF). Incubate at 50° for 1 h per wash (to inactivate residual proteinase K).

14. Decant the solution and wash four times with 50 ml of WBC at 50° for 1 h per wash (any of these washes can be carried out overnight at 4°, if desired). Store in WBC at 4°.

Partial Digestion

1. Before restriction enzyme digestion, buffer exchange two whole plugs into 1× restriction enzyme buffer B (New England Biolabs) supple-mented with 0.2 mg/ml BSA, 2 mM spermidine, and 1 mM DTT (REBS). Wash twice with 20 ml of buffer for 30 min/wash on ice.

2. Using a glass coverslip, cut each plug into six pieces and transfer nine of these pieces each to a 1.5-ml microfuge tube on ice. Label the tubes 1–9 and add 250 μl of REBS to each tube.

3. Starting at tube no. 2, add a specific quantity of *Hin*d III to each tube on ice, adding 2, 5, 10, 20, 40, 80, 160, and 200 units of enzyme to tubes 2–9 respectively (Fig. 1). *Note*: Use a 1:10 dilution of the *Hin*d III stock (New England BioLabs) on ice in 1× REBS for 20 units and below to ensure accuracy.

4. Gently mix the contents of each tube and incubate tubes on ice for 2 h to allow the enzyme to diffuse into the plugs.

5. Place the tubes in a 37° waterbath and incubate for *exactly* 30 min, after which immediately place on ice and add 30 μl of 0.5 M EDTA (pH 8.0) to each tube (to inhibit further enzyme activity). Agitate gently and keep on ice.

6. Make a 1.0% agarose gel in 0.5× TBE using the small Bio-Rad CHEF gel casting stand and a 15-tooth gel comb.

7. Place a high-molecular-weight DNA marker in the first well of the gel, then transfer plugs into consecutive wells and seal each with molten 0.5% TBE low-melting point agarose (preequilibrated to 42°). Run the gel using the following parameters: buffer temperature = 12°, volts/cm = 6.0, included angle = 120 degrees, initial switch time = 1.0 s, final switch time = 40.0 s, ramping = linear, running time = 18 h.

8. Place the gel in a staining dish, cover with MBG water, add 10 μl of ethidium bromide (10 mg/ml), and rotate slowly for 20 min. Examine the gel on an ultraviolet (UV) transilluminator, and photograph. Figure 1 illustrates how the optimal enzyme concentration is determined.

Interpreting the Results. The bulk of the DNA in the undigested sample should be more than 600 kb in length and lie in the compression zone, with no more than a faint smear of smaller fragments visible. If the DNA is smaller than this, it has degraded during either processing or storage and is not suitable for library construction. In the other samples, fragment size should be seen to decrease with increasing enzyme concentration (see Fig. 1); if not, several possibilities should be investigated: (1) The *Hin*d III is inactive; (2) the DNA is not digestible; (3) one or more reagents is inactive/contaminated; or (4) EDTA was not sufficiently washed from the plugs. The optimal size range of restriction fragments for cloning is 100–350 kb, and one or more of the partial digest conditions should produce DNA fragments with a mean length in this range. Those conditions that produce the largest percentage in the desired range will be used to perform a large-scale digest.

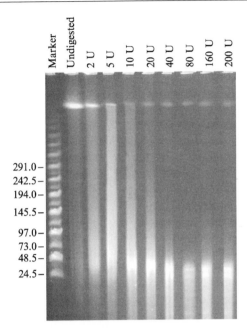

FIG. 1. Pulse field gel separation of partial digests of high-molecular-weight DNA prepared from *Carex lupulina*. Nuclei plugs were prepared and subjected to partial digestion as described in the text. Partial digests were carried out using *Hin*d III, the number of units of enzyme used in each treatment is indicated above each lane. The lane in which 5 units of enzyme was used for partial digestion can be seen to generate the most fragments in the 100–350 kb range and for this reason was selected for bulk digestions.

Size Fractionation

1. After determination of optimal digestion conditions, buffer exchange six to eight whole plugs (equivalent to 1.2–1.6 ml) into REBS and digest as described above with the conditions chosen, and cut each plug in half (along the long axis), rather than in six pieces as in the previous test.

2. Cast a 100-ml 0.5× TBE SeaKem Gold agarose (FMC) PFG and cut a single preparative well using a glass coverslip. Well size is dependent on the volume of plugs used but should not be more than about 5–8 mm in height (increasing height leads to greater size variation at any point in the gel after electrophoresis). Stack the 1/2 plugs as compactly as possible into the well and place a suitable PFG marker (e.g., Mid Range PFG marker II, NEB) lane on each side. Seal with 1% low melting point agarose in 0.5× TBE (preequilibrated to 42°). Run the gel as described in step 7 of "Partial Digestion" (p. 392).

3. After electrophoresis, excise the edges of the gel (including the marker lanes and ~5 mm of the sample lanes) and stain with ethidium bromide. Visualize on a UV transilluminator and precisely measure the distance from the edge of the gel to each size marker.

4. Using these measurements, return to the unstained portion of the gel and excise size fractions of approximately 100–150 kb, 150–200 kb, and 200–250 kb, marking the orientation of each slice (slices can be stored at 4° in TE50 [10 mM Tris–HCl, 50 mM EDTA] if desired).

5. Fractionate gel slices a second time on a 0.5X TBE, LMP agarose gel (FMC) under identical conditions except that the excised gel slices are loaded in the opposite orientation (turned horizontally through 180°) to the original gel. Again, remove and stain the edges of each track with ethidium bromide.

Note: This second fractionation simultaneously removes DNA that migrated anomalously in the first separation and focuses the DNA back into a zone approximately equal to the size of the original preparative well, helping to maintain a DNA concentration suitable for cloning.

6. Place the stained gel edges on a UV transilluminator and measure the precise location of the most concentrated part of each sample with a ruler. Return to the unstained portion of the gel, precisely excise each sample from the gel, and store at 4° in TE50 (~10 ml) in sterile 15-ml polypropylene tubes.

Isolation of DNA from Agarose

Isolation of high-molecular-weight DNA from agarose can be achieved effectively by either electroelution or using β-agarase. In our experience, β-agarase digestion is highly effective and does not require additional equipment. β-Agarase is active in T4 DNA ligase buffer and performing this step in this buffer helps to reduce dilution of the DNA in the ligation reaction.

1. Wash the gel slices five times with 1× T4 DNA ligase buffer using approximately 5 ml/wash (30 mM Tris–HCl [pH 7.8], 10 mM MgCl$_2$, 10 mM DTT, 1 mM ATP) on ice, for 10 min/wash.
2. Decant the final wash, carefully removing as much liquid as possible using a pipette.
3. Melt the slices at 70° for 5 min and then quickly transfer the tubes to a 45° waterbath and equilibrate for 10 min.
4. Add 1 unit of β-agarase (New England Biolabs) for every 100 μl of agarose to each tube.

5. Gently mix the contents of each tube by gently stirring with a 200 μl pipette tip and incubate at 45° for 2 h. Store at 4°.

Quantitation of DNA

1. Prepare a 1% agarose submarine minigel in 1× TAE.

2. Place a 5-μl aliquot of each DNA sample in microcentrifuge tubes on ice and add 5 μl of MBG water and 1 μl of 10× DNA loading dye to each tube.

3. Make up a series of standards of uncut lambda DNA (2, 5, 7.5, 10, 20, and 40 ng), each in 10 μl, in separate 1.5-ml microcentrifuge tubes. Add 1.0 μl of 10× DNA loading dye to each.

4. Load the gel and run in 1× TAE buffer in an appropriate apparatus at 100 V for about 15 min. Based on comparison of the relative fluorescence in the sample and standard lanes, estimate the concentration of each sample (Fig. 2).

Note: In our experience, as long as the DNA concentration is more than 1.0 ng/μl, it can be successfully used for ligation. High-molecular-weight (insert) DNA is usually stable for at least a week at 4°, but it is recommended that test ligations are set up immediately.

Ligation

Optimal ligation conditions are somewhat variable, depending on both vector and insert DNA quality (plus some inherent error in visually estimating DNA concentrations). A 10:1 vector/insert molar ratio most frequently provides the best compromise between total clone number and the percentage of noninsert clones, but it is advisable to test vector/insert molar ratios of 5:1, 10:1, and 15:1 using small-scale ligations.

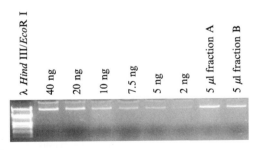

FIG. 2. Quantitation of genomic DNA after size fraction. Five-microliter samples of *Carex luplina* genomic DNA (fraction A = 100–150 kb; fraction B = 150–200 kb), compared with a series of undigested λ DNA standards from 40–2 ng as indicated. From this gel, fraction A was estimated to contain 40 ng (8 ng/μl) of DNA and fraction B about 20 ng (4 ng/μl).

1. Calculate how many nanograms of DNA are present in 27 μl of insert DNA, then based on the size (in kilobases) of the genomic DNA (average size of the fraction) and the vector, extrapolate to determine how much digested dephosphorylated vector DNA is required to give molar ratios of 5:1, 10:1, and 15:1 (vector:insert).

2. For each ligation condition, mix the size fractionated genomic DNA with the appropriate quantity of dephosphorylated *Hin*d III cut vector, heat to 60° for 10 min and then allow to cool to room temperature.

3. Add 1 μl of 10 mM ATP and 1 μl (3 units) of T4 DNA ligase and mix gently by stirring with a pipette tip. Incubate the ligation mix at 4° for 18 h.

4. Desalt each ligation mixture by drop dialysis, transform into *E. coli* by electroporation, and culture as described in the section "*E. coli* Transformation: Electroporation."

Note: Addition of supplemental ATP is to compensate for breakdown of ATP in the ligase buffer during β-agarase digestion.

Interpreting the Results. The limits of acceptability of a ligation mixture are somewhat dependent on the size of library under construction and how many electroporations one is prepared to perform. A good ligation can generate 1000 cfu/μl of ligation mixture, with less than 10% blue colonies. It is not uncommon for the figure to be closer to 250 cfu/μl of ligation mixture with about 20% blue colonies; for smaller libraries, this may still be acceptable for library construction.

1. Inoculate cultures with 20 white clones from each ligation, perform BAC minipreps and roughly assess the average insert size of the clones using the procedures in the section "Isolation of BAC DNA," later in this chapter.

2. Once optimum ligation conditions and average insert size have been established, perform a large volume ligation on a scale that will provide the number of clones desired for the library. Desalt the ligation mixture, electroporate into *E. coli* and plate as described earlier.

Library Arraying and Storage

1. Fill 384-well plates (Genetix) with 50 μl/well LB freezing buffer (36 mM K$_2$HPO$_4$, 13.2 mM KH$_2$PO$_4$, 1.7 mM sodium citrate, 0.4 mM MgSO$_4$, 6.8 mM (NH$_4$)SO$_4$, 4.4% glycerol, LB broth) supplemented with the appropriate antibiotic, using a multichannel pipette.

2. Pick the white colonies into 384-well plates. This can be carried out by hand with sterile toothpicks in a laminar flow hood or using one of various robots that have been designed for this purpose. If picking by

hand, it is advisable to leave the toothpicks in the well being inoculated until each entire row has been completed to avoid confusion.

3. Seal the plates with parafilm and incubate plates at 37° overnight. After the incubation, cool to room temperature and then store at −80°.

Note: It is highly advisable to make a duplicate copy of the library as soon as possible, and use one as a working copy and the other as a backup copy in a separate −80° freezer.

Library Replication

1. Fill the desired number of 384-well plates with LB freezing buffer supplemented with the appropriate antibiotic.
2. Sterilize a 384-pin stainless-steel replicator by dipping it sequentially into baths of 10% bleach, water, and 70% ethanol and finally flaming in a Bunsen burner. Allow the pins to cool for 2 min.
3. Carefully insert the replicator into the plate to be copied, remove, and insert into a fresh plate (taking care to maintain the orientation between the two plates).
4. Remove the replicator and repeat the process with the next plate.
5. Incubate freshly inoculated plates overnight at 37° and then store at −80°.

Isolation of BAC DNA

An accurate figure for genome coverage should be determined by assessing the average insert size and percentage of non-insert–containing clones in the library. The average insert size is determined by performing digests with a restriction enzyme that will either excise the insert from the vector or linearize each clone, with the 8-base cutter *Not* I being most frequently used. BAC minipreps should be prepared from at least 100 random insert–containing clones, digested, and analyzed on a 5–250 kb PFG separation. This will also provide an estimate of the number of non-insert–containing clones. BAC DNA is prepared using a modified alkaline lysis procedure. This can be performed using homemade solutions or with those from a variety of commercial sources, but we find that Wizard Plus Miniprep solutions (Promega) provide the most consistent results.

1. 2-ml cultures are grown in LB supplemented with 12.5 μg/L of the appropriate antibiotic for 16 h.

2. Transfer 1.7 ml of each culture into Microfuge tubes and pellet by centrifugation at 8000 rpm for 2 min. Pour off supernatant and repeat

centrifugation. Using a pulled Pasteur pipette (made by heating the end of the pipette in a Bunsen burner to red heat, then removing from the flame and pulling to create a fine tip), remove *all* residual media.

3. Resuspend cells in 200 μl of cell resuspension buffer (Wizard Plus Miniprep, Promega) by vortexing. Add 200 μl of lysis buffer, mix by gentle inversion, and incubate for 5 min at room temperature. Neutralize by adding 200 μl of neutralization buffer, again mix by gentle inversion, and centrifuge for 20 min at 16,000g (4°).

4. After centrifugation, transfer 570 μl of supernatant to a fresh tube (taking care to avoid the white flocculent material) and precipitate by adding 400 μl of isopropanol.

5. Pellet precipitated DNA by centrifugation for 15 min at 16,000g (4°) and decant supernatant. Spin again for 5 min at 16,000g (4°) and carefully remove *all* liquid with a pulled Pasteur pipette. Wash pellets with 750 μl of 70% ethanol and air-dry. Pellets can be resuspended directly in enzyme mix for subsequent digestion.

6. *Not* I digests are performed using 10 units of *Not* I per digest in the manufacturer's recommended buffer with the addition of 1 mM DTT, 100 μg/μl RNase A, 1 mM spermidine, and 100 μg/μl BSA (final concentrations). Incubate for at least 4 h at 37°.

7. Analyze *Not* I digests using 5–250 kb PFG separations (1% SeaKem Gold agarose gel, run at 6 V/cm, 0.1–22.0 s switching interval, 120-degree angle, 11–14° in 0.5× TBE, for 15 h).

Concluding Remarks

BAC libraries are a powerful resource for genome analysis. The largest hurdles to overcome are preparation of digested, dephosphorylated vector of sufficient quality and extraction of high-molecular-weight, readily digestable genomic DNA. The former now can be solved easily by purchasing a commercial vector preparation. By taking care to keep genomic DNA preparations in a reducing environment as described here, it should be possible to avoid the oxidative cross-linking that renders DNA undigestible. Following the protocols described, it should be possible for individual groups to generate BAC resources for most plant species. For groups working with animal species, nuclei preparation does not generally require a protocol as involved as that provided in this chapter, which is designed specifically for removal of plant cell walls and secondary compounds. A protocol for nuclei preparation from animal cells can be found in Osoegawa and de Jong (2004). The procedures downstream of nuclei preparation that we have provided are applicable to all species.

References

Bernacchi, D., and Tanksley, S. D. (1997). An interspecific backcross of *Lycopersicon esculentum* X *L. hirstutum*: Linkage analysis and a QTL study of sexual compatibility factors and floral traits. *Genetics* **147**, 861–877.

Bradshaw, H. D., Otto, K. G., Frewen, B. E., McKay, J. K., and Schemske, D. W. (1998). Quantitative trait loci affecting differences in floral morphology between two species of monkeyflower (*Mimulus*). *Genetics* **149**, 367–382.

Burke, D. T., Carle, G. F., and Olson, M. V. (1987). Cloning of large segments of exogenous DNA into yeast by means of artificial chromosome vectors. *Science* **236**, 806–812.

Choi, S., and Wing, R. A. (1999). The construction of bacterial artificial chromosome (BAC) libraries. *In* "Plant Molecular Biology Manual" (S. Gelvin and R. Schilperoort, eds.), 2nd Ed., Suppl. IV, pp. 1–32. Kluwer Academic Publishers, The Netherlands.

Gale, M. D., and Devos, K. M. (1998). Comparative genetics in the grasses. *Proc. Natl. Acad. Sci. USA* **95**, 1971–1974.

Ku, H.-M., Vision, T., Liu, J., and Tanksley, S. D. (2000). Comparing sequenced segments of the tomato and *Arabidopsis* genomes: Large-scale duplication followed by selective gene loss creates a network of synteny. *Proc. Natl. Acad. Sci. USA* **97**, 9121–9126.

Liu, Y. G., and Whittier, R. F. (1994). Rapid preparation of megabase plant DNA from nuclei in agarose plugs anad microbeads. *Nucleic Acids Res.* **22**, 2168–2169.

Nadeau, J. H., and Sankoff, D. (1998). Counting on comparative maps. *Trends Genet.* **14**, 495–501.

Noor, M. A. F., Grams, K. L., Bertucci, L. A., and Reiland, J. (2001). Chromosomal inversions and the reproductive isolation of species. *Proc. Natl. Acad. Sci. USA* **98**, 12084–12088.

Osoegawa, K., and de Jong, P. J. (2004). BAC library construction. *Methods Mol. Biol.* **255**, 1–46.

Paterson, A. H., Lan, T. H., Reischmann, K. P., Chang, C., Lin, Y. R., Liu, S. C., Burow, M. D., Kowalski, S. P., Katsar, C. S., DelMonte, T. A., Feldmann, K. A., Schertz, K. F., and Wedel, J. F. (1996). Toward a unified genetic map of higher plants, transcending the monocot-dicot divergence. *Nat. Genet.* **14**, 380–382.

Paterson, A. H., Bowers, J. E., Burow, M. D., Drayer, X., Elsik, C. G., Jiang, C. -X., Katsar, C. S., Lan, T.-H., Lin, Y.-R., Ming, R., and Wright, R. J. (2000). Comparative genomics of plant chromosomes. *Plant Cell* **12**, 1523–1539.

Peterson, D. G., Tomkins, J. P., Frisch, D. A., Wing, R. W., and Paterson, A. H. (2000). Construction of plant bacterial artificial chromosome (BAC) libraries: An illustrated guide. *J. Agr. Genomics* **5**, 1–100. http://www.ncgr.org/jag/.

Rieseberg, L. H. (2001). Chromosomal rearrangements and speciation. *Trends Ecol. Evol.* **16**, 351–358.

Rieseberg, L. H., Sinervo, B., Linder, C. R., and Arias, M. C. (1996). Role of gene interactions in hybrid speciation: Evidence from ancient and experimental hybrids. *Science* **272**, 741–745.

Shizuya, H., Birren, B., Kim, U.-J., Mancino, V., Slepak, T., Tachiiri, Y., and Simon, M. (1992). Cloning and maintenance of 300-kilobase-pair fragments of human DNA in *Escherichia coli* using an F-factor–based vector. *Proc. Natl. Acad. Sci. USA* **89**, 8794–8797.

Tanksley, S. D., Bernatzky, R., Lapitan, N. L., and Prince, J. P. (1988). Conservation of gene repertoire but not gene order in pepper and tomato. *Proc. Natl. Acad. Sci. USA* **85**, 6419–6423.

Tanksley, S. D., Ganal, M. W., Prince, J. P., Vicente, M. C. D., Bonierbale, M. W., Broun, P., Fulton, T. M., Giovannoni, J. J., Grandillo, S., and Martin, G. B. (1992). High density molecular linkage maps of the tomato and potato genomes. *Genetics* **132**, 1141–1160.

Westerbergh, A., and Doebley, J. (2002). Morphological traits defining species differences in wild relatives of maize are controlled by multiple quantitative trait loci. *Evolution* **56,** 273–283.

Wild, J., Hradecna, Z., and Szybalski, W. (2002). Conditionally amplifiable BACs: Switching from single-copy to high-copy vectors and genomic clones. *Genome Res.* **12,** 1434–1444.

Wu, C.-I., and Hollocher, H. (1998). Subtle is nature: The genetics of species differentiation and speciation. *In* "Endless Forms: Species and Speciation" (D. J. Howard and S. H. Berlocher, eds.), pp. 339–351. Oxford University Press, Oxford.

[22] Comparative EST Analyses in Plant Systems

By Qunfeng Dong, Lori Kroiss, Fredrick D. Oakley,
Bing-Bing Wang, and Volker Brendel

Abstract

Expressed sequence tag (EST) data are a major contributor to the known plant sequence space. Organization of the data into non-redundant clusters representing tentative unique genes provides snapshots of the gene repertoires of a species. This chapter reviews availability of sequences and sequence analysis results and describes several resources and tools that should facilitate broad-based utilization of EST data for gene structure annotation, gene discovery, and comparative genomics.

Introduction

Expressed sequence tags (ESTs) are generated by high-throughput single-pass sequencing of complementary DNA (cDNA) clones (Adams *et al.*, 1991). A cDNA library is a snapshot representation of the messenger RNA (mRNA) population within a given cell type, tissue, or organism, under a certain set of conditions at a particular moment. Over the past decade, ESTs have accumulated at an exponential rate and have become a major source of plant sequence data. Today more than 2 million plant-derived ESTs from various species are available at public databases. These data have provided a rich resource for gene discovery and annotation (Rudd, 2003). Additional value can be leveraged from these collections by comparing ESTs from multiple species (Fulton *et al.*, 2002; Vincentz *et al.*, 2004). Despite this potential, the resources available for comparative EST analyses in plants remain fairly limited and dispersed. The PlantGDB database (Dong *et al.*, 2004; http://www.plantgdb.org/) was developed to provide informatics tools for comparative plant genomics. Drawing on our experiences with this project, we discuss various data sources and tools for plant EST analyses and then describe a few case studies on ways EST data may be mined.

Getting the Data

Data Sources

dbEST (http://www.ncbi.nlm.nih.gov/dbEST/) (Boguski *et al.*, 1993), a division of GenBank, is a central repository for all the publicly available EST sequences. Users can search ESTs from plants (as well as other king-doms) with NCBI's Entrez system. For example, all the maize ESTs can be retrieved by typing "Zea mays [ORGN]" into the search text box on the dbEST front page. The resulting sequences can subsequently be down-loaded in a variety of formats (e.g., FASTA) for users to analyze. The complete Entrez syntax is described at http://www.ncbi.nlm.nih.gov/entrez/ query/static/help/helpdoc.html. A convenient shortcut to the data from large-scale plant EST sequencing projects is provided at the NCBI Plant Genomes Central home page (http://www.ncbi.nlm.nih.gov/genomes/ PLANTS/PlantList.html). In addition to web retrieval through Entrez, EST data can be obtained from NCBI by anonymous ftp (ftp://ftp.ncbi. nih.gov/repository/dbEST/). Unfortunately, all ESTs in the dbEST ftp repository are bundled, irrespective of sequence origin. Customized pars-ing of the bulk files is required if users are only interested in a subset of plant species. An alternative data resource is the PlantGDB download site (http://www.plantgdb.org/download.php), which provides EST sequences in FASTA format organized by species. As discussed later in this chapter, PlantGDB also displays EST assembly and annotation results based on the regularly downloaded sets from dbEST. Because of the complexity of that analysis, there will typically be a time lag between the most up-to-date dbEST collections and the subsets of those sequences that were processed at PlantGDB.

In addition to the sequence data, for some applications it is critical to have the quality scores associated with the sequence entry. A quality score is a number between 1 and 100 that gives the relative confidence that a particular base was determined correctly. The scores are generated by the sequencing centers at the time the bases are called. Because of the single-pass nature of EST sequencing, the sequencing error rate is relatively high, estimated in one study at about 3% (Hillier *et al.*, 1996). Sequencing accuracy is probably higher with modern sequencing machines; however, for applications such as single nucleotide polymorphism (SNP) analysis (Garg *et al.*, 1999), the accuracy of each base is critical and must be assessed from the quality scores. The NCBI Trace Archive (http://www.ncbi.nlm. nih.gov/Traces/trace.cgi) intends to be a central repository to store trace data generated by sequencing centers. At this point, the vast majority of the trace data posted are from genomic sequencing rather than ESTs. Therefore, users who want to obtain EST trace files will generally have to contact the individual sequence providers for availability. For some plant species, their community database stores trace files. For example, the EST trace files from the Maize Gene Discovery Project (Lunde *et al.*, 2003) are available through MaizeGDB (http://www.maizegdb.org) (Lawrence *et al.*, 2003).

Filtering Out Contaminants

Although various sequence processing suites are widely available to prepare sequences for submission to the databases (Chou and Holmes, 2001; Parkinson *et al.*, 2004; Scheetz *et al.*, 2003), our advice is to still screen the EST sequences deposited at dbEST for contaminating nonnative sequences before conducting survey studies using the ESTs. Some contaminants, such as sequences from plant-associated fungi, may be integral to an EST library preparation because they are not separable at the time of RNA extraction. More typically, contaminations are artifacts of the sequencing process itself. This type of contamination includes cloning vectors, bacterial host, and even human sequences presumably derived from the library preparation team. Such contaminants must be removed before sequence analysis because their presence poses various problems. For instance, if a contaminated sequence was used for primer design, the primer would not allow the intended target region to be amplified. Contaminant sequences found in multiple species might suggest an evolutionary relationship that does not exist. Approaches to identify contaminants for removal are based on matching EST sequences of interest to an annotated database of common contaminants. For example, the UniVec database at NCBI (http://www.ncbi.nlm.nih.gov/VecScreen/UniVec.html) collects a large number of commonly used cloning vector, adapters, and primers.

In addition to the nonnative sequences discussed earlier, analysis of EST sets may be confounded if the ESTs contain sequences that are derived from native repetitive elements. The TIGR Plant Repeat Database (http://www.tigr.org/tdb/e2k1/plant.repeats/) (Ouyang and Buell, 2004) is commonly used to screen plant EST sequences for the presence of sequences derived from retrotransposons and other repetitive elements. To compare ESTs with repeat databases, similarity-search–based programs such as BLAST (Altschul *et al.*, 1997) and cross_match (http://www.phrap.org/phredphrapconsed.html) may be used. Vmatch (Abouelhoda *et al.*, 2004;

http://www.vmatch.de/) is a less known but highly versatile and efficient
sequence comparison program that implements a suffix-array algorithm
for fast string comparison. As part of the PlantGDB EST processing pipeline, we apply Vmatch routinely to remove vector contaminations and
identify repetitive elements. The percentage of contaminated sequences
overall is not large; for example, when processing the maize ESTs at
PlantGDB, less than 0.4% of the sequences were found to contain vector
sequences. However, because more than 400,000 sequences are in the maize
collection, it is still a matter of thousands of sequences, which, if not
removed, could confound a whole set analysis. Even when an EST dataset
does not match databases of known contaminants or repeats, this does not
mean the sequences are free of these types of sequences, because the
screening databases, for example, the TIGR Plant Repeat Database, are
being added to continually, and will not anytime soon consist of a comprehensive list of all repetitive plant sequences.

Use of Library Information

Information about the source of ESTs (e.g., the tissue type, the stage of
development of the plant, and the cultivar) allows the sequence information to be more fully used. For example, for a researcher interested in the
initiation of flowering in maize, it would be desirable to compare ESTs
derived from floral meristematic tissue in maize to ESTs from other types
of tissue. In this way, candidate flowering-specific genes might be identified.
Making the ability to perform these kinds of queries easily accessible to
plant biologists is an ongoing challenge. Although library information is
embedded in the GenBank EST entries, it is very complicated for a user to
retrieve sets of ESTs from multiple libraries with specific characteristics
from the NCBI site. However, a user can easily do such a search at
PlantGDB. This is made possible by organizing the library information
(taken from GenBank) into a relational database in PlantGDB, which
allows more specific searches than are possible with NCBI-Entrez. For
example, even though the requisite information is part of the GenBank
records, the execution of a query like "display all the root ESTs generated from maize a particular inbred line" is currently impossible at Gen-
Bank. The TableMaker tool at PlantGDB makes this kind of query
straightforward (http://www.plantgdb.org/TableMaker.php).
Another consideration when using ESTs to infer expression under
specific conditions is how the library was generated. Some libraries are
constructed to reflect the relative abundance of the mRNA under certain
conditions, at a particular stage, and/or tissue type. Other libraries, referred
to as *normalized libraries*, attempt to capture a higher percentage of lower

copy mRNA by filtering out redundant copies of more abundant mRNAs before constructing the library. If downloaded EST sequence data are used to explore expression profiles, expression level ought only to be inferred from nonnormalized expression libraries. When multiple EST libraries made under different, well-documented conditions exist, they can be used to make inferences about expression differences among the represented tissue sources or developmental stages. For example, Ronning *et al.* (2003) surveyed the potato transcriptome using ESTs from diverse tissues, in particular tissues challenged with late-blight pathogen. The authors were able to identify a number of candidate genes that were specifically expressed during incompatible interactions with the late-blight pathogen. Results gleaned from comparison between multiple libraries must be statistically supported. Stekel *et al.* (2000) describe a likelihood ratio method for comparing the level of gene expression from multiple EST libraries to identify differentially expressed genes. The author also provided a nice review of alternative methods for comparing expression levels between two EST libraries.

EST Clustering and Unigene Assembly

Because ESTs usually correspond to only partial cDNA sequences, and because EST samples typically are highly redundant, ESTs are commonly clustered to derive a set of unique putative genes. During this clustering process, ESTs are grouped based on mutual percent identity over a minimum number of overlapping bases. Consensus sequences are derived from the multiple-sequence alignment from each group to provide a tentative "unigene." EST clustering is computationally intensive given a large redundant set of ESTs, and for this reason specialized databases, listed in Table I, have evolved to make the resulting clusters available to the public.

TABLE I
MAJOR RESOURCES FOR PLANT EST DATA

Organization	URL
Génoplante	http://genoplante-info.infobiogen.fr/
NCBI dbEST	http://www.ncbi.nlm.nih.gov/dbEST/
NCBI Unigenes	http://www.ncbi.nlm.nih.gov/entrez/query.fcgi?db=unigene
PlantGDB	http://www.plantgdb.org
Sputnik	http://sputnik.btk.fi
TIGR	http://www.tigr.org/tdb/tgi/plant.shtml

PlantGDB is an example of this type of resource. At PlantGDB, for each species, ESTs are organized in terms of tentative unique genes (TUGs), which are either contigs (EST clusters of two or more sequences; see Fig. 1 for a sample display of an EST contig) or singlets (ESTs that are not significantly similar to any other ESTs). This is achieved by a routine data analysis pipeline. After processing EST sequences with Vmatch to remove contaminants and repeats, the ESTs are clustered with the PaCE program (Kalyanaraman et al., 2003). PaCE employs parallel computing technology to allow fast processing of huge datasets. Therefore, the TUGs can be updated in a timely fashion to reflect the current status of known expressed genes. The program groups sequences that overlap to a certain degree. Then a consensus sequence for each cluster is generated with CAP3 (Huang and Madan, 1999).

In addition to PlantGDB, other databases provide collections of clustered ESTs using different software. The Sputnik project (http://sputnik.btk.fi/) aims to place paralogous sequences in separate clusters using proprietary software. Each EST cluster and the associated peptide are functionally and structurally annotated in an automatic pipeline (Rudd et al., 2003). The TIGR gene indices (http://www.tigr.org/tdb/tgi/plant.shtml) are constructed with the TIGR assembler (Pertea et al., 2003; Quakenbush et al., 2001). In addition to the raw EST data available at NCBI or EBI, used by PlantGDB and Sputnik, the TIGR gene indices are constructed from NCBI gene sequences and expressed transcript (ET) sequences from the TIGR Expressed Gene Anatomy Database (http://www.tigr.org/tdb/egad/egad.shtml). A third collection is the Unigene set maintained by NCBI at GenBank, http://www.ncbi.nlm.nih.gov/entrez/query.fcgi?db=unigene. This collection differs significantly from that of PlantGDB, the Sputnik project, and the TIGR gene indices, as the Unigene clusters are not assembled. Furthermore, all Unigene clusters include sequences providing evidence of the 3' end of the transcription unit. As a result, not all of the dbEST sequences are represented by a Unigene (Pontius et al., 2003).

Different assembly programs or parameters used for assembly largely explain the differences between the EST clusters provided by different databases, in addition to the differences due to the scope of ESTs analyzed. Efforts to compare different assemblies are currently under way (http://www.phytome.org/). For species with large amounts of genome sequence available, the alignment between ESTs and genomic data can be used as a "gold standard" to calibrate the assembly parameters, that is, ESTs aligned to the same locus should have been clustered (Zhu et al., 2003). However, users of Unigene sets should realize that the parameters being used for assembly were usually developed by trial and error. Thus, when looking at

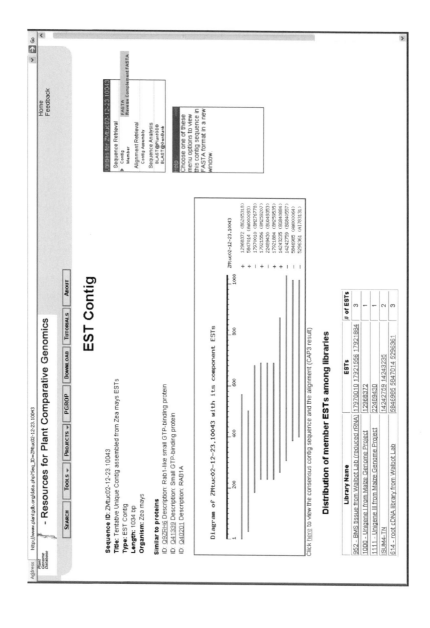

the EST clusters provided by the databases, they should be considered heuristic results. Although reclustering the entire EST dataset is typically out of the computational capabilities of general biologists, users are encouraged to recluster subsets of their ESTs of interest with different assembly programs or parameters to evaluate the robustness of results for themselves.

EST Annotation

If a genome were to be completely annotated, the function, gene structure, location, and evolutionary history of each gene would have to be determined. Gene structure annotation, via the spliced alignment of ESTs to genomic DNA, is covered in the last section of this chapter. To facilitate physical mapping in other species, their ESTs were compared to the *Arabidopsis* genome and results displayed on maps of the *Arabidopsis* chromosomes (http://genopole.toulouse.inra.fr/bioinfo/Iccare/) (Muller *et al.*, 2004). Similar attempts to leverage whole genome information for model species can be anticipated as more genomes are being finished.

As a step toward the elusive goal of whole genome annotation, ESTs themselves can be functionally annotated, and this is commonly done to give a general idea of what kinds of genes are expressed by an organism. Ideally all annotation would be either done or checked manually by expert curators. Because it is impossible to functionally annotate tens of thousands of ESTs this way, major EST databases rely on automated BLAST searches to assign putative function based on sequence similarity to gene products that have previously been functionally annotated. For example, at PlantGDB, contig consensus sequences and singleton sequences are compared to the protein database GenPept at NCBI using BLASTX. The top three hits below the default E-20 cutoff value are assigned as putative functional homologues of the protein products coded for in the EST contig. The NCBI Unigene collection is annotated by providing protein similarity data for one representative sequence from each cluster to proteins from a group of eight model organisms (including *Arabidopsis*). Obviously, such functional annotation provided at any EST

FIG. 1. A screenshot of an expressed sequence tag (EST) contig page displayed at PlantGDB. On the left from top to bottom: contig identification, links to similar proteins that were assigned by BLASTX searches against protein databases, and a schematic diagram of the EST assembly into the contig. Each member EST is linked to its sequence and information about the library in which the EST was generated. On the right are links for downloading either the consensus sequence of the contig or all the sequences of the component ESTs, as well as links to analysis tools such as BLAST@PlantGDB.

database, often retrieved through general text search tools, is just a sugges-
tion, suffering from potential inaccurate transfer of functional assignments
and cascading of annotation errors. However, automated annotation provides a starting point, and users can and should always reannotate their
sequences of interest. Different programs or databases can be used, parameters can be changed, and alignments can be manually examined. This
may reduce the propagation of errors due to the limitations of automated
annotation. At PlantGDB, end-user functional annotation is facilitated by
links to analysis tools from each record display page (Fig. 1).

A potential barrier to the custom annotation of user-derived EST datasets is the cumbersome nature of the BLAST output generated from running searches for functional annotation purposes. Several programs can be
used to process BLAST output, such as BEAUTY (Worley *et al.*, 1998),
PhyloBLAST (Brinkman *et al.*, 2001), MuSeqBox (Xing and Brendel,
2001), and Zerg (Paquola *et al.*, 2003). MuSeqBox was designed specifically
to meet the need for a convenient way of examining multiquery sequence
BLAST output during large-scale EST annotation projects. It parses the
informative parameters of BLAST hits into a tabular form, from which
subsets of BLAST hits can be selected according to user-specified criteria.
For example, MuSeqBox can screen the BLASTX output from an EST
versus protein database search to indicate sequences that potentially represent full-length coding sequences. Extensive tutorials for the use of MuSeq-
Box are available at http://bioinformatics.iastate.edu/bioinformatics2go/
mb/MuSeqBox.html.

Often biologists want to derive open reading frames (ORFs) from ESTs
so that protein sequences can be used for further analysis (e.g., domain
analysis at protein motif databases). In addition to using alignments with
protein sequence to determine ORFs, *ab initio* programs can be used. For
example, ESTscan2.0 (Lottaz *et al.*, 2003) can be used to identify coding
regions in EST contigs. The efficacy of ESTscan2.0 and similar programs
for translation initiation site determination has been assessed (Nadershahi
et al., 2004). For domain analysis, there are also programs using nucleotide
sequences directly to search for protein motifs. For example, InterProScan
(http://www.ebi.ac.uk/interpro/) will read the input nucleotide sequences,
translate them into protein sequences internally, and then search the
domain databases (Zdobnov and Apweiler, 2001).

Tools for Comparative EST Analyses

A number of tools are available for comparative genomics; however,
most of the tools are not used by a wide audience, either because of barriers,
such as an absence of a user-friendly interface, or because of a lack of

understanding for how to distinguish the underlying algorithms. The most popular tools remain some of the most "simple," such as BLAST. Here, we discuss a few tools for comparative EST analysis. First is the multispecies BLAST at PlantGDB. Although BLAST is available on many servers, the user is generally limited to searching against either very general groups or one species or data type at a time. The unique feature of BLAST@PlantGDB (http://www.plantgdb.org/cgi-bin/PlantGDBblast) is the ability for the user to search against any combination of species and EST or other sequence types. Searches can be conducted against user-designated single or multiple BLAST databases simultaneously (e.g., only rice or maize ESTs or both; or all monocot ESTs; or rice ESTs and all cereal EST contigs). The selection of specific databases for comparison can be critical to correctly assess statistical significance of observed sequence matches.

Another common task is to align ESTs to genomic DNA to determine exon–intron gene structure. BLAST is not the most suitable program for this task, because it was not designed to perform the spliced alignments necessary to overlay exons, as represented by the EST sequence, onto genomic DNA. Several spliced-alignment programs have been developed, such as sim4 (Florea et al., 1998), Spidey (Wheelan et al., 2001), BLAT (Kent, 2002), and GeneSeqer (Brendel et al., 2004; Usuka et al., 2000). GeneSeqer is perhaps the most versatile tool for annotating plant genomic DNA. It can produce plant gene structure models based on spliced alignment to genomic sequences of both native and homologous ESTs, cDNAs, and protein sequences (Usuka and Brendel, 2000). The GeneSeqer@PlantGDB server (Schlueter et al., 2003) allows users to run the GeneSeqer program over the web. As with BLAST @PlantGDB, users can select any combination of EST species collections from the PlantGDB database, or they can supply their EST sequences. Figure 2 displays a typical spliced alignment generated from GeneSeqer@ PlantGDB.

In addition to using analysis programs such as BLAST and GeneSeqer, users may want to selectively retrieve information from databases for more detailed analyses and comparisons. Simple text search capabilities and fixed table reporting, typically implemented at biological databases, do not always meet the needs of biological researchers. The PlantGDB TableMaker tool (http://www.plantgdb.org/TableMaker.php) allows for criteria to be specified for different columns and for the selection of columns for inclusion in a tabular report format. Executing queries such as "Show me all the maize ESTs, library name, tissue type, and EST contigs they belong to" can be accomplished by filling out web forms.

```
ATGGTGCGGA TTTT...... .......... .......... .......... ..........          814


GTTAGATGAT AGCAACTGCA GAATTCAATG GTTCATCAAT AGCAAGAGAA GAGGAAGATT      259562

.......... .......... .......... .......... .......... ..........          814


TGTACTGACG AAATCAACAC CGATGGTACT GATGTAAGTG TCGATGTACG AATCGTCCTG      259622
           | ||||||||| |  ||||||||||||  |||||| ||| || | ||||| |||||||
.........G AAATCAACGC CGATGGTACT GATGTAGGTG TCAACGTACG AATCGTCGG.          864


CATGAATCAA ACGAACCGGA ACTTGTCAGG ACGAGAAGTT AATTAAACAC CGAACTGTGC      259682

.......... .......... .......... .......... .......... ..........          864


TACTCTATTG AAATAAAGAA GAGAAGTTTC CGAGCGAAAT TAAGATGGAG GCTTTCTTAC      259742

.......... .......... .......... .......... .......... ..........          864


AGCAAATCGG AGGAGGAGGC AGGACTTGCC GACGGAGGAG TCGCCGATGA GGAGCAGCTT      259802
 | || ||| |||||||||| |||||||||| |||||||||  |||||||||| ||||||||||
..CGAAACGG AGGAGGAGGC AGGACTTGCC GACGGAGGAA TCGCCGATGA GGAGCAGCTT          922


GAACAGGTAA TCGCTGCATC AGAAAACGCA AACACCATTA ATCGCAGTTT CACAAAGAGG      259862
||| || || |||
GAAGAGATAG TCG...... .......... .......... .......... ..........          935
```

Fig. 2. A typical spliced alignment generated by the GeneSeqer program. For each row of the alignment, the upper line represents the genomic sequence, and the bottom line represents the expressed sequence tag (EST) sequences. The dot indicates the gap in the alignment between the genomic DNA and the EST, referring to the intron positions. The same task performed by BLAST program, as used by many biologists to annotate exon–intron positions, will only result in obtaining approximately the aligned parts (exons) while missing the display of intron positions. Thus, the results of BLAST often require manual editing to recover a simple spliced alignment as shown here.

Using EST Collections

Plant ESTs can be used for various investigations. The primary usage of ESTs is for gene discovery. A typical biologist is usually interested in questions such as whether the gene of interest exists and is expressed in certain organisms or tissues. Such questions are typically answered by simply searching EST databases using the gene of interest as a query in a BLAST run. Here we discuss the application of plant ESTs for more complex tasks: identification of gene families that are conserved among species considered, discovery of splicing variants, and delineating gene structure via spliced alignments.

Identification of Gene Families Conserved across Species

Gene families conserved across plants can be used in phylogenetic studies and in the identification of conserved noncoding regions or to make comparative maps of major crop species. Although the amount of genetic data available for plants is increasing exponentially, most of the work is being done in just a few species. The identification of a set of gene families common to a large variety of plants (e.g., all angiosperms) would allow researchers studying less well-characterized plants to capitalize more fully on gains being made in *Arabidopsis,* rice, *Medicago,* and other model or reference plant species. Because ESTs give an idea of the genes expressed in an organism, and because EST data are abundant for a variety of species, ESTs are ideal starting material for the identification of genes conserved among species. There are two general approaches. The first is aligning ESTs or protein sequences from one species to the genomic DNA of another species. Fulton *et al.* (2002) developed a Conserved Ortholog Set (COS) of 1025 single-copy markers by matching tomato EST-derived Unigene sequences to the *Arabidopsis* genomic sequence with BLAST. A second common approach is to compare ESTs from one species to ESTs from another species, usually also based on BLAST-type sequence comparisons. For example, Lee *et al.* (2002) identified Tentative Ortholog Groups (TOGs) by selecting reciprocal BLASTN best-hit pairs.

Identification of Alternative Splicing

There are usually two approaches to detect alternative splicing by EST alignments. For species that have nearly completed genome sequences, such as *Arabidopsis* and rice, all EST and cDNA sequences can be aligned against the genome sequence using spliced alignment programs. Alternative splice cases can be inferred from the coordinates of predicted introns and exons. This method has been widely used in animal system including *Caenorhabditis elegans*, mouse, and human (Kan and Gish, 2002; Kent and Zahler, 2000; Thanaraj *et al.*, 2003). For species lacking genome sequences, ESTs can be aligned against themselves. Any reliably predicted insertions in one sequence relative to another would indicate alternative splicing cases. This method is applicable only when the insertions are long enough to be distinguished from polymorphisms and sequencing errors. Different types of alternative splicing, however, cannot always be reliably distinguished from the EST–EST alignments. A study by Kan *et al.* (2004)

revealed a possible third approach: aligning ESTs of one species to the genome sequence of close relatives.

Because *Arabidopsis* and rice have nearly completed genome sequences and large EST and cDNA collections, we deployed the first approach to detect alternative splicing in plants (see also Zhu *et al.*, 2003). A total of 4161 *Arabidopsis* genes and 5378 rice genes showed five types of alternative splicing: alternative donor sites, alternative acceptor sites, alternative position (both donor and acceptor sites are different), exon skipping, and intron retention. Updates and details about these cases are available at the Alternative Splicing In Plants (ASIP) web site, http://www.plantgdb.org/prj/SiP/ASIP/EnterDB.php.

An example from the ASIP database is shown in Fig. 3. At3g01150 (atPTB2a) is one of the three polypyrimidine tract binding proteins in *Arabidopsis* (Wang and Brendel, 2004). Its human homologue functions as a splicing regulator by binding to the polypyrimidine tract of introns (Lin

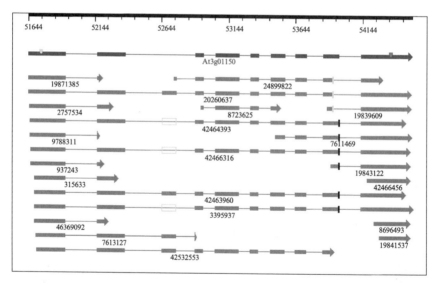

FIG. 3. Visualization of expressed sequence tag (EST)/complementary DNA (cDNA) alignments and alternative splicing. Arabidopsis ESTs and cDNAs were aligned against the genome sequence using GeneSeqer. The black scale on top of the picture indicates the chromosome three coordinates of the alignments. Thick horizontal bars represent exons, and thin lines represent introns. Red indicates EST/cDNAs alignments. Blue indicates GenBank annotation. The green and red triangles above the blue bars indicate the translation start and stop positions, respectively. For alternative splicing, the green box on the red lines indicates an exon-skipping case. For the alternative donor site in the last intron sites, a black vertical bar indicates the most prevalent splice site, and a green vertical bar represents the alternative splice site. (See color insert.)

and Patton, 1995). As shown in the figure, a 102-bp exon was skipped in three full-length cDNAs (gi3395937, gi42466316, and gi42464393). The skipped exon is indicated by a green box. Inclusion of the exon in two full-length cDNAs (gi20260637, gi42463960) and ESTs (gi7613127, gi42532553) will introduce an in-frame premature stop codon. Thus, the exon skipping transcript is likely the wild-type splice form. In addition, the last intron of the gene can also be alternatively spliced. As shown in Fig. 3, six ESTs/cDNAs use a canonical donor site, which is marked by a black bar. Three EST/cDNAs, however, use an alternative donor site located 47 bp upstream of the canonical site, marked by a green bar. Both the constitutive donor site and the upstream donor site have conserved splice signals, as indicated by high scores in the GeneSeqer alignment. Detailed alignment of representative ESTs/cDNAs is displayed in Fig. 4. Use of the upstream donor site introduces a frameshift and premature stop codon in the transcript.

Genome Annotation

The annotation accompanying the whole genome sequences of *Arabidopsis* and rice as stored at NCBI was mostly generated by TIGR (http://www.tigr.org/). Much of their structural annotation was based on *ab initio* gene prediction. Individual biologists could reannotate their regions of interest using spliced alignment programs to align ESTs to the sequence. A recommended starting point is to visit genome browsers provided by major databases. At PlantGDB, for example, for *Arabidopsis thaliana*, the most recent EST and full-length cDNA collections are periodically threaded onto the model organism's five established chromosome sequences. The AtGDB genome browser (http://www.plantgdb.org/AtGDB/) provides a means for the visual assessment of the resulting gene structure annotation (Zhu *et al.*, 2003). Specifically, at any chosen region of the *Arabidopsis* genome, users can view the original annotation downloaded from NCBI (provided by TIGR), spliced alignments of all ESTs and full-length cDNAs, and exon–intron structural annotation based on the spliced alignment results. Currently 206,672 EST sequences, 58,679 full-length cDNA sequences, and 28,952 TIGR-predicted gene models are incorporated into AtGDB. Approximately 60% of the predicted gene models (protein coding regions) are fully supported by EST or cDNA evidence, whereas 25% are based solely on computational gene structure prediction. In addition, 70% of the predicted gene models are found in a genomic context so that corresponding EST and cDNA alignments display some level of incongruence, including incompletely annotated noncoding regions, alternative splicing, and erroneous gene predictions.

```
Genome     GCAGCAATGG CCAAAGAAGC ACTGGAAGGA CACTGCATTT ATGACGGTGG CTACTGTAAG    53923
           |||||||||| |||||||||| |||||||||| |||||||||| |||||||||| ||||||||||
7611469    GCAGCAATGG CCAAAGAAGC ACTGGAAGGA CACTGCATTT ATGACGGTGG CTACTGTAAG    224
           |||||||||| |||||||||| |||||||||| |||||||||| |||||||||| |||||
2026037    GCAGCAATGG CCAAAGAAGC ACTGGAAGGA CACTGCATTT ATGACGGTGG CTACT.....    1136


Genome     CTTCGACTAT CATACTCTCG TCATACTGAT CTGAATGTTA AGGTACATAA GATTTGTTGC    53983
           |||||||||| |||||||||| |||||||||| |||||||||| ||
7611469    CTTCGACTAT CATACTCTCG TCATACTGAT CTGAATGTTA AG....... ..........    266

2026037    .......... .......... .......... .......... .......... ..........    1136


Genome     TTTTTTGAGA TGTGTACCTA AAAAAAAAAA CAATTGTGAG GGGAAGAGAA TAGTAACGGA    54043

7611469    .......... .......... .......... .......... .......... ..........    266

2026037    .......... .......... .......... .......... .......... ..........    1136


Genome     ATTGCGTATG CTTCTTTCTT TTCTACATTT CGCAGCAAAA TATTAAATCA ACGTCTAATT    54103

7611469    .......... .......... .......... .......... .......... ..........    266

2026037    .......... .......... .......... .......... .......... ..........    1136


Genome     ATCTTCTTCA ATTTTCCTGT GATTGAATGA AGGCTTTTAG CGACAAAAGC AGAGACTATA    54163
                                          |||||||||| |||||||||| ||||||||||
7611469    .......... .......... .......... ..GCTTTTAG CGACAAAAGC AGAGACTATA    294
                                          |||||||||| |||||||||| ||||||||||
2026037    .......... .......... .......... ..GCTTTTAG CGACAAAAGC AGAGACTATA    1164
```

FIG. 4. Alternative donor sites for the last intron in gene At3g01150 (atPTB2a). The displayed alignment was constructed manually from GeneSeqer alignment of expressed sequence tag (EST)/complementary DNAs (cDNAs) gi7611469 and gi2026037. The top sequence for each alignment block is the genome sequence, with sequence of gi7611469 in the middle and gi2026037 at the bottom. Intron positions are indicated by dots. Vertical bars indicate base matches. Numbers on the right indicate the coordinates of the last base in each displayed sequence. The EST/cDNA gi2026037 uses an alternative donor site, which is located 47 bp upstream of the donor site used by gi7611469. The upstream donor site will cause a frameshift and introduce a premature stop codon, indicated by the black box on the sequence.

A similar browser has been created for rice (http://www.plantgdb.org/ OsGDB), although the framework here is the complete rice bacterial artificial chromosome (BAC) sequences, rather than pseudo-chromosomes, the assembly of which is still in flux. Currently, 284,006 EST sequences, 32,128 full-length cDNA, 3572 gene models defined as GenBank file features, and 59,712 transcription unit (TU) models defined by TIGR (http://www. tigr.org/tdb/e2k1/osa1) are on display at OsGDB. Of the more current TU gene model predictions, 43% are fully confirmed by EST and cDNA evidence, whereas 48% are based on computational gene prediction alone. In addition, 61% of the TUs display some level of incongruence with local EST and cDNA alignments, as discussed for *Arabidopsis*.

At both AtGDB and OsGDB, users can contribute updated annotations of their own to a shared community annotation system through the use of web-based annotation tools (Schlueter *et al.*, 2005). These tools were developed to allow a user to easily access *ab initio* predictions, native and homologous sequence alignments, ORF estimations, and other useful analyses for gene structure determination. The results of these individual analyses are incorporated into the annotation system so a user can simply select the custom gene structure they desire. This "User-Contributed Annotation" is then credited to the user but after validation is owned by the community.

Concluding Remarks

Plant ESTs are a rich resource. The examples discussed here might indicate how this resource can still be used more effectively by a broader community. Both for genome annotation in species with whole genome sequencing efforts and for gene discovery in species without substantial genome sequencing, EST sequencing seems to be a highly cost-effective approach.

Acknowledgments

This work was supported in part by NSF Plant Genome Research Projects grants DBI-0110189 and DBI-0321600. Some of the work was initiated while L. K. was a participant of the NIH-NSF 2003 Summer Institute in Bioinformatics and Computational Biology at ISU (supported by NSF grant EEC-0234102).

References

Abouelhoda, M. I., Kurtz, S., and Ohlebusch, E. (2004). Replacing suffix trees with enhanced suffix arrays. *J. Discrete Algorithms* **2**, 53–86.
Adams, M. D., Kelley, J. M., Gocayne, J. D., Dubnick, M., Polymeropoulos, M. H., Xiao, H., Merril, C. R., Wu, A., Olde, B., Moreno, R. F., Kerlavage, A. R., McCombie, W. R., and Venter, J. C. (1991). Complementary DNA sequencing: Expressed sequence tags and human genome project. *Science* **252**, 159–164.
Altschul, S. F., Madden, T. L., Schäffer, A. A., Zhang, J., Zhang, Z., Miller, W., and Lipman, D. J. (1997). Gapped BLAST and PSI-BLAST: A new generation of protein database search programs. *Nucleic Acids Res.* **25**, 3389–3402.
Boguski, M. S., Lowe, T. M., and Tolstoshev, C. M. (1993). dbEST—database for "expressed sequence tags." *Nat. Genet.* **10**, 369–371.
Brendel, V., Xing, L., and Zhu, W. (2004). Gene structure prediction from consensus spliced alignment of multiple ESTs matching the same genomic locus. *Bioinformatics* **20**, 1157–1169.
Brinkman, F. S., Wan, I., Hancock, R. E., Rose, A. M., and Jones, S. J. (2001). PhyloBLAST: Facilitating phylogenetic analysis of BLAST results. *Bioinformatics* **17**, 385–387.
Chou, H.-H., and Holmes, M. H. (2001). DNA sequence quality trimming and vector removal. *Bioinformatics* **17**, 1093–1104.

Dong, Q., Schlueter, S. D., and Brendel, V. (2004). PlantGDB, plant genome database and analysis tools. *Nucleic Acids Res.* **32,** 354–359.

Florea, L., Hartzell, G., Zhang, Z., Rubin, G. M., and Miller, W. (1998). A computer program for aligning a cDNA sequence with a genomic DNA sequence. *Genome Res.* **8,** 967–974.

Fulton, T. M., Van der Hoeven, R., Eannetta, N. T., and Tanksley, S.D. (2002). Identification, analysis, and utilization of conserved ortholog set markers for comparative genomics in higher plants. *Plant Cell* **14,** 1457–1467.

Garg, K., Green, P., and Nickerson, D. A. (1999). Identification of candidate coding region single nucleotide polymorphisms in 165 human genes using assembled expressed sequence tags. *Genome Res.* **9,** 1087–1092.

Hillier, L.D., Lennon, G., Becker, M., Bonaldo, M.F., Chiapelli, B., Chissoe, S., Dietrich, N., DuBuque, T., Favello, A., Gish, W., Hawkins, M., Hultman, M., Kucaba, M., Lacy, M., Le, M., Le, N., Mardis, E., Moore, B., Morris, M., Parsons, J., Prange, C., Rifkin, L., Rohlfing, T., Schellenberg, K., and Marra, M. (1996). Generation and analysis of 280,000 human expressed sequence tags. *Genome Res.* **6,** 807–828.

Huang, X., and Madan, A. (1999). CAP3: A DNA sequence assembly program. *Genome Res.* **9,** 868–877.

Kalyanaraman, A., Aluru, S., Kothari, S., and Brendel, V. (2003). Efficient clustering of large EST data sets on parallel computers. *Nucleic Acids Res.* **31,** 2963–2974.

Kan, Z., Castle, J., Johnson, J. M., and Tsinoremas, N. F. (2004). Detection of novel splice forms in human and mouse using cross-species approach. *Pac. Symp. Biocomput.* **9,** 42–53.

Kan, Z., States, D., and Gish, W. (2002). Selecting for functional alternative splices in ESTs. *Genome Res.* **12,** 1837–1845.

Kent, W. J. (2002). BLAT–the BLAST-like alignment tool. *Genome Res.* **12,** 656–664.

Kent, W. J., and Zahler, A. M. (2000). The intronerator: Exploring introns and alternative splicing in *Caenorhabditis elegans. Nucleic Acids Res.* **28,** 91–93.

Lawrence, C. J., Dong, Q., Polacco, M. L., Seigfried, T. E., and Brendel, V. (2003). MaizeGDB, the community database for maize genetics and genomics. *Nucleic Acids Res.* **32,** D393–D397.

Lee, Y., Sultana, R., Pertea, G., Cho, J., Karamycheva, S., Tsai, J., Parvizi, B., Cheung, F., Antonescu, V., White, J., Holt, I., Liang, F., and Quackenbush, J. (2002). Cross-referencing eukaryotic genomes: TIGR Orthologous Gene Alignments (TOGA). *Genome Res.* **12,** 493–502.

Lin, C. H., and Patton, J. G. (1995). Regulation of alternative 3′ splice site selection by constitutive splicing factors. *RNA* **1,** 234–245.

Lottaz, C., Iseli, C., Jongeneel, C. V., and Bucher, P. (2003). Modeling sequencing errors by combining hidden markov models. *Bioinformatics* **19,** ii103–ii112.

Lunde, C. F., Morrow, D. J., Roy, L. M., and Walbot, V. (2003). Progress in maize gene discovery: A project update. *Funct. Integr. Genomics* **3,** 25–32.

Muller, C., Denis, M., Gentzbittle, L., and Faraut, T. (2004). The Iccare web server: An attempt to merge sequence and mapping information in plant and animal species. *Nucleic Acids Res.* **32,** W429–W434.

Nadershahi, A., Fahrenkrug, S. C., and Ellis, L. B. (2004). Comparison of computational methods for identifying translation initiation sites in EST data. *BMC Bioinformatics* **5,** 14.

Ouyang, S., and Buell, C. R. (2004). The TIGR Plant Repeat Databases: A collective resource for the identification of repetitive sequences in plants. *Nucleic Acids Res.* **32,** D360–D363.

Paquola, A. C., Machado, A. A., Reis, E. M., Da Silva, A. M., and Verjovski-Almeida, S. M. (2003). Zerg: A very fast BLAST parser library. *Bioinformatics* **19**, 1035–1036.

Parkinson, J., Anthony, A., Wasmuth, J., Schmid, R., Hedley, A., and Blaxter, M. (2004). PartiGene – constructing partial genomes. *Bioinformatics* **20**, 1398–1404.

Pertea, G., Huang, X., Liang, F., Antonescu, V., Sultana, R., Karamycheva, S., Lee, Y., White, J., Cheung, F., Parvizi, B., Tsai, J., and Quackenbush, J. (2003). TIGR Gene Indices clustering tools (TGICL): A software system for fast clustering of large EST datasets. *Bioinformatics* **19**, 651–652.

Pontius, J. U., Wagner, L., and Schuler, G. D. (2003). UniGene: A unified view of the transcriptome. *In* "The NCBI Handbook." Bethesda, IL. http://www.ncbi.nlm.nih.gov/books/bv.fcgi?rid=handbook.chapter.857.

Quakenbush, J., Cho, J., Lee, D., Liang, F., Holt, I., Karamycheva, S., Parvizi, B., Pertea, G., Sultana, R., and White, J. (2001). The TIGR Gene Indices: Analysis of gene transcript sequences in highly sampled eukaryotic species. *Nucleic Acids Res.* **29**, 159–164.

Ronning, C. M., Stegalkina, S. S., Ascenzi, R. A., Bougri, O., Hart, A. L., Utterbach, T. R., Vanaken, S. E., Riedmuller, S. B., White, J. A., Cho, J., Pertea, G. M., Lee, Y., Karamycheva, S., Sultana, R., Tsai, J., Quackenbush, J., Griffiths, H. M., Restrepo, S., Smart, C. D., Fry, W. E., Van Der Hoven, R., Tanksley, S., Zhang, P., Jin, H., Yamamoto, M. L., Baker, B. J., and Buell, C. R. (2003). Comparative analyses of potato expressed sequence tag libraries. *Plant Physiol.* **131**, 419–429.

Rudd, S. (2003). Expressed sequence tags: Alternative or complement to whole genome sequences? *Trends Plant Sci.* **8**, 321–329.

Rudd, S., Mewes, H.-W., and Mayer, K. F. (2003). Sputnik: A database platform for comparative plant genomics. *Nucleic Acids Res.* **31**, 128–132.

Scheetz, T. E., Trivedi, N., Roberts, C. A., Kucaba, T., Berger, B., Robinson, N. L., Birkett, C. L., Gavin, A. J., O'Leary, B., Braun, T. A., Bonaldo, M. F., Robinson, J. P., Sheffield, V. C., Soares, M. B., and Casavant, T. L. (2003). ESTprep: Preprocessing cDNA sequence reads. *Bioinformatics* **19**, 1318–1324.

Schlueter, S. D., Dong, Q., and Brendel, V. (2003). GeneSeqer@PlantGDB: Gene structure prediction in plant genomes. *Nucleic Acids Res.* **31**, 3597–3600.

Schlueter, S. D., Wilkerson, M. D., Huala, E., Rhee, S. Y., and Brendel, V. (2005). Community-based gene structure annotation for the *Arabidopsis thalania* genome. *Trends Plant Sci.* **10**, 9–14.

Stekel, D. J., Git, Y., and Falciani, F. (2000). The comparison of gene expression from multiple cDNA libraries. *Genome Res.* **10**, 2055–2061.

Thanaraj, T. A., Clark, F., and Muili, J. (2003). Conservation of human alternative splice events in mouse. *Nucleic Acids Res.* **31**, 2544–2552.

Usuka, J., and Brendel, V. (2000). Gene structure prediction by spliced alignment of genomic DNA with protein sequences. *J. Mol. Biol.* **297**, 1075–1085.

Usuka, J., Zhu, W., and Brendel, V. (2000). Optimal spliced alignment of homologous cDNA to a genomic DNA template. *Bioinformatics* **16**, 203–211.

Vincentz, M., Cara, F. A., Okura, V. K., da Silva, F. R., Pedrosa, G. L., Hemerly, A. S., Capella, A. N., Marins, M., Ferreira, P. C., Franca, S. C., Grivet, L., Vettore, A. L., Kemper, E. L., Burnquist, W. L., Targon, M. L., Siqueira, W. J., Kuramae, E. E., Marino, C. L., Camargo, L. E., Carrer, H., Coutinho, L. L., Furlan, L. R., Lemos, M. V., Nunes, L. R., Gomes, S. L., Santelli, R. V., Goldman, M. H., Bacci, M. Jr., Giglioti, E. A., Thiemann, O. H., Silva, F. H., Van Sluys, M. A., Nobrega, F. G., Arruda, P., and Menck, C. F. (2004). Evaluation of monocot and eudicot divergence using the sugarcane transcriptome. *Plant Physiol.* **134**, 951–959.

Wang, B.-B., and Brendel, V. (2004). The ASRG database: Identification and survey of *Arabidopsis thaliana* genes involved in pre-mRNA splicing. *Genome Biol.* **5**, R102.

Wheelan, S. J., Church, D. M., and Ostell, J. M. (2001). Spidey: A tool for mRNA-to-genomic alignments. *Genome Res.* **11,** 1952–1957.

Worley, K. C., Culpepper, P., Wiese, B. A., and Smith, R. F. (1998). BEAUTY-X: Enhanced BLAST searches for DNA queries. *Bioinformatics* **14,** 890–891.

Xing, L., and Brendel, V. (2001). Multi-query sequence BLAST output examination with MuSeqBox. *Bioinformatics* **17,** 744–745.

Zdobnov, E.M., and Apweiler, R. (2001). InterProScan–an integration platform for the signature-recognition methods in InterPro. *Bioinformatics* **17,** 847–848.

Zhu, W., Schlueter, S.D., and Brendel, V. (2003). Refined annotation of the *Arabidopsis thaliana* genome by complete EST mapping. *Plant Physiol.* **132,** 469–484.

[23] Isolation of Genes from Plant Y Chromosomes

By DMITRY A. FILATOV

Abstract

Few plant species are dioecious and only a small fraction of these species are known to have sex chromosomes. Considerable efforts to isolate sex-linked genes from dioecious *Silene latifolia* (*Caryophillaceae*) have resulted in the isolation of surprisingly few sex-linked genes, suggesting that the methods used previously were not efficient in plants. This chapter analyzes the methods that have been and can be used for isolation of genes from plant sex chromosomes. The most successful method used for the isolation of Y-linked genes included the screening of a male complementary DNA (cDNA) library with the probe obtained by degenerate oligonucleotide-primed polymerase chain reaction (PCR) of the microdissected Y chromosomes. However, chromosome microdissection requires sophisticated equipment and is difficult to apply to species with cytologically indistinguishable sex chromosomes. Genome and cDNA library subtraction methods were surprisingly unsuccessful, probably because of low divergence between the homologous X- and Y-linked genes in plants. Segregation testing and genomics-based methods are increasingly popular and are the most promising approaches for isolation of multiple genes from plant sex chromosomes.

Introduction

Sex chromosomes are probably quite a rare phenomenon in plants. Only about 50% of plant species are dioecious (having separate male and female individuals), and only a fraction of these dioecious species possess morphologically distinguishable sex chromosomes (Westergaard, 1958;

METHODS IN ENZYMOLOGY, VOL. 395

Yampolsky and Yampolsky, 1922). Only a few cases of plant sex chromosomes have been described. Among the higher plant genera, *Silene, Rumex, Humulus, Cannabis, Coccinia* (Ainsworth, 1999), and *Carica papaya* (Liu *et al.*, 2004) are known to have sex chromosomes, with males being the heterogametic sex. The haploid liverwort *Marchantia* also has sex chromosomes, with the Y chromosome confined to males and the X to females (Bischler, 1998). Because plant sex chromosomes are probably quite young, the X and Y may be indistinguishable under the microscope (e.g., in papaya), and only a detailed genetic study may demonstrate the presence of a region where X and Y do not recombine and diverge (Liu *et al.*, 2004). Thus, functional sex chromosomes may be present in many dioecious plants, but they have been discovered mostly in agriculturally important species: hop (*Humulus*), hemp (*Cannabis*), and papaya (*Carica papaya*).

The most prominent feature of Y chromosomes is that they do not recombine and are usually genetically degenerate, containing only few genes (e.g., in humans, Skaletsky *et al.*, 2003). Genetic degeneration is thought to occur as a result of lack of recombination, which results in reduced efficacy of natural selection. As a result, Y chromosomes accumulate deleterious mutations and Y-linked genes gradually become nonfunctional (Charlesworth and Charlesworth, 2000). This scenario is consistent with the substantial lack of DNA diversity on *Silene latifolia* Y chromosome (Filatov *et al.*, 2000, 2001) and on the *Drosophila* neo-Y chromosome (Bachtrog and Charlesworth, 2002; Yi and Charlesworth, 2000), which is predicted by theory (Charlesworth, 1996; Rice, 1987).

Plant sex chromosomes are usually fairly young. Recently characterized papaya sex chromosomes might be very young: They are indistinguishable under the microscope and the non- recombining region is only a few megabases long, accounting for only 10% of the entire chromosome (Liu *et al.*, 2004). Yet, the density of the genes on the Y chromosome was reported to be significantly lower compared to that of those on the X (Liu *et al.*, 2004). In *Silene*, the non- recombining region is much larger, making up more than 90% of the entire length of the sex chromosomes, and they are easily distinguishable cytologically (Westergaard, 1958), suggesting that they are older than in papaya. Sex chromosomes were found only in a cluster of dioecious *Silene* species (section *Elisanthe: S. latifolia, Silene dioica, Silene diclinis, Silene heuffelii,* and *Silen marizii*), whereas the rest of the genus is nondioecious (exept *Silene otites,* which apparently evolved dioecy independently from *Elisanthe*) and lacks sex chromosomes. Silent divergence between dioecious *S. latifolia* and hermaphroditic *Silene conica* is about 15%, suggesting that *S. latifolia* sex chromosomes are probably not older than 15 million years (Filatov and Charlesworth,

2002). Because the YY plants (having no X) are usually inviable (Ye et al., 1990), the S. latifolia Y chromosome has probably degenerated to some extent. Rumex sex chromosomes are probably more ancient than in Silene, because the Y chromosome is almost completely heterochromatic (Westergaard, 1958).

The genetic degeneracy of the Y chromosomes may be a serious obstacle for the successful isolation of Y-linked genes. Few active genes have been isolated from plant Y chromosomes. We may expect ongoing genomic work to result in the isolation of many Y-linked genes from *Marchantia* (Okada et al., 2000) and *Carica papaya* (Liu et al., 2004) Y chromosomes. However, for the moment plant Y-linked genes have been isolated only from S. latifolia Y chromosomes. The hunt for the sex-determination loci in S. latifolia has yielded several dozen genes, many of which are specifically expressed at the early stages of the male flower bud development (Barbacar et al., 1997; Matsunaga et al., 1996; Robertson et al., 1997; Scutt et al., 1997a,b). Surprisingly, none of the genes isolated in these studies was originally identified as sex linked, resulting in speculation that most genes involved in sex determination are autosomal. The segregation tests of the genes identified by Matsunaga et al. (1996) revealed that one of the genes MROS3 is X-linked and that it has a nonfunctional Y homolog, which was interpreted as a sign of genetic degeneration of the MROS3Y gene originally present on the proto-Y chromosome (Guttman and Charlesworth, 1998). A screen of a S. latifolia cDNA library, using a Y-derived probe resulted in the isolation of two pairs of sex-linked genes, SlX1/SlY1 and SlX4/SlY4 (Atanassov et al., 2001; Delichère et al., 1999). Another search for male-specific S. latifolia genes using fluorescent differential display resulted in the isolation of several cDNA clones, two of which, Men-153 and Men-470, showed male-specific bands when hybridized with genomic Southern blots (Scutt et al., 2002). However, no conclusive evidence for Y linkage of these genes was presented. Using a similar approach, followed by segregation analysis, Moore et al. (2003) isolated and characterized a pair of homologous X- and Y-linked genes, DD44X and DD44Y, respectively. Matsunaga et al. (2003) used degenerate primers to isolate S. latifolia MADS box genes. One of the isolated genes SlAP3Y gave a male-specific band on Southern blots and appeared to be Y linked. Finally, Nakao et al. (2002) reported a Y-specific open reading frame (ORF), but no evidence for this ORF being active was provided.

Given the amount of effort invested into the search for sex-linked genes in S. latifolia, it is surprising that only four active Y-linked genes (Atanassov et al., 2001; Delichère et al., 1999; Matsunaga et al., 2003; Moore et al., 2003) and a few more nonactive (Guttman and Charlesworth, 1998; Nakao et al., 2002), or active, but not definitely Y-linked genes (Scutt et al.,

2002) have been isolated. In this chapter, I describe and analyze the methods that have been and could be used for isolation of active Y-linked genes and Y-specific sequences in plants.

The approaches used to isolate the sex-linked genes can be divided into the methods based on (1) cDNA or genome subtraction, (2) isolation of the Y (and/or X) chromosome sequences via microdissection or flow sorting, (3) localization of random or preselected clones on the sex chromosomes using *in situ* hybridization and/or segregation analysis, and (4) "chromosome walking" from already known Y-linked sequences. The methods belonging to the first two groups are not completely reliable. Although they yield a fraction enriched for sequences from a specific chromosome, the sequences from the other chromosomes may also be present. Thus, it is difficult to be certain of the sex linkage of a gene isolated by these methods without checking it further for sex linkage by segregation analysis. The first two groups of approaches may, therefore, be viewed as preliminary screening methods, which should be followed by more accurate segregation analysis and *in situ* hybridization.

Subtraction Methods

Y chromosomes are confined to males (in case of male heterogamety), and in principle, Y-linked sequences can be isolated through a comparison of male and female gene pools or genomes. Various genome subtraction methods proved successful in the isolation of several kilobase long regions from relatively small ($\sim 10^8$ bp) genomes (Straus, 1995). However, it is not clear whether such methods will be effective in complex plant genomes (e.g., $\sim 3 \times 10^9$ bp in *S. latifolia*) for isolation of Y-specific sequences. Another potential problem with plant Y chromosomes is the low divergence between the homologous X and Y chromosomes (Filatov and Charlesworth, 2002), which may render the subtraction methods ineffective.

Several studies have used cDNA library subtraction or differential display methods to isolate genes expressed specifically in males (Barbacar *et al.*, 1997; Matsunaga *et al.*, 1996; Moore *et al.*, 2003; Scutt *et al.*, 2002). As such genes are not necessarily Y linked, these methods do not specifically select for Y-linked genes. The methodology of cDNA subtraction and differential display is described in detail elsewhere (Martin and Pardee, 1999; Sambrook *et al.*, 1989; Straus, 1995), so they are not discussed in detail here.

In contrast to the cDNA library subtraction and the differential display methods, representational difference analysis (RDA) (Lisitsyn *et al.*, 1993) allows subtraction of the female genome from the male genome, resulting in isolation of male-specific Y-linked sequences. This method allows efficient selection of Y-specific sequences, and Donnison et al., (1996) successfully applied this method to isolate Y-specific sequences from the S. latifolia genome (see Appendix A). None of the ten isolated male-specific sequences corresponded to coding regions and all the sequences were represented by multiple copies on the S. latifolia Y chromosome. This is, probably, not surprising because this method does not specifically select for coding sequences and the substantial proportion of the genome (and especially of the Y chromosome) is represented by repetitive noncoding sequences. Thus, RDA is not a very efficient method to search for expressed Y-linked genes. However, it may be useful for the isolation of Y-specific noncoding sequences.

In principle, it would be very efficient to use microarray technology to screen for Y-specific genes. A microarray of cDNA clones from a male plant can be hybridized to a mixture of differentially fluorescently labeled male (e.g., red labeled) and female (e.g., blue labeled) fragmented genomic DNA (or random amplified DNA). Autosomal genes are present in equal proportions in males and females, giving an intermediate ratio of red/blue fluorescence, whereas Y-linked genes should preferentially hybridize with male DNA, giving a much higher red/blue fluorescence ratio. However, to my knowledge, this approach has never been used to search for Y-linked genes in plants.

Surprisingly, only two genes, MROS3X (Guttman and Charlesworth, 1998) and DD44X/Y (Moore et al., 2003), out of several dozens isolated by subtraction and differential display methods appeared to be sex linked, and only one actively expressed Y-linked gene, DD44Y (Moore et al., 2003), was found. This may suggest that subtraction methods do not select for or even select against the sex-linked genes in plants. Indeed, these methods select against Y-linked genes, which have a homologous gene on the X (e.g., SlX1/SlY1, SlX4/SlY4, DD44X/DD44Y) (Atanassov et al., 2001; Delichère et al., 1999; Moore et al., 2003) or on the autosomes (e.g., SlAP3Y) (Matsunaga et al., 2003). As S. latifolia sex chromosomes are fairly young, total nucleotide divergence between the homologous Y- and X-linked (or autosomal) genes does not exceed 10% in coding regions (Filatov and Charlesworth, 2002), allowing for cross-hybridization between the homologous Y- and X-linked (or autosomal) genes. As a result, any Y-linked gene that has an X-linked or autosomal homolog will be selected against by genome subtraction or differential display. Thus, the subtraction-based methods are not suitable for isolation of young Y-linked genes (e.g., S. latifolia SlY1 gene) (Delichère et al., 1999), but they may work for the old Y-linked genes (e.g., S. latifolia SlY4 gene) (Atanassov et al., 2004) or old sex chromosomes. In particular, subtraction methods might be suitable for isolation of Y-specific gene families (NRY class 2 genes in classification of Lahn *et al.*, 2001). Overall, the subtraction-based methods may be used to isolate only a small fraction of plant Y-linked genes, which substantially limits their applicability and success rate.

Chromosome Isolation Methods

Isolation of the Y chromosomes is probably the most direct approach to the cloning of Y-linked genes and noncoding sequences. Y chromosomes can be isolated from the rest of the genome by flow sorting or chromosome microdissection. A serious drawback of these methods is that they require sophisticated and expensive equipment. Another potential problem is the tiny amounts of DNA yielded by these methods, making it necessary to conduct polymerase chain reaction (PCR) amplification of the sequences of the entire Y chromosome using degenerate primers. This procedure is extremely sensitive to contamination, making these methods technically quite challenging. It may also be problematic to use flow sorting and chromosome microdissection with species in which Y chromosomes are difficult to distinguish from the autosomes. This is not a problem with *S. latifolia*, where the sex chromosomes are substantantially larger than the autosomes, and consequently, several studies attempted to isolate Y-specific (Matsunaga *et al.*, 1999b) or X-specific (Buzek *et al.*, 1997) sequences from this species. Similar to the RDA method discussed earlier, the sequences isolated by the chromosome isolation (CI) methods are of noncoding repetitive nature, probably because such sequences are very abundant on the Y chromosome. However, unlike the RDA results, the CI methods yield nonspecific Y-linked sequences that are also present on the autosomes (Buzek *et al.*, 1997; Matsunaga *et al.*, 1999b).

Chromosome flow sorting is often used for isolation of plant chromosomes (Dolezel *et al.*, 1999; Lucretti *et al.*, 1993) (see Appendix B). *S. latifolia* may be thought of as quite a convenient species for this method because the sex chromosomes are larger than any of the autosomes. Indeed, flow sorting was successfully used for isolation of the X chromosomes from this species (Kejnovsky *et al.*, 2001). However, the attempt to isolate *S. latifolia* Y chromosomes by this method was unsuccessful because of substantial contamination of the sorted Y chromosome preparations with clumped autosomes (Kejnovsky *et al.*, 2001).

Several authors used micromanipulator microdissection to isolate sequences from the sex chromosomes of *S. latifolia* (Atanassov *et al.*, 2001; Buzek *et al.*, 1997; Delichère *et al.*, 1999) and *Rumex acetosa* (Shibata *et al.*, 1999). In this method, a steel needle is used to scratch bits of the Y chromosomes from a chromosomal preparation under an inverted micro-scope (see Appendix C). Another way to microdissect Y chromosomes is to use ultraviolet (UV) laser microbeam to destroy all the autosomes and X chromosomes on the chromosomal preparation (Matsunaga *et al.*, 1999b; Scutt *et al.*, 1997a). The existing technology does not allow the recognition of the Y chromosomes automatically, so the Y chromosomes have to be found by eye. Moreover, although the system can burn everything on the chromosome preparation apart from designated small squares around the Y chromosomes, the remaining material in the proximity of the Y chromo-somes has to be destroyed under manual control, limiting the number of Y chromosomes remaining on the preparation after the UV laser microbeam microdissection (see Appendix D).

Both chromosome microdissection and flow sorting yield tiny amounts of DNA, which may be problematic even for highly sensitive PCR amplifi-cation. To increase the amount of genetic material, whole genome ampli-fication (WGA) methods are used (Dietmaier *et al.*, 1999; Zhang *et al.*, 1992). With WGA the entire genome (or the entire Y chromosome in our case) can be amplified using degenerate primers and then can be analyzed in multiple specific PCRs. One WGA approach is degenerate oligonucleotide-primed PCR (DOP-PCR), (Telenius *et al.*, 1992; see Appendix E). The DOP-PCR approach was demonstrated to be inferior, compared to primer-extention-preamplification PCR (PEP-PCR) (Zhang *et al.*, 1992) and to the improved PEP-PCR (I-PEP-PCR) (Dietmaier *et al.*, 1999; see Appendix F). Because WGA methods use degenerate primers, they are quite sensitive to contamination with any other DNA. An inter-esting approach to overcome the contamination problem was proposed by Matsunaga *et al.* (1999a). They used the pressure of a laser beam to manipulate single pollen grains without any physical contact (so-called "laser-pressure catapulting") to reduce the risk of contamination. Matsunaga *et al.* (1999a) used this method for the amplification of the whole *Silene* genome from a single pollen grain.

The attempts to isolate Y- or X-specific sequences from *Silene* and *Rumex* using the methods described earlier yielded only repetitive DNA, suggesting that a substantial proportion of the sex chromosomes consist of repetitive elements. The repetitive sequences isolated from *S. latifolia* X and Y chromosomes hybridized unspecifically to the sex chromosomes and to autosomes (Buzek *et al.*, 1997; Matsunaga *et al.*, 1999b; Scutt *et al.*, 1997a), while *Rumex acetosa* sequences isolated by these methods were Y specific (Shibata *et al.*, 1999), which is consistent with the view that Y chromosomes in *Rumex* are older than those in *Silene*.

Failure to isolate coding sex-linked genes using chromosome isolation or genome subtraction methods is probably due to the lack of a specific coding sequence selection step in these methods. The attempt to introduce such a selection step was undertaken by Delichère *et al.* (1999) and Atanassov *et al.* (2001), who used chromosome microdissection and degenerate oligonucleotide-primed PCR (DOP-PCR) to isolate the pool of *S. latifolia* Y-linked sequences (see Appendix G). They then used this pool of Y-linked sequences as a probe to screen a male flower bud cDNA library for Y-linked genes. Unlike the genome or cDNA library subtraction methods discussed earlier, this approach does not select against the Y-linked genes, which have X-linked or autosomal homologs, and is suitable for isolation of any expressed Y-linked genes. This strategy was quite successful, resulting in the isolation of two expressed *S. latifolia* Y-linked genes, *SlY1* (Delichère *et al.*, 1999) and *SlY4* (Atanassov *et al.*, 2002).

In Situ Hybridization with Plant Chromosomes

A genomic clone can be localized on the chromosomes using fluorescence *in situ* hybridization (FISH) (de Jong *et al.*, 1999). The method is based on hybridization of a fluorescently labeled DNA probe to mitotic or meiotic chromosomes spread on a glass slide. To visualize the chromosomes, they are usually stained with blue-fluorescing 4′,6-diamino-2-phenylindole (DAPI) before the analysis. Using several fluorochromes, it is possible to hybridize several probes to the same slide (multicolor FISH) (Speicher *et al.*, 1996). This is particularly useful if the sex chromosomes are not readily distinguishable under the microscope, as the chromosomes may be identified by hybridization of a probe of known location, or even a specific "chromosome paint," the DNA probe specifically hybridizing to the whole chromosome. Such "chromosomal paints" can be developed using DOP-PCR with flow-sorted or microdissected Y chromosomes or simply using a mixture of BAC clones from the same chromosome, if available. Transposable elements are often very abundant on the Y chromosomes and can be used as probes preferentially hybridizing with the Y chromosome.

The difficulty in distinguishing the arms of the sex chromosomes in *S. latifolia* stimulated the search for FISH markers in this species. One of the first FISH markers was identified by Buzhek *et al.* (1997). Isolation of further X- and Y-linked sequences allowed more precise FISH analysis, which resulted in a revision of the map of the *S. latifolia* sex chromosomes (Lengerova *et al.*, 2003). Thus, FISH is a valuable method that allows for fairly accurate localization of the genomic clones on the chromosomes. We have successfully used the FISH protocol to localize the *SlX1/SlY1* and *SlX4/SlY4* genes on the *S. latifolia* sex chromosomes (Appendix H).

In principle, FISH can be used to identify new Y-linked sequences. In practice, however, several complications make it difficult to use this approach to search for new Y-linked sequences. One serious limitation of FISH is the size of the probe used for hybridization. According to our experience, it is fairly difficult to get a good signal with probes smaller than 5 Kb. In practice, much larger probes (cosmids and BAC clones) are often being used (de Jong et al., 1999). Such long genomic DNA fragments usually contain repetitive sequences, resulting in hybridization to several chromosomes (Lengerova et al., 2004). Duplications of the genomic regions may have similar effects, creating problems with interpretations of the FISH results. If the sex chromosomes or chromosome arms are difficult to recognize, multiprobe FISH has to be used to localize the new FISH probe relative to the positions of the known FISH markers (Lengerova et al., 2003). With multiple probes, the probe cross-hybridization may be a serious problem, which we found difficult to overcome. Overall, although FISH is very useful for localization of plant genes, we found it extremely time consuming and labor intensive to be used in a search for new Y-linked genes, when many clones have to be tested.

Segregation-Based Tests

Segregation analysis represents the most reliable approach to test the sequences for sex linkage. Regardless of the nature of the molecular variants, father-to-son inheritance demonstrates Y linkage, whereas inheritance of paternal variants by daughters only indicates X linkage of the gene (Guttman and Charlesworth, 1998). This approach requires genotyping of parents and offspring, which may be labor intensive for large numbers of markers. If it is required to isolate only Y-linked markers, the use of the family is not necessary and the procedure can be simplified: The male-specific markers in a population of several unrelated males and females should be Y linked. Isolation of X-linked markers, however, requires segregation analysis in a family. If X-linked markers are of interest, but the separate analysis of every individual in the progeny is too costly, the bulked segregant analysis (BSA) (Michelmore et al., 1991) may be used. BSA screen for sex-linked markers involves a comparison of two pooled DNA samples, of females and of males from a segregating population originating from a single cross. If a marker present in the father is inherited by the male bulk, but not by the female bulk, it should be Y linked, whereas the inheritance of paternal variants only by the female bulk should indicate X linkage. These straightforward analyses can be applied to any kind of genetic markers: random amplified polymorphic DNA (RAPD) (Williams et al., 1990), amplified fragment length polymorphism (AFLP) (Vos et al.,

1995), sequence characterized amplified regions (SCARs), cleaved amplified polymorphic sequences (CAPS) (Konieczny and Ausubel, 1993), single nucleotide polymorphisms (SNPs), and so on.

RAPDs and AFLPs

RAPDs are the simplest and the cheapest markers to develop (Williams et al., 1990). The protocol uses short (~10 bases) primers in a low-stringency PCR to generate a number of randomly primed PCR products, some of which may be male specific. The number of resolved PCR products depends on the electrophoresis conditions. With agarose gels, it is difficult to resolve more than 10 PCR products. Thus, to screen a sufficient number of bands for sex linkage, it is necessary to use many combinations of 10-mer PCR primers. The development of AFLP markers is somewhat more laborious, including digestion of genomic DNA by restriction enzymes, ligation of oligonucleotide adapters to the ends of digested genomic fragments, and PCR amplification with primers homologous to the ligated adapters. As the number of PCR products obtained this way may be very large, several additional "selective" nucleotides are added to the 3' end of the PCR primers to reduce the number of PCR products (Vos et al., 1995). PCR products are resolved on the sequencing gels. This approach is more powerful, compared to the RAPD, because it allows up to 100 PCR products to be resolved on a single gel. Sex-specific PCR products detected by RAPD or AFLP may be excised from the gel, purified, and sequenced to produce more reliable and informative SCARs and sequence-tagged sites (STSs). This approach has been used successfully to isolate sequences from the Y chromosomes of *Rumex nivalis* (Stehlik and Blattner, 2004) and *S. latifolia* (Nakao et al., 2002; Obara et al., 2002; Zhang et al., 1998). In most cases, the isolated sequences appeared to be repetitive DNA and only Nakao et al. (2002) succeeded in isolating a region containing a Y-specific ORF using RAPDs. Repetitive DNA (e.g., transposable elements) is often present on several chromosomes, and in many cases, the Y-linked sequences isolated by RAPD or AFLP gave FISH hybridization to several autosomes as well as to the Y chromosome (Zhang et al., 1998). The obvious way to avoid isolation of repetitive DNA in a screen for sex-linked sequences is to test the individual coding genes for sex linkage.

Single-Strand Conformation Polymorphism (SSCP)

Guttman and Charlesworth (1998) successfully used "cold SSCP" analysis to demonstrate X linkage of the *S. latifolia* MROS3 gene. The method is based on the effect of nucleotide polymorphisms on conformation of the denatured single-stranded DNA and its mobility in polyacrylamide gels (Orita *et al.*, 1989). Unlike the original SSCP protocol, the "cold SSCP" does not require the use of radiolabeled PCR primers and long autoradiography exposures to visualize the SSCP bands. Instead, ethidium bromide staining is used, reducing the time of the SSCP analysis to approximately 3 h, and simplifying the excision of the SSCP bands from the gel for sequence analysis (Hongyo *et al.*, 1993).

Although this method is suitable for segregation analysis of plant genes, it requires careful fine-tuning of the conditions of the SSCP analysis. The SSCP banding pattern is quite sensitive to temperature during electrophoresis, concentration of the PCR product, and the nature of the chemicals used to denature the DNA, resulting in occasional appearance of metastable "ghost" bands, double-stranded, or blurred SSCP bands (Hongyo *et al.*, 1993). In addition, SSCP analysis is limited to fairly short PCR products (<0.4 kb). If the size of the PCR products is not known *a priori* (e.g., if genomic sequence is not known and cDNA sequences are used to design PCR primers), it is necessary to analyze the size and the quality of the PCR product before the SSCP analysis. If the product is too long, restriction digestion of the PCR products has to be conducted. To simplify the interpretation of the SSCP banding pattern, it is advisable to analyze a single PCR or restriction digestion product, often making it essential to excise individual bands from agarose gels and analyze these bands in separate SSCP reactions. In principle, it is possible to analyze several bands in the same SSCP reaction, but in this case the banding pattern may be quite complex and difficult to interpret. As only the presence/absence of the particular SSCP band can be scored (and not always reliably), such markers have to be treated as dominant.

In practice, SSCP analysis of every gene requires time-consuming individual adjustments of the conditions, making it unsuitable for parallel analysis of several genes. Direct sequencing of the PCR products may be a more reliable, faster, and cheaper way to find markers suitable for segregation analysis (see below).

Restriction Cleavage-Based Polymorphisms (CAPS)

Using the published sequences of unlocalized genes or random cDNA clones, it is possible to design PCR primers to amplify short regions of such genes from parents and offspring and to analyze the segregation of the amplified region using restriction enzymes (the CAPS markers). The cheapest but fairly time-consuming aproach to find CAPS markers segregating in a family is to use a panel of restriction enzymes to digest the PCR products from parent individuals. Polymorphisms identified this way may be used to analyze the segregation in the progeny. A somewhat more expensive but much faster way to identify polymorphic restriction sites is to compare the directly sequenced PCR products from the parents. This approach was undertaken by Laporte and Charlesworth (2001), to analyze segregation of six *S. latifolia* genes. None of the analyzed genes appeared to be sex linked.

Sequence Analysis

The use of the CAPS markers limits the choice of polymorphisms to those that occur at restriction sites. With decreasing sequencing costs, it is now feasible to sequence PCR products from parents and progeny to trace the segregation of polymorphisms at the DNA level (SNPs, insertions, and deletions). As the DNA diversity in plant species is usually quite high (e.g., ~1.5% in *S. latifolia*; Filatov *et al.*, 2000, 2001), sequencing 300–500 bp (less than a single read on modern sequencing machines) usually results in detection of several SNPs segregating in a family. The segregation of these SNPs allows rapid testing of genes for sex linkage. A segregation pattern incompatible with sex linkage usually allows a gene to be identified as non–sex-linked based on the analysis of only few individuals. To minimize the costs and allow high throughput, it is possible to PCR amplify and directly sequence a gene from the two parents, four male and two female progeny, making this approach compatible with the 96-well plate format: Every column (eight wells) in a plate corresponds to a single gene, allowing up to 12-genes to be tested per plate. Those genes demonstrating segregation compatible with X (from father to daughters, but not sons) or Y linkage (from father to sons, but not daughters) should be tested further using more male and female progeny. This approach was recently used for isolation of new *S. latipolia* sex-linked genes, s/ssX and s/ssY (Filatov, 2005). In principle, the efficiency of this approach can be increased using bulked segregant analysis (BSA) (see above), using only four wells per gene (24 genes per 96-well plate) with PCR products from the two parents and the male and female bulks. However, in practice, the indel differences between the segregating alleles often result in unreadable sequences in the bulks, making the sequence analysis of bulked progeny problematic.

Sequencing-based segregation analysis is compatible with high throughput and is fairly robotics friendly. The use of 96-well plates for PCR and sequencing allows most stages of the process to be automated. The only stage that is difficult to automate is separation of the PCR products on the agarose gels and excision of individual PCR bands from the gel. This stage is necessary because the PCR primers designed using cDNA sequence often yield several bands because of nonspecific annealing or the presence of several gene copies in the genome, and occasionally fail to produce a PCR product, possibly because of presence of introns in the priming sites or long introns between the primers. The gel-based separation of PCR products allows for simultaneous control of the specificity and yield of the PCRs and elimination of nonspecific PCR products. In principle, it might be possible to replace gel purification by much more robotics-friendly analytical high-performance liquid chromatigraphy (HPLC) separation of PCR products (Wages et al., 1995). However, the existing HPLC equipment is quite expensive and is not suitable for the high-throughput 96-well format, making this option impractical.

Data analysis presents another problem with automation of segregation analysis. Although existing programs allow for automatic SNP detection (e.g., PHRED; Ewing et al., 1998), the task is more complex: The system has to track the relationship between the particular individuals in the family to analyze the segregation of SNPs. The optimal solution would be to have an integrated system that stores the information about all the individuals, genes, and SNPs in a relational database, analyze sequence data to find SNPs, and trace the segregation of these SNPs automatically. It is possible to assemble such a system on a UNIX machine using existing programs to assess the sequence quality (e.g., PHRED), find SNPs, store them in a relational database (e.g., MySQL), and to use perl scripts to "glue" all of the components together. However, such a solution requires programming skills and would be fairly time consuming to implement and difficult to deploy. To our knowledge, there is no software that can perform all the tasks required for SNP segregation analysis and that is sufficiently user friendly, easy to use, and cheap.

Thus, for the moment, complete "walk-away" automation of the sequencing-based segregation analysis is not possible; however, the automation of certain stages of this process makes it substantially less time consuming and labor intensive.

SNP Genotyping

Sequence analysis requires sequencing of the same region of a gene from both parents and progeny to find SNPs and to trace their segregation. Although the cost of sequencing is reasonably low, it may be more cost effective to use SNP genotyping for segregation analysis. An additional advantage of this approach is highly parallel design, allowing many genes to be genotyped simultaneously. Depending on the technology used, it is possible to analyze hundreds or even thousands of genes simultaneously (reviewed in Syvänen, 2002). The major difficulty in the application of these methods to plants is the lack of already characterized SNPs (except

SNPs in *Arabidopsis*, which are not of much use for the analysis of plant sex chromosomes).

Various SNP-hunting stratagies (Syvänen, 2002) can be employed to search for SNPs in the plant of interest. The simplest strategy to follow is to sequence a short region of a gene of interest in parent individuals to discover the nucleotide differences suitable for segregation analysis and to use SNP genotyping methods to trace the segregation of these polymorphisms in the progeny. Once the way to genotype a certain SNP was developed, it became very fast and easy to genotype a large number of individuals. The discussion of various SNP genotyping techniques is beyond the scope of this chapter and only the simplest SNP genotyping technique, allele-specific oligonucleotide (ASO) primed PCR, is discussed. This technique is easy and inexpensive to implement in any laboratory without specialized equipment.

The principle of the ASO-PCR is simple (Fig. 1A): As the PCR is sensitive to the correct match of the very 3′ end nucleotide of the PCR primers, making one of the PCR primer with the 3′ end nucleotide matching one of the alleles at the polymorphic site would make it allele specific (ASO primer). Such a marker is dominant—ASO primer will work in individuals heterozygous or homozygous for the matching nucleotide, but not in homozygotes for the mismatching nucleotide (Fig. 1B). This makes the method suitable for testing for sex linkage using a single ASO-PCR per individual. If the paternal individual is homozygous (or hemizygous) for, say, "G" at a certain SNP and the maternal individual does not have "G" at the same position (homozygous or heterozygous for some other nucleotide), then using a primer pair with one of the primers being allele specific, it is possible to trace the segregation of the paternal allele from the patterns of successful and unsuccessful PCR amplification in the progeny. X-linked genes will give successful PCR amplification only in the female progeny, but not in the male progeny, whereas Y linkage should result in successful PCR amplification in the male progeny but not in the female progeny. In practice, PCR is not 100% successful even if the primers match. Thus, it is more reliable to conduct two separate ASO-PCR for each individual to genotype the particular SNP: one reaction with the primers specific to one allele and the other reaction with the primers specific to the other allele. Both PCR will be successful in a heterozygous individual, wherease one successful and one failed reaction would indicate a homozygote (Fig. 1B). Testing both SNP alleles provides a positive control against PCR failure.

Because this approach allows for parallel genotyping of multiple SNPs, segregation (and co-segregation) of many genes can be tested simultaneously. Conducting PCR in 96-well plates makes it is possible to genotype

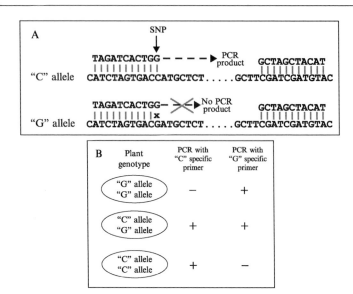

Fig. 1. SNP genotyping with allele-specific oligonucleotide primed polymerase chain reaction PCR product is synthesized only if the very 3′ nucleotide of the ASO primer corresponds to the nucleotide at the SNP site. (B) The possible outcomes of the ASO-PCR to genotype the C/G SNP. The PCR with ASO primers designed to specifically amplify the "C" allele should be successful in heterozygotes and homozygotes for this allele, and ASO-PCR with "G" allele-specific primers should be successful in heterozygotes and homozygotes for the "G" allele. Thus, both successful reactions indicate heterozygotes, whereas one of the reactions being successful indicates one of the homozygotes.

up to 48 SNPs in the same individual simultaneously (testing for both alleles in every SNP). As the success of every PCR has to be tested on a gel, this stage is a serious limitation for the ASO-PCR approach, making it unsuitable for high-throughput applications. To avoid this limitation, the success/failure of the ASO-PCR can be checked automatically using Taq-Man (or similar) probes in a real-time PCR machine or using a spectro-photometer (Lee et al., 1993). Such probes contain fluorescent dye and a quencher. If PCR is successful, quencher is released and fluorescence gradually increases over time, which can be detected by the real-time PCR machine or a spectrophotometer. Although TaqMan probes are quite expensive, this approach is very efficient because it eliminates the time-consuming stage of gel electrophoresis and isolation of PCR products from the gel.

According to our experience, the following ASO-PCR–based strategy is optimal to search for plant sex-linked genes. The approach below assumes that at least partial sequences of the genes to be tested are known. Such sequences can be obtained by sequencing random clones from a cDNA library. These sequences can be used to design PCR primers for the amplification of short (up to 2 kb) regions from these genes. PCR amplification and agarose gel electrophoresis of a region from maternal and paternal individuals may allow the detection of size differences between the paternal and the maternal PCR products. Such size differences can be used as molecular genetic markers, which allows more costly SNP genotyping analysis to be avoided. If the size of the maternal and paternal PCR products is the same, then the bands should be cut from the gel and extracted from agarose. For this purpose, we usualy use the Qiagen gel extraction kit, which yields sufficiently clean DNA to be sequenced directly. The same PCR primers usually can be used for sequencing. The comparison of the paternal and maternal sequences often allows to find nucleotide polymorphisms (SNPs), which have different alleles in the maternal and paternal genotypes. Segregation of these SNPs in the progeny can be tested using ASO-PCR. The genomic DNA from each individual in the progeny can be used in two PCR amplifications, each with one of the two ASO primers, working in pair with the primer used in the original PCR. To increase the throughput of the method, we use the bulked segregant analysis (Michelmore et al., 1991), pulling the genomic DNA from 10 male and 10 female progeny into the male and the female bulks, respectively. Although the genotypes of the parents were detected by direct sequencing, it is essential to conduct ASO-PCRs for the parents and the progeny bulks simultaneously to test the specificity of the ASO-PCR. As two ASO-PCRs have to be conducted for each of the parents and for the male and female bulks, in total, eight ASO-PCRs are needed to study the segregation of a single gene. Thus, it is very convenient to conduct the genotyping in the 96-well plates, testing 1 gene in eight "column" wells, allowing 12 genes per plate to be tested.

Genomics-Based Approaches

Genomics approaches have become increasingly common in isolation of Y-specific sequences (Liu et al., 2004; Okada et al., 2000). Isolation of large regions of the Y chromosome in the form of cosmids, BAC, or PAC clones can be achieved by screening a genomic (e.g., BAC) library using Y-specific markers, isolated in a preliminary step using one of the approaches described in this chapter (microdissection, flow sorting, FISH, and segregation analysis). Alternatively, it is possible to extend already cloned region(s) of the Y chromosome (e.g., sequenced Y-linked genes, if available) by "chromosome walking," using ends of the already cloned

Y-linked sequences to screen for clones that extend further. This approach was successfully employed to construct a contig covering the male-specific region on the *Carica papaya* primitive sex chromosomes (Liu *et al.*, 2004). For the chromosome-walking approach to be successful, a representative (at least x10 coverage) genomic library is required. This may be prohibitively expensive for species with large genomes, such as *S. latifolia*.

Regardless of the approach used, the Y linkage of the candidate clones should be tested by segregation analysis. Segregation analysis of individual random BAC clones by itself may be an efficient strategy to isolate large regions of the Y chromosome. To test the segregation of a BAC clone, it is usually sufficient to sequence short regions at the ends of the clone, design PCR primers to these regions, amplify them from the parents and the progeny in a family, and use one of the approaches described earlier to find the genetic markers in these regions and trace their segregation in the progeny. The isolated Y-linked genomic clones may be used further as a probe to screen a male cDNA library for expressed Y-linked genes.

Clearly, the genomics "brute-force" approach is fairly expensive; however, it may be very successful for agriculturally important organisms with small genomes, such as papaya (Liu *et al.*, 2004). Y chromosomal contigs of less agriculturally important plants with large genomes, such as *Silene* or *Rumex*, will probably never be constructed and the other approaches, such as segregation analyses, have to be used.

Acknowledgments

I thank Joe Ironside and Dave Gerrard for critical reading of the manuscript and the BBSRC for funding.

Appendix A

Representational difference analysis (RDA) (Lisitsyn *et al.*, 1993): Digest male and female DNA with the same rare cutting restriction enzyme (e.g., *Bgl*II, *Eco*RI, *Bam*HI) and ligate adapters to the genomic fragments in such a way to ensure reconstruction of the original restriction site. PCR-amplify the male and female genomic DNA using oligonucleotide primers to the adapters. Digest the amplification products with the same restriction enzyme to remove the adapters. Ligate new unphosphorylated oligonucleotide adapters to the amplified tester (male) DNA only. As the oligonucleotides lack 5'-phosphate, they will only be ligated to the 5' end of the genomic DNA fragments. Denature male DNA and anneal it to an excess of denatured female genomic DNA. Because of excess female genomic DNA, most male genomic fragments having homology in the female genome will anneal to female genomic fragments, and male/male double-strand fragments will be formed mostly for male-specific sequences. As male fragments have adapters at the 5' end only, after annealing, only male/male double-strand fragments will have adapters at both ends. Thus, in the PCR with primers homologous to the adapters, only such male/male fragments with adapters at both ends will be amplified, resulting in massive enrichement for male-specific fragments in the resulting PCR products. The pool of DNA fragments after the first subtraction can be used as tester DNA in further rounds of RDA to remove all the non–male-specific DNA fragments, which can still be present after the first round of subtraction.

Appendix B

Chromosome flow sorting (Dolezel *et al.*, 1989; Kejnovsky *et al.*, 2001): Incubate root tips for 2–5 h in a 0.05% solution of colchicine for metaphase synchronization (Pan *et al.*, 1993). Cut synchronized root tips (~1 mm), rinse them in distilled water, fix for 20 min at 5° in 2% formaldehyde made in Tris buffer (10 m*M* Tris, 10 m*M* Na$_2$EDTA, 100 m*M* NaCl pH 7.5), supplemented with 0.1% Triton X-100, and wash (three times for 5 min) in the Tris buffer. Transfer approximately 200 root tips into 1 ml of LB01 lysis buffer (Dolezel *et al.*, 1989). Release the chromosomes by mechanical homogenization (e.g., with a Polytron PT1200 homogenizer, Kinematika AG, Littau, Switzerland) at 15,000 rpm for 10 s. Pass the suspension through a 50-μm pore size nylon mesh and stain with 4'6-diamidino-2-phenylinodole (DAPI) at a final concentration of 2 μg/ml. Use a flow cytometer (e.g., FACSVantage, Becton Dickinson, San Jose, CA) equipped with a chromosome sorter to sort approximately 25,000 chromosomes at rates of 5–10/s into an Eppendorf tube containing deionized water. To achieve the highest purity in sorted fractions, the second sorting step can be added. In this case, during the first step sort the chromosomes into a tube with 1.5X LB01 buffer and add DAPI before the second sorting step. The sorted chromosomes can be storred in deionized water at −70° or used directly for DOP-PCR (see Appendix E).

Appendix C

Micromanipulator microdissection (Buzek *et al.*, 1997): Synchronize root tip meristems incubating root tips for 2–5 h in a 0.05% solution of colchicine or 2 m*M* 8-hydroxyquinoline (Pan *et al.*, 1993). Fix them in 45% acetic acid, or 1:3 acetic acid:ethanol. Digest the fixed root tips for about 30 min at 37° by a mixture of 2.5% pectolyase Y-23 and 2.5% cellulase

Ozonuka R-10 in 2X SSC pH 4.5 and squash them under coverslips after fixation in 45% acetic acid. Remove the coverslips using the dry-ice or liquid nitrogen method. Microdissect the Y chromosomes under an inverted microscope using a micromanipulator and collect the micro-dissected chromosomes in a droplet of distilled water or a collection solution (10 mM Tris–HCl pH 8.0, 10 mM NaCl, 0.1% sodium dodecyl sulfate) overlaid with water-saturated paraffin oil. The microdissected Y chromosomes can be used for DOP-PCR (see Appendix E).

Appendix D

Laser microdissection (Matsunaga *et al.*, 1999b): Treat the harvested roots with 15 mg/ml of aphidicolin at 26° for 23 h, rinse, and incubate in distilled water for 5 h. Synchronize the root tips with 0.05% colchicine (2 h in the dark), incubate in distilled water at 4° for 24 h, and fix for 5 min in ice-cold 1:3 glacial acetic acid: ethanol. For tissue maceration, an incubation with 2% cellulase and 2% pectolyase (in 75 mM KCl and 7.5 mM EDTA, pH 4) at 37° for 40 min can be used, followed by a rinse in distilled water. For laser microdissection, transfer the tissues into a drop of 45% acetic acid on a laser-penetrable film (Dura Seal, Diversefield Biotech, Boston, MA), stuck on a polilysin-coated cover glass in a Glass Bottom Culture Dish (Meridian, Tokyo, Japan). After the acetic acid evaporates at room temperature, cut the film into 0.5- by 0.5-mm squares. Stain the chromosomes with 5% Giemsa solution (Merck) in PBS buffer (137 mM NaCl, 2.7 mM KCl, 4.3 mM Na$_2$HPO$_4$, 1.4 mM KH$_2$PO$_4$) for 10 min and microdissect with an ultraviolet laser microbeam, sublimating all the regions of the preparation except the Y chromosomes. This can be done using a laser microscope, such as the PALM laser microscope system (PALM, GmbH, Wolfrastshausen, Germany), coupled with an Axiovert 135 inverted microscope (Carl Zeiss, Oberkochen, Germany). After the microdissection, place the film with the remaining Y chromosomes in a PCR tube contaning PCR buffer (40 mM Tris–HCl pH 7.5, 20 mM MgCl$_2$, 50 mM NaCl), vortex for 10 min and incubate overnight at 4°. Prior to PCR, treat the preparation with topoisomerase I (1 unit, 37° for 30 min) and hit-inactivate it at 96° for 10 min. This preparation can be used for DOP-PCR (see Appendix E).

Appendix E

Degenerate oligonucleotide-primed PCR (DOP-PCR) (Cheung and Nelson, 1996; Telenius *et al.*, 1992): DOP-PCR is conducted with one or two primers defined at both ends and a random hexamer or heptamer sequence in the middle. For example, Matsunaga et al. (1999a) used the primers 7-MW1, 5′-CCGACTCGAGNNNNNNNCCCT-3′, and 7-MW2, 5′-CCGACTCGAGNNNNNNNCCTCC-3′, and Delichère et al. (1999) used the primers 5′-CCGACTCGAGNNNNNNNATGTGG-3′, 5′-CCGACTCGAGNNNNAANNATGG-3′. The following PCR buffer can be used: 40 mM Tris–HCl pH 7.5, 20 mM MgCl$_2$, 50 mM NaCl, and 200 μM dNTPs. The PCR is performed at low stringency conditions for the first five cycles and with more stringent conditions for the following 35 cycles.

Appendix F

Improved primer-extention-preamplification PCR (I-PEP-PCR) (Dietmaier et al., 1999): A single totally degenerate 15-mer primer is used in PEP-PCR (and I-PEP-PCR). Dietmaier et al. (1999) used the following 50-μl I-PEP-PCR mixture: final concentration 0.05 mg/ml gelatine, 16 μmol/L totally degenerate 15-mer primer, 0.1 mmol/L dNTPs, 3.6 units Expand High Fidelity polymerase (Roche), 2.5 mmol/L MgCl$_2$ and the PCR buffer 3 from the Expand High Fidelity PCR system (Roche). In each of 50 cycles of the I-PEP-PCR, the denaturing step (94°, 1 min) is followed by low-stringent annealing at 37° (2 min), which is then continuously increased to 55° (0.1°/s), high-stringent annealing and extention at 55° (4 min), and an extension step at 68° (30 s). The improvements introduced to I-PEP-PCR, compared to PEP-PCR, are that Expand High Fidelity polymerase (Roche) is used instead of normal Taq polymerase and an extra extension step at 68° is added at the end of each cycle, which substantially increased the yield and sensitivity of the reaction (Dietmaier et al., 1999).

Appendix G

cDNA library screen using Y-specific probe (Delichère et al., 1999): Isolate Y chromosomes by flow sorting or microdissection as described in Appendixes C and D. Use at least 10 microdissected Y chromosomes as a template in DOP-PCR or I-PEP-PCR. The products of this PCR amplification can be used as the Y-derived probe to screen a male flower bud cDNA library. Radiolabel 2 μg of the Y-specific probe using 500 mCi of [α-^{32}P]dCTP and a Random Priming DNA kit (Roche). Use this probe to screen about 35,000 phage plates as discribed by Sambrook et al. (1989). Perform hybridization at 65° in 5X SSPE (0.75 M NaCl, 50 mM NaH$_2$PO$_4$, 5 mM EDTA), 5X Denhardt solution (Sambrook et al., 1989), 1% Sodium dodexyl sulfatt (SDS), and 0.1 mg/ml of sperm DNA. Wash four times (or more if needed) with the decreasing concentration of SSC and SDS

(Sambrook *et al.*, 1989): 15 min at 65° in 2X SSC, 0.1% SDS, 15 min at 65° in 1X SSC, 0.1% SDS, 15 min at 65° in 0.5X SSC, 0.1% SDS, and 0.1X SSC, 0.1% SDS. Each positive clone selected in this screen should be further tested for sex linkage using segregation analysis.

Appendix H

FISH with *S. latifolia* meiotic chromosomes (Filatov and Armstrong, unpublished): Fix anthers in Carnoy's fixative (6:3:1 ethanol:chloroform: acetic acid) for 15–30 min at room temperature, changing fixative every 10 min. Rinse in 3:1 (ethanol:acetic acid) and wash three times for 2 min to remove chloroform. Wash in citrate buffer (pH 4.5) for 2 min at room temperature. Incubate anthers with enzyme mixture (0.3% pectolyase and 0.3% cellulase in citrate buffer) at 37° for 1 h and stop the reaction transferring the tissue into ice-cold distilled water. Place anther on a glass slide and tap the tissue out in a small drop of water. Add 10 μl 60% acetic acid, heat the slides to 45° in a hot block, add 20 μl 60% acetic acid, and stir gently with a needle for about 1 min. Cool the slide and fix with cold 3:1 (ethanol:acetic acid) and air-dry the preparation with a hair dryer. Rinse the slide in 2X SSC for 10 min at room temperature and place it in a 0.00001% solution of proteinase K for 5 min at room temperature. Rinse the slide in 2X SSC and fix in freshly prepared 4% paraformaldehyde (pH 8) for 10 min at room temperature. Rinse in distilled water, dehydrate in 70%, 85%, and 100% ethanol and air-dry the slide. Denature the hybridization mixture (the biotinilated probe in 2X SSC, 50% deionized formamide, and 10% dextran sulfate) at 96° for 5 min and cool it down on ice. Apply about 20 μl of the hybridization mixture to the slide, cover with a coverslip, surround with rubber glue to prevent the slide from drying out, and denature at 75° for 2 min. Transfer the slides to a moist chamber and leave overnight at 37°. After hybridization, remove the coverslips with a razor blade, wash three times in a 50% solution of formamide in 2X SSC at 45° for 5 mins, once in 2X SSC at 45° for 5 mins, once in 4X SSC and 0.05% Tween 20 at 45° for 5 mins and once in 4X SSC and 0.05% Tween 20 at room temperature. Incubate the slide with 5 μg/ml streptavidin-Cy3 conjugate (Cambio, UK) in the blocking buffer (4X SSC, 0.05% Tween 20, 0.05% fat-free dry milk) for 20 min. Wash three times in 4X SSC, 0.05% Tween 20 solution for 5 min. Incubate the slide with 5 μg/ml biotinilated anti-avidin (Vector Labs) in the blocking buffer for 20 mins, and wash three times in 4X SSC, 0.05% Tween 20 solution for 5 mins. Incubate the slide with 5 μg/ml streptavidin-Cy3 conjugate (Cambio, UK) in the blocking buffer (4X SSC, 0.05% Tween 20, 0.05% fat-free dry milk) for 20 min. Wash three times in 4X SSC, 0.05% Tween 20 solution for 5 min. Add 7 μl of the 10 μg/ml DAPI solution in Vectorshield mounting medium (Vector Labs), cover with a coverslip, and analyze the slides under a fluorescent microscope.

References

Ainsworth, C. C. (1999). "Sex Determination in Plants." Bios Scientific Publishers, Oxford.

Atanassov, I., Delichère, C., Filatov, D. A., Charlesworth, D., Negrutiu, I., and Moneger, F. (2001). Analysis and evolution of two functional Y-linked loci in a plant sex chromosome system. *Mol. Biol. Evol.* **18,** 2162–2168.

Bachtrog, D., and Charlesworth, B. (2002). Reduced adaptation of a non-recombining neo-Y chromosome. *Nature* **416,** 323–326.

Barbacar, N., Hinnisdaels, S., Farbos, I., Moneger, F., Lardon, A., Delichère, C., Mouras, A., and Negrutiu, I. (1997). Isolation of early genes expressed in reproductive organs of the dioecious white campion (*Silene latifolia*) by subtraction cloning using asexual mutant. *Plant J.* **12,** 805–817.

Bischler, H. (1998). Systematics and evolution in the genera of the *Marchantiales. Bryophyt. Biblioth.* **51,** 1–201.

Buzek, J., Koutnikova, H., Houben, A., Riha, K., Janousek, B., Siroky, J., Grant, S., and Vyskot, B. (1997). Isolation and characterisation of X chromosome-derived DNA sequences from a dioecious plant *Melandrium album. Chromosome Res.* **5,** 57–65.

Cheung, V., and Nelson, S. F. (1996). Whole genome amplification using a degenerate oligonucleotide primer allows hundreds of genotypes to be performed on less than one nanogram of genomic DNA. *Proc. Natl. Acad. Sci. USA* **93,** 14676–14679.

Charlesworth, B. (1996). Background selection and patterns of genetic diversity in *Drosophila melanogaster. Genet. Res. Cambs.* **68,** 131–150.

Charlesworth, B., and Charlesworth, D. (2000). The degeneration of Y chromosomes. *Philos. Trans. R. Soc. Lond. B Biol. Sci.* **355,** 1563–1572.

Delichère, C., Veuskens, J., Hernould, M., Baarbacar, N., Mouras, A., Negrutiu, I., and Monéger, F. (1999). *SlY1,* the first active gene cloned from a plant Y chromosome, encodes a WD-repeat protein. *EMBO J.* **18,** 4169–4179.

Dietmaier, W., Hartmann, A., Wallinger, S., Heinmoller, E., Kerner, T., Endl, E., Jauch, K.-W., Hofstadter, F., and Ruschoff, J. (1999). Multiple mutation analyses in single tumor cells with improved whole genome amplification. *Am. J. Pathol.* **154,** 83–95.

Dolezel, J., Binarova, P., and Lucretti, S. (1989). Analysis of nuclear DNA content in plant cells by flow cytometry. *Biol. Plant* **31,** 113–120.

Dolezel, J., Macas, J., and Lucretti, S. (1999). Flow analysis and sorting of plant chromosomes. *In* "Current Protocols in Cytometry" (J. P. Robinson, Z. Darzynkiewicz, P. N. Dean, L. G. Dressler, and H. A. Crissman, eds.), pp. 5.3.1–5.3.33. John Wiley and Sons, New York.

Donnison, I. S., Siroky, J., Vyskot, B., Saedler, H., and Grant, S. R. (1996). Isolation of Y chromosome-specific sequences from *Silene latifolia* and mapping of male sex-determining genes using representational difference analysis. *Genetics* **144,** 1893–1901.

Ewing, B., Hillier, L., Wendl, M. C., and Green, P. (1998). Base-calling of automated sequencer traces using *phred.* I. Accuracy assessment. *Genome Res.* **8,** 175–185.

Filatov, D. A. (2005). Substitution rates in a new silene latipolia sex-linked gene, s/ssX/Y. *Mol. Biol. Evol.* **22,** 402–408.

Filatov, D. A., Moneger, F., Negrutiu, I., and Charlesworth, D. (2000). Low variability in a Y-linked plant gene and its implications for Y-chromosome evolution. *Nature* **404,** 388–390.

Filatov, D. A., Laporte, V., Vitte, C., and Charlesworth, D. (2001). DNA diversity in sex linked and autosomal genes of the plant species *Silene latifolia* and *S. dioica.*. *Mol. Biol. Evol.* **18**, 1442–1454.

Filatov, D. A., and Charlesworth, D. (2002). Substitution rates in the X- and Y-linked genes of the plants, *Silene latifolia* and *S. dioica.*. *Mol. Biol. Evol.* **19**, 898–907.

Guttman, D. S., and Charlesworth, D. (1998). An X-linked gene with a degenerate Y-linked homologue in a dioecious plant. *Nature* **393**, 263–266.

Hongyo, T., Buzard, G. S., Calvert, R. J., and Weghorst, C. M. (1993). "Cold SSCP": a simple, rapid and non-radioactive method for optimized single-strand conformation polymorphism analyses. *Nucleia Acids Res.* **21**, 3637–3642.

de Jong, J. H., Fransz, P., and Zabel, P. (1999). High resolution FISH in plants—techniques and applications. *Trends Plant Sci.* **4**, 258–263.

Kejnovsky, E., Vrana, J., Matsunaga, S., Soucek, P., Siroky, J., Dolezel, J., and Vyskot, B. (2001). Localisation of male-specifically expressed MROS genes of *Silene latifolia* by PCR and flow-sorted sex chromosomes and autosomes. *Genetics* **158**, 1269–1277.

Konieczny, A., and Ausubel, F. (1993). A procedure for mapping *Arabidopsis* mutations using codominant ecotype-specific PCR-based markers. *Plant J.* **4**, 403–410.

Lahn, B. T., Pearson, N. M., and Jegalian, K. (2001). The human Y chromosome, in the light of evolution. *Nat. Rev. Genet.* **2**, 207–216.

Laporte, V., and Charlesworth, D. (2001). Non-sex–linked, nuclear cleaved amplified polymorphic sequences in *Silene latifolia*. *J. Heredity* **92**, 357–359.

Lee, L. G., Connell, C. R., and Bloch, W. (1993). Allelic discrimination by rick-translation PCR with fluorogenic probes. *Nucleic Acid Res.* **21**, 3761–3766.

Lengerova, M., Moore, R. C., Grant, S. R., and Vyskot, B. (2003). The sex chromosomes of *Silene latifolia* revisited and revised. *Genetics* **165**, 935–938.

Lengerova, M., Kejnovsky, E., Hobza, R., Macas, J., Grant, S. R., and Vyskot, B. (2004). Multicolor FISH mapping of the dioecious model plant, *Silene latifolia*. *Theor. Appl. Genet.* **108**, 1193–1199.

Lisitsyn, N., Lisitsyn, N., and Wigler, M. (1993). Cloning of the difference between two complex genomes. *Science* **259**, 946–951.

Liu, Z., Moore, P. H., Ma, H., Ackerman, C. M., Ragiba, M., Yu, Q., Pearl, H. M., Kim, M. S., Charlton, J. W., Stiles, J. I., Zee, F. T., Paterson, A. H., and Ming, R. (2004). A primitive Y chromosome in papaya marks incipient sex chromosome evolution. *Nature* **427**, 348–352.

Lucretti, S., Dolezel, J., Schubert, J. I., and Fuchs, J. (1993). Flow cariotyping and sorting of *Vicia faba* chromosomes. *Theor. Appl. Genet.* **85**, 665–672.

Martin, K. J., and Pardee, A. B. (1999). Principles of differential display. *Methods. Enzymol.* **303**, 234–258.

Matsunaga, S., Kawano, S., Takano, H., Uchida, H., Sakai, A., and Kuroiwa, T. (1996). Isolation and developmental expression of male reproductive organ-specific genes in a dioecious campion, *Melandrium album*. *Plant J.* **10**, 679–689.

Matsunaga, S., Schutze, K., Donnison, I. S., Grant, S. R., Kuroiwa, T., and Kawano, S. (1999a). Single-pollen typing combined with laser-mediated manipulation. *Plant J.* **20**, 371–378.

Matsunaga, S., Kawano, S., Michimoto, T., Higashiyama, T., Nakao, S., Sakai, A., and Kuroiwa, T. (1999b). Semi-automatic laser beam microdissection of the Y chromosome and analysis of Y chromosome DNA in a dioecious plant, *Silene latifolia*. *Plant Cell Physiol.* **40**, 60–68.

Matsunaga, S., Isono, E., Kejnovsky, E., Vyskot, B., Dolezel, J., Kawano, S., and Charlesworth, D. (2003). Duplicative transfer of a MADS box gene to a plant Y chromosome. *Mol. Biol. Evol.* **20**, 1062–1069.

Michelmore, R. W., Paran, I., and Kesseli, R. V. (1991). Identification of markers linked to disease-resistance genes by bulked segregant analysis: A rapid method to detect markers in specific genomic regions by using segregating populations. *Proc. Natl. Acad. Sci. USA* **88**, 9828–9832.

Moore, R. C., Kozyreva, O., Lebel-Hardenack, S., Siroky, J., Hobza, R., Vyskot, B., and Grant, S. R. (2003). Genetic and functional analysis of DD44, a sex-linked gene from the dioecious plant *Silene latifolia*, provides clues to early events in sex chromosome evolution. *Genetics* **163**, 321–334.

Nakao, S., Matsunaga, S., Sakai, A., Kuroiwa, T., and Kawano, S. (2002). RAPD isolation of a Y chromosome specific ORF in a dioecious plant, *Silene latifolia*. *Genome* **45**, 413–420.

Obara, M., Matsunaga, S., Nakao, S., and Kawano, S. (2002). A plant Y chromosome-STS marker encoding a degenerate retrotransposon. *Genes Genet. Syst.* **77**, 393–398.

Okada, S., Fujisawa, M., Sone, T., Nakayama, S., Nishiyama, R., Takenaka, M., Yamaoka, S., Sakaida, M., Kono, K., Takahama, M., Yamato, K. T., Fukuzawa, H., Brennicke, A., and Ohyama, K. (2000). Construction of male and female PAC genomic libraries suitable for identification of Y-chromosome–specific clones from the liverwort, *Marchantia polymorpha*. *Plant J.* **24**, 421–428.

Orita, M., Iwahana, H., Kanazawa, H., Hayashi, K., and Sekiya, T. (1989). Detection of polymorphisms of human DNA by gel electrophoresis as single-strand conformation polymorphisms. *Proc. Natl. Acad. Sci. USA* **86**, 2766–2770.

Pan, W. H., Houben, A., and Schlegel, R. (1993). Highly effective cell synchronisation in plant roots by hydroxyurea and amiprophos-methyl or colchicine. *Genome* **36**, 387–390.

Rice, W. R. (1987). Genetic hitchhiking and the evolution of reduced genetic activity of the Y sex chromosome. *Genetics* **116**, 161–167.

Robertson, S. E., Li, Y., Scutt, C. P., Willis, M. E., and Gilmartin, P. M. (1997). Spatial expression dynamics of *Men-9* delineate the third floral whorl in male and female flowers of dioecious *Silene latifolia*. *Plant J.* **12**, 155–168.

Sambrook, J., Fritsch, E. F., and Maniatis, T. (1989). "Molecular Cloning: A Laboratory Manual," 2nd Ed., Cold Spring Harbor Laboratory Press, New York.

Scutt, C. P., Kamisugi, Y., Sakai, F., and Gilmartin, P. M. (1997a). Laser isolation of plant sex chromosomes: studies on the DNA composition of the X and Y sex chromosomes of *Silene latifolia*. *Genome* **40**, 705–715.

Scutt, C. P., Li, T., Robertson, S. E., Willis, M. E., and Gilmartin, P. M. (1997b). Sex determination in dioecious *Silene latifolia*. Effects of the Y chromosome and the parasitic smut fungus (*Ustilago violacea*) on gene expression during flower development. *Plant Physiol.* **114**, 969–979.

Scutt, C. P., Jenkins, T., Furuya, M., and Gilmartin, P. M. (2002). Male specific genes from dioecious white campion identified by fluorescent differential display. *Plant Cell Physiol.* **43**, 563–572.

Shibata, F., Hizume, M., and Kuroki, Y. (1999). Chromosome painting of Y chromosomes and isolation of a Y chromosome-specific repetitive sequences in the dioecious plant *Rumex acetosa*. *Chromosoma* **108**, 266–270.

Skaletsky, H., Kuroda-Kawaguchi, T., Minx, P. J., Cordum, H. S., Hiller, L., Brown, L. G., Repping, S., Pyntikova, T., Ali, J., Bieri, T., Chinwalla, A., Delehaunty, A., Delehaunty, K., Du, H., Fewell, G., Fulton, L., Fulton, R., Graves, T., Hou, S. F., Latrielle, P., Leonard, S., Mardis, E., Maupin, R., McPherson, J., Miner, T., Nash, W., Nguyen, C., Ozersky, P., Pepin, K., Rock, S., Rohlfing, T., Scott, K., Schultz, B., Strong, C., Tin-Wollam, A., Yang, S. P., Waterson, R. H., Wilson, R. K., Rozen, S., and Page, D. C. (2003). The male-specific region of the human Y chromosome is a mosaic of discrete sequence classes. *Nature* **423**, 825–837.

Speicher, M. R., Ballard, S. G., and Ward, D. C. (1996). Karyotyping human chromosomes by combinatorial multi-fluor FISH. *Nat. Genet.* **12,** 368–375.

Stehlik, I., and Blattner, F. R. (2004). Sex-specific SCAR markers in the dioecious plant *Rumex nivalis* (*Polygonaceae*) and implications for the evolution of sex chromosomes. *Theor. Appl. Genet.* **108,** 238–242.

Straus, D. (1995). Genomic subtraction. *In* "PCR Strategies" (M. A. Innis, D. H. Gelfand, and J. J. Sninsky, eds.), pp. 220–236. Academic Press.

Syvänen, A. C. (2002). Accessing genetic variation: genotyping single nucleotide polymorphisms. *Nat. Rev. Genet.* **2,** 930–942.

Telenius, H., Carter, N. P., Bebb, C. E., Nordenskjold, M., Ponder, B. A., and Tunnacliffe, A. (1992). Degenerate oligonucleotide-primed PCR: General amplification of target DNA by a single degenerate primer. *Genomics* **13,** 718–725.

Vos, P., Hogers, R., Bleaker, M., Reijans, M., van de Lee, T., Hornes, M., Frijters, A., Pot, J, Peleman, J., and Kuiper, M. (1995). AFLP: A new technique for DNA fingerprinting. *Nucl. Acids Res.* **23,** 4407–4414.

Wages, J. M., Zhao, X., and Katz, E. D. (1995). High-performance liquid chromatography analysis of PCR products. *In* "PCR Strategies" (M. A. Innis, D. H. Gelfand, and J. J. Sninsky, eds.), pp. 140–153. Academic Press.

Westergaard, M. (1958). The mechanism of sex determination in dioecious flowering plants. *Adv. Genet.* **9,** 217–281.

Williams, J. G. K., Kubelik, A. R., Livak, K. J., Rafalski, J. A., and Tingey, S. V. (1990). DNA polymorphisms amplified by arbitrary primers are useful as genetic markers. *Nucleic Acids Res.* **18,** 6531–6535.

Yampolsky, C., and Yampolsky, H. (1922). Distribution of sex forms in the phanerogamic flora. *Biblioth. Genet.* **3,** 1–62.

Ye, D., Installé, P., Ciuperescu, C., Veuskens, J., Wu, Y., Salesses, G., Jacobs, M., and Negrutiu, I. (1990). Sex determination in the dioecious *Melandrium*. I. First lessons from androgenic haploids. *Sex Plant Reprod.* **3,** 179–186.

Yi, S., and Charlesworth, B. (2000). Contrasting patterns of molecular evolution of the genes on the new and old sex chromosomes of *Drosophila miranda*. *Mol. Biol. Evol.* **17,** 703–717.

Zhang, L., Cui, X., Schmitt, K., Hubert, R., Navidi, W., and Arnheim, N. (1992). Whole genome amplification from a single cell: Implications for genetic analysis. *Proc. Natl. Acad. Sci. USA* **89,** 5847–5851.

Zhang, Y. H., Di Stilio, V. S., Rehman, F., Avery, A., Mulcahy, D., and Kesseli, R. (1998). Y chromosome specific markers and the evolution of dioecy in the genus. *Silene. Genome* **41,** 141–147.

[24]   Preparation of Samples for Comparative Studies of Plant Chromosomes Using *In Situ* Hybridization Methods

By Jason G. Walling, J. Chris Pires, and Scott A. Jackson

Abstract

The development of fluorescence *in situ* hybridization (FISH) has led to the advancement of chromosome studies not only for physical mapping and genome analyses but also as a tool for evolutionary studies. Isolated repetitive DNA sequences have been useful cytological markers, but large-insert genomic libraries (e.g., bacterial artificial chromosomes [BACs], yeast artificial chromosomes [YACs], and cosmids) are being increasingly used to serve as probes for large segments of DNA across related genomes. Although FISH is usually applied to metaphase chromosomes, fiber FISH, a variation of FISH using extended DNA fibers, is now used to measure loci at the resolution of a few kilobases to compare orthologous genome segments across related genomes. A generalized set of protocols for chromosome preparation, FISH, and fiber FISH are presented; however, it is often necessary to experiment with techniques for different plant taxa for successful molecular cytogenetic studies.

Introduction

Chromosome changes occur frequently in evolution, and even the most closely related species can show differences in their karyotypes. In both animals and plants, lineages with high rates of chromosomal change also have high rates of speciation (Bennett *et al.*, 2000; Greilhuber, 1998; Navarro and Barton, 2003). Polyploidization and chromosomal rearrangements are thought to generate reproductive isolation or prevent recombination in linkage groups that contain ecologically important loci, thereby causing speciation (Rieseberg, 2001). Fluctuations in chromosome number and genome size have also been correlated with a number of novel morphological and environmental traits, which can lead to habitat divergence (Bennett, 1987; Bennett *et al.*, 2000; Grime, 1998; Levin, 1983; Ohri *et al.*, 1998; Osborn *et al.*, 2003).

Classic studies in karyotype analysis involved the time-consuming activity of counting chromosomes and examining gross structural changes. Although chromosome counting for its own sake has done very little, cytogenetics still plays an essential role in answering several biological questions (Heslop-Harrison, 2000; Stace, 2000). These questions can range from basic species identification to determining the extent of evolutionary change between species at the molecular and organismic levels. Because interspecific and intergeneric hybridization forms an important part of evolution in many plant species, it is often necessary to determine the ancestors of polyploids or the origin of alien chromosomes or chromosomal segments in natural and cultivated hybrids.

In the early 1980s, the development of fluorescence *in situ* hybridization (FISH) led to the advancement of chromosome studies not only for physical mapping and genome analyses but also as a tool for evolutionary studies (Heslop-Harrison, 2000; Jiang and Gill, 1994, 1996; Schwarzacher and Heslop-Harrison, 2000). With FISH, it is possible to map sequences across related species and genera to show not only conservation but also the location of sequences on chromosomes. Repetitive DNA sequences, including dispersed and tandem repeats, have been useful cytological markers because they constitute the major fraction of most plant genomes (Heslop-Harrison, 2000). Tandem repeats include ribosomal DNA (rDNA) and other satellite DNA (satDNA) that are localized at particular regions of chromosomes as blocks. All three types of repetitive sequences (dispersed, rDNA, satDNA) can evolve dramatically after hybridization or polyploidization. For example, Hanson *et al.* (1998) used FISH to show that dispersed repetitive sequences in *Gossypium* specific to the A-genome at the diploid level have colonized the D-genome at the polyploid level. Similarly, the copy number of 18S-5.8S-26S and 5S rDNA loci can change in allopolyploids. FISH of rDNA loci in *Sanguisorba* (Rosaceae) showed that some duplicated 5S rDNA sites were evidently lost after polyploidization (Mishima *et al.*, 2002), whereas another study showed that duplicated 18S-5.8S-26S rDNA sites were lost in polyploid *Zingeria trichopoda* (Poaceae) (Kotseruba *et al.*, 2003). Finally, the evolution of satDNA is dynamic in the polyploid genomes of *Nicotiana* (Lim *et al.*, 2000) and other taxa. Evolutionary applications of FISH in *Paeonia* (Zhang and Sang, 1999), *Nicotiana* (Lim *et al.*, 2000), and *Tragopogon* (Pires *et al.*, 2004) demonstrate the utility of using specific repetitive DNA sequences as probes.

With the advent of large-insert genomic libraries (e.g., bacterial artificial chromosomes [BACs], yeast artificial chromosomes [YACs], and cosmids), a new substrate became available to sample large segments of DNA across related genomes (Jackson *et al.*, 2000; Koumbaris and Bass, 2003). For example, Jackson *et al.* (2000) used six BACs as FISH probes to identify a 431-kb region of *Arabidopsis thaliana* chromosome 2 and then used the same BACs to find the homologous region in four to six areas of the *Brassica rapa* genome. This result is consistent with the hypothesis that the "diploid" *B. rapa* has experienced ancient polyploid events relative to the *Arabidopsis* genome. Similarly, the small genome of *Sorghum* has been used as a foundation for integrating genetic and physical maps across grass genera with larger genomes (Draye *et al.*, 2001; Kim *et al.*, 2002; Koumbaris and Bass, 2003). The key genomics tool in both of these sets of studies was a BAC library made at a reasonable cost from a small genome species (*Arabidopsis* and *Sorghum*). BAC libraries have also been used to identify genomic clones containing repetitive DNA sequences that can be used as cytological markers (Fig. 1A).

Fiber FISH, a variation of FISH using extended DNA fibers, has been used to examine the structure at high resolution of orthologous genome segments across related genomes. For example, Jackson *et al.* (2000) used fiber FISH in their comparative analyses of *Arabidopsis* and *B. rapa* genomes to show that there was very little change in size (measured in kilobases [kb]) between the homologous regions (Fig. 1B). Fiber FISH allows one to measure loci at the resolution of a few kilobases, but because the fibers are spread out, chromosomal orientation is lost. However, when combined with metaphase or pachytene FISH to show chromosomal location, fiber FISH is a powerful tool to measure the size of loci, the number of copies of a tandem repeat, or the distance between adjacent sequences (Fransz *et al.*, 1996; Jackson *et al.*, 1998, 2000).

Another variation of FISH involves the ability to track individual genomes via genome *in situ* hybridization (GISH) to identify the progenitor species of hybrids or allopolyploids (Stace and Bailey, 1999). In GISH, total genomic DNA is used as a molecular probe to differentiate related species and to establish relationships between them. The method has been used successfully across a wide range of plant species including the grasses, *Brassica, Crocus, Solanum*, and *Nicotiana*.

To more fully exploit these molecular cytogenetic tools for evolutionary studies, a generalized set of protocols for chromosome preparation, FISH and fiber FISH, is presented (other protocols can be found in Leitch *et al.*, 1994; Schwarzacher and Heslop-Harrison, 2000; Singh, 2003). In essence, FISH is similar to Southern blotting in that it employs DNA–DNA hybridization and, as with Southern blotting, there are standard issues of controlling stringency and blocking DNA.

Isolation and Preparation of Chromosomes

Following is a generalized protocol for preparing mitotic and meiotic chromosomes for FISH and/or traditional karyotypic analyses. For any specific species, the protocol will have to be optimized for the peculiarities

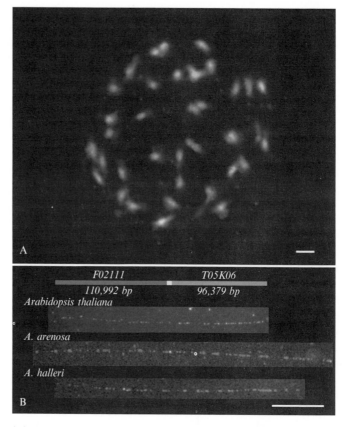

Fig. 1. (A) A soybean (*Glycine max*) bacterial artificial chromosome (BAC) composed of repetitive sequences was identified and mapped to chromosomes of soybean. This BAC (76J21) was found in the pericentromeric regions of all 40 chromosomes (bar = 5 μm). (B) Two BACs from *Arabidopsis thaliana* were used as probes for fiber fluorescent *in situ* hybridization (FISH) to genomic fibers from *A. thaliana* (top), *Arabidopsis arenosa* (middle), and *Arabidopsis halleri* (bottom). The sequences contained in these BACs were conserved in all three species, although the size of this region varied between species. A schematic of the BACs based on sequencing data is shown across the top (bar = 20 μm).

inherent to the species of interest. As with any technique, but perhaps more so for chromosome preparation, the technique has to be practiced to gain proficiency. For many species, the chromosomes will be very small (e.g., *Arabidopsis* and rice) and it takes a practiced eye to recognize the chromosomes from the rest of the cellular debris on the slide. It may be useful when starting out to practice with a standardized species with large chromosomes such as wheat (Singh, 2003).

Harvesting and Treatment of Root Tips for Cytological Preparations

Root tips can be harvested from a variety of sources including greenhouse plants, field samples, or germinating seeds. The quality of chromosomes varies by source. In general, root tips collected from field-grown material are superior; however, this can vary and the best source of chromosomes has to be determined on a per-species basis. In addition to the tissue source of chromosomes, other variables come into play such as the time of day collected, season, temperature, and freshness of the spindle-fiber inhibitor.

To harvest root tips from greenhouse-grown plants, empty the soil plug from the pot by first inverting the pot and tapping smartly on the bottom. The root system from each plant should be carefully dissected away from the potting mix. Dipping the root system in water can aid in this. Dissect roots approximately 2–3 cm proximal to the root tip. The root tips can be excised closer to the end; however, this unused portion acts as a buffer zone for subsequent manipulation of the root tip. The root tips should be placed directly into ice-cold ddH$_2$O bath in a Petri dish until they are further processed or placed directly into a spindle-fiber inhibitor.

Spindle-fiber inhibitors are used to facilitate the accumulation of metaphase chromosomes in the meristematic portion of the root tip. Many spindle-fiber inhibitors exist: cold treatment, 1-bromonaphthalene, colchicine, and hydroxyquinoline. In our hands, (2 mM) 8-hydroxyquinoline produces high-quality chromosomes in many species. The incubation time in the spindle inhibitor is species dependent and empirical data should be taken from time-course experiments to find an appropriate incubation period—usually between 2 and 6 h at room temperature (RT). The treatment should be performed in the dark because hydroxyquinoline is light sensitive.

To preserve chromosome samples, it is necessary to fix the cells. Most fixatives include a short-chain organic acid such as acetic or propionic acid mixed with an alcohol such as methanol or ethanol (EtOH). The most commonly used fixative is a 3:1 solution of absolute EtOH to glacial acetic acid (Carnoy's solution). Root tips can be stored in fixative at RT or at $-20°$ up to a year. Many cytologists claim that the best chromosome preparations come from samples that have not been fixed for a lengthy period, but samples should be fixed for at least 2–3 days.

Chromosome Pretreatment

To remove the fixative, the root tips are washed in ddH$_2$O. In many plant species, the cell wall is an impediment to good chromosome preparations, and enzymatic treatment with cellulase and pectinase is necessary. These enzymes are typically used at a 1.0%/1.0% concentration and incubated with the root tips at 37° to hydrolyze components of the cell wall. The length of this enzymatic treatment is species specific. If the species has tips that are large and fibrous, it may be necessary to treat longer; however, many species, such as wheat, rice, and *Arabidopsis*, require no treatment. This treatment will greatly soften the root tips so care must be taken in subsequent steps.

Briefly rinse the roots tips in water to remove excess enzymes. Many cytologists "clear" their specimens before mounting by immersing them in 45% acetic acid (v/v) for approximately 8–10 min. This step turns most cell debris translucent and reduces the appearance of cell debris in preparations. To assist in visualizing cytological preparations, the nuclear material can be stained with either acetocarmine (Carolina Biological Supply) or acetoorcein (Carolina Biological Supply) for 10–30 min, depending on the level of staining desired. When a quality phase-contrast scope is used to inspect preparations, the use of stains may not be necessary.

Root Tip Squashing Procedures

Root tips should be removed from the staining solution (if used) and placed on a clean slide. A dissecting (stereo) microscope can be used during the following steps to aid in dissection. To rinse the roots of excess stain and prevent drying, a drop of 45% acetic acid is applied to the root tip. The meristem tissue should be easily recognized from the previous staining treatment because it stains a darker red than the rest of the root material. If no stain is used, the meristem tissue typically appears as a white opaque portion of the root tip (Fig. 2A). The root cap should first be dissected from the specimen. The root cap appears as the minute portion at the apex of the root tip, just distal to the meristem tissue. Using a razor blade, a small section of the meristem tissue should be dissected away from the root tip and placed on a clean microscope slide in a drop of 45% acetic acid. In general, the best chromosome preparations are obtained with very small amounts of tissue. Often the rest of the root tip can be saved because it is feasible to make several slides from one root tip.

Place an 18- by 18-mm or 22- by 22-mm coverslip over the root tip section and acetic acid. A double-edged razor blade is used to separate one corner of the coverslip from the slide. Tightly press a finger down on one corner of the coverslip opposite the corner with the razor blade (Fig. 2B). Tap the coverslip with a semiblunt instrument (we use sharpened dowel rods) to spread out the tissue between the coverslip and the slide. The objective is to spread out the cells, but care should be taken not to push the cell material out from under the coverslip. If the tissue was stained with acetocarmine, the tissue should be tapped until it is a "pink haze." It is

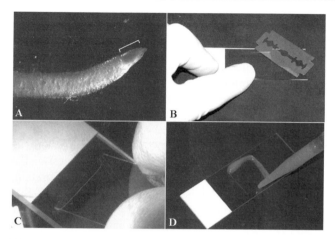

Fig. 2. (A) Microphotograph of root tip. Line indicates meristematic portion of root tip distal to which is the root cap. (B) Example of coverslip and razor blade inserted at one corner to create space between coverslip and slide to squash tissue. (C) Extension of lysed nuclei to create extended DNA fiber on slide. Slow, steady motion is used to drag the fibers down the slide using the coverslip. Note that coverslip does not contact the slide. (D) Rubber cement is pipetted onto the edges of the coverslip to seal the probe mixture between the coverslip and slide.

a learned skill to make good chromosome preparations. There are many variables in the squashing process and one should not give up easily; persistence is required. When the tissue is sufficiently broken up, the razor blade can be removed, and several additional taps on top of the coverslip will complete the spread. It is imperative to minimize any movement of the coverslip during the squash because chromosomes will "roll," ruining chromosome morphology.

After the squash, the entire slide should be heated over a small flame without boiling. Using a piece of Whatman filter paper, hand pressure is then applied to the coverslip to remove excess liquid. To determine the extent and success of the chromosome spread, the slide can be briefly scanned under phase contrast for presence and quality of chromosomes. A good slide is relatively free of cell debris and consists of many chromosome spreads exhibiting typical metaphase morphology. Slides should be placed at $-80°$ as soon as possible until used for FISH.

Preparation of Plant Nuclei for Fiber FISH

Fiber FISH increases the resolution of FISH analyses down to a few kilobases. It can be used to gauge distances between adjacent clones from a few kilobases up to several hundred kilobases and measure repetitive loci up to several megabases (Mb). Fiber FISH is basically FISH but on extended DNA fibers. Combined with metaphase analysis, this tool allows molecular cytogeneticists to map loci to specific chromosomes and analyze at high resolution (the Watson-Crick level) the structure of a locus. This protocol details the isolation of plant nuclei needed for fiber FISH and the preparation of extended DNA fibers on glass slides. Isolation of plant nuclei is similar to the preparation of high-molecular-weight DNA (Peterson *et al.*, 2000) except that isolated nuclei are suspended in 50% glycerol and stored at −20°.

Isolation of Plant Nuclei. Snap-freeze 2–5 g of fresh leaf tissue in liquid nitrogen and grind to fine powder using a mortar and pestle. Transfer the powder to a 50-ml centrifuge tube on ice and add 20 ml of prechilled nuclei isolation buffer (NIB: 10 mM Tris–HCl [pH 9.5], 10 mM EDTA, 100 mM KCl, 0.5 M sucrose, 4.0 mM spermidine, 1.0 mM spermine, 0.1% mercaptoethanol added just before use). NIB can be stored at 4° without mercaptoethanol.

Sequentially filter the emulsion through 220, 148, and 48 μm nylon mesh into prechilled 50-ml conical tubes, always on ice. Add 1 ml of NIB supplemented with 10% (v/v) of Triton X-100 to the filtrate and gently invert several times. Centrifuge at 2000g for 10 min at 4° and decant the supernatant. Resuspend the pellet in 200 μl to 1 ml of 1:1 NIB (without mercaptoethanol) and glycerol; the amount depends on the size of the pellet. The nuclei can be stored at −20° for up to a year without noticeable deterioration.

Extension of DNA Fibers. In the nuclei stocks, nuclei settle just above a darker layer of cellular debris. Although light mixing may be necessary, vigorous shaking should be avoided. The concentration of nuclei can be determined by placing a drop of the nuclei suspension on a slide, staining with DAPI (4′-6′-diamidino-2-phenylindole) and observing under an epifluorescence microscope.

Using a cut pipette tip, extract 1–10 μl of the nuclei suspension per slide, place into a clean Microfuge tube containing about 100 μl of NIB (without mercaptoethanol), and mix gently. Centrifuge sample at 3600 rpm for 5 min and then carefully remove supernatant and resuspend pellet in 2.5 μl of 1× phosphate-buffered saline (PBS) per slide. Pipette 2.5 μl of nuclei solution across a poly-L-lysine (Sigma, Poly-Prep, catalog no. P0425) slide and allow to air-dry for 5–10 min. The nuclei solution should be allowed to dry to the point at which it appears "sticky," neither wet nor completely dry. Pipette 10 μl of STE lysis buffer (0.5% sodium dodecyl sulfate [SDS], 5 mM ethylenediaminetetraacetic acid [EDTA], 100 mM Tris, pH 7.0) on top of the nuclei and incubate at RT for 5 min. To extend the DNA fibers, use a 22- by 40-mm coverslip held at an angle to slowly drag the solution down the slide (Fig. 2C). The coverslip should not touch the slide but should be held just above the slide while still in contact with the solution to prevent scraping the nuclei off the slide. Air-dry the slide(s) for 5 min, fix in fresh 3:1 EtOH:acetic acid for 2 min, and then bake the slide(s) at 60° for 30 min. The preparations perform best when used immediately, but successful *in situ* experiments have been performed using slides that have been stored at RT for several weeks.

Isolation and Preparation of Meiocytes for Meiotic FISH

Meiotically derived chromosomes at pachytene stage have been exploited for their high resolving power (Cheng *et al.*, 2001; Zhong *et al.*, 1999). Meiotic chromosomes can be 10–40 times longer than mitotic chromosomes, and in normal meiosis, they are grouped into a haploid number of chromosomes. Thus, meiotic preparations provide superior FISH targets.

Meiocytes are most commonly obtained from young flowers with the purpose of targeting pollen mother cells (PMCs) at the pachytene stage. Anthers are the preferred source of meiocytes because they usually contain many PMCs. Furthermore, in many plant species, the PMCs within an anther are developmentally synchronous and the meiotic stage is correlated with anther development.

Flowers are excised and fixed directly in Carnoy's solution without the need for a spindle-fiber inhibitor. After 2–3 days of fixation, the anthers of one flower are carefully dissected away from the flower in 45% acetic acid using a stereomicroscope if necessary. A small section of an anther can be squashed in 45% acetic acid and viewed under phase contrast to determine the stage of pollen development. If tetrads or fully developed pollen is seen, it will be necessary to sample anthers from florets at an earlier developmental stage. If meiotic stages near pachytene are seen, it may be helpful to sample anthers from different flowers within a floral cluster. In grasses, for instance, early meiotic stages are found toward the end of the panicle and later stages are found in the middle of the panicle.

Preparations of pachytene chromosomes can be done using the same squashing technique as described for mitotic preparations; however, a small cross-section of anther material is used rather than root tips. For the best preparations, the tip of an anther can be cut and the meiocytes gently squeezed from the anther into a drop of 45% acetic acid on a slide and then squashed.

Probe Preparation and Labeling for In Situ Detection

The following protocol describes a nick translation method for labeling genomic DNA and both large- and small-insert clones such as plasmids, BACs, and PACs for use in FISH experiments. Numerous labeling kits can be used in place of the following procedure (e.g., Molecular Probes, Roche Biochemical).

Incorporation of fluorescently labeled nucleotides is done using the DNA nicking of DNAse I coupled with DNA polymerase I. If not using an off-the-shelf kit, the DNAse activity should be determined by digesting 1 μg lambda DNA with a serial dilution of DNAse I at 15° for 2 h in polymerase I buffer. The reaction is stopped by adding 5 μl of 0.2 M EDTA (pH 8.0). Reaction products are analyzed using gel electrophoresis. An appropriate dilution should be selected that yields 300–600 bp DNA fragments. The nick translation reaction is set up as follows, incubated for 2 h at 15°, and stopped by adding 5 μl of 0.2 M EDTA. Unincorporated nucleotides are removed with a Sephadex column or a commercial nucleotide removal kit (e.g., Qiagen).

Nick Translation Mastermix

	(μl/rxn)
DNAse I[*]	1
Polymerase I buffer	5
Polymerase I	1
0.5 mM ACG nucleotides	5
Labeled dUTP/TTP[**]	5
DNA (1 μg)	x
QS to 50 μl w/ddH$_2$O	x
Total:	50 μl

[*] Concentration determined empirically.
[**] Digoxigenin–dUTP or biotin–dUTP (Roche Biochemical). The ratio of UTP/TTP follows manufacturers instructions.

Blocking DNA

The size and complexity of plant genomes vary greatly across genera and can be problematic for FISH. In particular, FISH analysis of cloned DNA (i.e., BACs) on large plant genomes with high repetitive fractions can be very difficult because of the dispersed repeats throughout the genome. Cot DNA fractions can be used to block repetitive DNA sequences during FISH. Briefly, in a kinetically driven experiment, total DNA is denatured and allowed to reanneal under controlled conditions with the purpose of selectively recovering only highly (Cot-1) or moderately (Cot-100) repetitive sequences. In some cases, such as rice (Jiang et al., 1996), Cot-1 works fine, whereas in other species, such as tomato (Budiman *et al.*, 2004), Cot-100 fractions are necessary. The appropriate Cot fraction for any species has to be determined on a case-by-case basis.

Before starting the Cot extractions, the DNA concentration should be determined and the genomic DNA should be sheared to an average size of 300 bp. Shearing can be done using NaOH, sonication, or autoclaving.

Cot-1 (Jiang et al., *1996).* Determine the annealing time using the concentration of the sheared DNA in the following equation: time to anneal (in min) = 5.92/DNA concentration (mg/ml).

Denature the previously sheared DNA by incubating at 90° in a heat block or in boiling water for 5 min. Anneal to Cot-1 by first equilibrating the sheared DNA in a 65° water bath by adding prewarmed (65°) 10 M NaCl solution to a final concentration of 0.3 M and incubating for the calculated reannealing time. Digest the residual single-stranded DNA using S1 nuclease (1 U/mg starting material) and 1 volume of ice-cold 2× S1 nuclease buffer (0.5 M NaCl, 2 mM ZnSO$_4$, 0.06 M NaAc [pH 4.2], 10% glycerol) and incubate the solution at 37° for 30 min. The reaction is stopped and purified simultaneously via a phenol chloroform extraction. Determine the final concentration of the Cot-100 DNA.

Cot-100 (Budiman et al., *2004).* Annealing to Cot-100 will recover more of the moderately repetitive DNA sequences, which should result in the blocking of many repetitive sequences found in large, complex genomes. The procedure is identical to that of Cot-1 with the exception of the incubation time and temperature, which can be calculated using the following equations:

$$\text{Melting Temperature} = 81.5 + 16.6\text{LogM} + 41(\%G + C) - 500/L - 0.62F,$$

where M = molar concentration of monovalent cations; %G + C = G + C fraction of genome; L = length of sheared DNA; F = molar concentration of formamide; Reassociation Temp = melting temperature-25; Reassociation Time = 100/([DNA μg/ml]/339) = time in s.

In Situ Hybridization Procedure

The following procedure can be used to hybridize probes of differing sizes to chromosomes or extended DNA fibers from previous steps. Depending on the target, the target and the probe may be denatured together or separately. It is especially important in the following steps to not let the previously prepared mitotic/meiotic slides thaw before dehydrating because it will destroy the chromosome morphology. The probe solution is prepared as follows:

Probe Solution

	μl/slide	Final concentration
Deionized formamide	5	50%[**]
50% dextran sulfate	2	10%
20× SSC	1	2×
10 mg/ml salmon sperm	1	1 mg/ml
Probe DNA	1[*]	~17 ng
Total:	10 μl	

[*] If Cot-1/100 blocking DNA and/or multiple probes are to be used, the DNA (probe and salmon sperm) should be dried down using a vacuum centrifuge before the rest of the components are added. ddH$_2$O should be used to make up the difference in volume.

[**] In instances when lower stringency hybridization is required, such as using heterologous probes, the formamide concentration can be decreased from 50% (5 μl) to 30% (3 μl) and ddH$_2$O used to make up the difference in volume.

Depending on the target DNA used, the probe and target DNA can be denatured separately or simultaneously. In the case of meiotic and mitotic preparations, they should be denatured separately and combined for hybridization. For fiber FISH, the probe and fibers can be denatured together.

Denaturation/Hybridization for Meiotic and Mitotic Preparations

Cytological preparations stored at −80° should be removed just before hybridization experiments without letting them thaw. Immediately upon removal from −80°, the coverslip should be flicked off by inserting a single-edged razor blade under a corner of the coverslip and rapidly lifting straight up. The slide should then be placed immediately into a Coplin jar containing 70% EtOH. To maintain the appearance of the chromosomes, it is important to prevent the slide from thawing before placing in EtOH. The EtOH dehydration series is as follows: 70%, 90%, and then 100% EtOH rinses at RT for 5 min each. Allow the slides to air-dry on the benchtop at RT.

Slide denaturation: Place 100 μl of 70% formamide in 2× SSC over the sample on the slide and cover with a 22- by 40-mm coverslip. Incubate the slides on a hot plate at 90° for exactly 90 s. Immediately immerse the slide(s) in 70% EtOH (prechilled to −20°) for 5 min, allowing the coverslip to fall off in the EtOH bath. In the same fashion as the previous EtOH series, this step should be followed by 90% EtOH and 100% EtOH washes, all prechilled to $-20°$. Allow the slide to dry at RT on a paper towel.

Probe: The probe DNA is denatured by incubating the sample at 80–90° for 10 min, followed by immediate submersion in ice bath to prevent reannealing.

Hybridization: Apply 10 μl of denatured probe mixture to the area of the slide containing the chromosome spread and cover with an 18- by 18-mm coverslip. Seal the edges of the coverslip with rubber cement to prevent evaporation (Fig. 2D). Hybridize for 6–24 h at 37° in a sealed humid chamber.

Denaturation/Hybridization for Fiber FISH

Add 10 μl of probe solution to the area of the slide in which the DNA fibers are fixed and cover with an appropriately sized (22- by 22-mm) coverslip. Seal the edges of the coverslip with rubber cement and allow it to dry (Fig. 2D). Place the slide on an 80° hot plate, denature for 3–5 min, and then place the slides directly into moist chamber and hybridize for 6–24 h at 37°.

Antibody Detection

After hybridization, the slides should be removed from the humid incubation chamber. Using forceps, peel away the line of rubber cement, taking care not to move the coverslip. The slides are then placed in a series of washes, as outlined below. During the first wash, the coverslips should be allowed to fall off.

Washes

Buffer	Min	Temp	Notes
2× SSC[*]	–	RT	Allow coverslips to fall off
2× SSC	5	RT	
2× SSC	10	42°	(50% formamide in 2× SSC is often used)
2× SSC	5	RT	
1× PBS[**]	5	RT	

[*] 20× SSC: (3 M NaCl, 0.3 M sodium citrate).
[**] 10× PBS: (0.13 M NaCl, 0.0007 M Na$_2$HPO$_4$, 0.003 M NaH$_2$PO$_4$).

For meiotic and mitotic FISH, single-layer antibodies or fluorescently labeled nucleotides are generally sufficient, but for fiber FISH, multiple antibody layers are used to increase the fluorescence signal. When using a multiple layer antibody scheme, it is important to avoid cross-reactivity between antibodies. Examples of single-layer and multilayer detection schemes are given.

Single-Layer Detection for Meiotic and Mitotic FISH. To detect biotin-labeled probes, streptavidin-conjugated fluorophore is used, whereas for digoxigenin-labeled probes, an anti-digoxigenin–conjugated fluorophore is used. The most commonly used fluorophores include rhodamine, AlexaFluor 568 (Molecular Probes), Texas Red, and Cy3 for the red channel and fluorescein (FITC), AlexaFluor 488 (Molecular Probes), and Cy2 for the green channel. The antibodies are allowed to form conjugates according to the following protocol:

Antibody Solution

	μl/slide
5× buffer[*]	20
ddH$_2$O	80
1 Antibody	1–2 (per manufacturer's directions)
Total:	~100 μl/slide

[*]5× antibody buffer: 5× PBS, 25 mM EDTA, 5% BSA.

Pipette 100 μl of the antibody solution on each slide and cover with a 22- by 40-mm coverslip. Incubate the slides in a moist chamber in the dark for 30 min at 37°. After incubation, the coverslips are allowed to fall off by holding the slide vertically and then the slides are washed three times in 1× PBS for 5 min at RT in the dark. After washing, the slides are removed from the 1× PBS and coverslips mounted in about 10 μl VectaShield solution (Vector Labs) supplemented with either DAPI or propidium iodide (PI). For slides containing only a single probe detected with a fluorophore in the green channel, PI can be used to counterstain the chromosomes red. For all red fluorophores, DAPI is used to counterstain the chromosomes in the blue channel. Slides are visualized immediately under fluorescence microscopy or stored at −20°.

Multilayer Antibody Detection for Fiber FISH. The following is a scheme that employs an antibody layering system used to amplify signal (Zhong *et al.*, 1996). This protocol has been designed for use with two separate probes, one of which is labeled with biotin and the other with digoxigenin.

Wash the slide(s) in $1\times$ 4T ($4\times$ SSC, 0.05% Tween 20) for 5 min with constant shaking. An optional blocking step to reduce background noise can be done by pipetting 100 μl of $1\times$ 4 M (3–5% nonfat dry milk in $4\times$ SSC) on each slide, adding a coverslip, and incubating at 37° for 30 min. This blocking step is followed by an additional 5-min wash in $1\times$ 4T.

Antibody Mastermix I (AMI)—Layer I

	μl/slide
$5\times$ TNB[*]	20
ddH$_2$O	80
Streptavidin-FITC	1
Total:	~100 μl/slide

[*]TNB: 0.1 M Tris pH 7.5, 0.15 M NaCl, 0.5% blocking reagent (Amersham Biosciences).

Place 100 μl of AMI on each slide, cover with a 22- by 40-mm coverslip and incubate in moist chamber for 30 min. Allow the coverslip to fall off by holding vertically and wash three times in $1\times$ TNT (0.1 M Tris–HCl, 0.15 M NaCl, 0.05% Tween 20, pH 7.5) for 5 min each.

Antibody Mastermix II (AMII)—Layer 2

	μl/slide
$5\times$ TNB	20
ddH$_2$O	80
Biotin anti-Streptavidin	0.5
Mouse anti-digoxigenin	1
Total:	~100 μl/slide

Place 100 μl of AMII layer onto each slide, cover with a 22- by 40-mm coverslip, incubate for 30 min, and then wash three times in $1\times$ TNT for 5 min each.

Antibody Mastermix III (AMIII)—Layer 3

	μl/slide
$5\times$ TNB	20
ddH$_2$O	80
Streptavidin-FITC	1
Rhodamine antimouse	1
Total:	~100 μl/slide

Place 100 μl of AMIII antibody layer onto each slide and repeat the incubation and washing procedures explained above. The procedure can be taken through the same final washes and mounting techniques as the single-layer FISH protocol except that neither DAPI nor PI is added to the VectaShield. Some labs dehydrate the slides with and EtOH series before mounting the coverslip. Slides can be viewed immediately or stored at $-20°$. In general, it is necessary to use a $60\times$ or greater oil immersion lens to view fiber FISH slides.

Concluding Remarks

The significant amount of time, resources, and specialized expertise needed for molecular cytogenetic studies necessitates that only biological questions that cannot be addressed with more straightforward molecular mapping techniques should be undertaken. Although a generalized set of protocols for chromosome preparation, FISH, and fiber FISH have been presented here, successful molecular cytogenetic techniques must be crafted for each plant taxa to be studied. We encourage researchers to look at various other protocols that may be more relevant to their particular group of model organisms (Leitch et al., 1994; Schwarzacher and Heslop-Harrison, 2000; Singh, 2003). Despite these caveats, the impressive amount of genomic resources (e.g., BAC libraries) becoming available for a wide variety of organisms ensures a promising future for plant molecular cytogenetics.

References

Bennett, M. D. (1987). Variation in genomic form in plants and its ecological implications. N. Phytol. **106**, 177–200.

Bennett, M. D., Bhandol, P., and Leitch, I. J. (2000). Nuclear DNA amounts in angiosperms and their modern uses—807 new estimates. Ann. Botany **86**, 859–909.

Budiman, M. A., Chang, S.-B., Lee, S., Yang, T. J., Zhang, H-B., de Jong, H., and Wing, R. A. (2004). Localization of jointless-2 gene in the centromeric region of tomato chromosome 12 based on high resolution genetic and physical mapping. Theor. Appl. Genet **108**, 190–196.

Cheng, Z. K., Presting, G. G., Buell, C. R., Wing, R. A., and Jiang, J. (2001). High-resolution pachytene chromosome mapping of bacterial artificial chromosomes anchored by genetic markers reveals the centromere location and the distribution of genetic recombination along chromosome 10 of rice. Genetics **157**, 1749–1757.

Draye, X., Lin, Y. R., Qian, X. Y., Bowers, J. E., Burow, G. B., Morrell, P. L., Peterson, D. G., Presting, G. G., Ren, S. X., Wing, R. A., and Paterson, A. H. (2001). Toward integration of comparative genetic, physical, diversity, and cytomolecular maps for grasses and grains, using the Sorghum genome as a foundation. Plant Physiol. **125**, 1323–1341.

Fransz, P. F., Alonso-Blanco, C., Liharska, T. B., Peeters, A. J. M., Zabel, P., and de Jong, J. H. (1996). High-resolution physical mapping in Arabidopsis thaliana and tomato by fluorescence in situ hybridization to extended DNA fibers. Plant J. **9**, 421–430.

Greilhuber, J. (1998). Intraspecific variation in genome size: A critical reassessment. *Ann. Botany* **82**(suppl. A), 27–35.

Grime, J. P. (1998). Plant classification for ecological purposes: is there a role for genome size. *Annals of Botany* **82**(suppl. A), 117–120.

Hanson, R. E., Zhao, X.-P., Islam-Faridi, M. N., Paterson, A. H., Zwick, M. S., Crane, C. F., McKnight, T. D., Stelly, D. M., and Price, H. J. (1998). Evolution of interspersed repetitive elements in *Gossypium* (Malvaceae). *Am. J. Botany* **85**, 1364–1368.

Heslop-Harrison, J. S. (2000). Comparative genome organization in plants: From sequence and markers to chromatin and chromosomes. *Plant Cell* **12**, 617–635.

Jackson, S. A., Cheng, Z., Wang, M.-L., Goodman, H. M., and Jiang, J. (2000). Comparative FISH mapping of a 431-kb *Arabidopsis thaliana* BAC contig reveals the role of chromosomal duplications in the expansion of the *Brassica rapa* genome. *Genetics* **156**, 833–838.

Jackson, S. A., Wang, M., Goodman, H., and Jiang, J. (1998). Fiber-FISH analysis of repetitive DNA elements in *Arabidopsis thaliana*. *Genome* **41**, 566–572.

Jiang, J., and Gill, B. S. (1994). Nonisotopic *in situ* hybridization and plant genome mapping: The first 10 years. *Genome* **37**, 717–725.

Jiang, J., and Gill, B. S. (1996). Current status and potential of fluorescence *in situ* hybridization in plant genome mapping. *In* "Genome Mapping in Plants" (A. H. Paterson, ed.), pp. 127–135. Landes Co., Austin, TX.

Jiang, J., Hulbert, S. H., Gill, B. S., and Ward, D. C. (1996). Interphase fluorescence *in situ* hybridization mapping: A physical mapping strategy for plant species with large complex genomes. *Mol. Gen. Genet.* **252**, 497–502.

Kim, J. S., Childs, K. L., Islam-Faridi, M. N., Menz, M. A., Klein, R. R., Klein, P. E., Price, H. J., Mullet, J. E., and Stelly, D. M. (2002). Integrating karyotyping of *Sorghum* by *in situ* hybridization of landed BACs. *Genome* **45**, 402–412.

Kotseruba, V., Gernand, D., Meister, A., and Houben, A. (2003). Uniparental loss of ribosomal DNA in the allotetraploid grass *Zingeria trichopoda* (2n = 8). *Genome* **46**, 156–163.

Koumbaris, G. L., and Bass, H. W. (2003). A new single-locus cytogenetic mapping system for maize (*Zea mays* L.): Overcoming FISH detection limits with marker-selected *Sorghum* (*S. propinquum* L) BAC clones. *Plant J.* **35**, 647–659.

Leitch, A. R., Schwarzacher, T., Jackson, D., and Leitch, I. J. (1994). "*In Situ* Hybridization: A Practical Guide," p. 118. Bios, Oxford, UK.

Levin, D. A. (1983). Polyploidy and novelty in flowering plants. *Am. Naturalist* **122**, 1–25.

Lim, K. Y., Matyásek, R., Lichtenstein, C. P., and Leitch, A. R. (2000). Molecular cytogenetic analyses and phylogenetic studies in the *Nicotiana* section *Tomentosae*. *Chromosoma* **109**, 245–258.

Mishima, M., Ohmido, N., Fukui, K., and Yahara, T. (2002). Trends in site-number change of rDNA loci during polyploid evolution in *Sanguisorba* (Rosaceae). *Chromosoma* **110**, 550–558.

Navarro, A., and Barton, N. H. (2003). Chromosomal speciation and molecular divergence— accelerated evolution in rearranged chromosomes. *Science* **300**, 321–324.

Ohri, D., Fritsch, R. M., and Hanelt, P. (1998). Evolution of genome size in *Allium* (Alliaceae). *Plant Syst. Evol.* **210**, 57–86.

Osborn, T. C., Pires, J. C., Birchler, J. A., Auger, D. L., Chen, Z. J., Lee, H.-S., Comai, L., Madlung, A., Doerge, R. W., Colot, V., and Martienssen, R. A. (2003). Understanding mechanisms of novel gene expression in polyploids. *Trends Genet.* **19**, 141–147.

Peterson, D. G., Tomkins, J. P., Frisch, D. A., Wing, R. A., and Paterson, A. H. (2000). Construction of plant bacterial artificial chromosome libraries: An illustrated guide. *J. Agric. Genom.* **5**, http://www.ncgr.org/jag.

Pires, J. C., Lim, K. Y., Kovařík, A., Matyásek, R., Boyd, A., Leitch, A. R., Leitch, I. J., Bennett, M. D., Soltis, P. S., and Soltis, D. E. (2004). Molecular cytogenetic analysis of recently evolved *Tragopogon* (Asteraceae) allopolyploids reveal a karyotype that is additive of the diploid progenitors. *Am. J. Botany* **91,** 1022–1035.

Rieseberg, L. H. (2001). Chromosomal rearrangements and speciation. *Trends Ecol. Evol.* **16,** 351–358.

Schwarzacher, T., and Heslop-Harrison, J. S. (2000). "Practical *In Situ* Hybridization." Oxford Press, New York, NY.

Singh, R. J. (2003). "Plant Cytogenetics, Second Edition." CRC Press, Boca Raton, FL.

Stace, C. A. (2000). Cytology and cytogenetics as a fundamental taxonomic resource for the 20th and 21st centuries. *Taxon* **49,** 451–477.

Stace, C. A., and Bailey, J. P. (1999). The value of genomic *in situ* hybridization. (GISH) in plant taxonomic and evolutionary studies. *In* "Molecular Systematics and Plant Evolution" (P. M. Hollingsworth, R. M. Bateman, and R. J. Gornall, eds.), pp. 199–210. London.

Zhang, D., and Sang, T. (1999). Physical mapping of ribosomal RNA genes in peonies (*Paeonia*, Paeoniaceae) by fluorescent *in situ* hybridization: Implications for phylogeny and concerted evolution. *Am. J. Botany* **86,** 735–740.

Zhong, X., Fransz, P. F., van Wennekes, E. J., Zabel, P., van Kammen, A., and de Jong, H. J. (1996). High-resolution mapping on pachytene chromosomes and extended DNA fibres by fluorescence *in-situ* hybridization. *Plant. Mol. Biol. Rep.* **14,** 232–242.

Zhong, X. B., Bodeau, J., Fransz, P. F., Williamson, V. M., van Kammen, A., deJong, J. H., and Zabel, P. (1999). FISH to meiotic chromosomes of tomato locates the root knot nematode resistance gene Mi-1 and the acid phosphatase gene Aps-1 near the junction of euchromatin and pericentromeric heterochromatin of chromosome arms 6S and 6L, respectively. *Theor. Applied Genet.* **98,** 365–370.

[25] Preparation of Samples for Comparative Studies of Arthropod Chromosomes: Visualization, In Situ Hybridization, and Genome Size Estimation

By Rob DeSalle, T. Ryan Gregory, and J. Spencer Johnston

Abstract

The ability to obtain large amounts of genomic sequence for organisms and high throughput technology has led to a change in the thrust of research at the level of chromosomes in animals. In the past chromosomal analysis of animals was focused on gross changes such as inversions, translocations, and deletions for both genetic and evolutionary studies. The advent of *in situ* hybridization technology and the ability to measure genome content size changed both the precision and the scale of studies addressing chromosomal change as a tool in evolutionary biology. This chapter addresses two of the major areas of change that have occurred in chromosomal studies in the past decade - examination of more refined and genome enabled structural changes in chromosomes and genome size measure. This cahpter describes some of the chromosome structure approaches such as fluorescent *in situ* hybridization (FISH), comparative genomic hybridization (CGH) and other techniques. As well, advances in Genome size measurement and theory are described herein.

Introduction

In the past, chromosome-based approaches to studying the evolution of insects and other arthropods have been limited to using gross and "visible" chromosomal changes as genetic and phylogenetic markers. Although such markers are useful for following the population genetic or long-term evolutionary history of specific arthropod groups, their resolution is rather limited. It is, therefore, not surprising that so much excitement has been generated with the sequencing of entire genomes and from taking a whole genome approach to chromosomal evolution (Eichler and Sankoff, 2003). Indeed, it is safe to say that arthropod cytogenetics is on the cusp of a revolution.

Even though only a handful of arthropod genomes have been sequenced (see below), it already is obvious that the large-scale study of whole genomes holds immense promise for evolutionary biology. This arises from the newfound ability to determine with precision the breakpoints of changes in chromosomal morphology (down to the level of the base pair) and the ability to assess with a high degree of accuracy the amount of DNA contained in a species' genome.

This chapter is concerned with two of the most important aspects of modern research in arthropod cytogenetics: the analysis of structural chromosome changes and genome size estimation. This includes a discussion of the rationale for carrying out such studies, as well as providing the technical background information and recommended protocols required for their implementation. Because the kinds of information sought in chromosome counting and morphological studies differ from those in genome size estimates, this chapter is separated into two roughly equal parts, one on "chromosomes" and the other on "genome size," which are treated in parallel.

Chromosomes

Polytene and Nonpolytene Types

Detailed examinations of the cytogenetics of specific arthropod groups can be found, for example, in the reviews of the *Animal Cytogenetics* series (for which, unfortunately, not all of the intended volumes were published) and in the *Handbook of Genetics* (King, 1975; specifically Volume 3, *Invertebrates of Genetic Interest*). As such works make clear, chromosomes have been examined for decades as a source of transmission genetic, population genetic, and phylogenetic information, all falling under the general heading of cytogenetics.

Dipteran insects, in particular, have a long tradition of use in cytogenetic research. Insect chromosomes come in two basic kinds: polytene and non-polytene, the former of which have been especially useful in cytogenetic research. Polytene chromosomes are giant structures that contain upwards of 1024 copies of the genome (10 doublings from the initial diploid state). What makes polyteny special, as compared with polyploidy, is that these 1000 or so copies of the genome remain conjoined, aligning perfectly into thick "cables" of DNA. Balbiani (1881) was perhaps the first to note the presence of these large, chromatin-dense structures in the nuclei of *Chironomus* salivary gland cells, but it was Thomas Hunt Morgan and his famous *Drosophila* group who capitalized on the utility of polytene chromosomes to study dipteran genetics. To be sure, Morgan and his group played an important role in making *Drosophila* a key model organism (and by extension, one of the first animals to have its genome sequenced) by taking advantage of these kinds of chromosomes in both genetic and evolutionary studies.

As noted by Wagner and Crow (2001), Painter was the real driving force behind the development of polytene chromosomes as genetic and evolutionary tools. In his initial work in Morgan's lab (done in collaboration with Muller), Painter examined the effects of radiation on chromosomes and soon found that nonpolytene chromosome spreads were suitable for studying gross chromosomal rearrangements. However, he quickly turned to the large, banded salivary gland chromosomes to obtain a more detailed view of chromosomal breaks (Painter, 1933). These conspicuous bands, which can be revealed by different staining techniques (see later discussion), result from the interspersion of highly condensed and relatively open regions of the chromosome. Specifically, the darker regions form because of the high degree of condensation of heterochromatic regions, whereas the less condensed lighter regions correspond to potentially transcribed euchromatic regions. In *Drosophila melanogaster*, there are more than 5000 of these lighter regions, and comparable numbers of bands are found in other dipteran species.

Basic Visualization

The method of choice for viewing a particular set of chromosomes is dependent on their size. In rapidly replicating polytene cells, gross chromosome structure sometimes can be viewed directly by the naked eye, but this is far from typical for most chromosomes. Because nonpolytene chromosomes are usually tiny, it is best to view them at their highest level of condensation, namely during metaphase. At other stages of the cell cycle, the DNA may be far too dispersed to be visualized as distinct chromosomes. To maximize the likelihood of finding nicely spread metaphase chromosomes, samples usually are taken from rapidly dividing tissues. The addition of chemicals that inhibit spindle-fiber formation also is used to enhance the probability of finding suitable chromosomes, and bathing in hypotonic solution to swell the cells may be done to improve the visualization of chromosomes in the metaphase.

Physical squashing onto slides (usually with a coverslip) further promotes the even spreading of the chromosomal preparation to facilitate enumeration and morphological analysis. In addition, binding certain optically dense chemicals to chromosomes enhances their observation. So treated, the banding patterns of the chromosomes can usually be viewed at low magnification ($10\times$ to $100\times$) on many kinds of microscopes. Of course, high-resolution observation of chromosome spreads requires viewing at high magnification ($1000\times$). On the other hand, phase contrast or Nomarski optics using an appropriately high magnification may allow one to assess the quality of a chromosomal spread, count chromosome arms, and even identify banding patterns of the different chromosome arms without staining. The choice to stain and the magnification employed must be based on the intended use of the image of the spread chromosomes. If one simply is trying to view chromosome morphology, banding pattern, or arm number, staining immediately after spreading is appropriate. However, if additional manipulations (e.g., chromosome dissection or chromosomal painting via *in situ* hybridization; see later discussion) are planned after confirmation of spread quality, then obviously staining of the chromosomes is not a desired operation.

Charge group on the chromosomes is taken advantage of in chromosome staining. DNA is negatively charged, so positively charged stains are preferred. Many chromosome stains that take advantage of charge to bind to chromosomes are the classically used acetoorcein, acetocarmine, gentian violet, hematoxylin, Leishman's stain, and Wright's stain. These standard stains uniformly stain chromosomes and leave the centromere constricted, which enhances the ability of cytogeneticists to count chromosomes and assess chromosomal morphology more accurately (Petitpierre, 1996). Binding of proteins to chromosomes (most notably, histone and nonhistone chromosome-specific proteins) complicates chromosome staining. An interesting (though still unexplained) exception to this is orcein stain, one of the first used in cytogenetics, which actually binds better to chromosomes associated with proteins.

An important feature of orcein and some of the other early stains is that once bound to chromosomes, they cannot be washed off. This fact allowed the development of the family of chromosome stains known as Romanovsky dyes, which include Giemsa, Leishman's, and Wright's stains. These bind differentially under certain conditions and, thus, can be used together in a process of "counterstaining" to highlight different structural features of chromosomes.

Three major staining techniques are used to reveal banding patterns in metaphase chromosomes. G-banding is a technique that starts with digestion of chromosome spreads with trypsin. Giemsa stain, which binds to DNA in AT-rich regions of the chromosomes, then is used to stain the digested chromosomes. The chromosomes then show dark (G-bands) and light (G-negative) patterns under light microscopy. Q-banding uses fluorescent dyes such as Quinacrine, DAPI (4′, 6-diamidino-2-phenylindole dihydrochloride), or Hoechst 33258 to stain spread or condensed chromosomes, which then can be viewed under ultraviolet (UV) fluorescence microscopy. The resulting Q-bands directly mimic G-bands, which also indicate AT-rich regions. R-banding requires preliminary heat denaturation in saline solution of the chromosome spreads. Because AT-rich regions will melt at lower temperatures than GC-rich ones, the heat treatment preferentially denatures the AT-rich regions. The GC-rich areas then can be observed as darkly staining bands using Giemsa or a stain specific to these regions (e.g., chromomycin A3, olivomycin, or mithramycin). R-bands mimic the Q-negative bands in Q-stained chromosome spreads. Of these three spreads, G- and Q-banding appear to be the methods of choice (Steiniger and Mukherjee, 1975). Chromosome spreads that have been treated with restriction enzymes that recognize sequences in satellite DNA regions of chromosomes also have been used in concert with these standard stains, yielding more information about the location of specific kinds of sequences on chromosomes.

Other kinds of chromosome manipulation are used as genetic and evolutionary tools. Silver stains can be used to enhance the visualization of nucleolus organizer regions (NORs), and hence to identify the karyotypic location of the ribosomal DNA cistrons. Other kinds of procedures similarly take advantage of specific sequences in well-defined positions on chromosomes, which may be in either genic or noncoding regions. For instance, telomeric regions can be detected on spread chromosomes (Sahara et al., 2002) using the techniques described later in this chapter.

By far the method of choice for detecting these stained regions is in situ hybridization, which began as a method for studying polytene chromosomes, but which subsequently was used with success in studies of metaphase chromosomes. Initially, this method used radioactively labeled (usually ^{3}H or ^{125}I) DNA probes complementary to specific genes or gene families. However, because the radioactively labeled probe method was cumbersome (requiring the coating of slides with emulsion after *in situ* hybridization, long exposure times, and photographic techniques for developing the final product), more accessible procedures have been developed. These include the labeling of DNA or RNA probes with chemical agents that can be detected using immunological procedures, followed by *in situ* hybridization of the labeled probe to chromosome spreads. At first, nonfluorescent labeling of probes was preferred, but fluorescence *in situ* hybridization (FISH) has become a popular method for locating genes on both polytene and metaphase chromosomes. The availability of a large number of fluorescent dyes that can be used to label probes makes it possible to stain chromosomes simultaneously with several different probes. This approach, known as "chromosome painting," because it generates a karyotype marked by dyes of several colors, bears notable similarities to the older method of counterstaining described earlier in this chapter.

Cytogenetics in Evolutionary Studies

Chromosome spreads have had broad usage in evolutionary studies. In general, the gross structures of nonpolytene metaphase chromosomes change relatively slowly and, therefore, are useful for comparisons across species but not at the population level. Polytene chromosomes, by contrast, provide more precise information about changes at the chromosomal level, making them good indicators of variation both at the population level and among closely related species. The classic examples of this approach are provided by the work of Theodosius Dobzhansky and colleagues on *Drosophila*. These studies took advantage of naturally occurring chromosomal inversions to follow lethal genes and other traits in these populations and were followed by many other studies that examined the correlation of environmental clines and chromosomal inversion frequencies. Occasionally, salivary chromosomes also may retain enough information to make broader evolutionary studies possible. *In situ* hybridizations of DNA sequence markers, such as transposable elements, to polytene chromosomes have been used to follow population genetic phenomena. *In situ* studies of polytene chromosomes also have aided in pinpointing the breakpoints of important chromosomal inversions associated with lethals and other traits. So, although the techniques have advanced considerably since Dobzhansky's time, many of the fundamental questions being addressed remain the same.

Various aspects of chromosome morphology can be observed with metaphase spreads, including chromosome number and size, and structural features such as the relative position of the centromere. With regards to the latter, it is helpful to determine whether a metaphase chromosome is metacentric (centromere in the middle of the chromosome), acrocentric (centromere off middle of chromosome, but not at the end), or telomeric (centromere placed at the end of a chromosome arm). Centromeric position also is useful in describing chromosomal morphology with respect to inversions, which can be either pericentric (an inversion including the centromere) or paracentric (not including the centromere). Other important evolutionary events, such as translocations and chromosomal fusions, can be detected using both polytene and nonpolytene metaphase chromosomes.

Several authors realized the importance of polytene chromosome inversions in elucidating phylogenetic relationships among arthropods. Again, *Drosophila* serves as the prime example of this approach, as exemplified by the classic chromosome inversion phylogeny developed by Carson in which he established the phylogenetic relationships of over 100 Hawaiian *Drosophila* species (Carson, 1970, 1972; Carson and Stalker, 1968a,b,c, 1969). Importantly, molecular and cladistic analyses have been congruent with most of the species relationships defined by Carson (O'Grady *et al.*, 2002). Other examples of dipteran groups subjected to polytene chromosome phylogenetics include important disease vectors like *Anopheles* mosquitoes, as well as species in the families Chironomidae, Sciariidae, and Simulidae. As summarized by White (1873) and Marin and Baker (1998), the study of sex chromosomes also has provided substantial insights into insect (especially dipteran) biology. In the Diptera, such work was crucial in revealing the vast range in sex determination mechanisms exhibited by the members of this highly diverse group.

Metaphase spread chromosome methods that incorporate FISH techniques have been used to enhance the evaluation of chromosome arms in addition to chromosome number and shape. The breadth of arthropod species examined is extensive, but the primary focus has been on the insects. Disease vectors and agricultural pests comprise a large proportion of the subjects of these studies, but it is evident that FISH and chromosome painting methods are becoming regular tools for broader evolutionary cytogenetic studies as well.

In addition, a clever technique known as comparative genomic hybridization (CGH) has been developed (Kalliniemi *et al.*, 1992; Struski *et al.*, 2002; Traut *et al.*, 1999) that allows assessments of chromosomal synteny (conservation of gene or other chromosomal segment order in different species). This approach works by using *in situ* hybridizations to metaphase chromosomes of two differentially labeled genomic DNAs, one representing a test genome and the other a reference genome. Fluorescence microscopy is then used to detect the positions of hybridization of the two genomic DNAs simultaneously. Much of the visualization process in this approach has been digitized, thereby facilitating the rapid assessment of hybridization patterns (Joos *et al.*, 1994; http://amba.charite.de/cgh/). (Indeed, digital imaging can be a useful component of most types of modern cytogenetic analyses, making it possible to perform highly accurate computerized measurements and providing a permanent record of the chromosome preparations examined.) The patterns obtained from CGH indicate where chromosomal breaks or breaches in synteny occur in the test chromosomes relative to the reference genome. This novel technique was developed primarily to detect gross chromosomal rearrangements in cancer studies (Kallioniemi *et al.*, 1992), but it clearly also has exciting potential for comparative genomics and evolutionary biology. Once again, the development of a new technology has improved greatly the capacity to address old questions. In this case, the detailed synteny information gleaned from techniques like CGH can be likened to the original polytene chromosome studies in which banding patterns were used to determine inversions and, hence, gross gene order changes in the genomes of Diptera.

Chromosome Protocols

Preparations for Chromosome Morphology Studies

The field of arthropod cytogenetics is a well-established one, and consequently, hundreds of protocols exist for the preparation of both polytene and metaphase chromosomes. These preparations are most suitably summarized in cytogenetics textbooks, as listed in Table IA. In addition, because *Drosophila* has been the focus of intensive cytogenetic research over the past century, many excellent cytology protocol books have appeared in the past decade (Table IB). These protocols, though specific to *Drosophila*, can be modified to accommodate other arthropod species. Finally, there are a number of web sites that include laboratory protocols for the preparation of animal chromosomes and arthropod chromosomes in particular (Table IC). Because so many specific protocols are available in the literature and on the web, only generic protocols are presented here for use in direct observation and *in situ* hybridization. This applies to the preparation of both polytene salivary gland chromosomes (Box I) and nonpolytene metaphase chromosomes (Box II). The reader is referred to standard *in situ* hybridization protocols for further treatment of the

TABLE I
Texts and Web Sites with Manuals and Protocols for Arthropod Chromosome
Manipulation

Title	Author	Publisher
A. General Animal Cytogenetics Manuals and Protocols		
Chromosome Analysis Protocols	Gosden, 1994	Humana
Analyzing Chromosomes	Czepulkowski, 2001	Oxford
Cytogenetics Laboratory Manual	Barch *et al.*, 1997	Lippincott Williams & Wilkins
Molecular Cytogenetics: Protocols and Applications	Fan, 2003	Humana
Insect Cytogenetics	Blackman, 1980	Halsted Pub
Animal Cytogenetics: Techniques in Animal Cytogenetics	Popescu *et al.*, 2000	INRA
B. *Drosophila*-Specific Manuals and Protocols		
Drosophila: A Practical Approach	Roberts, 1998	IRL-PR
Drosophila Protocols	Sullivan *et al.*, 2000	*CSHL*
Drosophila Cytogenetics Protocols	Henderson, 2004	Humana
Drosophila: A Laboratory Manual	Ashburner, 1989	CSHL

C. Animal Cytogenetics Techniques Web Sites

General Cytogenetic Protocols
   http://www.ksu.edu/wgrc/Protocols/labbook.html
   http://hdklab.wustl.edu/lab_manual/cam/camcontents.html
   http://amba.charite.de/cgh/protocol/02/prot02.html
Drosophila Protocols
   http://www.fruitfly.org/about/methods/cytogenetics.html
   http://www.biochem.northwestern.edu/carthew/manual/Chrom.html
   http://www.ceolas.org/VL/fly/protocols.html
   http://www.ou.edu/journals/dis/Techniques/Techhome.html
Specific protocols
   http://iprotocol.mit.edu/protocol/57.htm (Salivary gland preparation)
   http://biology.clc.uc.edu/fankhauser/Labs/Genetics/Drosophila_chromosomes/
      Drosophila_Chromosomes.htm (salivary gland preparation)
   http://www.biovisa.net/protocol/protocol_read.php3?pid = 1141 (metaphase preparation)
   http://grimwade.biochem.unimelb.edu.au/bfjones/ish1.htm (*in situ* preparations)
   http://www.utoronto.ca/krause/FISH.html (fluorescence *in situ* hybridization [FISH]
      protocols)
   http://www.biovisa.net/protocol/protocol_read.php3?pid = 1215 (G-band staining)
Microscopy
   http://iprotocol.mit.edu/protocol/44.htm
   http://iprotocol.mit.edu/protocol/46.htm
   http://iprotocol.mit.edu/protocol/45.htm
Confocal microscopy
   http://helios.mol.uj.edu.pl/conf_c/chapters/chap_3.htm
   http://www.acs.ucalgary.ca/~phillips/users/schoel/chromosome.html
   http://www.soton.ac.uk/~kpa/molecol/60.html
Comparative genomic hybridization (CGH)
   http://amba.charite.de/cgh/

BOX I

POLYTENE CHROMOSOME PREPARATION FROM SALIVARY GLAND TISSUE FOR USE IN DIRECT VISUALIZATION OR *IN SITU* HYBRIDIZATION METHODS

Materials Required:

- Ringer's solutions (in 100 ml dH_2O, mix: 860 mg NaCl, 30 mg KCl, and 35 mg $CaCl_2$)
- Acetic acid orcein: (1–2%): mix 1 g orcein in 50 ml of 45% acetic acid, heat dissolve, then cool and filter through Whatman no. 1 paper
- 3:2:1 solution of acetic acid:lactic acid:dH_2O
- 95% ethanol; absolute ethanol
- Dry ice, razor blades, slides, coverslip, Permamount

Protocol Steps:

1. Dissect plump third instar larvae in which the cuticle has not yet hardened in a drop of saline solution (preferably Ringer's, as above) on a slide. The closer the larva is to puparium formation, the more replicated the polytene chromosomes will be, but too close will give mediocre polytene chromosomes. This dissection can best be accomplished under a dissecting microscope using two fine-tipped forceps. With one forceps, either grasp or pin down the larva at its posterior end. With the second forceps in the other hand, grasp the head between its mouth parts and pull gently.

2. Discard the posterior carcass of the larva and dissect the salivary glands from the mass of tissue left on the slide, which contains mostly fat bodies, digestive tract, and salivary glands. The digestive tract and fat bodies are not opaque like the salivary glands.

3. Remove dissected fat bodies and digestive tract and leave the dissected salivary gland on the slide. It is best to include as many salivary glands from as many larvae as possible on the slide, to enhance the probability that several good spreads will be obtained. Place a drop of acetoorcein on the slide near the saline drop. Transfer the salivary glands to the acetoorcein. If an *in situ* hybridization is desired, a 3:1:2 mixture of acetic acid:lactic acid:water should be used instead of the acetoorcein.

4. Place a coverslip over the salivary gland(s) and fold a portion of lab paper towel so that it is a little larger than the coverslip. Place the folded paper towel over the coverslip and press gently. This step generally requires considerable practice to execute correctly and is critical. Alternatively, excellent spreads can be obtained by gently tapping the coverslip with a pencil using the eraser end. In either case, the purpose of the squash is to flatten the material into a single focal plane.

5a. *For direct observation of polytene chromosomes.* Once the glands are squashed, the excess solution on the slide of acetoorcein stain can be blotted off with a paper towel. The slide is now ready for viewing under a compound microscope. Spread polytene chromosomes should be easily visible at $10\times$ and the fine details of the banding patterns on the chromosomes begin to become visible at $20\times$, with the best definition found under immersion oil at $100\times$. To make the spread permanent, wait 24 h and place the slide on a block of dry ice. When the slide freezes, flip the coverslip off with a razor blade and dehydrate the chromosomes on the slide first in 95% ethanol for 5 min and then in absolute ethanol for 10 min. Permamount or some other suitable mounting solution then can be used to mount the slide.

5b. *For* in situ *hybridization of polytene chromosomes.* Once the glands are squashed, the excess solution on the slide, of acetic acid:lactic acid:water, can be blotted off with a paper towel. At this point, it is advisable to view the slide under phase contrast to assess whether an adequate number of polytene chromosomes have spread on the slide. Once this has been confirmed, the slide and coverslip are frozen on a block of dry ice, and the coverslip flipped off with a razor blade. After the coverslip has been removed, the slide is immediately dipped into 95% ethanol for 10 min followed by two similar washes in fresh 95% ethanol. After the second 95% wash, the slides are air-dried and stored at 4° until the *in situ* hybridization steps are performed.

BOX II

METAPHASE CHROMOSOME PREPARATION FROM LARVAL BRAIN TISSUE FOR USE IN DIRECT VISUALIZATION OR *IN SITU* HYBRIDIZATION METHODS

Materials Required:

- Ringer's solutions (in 100 ml dH$_2$O, mix: 860 mg NaCl, 30 mg KCl, and 35 mg CaCl$_2$)
- Colcemid solution; 0.01 mM Colcemid in Ringer's solution
- Acetic acid orcein (4%): Mix 2 g orcein in 50 ml 45% acetic acid, heat dissolve, then cool and filter through Whatman no. 1 paper *or* 3% Giemsa solution: mix 3 g Giemsa in 100 ml of 45% acetic acid
- Depression slide to hold Colcemid and fixative solutions
- 1% sodium citrate
- 3:1 by volume solution of methanol:acetic acid, made fresh
- 60% acetic acid
- Dry ice, razor blades, slides, coverslip, slide warmer, Permamount

Protocol Steps:

1. Place third instar larvae on a slide in a drop of Ringer's solution. Dissection of the larvae best can be accomplished under a dissecting microscope using two fine-tipped forceps. With one forcep, either grasp or pin down the larva at its posterior end. With the second forcep in the other hand, grasp the head between its mouth hooks and pull gently.

2. The larval brain will most often be associated with tissues connected to the mouth hooks. It is a small inconspicuous bit of tissue that is best recognized by identifying the two hemispheres of the brain as small rounded white organs with stringy fibers (neural tissue) hanging off of them.

3. Place the larval brain in the Colcemid solution for at least 45 min at room temperature.

4. Swell the brain cells by room temperature incubation in 1% sodium citrate for 10–15 min.

5. Transfer the brain tissue to fixative (3:1 methanol:acetic acid) and change the fixative several times.

6. Pick brains out of the fixative and place on a slide. With a small piece of a paper towel or tissue, soak away excess fixative by capillary action, making certain that the brain tissue remains on the slide.

7. Add a drop of 60% acetic acid and place the slide on the slide warmer; set at 45°. As the acetic acid begins to evaporate, remove the slide from the warmer and roll the drop of acetic acid with brain tissue in it around the slide in an area about the size of the coverslip. This action will complete the evaporation of the acetic acid and spread the brain cells in a monolayer on the slide. Let the slides dry further for a few minutes.

8. Store the chromosome spreads dry and use them as soon as possible for either staining with Giemsa or orcein, or as targets for *in situ* hybridization (fluorescence *in situ* hybridization [FISH], etc.).

9. When the secondary manipulation of the chromosomes (direct staining or FISH) is completed, the chromosomes spreads can be made permanent by mounting in Permamount.

---

chromosomes for this technique. The preparations described in Boxes I and II are suitable for visualization using confocal laser microscopy, but further fixation of the slide preparation and RNase treatment is required in this case (see Table IC for web sites listing fixation protocols for confocal imaging of chromosomes).

Genome Size

A Brief Introduction

The genome size (or "C-value") of an organism is defined as the total amount of DNA contained within a single (i.e., haploid) set of its chromosomes. In general, but with some interesting exceptions and minor fluctuations, nuclear genome size is constant within a given species. Genome size typically is measured in terms of either mass (usually in picograms, where 1 pg = 10^{-12}g) or the number of nucleotide base pairs (bp), the latter of which can be modified with metric prefixes to cover larger scales (e.g., kilobases [kb], megabases [Mb], or even gigabases [Gb]). The conversion between these units is simple, with 1 pg being roughly equivalent to 1 billion bases (i.e., 1000 Mb or 1 Gb). More precisely (Dolezel *et al.*, 2003),

$$\text{number of base pairs} = \text{mass in pg} \times 0.978 \times 10^9$$

and

$$\text{mass in pg} = \text{number of base pairs} \times 1.022 \times 10^{-9}$$

The first few genome size measurements were made in the late 1940s (Boivin *et al.*, 1948), leading to broader surveys a few years later (Mirsky and Ris, 1951). To date, about 3800 animal genome sizes have been published, most ($\sim$2500) of which are from vertebrates (Gregory, 2001a). Despite their enormous abundance, arthropods are represented in the current dataset by a mere 400 insects, 230 crustaceans, and 115 spiders (Gregory, 2001a). Other arthropods such as ticks, mites, scorpions, centipedes, millipedes, and smaller groups are only scarcely represented, if at all. Moreover, many of the arthropod values that *are* available have been published only within the past few years, further emphasizing the lack of attention paid to them through much of the 50-year history of genome size study. Clearly, there is a major taxonomic discrepancy in the existing genome size dataset, one that would require a substantial effort to correct. The question is, would such an investment be justified? Or, put more bluntly, why should anyone care about arthropod genome sizes?

Relevance for Sequence and Structure

Undoubtedly, genome sequencing projects are the most obvious area in which genome size has direct relevance. This is particularly true of insects, which are some of the most important targets for complete sequencing because of their roles as laboratory models, disease vectors, domesticated producers, and pests. The publication of the *Drosophila melanogaster* genome sequence in 1998 and that of the malaria mosquito, *Anopheles gambiae*, in 2002 were only the beginning. The honeybee (*Apis mellifera*) genome has been completed in draft form, and several more insect projects are underway, including additional species of *Drosophila* and mosquitoes (*Aedes* spp. and *Culex pipiens*), the silkworm moth (*Bombyx mori*), and the tobacco budworm (*Heliothis virescens*) (Genomes OnLine Database: http://wit.integratedgenomics.com/GOLD/). Additional insects such as the flour beetle *Tribolium castaneum*, the medfly *Ceratitis capitata*, and the tsetse fly *Glossina morsitans*, as well as a few other arthropods like ticks (*Amblyomma* spp.) now are being studied from a large-scale genomic perspective, for example, in expressed sequence tag (EST) or broader sequencing studies.

Although the choice of arthropod models remains largely restricted to species having direct importance to human medicine or agriculture, this list is likely to expand to include species of more general evolutionary, ecological, and/or taxonomic interest. So long as only a limited number of species can be sequenced (because of funding constraints or limitations of technology)—that is, for the foreseeable future—genome size will remain a key consideration in the choice of sequencing projects. Indeed, in their recent discussion of the future of insect genome sequencing, Evans and Gundersen-Rindal (2003) list genome size first among their proposed criteria for selecting the next series of target species. The reason is obvious, namely that both cost of sequencing and difficulty of assembly and analysis increase with genome size.

Fortunately, those species chosen for practical reasons have tended to have small genomes, but this is not true of all arthropods. Known insect genome sizes range about 180-fold, from roughly 0.1 pg in Strepsiptera, some flies, and parasitic wasps to about 17 pg in certain grasshoppers. Arachnid genome sizes range 70-fold, from 0.08 pg in the two-spotted spider mite (*Tetranychus urticae*) to 5.7 pg in jumping spiders in the genus *Habronattus*. In crustaceans, the variation is 240-fold, from 0.16 pg in the smallest water flea to 38 pg in the largest deep-sea shrimp (Gregory, 2001a). Of course, these ranges are minimum estimates, because the vast majority of species have yet to be studied. Nevertheless, it is already clear that arthropod species cannot be chosen at random for genome sequencing projects, and that it is very important that their genome sizes be evaluated before such work. More generally, genome size will remain an important parameter for *any* methods involving whole-scale genome manipulation, such as the construction of genetic libraries.

Comparative genomics involves far more than evaluations of small-scale differences in base pair sequences. Large-scale features of genomic "anatomy," specifically structure and organization, are also worthy of detailed analysis (Eichler and Sankoff, 2003). Size is perhaps the most fundamental aspect of anatomy for any biological entity, genomes included. It is simply impossible to achieve a full understanding of genome structure without data on genome size. Information on genome size also provides the larger context for analyses of subgenomic structure at all scales, from chromosome organization down to the relative abundance of the different types of sequences that make up a given genome.

Relevance for Arthropod Morphology, Development, and Evolution

The bulk mass of nuclear DNA is known to correlate positively with cell size and negatively with cell division rate across a wide range of taxa (Gregory, 2001b). By extension, genome size has the potential to influence any features at the organism level dependent on these cellular parameters, including metabolism, morphology, and development.

Although most of the investigations of such relationships have come from vertebrates and plants, it is becoming increasingly apparent that arthropod biology likewise is affected by variation in genome size. For example, genome size correlates negatively with developmental rate and positively with body size in copepod crustaceans (Gregory et al., 2000; McLaren et al., 1988, 1989; White and McLaren, 2000) and positively with egg volume in water fleas (Beaton, 1995). It has been noted that metamorphosis, a time-limited period of intensive tissue differentiation, may impose limits on genome size (and vice versa) in insects, with those species exhibiting holometabolous development (complete metamorphosis) having genomes less than 2 pg, and those with ametabolous or hemimetabolous (no or incomplete metamorphosis) often exceeding this threshold by a wide margin (Gregory, 2002a). In spiders and other arthropods, the existing data are insufficient to identify such patterns (Gregory and Shorthouse, 2003), but it is very likely that additional surveys will pay large dividends in this regard.

In their pioneering survey—published 2 years before the elucidation of the structure of DNA, and thus before the advent of modern genome biology—Mirsky and Ris (1951) noted a disconnect between genome size and organismal complexity. This raised a significant challenge for theorists over the following two decades, given that C-values were assumed to be constant because DNA is the stuff of genes, and yet the presumed number of coding genes was clearly unrelated to DNA content. The discovery that most eukaryotic DNA is noncoding resolved this "C-value paradox" but also generated several new questions regarding the sources, mechanisms of spread and loss, phenotypic impacts, and differential representation of this nongenic majority. Taken together, these questions make up the much more complex "C-value enigma" (Gregory, 2001b), which has yet to be fully unraveled.

The C-value enigma represents one of the longest running puzzles in genetics and is of substantial significance for evolutionary theory. For example, the abundance of "parasitic" transposable elements (TEs) within genomes implies the operation of natural selection at the subgenomic level, while it increasingly is becoming apparent that genome size, which is influenced strongly by the abundance of TEs, can affect (and be influenced by) features at the organism level (Gregory, 2004). For the most part, this type of multilevel selection process has been discussed only with regards to higher levels (e.g., groups and species), but it seems very appropriate for application to the genome (Gregory, 2004). Because of their unmatched diversity, it is a simple truism that most animal genomes reside within arthropod cells. As such, arthropods must play a central role in the ongoing study of the C-value enigma. Unfortunately, in spite of being relatively inexpensive once the measuring equipment has been purchased, the accumulation of arthropod genome size data has lagged far behind that for vertebrates, meaning that truly general conclusions about animal genome size evolution remain out of reach.

Genome Size Protocols

Nuclear Preparations for Genome Size Estimates

As it stands, the only reliable way to estimate a species' genome size is to use a relative measurement technique—that is, one involving an indirect comparison against a "known" standard. This may seem counterintuitive in the age of complete genome sequencing, but the reality is that sequencing rarely is fully complete, leaving a significant margin of error in the total genome size estimate, particularly in larger genomes containing substantial amounts of heterochromatin. Moreover, "complete" genome sequencing remains a highly inefficient means of estimating genome size, especially in comparison to the rapid and comparatively inexpensive relative measurement techniques typically employed.

Two methods are commonly used to estimate genome size: Feulgen densitometry and flow cytometry. These methodologies are opposite in terms of the underlying physics, given that the former is based on light absorption (optical density) and the latter on emission (fluorescence), but nonetheless are similar by relying on the aforementioned comparison between an unknown and a standard. Likewise, the techniques of tissue preparation differ substantially for the two methods, but the common goal in either case is the production of a set of stained nuclei of known ploidy that can be analyzed individually. In Feulgen densitometry, this involves preparing monolayers of cells fixed onto microscope slides. For flow cytometry, it is necessary to obtain free nuclei in suspension.

Slide Preparation for Feulgen Densitometry

The preparation of slides for most vertebrates is very simple, because an air-dried smear of a single drop of blood provides more than enough diploid nuclei for measurement (Hardie *et al.*, 2002). Invertebrates lack such a ready-made source of individual nuclei, and the techniques for preparing slides are, therefore, more complicated. In larger arthropods, such as spiders, myriapods, decapod crustaceans, and large insects, it is possible to prepare air-dried droplets of hemolymph by sampling with a needle and syringe or by removing a leg with tweezers or scissors and touching the wound to the slide so that the fluid will exude by capillary action. Arthropod hemocytes also are diploid but are far more sparse than vertebrate erythrocytes, and a balance must be struck between too little fluid (with too few cells) and too much (which results in background staining and/or cracking upon drying, thereby damaging the cells).

In most insects, particularly small-bodied species, it may be desirable to obtain dispersed samples of spermatozoa (or spermatids). This usually can be accomplished by dissecting the testes and associated tubules under insect Ringer's saline (1 L distilled H_2O + 7.5 g NaCl + 0.35 g $CaCl_2$ + 0.21 g KCl) and mixing them vigorously with dissecting pins to free individual sperm. In cases in which the insect is too small to permit visualization of the gonads even under a dissecting microscope, it is usually sufficient to macerate the entire abdomen with pins under saline to free some sperm. The entire solution can be allowed to air-dry for later staining (at which time, any salt crystals that form will simply wash off). In general, insect sperm nuclei appear thin and hairlike when stained, although in the Orthoptera (crickets and grasshoppers), Phasmida (stick insects), and Embiidina (webspinners), the sperm more closely may resemble those of vertebrates with a distinct head and tail (Fig. 1) (see Gregory [2002b] and Gregory and Hebert [2003] for additional photos). In some groups, such as the family Coccinellidae in the Coleoptera (ladybird beetles), the sperm may tend to be very thin and highly elongated, thereby greatly complicating or even precluding their measurement (Gregory *et al.*, 2003). In others, like the Lepidoptera (moths and butterflies), the sperm may form tight bundles that can only be dispersed with some difficulty (see Gregory and Hebert [2003] for suggested techniques and photos).

When working with very small arthropods, it may be necessary to stain whole organisms *en bloc* and to dissect specific tissues or to prepare

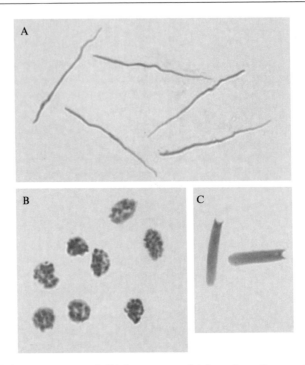

Fig. 1. (A) Spermatozoa and (B) hemocyte nuclei from the yellow mealworm beetle (*Tenebrio molitor*), and (C) sperm nuclei from the northern walkingstick (*Diapheromera femorata*). In many cases, insect hemocytes and sperm give genome size estimates in good agreement with each other, but in beetles, the two cell types may not be compatible. Walkingsticks (order: Phasmida), webspinners (order: Embiidina), and crickets (order: Orthoptera) have compacted sperm nuclei, whereas the sperm nuclei of most other insects, though varying greatly in length, take the generally elongated form illustrated by the beetle example (A).

squashes after staining. With copepod crustaceans, for example, either the epidermal layers can be dissected (Gregory *et al.*, 2000) or the entire animal can be squashed in a drop of acetic acid under a coverslip, which can then be removed by freezing in liquid nitrogen or on dry ice and "flipping" it off with the edge of a razor blade (Wyngaard and Rasch, 2000). In the former case, the cells will be diploid (McLaren *et al.*, 1989), but in the latter, a range of ploidy levels will be found, so diploid cells must be selected carefully. In cladoceran crustaceans, the exopodites of the thoracic limbs usually are dissected after *en bloc* staining because these represent some of the few reliably diploid cells found in adult specimens (Beaton and Hebert, 1988, 1989). Methods such as these almost certainly can be adapted to other small crustaceans to obtain the necessary monolayer of undamaged nuclei of known ploidy. Generally speaking, the preparation methods are flexible, so long as a set of nonoverlapping, undamaged nuclei can be prepared for both the unknown and the chosen standard.

Genome Size by Feulgen Densitometry

The historical development and background chemistry and physics of Feulgen densitometry methods have been reviewed in detail by Hardie *et al.* (2002). In short, densitometric methods involve determining the absorbances (optical densities) of nuclei stained by the well-known Feulgen reaction (Feulgen and Rossenbeck, 1924). This involves hydrolyzing fixed nuclei in strong acid to split off the purine bases and expose aldehyde groups in the DNA molecule, followed by staining with the aldehyde-specific Schiff reagent, which turns pink upon binding and absorbs light in proportion to the amount of DNA present (see Hardie *et al.* [2002] and Rasch [2003] for reagent recipes and recommended procedures).

Three factors complicate the densitometric analysis of stained nuclei: (1) It is not possible to measure absorbance directly, because this is the amount of light *not* passing through an object and is, therefore, "noninformation"; (2) the DNA–stain complex in nuclei is not homogeneous because of intranuclear variation in DNA compaction; and (3) individual nuclei differ in size and shape. The first problem is solved by calculating optical density (OD) from measures of transmittance (T, the amount of light passing through the nucleus subtracted from a sample of uninterrupted "background" light), which are related as follows:

$$OD = \log_{10}(1/T)$$

The second and third issues are dealt with by dividing the nucleus into individual "point densities," which are then summed to give an "integrated optical density" (IOD). Traditionally, this has been accomplished either by passing the nucleus through a fixed light beam ("scanning stage densitometry") or by scanning a moveable light beam through a static nucleus ("flying spot densitometry"). The latter of these has been reviewed in detail by Rasch (1985, 2003). The major shortcoming of both methods is that they are time consuming because they allow only a single nucleus (and indeed, one point density) to be measured at a time. It has also become difficult to acquire and repair the equipment needed for these techniques. The new method of Feulgen image analysis densitometry, which uses the same time-tested staining procedures combined with modern computer-based equipment, greatly accelerates such measurements by capturing a digital image of the microscope field (which contains many nuclei), taking each image pixel as a point density using its grayscale value from 0 (black) to 255 (white), and calculating IODs for all nuclei simultaneously. Although it deals most explicitly with the measurement of vertebrate tissue, the detailed discussion of image analysis equipment and protocols provided by Hardie *et al.* (2002) is also of use in arthropod studies. Indeed, the methods outlined therein have since been used in studies of insects (Gregory, 2002b; Gregory and Hebert, 2003; Gregory *et al.*, 2003) and spiders (Gregory and Shorthouse, 2003) involving sperm and/or hemocytes.

Again, it is only possible to translate the arbitrary units of IOD to absolute genome size (in picograms or base pairs) by comparison of the unknown with the IOD of a reliable standard. This makes the choice of standard an important component of the densitometric protocol, which can lead to significant errors if not done carefully. In particular, the level of DNA compaction and, therefore, the capacity for stain uptake vary widely among cell types, making it highly advisable to select a standard of the same cell type as the unknown. For example, sperm from *Drosophila melanogaster* (1 C = 0.18 pg) or hemocytes from *Tenebrio molitor* (2 C = 1.04 pg) have been used in insect studies in which these cell types were sampled from the unknowns. In cases in which a fully compatible standard is unavailable, a commonly used one like chicken erythrocytes (see below) should be employed so the values can be easily corrected at a later date if necessary.

Nuclear Suspension Preparation for Flow Cytometry

As with densitometry methods, vertebrate blood represents an ideal tissue for genome size analysis by flow cytometry. Unfortunately, arthropod "blood" generally contains too few cells and may cause problems with melanization and clumping during passage through the flow cytometry equipment. As a result, it is usually necessary to generate a suspension of isolated nuclei from solid tissues of known or identifiable ploidy for use in calculating haploid genome size. Also, unlike with Feulgen densitometry, nuclei are treated with RNase to remove RNA before staining in flow cytometric protocols (even though not all dyes will bind to RNA).

Various protocols have been used in the preparation of insect tissues for flow cytometry (Bennett *et al.*, 2003; Marescalchi *et al.*, 1990, 1998; Vieira *et al.*, 2002). The protocol recommended here is outlined in Box III using the example of brain tissue from *Drosophila*. The same protocol can be applied to brain and other tissues from larger insect species (as well as spiders and other arthropods), which can be dissected out before grinding if necessary. Dissecting the brain is advisable for blood-sucking insects, because their salivary glands may contain potent DNA-degrading enzymes. It

---

BOX III

RECOMMENDED PROTOCOL FOR THE PREPARATION AND STAINING OF NUCLEAR SUSPENSIONS FOR FLOW CYTOMETRY

Materials Required:

- Galbraith buffer (per liter: 4.26 g $MgCl_2$, 8.84 g sodium citrate, 4.2 g 3-[N-morpholino]-propane sulfonic acid, 1 ml Triton X-100, 20 μg/ml boiled ribonuclease A, pH 7.2; Johnston et al., 1999)
- Kontes Dounce tissue grinder
- 20 μm nylon filtration mesh
- Propidium iodide

Protocol Steps:

1. Place the heads from individual specimens (preferably adults) in 1–2 ml of modified Galbraith buffer. For *Drosophila* and other small insects, entire heads can be used. For larger specimens, it may be desirable to dissect out the brains before grinding. Tissue from unknowns and standards should be prepared, stained, and measured together.

2. Stroke the heads 15 times in about 10 s with the "A" pestle in a Kontes Dounce tissue grinder.

3. Filter the mixture through 20-μm nylon mesh into a Microfuge tube and store it at 4° until staining.

4. Stain the nuclear suspensions with propidium iodide set to a final concentration of 50 parts per million for up to 24 (usually 1–9 h) in the dark and cold. See text for a discussion of flow cytometer settings to be used during the measurement phase.

Note: If the preparations must be shipped or stored (for up to 6 mo), the grinding buffer described in Thindwa et al. (1994) is recommended instead of Galbraith buffer (380 ml H_2O, 50 ml 2-methyl-2,4 penetadiol [Eastman], 50 ml glycerol, 125 g sucrose, 0.15 g PIPES, 0.55 g $CaCl_2$, 50 mg RNase A [Sigma R5000], adjusted to pH 7.4).

In this case, the storage buffer must be replaced because the sugar it contains will quench fluorescence. To do this, pellet the sample by centrifugation for 1 min (at between 1500 and 2500 rpm, with the appropriate speed determined by trial and error according to the size of the nuclei and the need to balance the number of nuclei with the amount of debris maintained in the sample), pour out the supernatant, and add staining buffer (per liter: 50 mg propidium iodide [Sigma R4170], 50 mg RNase A [Sigma R5000], 9 g NaCl, 3 g PIPES, pH 7.2).

---

is important not to include stomach contents from such insects or ticks, because this will often contain vertebrate blood cells.

Because many insects have one or more sex chromosomes or exhibit haplodiploidy, it is important to either measure the sexes independently or report which sex was measured. This is best practiced not only for the sample, but also for the standard (e.g., females in the *D. melanogaster* Iso-1 line have ~1.5% larger, though slightly less variable, genomes than males). It is recommended that abdominal tissue not be used for insect flow cytometry, because this contains enzymes and gut contents that may interfere with the analysis. Unlike with densitometric techniques, insect sperm is not recommended for use in flow cytometry, because it takes up stain slowly or not at all.

Crustaceans have not been as well studied as insects using flow cyto-metry, but there are a few examples dealing with both large- and small-bodied taxa. For example, in their study of the small crustacean *Daphnia*, Korpolainen *et al.* (1997) ground whole specimens on ice in 0.5 ml of buffer (10 mM Tris–HCl at pH 7.4, 10 mM CaCl$_2$, 3 mM MgCl$_2$, 0.5% Nonidet P-40) and then passed the suspension through nylon filters of 200 and 50 μm mesh sizes to obtain the nuclei. Similar whole-body protocols also have been used for aphids (Thindwa *et al.*, 1994). For larger decapods, Deiana *et al.* (1999) used antennal gland cells minced with fine scissors in saline buffer, mixed with vertebrate blood standards, and then filtered through a 30μm mesh.

In all cases, caution must be exercised when choosing tissues for analy-sis, because many arthropod tissues tend to be composed of polyploid cells. Although this convention is not always observed, it is strongly recom-mended that aliquots of suspensions from standard and unknown be prepared the same way they are mixed, stained, and measured together rather than in separate runs. Although variations between runs and biases due to effects of eye pigments and possible inhibitors of staining cannot be avoided entirely, they can be minimized by co-preparing, staining, and measuring the standard and sample together in the same tube. It bears noting in this context that because inhibitors and competitors for staining within the organism, dye concentration, stain time, tissue type, and daily machine fluctuations all can produce small variations in genome size esti-mates, and small observed differences between (or within) species cannot be considered completely reliable unless it can be shown that the two sets of nuclei produce separate 2 C peaks when co-prepared, stained, and run as a single sample.

Female *D. melanogaster* (Iso-1 strain) should be used for calibration, as their genome size is reliably known (2 C = 0.35 pg). The Canton-S strain, by contrast, is problematic, because a 15% difference may be found be-tween two females in the same culture because of the occurrence in this strain of aneuploids that contain three copies of the fourth chromosome. The nematode *Caenorhabditis elegans* (2 C = 0.204 pg) provides another low-end standard whose genome size is well established. For insects with large genomes, chicken blood added to the sample before preparation serves as a suitable standard. As with *Drosophila*, the strain (e.g., "white leghorn") and sex of the chicken should be provided because the DNA contents of males (2 C = 2.54 pg) and females (2 C = 2.48 pg) are slightly different. For both flow cytometry and Feulgen densitometry, it would be desirable to develop a series of reliable and readily available standards for use in studies of arthropods and other animals, as has been done for plants (Dolezel *et al.*, 1998; Johnston *et al.*, 1999).

Genome Size by Flow Cytometry

Several fluorochromes are available for use in flow cytometry. Some of these are base-pair specific, such as DAPI and Hoechst 33258, which bind to AT-rich regions, and mithramycin, which binds to GC-rich regions. A DNA intercalating dye, which is not base-pair specific, is greatly preferable. Examples include propidium iodide (PI) and ethidium bromide (EtBr), the former of which is recommended here (Box III).

PI has been used in studies of *Drosophila* and stick insects (Bennett *et al.*, 2003; Marescalchi *et al.*, 1990, 1998; Vieira *et al.*, 2002). The protocol recommended here follows that of Bennett *et al.* (2003) and involved staining with PI (50 ppm) for a known duration of up to 24 h (usually 1–9 h) in the dark and cold. Although the stain will begin to saturate within 15 min with only a 10–15% increase in 24 h, and although staining tends to increase at the same rate in the sample and standard, it is best practice to use the same stain time for all samples that are to be compared one to the other.

In their study of *Daphnia*, Korpelainen *et al.* (1997) chose EtBr and stained by adjusting the volume of the filtered nuclear suspension to 2 ml by adding buffer and then by adding 40 μg of RNase and 100 μg of EtBr and allowing the suspension to sit in the dark and cold for 5 h before flow cytometric analysis. For their study of larger decapods, Deiana *et al.* (1999) centrifuged the tissue–saline mixture and then suspended the nuclei in 1 ml of solution 1 ml containing 0.12% sodium citrate, 0.005% PI, and 0.1% RNase. RNA digestion lasted 30 min at 25°, and staining was performed overnight at 4° in this case.

The actual measurements in flow cytometry are produced by passing the stained nuclei through a laser beam and measuring the intensity of the fluorescence emitted by the stimulated fluorochrome. All of this takes place within specialized flow cytometers, with the output being a histogram of nuclear intensities compiled from the measurement of large numbers of individual nuclei (Fig. 2). Samples prepared using Galbraith buffer may be run 20 min after staining. It is recommended that samples prepared in the grinding storage buffer stain at least 40 min. For small genomes, such as *D. melanogaster*, 514-nm (green) illumination produces the best signal-to-noise ratio; however, reliable results can also be obtained using 488-nm (blue excitation) provided gates are used to exclude all but clean 2 C nuclei and Dounce tissue grinding is gentle to minimize broken cells and debris. Best results are obtained with a flow rate of 1000 nuclei/min.

Insect samples prepared as described earlier are usually relatively "dirty" (i.e., contain broken cells and cuticular debris), so it is important to count ("trigger on") only fluorescent cells. Insect nuclei scatter little

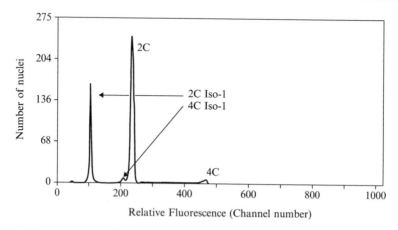

FIG. 2. Sample output of a flow cytometry run showing the relative propidium iodide (PI) fluorescence peaks for an unknown and a standard. The peaks labeled "Iso-1" represent nuclei from the head of a single female *Drosophila melanogaster* specimen from the sequenced Iso-1 strain. The others are from co-prepared brain tissue of a male warning color butterfly *Heliconius erato*. Diploid (2C) and tetraploid (4C) nuclei can be found in both sets of tissue.

light, so a gate that selects only the nuclei with the least amount of forward and/or side scatter will select the subpopulation of nuclei that are intact and free of cytoplasmic tags. Similarly, insect cells so prepared are dilute when compared to most tissues that are prepared for flow cytometry. As a consequence, the run time for an insect sample should be 2–20 min. The slowest possible flow rate ensures optimal results and extends the time required per sample but ensures best results.

Because few arthropods have been examined, it is difficult to predict the genome size of most new samples and thus difficult to predict where the 2-C sample peak will appear for any given voltage and gain setting. As a consequence, it is important that the sample and the standard be run individually before the measurement runs containing the co-prepared samples of unknown and standard. This permits the operator to ensure that the instrument is adjusted with the lowest (usually 2-C) sample peak on screen. It also permits the operator to predict (and adjust as necessary) the relative positions of the 2-C sample peak and the 2-C standard peaks so that they both appear in the cytogram.

Flow cytometry has traditionally involved the use of large, expensive machines. However, recent advances in laser technology have allowed the development of compact inexpensive models. It is likely that these will become more widespread, thereby increasing the worldwide capacity to survey arthropod genome sizes and to help correct the current dataset's severe taxonomic imbalance.

Web-Based Information

Arthropod Cytogenetic and Genome Databases

The *Animal Cytogenetics* series mentioned in the introduction provides detailed (though in some cases, by now dated) information for specific groups of insects. Unfortunately, this information has not been made available online in any comprehensive way. Indeed, although the list of insect species that have been examined cytogenetically is large, few web sites are dedicated to archiving chromosome number or morphology in insects or any other arthropods. Furthermore, the overriding trend is for web sites that do present arthropod cytogenetic information to focus on model insects whose genomes have been or are being sequenced. These sites and others providing additional bibliographic material on arthropod cytogenetics are listed in Table II. The latter can be used to good effect by searching their content for cytogenetic keywords.

Genome Size Databases

In 1972, Sparrow and colleagues published a compilation of genome sizes from plants, animals, fungi, and bacteria, and a few years later, the first in a series of large angiosperm C-value collections was published by Bennett and Smith (1976). In 1997, the *Angiosperm DNA C-Values Database* was made available online and has since been expanded to include gymnosperms and other groups as part of the broader *Plant DNA C-Values Database* (Bennett and Leitch, 2003), which contains data for about 4000 species of plants.

In the animal cytogenetics community, by contrast, little effort was made to follow up on the initial survey of Sparrow *et al.* (1972). In fact, no comprehensive compilation of animal genome sizes was available before 2001, when the online *Animal Genome Size Database* was launched (Gregory, 2001a). As noted previously, the database contains genome size estimates for roughly 3800 species, most from vertebrates and only about 20% from arthropods. The database can be freely accessed and contains information about the methods, cell types, and standards used in each measurement, as well as summary statistics for all the major animal groups represented. Chromosome numbers also are provided for some arthropod groups, although this is limited to species for which genome sizes are also available and is, therefore, not comprehensive.

TABLE II
WEB-BASED ARTHROPOD CHROMOSOMAL AND GENOMIC DATABASES

Insect Cytogenetics
Drosophila melanogaster cytological information
   http://flybase.bio.indiana.edu:82/maps/fbgrmap.html
FlyView
   http://pbio07.uni-muenster.de/FlyView/Home.html
Bridges drawings
   http://www.hawaii.edu/bio/Chromosomes/poly/poly.html
   Anopheles gambiae (AnoBase)
     http://skonops.imbb.forth.gr/cgi-bin/insitu.pl
   Anopheles chromosome bibliography
     http://wrbu.si.edu/www/culicidae/anophelinae/an/An.catalog.html

Arthropod Genome Web Sites
Apis
   http://www.genome.clemson.edu/projects/stc/bee/AM-Ba/
   http://www.ncbi.nlm.nih.gov/genome/guide/bee/
BombMap
   http://www.nises.affrc.go.jp/sgp/BombMap.htm
SilkBase
   http://www.ab.a.u-tokyo.ac.jp/silkbase/
FlyBase
   http://www.flybase.org/
Mosquito genomics
   http://mosquito.colostate.edu/tikiwiki/tiki-page.php?pageName=Mosquito_genomics
   Insect genetic resources
   http://www.ars-grin.gov/nigrp/

Miscellaneous
Animal Genome Size Database
   http://www.genomesize.com/
Insect bibliographic databases
   http://www.sel.barc.usda.gov/selhome/database.htm

The creation of the *Animal Genome Size Database* has allowed broad analyses of the relationships between genome size and cell size, metabolic rate, developmental rate, and other life-history parameters in vertebrates. The equivalent studies in arthropods have thus far been hampered by lack of data. Of course, as improvements in technology continue to facilitate new measurements, as the importance of collecting arthropod genome size data becomes appreciated more widely, and as the resulting data are incorporated into the growing database, such broad comparisons finally will be extended to the most diverse of all animal phyla.

Concluding Remarks

The study of arthropod chromosomes has played an important part in the history of evolutionary biology and genetics, not least by paving the way for the rise of modern genomics. The development of advanced comparative techniques as part of the genomic revolution has, in turn, expanded the possibilities for cytogenetic study. This involves both the application of new methodologies to long-standing questions and the opening of entirely new avenues of research. In both cases, the prospects for an improved understanding of large-scale genomic properties are very exciting. Far from being rendered obsolete by detailed sequence-based analysis, cytogenetic analyses are likely to play an increasingly important role as a more holistic approach continues to develop in genomics.

Acknowledgments

R. D. thanks the Lewis and Dorothy Culman Program for Molecular Systematics Studies at the AMNH. TRG was supported by a Natural Sciences and Engineering Research Council of Canada (NSERC) postdoctoral fellowship and the NSERC Howard Alper Postdoctoral Prize.

References

Ashburner, M. (1989). "Drosophila: A Laboratory Manual." CSHL Press, Cold Spring Harbor, NY.

Balbiani, E. G. (1881). Sur la structure de noyau des cellules salivaires les larves de Chironomus. *Zool. Anz.* **4,** 637–641, 662–666.

Barch, M. J., Knutsen, T., and Spurbeck, J. L. (1997). "The AGT Cytogenetics Laboratory Manual," 3rd Edn., Lippincott–Raven Press, Philadelphia, PA.

Beaton, M. J., and Hebert, P. D. N. (1988). Geographic parthenogenesis and polyploidy in *Daphnia pulex. Am. Naturalist* **132,** 837–945.

Beaton, M. J., and Hebert, P. D. N. (1989). Miniature genomes and endopolyploidy in cladoceran crustaceans. *Genome* **32,** 1048–1053.

Beaton, M. J. (1995). Patterns of endopolyploidy and genome size variation in *Daphnia.* PhD Thesis. University of Guelph, Guelph.

Bennett, M. D., and Smith, J. B. (1976). Nuclear DNA amounts in angiosperms. *Philos. Trans. R. Soc. Lond. Series B* **274,** 227–274.

Bennett, M. D., and Leitch, I. J. (2003). Plant DNA C-values database. Royal Botanic Gardens, Kew, UK. http://www.rbgkew.org.uk/cval/homepage.html.

Bennett, M. D., Leitch, I. J., Price, H. J., and Johnston, J. S. (2003). Comparisons with *Caenorhabditis* (~100 Mb) and *Drosophila* (~175 Mb) using flow cytometry show genome size in *Arabidopsis* to be ~157 Mb and thus ~25 % larger than the *Arabidopsis* Genome Initiative estimate of ~125 Mb. *Ann. Botany* **91,** 547–557.

Blackman, R. L., Hewitt, G. M., and Ashburner, M. (1980). "Insect Cytogenetics." Blackwell Scientific Publications, London.

Boivin, A., Vendrely, R., and Vendrely, C. (1948). L'acide désoxyribonucléique du noyau cellulaire dépositaire des caractères héréditaires; arguments d'ordre analytique. *Comptes Rendus de l'Académie des Sciences* **226,** 1061–1063.

Carson, H. L., and Stalker, H. D. (1968a). Polytene chromosome relationships in Hawaiian species of Drosophila. I. The D. grimshawi group. *In* "Studies in Genetics" (M. R. Wheeler, ed.), pp. 335–354, IV. Research reports. The University of Texas, Austin.

Carson, H. L., and Stalker, H. D. (1968b). Polytene chromosome relationships in Hawaiian species of *Drosophila.* II. The D. planitibia subgroup. *In* "Studies in Genetics" (M. R. Wheeler, ed.), pp. 355–365, IV. Research reports. The University of Texas, Austin.

Carson, H. L., and Stalker, H. D. (1968c). Polytene chromosome relationships in Hawaiian species of *Drosophila.* III. The *D. adiastola* and *D. punalua* subgroups. *In* "Studies in Genetics" (M. R. Wheeler, ed.), pp. 367–380, IV. Research reports. The University of Texas, Austin.

Carson, H. L., and Stalker, H. D. (1969). Polytene chromosome relationships in Hawaiian species of *Drosophila.* IV. The D. primaeva subgroup. *In* "Studies in Genetics" (M. R. Wheeler, ed.), pp. 85–94, V. The University of Texas, Austin.

Carson, H. L. (1970). Chromosome tracers of the origin of species. *Science* **168,** 1414–1418.

Carson, H. L. (1972). Inversions in Hawaiian *Drosophila. In* "Drosophila Inversion Polymorphism" (C. B. Krimbas and J. R. Powell, eds.), pp. 407–439. CRC Press, Boca Raton, FL.

Czepulkowski, B. H. (2001). "Analyzing Chromosomes." Oxford Press, Oxford.

Deiana, A. M., Cau, A., Coluccia, E., Cannas, R., Milia, A., Salvadori, S., and Libertini, A. (1999). Genome size and AT-DNA content in thirteen species of decapoda. *In* "Crustaceans and the Biodiversity Crisis" (F. R. Schram and J. C. von Vaupel Klein, eds.), pp. 981–985. Koninklijke Brill NV, Leiden, The Netherlands.

Dolezel, J., Greilhuber, J., Lucretti, S., Meister, A., Lysak, M. A., Nardi, L., and Obermayer, R. (1998). Plant genome size estimation my flow cytometry: Inter-laboratory comparison. *Ann. Botany* **82**(suppl. A), 17–26.

Dolezel, J., Bartos, J., Voglmayr, H., and Greilhuber, J. (2003). Nuclear DNA content and genome size of trout and human. *Cytometry* **51A,** 127–128.

Eichler, E. E., and Sankoff, D. (2003). Structural dynamics of eukaryotic chromosome evolution. *Science* **301,** 793–797.

Evans, J. D., and Gundersen-Rindal, D. (2003). Beenomes to *Bombyx*: Future directions in applied insect genomics. *Genome Biol.* **4,** 101–107.

Fan, Y. (2003). "Molecular Cytogenetics and Applications." Humana Press, Totawa, NJ.

Feulgen, R., and Rossenbeck, H. (1924). Mikroskopisch-chemischer Nachweis einer Nucleinsäure vom Typus der Thymonucleinsäure und die darauf beruhende elektive Färbung von Zellkernen in mikroskopischen Präparaten. *Hoppe-Seyler's Zeitschrift für Physiologische Chemie* **135,** 203–248.

Frydrychová, R., and Morec, F. (2002). Repeated losses of TTAGG telomere repeats in evolution of beetles (Coleoptera). *Genetica* **115,** 179–187.

Gosden, J. R. (1994). "Chromosome Analysis Protocols." Humana Press, Totawa, NJ.

Gregory, T. R. (2001a). Animal Genome Size Database. http://www.genomesize.com.

Gregory, T. R. (2001b). Coincidence, coevolution, or causation? DNA content, cell size, and the C-value enigma. *Biol. Rev.* **76,** 65–101.

Gregory, T. R. (2002a). Genome size and developmental complexity. *Genetica* **115,** 131–146.

Gregory, T. R. (2002b). Genome size of the northern walkingstick, *Diapheromera femorata* (*Phasmida: Heteronemiidae*). *Can. J. Zool.* **80,** 1303–1305.

Gregory, T. R., and Shorthouse, D. P. (2003). Genome sizes of spiders. *J. Heredity* **94**, 285–290.

Gregory, T. R., and Hebert, P. D. N. (2003). Genome size variation in lepidopteran insects. *Can. J. Zool.* **81**, 1399–1405.

Gregory, T. R., Hebert, P. D. N., and Kolasa, J. (2000). Evolutionary implications of the relationship between genome size and body size in flatworms and copepods. *Heredity* **84**, 201–208.

Gregory, T. R., Nedved, O., and Adamowicz, S. J (2003). C-value estimates for 31 species of ladybird beetles (*Coleoptera: Coccinellidae*). *Hereditas* **139**, 121–127.

Gregory, T. R. (2004). Macroevolution, hierarchy theory, and the C-value enigma. *Paleobiology* **30**(2), 179–202.

Hardie, D. C., Gregory, T. R., and Hebert, P. D. N. (2002). From pixels to picograms: A beginners' guide to genome quantification by Feulgen image analysis densitometry. *J. Histochem. Cytochem.* **50**, 735–749.

Henderson, D. (2004). "*Drosophila* Cytogenetics Protocols." Humana Press, Stony Brook, NY.

Johnston, J. S., Bennett, M. D., Rayburn, A. L., Galbraith, D. W., and Price, H. J (1999). Reference standards for determination of DNA content of plant nuclei. *Am. J. Botany* **86**, 609–613.

Joos, S., Fink, T. M., Rätsch, A., and Lichter, P. (1994). Mapping and chromosome analysis: The potential of fluorescence *in situ* hybridization. *J. Biotechnol.* **35**, 135–153.

Kallioniemi, A., Kallioniemi, O.-P., Sudar, D., Rudovitz, D., Gray, J. W., Waldman, F. M., and Pinkel, D. (1992). Comparative genomic hybridization for molecular cytogenetic analysis of solid tumors. *Science* **258**, 818–821.

King, R. (1975). "Handbook of Genetics," "Invertebrates of Genetic Interest." Vol. 3. Plenum Press, New York.

Korpelainen, H., Ketola, M., and Hietala, J. (1997). Somatic polyploidy examined by flow cytometry in *Daphnia. J. Plankt. Res.* **19**, 2031–2040.

Marescalchi, O., Scali, V., and Zuccotti, M. (1990). Genome size in parental and hybrid species of *Bacillus* (*Insecta, Phasmatodea*) from southeastern Sicily: A flow cytometric analysis. *Genome* **33**, 789–793.

Marescalchi, O., Scali, V., and Zuccotti, M. (1998). Flow-cytometric analyses of intra-specific genome size variations in *Bacillus atticus* (Insecta, Phasmatodea). *Genome* **41**, 629–635.

Marin, I., and Baker, B. S. (1998). The evolutionary dynamics of sex determination. *Science* **281**, 1990–1994.

McLaren, I. A., Sévigny, J.-M., and Corkett, C. J. (1988). Body size, development rates, and genome sizes among *Calanus* species. *Hydrobiologia* **167/168**, 275–284.

McLaren, I. A., Sévigny, J.-M., and Frost, B. W. (1989). Evolutionary and ecological significance of genome sizes in the copepod genus *Pseudocalanus. Can. J. Zool.* **67**, 565–569.

Mirsky, A. E., and Ris, H. (1951). The desoxyribonucleic acid content of animal cells and its evolutionary significance. *J. Gen. Physiol.* **34**, 451–462.

O'Grady, P. M., Baker, R. H., Durando, C. M, Etges, W. J., and DeSalle, R. (2002). Polytene chromosomes as indicators of phylogeny in several species groups of *Drosophila. BMC Evol. Biol.* **1**, 6.

Painter, T. S. (1933). A new method for the study of chromosome rearrangements and the plotting of chromosome maps. *Science* **78**, 585–586.

Petitpierre, E. (1996). Molecular cytogenetics and taxonomy of insects, with particular reference to the coteoptera. *Int. J. Insect morph. Embriol.* **25**, 115–134.

Popescu, P., Hayes, H., and Dutrillaux, B. (2000). "Techniques in Animal Cytogenetics." Springer, New York.

Rasch, E. M. (1985). DNA "standards" and the range of accurate DNA estimates by Feulgen absorption microspectrophotometry. In "Advances in Microscopy" (R. R. Cowden and S. H. Harrison, eds.), pp. 137–166. Alan R. Liss, New York.

Rasch, E. M. (2003). Feulgen-DNA cytophotometry for estimating C values. In "Methods in Molecular Biology," (D. S. Henderson, ed.), Vol. 247, pp. 163–201. Humana Press, Totowa, NJ.

Roberts, D. B. (1998). "Drosophila. A Practical Approach." IRL, Oxford.

Smith, S. G., and Virkki, N. (1978). "Insecta 5. Coleoptera. Animal Cytogenetics," Vol. 3. Borntraeger Verlagsbuchhandlung, Science Publishers, Stuttgart.

Sparrow, A. H., Price, H. J., and Underbink, A. G. (1972). A survey of DNA content per cell and per chromosome of prokaryotic and eukaryotic organisms: some evolutionary considerations. In "Evolution of Genetic Systems" (H. H. Smith, ed.), pp. 451–494. Gordon and Breach, New York.

Steiniger, G. E., and Mukherjee, A. B. (1975). Insect chromosome banding: technique for G- and Q-banding patterns in the mosquito Aedes albopictus. Can. J. Genet. Cytol. 17, 241–244.

Struski, S., Doco-Fenzy, M., and Cornillet-Lefebvre, P. (2002). Compilation of published comparative genomic hybridization studies. Cancer Genet. Cytogenet. 135, 63–90.

Sullivan, W., Ashburner, M., and Hawley, R. S. (2000). "Drosophila Protocols." CSHL Press, Cold Spring Harbor.

Thindwa, H. P., Teetes, G. L., and Johnston, J. S. (1994). Greenbug DNA content. Southwestern Entomol. 19, 371–378.

Traut, W., Sahara, K., Otto, T. D., and Marec, F. (1999). Molecular differentiation of sex chromosomes probed by comparative genomic hybridization. Chromosoma 108, 173–180.

Vieira, C., Nardon, C., Arpin, C., Lepetit, D., and Biémont, C. (2002). Evolution of genome size in Drosophila. Is the invader's genome being invaded by transposable elements? Mol. Biol. Evol. 19, 1154–1161.

Wagner, R. P., and Crow, J. F. (2001). The other fly room: J. T. Patterson and Texas genetics. Genetics 157, 1–5.

White, M. J. D. (1873). "Animal Cytology and Evolution," 3rd Edn., University of Cambridge Press, Cambridge.

White, M. M., and McLaren, I. A. (2000). Copepod development rates in relation to genome size and 18S rDNA copy number. Genome 43, 750–755.

Wyngaard, G. A., and Rasch, E. M. (2000). Patterns of genome size in the Copepoda. Hydrobiologia 417, 43–56.

Section II

Comparing Macromolecules: Functional Analyses

[26]   Experimental Methods for Assaying Natural Transformation and Inferring Horizontal Gene Transfer

By JESSICA L. RAY and KAARE M. NIELSEN

Abstract

The observation of frequent lateral acquisitions of genes in sequenced bacterial genomes has spurred experimental investigations to elucidate the factors governing ongoing gene transfer processes in bacteria. The uptake of naked DNA by natural transformation is known to occur in a wide range of bacterial species and in some archaea. We describe a series of protocols designed to dissect the natural genetic transformability of individual bacterial strains under conditions that progress from standard *in vitro* conditions to purely *in situ*, or natural, conditions. One of the most important factors in ensuring the success of any transformation assay system is the use of a sensitive, effective, and distinguishable selection regimen. Detailed template protocols for assaying bacterial transformation *in vitro* are presented using the naturally competent bacterium *Acinetobacter baylyi* strain BD413 as a model. Factors increasing the complexity of the assay systems are included in the following section describing the incorporation of components of natural systems to the *in vitro* models, such as in soil and water microcosm experiments. We then present template protocols for the transformation of bacteria in modified natural systems, such as in the presence of host tissues and extracts or in the greenhouse. Clear and ecologically meaningful demonstrations of *in situ* natural transformation are most desirable but are also the most complex and challenging. Because of the highly variable nature of these experiments, we include a discussion of important factors that should be considered when designing such experiments. Some advantages and disadvantages of the experimental systems with regard to resolving the hypotheses tested are included in each section.

Introduction

A growing amount of data suggests that phylogenetic incongruencies are due to horizontal gene transfer (HGT) and drives the initiative to accumulate empirical data assessing the relevance, frequency, and importance of HGT in bacterial evolution. Phylogenetics has provided enough information from available genome sequences to assert with a high degree of certainty that the horizontal transfer of entire genetic traits between bacteria and even between kingdoms has made significant contributions to the genetic diversity observed in bacteria today. Genetic exchange in prokaryotes can be mediated via conjugation, transduction, or natural transformation (also called *natural genetic transformation*). The number of culturable bacteria that are amenable to natural genetic transformation in the laboratory is rapidly increasing (Lorenz and Wackernagel, 1994; Wackernagel, 2003), providing evidence that uptake of naked DNA from the environment (natural transformation) represents a potentially significant mechanism by which bacterial populations acquire novel genetic material and evolve. A universal function for natural transformation has not been identified. It has been hypothesized that natural transformation exists to facilitate the exchange of genetic traits that may promote fitness adaptation, to acquire genetic information that can serve as a template for DNA repair, or to provide nutrition during periods of starvation (Dubnau, 1999).

To assess the relevance of HGT (via natural transformation) in contemporary bacterial populations, we can first determine whether the bacterium of interest is amenable to natural transformation. A clear understanding of genetic transformation in natural habitats presents a technical challenge but has become facilitated by methods that in our approximation more closely resemble conditions found in nature. Empirical data could then ideally be applied to the interpretation of phylogenetic analyses.

The goal of this chapter is to present methods that can be used to obtain knowledge about the amenability of a bacterium to natural transformation and to further understand the environmental and cellular factors important for the process to occur. We have chosen to arrange methods for assaying natural transformation in bacteria along an arbitrary spectrum from purely *in vitro* to purely *in situ* (Fig. 1) based on the estimated fidelity with which

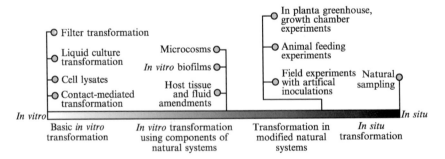

FIG. 1. Arbitrary spectrum of biological relevance of the various transformation methods, ranging from purely *in vitro* to purely *in situ* (field conditions).

experimental designs simulate natural conditions. Following this scheme, we describe four arbitrary classes of methods for assaying natural transformation. The underlying structure of each method is simple: exposure of recipient bacteria to a donor DNA source followed by the application of selection to identify transformants from the recipient population. To build a foundation toward the assessment of *in situ* transformation, the first section begins with basic *in vitro* transformation protocols. The second section describes protocols for *in vitro* transformation using modified components of real systems. In the last two sections, we have chosen to discuss methods for assaying transformation in modified natural systems and *in situ* in the context of published studies (the complexity of such experiments precludes the practicality of a single generic protocol). This progression reflects incorporation of additive conditional modifications (inclusion of soil in transformation media, adaptation of the transformation assay to a living system, etc.) that gradually increase the similarity of the reaction to natural conditions. Practical application of the methods in published studies of natural transformation are discussed in each section and are summarized in Table I.

The culture-based methods provided here are intended as means to assay the outcome of a natural transformation event, namely a predictable and detectable change in the genotype and hence, phenotype of the recipient organism. We do not include protocols for the dissection of the molecular mechanisms governing genetic competence or physiology of DNA uptake in recipient organisms. Dubnau (1999) provides an excellent review of research on the mechanisms and development of genetic competence in bacteria. Although this chapter refers explicitly to the transformation of bacteria, the protocols may be modified to assay transformation in other prokaryotes or single-celled eukaryotes (Nevoigt *et al.*, 2000).

Selection Markers

Good selection markers should confer a dominant phenotypic alteration that would otherwise occur (due to genetic mutation) at negligible frequency in that population of recipient bacteria. The most popular selection markers employed in recombinant DNA technologies and in assaying gene transfer events are antibiotic resistance genes. These include the *bla* gene family, conferring resistance to β-lactam antibiotics, the *nptII* gene, conferring resistance to aminoglycoside antibiotics, and members of the *aad* gene family, which confer resistance to various additional aminoglycoside antibiotics. A less commonly used selection strategy in gene transfer assays is the rescue of a mutant phenotype that occurs via chromosomal marker exchange (Juni, 1978; Williams *et al.*, 1996). Genes conferring

TABLE I

OVERVIEW OF METHODS TO ASSAY NATURAL TRANSFORMATION

Biological relevance	Donor DNA source	Transformation medium	Advantages	Disadvantages	Example references
In vitro	Genomic DNA	Filter or liquid culture	Cost-effective, reproducible, fine-tuning, high detection sensitivity	Artificial conditions, results do not always translate to *in situ*	Harris-Warrick and Lederberg, 1978; Majewski et al., 2000; Palmen et al., 1993
	Plasmid DNA	Filter or liquid culture	Cost-effective, reproducible, broad host range	Less efficient than genomic donor DNA	Nielsen et al., 1997; Sikorski et al., 1998, 2002
	Cell lysates	Filter or liquid culture	No DNA purification required, more representative of DNA in the environment	Dilute donor DNA, presence of DNA-degrading factors	Nielsen et al., 2000
	Intact donor cells	Filter	More representative of natural bacterial interactions	Difficult to verify mechanism using DNAse-sensitivity assay	Albritton et al., 1982; Stewart et al., 1983
In vitro using components of natural systems	Genomic or plasmid DNA, cell lysates or intact donor cells	Soil microcosm	More representative of soil bacterial dynamics, cost-effective, medium throughput	Higher variability depending on soil type and treatment, decreased detection sensitivity	Demanèche et al., 2001; Graham and Istock, 1978; Nielsen et al., 1997; Sikorski et al., 1998

System	DNA source	Setting	Advantages	Disadvantages	References
Plant tissue microcosm			More representative of environmental conditions, cost-effective, sterility can be controlled	Decreased detection sensitivity, interference from indigenous plant microbes when using nonsterile plant tissue	Becker et al., 1994; Tepfer et al., 2003
Biofilms			Enable both enumeration and visualization of transformation events, cost-effective, more realistic for many aquatic bacteria	Do not reflect microbial or environmental fluctuations in aquatic habitats	Hendrickx et al., 2000; Li et al., 2001; Williams et al., 1996
Biological fluids and/or tissues			Assay transformation in a more biological context, good preparation for in vivo experiments	Materials are often artificially processed prior to use, variability in material composition between samples	Mercer et al., 1999a,b, 2000
Modified natural systems	Genomic or plasmid DNA, cell lysates, or intact donor cells	in planta, in the greenhouse, or growth chamber	More representative of in situ conditions, amenable to manipulations, sterility can be controlled	Less cost-effective, decreased detetion sensivity, variability	Bertolla et al., 1999; Kay et al., 2002, 2002b

(continued)

TABLE I (continued)

Biological relevance	Donor DNA source	Transformation medium	Advantages	Disadvantages	Example references
		Animal feeding experiments	Analysis of DNA and bacterial dynamics *in vivo*	Expensive, time-consuming, ethical considerations, decreased detection sensitivity, use of nonnatural hosts, artificial introduction of recipients and donor bacteria, inability to control for conjugation or transduction as gene transfer mechanisms	Dahl *et al.*, 1999; Flint *et al.*, 2002
		Field experiments with artificial donor and/or recipient inoculations	Demonstration of competence development/ transformation *in situ*		Williams *et al.*, 1996
In situ	Natural source	Field study	Identification of gene transfer events from natural donor DNA sources	Detection sensitivity insufficient, inability to distinguish transformation from conjugation or transduction, low reproducibility, time-consuming, requires large sampling size, inability to account for all external factors	Gebhard and Smalla, 1999; Paget *et al.*, 1998

resistance to metal ions are also becoming popular selection alternatives to antibiotic resistance genes (Baulard *et al.*, 1995). When transformants are to be identified by direct visualization, genes encoding fluorescent proteins, most notably the green fluorescent protein (GFP), are convenient for use as selection markers (Errampalli *et al.*, 1999). Because fluorescent protein markers provide no selective advantage by themselves, they are often used as secondary selection markers in combination with genes encoding antibiotic resistance as primary selection markers.

Before transformation assays, the effectiveness of selection markers in distinguishing transformants from a population of recipients should be tested. Background resistance (to the selective pressure) that is present in recipient populations will undermine the detection sensitivity of the selection strategy. Commercially available diagnostic tests such as Etest or PDM Antibiotic Sensitivity Disks (AB Biodisk, Sweden) are recommended for measuring antibiotic resistance levels in individual recipients. In natural samples where intrinsic phenotypic resistance to the antibiotics used for selection is abundant (Heuer *et al.*, 2002; van Overbeek *et al.*, 2002), the researcher must possess the means to distinguish the target selection marker gene from the indigenous population of intrinsic or *a priori* resistant bacteria that will confer the same selectable phenotype. Selection for auxotrophy to prototrophy conversions and for antibiotic resistance that can arise by mutation in housekeeping genes requires that the frequency of spontaneous mutation in the recipient strain are below the limit of detection under negative control transformation conditions. Donor DNAs containing multiple selectable markers should be tested to determine the best marker to use in further transformation assays. For all transformations, we recommend additional genetic or molecular tests to verify the presence of the specific genetic trait in transformants.

Basic *In Vitro* Transformation

In vitro transformation assays allow the dissection of physical and chemical parameters that govern a bacterium's amenability to natural genetic transformation to establish so-called "optimal" conditions. The setup of a basic transformation assay is quite simple: Cells are cultured, exposed to donor DNA, and then plated on a selective medium to identify transformants. Obtaining maximum or even detectable transformation events depends on several factors: The conditions for DNA uptake must be met, donor DNA must be taken up and expressed by recipients, and the selection applied must be specific and sufficient to increase the limit of detection above the background and the rate of spontaneous mutation. When assaying a bacterium that has not previously been transformed in the laboratory, manipulation of various parameters within the transformation reaction may ultimately identify permissible transformation conditions (Fig. 2). Physical and chemical factors that can influence the ability of bacterial cells to achieve competence in culture include monovalent and divalent cation concentration, temperature, pH, nutrient starvation, and water availability and have been discussed in detail previously (Lorenz and Wackernagel, 1994).

Controls to test the sterility of donor DNA, buffer/saline, nutrient media, and filters, as well as the purity of bacterial cultures should always be included in transformation assays. Both recipient cells and transformants should be enumerated from each treatment to verify recipient growth and/or survival during the transformation period. Once permissible conditions are identified, positive control donor DNA treatments can also be included to control for variations in the system that may have affected recipient growth.

Cultivation and Filter Transformation of Acinetobacter sp. BD413

Acinetobacter sp. BD413 is a gram-negative γ-proteobacterium found in soil and water and is naturally competent for nondiscriminatory uptake of both chromosomal and plasmid DNA (Juni, 1978; Nielsen *et al.*, 1997a; Palmen *et al.*, 1993). For the following protocols using *Acinetobacter* sp. BD413 as recipient, the denotation *BD413* refers to a spontaneous rifampicin-resistant mutant generated by Nielsen *et al.* (1997b).

BD413 is cultured at 30° in Luria-Bertani medium (Sigma-Aldrich, Inc., St. Louis, MO) as broth (LB) or agar (LA) plates, both amended with 50 μg/ml rifampicin (LBR or LAR, respectively). Liquid cultures should be well aerated, and cultures on solid media should be incubated in high relative humidity conditions. BD413 will form large, diffuse colonies on solid media if excess moisture is present, so moisture on solid media surfaces should be allowed to dry before incubation.

We routinely use Millipore White GSWP, 47-mm, 0.22-μm filters (Millipore, Bedford, MA). When different filter types are available, control experiments can be performed to compare their function in transformation assays. The size of these filters allows easy retention and recovery of bacterial cells for enumeration after transformation. The filters are obtainable presterilized. Nonsterile filters are sterilized in batches of 25–50 in dH$_2$O, because dry sterilization causes the filters to become brittle.

Protocol for the Preparation of Competent Cells

1. Prepare an overnight culture of BD413 in liquid medium by inoculating a single BD413 colony into 3–5 ml of LBR and incubating at 30° with good aeration.

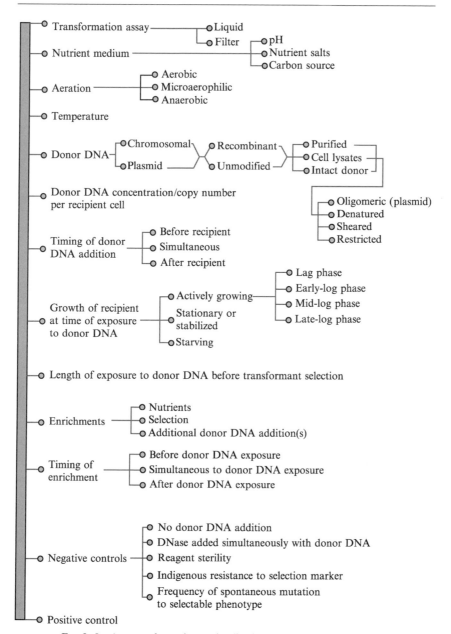

FIG. 2. *In vitro* transformation optimalization parameters and controls.

2. The following day, subculture BD413 1:100 into 100 ml of fresh LBR. Incubate at 30° with good aeration for about 4–5 h or until the $OD_{600\ nm}$ equals about 0.6.

3. Measure the $OD_{600\ nm}$ of the culture and calculate the number of cells in 100 ml of culture (NB: This step requires that a growth experiment has been performed in advance to correlate CFU milliliter^{-1} with $O.D._{600nm}$). From this number, calculate the volume of buffer required to resuspend the cell pellet so the final concentration is 1.0×10^9 cells/ml.

4. Pellet the cells by spinning at 4000 rpm for 10 min at 4° in a Sorvall SS-34 rotor. Competent cells should not be centrifuged at speeds faster than 2000g. Discard the supernatant.

5. Resuspend the cell pellet in the appropriate volume (calculated in step 3) of cryostorage buffer (recommended 15% [w/v] glycerol in LB). Transfer 1-ml aliquots of the cell suspension into sterile 1.5-ml Microfuge tubes (each tube contains 1.0×10^9 cells). Place cells immediately in −80° freezer.

Each transformation reaction is performed using 1.0×10^8 competent recipient cells. Thus, each tube contains competent cells for 10 transformation reactions.

Protocol for Transformation on Filters

Day 1

1. Determine the number of donor DNA samples and controls to be used, and then from this number, determine the number of filters and cell resuspension tubes necessary. Triplicate transformation reactions are recommendable for each donor DNA sample and control.

2. Prepare cell resuspension tubes by aseptically aliquotting 4 ml of sterile 0.85% (w/v) NaCl (saline) into sterile 50-ml Falcon tubes. These tubes can be stored at 4° until needed for step 11 on day 2 of the transformation experiment.

3. Prepare serial dilution tubes (∼6–7 dilution tubes per filter) by aseptically aliquotting 0.9 ml of sterile saline into sterile 1.5-ml Microfuge tubes. These can also be prepared in advance and stored at 4° until needed for step 13 on day 2 of the transformation experiment.

4. Working in a laminar-flow bench and using flame-sterilized forceps, place a sterile filter on the surface of an LA plate. Use a flame-sterilized glass spreader to smooth the filter and remove any air bubbles between the filter and agar.

5. Thaw the competent cells at room temperature.

6. Centrifuge tubes at 4000 rpm for 5 min in a tabletop Microfuge to pellet the cells. Discard the supernatant and resuspend each pellet

(containing 1.0×10^9 cells) in 800 μl of sterile saline. Each of the ten 80-μl aliquots in the tube now contains 1.0×10^8 competent cells.

7. Pipette 3 volumes of each 0.5 μg/ml donor DNA sample (1–20 μg donor DNA per volume) into sterile 1.5-ml Microfuge tubes. We routinely use 10-μg donor DNA per reaction once optimal transformation conditions have been determined. Where the total volume of donor DNA added to each tube is less than 60 μl, add sterile water to a final volume of 60 μl.

8. To each of these tubes containing 60 μl of donor DNA, add 240 μl (3×80 μl) competent cells. Each sample tube now contains 30 μg donor DNA and 3.0×10^8 recipient cells in 300 μl. Mix by vortexing briefly.

9. Label three LA plates containing sterile filters with each donor DNA sample. To the center of each filter, apply 100 μl donor DNA/ recipient cell mixture. Once the mixture is applied, avoid agitating the plates, which may cause the mixture to run off the filter. Repeat this step until all donor DNA/recipient cell mixtures have been applied to filters. Do *not* invert the plates for incubation.

10. Incubate the transformation reactions for 24 h at 30°.

Day 2/3: Bacterial growth should be visible on filters.

11. Using flame-sterilized forceps, remove each filter from the plate and place in the bottom of a cell resuspension tube labeled with the name of the donor DNA sample and replicate number. We recommend transferring filters to resuspension tubes only three at a time to ensure that filters do not dry before processing.

12. Vortex the resuspension tube for 60 s to remove all cells from the filter.

13. Make a dilution series of this cell resuspension in 10-fold steps from 10^{-1} to 10^{-6} or 10^{-7} using the previously prepared dilution tubes.

14. To enumerate recipients, plate 100 μl of either the 10^{-6} or the 10^{-7} dilution onto three LAR plates and spread cells evenly using a flame-sterilized glass spreader. Dilution plating should be optimized to obtain 30–300 colonies/plate.

15. To enumerate transformants, plate 100 μl of the undiluted, 10^{-1}, 10^{-2}, 10^{-3}, or 10^{-4} dilutions onto three LAR + appropriate selection-containing plates. The amount of dilution required to achieve a detectable or countable number of transformant colonies depends on the frequency of transformation. When assaying transformation in a new bacterium, using previously untested donor DNA, or when plating negative controls (e.g., saline, heterologous DNA, or control DNA containing no selectable marker), plate undiluted cell suspensions to maximize the lower limit of detection of transformation.

16. Once all filters have been processed and all plating is completed, allow the plates to dry on the bench top (agar-side down) to ensure that excess moisture is reabsorbed into the agar.

17. Invert plates and incubate agar-side up at 30° for 2–3 days or until colonies are sufficiently large for enumeration.

Day 4

18. Count both recipient and transformant colonies for all filters and replicates to obtain the number of colony-forming units (CFUs) per plate. Plates containing 30–300 colonies are optimal for enumeration. Calculate the transformation frequency for each sample:

transformation frequency = (no. of transformants)/(no. of recipients)

19. If desired, the transformation efficiency may also be calculated:

transformation efficiency = (no. of transformants)/(μg DNA)

Calculation of transformation efficiency is used to determine the amount of donor DNA required to achieve optimal transformation frequencies for each set of experimental conditions.

Transformation in Liquid Culture Media

Variants of the following protocol for transformation in liquid culture media have been used with success to optimize the *in vitro* transformation of, for example, *Acinetobacter* sp. BD413 (de Vries and Wackernagel, 2002; Palmen *et al.*, 1993), *Streptococcus pneumoniae* (Frischer *et al.*, 1994; Majewski *et al.*, 2000), *Synechococcus* and *Synechocystis* spp. (Frischer *et al.*, 1994), and *Agmenellum quadriplatum* (Essich *et al.*, 1990).

Day 1

1. Subculture an overnight liquid culture of recipient bacteria by diluting 1:100 in enough prewarmed fresh nutrient broth to provide 2 ml of culture for each donor DNA treatment to be tested. Incubate at the appropriate temperature and aeration until the desired phase of growth is achieved, usually about 2–4 h.

2. Transfer 1.0-ml aliquots of the culture into sterile 1.5-ml microcentrifuge tubes. Label sets of two tubes with a description of the donor DNA treatment.

3. Add 10 μl of a 0.1 μg/μl donor DNA solution to one tube in each set. To the second tube in each set, add 10 μl of sterile saline.

4. Incubate the tubes at the desired temperature and aeration for 1 h to allow donor DNA uptake.

5. If desired, add DNase to each tube to a final concentration of 100 μg/ml and continue incubating for 60 min to allow digestion of extracellular donor DNA and expression of the selection marker before selective plating.

6. Pellet cells by centrifugation at 4000 rpm for 10 min in a tabletop microfuge. Discard supernatant and resuspend cells in 500 μl of sterile saline by vortexing.

7. Use 100 μl of this suspension to make a 10-fold dilution series in sterile saline for the enumeration of recipient cells. Plate three 100-μl aliquots of the desired dilution (usually 10^{-6} or 10^{-7}) on nonselective plates using a sterile glass spreading bar.

8. Plate three 100-μl aliquots of the desired dilution(s) (usually undiluted, 10^{-1} or 10^{-2}) onto selection-containing plates using a sterile glass spreading bar.

9. Allow the plated liquid to be absorbed into the agar before inverting plates and incubating until colonies are large enough to count.

Day 2/3

10. Count colonies and calculate transformation frequency for each donor DNA treatment using Equations 1 and 2.

Cell Lysate or Culture Supernatant Transformation

Donor cell lysates may be substituted for purified donor DNA in any transformation protocol. Sterile cell lysates are prepared according to the method of Nielsen *et al.* (2000):

1. Prepare a liquid culture of donor bacteria by inoculating a single colony into 2–3 ml of liquid media. Incubate overnight at appropriate temperature and aeration.

2. Pellet cells by centrifugation at 6000 rpm for 10 min in a tabletop Microfuge. The supernatant may be saved and stored at $-80°$ if its transforming activity is to be tested. If not, then discard the supernatant.

3. Resuspend cells in 1.0-ml of sterile filtered water and pellet by centrifugation at 6000 rpm for 5 min in a tabletop Microfuge. Discard the water wash and resuspend pellets in 500 μl of sterile filtered water.

4. Heat the suspension for 15 min at $80°$ to lyse cells.

5. Plate a small aliquot of each lysate on nonselective nutrient agar to verify sterility. The concentration of DNA present in the lysate can be roughly estimated using the calculations of Torsvik and Goksøyr (1978) or

by multiplying the number of CFUs in the lysed cell suspension by the genome weight.

6. When not in use, lysates should be frozen at −80°.

7. To test the supernatant saved from step 2 for transforming activity, substitute sterile (0.22 μm-filtered) donor culture supernatant for donor DNA in the transformation protocol. Sterilized supernatants should be stored at −80°.

Cell-Contact Transformation

It has been proposed that some bacteria express DNA on their cell surface during periods of competence (Lorenz and Wackernagel, 1994). Surface-expressed DNA may have transforming activity for competent bacteria that come in contact with donor cells (Albritton *et al.*, 1982; Stewart and Sinigalliano, 1989; Stewart *et al.*, 1983). This mechanism of transfer is considered a type of transformation rather than conjugation because DNA is taken up into the recipient cell in the absence of mobilization or transfer functions on the part of either donor or recipient cell. A possible advantage of cell-contact transformation is that it may eliminate the need for passage of donor DNA between cells in the environment, where its transforming activity may be diminished by external factors such as DNases (Graham and Istock, 1978; Stewart *et al.*, 1983).

Cell-contact transformation can be tested by including intact donor cells instead of purified DNA as the donor DNA source in the filter transformation protocol described above or in the modified *in vitro* experiments described below. A strong counterselection must be available to either prevent the growth of donor cells or distinguish donor from recipient cells on selective media after the transformation period. A note should be made that gene transfer in such experiments can be bidirectional between donor and recipient, which must be taken into account when selecting for transformants. Although it is desirable to perform DNase controls to confirm transformation as the mechanism of gene transfer, the DNase sensitivity of cell-contact transformation cannot always be demonstrated (Albritton *et al.*, 1982). Verification of the absence of any intrinsic mobilization or transfer elements in either the donor or the recipient would, in such cases, provide more solid evidence of transformation as the mechanism of gene transfer.

The optimal ratio of donor/recipient cells used will vary between donor and recipient bacteria and experimental conditions; however donor/recipient ratios of 1000:1 to 1:1000 in 10-fold increments may be appropriate for preliminary tests. DNase should be added after the transformation period to exclude the detection of transformation events that occur after plating. Transformation assays are otherwise processed according to the aforementioned procedures.

Application of the Methods

Basic *in vitro* transformation assays have proved widely used and powerful tools for assessing the amenability of bacteria to genetic transformation. Determining which transformation method to use when testing an unknown bacterial recipient is somewhat arbitrary and may be dictated by optimal growth conditions in the laboratory. *Acinetobacter* sp. BD413 has demonstrated efficient transformation using both filter (Nielsen *et al.*, 1997b) and liquid culture (de Vries and Wackernagel, 2002; Palmen *et al.*, 1993) transformation protocols. On filters, average transformation frequencies of about 10^{-3} transformants per recipient have been obtained using purified homologous chromosomal DNA as donor DNA source (Nielsen *et al.*, 2000; Ray and Nielsen, in preparation), while plasmid DNA transforms *Acinetobacter* approximately 2 orders of magnitude less frequently using the same concentration of DNA (Nielsen *et al.*, 1997b). Transformation frequencies using cell lysates as donor DNA were even lower ($\sim10^{-5}$ transformants per recipient using homologous donor cell lysates), presumably because of decreased concentration of donor DNA molecules per unit volume in comparison to purified donor DNA (Nielsen *et al.*, 2000). For bacteria such as *S. pneumoniae,* in which competence is induced via a quorum-sensing mechanism, transformation frequencies of nearly 10^{-2} transformants per recipient are obtained using homologous donor DNA in liquid culture transformations (Majewski *et al.*, 2000). Strains of the oral bacterium *Streptococcus mutans* are transformed by homologous chromosomal DNA at frequencies as high as 10^{-3} transformants per recipient in biofilms. Transformation of planktonic cells by the same donor DNA occurred typically 10- to 100-times less frequently, indicating that this species may require a solid surface and/or biofilm matrix for efficient transformation (Li *et al.*, 2001).

In Vitro Transformation Using Components of Natural Systems

Moving one step closer to understanding transformation in natural habitats, we discuss methods for assaying transformation using experimental systems that assimilate environmental conditions by the inclusion of naturally occurring components in the *in vitro* transformation reaction. The section begins with a protocol for the transformation of *Acinetobacter* sp. BD413 in soil microcosms and continues with assays for natural transformation using plant tissue as the donor DNA source, in biofilms, in simulated *in vivo* conditions by using animal tissues or fluids in the transformation medium. An important modification in all remaining transformation protocols is the necessary inclusion of a posttransformation DNase I degradation step to prevent the detection of transformation events that may hypothetically occur on the selective media after plating. When using antibiotic resistance genes as selection markers for transformation in soil microcosms, a thorough analysis of the types and frequency of similar antibiotic resistance genes present in the soil matrix should be done before their use in the transformation assay. This applies most particularly for transformation in nonsterile soil microcosms. When sterile components are included, controls to verify their sterility should be included. Continued identification of conducive conditions in these protocols will most probably be required (Fig. 2), as changes in the physical or chemical aspects of any transformation reaction will likely result in changes in observed transformation frequencies.

Transformation of Acinetobacter sp. BD413 in Sterile Soil Microcosms

Because of its versatility, cost effectiveness, and similarity to natural soil conditions, the soil microcosm is suitable for the *in vitro* investigation of parameters that may influence natural transfer events in a soil matrix (Nielsen *et al.*, 2004). Although the matrices used in soil microcosms include various mixtures of sand, minerals, and other organic matter, they are referred to as "soil" in the following protocols for simplicity. Lorenz and Wackernagel (1994) provide an excellent review of published studies on soil type and its influence on both DNA stability and natural transformation. The protocol given here includes a sterile soil matrix and the exogenous addition of nutrients to a stabilized recipient population of *Acinetobacter* sp. BD413 cells shortly before the introduction of donor DNA.

Natural transformation of *Acinetobacter* sp. BD413 in sterile soil microcosm (Nielsen *et al.*, 1997a,b):

Day 1

1. Sieved (4 mm) and air-dried soil samples are sterilized by gamma-radiation (4 Mrad) in small polypropylene containers.

2. Pack the sterilized soil under sterile conditions into polypropylene cylinders of approximately 7 mm in height. The cylinders are made from 15-ml centrifuge tubes and are sterilized in water by autoclaving and subsequently dried. Cylinders typically contain 1.1–1.5 g of soil, depending on the soil type used.

3. Place the cylinders in sterile Petri dishes and add 100 μl (containing 1.0 × 10⁸ cells) of *Acinetobacter* sp. BD413 in water (prepared as described earlier) to the center of the soil matrix using a micropipette.

4. Incubate the soil microcosms at room temperature for 24 h to allow the recipient bacteria to adjust to the soil matrix.

5. Add 100 μl of nutrient solution (5× M9 salts with 25× M9 phosphate salts and 2% lactose) (Nielsen *et al.*, 1997a) to the soil microcosm.

6. Because of the nutrient addition, the bacteria will start growing and hence develop competence. After 1–6 h, add 10 μl of the DNA source (purified or cell lysates, concentration variable) to the soil microcosm.

7. After 24 h of incubation, the central core of the soil microcosm is sampled with the wide end of a sterile 1-ml pipette tip and transferred to a 1.5 ml microfuge tube containing 1-ml of 0.1% sodium pyrophosphate (tetrasodium diphosphate decahydrate) and four to five pieces of small sterile gravel to enable complete resuspension of the soil sample.

8. Vortex the Microfuge tube until a fine soil suspension is obtained. Further dilutions, plating, and enumeration are performed as described in the section "Basic *In Vitro* Transformation," earlier in this chapter.

In addition to the controls mentioned earlier, controls of soil sterility is essential. Putative transformants should be verified as the intended recipient through, for instance, metabolic profiling with BIOLOG (Hayward) or other methodology allowing the introduced recipient and the transformants to be shown to be of the same strain. For work with nonsterile samples, 100 μg/ml cycloheximide is added to all agar plates to suppress fungal growth.

Transformation Using Plant Tissue as the Source of Donor DNA

The transfer of marker genes from transgenic plants to bacteria can be detected using the modified *in vitro* assays described in the first section of this chapter. In short, intact or disrupted transgenic plant tissue can be used directly as the source of donor DNA (Kay *et al.*, 2002a; Tepfer *et al.*, 2003) in the transformation experiment. It is included here because of the interest in the transforming activity of plant DNA present in the phytosphere and in soil litter. Sterile plant tissue may be desirable to test transformation in the absence of indigenous plant-associated microorganisms. Plant tissue, intact or disrupted, can be blended with the soil matrix before forming microcosms or added after the establishment of the recipient population in the soil. In this case, plant tissue homogenates are added to soil microcosms in the same way that purified donor DNA or donor cell lysates are added. When nonsterile soil or plant tissue is used, cycloheximide, an inhibitor of eukaryotic protein synthesis, is included as an amendment in solid media (100 μg/ml) to inhibit fungal growth.

Transformation in Biofilms and Simulated Aquatic Environments

A surge of literature has provided clear evidence that many aquatic bacteria are not only genetically competent, but also that natural transformation occurs at high rates in simulated aquatic habitats and potentially within biofilms that nonplanktonic aquatic bacteria form. This protocol describes a simple laboratory method for assaying transformation of recipient cells in a monoculture biofilm formed on a glass slide in a noncirculating liquid bath (Hendrickx et al., 2000; Williams et al., 1996). The protocol can be easily modified to use sterile microtiter plate wells as surfaces for biofilm formation for more high-throughput applications (Li et al., 2001).

Day 1

1. Aseptically transfer a sterile glass microscope slide (or other sterile portable surface) into the bottom of a sterile 1-L glass beaker in a laminar flow hood.

2. Add 200 ml of nutrient broth to the beaker.

3. Subculture a fresh overnight culture of recipient bacteria 1:10,000 into the beaker. Seal the beaker to retain axenic conditions within the culture during incubation.

4. Incubate the beaker at the appropriate temperature on a tilt shaker on low speed overnight to allow biofilm to form on the glass slide.

Day 2

5. Working in the laminar flow hood, lift the glass slide out of the culture liquid using sterile forceps and handling the slide along on the edges.

6. Wash the slide by gently pipetting sterile saline across the surface of the slide.

7. Replace the glass slide to a new sterile 1-L glass beaker containing 100 ml of nutrient broth amended with 1 μg/ml salmon sperm DNA (blocking DNA).

8. Incubate with tilt rotation for 1 h at the appropriate temperature to allow saturation of nonbacterial-coated surfaces by the blocking DNA.

9. Pour off the blocking DNA solution and replace with 100 ml of nutrient broth amended with donor DNA to a final concentration of 1 μg/ml.

10. Continue incubation with tilt rotation for 2 h (up to 2 days) at the appropriate temperature to allow transformation.

11. Pour off the transforming DNA solution and replace with 20 ml of sterile saline amended with 100 μg/ml of DNase I.

12. Sonicate for 10 s at 20% power to remove cells from the slide and break up aggregates. Vortex briefly.

13. Make 10-fold dilutions of the cell suspension in sterile saline and plate appropriate dilutions onto nonselective nutrient media for recipient enumeration.

14. Plate either undiluted or diluted cell suspension onto selective nutrient media for transformant enumeration.

15. Allow all plated liquid to be reabsorbed into the agar before inverting plates and incubating at the appropriate temperature until colonies are large enough to count.

Day 3/4

16. Count all transformant and recipient colonies and calculate transformation frequencies/efficiencies using Equations 1 and 2 above.

Transformation in the Presence of Host Tissues or Fluids

Of the approximately 81 culturable bacterial species that are known to be naturally transformable in the laboratory, approximately 50% are human pathogens (Wackernagel, 2003). Frequent lateral gene exchange has been shown to occur in many human pathogens and is responsible for the generation of the tremendous genetic diversity found among sexually recombining bacterial isolates. *In vitro* optimal transformation is an appropriate method for demonstrating the basic ability of a bacterium to take up and express environmental DNA, but this method does not approximate the *in situ* conditions that a bacterium encounters in its host, where gene exchange is most likely facilitated by proximity and by constant selective pressure for genetic variation and adaptability. The following protocol is adapted from the methods provided in previous studies of transformation in biological fluids (Mercer *et al.*, 1999a,b, 2001) and allows the researcher to assess the transformability of a recipient bacterial strain in simulated *in situ* conditions by including biological fluids or tissues in the transformation mixture. Here, we describe a method for assaying transformation of bacteria in the presence of human saliva. The saliva is collected in the morning before eating or brushing to minimize any contamination by food components. This method can also be used to test the influence of, for example, blood, serum, stomach contents, rumen fluid, fecal slurry, and others on transformation. Filter sterilize saliva by passing through a 0.22-μm filter.

Store at $-80°$ when not in use. When possible, a positive transformation control strain is included in the experimental design to account for any decrease in transformation frequency due to donor DNA degradation in the presence of the biological fluids. This is particularly important when assaying transformation in an "unknown" bacterium.

1. Prepare a suspension of competent cells in sterile saline and subculture 1:100 into fresh prewarmed nutrient medium amended with 10% (v/v) filter-sterilized human saliva. Prepare a large enough culture to provide 1-ml aliquot per donor DNA sample.

2. Incubate at appropriate temperature and aeration until desired growth phase is achieved.

3. Transfer 1-ml aliquots of the competent cell/saliva culture to sterile 1.5-ml Microfuge tubes.

4. Add donor DNA to a final concentration of 1 μg/ml and incubate tubes with aeration for 1 h.

5. Add DNase I to a final concentration of 100 μg/ml and continue incubating for 30 min.

6. Pellet cells by centrifugation at 4000 rpm for 10 min in a tabletop microfuge. Discard the supernatant.

7. Resuspend pellets in 500 μl of sterile saline by vortexing.

8. Make desired 10-fold dilutions in sterile saline and plate on selective and nonselective media in triplicate to enumerate transformants and recipients, respectively.

9. Count recipient and transformant colonies and calculate transformation frequencies/efficiencies using Equations 1 and 2 above.

Application of the Methods

Knowledge of optimal *in vitro* transformation conditions can theoretically be applied to *in situ* simulations to assess the likelihood of the conditions necessary for transformation arising in the natural habitat of the bacterium. *Acinetobacter* sp. (formerly named *Acinetobacter calcoaceticus*) is transformed at the highest frequencies when cells are actively growing, suggesting a dependence of competence development on growth (Palmen *et al.*, 1993). *Acinetobacter* grows poorly in sterile soil microcosms and hence is inefficiently transformed under such conditions. The exogenous addition of nutrients to microcosms shortly before or after the introduction of donor stimulates growth sufficiently to restore temporary genetic competence and the detection of transformants (Nielsen and van Elsas, 2001). The *in situ* transformation of *Acinetobacter*, therefore, would be dependent on the simultaneous availability of nutrients required for growth and competence. The plant rhizosphere is thought to be one nutrient-rich environment that supports active bacterial growth (Jaeger *et al.*, 1999). Indeed, Nielsen and van Elsas (2001) demonstrated that transformation of *Acinetobacter* in soil microcosms was stimulated by the addition of soluble compounds commonly found in root exudates.

The introduction of natural materials into transformation assays can, however, introduce unaccountable factors that affect reproducibility. A study by Baur *et al.* (1996), for example, demonstrated that Ca^{2+} may be necessary for *in situ* and *in vitro* transformation of *Escherichia coli*. Variation between different spring water samples tested, however, shows clearly that either Ca^{2+} alone is insufficient for inducing competence or that unknown factors present in natural spring water negatively affect competence even in the presence of Ca^{2+}.

A study of natural transformation of *Pseudomonas fluorescens* and *Agrobacterium tumefaciens* (Demanèche *et al.*, 2001) in soil microcosms demonstrated that *P. fluorescens* was transformable in soil microcosms but not *in vitro* despite a wide range of *in vitro* conditions tested. The inability to detect transformants *in vitro* may, therefore, be caused by the absence of natural factors that are present in the environment (but lacking *in vitro*), and which promote natural transformation. In the same study, Demanèche *et al.* (2001) demonstrated that transformation can occur between intact donor and recipient bacterial cells in soil microcosms, though at low levels. Interestingly, they obtained similar transformation frequencies for recipient *P. fluorescens* cells when the donor DNA source was purified plasmid DNA or intact *E. coli* plasmid donor cells. In contrast, they obtained almost 10-fold lower transformation frequencies when *A. tumefaciens* was the recipient and intact *P. fluorescens* plasmid donor cells were used as donor DNA source. In the *P. fluorescens/E. coli* system, the rapid decline of the donor *E. coli* strain, which is not a natural soil inhabitant, in microcosms may have resulted in the release of free plasmid DNA, which would be available for recipient uptake. Free DNA may not have been comparably available in the *P. fluorescens/A. tumefacium* system because *P. fluorescens* is a natural soil inhabitant and survived well during the experimental period. This study emphasizes the importance of including ecologically relevant strains in various transformation systems. Whether the transformation between donor and recipient cells truly occurred via cell–cell contact (as opposed to release and subsequent uptake of DNA) was not clarified. The distinction between free DNA uptake and cell contact–mediated DNA uptake may represent a trivial subdivision of natural transformation mechanisms if the researcher only desires to demonstrate transformation. The introduction of intact donor cells into transformation systems, however, is relevant because these conditions most likely represent those encountered in the environment. It was not demonstrated that plasmid transfer in all cases was sensitive to DNase treatment and, therefore, occurred by natural transformation.

The decision to incorporate natural components into *in vitro* systems is accompanied by the choice to use them in either a sterile or nonsterile context. The use of irradiation, fumigation, filtration, or autoclaving to sterilize natural reagents can dramatically change their chemical and/or biological properties (Sikorski *et al.*, 1998), making a clear extrapolation of the results in a natural context more difficult. The effect of soil sterility on transformation in soil microcosms, for example, is highly variable. Nielsen *et al.* (2000) observed a decrease in transformation frequencies of *Acinetobacter* in nonsterile soil relative to that in sterile soil, whereas Demanèche *et al.* (2001) obtained evidence that suggests an increased transformation frequency of *P. fluorescens* in nonsterile versus sterile soil. The choice of sterile or nonsterile in transformation assays depends in part on the strength and sensitivity of the detection methods used and in part on the hypothesis being tested. Nonsterile soils, for example, are recommended for soil microcosm experiments whose purpose is to determine the transformability of indigenous microbes or the effect of indigenous microbes on the transformability of an inoculated recipient. In the latter case, the use of nonsterile soils requires a selection regimen that can clearly distinguish transformants of the desired recipient strain from indigenous bacteria present in the soil. The use of antibiotic selection markers to assay transformation in nonsterile soils can be problematic as intrinsic resistance to antibiotics is often found in natural samples, producing high background levels of resistant colonies and severely limits the ability to detect true transformants.

In vitro biofilm transformation assays are useful for determining the likelihood that transformable aquatic bacteria can take up DNA *in situ*. Modified *in vitro* assays for the transformation of biofilm bacteria can help predict the likelihood at which bacteria naturally present in aquatic ecosystems might be transformed by DNA from, for example, antibiotic resistant or chemical-pollutant isolates present in the same habitat. Hendrickx *et al.* (2000) developed an interesting *in vitro* biofilm system to assess the transformation of biofilm-resident *Acinetobacter* cells via direct visualization of expression of a marker gene. Their results suggest the importance of physical proximity between recipient cells and donor DNA in the frequency of transformation, namely that biofilm cells located near the biofilm–nutrient medium interface and the surface of the glass slide were transformed at the highest frequencies because the highest concentrations of DNA are thought to be found at those locations. The reported transformation frequencies here, however, may be underestimations of optimal transformation frequencies as the donor DNA may have bound to regions of the glass slide not in proximity to bacteria, thereby reducing the total amount of donor DNA available for transformation of the biofilm *Acinetobacter* cells. The inclusion of blocking DNA in transformation assays is to saturate all nonbiological surfaces where donor DNA could bind and thereby lose its transforming activity (Sikorski *et al.*, 1998).

The use of sterile or simulated biological fluids and tissues in transformation experiments with bacteria that normally colonize host tissues similarly may be used to extrapolate the likelihood of transformation in *in vivo* conditions. In a study of *Streptococcus bovis* transformation, Mercer *et al.* (1999b) obtained transformation frequencies of about 1.3×10^{-3} transformants per recipient *in vitro* using a standard liquid transformation protocol and assaying the integration of a nonreplicative plasmid into an introduced site of homology on the bacterial chromosome. In the presence of nonsterile saliva, serum, or rumen fluid, transformation frequencies decreased to below detection, suggesting that natural transformation of this bacterium in its natural habitat (the rumen of livestock) is not likely. In a separate study examining the transformation of *Streptococcus gordonii* in human saliva, they obtained results that would indicate that conditions in the human oral cavity may be conducive to natural transformation in this bacterium (Mercer *et al.*, 1999a).

Transformation in Modified Natural Systems

This section highlights methods used to assess transformation *in planta* and *in vivo* using systems that are slightly modified from conditions found in nature. The advantage of using such systems is obvious in allowing even more accurate simulation of conditions that bacteria likely experience in nature. However, the mere scale of the systems is often large enough that very sensitive detection methods are required to detect transformation events that almost invariably occur at frequencies several orders of magnitude lower than in optimal laboratory conditions. Despite technical complications, *in planta* and *in vivo* transformation systems remain powerful tools for learning more about the frequency of and factors important for bacteria to undergo genetic transformation *in situ*. Because of the high variability of such methods, the dependency on the system used and the hypotheses being tested, we present a brief description of the basic design of such experiments rather than detailed protocols.

In Planta *Greenhouse and Growth Chamber Experiments*

Little is known about the occurrence or frequency of gene transfer events in plant-associated bacterial communities. Given that many bacteria require nutrients to develop competence for natural transformation

(Lorenz and Wackernagel, 1994), it seems reasonable to infer that bacteria associated with plants could be transformed if they were located in areas of sufficient nutrient concentration (Lindow, 2003). Some pathogenic bacteria are able to promote the enzymatic breakdown of their plant hosts during colonization, thereby releasing nutrients that may support growth and, hence, potential transformation of co-inhabitant bacteria (Bertolla *et al.*, 1999). In a nonsterile growth chamber experiment, Bertolla *et al.* (1999) demonstrated that the tomato pathogen *Ralstonia solanacearum* can be transformed by purified donor plasmid DNA injected into the stems of *Ralstonia*-infected tomato, but only during periods of active bacterial growth and then at a maximum frequency of 3.3×10^{-7} transformants/ recipient. A water control confirmed that the transformants obtained in this experiment were due to uptake of inoculated plasmid DNA (as opposed to spontaneous mutation or background antibiotic marker resistance). In addition, they were also able to demonstrate exchange of genetic markers between two co-infecting *Ralstonia* strains *in planta* at frequencies as high as 1.34×10^{-7} transformants/recipient. They were unable to detect transformants of bacteria other than the inoculated *Ralstonia* species, which may be because of outcompetition of the resident bacteria by the *Ralstonia* so that any indigenous transformants would be indistinguishable.

Another growth chamber experiment conducted by Kay *et al.* (2002a) demonstrated that the *in planta* transformation of *Acinetobacter* sp. BD413 by transgenic plant DNA in transplastomic tobacco plants co-infected with *R. solanacearum*. Detectable transformation in this system required the presence of chloroplast DNA sequences on a plasmid in the recipient *Acinetobacter* cells to provide homology for recombination with the transgenic plant DNA. They were unable to calculate exact transformation frequencies in this experiment because of variability between the different replicates; however, they were able to demonstrate that *Acinetobacter* can be transformed in a semi-natural habitat if favorable conditions develop.

Bertolla *et al.* (1999) demonstrated that *R. solanacearum* developed genetic competence during infection of tomato plants and could be transformed by plasmid DNA that was artificially inoculated into infected plants. Kay *et al.* (2002b) also showed that *Acinetobacter,* which normally does not survive well in plants, was able to grow and achieve a genetically competent state when co-inoculated into tomato plants with the plant pathogen *R. solanacearum*. Although this study did not test the *Acinetobacter* cells for *in planta* transformation, the development of competence suggested that *Acinetobacter* could be transformed by DNA from the plant or from co-inhabiting bacteria. Indeed, in a follow-up study, Kay *et al.* (2002a) demonstrated the transfer of plant chloroplast DNA to artificially inoculated

Acinetobacter cells in *Ralstonia*–co-infected transplastomic tobacco plants. Homology between the plant transgenic construct and the recipient *Acinetobacter* chromosome was required for the detection of gene transfers in this experiment.

Animal Feeding Experiments

The digestive tract of higher eukaryotes provides an enclosed environment in which both resident and ingested bacteria can interact and theoretically exchange genetic material at high rates (Salyers, 1993). Although genetic exchange in this context is thought to occur most often via the direct transfer of mobile elements (conjugation) (Mercer *et al.*, 1999a), the question about whether transfer of chromosomal genes or nonconjugative plasmids present in the digestive tract can occur by natural transformation mechanisms remains. Protocols to assay the transformation of bacteria *in vivo* require controls to verify that the recipient bacteria are present in the digestive tract of host animals, that they achieve competence, that the donor DNA is successfully introduced into the digestive tract, that the bacteria are exposed to donor DNA, and that sufficient numbers of recipient bacteria are recovered from various organs and tissues of the test animals to increase the sensitivity of detection for what will most likely be rare transformation events. All transformants recovered should also be screened for the presence of the particular donor DNA construct using molecular methods.

To our knowledge, only a few studies of an animal feeding experiment to assay *in vivo* natural transformation has been published. Flint *et al.* (2002) fed purified plasmid donor DNA to laboratory rats (not germ free) and then attempted to detect transformation of bacteria within the rats' natural bacterial microflora by sampling feces and the contents of the small intestine, cecum, and colon. Although they detected a higher number of resistant bacteria in plasmid-treated rats, further PCR analysis of random putative transformants was unable to amplify the plasmid-encoded resistance genes.

Field Experiments with Artificial Inoculations

Williams *et al.* (1996) attempted to demonstrate the *in situ* transformation of *Acinetobacter* cells in river epilithon by inoculating donor and recipient cells on to sterilized river stones and then incubating the stones in a natural river before recipient and transformant enumeration. This study clearly demonstrates that genetic exchange in *Acinetobacter* can occur in natural conditions. This study exemplifies the many challenges that arise when conducting open field investigations. First, the number of replicates for each transformation was low, with each *in situ* experiment repeated between one and six times. Second, although the mean transformation frequencies are reported in the experimental summaries, the authors mention that variability in transformation frequencies between the different replicates was sometimes as high as 100-fold. Third, the donor and recipient cells were artificially inoculated onto the river stones before incubation *in situ*. Although *Acinetobacter* is a natural inhabitant of water, the high density of donor and recipient populations in close physical proximity may have yielded artificially high transformation frequencies. Fourth, the unidirectional transfer of donor plasmid from donor cells to recipient cells cannot be verified. Fifth, the plasmid used in this study was isolated from *Acinetobacter* and may, therefore, transform recipient *Acinetobacter* cells at frequencies that are not representative of those obtained using other plasmids. Finally, the transfer of the donor plasmid from donor cells to recovered transformants via conjugation or transduction cannot be excluded using the aforementioned experimental system.

Transformation *In Situ*

The ultimate goal of understanding transformation in laboratory conditions is to extrapolate the likelihood or conditions in which transformation may occur in natural settings. Physiological stresses encountered by bacteria *in situ* have not yet been adequately quantified or described in terms of the corresponding variation in selection pressure to uptake of environmental DNA for the purposes of adaptation, repair, or nutrition. Because of the complexity of any experiment designed to detect transformation *in situ*, we have chosen to include here a practical consideration of the methods used in published studies rather than discuss the design of individual experiments. Each experiment should be designed in consideration of the system being tested, the hypothesis of the researcher, and the resources available.

The first published field study attempting to detect gene transfer *in situ* was conducted by Paget *et al.* (1998), in which soil bacteria from a field that had been planted with transgenic tobacco plants were screened for the incorporation of transgenic plant DNA into their own genomes. A high level of background antibiotic resistance resulted in the isolation of high numbers of colonies on transformant-selective media. PCR analysis of 600 of these colonies was unable to confirm the transfer of plant DNA into resident bacteria. The second study (Gebhard and Smalla, 1999) similarly attempted to screen bacteria isolated from a field to which transgenic sugar beet plant material was introduced. Using a similar detection strategy, they first isolated kanamycin-resistant bacteria from soil samples and then screened these isolates for the presence of the transgenic plant DNA using dot-blot hybridization and PCR. They isolated approximately 4.0×10^7 kanamycin-resistant colonies from of soil, and 4000 of these isolates were subjected to dot-blot hybridization analysis. They were also unable to confirm the incorporation of transgenic plant DNA into the bacterial isolates examined. Thus, neither of the two published studies examining the transfer of engineered plant transgenes into resident field bacterial populations detected transformants from among the bacteria screened. This is presumably because the detection methods and sample sizes were insufficient to identify rare bacterial transformants among the approximately 10^8 bacteria found in every gram of soil (Pickup, 1995).

Although a critical consideration of the data and conclusions is necessary, the *in situ* transformation experiments described nonetheless attempt to answer important biological questions on the evolutionary dynamics of bacteria in their natural environment and the impact that human-made technologies may have on those dynamics. The scarcity of published studies on *in situ* transformation is good evidence that we simply lack the detection methods to detect rare transformation events that may occur in those systems. Progress in the field of *in situ* transformation assessment, therefore, requires the development of higher resolution selection regimens and higher throughput sampling methods. These factors are essential if empirical data are to withstand statistical scrutiny. This will, in turn, allow the researcher to make more inferences about natural transformation in bacteria in the environment.

Concluding Remarks

The methods presented here are intended to facilitate the elucidation of optimal conditions for transformation of bacteria *in vitro,* with the ultimate goal of understanding the importance or relevance of natural transformation to bacteria *in situ.* As evidenced by the studies discussed earlier, optimal conditions for transformation *in vitro* do not always translate to optimal conditions for transformation *in situ* because of the unavoidable presence of unaccounted factors present in the natural habitats of the bacteria under investigation. The results from optimal experiments, however, do provide information about the transformation biology of the recipients, which will ideally be applied to inform the design of more ecologically relevant experiments. Speaking in terms of transformation frequencies, however, may be irrelevant to understanding the long-term effect of natural transformation events in the environment, because theoretically even a single, and thus extremely rare, transformation event may be sufficient to set off a chain reaction of ecological consequences.

Acknowledgments

The authors thank Kristin Dahl, Derry Mercer, Pascal Simonet, and Sasha Shafikhani for helpful discussions and critical reading of the manuscript. The authors received financial support from the Research Council of Norway (J. Ray, Project 153587/432, and K. Nielsen, Project 140890/720).

References

Albritton, W. L., Setlow, J. K., and Slaney, L. (1982). Transfer of *Haemophilus influenzae* chromosomal genes by cell-to-cell contact. *J. Bacteriol.* **152,** 1066–1070.

Baulard, A., Escuyer, V., Haddad, N., Kremer, L., Locht, C., and Berche, P. (1995). Mercury resistance as a selective marker for recombinant mycobacteria. *Microbiology* **141,** 1045–1050.

Baur, B., Hanselmann, K., Schlimme, W., and Jenni, B. (1996). Genetic transformation in freshwater. *Escherichia coli* is able to develop natural competence. *Applied Environ. Microbiol.* **10,** 3673–3678.

Bertolla, F., Frostegard, A., Brito, B., Nesme, X., and Simonet, P. (1999). During infection of its host, the plant pathogen *Ralstonia solanacearum* naturally develops a state of competence and exchanges genetic material. *Mol. Plant Microbe. Int.* **12,** 467–472.

de Vries, J., and Wackernagel, W. (2002). Integration of foreign DNA during natural transformation of *Acinetobacter* sp. by homology-facilitated illegitimate recombination. *Proc. Natl. Acad. Sci. USA* **99,** 2094–2099.

Demanéche, S., Kay, E., Gourbiere, F., and Simonet, P. (2001). Natural transformation of *Pseudomonas fluorescens* and *Agrobacterium tumefaciens* in soil. *Applied Environ. Microbiol.* **67,** 2617–2621.

Dubnau, D. (1999). DNA Uptake in Bacteria. *Annual Reviews of Microbiology.* **53,** 245–281.

Errampalli, D., Leung, K., Cassidy, M. B., Kostrzynska, M., Blears, M., Lee, H., and Trevors, J. T. (1999). Applications of the green fluorescent protein as a molecular marker in environmental microorganisms. *J. Microbiol. Methods* **35,** 187–199.

Essich, E., Stevens, S. E. J., and Porter, R. D. (1990). Chromosomal transformation in the cyanobacterium *Agmenellum quadruplicatum. J. Bacteriol.* **172,** 1916–1922.

Flint, H. J., Mercer, D. K., Scott, K. P., Melville, C., and Glover, A. L. (2002). Survival of Ingested DNA in the Gut and the Potential for Genetic Transformation of Resident Bacteria. Report FSG01007. Food Standards Agency Report, United Kindgom.

Frischer, M. E., Stewart, G. J., and Paul, J. H. (1994). Plasmid transfer to indigenous marine bacterial populations by natural transformation. *FEMS Microbiol. Ecol.* **15,** 127–135.

Gebhard, F., and Smalla, K. (1999). Monitoring field releases of genetically modified sugar beets for persistence of transgenic plant DNA and horizontal gene transfer. *FEMS Microbiol. Ecol.* **28,** 261–272.

Graham, J. B., and Istock, C. A. (1978). Genetic exchange in *Bacillus subtilis* in soil. *Mol. Gen. Genet.* **166,** 287–290.

Hendrickx, L., Hausner, M., and Wuertz, S. (2000). *In situ* monitoring of natural genetic transformation of *Acinetobacter calcoaceticus* BD413 in monoculture biofilms. *Water Sci. Technol.* **41,** 155–158.

Heuer, H., Krogenecklenfort, E., Wellington, E. M. W., Egan, S., van Elsas, J. D., van Overbeek, L. S., Collard, J. M., Guillaume, G., Karagouni, A., Nikolakopoulou, D., and Smalla, K. (2002). Gentamicin resistance genes in environmental bacteria: Prevalence and transfer. *FEMS Microbiol. Ecol.* **42,** 289–302.

Jaeger, C. H., III, Lindow, S. E., Miller, W., Clark, E., and Firestone, M. K. (1999). Mapping of sugar and amino acid availability in soil around roots with bacterial sensors of sucrose and tryptophan. *Applied Environ. Microbiol.* **65,** 2685–2690.

Juni, E. (1978). Genetics and physiology of acinetobacter. *In* "Annual Review of Microbiology" (M. P. J. L. I. Starr and R. Sidney, eds.), Vol. 32, pp. 349–371. Annual Reviews, Inc., Palo Alto, CA.

Kay, E., Vogel, T. M., Bertolla, F., Nalin, R., and Simonet, P. (2002a). *In situ* transfer of antibiotic resistance genes from transgenic (transplastomic) tobacco plants to bacteria. *Applied Environ. Microbiol.* **68,** 3345–3351.

Kay, E., Bertolla, F., Vogel, T. M., and Simonet, P. (2002b). Opportunistic colonization of Ralstonia solanacearum-infected plants by *Acinetobacter* sp. and its natural competence development. *Microb. Ecol.* **43,** 291–297.

Li, Y.-H., Lau, P. C. Y., Lee, J. H., Ellen, R. P., and Cvitkovitch, D. G. (2001). Natural genetic transformation of *Streptococcus mutans* growing in biofilms. *J. Bacteriol.* **183,** 897–908.

Lindow, S. E. (2003). Overview of phyllosphere microbial communities. *In* "Proceedings of the Impact of Genetically Modified Plants (GMPs) on Microbial Communities" (K. M. Nielsen, ed.), pp. 16–17. European Science Foundation Programme: "Assessment of the Impact of Genetically Modified Plants. Tromsø, Norway, 24–28 May.

Lorenz, M. G., and Wackernagel, W. (1994). Bacterial gene transfer by natural genetic transformation in the environment. *Microbiol. Rev.* **58,** 563–602.

Majewski, J., Zawadzki, P., Pickerill, P., Cohan, F. M., and Dowson, C. G. (2000). Barriers to genetic exchange between bacterial species: *Streptococcus pneumoniae* transformation. *J. Bacteriol.* **182,** 1016–1023.

Mercer, D. K., Scott, K. P., Bruce-Johnson, W. A., Glover, L. A., and Flint, H. J. (1999a). Fate of free DNA and transformation of the oral bacterium *Streptococcus gordonii* DL1 by plasmid DNA in human saliva. *Applied Environ. Microbiol.* **65,** 6–10.

Mercer, D. K., Melville, C. M., Scott, K. P., and Flint, H. J. (1999b). Natural genetic transformation in the rumen bacterium *Streptococcus bovis* JB1. *FEMS Microbiol. Lett.* **179,** 485–490.

Mercer, D. K., Scott, K. P., Melville, C. M., Glover, L. A., and Flint, H. J. (2001). Transformation of an oral bacterium via chromosomal integration of free DNA in the presence of human saliva. *FEMS Microbiol. Lett.* **200,** 163–167.

Nevoigt, E., Fassbender, A., and Stahl, U. (2000). Cells of the yeast *Saccharomyces cerevisiae* are transformable by DNA under non-artificial conditions. *Yeast* **16,** 1107–1110.

Nielsen, K. M., Bones, A. M., and Van Elsas, J. D. (1997a). Induced natural transformation of *Acinetobacter calcoaceticus* in soil microcosms. *Applied Environ. Microbiol.* **63,** 3972–3977.

Nielsen, K. M., Van Weerelt, M. D. M., Berg, T. N., Bones, A. M., Hagler, A. N., and Van Elsas, J. D. (1997b). Natural transformation and availability of transforming DNA to *Acinetobacter calcoaceticus* in soil microcosms. *Applied Environ. Microbiol.* **63,** 1945–1952.

Nielsen, K. M., Smalla, K., and van Elsas, J. D. (2000). Natural transformation of *Acinetobacter* sp strain BD413 with cell lysates of *Acinetobacter* sp, *Pseudomonas fluorescens,* and *Burkholderia cepacia* in soil microcosms. *Applied Environ. Microbiol.* **66,** 206–212.

Nielsen, K. M., and van Elsas, J. D. (2001). Stimulatory effects of compounds present in the rhizosphere on natural transformation of *Acinetobacter* sp BD413 in soil. *Soil Biol. Biochem.* **33**, 345–357.

Nielsen, K. M., Ray, J., and van Elsas, J. D. (2004). Natural transformation in soil: Microcosm studies. *In* "Molecular Microb. Ecol. Manual," 2nd Ed., Vol. 5, pp. 1069–1080. Kluwer, The Netherlands.

Paget, E., Lebrun, M., Freyssinet, G., and Simonet, P. (1998). The fate of recombinant plant DNA in soil. *Eur. J. Soil Biol.* **34**, 81–88.

Palmen, R., Vosman, B., Buijsman, P., Breek, C. K. D., and Hellingwerf, K. J. (1993). Physiological characterization of natural transformation in *Acinetobacter calcoaceticus*. *J. Gen. Microbiol.* **139**, 295–305.

Pickup, R. W. (1995). "Sampling and Detecting Bacterial Populations in Natural Environments." Soc. Gen. Microbiol. Symp., pp. 298–315. Cambridge University Press, Cambridge.

Ray, J. L., and Nielsen, K. M. (in preparation). Effect of sequence divergence on transformation in *Acinetobacter* sp. BD413.

Salyers, A. A. (1993). Gene transfer in the mammalian intestinal tract. *Curr. Opin. Biotechnol.* **4**, 294–298.

Sikorski, J., Graupner, S., Lorenz, M. G., and Wackernagel, W. (1998). Natural genetic transformation of *Pseudomonas stutzeri* in a non-sterile soil. *Microbiology* **144**, 569–576.

Stewart, G. J., Carlson, C. A., and Ingraham, J. L. (1983). Evidence for an active role of donor cells in natural transformation of *Pseudomonas stutzeri*. *J. Bacteriol.* **156**, 30–35.

Stewart, G. J., and Sinigalliano, C. D. (1989). Detection and characterization of natural transformation in the marine bacterium *Pseudomonas stutzeri* strain ZoBell. *Arch. Microbiol.* **152**, 520–526.

Tepfer, D., Garcia-Gonzales, R., Mansouri, H., Seruga, M., Message, B., Leach, F., and Perica, C. (2003). Homology-dependent DNA transfer from plants to a soil bacterium under laboratory conditions: implications in evolution and horizontal gene transfer. *Transgenic Res.* **12**, 425–437.

Torsvik, V. L., and Goksøyr, J. (1978). Determination of bacterial DNA in soil. *Soil Biol. Biochem.* **10**, 7–12.

van Overbeek, L. S., Wellington, E. M. W., Karagouni, A., Smalla, K., Collard, J. M., and van Elsas, J. D. (2002). Prevalence of streptomycin resistance genes in European habitats. *FEMS Microbiol. Ecol.* **42**, 277–288.

Wackernagel, W. (2003). Methods to detect horizontal gene transfer. *In* "Proceedings of the Impact of Genetically Modified Plants (GMPs) on Microbial Communities" (K. M. Nielsen, ed.), pp. 39. European Science Foundation Programme: "Assessment of the Impact of Genetically Modified Plants." Tromsø, Norway, 24–28 May.

Williams, H. G., Day, M. J., Fry, J. C., and Stewart, G. J. (1996). Natural transformation in river epilithon. *Applied Environ. Microbiol.* **62**, 2994–2998.

[27]   Use of Confocal Microscopy in Comparative Studies of Vertebrate Morphology

By ANDRES COLLAZO, OLIVIER BRICAUD, and KALPANA DESAI

Abstract

Laser scanning confocal microscopy provides a means to acquire and analyze images of complex morphological structures and to help place molecules or cells of interest in their proper morphological context. Confocal microscopy is a form of fluorescence microscopy that sharpens the images collected by visualizing the light from only one plane of focus. This allows for the collection of multiple focal planes in what is called a *z-stack,* which provides three-dimensional data. Five steps that any investigator using a confocal microscope should follow are described: (1) labeling and (2) mounting of specimens for viewing, (3) optimizing the image on the confocal, and (4) collecting and (5) analyzing of confocal image data. We describe three specific protocols incorporating these steps from our work on vertebrate inner ear development. The first two describe a collection of z-stacks in living, fluorescently labeled, and intact embryos. The second protocol is for time-lapse imaging of multiple focal planes at each time point. The third protocol describes confocal imaging of preserved material double labeled with antibodies and by retrograde labeling of neurons via axonal uptake. Finally, three alternative or complementary approaches to standard confocal microscopy are described and discussed.

Introduction

Molecular evolutionary studies, concentrating on molecules common to many taxa, have had great success in determining phylogenetic relationships. Often little is known about the morphological context of these molecules. For some, this is because they are basic to the function of the cellular machinery, such as ribosomal RNA (rRNA), and so are expressed ubiquitously in an organism (Field *et al.*, 1988; Hedges *et al.*, 1990). For other molecules being used for phylogenetic analysis such as the Hox genes (de Rosa *et al.*, 1999), their restricted morphological expression is part of the reason they were chosen and the basis for understanding their role in evolution. Hox genes are homeodomain containing transcription factors of evolutionary interest because of their dramatic and conserved role in the development of many metazoan taxa (Krumlauf, 1992). Developmental data provide not just a molecular context for the formation of the varied organs or body structures of interest to evolutionary biologists but also cellular and tissue level contexts. Understanding the molecules, cells, and tissues that contribute to the development of a given organ or structure and how these might change during the course of evolution is a potentially powerful tool for comparative organism studies. One difficulty such studies have is visualizing the molecules and/or cells of interest, which is addressed in this chapter through the technique of fluorescence microscopy—more specifically confocal fluorescence microscopy.

Light microscopy has been a powerful tool for biology since the first studies by Robert Hooke and Anton van Leeuwenhoek in the seventeenth century. The visible spectrum of light ranges from wavelengths of 400–700 nm. Fluorescence microscopy uses a restricted part of this spectrum and the luminescent property of molecules (fluorophores) to maximize sensitivity and specificity (Berland, 2001). This type of microscopy helps address a major challenge to biological imaging: how to visualize and localize specific cell types or molecules against a background of many. Fluorescence microscopy achieves this by maximizing the ratio of signal (what we are interested in seeing) to noise (background we do not want to see), helped by looking at a restricted portion of the visible spectrum.

What advantage does confocal fluorescence microscopy offer over standard fluorescence microscopy? Mainly, confocal microscopy eliminates the light above or below the focal plane, which allows for subcellular localization of a molecule of interest, such as whether that molecule is located in the cell membrane, nucleus, cytoplasm, or some other specific cell organelle. Although subcellular localization requires high magnifications, confocal microscopy benefits images collected at any magnification by sharpening the fluorescent image and eliminating haze. Multiple two-dimensional images obtained sequentially at different planes of focus, called a *z-stack* because they are typically collected along the z- or depth axis, provide the data needed to generate sharp three-dimensional (3D) images and for analyses ranging from localization to quantitation.

Basics of Confocal Fluorescence Microscopy

Laser scanning confocal microscopy is a form of fluorescence microscopy that differs from the more typical form used in laboratories, in three important ways: light source used, presence of a pinhole aperture, and the need for a computer to reconstruct the image (Fig. 1). First, instead of using halogen, mercury, or xenon light bulbs, which are broad spectrum, confocal microscopy uses lasers with monochromatic lines. Bulbs usually illuminate the whole field of view while lasers have a much narrower field of illumination.

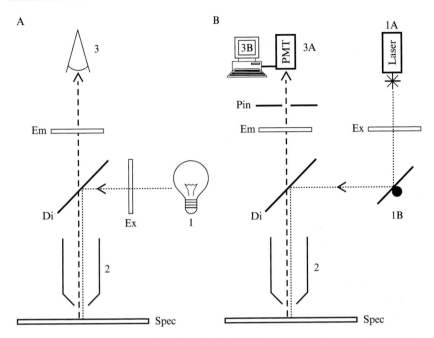

FIG. 1. Comparisons of the light paths of typical (A) fluorescence microscopy with (B) laser scanning confocal microscopy. (A) Fluorescence microscopy typically uses some type of broad-spectrum bulb for a light source (1). The light (dotted line) passes through an excitation filter (Ex) that restricts the wavelength and is reflected to the specimen (Spec), through an objective (2) by a dichroic beam splitter (Di). The light emitted by the specimen (dashed line) passes through an objective (2), the Di, and an emission filter (Em) until it reaches the eye of the observer (3) or a camera. (B) Confocal microscopy uses a laser as a light source (1A), which is scanned across the specimen (Spec) by a motorized mirror (1B). Laser light (dotted line) may or may not pass through an excitation filter (Ex). Light is reflected to the specimen, through an objective (2) by a Di. Light emitted by the specimen (dashed line) passes through an objective (2), a Di, an Em, and the pinhole aperture (Pin). Unlike what is shown, the Em is commonly located after the Pin. Emitted light is detected by a photo multiplier tube (PMT), (3A) and images are reconstructed by computer (3B).

For typical fluorescence microscopy, light from broad-spectrum bulbs (mercury being the most common) needs to be restricted to the wavelengths needed to excite the fluorophore. This is done by passing light through an excitation filter, one of three glass elements making up a typical fluorescence microscope cube, the other two being the dichroic beam splitter and emission filter (Fig. 1A). The dichroic beam splitter is located at a 45-degree angle to the incoming light path and has some complex optical properties, reflecting light of one wavelength while allowing longer wavelength light to pass through. The emission filter restricts the light visualized to just the wavelength of interest, eliminating any extraneous light of the excitation wavelength. It is important to note that the excitation wavelength that is absorbed by the fluorophore is always shorter than the emission wavelength and this difference, called the *Stokes shift,* needs to be maximized for a higher signal-to-noise ratio. Of these glass elements, typical confocal microscopes use all but the excitation filter, unless the laser produces more than one wavelength.

The second difference is the pinhole or iris aperture placed at the confocal plane of focus, which basically limits the light seen by the detector to just that light coming from the plane of focus (Fig. 1B). The size of the aperture determines the thickness of the section being viewed: the smaller the aperture, the thinner the section. Although it would seem that thinner is better, it is important to note that an image through too small a pinhole will be too dim to see. Also, the resolution along the z-axis will always be less than that along the x- and y-axes.

Finally, confocal microscopy uses a computer to reconstruct the image as the laser scans across the specimen, which makes confocals slower than video rate or live imaging, although we discuss exceptions later. Confocal microscopes detect the emitted light through a photo multiplier tube (PMT), which differs from other digital detectors such as charged-coupled devices (CCDs) in being basically a point source detector and more light sensitive. This is where the *scanning* in *laser scanning confocal microscopy comes* from; a mirror moves the laser across the field of view and a computer is used to reconstruct the image from the light intensity values determined by the PMT (Fig. 1B). The original confocal microscopes moved the specimen, but all modern systems move just the laser light. It is the advent of personal computers powerful enough to process the collected image data that allowed confocals to become more cost effective and accessible in the 1980s. Although the confocal microscope shows only the light coming from the plane of focus, the laser is still exciting above and below this plane of focus, potentially causing bleaching and, in living specimens, phototoxic effects, even in regions that are not being viewed. Also, the depth that a confocal microscope can penetrate into a specimen is not any deeper than that of typical fluorescence microscopy, although the images provided by the confocal will be sharper, giving the appearance of being able to penetrate deeper.

Methods

With so many possible fluorophores, brands of confocal microscopes, and methods to prepare samples for viewing, it is impossible to describe all possible uses for a confocal microscope. Good reviews are available that list and explain the bewildering array of fluorophores available for fluorescence microscopy (Harper, 2001). Our research focuses on vertebrate development, so our examples were chosen with this in mind. We show examples that use both living and preserved material. We begin with five general steps that every investigator should follow and then place these into context using specific examples.

1. *Labeling specimens for viewing.* Confocal microscopes require a fluorescent label or fluorophore for viewing. There are many different ways to label specimens for viewing. Living animals can be labeled with vital fluorescent dyes, whereas preserved specimens can be labeled with antibodies to specific epitopes of interest. Some specimens may not need to be labeled; viewing the autofluorescence available at certain wavelengths may provide sufficient information. For example, light-reflecting pigment cells called iridophores can be seen without labeling. Ideally, one should optimize the fluorophore to be used with the laser excitation wavelength. This is important because some commonly used fluorophores (e.g., the nuclear stains DAPI and Hoechst) have an ultraviolet (UV) excitation wavelength (Harper, 2001), which is often not available in standard configurations of confocal microscopes. UV lasers have been available, but they have been large and expensive and they generate a great deal of heat. This is one great advantage of two-photon, as opposed to the single-photon, confocal microscopy that is described for most applications in this chapter, although we discuss this alternative in the section "Alternative and Complementary Approaches," later in this chapter.

2. *Mounting specimens for viewing.* The type of compound microscope, inverted or upright (Foster, 1997), to which the remaining confocal hardware is attached determines how the specimen needs to be mounted. Inverted microscopes are most commonly used in biological applications. In these, the objective comes from the bottom, and living specimens can be kept covered avoiding contamination of cultured cells or materials. Although this allows viewing through such large containers as plastic multiwell plates or tissue culture flasks, for higher magnifications or resolutions, it is best to look through a glass coverslip, which usually requires a special holder for mounting the specimen. Confocals attached to upright microscopes were mostly used for material sciences and engineering applications but are becoming more commonly used for biological applications as well. Upright microscopes allow the use of high-magnification water immersion objectives with long working distances that are optically corrected for use without a coverslip, allowing for imaging of living aquatic specimens. Although contamination of specimens in sterile media is a problem with upright microscopes, there are ways of

mounting specimens to eliminate this concern (Potter, 2000). If the specimen is alive and moving, it needs to be anesthetized or immobilized during imaging. We provide specific examples of how to anesthetized aquatic vertebrate embryos in the section "Protocols," later in this chapter.

Deciding on the optimum objective is difficult because an objective's design is a compromise between resolution and working distance. The resolution of an objective is a function of the numerical aperture (NA), which is written on the objective and cannot be higher than the refractive index of the media the light is traveling through from the specimen or coverslip to the lens (Foster, 1997). Air has a refractive index of 1.00, whereas those of water and oil are 1.33 and 1.51, respectively, so the highest resolution objectives are generally oil immersion. Ideally one wants to match the type of objective (air, water, or oil) with the media in which the specimen is mounted. Histological sections mounted in Permount have a similar refractive index to oil and so can benefit from the high NA of these objectives. Living biological material is often in an aqueous media, so water immersion objectives are a logical but not necessary choice. Unfortunately, high NA objectives generally have a short working distance, particularly short in the case of oil immersion objectives, limiting their usefulness for thicker specimens. The high NA water immersion objectives specifically designed for use with confocal microscopes are expensive because they require so many glass elements inside to maintain a relatively long working distance. A high NA is important for resolution along the x- and y-axes, but it is particularly critical for z-axis or depth resolution.

3. *Optimizing image on the confocal.* There are many different brands of scanning laser confocal microscopes, and each manufacturer uses its own proprietary software for controlling the instrument and analyzing the data collected. The major vendors, in alphabetical order, are Bio-Rad Laboratories Life Sciences Group, Carl Zeiss, Leica, Nikon, and Olympus. Given the variety of instruments and software, often with differing nomenclature for the same component (e.g., the pinhole could be called the *iris*), it is difficult to provide detailed descriptions for how to optimize an image. Nevertheless, some general tips can be provided. The z-section thickness can be adjusted by closing down the pinhole or iris to make it thinner or opening the pinhole for thicker and brighter images. Deciding on the optimal z-section or depth thickness is difficult because it depends on so many variables: the specimen, objective used, and resolution desired. Some confocal software comes with tools that provide an "optimal" pinhole adjustment for the objective used, but this does not take into account unusual characteristics of a particular specimen. Still this software can be useful. See the manufacturer of the instrument being used for more details.

Besides changing the pinhole, image brightness can be adjusted in two ways. The laser power being used can be increased or decreased depending on the brightness desired. Typically, one wants to keep the laser power as low as possible because this minimizes bleaching (or fading) of the fluorophore and phototoxic effects in living samples. Phototoxic effects are typically due to the generation of free radicals. Brightness can also be increased digitally by increasing the sensitivity of the PMT in an adjustment called *gain* or *brightness*. Increasing brightness this way has no bleaching or phototoxic effects, but it can decrease the signal-to-noise ratio by increasing the background detected. Increasing the length of time the laser scans across the specimen can minimize the noise problem, but again this can lead to bleaching or phototoxic effects. Maximizing the signal-to-noise ratio of the light collected is helped by viewing a histogram of the different light levels or using software that points out light of maximum and minimum value in the image. Ideally the plot of light intensities revealed by the histogram should appear as a bell curve spread across the entire range detected. Too much brightness in the area of interest in the image can lead to saturation, which limits resolving of details and any future quantification. When adjusting brightness, one should adjust the contrast (also called the "black level" or "gamma" setting), which is also important for maintaining image quality. All these adjustments should be made as quickly as possible to minimize bleaching or phototoxic effects.

Older confocal microscopes operate in 8-bit mode, which only resolves 256 shades of gray. Most newer instruments operate in 12- (4096 shades of gray) or 16-bit mode (65,280 shades of gray), and with their increased bit depth, the need for a bell curve of light intensities spread across the full range of the histogram is less critical. However, when images are transferred into software such as Adobe Photoshop, which operates in 8-bit mode, the images need to be readjusted to cover the full range of the histogram to compensate for the decreased bit depth.

4. *Collecting image data.* All laser scanning confocal microscopes motorize the focus so they can collect multiple images at fixed and repeatable intervals, into what are called *z-series stacks*. This is one of the more powerful features of the instrument, because it allows for automation of image collection and localization in three dimensions. It is difficult to decide how many sections to collect in a z-series stack. This often depends on the z-resolution desired. Of course, single sections can be collected as well, and this is often sufficient for the question being addressed.

In discussing objectives in step 2, we did not emphasize magnification. This is because magnification is not as important as resolution and working distance. Confocal microscopes, by scanning with the laser for the same number of points but in a smaller area, can provide a magnification or zoom setting adjustable through the software that is limited by the mechanics of the scanning mirror and resolution of the objective. All confocal microscopes allow for the collection of images of different file sizes (e.g., 256 × 256, 515 × 512, and 1024 × 1024, although they do not have to be square), with larger images taking longer to acquire. Although one wants to use magnifications high enough to fill the field of view with the labeled area of interest, qualities of a specimen such as thickness may limit the choice of objective to one with a lower magnification if it has the working distance necessary. The highest NA objective that provides the working distance needed to view the specimen should be used.

Confocal microscopes can also collect bright field images using optical techniques such as phase or differential interference contrast microscopy (Davidson and Keller, 2001; Foster, 1997), but these are not optical sections, as are collected with fluorescence microscopy. It is often useful to collect a bright field image or even a z-stack of bright field images at the same positions that a fluorescent z-series stack is acquired, to help register the fluorescent images with morphological details in the specimen, particularly when the fluorescent labeling is highly restricted. A final feature of particular use with living material is the ability to collect z-series stacks at multiple time points. These four-dimensional (4D) time-lapse datasets can be quite large in terms of file size and during collection require a stable mounting of the specimen.

5. *Analyzing confocal data.* Analysis of the confocal images collected, especially z-series stacks, can be done with the software provided with the confocal microscope or through some third-party software. Quantitative analyses of the images can entail measurements of structures within the field of view or of light intensity, as an indicator of the amount of fluorophore. It is often useful to take z-series stacks and project them down to one plane of focus so that all the planes of focus are visible. This is one of the more common ways to present confocal microscope data, and although x-y projections are the most common, x-z, y-z, and even projections at an angle are possible depending on the software. The 3D nature of z-series stacks is one of the most powerful features of these data, and they are often best analyzed by stepping through the individual z-sections. However, it can be difficult to visualize complex 3D structures this way. All confocal software also provides a means to make 3D images viewed as stereo pairs or through 3D glasses if they are red-green anaglyph images. In addition, 4D datasets are particularly difficult to analyze because of the computer resources necessary to analyze such generally large file sizes, but software that allows the user to step quickly through both the z and the time dimensions is the most powerful.

Protocols

Imaging Mechanosensory Hair Cells in the Inner Ear of Living Tadpoles

In our laboratory we are interested in the development of the sensory organs of the inner ear. The vertebrate inner ear arises from an epidermal thickening called the *otic placode*. The cranial placodes are a tissue unique to vertebrates and contribute to all the paired sensory structures (Baker and Bronner-Fraser, 2001). The sensory organs consist of mechanosensory hair cells and surrounding supporting cells. The number of sensory organs in the inner ear varies across vertebrates, ranging from 6 to 9 (Fritzsch *et al.*, 2002), and these need to be precisely positioned for proper function in hearing and balance. We have previously described our methods for labeling these cells in living tadpoles of the frog *Xenopus laevis* (Kil and Collazo, 2001) but did not discuss viewing labeled tadpoles with the confocal microscope.

1. *Labeling specimens for viewing.* The vital dye and the method used to inject it into the inner ear of living tadpoles to label hair cells has been described (Collazo and Fraser, 1996; Kil and Collazo, 2001). Briefly, tadpoles were anesthetized with a 1:2500 solution of MS-222 (Argent Chemical Laboratories) in 0.1× Marc's modified Ringer's (MMR) solution (Sive *et al.*, 2000) and were gently held in place on a 2% bacto-agar bed (DIFCO Laboratories) with tungsten staples. The fluorophore used was the vital dye 4-(4-[diethylamino] styryl)-*N*-methylpyridinium iodide (4-Di-2-ASP; Molecular Probes, Inc., D-289), of which 2 mg was dissolved in 10 ml of 0.1× MMR solution. This was injected into the frog inner ear where the fluorophore is preferentially taken up by the mechanosensory hair cells.

2. *Mounting specimens for viewing.* The living specimen was imaged on an inverted microscope. The tadpole was anesthetized as described earlier and placed upside down in a stainless-steel chamber (Atto Bioscience, Rockville, MD) with a glass coverslip on the bottom (Fig. 2A). The inner ear of the tadpole is relatively superficial at this stage and located on the dorsal surface, which is laterally compressed in this species. The objective used was a water immersion Carl Zeiss C-Apochromat 40× with a 1.2 NA and a working distance of 225 microns. This working distance was insufficient to image through the whole inner ear, but it did allow for the visualization of the three most dorsal sensory organs, the cristae, at the base of the semicircular canals, which sense rotational acceleration (Fig. 2B) (Goldberg and Hudspeth, 2000).

3. *Optimizing image on the confocal.* We used an older model Zeiss scanning laser confocal microscope called the *LSM 410*. The 4-Di-2-ASP–

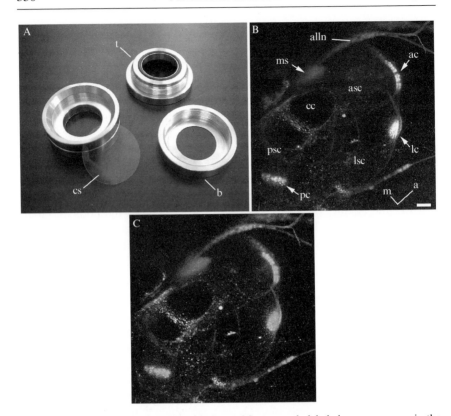

Fig. 2. *In vivo* confocal microscope images of fluorescently labeled sensory organs in the inner ear of the frog *Xenopus laevis* and the chamber used to hold the specimen. (A) Stainless-steel chamber used to view living, labeled tadpole. On the left side is the intact chamber with a 25-mm coverslip (cs) resting on it for scale. Chamber unscrews into top (t) and bottom (b) pieces with the coverslip placed in the bottom. Top piece's black gasket provides a seal when it is flipped and screwed into bottom. (B) Maximum projection of confocal z-stack of the dorsal sensory organs of *Xenopus* inner ear. Arrows point to the sensory organs visible. Three most dorsal are the anterior crista (ac), lateral crista (lc), and posterior crista (pc) at the base of the three semicircular canals: anterior (asc), lateral (lsc), and posterior (psc). A fourth sensory organ, macula sacculus (ms), is barely visible because it is too ventral for the working distance of our objective. Anterior and posterior semicircular canals meet in a structure called the *common crus* (cc). The anterior lateral line nerve (alln) outside of the inner ear also labeled with 4-Di-2-ASP that leaked out of the injected ear. Dorsal view with the anterior (a) and medial (m) axes indicated by lines. Scale bar is 50 μm for this and the following image. (C) Three-dimensional view of the same ear shown in (B) is visible with red-green three-dimensional glasses (green over left eye). The positions of the semicircular canals and the relative heights of the cristae are particularly clear in this view with the appropriate glasses. (See color insert.)

labeled hair cells were extremely bright so the laser power could be reduced to 10%, and because this fluorophore has a broad range of excitation, we could visualize it with two different laser lines. In our image, we saturated the hair cells, making individual cells difficult to visualize in order to increase the background so the walls of the semicircular canals could be seen (Fig. 2B, C).

4. *Collecting image data.* The laser excitation wavelength used was 568 and the emitted light was passed through a 570 long-pass (LP) (allows light >570 nm to pass through) emission filter. The pinhole aperture was set at 20 (out of a range of 255) and the image size was 512×512 pixels with a scan time of 8 s. The software used was Carl Zeiss LSM version 3.993 for Windows. The z-series stack from which the image was generated consisted of 40 sections collected at 2-micron intervals. The working distance of the objective limited how deep the stack went into the tadpole. The two ventral chambers of the inner ear and their sensory organs could not be seen.

5. *Analyzing confocal data.* The z-series stack was projected with a maximum summation, which adds all the intensity values for all sections (Fig. 2B). The projection shown was done by transferring the confocal stack into another image analysis software package, Metamorph 4.1 (Universal Imaging Corporation), which could read the files. The projection was also constructed as a red-green anaglyph, which allows for 3D viewing with red-green 3D glasses (green over left eye to see Fig. 2C in 3D).

Time-Lapse Confocal Microscopy of the Developing Otic Placode in Zebrafish Embryos

The zebrafish *Danio rerio* is a teleost fish that has become a powerful model system for vertebrate genetics and developmental studies (Haffter *et al.*, 1996). The transparency and rapid development of the embryo make it an excellent system for confocal *in vivo* imaging. One of the most common applications for *in vivo* imaging of zebrafish embryos is to document the cell movements occurring during the principal morphogenetic events of embryogenesis. One common method is to use green fluorescent protein (GFP) or more accurately a modified form called enhanced GFP (EGFP) transgenic embryos (Amsterdam *et al.*, 1995; Shafizadeh *et al.*, 2002). GFP, a naturally fluorescent protein from a jellyfish, is expressed in a tissue or group of cells under the control of a known promoter active within this tissue or group of cells. Because of the high stability of GFP, the cells where its expression has been switched on can be followed even after GFP expression has been switched off. This method, though elegant and powerful, requires the availability of a promoter driving expression of GFP in the cells of interest and generation of transient or stable transgenic embryos.

We chose to describe a method routinely used in the lab to study morphogenetic events during inner ear formation in the zebrafish embryo. This method is applicable to many systems or tissues as the boundaries of all cells are labeled. It uses Bodipy-ceramide (full name: Bodipy FL C_5-Cer/C_5-DMB-Cer [N-(4, 4-difluoro-5, 7-dimethil-4-bora-3a, 4a-diaza-s-indacene-3-pentanoyl) sphingosine; MW 631; Molecular Probes, Inc., no. D-3521], a fluorescent sphingolipid that has been used for many years as a vital stain for the Golgi apparatus in cultured vertebrate tissue cells. When applied to intact zebrafish embryos, Bodipy-ceramide stains cytoplasmic particles within the cells of the superficial and non–embryonic enveloping layer (EVL). However, after the fluorescent lipid diffuses through the EVL epithelium, the fluorescent lipid remains localized within the interstitial fluid of the embryo and freely diffuses between cells, surrounding and outlining them (Cooper *et al.*, 1999a,b). Furthermore, because there is a large reservoir of mobile fluorescent lipid present in the embryo, bleached molecules are quickly exchanged with unbleached molecules, allowing for acquisition of images over extended periods (up to 10 h). Because of the broad extent of the labeling, optical sections as acquired by confocal microscopy are necessary for imaging and resolving individual cells.

1. *Labeling specimens for viewing.* The embryos were vitally stained at mid-gastrula stage by placing them in staining solution for 30 min. Staining was done in a 35-mm Petri dish in which the bottom was coated with agarose. The staining solution consists of a 100-μM Bodipy-ceramide in HEPES-buffered embryo rearing medium [10 mM HEPES in embryo medium (Westerfield, 1993), pH 7.2] and 2% DMSO. After staining, the embryos were thoroughly rinsed in successive HEPES-buffered embryo rearing medium washes and then allowed to grow at 28° to the stage of interest.

2. *Mounting specimens for viewing.* The viewing chamber consists of an embedding mold in which the bottom has been cut off and glued to a glass coverslip using vacuum silicone grease to create a little pool. The whole apparatus was then placed in a stainless-steel chamber on an inverted microscope with a Petri dish lid on top. When the embryos reach the stage of interest (i.e., 16 hours postfertilization [hpf] in our case), they were anesthetized in 0.4 mg/ml solution of MS-222 (Argent Chemical Laboratories) in embryo rearing medium. They were then mounted in the viewing chamber and oriented individually in a solution of 0.7% agarose in

embryo rearing medium containing 0.4 mg/ml of MS-222 to prevent them from moving but to allow for growth. The viewing chamber was then flooded with embryo rearing medium containing 0.4% of MS-222 to prevent dehydration and movement of the specimen during image acquisition. The objective used was a water immersion Carl Zeiss C-Apochromat 40× with a 1.2 NA and a working distance of 225 μm. This working distance was sufficient at these stages to image through the whole embryo either dorsally to ventrally or from one side to the other.

3. *Optimizing image on the confocal.* We used an LSM 410 Zeiss scanning laser confocal microscope to acquire the 4D datasets. Bodipy-ceramide staining is very bright and images can be video enhanced after acquisition, which allowed us to reduce the power of the scanning laser to only 10% of its full power to avoid both photobleaching and phototoxic damage to the embryos.

4. *Collecting image data.* Our labeled zebrafish embryos were excited with a 488-nm laser line and the emitted light went through a 500 long-pass (allows light >500 nm to pass through) emission filter. The pinhole aperture was set at 20 (out of 255) and the image size was 512 by 512 pixels with a scanning time of 16 s. The software used to collect the images was Carl Zeiss LSM version 3.993 for Windows, which has a macro command to automatically collect a z-series stack at multiple user-specified times. The Zeiss software saves the image data as individual, relatively generic TIF files readable by many programs. We fixed the number of sections at 10, with an interval of 10 μm between each section and with the median section focused at the middle of the otic placode. This setting allowed us to image all the way through the developing inner ear at these stages. We chose a time interval of 5 min, which was long enough to let the microscope scan the 10 sections and short enough to capture any cell rearrangements.

5. *Analyzing confocal data.* After acquisition, all images at every time point were standardized in terms of gray level, orientation, and size using the Macintosh software NIH-Image (freely downloadable at http://rsb.info.nih.gov/nih-image/index.html). At the same time, the name of each individual file was changed to be compatible with the subsequent steps of the analysis. This tedious process was made easier by the already available macro commands in NIH-Image. Once all the images were formatted and named properly, we used the program's 4D-Turnaround and 4D-Viewer, freely available at the Integrated Microscopy Resource's 4-D Microscopy web site (http://www.loci.wisc.edu/4d/) to process the data. 4D-Turnaround converts the time-lapse data for each focal plane into a Quicktime movie. 4D-Viewer allows the set of Quicktime movies generated for each focal plane to be viewed as a multilayered movie. What

makes 4D-Viewer particularly useful is that it allows one to easily move through the z-sections at one time point as well as one section through different time points. Cells often move above or below the plane of focus through time, so the ability to quickly change the level in any stack at any time point provides the flexibility required to follow cell movements (Fig. 3).

Antibody Staining and Retrograde Labeling of Neurons in the Hindbrain of Zebrafish Embryos

The morphological structure of the developing hindbrain of zebrafish has been the subject of much research that has identified many of the earliest neurons forming in this region of the nervous system (Kimmel, 1982; Metcalfe *et al.*, 1986). The hindbrain contains nuclei for processing of auditory and vestibular information (McCormick, 1999) and our lab is interested in studying the development of neural projections between inner ear sensory organs and the hindbrain, because the embryonic origin of these nuclei in the zebrafish is not well understood. During development, transient bulges appearing shortly after neurulation, called *rhombomeres,* segment the hindbrain (Lumsden, 1990). Imaging a select set of stereotypically developing neurons in the rhombomeres allows for a simple way to find landmarks against which to assay the position of auditory and vestibular projections. One set of functionally distinct neurons is termed the *reticulospinals,* and these have been described in the developing zebrafish larvae (Kimmel, 1982; Metcalfe *et al.*, 1986). A reticulospinal neuron in the brain possesses an unbranched axon that descends into the spinal cord. Among the reticulospinals, the Mauthner neuron located in rhombomere 4 is one of the most readily distinguishable neurons (Fig. 4).

We can assay hindbrain development and inner ear development by imaging neurons within these regions using two methods. The first uses retrograde labeling or back-filling of a neuron with a fluorophore. In this method, the axonal process from a neuron or group of neurons in living anesthetized animals is severed. Severed axons have a tendency to heal and draw up surrounding material. Certain fluorophores placed at the site of damage, such as fluorescent dextrans, will be taken up and passively transported to the soma of the neuron, rendering that cell visible. The second method for imaging neurons employs immunohistochemistry, in which the primary antibody recognizes a specific epitope, unique to the cell type of interest. To visualize the primary antibody, a secondary antibody specific to it and coupled to a fluorophore is used. For comparative studies, often an antibody raised specifically to your species of interest is not available. Fortunately, some antibodies will recognize the same epitope across many species, even across major clades of metazoans. These two

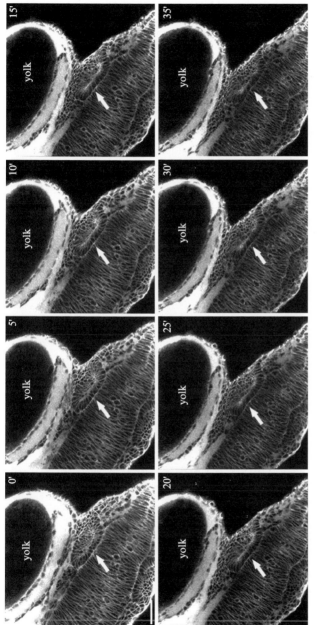

FIG. 3. Time-lapse confocal microscopy of the developing otic placode in zebrafish embryos. A time-lapse recording of an 18-hpf zebrafish embryo vitally stained with the fluorescent lipid, Bodipy-ceramide (anterior to right bottom corner; all images are dorsal view). The white arrow indicates the otic placode undergoing hollowing by cavitation. These pictures are extracted from a multilevel time-lapse recording of the otic region over a several hour period. The time interval between each picture is 5 min, and the focal plane of the series shown here goes right through the center of the otic placode. Scale bar, 20 μm.

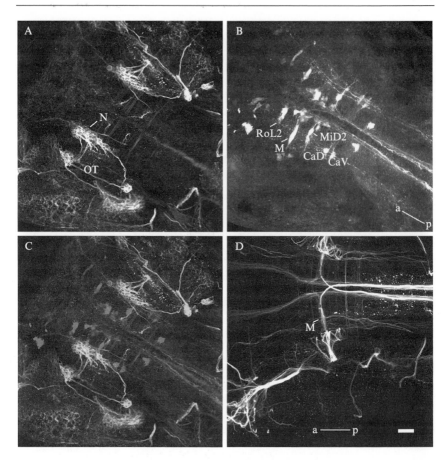

Fig. 4. Antibody staining and retrograde labeling of neurons in the hindbrain of zebrafish embryos 3 days post-fertilization (dpf). Ventral view of inner ear and hindbrain. Images in panels A to C are of the same embryo, taken with two different wavelengths as indicated in the text. The anterior/posterior axis for panels A, B, and C is indicated in B. (A) Immunostaining with Zn-8 antibody and Alexa 488 secondary antibody reveals a yet to be identified nucleus (N) just medial to the developing saccule of the zebrafish inner ear. The central part of the otic vesicle (OT) is indicated. (B) Retrograde labeling with lysinated rhodamine dextran of the reticulospinal neurons. Five of the neurons are identified. Four of the names (RoL2, MiD2, CaD, CaV) incorporate their hindbrain position: rostral (Ro), middle (Mi), and caudal (Ca). The cell body of the Mauthner (M) neuron resides in rhombomere 4, in the middle of the hindbrain. (C) A merged pseudo-colored image in which the immunostaining of Zn-8 is green and the retrograde labeling of the reticulospinal neurons is red. This allows for determination of hindbrain position of our Zn-8 antibody labeling. (D) Immunostaining with 3A10 antibody (recognizes neurofilaments). The cell body of the Mauthner (M) neuron is indicated. 3A10 is not a zebrafish-specific antibody, but it labels specific neurons in the central nervous system. Note the similarity to the retrograde labeling in (B) Scale bar, 25 μm (same for all panels). (See color insert.)

labeling techniques often can complement each other by providing land-marks for locating the anatomical position of the unknown labeling. The following example uses preserved material, which for many comparative studies is the most commonly available material.

1. *Labeling specimens for viewing.* Three days postfertilization (3 dpf) zebrafish embryos underwent retrograde labeling of some reticulospinal neurons followed by immunohistochemistry with the monoclonal antibody Zn-8 (Institute of Neuroscience, University of Oregon, Zn standing for zebrafish neural) (Westerfield, 1993). The reticulospinal neurons were back-filled by retrograde labeling. In this procedure, a tip of a tungsten needle laden with a sticky paste of fluorescent rhodamine dextran (full name: dextran tetramethylrhodamine lysine fixable; MW 10,000; Molecular Probes, Inc., no. D-1817) was used to transect the ventral part of a 3-dpf zebrafish spinal cord (Hill *et al.*, 1995). After recovery, the zebrafish embryos were post-fixed in 4% paraformaldehyde (PFA). These embryos then underwent antibody staining using a monoclonal antibody Zn-8 (1:10) with a 488 Alexa Fluor secondary (1:200) (Molecular Probes A-11001). Another set of 3-dpf embryos was immunostained with a neurofilament antibody (raised against the chick) 3A10 (1:10) (Developmental Studies Hybridoma Bank, University of Iowa), which labels the Mauthner neuron (Hatta, 1992). Again, the 488 Alexa Fluor secondary antibody (1:200) (Molecular Probes A-11001) was employed for imaging. During the immunohistochemistry procedure, the embryos were washed in a wash buffer of phosphate-buffered saline (PBS) + 0.1% Tween + 1% dimethylsulfoxide (DMSO) + 0.5% Triton for several hours between steps. The nonspecific binding of the primary antibody was blocked by exposing the embryos to wash buffer + 10% sheep serum for 2–4 h. The embryos were incubated in primary antibody or secondary antibody overnight at 4° in wash buffer + 1% sheep serum.

2. *Mounting specimens for viewing.* The embryos were placed on watchmaker glass, and the yolk and the branchial cartilages were removed with forceps. After clearing the embryos in glycerol, they were mounted in Fluoromount-G (Southern Biotechnology Associate, Inc.). The embryos were placed on a 60- by 24-mm coverslip. Four small dabs of vacuum grease were placed around the embryo and an 18- by 18-mm coverslip was set atop the grease. After correctly positioning the embryo, the smaller coverslip was gently pushed down to hold the embryo in place. A brass holder (Westerfield, 1993) was employed to hold the coverslip in place on the microscope stage.

3. *Optimizing image on the confocal.* Using the LSM 410 Zeiss confocal and the Carl Zeiss C-Apochromat 40× objective, we imaged the

embryos from the ventral side. Because the confocal used had an inverted microscope, the embryos were mounted with their dorsal side against the smaller coverslip and the ventral side against the 60- by 24-mm coverslip. To visualize images from the fluorescent rhodamine dextran retrograde labeling, the confocal was set to excite at 568 nm with the light collected going through a 570 LP emission filter. To visualize the antibody labeling, the confocal was set to excite at 488 nm with the light collected going through a 500 LP emission filter.

4. *Collecting image data.* Each image was acquired with the pinhole set at 20 (out of 255) over 16 s and measured 1024 by 1024 pixels. To visualize the retrogradely labeled reticulospinal neurons, a z-stack of 58 images, restricted to the dorsal region of the embryo, was taken at a step size of 0.5 μm (Fig. 4B). The same embryo was also stained with Zn-8 antibody using a different and distinguishable fluorophore (Fig. 4A). For this staining, a stack of 106 images were taken, also at a step size of 0.5 μm. To visualize immunohistochemistry using the 3A10 antibody, 64 images were taken from a second embryo at a step size of 1 μm (Fig. 4D).

5. *Analyzing confocal data.* The stacks from each set of images were projected along the z-axis using a maximum summation algorithm, into one image, with the confocal software Carl Zeiss LSM 3.993 for Windows. The projections were optimized under Adobe Photoshop 5.5 and pseudo-colored to make rhodamine dextran back-filling red and Zn-8 green. The overlay of these two projections represented the relationship of each projection along the anterior/posterior axis relative to each other (Fig. 4C). The double labeling allows us to place an as yet unidentified hindbrain nucleus, adjacent to the otic (inner ear) vesicle, in rhombomeres 4 and 5 (Fig. 4).

Alternative and Complementary Approaches

Confocal microscopy provides a powerful tool for biological studies, but there are related and alternative instruments or techniques for producing comparable images. An addition to laser scanning confocal microscopy is the ability to scan the spectral properties of the sample, allowing the separation of fluorophores that have overlapping emission spectra for which glass filters were never effective (Dickinson et al., 2001). Because the entire emission spectra is analyzed, fluorophores that emit at the same wavelength, such as EGFP and fluorescein isothiocyanate (FITC) (Harper, 2001), can be distinguished in the same sample once the system is calibrated for each individually. In this section, we discuss three methods that can complement and provide alternatives to confocal microscopy, and although all these methods could benefit from the spectral scanning just described, because of the software needed, they are only implemented in a few brands of confocal microscopes.

Two-Photon Microscopy

The confocal microscopy discussed earlier in this chapter uses a single photon of light, but another approach with great promise uses two photons produced from a rapidly pulsed (femtoseconds) infrared laser (Potter, 1996). Two-photon microscopy offers several advantages over typical confocal microscopy. Fluorophores are excited only at the plane of focus because that is the only place where two photons meet. This means that unlike with single-photon confocal microscopy, there is no excitation above or below the plane of focus. This restricts any bleaching and phototoxic effects to just that plane. The lack of any emission except from the plane of interest means that all the light emitted can be collected, eliminating the need for the pinhole, which simplifies the mechanics of the instrument. Infrared wavelengths penetrate deeper into most biological tissues, allowing two-photon confocal microscopes to image deeper than standard confocal microscopes. Two-photon lasers are typically tunable, allowing for the excitation of many wavelengths. The two-photon excitation wavelength differs from that of a typical confocal (Xu et al., 1996) (wavelength indicated in parentheses); for example, at 810 nm on our two-photon system, we can excite both DAPI (358 nm) and EGFP (488 nm) (Harper, 2001). A particular advantage of two-photon microscopy is the ability to excite fluorophores that are excited by UV wavelengths. Such fluorophores are widely used, yet most confocal systems do not have a UV laser.

As with any technology, there are some disadvantages to two-photon microscopy. Melanin-containing pigment cells absorb so much of the infrared light that they explode, limiting the viewing of samples with such pigment cells such as the eye's retina (surrounded by the pigmented epithelium) and whole nonalbino larvae. The exotic titanium sapphire laser used for two-photon microscopy is very expensive (ranging from $150,000 to $200,000). Finally, to view multiple fluorophores at a given two-photon wavelength, high-quality emission filters must be used.

Real-Time Confocal Microscopy

Laser scanning confocal microscopy cannot collect images in milliseconds without reducing image size and quality to a generally unusable degree. Two-photon microscopy shares this limitation. An alternative instrument is the spinning disk confocal microscope, which uses a Nipkow spinning disk with small, spirally arranged holes in front of the light source to produce a live confocal image (Maddox *et al.*, 2003). The main advantage of these instruments is the fast acquisition of confocal images, which can be crucial for time-dependent experiments such as calcium flux imaging (Yuste *et al.*, 2000). These systems are typically much less expensive than a standard confocal microscope because they are generally attached to a standard compound fluorescence microscope, use its optics, and do not use a laser. Images are typically acquired with a digital camera, which limits acquisition speed to that of the camera, although this can be quite fast given the right model camera. The greatest limitation of these systems is the need for a brightly labeled sample because relatively little light passes through the small holes in the spinning disk. Modifications that place lenses in front of each of the holes can help (Nakano, 2002), but even these systems are not as sensitive as a standard confocal microscope.

Deconvolution Software

All the methods discussed to this point use optical methods to eliminate the fluorescence outside the plane of interest. It is also possible to process standard fluorescence microscopy images acquired digitally, using sophisticated mathematical algorithms to computationally remove the out-of-focus fluorescence (Shotton, 1995). Such deconvolution software is less expensive than a confocal microscope and uses a standard compound fluorescence microscope for image acquisition. Deconvolution software is an option or a standard feature of many imaging systems (Davidson and Keller, 2001). The problems with deconvolution software include the potential to introduce artifacts into images, although for most samples, this is highly unlikely and newer sophisticated algorithms further minimize this problem. Also, depending on the power of the computer used, it can take minutes to hours or more to process stacks of multiple images. Advances in deconvolution software have minimized these problems, but careful controls are required.

These alternative methods and laser scanning confocal microscopy are not mutually exclusive. Current marketing practices do not sell separate two-photon systems but generally include these as an upgrade to typically high-end confocal systems. It is interesting to note that images collected with any of the two-photon or confocal microscopy systems discussed can be enhanced with deconvolution software, especially when the acquired images were not fully optimized to avoid photobleaching or quicken acquisition time.

Concluding Remarks

Confocal microscopy allows for the visualization of molecular expression patterns and morphological structures. Five steps are required for collecting confocal image data: The first two are labeling and mounting of specimens for viewing, and the next three are image optimization, collection, and analysis. In this chapter, we have used examples from our studies of vertebrate development. These examples draw from a narrow taxonomic sample, an amphibian and a teleost fish. These species are model systems for developmental biology studies and as such may represent taxa of limited evolutionary interest (Bolker, 1995; Metscher and Ahlberg, 1999). As comparative biologists, we are interested in a breadth of species for which the manipulations for viewing with confocal microscopy will be entirely different than for the two species discussed. Still, the protocols discussed are applicable to many other, though by no means all, species of amphibians and teleost fish with some modification. Although teleosts represent nearly half of all vertebrate species (Nelson, 1984), it is important to emphasize that these methods also can be effectively used to study amniote vertebrates (reptiles, birds, and mammals). The biggest challenge to studying amniote development *in vivo* is that embryos develop in an egg or inside the mother, in the case of placental mammals. Even so, methods have been developed to do confocal microscopy, even time-lapse confocal microscopy, on living avian embryos inside the egg (Kulesa and Fraser, 1999) and cultured mouse embryos (Jones *et al.*, 2002). These powerful techniques raise the possibility of studying the development *in vivo* of many other species of amniotes. Although confocal microscopy is a powerful tool for studying development, it has much more general applicability for studies across eukaryotes and prokaryotes.

Acknowledgments

This work was supported by a grant (RO1 DC04061) from the National Institutes of Health and the National Institute for Deafness and Other Communication Disorders. This work was also supported by the Oberkotter Foundation.

References

Amsterdam, A., Lin, S., and Hopkins, N. (1995). The Aequorea victoria green fluorescent protein can be used as a reporter in live zebrafish embryos. *Dev. Biol.* **171**, 123–129.

Baker, C. V., and Bronner-Fraser, M. (2001). Vertebrate cranial placodes I. Embryonic induction. *Dev. Biol.* **232**, 1–61.

Berland, K. (2001). Basics of fluorescence. *In* "Methods in Cellular Imaging" (A. Periasamy, ed.), pp. 5–19. Oxford University Press, Oxford.

Bolker, J. A. (1995). Model systems in developmental biology. *Bioessays* **17**, 451–455.

Collazo, A., and Fraser, S. E. (1996). Integrating cellular and molecular approaches into studies of development and evolution: The issue of morphological homology. *Aliso* **14**, 237–262.

Cooper, M. S., D'Amico, L. A., and Henry, C. A. (1999a). Analyzing morphogenetic cell behaviors in vitally stained zebrafish embryos. *Methods Mol. Biol.* **122**, 185–204.

Cooper, M. S., D'Amico, L. A., and Henry, C. A. (1999b). Confocal microscopic analysis of morphogenetic movements. *Methods Cell Biol.* **59**, 179–204.

Davidson, L., and Keller, R. (2001). Basics of a light microscopy imaging system and its application in biology. *In* "Methods in Cellular Imaging" (A. Periasamy, ed.), pp. 53–65. Oxford University Press, Oxford.

de Rosa, R., Grenier, J. K., Andreeva, T., Cook, C. E., Adoutte, A., Akam, M., Carroll, S. B., and Balavoine, G. (1999). Hox genes in brachiopods and priapulids and protostome evolution. *Nature* **399**, 772–776.

Dickinson, M. E., Bearman, G., Tille, S., Lansford, R., and Fraser, S. E. (2001). Multi-spectral imaging and linear unmixing add a whole new dimension to laser scanning fluorescence microscopy. *Biotechniques* **31**, 1272.

Field, K. G., Olsen, G. J., Lane, D. J., Giovannoni, S. J., Ghiselin, M. T., Raff, E. C., Pace, N. R., and Raff, R. A. (1988). Molecular phylogeny of the animal kingdom. *Science* **239**, 748–753.

Foster, B. (1997). "Optimizing Light Microscopy for Biological and Clinical Laboratories." Kendall/Hunt Publishing Company, Dubuque, IA.

Fritzsch, B., Beisel, K. W., Jones, K., Farinas, I., Maklad, A., Lee, J., and Reichardt, L. F. (2002). Development and evolution of inner ear sensory epithelia and their innervation. *J. Neurobiol.* **53**, 143–156.

Goldberg, M. E., and Hudspeth, A. J. (2000). The vestibular system. *In* "Principles of Neural Science" (E. R. Kandel, J. H. Schwartz, and T. M. Jessell, eds.), pp. 801–815. McGraw-Hill, New York.

Haffter, P., Granato, M., Brand, M., Mullins, M. C., Hammerschmidt, M., Kane, D. A., Odenthal, J., Vaneeden, F. J. M., Jiang, Y. J., Heisenberg, C. P., Kelsh, R. N., Furutaniseiki, M., Vogelsang, E., Beuchle, D., Schach, U., Fabian, C., and Nussleinvolhard, C. (1996). The identification of genes with unique and essential functions in the development of the zebrafish, Danio-Rerio. *Development* **123**, 1–36.

Harper, I. (2001). Fluorophores and their labeling procedures for monitoring various biological signals. *In* "Methods in Cellular Imaging" (A. Periasamy, ed.), pp. 20–39. Oxford University Press, Oxford.

Hatta, K. (1992). Role of the floor plate in axonal patterning in the zebrafish CNS. *Neuron* **9**, 629–642.

Hedges, S. B., Moberg, K. D., and Maxson, L. R. (1990). Tetrapod phylogeny inferred from 18S and 28S ribosomal RNA sequences and a review of the evidence for amniote relationships. *Mol. Biol. Evol.* **7**, 607–633.

Hill, J., Clarke, J. D., Vargesson, N., Jowett, T., and Holder, N. (1995). Exogenous retinoic acid causes specific alterations in the development of the midbrain and hindbrain of the zebrafish embryo including positional respecification of the Mauthner neuron. *Mech. Dev.* **50**, 3–16.

Jones, E. A., Crotty, D., Kulesa, P. M., Waters, C. W., Baron, M. H., Fraser, S. E., and Dickinson, M. E. (2002). Dynamic *in vivo* imaging of postimplantation mammalian embryos using whole embryo culture. *Genesis* **34**, 228–235.

Kil, S. H., and Collazo, A. (2001). Origins of inner ear sensory organs revealed by fate map and time-lapse analyses. *Dev. Biol.* **233**, 365–379.

Kimmel, C. B. (1982). Reticulospinal and vestibulospinal neurons in the young larva of a teleost fish, *Brachydanio rerio*. *Prog. Brain Res.* **57**, 1–23.

Krumlauf, R. (1992). Evolution of the vertebrate Hox homeobox genes. *Bioessays* **14**, 245–252.

Kulesa, P. M., and Fraser, S. E. (1999). Confocal imaging of living cells in intact embryos. *Methods Mol. Biol.* **122**, 205–222.

Lumsden, A. (1990). The cellular basis of segmentation in the developing hindbrain. *Trends Neurosci.* **13**, 329–335.

Maddox, P. S., Moree, B., Canman, J. C., and Salmon, E. D. (2003). Spinning disk confocal microscope system for rapid high-resolution, multimode, fluorescence speckle microscopy and green fluorescent protein imaging in living cells. *Methods Enzymol.* **360**, 597–617.

McCormick, C. A. (1999). Anatomy of the central auditory pathways of fish and amphibians. *In* "Comparative Hearing: Fish and Amphibians," (R. R. Fay and A. N. Popper, eds.), Vol. 11, pp. 155–217. Springer, New York.

Metcalfe, W. K., Mendelson, B., and Kimmel, C. B. (1986). Segmental homologies among reticulospinal neurons in the hindbrain of the zebrafish larva. *J. Comp. Neurol.* **251**, 147–159.

Metscher, B. D., and Ahlberg, P. E. (1999). Zebrafish in context: Uses of a laboratory model in comparative studies. *Dev. Biol.* **210**, 1–14.

Nakano, A. (2002). Spinning-disk confocal microscopy—a cutting-edge tool for imaging of membrane traffic. *Cell Struct. Funct.* **27**, 349–355.

Nelson, J. S. (1984). "Fishes of the World." John Wiley & Sons, New York.

Potter, S. M. (1996). Vital imaging: Two photons are better than one. *Curr. Biol.* **6**, 1595–1598.

Potter, S. M. (2000). Two-photon microscopy for 4D imaging of living neurons. *In* "Imaging Neurons: A Laboratory Manual" (R. Yuste, F. Lanni, and A. Konnerth, eds.), pp. 20.1–20.16. Cold Spring Harbor Laboratory Press, New York.

Shafizadeh, E., Huang, H., and Lin, S. (2002). Transgenic zebrafish expressing green fluorescent protein. *Methods Mol. Biol.* **183**, 225–233.

Shotton, D. M. (1995). Robert Feulgen Prize Lecture 1995. Electronic light microscopy: Present capabilities and future prospects. *Histochem. Cell. Biol.* **104**, 97–137.

Sive, H. L., Grainger, R. M., and Harland, R. M. (2000). "Early Development of *Xenopus laevis*: A Laboratory Manual." Cold Spring Harbor Laboratory Press, Cold Spring Harbor, NY.

Westerfield, M. (1993). "The Zebrafish Book: A Guide for the Laboratory Use of Zebrafish (*Brachydanio rerio*)." M. Westerfield, Eugene, OR.

Xu, C., Zipfel, W., Shear, J. B., Williams, R. M., and Webb, W. W. (1996). Multiphoton fluorescence excitation: New spectral windows for biological nonlinear microscopy. *Proc. Natl. Acad. Sci. USA* **93**, 10763–10768.

Yuste, R., Lanni, F., and Konnerth, A. (2000). "Imaging Neurons: A Laboratory Manual." Cold Spring Harbor Laboratory Press, New York.

[28] PrimerSelect: A Transcriptome-Wide Oligonucleotide Primer Pair Design Program for Kinetic RT-PCR–Based Transcript Profiling

By KENNETH J. GRAHAM and MICHAEL J. HOLLAND

Abstract

We describe PrimerSelect, a program capable of transcriptome-wide design of primer pairs for optimal performance in kinetic reverse-transcriptase polymerase chain reaction (RT-PCR). For the yeast *Saccharomyces cerevisiae,* PrimerSelect designs primer pairs for 86% of genomic open reading frames (ORFs) using design criteria we previously established to be optimal for kinetic RT-PCR (kRT-PCR)–based transcript quantitation. Primer pairs designed by PrimerSelect for 230 yeast ORFs were evaluated for primer dimer potential, PCR cyclewise yield, and cross-priming. Performance of 95% of these primer pairs is optimal with respect to primer dimer potential and PCR cyclewise yield for quantitating even the rarest yeast transcript. All of the primer pairs produced a single amplicon of the expected size from yeast genomic DNA template. The utility of PrimerSelect for designing primer pairs complementary to ORF sequences defined for multiple isolates of the human bacterial pathogens *Helicobacter pylori* and *Staphylococcus aureus* is also demonstrated.

Introduction

For many years, polymerase chain reaction (PCR) was only analyzed after completion of thermal cycling, typically by examination of the amplicon product(s) after gel electrophoresis. The first attempts to monitor the kinetics of amplicon accumulation during PCR were performed by Watson *et al.* in the late 1980s (Higuchi *et al.*, 1993). They were interested in the origin and kinetic behavior of competing reactions such as cross-priming and the primer dimer reaction, which interfere with amplification of the intended target sequence during PCR. Perhaps the most important result obtained from these initial studies was the realization that the primer dimer reaction can abolish or significantly alter the kinetics of target amplicon accumulation if the target sequence is present at a low concentration in the sample.

Primer pairs can form a double-stranded DNA product, referred to as "primer dimer," in the absence of an exogenous template. This product forms in a time-dependent fashion at low temperature and is amplified in PCR. The potential to form primer dimer is an inherent property of each primer pair. The primer dimer reaction is substantially delayed or eliminated by the features of the PrimerSelect program described here. Because the primer dimer reaction competes with the templated reaction for oligonucleotides, it is important for quantitative kPCR-based analyses that the primer dimer reaction is delayed relative to the templated reaction.

Kinetic monitoring of PCRs is now used widely to quantitate the amount of a target sequence in a DNA (kPCR) or RNA (kRT-PCR) sample. The basic principle of these analytical methods is that the kinetics of amplicon accumulation in PCR is directly related to the initial concentration of the target sequence in the sample to be analyzed. Thus, the earlier the amplicon accumulation growth curve, the more abundant the target sequence in the sample. Higuchi and Watson (1999) determined that the point at which the growth curve crosses a predetermined normalized fluorescence produces the most reliable descriptor of each growth curve. The fractional PCR cycle in which the growth curve crossed this arbitrary fluorescence level (AFL) was determined by linear regression analysis of the growth curve and was designated the *fractional PCR cycle* at threshold or Ct.

The overwhelming cause of failed kPCR is poor primer pair design. A deficiency in primer pair design contributes to early primer dimer formation, poor selectivity, and delayed kinetics of amplicon accumulation. The primer dimer reaction, as does cross-priming to form a secondary amplicon, consumes oligonucleotide primers, substrates for the synthesis of target amplicon in PCR. As a consequence, these secondary reactions can profoundly affect the reliability of kPCR or kRT-PCR–based quantitation of target sequence in a sample. These latter effects occur independent of whether the kinetics of amplicon accumulation is monitored with a dye or a hybridization probe.

Control of the Primer Dimer Reaction

The most efficient way to minimize primer dimer potential is through primer design. Typically, a poorly designed primer pair cannot be rescued by incorporation of one of the "hot-start" reaction features described later in this chapter. Minimizing intramolecular and intermolecular priming potential is easily achieved using most primer design programs. If possible, all oligonucleotides should contain an AA dinucleotide at their 3' termini. The importance of this latter feature for controlling the primer dimer reaction was discovered by Randy Saiki (Roche Molecular Systems, Inc.) and has been repeatedly demonstrated through the characterization of thousands of primer pairs in kinetic PCR. At some point, the notion of a G:C "clamp" at the 3′ end of PCR oligonucleotides was popularized. Suffice it to say that with the exception of complementarily at the 3′ termini of primers, a 3′ G:C clamp is the worst possible oligonucleotide feature with respect to encouraging early primer dimer accumulation in PCR.

Specialized reaction conditions for controlling the primer dimer reaction in PCR and RT-PCR employ a "hot-start" feature to prevent primer dimer formation in the foundation stages of PCR at low temperature (Innis and Gelfand, 1999). A hot start can be achieved with antibody against the DNA polymerase that inhibits activity at low temperature but not high temperature, temperature-sensitive aptamer inhibitors of DNA polymerase or the use of a thermostable DNA polymerase, which is chemically modified to render the enzyme inactive at low temperatures. These latter gold or gilded versions of DNA polymerases are activated at 95°, at which the chemical modification is hydrolyzed to generate a hot start (Birch, 1996). These derivatized DNA polymerases are not compatible with single enzyme RT-PCR, however, because of hydrolysis of the RNA template at the high activation temperature.

For kRT-PCR–based transcript quantitation, we include uracil N-glycosidase (UNG) to hydrolyze amplicon carryover (all kPCRs are performed with an excess of dUTP over dTTP to generate dU-containing amplicons) (Kang and Holland, 1999; Kang et al., 2000). A 2-min incubation at 50° with 2 units of UNG completely abolishes PCR from dU-containing amplicon template. Consequently, UNG provides a hot-start feature because the enzyme is active at 50° but is inactivate at 60°, at which reverse transcription occurs. Thus, primer dimer generated at temperatures below 60° is hydrolyzed.

Control of Cross-Priming Reactions

Cross-priming potential of each primer pair (upper and lower oligonucleotide; average length 25 nucleotides) is effectively evaluated by comparison with a complete genome sequence using BLAST. For primer pairs with four or more mismatches to the closest secondary target, the only secondary amplicons we have ever detected after PCR are authentic splice variants of human transcripts generated from the same open reading frame (ORF). Oligonucleotides with fewer than four mismatches each to a secondary target that can make an amplicon of 1000 of fewer base pairs are problematic. These secondary targets typically correspond to closely related ORFs or gene families, and the PrimerSelect program reports them as gene families. We emphasize that selectivity in RT-PCR is solely a property of the forward and reverse primer pair. Monitoring amplicon accumulation with a third hybridization probe does not influence the effect of a cross-priming event on the observed target amplicon growth curve, or Ct.

Normalized Primer Pair Performance in the Reverse Transcription and PCR Phases of Kinetic RT-PCR

We showed previously that the reverse-transcription efficiency of multiple primer pairs designed against the same transcript varied within a factor of 2 when amplicon size is restricted to 100–300 bp (Kang *et al.*, 2000). We have obtained similar results for multiple primer pairs designed for more than 100 different yeast, bacterial, and human transcripts. Furthermore, the average PCR cyclewise efficiency or yield for more than 2000 primer pairs tested is 95% for amplicons ranging from 100 to 300 bp. For these reasons, we have restricted amplicon size to 100–300 bp.

Automated Transcriptome-Wide Primer Pair Design Using the PrimerSelect Program

Using the primer design program Oligo 5.0 and an online BLAST program, an experienced investigator can design a primer pair for a cellular transcript in approximately 20 min. Because of this relatively slow throughput and fatigue factors, manual primer pair design is not practical for large-scale analyses. The program described here, PrimerSelect, incorporates all of the design features we have determined to be important for optimal performance in kinetic PCR applications.

The PrimerSelect program runs under the Linux operating system using Mandrake, Red Hat, or SuSE distributions. We choose this platform because the Primer3 program (Rozen and Skaletsky, 2000) is written for Unix-based operating systems. Perl scripting language (version 5.6.1) was used to join the Primer3 and BLAST programs and to perform other functions as noted. Using a Pentium IV, 1.6-GHz processor with 512 MB of RAM, the transcriptome-wide analyses of the bacterial and yeast genomes described here are completed in 1–3 h. The PrimerSelect program and instructions for its use are available at http://biochem.ucdavis.edu/faculty/holland/ReviewEnzData.htm.

The PrimerSelect program designs primer pairs for a complete transcriptome without manual intervention. In the first phase of the program, a set of parameters for oligonucleotide primer pair design is used to minimize the potential for primer dimer reaction. These latter parameters were identified through trial-and-error design and evaluation of thousands of primer pairs for *Saccharomyces cerevisiae, Helicobacter pylori,* and humans. In the second phase of the design, a genome-wide BLAST analysis is used to evaluate the potential of each primer pair for cross-priming.

A schematic of the PrimerSelect program is shown in Fig. 1. The details of PrimerSelect are summarized as follows: A tab-delimited file with the coordinates for all of the ORFs predicted for the transcriptome of interest is imported into the program (Fig. 1, "ORF coordinates") together with the complete genomic sequence for that organism in FASTA format (Fig. 1; "Reference genomic sequence"). A Perl subroutine then extracts all of the ORF sequences sequentially (Fig. 1; "ORF sequence"). Primer pairs are then designed sequentially for all ORFs (≥100 bases) using the Primer3 program (Rozen and Skaletsky, 2000). Key parameters for primer pair design are balanced Tm (68° minimum, 72° maximum, 68° preferred), no

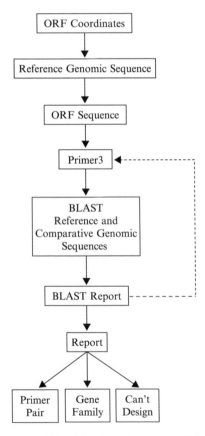

Fig. 1. Schematic outline of the PrimerSelect program. Details of this program are provided in the text.

more than four consecutive identical bases, and amplicon sizes between 100 and 300 bp in length. The Primer3 program was modified to ensure that each primer contains AA at its 3′ terminus. The last five nucleotides at the 3′ end of both primers are evaluated for calculated stability of both intra-molecular and intermolecular base pairing. A level 6 stability (−6 Kcal/mole) corresponds to the lowest primer dimer potential. If no primer pairs can be designed with a level 6 stability, primer pairs with a level 7 stability are designed. Once primer pairs are designed at the best stability level, the search stops. If two or more primer pairs are designed at a given stability level, a Perl subroutine sorts them relative to the 3′ end of the ORF. An additional Perl subroutine rejects primer pairs when one or both oligonu-cleotide contains more than three consecutive G nucleotides. This latter feature of the program ensures optimal repetitive yield during oligonucleo-tide synthesis. If no primer pair can be designed with a level 9 stability, then failure is reported (Fig. 1; "Cannot design"). Typically, this occurs when the transcript is short.

The first primer pair from the sorted list at the best stability level is compared to the entire genomic reference sequence using BLAST (version 2.2.5) with an expectation value of 0.1. If the upper or the lower oligonu-cleotide displays four or more mismatches to the best secondary genomic target, that primer pair is reported and the search is complete (Fig. 1; "Primer pair"). This ensures that the primer pair nearest the 3′ end of the ORF at the lowest primer dimer stability level is reported.

If the upper and lower oligonucleotides each display fewer than four mismatches to a secondary genomic sequence and this secondary target generates an amplicon of 1000 bp or less, the primer pair is rejected. If all of the primer pairs designed at a particular stability level are rejected on this basis, Primer3 generates a list of primer pairs at the next stability level (Fig. 1, dashed line) and the BLAST analysis is repeated. If all of the primer pairs designed up to a level 9 stability are rejected because of their potential to amplify from a secondary target, that ORF is reported to be a member of a gene family (Fig. 1; "Gene family").

Performance of Primer Pairs Designed by the PrimerSelect Program

All of the ORFs (≥100 bases) for *S. cerevisiae* strain S288c were analyzed using PrimerSelect and the results are shown in Table IA. Primer pairs were designed for 86% of these yeast ORFs. The remaining ORFs fell into either the "gene family" or the "cannot design" category.

We evaluated 230 of the yeast primer pairs designed by the current version of PrimerSelect for primer dimer potential, PCR cyclewise yield, and cross-priming. kPCR was performed in 20-μl reactions containing:

TABLE I
DESIGN OF PRIMER PAIRS FOR *S. CEREVISIAE, H. PYLORI,* AND *S. AUREUS* USING PRIMERSELECT

A

Reference genome[a]	ORFs[b]	Primer pair[c]	Gene family[d]	Cannot design[e]
Saccharomyces cerevisiae S288C	5842	5054	284	504
Helicobacter pylori 26695	1386	1212	8	166
Staphylococcus aureus Mu50	707	569	0	138

B

Reference genome[a]	Comparative genome(s)[f]	Common primer pair[g]
Helicobacter pylori 26695	*H. pylori* J99	562
Staphylococcus aureus Mu50	*S. aureus* N315, *S. aureus* MW2	534

[a] *Reference genome* refers to the genomic nucleotide sequence used to design primer pairs.

[b] For *S. cerevisiae* strain S288c, open reading frames (ORFs) is the total number of annotated ORFs that are 100 bases or larger in the reference genome. For *H. pylori* strain 26695, ORFs is the total number of annotated ORFs shared by strains 26695 and J99 that are 100 bases or larger in the reference genome. For *S. aureus* strain Mu50, ORFs is the total number of fully annotated ORFs shared by strains Mu50, N315, and MW2 that are 100 bases or larger in the reference genome.

[c] *Primer pair* is the total number of primer pairs that were successfully designed using PrimerSelect.

[d] *Gene family* is the number of ORFs determined by PrimerSelect to be present two or more times in the reference genome.

[e] *Cannot design* is the number of ORFs in the reference genome for which PrimerSelect cannot design a primer pair.

[f] *Comparative genome(s)* is the genome of additional isolates of the indicated pathogen.

[g] *Common primer pair* is the total number of primer pairs designed by PrimerSelect to the reference genome that are perfect matches to the comparative genome(s).

50 mM Tricine buffer, pH 8.3, 110 mM KOAc, 13% glycerol, 0.3 mM dATP, dCTP, and dGTP, 0.05 mM dTTP, 0.5 mM dUTP, 2.4 mM Mn(OAc)$_2$, 2.5 μM ethidium bromide, 0.25 μM primers, 4 units rTth DNA polymerase, provided by Roche Molecular Systems, 2 U uracil N-glycosylase (UNG). Primer dimer potential was evaluated for each primer pair in reactions containing no template. PCR cyclewise yield was

evaluated in triplicate reactions containing 10^5 genome equivalents of highly purified genomic DNA isolated from strain S288c. Amplicon product(s) produced with each primer pair in these latter genomic DNA-templated reactions was analyzed by agarose gel electrophoresis to test for putative secondary amplicons produced by cross-priming. kPCR was performed under the same conditions used for kRT-PCR (Holland, 2002) in the following steps: 2 min at 50° for UNG-dependent "hot start" and digestion of any dU-containing amplicon carryover, 30 min at 60° for UNG inactivation and reverse transcription, and 50 cycles of PCR (anneal/extend, 40 s at 60°; denaturation, 20 s at 95°). Amplicon accumulation was monitored in a kinetic thermal cycler using ethidium bromide dye as previously described (Kang et al., 2000).

Of the 230 primer pairs tested in reactions containing no template, 81 failed to produce primer dimer after 50 cycles of PCR and only 7 (3%) produced primer dimer earlier than PCR cycle 35. Under the same experimental conditions, the rarest yeast transcripts we have measured display Ct values before cycle 35. The average PCR cyclewise yield determined from genomic DNA templated reactions with all 230 primer pairs was 95%. Only five primer pairs (2%) displayed PCR cyclewise yields less than 85%. Finally, all of the primer pairs directed the synthesis of a single amplicon of the expected size from yeast genomic DNA template. Thus, the properties of 95% of the primer pairs tested were in the optimal range for kRT-PCR–based transcript quantitation.

Primer Pair Design for Multiple Isolates of Pathogenic Organisms

The PrimerSelect program is capable of designing primer pairs to conserved ORFs within the genomes of multiple clinical isolates of the same bacterial pathogen. This feature of the program is illustrated using the genomic sequences of two H. pylori isolates and 707 fully annotated ORFs shared by three isolates of S. aureus. H. pylori strain 26695 (Tomb et al., 1997) and J99 (Alm et al., 1999) share 1386 common ORFs. When strain 26695 is used as the reference genomic sequence and the BLAST analysis is performed only with the reference genomic sequence, 1212 primer pairs are designed corresponding to 87% of these conserved ORFs (Table IA).

To design primer pairs that are perfect matches to the corresponding ORF sequences in H. pylori strains 26695 and J99, the BLAST analysis of primer pairs designed by the Primer3 program is extended to the strain J99 comparative genomic sequence. Specifically, each primer pair designed against the strain 26695 reference genomic sequence is subjected to a BLAST analysis as described earlier for yeast. Primer pairs that are successfully designed for strain 26695 are then compared to the strain J99 comparative genomic sequence to identify perfect sequence matches to both oligonucleotide primers. If no perfect match is found, the next primer pair designed for strain 26695 is compared to the J99 sequence. This search will continue through all of the primer pairs designed up to and including those designed at level 9 stability. For the two *H. pylori* transcriptomes that share approximately 85% sequence identity, primer pairs for 562 common ORFs were identified with perfect nucleotide sequence matches to both genomes (Table IB).

We evaluated 707 fully annotated ORFs shared by *S. aureus* strains Mu50, N315, and MW2 (Baba *et al.*, 2002; Kuroda *et al.*, 2001). Primer pairs were successfully designed for 569 (80%) of these shared ORFs using only the *S. aureus* strain Mu50 reference genomic sequence (Table IA). Because the three *S. aureus* genomic sequences are highly conserved, 534 ORF-specific primer pairs were identified that matched all three genomes (Table IB).

Additional Considerations

The version of PrimerSelect described here requires an AA dinucleotide at the 3' terminus of each oligonucleotide primer, tightly balanced primer pair Tm, and for multiple genomes, a perfect nucleotide sequence match between the reference genome and the comparative genome. We have successfully designed hundreds of primer pairs for transcripts that fall into the "gene family" or the "cannot design" categories described earlier. This is accomplished by relaxing the requirement for a 3' AA, allowing a wider range of Tm differential or allowing a mismatch between the reference and comparative genome within the first five nucleotides at the 5' end of an oligonucleotide primer. As expected, when these parameters are relaxed, average performance of the primer pairs declines. Nevertheless, we have successfully designed primer pairs to discriminate among yeast glycolytic gene families that share greater than 95% sequence identity (Kang *et al.*, 2000). Similarly, we have designed a number of primer pairs containing a single mismatch near the 5' end of an oligonucleotide primer that are suitable for comparative transcript analysis with the two *H. pylori* isolates. PrimerSelect is an open program and can be easily modified to allow relaxation of any of these parameters. Thus, a stratified approach can be used to design primer pairs to a complete cellular transcriptome beginning with the most stringent design criteria.

Acknowledgments

The authors thank Robert Watson and David Gelfand (Program in Core Research group at Roche Molecular Systems, Inc.) for advice, thoughtful comments, and support. This work was supported by National Institutes of Health grants HG1736 and AI47400.

References

Alm, R. A., Ling, L. S., Moir, D. T., King, B. L., Brown, E. D., Doig, P. C., Smith, D. R., Noonan, B., Guild, B. C., deJonge, B. L., Carmel, G., Tummino, P. J., Caruso, A., Uria-Nickelsen, M., Mills, D. M., Ives, C., Gibson, R., Merberg, D., Mills, S. D., Jiang, Q., Taylor, D. E., Vovis, G. F., and Trust, T. J. (1999). Genomic-sequence comparison of two unrelated isolates of the human gastric pathogen Helicobacter pylori. *Nature* **397,** 176–180.

Baba, T., Takeuchi, F., Kuroda, M., Yuzawa, H., Aoki, K., Oguchi, A., Nagai, Y., Iwama, N., Asano, K., Naimi, T., Kuroda, H., Cui, L., Yamamoto, K., and Hiramatsu, K. (2002). Genome and virulence determinants of high virulence community-acquired MRSA. *Lancet* **359,** 1819–1827.

Birch, D. E. (1996). Simplified hot start PCR. *Nature* **381,** 445–446.

Higuchi, R., Fockler, C., Dollinger, G., and Watson, R. (1993). Kinetic PCR analysis: real-time monitoring of DNA amplification reactions. *Biotechnology (NY)* **11,** 1026–1030.

Higuchi, R., and Watson, R. (1999). Kinetic PCR Analysis Using a CCD Camera and without Using Oligonucleotide Probes. *In* "PCR Applications: Protocols for Functional Genomics" (M. A. Innis, D. H. Gelfand, and J. J. Sninsky, eds.), pp. 263–284. Academic Press, San Diego, CA.

Holland, M. J. (2002). Transcript abundance in yeast varies over six orders of magnitude. *J. Biol. Chem.* **277,** 14363–14366.

Innis, M., and Gelfand, D. (1999). Optimization of PCR: Conversations Between Michael and David. *In* "PCR Applications: Protocols for Functional Genomics" (M. A. Innis, D. H. Gelfand, and J. J. Sninsky, eds.), pp. 3–22. Academic Press, San Diego, CA.

Kang, J. J., and Holland, M. J. (1999). Cellular Transcriptome Analysis Using a Kinetic PCR Assay. *In* "PCR Applications: Protocols for Functional Genomics" (M. A. Innis, D. H. Gelfand, and J. J. Sninsky, eds.), pp. 429–444. Academic Press, San Diego, CA.

Kang, J. J., Watson, R. M., Fisher, M. E., Higuchi, R., Gelfand, D. H., and Holland, M. J. (2000). Transcript quantitation in total yeast cellular RNA using kinetic PCR. *Nucleic Acids Res.* **28,** e2.

Kuroda, M., Ohta, T., Uchiyama, I., Baba, T., Yuzawa, H., Kobayashi, I., Cui, L., Oguchi, A., Aoki, K., Nagai, Y., Lian, J., Ito, T., Kanamori, M., Matsumaru, H., Maruyama, A., Murakami, H., Hosoyama, A., Mizutani-Ui, Y., Takahashi, N. K., Sawano, T., Inoue, R., Kaito, C., Sekimizu, K., Hirakawa, H., Kuhara, S., Goto, S., Yabuzaki, J., Kanehisa, M., Yamashita, A., Oshima, K., Furuya, K., Yoshino, C., Shiba, T., Hattori, M., Ogasawara, N., Hayashi, H., and Hiramatsu, K. (2001). Whole genome sequencing of meticillin-resistant Staphylococcus aureus. *Lancet* **357,** 1225–1240.

Rozen, S., and Skaletsky, H. (2000). Primer3 on the WWW for general users and for biologist programmers. *Methods Mol. Biol.* **132,** 365–386.

Tomb, J. F., White, O., Kerlavage, A. R., Clayton, R. A., Sutton, G. G., Fleischmann, R. D., Ketchum, K. A., Klenk, H. P., Gill, S., Dougherty, B. A., Nelson, K., Quackenbush, J., Zhou, L., Kirkness, E. F., Peterson, S., Loftus, B., Richardson, D., Dodson, R., Khalak, H. G., Glodek, A., McKenney, K., Fitzegerald, L. M., Lee, N., Adams, M. D., Venter, J. C., *et al.* (1997). The complete genome sequence of the gastric pathogen Helicobacter pylori. *Nature* **388,** 539–547.

[29]   Detecting Differential Expression of Parental or Progenitor Alleles in Genetic Hybrids and Allopolyploids

By CRAIG S. PIKAARD, SASHA PREUSS, KEITH EARLEY,
RICHARD J. LAWRENCE, MICHELLE S. LEWIS, and
Z. JEFFREY CHEN

Abstract

Three assays useful for detecting specific RNA transcripts are primer extension, S1 nuclease protection, and reverse-transcription–cleaved amplified polymorphic sequence (RT-CAPS) analysis. All three of these techniques are used routinely for gene expression analyses and allow insights not possible by RNA blot (northern blot) hybridization. In this chapter, we describe how the primer extension, S1 nuclease protection, and RT-CAPS methods can be used to discriminate one or more parental or progenitor alleles in hybrids or allopolyploids. We discuss the rationale for using the different techniques and provide examples of the data generated.

Introduction

A molecular understanding of genetic or epigenetic traits in diploid hybrids and allopolyploids (polyploid hybrids) requires measuring the expression state of alleles inherited from the different parents or progenitor species. Indeed, uniparental gene expression is the basis for a number of intriguing epigenetic phenomena, including gametic imprinting (Brannan and Bartolomei, 1999; Sleutels *et al.*, 2000), mammalian X-chromosome inactivation (Avner and Heard, 2001; Huynh and Lee, 2001), and nucleolar dominance (Pikaard, 2000a,b; Reeder, 1985).

Typically, transcripts from orthologous alleles are too similar in sequence to be discriminated from one another using RNA blot hybridization. Therefore, more precise methods are needed. The primer extension, S1 nuclease protection, and reverse-transcription–cleaved amplified polymorphic sequence (RT-CAPS) assays can all be designed to provide the needed precision, because all three methods can exploit relatively minor sequence differences among allelic RNAs. Primer extension (Boorstein and Craig, 1989) is useful for discriminating allele transcripts that share regions of sequence identity interrupted by insertions/deletions in the region near the transcription initiation site. S1 nuclease protection

(Berk and Sharp, 1977) is useful for discriminating transcripts that may be identical in size but have at least several contiguous nucleotides that are polymorphic. RT-CAPS is a variation of the CAPS method used for generating polymerase chain reaction (PCR)–based markers for genetic mapping (Konieczny and Ausubel, 1993). This method exploits one or more restriction endonuclease site polymorphisms following reverse transcription of RNA into first-strand complementary DNA (cDNA) and subsequent PCR amplification. All three techniques require knowledge of the allele sequences to design the appropriate primers or probes. Different alleles can only be discriminated if there is sequence variation between them. The type of sequence differences among the alleles then dictates which methods are viable options. Examples of primer extension, S1 nuclease protection, and RT-CAPS data are presented based on published (Chen and Pikaard, 1997; Lewis and Pikaard, 2001) and unpublished studies of nucleolar dominance, the epigenetic phenomenon that describes the expression of only one parental set of ribosomal RNA (rRNA) genes in a genetic hybrid (Pikaard, 2000a; Reeder, 1985; Viegas et al., 2002).

Primer Extension Assay for Detection of Allele-Specific Transcripts

Primer extension is a method typically used to map the 5′ end(s) of an RNA, thus defining the transcription start site and providing initial evidence for where the promoter is located within a cloned gene. The method uses a single-stranded DNA primer that is designed to hybridize to the RNA at a position downstream of the transcription start site (defined as +1), preferably within about 200 bp of +1. One then adds RNA-dependent DNA polymerase (reverse transcriptase [RT]) and deoxyribonucleotide triphosphates. RT catalyzes the sequential addition of deoxyribonucleotides to the 3′ hydroxyl group supplied by the oligonucleotide. The oligonucleotide thus acts as a primer for the synthesis of first-strand cDNA, which continues to extend until the 5′ end of the RNA template is reached. As a consequence of this reaction, the relatively short (25–40 nt) primer becomes extended into a longer DNA molecule. By labeling the primer radioactively at its 5′ end, both the initial primer and the final extended product can be resolved and visualized after denaturing gel electrophoresis and exposure to x-ray film or a PhosphorImager screen. The size of the extended product corresponds to the precise distance from the labeled nucleotide to the 5′ end of the RNA. This size can be estimated by comparison to end-labeled size markers run on the same gel. Better yet is to perform a dideoxynucleotide sequencing reaction using the 5′ end–labeled primer and genomic DNA as template. In this manner, one generates a sequencing ladder that is run on the gel adjacent to the primer extension product. The fragment in the sequencing ladder that matches the size of the extension product defines the precise nucleotide that corresponds to +1. In addition to being useful for mapping transcription start sites, primer extension results are quantitative so long as the primer is in excess of the RNA to which it hybridizes, which is easily accomplished. Thus, the amount of radioactive primer extension signal on the gel is directly proportional to the abundance of the target RNA.

Primer extension can be used as an assay to discriminate alleles of two progenitors if the sequences of the two RNAs are different enough that two primers can be designed, each specific for the allele of only one progenitor. However, we find primer extension to be most useful when the sequences of the progenitors' alleles are very similar but have insertions/deletions relative to one another. In this latter scenario, a single primer is used, but distinct primer extension products are generated. The use of a single primer is ideal, because there is no need to worry about possible differences in the specific activity or hybridization efficiency of two different primers.

Figure 1 provides an example of the use of primer extension to map the transcription start sites for rRNA genes of *Brassica oleracea* and *Brassica nigra* and to show that only *B. nigra* rRNA genes are expressed in *Brassica carinata*, the allotetraploid hybrid of *B. oleracea* and *B. nigra*. The rRNA genes of the two progenitor species are very similar in sequence, but at a location 87 bp downstream of the transcription start site, *B. oleracea* has 10 bp inserted relative to *B. nigra* (see diagram in Fig. 1). A 30-nt primer corresponding to positions 158–187 (numbered relative to the *B. nigra* start site) was designed because there are only two nucleotide differences between *B. nigra* and *B. oleracea* in this interval. Both polymorphic nucleotides are located near the 5′ end of the primer where they are unlikely to interfere with hybridization near the 3′ end of the primer where the extension reaction takes place. Indeed, this primer works with equal efficiency for primer extension of both *B. oleracea* and *B. nigra* RNA but yields a longer extension product with *B. oleracea* RNA, as expected (Fig. 1, compare lanes 5 and 7). Side-by-side sequencing reactions performed using the primer with genomic DNA clones of the promoter regions for the two species shows that the sequences at the transcription start sites of *B. oleracea* and *B. nigra* are identical (shown to the side of the photo). Primer extension of RNA isolated from the allotetraploid hybrid *B. carinata* shows that only *B. nigra* rRNA gene transcripts are detected (Fig. 1, lane 6). Collectively, the data indicate that *B. nigra* rRNA genes are expressed and *B. oleracea* rRNA genes are silenced in *B. carinata*, an example of the epigenetic phenomenon known as *nucleolar dominance*.

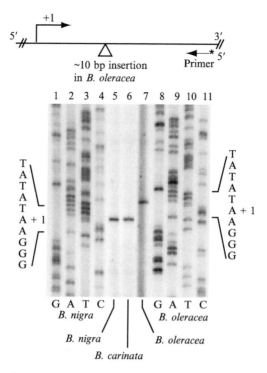

FIG. 1. Use of the primer extension assay to reveal nucleolar dominance in *Brassica carinata*, the allotetraploid hybrid of *Brassica nigra* and *Brassica oleracea*. The diagram illustrates the fact that a single end-labeled primer can hybridize to ribosomal RNA (rRNA) transcripts of either *B. nigra* or *B. oleracea*. Because of an estimated 10-bp insertion in *B. oleracea* relative to *B. nigra,* extension of the primer by reverse transcriptase results in primer extension product that is longer for *B. oleracea* (compare lanes 5 and 7). Primer extension of RNA isolated from *B. carinata* reveals abundant rRNA transcripts from the *B. nigra* genes, but no *B. oleracea* rRNA transcripts can be detected (lane 6). The same end-labeled primer was used to generate sequencing ladders from promoter DNA clones (lanes 1–4, 8–11) allowing the transcription initiation sites (+1) to be accurately mapped to the same sequence motif. Reprinted, with permission, from Chen and Pikaard (1997).

Solutions Needed

A. primer extension reaction mix: 20 mM Tris–HCl (pH 8.7), 10 mM MgCl$_2$, 5 mM DTT (added just before use), 50 μg/ml actinomycin D, 0.5 mM each dNTP (dATP, dCTP, dGTP, TTP), 2–5 units/μl MMLV RT.

B. Formamide gel-loading dye: 80% formamide (deionized ultrapure), 10 mM NaOH (to help degrade RNA), 10 mM EDTA, 0.5 mg/ml xylene cyanol FF dye, 0.5 mg/ml bromophenol blue dye.

Procedure

Primer extension uses a radioactively labeled primer, so take the appropriate precautions for handling radioisotopes. These include wearing gloves, a lab coat, and eye protection and keeping your work behind a radiation shield at all times. Keep a radiation monitor (Geiger counter) at your side and check your hands and pipetter often to ensure they do not become contaminated. As for all assays involving RNA, use sterilized tubes, pipette tips, and solutions to prevent introduction of RNases into the reactions.

1. Radioactive 5′ end labeling of the primer:
   To a 1.7-ml or 0-5 ml plastic Microfuge tube add the following:
   5 μl water (distilled and/or Milli-Q treated)
   1 μl of primer DNA ($\sim$50–70 ng; 25–30 nt long)
   2 μl 660 mM Tris–HCl pH 7.5
   2 μl 100 mM DTT
   2 μl 100 mM MgCl$_2$
   2 μl 10 mM spermidine
   1 μl T4 polynucleotide kinase ($\sim$5 units)
   5 μl gamma-labeled ^{32}P-ATP (10 μCi/μl; 6000 Ci/mmol)

Incubate labeling reaction at 37° for 60 min. Heat reaction at 65° for 15 min to inactivate the kinase. Remove unincorporated ^{32}P-ATP by centrifugation through a Sephadex G50 spin column (these are often marketed for removing unincorporated nucleotide triphosphates from DNA sequencing reactions). Alternatively, one can use ethanol precipitation using ammonium acetate, rather than sodium acetate, as the salt according to standard methods (Sambrook and Russell, 2001).

2. In a fresh Microfuge tube, add the RNA to be tested, dissolved in sterile water (10–40 μg depending on the abundance of the transcript). Add about 10^5 cpm ($\sim$1 ng) of labeled primer. Ethanol precipitate the RNA and primer together by adding 1/10 volume of 3 M sodium acetate pH 5.2 and 2.5 volumes of cold (−20°) ethanol. Incubate on ice 5–10 min, then centrifuge at 14,000g for 15 min. Wash the pellet with 70% ethanol and air-dry the pellet.

3. Resuspend the RNA and primer in 8 μl TE (10 mM Tris, 1 mM EDTA), pH 8.4 by pipetting up and down repeatedly. Add 2 μl of 1.25 M

KCl and mix. Spin tube briefly in a microcentrifuge to collect all of the solution at the bottom of the tube.

4. Bring a water bath to boiling. Turn off the gas and put the tube containing the RNA and primer into the bath to denature the RNA and primer. After about 1–2 min, transfer the tube to a 42° waterbath for 30 min to allow the primer and RNA to anneal and hybridize.

5. Add 24 μl of primer extension reaction mix. Mix the reaction and spin briefly in a microcentrifuge to collect all of the solution at the bottom of the tube.

6. Incubate 1 h at 42° to allow primer extension (cDNA synthesis) to occur.

7. Add 70 μl TE, pH 8.0. Precipitate nucleic acids by addition of 50 μl 7.5 M ammonium acetate and 375 μl cold ($-20°$) ethanol. Incubate on ice 5–10 min and then centrifuge at 14,000g for 15–20 min. Wash pellet with 70% ethanol and air-dry.

8. Dissolve pellet in 6-8 μl of formamide gel-loading dye. Place tubes into a boiling waterbath just after turning off the heat and incubate for 5 min. Chill tubes on ice.

9. Load sample onto a 6–8% polyacrylamide, 8 M urea sequencing gel (Sambrook and Russell, 2001). As controls, load a small aliquot of the labeled primer and load end-labeled size markers or dideoxynucleotide sequencing reaction products.

10. Run the gel until the bromophenol dyes have migrated the desired distance, then pry apart the gel plates, transfer the gel to filter paper, and dry, using a vacuum gel dryer.

11. Expose the dried gel to X-ray film or a PhosphorImager screen to obtain an image analogous to that shown in Fig. 1.

S1 Nuclease Protection Assay for Detecting Allele-Specific Transcripts

The S1 nuclease protection assay allows accurate estimations of transcript abundance and has the ability to distinguish RNA transcripts that sometimes differ by only a few nucleotide substitutions (Berk and Sharp, 1977). In this method, equal aliquots of RNA are hybridized with radioactively 5' end–labeled DNA probes specific for the parental/progenitor alleles to be assayed. The probes are typically made from genomic clone restriction fragments spanning the transcription start site (as diagrammed in Fig. 2), but long oligonucleotides (typically >50 nt) can also be used. The DNA probes are designed so they include short regions, often only several nucleotides long, where the nucleotide sequence is different in the alleles to be discriminated. A probe molecule that hybridizes to a

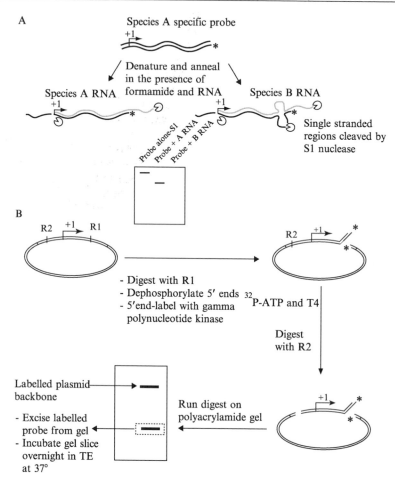

FIG. 2. The S1 nuclease protection assay and probe preparation. (A) Diagrammatic representation of the specificity of the S1 nuclease protection assay. A double-stranded DNA fragment from species A is 5′ end–labeled on the bottom strand (*), complementary to the RNA, at a position of about 50–300 bp downstream from the transcription start site (+1). The probe is then denatured in the presence of formamide and RNA from species A or species B. Upon hybridization with RNA from species A, a perfect match between the probe and RNA occurs from the 5′ end of the RNA to the labeled end of the probe. Single-stranded portions of the DNA probe extend upstream of the RNA 5′ end, and single-stranded RNA extends beyond the labeled end of the probe. S1 nuclease digests these single-stranded nucleic acids, leaving the RNA–DNA duplex intact. As a result, the initial labeled strand of the probe becomes shorter and migrates faster on a denaturing sequencing gel, depicted at the bottom of panel A. By contrast, hybridization of the probe from species A with RNA from species B, which does not result in a perfect hybrid, results in looped out single-stranded regions that are digested with S1 nuclease to produce labeled fragments not detected on the subsequent gel.

transcript for which it is an exact match forms a perfectly base-paired RNA–DNA duplex that is resistant to digestion by S1 nuclease (Fig. 2A). By contrast, hybridization of the probe to a homologous but nonidentical RNA transcribed from a different allele will form a duplex that leaves single-stranded loops at sites of sequence mismatch. These single-stranded regions are cleaved by S1 nuclease (Fig. 2A). Placement of the 5' radioactive end-label near such regions of nonhomology ensures that any resulting labeled digestion products are small enough that they migrate off the bottom of a sequencing gel.

DNA probes used for S1 nuclease protection are designed such that they begin upstream of the transcription start site and extend 50–300 nucleotides downstream of the start site. Upon hybridizing to RNA initiated at the start site, the DNA probe overhangs the 5' end of the RNA and the RNA extends beyond the downstream end of the probe (see Fig. 2A). S1 nuclease cleaves these single-stranded DNA and RNA overhangs, and any single-stranded bubbles within a non-perfect hybrid but does not digest the double-stranded regions. As a result, the initial labeled DNA probe becomes shorter in a successful S1 nuclease protection assay because of digestion of the single-stranded DNA that extends upstream of the RNA's 5' end.

We employ two methods for generating probes for the S1 nuclease protection assay. The simplest method is to use long (55–70 nt) gel-purified DNA oligonucleotides as probes. These oligos are designed to be complementary to the RNA and extend upstream of the transcription start site by about 10 nt, leaving about 45–60 nt to form a DNA–RNA hybrid that is quite stable. The oligos are 5' end–labeled using T4 kinase and can be used directly in the hybridization reaction, although gel purification is generally a good idea to eliminate smaller oligonucleotides in the synthesis reaction that will also hybridize to the target RNA. Despite the ease of labeling oligonucleotides to generate probes, our preferred method is to use restriction

---

(B) Probe preparation from a genomic DNA fragment spanning the promoter region, cloned within a plasmid. Based on the known sequence of the genomic clone and plasmid, two restriction sites (R1 and R2) are chosen which, ideally, are unique in the plasmid clone. Site R1 is typically located about 50–300 bp downstream from the transcription start site (+1). The plasmid is linearized at site R1, dephosphorylated, and then 5' end–labeled on both the top and bottom strands (*). The plasmid is then cut at site R2, which liberates the desired promoter fragment from the remainder of the plasmid. The promoter fragment is labeled only on the bottom strand, which is complementary to the RNA strand. The desired promoter fragment is then gel purified and allowed to diffuse out of a gel slice into TE for use in the S1 nuclease protection assay.

fragments that include the transcription start site and are isolated from genomic clones in plasmid vectors (see Fig. 2B). We find that the longer probes generated from a cloned fragment tend to generate cleaner results than from a synthetic oligonucleotide.

To make probes from cloned genomic DNA, the plasmid is first cut at a restriction endonuclease site typically chosen to be 50–300 bp downstream from the transcription start site (see Fig. 2B). Ideally, this site is unique in the clone and leaves a 5′ overhang (sticky end). The cut DNA is then dephosphorylated with alkaline phosphatase and is 5′ end–labeled using T4 polynucleotide kinase and γ-^{32}P-ATP. The plasmid is then cut with a second restriction enzyme at a site upstream of the transcription start site, thus liberating the labeled fragment from the plasmid, which is also labeled. The desired fragment that includes the transcription start site is then gel purified, using a 5% non-denaturing polyacrylamide gel (Fig. 2B). Although this probe fragment is double stranded, only the strand complementary to the RNA is labeled. If desired, one can isolate only this labeled strand after NaOH denaturation of the duplex and loading onto a long, strand-separating gel (Sambrook and Russell, 2001). However, we typically find the use of strand-separating gel electrophoresis to be unnecessary because by denaturing the double-stranded DNA fragment in the presence of the RNA to be tested and allowing hybridization to take place in the presence of a high concentration of formamide, RNA–DNA hybrid molecules are favored over the DNA–DNA duplexes that would result from reannealing of the probe fragment. Resulting RNA–DNA hybrids are then treated with S1 nuclease and the digestion products are resolved on a sequencing gel and visualized by autoradiography. As with the primer extension assay, end-labeled size markers are run on the gel as controls adjacent to the S1 digestion products. Alternatively, an end-labeled primer (25–30 nt) whose labeled 5′ end corresponds precisely to the labeled 5′ end of the S1 probe can be used to generate a dideoxynucleotide sequencing ladder. The sequence in the ladder corresponding to the size of the S1-protected probe fragment defines the transcription start site.

An example of the use of the S1 nuclease protection assay to discriminate rRNA transcripts from two parents of a hybrid is shown in Fig. 3. rRNAs of *A. lyrata* and *A. thaliana* are highly conserved in sequence but have occasional nucleotide substitutions, including short intervals of 2–3 polymorphic nucleotides. Thus, the S1 nuclease protection assay can be used to discriminate and quantify the relative expression levels of parental rRNA genes in F1 hybrids of *A. thaliana* and *A. lyrata*. Using probes derived from PCR-generated genomic clones, one specific for *A. thaliana* and one specific for *A. lyrata*, the experiment revealed that F1 hybrids

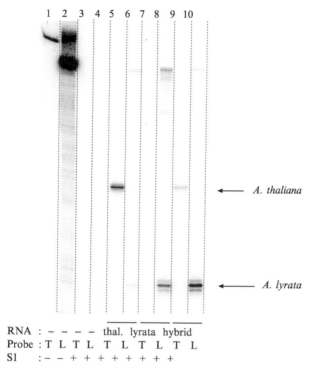

FIG. 3. Use of the S1 nuclease protection assay to reveal nucleolar dominance in F1 hybrids of *Arabidopsis thaliana* × *Arabidopsis lyrata*. Twenty micrograms of *A. thaliana* (T), *A. lyrata* (L), or F1 hybrid total RNA was hybridized with S1 probes specific for *A. thaliana* or *A. lyrata* ribosomal RNA (rRNA) transcripts. After S1 nuclease digestion and electrophoresis through a sequencing gel, the gel was dried and exposed to a PhosphorImager screen to produce the image shown. Controls include aliquots of the probes, to help recognize bands corresponding to undigested probe in the experimental lanes (lanes 1 and 2); mock reactions that include probe but not RNA, to control for possible S1 digestion products not derived from DNA–RNA hybrids (lanes 3 and 4); and tests of the two probes with RNA from the two parents, to demonstrate the parent specificity of the probes and to determine whether their specific activities are similar (lanes 5–8). The two experimentally meaningful lanes (lanes 9 and 10) test RNA purified from an *A. thaliana* × *A. lyrata* F1 hybrid. Comparison of these two lanes reveals low levels of *A. thaliana* rRNA and 10-fold higher levels of *A. lyrata* rRNA transcripts in the hybrid.

express the *A. lyrata* transcripts at about 10-fold higher levels than *A. thaliana* transcripts (Fig. 3, compare lanes 9 and 10), providing another example of nucleolar dominance.

Solutions Needed

   A. Probe-RNA hybridization buffer: 40 mM PIPES (pH 6.4), 400 mM NaCl, 1 mM EDTA, 80% deionized formamide

   B. S1 digestion buffer: 5% glycerol, 1 mM zinc sulfate, 30 mM sodium acetate (pH 4.5), 50 mM NaCl

   C. Formamide-based gel loading buffer (see the section "Primer Extension Assay for Detection of Allele-Specific Transcripts," earlier in this chapter).

S1 Probe Preparation

The S1 nuclease protection assay uses radioactive probes, so take the appropriate precautions for the safe handling of radioisotopes, some of which were mentioned previously in the section "Primer Extension Assay for Detection of Allele-Specific Transcripts."

1. Cut about 2 μg of plasmid DNA containing the cloned promoter region of interest, using 5–10 units of a restriction endonuclease that cleaves downstream from the transcription start site.

2. Dephosphorylate the 5′ ends by addition of 5–10 units of shrimp alkaline phosphatase and incubation at 37° for 60 min. Heat-inactivate the shrimp alkaline phosphatase at 65° for 20 min.

3. Precipitate the dephosphorylated DNA by addition of 1/10 volume of 3 M sodium acetate (pH 5.2) and 2 volumes of cold ($-20°$) absolute ethanol. Mix by vortexing, store on ice 5 min, and centrifuge at 14,000g for 15 min. Wash the DNA pellet with 70% ethanol, dry the pellet, and resuspend in 6 μl of sterile water or TE, pH 8.0. (Note that steps 1–3 can be scaled up 10-fold to yield enough dephosphorylated DNA for 10 or more experiments.)

4. End-label the DNA using T4 polynucleotide kinase and γ-^{32}P-ATP.

     6 μl dephosphorylated DNA (2 μg)

     2 μl 660 mM Tris–HCl pH 7.5

     2 μl 100 mM DTT

     2 μl 100 mM MgCl$_2$

     2 μl 10 mM spermidine

     1 μl T4 polynucleotide kinase ($\sim$5 units)

     5 μl gamma-labeled ^{32}P-ATP (10 μCi/μl; 6000 Ci/mmol)

Incubate the labeling reaction at 37° for 60 min. Heat the reaction at 65° for 15 min to inactivate the kinase. Precipitate the labeled DNA by addition of 1/10 volume 3 M sodium acetate (pH 5.2) and 2 volumes of cold ($-20°$) absolute ethanol. Mix by vortexing, store on ice 5 min, and centrifuge at 14,000g for 15 min. Wash the DNA pellet with 70% ethanol, dry the pellet, and resuspend in 20 μl of 1× restriction endonuclease buffer (the buffer depends on the enzyme that will be used to cut the plasmid upstream of the transcription start site). Add 5–10 units of the restriction endonuclease and incubate at the appropriate temperature for 60 min.

5. Gel purify the probe DNA fragment by electrophoresis through a 5% native polyacrylamide gel. Pry apart the glass plates, cover the gel with plastic wrap, and expose the gel to X-ray film for 5–10 min, being sure to align one corner of the film with one corner of the gel plate inside the film holder. Lay the exposed film back down onto the gel, aligning the film as during the exposure step, and use a razor blade to trace the outline of the desired probe fragment, slicing through the film and the gel. Using forceps, transfer the radioactive gel slice to a 1.7-ml microcentrifuge tube and add 500 μl TE, pH 8.0. Incubate at 65° for 4 h, or overnight at 37°, to allow the labeled DNA to diffuse out of the gel slice.

6. Vortex the tube containing the gel slice in TE. Remove about 1/10 of the TE for each S1-protection reaction to be performed and pipette this into a fresh tube. Add 20–50 μg total RNA. Precipitate the RNA and probe together by addition of 1/10 volume 3 M sodium acetate (pH 5.2) and 2.5 volumes of cold (−20°) absolute ethanol. Vortex, store on ice 5–10 min, and centrifuge at 14,000g for 15–20 min. Wash the DNA pellet with 70% ethanol and then dry the pellet.

7. Resuspend the RNA and probe thoroughly in 30 μl of probe–RNA hybridization buffer. Add 50 μl of light mineral oil to prevent the reaction from evaporating during hybridization. Incubate the hybridization reaction at 90° in dry-bath incubator (or in a boiling waterbath after turning off the heat) for 15 min to denature the probe and RNA.

8. Carry the hot metal block of the dry-bath incubator to a waterbath set at the proper hybridizing temperature (generally 37–55°, depending on GC content and the length of hybrid that can form; usually, 37° works fine). Quickly move the tube(s) from the hot block to the waterbath.

9. Allow hybridization to proceed 2 h to overnight.

10. Open the tubes without removing them from the waterbath and add 270 μl of S1 digestion buffer containing about 200 units S1 nuclease/ml (S1 is added just before use). The amount of S1 nuclease can be increased in future experiments if necessary to achieve complete digestion.

11. Spin 5 s in a Microfuge tube to get the S1 digestion mix below the layer of mineral oil. Vortex briefly to mix and repeat the spin to get the thin layer of oil at the top again.

12. Incubate the S1 digestion reaction 30–45 min at 37°.

13. Stop the reactions by removing 280 μl from the bottom of each reaction tube (i.e., avoid paraffin oil) to a fresh tube containing 10 μl of 10%

sodium dodecyl sulfate (SDS) and 5 μl of 0.5 M ethylenediaminetetraacetic acid (EDTA) (the EDTA chelates the zinc required for S1 nuclease activity). Vortex briefly to mix.

14. Add 30 μl of 7.5 M ammonium acetate and vortex to mix. Add 1 ml cold ($-20°$) absolute ethanol to precipitate the DNA–RNA hybrid and vortex. Store on ice 5–10 min and then centrifuge at 14,000g for 15–20 min.

15. Carefully remove supernatant (pellets may not be visible) and discard in radioactive waste container. Wash pellet with 70% ethanol and dry pellets under vacuum (or in 65° waterbath with tube caps open).

16. Resuspend the pellets in 6–8 μl of formamide-based sequencing gel-loading buffer and conduct denaturing gel electrophoresis and autoradiography as described for primer extension.

RT-CAPS Assay of Allele-Specific Transcripts

Reverse transcription of RNA, followed by CAPS analysis, exploits polymorphic restriction sites within the transcribed portions of different alleles to discriminate their transcripts. Because the only difference needed to differentiate sequences is a single nucleotide polymorphism, this method can sometimes be used when there are insufficient sequence differences for S1 nuclease protection or primer extension assays to be employed. RT-CAPS requires smaller quantities of RNA than S1 nuclease protection or primer extension and is generally less time consuming, especially when the number of samples to be assayed is large.

An example of the use of RT-CAPS is the analysis of nucleolar dominance in *A. suecica*, the allotetraploid hybrid of *A. thaliana* and *A. arenosa*. A single nucleotide polymorphism in internal transcribed spacer sequence 1 (ITS1) results in a polymorphic *Hha* I restriction endonuclease site (Fig. 4) that is present in *A. arenosa* rRNA genes but missing in *A. thaliana* rRNA genes. If one performs simple CAPS analysis using genomic DNA as the template for PCR, the *Hha* I restriction fragment patterns for the two parental species are different (Fig. 4, lanes 1 and 2). The *Hha* I fragment pattern for the hybrid is the sum of the parental patterns, indicating that both parental sets of rRNA genes are present in *A. suecica* (lane 3). If RNA from the parental species is converted into cDNA using RT and then subjected to CAPS, the same fragment patterns seen using genomic DNA are observed (lanes 4 and 5), as expected. However, RT-CAPS performed using RNA of the allotetraploid hybrid, *A. suecica* reveals the banding pattern expected for *A. arenosa* rRNA transcripts but only trace amounts of the *A. thaliana* specific band is detected (lane 6), indicating

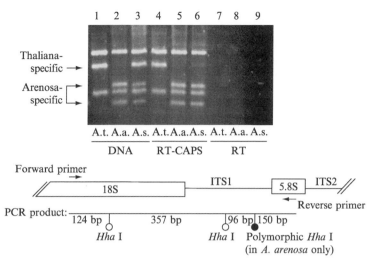

Fig. 4. Use of the reverse-transcription–cleaved amplified polymorphic sequence (RT-CAPS) assay to show that *Arabidopsis thaliana* ribosomal RNA (rRNA) genes are repressed and *Arabidopsis arenosa* rRNA genes are dominant in the allotetraploid hybrid *Arabidopsis suecica*. Genomic DNA (lanes 1–3) or reverse-transcribed (RT) total RNA (lanes 4–6) was amplified by polymerase chain reaction (PCR) using primers flanking internal transcribed spacer 1 (ITS1; see diagram) and then cleaved using the restriction endonuclease *Hha* I. An extra (polymorphic) *Hha* I site in ITS1 of *A. arenosa* rRNA genes allows *A. arenosa* (A.a.) and *A. thaliana* (A.t.) genes and their transcripts to be discriminated in *A. suecica* (A.s.). Both progenitors' rRNA genes are present in *A. suecica* (lane 3), but only *A. arenosa* rRNA gene transcripts are abundant in the hybrid (lane 6). Controls in which RNA samples were not incubated with reverse transcriptase before PCR show that the RNA samples are free of contaminating DNA (lanes 7–9). Reprinted, with permission and minor modifications, from Lewis and Pikaard (2001).

that the *A. thaliana* rRNA genes are mostly silenced. As controls, mock reactions that include RNA but no reverse transcriptase are performed to be sure that contaminating genomic DNA is not contributing to the signals (lanes 7–9).

Reagents Needed

   A. RQ1 DNAse reaction buffer and stop solution (Promega).
   B. RT kit (e.g., Invitrogen Superscript kits).
   C. 10X PCR buffer (various commercial sources).
   D. Restriction endonuclease and buffer (various commercial sources).

Procedure

1. Isolate total RNA from both the hybrid (or allopolyploid) and the parental species. Treat 500 ng of total RNA from each species with RNase-free DNAse (we use Promega RQ1 DNAse) in a 10-μl reaction to eliminate contaminating genomic DNA. Incubate for 30 min at 37°. Stop the reaction with 1 μl of RQ1 stop solution. Incubate for 15 min at 65° to inactivate the enzyme.

2. Add 5 μl of the DNAse-treated RNA to a 20 μl RT reaction using 125 ng of random hexamer oligonucleotide primers (New England BioLabs) to prime cDNA synthesis. For the RT reaction itself, we use the reagents supplied in the Superscript Reverse Transcriptase kit (Invitrogen). As a negative control, add 5 μl of DNAse-treated RNA to a 20 μl reaction but omit the RT. Incubate the RT reaction for 50 min at 42°. Stop the reaction by heating at 85° for 5 min.

3. Perform a 50 μl PCR containing the following:
   1-2 μl of RT reaction product (or –RT control)
   5 μl 10× PCR buffer
   0.2 mM each dNTP (dATP, dCTP, dGTP, TTP)
   2.5 mM MgCl$_2$
   10–20 pmol each of the forward and reverse primers
   2 units Taq polymerase
Perform 25–35 cycles of PCR using conditions appropriate for the primers chosen.

4. Digest 20 μl of the PCR with the restriction enzyme that will reveal the polymorphism. It may be necessary to first ethanol precipitate the PCR products if the PCR buffer is incompatible with the restriction endonuclease to be used.

5. Analyze digestion products on a 1–2% agarose gel and visualize DNA bands by staining with ethidium bromide or an alternative reagent.

Concluding Remarks

Of the three techniques described in this chapter, we tend to use S1 nuclease protection most frequently. Despite being the most laborious of the three assays, the S1 nuclease protection assay consistently yields the most reliable quantitative data. Presumably, this is due to the longer, more stable probe–RNA hybrids that are formed, compared to the primer–RNA hybrids formed for the primer extension assay. Furthermore, the S1 nuclease protection assay does not require signal amplification as in the PCR-based RT-CAPS method, which can lead to skewed results. In our hands, RT-CAPS analysis typically makes differences in allele expression appear smaller than they really are. For this reason, we strongly advise that any differences in allele expression found between two alleles using the RT-CAPS method be confirmed using a second technique.

For additional discussion of the S1 nuclease protection assay, primer extension assay, and RT-PCR, including detailed protocols that may vary slightly from ours, we highly recommend the *Molecular Cloning Laboratory Manual* (Sambrook and Russell, 2001).

Acknowledgments

The experiments and assays described in this chapter were made possible by grants from the National Institutes of Health (R01-GM60380), the National Science Foundation (DMB-9018428), and the USDA National Research Initiative Competitive Grants Program (94-373012-0658). Any opinions, findings, and conclusions or recommendations expressed in this material are those of the author(s) and do not necessarily reflect the views of the National Institutes of Health, National Science Foundation, or USDA. S. P. is supported by a National Research Service Award from the National Institutes of Health (1 F32 GM69206).

References

Avner, P., and Heard, E. (2001). X-chromosome inactivation: Counting, choice, and initiation. *Nat. Rev. Genet.* **2,** 59–67.

Berk, A. J., and Sharp, P. A. (1977). Sizing and mapping of early adenovirus mRNAs by gel electrophoresis of S1 endonuclease-digested hybrids. *Cell* **12,** 721–732.

Boorstein, W. R., and Craig, E. A. (1989). Primer extension analysis of RNA. *Methods Enzymol.* **180,** 347–369.

Brannan, C. I., and Bartolomei, M. S. (1999). Mechanisms of genomic imprinting. *Curr. Opin. Genet. Dev.* **9,** 164–170.

Chen, Z. J., and Pikaard, C. S. (1997). Transcriptional analysis of nucleolar dominance in polyploid plants: Biased expression/silencing of progenitor rRNA genes is developmentally regulated in *Brassica. Proc. Natl. Acad. Sci. USA* **94,** 3442–3447.

Huynh, K. D., and Lee, J. T. (2001). Imprinted X inactivation in eutherians: A model of gametic execution and zygotic relaxation. *Curr. Opin. Cell Biol.* **13,** 690–697.

Konieczny, A., and Ausubel, F. M. (1993). A procedure for mapping *Arabidopsis* mutations using co-dominant ecotype-specific PCR-based markers. *Plant J.* **4,** 403–410.

Lewis, M. S., and Pikaard, C. S. (2001). Restricted chromosomal silencing in nucleolar dominance. *Proc. Natl. Acad. Sci. USA* **98,** 14536–14540.

Pikaard, C. S. (2000a). The epigenetics of nucleolar dominance. *Trends Genet.* **16,** 495–500.

Pikaard, C. S. (2000b). Nucleolar dominance: Uniparental gene silencing on a multi-megabase scale in genetic hybrids. *Plant Mol. Biol.* **43,** 163–177.

Reeder, R. H. (1985). Mechanisms of nucleolar dominance in animals and plants. *J. Cell Biol.* **101,** 2013–2016.

Sambrook, J., and Russell, D. R. (2001). "Molecular Cloning: A Laboratory Manual," 3rd Ed. Cold Spring Harbor Laboratory Press, Cold Spring Harbor, NY.

Sleutels, F., Barlow, D. P., and Lyle, R. (2000). The uniqueness of the imprinting mechanism. *Curr. Opin. Genet. Dev.* **10,** 229–233.

Viegas, W., Neves, N., Caperta, A., Silva, M., and Morais-Cecílio, L. (2002). Nucleolar dominance: A 'David and Goliath' chromatin imprinting process. *Curr. Genomics.* **3,** 563–576.

[30] Methods for Genome-Wide Analysis of Gene Expression Changes in Polyploids

By Jianlin Wang, Jinsuk J. Lee, Lu Tian, Hyeon-Se Lee,
Meng Chen, Sheetal Rao, Edward N. Wei, R. W. Doerge,
Luca Comai, and Z. Jeffrey Chen

Abstract

Polyploidy is an evolutionary innovation, providing extra sets of genetic material for phenotypic variation and adaptation. It is predicted that changes of gene expression by genetic and epigenetic mechanisms are responsible for novel variation in nascent and established polyploids (Liu and Wendel, 2002; Osborn *et al.*, 2003; Pikaard, 2001). Studying gene expression changes in allopolyploids is more complicated than in autopolyploids, because allopolyploids contain more than two sets of genomes originating from divergent, but related, species. Here we describe two methods that are applicable to the genome-wide analysis of gene expression differences resulting from genome duplication in autopolyploids or interactions between homoeologous genomes in allopolyploids. First, we describe an amplified fragment length polymorphism (AFLP)–complementary DNA (cDNA) display method that allows the discrimination of homoeologous loci based on restriction polymorphisms between the progenitors. Second, we describe microarray analyses that can be used to compare gene expression differences between the allopolyploids and respective progenitors using appropriate experimental design and statistical analysis. We demonstrate the utility of these two complementary methods and discuss the *pros* and *cons* of using the methods to analyze gene expression changes in autopolyploids and allopolyploids. Furthermore, we describe these methods in general terms to be of wider applicability for comparative gene expression in a variety of evolutionary, genetic, biological, and physiological contexts.

AFLP-cDNA Display

Introduction

Amplified fragment length polymorphism (AFLP) (Vos *et al.*, 1995) is used to develop DNA markers for genome mapping and genetic diversity analysis. The method can be used to analyze simultaneously different DNA regions throughout the entire genome regardless of its origin or complexity

(Mei *et al.*, 2004). A modification of the method using complementary DNA (cDNA) samples in the analysis, known as AFLP-cDNA or cDNA-AFLP display, allows the characterization of tissue-specific gene expression patterns (Bachem *et al.*, 1996) and detection of gene expression differences in allopolyploids (Comai *et al.*, 2000; Lee and Chen, 2001; Madlung *et al.*, 2002). The technique employs the use of restriction polymorphism in analyzing cDNA samples (Fig. 1A). After the cDNA fragments are digested, they are ligated onto adapters with compatible restriction ends. Ligated products are subjected to polymerase chain reaction (PCR) analysis using two steps of preselective and selective amplification so that subsets of transcripts are amplified and resolved in a sequencing gel (Sambrook *et al.*, 1989). Any fragments that are present in one sample but not the other are isolated and sequenced (Lee and Chen, 2001). The method is particularly useful to detect expression changes in orthologous and homoeologous loci that display polymorphism in the restriction sites of the enzymes used in the assay. Compared with the traditional messenger RNA (mRNA) display (Liang and Pardee, 1992) and random arbitrarily primed PCR (RAP-PCR) (Welsh *et al*, 1992), the AFLP-cDNA display is relatively sensitive and reproducible. The technique is sensitive enough to detect weakly expressed transcripts, such as genes encoding transcription factors that are undetectable by Northern blot analysis (Lee and Chen, 2001). Moreover, it does not detect as many false positives as observed in regular mRNA display. Finally, it can be used for quantitative detection because the intensity of each fragment reflects the expression level of the gene relative to internal controls (e.g., constantly expressed cDNAs), allowing measurement of dosage-dependent gene regulation in autoploidy series.

Another advantage of using AFLP-cDNA display is that there is no prerequisite for the DNA sequence information. Novel transcripts are readily detectable. The origin of the expressed homoeologous transcripts may be determined by comparing the polymorphic cDNA fragments between the allotetraploids and their progenitors. Moreover, the method can be applied to any organisms, especially those whose genomic resources are underdeveloped. By comparing the sequences of the cloned fragments with known sequences in the databases, one might identify homologous genes in the related species. However, the efficiency of AFLP-cDNA display is limited by *post hoc* cloning and sequencing of individual cDNA fragments. False positives may be detected because of PCR artifacts or heterogeneous populations (e.g., outcrossing plants). The comparison of gene expression patterns (absence or presence of a fragment) between an allopolyploid and its progenitors may fail to distinguish gene expression differences because of allelic variation among individual plants within a progenitor species.

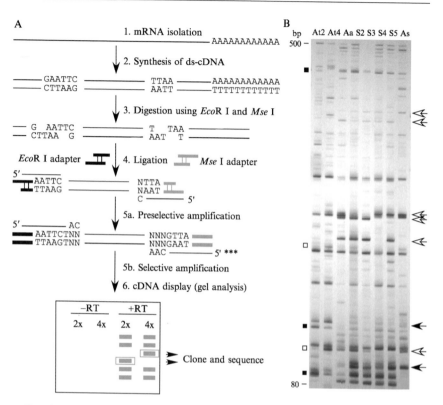

Fig. 1. Amplified fragment length polymorphic complementary DNA (AFLP-cDNA) display analysis of gene expression. (A) A simple diagram of the AFLP-cDNA display procedures including messenger RNA (mRNA) isolation, cDNA synthesis, restriction digestion, ligation, preselective and selective amplification, and gel electrophoresis. The detailed protocols are described in the text. Only a few hypothetical fragments are shown to illustrate how *EcoR* I (GAATTC)- and *Mse* I (TTAA)–digested fragments are ligated and amplified during the analysis. (B) AFLP-cDNA display analysis using leaves of *Arabidopsis thaliana* diploid (At2), autotetraploid (At4), *Arabidopsis arenosa* (Aa), and generations 2–5 (S2-5) of new allotetraploids, and a natural *Arabidopsis suecica* (As) line (Wang *et al.*, 2004). The primers used in selective amplification are E-TA and M-CTG. Open and filled squares indicate that the genes are up-regulated and down-regulated, respectively, in the isogenic autotetraploid (At4). Arrows indicate gene silencing or activation (filled) and random changes in gene expression (open) in different generations of the allotetraploids, respectively. Negative controls using mRNA samples without reverse transcriptase in the reaction (Lee and Chen, 2001; Wang *et al.*, 2004) are not shown because of space limitations.

Therefore, to reduce the number of false positives, it is recommended to use biological replications and to verify the gene expression patterns detected in AFLP-cDNA display by other methods such as reverse-transcriptase PCR (RT-PCR) and cleaved amplified polymorphic sequence (CAPS), quantitative RT-PCR, sequencing, and/or SSCP (Adams *et al.*, 2003; Lee and Chen, 2001; Lee *et al.*, 2004), and other methods as described in Chapter 29.

Protocol

The protocol for AFLP-cDNA display is modified from the published methods (Bachem *et al.*, 1996; Comai *et al.*, 2000; Lee and Chen, 2001) and recommendations from the manufacturer (Invitrogen, Carlsbad, CA). The protocol includes six steps (Fig. 1A). They are (1) isolate mRNA; (2) synthesize double-stranded (ds) cDNA; (3) generate cDNA fragments by restriction digestion; (4) ligate the cDNA fragments onto adapters that contain the restriction enzyme sites; (5) amplify subsets of cDNA fragments using two-step (preselective and selective) amplification; and (6) display the cDNA fragments in a sequencing gel, clone, and sequence the cDNA fragments that are differentially expressed between the two samples. In this experiment (Fig. 1B), we prepared RNA samples from an *Arabidopsis thaliana* diploid (At), an autotetraploid (At4), *Arabidopsis arenosa* autotetraploid, four selfing generations (S2–S5) of a new allotetraploid (745), and natural allotetraploid *Arabidopsis suecica* (As), to demonstrate the detection of changes of gene expression in diploid and isogenic autotetraploid and allotetraploid lines. In addition to the results shown in Fig. 1B, detailed analyses of gene expression patterns in these lines are described elsewhere (Wang *et al.*, 2004).

Isolation of Total RNA

All reagents are purchased from Invitrogen, unless noted otherwise. Every solution is made by diethylpyrocarbonate (DEPC)-treated double-deionized autoclaved water (ddH$_2$O) (Sambrook *et al.*, 1989). For each line, total RNA is extracted from fresh leaves or flower buds of pools of 10–20 plants (3–4 weeks old) using the TRIzol reagent (Invitrogen). In a typical preparation, about 500 μg of total RNA can be obtained from 1 g of fresh leaves or flower buds in *Arabidopsis* and *Brassica*.

Grind the plant samples (1 g) in liquid nitrogen and transfer the fine powders to a 30-ml centrifuge tube. Add 10 ml of TRIzol reagent, mix well by vortexing for 20 s, and incubate in a waterbath at 65° for 5 min. After cooling, add 2 ml of chloroform and mix vigorously for 20 s. Incubate at room temperature (RT) for 5 min followed by centrifugation at 10,000g for 15 min at 4°. Transfer the supernatant to a 50-ml polypropylene conical tube, add about 1 volume of isopropyl alcohol, and incubate at RT for 45 min. Centrifuge at 3000 rpm for 15 min at 4° to precipitate RNA. Wash the RNA pellet twice using 70% ethanol, briefly dry the RNA pellet, and dissolve the pellet in 300 μl of DEPC-treated ddH$_2$O or RNase-free TE buffer (10 mM Tris–HCl and 1 mM ethylenediaminetetraacetic acid [EDTA], pH 8.0).

Isolation of mRNA

mRNA is isolated using the Micro-FastTrack 2.0 mRNA isolation kit (Invitrogen). Approximately 5 μg of poly(A) RNA can be purified from about 500 μg of total RNA using 25 mg of oligo-(dT) cellulose.

Add 1 ml of binding buffer (500 mM NaCl, 10 mM Tris–Cl, pH 7.5) to the total RNA solution, heat it at 65° for 5 min, and place it immediately on ice for exactly 1 min.

Transfer the solution to a 2-ml microcentrifuge tube containing 25 mg of oligo-(dT) cellulose and incubate at RT for 2 min. Then gently rotate the tube for 0.5–1.0 h at RT to increase the efficiency of mRNA binding to oligo-(dT) cellulose. Centrifuge at 4000g for 5 min at RT and carefully remove the supernatant (do not disturb the cellulose pellet).

Resuspend the oligo-(dT) cellulose in 1.3 ml of binding buffer and centrifuge at 4000g for 5 min and carefully remove the supernatant. This step is repeated (at least three times) until the buffer is no longer cloudy.

Resuspend the oligo-(dT) cellulose in 0.3 ml of binding buffer and transfer it to a spin column. Centrifuge at 4000g for 10 s at RT and carefully remove the supernatant. Repeat this washing procedure using 0.5 ml of binding buffer for at least three times until the OD$_{260}$ of the flow through is 0.05 or less.

To the oligo-(dT) cellulose, add 200 μl of low-salt washing buffer (250 mM NaCl, 10 mM Tris–Cl, pH 7.5) and gently mix. Centrifuge the column and remove the liquid as described earlier. Repeat washing with additional 200 μl of low-salt washing buffer for at least three times. Washing using low-salt buffer helps remove sodium dodecyl sulfate (SDS) and non-polyadenylated RNAs such as rRNAs.

Place the spin column into a new microcentrifuge tube, add 100 μl of elution buffer, and mix gently by pipetting. Centrifuge 10 s at RT. Repeat the procedure using another 100 μl of elution buffer. The final volume of mRNA in the elution buffer is 200 μl.

Quantify mRNA in a 1:1 dilution (5 μl of mRNA sample and 5 μl of elution buffer) using a spectrophotometer (GeneQuant *pro*, Amersham Bioscience, Piscataway, NJ). Use 10 μl of elution buffer to calibrate the spectrophotometer.

To the about 200 μl of mRNA solution, add 10 μl of 2 mg/ml of glycogen, 30 μl of 3 M Na-acetate, 600 μl of 100% ethanol, and mix. Store at $-70°$ until use.

Place the samples at RT for 15 min. Centrifuge at 14,000 rpm for 20 min at 4°. Remove the ethanol and wash the pellet with 80% ethanol. Centrifuge briefly, remove ethanol, and dissolve the mRNA pellet in the elution buffer to a final concentration of 100 ng/μl. The mRNA can be used immediately or stored at $-70°$.

First-Strand cDNA Synthesis

The cDNAs are synthesized using a slightly modified procedure from that provided by the universal RiboClone cDNA synthesis system (Promega, Madison, WI). To synthesize the first-strand cDNA, add 2 μl oligo-(dT)$_{12-18}$ (500 μg/ml) and 0.5–1.0 μg of mRNA to a 0.5-ml RNase-free microcentrifuge tube. Adjust the volume to 15 μl using nuclease-free ddH$_2$O. Heat the reaction mixture at 70° in a thermal cycler for 10 min and quickly chill the tube on ice. After brief centrifugation, add 5 μl of first-strand 5× buffer, 1 μl of RNasin ribonuclease inhibitor (40 U/μl), 2.5 μl of sodium pyrophosphate (40 mM), and 2.5 μl of AMV reverse transcriptase (24 U/μl) to each tube. Mix the solution by gently pipetting up and down several times and incubate at 42° for 1.5 h.

Second-Strand cDNA Synthesis

The procedure is performed immediately after the previous reaction. Add 40 μl of second-strand 2.5× buffer, 2.5 μl of DNA polymerase I (9 U/μl), 0.8 μl of RNase H (2 U/μl), 5 μl of acetylated bovine serum albumin (BSA) (1 mg/ml), and 26.7 μl of nuclease-free ddH$_2$O. Gently mix the solution in a total volume of 100 μl and incubate at 14° for 2 h. Heat this reaction to 70° for 10 min and place it on ice. Add 1 μl of T4 DNA polymerase (7.7 U/μl) and incubate it at 37° for 10 min. The reaction is terminated by the addition of 4 μl of 200 mM EDTA. Place the tube on ice or store at $-20°$.

The size of cDNA should be examined after the second-stranded cDNA synthesis. Take 5 μl of the reaction solution (and 0.5 μl 10× loading buffer) and load it onto a 1% agarose gel. After electrophoresis, the DNA can be stained using SYBR gold dye (Molecular Probes, Inc., Eugene, OR) and visualized under ultraviolet (UV) light. A successful reaction generates a smear of DNA with a range of 500–4000 bp in length.

The cDNAs are purified by phenol–chloroform extraction and ethanol precipitation (Sambrook *et al.*, 1989). To the sample tube, add an equal volume (~100 μl) of phenol:chloroform:isoamyl alcohol (25:24:1) saturated with TE buffer (pH 8.0) and mix it by vortexing. Centrifuge the tubes at 14,000 rpm in a microcentrifuge tube for 2 min and transfer the aqueous phase of the sample to a new tube. Add 0.1 volume of 2.5 M sodium acetate (pH 5.2) and 2 volumes of chilled ($-20°$) 100% ethanol, and then place the tube at $-20°$ for 30 min. After centrifugation at 14,000 rpm for 20 min, the supernatant of the sample is carefully removed and the pellet is rinsed with 0.5 ml of cold ($-20°$) 70% ethanol. After a brief centrifugation, the 70% ethanol is carefully removed and the pellet is dried at RT. The cDNA pellet is dissolved in 10–50 μl of TE buffer and ready for restriction digestion or stored at $-20°$ for future use.

Restriction Digestion

The double-stranded cDNAs are digested using two restriction endonucleases together to generate DNA fragments for ligation and amplification. In general, two restriction enzymes with 6- and 4-bp recognition sites, respectively (e.g., *Eco*R I and *Mse* I) are used to digest cDNA so that a relatively large number of polymorphic DNA fragments can be generated. Customized primers for adapter ligation and PCR amplification can be synthesized if other restriction enzymes (e.g., *Hind* III instead of *Eco*R I) are used. The combination of these two restriction endonucleases generates an optimum spectrum of fragments (<1 kb) for amplification and separation on a polyacrylamide gel. The completion of restriction enzyme digestion is very important as partial digestion of the cDNA generates ambiguous polymorphism that is not caused by the differences in cDNA or transcript profiles.

To digest cDNA, add 5 μl of 5× reaction buffer, about 18 μl of cDNA (~250 ng), 2 μl of *Eco*R I/*Mse* I (1.25 U/μl each) as supplied by the manufacturer (Invitrogen) to a final volume of 25 μl. Mix gently, centrifuge briefly, and incubate the reaction at 37° for at least 2 h. The reaction is inactivated by heating at 70° for 15 min.

Ligation of Adapters

The digested cDNA fragments are ligated onto double-stranded oligo DNA fragments that contain *Eco*R I and *Mse* I recognition sites (Fig. 1A). These short DNA fragments are called *Eco*R I and *Mse* I adapters. The cDNA fragment ends with specific sequences can be used to design primers for PCR amplification, which prevents the nonspecific amplifications observed in regular mRNA displays (Liang and Pardee, 1992). Thus, a high efficiency of ligation is as important as the completion of restriction enzyme digestion so that the amplified fragments represent the expression profile differences among the biological samples tested.

To the reaction containing digested cDNAs, add 24 μl of adapter ligation solution (*Eco*R I adapter, 5 μM and *Mse* I adapter, 50 μM), 0.4 mM ATP, 10 mM Tris–HCl, pH 7.5, 10 mM magnesium acetate, and 50 mM potassium acetate) and 1 μl of T4 DNA ligase (1 U/μl, Invitrogen). Centrifuge briefly and incubate at 20° for 2 or more hours. After ligation, the reaction mixture can be diluted to 5, 10, or 15 times for PCR amplification. Store the unused portion of the reaction at -20°.

PCR Amplification

PCR is performed using two consecutive steps, namely, preselective and selective amplification. Ligated cDNA template is used for preselective amplification, in which the *Mse* I primer has one selective nucleotide at the 3′ end, whereas the *Eco*R I primer contains no selective nucleotide, so only a fraction of ligated cDNA fragments are amplified. The PCR products from preselective amplification are diluted and used for selective amplification using another primer pair, an *Mse* I primer containing three selective nucleotides and an *Eco*R I primer containing two selective nucleotides. The combination of primer pairs used in each amplification should be tested empirically or according to the recommendations of the manufacturer (Gibco BRL/Invitrogen). The two-step amplification strategy results in a high-resolution sequencing gel with reproducible cDNA fragments produced from thousands of AFLP reactions. Moreover, the primer design and selective amplification strategy preferentially amplify *Mse* I–*Eco*R I fragments rather than *Mse* I–*Mse* I and *Eco*R I–*Eco*R I fragments.

Preselective Amplification

To a 0.2-ml thin-walled microcentrifuge tube, add 4 μl of diluted template DNA (from the ligation reaction), 40 μl of preselective primer mix (supplied by the manufacturer), 5 μl of 10× PCR buffer (including 15 mM MgCl$_2$), and 1 μl of *Taq* DNA polymerase (5 U/μl) in a final volume of 50 μl. Mix and centrifuge briefly. PCR is performed for 20 cycles of 94° for 30 s, 56° for 60 s, and 72° for 60 s, with a final extension at 72° for 10 min. The PCR products are diluted by 1:25 (transfer 5 μl of the PCR products into a new 1.5-ml tube containing 120 μl of ddH$_2$O).

An optional step monitors the results of preselective amplification. Label the *Eco*R I primer (without selective nucleotide) with [γ-^{33}P]ATP (see later discussion) and add 5 μl of the labeled primer and 35 μl pre-selective primer mix to the PCR. After PCR amplification, an aliquot of the PCR products can be electrophoresed in a polyacrylamide gel. The gel is dried in a vacuum gel dryer (Bio-Rad, Hercules, CA) at 80° for 2 h and exposed to an X-ray film or a PhosphorImager (Fujifilm Medical Systems, Inc., Stamford, CT). Successful amplification from the ligated cDNA templates generates a smear of DNA fragments ranging from 50 to 700 bp. No visible smearing above 2 kb or around the well should be observed. However, the products of abundant transcripts may be visible as prominent bands. Therefore, it is recommended to dilute (1:20–50) the amplified products to about 1 ng/μl for selective amplification.

Selective AFLP Amplification

Selective amplification is performed using primer pairs that contain two to three selective nucleotides in the 3′ end of each primer. The introduction of selective bases reduces the number of EcoR I–Mse I fragments that can be amplified during PCR. The number of all possible primer combinations is 1024 (16 × 64) if two selective bases are used in one primer and three in another. However, not every primer pair is useful for PCR amplification. Among all primer pairs tested by the manufacturer (Gibco BRL), 64 pairs are recommended for selective amplification in small genome-size species like Arabidopsis and Brassica. These include (1) eight EcoR I primers, each with two selective nucleotides in the 3′ end: 5′-GAC TGC GTA CCA AAT CAA-3′ (E-AA), CAC-3′ (E-AC), CAG-3′ (E-AG), CAT-3′ (E-AT), CTA-3′ (E-TA), CTC-3′ (E-TC), CTG-3′ (E-TG), and CTT-3′ (E-TT); and (2) eight Mse I primers, each with three selective nucleotides: 5′-GAT GAG TCC TGA GTA ACA A-3′ (M-CAA), ACAC-3′ (M-CAC), ACAG-3′ (M-CAG), ACAT-3′ (M-CAT), ACTA-3′ (M-CTA), ACTC-3′ (M-CTC), ACTG-3′ (M-CTG), and ACTT-3′ (M-CTT). For large-genome species such as cotton and maize, a different set of EcoR I primers is recommended. The eight EcoR I primers are as follows: 5′-GAC TGC GTA CCA AAT CAA C-3′ (E-AC), AAG-3′ (E-AG), ACA-3′ (E-CA), ACT-3′ (E-CT), ACC-3′ (E-CC), ACG-3′ (E-CG), AG C-3′ (E-GC), and AGG-3′ (E-GG). The Mse I primers are the same for both small- and large-genome species.

Primer Labeling

In a 1.5 microcentrifuge tube, add 9 μl of an EcoR I primer (27.8 ng/μl), 5 μl of ddH$_2$O, 5 μl of 5× kinase buffer, 1 μl of T4 polynucleotide kinase (10 U/μl, Invitrogen), and 5 μl of [γ-^{33}P] ATP to a final volume of 25 μl.

Mix and centrifuge briefly and incubate the reaction at 37° for 1 h. The reaction is inactivated by heating at 70° for 10 min.

For "mix 1," add 2.5 μl of the [γ-^{33}P] ATP-labeled EcoR I–primer and 22.5 μl of the Mse I–primer mix (6.7 ng/μl) that contains deoxyribonucleic triphosphates (dNTPs) to the final volume of 25 μl.

For "mix 2," add 39 μl of ddH$_2$O, 10 μl of 10× PCR buffer (including 15 mM MgCl$_2$), and 1 μl of Taq DNA polymerase (5 U/μl) to the final volume of 50 μl.

Add 5 μl of mix 1 (primers, dNTPs) and 10 μl of mix 2 to a 0.2-ml thin-walled microcentrifuge tube containing 5 μl of diluted PCR products from preselective amplification. Mix gently and centrifuge briefly before the tubes are placed in a thermal cycler. PCR is performed for one cycle of 94° for 30 s, 65° for 60 s, and 72° for 60 s plus 12 cycles of amplification with a descending annealing temperature of −0.7° in each cycle, which is followed by additional 23 cycles of 94° for 30 s, 56° for 60 s, and 72° for 60 s, with a final extension at 72° for 10 min. A high annealing temperature (at 65°) in the first PCR cycle and a gradient decrease in the annealing temperature during the subsequent 12 cycles provide a high level of selectivity for the primers containing selective nucleotides during PCR amplification.

AFLP-cDNA Analysis in a Polyacrylamide Gel

Many sequencing gel systems are appropriate for AFLP-cDNA display. We use a Gibco BRL (Invitrogen) or Bio-Rad sequencing apparatus, in which 49 or 97 samples can be loaded using 0.4-mm spacers and shark-tooth combs.

Make a 5% polyacrylamide gel solution containing 5% acrylamide:bis-acrylamide (19:1), 7.5 M urea, and 1× TBE buffer (89 mM Tris, 89 mM Borate, 2 mM EDTA, pH 8.0) (Sambrook et $al.$, 1989). Add 200 μl of fresh 10% ammonium persulfate (APS) and 50 μl of N,N,N′,N′-tetramethyl-ethylenediamine (TEMED) to 80 ml of gel solution for polymerization. Pour the gel solution quickly into a sequencing gel unit.

Add 20 μl of formamide loading dye (98% formamide, 10 mM EDTA, bromophenol blue, xylene cyanol) to 20 μl of the PCR-amplified products. Heat at 90° for 3 min and chill the tube on ice before loading.

Pre-electrophorese the gel until the gel temperature is about 45° (45 Watts for 30 min to 1 h). Load 5 μl of each sample to the gel. Electrophorese the gel at 80 Watts (constant wattage) until the xylene cyanol dye is 1/2–2/3 down the length of the gel (when the bromophenol blue runs off the gel), which usually results in a display of fragment sizes from 50 to 1000 bp.

After electrophoresis, transfer the gel onto a Whatman 3MM paper and dry the gel in a vacuum gel dryer (Bio-Rad) at 80° for 2 h. Expose the gel to a Kodak X-ray film at −80° overnight and develop the film in an automatic film developer (SRX-101A, Konica) or expose the gel to a screen for 3–8 h and detect the images in a PhosphorImager (Fujifilm Medical Systems, Inc.). The AFLP-cDNA profiles are clearly visible in the autoradiogram (Fig. 1B) or in a TIFF file (data not shown).

Reamplification of the Differentially Expressed cDNA Fragments

To recover the DNA fragments from the gel, completely dry the DNA sequencing gel and clearly mark each corner of the gel using a permanent marker. Alternatively, staple the gel to the film or poke a few "guide holes" using a needle.

After the film is developed, the autoradiogram is "perfectly" aligned with the gel using the marks in each corner, the staple, or guide holes. Use sturdy (bulldog) clips to hold the gel and film and place the film directly towards a light box. The bands should be visible through the gel and Whatman 3M paper. Use a sharp razor blade to slice out the bands of interest. Clearly mark the band locations on the autoradiogram.

Place the gel slice in 10 μl of TE on a piece of parafilm for 3–5 min and carefully remove the 3M paper attached to the gel using a pair of fine tweezers. Transfer the gel slice and TE into a 1.5-ml microcentrifuge tube containing 90 μl of TE. Pierce a hole in the lid and boil the tube for 15 min. Centrifuge at 14,000 rpm for 10 min and transfer the supernatant to a new tube. Add 10 μl of 3M sodium acetate, 5 μl of glycogen (10 mg/ml), and 300 μl of 95% ethanol to the solution and incubate at −20° for 2 h. Centrifuge at 14,000 rpm for 20 min at 4°, remove the supernatant, and wash the DNA pellet using 500 μl of cold 70% ethanol. The DNA pellet is dried at RT and dissolved in 20 μl of ddH$_2$O. An aliquot of 5 μl is used for reamplification of the DNA fragment using PCR. The protocol for cloning and sequencing the amplified PCR products is described elsewhere (Lee and Chen, 2001). After the sequence and its coding genes are determined, the expression levels of the gene are verified using RT-PCR, CAPS, and/or SSCP analysis (Adams et al., 2003; Lee and Chen, 2001; Lee et al., 2004).

Notes

1. It is extremely important to collect the samples or tissues from plants growing in the same conditions for gene expression analysis because expression profiles may change in variable growth conditions. Biological

replications (RNAs isolated from two different pools of plants of the same genotype or treatment) should be used whenever possible. Protocols for germinating seeds and growing plants are described elsewhere (Comai et al., 2000; Lee and Chen, 2001; Lee et al., 2004; Madlung et al., 2002).

2. The completion of restriction digestion is critical for AFLP analysis, which can be monitored using controls in which genomic and plasmid DNAs are digested using the same enzymes.

3. Adapter ligation is another critical process that should be monitored using an aliquot of ligation mix for PCR amplification. A successful ligation will result in a clear smear of DNA fragments between 50 and 600 bp.

4. Primer pairs used for preselective and selective amplification may be empirically tested. Alternatively, EcoR I and its primers may be replaced with another restriction enzyme such as Hind III or Dra I. EcoR I activity is sensitive to DNA methylation, which may affect the genomic AFLP results. However, data obtained from our lab and others indicate that similar polymorphism levels of the fingerprints may be obtained using restriction enzymes other than EcoR I (Mei et al., 2004).

5. Glassware, disposable containers, and solutions for RNA samples should be RNase free as recommended (Sambrook et al., 1989). Some plasticware that cannot be autoclaved may be treated using special detergents (e.g., RNase Away, Molecular Bioproducts, San Diego, CA).

6. AFLP analysis can be automated using a LI-COR sequencer (LI-COR Biosciences, Lincoln, NE) if the primers are labeled using IRDyes or other fluorescent dyes (Klein et al., 2000) instead of $[\gamma-^{33}P]ATP$.

Spotted 70-mer Oligonucleotide (Oligo) Microarrays

Introduction

Although AFLP-cDNA display has the advantage of discriminating expressed homoeologous transcripts between the allotetraploids and their progenitors (Comai et al., 2000; Lee and Chen, 2001; Madlung et al., 2002; Wang et al., 2004), the efficiency is limited by *post hoc* cloning and sequencing of individual cDNA fragments. Other methods such as serial analysis of gene expression (SAGE) (Velculescu et al., 1995) and massively parallel signature sequencing (MPSS) (Brenner et al., 2000) also can be used for genome-wide analysis of transcription profiles. However, the techniques require a series of molecular manipulations.

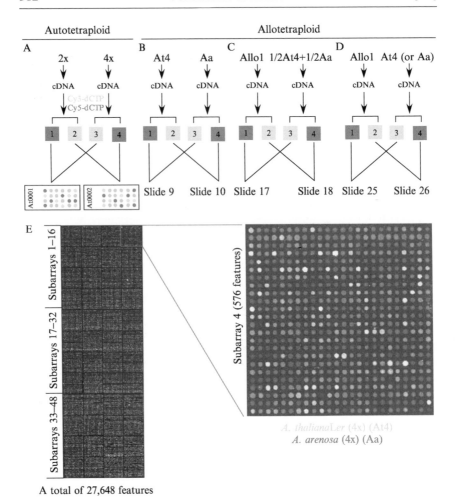

FIG. 2. Microarray experimental design and analysis for detecting gene expression changes in autotetraploids and allotetraploids. (A) A dye-swap experiment using messenger RNA (mRNA) from a diploid and an isogenic autotetraploid to study changes of gene expression in response to ploidy increase. Four labeling reactions using Cy3- and Cy5-dCTP are prepared and mixed reciprocally (e.g., 1 + 3 and 2 + 4) to hybridize two slides in one dye-swap experiment. Four dye-swap experiments (or eight slides) are used in each comparison. (B–D) To detect genes that are differentially expressed in an allotetraploid, four sets of dye-swap experiments are performed for each comparison between the two parents (B), between the allotetraploid and two progenitors (C), or between the allotetraploid and each progenitor (D). Results from the experiments in (B) and (C) determine the overall changes of gene expression in the allotetraploid relative to the progenitors but may underestimate the number of homoeologous genes that are silenced (see text). (E) A hybridization image shows 27,648 features in the *Arabidopsis* spotted 70-mer oligogene microarray, as previously described

DNA microarrays (Fodor *et al.*, 1991; Hughes *et al.*, 2001; Lipshutz *et al.*, 1999; Schena *et al.*, 1998; Singh-Gasson *et al.*, 1999) are applicable to transcriptome analysis in polyploid series. They offer advantages over AFLP-cDNA display and other methods because all gene sequences can be spotted simultaneously and the sequences of the spots on the microarrays are known. Although the technology is dependent on which genes or sequences are spotted on the microarrays, the identities of all genes displaying expression differences are known immediately. DNA microarrays are readily applicable to the analysis of gene expression in autoploidy series such as yeast (Galitski *et al.*, 1999). However, DNA microarrays may not be able to distinguish expression between homologous and/or homoeologous genes in allopolyploids. It requires appropriate experimental design (Fig. 2) and statistical analysis for the detection of transcriptome differences in *Arabidopsis* allotetraploids (Chen *et al.*, 2004) and within and among species (Townsend and Taylor, Chapter 31).

Various microarray platforms and their utilities have been reviewed and discussed (Fodor *et al.*, 1991; Hughes *et al.*, 2001; Johnston, 1998; Lee *et al.*, 2004; Schena *et al.*, 1995, 1998; Singh-Gasson *et al.*, 1999). Here we describe a spotted microarray technology using synthesized long (70-mer) oligos (Lee *et al.*, 2004). We spot about 26,000 oligos designed from all approximately 26,000 annotated genes in the *Arabidopsis* genome available in the National Center of Biological Institutes (NCBI). (http://www.ncbi.nlm.nih.gov/entrez/query.fcgi?db=unigene) and The Institute for Genomic Research (TIGR) (http://www.tigr.org/tdb/agi/) (February 2002) (Arabidopsis Genome Initiative 2000). Oligos with a uniform melting temperature ($75 \pm 5°$) are selected within 1000 nucleotides of the 3′ end of predicted coding sequences (Lee *et al.*, 2004). Gene names and GenBank accession numbers of the 26,090 70-mer oligos and their corresponding cDNA sequences can be found at http://www.operon.com/arrays/omad.php. The slide design, printing, and hybridization of our microarrays are described elsewhere (Lee *et al.*, 2004), except that now each slide is printed with 27,648 features (26,090 oligo-genes plus controls) (Fig. 2E).

(Chen *et al.*, 2004). The slide is hybridized to complementary DNA (cDNA) prepared from *Arabidopsis thaliana* and *Arabidopsis arenosa*. The red, green, and yellow-white features in the subarray indicate low, high, and equal levels of gene expression, respectively, in *A. thaliana*. The data obtained from four sets of dye-swap experiments in each comparison are analyzed using a linear model (Lee *et al.*, 2004) to detect differentially expressed genes at a significance level ($\alpha = 0.05$, see text). (See color insert.)

Microarray Experimental Design

There are many sources (both biological and technical) of variation that are involved in the microarray experiments; therefore, appropriate experimental design is required for the purpose of accurately estimating biological and technical variation (Churchill, 2002; Kerr and Churchill, 2001; Lee *et al.*, 2004; Quackenbush, 2001). Designing a microarray experiment for the analysis of gene expression is essentially equivalent to planning yield trials in multiple locations and years, as is done by plant breeders. The basic principles are to replicate the experiment so that one can estimate the sources of variation for every (biological or technical) parameter in a statistical model. We employ a dye-swap experimental design for microarray analysis (Fig. 2A–D) (Lee *et al.*, 2004). The advantage of the dye swap is that by definition it includes technical replication, and it allows the estimation of all parameters in a linear model, because every possible combination of factors is observed. In fact when multiple samples are involved in a comparative analysis of gene expression, other methods such as loop and reference designs can be implemented (Churchill, 2002; Kerr and Churchill, 2001). We use a linear model analysis that is equivalent to an analysis of variance (ANOVA) (i.e., the treatments involved in array experiments are typically qualitative). Therefore, each gene can be tested readily for differential expression, and a conclusion drawn that pertains to the statistical significance for that gene. Because so many genes are usually represented on the array, a statistical complication arises due to the very large number of hypothesis tests that are made (commonly referred to as the *multiple comparisons problem* in statistics). To control the type I error across the family of tests made, one can use a variety of multiple comparison corrections (e.g., false-discovery rate [FDR] and family-wise error rate [FWER]) (Hochberg and Benjamini, 1990; Hochberg and Tamhane, 1987).

Using the described dye-swap experimental design, the comparison of gene expression differences between a diploid and isogenic autopolyploid is relatively straightforward (Fig. 2A). However, detecting gene expression differences between an allotetraploid and its two progenitors is complicated. Several comparisons are needed to make the data interpretation meaningful. First, a comparison of gene expression differences between two progenitors determines the levels of divergence between the two parents (Fig. 2B). Second, a comparison between the allotetraploid and midvalue of the two parents (an equal mixture of RNAs from two parents) determines overall changes of gene expression relative to the two parents. However, it does not reveal the parental origin of differentially expressed genes. Silencing or activation of one of the two homoeologous genes may not be detected. Third, to determine the origin of gene expression changes, the allotetraploid is compared with one parent at a time so changes of gene expression in every homoeologous locus may be detected. Because each experiment requires four dye swaps or eight replicates (Chen *et al.*, 2004), a total of 32 slide hybridizations are needed for each experiment involving an allotetraploid and two progenitors (Fig. 2B–D). A microarray image of one slide hybridization between *A. thaliana* and *A. arenosa*, progenitors for *Arabidopsis* allotetraploids, is shown in Fig. 2E. The genes that are highly expressed in *A. thaliana*, *A. arenosa*, or are equally expressed are shown in red, green, and yellow, respectively. In this protocol, we describe microarray manufacturing, probe preparation, slide hybridization and washing, image scanning, and data collection and processing. Statistical models and data analysis tools are described elsewhere (Black and Doerge, 2002; Kerr and Churchill, 2001; Lee *et al.*, 2004).

Oligo Design and Distribution

Each 70-mer oligo is designed from 1000 bp with the 3' end of an annotated gene (Lee *et al.*, 2004). The oligo contains no significant secondary structures, has a melting temperature of $70 \pm 2°$, and is coupled with C_6 amino linker that helps DNA bind to covalent substrate coated on the slides. The database for oligo sequences and corresponding annotated genes are maintained by Qiagen at http://oligos.qiagen.com/arrays/omad.php. After synthesis and appropriate quality control, an equal amount of each oligo (600 pmol) is dispensed into 384-well plates (Genetix, Ltd, Hampshire, UK). Contact Qiagen if the oligos need to be distributed in customized plates. The oligos are lyophilized and stored at $-20°$.

Array Fabrication

Spotting high-quality slides is the first step toward a successful microarray experiment. We print microarray slides using the OmniGrid Accent microarrayer (GeneMachines, San Carlos, CA) and microspotting Stealth 3 pins (SMP3) or 946 microspotting pins (946MP3) (TeleChem International, Inc., Sunnyvale, CA). The Accent benchtop model is relatively easy to operate and spots 50 slides at a time, while an OmniGrid 300 microarrayer can print 300 slides. We tested various slides coated with different substrates including poly-L-lysine, SuperAmine, aldehyde, phenylene diisocyanate (PDC), and nitrocellulose. The SuperAmine (SMM) (Telechem International, Inc., Sunnyvale, CA) and poly-L-lysine (PLL) (Cel Associates, Inc., Houston, TX) slides are inexpensive and produce the best and uniform spots.

Place the slides carefully onto the microarrayer station. Use an air duster (from any convenience store) to remove debris on the surface of the slides. Adjust the distance between the pins and each slide by calibrating the slide Z-offsets (one motor step is equivalent to 2.5 μm) so that each slide is spotted by all 16 pins spontaneously with no missing spots.

Set the speed to "low" with a velocity of 5 cm/s and an acceleration of 30 cm/s^2 for both "dipping" and "spotting." The pins are blotted (spotting on a clean plastic plate) five times before spotting onto the slides. Between two spotting steps, the pins are cleaned by sonication at 2000 ms followed by three cycles of wash and dry (1000 ms each).

Before printing, the pins should be tested using a blue food dye (McCormick & Company, Inc., Hunt Valley, MD) dissolved in 3× SSC. Each pin should print more than 100 consistent spots. If printed spots are inconsistent, clean the pins using the microspotting pin and printhead cleaning kit (Telechem International). We routinely clean the pins (in addition to a regular clean step during spotting as just mentioned) after each spotting operation or after a long period of storage, to remove surface contaminants such as salts and organic films that may diminish printing performance. Also, clean the pins after about 12 h of continuous spotting (or each of three subsets of spotting; see later discussion).

Fill the ultrasonic bath with 1 liter of 5% microcleaning solution and place pins in a floatable pin rack. Place the rack with pins in the ultrasonic bath and sonicate for 5 min. Rinse the rack and pins exhaustively with tap water. Sonicate for 5 min in ddH$_2$O, remove pins, and dry. Place clean pins in a printhead and set up for printing.

Centrifuge the plates at 1000 rpm for 2 min to ensure that the oligo pellets fall to the bottom before unsealing the plates.

Make 1 liter of 3× SSC and attach the bottle to a μFill microplate bulk reagent dispenser (Bio-Tek Instruments, Inc., Winooski, VT). Aliquot 20 μl of 3× SSC into each well (600 pmol/oligo) of the plates. Seal the plates using a manual sealer (Corning Incorporated, New York, NY), place them on a shaker, and dissolve the oligos overnight. Centrifuge the plates at 1000 rpm before spotting. The final concentration of each oligo is 30 μM.

A total of 26,090 oligo samples are distributed in sixty-nine 384-well plates. A control plate contains 16 replicates of 12 negative controls (random 70-mer oligos that do not match any sequences in the databases) and 12 positive controls (70-mer oligos designed from constitutive genes). For spotting, we use 16 pins in 4-by-4 configuration. Each pin prints 576 (or 24 by 24) features per subarray (Fig. 2E). The 16 pins spot three subsets of 16 subarrays or a total of 48 subarrays in each slide, producing a total of 27,648 (16 by 3 by 576) features (Fig. 2E). The 16 pins spot samples from 24 plates (23 sample plates plus the control plate) so that the control plate is spotted in each subset. We set the parameters of low velocity and acceleration for both dipping (picking up the oligos from 384-well plate) and printing (spotting the oligos onto the slide). Each spot is about 100 μm in diameter with 180- by 180-μm spacing between spots. The printing quality is dependent on various mechanical and human factors. We can routinely print high-quality slides (Fig. 2E) using Telechem pins, about 40–65% relative humidity in the printing chamber, stable printing buffer (3× SSC), low salt concentration, and PLL or SMM slides. "Double dipping" reduces the number of missing spots during the printing. It takes about 36 h to print 50 slides each with 27,648 features.

To estimate the variation among different slides, we also generate different printing patterns by changing the order of the plates during printing or rotating each plate to 180 degrees so spots in different printing patterns are generated by different pin-sample combinations.

After each spotting operation, we dry the oligo samples in a laminar flow hood, seal the plates using manual dealers (Corning, Inc.), and store at $-20°$. The oligos can be re-dissolved in 20 μl of ddH$_2$O overnight before a subsequent printing.

Posttreatment of the Slides

Using a diamond scriber, mark the corners of the printing area that will be covered during hybridization. Arrange 10–15 slides in the same orientation and place them in a slide rack (VWR Scientific Products, West Chester, PA). Hold the slides with the printed side toward open plasticware containing 1× SSC heated at 37°. After about 20 s, the surface of the slides becomes foggy. Place the slides on a heating block at 100° and snap-dry the slides for 15 s. Repeat the procedure twice. This procedure allows the DNA spots to "swell" (redistributing the DNA uniformly throughout the spots) but not spread into each other, which results in rounded spot patterns fixed by the heating step.

Place the slides in a Stratalinker UV Crosslinker (Stratagene, La Jolla, CA) and cross-link the DNA to the glass substrate using the setting of 250 mJ.

Prepare the blocking solution: Dissolve 9 g of succinic anhydride in approximately 500 ml of 1-methyl-2-pyrrolidinone (Sigma, St. Louis, MO). Immediately after the last flake of the succinic anhydride is dissolved, add 22.4 ml of 1 M of sodium borate (pH 8.0).

To make 1 M of sodium borate, dissolve 6.2 g of boric acid (Fisher Scientific International, Inc., Fair Lawn, NJ) in a final volume of 100 ml and adjust pH to 8.0 using NaOH.

For blocking the slides, place the slides in a slide rack and quickly plunge the slide rack up and down in the blocking solution for 1 min. Then place the slides in the blocking solution and mix on an orbital shaker at 60 rpm for 15–20 min.

Carefully transfer the slide rack to a slide dish containing ddH_2O and plunge the rack up and down several times to rinse off the blocking solution. Then immerse the slides in ddH_2O for 1 min. (Note: It is critical to remove the residual blocking solution as much as possible to prevent background problems during slide hybridization.)

Transfer the slide rack to plasticware containing 95% ethanol and plunge up and down five to eight times. Dry the slides by centrifugation in a tabletop centrifuge Savant SpeedVac (TeleChem International) at 850 rpm for 5 min. The slides can be used immediately for hybridization or stored in a humidity-controlled Dry Keeper (Sanplatec Corp., Japan) with a relative humidity of 10–20% at RT for up to a year.

Probe Labeling (Direct Dye Incorporation Method)

The procedures for isolating total RNA and purifying mRNA are described in the section "AFLP-cDNA Display," earlier in this chapter. We use 500 ng of mRNA in each labeling reaction using Cy3- or Cy5-dCTP (Amersham Biosciences). For one dye-swap experiment, we use two sets or four labeling reactions (Fig. 2A). After labeling, the Cy3-dCTP reaction is mixed with the Cy5-dCTP reaction to make one probe. Therefore, two "identical" probes, each containing an equal amount of Cy3- and Cy5-labeled cDNAs, are hybridized with two slides, which constitute one dye-swap experiment.

Primer Annealing

Add 16 μl of mRNA (~500 ng), 1 μl of anchored oligo (dT) (2 $\mu g/\mu l$), and 1 μl of random nonamer (2 $\mu g/\mu l$, Gene Link, Hawthorne, NY) to a 1.5-ml microcentrifuge tube on ice. Mix gently by pipetting. Incubate the reaction at 65° for 5 min. Cool reactions at RT for 10 min to allow the primers and the RNA template to anneal. Centrifuge briefly.

Reverse Transcription

Place reactions on ice and add 6 μl of reverse-transcriptase buffer (5×), 3 μl of DTT (0.1 M), 1 μl of dNTP (10 mM dATP, dTTP, dGTP, 2.5 mM dCTP), 1 μl of Cy3- or Cy5-dCTP, and 1 μl of Superscript II RT (Invitrogen) to a final volume of 30 μl.

Mix very gently by pipetting and centrifuge briefly. Quickly wrap the tube with aluminum foil and incubate the reactions at 42° for 2 h. From this step on, minimize the direct exposure of the labeling reactions to light.

RNA Degradation

Add 3 μl of 2.5 M NaOH, mix well by pipetting, and incubate the reactions at 37° for 15 min in the dark (wrap the tube with aluminum foil). Add 15 μl of 2 M 4-(2-hydroxyethyl)-1-piperazineethanesulfonic acid (HEPES) to neutralize the solution.

To make 10 ml of 2 M HEPES, dissolve 4.77 g of HEPES (Sigma) in ddH$_2$O to a final volume of 10 ml, and filter-sterilize the solution using a 0.45-μm filter (Millipore, Billerica, MA).

Probe Purification

The labeled probes are purified using a PCR purification kit according to the recommendation of the manufacturer (Qiagen). Briefly, add 300 μl of binding buffer (see earlier discussion) to the reaction and elute twice using 30 μl of ddH2O, resulting in 60 μl. Dry the samples using a Savant SpeedVac (TeleChem International) covered with a black cloth wrap.

Probes can be labeled using two other methods known as (1) amino-allyl dye coupling or postlabeling (Randolph and Waggoner, 1997) (Amersham Biosciences) and (2) 3D labeling with dendritic nucleic acid structure (Nilsen et al., 1997; Stears et al., 2000) (Genisphere, Hatfield, PA). In our experience, the postlabeling method works as well as the direct labeling method. However, for unknown reasons, the dendrimer molecules from the products we tested in 2002 hybridized randomly to some 70-mer oligos (Lee et al., 2004). We have not tested a new product that is claimed to be suitable for labeling probes used in hybridization to spotted oligo microarrays (Genisphere).

RNA Labeling Using CyScribe PostLabeling Kit (RPN5660) (Amersham Biosciences)

The indirect labeling procedure requires two steps. The first step involves incorporation of amino allyl-dUTP (AA-dUTP) during cDNA synthesis. The second step involves conjugation of chemically amino allyl-modified cDNA and cyanine (Cy3 or Cy5) dye. This two-step incorporation of cyanine dyes presumably improves the labeling efficiency by reducing the bias of direct Cy3- and Cy5-dye incorporation during cDNA synthesis.

We use about 500 ng of mRNA for each indirect labeling reaction according to the recommendations of the manufacturer (Amersham Biosciences). Synthesize the first-strand cDNA using AA-dUTP and Superscript II RT (Invitrogen). The Superscript II RT works better than the Cyscript RT provided in the kit. Incorporate the Cy3 or Cy5 dye to amino allyl–modified cDNA through the reaction of the amino allyl group with the monofunctional N-hydroxysuccinimide (NHS)-ester Cy dye (Amersham Biosciences) according the manufacturer's recommendations. The labeled cDNA is then purified using a PCR purification kit (Qiagen) as described earlier.

Hybridization

Resuspend the dried probe in 56 μl of ddH$_2$O and denature at 95° for 3 min. Add 12 μl of 20× SSC, 2 μl of 10% SDS, and 8 μl of 10% BSA to the probe, and mix well in a final volume of about 80 μl.

Place the posttreated microarray slide on a heating plate (at 40–50°), add the probe as one drop of liquid to an area that is 5 mm from the marked edge of the printed area, and slowly lay down a 22-by-26 LifterSlip (Erie Scientific Company, Portsmouth, NH) over the hybridization solution so that the probe is evenly distributed without bubbles.

Alternatively, lay down the LifterSlip and add the 2/3 of labeled probe very slowly from one side of the coverslip. The solution flows to completely cover the hybridization area so no bubbles are formed. Add the rest of the probe from the opposites side of coverslip. (Do not let the pipette tip move the LifterSlip.)

Place the slide in a sealed hybridization cassette (TeleChem International) and add 20 μl of 3× SSC to both ends of the slide in the cassette. Place the hybridization cassette in a hybridization oven at 65° for 12–16 h. We normally hybridize four or eight slides each time. The hybridization and washing should be performed in dark conditions whenever possible.

Remove each slide from the hybridization cassette without disturbing the coverslip and place the slides in a slide rack. Place the slide rack in a washing station (HTW) (Telechem International) containing low-stringency washing buffer (2× SSC, 0.2% SDS). The coverslip gradually slips off the slide into washing buffer in less than 1 min. Gently remove the coverslip and continue washing slides for 4 min at RT.

Wash the slides in 0.2× SSC buffers for 2 min at RT and repeat once. Wash the slides at high-stringency in another washing station containing 0.05× SSC for 4 min at RT. Remove the slide rack and plunge the slides up and down several times in the washing station containing fresh 0.05× SSC buffer. Dry the slides immediately by centrifugation in an Allegra X-12 tabletop centrifuge (Beckman Coulter, Inc., Fullerton, CA) at 850 rpm for 10 min. The slides then are assembled in a slide rack and quickly placed in a dark container.

Microarray Scanning (Parameters) and Data Analysis

Take one slide at a time, carefully place it in the scanning chamber of a GenePix 4000B scanner (Axon Instruments, Inc., Union City, CA), and gently close the lid. The images can be scanned using "preview scan" function to estimate the relative intensities of the two-color images. The red and green signals need to be balanced during array scanning by adjusting the photomultiplier tube (PMT, maximum 1000) gain and 100% laser power for each wavelength. We normally set the power to 100% and the PMT gain to 900–990 for the 635-nm (red) laser and 800–900 for the 532-nm (green) laser. Each microarray spot is about 100 μm in diameter. When the slide is scanned using the setting of 10-μm pixel size, each spot contains about 78 pixels, which is sufficient for an accurate intensity measurement using GenePix Pro 5.0 software. A pixel with an intensity of 65,535 or more is saturated and seen as a "white spot" on the slide. The minimum intensity that can be detected is 500. While scanning, we adjust the PMT gain in each channel using a "count ratio" until the ratio is about 1.0. After "preview scan," we select the area and scan it again using the "data scan" function.

After the scanning is completed, we save the image as a multi-image TIFF file that can be exported as a JPEG file. The PMT values and probe/hybridization information for each file are recorded according to the requirement for maintaining minimum information about microarray experiments (MIAME) (Brazma *et al.*, 2001). For the quantification of each hybridized feature, the "deconvolution blocks" are generated by the GeneMachines microarrayer using "generate grid" (384 by 72), which is exported to the GenePix Pro 5.0 as a GAL file. In the GenePix Pro 5.0, we use "loading array list" to overlay the "grid" to the scanned images. We then adjust the grid to fit the features automatically, followed by manual corrections so that the quantification area is localized in the center of every feature. The intensities of each feature at 532 and 635 nm are generated in a GPR file for further analysis. GenePix Pro 5.0 has a few analysis functions to display scatter plots and intensity ratios.

Data Normalization and Analysis

The GPR files generated from each experiment are imported from GenePix Pro 5.0 and analyzed using both Matlab (MathWorks, Inc., Natick, MA) (Ihaka and Gentleman 1996) and custom statistical programs written with the statistical package known as R (http://www.r-project.org). Data normalization can be used to correct some technical variations involved in microarray experiments, which includes unequal amounts of RNA samples, different spotting pins, variable surface on the slides, different labeling and detection efficiencies between the cyanine dyes used, and systematic biases in the measured expression levels (Quackenbush, 2002). Normalization of microarray data is equivalent to the use of a reference or control to estimate the expression levels of the genes of interest relative to the control. Typically, data normalization is not necessary for the ANOVA model (Lee *et al.*, 2004). The data may be collected directly from scanned images and transformed using $\log_2$(ratio) total intensities. However, in many microarray experiments, local variation is commonly observed because of factors such as unequal hybridization, biased dye incorporation, and block (or pin)-to-block variation. A robust locally-weighted linear regression (lowess) (Cleveland, 1979) is recommended for two-color microarray experiments. This normalization process allows the correction for the bias caused by nonlinear rates of dye incorporation and other systematic technical factors such as pins. Lowess normalization can be applied either globally or locally. Local normalization has the advantage of correcting systematic spatial variation that is often observed in large array experiments. In our experiments, we use the local (pin-based) lowess normalization and analyze data using the ANOVA model (Chen *et al.*, 2004). The multiple comparison tests using a per-gene variance and a common gene variance are employed to select differentially expressed genes. Holm's FWER and/or FDR (Benjamini and Hochberg, 1995; Hochberg and Tamhane, 1987) are applied to control multiple comparison testing errors at a significant level $\alpha = 0.05$.

Commercial software packages such as GeneSpring (Silicon Genetics, Redwood City, CA) and Spotfire (Spotfire, U.S., Somerville, MA) are available for the data normalization, quality control, and data analysis. A caveat for using commercial software is that underlying statistical models (how to calculate variance, multiple comparisons corrections, etc.) are not clear. Biologists should consult with a statistician/bioinformatician to be sure that their experimental design is compatible with the underlying models in these computer programs.

The differentially expressed genes detected (Fig. 2) are confirmed and analyzed using other methods, such as quantitative RT-PCR, RT-PCR and CAPS, and SSCP (Adams *et al.*, 2003; Lee and Chen, 2001; Lee *et al.*, 2004).

Notes

1. Current printing patterns do not contain replications of features within a slide. If the statistical model requires replications, every feature can be duplicated in each slide. However, spotting about 60,000 features in one slide may decrease spacing between spots and resolution in subsequent hybridization and detection. Alternatively, a set of 26,000 oligos can be printed on two slides, although this increases variation between slides.

2. Slide hybridization using a LifterSlip cover is one of the best choices for us, because the printed area occupies the maximum amount of space on the slide and leaves little room for sealers that are used in many automated systems. It is recommended that before using probe-containing hybridization solutions, one should practice several times using the LifterSlip to cover hybridization solutions without probes on a regular slide until no bubbles are formed.

3. We recommend using direct or indirect labeling methods for probe preparation. The direct incorporation method is routinely used because the Superscript II RT (Invitrogen) has a similar efficiency for Cy3- and Cy5-dCTP incorporation. However, when mRNA is limiting (e.g., from cotton fiber tissues in early stages), the Amino Allyl MessageAmp aRNA kit (Ambion, Austin, TX) can be used to label probes for microarray hybridization. Only 5 μg of total RNA is required for each labeling reaction (Lee and Chen, unpublished data).

4. For genomic DNA analysis, we label genomic DNA using a modified protocol (Pollack et al., 1999). For each labeling, 1 μg of genomic DNA is digested by Mse I and purified using a QIAquick PCR kit, which is then random-primer labeled using a Bioprime labeling kit (Gibco BRL). The labeling reaction includes 1 μg of Mse I–digested genomic DNA, dATP, dGTP, and dTTP (120 μM each), dCTP (90 μM), and Cy3-dCTP or Cy5-dCTP (30 μM) in 50 μl of reaction. The labeled probe is purified as described earlier. The same hybridization and washing procedures are used except that 20 μg of human cot-1 DNA and 100 μg of tRNA (Invitrogen) are included in about 80 μl of hybridization solution.

5. We have shown that spotted 70-mer oligo microarrays can be used to hybridize homoeologous and orthologous genes in Arabidopsis and related species including Brassica (Lee et al., 2004).

6. Although the spotted oligo microarray has many advantages (Lee et al., 2004), the oligo design is dependent on accuracy of gene annotation. Therefore, gene indices should be updated periodically according to the new release of annotated databases from research groups such as TAIR (http://arabidopsis.org/) and TIGR (http://www.tigr.org/).

7. PLL slides need to be freshly made. High levels of hybridization background are often associated with old PLL slides. We prefer SMM to PLL slides.

Acknowledgments

We thank James A. Birchler, Robert Martienssen, Thomas C. Osborn, and J. Chris Pires for comments on the manuscript and insightful discussions leading to the development of spotted oligo gene microarrays in *Arabidopsis*. The research is supported by a grant from the National Science Foundation (DBI0077774). Work in the Chen laboratory is supported in part by the National Institutes of Health (GM067015). The opinions expressed are those of the authors and do not reflect the official policy of the National Institutes of Health, the National Science Foundation, or the U.S. government.

References

Adams, K. L., Cronn, R., Percifield, R., and Wendel, J. F. (2003). Genes duplicated by polyploidy show unequal contributions to the transcriptome and organ-specific reciprocal silencing. *Proc. Natl. Acad. Sci. USA* **100,** 4649–4654.

Arabidopsis Genome Initiative. (2000). Analysis of the genome sequence of the flowering plant *Arabidopsis thaliana. Nature* **408,** 796–815.

Bachem, C. W., van der Hoeven, R. S., de Bruijn, S. M., Vreugdenhil, D., Zabeau, M., and Visser, R. G. (1996). Visualization of differential gene expression using a novel method of RNA fingerprinting based on AFLP: Analysis of gene expression during potato tuber development. *Plant J.* **9,** 745–753.

Benjamini, Y., and Hochberg, Y. (1995). Controlling the false discovery rate: A practical and powerful approach to multiple testing. *J. R. Stat. Soc. Series B* **57,** 289–300.

Black, M. A., and Doerge, R. W. (2002). Calculation of the minimum number of replicate spots required for detection of significant gene expression fold change in microarray experiment. *Bioinformatics* **18,** 1609–1616.

Brazma, A., Hingamp, P., Quackenbush, J., Sherlock, G., Spellman, P., Stoeckert, C., Aach, J., Ansorge, W., Ball, C. A., Causton, H. C., Gaasterland, T., Glenisson, P., Holstege, F. C., Kim, I. F., Markowitz, V., Matese, J. C., Parkinson, H., Robinson, A., Sarkans, U., Schulze-Kremer, S., Stewart, J., Taylor, R., Vilo, J., and Vingron, M. (2001). Minimum information about a microarray experiment (MIAME)-toward standards for microarray data. *Nat. Genet.* **29,** 365–371.

Brenner, S., Johnson, M., Bridgham, J., Golda, G., Lloyd, D. H., Johnson, D., Luo, S., McCurdy, S., Foy, M., Ewan, M., Roth, R., George, D., Eletr, S., Albrecht, G., Vermaas, E., Williams, S. R., Moon, K., Burcham, T., Pallas, M., DuBridge, R. B., Kirchner, J., Fearon, K., Mao, J., and Corcoran, K. (2000). Gene expression analysis by massively parallel signature sequencing (MPSS) on microbead arrays. *Nat. Biotechnol.* **18,** 630–634.

Chen, Z. J., Wang, J., Tian, L., Lee, H.-S., Wang, J. J., Chen, M., Lee, J. J., Josefsson, C., Madlung, A., Watson, B., Lippman, Z., Vaughn, M., Pires, J. C., Colot, V., Doerge, R. W., Martienssen, R. A., Comai, L., and Osborn, T. C. (2004). The development of an *Arabidopsis* model system for genome-wide analysis of polyploidy effects. *Biol. J. Linn. Soc.* **82,** 689–700.

Churchill, G. A. (2002). Fundamentals of experimental design for cDNA microarrays. *Nat. Genet.* **32,** 490–495.

Cleveland, W. S. (1979). Robust locally weighted regression and smoothing scatterplots. *J. Am. Stat. Assoc.* **74**, 829–836.

Comai, L., Tyagi, A. P., Winter, K., Holmes-Davis, R., Reynolds, S. H., Stevens, Y., and Byers, B. (2000). Phenotypic instability and rapid gene silencing in newly formed *Arabidopsis* allotetraploids. *Plant Cell* **12**, 1551–1568.

Fodor, S. P., Read, J. L., Pirrung, M. C., Stryer, L., Lu, A. T., and Solas, D. (1991). Light-directed, spatially addressable parallel chemical synthesis. *Science* **251**, 767–773.

Galitski, T., Saldanha, A. J., Styles, C. A., Lander, E. S., and Fink, G. R. (1999). Ploidy regulation of gene expression. *Science* **285**, 251–254.

Hochberg, Y., and Benjamini, Y. (1990). More powerful procedures for multiple significance testing. *Stat. Med.* **9**, 811–818.

Hochberg, Y., and Tamhane, A. C. (1987). "Multiple Comparison Procedures." Wiley, New York.

Hughes, T. R., Mao, M., Jones, A. R., Burchard, J., Marton, M. J., Shannon, K. W., Lefkowitz, S. M., Ziman, M., Schelter, J. M., Meyer, M. R., Kobayashi, S., Davis, C., Dai, H., He, Y. D., Stephaniants, S. B., Cavet, G., Walker, W. L., West, A., Coffey, E., Shoemaker, D. D., Stoughton, R., Blanchard, A. P., Friend, S. H., and Linsley, P. S. (2001). Expression profiling using microarrays fabricated by an ink-jet oligonucleotide synthesizer. *Nat. Biotechnol.* **19**, 342–347.

Ihaka, R., and Gentleman, R. (1996). A language for data analysis and graphics. *J. Comput. Graphical Statist.* **5**, 299–314.

Johnston, M. (1998). Gene chips: Array of hope for understanding gene regulation. *Curr. Biol.* **8**, R171–R174.

Kerr, M. K., and Churchill, G. A. (2001). Experimental design for gene expression microarrays. *Biostatistics* **2**, 183–201.

Klein, P. E., Klein, R. R., Cartinhour, S. W., Ulanch, P. E., Dong, J., Obert, J. A., Morishige, D. T., Schlueter, S. D., Childs, K. L., Ale, M., and Mullet, J. E. (2000). A high-throughput AFLP-based method for constructing integrated genetic and physical maps: Progress toward a sorghum genome map. *Genome Res.* **10**, 789–807.

Lee, H. S., and Chen, Z. J. (2001). Protein-coding genes are epigenetically regulated in *Arabidopsis* polyploids. *Proc. Natl. Acad. Sci. USA* **98**, 6753–6758.

Lee, H. S., Wang, J., Tian, L., Jiang, H., Black, M. A., Madlung, A., Watson, B., Lukens, L., Pires, J. C., Wang, J. J., Comai, L., Osborn, T. C., Doerge, R. W., and Chen, Z. J. (2004). Sensitivity of 70-mer oligonucleotides and cDNAs for microarray analysis of gene expression in *Arabidopsis* and *Brasssica*. *Plant Biotechnol. J.* **2**, 45–57.

Liang, P., and Pardee, A. B. (1992). Differential display of eukaryotic messenger RNA by means of the polymerase chain reaction. *Science* **257**, 967–971.

Lipshutz, R. J., Fodor, S. P., Gingeras, T. R., and Lockhart, D. J. (1999). High density synthetic oligonucleotide arrays. *Nat. Genet.* **21**, 20–24.

Liu, B., and Wendel, J. (2002). Non-Mendelian phenomenon in allopolyploid genome evolution. *Curr. Genomics* **3**, 489–505.

Madlung, A., Masuelli, R. W., Watson, B., Reynolds, S. H., Davison, J., and Comai, L. (2002). Remodeling of DNA methylation and phenotypic and transcriptional changes in synthetic *Arabidopsis* allotetraploids. *Plant Physiol.* **129**, 733–746.

Mei, M., Syed, N. H., Gao, W., Thaxton, P. M., Smith, C. W., Stelly, D. M., and Chen, Z. J. (2004). Genetic mapping and QTL analysis of fiber-related traits in cotton (*Gossypium*). *Theor. Applied Genet.* **108**, 280–291.

Nilsen, T. W., Grayzel, J., and Prensky, W. (1997). Dendritic nucleic acid structures. *J. Theor. Biol.* **187**, 273–284.

Osborn, T. C., Pires, J. C., Birchler, J. A., Auger, D. L., Chen, Z. J., Lee, H. S., Comai, L., Madlung, A., Doerge, R. W., Colot, V., and Martienssen, R. A. (2003). Understanding mechanisms of novel gene expression in polyploids. *Trends Genet.* **19**, 141–147.

Pikaard, C. S. (2001). Genomic change and gene silencing in polyploids. *Trends Genet.* **17**, 675–677.

Pollack, J. R., Perou, C. M., Alizadeh, A. A., Eisen, M. B., Pergamenschikov, A., Williams, C. F., Jeffrey, S. S., Botstein, D., and Brown, P. O. (1999). Genome-wide analysis of DNA copy-number changes using cDNA microarrays. *Nat. Genet.* **23**, 41–46.

Quackenbush, J. (2001). Computational analysis of microarray data. *Nat. Rev. Genet.* **2**, 418–427.

Quackenbush, J. (2002). Microarray data normalization and transformation. *Nat. Genet..* **32**, 496–501.

Randolph, J. B., and Waggoner, A. S. (1997). Stability, specificity and fluorescence brightness of multiply-labeled fluorescent DNA probes. *Nucleic Acids Res.* **25**, 2923–2929.

Sambrook, J., Fritsch, E. F., and Maniatis, T. (1989). "Molecular Cloning: A Laboratory Manual." Cold Spring Harbor Laboratory Press, Cold Spring Harbor, NY.

Schena, M., Heller, R. A., Theriault, T. P., Konrad, K., Lachenmeier, E., and David, R. W. (1998). Microarrays: Biotechnology's discovery platform for functional genomics. *Trends Biotechnol.* **16**, 301–306.

Schena, M., Shalon, D., Davis, R. W., and Brown, P. O. (1995). Quantitative monitoring of gene expression patterns with a complementary DNA microarray. *Science* **270**, 467–470.

Singh-Gasson, S., Green, R. D., Yue, Y., Nelson, C., Blattner, F., Sussman, M. R., and Cerrina, F. (1999). Maskless fabrication of light-directed oligonucleotide microarrays using a digital micromirror array. *Nat. Biotechnol.* **17**, 974–978.

Stears, R. L., Getts, R. C., and Gullans, S. R. (2000). A novel, sensitive detection system for high-density microarrays using dendrimer technology. *Physiol. Genomics* **3**, 93–99.

Velculescu, V. E., Zhang, L., Vogelstein, B., and Kinzler, K. W. (1995). Serial analysis of gene expression. *Science* **270**, 484–487.

Vos, P., Hogers, R., Bleeker, M., Reijans, M., van de Lee, T., Hornes, M., Frijters, A., Pot, J., Peleman, J., and Kuiper, M. (1995). AFLP: A new technique for DNA fingerprinting. *Nucleic Acids Res.* **23**, 4407–4414.

Wang, J., Tian, L., Madlung, A., Lee, H. S., Chen, M., Lee, J. J., Watson, B., Kagochi, T., Comai, L., and Chen, Z. J. (2004). Stochastic and epigenetic changes of gene expression in *Arabidopsis* polyploids. *Genetics* **167**, 1961–1973.

Welsh, J., Chada, K., Dalal, S. S., Cheng, R., Ralph, D., and McClelland, M. (1992). Arbitrarily primed PCR fingerprinting of RNA. *Nucleic Acids Res.* **20**, 4965–4970.

[31] Designing Experiments Using Spotted Microarrays to Detect Gene Regulation Differences Within and Among Species

By JEFFREY P. TOWNSEND and JOHN W. TAYLOR

Abstract

Comparative studies of genome-wide gene expression must account for variation not only among species, but also within species. Such studies are necessarily large in scale, because they incorporate experiments on multiple individuals of multiple species in multiple developmental stages in multiple environmental conditions. If the experiments are carefully designed and performed, the data they provide are worth the effort. We describe the utility of spotted microarrays for these studies and highlight experimental design criteria that will maximize inferential and statistical power. We conclude with a discussion of experimental protocols that are designed for investigations of differential gene expression and their pitfalls.

Introduction

DNA microarrays have proved their worth in studies of development and mutation using single experimental strains and will be even more useful for the examination of natural genetic variation by the analysis of multiple individuals from populations and species. DNA microarray technology has particularly strong potential to illuminate studies of the molecular origin of phenotype (Singh, 2003). Where there is variation in phenotype, there will be underlying molecular correlates that will lead to a better understanding of the phenotype. Even complex phenotypes that may involve multiple metabolic and developmental pathways are approachable with appropriate experimental design. Where there is variation within populations or among species, questions about adaptation may be clearly framed. These questions will be best addressed when fitness can be assayed at the same time point for which transcription is profiled. For example, successful spore germination may be assayed among different individuals.

Well-studied model organisms such as yeast or *Escherichia coli* are ideal for this type of research because their genomes are sequenced, microarrays are available, their biochemical metabolism and their cell and molecular biology are well understood, and hypotheses about genes can be tested using molecular genetic experiments. However, non–model

TABLE I
FEATURES AND DRAWBACKS OF TYPES OF PROBES FOR MICROARRAYS

Type of probe	Genome[a]	Annotated[b]	Sensitivity to divergence	Completeness[c]	Expense
Random genomic clones	No	No	Depends on fragment size	Incomplete	Low
cDNAs from ESTs	No	No	Sensitive	Incomplete	Low
PCR products of ORFs	Yes	Yes	Sensitive	Complete	Laborious, 2 primers/ sequence
70-mers designed for ORFs	Yes	Yes	Very sensitive	Complete, low cross-hybridization	Intermediate
25-mer whole genome tiling arrays	Yes	No	Extremely sensitive	Complete, low cross-hybridization	Expensive

Note: cDNA, complementary DNA; EST, expressed sequence tag; ORF, open reading frame; PCR, polymerase chain reaction.
[a] Requires a known complete sequence for the organismal genome.
[b] Requires a well-annotated genome to design probes.
[c] Results in a complete description of genomic gene expression levels.

systems are not excluded from microarray research. In fact, a useful microarray can be constructed from fragments of the genome of any organism and used for competitive hybridization experiments to identify a pool of potentially interesting genes. The fragments of interest may then be sequenced, and their identification attempted through comparison with known genes. Within fungi, this approach has been used to find genes involved in *Histoplasma capsulatum* pathogenicity (Hwang *et al.*, 2003). There are many ways in which microarrays may be created, and each way has its advantages and disadvantages (Table I). Here, we present guidelines for experimental design of spotted DNA microarray studies, with particular focus on the examination of variation among individuals or species.

Experimental Design

Use of Comparative Hybridizations

Microarray experimental designs are highly influenced by the particular technology employed, and here we emphasize spotted microarray design. The most salient feature of experimental design for spotted microarrays is their comparative nature. In each use, two samples of messenger RNA (mRNA) are competitively hybridized to the deposited DNA, and the comparison allows for the elimination of many spot-specific confounding factors (Eisen and Brown, 1999; Townsend, 2004). In any experimental use of spotted DNA microarrays, the first question to consider is "how should samples of interest be compared?" Experimental design in this context may be approached by the use of universal reference samples (Novoradovskaya et al., 2004; Puskás et al., 2004), pooled samples, or circuits of comparisons. For most purposes, circuits of comparisons have the greatest statistical power to detect differences in gene expression among all samples examined (Kerr and Churchill, 2001a; Townsend, 2003; Townsend and Hartl, 2002; Wolfinger et al., 2001; Yang and Speed, 2002); thus, this method of experimental design is described here.

When comparisons are made in circuits, complex experiments can be designed to accommodate three major axes of biological variation. The first axis is that of genotype and is typically manifested in multiple individuals, $G_1 \ldots G_i$, where i is the number of individuals and each individual is a natural or mutational variant. The second axis includes any investigated aspects of the environment, $E_1 \ldots E_j$, where j is the number of environments, typically comprising several experimental treatments, microhabitats, or geographic locations. Lastly, a third axis along which samples are frequently compared is an axis of developmental states, typically time points in a developmental course or cycle, for example, $T_1 \ldots T_k$, where k is the number of time points. Experimental designs incorporating one, two, or three of these dimensions may be most clearly planned and depicted using design graphs.

Graph Theoretical Depiction of Experimental Designs

Spotted microarray experimental designs are commonly depicted using directed multigraphs (e.g., Fig. 1, a simple design incorporating all three major axes of variation; see also Figs. 2A–C and 3A–C). In these graphs, circular nodes represent samples of mRNA that have been harvested from a particular genotype, environment, and developmental state. Arcs between nodes represent competitive hybridizations between the mRNAs. This depiction is a *multi*graph because more than one arc can connect any two nodes. The number of these replicated hybridizations, or arcs, is vital to the degree of statistical significance for both large and small differences in expression level. Simultaneous comparison of two mRNAs on a single microarray is made possible by labeling the samples with two fluorescent dyes. The direction of the arc in the diagram conventionally indicates which sample is labeled with which fluorophore, (e.g., the pointed end

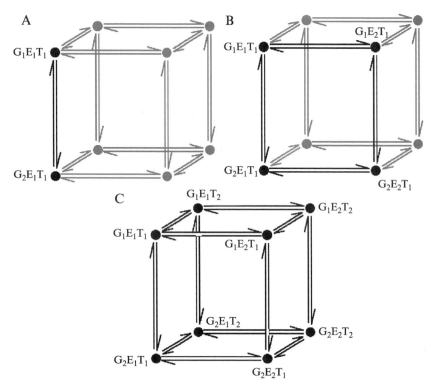

FIG. 1. A general example of spotted complementary DNA (cDNA) microarray experimental design that incorporates interrogation of eight messenger RNA (mRNA) samples from (A) two genotypes (G_1 and G_2), (B), two environments (E_1 and E_2) and (C) two developmental states (T_1 and T_2). Nodes represent the mRNA samples, and arcs represent competitive hybridizations. The direction of the arc indicates which mRNA sample was labeled with which fluorophore (e.g., the blunt end indicates the node labeled with Cy5, and the pointed end indicates the node labeled with Cy3). With eight mRNA samples, 24 competitive hybridizations are proposed. An additional 24 competitive hybridizations could be made symmetrically by including direct comparisons across the diagonals of the cube graph faces (i.e., $G_1E_1T_1$ to $G_2E_2T_1$, $G_2E_1T_1$ to $G_1E_2T_1$, etc.).

indicates the complementary DNA [cDNA] labeled with Cy5, and the blunt end indicates the cDNA labeled with Cy3). Between the fluorophore and the DNA spotted, there is a small interaction effect on the hybridization intensity, so it is recommended to perform two hybridizations with "flipped fluorophores" between any two samples that are to be directly compared. This pair of replicates is indicated in a directed multigraph by a pair of oppositely directed arcs (Fig. 1A).

For a given mRNA sample, or node, the number of adjacent nodes is the unweighted valence and the number of arcs is the weighted valence. In Fig. 1A, each mRNA is to be competitively hybridized two times against each of three other mRNAs, for a total of six hybridizations. Each node has an unweighted valence of three and a weighted valence of six. Thus, the weighting of the valence of a node in microarray experimental design is the degree of experimental replication applied to that particular mRNA sample. To increase replication in the experimental design in Fig. 1, for example, flipped-fluorophore cross-comparisons could be added to span from corner to corner of each face of the cubic graph. These added comparisons would increase the weighted valence of each node from 6 to 12 and allow one to resolve differences in gene expression as low as 1:1.2 instead of 1:1.4 (see discussion of statistical significance).

If variation along an axis of interest is continuous or ordinal (e.g., a quantitative genetic trait, a temperature gradient, a variable reagent concentration, or a time course), the experiment should be designed to compare each node with its nearest neighbors. This guideline ensures that the greatest power is applied to the detection of differences between the most similar samples. It might be argued that differences between the most similar samples are not of as much interest; in that case, the prudent action would be to pare down the number of nodes in the design until all samples are of interest.

If variation along an axis of interest is discrete (e.g., wild type vs. mutant, presence vs. absence of environmental contaminants, or a series of developmentally plastic phenotypes), it is useful to compare each node with the others in as symmetrical a fashion as possible. This guideline ensures that there will be equivalent power to detect differences among all nodes in your design. Of course, if one experimental node is of greater interest, then it is sensible for your experimental design to feature a higher valence for that node.

Generally, it is easy to design a microarray experiment that examines many factors for their influence on gene expression in organisms. The difficult part is carefully performing such an experiment. Our ability to conceive of interesting variables to manipulate far exceeds our ability to examine them. Figure 2 depicts a more complex example of an experiment along the lines of Fig. 1, but with greater inferential power. However, this experiment would require more than four times as many microarray hybridizations as the experiment depicted in Fig. 1. Figure 3 depicts an example of an experiment to jointly examine population and species variation in gene expression, and this design would need almost as many microarrays as in the experiment depicted in Fig. 2. The technical requirements of these fairly simple-to-conceive designs make clear how rapidly the multiplicative examination of various influences on gene expression can

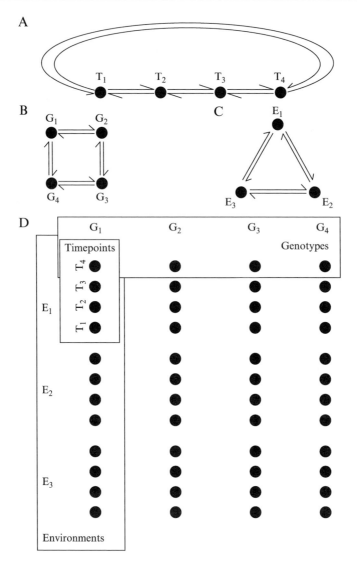

FIG. 2. A general example of an extensive spotted cDNA microarray experimental design, incorporating factors including developmental state ($T_1 \ldots T_4$), genotype ($G_1 \ldots G_4$), and environment ($E_1 \ldots E_3$). Nodes (filled circles) represent 48 mRNA samples, and arcs represent competitive hybridizations. The direction of the arc indicates which mRNA sample was labeled with which fluorophore (e.g., the blunt end indicates the node labeled with Cy5, and the pointed end indicates the node labeled with Cy3). (A) Circuit design for a time course. Each ordinal time point has been compared to its neighbor, and additionally the last time point has been compared to the initial time point. This design comprises eight microarrays and

make experimental designs costly and time consuming. Fortunately, spotted microarrays are not as expensive as other high-throughput expression analyses, and large numbers of arrays can be used. Completing all parts of the experiments diagrammed in Fig. 1A–C or Fig. 2A–C as depicted in Fig. 1D or Fig. 2D results in every experimental node possessing a weighted valence of 12, as well as resolution of the significance of most differences in gene expression of a factor of 1.2 or more. A key step after brainstorming an interesting set of experiments lies in paring down those experiments so they are feasible in both time and expense while retaining the essential comparisons to answer the questions of importance in your system. This stage of the design cannot be prescribed but will always belong to the experimentalist, who must apply his or her expertise and insight to narrow down the biological factors examined to those that are truly relevant.

Replication and Resolution of Differences in Gene Expression

Replication of microarray hybridizations is essential to precise inference of gene expression level (Lee *et al.*, 2000; Pan *et al.*, 2002). Flipping fluorophores is one important method of replication. However, just two hybridizations between a pair of nodes frequently are insufficient for the resolution of small differences in gene expression. Closed circuit designs like those in Figs. 1–3 help to increase the resolution of differences in gene expression, because data from alternate longer paths in the design also provide moderate inferential power (Townsend, 2003), Performing additional hybridizations beyond the minimum of two between two nodes is another option. Additional replicate comparisons increase the probability that the experiment performed will resolve large and small differences in gene expression as statistically significant. A useful summary statistic of the power of an experiment to resolve the statistical significance of small

---

should detect as significant most differences greater than about 1.6-fold. Cross-comparisons from T_2 to T_4 and from T_1 to T_3 could be added to increase resolution. (B) Circuit design for comparison of four genotypes; note that the graph is topologically the same as in panel A. (C) Circuit design for comparison of three environments. This design comprises six microarrays and should detect as significant most differences greater than about 1.6-fold. (D) Diagram showing all 48 mRNA samples that would have to be collected to examine all of these factors. Boxes indicate sets of mRNA samples that correspond to nodes in graphs A–C. An examination of all of these factors in one experiment could be conducted by performing the comparisons in panel A for all combinations of the four genotypes and three environments, by performing the comparisons in panel B for all combinations of the four time points and three environments, and by performing the comparisons in panel C for all combinations of the four time points and the four genotypes. Combined into a single dataset and analyzed together, these 288 microarrays should yield a net resolution of the statistical significance of most differences greater than about 1.2-fold.

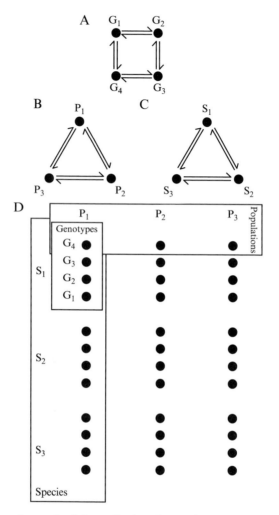

Fig. 3. A general example of the application of spotted cDNA microarray experimental design applied to the study of variation within populations and among species. In this figure, factors examined include genotype ($G_1 \ldots G_4$), population ($P_1 \ldots P_3$), and species ($S_1 \ldots S_3$). Nodes represent mRNA samples, and arcs represent competitive hybridizations. The direction of the arc indicates which mRNA sample was labeled with which fluorophore (e.g., the blunt end indicates the node labeled with Cy5, the pointed end indicates the node labeled with Cy3). (A) Circuit design comparing genotypes G_1 through G_4. This design comprises eight microarrays and should detect as significant most differences greater than about 1.6-fold. Cross-comparisons from G_2 to G_4 and from G_1 to G_3 could be added to increase resolution. (B) Circuit design comparing localities a, b, and c. This design comprises six microarrays and should detect as significant most differences greater than about 1.6-fold. (C) Circuit design comparing species S_1, S_2, and S_3, graph theoretically the same as in panel B.

differences in gene expression is the gene expression level (GEL) at which there is an empirical 50% probability of a statistically significant call (GEL_{50}). For any pair of samples, this statistic may be calculated across all genes present on a microarray by logistic regression of the statistical significance call on estimated GEL (Fig. 4). Additional replicates increase the precision of estimates of GEL, decrease the GEL_{50} for an experiment, and thus increase the power to resolve small differences in gene expression.

Statistical Significance and Its Importance

Early transcriptional profiling experiments identified genes as differentially expressed by a "twofold threshold," so genes whose expression level was greater or lesser by a factor of 2 in a comparison of an experimental to a reference sample were considered differentially expressed (Alexandre *et al.*, 2001; DeRisi *et al.*, 1997; Lyons *et al.*, 2000; Sudarsanam *et al.*, 2000). The twofold threshold has no theoretical basis and often, inappropriately, serves double duty as a signifier of both statistical and biological significance.

Disentangling statistical and biological significance is essential to understanding the power of a study to reveal biological differences among samples (Townsend and Hartl, 2002; Tseng *et al.*, 2001; Wolfinger *et al.*, 2001). The finer resolution of smaller and smaller differences in gene expression with increased replication demonstrated in several empirical studies (Townsend, 2004) and in simulations (see Fig. 4) shows that a particular transcriptional profiling experiment cannot reveal all of the differential expression among treatments or genotypes. Rather, a transcriptional profiling experiment reveals those genes whose differences in expression are sufficiently large and sufficiently consistent in measurement to be statistically different. The number of such genes is strongly influenced by the replication present in the experimental design.

(D) Diagram showing all 36 mRNA samples that would have to be collected to examine all of these factors. Each mRNA sample is represented by a filled circle. Boxes indicate sets of mRNA samples that comprise comparison diagrams A–C. An examination of all of these factors in one experiment could be conducted by performing the comparisons in panel A for all combinations of the three localities and three species, by performing the comparisons in panel B for all combinations of the four genotypes and three species, and by performing the comparisons in panel C for all combinations of the four genotypes and the three localities. Combined into a single dataset and analyzed together, these 216 microarrays should yield resolution of the statistical significance of most differences greater than about 1.2-fold.

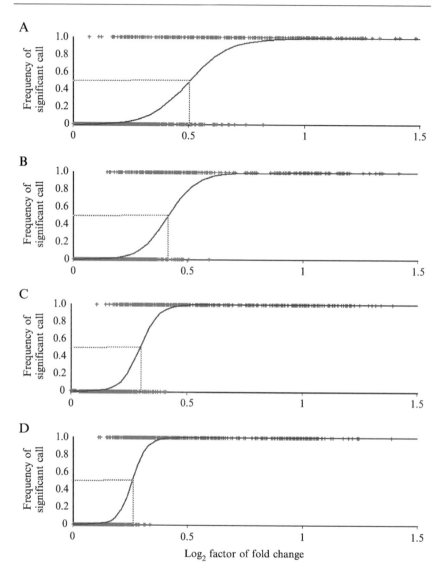

FIG. 4. Logistic regressions of the frequency of affirmative significance call over $\log_2$ factor of difference in gene expression. The logistic model plotted is that $\log_e p/(1 - p) = mx + b$, where x is the $\log_2$ factor of difference in gene expression. Cross symbols represent estimated expression levels from simulated data using a Bayesian analysis of gene expression level with additive small error terms. Each cross is placed on the abscissa at the estimated expression level, either at the top of the plot (significant, S) or at the bottom (not significant, NS). Ratio data were simulated using the probability distribution of (Fieller, 1932) assuming a constant coefficient of variation across samples. Logistic regressions are on the factors of difference

Furthermore, restricting analysis to only those genes showing statistically significant changes yields much more meaningful results. Genes whose expression level measurements were highly variable in a given study should be filtered from large-scale comparisons because of their lack of statistical significance (i.e., large credible or confidence intervals for expression level). The filtered subset of expression level measurements demonstrate vastly increased biological correlation (Townsend, 2003; Townsend and Hartl, 2002; Townsend et al., 2003), because the well-measured genes are not swamped by a morass of poorly measured genes. For instance, large numbers of poorly resolved genes will by chance be clustered within small clusters of genes that are well measured and have true biological association, obscuring otherwise clear functional groupings. The best verification of the results of a DNA microarray study, in the end, is the concordance of expression data with known biology, in particular with molecular biological data on transcription factor recruitment, metabolic pathways, and protein–protein interaction. For DNA microarray studies on population and species differences in expression, it should be kept in mind that these molecular data are well known only for model organisms. Thus, using prior knowledge of organismal biology to troubleshoot the development of DNA microarray technology and protocols is much easier when model organisms and their close relatives are the initial organisms of study.

---

estimated from simulated data comparing two samples to each other with different numbers of replicates. (A) An example with four replicates. The model has a highly significant fit ($\chi^2 = 1593.1$, $P < .0001$). The estimated intercept for the log odds, b, of a significant call versus no significant call is -5.4 (significant, $P < .0001$), and the estimated slope with $\log_2$ factor of difference in gene expression, m, is 11.5 (significant, $P < .0001$). The factor of gene expression at which 50% of estimated differences were identified as significant (GEL_{50}) was 1.4-fold. (B) An example with six replicates. The model has a highly significant fit ($\chi^2 = 1186.8$, $P < .0001$). The estimated intercept for the log odds, b, of a significant call versus no significant call is -6.1 (significant, $P < .0001$), and the estimated slope with $\log_2$ factor of difference in gene expression, m, is 16.5 (significant, $P < .0001$). The factor of gene expression at which 50% of estimated differences were identified as significant (GEL_{50}) was 1.29-fold. (C) An example with eight replicates. The model has a highly significant fit ($\chi^2 = 1186.8$, $P < .0001$). The estimated intercept for the log odds, b, of a significant call versus no significant call is -7.3 (significant, $P < .0001$), and the estimated slope with $\log_2$ factor of difference in gene expression, m, is 26.7 (significant, $P < .0001$). The factor of gene expression at which 50% of estimated differences were identified as significant (GEL_{50}) was 1.21-fold. (D) An example with 10 replicates. The model has a highly significant fit ($\chi^2 = 1307.2$, $P < .0001$). The estimated intercept for the log odds, b, of a significant call versus no significant call is -8.6 (significant, $P < .0001$), and the estimated slope with $\log_2$ factor of difference in gene expression, m, is 35.0 (significant, $P < .0001$). The factor of gene expression at which 50% of estimated differences were identified as significant (GEL_{50}) was 1.18-fold. (See color insert.)

Comparing within and among Species

Comparisons among different individuals are possible within and among species. Interpretation of such comparisons will be most convincing if information is already available on the circumscription of populations and species, preferably by phylogenetic and population genetic studies using nucleic acid variation and appropriate methods of statistical analysis. Typically, arrays of any kind will be designed based on sequence of one individual in one species. To incorporate natural variation into microarray studies, it will be necessary to use arrays designed from the sequence of one individual for competitive hybridizations among genetically different individuals or species. The utility of the array will decrease as the genetic distance increases between the "design" individual and the "experimental" individuals. Detailed knowledge of populations and species will allow researchers to select individuals with increasing genetic distances to test the range of a microarray to aid experimental design. Fragmented DNA can be used as the probe in these experiments.

In terms of design, and in keeping with the aforementioned guidelines, it is ideal to make direct comparisons among the most closely related populations and/or species and then work toward more divergent comparisons. This approach is needed because sequence divergence will increase with phylogenetic distance and confound the hybridizations at each spot on a microarray (Bozdech *et al.*, 2003; Letowski *et al.*, 2004; Nagpal *et al.*, 2004). In principle, a consistent low level of divergence across genes should not present a serious problem, because any decrease in hybridization due to that divergence will then be approximately constant across genes in that sample and a global normalization will compensate appropriately. However, there is considerable variation in the rate of divergence of genes (Graybeal, 1994). Thus, experimentalists must be wary of interpreting differential hybridization as representative of differential expression when mRNA is derived from divergent organisms.

Although divergence in DNA sequence creates a difficulty for the study of gene expression in closely to moderately diverged species, it is an advantage for the study of gene expression in very distantly related species that live in close association. For instance, many of the socially important and evolutionarily interesting questions concerning fungi involve mutualisms with other organisms, ranging from pathogenesis to symbiosis. Microarrays containing neighboring spots of DNA sequence for genes of both species can be used to simultaneously measure gene expression in the two organisms with little concern for cross-hybridization. Thus, single arrays composed of sequences homologous to both partners may be used to identify pools of candidate genes that are coordinately regulated in both partners in the mutualism (Johansson *et al.*, 2004). Microarrays are designed to measure mRNA level and profile transcription throughout the genome, but they also can be used to discriminate between alleles at single nucleotide polymorphism (SNP) loci throughout the genome (Steinmetz *et al.*, 2002). Therefore, experiments can be designed to combine QTL analysis and transcription profiling in progeny from parents with phenotypes important to adaptation: pathogenicity, industrial production, and so on.

Disturbance Involved in Harvesting mRNA

A key element to good competitive hybridization experiments is to minimize disturbance of the organisms in the process of harvesting RNA, to avoid accidentally studying the effects of the disturbance. Moreover, it is absolutely vital to confine differences among experimental treatments to those being tested in the experiment; any other differences will confound analysis and interpretation. Consider with extreme care how samples will be handled immediately before RNA harvest. An incautiously designed microarray experiment may easily reveal only the manifold effects of centrifugation, filtering, or other lab manipulations. Of course, the same experimental protocols cannot be applied to all organisms, for example, contrast the swift and uniform arresting of bacterial cell activity at $-80°$ or with ethanol (EtOH)–phenol treatment (Zimmer *et al.*, 2000) with the quite variable treatments of human and other primate organs before RNA extraction (Enard *et al.*, 2002). Even routine experimental manipulations can have a strong effect on transcription, as exemplified by a study of gravitropism in plants that showed significant changes in transcriptional profile due solely to the minimal handling of plants before RNA extraction (Moseyko *et al.*, 2002).

Technical Protocols

DNA Microarray Construction

DNA microarrays are multiplexed Southern hybridizations (Fig. 5). Using terminology that mirrors the functional roles of the nucleic acids in Southern hybridizations, the "probe" in a microarray experiment is unlabeled and affixed to a solid substrate (the coated glass slide). The "target," counterintuitively, is washed over the multiple affixed probes and consists of fluorescently labeled cDNA made from mRNA harvested from the organism. Spotted microarrays can be constructed by deposition of polymerase chain reaction (PCR) products for each and every open reading

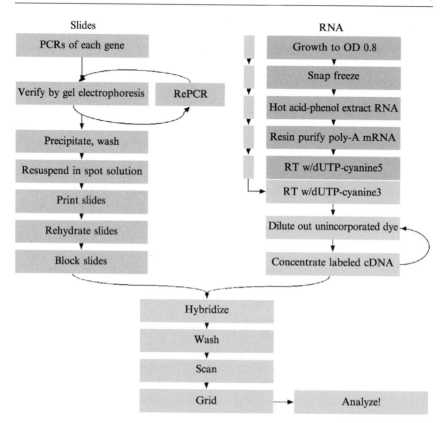

Fig. 5. Schematic diagram of the steps employed to create spotted microarrays from polymerase chain reaction products and to perform a comparative microarray hybridization. (See color insert.)

frame (ORF) (Eisen and Brown, 1999), by deposition of oligomers (70 or 50 nucleotides) designed from predicted ORF sequences, or by deposition of random clones from a cDNA library or even from random clones of genomic DNA fragments. Oligomer arrays and cDNA arrays appear to have similar sensitivities for differential gene expression (Lee *et al.*, 2004), but there are advantages and disadvantages to each method (Table I). For example, it is easier to design oligomer probes that minimize cross-hybridization, but 10% sequence divergence randomly located within a 70-mer long oligo may reduce hybridization intensity by 64% (Bozdech *et al.*, 2003). Conversely, whole-ORF PCR products do cross-hybridize but appear not to be very sensitive to SNPs (Ranz *et al.*, 2003).

ORFs may be amplified from clones or from genomic DNA by means of the PCR. Longer ORFs may be amplified with any of a variety of special polymerase kits that have high processivity. Each amplified product should be confirmed for correct length by agarose gel electrophoresis. Amplified DNA may be precipitated in 96-well format with isopropanol, washed with 70% EtOH, and resuspended in a salt spotting solution such as 3× SSC. DNA is commonly spotted on polylysine-coated glass slides (Eisen and Brown, 1999) or γ-aminopolysilane (GAPS)–coated glass slides (Corning, Corning, NY), using a microarraying robot (e.g., http://cmgm.stanford.edu/pbrown/mguide/index.html). GAPS slides are more expensive but have a better shelf life and tend to be less variable in quality.

Extraction of RNA and Reverse Transcription

The key reagent in transcriptional profiling is the RNA harvested from an organism. This RNA is used as a template for reverse transcription to make cDNA for competitive hybridization against the affixed probes on the microarray. RNA is best extracted from flash-frozen pellets of tissue or culture grown in meticulously maintained common garden conditions. The flash-frozen matter should have its RNA extracted in a manner that will not result in mRNA degradation (i.e., performed rapidly upon or while thawing). Nucleic acids may be EtOH precipitated, washed, dried, and redissolved in TE buffer. Yield ranges from organism to organism, but a spectrophotometric ratio of absorption (260 nm/280 nm) of about 2.0 indicates a clean preparation without much protein contaminant. mRNA may be purified easily using a Qiagen Extraction Kit (Valencia, CA), which contains columns that retain poly-A RNA and allow much of the tRNA and rRNA to pass through. For eukaryotes, reverse transcription of eluted mRNA may be performed with oligo-dT primers of an appropriate length to bind with the poly-A tails of mRNA from the organism of study. For both eukaryotes and prokaryotes, the reactions may be primed with random hexamer primers, supplied with deoxyribonucleic triphosphates (dNTPs), performed by a reverse transcriptase such as Superscript II. To provide a ligand for dye labeling, amino-allyl-dUTP is incorporated into the cDNA along with the dNTPs. After at least 2 h of reverse transcription, the approximately 20-μl reaction should be stopped with 10 μl of 1 M NaOH and 10 μl of 0.5 M of ethylenediaminetetraacetic acid (EDTA), and the mix is incubated at 65° for 15 min. Then 25 μl of 1 M HEPES pH 7.5 is added to stabilize the solution.

Both total RNA and purified mRNA have been successfully used as templates for the production of labeled cDNA for microarray hybridization.

Also, both oligo-dT alone and a mixture of oligo-dT and oligo-dN primer have been used as primers to reverse transcribe mRNA. For the most part, both purifying the mRNA and using oligo-dT alone help decrease noise caused by errant cDNAs from tRNA and rRNA. However, using total RNA is cheaper and faster, and using oligo-dN primers can dramatically increase signal. The tradeoff must be examined in each particular microarray experimental context, because these factors interact with other aspects of experimental setup such as spotted DNA fragment size, slide chemistry, deposition solution, and hybridization conditions. Figure 6 may be used as a guide for optimization.

Cyanine Dye Coupling

To retain the cDNA and discard unincorporated nucleotides, the products of reverse transcription are diluted and filtered in a Microcon-30 microconcentrator, which retains the long polymers of cDNA but not unincorporated nucleotides. Typically, an initial 10-fold dilution of the reverse transcription reaction product is followed by a 20-fold concentration. Two more rounds of 20-fold dilution and concentration complete the cleanup. NHS dye may be bound to cDNA via amino-allyl-dUTP residues by raising the pH. To 10–13 μl of purified concentrate, 0.8 μl of 1 M NaHCO$_3$ pH 9 can be added, with an appropriate NHS-cyanine dye aliquot. This coupling reaction is incubated in the dark at 25° for 75 min and then stored in the dark at 4° and used in less than 24 h. The labeled cDNA may then be purified with a QIAquick column. This elution of about 55 μl of purified cyanine-labeled cDNA may also be stored at 4° and should be used in less than 24 h.

	TTT... primers	TTT... primers and NNN... primers
Total RNA	Low signal Intermediate noise	High signal High noise
Poly-A RNA	Low signal Low noise	High signal Intermediate noise

FIG. 6. Diagram relating signal and noise characteristics for microarray hybridizations conducted using messenger RNA (mRNA) or total RNA and poly-T or polyT + polyN primers used in the reverse-transcriptase reaction for production of the labeled mRNA target to be washed over the microarray of fixed probe spots.

Hybridization

For each competitive hybridization, the labeled target cDNAs from two samples are used, i.e., one labeled with cyanine-3 and one labeled with cyanine-5. The labeled cDNA is concentrated to 20 μl in Microcon-30 microconcentrators, combining appropriate cyanine-3– and cyanine-5– labeled paired samples. Note that 1.5 μl of poly-dA oligomers of appropriate length for the organism of study may be added to block poly-T tails of the cDNA. Next, 3 μl of 20× SSC and 0.5 μl of 1 M HEPES pH 7.0 are added. The mix can then be filtered of any dust or residues with a wetted (10 μl ddH$_2$O) Millipore-0.45 μm filter; 10% sodium dodecyl sulfate (SDS) is added, and the mix is then boiled for 2 min to denature the nucleic acids. It should then be cooled at 27° for 10 min. Hybridizations using labeled target at temperatures above room temperature can result in extremely high background fluorescence. A microarray slide is set in a hybridization chamber. To keep the slide stable within the chamber, deposit drops of 3× SSC on the underside of the slide, allowing them to adsorb to the slide corners and the chamber bottom. To prevent dehydration of the labeled cDNA solution from beneath the coverslips, 3× SSC is added to the hybridization chamber wells. A coverslip (LifterSlips are very convenient for this purpose) should be cleaned with EtOH and then placed over the printed microarray. The labeled cDNA mix is then injected at the corners of the coverslip, and the chamber is sealed and then placed level in a 60° waterbath, to be incubated at 60–63° for 12–15 h to reach equilibrium (Sartor et al., 2004).

Array Wash

Hybridized microarray slides can be washed by repeated plunging in a solution of 387 ml of purified water, 12 ml of 20× SSC, and 1 ml of 10% SDS, and rinsed by repeated plunging in a solution of 399 ml of purified water and 1 ml of 20× SSC. The array should be scanned as soon as possible; if there must be a delay, the array may be stored in the dark, but for no more than 2 h.

Data Acquisition and Analysis

Fluorescent DNA bound to the microarray may be detected with a GenePix 4000 microarray scanner (Axon Instruments, Foster City, CA), using the GenePix 4000 software package to locate spots in the microarray. Other scanners are available, as are alternative, open source, and freely downloadable scanning software, such as TIGR's Spotfinder (http://www.tigr.org/software).

Normalization

Fluorescence intensity values are commonly adjusted in each channel by subtracting the average background intensity observed surrounding the spot from the observed foreground intensity of the spot. Typical foreground intensities for quality microarray hybridizations should be 10 or more times as intense as background. However, some spots do not achieve this ideal even in excellent hybridizations. To eliminate signals that are most prone to estimation error, any spot can be excluded from analysis if both the Cy3 and the Cy5 fluorescence intensities for that spot are within 3 standard deviations of the distribution of intensities of the background pixels for that spot. This procedure avoids artificially inflated measurements of relative expression for the competing mRNA samples to that spot due to near-zero background-subtracted intensity values in one fluorescence channel.

Relative expression levels for the two competing dye-labeled samples may be normalized by linear scaling of the cyanine-5 values so the mean cyanine-5 and cyanine-3 background-corrected intensity values of nonexcluded spots are equal when hybridizations are of uniformly high quality. This straightforward method should then yield a linear log-log cyanine-3–cyanine-5 intensity and no further normalization will be necessary. If the relationship is not linear because of systemic technical problems, it may be useful to use LOWESS smoothing or other statistical approaches. However, curved or abnormally shaped log-log regressions generally indicate poor and therefore misleading data, regardless of applied statistical sophistication. It should be considered by the experimentalist whether extra analytical efforts to glean the most information from poor experiments are more fruitful than repeating hybridizations and refining experimental technique.

Data Analysis

Multifactorial experimental designs, as described earlier, should be analyzed statistically. Two major methods for analysis of such designs are analysis of variance (ANOVA) methods (Kerr and Churchill, 2001a,b; Wolfinger *et al.*, 2001) and Bayesian methods (Townsend, 2004; Townsend and Hartl, 2002). The details of these approaches are adequately covered in the primary literature. The two approaches yield consistent results when used to analyze the same dataset (Whitfield *et al.*, 2003), so the choice of method may come down to a combination of philosophical preference and practicality. ANOVA presents a powerful method that can, with a fair degree of statistical savvy on the part of the user, be tailored tightly to incorporate normalization and downstream analysis (Jin *et al.*, 2001) into a single cohesive whole. A powerful and flexible method for the Bayesian analysis of gene expression levels (BAGEL), from multifactorial experiments, has been implemented to analyze normalized data on multiple platforms in a straightforward and freely available software package (http://web.uconn.edu/townsend/software.html).

Acknowledgments

We thank the Miller Institute for Basic Research and NIH grant GM068087 for supporting our research.

References

Alexandre, H., Ansanay-Galeote, V., Dequin, S., and Blondin, B. (2001). Global gene expression during short-term ethanol stress in *Saccharomyces cerevisiae*. *FEBS Lett.* **498**, 98–103.

Bozdech, Z., Zhu, J., Joachimiak, M. P., Cohen, F. E., Pulliam, B., and DeRisi, J. L. (2003). Expression profiling of the schizont and trophozoite stages of *Plasmodium falciparum* with a long-oligonucleotide microarray. *Genome Biol.* **4**, R9.

DeRisi, J. L., Iyer, V. R., and Brown, P. O. (1997). Exploring the metabolic and genetic control of gene expression on a genomic scale. *Science* **278**, 680–686.

Eisen, M. B., and Brown, P. O. (1999). DNA arrays for analysis of gene expression. *Methods Enzymol.* **303**, 179–205.

Enard, W., Khaitovich, P., Klose, J., Zollner, S., Heissig, F., Giavalisco, P., Nieselt-Struwe, K., Muchmore, E., Varki, A., Ravid, R., Doxiadis, G. M., Bontrop, R. E., and Paabo, S. (2002). Intra- and interspecific variation in primate gene expression patterns. *Science* **296**, 340–343.

Fieller, E. C. (1932). The distribution of the index in a normal bivariate population. *Biometrika* **24**, 428–440.

Graybeal, A. (1994). Evaluating the phylogenetic utility of genes: A search for genes informative about deep divergences among vertebrates. *Syst. Biol.* **43**, 174–193.

Hwang, L., Hocking-Murray, D., Bahrami, A. K., Andersson, M., Rine, J., and Sil, A. (2003). Identifying phase-specific genes in the fungal pathogen *Histoplasma capsulatum* using a genomic shotgun microarray. *Mol. Biol. Cell* **14**, 2314–2326.

Jin, W., Riley, R. M., Wolfinger, R. D., White, K. P., Passador-Gurgel, G., and Gibson, G. (2001). The contributions of sex, genotype and age to transcriptional variance in *Drosophila melanogaster*. *Nat. Genet.* **29**, 389–395.

Johansson, T., Le Quere, A., Ahren, D., Soderstrom, B., Erlandsson, R., Lundeberg, J., Uhlen, M., and Tunlid, A. (2004). Transcriptional responses of Paxillus involutus and Betula pendula during formation of ectomycorrhizal root tissue. *Mol. Plant Microbe Interact.* **17**, 202–215.

Kerr, M. K., and Churchill, G. A. (2001a). Experimental design for gene expression microarrays. *Biostatistics* **2**, 183–201.

Kerr, M. K., and Churchill, G. A. (2001b). Statistical design and the analysis of gene expression microarray data. *Genet. Res.* **77**, 123–128.

Lee, H. S., Wang, J. L., Tian, L., Jiang, H. M., Black, M. A., Madlung, A., Watson, B., Lukens, L., Pires, J. C., Wang, J. J., Comai, L., Osborn, T. C., Doerge, R. W., and Chen, Z. J. (2004). Sensitivity of 70-mer oligonucleotides and cDNAs for microarray analysis ofgene expression in Arabidopsis and its related species. *Plant Biotechnol. J.* **2**, 45–57.

Lee, M.-L. T., Kuo, F. C., Whitmore, G. A., and Sklar, J. (2000). Importance of replication in microarray gene expression studies: Statistical methods and evidence from repetitive cDNA hybridizations. *Proc. Natl. Acad. Sci. USA* **97,** 9834–9839.

Letowski, J., Brousseau, R., and Masson, L. (2004). Designing better probes: Effect of probe size, mismatch position and number on hybridization in DNA oligonucleotide arrays. *J. Microbiol. Methods.* **57,** 269–278.

Lyons, T. J., Gasch, A. P., Gaither, L. A., Botstein, D., Brown, P., and Eide, D. (2000). Genome-wide characterization of the Zap1p zinc-responsive regulon in yeast. *Proc. Natl. Acad. Sci.* **97,** 7957–7962.

Moseyko, N., Zhu, T., Chang, H. S., Wang, X., and Feldman, L. J. (2002). Transcription profiling of the early gravitropic response in Arabidopsis using high-density oligonucleotide probe microarrays. *Plant Physiol.* **130,** 720–728.

Nagpal, S., Karaman, M. W., Timmerman, M. M., Ho, V. V., Pike, B. L., and Hacia, J. G. (2004). Improving the sensitivity and specificity of gen expression analysis in highly related organisms through the use of electronic masks. *Nucleic Acids Res.* **32,** e51.

Novoradovskaya, N., Whitfield, M. L., Basehore, L. S., Novoradovsky, A., Pesich, R., Usary, J., Karaca, M., Wong, W. K., Aprelikova, O., Fero, M., Perou, C. M., Botstein , D., and Braman, J. (2004). Universal reference RNA as a standard for microarray experiments. *BMC Genomics* **5,** 20.

Pan, W., Lin, J., and Le, C. T. (2002). How many replicates are required to detect gene expression changes in microarray experiments? A mixture model approach. *Genome Biol.* **3,** research 0022.1–0022.10.

Puskás, L. G., Zvara, Á., Hackler, L., Micsik, T., and van Hummelen, P. (2004). Production of bulk amounts of universal RNA for DNA microarrays. *Biotechniques* **33,** 898–904.

Ranz, J. M., Castillo-Davis, C. I., Meiklejohn, C. D., and Hartl, D. L. (2003). Sex-dependent gene expression and evolution of the *Drosophila* transcriptome. *Science* **300,** 1742–1745.

Sartor, M., Schwanekamp, J., Halbleib, D., Mohamed, I., Karyala, S., Medvedovic, M., and Tomlinson, C. R. (2004). Microarray results improve significantly as hybridization approaches equilibrium. *Biotechniques* **36,** 790–796.

Singh, R. S. (2003). Darwin to DNA, molecules to morphology: The end of classical population genetics and the road ahead. *Genome* **46,** 938–942.

Steinmetz, L. M., Sinha, H., Richards, D. R., Spiegelman, J. I., Oefner, P. J., McCusker, J. H., and Davis, R. W. (2002). Dissecting the architecture of a quantitative trait locus in yeast. *Nature* **416,** 326–330.

Sudarsanam, P., Iyer, V. R., Brown, P. O., and Winston, F. (2000). Whole-genome expression analysis of *snf/swi* mutants of *Saccharomyces cerevisiae*. *Proc. Natl. Acad. Sci. USA* **97,** 3364–3369.

Townsend, J. P. (2003). Multifactorial experimental design and the transitivity of ratios with spotted DNA microarrays. *BMC Genomics* **4,** 41.

Townsend, J. P. (2004). Resolution of large and small differences in gene expression using models for the Bayesian analysis of gene expression levels and spotted DNA microarrays. *BMC Bioinformatics* **5,** 54.

Townsend, J. P., Cavalieri, D., and Hartl, D. L. (2003). Population genetic variation in genome-wide gene expression. *Mol. Biol. Evol.* **20,** 955–963.

Townsend, J. P., and Hartl, D. L. (2002). Bayesian analysis of gene expression levels: Statistical quantification of relative mRNA level across multiple treatments or samples. *Genome Biol.* **3,** research 0071.1–0071.16.

Tseng, G. C., Oh, M.-K., Rohlin, L., Liao, J. C., and Wong, W. H. (2001). Issues in cDNA microarray analysis: Quality filtering, channel normalization, models of variations and assessment of gene effects. *Nucleic Acids Res.* **29,** 2549–2557.

Whitfield, C. W., Cziko, A.-M., and Robinson, G. E. (2003). Gene expression profiles in the brain predict behavior in individual honey bees. *Science* **302,** 296–299.

Wolfinger, R. D., Gibson, G., Wolfinger, E. D., Bennett, L., Hamadeh, H., Bushel, P., Afshari, C., and Paules, R. S. (2001). Assessing gene significance from cDNA microarray expression data via mixed models. *J. Comput. Biol.* **8,** 625–637.

Yang, Y. H., and Speed, T. (2002). Design issues for cDNA microarray experiments. *Nat. Rev.* **3,** 579–588.

Zimmer, D. P., Soupene, E., Lee, H. L., Wendisch, V. F., Khodursky, A. B., Peter, B. J., Bender, R. A., and Kustu, S. (2000). Nitrogen regulatory protein C-controlled genes of *Escherichia coli*: Scavenging as a defense against nitrogen limitation. *Proc. Natl. Acad. Sci. USA* **97,** 14674–14679.

[32] Methods for Studying the Evolution of Plant Reproductive Structures: Comparative Gene Expression Techniques

By ELENA M. KRAMER

Abstract

A major component of evolutionary developmental (evo-devo) genetics is the analysis of gene expression patterns in nonmodel species. This comparative approach can take many forms, including reverse-transcriptase polymerase chain reaction, Northern blot hybridization, and *in situ* hybridization. The choice of technique depends on several issues such as the availability of fresh tissue, as well as the expected expression level and pattern of the candidate gene in question. Although the protocols for these procedures are fairly standard, optimization is often required because of the specific characteristics of the species under analysis. This chapter describes several methods commonly used to determine gene expression patterns in angiosperms, particularly in floral tissues. Suggestions for adapting basic protocols for diverse taxa and troubleshooting are also extensively discussed.

General Considerations for Working with RNA

RNA is, by nature, a less stable molecule than DNA and there are many sources of RNase in the environment, which can lead to its rapid degradation. These facts often lead to trepidation on the part of researchers who are not experienced with RNA work. However, such concerns are largely unnecessary because simple precautions can effectively prevent

RNA degradation. First, researchers must always wear sterile gloves when working with RNA. Gloves should be changed if the skin or items generally handled without gloves, such as doorknobs, are touched. Many labs go so far as to have specific RNA-use benches or even rooms, but this is not an absolute requirement as long as the bench space is kept clean. Second, all plasticware (tips, pipets, tubes) should be purchased RNase-free and kept that way by maintaining them as separate lab stocks that are handled only with gloves. For instance, RNase-free supplies can be stored in a separate cabinet that is kept locked to prevent unintentional contamination. Third, glass or metalware should be made RNase-free either by baking at 235° for more than 2 h (careful baking bottle caps, most will melt!) or by treating with 0.1 M NaOH overnight followed by thorough rinsing with sterile water. Before baking, wrap bottle mouths and metal items, such as slide racks or stir bars, in aluminum foil to help keep them RNase-free after baking. Fourth, all solutions should be made with water, chemicals, stir bars, cylinders, bottles, and others, which are all RNase-free. RNase-free water can be prepared using diethylpyrocarbonate (DEPC), which is highly toxic and should be used only under a fume hood. DEPC is typically diluted in double-deionized water (ddH$_2$O) to produce a 0.1% solution. Because DEPC is unstable once the bottle seal is broken, it is best to prepare large batches of DEPC water at one time in order to use the entire volume. Take clean bottles of various sizes (100 ml to 2 L), fill with fixed amounts of ddH$_2$O, and place under a fume hood. Aliquot the appropriate amounts of DEPC into each bottle, tightly secure the bottle caps, and shake. Let the bottles sit in the fume hood for 2–24 h. DEPC has only a 30 min half-life in ddH$_2$O, but autoclaving for 15–30 min will ensure that all DEPC is inactivated. It is important to note that DEPC can be added to many solutions directly, but it is not compatible with Tris or MOPS buffers. Tris and MOPS solutions can be prepared using DEPC water, and RNase-free bottles and chemicals so treatment is not necessary after solution preparation. Also, remember that the bottles that are used to make DEPC water can also be treated as RNase-free themselves after they are emptied. See http://www.ambion.com/techlib/basics/rnasecontrol/ for more information on DEPC water and RNase control methods.

Methods Based on RNA Extraction

Collecting and Storing Tissue

Given the instability of RNA, the collection and storage of tissue becomes a much greater concern than when the goal is DNA extraction. There are two primary methods for storing tissue for future RNA extraction:

freezing at $-80°$ or infiltration with a special RNA preservative such as RNA*later* (Ambion). In either case, it is critical to quickly collect the material and complete the selected treatment. Some tissues may be more sensitive to RNA degradation than others, but no more than 15 min at ambient temperature or 1 h on ice is ideal. For best results, tissue should be treated immediately. In cases in which dissection is required, material can be kept on ice during the dissection and the separated tissues treated as soon as the desired amount is obtained (but observing the time frame mentioned earlier).

For freezing, it is best to collect tissue in plastic screw-top or snap-top tubes. Plastic or paper bags perform poorly during long-term storage at $-80°$. Make sure to mark all samples with ink or labels that are compatible with cold storage. Once material is frozen, it is critical that it be maintained at temperatures less than $-50°$; RNA will not tolerate freeze–thaw cycles of more than $30°$. Material can be stored at $-80°$ for very long periods, however, as long as the temperature is controlled.

As a general rule, tissue that has been fixed in formaldehyde or similar chemical fixatives is not suitable for RNA extraction. Likewise, silica-preserved tissues will not yield usable RNA. These issues must be taken into consideration when planning any experiment that will use RNA and, unfortunately, often mean that remote field collections are not compatible with RNA-based techniques.

Methods of RNA Extraction

Many expression analysis techniques, such as Northern blot hybridization and reverse-transcriptase polymerase chain reaction (RT-PCR), start with the extraction of total RNA. Although numerous plants are perfectly amenable to this process, it is not uncommon for the presence of polyphenols, polysaccharides, mucilage, and other compounds to inhibit RNA extraction from plant tissues. A number of specialized protocols have been developed to deal with particularly recalcitrant species or tissue types (Chang *et al.*, 1993; Suzuki *et al.*, 2003; Wang *et al.*, 2000). As a general rule of thumb, younger tissues, such as flower buds or immature leaves, are easier to work with. Many kits and products for RNA extraction are available (see http://www.ambion.com, http://www. qiagen.com, and http://www.invitrogen.com, among others). We prefer to use Concert Plant RNA Reagent by Invitrogen (Carlsbad, CA) because of the simple scalability of the protocol for various amounts of tissue (100 mg to 5 g). This protocol is not reproduced here because we follow the manufacturer's instructions quite closely, but a number of general considerations are noteworthy.

• It is unnecessary to treat items such as mortars and pestles or spatulas to make them RNase-free. Although they should be clean, there is much more RNase in the tissue than there is on these tools.

• The best way to prevent degradation of RNA during the grinding process is to keep all utensils very cold. This will also aid in grinding. Prechill mortars and pestles with liquid nitrogen for 5 min before grinding the tissues in liquid nitrogen. Grind small amounts of tissue in microtubes using disposable micropestles (e.g., VWR KT49521-1590) prechilled in liquid nitrogen.

• Best results will be obtained by grinding the frozen tissue to a dustlike consistency. This should be rapidly transferred to a large tube containing the appropriate amount of extraction buffer using a prechilled spatula. For small amounts of tissue, further maceration of material in the extraction buffer can greatly increase yield.

• Once an RNA pellet has been obtained, rinse it carefully in 70% ethanol made with diethylpyrocarbonate (DEPC) water. The dry pellet should become translucent and glassy.

• Resuspension of the RNA is aided by the use of prewarmed (50–60°) RNase-free water or 2 mM of ethylenediaminetetraacetic acid (EDTA) pH 8.0. Resuspend the pellet in as small a volume as possible to facilitate downstream protocols. Inability to resuspend the pellet may result from overdrying (never dry RNA in a SpeedVac) but is also often indicative of the presence of secondary compounds or starch. This may indicate that alternative preparation protocols will be necessary.

• RNA should be stored at −80°.

Northern Blot Hybridization

Northern blots can be useful to examine the expression of genes in various tissue types or over a range of developmental stages. They can be prepared using either total RNA or poly(A) RNA. Several standard protocols for Northern blot preparation are available, including that of Sambrook and Russel (2001a). General tips for performing Northern blot hybridization are as follows:

• Prepare the electrophoresis apparatus, gel casting tray, and gel comb by treatment with 0.1 M NaOH for 12 h, followed by thorough rinsing in water.

• The RNA must be suspended at a reasonably high concentration to allow loading of a large amount of RNA (5–10 μg) in 1–5 μl of sample. Care should be taken to equalize the loading amount of each sample so direct comparisons can be made across different lanes.

• Once the RNA transfer to the membrane is complete, the transfer efficiency can be checked by illuminating the blot with a handheld ultraviolet (UV) light source. The nucleic acids must be fixed to the membrane surface, which can be done using commercially available UV cross-linkers or by baking in a vacuum oven for at least 1 hr at 80°. Using both methods increases the stability of the blot and facilitates repeated probing and stripping.

• I recommend a hybridization solution for Northerns composed of 50% formamide, 3× SSC (3.0 M NaCl, 0.3 M sodium acetate), 0.5% sodium dodecyl sulfate (SDS), 0.1 mg/ml of herring sperm DNA, 5X Denhardt's solution (1% Ficoll, 1% polyvinylpyrrolidone, 1% bovine serum albumin [BSA], filter sterilized), and 25 mM of ethylenediaminetetraacetic acid (EDTA) pH 8. This increases the stringency of the hybridization.

Blots prepared in this manner can be stripped and hybridized with different probes up to six times. Relative lane loading can also be assessed by hybridizing the blot with a control probe, such as ubiquitin or actin, in the final hybridization. When combined with phospho imaging, this can allow more accurate quantification of relative expression levels.

PCR-Based Methods

To an increasing degree, PCR-based approaches are replacing the use of Northern blots to assess gene expression. These techniques include "semiquantitative" RT-PCR and quantitative or real-time RT-PCR (RT-qPCR). Before PCR can be conducted, first-strand cDNA must be synthesized from the RNA. Although total RNA can be used as the template for this reaction, I have found that using poly(A) RNA yields the best results. A caveat to this statement, however, is that poly(A) RNA constitutes only about 10% of a total RNA extraction, and typical yields are in the range of 1–5%. This means that extraction of poly(A) from total RNA is worth pursuing only when starting with a fairly large amount of total RNA (at least 50–100 μg). Many poly(A) extraction kits are available, including some that allow direct preparation of poly(A) from tissue (see Ambion, Qiagen, Invitrogen, Clontech). I prefer to use magnet-based methods (Dynal, Ambion, Novagen), which allow the elution of poly(A) RNA in very small volumes.

Before cDNA synthesis, RNA should be treated with RNase-free DNAse, particularly when working with total RNA. The stop solutions typically used with DNAse, however, contain concentrations of EDTA that may inhibit further enzymatic reactions. This means that treated RNA must be column cleaned using a product such as RNAqueous (Ambion, Austin, TX) or RNeasy (Qiagen, Valencia, CA) and eluted in DEPC water.

RNA can be easily quantified with UV spectrophotometry; high-quality RNA should give an A_{260}/A_{230} reading of more than 1.8. Electrophoresis in a typical TBE or TAE gel is also useful to assess RNA quality. Keeping a small electrophoresis rig, gel casting tray, and gel comb RNase-free is useful if this is going to be performed frequently. Good quality total RNA will appear as a long, bright smear punctuated by two to four distinct bands of moderate molecular weight, which represent the rRNAs and sometimes *rbcL* transcripts if the tissue is photosynthetic. Poly(A) RNA does not look like much on a gel, just a bright smear in the range of 400–2000 bp. A faint smear with a bright spot at low molecular weight (<100 bp) indicates that the RNA is degraded.

Preparation of first-strand cDNA can be done in a separate reaction or at the same time as the PCR (one-step RT-PCR). Again, many manufacturers supply products for both types of procedure (Ambion, Invitrogen, Stratagene, Clontech, Promega, etc.). Because my laboratory is typically interested in the expression of multiple genes, my colleagues and I prefer to prepare cDNA in a separate reaction using an anchored poly-T primer [see Kramer *et al.* (1998)]. This type of cDNA is stable at $-20°$ for several months and can be used with any combination of degenerate or specific PCR primers. Following cDNA synthesis, the reaction can be treated with RNase to remove all remaining RNA, and the cDNA can be column purified using any available PCR cleanup kit. The best way to assess the success of the cDNA reaction is to use it as a template in a PCR with primers for control loci such as ubiquitin or actin.

Semiquantitative RT-PCR simply involves performing PCR on cDNA using primers that are designed to be specific to the gene of interest. It is advisable to design the primers so that they span at least one intron. This allows the amplification of contaminating genomic DNA to be simply detected because the product derived from genomic DNA will be larger than that derived from cDNA. In addition to the typical concerns for designing PCR primers (annealing temperature, GC content, etc.), primer specificity should be carefully considered so that only the gene of interest is amplified, particularly when working in large gene families. Every effort should be made to use equivalent amounts of cDNA in each reaction, and it is important to run positive control reactions using primers for a gene that is expected to be expressed fairly uniformly, such as actin.

Several caveats are important to consider when performing semiquantitative RT-PCR and interpreting the results. The highly sensitive nature of PCR means that even samples that have vanishingly small amounts of target can yield bands if subjected to enough amplification cycles. For this reason, it is advisable to use cycle numbers in the 20–25 range to increase confidence that the detection of cDNA is significant. Of course, some loci are expressed at such low levels that additional cycles are necessary. In this case, it is preferable to show the results for several experiments using both low and high cycle numbers. Real-time or quantitative RT-PCR is rapidly becoming a standard method for more accurate analysis of relative gene expression levels. These methods require specialized equipment, including light-detecting PCR machines. The technical considerations and data analyses involved in this procedure are quite complex and beyond the scope of this chapter. Several excellent web sites on this topic exist, including http://www.wzw.tum.de/gene-quantification/. For both semiquantitative and RT-qPCR, however, several limitations remain. Early stages of development are often difficult to analyze, because of the inability to dissect separate organs for RNA extraction. In addition, important aspects of gene expression, such as spatial patterns of RNA distribution within an organ, are not observable. For these reasons, the best tool for assessing gene expression patterns remains *in situ* hybridization.

In Situ Hybridization

In situ hybridization is, as already mentioned, the best available method for obtaining information on the specific spatial and temporal patterns of gene expression in developing tissues and organs. At the same time, the method is complex and requires considerable preparation and expertise. Commonly used protocols involve hybridization of radioactive or nonradioactive RNA probes to sectioned tissue. Radioactive probes are thought to be more sensitive but have many drawbacks, including poor localization of the signal, instability of the probe, and the requirement to process slides in complete darkness. Several protocols for labeling and hybridization of radioactive probes are available, including that of Weigel and Glazebrook (2002). Nonradioactive *in situ* hybridization is generally preferable and is described here. Many variations on the protocol exist, but all are derived from Jackson (1991). The following protocol has been successfully used with magnoliid dicot, monocot, and eudicot floral tissue. The procedure is described in six parts: fixation and embedding of tissue, preparation of RNA probes, preparing for *in situ* hybridization, sectioning and prehybridization, posthybridization, and imaging.

Fixation and Embedding of Tissue

As with tissue collected for RNA extraction, tissue intended for ultimate use in *in situ* hybridization must be carefully processed.

1. Dissect or collect tissue and immediately submerge in ice cold, freshly prepared FAA (50% ethanol, 10% formalin, 5% acetic acid). It is critical to use freshly prepared FAA and preferable to use formalin that is not more than a year old.

2. Place the tube containing the FAA and tissue into a beaker filled with ice, and place this into a desiccator. Vacuum infiltrate the tissue for 1.0–1.5 h. Hold the vacuum for 15 min and then release slowly. Repeat. Tissue may sink, and many protocols suggest that this is a necessity. However, it is common for plant materials that are very pubescent (hairy) to remain buoyant in FAA. *It is more important not to overfix your tissue than to wait for it to sink.*

3. Following vacuum infiltration, refill the tube with fresh FAA as necessary and place it on an oscillating shaker at 4°. Protocols vary as to how long tissue should be incubated in fixative after vacuum infiltration. Very small herbaceous tissue can be immediately removed from the fixative following vacuum infiltration and dehydrated. Larger, dense tissue typically requires 4–12 h of incubation in FAA for proper fixation. Never fix tissue for more than 16 h total.

Note: Many people are not aware that tissue can be overfixed. This is a serious problem that must be avoided for optimal *in situ* hybridization results. Tissue that has been in fix for more than 24 h is unlikely to be useful for *in situ* hybridization.

4. Dehydrate the samples through the following ethanol series at 4° with agitation for 30-90 min each: 50% ethanol, 70% ethanol, 85% ethanol, 90% ethanol, 100% ethanol. The duration of the incubation depends on the nature of the tissue. Similar to the case with fixation, small herbaceous tissue requires shorter periods than denser tissues. Tissue can be stored overnight or for longer periods at 4° during the 70% ethanol stage. After the 100% ethanol step, exchange the solution for fresh 100% ethanol and leave overnight at 4°.

5. Exchange the solution for fresh 100% ethanol and incubate at room temperature (RT) for 1 h. Follow this by subsequent RT incubations in 50% ethanol/50% Citrisolv (Fisher Scientific 22-143-975) for 2 h, then 100% Citrisolv for 2 h.

6. Transfer tissue to a small glass beaker or scintillation vial and add just enough fresh Citrisolv to cover the tissue. Fill the beaker with Paraplast Plus chips and incubate overnight at 55–60°. For the next 2 days, exchange the Paraplast with fresh molten Paraplast two to three times each day. Material must be maintained at 55–60°. With dense tissue or organs that tend to retain air bubbles (such as spurs), vacuum infiltration of the molten Paraplast may be necessary. This can be accomplished using a vacuum oven set to 60°, applying moderate vacuum for 30–60 min.

7. Embed the tissue by filling molds (such as Peel-A-Way or Tissue-Tek) with molten Paraplast and placing individual samples in each. This can be done on a hot plate to allow time to properly orient the tissue. Allow the molds to cool in a RT waterbath.

8. Embedded tissue can be stored for long periods at 4°.

Preparation of RNA Probes

For nonradioactive probes, antisense and sense RNA is typically labeled using digoxigenin (DIG), which is then detected using an anti-DIG antibody conjugated to alkaline phosphatase. All DIG-labeling supplies mentioned below are available from Roche Applied Science (http://www.roche-applied-science.com). All solutions are prepared with DEPC water and RNase-free chemicals or treated with DEPC and autoclaved after preparation.

1. Chose a region of your gene of interest to serve as the probe template. Generally, a 200-500 bp fragment is optimal. Although *in situ* hybridization is typically quite stringent in its specificity, it is preferable to use regions that do not show high conservation across members of a gene family, such as DNA-binding domains. However, using solely 3' UTR sequence as template is not advisable because these regions often form secondary structures that can interfere with probe hybridization.

2. Generate a linearized version of the template either by restriction digestion or by PCR. If using PCR, the RNA polymerase binding site can be incorporated into the fragment in one of the oligonucleotide primers. PCR products should be cleaned and concentrated using a spin column. In the case of plasmid linearization, it is important to use a restriction enzyme that does *not* create a 3' overhang, which can result in nonspecific polymerase initiation. The chosen enzyme should cut at the end of the template insert opposite from the RNA polymerase binding site. Confirm complete digestion by running linearized plasmid on an agarose gel.

3. Extract the linearized DNA with an equal volume of phenol/chloroform, then chloroform. Use RNase-free tubes and tips following the first extraction. Precipitate the DNA by adding 2 volumes 100% ethanol and 0.1 volume of 3 M NaAC, incubating at $-20°$ for 2 h, and centrifuging at high speed for 10 min. Wash the pellet with ice cold 70% ethanol, dry, and resuspend in DEPC water at a concentration of about 1 $\mu g/\mu l$.

4. Set up a runoff transcription reaction as follows:

1 μl DNA at 1 $\mu g/\mu l$
2 μl NTP/DIG–UTP mix
2 μl 10X transcription buffer

2 μl T7, T3, or SP6 RNA polymerase
1 μl RNase inhibitor
12 μl DEPC water
Mix by gently pipetting and incubate at 37° for 2 h.

5. Set aside a 1-μl sample for later use (see step 7).

6. Add 2 μl of RNase-free DNAse. Mix and incubate at 37° for 15 min.

7. Take another 1-μl sample. Run this sample side by side with the pre-DNAse sample (step 5) on a small agarose gel (use RNase-free loading buffer to run the samples). Bright RNA bands should be visible in both lanes. RNA probe often runs as multiple bands. DNA template should be visible in the pre-DNAse sample but not in the post-DNAse. If the DNA is not eliminated, add 1 μl of DNAse and incubate another 30 min at 37°. Repeat analysis as needed.

8. Once DNAse completion has been confirmed, stop the reaction by adding 4 μl 200 mM EDTA pH 8.0. Precipitate the RNA by adding 5 μl 4 M LiCl and 150 μl ethanol. Incubate at −20° for 2 h before precipitation by centrifugation for 10 min at maximum speed.

9. Wash pellet with ice cold 70% ethanol and allow to air dry.

10. Hydrolize RNA probe to desired length by resuspending pellet in 50 μl of 0.1 M NaHCO$_3$ pH 10.2. Incubate at 60° for an amount of time determined by the formula $t = (L_i - L_f)/K(L_i)(L_f)$, where t is the time in min, K is 0.11 breaks/min, L_i is the initial length of the probe in kilobases, and L_f is the final desired length in kilobases. For instance, to hydrolyze an initial probe of 0.5 kb to 0.15 kb (the typically recommended length), $t = (0.5–0.15)/(0.11 \times 0.5 \times 0.15) = 42$ min. See *Trouble-shooting in situ hybridization* section for further discussion of probe hydrolysis.

11. Stop the hydrolysis reaction by adding 5 μl 5% acetic acid, 5 μl 3 M sodium acetate, and 125 μl ethanol. Incubate at −20° for two h. Precipitate by centrifugation for 10 min at maximum speed, wash pellet in ice cold 70% ethanol, dry, and resuspend in 20 μl deionized formamide.

12. Quantify probe concentration using the Roche protocol, available at http://www.roche-applied-science.com under the title "Estimating the Yield of DIG-labeled Nucleic Acids." Probe can be stored at −20° for several months.

Obtaining a good-quality probe is very important for the success of the hybridization. Reactions that yield poor concentrations are unlikely to perform well. In my experience, T7 or T3 RNA polymerase gives better results than SP6. For probe templates that are particularly GC rich, it may be necessary to prepare a special NTP/DIG–UTP mix. The DIG–RNA labeling mix supplied by Roche has 3.5 mM DIG-11–UTP/6.5 mM dTTP and 10 mM of each remaining NTP. DIG-11–UTP and NTP solutions can be purchased separately in order to prepare alternate concentrations, such as 6.5 mM DIG-11–UTP/3.5 mM TTP. Higher DIG-11–UTP concentrations will result in more DIG incorporation, but it also will reduce the efficiency of the transcription reaction. It is recommended to prepare both sense and anti-sense probes. Anti-sense probe will be produced by transcription from an RNA polymerase binding site at the 3′ end of the template fragment, whereas the sense probe is made by transcription from the 5′ end.

Preparing for In Situ *Hybridization*

Preparing to perform *in situ* hybridization takes 3–5 days, depending on how many solutions need to be made and your experience with the process. The following basic RNase-free stock solutions are required [refer to Sambrook and Russel (2001b) for details]:

10× PBS
5× NTE
20× SSC
10× PBS with 20 mg/ml glycine (store at 4°)
1 M Tris pH 9.5
1 M pH Tris 8.0
1 M Tris 7.5
0.5 M EDTA pH 8.0
5 M NaCl
1 M MgCl$_2$

Additional required RNase-free solutions include

10× pronase buffer: .5 M Tris 7.5, 50 mM EDTA (see the discussion on pronase, later in this chapter)

10× *in situ* salts: 3 M NaCl, 100 mM Tris pH 8, 100 mM NaH$_2$PO$_4$ pH 6.8, 50 mM EDTA

Hybridization solution (800 μl): 100 μl 10× *in situ* salts, 400 μl deionized formamide, 200 μl 50% dextran sulfate (this will require heating to dissolve), 20 μl 50× Denhardt's solution, 10 μl tRNA (100 mg/ml in DEPC water), and 70 μl DEPC water

Note on the hybridization solution: It is best to prepare a relatively large volume (10–15 ml) and aliquot this into working amounts (1 ml) in RNase-free microtubes. It is absolutely critical that this solution be RNase free, so prepare it with care.

In addition to preparing these solutions, all glassware (additional bottles, graduated cylinders, Coplin jars), slide racks, and stir bars must be baked. Other plasticware, such as bottle caps, can be treated with 0.1 M

NaOH as described earlier. I recommend using square polyethylene (VWR 36318-045) and flat Nalgene (VWR 36212-204) boxes for incubation and washing steps. These can be thoroughly washed twice in an automated dishwasher, once with detergent and once without. Dry upside down on paper towels. Glass staining dishes can be similarly treated. Prepare at least 8 L of ddH$_2$O by autoclaving for 30 min in baked bottles. This "clean" water will be used for preparation of the hydration series and all diluted working solutions.

Sectioning and Prehybridization

Place five to six ProbeOn Plus slides (Fisher 15-188-52) onto a slide warmer set at about 40° and cover the unfrosted surface of each slide with a pool of DEPC water. It is important to use ProbeOn Plus slides to promote proper adhesion of the sections and allow sandwiching of slides (see below). Paraplast Plus embedded tissue should be sectioned at 8-μm thickness on a rotary microtome. Separate the ribbon of sections into strips of two to three sections using a clean razor blade and place sequentially onto the slides. Small paint brushes or wooden applicator sticks can be used to manipulate the sections. The sections must be allowed to flatten on the surface of the DEPC water. After about 15 min, carefully remove excess water using a Pasteur pipette around the edges of each slide. After another period of about 15–30 min, remaining water can be removed by gently tipping the edge of slide against a stack of Kim wipes or paper towels. Repeat this process as needed in order to obtain 20–25 slides.

Sectioning can be done the day before the prehybridization, but sectioning on the same day generally yields better results. However, for the sections to adhere properly, it requires at least 4 h of incubation from the time the slides are completely drained. During this period, slides should remain on the slide warmer with the lid closed. If you choose to section the day before, do so late in the afternoon and turn the temperature of the slide warmer down below 30° before leaving the slides overnight.

Before starting the prehybridization, several solutions need to be prepared. First, the hydration series should be arranged using square polyethylene boxes, which typically hold 300 ml each. Second, prepare all the diluted solutions, including 1.2 L of 1× PBS, 300 ml of 1× pronase buffer, and 300 ml of 1× PBS with 0.2% glycine. These can be made by dilution of the stocks (see earlier discussion) with the prepared "clean" water. It is also necessary to make a fresh solution of 4% paraformaldehyde in 1× PBS. Several methods are available for preparing this solution, but I prefer the following: Put 300 ml of 1× PBS in an RNase-free flask with a treated stir bar and bring to an active boil in the microwave. Immediately place the flask on a magnetic stir plate under a fume hood and add 12 g of paraformaldehyde to the hot solution. Stir. The powder should go into solution quickly. Allow the flask to cool and cover the mouth with parafilm or foil. Place the flask on ice under the hood to cool completely. This must be done before starting the prehybridization to allow the solution to cool but should not be performed a day in advance.

The following is the prehybridization protocol. Unless otherwise noted, the steps are performed in square polyethylene boxes in volumes of 300 ml.

1. Place slides into a metal slide rack, leaving every other slot empty to allow circulation of the solutions.

2. Incubate 10 min in about 300 ml of Citrisolv in a glass staining box. Repeat with fresh Citrisolv.

3. Rehydrate the slides by processing through an ethanol series as follows: 2 min in 100% ethanol, 1 min in 100%, and 1 min each in 95%, 85%, 70%, 50%, and 30%. Follow by 2 min in 150 mM NaCl, and then 2 min in 1× PBS.

4. Incubate the slides for 20 min at 37° in 1× pronase buffer (50 mM Tris pH 7.5, 5 mM EDTA) with 10 μg/ml pronase.

Note on pronase treatment: Either pronase or proteinase K can be used for digestion, but they use different buffers, so care must be taken to ensure use of the correct buffer. This step requires considerable optimization. First, the activities of both enzymes vary between preparations. When preparing the enzyme stock, follow the manufacturer's instructions to make a large volume (10–20 ml) at 10 mg/ml in DEPC water. This can be aliquoted into small working volumes in RNase-free microtubes, which can be stored long term at −20°. This will allow the use of the same stock (and hence, the same activity level) throughout repeated experiments. Second, once the stock is prepared, the specific activity must be assessed for the particular tissue being analyzed. The best way to do this is to run test digestions on sectioned tissue using a range of concentrations (5–20 μg/ml) or incubation lengths (20–30 min). For simplicity, vary only one component (concentration, time, or temperature) in the experiment. Ideally, the tissue should be digested as much as possible without affecting tissue/cell integrity. It is important to ensure that the cell contents are not affected by the digestion (the cell walls may look perfect even when the contents are gone). The use of stains can aid in the evaluation of the optimal enzyme treatment. Different tissue types or stages may require different treatments. The findings of this process will determine the enzyme concentration that will be used at this step in the actual prehybridization.

5. Following digestion, incubate in 1× PBS with 0.2% glycine for 2 min.

6. Rinse in 1× PBS for 2 min.

7. Incubate in 4% formaldehyde for 20 min under a fume hood.

8. Rinse in 1× PBS for 2 min (dispose of the PBS from this step in the formaldehyde waste).

9. Acetylation: Incubate slides for 10 min in 0.1 M triethanolamine pH 8.0 with 0.5% acetic anhydride.

Note: This solution must be prepared fresh but takes a while to complete because the triethanolamine is a thick liquid and is difficult to pipette. It is best to start preparing the 0.1 M of triethanolamine during the incubation period of step 7. Under a fume hood, place a glass staining dish on a magnetic stir plate. Place a stir bar and two pieces of plastic sterile 10-ml pipettes in the bottom. The pipettes should be broken so that they fit in the bottom of the container with room for the stir bar to spin. Fill the dish with 589.8 ml of "clean" water and add 7.8 ml of triethanolamine while stirring. Bring the pH to 8.0 with 2.4 ml of concentrated HCl. Check the pH using a pH strip. Carefully add 3.6 ml of acetic anhydride. Momentarily stop the stir bar and allow the pipette pieces to settle on either side of the stir bar. Place the slide rack into the glass box so that the rack is supported above the stir bar with sufficient room for the bar to move. Restart the stirring and incubate for 10 min.

10. Rinse in 1× PBS for 2 min (do this step in the fume hood and dispose of the PBS in the acetic anhydride waste).

11. Rinse in 150 mM NaCl for 2 min. Follow by processing through the ethanol dehydration series with 1 min of incubation per step: 30%, 50%, 70%, 85%, 95%, and 100%. It is not necessary to make a fresh dehydration series; in fact, the same solutions can be used for several experiments.

12. Incubate in fresh 100% ethanol for 2 min. Carefully remove slides from rack and place them section-side up onto fresh paper towel. Allow to dry completely.

13. Prewarm the necessary number of tubes of hybridization solution in a heat block at 80° (this should correspond to the number of slide pairs, see step 14). Set the hybridization oven to the appropriate temperature (see below).

14. Carefully examine the dry slides and choose those that have the best quality of tissue at the desired stage. Arrange the chosen slides into pairs. Mark the pairs as to which probe will be used on them (each pair will have the same probe). At least one pair should be hybridized with sense probe as a negative control.

15. Probe should be used at 0.5 ng/μl/kb. The amount of probe per slide pair is determined by $L_p \times 100\ \mu l \times (0.5\ \text{ng}/\mu l/\text{kb})$, where L_p is the length of the probe in kilobases. For instance, a probe hydrolyzed to

0.15 kb will require $0.15 \times 100 \times 0.5 = 7.5$ ng of probe. The concentration of the probe is determined in step 12 in the section "Preparation of RNA Probes." The first time that a given probe is hybridized, it should be tried at both the concentration calculated above (this is the $1\times$ concentration) and at $5\times$.

16. For each slide pair, the desired amount of probe should be added to 50% deionized formamide to make the volume up to 40 μl. Heat the probe to 80°.

17. Add 200 μl of preheated hybridization solution to each tube of heated probe and mix by pipetting. The hybridization solution is very viscous and is difficult to pipette accurately.

18. Apply the probe to each slide pair. The technique of probe application is a matter of taste and requires some practice to determine what works best for you. This is the method I use: Take the first slide of the pair and pipette 100 μl of probe mix onto the slide. Carefully use the side of an RNase-free pipette tip to spread the hybridization solution across the entire surface of the slide. Pipette the remaining probe solution onto the second slide along the narrow edge of the slide, opposite the frosted area. Take the spread slide and place its corresponding edge on the narrow edge of the second slide. The wet side of the first slide should be facing with the sections downward toward the second slide. Slowly, lower the spread slide onto the second slide, allowing the adhesion of the probe solution to pull the liquid across the entire surface of the slide. The final product will be a slide "sandwich" with the probe solution in between the surfaces of the two slides.

19. Elevate the slide sandwiches above wet paper towels in a flat plastic box. This can be achieved by breaking plastic sterile 1-ml pipettes into halves and placing them across the bottom of the box, which is lined with wet paper towels. Line the lid with Saran wrap and close tightly. Place the box into the preheated hybridization oven for 14–16 h. Hybridization temperatures generally range from 38 to 45°. Initial hybridizations should be performed at a lower temperature, such as 40°.

Make careful notes concerning the probes used and their concentration, the hybridization temperature, and any other variables such as the pronase concentration.

Posthybridization

At the end of the prehybridization protocol, several solutions for the posthybridization should be prepared 1 day in advance. Prepare 1.5 L of $0.2\times$ SSC and place at 48–55°. The temperature depends on the desired stringency of your wash. A good rule of thumb is to wash at a temperature 10°

above the hybridization temperature. Prepare 1.5 L of 1× NTE solution and place at 37°. In addition, you can prepare 300 ml of 1× PBS; 650 ml of 100 mM Tris 7.5, 150 mM NaCl; and 250 ml 100 mM Tris pH 9.5, 100 mM NaCl, 50 mM MgCl$_2$. The morning of the posthybridization, several additional solutions should be prepared using the solutions above: 120 ml 1.0% Roche Blocking Reagent (Roche 1 096 176) in 100 mM Tris 7.5, 150 mM NaCl (this requires warming for complete incorporation, so allow time for cooling before use), and 520 ml 1.0% BSA in 100 mM Tris 7.5, 150 mM NaCl, and 0.3% Triton. The following is the posthybridization protocol. Unless otherwise noted, the incubations are performed in square polyethylene boxes in volumes of 300 ml.

1. Fill a square plastic box with about 300 ml of preheated 0.2× SSC. Immerse each slide sandwich and gently separate by opening, not sliding. Place the slides in a slide rack, leaving an empty slot between each one for circulation.

2. Wash the slides for 1 h in fresh preheated 0.2× SSC with gentle agitation in a hybridization oven set to the chosen wash temperature.

3. Repeat 60-min wash with fresh preheated 0.2× SSC.

4. Incubate the slides in preheated 1× NTE for 5 min at 37° with gentle agitation.

5. Repeat 5-min 1× NTE wash.

6. Treat with 20 μg/ml RNase A in preheated 1× NTE for 20 min at 37°.

7. Incubate the slides in preheated 1× NTE for 5 min at 37° with gentle agitation.

8. Repeat 5-min 1× NTE wash.

9. Wash for 60 min in preheated 0.2× SSC at wash temperature with gentle agitation.

10. Incubate 5 min in 1× PBS at RT.

11. Place each slide on the bottom of a flat plastic box, section-side up with no slides overlapping each other. The flat boxes cited above will hold 10 slides. Fill each box with 100 ml of 1.0% Roche Blocking Reagent in 100 mM Tris 7.5, 150 mM NaCl. This should be just enough to cover the slides. Incubate for 45 min on a rocking platform at RT.

12. Replace block solution with 100 ml 1.0% BSA in 100 mM Tris 7.5, 150 mM NaCl, 0.3% Triton. Incubate for 45 min on a rocking platform at RT.

13. Dilute 8 μl of alkaline phosphatase-conjugated anti-DIG antibody (Roche 1 093 274) into 10 ml of 1.0% BSA in 100 mM Tris 7.5, 150 mM NaCl, 0.3% Triton. Make a puddle of the solution in a large plastic weigh dish. Place each slide pair in the puddle with their long edges in the

solution. Sandwich the slides together, with the sections on the inside, drawing up the antibody solution between them. Drain the slide pair on a stack of Kim wipes and place the edge back into the puddle. Solution will be drawn up by capillary action. Repeat once, avoiding air bubbles.

14. Elevate the slide sandwiches with antibody solution above wet paper towels as described above. Allow to sit at RT for 2 h.

15. Drain slides on Kim wipes and separate carefully. Place on the bottom of a flat plastic box as in step 11. Wash with 100 ml of 1.0% BSA in 100 mM Tris 7.5, 150 mM NaCl, 0.3% Triton for 15 min at RT on a rocking platform. Repeat three times for a total of 4 washes.

16. Replace BSA/Triton solution with 100 ml 100 mM Tris pH 9.5, 100 mM NaCl, 50 mM MgCl$_2$. Wash for 10 min with rocking.

17. Fill a Coplin jar with 100 mM Tris pH 9.5, 100 mM NaCl, and 50 mM MgCl$_2$ and dip each slide into the solution to ensure that all of the detergent is removed.

18. Prepare substrate solution for color detection by adding 22 μl NBT (Roche 1 383 213) and 16 μl BCIP (Roche 1 383 221) to 10 ml of 100 mM Tris pH 9.5, 100 mM NaCl, 50 mM MgCl$_2$.

19. Apply solution to slide sandwiches as described in step 13.

20. Elevate slide sandwiches above wet paper towels and seal box tightly. Carefully wrap box in aluminum foil and place in a dark drawer to prevent light contamination.

Again, make notes on all the variables, such as the wash temperature.

Imaging

Color development generally takes 12–48 h. Good signal is usually visible to the naked eye, although this depends on the size of the expression domain. Carefully examine the sandwiches under a microscope to assess staining. When adequate signal is detectable, immerse the sandwiches in TE to separate and stop the staining reaction. Place slides into a slide rack and incubate in 1× PBS for 5 min at RT. Counterstaining with calcofluor (also known as Fluorescent Brightener, Sigma F-3397) can significantly improve contrast and visualization. Incubate slides in 0.002% solution of calcofluor in 1× PBS. Follow by rinsing for 5 min in 1× PBS. Dehydration will significantly reduce staining intensity, so mount slides with water, glycerol, or other aqueous medium (I prefer water). Do not let the slides dry out at any step.

The slides can now be photographed. To take advantage of the calcofluor counterstaining, sections should be illuminated simultaneously with fluorescent and white light. Dial down the intensity of the white light until the fluorescence of the tissue is just visible. This will appear as a blue-white glow in the cell walls. Certain tissue absorbs calcofluor better than others and this may be inconsistent between sections. "Real" signal should be light brown or lilac to dark brown. Dark blue staining is generally not real signal (see below). Photodocument every section with informative staining because water-mounted slides will not be stable for more than a week or two (at 4° elevated above wet paper towels).

Troubleshooting In Situ *Hybridization Results*

As mentioned earlier, performing *in situ* hybridization can be a challenging process and the protocol requires optimization of many steps. The following are suggestions for troubleshooting:

No or low signal: This may be due to a number of factors including low probe concentration, hybridization, or wash temperatures that are too high, RNase contamination of reagents (particularly hybridization solution), overfixation of tissue, insufficient tissue digestion, or improper probe hydrolization (see below).

High background: This typically results from hybridization or wash temperatures that are too low. It is important to note that high background can be general, affecting all tissues, or disconcertingly specific, only affecting certain tissues. Distinguishing nonspecific staining from "real" staining is sometimes difficult. Tissues that are prone to nonspecific staining include stamens and pollen, vascular tissues, and very small dense meristematic cells. Nonspecific signal often takes the form of very dark blue staining that is associated with the cell walls. It is important to remember that *in situ* hybridization targets mRNA, which should be in the cytoplasm, not the nucleus or cell walls. The caveat to this statement is that in mature plant cells that are highly vacuolated, the cytoplasm may be closely pressed to the cell wall. The most commonly analyzed genes for comparative gene expression, such as MADS-box containing genes or *CYCLOIDIA* homologues, typically have high levels of expression that should be clearly discernible. Comparisons with sense control slides can sometimes help to distinguish the real signal from the nonspecific staining.

Tissue appears to be degraded or cell contents are gone: Tissue has been overdigested. Lower the enzyme concentration.

Note on probe hydrolysis: Protocols commonly call for probe to be hydrolyzed to 70–150 bp (Jackson, 1991; Weigel and Glazebrook, 2002). Although this works well for many probes, others perform better at different lengths. Optimization of probe length is a trial-by-error process; there are no good rules of thumb. Although 200 bp may work well for a gene in one species, the ortholog in another taxon may work best at 300 bp. If little signal is observed even at low hybridization temperatures and high probe concentrations, using a shorter probe is often successful. Alternatively, high background (either general or specific) can be eliminated by using longer probe.

Several alternatives to the type of *in situ* hybridization described earlier have been developed. For organs with complex three-dimensional structure, whole-mount *in situ* hybridization (Zachgo *et al.*, 2000) can yield more informative results. Another alternative that has yet to be broadly used is *in situ* RT-PCR. One drawback to the DIG-labeled form of this protocol (Johansen, 1997) is that it sometimes results in nonspecific labeling of all nuclei, which can make interpretation of the data difficult. A new method using Oregon green–labeled UTP (Ruiz-Medrano *et al.*, 1999) has yielded beautiful results, however (Kim *et al.*, 2003a,b).

Concluding Remarks

The generation of gene expression data is only the first step in the difficult process of assessing gene function. It has been demonstrated for many genes that expression is not necessarily a simple proxy for the spatial extent of gene function, both because of non–cell autonomous effects and posttranscriptional or posttranslational regulation. Therefore, any data resulting from the types of analyses described earlier should be interpreted with care. However, given that direct analyses of gene function using genetic tools are commonly unavailable in nonmodel species, gene expression is often our only comparative tool. Therefore, it is preferable to use a combination of all of the aforementioned techniques to obtain a detailed and clear picture of gene expression patterns.

References

Chang, S., Puryear, J., and Cairney, J. (1993). A simple and efficient method for isolating RNA from pine trees. *Plant Mol. Biol. Rep.* **11**, 113–116.

Jackson, D. (1991). *In situ* hybridisation in plants. *In* "Molecular Plant Pathology: A Practical Approach" (D. J. Bowles, S. J. Gurr, and P. McPherson, eds.), pp. 163–174. Oxford University Press, Oxford.

Johansen, B. (1997). *In situ* PCR on plant material with sub-cellular localization. *Ann. Bot.* **80**, 697–700.

Kim, M., McCormick, S., Timmermans, M., and Sinha, N. (2003a). The expression domain of PHANTASTICA determines leaflet placement in compound leaves. *Nature* **424**, 438–443.

Kim, M., Pham, T., Hamidi, A., McCormick, S., Kuzoff, R. K., and Sinha, N. (2003b). Reduced leaf complexity in tomato wiry mutants suggests a role for PHAN and KNOX genes in generating compound leaves. *Development* **130**, 4405–4415.

Kramer, E. M., Dorit, R. L., and Irish, V. F. (1998). Molecular evolution of genes controlling petal and stamen development: Duplication and divergence within the *APETALA3* and *PISTILLATA* MADS-box gene lineages. *Genetics* **149,** 765–783.

Ruiz-Medrano, R., Xoconostle-Cazares, B., and Lucsa, W. (1999). Phloem long-distance transport of CmNACP mRNA: Implications for supracellular regulation in plants. *Development* **126,** 4405–4419.

Sambrook, J., and Russel, D. W. (2001a). "Molecular Cloning," Vol. 2. Cold Spring Harbor Press, Cold Spring Harbor, NY.

Sambrook, J., and Russel, D. W. (2001b). "Molecular Cloning," Vol. 3. Cold Spring Harbor Press, Cold Spring Harbor, NY.

Suzuki, Y., Hibino, T., Kawazu, T., Wada, T., Kihara, T., and Koyama, H. (2003). Extraction of total RNA from leaves of Eucalyptus and other woody and herbaceous plants using sodium isoascorbate. *Biotechniques* **34,** 988–993.

Wang, S. X., Hunter, W., and Plant, A. (2000). Isolation and purification of functional total RNA form woody branches and needles of Sitka and white spruce. *Biotechniques* **28,** 292–296.

Weigel, D., and Glazebrook, J. (2002). "Arabidopsis: A Laboratory Manual." Cold Spring Harbor Press, Cold Spring Harbor, NY.

Zachgo, S., Perbal, M.-C., Saedler, H., and Schwarz-Sommer, Z. (2000). *In situ* analysis of RNA and protein expression in whole mounts facilitates detection of floral gene expression dynamics. *Plant J.* **23,** 697–702.

[33] Developing Antibodies to Synthetic Peptides Based on Comparative DNA Sequencing of Multigene Families

By Roger H. Sawyer, Travis C. Glenn, Jeffrey O. French, and Loren W. Knapp

Abstract

Using antisera to analyze the expression of specific gene products is a common procedure. However, in multigene families, such as the β-keratins of the avian integument where strong homology exists among the scale (ScβK), claw (ClβK), feather (FβK), and feather-like (FlβK) subfamilies, determining the cellular and tissue expression patterns of the subfamilies is difficult because polyclonal antisera produced from any one protein recognize all family members. Traditionally, researchers produced and screened multiple monoclonal antisera produced from the proteins of interest until an antiserum with sufficient specificity could be obtained. Unfortunately, this approach requires a lot of effort, and once obtained, such antisera may have limited applications. Here, we present procedures by which comparative DNA sequences of members from the β-keratin multigene family were translated and aligned to identify amino acid domains that were conserved within the FβK subfamily, but which were divergent from the other subfamilies. A synthetic 23-mer peptide with the conserved amino acid sequence was generated and used to produce a polyclonal antiserum that recognizes only the FβK subfamily of proteins. Western blot analysis and confocal microscopy with this antiserum are now providing valuable new insights concerning the developmental and evolutionary relationships between the scale, claw, and feather proteins found in birds. This represents a powerful new approach combining techniques from molecular evolution and developmental biology to study the expression and evolution of specific members of multigene families.

Introduction

Functional genomics encompasses many aspects of biology ranging from highly experimental systems in which variation in transcription and translation are monitored closely *in vitro*, to overarching relationships in regulation of multigene families across taxa of organisms with shared evolutionary histories (http://www.nsf.gov/pubs/2001/bio012/). Here, we present techniques that have allowed us to address the role of a multigene family in the evolution of a phenotypic innovation, namely the avian feather (Sawyer *et al.*, 2003a). Our model examines the role of the multigene family of epidermal β-keratins whose expression is restricted to the skin and its appendages in reptiles and birds (Alibardi and Sawyer, 2002; Sawyer *et al.*, 2000, 2003b). The expression of this family of genes is intimately associated with the morphogenesis of integumentary appendages (Haake *et al.*, 1984; Sawyer and Knapp, 2003), so these unique and highly insoluble protein products (Shames *et al.*, 1988) make up complex structural elements, such as the barbs, barbules, and hooklets of feathers (Haake *et al.*, 1984). Understanding the evolutionary origin of feathers depends on understanding the evolution of development, including an understanding of the evolution of the developmental expression of the members of the β-keratin gene family.

The overall approach that we describe is generic, uses common techniques, and can be applied to study many multigene families. Many of the specific methods that we have used draw on techniques from diverse fields of biology, and thus, the salient points of all the methods chosen may not be readily apparent to many readers. In this chapter, we describe the rationale for each method and the important goals that must be achieved and illustrate the method with specific techniques we have used to study β-keratins. By describing methods in this way, we hope that the overall approach will be accessible to a larger pool of developmental and evolutionary biologists.

Background on Example Used

Four subfamilies of β-keratins are known for the chicken: scale β-keratins (ScβK), claw β-keratins (ClβK), feather β-keratins (FβK), and feather-like β-keratins (FlβK) (Presland *et al.*, 1989a,b; Shames *et al.*, 1988). Because of their poor solubility, tendency to aggregate in gels, and complex expression patterns in embryonic epidermal appendages, generation of non–cross-reacting antisera has not been possible (Shames *et al.*, 1988). To date, polyclonal antisera generated against the ScβK isolated by one- and two-dimensional polyacrylamide gel electrophoresis have cross-reacted with the FβK on Western blots and with feather tissues (Carver and Sawyer, 1987; Haake *et al.*, 1984; O'Guin *et al.*, 1982; Sawyer *et al.*, 2000; Shames *et al.*, 1988). These antisera, such as "anti-β_1" (Sawyer *et al.*, 2000; Shames *et al.*, 1988) also react with the β-keratins found in the epidermal appendage of reptiles (Alibardi and Sawyer, 2002).

Here, we describe a comparative genomic approach, in which we have taken advantage of DNA and protein sequence data in GenBank to align the four subfamilies of the β-keratin family and design polymerase chain reaction (PCR) primers specific for the FβK subfamily. By aligning the inferred amino acid sequences of the FβK from several orders of birds, it is possible to identify highly conserved peptide sequences within the FβK subfamily, which are divergent from the other subfamilies. A synthetic peptide with the identified sequence can then be used to generate specific antisera. Such antisera do not cross-react with the other subfamilies of the β-keratin multigene family, thus allowing specific localization of the FβK subfamilies of proteins in other skin appendages of birds and possibly reptiles (Sawyer *et al.*, 2003a,b). Antisera generated in such a manner may also be used to analyze the expression of the FβKs during the development of various skin appendages using both Western blots of extracted proteins and indirect immunofluorescence of tissues (Sawyer *et al.*, 2003a,b).

Methods (Fig. 1)

Initial Characterization of the Multigene Family

Information on the expression of multigene families during embryogenesis has generally come from comparative analysis of proteins extracted from stage-specific tissues (Kemp and Rogers, 1972; Molloy *et al.*, 1982; O'Guin and Sawyer, 1982). Direct amino acid sequencing of such proteins provides preliminary sequence information (Inglis *et al.*, 1987; Walker and Bridgen, 1976). The generation of cDNA libraries from messenger RNA (mRNA)

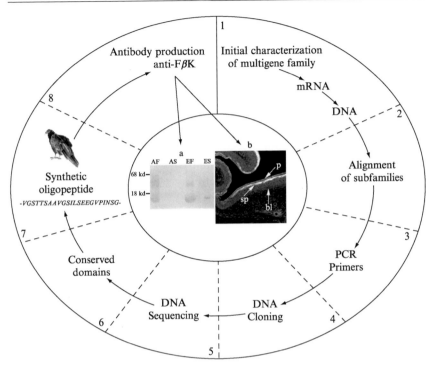

FIG. 1. A schematic representation of the procedures used to obtain antisera against specific conserved peptides, identified through genomic analysis. Each number around the circle, 1–8, represents a step in the procedure and corresponds to the eight headings in this manuscript. The letters *a* and *b* in Fig. 1 show a Western blot and a confocal image, respectively. These figures are described in detail in Figs. 3 and 4, respectively.

populations provides the initial sequence data (Presland *et al.*, 1989a,b; Wilton *et al.*, 1985), and analysis of genomic clones then provides comparative information on gene sequence and organization (Presland *et al.*, 1989a,b; Rogers *et al.*, 1998). Hopefully these data provide enough sequence information so that conserved regions can be identified and PCR primers can be designed.

Example, β-Keratins (Fig. 1)

Initially, the β-keratins of scales and feathers were analyzed by gel electrophoresis (Kemp and Rogers, 1972). Using direct sequencing of polypeptides, the amino acid sequences of a few β-keratins were published (Inglis *et al.*, 1987; Walker and Bridgen, 1976), yet it was not until cDNAs were generated from mRNA populations that DNA sequences for the FβK and ScβK were first described (Presland *et al.*, 1989a,b; Wilton *et al.*, 1985). Subsequently, using feather and scale cDNAs as probes, genomic libraries provided information on the genomic organization of the subfamilies of β-keratins (Molloy *et al.*, 1982; Presland *et al.*, 1989a,b; Rogers *et al.*, 1998; Whitbread *et al.*, 1991; Wilton *et al.*, 1985). The overall structure of the FβKs is presented in Fig. 2.

If a sufficient amount of information exists (i.e., can be found in GenBank), then steps 2–5 may be omitted. Here, we assume that a small, but limited, amount of initial information is available. If that is true, then it will be often desirable to increase the amount of comparative information available to identify conserved domains before attempting to design the oligopeptides (step 7). By using well-developed and widely available tools from molecular genetics and molecular evolution (steps 2–5), researchers can increase the amount of comparative information available relatively quickly and easily. In most situations, researchers will want to skip ahead to step 6, determine how much additional information is desired, and then come back to steps 2–5.

Aligning DNA Sequences

After obtaining DNA sequence information for some members of the multigene family, the next step is to align the sequences. A variety of

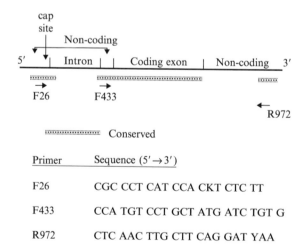

Primer	Sequence (5′→3′)
F26	CGC CCT CAT CCA CKT CTC TT
F433	CCA TGT CCT GCT ATG ATC TGT G
R972	CTC AAC TTG CTT CAG GAT YAA

FIG. 2. Feather β-keratin gene structure (summarizing Presland *et al.* [1989a]) and the locations of the primers used in polymerase chain raction (PCR) and sequencing. F26 and R972 were used in all successful PCRs. F433 and R972 were used in sequencing of the focused dataset (after Sawyer *et al.* [2003]).

software tools and protocols are available to aid in this process. One of the best and mostly widely used alignment tools is Clustal (Thompson *et al.*, 1994, 1997). Stand alone versions of ClustalX are freely available (ftp://ftp-igbmc.u-strasbg.fr/pub/ClustalX/), although some researchers may benefit from step-by-step guides (e.g., Hall, 2001; http://www.biozentrum. unibas.ch/~biophit/clustal/ClustalX_help.html). The Clustal algorithm is also incorporated into a variety of commercial (GCG, DS Gene, MacVector, etc.) and free (http://workbench.sdsc.edu/) software programs and web sites. Researchers with little experience in this area should keep in mind that software packages can be very sensitive to input parameters and variation in sequence length, especially inconsistencies in the homology of beginning and ending locations. Because the alignment produced in this situation is used simply to identify conserved regions of DNA sequence to be used in the subsequent steps of this protocol, many researchers may find manual alignment the quickest way to achieve this goal.

Example, the β-Keratin Multigene Family (Fig. 1). It is desirable to identify conserved sequences that will allow amplification of the entire coding region, especially when the coding region is small, as is the case for FβKs. Presland *et al.* (1989a) include alignments of the β-keratin data available at the time of publication. Initially, we supplemented the published alignments using GCG (Accelrys, Inc., San Diego, CA) to align all sequences of β-keratin available. Subsequently, we have used the ClustalW algorithm (Thompson *et al.*, 1994), implemented in Sequence Navigator 1.0.1 (Applied Biosystems). From these alignments, it was clear that a region including the cap site, 5′ to the coding region, and a 3′ untranslated region contained few substitutions among the sequences available.

Designing and Using PCR Primers

After identifying conserved regions of DNA sequence, primers for the polymerase chain reaction (PCR) need to be designed. Again, various software tools and protocols (Sambrook and Russell, 2001) are available to aid in this process. One commercial software package (Oligo 6 [Molecular Biology Insights, Cascade, CO]) includes an option to identify primer-binding regions from alignments of multiple sequences. Most researchers, however, will find it easier and more economical to simply import the consensus sequence or a representative sequence into the chosen primer design software. Primers can then be designed from conserved regions of the DNA sequence. If variable sites occur in potential primer-binding regions, then mixed bases can be integrated into the primers (Palumbi, 1996). Use of mixed bases should be minimized, especially on the 3′ ends of primers. Many software packages do not allow mixed bases to be analyzed automatically, but they will allow primers to be manually edited. By manually editing potential primers, the effects of different specific bases at variable sites can be investigated.

Example, Feather β-keratin Subfamily (Fig. 1). Using aligned feather β-keratin DNA sequences from chicken (Presland *et al.*, 1989a,b), primers for PCR were designed in two conserved noncoding regions, which span the entire coding region (Fig. 2). These primers were tested by using them in a PCR with chicken genomic DNA as the template. Primers F26 and R972 successfully amplify about 900–1000 bp fragments of DNA from a variety of birds (French, 2001).

Cloning Members of the Multigene Family

Because multiple members of the gene family are amplified during PCR, it is not possible to determine the sequences directly from the PCR products. First, the PCR products must be separated so that amplicons from individual members can be sequenced separately. The most straightforward way to accomplish this is to clone the PCR products. During the process of cloning, individual amplicons are incorporated into vectors. Each vector–insert (ideally the insert is an amplicon of interest) is then incorporated into a single bacterium, and the bacteria multiply each vector–insert pair. Many different strategies for cloning PCR amplicons have been described (Frohman, 1994; Sambrook and Russell, 2001). In most situations and laboratories, TA cloning (Holton and Graham, 1991; Marchuk *et al.*, 1991; Sambrook and Russell, 2001) will be easiest to implement. Several vendors (e.g., Invitrogen, Carlsbad, CA, and Promega, Madison, WI) offer products for such cloning experiments. In the example below, we describe some specific techniques that may be useful when working with amplicons from multigene families.

Following cloning, the individual bacterial colonies must be screened to ensure inserts of the appropriate size have been obtained. Various methods have been described to achieve this goal (Sambrook and Russell, 2001). In many labs, this goal is most easily accomplished by using PCR and primers that bind to the multiple cloning site of the vector (e.g., M13 forward and M13 reverse). The number of colonies that should be screened depends on the efficiency of cloning products of the appropriate size and the number that are needed for sequencing (see below).

Example: Feather β-Keratins from Individual Birds (Fig. 1). To reduce cloning small nonspecific PCR amplicons (i.e., primer dimers), PCR products were precipitated by adding an equal volume of 20% PEG with 2.5 *M* NaCl, centrifuging, washed twice with 80% EtOH (Kusukawa, 1990), and quantified on a 1.5% agarose gel using varying quantities of lambda DNA as a standard. To improve ligation efficiency, A-tailing reactions were carried out for 30 min at 70° in 10-μl volumes with final concentrations of 50 mM KCl, 10 mM Tris–HCl pH 9, 1% Triton X-100, 2.0 mM MgCl$_2$, 0.2 mM dATP, 2 units of *Taq* DNA Polymerase (Promega, Madison, WI), and 250 ng of precipitated PCR amplicons. Three microliters (75 ng) of the A-tailing reaction was used in the ligation reaction for each sample. Ligations were performed using pGEM-T Vector System kits (Promega, Madison, WI) according to the manufacturer's protocol. Following ligation, 2 volumes (20 μl) of TE were added to each reaction, and each sample was incubated at 65° for 20 min to dissociate the ligase from the DNA. Transformations were performed using Epicurian Coli XL-10 Gold^{TM} Ultracompetent Cells (Stratagene, La Jolla, CA) according to the manufacturer's protocol or by electroporation of this strain.

For each sample, 48–192 colonies, identified as potentially positive for inserts by color screening, were picked and transferred to 40 μl of water. These colony suspensions were used as templates to determine insert size by performing PCR using M13 Forward and Reverse primers (Stratagene, La Jolla, CA) and an initial denaturation period of 10 min. PCR amplicons were examined on 1.5% agarose gels (Sambrook and Russell, 2001); those that contained inserts of 800–1100 bp were used as templates for sequencing reactions.

Automated DNA Sequencing of Clones

Following identification of clones with inserts of interest, the DNA sequence needs to be determined. Various options are available (Sambrook and Russell, 2001). Researchers should choose the method that will be most efficient and economical for them. Most major research institutions have automated DNA sequencers available in core labs and/ or in the labs of genomics researchers. The newest generation of capillary sequencers allow read lengths of 1000 bases or more and run costs less than $1/lane. Outsourcing to commercial or core labs at other institutions that have these capillary instruments is a viable option, even when other automated sequencers are available at a researcher's own institution. Because it is straightforward to complete cycle sequencing reactions in any lab equipped to conduct PCR, many researchers will find it economical to do the reactions themselves and simply employ the commercial/core lab to run reactions on their instruments, sending the unedited chromatograms back to the researchers.

After the chromatograms are generated, they must be edited to ensure accuracy of the sequence obtained. A variety of commercial (Sequencher–Genecodes, Ann Arbor, MI; Lasergene–DNAStar, Inc., Madison, WI;

GeneTool–BioTools, Inc., Edmonton, Alberta, Canada; e-Seq–Licor, Lincoln, NE; Vector NTI–InforMax, Frederick, MD) and free (e.g., Sequence Viewer–Applied Biosystems, Foster City, CA; AutoEditor–The Institute for Genomic Research, Rockville, MD) software packages are available for this purpose. Many research institutions support free software and shared user licenses of the commercial software, often through the core sequencing facilities. Researchers should invest sufficient resources in determining the accuracy of DNA sequences to justify the considerable expenses incurred by subsequent steps of this approach.

Example: Sequencing Clones of Feather β-Keratins (Fig. 1). Two methods of automated DNA sequencing were used. To produce clean reads of more than 1000 bases, a LICOR 4000L (Lincoln, NE) automated sequencer with fluorescently labeled primers was used. Sequences from both DNA strands were determined from the products of the colony screening PCRs, which had been purified by PEG precipitation and quantified. Cycle sequencing (Amersham, Cleveland, OH) of purified products was conducted with M13 Forward and Reverse primers labeled with IRD41 (LICOR, Lincoln, NE). Cycle sequencing parameters were 95° for 2 min, followed by 25 cycles of 95° for 30 s, 55° for 30 s, and 70° for 1.25 min.

The second dataset was generated using an ABI 377XL automated sequencer (Applied Biosystems 1998), primer R972, and an internal primer F433 (Table I) designed specifically for the purpose of allowing us to focus on the protein coding region of the gene. Sequences from both DNA strands were determined directly from the products of the colony PCRs using Big-Dye terminator chemistry (Applied Biosystems, Foster City, CA). All cycle sequencing reactions were carried out in 10-μl volumes with 0.32 μM primer, about 50 ng of PCR product, and Big-Dye terminators using the manufacturer's specifications except that the terminator mix was diluted 1:1 with halfBD (GENPAK Ltd., Stony Brook, NY) or a homemade equivalent (400 mM Tris–HCl pH 9, 10 mM MgCl$_2$; see http://www.genome.ou.edu/proto.html for details).

Chromatograms of forward and reverse sequences from each sample were imported into Sequencher 3.1.1 (Gene Codes, Ann Arbor, MI), which was used to edit and align the individual sequences into contigs. Sequencher, using the universal translation codes, allowed inference of the amino acid sequence, which was then exported for further manipulation.

Aligning Amino Acid Sequences and Identifying Domains of Interest

After amino acid sequences of interest have been obtained (from steps 1 and 5), they need to be aligned. As with aligning DNA sequences, several software tools and protocols are available to assist with aligning amino acid

TABLE I

ALIGNMENT OF AMINO ACID SEQUENCES FOR β-KERATINS WAS DETERMINED DIRECTLY OR INFERRED FROM DNA SEQUENCES[a]

	Amino acid position								
	01	20	30	40	50	60	70	80	83
Turkey vulture	MSCYDLCRPC----GPTPLANSCNEPCVRQCQDSRV VIEPSPVVVTLPGPILSSFPQNTAVGSTTSAAVGSILSEEGVPINSG								
Wood stork	...........................I.............................A........S...								
Pigeon	...NP.L..QC...............Q.......................S..........C........								
Chicken FBK A	..........S...A..........Q...............................S..........S...								
Chicken feather-like	..FHV.Q--..-.............T..Q..................T...SA......A..AG.......S...								
Chicken claw	...SSLCA.ACVAT-.....D......P.T....Q.PAT...F.............YA.....AGVP......GMGGTFGRGAGF								
Chicken scale	......P.TSCISR.Q.I.D.G.....P..TT..Q.P...F...............DSV....SGAPIF.GSSLGY.GSSLGY								

Note: The feather β-keratin antiserum (anti-FβK) was generated using a synthetic polypeptide whose sequence was identical to the underlined amino acids of the turkey vulture sequence.

[a] Sequences correspond (from top to bottom) to accessions: AF308826, AF308827, **KRPYF4**, X17511, X17521, M37698, and X00315 of GenBank or GenPept.

sequences (http://us.expasy.org/tools/). The amount of effort required for amino acid alignment will likely range from nearly trivial to extraordinarily difficult. The goal is for the alignment to be of sufficient quality that regions of interest can be identified. In the overall approach we are describing, a region of interest is a string of amino acid sequence that is conserved within a subgroup of multigene family members, but variable among subgroups. Thus, researchers should choose a method that produces an alignment of sufficient quality to achieve that goal.

Example: The β-Keratin Multigene Family (Fig. 1). Amino acid sequences from GenBank and inferred from DNA sequences (described earlier) were obtained. We used the Clustal algorithm implemented in Sequence Navigator 1.0.1 (Applied Biosystems, Foster City, CA) to align the sequences. From this alignment (Table I), it is clear that all β-keratins share regions of high similarity (e.g., residues 44–57 have only one variable amino acid among all β-keratins), but regions of dissimilarity exist among the subgroups (e.g., residues 61–83 have few substitutions within FβKs, but several to many substitutions compared with other subgroups). Using information from an alignment that included many more FβK sequences, we chose a region of interest to synthesize as an oligopeptide.

Synthesizing an Oligopeptide

One of the advantages of knowing the specific sequences of families of related genes is that regions of high similarity within a subfamily and heterogeneity among subfamilies can be ascertained. However, other regions of high similarity among all members of the multigene family often exist within the same proteins. Thus, regions of varying similarity to other proteins are linked in any intact protein. If researchers rely on using intact proteins isolated from biological tissues to produce poly-clonal antibodies (as is traditionally done [see step 8 below]), then the antibodies will recognize all proteins with similarity to any of the domains. Such polyclonal antibodies are, therefore, not specific to subfamilies (e.g., FβKs vs. ScβKs).

A strategy to avoid the problems associated with using intact proteins isolated from biological tissue is to synthesize an oligopeptide that is homologous to unique regions of interest. This strategy allows antibodies (step 8) to be made, which will only react with proteins containing the unique region of interest, even though other regions may have amino acid sequences homologous to other proteins.

Example: The β-Keratin Multigene Family (Fig. 1). Alignment of amino acid sequences for β-keratins revealed distinct amino acid sequences of FβKs, which were not present in other subfamilies. We used this unique amino acid sequence, conserved in all FβKs, to synthesize a feather-specific peptide useful in the production of monospecific antiserum (Sawyer et al., 2003a). Based on sequence analysis and alignment of feather β-keratin amino acid sequences, the 23-mer VGSTTSAAVGSILSEEGVPINSG-CONH2 (a C-terminal amide) was synthesized on an applied Biosystems Pioneer automated peptide synthesizer according to the methods described in Sawyer et al. (2003a).

Producing an Antiserum

There are two convenient approaches to making high-specificity antibodies. One is through the technically demanding methods of producing hybridomas following fusion of activated murine lymphocytes and immortalized myeloma cells (Zola, 1987), and the other is through the use of oligopeptides in conventional methods involving injection of these antigens into mammals (O'Guin et al., 1982; Sawyer et al., 2003a). Monoclonal antibodies produced by hybridomas are useful but often tend to be less efficacious when epitopes in the original proteins are rare or conformationally cryptic. Oligopeptides used for injection into a mammal, on the other hand, have multiple but limited epitopes that are more likely available in the intact protein under a wide range of conditions (e.g., immunoblotting of proteins and immunostaining of cells and tissues). Antisera made in animals are much less technically demanding to produce because of the relative simplicity of animal use methods (O'Guin et al., 1982).

Example: The β-Keratin Multigene Family (Fig. 1). In the case of our work, a highly specific (FβK antiserum was produced in a male New Zealand white rabbit (Sawyer et al., 2003a). The synthetic peptide antigen was cross-linked to keyhole limpet hemocyanin (KLH) using glutaraldehyde (Sawyer et al., 2003a). The primary and secondary injections of dialyzed Freund's emulsified antigen contained 200 mg of the synthetic peptide cross-linked with KLH. Serum was collected and processed 14 days after the second injection. Initial screening of the serum against extracts of feather keratins demonstrated specificity for the feather β-keratins. Preimmune serum controls were negative. For convenience, the antiserum was divided into 1-ml aliquots and frozen at −70° for preservation and future use.

Step 8a: Immunoblotting (Western Blots)

To verify that the synthetic peptide did indeed present a unique set of epitopes for the production of an antiserum that would react specifically against FβKs, we analyzed its reactivity against the epidermal polypeptides extracted from feather and scale tissues (O'Guin et al., 1982; Sawyer *et al.*, 2003a). The epidermal keratins were extracted with Triton X-100 in 1.5 *M* KCl and separated by electrophoresis on a 10% polyacrylamide gel electrophoresis (PAGE). The peptides were stained with Coomassie brilliant blue (Fig. 3). Note that the α-keratins, which are also prevalent in epidermal tissues of all chordates (O'Guin and Sawyer, 1982; Sawyer *et al.*, 2000; Whitbread *et al.*, 1991) migrate as bands in the molecular weight range of 40–70 kd, while scale β-keratins migrate in the range of 17–20 kd. The FβKs migrate slightly lower at 10–14 kd (Haake *et al.*, 1984; O'Guin and Sawyer, 1982; Whitbread *et al.*, 1991).

Keratins separated by PAGE were transferred to nitrocellulose or *Immobulon* membranes (Sawyer *et al.*, 2003a) and incubated overnight in a 1:5000 dilution of anti-FβK in phosphate-buffered saline. Membranes were rinsed thoroughly and exposed to secondary antibodies (goat anti-rabbit IgG linked to horseradish peroxidase). An appropriate substrate

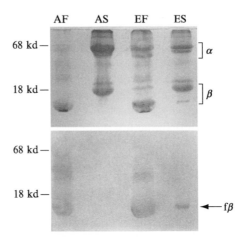

FIG. 3. The upper panel shows the epidermal keratins of adult feather (AF), adult scutate scale (AS), embryonic feather (EF), and embryonic scutate scale (ES) extracted with Triton X-100/1.5*M* KCl and separated by electrophoresis on a 10% polyacrylamide gel. The keratins are stained with Coomassie brilliant blue. The alpha (α) keratins migrate as bands in the range of 40–70 kd, whereas the scale-type β keratins (ScβKs) migrate to the range of 17–20 kd, and the feather-type β keratins (FβKs) migrate to the range of 10–14 kd. The lower panel shows the corresponding Western blot. The keratins separated as shown in the upper panel were transferred to a nitrocellulose membrane and incubated overnight with the FβK antiserum, and the FβK antiserum was localized with goat anti-rabbit horseradish peroxidase. The FβK is present in the AFs and EFs as evidenced by the broad band in the molecular weight range of the β keratins. The bands at higher molecular weight are the result of aggregation known to occur with β keratins. The FβK band is absent from the adult scale (AS) but is present in the embryonic scale (ES). The FβK antiserum does not react with the ScβKs present in both the adult and the ES (after Sawyer *et al.* [2003]).

(e.g., diaminobenzidine) was used to produce a precipitate on the membrane to mark the location of antigens (O'Guin *et al.*, 1982; Sawyer *et al.*, 2003a). Figure 3 shows the localization of FβK antiserum on extracts of epidermal β keratins of adult feather, adult scale, embryonic feather and embryonic scale.

Step 8b: Immunostaining of Histological Sections

As important as the specificity of this antiserum is in determining the biochemical identity of the FβKs in immunoblots, it is equally important in functional studies to determine the timing of expression and particularly the spatial distribution of the β-keratins in the cells of the structures themselves as they develop. For this aspect of the analysis, we employed immunohistochemistry of tissues that were embedded in paraffin and sectioned on a microtome (Sawyer *et al.*, 2003a). The antiserum was used as a probe for location of the protein in actual feather and scale epidermal tissue. Specifically, we used indirect immunofluorescence techniques in conjunction with confocal microscopy (Sawyer *et al.*, 2003a). The advantage of confocal microscopy is in the resolution attainable because of a dramatic reduction in flare and the ability to optically section thick sections (10–20 μm) of tissue. Combining paraffin sectioning with confocal analysis allows the collection of data on serial sets of sections. In Fig. 4, confocal microscope images of feather and scale show localization of feather type β-keratins using FITC-tagged anti-rabbit IgG antiserum to localize the binding of the FβK antiserum to its specific antigens.

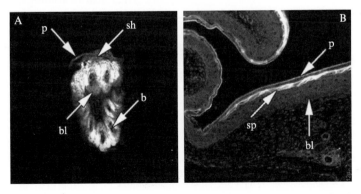

FIG. 4. (A) A confocal microscope image of a 17-day-old embryonic feather localizing feather-type β keratin using FITC-tagged anti-rabbit immunoglobulin G (IgG) antiserum, which localized the anti-FβK. (B) The location of the feather-type β keratins in a 17-day-old embryonic scutate scale. (After Sawyer *et al.* [2003].) b, barb ridge; sh, sheath cells; p, primary periderm cells; sp, subperidermal cells; bl, stratum basale region.

Unlike the results with the Western blots, which solely demonstrate reactivity to the electrophoretically separated proteins, the results of immunohistochemistry indicate where such proteins reside in the cells and tissues themselves. For example, in Fig. 4 individual cells of the barb ridges of an embryonic day 17 feather are easily defined by both position and intensity. In Fig. 4B, an embryonic day 17 scutate scale (found on the anterior surface of the bird's foot) displays two separate and distinct cell layers in which the FβK are localized. This approach adds a new dimension to the functional genomic analysis presented here, in that for the first time it can be unambiguously demonstrated that embryonic scutate scales express FβKs as demonstrated by both Western blots and immunohistochemistry.

Concluding Remarks

There are clear advantages to understanding developmental processes in light of a relatively complete sequence of events from transcription to translation to differentiation and histological organization. One advantage is that we are not left with much doubt about what the fate of such expression is in terms of functionality. Unlike many systems in which the expression of an enzyme or transcription factor is linked to many different pathways or is transient in presentation, the use of structural elements such as β-keratin has a fate that is less complicated. Another advantage for functional genomics is the ability to compare species with historical evolutionary relationships and determine what is unique about processes such as morphogenesis and those aspects of the process that are shared. With respect to feather development, it is now becoming clear that there are events in the morphogenesis of feathers that relate not only to birds, but also potentially to modern reptiles and their predecessors. The functional genomics approach employed in this methods chapter embraces the desire to uncover the common themes of evolutionary developmental biology while retaining our interest in the unique aspects of each species under study.

References

Alibardi, L., and Sawyer, R. H. (2002). Immunocytochemical analysis of beta (β) keratin in the epidermis of chelonians, lepidosaurians and archosaurians. *J. Exp. Zool.* **293,** 27–38.
Carver, W. E., and Sawyer, R. H. (1987). Development and keratinization of the epidermis in the common lizard, *Anolis carolinensis. J. Exp. Zool.* **243,** 435–443.
French, J. O. (2001). "Characterization and evolutionary relationships among copies of feather β-keratin genes." M.S. Thesis, University of South Carolina, Columbia.
Frohman, M. A. (1994). Cloning PCR products. *In* "The Polymerase Chain Reaction" (K. B. Mullis, F. Ferre, and R. A. Gibbs, eds.), pp. 14–37. Birkhauser, Boston, MA.

Haake, A. R., Koenig, G., and Sawyer, R. H. (1984). Avian feather development: Relationships between morphogenesis and keratinization. *Dev. Biol.* **106**, 406–413.

Hall, B. G. (2001). "Phylogenetic Trees Made Easy: A How-To Manual for Molecular Biologists. Sinauer Associates, Inc., Sunderland, MA.

Holton, T. A., and Graham, M. W. (1991). A simple and efficient method for direct cloning of PCR product using ddT-tailed vectors. *Nucleic Acids Res.* **19**, 1156.

Inglis, A. S., Gillespie, J. M., Roxburgh, C. M., Whittaker, L. A., and Casagranda, F. (1987). *In* "Proteins, Structure, and Function" (J. L. L'Italien, ed.), p. 757. Plenum Press, New York and London.

Kemp, D. J., and Rogers, G. E. (1972). Differentiation of avian keratinocytes: Characterization and relationship of the keratin proteins of adult and embryonic feather and scales. *Biochem.* **11**, 969–975.

Kusukawa, N., Uemori, T., Asada, K., and Kato, I. (1990). Rapid, reliable protocol for direct sequencing of material amplified by the PCR. *Biotechniques* **9**, 66–72.

Marchuk, D., Drumm, M., Saulino, A., and Collins, F. S. (1991). Construction of T-vectors, a rapid and general system for direct cloning of unmodified PCR products. *Nucleic Acids Res.* **19**, 1154.

Molloy, P. L., Powell, B. C., Gregg, K., Barone, E. D., and Rogers, G. E. (1982). Organization of feather keratin genes in the chick genome. *Nucleic Acids Res.* **10**, 6007–6021.

O'Guin, W. M., Knapp, L. W., and Sawyer, R. H. (1982). Biochemical and immunohistochemical localization of alpha and beta keratin in avian scutate scale. *J. Exp. Zool.* **220**, 371–376.

O'Guin, W. M., and Sawyer, R. H. (1982). Avian scale development: VIII. relationship between morphogenetic and biosynthetic differentiation. *Dev. Biol.* **89**, 485–492.

Palumbi, S. R. (1996). Nucleic acids III: The polymerase chain reaction. *In* "Molecular Systematics" (D. M. Hillis, C. Moritz, and B. K. Mable, eds.), 2 Ed., pp. 205–247. Sinauer Associates, Inc., Sunderland, MA.

Presland, R. B., Gregg, K., Molloy, P. L., Morris, C. P., Crocker, L. A., and Rogers, G. E. (1989a). Avian keratin genes. I. A molecular analysis of the structure and expression of a group of feather keratin genes. *J. Mol. Biol.* **209**, 549–559.

Presland, R. B., Whitbread, L. A., and Rogers, G. E. (1989b). Avian keratin genes. II. chromosomal arrangement and close linkage of three gene families. *J. Mol. Biol.* **209**, 561–576.

Rogers, G. E., Dunn, S., and Powell, B. (1998). Late events and the regulation of keratinocyte differentiation in hair and feather follicles. *In* "Molecular Basis of Epithelial Appendage Morphogenesis," pp. 315–340. R. G. Landes Company, Austin, TX.

Sambrook, J., and Russell, D. W. (2001). "Molecular Cloning: A Laboratory Manual." Cold Spring Harbor Laboratory Press, Cold Spring Harbor, NY.

Sawyer, R. H., Glenn, T. C., French, J. O., Mays, B., Shames, R. B., Barnes, G. L., Rhodes, W., and Ishikawa, Y. (2000). The expression of beta (β) keratins in the epidermal appendages of reptiles and birds. *Am. Zool.* **40**, 530–539. Wiley-Liss, New Jersey.

Sawyer, R. H., and Knapp, L. W. (2003). Avian skin development and the origin of feathers. *In* "Molecular and Developmental Evolution" (C.-M. Chuong, ed.), pp. 57–72.

Sawyer, R. H., Salvatore, B. A., Potylicki, T.-T. F., French, J. O., Glenn, T. C., and Knapp, L. W. (2003a). Origin of feathers: Feather beta (β) keratins are expressed in discrete epidermal cell populations of embryonic scutate scales. *J. Exp. Zool. (Mol. Dev. Evol.)* **295B**, 12–24.

Sawyer, R. H., Washington, L. D., Salvatore, B. A., Glenn, T. C., and Knapp, L. W. (2003b). Origin of archosaurian integumentary appendages: The bristles of the wild turkey beard express feather-type β keratins. *J. Exp. Zool. (Mol. Dev. Evol.)* **297B**, 1–8.

Shames, R. B., Knapp, L. W., Carver, W. E., and Sawyer, R. H. (1988). Identification, expression and localization of beta keratin gene products during development of avian scutate scales. *Differentiation* **38**, 115–123.

Thompson, J. D., Gibson, T. J., and Higgins, D. G. (1994). The ClustalW: Improving the sensitivity of progressive multiple sequence alignment through sequence weighting, position, specific gap penalties, and weight matrix choice. *Nucleic Acids Res.* **22**, 4673–4680.

Thompson, J. D., Gibson, T. J., Plewniak, F., Jeanmougin, F., and Higgins, D. G. (1997). The ClustalX-Windows interface: Flexible strategies for multiple sequence alignment aided by quality analysis tools. *Nucleic Acids Res.* **25**, 4876–4882.

Walker, I. D., and Bridgen, J. (1976). The keratin chains of avian scale tissue, sequence heterogeneity and the number of scale keratin genes. *Eur. J. Biochem.* **67**, 283–293.

Whitbread, L. A., Gregg, K., and Rogers, G. E. (1991). The structure and expression of a gene encoding chick claw keratin. *Gene* **101**, 223–229.

Wilton, S. D., Crocker, L. A., and Rogers, G. E. (1985). Isolation and characterization of keratin m RNA from the scale epidermis of the embryonic chick. *Biochim. Biophys. Acta* **824**, 201–201.

Zola, H. (1987). "Monoclonal Antibodies: A Manual of Techniques." CRC Press, Boca Raton, FL.

[34] Applications of Ancestral Protein Reconstruction in Understanding Protein Function: GFP-Like Proteins

By Belinda S. W. Chang, Juan A. Ugalde, and
Mikhail V. Matz

Abstract

Recreating ancestral proteins in the laboratory increasingly is being used to study the evolutionary history of protein function. More efficient gene synthesis techniques and the decreasing costs of commercial oligo-synthesis are making this approach both simpler and less expensive to perform. Developments in ancestral reconstruction methods, particularly more realistic likelihood models of molecular evolution, allow for the accurate reconstruction of more ancient proteins than previously possible. This chapter reviews phylogenetic methods of ancestral inference, strategies for investigating alternative reconstructions, gene synthesis, and design, and an application of these methods to the reconstruction of an ancestor in the green fluorescent protein family.

Introduction

Ancestral protein reconstruction allows for the recreation of protein evolution in the laboratory so that it can be studied directly. This approach is a natural extension of experimental studies that examine present-day protein function, from which the evolutionary history of function may then be extrapolated. Reconstructing ancestral proteins can provide additional information not easily obtainable from studies of extant proteins, which are necessarily limited to the range of function available today. This kind of experimental approach has been used as a window into the evolutionary past of proteins in an increasing number of systems (Chandrasekharan et al., 1996; Sun et al., 2002; Thornton et al., 2003; Zhang and Rosenberg, 2002; for reviews, see Chang and Donoghue, 2000; Stewart, 1995; Thornton, 2004). In addition to providing valuable information about the evolution of present-day molecular structure and function, it also can lead to the discovery of new aspects of biochemical function that have subsequently been lost in extant proteins or that exist only in obscure or difficult to obtain organisms (Adey et al., 1994; Jermann et al., 1995; Malcolm et al., 1990). It can lead to new insights into the biology, or even the environment, of extinct organisms (Chang et al., 2002b; Gaucher et al., 2003).

Traditional methods of studying the structure and function of proteins generally have employed mutagenesis methods to identify residues or regions of the protein that are important for function. Although the targets of mutagenesis may often be directed by knowledge of the three-dimensional structure of the protein of interest, researchers are none-theless faced with choosing from a vast number of possible mutations, either singly or in combination. Finding a number of mutations that produce properly folded and functional proteins, never mind those that show interesting and substantial effects in biochemical assays, can be difficult indeed. Multiple site mutants are, therefore, likely to be limited to combinations of single mutants that produce interesting phenotypes. Although this can be a reasonable approach given the expense and effort involved to make and express hundreds of mutants, it necessarily precludes combinations of mutations that by themselves do not produce any effect worth noting but together may create a novel or, better yet, interesting phenotype.

Reconstructing the evolutionary history of proteins in the laboratory offers several intriguing advantages compared to these more traditional approaches. Because the process of natural selection tends to eliminate the vast majority of mutations producing dysfunctional proteins, this approach effectively screens out mutations resulting in misfolded proteins and focus-es on those changes that may have altered protein function during its evolutionary history. Moreover, the problem of assessing which combina-tions of mutations may be interesting from a functional point of view is addressed nicely using this approach; ancestral proteins are, in effect, combinations of mutations that, if properly chosen, have been selected to produce marked shifts in evolutionary function.

Here, we review some of the methods of ancestral inference, as well as the design and synthesis of ancestral genes in the laboratory. We also touch on some of the issues that can arise, particularly if different methods do not agree in their inference of ancestral states, and how these issues can be overcome. Finally, we highlight an example of how these methods can be applied to the green fluorescent protein (GFP)–like family of proteins.

Ancestral Gene Inference Methods

Generally speaking, two types of phylogenetic methods are used to infer ancestral sequences: parsimony and likelihood/bayesian methods (Table I). Parsimony methods (Swofford, 2002) minimize the amount of evolutionary change along each branch, assuming slow and consistent rates of evolutionary change. Likelihood methods, on the other hand, incorporate an explicit model of substitution, which allows for statistical comparisons among models in order to determine which is a better fit to the data at hand, at least among nested models (likelihood ratio tests; see later discussion). Parsimony methods, with reference to the reconstruction of ancestral states, have been extensively discussed and reviewed elsewhere (Cunningham *et al.*, 1998; Maddison, 1995; Omland, 1999; Swofford *et al.*, 1996) and are not discussed further here.

TABLE I
STATISTICAL METHODS OF INFERENCE OF ANCESTRAL STATES

Method	Programs available	Computer programs	References
Maximum Parsimony	http://macclade.org/ macclade.html	MacClade (Maddison and Maddison, 1993)	(Maddison, 1995; Swofford *et al.*, 1996)
	http://paup.csit.fsu.edu/	PAUP* (Swofford, 2002)	
Maximum likelihood/ bayesian	http://abacus.gene.ucl. ac.uk/software/paml.html	PAML (Yang, 1997)	Empirical Bayes (Yang *et al.*, 1995)
	http://morphbank.ebc.uu.se/ mrbayes/	MrBayes (Huelsenbeck and Ronquist, 2001)	Hierarchical Bayes (Huelsenbeck and Bollback, 2001a)

Phylogenetic methods based on maximum likelihood analysis (implemented in programs such as PHYLIP [Felsenstein, 1991], MOLPHY [Adachi and Hasegawa, 1994], PAML [Yang, 1997], and NHML [Galtier and Gouy, 1998]) use a likelihood score as an optimality criterion. This likelihood score is calculated according to a specified model of evolution (Felsenstein, 1981). Optimizing the likelihood score can be used to infer the most likely tree topology, as well as parameters such as branch lengths, character state frequencies, and ancestral states. Bayesian methods are then used to calculate ancestral states with the highest posterior probability. This can be done by using the maximum likelihood topology, branch lengths, and model parameters as priors (empirical Bayes method [Yang et al., 1995]), or alternatively the posterior probabilities can be calculated by taking into account the uncertainty in the maximum likelihood topology and parameters using a Markov chain Monte Carlo approach (hierarchical Bayes method, (Huelsenbeck and Bollback, 2001a) (see Table I). This approach has some advantages over parsimony methods (Koshi and Goldstein, 1996; Lewis, 1998; Yang et al., 1995). In using an explicit model of molecular evolution, stochastic methods allow for the incorporation of knowledge of the mechanisms and constraints acting on coding sequences, as well as the possibility of comparing the performance of different models, ultimately resulting in the development of more realistic models (Goldman, 1993).

With stochastic methods such as maximum likelihood and Bayesian analysis, it is important to explore different models of molecular evolution to determine how robust the ancestral reconstruction results are (Huelsenbeck and Bollback, 2001b; Huelsenbeck et al., 2002). Oversimplified or unrealistic models have been shown in certain cases to yield spurious phylogenetic reconstructions (Buckley, 2002; Cao et al., 1994; Huelsenbeck, 1997), emphasizing the importance of model selection. Models can be grouped into different classes: nucleotide, amino acid, and codon. Nucleotide models range from the simplest (Jukes-Cantor, 1969), which assumes equal base frequencies and rates of transitions and transversions, to much more complex models allowing unequal base frequencies (Felsenstein, 1981), transition/transversion bias (Kimura, 1980), among-site rate heterogeneity (Yang, 1994), and/or nonstationary base composition (Galtier and Gouy, 1998).

The simplest amino acid model is the Poisson, which assumes equal amino acid frequencies and rates of substitution among amino acids. This model is clearly unrealistic and would not be expected to perform well. More realistic models have been developed that allow unequal amino acid frequencies (Hasegawa and Fujiwara, 1993), and among-site rate heterogeneity (Yang, 1994), in addition to a general time-reversible (GTR) model for amino acids, which allows for unequal numbers of substitutions in the rate matrix for all the different classes of amino acid substitutions (Yang, 1997). Fixed rate matrices have also been calculated for a number of datasets, including globular proteins (Cao *et al.*, 1994; Dayhoff, 1978; Jones *et al.*, 1992; Kishino *et al.*, 1990), and mitochondrial transmembrane proteins (Adachi and Hasegawa, 1996). The use of fixed or constant parameters in the rate matrix can be advantageous because it allows for a reduction in the number of parameters in the model of evolution being used. Recent developments include amino acid models that allow replacement rates to be proportional to the frequencies of both the replaced and the resulting residues (+*gwF* model; (Goldman and Whelan, 2002).

Codon-based models of molecular evolution show the most promise, as they have the potential to incorporate information about different types of nucleotide substitutions and whether they change the amino acid or not. The original codon-based models assumed equal nonsynonymous-to-synonymous rate ratios among sites and lineages (Goldman and Yang, 1994; Muse and Gaut, 1994). Subsequently, models were developed that allowed that ratio to vary across lineages in a phylogeny (Yang, 1998), across sites in a protein (Nielsen and Yang, 1998), and across both sites and lineages (Yang and Nielsen, 2002). Several models, each employing a different statistical distribution of nonsynonymous-to-synonymous rate ratios (dN/dS) across sites, have been developed as tools to detect positively selected sites using likelihood ratio tests ("random sites models" [Yang *et al.*, 2000]; for reviews see Bielawski and Yang [2003] and Yang and Bielawski [2000]). If, for example, there is some *a priori* information available based on tertiary protein structure that can be used to partition sites in the protein into different classes, then fixed-sites models may also be used (Yang and Swanson, 2002).

Given the diversity of models now available, the choice of a particular model for use in phylogenetic analysis and ancestral inference is critical. An inappropriate model of evolution can lead to inconsistency in the likelihood analysis and convergence to an incorrect result (Huelsenbeck, 1998). Ancestral inference methods are particularly sensitive to model choice. The possibility of an incorrect result can be reduced by selecting a model of evolution that displays the best fit to the sequence data at hand. To this end, likelihood ratio tests can be used to compare two models of evolution that are nested with respect to each other, to determine whether the more complex model fits the sequence data significantly better than the simpler model (Felsenstein, 1981; Huelsenbeck and Rannala, 1997; Yang *et al.*, 1994). For nested models, a more complex model (H_1) will contain all the parameters of the original model (H_0), as well as additional parameters. If the models are not nested, they cannot be directly compared using a likelihood ratio test, and other methods, such as the generation of the distribution of the test statistic using Monte Carlo simulation, must be used (Goldman, 1993).

Choosing an Ancestral Sequence to Reconstruct

What happens when different likelihood models or even methods of inferring ancestral sequences result in different amino acid reconstructions at particular sites in the protein? This may not happen if the sequences are closely related, where we would expect inferences about ancestral states to be fairly robust to changes in the particular model of evolution used. Additionally, it may be possible to compare likelihood models using likelihood ratio tests, if the models are nested, and then only the reconstructions from the model with the better fit would be considered. However, it is not always possible to compare models this way, and there usually will be a certain proportion of sites for which feasible alternative reconstructions exist, among which it may be difficult to choose.

There are different approaches that can be taken to address this problem. The simplest is to randomly choose one reconstruction (per site) among all the alternatives. This has obvious drawbacks if the true reconstruction is not among the alternatives chosen. In some cases, the probability that the ancestral sequence chosen will match the true ancestor across all sites may be quite small. This may have important consequences for subsequent interpretation of the results of functional assays of the ancestral protein. It seems more appropriate to incorporate at least some exploration of alternative methods and/or models in reconstructing the ancestral protein in the laboratory. This can be done in a number of ways. For the purpose of synthesizing ancestral proteins in the laboratory, different types of models may be most suitable, depending on how deep the ancestral node is in the tree. More recent ancestors that are not too diverged from existing sequences may be best reconstructed using nucleotide models, where parameters like transition/transversion rate ratios predominate. In contrast, as divergence increases, the more ancient nodes may be best reconstructed using the amino acid models, where factors such as side-chain properties predominate. For this reason, a useful approach is to synthesize several variants of the gene predicted to be the best reconstruction under different models, so the results of different models can be compared in functional assays in the laboratory (Chang and Donoghue, 2005).

Another approach entirely would be to incorporate degeneracies at sites where alternative reconstructions exist *during* the gene synthesis (degenerate reconstruction). In this way, it is possible to pool together the predictions of different models, which mitigates the problem of choosing the most appropriate model to a certain extent. As a result, a library of possible ancestral genes is obtained instead of just a single gene that is the most probable according to a particular model. Although the degenerate reconstruction approach is not really aimed at directly comparing the most likely ancestral reconstructions inferred using different likelihood models, it may be useful to compare models' performances by evaluating the probabilities of different phenotypes according to each of the models. If the different phenotypes are observed in the combinatorial library, the corresponding genes can be sequenced to determine the combinations of degenerate sites that are responsible for the phenotypes. It is then possible to go back and evaluate how probable these particular combinations are relative to each other according to the model's predictions. In this way, one can obtain the result in the form such as "at the node N, model A predicts phenotype X with probability P1, phenotype Y with probability P2, and so on." The necessary information concerning probabilities of each state at each site at each node can be extracted with the PAML 3.13 package by specifying "verbose = 2" along with "Rate Ancestor = 1" in a control file for *codeml* or *baseml*. The biggest consideration in estimating phenotypic probabilities is that it requires an efficient method for screening the combinatorial library for different phenotypes, which may not be feasible for many proteins. Therefore, the degenerate and targeted reconstruction approaches are equally valid and, in fact, somewhat complementary to each other in that they explore alternative reconstructions in different ways.

Ancestral Gene Design and Synthesis

Artificial Gene Design

Once an ancestral protein sequence or an array of possible sequences has been inferred, the degeneracy of the genetic code can be used to design artificial genes with properties useful in the synthesis and expression of the ancestral gene. Unique restriction sites and potential primer sites that later will aid in the characterization and construction of the gene can be incorporated. Codon usage bias can be optimized for a particular species or cell type (Sharp *et al.*, 1988). In many expression systems, rare codons are known to cause translational problems caused by limited tRNA availability, resulting in misincorporations, truncated proteins, and overall reduced translational efficiency (Kane, 1995). Conversely, although the goal of optimizing codon usage frequencies is usually increased expression levels, the incorporation of unpreferred codons is occasionally useful in slowing translation of signal sequences so that cellular membrane translocation systems are not saturated (Karnik *et al.*, 1987). The secondary structure of mRNA also has been implicated in lowered expression levels in *Escherichia coli* (Griswold *et al.*, 2003). GC content can affect levels of heterologously expressed proteins (Sinclair and Choy, 2002) and may need to be adjusted to minimize potential difficulties in later molecular biology manipulations such as cloning and sequencing. Epitopes for antibodies or other tags that would aid in protein purification also can be introduced in the design of an artificial gene.

Gene Synthesis Incorporating Degenerate Sites

The principle of the method described here is depicted in Fig. 1. It uses an array of overlapping oligonucleotides 30–35 bases long to assemble both strands of the synthesized gene by means of ligation, followed by PCR amplification of the target product using flanking oligonucleotides as primers. Note that degenerate sites, if they are to be incorporated into the gene synthesis, should be positioned as far as possible from the ligation points. Though very simple, the method has important advantages over previously reported techniques that rely on longer oligonucleotides (Chang *et al.*, 2002a; Ferretti *et al.*, 1986), because it minimizes the number of errors introduced during the chemical synthesis of the oligomers and allows for them to be ordered commercially instead of requiring an in-house oligonucleotide synthesizer. Long oligonucleotides tend to form secondary structures during synthesis, resulting in frequent errors that can include deletions and insertions of varying sizes. In contrast, shorter oligonucleotides are usually synthesized for a minimal cost with accuracy that is high enough to skip the expensive purification procedures, which are often necessary for long oligonucleotides. In addition, in our method, the oligonucleotides do not need to be modified (e.g., they do not require 5′ phosphates), which further decreases the cost of the project.

Finally, a gene synthesis strategy that incorporates so many ligation points per gene is particularly useful because the ligation efficiency is significantly diminished by the presence of mismatches in the vicinity of the ligation site. In our protocol, the separation between the ligation sites is only 16–17 bases, which means that almost three-fourths of the gene length is actually "proofread" at the ligation step because DNA ligase is sensitive to mismatches at least up to 6 bases from the ligation site (Roth *et al.*, 2004). For example, the mutated clones incorporating accidental errors at this step in our experiments made up less than 50% of the total number, and even in those clones, the mutations were likely to be PCR errors rather than gene assembly artifacts.

Mixture of all oligonucleotides comprising the gene except the two 5'-terminal ones

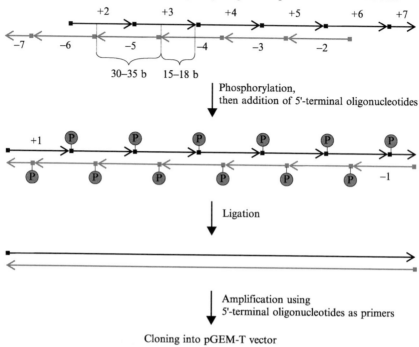

FIG. 1. Schematic outline of the described gene synthesis strategy. Oligonucleotides corresponding to plus and minus DNA strands are shown as black and gray arrows, respectively. Arrowheads correspond to free 3' termini, squares–to free 5' termini. For simplicity of representation, the scheme shows the synthesis of a short fragment about 210–250 bp in length; however, the strategy will work for the longer genes as well. In our experiment, as discussed in the text, the synthesized genes were 730 bp long.

Gene Synthesis Protocol

Oligonucleotides: The artificially designed ancestral gene should be divided into overlapping oligonucleotide fragments of about 30–35 bases in length. These oligonucleotides can be ordered from any reliable commercial service. No additional purification or modification is required; the smallest offered synthesis scale (usually 25 nmol) is sufficient.

Phosphorylation: In a 0.5-ml tube, combine 5 μl of 2× buffer for T4 ligase, 4 μl of the oligonucleotide mixture (all oligonucleotides that comprise the gene in a concentration of 0.1 μM each, except the two 5'-terminal ones that will not be ligated by their 5' ends, see Fig. 1) and 1 μl of T4 polynucleotide kinase (New England BioLabs, Beverly, MA). We used the buffer provided within the pGEM-T PCR cloning kit (Promega) because it is similar in composition to the standard T4 polynucleotide kinase buffer but already contains ATP in appropriate concentration. Incubate the reaction at 37° for 30 min, and then incubate it at 65° for 20 min to deactivate the enzyme.

Ligation: To the completed phosphorylation reaction, add 5 μl of 2× Ligation Buffer (Promega, Madison, WI), 4 μl of the terminal oligonucleotides mixture (Fig. 1, 0.1 μM each) and 1 μl of the T4 DNA ligase (New England BioLabs, Beverly, MA). Incubate the reaction for 2 h at 37°.

PCR amplification of the ligated products: It is important to use a polymerase or polymerase mixture exhibiting proofreading activity, to minimize PCR errors. In our experiments, we used Advantage 2 polymerase mixture (BD Biosciences Clontech, San Jose, CA) with the buffer provided. To perform the amplification, combine in an 0.5 thin-walled PCR tube: 2 μl of the ligation reaction, 2 μl of each of the 5′-terminal oligonucleotides diluted to 1 μM, 2 μl of the 10× reaction buffer, 2 μl of 5 mM dNTP mixture, 12 μl deionized water, and 0.5 μl of the Advantage 2 polymerase mix. Perform cycling according to the following program: 45 s at 94°, 1 min at the annealing temperature (depends on the sequence of the primers), 1 min at 72° (add 1 min per each 1000 bp of the synthesized gene over 1500 bp); 15–20 cycles. The accumulation of the PCR product should be monitored to keep the number of PCR cycles to the necessary minimum. The product should become visible on a standard agarose gel after 15–20 cycles, when 1/10 of the reaction volume is loaded into the well. The PCR product is then cloned into pGEM-T (Promega) in order to obtain bacterial expression libraries.

Example of Ancestral Protein Expression: GFP-Like Proteins

The primary function of the family of GFPs (and related colored proteins), first isolated from the jellyfish *Aequorea victoria*, is coloration and/or fluorescence. This is acquired by these proteins via autocatalytic synthesis of the chromophore moiety within its own globule, using its own side chains as substrates (Heim *et al.*, 1994; Matz *et al.*, 2002). GFP-like proteins are the only natural pigments in which both chromophore and protein are contained within a single gene, which has earned them great popularity as biotechnology tools (see Lippincott-Schwartz and Patterson [2003] for a recent review). They come in four basic colors, roughly corresponding to distinct types of the chromophore chemical structure: fluorescent colors including green, yellow, and red, and nonfluorescent purple-blue (Labas *et al.*, 2002). In many ways GFP-like proteins represent a convenient model for basic studies in the evolution of gene families (Matz *et al.*, 2002). The family contains many cases of gene duplication followed by diversification of function and may even present several cases of convergent evolution of complex features at the molecular level (Labas *et al.*, 2002; Shagin *et al.*, 2004). At the same time, the proteins are small (~230 amino acid residues long) and can be expressed easily in a functional form in a variety of heterologous systems including bacteria. The phenotype, which is simply the color of fluorescence, can already be precisely quantified in the bacterial colonies growing on a solid media. No further purification of the expressed proteins is necessary. This provides an excellent opportunity for high-throughput screening of expression libraries of mutants or, as we show in this chapter, reconstructed ancestral genes.

The experiment presented here is part of an ongoing study of the color evolution among paralogous lineages of GFP-like proteins found in the great star coral *Montastrea cavernosa* (Ugalde *et al.*, 2004). This species possesses several (at least four and maybe up to seven) genetic loci coding for GFP-like proteins comprising four paralogous groups corresponding to cyan (emission max at 480–495 nm), shortwave green (emission max at 500–510 nm), long-wave green (emission max at 515–525 nm), and red (emission max at 575–585 nm) colors (Kelmanson and Matz, 2003). These colors share a common ancestor sometime after the first diversification of corals in early Triassic (240 million years ago) and before mid-Jurassic (180 million years ago), according to the fossil record and the phylogenetic tree topology (Kelmanson and Matz, 2003). The most ancient ancestral phenotype of GFP-like proteins is likely to be shortwave green (Shagin *et al.*, 2004), while the phenotype of the common ancestor of *M. cavernosa* paralogs could be anything because more basal branches in the phylogeny lead to proteins of all colors (Kelmanson and Matz, 2003; Shagin *et al.*, 2004). We set out to reconstruct proteins in the nodes representing the common ancestor of all phenotypes (ALL ancestor), the common ancestor of all the red proteins (Red ancestor, or R), and two intermediate nodes, corresponding to the two possible common ancestors of reds and longwave greens (Red/Green ancestor, or RG; and pre-Red ancestor, or pre-R; see Fig. 4B).

For the prediction of the ancestral sequences, the dataset described in Shagin *et al.* (2004), comprising most of the cnidarian GFP-like proteins known, was used. Three alternative maximum likelihood models were applied: amino acid–based JTT (Jones *et al.*, 1992), codon-based M5 (Yang *et al.*, 2000), and nucleotide-based GTR+G3 (Tavare, 1986). The latter model was different from the more common GTR+G in that it assumed, not one, but three independent gamma-distributed rates of evolution at individual nucleotide sites, corresponding to their positions within codons. GTR+G3 performed best among nucleotide-based models compared in likelihood ratio tests using the Modeltest program (Posada and Crandall, 1998).

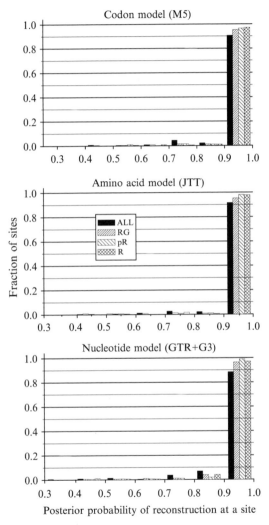

FIG. 2. Performance of different models in reconstruction of four ancestral sequences. Horizontal axis shows bin limits, so that, for example, the bars appearing between marks 0.8 and 0.9 show the fractions of sites that are predicted with posterior probability >0.8 and ≤0.9. In the legend, ALL, RG, pR, and R correspond to the ALL, red/green, pre-red, and red ancestors, respectively.

The reconstructions of all four ancestral sequences were quite robust with any model, although the GTR+G3 model was slightly less robust in its predictions compared to the other two (Fig. 2 and Table II). The least posterior probability at a reconstructed site was 0.328, observed in the reconstruction of the ALL-ancestor under the GTR+G3 model (the other two models predicted the same site with posterior probability 0.78). Comparison of the most probable reconstructions (Fig. 3) revealed that a small number of sites were predicted differently under different models. Apparently, these sites were poorly predictable by some or all of the three models, because no disagreement was observed between models when all three of them generated the site prediction with posterior probability exceeding 0.80. When planning ancestral gene synthesis, the codons corresponding to these ambiguous sites were designed to be degenerate to incorporate the alternative predictions. As a result, the designed genes for ALL, RG, pre-R, and R ancestors contained eight, six, four, and six degenerate codons, respectively.

A total of 500–1000 fluorescent clones from each of the four combinatorial libraries were visually surveyed using a fluorescent stereomicroscope (Leica MZ FLIII) with the optical filters providing excitation in the 400–450 nm range and emission from 475 nm and up (long-pass filter). Such a filter combination allows for easy discrimination of different fluorescent phenotypes by human eye, even such similar ones as long-wave and short-wave green. In two of the four cases (ALL ancestor and Red ancestor), no phenotypic diversity was observed, while the Red/Green and Pre-Red ancestors were represented by clones appearing in slightly different shades of yellow. Twenty-four clones from each library were sequenced and plated onto new plates for spectroscopy. The fraction of clones containing no additional mutations was 0.54–0.75. Among these "clean" clones, there were variations at all the degenerate sites. As predicted, the common ancestor of all colors (ALL ancestor) turned out to be shortwave green. Most interestingly, all clones corresponding to the two possible common ancestors of red and green proteins (Red/Green and pre-Red) showed an

TABLE II
AVERAGE POSTERIOR PROBABILITIES OF ANCESTRAL RECONSTRUCTION AT A SITE

Ancestor	JTT	Models GTRG3	M5
ALL	0.972	0.958	0.971
Red/Green	0.979	0.981	0.983
Pre-Red	0.987	0.987	0.988
Red	0.990	0.981	0.989

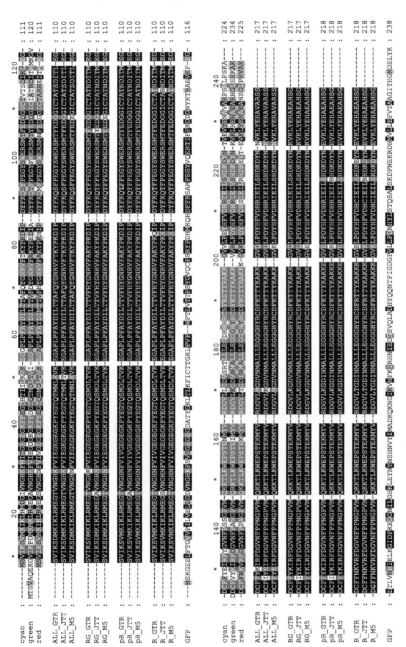

FIG. 3. Alignment of extant cyan, green, and red fluorescent proteins from *Montastrea cavernosa* (GenBank accession numbers AY181556, AY181554, and AY181552, respectively) (Kelmanson and Matz, 2003); and their ancestors predicted using three models (GTR+G3, JTT, and M5, see text for details). The ancestral sequences correspond to the nodes denoted on Fig. 2: ALL-ancestor (ALL), red/green ancestor (RG), pre-red ancestor (pR), and red ancestor (R). The green fluorescent protein (GFP) from *Aequorea victoria* (accession number M62539) is aligned below for reference.

intermediate long-wave green/red phenotype. Although the majority of the expressed protein bulk remained long-wave green, a small fraction was able to complete the third chromophore maturation stage resulting in a minor peak of red emission. Clones of the Red ancestor showed an "imperfect red" phenotype; although in them the red emission peak always dominated, the rate of green-to-red conversion during the last chromophore maturation stage was apparently still slower than in extant reds, resulting in a prominent minor peak of green fluorescence (Fig. 4A and C).

It is clear from these experiments that the evolution of red emission color, which corresponds to an increase of functional and structural complexity (Shagin *et al.*, 2004), progressed through a series of intermediate stages. Also, it was possible to establish the color of the common ancestor of all the *M. cavernosa* paralogs as shortwave green. The complete molecular analysis of the color evolution in fluorescent proteins that would include studies of selection pressure across individual sites and mutagenesis experiments will be published elsewhere.

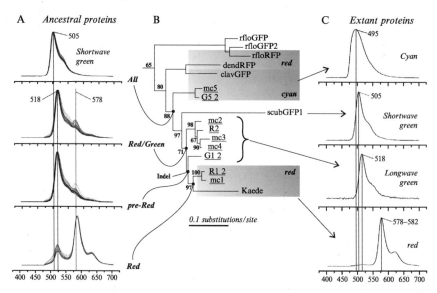

Fig. 4. Evolution of colors in a subset of cnidarian green fluorescent protein (GFP)–like proteins. (A) Fluorescence spectra of the reconstructed ancestral proteins. Multiple curves correspond to clones bearing variations at degenerate sites. (B) The portion of the phylogenetic tree of GFP-like proteins discussed here. The names of sequences originating from the great star coral *Montastrea cavernosa* are underlined. The values at the branches are nonparametric bootstrap support under the maximum likelihood criterion with GTR+G+I model. The sequence G1.2, which in an unconstrained bootstrap analysis has an uncertain affinity either to red or long-wave green cluster, has been forced to group with the red proteins on the basis of shared three-nucleotide (one codon) indel. (C) Fluorescence spectra of extant proteins.

Concluding Remarks

Laboratory synthesis of ancestral proteins is becoming a fast and fairly inexpensive method for studying the evolution of molecular structure and function. We no longer have to rely on inferences about the evolution of protein function based on amino acid substitutions thought to be historically important for functional shifts, which are then incorporated into site-directed mutagenesis studies of present day proteins. Ancestral protein synthesis offers a much more direct view of the evolutionary history of proteins, where the entire ancestral protein can be recreated and functionally assayed in the laboratory. This can shed light not only on the structure and function of present-day proteins, but also potentially on how evolution gave rise to the diversity of function seen today.

Acknowledgments

This work was supported by grants from the NSF and the National Sciences and Engineering Research Concil of Canada (B.S.W.C.), Grass Foundation (J.A.U.), and the U.S. Department of Defense and NIH (M.V.M.)

References

Adachi, J., and Hasegawa, M. (1994). "MOLPHY." Institute of Statistical Mechanics, Tokyo, Japan.

Adachi, J., and Hasegawa, M. (1996). Model of amino acid substitution in proteins encoded by mitochondrial DNA. *J. Mol. Evol.* **42**, 459–468.

Adey, N. B., Tollefsbol, T. O., Sparks, A. B., Edgell, M. H., and Hutchison, C. A., III (1994). Molecular resurrection of an extinct ancestral promoter for mouse L1. *Proc. Natl. Acad. Sci. USA* **91**, 1569–1573.

Bielawski, J. P., and Yang, Z. (2003). Maximum likelihood methods for detecting adaptive evolution after gene duplication. *J. Struct. Funct. Genomics* **3**, 201–212.

Buckley, T. R. (2002). Model misspecification and probabilistic tests of topology: Evidence from empirical data sets. *Syst. Biol.* **51**, 509–523.

Cao, Y., Adachi, J., Yano, T.-A., and Hasegawa, M. (1994). Phylogenetic place of guinea pigs: No support of the rodent-polyphyly hypothesis from maximum-likelihood analyses of multiple protein sequences. *Mol. Biol. Evol.* **11**, 593–604.

Chandrasekharan, U. M., Sanker, S., Glynias, M. J., Karnik, S. S., and Husain, A. (1996). Angiotensin II—forming activity in a reconstructed ancestral chymase. *Science* **271**, 502–505.

Chang, B. S., and Donoghue, M. J. (2005). Phylogenetic reconstruction of the origin of rod opsins from cone opsin ancestors. *submitted.*

Chang, B. S., Kazmi, M. A., and Sakmar, T. P. (2002a). Synthetic gene technology: Applications to ancestral gene reconstruction and structure-function studies of receptors. *Methods Enzymol.* **343**, 274–294.

Chang, B. S. W., and Donoghue, M. J. (2000). Recreating ancestral proteins. *Trends Ecol. Evol.* **15**, 109–114.

Chang, B. S. W., Jonsson, K., Kazmi, M., Donoghue, M. J., and Sakmar, T. P. (2002b). Recreating a functional ancestral archosaur visual pigment. *Mol. Biol. Evol.* **19,** 1483–1489.

Cunningham, C. W., Omland, K. E., and Oakley, T. H. (1998). Reconstructing ancestral character states: A critical reappraisal. *Trends Ecol. Evol.* **13,** 361–366.

Dayhoff, M. O. (1978). A model of evolutionary change in proteins. Matrices for detecting distant relationships. *In* "Atlas of Protein Sequence and Structure," (M. O. Dayhoff, ed.), Vol. 5(Suppl. 3), pp. 345–358. National Biomedical Research Foundation, Washington, DC.

Felsenstein, J. (1981). Evolutionary trees from DNA sequences: A maximum likelihood approach. *J. Mol. Evol.* **17,** 368–376.

Felsenstein, J. (1991). "PHYLIP: Phylogeny Inference Package." University of Washington, Seattle, WA.

Ferretti, L., Karnik, S. S., Khorana, H. G., Nassal, M., and Oprian, D. D. (1986). Total synthesis of a gene for bovine rhodopsin. *Proc. Natl. Acad. Sci. USA* **83,** 599–603.

Galtier, N., and Gouy, M. (1998). Inferring pattern and process: Maximum-likelihood implementation of a nonhomogeneous model of DNA sequence evolution for phylogenetic analysis. *Mol. Biol. Evol.* **15,** 871–879.

Gaucher, E. A., Thomson, J. M., Burgan, M. F., and Benner, S. A. (2003). Inferring the palaeoenvironment of ancient bacteria on the basis of resurrected proteins. *Nature* **425,** 285–288.

Goldman, N. (1993). Statistical tests of models of DNA substitution. *J. Mol. Evol.* **36,** 345–361.

Goldman, N., and Whelan, S. (2002). A novel use of equilibrium frequencies in models of sequence evolution. *Mol. Biol. Evol.* **19,** 1821–1831.

Goldman, N., and Yang, Z. (1994). A codon-based model of nucleotide substitution for protein-coding DNA sequences. *Mol. Biol. Evol.* **11,** 725–736.

Griswold, K. E., Mahmood, N. A., Iverson, B. L., and Georgiou, G. (2003). Effects of codon usage versus putative 5′-mRNA structure on the expression of Fusarium solani cutinase in the *Escherichia coli* cytoplasm. *Protein Exp. Purif.* **27,** 134–142.

Hasegawa, M., and Fujiwara, M. (1993). Relative efficiencies of the maximum likelihood, maximum parsimony, and neighbor-joining methods for estimating protein phylogeny. *Mol. Phylogenet. Evol.* **2,** 1–5.

Heim, R., Prasher, D. C., and Tsien, R. Y. (1994). Wavelength mutations and posttranslational autoxidation of green fluorescent protein. *Proc. Natl. Acad. Sci.* **91,** 12501–12504.

Huelsenbeck, J. P. (1997). Is the Felsenstein zone a fly trap? *Syst. Biol.* **46,** 69–74.

Huelsenbeck, J. P. (1998). Systematic bias in phylogenetic analysis: Is the Strepsiptera problem solved? *Syst. Biol.* **47,** 519–537.

Huelsenbeck, J. P., and Bollback, J. P. (2001). Empirical and hierarchical Bayesian estimation of ancestral states. *Syst. Biol.* **50,** 351–366.

Huelsenbeck, J. P., Larget, B., Miller, R. E., and Ronquist, F. (2002). Potential applications and pitfalls of Bayesian inference of phylogeny. *Syst. Biol.* **51,** 673–688.

Huelsenbeck, J. P., and Rannala, B. (1997). Phylogenetic methods come of age: Testing hypotheses in an evolutionary context. *Science* **276,** 227–232.

Huelsenbeck, J. P., and Ronquist, F. (2001). MRBAYES: Bayesian inference of phylogenetic trees. *Bioinformatics* **17,** 754–755.

Jermann, T. M., Opitz, J. G., Stackhouse, J., and Benner, S. A. (1995). Reconstructing the evolutionary history of the artiodactyl ribonuclease superfamily. *Nature* **374,** 57–59.

Jones, D. T., Taylor, W. R., and Thornton, J. M. (1992). The rapid generation of mutation data matrices from protein sequences. *Comp. Appl. Biosci.* **8,** 275–282.

Jukes, T. H., and Cantor, C. R. (1969). Evolution of protein molecules. *In* "Mammalian Protein Metabolism" (H. N. Munro, ed.), pp. 21–132. Academic Press, New York.

Kane, J. F. (1995). Effects of rare codon clusters on high-level expression of heterologous proteins in *Escherichia coli. Curr. Opin. Biotech.* **6,** 494–500.

Karnik, S. S., Nassal, M., Doi, T., Jay, E., Sgaramella, V., and Khorana, H. G. (1987). Structure-function studies on bacteriorhodopsin. II. Improved expression of the bacterio-opsin gene in *Escherichia coli. J. Biol. Chem.* **262,** 9255–9263.

Kelmanson, I. V., and Matz, M. V. (2003). Molecular basis and evolutionary origins of color diversity in great star coral *Montastrea cavernosa* (Scleractinia: Faviida). *Mol. Biol. Evol.* **20,** 1125–1133.

Kimura, M. (1980). A simple method for estimating evolutionary rate of base substitutions through comparative studies of nucleotide sequences. *J. Mol. Evol.* **16,** 111–120.

Kishino, H., Miyata, T., and Hasegawa, M. (1990). Maximum likelihood inference of protein phylogeny and the origin of chloroplasts. *J. Mol. Evol.* **31,** 151–160.

Koshi, J. M., and Goldstein, R. A. (1996). Probabilistic reconstruction of ancestral protein sequences. *J. Mol. Evol.* **42,** 313–320.

Labas, Y. A., Gurskaya, N. G., Yanushevich, Y. G., Fradkov, A. F., Lukyanov, K. A., Lukyanov, S. A., and Matz, M. V. (2002). Diversity and evolution of the green fluorescent protein family. *Proc. Natl. Acad. Sci.* **99,** 4256–4261.

Lewis, P. O. (1998). Maximum likelihood as an alternative to parsimony for inferring phylogeny using nucleotide sequence data. *In* "Molecular Systematics of Plants II: DNA Sequencing" (P. S. Soltis, D. E. Soltis, and J. J. Doyle, eds.), pp. 132–163. Kluwer, Boston.

Lippincott-Schwartz, J., and Patterson, G. H. (2003). Development and use of fluorescent protein markers in living cells. *Science* **300,** 87–91.

Maddison, W. P. (1995). Calculating the probability distributions of ancestral states reconstructed by parsimony on phylogenetic trees. *Syst. Biol.* **44,** 474–481.

Maddison, W. P., and Maddison, D. R. (1993). "MacClade." Sinauer Associates, Sunderland, MA.

Malcolm, B. A., Wilson, K. P., Matthews, B. W., Kirsch, J. F., and Wilson, A. C. (1990). Ancestral lysozymes reconstructed, neutrality tested, and thermostability linked to hydrocarbon packing. *Nature* **345,** 86–89.

Matz, M. V., Lukyanov, K. A., and Lukyanov, S. A. (2002). Family of the green fluorescent protein: Journey to the end of the rainbow. *Bioessays* **24,** 953–959.

Muse, S. V., and Gaut, B. S. (1994). A likelihood approach for comparing synonymous and nonsynonymous nucleotide substitution rates, with application to the chloroplast genome. *Mol. Biol. Evol.* **11,** 715–724.

Nielsen, R., and Yang, Z. (1998). Likelihood models for detecting positively selected amino acid sites and applications to the HIV-1 envelope gene. *Genetics* **148,** 929–936.

Omland, K. E. (1999). The assumptions and challenges of ancestral state reconstructions. *Syst. Biol.* **48,** 604–611.

Posada, D., and Crandall, K. A. (1998). MODELTEST: Testing the model of DNA substitution. *Bioinformatics* **14,** 817–818.

Roth, M. E., Feng, L., McConnell, K. J., Schaffer, P. J., Guerra, C. E., Affourtit, J. P., Piper, K. R., Guccione, L., Hariharan, J., Ford, M. J., Powell, S. W., Krishnaswamy, H., Lane, J., Intrieri, G., Merkel, J. S., Perbost, C., Valerio, A., Zolla, B., Graham, C. D., Hnath, J., Michaelson, C., Wang, R., Ying, B., Halling, C., Parman, C. E., Raha, D., Orr, B., Jedrzkiewicz, B., Liao, J., Tevelev, A., Mattessich, M. J., Kranz, D. M., Lacey, M., Kaufman, J. C., Kim, J., Latimer, D. R., and Lizardi, P. M. (2004). Expression profiling using a hexamer-based universal microarray. *Nat. Biotechnol.* **22,** 418–426.

Shagin, D. A., Barsova, E. V., Yanushevich, Y. G., Fradkov, A. F., Lukyanov, K. A., Labas, Y. A., Semenova, T. N., Ugalde, J. A., Meyers, A., Nunez, J. M., Widder, E. A., Lukyanov,

S. A., and Matz, M. V. (2004). GFP-like proteins as ubiquitous metazoan superfamily: Evolution of functional features and structural complexity. *Mol. Biol. Evol.* **21,** 841–850.

Sharp, P. M., Cowe, E., Higgins, D. G., Shields, D. C., Wolfe, K. H., and Wright, F. (1988). Codon usage patterns in *Escherichia-coli, Bacillus-subtilis, Saccharomyces-cerevisiae, Schizosaccharomyces-pombe, Drosophila-melanogaster* and Homo-Sapiens—a review of the considerable within-species diversity. *Nucleic Acids Res.* **16,** 8207–8211.

Sinclair, G., and Choy, F. Y. M. (2002). Synonymous codon usage bias and the expression of human glucocerebrosidase in the methylotrophic yeast, *Pichia pastoris. Prot. Exp. Purif.* **26,** 96–105.

Stewart, C.-B. (1995). Active ancestral molecules. *Nature* **374,** 12–13.

Sun, H. M., Merugu, S., Gu, X., Kang, Y. Y., Dickinson, D. P., Callaerts, P., and Li, W. H. (2002). Identification of essential amino acid changes in paired domain evolution using a novel combination of evolutionary analysis and *in vitro* and *in vivo* studies. *Mol. Biol. Evol.* **19,** 1490–1500.

Swofford, D. L. (2002). "PAUP*, Phylogenetic Analysis Using Parsimony (*and Other Methods)." Sinauer, Sunderland, MA.

Swofford, D. L., Olsen, G. J., Waddell, P. J., and Hillis, D. M. (1996). Phylogenetic Inference. *In* "Molecular Systematics" (D. M. Hillis, C. Moritz, and B. K. Mable, eds.), pp. 407–514. Sinauer, Sunderland, MA.

Tavare, L. (1986). Some probabilistic and statistical problems of the analysis of DNA sequences. *Lect. Math. Life Sci.* **17,** 57–86.

Thornton, J. W. (2004). Resurrecting ancient genes: Experimental analysis of extinct molecules. *Nat. Rev. Gen.* **5,** 366–375.

Thornton, J. W., Need, E., and Crews, D. (2003). Resurrecting the ancestral steroid receptor: Ancient origin of estrogen signaling. *Science* **301,** 1714–1717.

Ugalde, J. A., Chang, B. S., and Matz, M. V. (2004). Evolution of coral pigments recreated. *Science* **305**(5689), 1433.

Yang, Z. (1994). Maximum likelihood phylogenetic estimation from DNA sequences with variable rates over sites: Approximate methods. *J. Mol. Evol.* **39,** 306–314.

Yang, Z. (1997). PAML: A program package for phylogenetic analysis by maximum likelihood. *Comput. Appl. Biosci.* **13,** 555–556.

Yang, Z. (1998). Likelihood ratio tests for detecting positive selection and application to primate lysozyme evolution. *Mol. Biol. Evol.* **15,** 568–573.

Yang, Z., and Bielawski, J. P. (2000). Statistical methods for detecting molecular adaptation. *Trends Ecol. Evol.* **15,** 496–503.

Yang, Z., Goldman, N., and Friday, A. (1994). Comparison of models for nucleotide substitution used in maximum-likelihood phylogenetic estimation. *Mol. Biol. Evol.* **11,** 316–324.

Yang, Z., Kumar, S., and Nei, M. (1995). A new method of inference of ancestral nucleotide and amino acid sequences. *Genetics* **141,** 1641–1650.

Yang, Z., and Nielsen, R. (2002). Codon-substitution models for detecting molecular adaptation at individual sites along specific lineages. *Mol. Biol. Evol.* **19,** 908–917.

Yang, Z., and Swanson, W. J. (2002). Codon-substitution models to detect adaptive evolution that account for heterogeneous selective pressures among site classes. *Mol. Biol. Evol.* **19,** 49–57.

Yang, Z. H., Nielsen, R., Goldman, N., and Pedersen, A. M. K. (2000). Codon-substitution models for heterogeneous selection pressure at amino acid sites. *Genetics* **155,** 431–449.

Zhang, J. Z., and Rosenberg, H. F. (2002). Complementary advantageous substitutions in the evolution of an antiviral RNase of higher primates. *Proc. Natl. Acad. Sci.* **99,** 5486–5491.

Section III

Comparing Macromolecules: Phylogenetic Analysis

[35] Advances in Phylogeny Reconstruction from Gene Order and Content Data

By BERNARD M. E. MORET and TANDY WARNOW

Abstract

Genomes can be viewed in terms of their gene content and the order in which the genes appear along each chromosome. Evolutionary events that affect the gene order or content are "rare genomic events" (rarer than events that affect the composition of the nucleotide sequences) and have been advocated by systematists for inferring deep evolutionary histories. This chapter surveys recent developments in the reconstruction of phylogenies from gene order and content, focusing on their performance under various stochastic models of evolution. Because such methods are quite restricted in the type of data they can analyze, we also present research aimed at handling the full range of whole-genome data.

Introduction: Molecular Sequence Phylogenetics

A phylogeny represents the evolutionary history of a collection of organisms, usually in the form of a tree. Sequence data are by far the most common form of molecular data used in phylogenetic analyses. We begin by briefly reviewing techniques for estimating phylogenies from molecular sequences, with emphasis on the computational and statistical issues involved.

Model Trees and Stochastic Models of Evolution

Most algorithms for phylogenetic reconstruction attempt to reverse a *model of evolution*. Such a model embodies certain knowledge and assumptions about the process of evolution, such as characteristics of speciation and details about evolutionary changes that affect the content of molecular sequences. Models of evolution vary in their complexity; in particular, they require different numbers of parameters. For instance, the Jukes-Cantor model, which assumes that all sites evolve identically and independently and that all substitutions are equally likely, requires just one parameter per edge of the tree, viz., the expected number of changes of a random site on that edge. Overall, then, a rooted Jukes-Cantor tree with n leaves requires $2n - 2$ parameters. Under more complex models of evolution, the process operating on a single edge can require up to 12 parameters (for the General

Markov model), although these models still require $\Theta(n)$ parameters overall. If edge "lengths" are drawn from a distribution, however, the complexity can be reduced, because the evolutionary process operating on the model tree can then be described just by the parameters of the distribution.

These parameters describe how a single site evolves down the tree and so require additional assumptions in order to describe how different sites evolve. Usually the sites are assumed to evolve independently; sometimes they are also assumed to evolve identically. Moreover, the different sites are assumed either to evolve under the same process or to have rates of evolution that vary depending on the site. In the latter case (in which each site has its own rate), additional k parameters are needed, where k is the number of sites. However, if the rates are presumed to be drawn from a distribution (typically, the gamma distribution), then a single additional parameter suffices to describe the evolutionary process operating on the tree; furthermore, in this case, the sites still evolve under the *i.i.d.* assumption.

Tree generation models typically have parameters regulating speciation rates, but also inheritance characteristics, and others. For more on stochastic models of (sequence) evolution, see Felsenstein (1981), Kim and Warnow (1999), Li (1997), and Swofford *et al.* (1996); for an interesting discussion of models of tree generation, see Heard (1996) and Mooers and Heard (1997).

By studying the performance of methods under explicit stochastic models of evolution, one can assess the relative strengths of different methods and understand how methods can fail. Such studies can be theoretical, for instance, proving *statistical consistency*; given long enough sequences, the method will return the true tree with arbitrarily high probability. Others can use simulations to study the performance of the methods under conditions closely approximating practice. In a simulation, sequences are evolved down different model trees and then given to different methods for reconstruction; the reconstructions can then be compared against the model trees that generated the data. Such studies provide important quantifications of the relative merits of phylogeny reconstruction methods.

Phylogeny Reconstruction Methods from Molecular Sequences

Three main types of methods are used to reconstruct phylogenies from molecular sequences: *distance-based* methods, *maximum parsimony* heuristics, and *maximum likelihood* heuristics.

Distance-Based Methods. Of the three types of methods, only distance-based methods include algorithms that run in polynomial time. Distance-based methods operate in two phases:

1. Pairwise distances between every pair of taxa are estimated.
2. An algorithm is applied to the matrix of pairwise distances to compute an edge-weighted tree T.

The statistical consistency (if any) of such two-phase procedures rests on two assumptions: first, that a statistically consistent distance estimator is used in the first phase and, second, that an appropriate distance-based algorithm is used in the second phase. The requirements that the first phase be statistically consistent means that the distance estimator should return a value that approaches the expected number of times a random site changes on the path between the two taxa. Thus, the estimation of pairwise distances must be done with respect to some assumed stochastic model of evolution. As an example, in the Jukes-Cantor model of evolution, the estimated distance between sequences s_i and s_j is given by the formula

$$d_{ij} = -\frac{3}{4}\ln\left(1 - \frac{4}{3}\frac{H_{ij}}{k}\right),$$

where k is the sequence length and H_{ij} denotes the Hamming distance (the number of positions in which the two sequences differ, which is the edit distance under mutation operations).

Algorithms that attempt to reconstruct trees from distance matrices are guaranteed to produce accurate reconstructions of the trees only when the distance matrix entries approach very closely the actual number of changes between the pair of sequences. [In the context of estimating model trees, this requirement means that the estimated distances need to be extremely close to the model distances, defined to be the expected number of times a random site changes on a leaf-to-leaf path. See Atteson (1999) and Kim and Warnow (1999) for more on this issue.] Naïvely defined, distances, such as the Hamming distance, typically underestimate the number of changes that took place in the evolutionary history; thus, the first step of a distance-based method is to *correct* the naïvely defined distance into one that accurately accounts for the expected number of unseen back-and-forth changes in a site. Such corrections are not without problems: As the measured distance grows larger, the variance in the estimator increases, so that although the values produced by the correction may be statistically correct, the reconstructions produced by using such estimates are often erroneous.

The most commonly used, and simplest, distance-based method is the *neighbor-joining (NJ)* algorithm of Saitou and Nei (1987); improved versions of this basic method include BioNJ (Gascuel, 1997) and a version known as *Weighbor*, which requires an estimate of the variance of the distance estimator (Bruno *et al.*, 2000). NJ is known to be statistically consistent under most models of evolution.

Maximum Parsimony. Parsimony-based methods seek the tree and sequences labeling its internal nodes that together minimize the total number of evolutionary changes (viewed as distances summed along all edges of the tree). Put formally, the problem is as follows: Given a set S of sequences in a multiple alignment, each of length k, find a tree T and a set of additional sequences S_0, all also of length k, so that with the leaves of T are labeled by S and its internal nodes by S_0, the value $\sum_{e \in E(T)}$ *Hamming*(e) is minimized, where *Hamming*(e) denotes the Hamming distance between the sequences labeling the endpoints of e. (Weighted or distance-corrected versions can also be defined.)

The maximum parsimony (MP) problem is thus an optimization problem—and a hard one: Finding the best tree is probably NP-hard (Day, 1983). This property effectively rules out exact solutions for all but the smallest instances; indeed, in practice, exact solvers run within reasonable times on at most 30 taxa. Thus, heuristics is the normal approach to the problem; most are based on iterative improvement techniques and appear to return very good solutions for up to a few hundred taxa. Many software packages implement such heuristics, among them MEGA (Kumar *et al.*, 2001), PAUP* (Swofford, 2001), Phylip (Felsenstein, 1993), and TNT (Goloboff, 1999).

Maximum Likelihood. Like MP, maximum likelihood (ML) is an optimization problem. ML seeks the tree and associated model parameter values that maximizes the probability of producing the given set of sequences. ML thus depends explicitly on the assumed model of evolution. For example, the ML problem under the Jukes-Cantor model needs to estimate one parameter (the substitution probability) for each edge of the tree, whereas under the General Markov model, 12 parameters must be estimated on each edge. ML is much more computationally expensive than MP; even the point estimation (how to score a fixed tree) cannot be done in polynomial time for ML [see Steel (1994) for a discussion], whereas it is easily accomplished in linear time for MP using Fitch's algorithm (Fitch, 1977). In consequence, exact solutions for ML are limited to fewer than 10 taxa and even heuristic solutions are typically limited to fewer than 100 taxa. Various software packages provide heuristics for ML, including PAUP* (Swofford, 2001), Phylip (Felsenstein, 1993), FastDNAml (Olsen *et al.*, 1994), and PhyML (Guindon and Gascuel, 2003).

Performance Issues

Methods can be compared in terms of their performance guarantees, in terms of their resource requirements, and in terms of the quality of the trees they produce. Very few methods offer any performance guarantees, except in purely theoretical terms. For instance, while ML is known to be statistically consistent under most models, the same cannot be said of its heuristic implementation; and even NJ, which is statistically consistent and is implemented exactly, may return very poor trees; the guarantee of statistical consistency only implies good performance in the limit, as sequence lengths become sufficiently large. In terms of computational requirements, the comparison is easy: Distance-based methods are efficient (running in polynomial time with low coefficients); parsimony is much harder to solve (systematists are accustomed to running MP for weeks on a dataset of modest size); and ML is much harder again than MP. These comparisons, however, all have limited value: As we saw, statistical consistency is a very weak guarantee, while a guarantee of fast running times is worthless if the returned solution is poor. Thus, experimental studies are our best tool in the study of the relative performance of methods. Simulation studies, in particular, can establish the absolute accuracy of methods (whereas studies conducted with biological datasets can only assess relative performance in terms of the optimization criterion). Such studies have shown that MP methods can produce reasonably good trees under conditions where NJ can have high topological error (significantly worse than MP); this possibly surprising performance holds under many realistic model conditions—in particular, when the model tree has a high evolutionary diameter (Moret *et al.*, 2002b; Nakhleh *et al.*, 2001a,b, 2002; Roshan *et al.*, 2004).

Limitations for Molecular Sequence Phylogenetics

Although existing methods often yield good estimates of phylogenies on datasets of small to medium size, all methods based on molecular sequences suffer from similar limitations. Perhaps most seriously, deep evolutionary histories can be hard to reconstruct from molecular sequence data; the further back one goes in time, the harder the alignment of sequences becomes and the greater the impact of *homoplasy* (multiple point mutations at the same position). Under these conditions, we have established, through extensive simulation studies, that most major methods (heuristics for MP and NJ) have poor topological accuracy (Moret *et al.*, 2002b; Nakhleh *et al.*, 2001b, 2002). The problem accrues from a combination of the small state space (only four nucleotides for DNA or RNA sequences and only 20 amino acids for protein sequences), the relatively high frequency of point mutations, and the limited amount of data. Concatenating—also called combining—gene sequences to obtain longer sequences may provide more data but brings its own problems. Different genes may follow different evolutionary paths (each potentially different from that of the organism as a whole)—a problem known as the *gene tree/species tree* problem (Ma *et al.*, 1998; Maddison, 1997; Page and Charleston, 1997a; Pamilo and Nei, 1998)—while reticulation events [such as hybridization, lateral gene transfer, gene conversion, etc; see Linder *et al.* (2004)] create convergent paths, with the result that tree-based analyses may run into contradictions and yield poor results. Research on resolving the gene tree/species tree problem [a process known as *reconciliation* (Page, 1998; Page and Charleston, 1997a,b)] and on identifying and properly handling reticulation events has not yet produced reliably accurate and scalable methods. Thus, phylogeny reconstruction based on site-evolution models will continue to suffer problems, for at least the near future, when attempting to infer deep evolutionary histories.

Whole-Genome Evolution

Systematists are interested in whole genomes because they have the potential to overcome two of the main problems afflicting sequence data. Because the entire genome is used, the data reflect organismal evolution, not the evolution of single genes, thereby avoiding the gene tree/species tree problem. (Naturally, however, if the phylogeny is based on organellar genomes, it need not coincide with the organismal phylogeny, nor will two phylogenies based on different organelles or plasmids necessarily agree.) Moreover, the events that affect the whole genome by altering its gene content or rearranging its genes are so-called "rare genomic events" (Rokas and Holland, 2000). They occur rarely and come from a very large set of choices (e.g., there can be a quadratic number of distinct inversion events), so they are unlikely to give rise to homoplasy, even in deep branches of the tree. Thus, genome rearrangements, in particular, have been increasingly used in phylogenetic analysis (Boore *et al.*, 1995; Cosner *et al.*, 2000a,b; Downie and Palmer, 1992; Jansen and Palmer, 1987; Stein *et al.*, 1992).

The main approach to whole-genome phylogenetic analysis has used the ordering of the genes along the chromosomes as its primary data. That is, each chromosome is considered as a linear (or circular) ordering of genes, with each gene represented by an identifier that it shares with its homologues on other chromosomes (or, for that matter, on the same chromosome, in the case of gene duplications). The genome is thus simplified in the sense that point mutations are ignored and evolutionary history is inferred on the basis of the gene content and gene order within each chromosome. A typical single circular chromosome for the chloroplast organelle of a *Guillardia* species (taken from the NCBI database) is shown in Fig. 1. Plant chloroplast genomes such as this have around 120 DNA

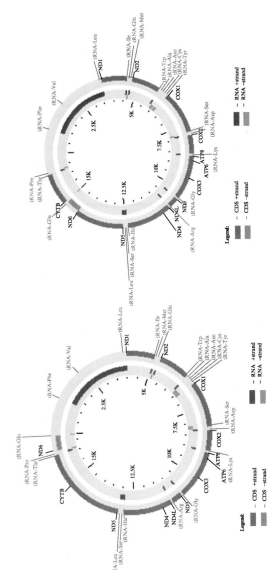

FIG. 1. The chloroplast chromosome of *Guillardia* (from NCBI). (See color insert.)

genes; in contrast, nuclear chromosomes in eukaryotes and free-living bacteria typically have several thousand genes.

Evolution of Gene Order and Content

Events that change gene order (but not content) along a single chromosome include inversions, which are well documented (Jansen and Palmer, 1987; Palmer, 1992), *transpositions*, which are strongly suspected in mitochondria (Boore and Brown, 1998; Boore *et al.*, 1995), and *inverted transpositions*; these three operations are illustrated in Fig. 2. In a multichromosomal genome, additional operations that do not affect gene content include *translocations*, which move a piece of one chromosome into another chromosome (in effect, a transposition between chromosomes), and *fissions* and *fusions*, which split and merge chromosomes without affecting genes. Finally, a number of events can affect the gene content of genomes: *insertions* (of genes without existing homologues), *duplications* (of genes with existing homologues), and *deletions*. In multichromosomal organisms, colocation of genes on the same chromosome, or *synteny*, is an important evolutionary character and has been used in phylogenetic reconstruction (Nadeau and Taylor, 1984; Sankoff and Nadeau, 1996; Sankoff *et al.*, 1997).

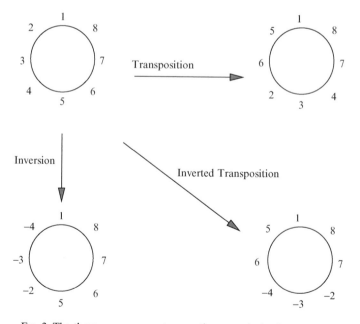

FIG. 2. The three rearrangements operating on a single chromosomo.

In this model, one whole chromosome forms a single character, whose state is affected by all of the operations just described. This one character can assume any of an enormous number of states—for a chromosome with n distinct single-copy genes, the number of states is 2^{n-1} $(n-1)!$, in sharp contrast to the 4 or 20 states possible for a sequence character. Even for a simple chloroplast genome, with a single small circular chromosome of 120 genes, the resulting number of states is very large—on the order of 10^{235}.

The use of gene-order and gene-content data in phylogenetic reconstruction is the subject of much current research. As mentioned earlier, such data present many advantages: (1) the identification of homologies can rest on a lot of information and thus tends to be quite accurate; (2) because the data capture the entire genome, there is no gene tree versus species tree problem; (3) there is no need for multiple sequence alignment; and (4) gene rearrangements and duplications are much rarer events than nucleotide mutations and thus enable us to trace evolution farther back in time—by two or more orders of magnitude.

However, there remain significant challenges. First is the lack of data: Mapping a full genome, though easier than sequencing it, remains far more demanding than sequencing a few genes. Table I gives a rough idea of the state of affairs around 2003; for obvious reasons, most of the bacteria sequenced are human pathogens, while the few eukaryotes are the model species chosen in genome projects: human, mouse, fruit fly, worm, mustard plant, yeast, and so on. Although the number of sequenced eukaryotic genomes is growing quickly, coverage at this level of detail will not, for the foreseeable future, exceed a small fraction of the total number of described organisms.

This lack of data has so far prevented us from understanding very much about the relative probabilities of different events that modify gene order and content; consequently, the stochastic models proposed (see the next section) remain fairly primitive.

Finally, the extreme (at least in comparison with sequence data) mathematical complexity of gene orders means that all reconstruction methods

TABLE I
EXISTING WHOLE-GENOME DATA (~2003) (APPROXIMATE VALUES)

Type	Attributes	Numbers
Animal mitochondria	1 chromosome, 40 genes	200
Plant chloroplast	1 chromosome, 140 genes	100
Bacteria	1–2 chromosomes, 500–5000 genes	50
Eukaryotes	3–30 chromosomes, 2000–40,000 genes	10

TABLE II
MAIN ATTRIBUTES OF SEQUENCE AND GENE-ORDER DATA

	Sequence	Gene order
Evolution	Fast	Slow
Errors	Small	Negligible
Data type	A few genes	Whole genome
Data Quantity	Abundant	Sparse
No. character states	Tiny	Huge
Models	Good	Primitive
Computation	Easy	Hard

(even the distance-based ones) face major computational challenges even on small datasets (containing only 10 or so genomes).

Table II summarizes the salient characteristics of sequence data and gene-order data.

Stochastic Models of Evolution

Models of genome evolution have been largely limited to simple combinations of the main rearrangement operations: inversion, transposition, and inverted transposition. The original model was proposed by Nadeau and Taylor (1984); it uses only inversions and assumes that all inversions are equally likely. We extended this model to produce the generalized Nadeau-Taylor (GNT) model (Wang and Warnow, 2001), which includes transpositions and inverted transpositions. Within each of the three types of events, any two events are equiprobable, but the relative probabilities of each type of event are specified by the two parameters of the model: α, the probability that a random event is a transposition and β, the probability that a random event is an inverted transposition. (The probability of an inversion is, thus, given by $1 - \alpha - \beta$.) The GNT model contains the Nadeau-Taylor model as a special case: just set $\alpha = \beta = 0$. Extensions of this simple model in which the probability of an event depends on the segment affected by the event have been proposed (Bender *et al.*, 2004), but no solid data exist to support one model over another; all that appears certain (Lefebvre *et al.*, 2003) is that short inversions are more likely than long ones in prokaryotes. Similarly, multichromosomal rearrangements such as translocations and events that affect the gene content of chromosomes, such as duplications, insertions, and deletions, have all been considered, but once again we have insufficient biological data to define any model with confidence.

Genomic Distances

The distance between two genomes (each represented by the order of its genes) can be defined in several ways. First, we have the *true evolutionary distance*, that is, the actual number of evolutionary events (mutations, deletions, etc.) that separate one genome from the other. This is the distance measure we would really want to have, but of course it cannot be inferred; as our earlier discussion of homoplasy made clear, we cannot infer such a distance even when we know the correct phylogeny and have correctly inferred ancestral data (at internal nodes of the tree). What we can define precisely and, in some cases, compute, is the *edit distance*, the minimum number of permitted evolutionary events that can transform one genome into the other. Because the edit distance invariably under-estimates the true evolutionary distance, we can attempt to *correct* the edit distance according to an assumed model of evolution in order to produce the *expected true evolutionary distance*. Finally, we can attempt to estimate the true evolutionary distance directly through various heuristic techniques.

We begin by reviewing distance measures between two chromosomes with equal gene contents and no duplications—the simplest possible case—then discuss research on distance measures between multichromosomal genomes and between genomes with unequal gene content.

Distances between Two Chromosomes with Equal Gene Content and No Duplications

Here we consider chromosomes that have identical gene content and exactly one copy of each gene, so that each chromosome can be viewed as a signed permutation of the underlying set of genes. Two distance metrics have been used in this context, one based on observed differences and one based on allowable evolutionary operations. Assume our two chromosomes each have seven genes, numbered 1 through 7, and are circular. Genome G_1 is given by $(1, 2, -4, -3, 5, 6, 7)$ and genome G_2 by $(1, 2, 3, 4, 5, 6, 7)$.

• The *breakpoint distance* simply counts the number of gene adjacencies (read on either strand) present in one chromosome, but not in the other. In our example, we have two breakpoints: The adjacencies 2, 3 and 4, 5 are present in G_2, but not in G_1. (Note that the adjacency 3, 4 is present on the forward strand on G_2 and on the reverse complement strand, as $-4, -3$, on G_1. Note also that the definition is symmetric: The adjacencies 2, -4 and -3, 5 are present in G_1, but not in G_2.) The breakpoint distance is, thus, not based on evolutionary events, only on the end result of such events.

- The *inversion distance* is the edit distance under the single allowed event of inversion. We need only one inversion to transform G_1 into G_2: We invert the two-gene segment $-4, -3$.

Every inversion clearly creates (or, equivalently, removes) at most two breakpoints, so that the inversion distance is at least half the breakpoint distance. The number of breakpoints is also clearly at most n in a circular chromosome of n genes (and $n + 1$ in a linear chromosome); less obviously, the same bound holds for the inversion distance (Meidanis *et al.*, 2000). Computing the breakpoint distance is trivially achievable in linear time. Computing the inversion distance, however, is a very complex problem. Indeed, it is computationally intractable (technically, it is NP-hard) for unsigned permutations (when we cannot tell on which strand each gene lies). For signed permutations, we showed that it can be computed in linear time (Bader *et al.*, 2001), but this result is the culmination of many years of research and rests on the very elaborate and elegant theory of Hannenhalli and Pevzner (1995a,b). Obtaining an actual sequence of inversions (as opposed to just the number of required inversions) is computationally more demanding. The classic algorithm of Kaplan *et al.* (1999) takes $O(dn)$ time, where d is the distance and n the number of genes; thus, for large distances, this algorithm takes quadratic time but was improved to $O(n\sqrt{n \log n})$ time (Tannier and Sagot, 2004).

Transpositions are also of significant interest in biology. However, although some of the same theoretical framework can be used (Bafna and Pevzner, 1995), results here are disappointing. No efficient algorithm has yet been developed to compute the transposition distance. The best result remains an approximation algorithm that could suffer from up to a 50% error (Bafna and Pevzner, 1995; Hartman, 2003); this result was extended, with the same error bound, to the computation of edit distances under a combination of inversions and inverted transpositions, with equal weights assigned to each (Hartman and Sharan, 2004).

Distance Corrections

None of these distances produces the true evolutionary distance; indeed, because all are bounded by n, all can produce arbitrary underestimates of the true evolutionary distance. To estimate the latter, we must use correction methods. Such methods are widely used in distance estimation between DNA sequences (Huson *et al.*, 1999b; Swofford *et al.*, 1996). We designed two techniques for "correcting" gene-order distances under the GNT model, one (IEBP) based on the breakpoint distance (Wang and Warnow, 2001) and the other empirically designed estimator [EDE] based on the inversion distance (Moret *et al.*, 2001a, 2002d).

• The *IEBP* estimator takes as input the breakpoint distance and the values of the two GNT parameters, α and β, and returns the expected number of inversions, transpositions, and inverted transpositions under the specified relative probabilities of the different events. The method is analytical and mathematically exact and can be implemented to run in cubic time.

• The *EDE* takes as input the inversion distance and produces an estimate of the expected number of inversions, under a model in which all inversions are equally likely. The formula used was derived with numerical techniques from a large series of simulations and can be computed in quadratic time.

Figure 3 (Moret *et al.*, 2002d) illustrates the EDE correction and compares it with the breakpoint and inversion distances, under an inversion-only scenario. The EDE offers no theoretical guarantee and only takes inversions into account but is easier to compute than the IEBP estimator, which also requires an accurate guess of the values of the two GNT parameters. Both estimators suffer from the simple fact that the variance of the true distance (as a function of the breakpoint or inversion distance) grows rapidly as the distance grows—as is clearly visible in Fig. 3. The most important aspect of EDE's performance is that trees reconstructed by distance-based methods using EDE are generally more accurate than those reconstructed by the same methods using any of the other distances (including IEBP), even when the evolutionary model involves many (or even only) transpositions (Moret *et al.*, 2001a, 2002d).

Distances between Two Chromosomes with Unequal Gene Content

Unequal gene content introduces new evolutionary events: insertions (including duplications) and deletions. If we first assume that each gene exists in, at most, one copy (no duplication), then the problem is to handle insertions of new genes and deletions of existing ones. (Note that the same event can be viewed as a deletion or as a nonduplicating insertion, depending on the direction of time flow.) In a seminal paper, El-Mabrouk (2000) showed how to extend the theory of Hannenhalli and Pevzner for a single chromosome without duplicated genes to handle both inversions and deletions exactly; we showed that the corresponding distance measure can also be computed in linear time (Liu *et al.*, 2003). In the same paper, El-Mabrouk also showed how to approximate the distance between two such genomes in the presence of both deletions and nonduplicating insertions along the same time flow.

Duplications are considerably harder from a computational standpoint, as they introduce a matching problem: Which homologue in one

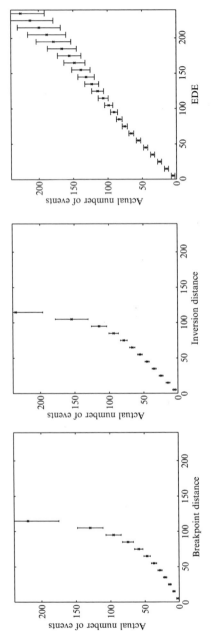

FIG. 3. Edit distances versus true evolutionary distances and the EDE correction.

genome corresponds to which in the other genome? Sankoff (1999) proposed to sidestep the entire issue by reducing the problem to one with no duplications using the *exemplar* approach. In that approach, a single copy is chosen from each gene family, and all other homologues in each family are discarded, in such a way to minimize the breakpoint or inversion distance between the two genomes. Unfortunately, choosing the exemplars themselves is an NP-hard problem (Bryant, 2000); moreover, the loss of information in nuclear genomes (which can have tens or hundreds of genes in each gene family) is very large—large enough to cause problems in reconstruction (Tang and Moret, 2003a; Tang *et al.*, 2004).

We proposed an approximation with provable guarantees (in terms of the edit distance) for the distance between two genomes under duplications, insertions, and deletions (Marron *et al.*, 2003). We later refined the approach to estimate true evolutionary distances directly (Swenson *et al.*, 2004) and used the new measure in a pilot study of a group of 13 γ-proteobacteria with widely differing gene contents, varying from 800 to more than 5,000 genes in their single nuclear chromosome (Earnest-DeYoung *et al.*, 2004). The latter study indicates that simple distance-based reconstructions can be very accurate even in the presence of enormous evolutionary distances: Many of the edges in the final tree have several hundred evolutionary events along them. Our simulations show that the distance computation tracks the true evolutionary distance remarkably well until saturation, which only occurs at extremely high levels of evolution (250 events on a genome of 800 genes, for instance). Figure 4 (Swenson *et al.*, 2004) shows typical results from these simulations: Genomes of 800 genes were generated in a simulation on a balanced tree of 16 taxa and all 120 pairwise distances computed and compared to the true evolutionary distances (the sum of the lengths of the edges in the true tree on the path connecting each pair). The figure shows the calculated edit lengths as a function of the generated edit lengths (the true evolutionary distances) on the left and an error plot on the right. The data used for this figure were generated using an expected edge length of 40, so the pairwise distance between leaves whose common ancestor is the root has an expected value of 320 events. Events were a mix of 70% inversions, 16% deletions, 7% insertions, and 7% duplications; the inversions had a mean length of 20 and a standard deviation of 10, while the deletions, insertions, and duplications all had a mean length of 10, with a standard deviation of 5. The figure shows excellent tracking of the true evolutionary distance for up to about 250 events, after which the computation consistently returns values less than the true distance; of course, distance corrections could be devised to remedy the latter situation.

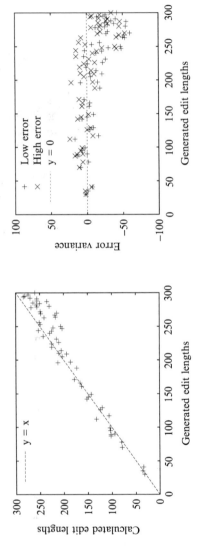

FIG. 4. Experimental results for 800 genes with expected edge length 40. Left: generated edit length vs. reconstructed length; right: the variance of computed distances per generated distance.

Distances between Multichromosomal Genomes

With multiple chromosomes, we can still make the same distinction as for single chromosomes, by first addressing genomes with equal gene content and no duplicate genes. The second of the two papers by Hannenhalli and Pevzner (1995b) showed how to handle a combination of inversions and translocations for such genomes, a result later improved by Tesler (2002) to handle any combination of inversions, translocations, fissions, and fusions. Using the same approach pioneered by Bader *et al.* (2001), one can devise linear time algorithms to compute other distances covered by the Hannenhalli-Pevzner theory, such as the translocation distance (Li *et al.*, 2004). However, multichromosomal genomes typically have large gene families, so the handling of duplications is crucial; moreover, few such genomes will have identical gene content—even such closely related genomes as two species of the nematode worm Caenorhabditis, *C. elegans* and *C. briggsae*, have different gene content. Thus, no distance method exists that could be used in the reconstruction of phylogenies for multichromosomal organisms.

Phylogenetic Reconstruction from Whole Genomes

The same three general types of approaches to phylogenetic reconstruction that we just reviewed for sequence data can be used for whole-genome data.

Distance-Based Methods

We have run extensive simulations under a variety of GNT model conditions (Moret *et al.*, 2001a, 2002d,e; Wang *et al.*, 2002), using breakpoint distance, inversion distance, and both EDE and IEBP corrections, all using the standard NJ (Saitou and Nei, 1987) and Weighbor (Bruno *et al.*, 2000) distance-based reconstruction algorithms, as well as within our own "DCM-boosted" extensions of the same (Huson *et al.*, 1999a). Figure 5 (Moret *et al.*, 2002d) shows typical results from these simulation studies: NJ is run with four distances (*BP* stands for breakpoint and *INV* for inversion) on datasets of various sizes (here, 10, 20, 40, 80, and 160 taxa), each taxon being given by a signed permutation of the same set of 120 genes, then the reconstructed trees are compared with the model trees. The figure shows the false-negative rate of the trees reconstructed with each distance measure, as a function of the diameter of the dataset. At low rates of evolution, the results are much the same for all four distance measures, but the corrected measures perform much better than the uncorrected ones when the diameter is large.

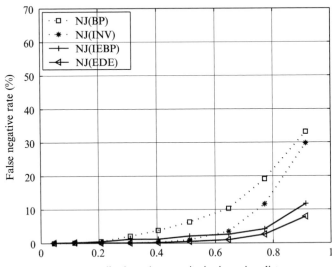

Fig. 5. False-negative rates for neighbor joining run with four genomic distance measures. The datasets have 10, 20, 40, 80, and 160 taxa; each taxon is a signed permutation of the same set of 120 genes, generated from the identity permutation at the root through an equal-weight mix of inversions, transpositions, and inverted transpositions. BP, breakpoint; INV, inn.

Our simulations have established the following:

• IEBP distances are generally the most accurate, even when given incorrect estimates of the parameters α and β. Although EDE is based on an inversion-only scenario, it produces highly accurate estimates of the actual number of events even under model conditions in which inversions play a minor role. Both EDE and IEBP are thus very robust to model violations.

• Phylogenies estimated using EDE are more accurate than phylogenies estimated using any other distance estimator (including IEBP), under all conditions, but especially when the model tree has a high evolutionary diameter. Phylogenies based on either EDE or IEBP are better than phylogenies based on inversion distances, which in turn are far better than phylogenies based on breakpoint distances.

Parsimony-Based Methods

Given a way of defining a distance between two genomes, we can define the "length" of a tree in which every node is labeled by a genome to be the sum of the lengths of the edges. This measure of length is thus similar to that used in sequence-based phylogenetic analyses, enabling us to define parsimony problems for gene-order data. Using the breakpoint distance, for instance, we would seek a tree and a collection of new genomes (for internal nodes) so that this tree, leaf-labeled by the given set of genomes and with internal nodes labeled by the new genomes, minimizes the breakpoint length of the tree—this problem is called the *breakpoint phylogeny* (Sankoff and Blanchette, 1998). Similarly we can define the *inversion phylogeny* by replacing breakpoint distances with inversion distances. Both problems are NP-hard; indeed, they are harder than MP for DNA sequences, as they remain NP-hard even for just three genomes (Caprara, 1999; Pe'er and Shamir, 1998). This last version is known as the *median problem*: given three genomes, find a fourth genome (to label the internal node connecting the three leaves) that minimizes the sum of its distances to the three given genomes. Exact solutions to the problem of finding a median of three genomes can be obtained for both the inversion and the breakpoint distance (Caprara, 2001; Moret *et al.*, 2002d; Siepel and Moret, 2001) and are implemented in our GRAPPA software suite. However, they take time exponential in the overall length and are thus applicable only to instances with modest pairwise distances. It should also be noted that all of these approaches are limited to unichromosomal genomes with equal gene content and no duplication.

Solving the parsimony problem for gene-order data requires the inference of "ancestral" genomes at the internal nodes of the candidate trees. Sankoff proposed to use the median problem in an iterative manner to refine rough initial guesses for these genomes (Sankoff and Blanchette, 1998), an approach that we implemented in GRAPPA with good results on small genomes of a few hundred genes (Moret *et al.*, 2001b, 2002a; Tang and Moret, 2003a,b; Tang *et al.*, 2004) and that was also implemented, with less accurate heuristics, to handle a few larger genomes by Bourque and Pevzner (2002). Identifying good candidate trees, however, is a much more expensive proposition; in that same paper, Sankoff proposed generating and scoring each possible tree in turn; the resulting BPAnalysis software is limited to breakpoint medians and to trees of eight or fewer leaves. We reimplemented Sankoff's algorithm and optimized its components to improve its speed, gradually, by 7 orders of magnitude (Moret *et al.*, 2001b, 2002a,d; Siepel and Moret, 2001), enabling the analysis of up to 16 taxa; we successfully coupled it with the Disk-Covering Method [the "DCM1" technique of Huson *et al.* (1999a)] to make it applicable to large datasets of several thousand taxa (Tang and Moret, 2003b).

Evidence from simulation studies and from the analysis of biological datasets indicates that even when the mechanism of evolution is based entirely on transpositions, solving the inversion phylogeny yields more accurate reconstructions than solving the breakpoint phylogeny (Moret *et al.*, 2002c; Tang *et al.*, 2004). Solving the inversion phylogeny was also found to yield better results than using distance-based methods. Figure 6 (Tang *et al.*, 2004) shows one example of such findings, on a small dataset of seven chloroplast genomes from green plants. Note that the phylogeny produced by NJ (using inversion distances after equalizing the gene contents) has false positives (edges in the inferred tree that are not present in the reference tree), while the breakpoint phylogeny (computed on the same input) resolves only one single edge. In contrast, the inversion

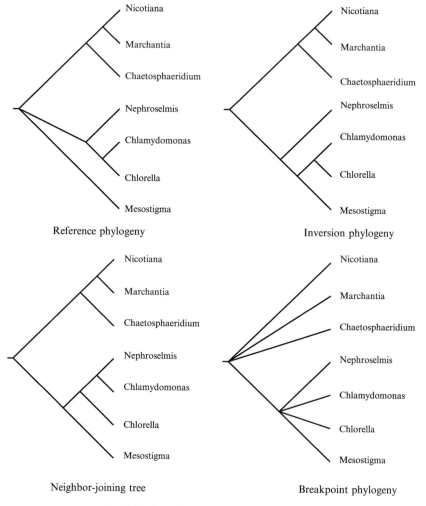

Fig. 6. Phylogenies on a 7-taxa cpDNA dataset

phylogeny matches the reference phylogeny, modulo the placement of *Mesostigma* (which remains in doubt).

A 2003 study (Lefebvre *et al.*, 2003) indicates that at least in prokaryotic genomes, short inversions are much more likely than long ones; work on developing new models of evolution that take such results into account is in progress. Applying these methods to large eukaryotic genomes may yield some surprises, however, because there is evidence that certain break-points in chromosomes are "hot spots" for rearrangements (Pevzner and Tesler, 2003); if an inversion is thus "anchored" at a fixed position in the chromosome, its net effect over several events is to produce one breakpoint per event, which might be better modeled with breakpoints than with inversions.

To date, our GRAPPA code remains the only parsimony-based recon-struction tool that can handle datasets with unequal gene content, albeit with only modest changes in gene content. In its current version (Tang *et al.*, 2004), it has been used to reconstruct chloroplast phylogenies of green plants in which gene content differs by a few genes at most. The algorithm proceeds in two phases: First, it computes gene contents for all internal nodes, then it proceeds (much as in the equal gene-content case) to estab-lish an ordering for these contents based on median computations. Gene content is determined from the leaves inward, under standard biological assumptions such as independence of branches; under such assumptions, the likelihood of simultaneous identical changes on two sibling edges is vanishingly small compared to the reverse change on the third edge (Maddison, 1990; McLysaght *et al.*, 2003).

Likelihood-Based Methods

Likelihood methods are based on a specific model of evolution. For a given tree T, they estimate the parameter values that maximize the proba-bility that T would produce the observed data; over all trees, they return that tree (and its associated parameter values) with the largest probability of producing the observed data. Because current statistical models for whole-genome evolution are so primitive (but also because any such mod-els will involve enormous complications due to the global nature of changes caused by a single event), no method exists to compute ML trees for gene-order data. The simpler Bayesian approach turns the tables; instead of seeking the distribution for the probability of producing the data given the tree, it seeks the probability of each tree given the data. The two approaches are similar in that they both rely on a model of evolution, but the Bayesian approach need not estimate all parameter values in order to produce high-scoring trees. In its standard implementation using

Markov chain Monte Carlo (MCMC) methods, the Bayesian approach uses a biased random walk through the space of trees and so depends mostly on a good choice of possible moves within tree space (and on the efficient computation of the effect of these moves on the current state and on the probability estimates). A preliminary implementation of such a method for equal gene content has yielded some promising results (Larget *et al.*, 2002).

Open Problems and Future Research

Nearly everything remains to be done! The work to date has convincingly demonstrated that gene-content and gene-order data can form the basis for highly accurate phylogenetic analyses; the demonstration is all the more impressive given the primitive state of knowledge in the area. We cannot give a detailed list of interesting open problems, because it would be far too long, but content ourselves with a short list of what, from today's perspective, seem to be the most promising or important avenues of exploration, from both computational and modeling perspectives.

- Solving the transposition distance problem.
- Handling transpositions along with inversions, preferably in a weighted framework.
- Adding length and location dependencies to the rearrangement framework.
- Formulating and providing reasonable approaches to solving the median problem in the above contexts.
- Developing a formal statistical model of evolution that includes all rearrangements discussed here, takes into account location within the chromosomes and length of affected segments, and obeys basic biological constraints (such as the need for telomeres and the presence of a single centromere).
- Designing a Bayesian approach to reconstruction within the framework just sketched.
- Combining DNA sequence data and rearrangement data—the sequence data may be used to rule out or favor certain rearrangements.
- Using rearrangement data in the context of network reconstruction (i.e., in the presence of past hybridizations, gene conversions, or lateral transfers).

The reader interested in more detail on the topics presented here and some of the research problems suggested should consult the survey articles of Wang and Warnow (2004) on distance corrections and distance-based methods and Moret *et al.* (2004) on parsimony-based methods.

Acknowledgments

Research on this topic at the University of New Mexico is supported by the National Science Foundation under grants ANI 02-03584, EF 03-31654, IIS 01-13095, IIS 01-21377, and DEB 01-20709 (through a subcontract to the University of Texas), by the NIH under grant 2R01GM056120-05A1 (through a subcontract to the University of Arizona), and by IBM Corporation, under contract NBCH30390004 from the U.S. Defense Advanced Research Projects Agency (the HPCS initiative). Research on this topic at the University of Texas is supported by the National Science Foundation under grants EF 03-31453, IIS 01-13654, IIS 01-21680, and DEB 01-20709, by the Program for Evolutionary Dynamics at Harvard, by the Institute for Cellular and Molecular Biology at the University of Texas at Austin, and by the David and Lucile Packard Foundation.

References

Atteson, K. (1999). The performance of the neighbor-joining methods of phylogenetic reconstruction. *Algorithmica* **25**(2/3), 251–278.

Bader, D., Moret, B., and Yan, M. (2001). A linear-time algorithm for computing inversion distance between signed permutations with an experimental study a preliminary version appeared in WADS '01, 365–376. *J. Comput. Biol.* **8**(5), 483–491.

Bafna, V., and Pevzner, P. (1995). Sorting permutations by transpositions. *In* "Proceedings of the 6th Annual ACM/SIAM Sympasium Discrete Algs. (SODA'95)," pp. 614–623. SIAM Press, Philadelphia.

Bender, M., Ge, D., He, S., Hu, H., Pinter, R., Skiena, S., and Swidan, F. (2004). Improved bounds on sorting with length-weighted reversals. *In* "Proceedings of the 15th Annual ACM/SIAM Symposium on Discrete Algs. (SODA'04)," pp. 912–921. SIAM Press, Philadelphia.

Boore, J., and Brown, W. (1998). Big trees from little genomes: Mitochondrial gene order as a phylogenetic tool. *Curr. Opini. Genet. Dev.* **8**(6), 668–674.

Boore, J., Collins, T., Stanton, D., Daehler, L., and Brown, W. (1995). Deducing the pattern of arthropod phylogeny from mitochondrial DNA rearrangements. *Nature* **376,** 163–165.

Bourque, G., and Pevzner, P. (2002). Genome-scale evolution: Reconstructing gene orders in the ancestral species. *Genome Res.* **12,** 26–36.

Bruno, W. J., Socci, N. D., and Halpern, A. L. (2000). Weighted neighbor joining: A likelihood-based approach to distance-based phylogeny reconstruction. *Mol. Biol. Evol.* **17**(1), 189–197.

Bryant, D. (2000). The complexity of calculating exemplar distances. *In* "Comparative Genomics: Empirical and Analytical Approaches to Gene Order Dynamics, Map Alignment, and the Evolution of Gene Families" (D. Sankoff and J. Nadeau, eds.), pp. 207–212. Dordrecht, Netherlands, Kluwer Academic.

Caprara, A. (1999). Formulations and hardness of multiple sorting by reversals. *In* "Proceedings of the 3rd Annual International Conference Comput. Mol. Biol. (RECOMB'99)," pp. 84–93. ACM Press, New York.

Caprara, A. (2001). On the practical solution of the reversal median problem. *In* "Proceedings of the 1st International Workshop Algs. in Bioinformatics (WABI'01)," Vol. 2149, of Lecture Notes in Computer Science, pp. 238–251. Springer Verlag.

Cosner, M., Jansen, R., Moret, B., Raubeson, L., Wang, L., Warnow, T., and Wyman, S. (2000a). An empirical comparison of phylogenetic methods on chloroplast gene order data in Campanulaceae. *In* "Comparative Genomics" (D. Sankoff and J. Nadeau, eds.), pp. 99–122. Kluwer Academic Publishers, Dordrecht Netherlands.

Cosner, M., Jansen, R., Moret, B., Raubeson, L., Wang, L., Warnow, T., and Wyman, S. (2000b). A new fast heuristic for computing the breakpoint phylogeny and experimental phylogenetic analyses of real and synthetic data. *In* "Proceedings of the 8th International Conference on Intelligent Systems for Molecular Biology (ISMB'00)," pp. 104–115. AAAI (American Association for Artificial Intelligence), Menlo Park, CA.

Day, W. (1983). Computationally difficult parsimony problems in phylogenetic systematics. *J. Theoret. Biol.* **103**, 429–438.

Downie, S., and Palmer, J. (1992). Use of chloroplast DNA rearrangements in Reconstructing plant phylogeny. *In* "Plant Molecular Systematics" (P. Soltis, D. Soltis, and J. Doyle, eds.), pp. 14–35. Chapman and Hall.

Earnest-DeYoung, J., Lerat, E., and Moret, B. (2004). Reversing gene erosion: Reconstructing ancestral bacterial genomes from gene-content and gene-order data. *In* "Proceedings of the 4th International Workshop Algs. in Bioinformatics (WABI'04)," Lecture Notes in Computer Science. Springer Verlag.

El-Mabrouk, N. (2000). Genome rearrangement by reversals and insertions/deletions of contiguous segments. *In* "Proceedings of the 11th Annual Symposium Combin. Pattern Matching (CPM'00)," Vol. 1848, of Lecture Notes in Computer Science, pp. 222–234. Springer Verlag.

Felsenstein, J. (1981). Evolutionary trees from DNA sequences: A maximum likelihood approach. *J. Mol. Evol.* **17**, 368–376.

Felsenstein, J. (1993). Phylogenetic Inference Package (PHYLIP), Version 3.5. University of Washington, Seattle.

Fitch, W. M. (1977). On the problem of discovering the most parsimonious tree. *Am. Naturalist* **111**, 223–257.

Gascuel, O. (1997). BIONJ: An improved version of the NJ algorithm based on a simple model of sequence data. *Mol. Biol. Evol.* **14**(7), 685–695.

Goloboff, P. (1999). Analyzing large datasets in reasonable times: Solutions for composite optima. *Cladistics* **15**, 415–428.

Guindon, S., and Gascuel, O. (2003). PHYML—a simple, fast, and accurate algorithm to estimate large phylogenies by maximum likelihood. *Syst. Biol.* **52**(5), 696–704.

Hannenhalli, S., and Pevzner, P. (1995a). Transforming cabbage into turnip (polynomial algorithm for sorting signed permutations by reversals). *In* "Proceedings of the 27th Annual ACM Symposium Theory of Comput. (STOC'95)," pp. 178–189. ACM Press, New York.

Hannenhalli, S., and Pevzner, P. (1995b). Transforming mice into men (polynomial algorithm for genomic distance problems). *In* "Proceedings of the 36th Annual IEEE Symposium Foundations of Comput. Sci. (FOCS'95)," pp. 581–592. IEEE Press, Piscataway, NJ.

Hartman, T. (2003). A simpler 1.5-approximation algorithm for sorting by transpositions. *In* "Proceedings of the 14th Annual Symposium Combin. Pattern Matching (CPM'03)," Vol. 2676, of Lecture Notes in Computer Science, pp. 156–169. Springer Verlag.

Hartman, T., and Sharan, R. (2004). A 1.5-approximation algorithm for sorting by transpositions and transreversals. *In* "Proceedings of the 4th International Workshop Algs. in Bioinformatics (WABI'04)," Vol. 3240, of Lecture Notes in Computer Science, pp. 50–61. Springer Verlag.

Heard, S. (1996). Patterns in phylogenetic tree balance with variable and evolving speciation rates. *Evol.* **50**, 2141–2148.

Huson, D., Nettles, S., and Warnow, T. (1999a). Disk-covering, a fast converging method for phylogenetic tree reconstruction. *J. Comput. Biol.* **6**(3), 369–386.

Huson, D., Smith, K., and Warnow, T. (1999b). Correcting large distances for phylogenetic reconstruction. *In* "Proceedings of the 3rd International Workshop Alg. Engineering (WAE'99)," Vol. 1668, of Lecture Notes in Computer Science, pp. 273–286. Springer Verlag.

Jansen, R., and Palmer, J. (1987). A chloroplast DNA inversion marks an ancient evolutionary split in the sunflower family (Asteraceae). *Proc. Natl. Acad. Sci. USA* **84,** 5818–5822.

Kaplan, H., Shamir, R., and Tarjan, R. (1999). Faster and simpler algorithm for sorting signed permutations by reversals. *SIAM J. Computing* **29**(3), 880–892.

Kim, J., and Warnow, T. (1999). Phylogenetic tree estimation. In "Proceedings of the 7th International Conference on Intelligent Systems for Moleculor Biology (ISMB'99)," Tutorial.

Kumar, S., Tamura, K., Jakobsen, I. B., and Nei, M. (2001). MEGA2: Molecular evolutionary genetics analysis software. *Bioinformatics* **17**(12), 1244–1245.

Larget, B., Simon, D., and Kadane, J. (2002). Bayesian phylogenetic inference from animal mitochondrial genome arrangements. *J. R. Stat. Soc. B* **64**(4), 681–694.

Lefebvre, J.-F., El-Mabrouk, N., Tillier, E., and Sankoff, D. (2003). Detection and validation of single gene inversions. In "Proceedings of the 11th International Conference on Intelligent Systems for Mol. Biol. (ISMB'03)," Vol. 19, of Bioinformatics, pp. i190–i196. Oxford University Press.

Li, G., Qi, X., Wang, X., and Zhu, B. (2004). A linear-time algorithm for computing translocation distance between signed genomes. In "Proceedings of the 15th Annual Symposium Combin. Pattern Matching (CPM'04)," Vol. 3109, of Lecture Notes in Computer Science. Springer Verlag.

Li, W.-H. (1997). "Molecular Evolution." Sinauer Associates.

Linder, C., Moret, B., Nakhleh, L., and Warnow, T. (2004). Network (reticulated) evolution: Biology, models, and algorithms. In "Proceedings of the 9th Pacific Sympoliyin on Biocomputing (PSB'04)." Tutorial.

Liu, T., Moret, B., and Bader, D. (2003). An exact, linear-time algorithm for computing genomic distances under inversions and deletions. *Tech. Rep.* TR-CS-2003-31, University of New Mexico.

Ma, B., Li, M., and Zhang, L. (1998). On reconstructing species trees from gene trees in terms of duplications and losses. In "Proceedings of the 2nd Annual International Conference Comput. Molecular Biology (RECOMB'98)," pp. 182–191. ACM Press, New York.

Maddison, W. (1990). A method for testing the correlated evolution of two binary characters: Are gains or losses concentrated on certain branches of a phylogenetic tree? *Evol.* **44,** 539–557.

Maddison, W. (1997). Gene trees in species trees. *Syst. Biol.* **46**(3), 523–536.

Marron, M., Swenson, K., and Moret, B. (2003). Genomic distances under deletions and insertions. In "Proceedings of the 9th International Conf. Computing and Combinatorics (COCOON'03)," Vol. 2697, of Lecture Notes in Computer Science, pp. 537–547. Springer Verlag.

McLysaght, A., Baldi, P., and Gaut, B. (2003). Extensive gene gain associated with adaptive evolution of poxviruses. *Proc. Natl. Acad. Sci. USA* **100,** 15655–15660.

Meidanis, J., Walter, M., and Dias, Z. (2000). Reversal distance of signed circular chromosomes, unpublished.

Mooers, A., and Heard, S. (1997). Inferring evolutionary process from phylogenetic tree shape. *Q. Rev. Biol.* **72,** 31–54.

Moret, B., Bader, D., and Warnow, T. (2002a). High-performance algorithm engineering for computational phylogenetics. *J. Supercomputing* **22,** 99–111.

Moret, B., Roshan, U., and Warnow, T. (2002b). Sequence length requirements for phylogenetic methods. In "Proceedings of the 2nd International Workshop Algs. in Bioinformatics (WABI'02)," Vol. 2452, of Lecture Notes in Computer Science, pp. 343–356. Springer Verlag.

Moret, B., Siepel, A., Tang, J., and Liu, T. (2002c). Inversion medians outperform break-point medians in phylogeny reconstruction from gene-order data. *In* "Proceedings of the 2nd International Workshop Algs. in Bioinformatics (WABI'02)," Vol. 2452, of Lecture Notes in Computer Science, pp. 521–536. Springer Verlag.

Moret, B., Tang, J., Wang, L.-S., and Warnow, T. (2002d). Steps toward accurate reconstructions of phylogenies from gene-order data. *J. Comput. Syst. Sci.* **65**(3), 508–525.

Moret, B., Tang, J., and Warnow, A. T. (2004). Reconstructing phylogenies from gene-content and gene-order data. *In* "Mathematics of Evolution and Phylogeny" (O. Gascuel, ed.). Oxford University Press.

Moret, B., Wang, L.-S., and Warnow, T. (2002e). New software for computational phylogenetics. *IEEE Computer* **35**(7), 55–64.

Moret, B., Wang, L.-S., Warnow, T., and Wyman, S. (2001a). New approaches for reconstructing phylogenies from gene-order data. *In* "Proceedings 9th International Conference on Intelligent Systems for Molecular Biology (ISMB'01)," Vol. 17, of Bioinformatics, pp. S165–S173.

Moret, B., Wyman, S., Bader, D., Warnow, T., and Yan, M. (2001b). A new implementation and detailed study of breakpoint analysis. *In* "Proceedings of the 6th Pacific Symposium on Biocomputing (PSB'01)," pp. 583–594 World Scientific Pub.

Nadeau, J., and Taylor, B. (1984). Lengths of chromosome segments conserved since divergence of man and mouse. *Proc. Natl. Acad. Sci. USA* **81**, 814–818.

Nakhleh, L., Moret, B., Roshan, U., John, K. S., and Warnow, T. (2002). The accuracy of fast phylogenetic methods for large datasets. *In* "Proceedings of the 7th Pacific Symposium on Biocomputing (PSB'02)," pp. 211–222. World Scientific Pub.

Nakhleh, L., Roshan, U., St. John, K., Sun, J., and Warnow, T. (2001a). Designing fast converging phylogenetic methods. *In* "Proceedings of the 9th International Conf. on Intelligent Systems for Mol. Biol. (ISMB'01)," Vol. 17, of Bioinformatics, pp. S190–S198. Oxford University Press.

Nakhleh, L., Roshan, U., St. John, K., Sun, J., and Warnow, T. (2001b). The performance of phylogenetic methods on trees of bounded diameter. *In* "Procedings 1st International Workshop Algs. in Bioinformatics (WABI'01)," Vol. 2149, of Lecture Notes in Computer Science, pp. 214–226. Springer Verlag.

Olsen, G., Matsuda, H., Hagstrom, R., and Overbeek, R. (1994). FastDNAml: A tool for construction of phylogenetic trees of DNA sequences using maximum likelihood. *Comput. Applied Biosci.* **10**(1), 41–48.

Page, R. (1998). GeneTree: Comparing gene and species phylogenies using reconciled trees. *Bioinformatic* **14**(9), 819–820.

Page, R., and Charleston, M. (1997a). From gene to organismal phylogeny: Reconciled trees and the gene tree/species tree problem. *Mol. Phyl. Evol.* **7**, 231–240.

Page, R., and Charleston, M. (1997b). Reconciled trees and incongruent gene and species trees. *In* "Mathematical Hierarchies in Biology" (B. Mirkin, F. R. McMorris, F. S. Roberts, and A. Rzehtsky, eds.), Vol. 37. American Mathematical Society.

Palmer, J. (1992). Chloroplast and mitochondrial genome evolution in land plants. *In* "Cell Organelles" (R. Herrmann, ed.), pp. 99–133. Springer Verlag.

Pamilo, P., and Nei, M. (1998). Relationship between gene trees and species trees. *Mol. Biol. Evol.* **5**, 568–583.

Pe'er, I., and Shamir, R. (1998). The median problems for breakpoints are NP-complete. *Elec. Colloq. Comput. Complexity* **71**.

Pevzner, P., and Tesler, G. (2003). Human and mouse genomic sequences reveal extensive breakpoint reuse in mammalian evolution. *Proc. Natl. Acad. Sci. USA* **100**(13), 7672–7677.

Rokas, A., and Holland, P. W. H. (2000). Rare genomic changes as a tool for phylogenetics. *Trends Ecol. Evol.* **15**, 454–459.

Roshan, U., Moret, B., Williams, T., and Warnow, T. (2004). Rec-I-DCM3: A fast algorithmic technique for reconstructing large phylogenetic trees. *In* "Proceedings 3rd IEEE Computational Systems Bioinformatics Conf. CSB'04." IEEE Press, Piscataway, NJ.

Saitou, N., and Nei, M. (1987). The neighbor-joining method: A new method for reconstructing phylogenetic trees. *Mol. Biol. Evol.* **4**, 406–425.

Sankoff, D. (1999). Genome rearrangement with gene families. *Bioinformatics* **15**(11), 990–917.

Sankoff, D., and Blanchette, M. (1998). Multiple genome rearrangement and breakpoint phylogeny. *J. Comput. Biol.* **5**, 555–570.

Sankoff, D., Ferretti, V., and Nadeau, J. (1997). Conserved segment identification. *J. Comput. Biol.* **4**(4), 559–565.

Sankoff, D., and Nadeau, J. (1996). Conserved synteny as a measure of genomic distance. *Disc. Appl. Math.* **71**(1–3), 247–257.

Siepel, A., and Moret, B. (2001). Finding an optimal inversion median: Experimental results. *In* "Proccedings 1st International Workshop Algs. in Bioinformatics (WABI'01)," Vol. 2149, of Lecture Notes in Computer Science, pp. 189–203. Springer Verlag.

Steel, M. A. (1994). The maximum likelihood point for a phylogenetic tree is not unique. *Syst. Biol.* **43**(4), 560–564.

Stein, D., Conant, D., Ahearn, M., Jordan, E., Kirch, S., Hasebe, M., Iwatsuki, K., Tan, M., and Thomson, J. (1992). Structural rearrangements of the chloroplast genome provide an important phylogenetic link in ferns. *Proc. Natl. Acad. Sci. USA* **89**, 1856–1860.

Swenson, K., Marron, M., Earnest-DeYoung, J., and Moret, B. (2004). Approximating the true evolutionary distance between two genomes. *Tech. Rep.* TR-CS-2004-15, University of New Mexico.

Swofford, D. (2001) PAUP*: Phylogenetic analysis using parsimony (*and other methods), version 4.0b8. Sunderland, MA.

Swofford, D., Olsen, G., Waddell, P., and Hillis, D. (1996). Phylogenetic inference. *In* "Molecular Systematics" (D. M. Hillis, B. K. Mable, and C. Moritz, eds.), pp. 407–514. Sinauer Assoc.

Tang, J., and Moret, B. (2003a). Phylogenetic reconstruction from gene rearrangement data with unequal gene contents. *In* "Procdings 8th International Workshop on Algs. and Data Structures (WADS'03)," Vol. 2748, of Lecture Notes in Computer Science, pp. 37–46. Springer Verlag.

Tang, J., and Moret, B. (2003b). Scaling up accurate phylogenetic reconstruction from gene-order data. *In* "Proceedings 11th International Conf. on Intelligent Systems for Mol. Biol. (ISMB'03)," Vol. 19, of Bioinformatics, pp. i305–i312. Oxford University Press.

Tang, J., Moret, B., Cui, L., and dePamphilis, C. (2004). Phylogenetic reconstruction from arbitrary gene-order data. *In* "Proceedings 4th IEEE Symp. on Bioinformatics and Bioengineering BIBE'04," pp. 592–599. IEEE Press, Piscataway, NJ.

Tannier, E., and Sagot, M. (2004). Sorting by reversals in subquadratic time. *In* "Proceedings 15th Ann. Symp. Combin. Pattern Matching (CPM'04)," Vol. 3109 of Lecture Notes in Computer Science. Springer Verlag.

Tesler, G. (2002). Efficient algorithms for multichromosomal genome rearrangements. *J. Comput. Syst. Sci.* **65**(3), 587–609.

Wang, L.-S., Jansen, R., Moret, B., Raubeson, L., and Warnow, T. (2002). Fast phylogenetic methods for genome rearrangement evolution: An empirical study. *In* "Proceedings 7th Pacific Symp. on Biocomputing (PSB'02)" pp. 524–535. World Scientific Pub.

Wang, L.-S., and Warnow, T. (2001). Estimating true evolutionary distances between genomes. *In* "Proceedings 33rd Ann. ACM Symp. Theory of Comput. (STOC'01)," pp. 637–646. ACM Press, New York.

Wang, L.-S., and Warnow, T. (2004). Distance-based genome rearrangement phylogeny. *In* "Mathematics of Evolution and Phylogeny" (O. Gascuel, ed.). Oxford University Press.

[36] Analytical Methods for Detecting Paralogy in Molecular Datasets

By JAMES A. COTTON

Abstract

Paralogy (common ancestry through gene duplication rather than speciation) is widely recognized as an important problem for molecular systematists. This chapter introduces the concepts of paralogy and orthology and explains why paralogy can complicate both systematic work and other studies of molecular evolution. The definition of paralogy is explicitly phylogenetic, and phylogenetic methods are crucial in elucidating the pattern of paralogy. In particular, knowledge of the species phylogeny is key. I introduce the theory behind methods for detecting paralogy and briefly discuss two particular software implementations of phylogenetic methods to detect paralogy from molecular data. I also introduce a statistical method for detecting paralogy and some future directions for work on paralogy detection.

Introduction

What is Paralogy?

Since Darwin, for most biologists (or at least, for evolutionary biologists) *homology* has come to mean something like "similarity due to common descent," to distinguish it from similarity due to convergent evolution. An accurate understanding of the relationships between living things depends on correctly identifying homologous characteristics of an organism from other similarity. A classic example would be the wings of bats and birds, which do not share a common evolutionary origin as wings. These wings and the limbs of other terrestrial vertebrates look very different but are truly homologous. Similarly, the genes of an organism can be homologues; genes share a common ancestor as features of an organism, and all mammal hemoglobins are descended from a hemoglobin gene

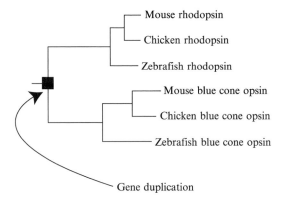

FIG. 1. Orthology and paralogy. The three rhodopsin genes are all orthologous to each other, as are the three blue cone opsin genes. The rhodopsin genes are paralogous to any of the cone opsins, and vice versa.

present in the ancestor of mammals, just as mammal limbs are descended from the limbs of this ancestor.

In genetics, however, homology can be a more complex phenomenon, because genes can be homologous in at least two distinct ways (Fitch, 2000). As well as descending from an ancestral species, genes also share a common ancestor as genes, in that related genes have arisen by duplication and gradual mutation. For example, all globin genes descend from a common ancestral globin gene. Fitch (1970) proposed new terms for these two classes of homology among genes. If the most recent common ancestor of two genes is a gene duplication event, the genes are *paralogous*, otherwise they are *orthologous* (Fig. 1). To use Fitch's original example, α and β hemoglobin are *paralogs,* whereas α hemoglobin in humans and mice are *orthologs*.

Why Does Paralogy Matter?

Fitch's introduction of the two terms makes it clear why the distinction between paralogs and orthologs is important: Only for orthologous genes does the "history of the gene reflect the history of the species." An organismal phylogeny based on a mixture of paralogous genes would be "biological nonsense" (Fitch, 1970, p.113). This realization that inference of species relationships should be based on orthologous genes alone dates back to the earliest days of molecular systematics (Fitch and Margoliash, 1967). If, for example, we (unknowingly) sampled just the chicken and zebrafish rhodopsin genes and the mouse cone opsin gene from Fig. 1, we would mistakenly conclude that chickens and zebrafish were more closely related to each other than either is to mice. We can think of the gene tree (a phylogeny constructed from some particular molecular data) and the species tree (the phylogeny representing the evolution of the organisms the genes have been sampled from) as distinct trees. The process of gene duplication is one of a number of reasons gene trees may not match the correct species tree (Maddison, 1997; Martin and Burg, 2002; Page, 1994).

This *problem of paralogy* is widely recognized by systematists (Sanderson and Shaffer, 2002), and many discussions of suitable genes for molecular phylogenetics suggest that the ideal molecular marker should be "single copy" (Cruickshank, 2002). The fear of paralogy has been one of the major reasons for the popularity of organelle genes (which are, perhaps wrongly, generally assumed to be single copy) and of ribosomal RNA genes (which are largely homogenized by gene conversion). This advice may, however, restrict systematists to relatively few loci, because most nuclear genes seem to be parts of families of related genes (Henikoff *et al.*, 1997; Kunin *et al.*, 2003; Slowinski and Page, 1999). Restricting work to these loci alone would mean rejecting the great possibilities opened up by genomic-level data becoming available for a widening range of organisms (Rokas *et al.*, 2003). In any case, even when a gene is single copy in known genomes, it cannot be certain that the gene is single copy in all organisms and has been single copy throughout evolutionary history. Standard molecular systematic studies involving just polymerase chain reaction (PCR) amplification of a locus and sequencing of the product are not readily capable of detecting multiple copies of a gene in a sample. Unfortunately, relatively few concrete suggestions for dealing with the problem have been put forward; if potentially multicopy genes must be used, most authors suggest only a vague hope that paralogy might be recognized by differences in molecular architecture. These might include differences in intron structure or size, as well as changes in codon usage or base composition. Although such molecular approaches may help in recognizing paralogy in specific cases, it seems by no means inevitable that paralogous copies will show such differences.

Molecular biologists may have other reasons for wanting to detect paralogous genes. Gene duplication is probably the most important mechanism by which genes evolve new functions (Long and Thornton, 2001; Ohno, 1970), so that genes that are *orthologs* are more likely to share a common function than *paralogs*. Functional characterization of a gene is, thus, best made by comparison with orthologous sequences. Gene duplication may be the only common mechanism; it is hard to imagine how an arbitrary sequence can evolve a useful function (although see Hayashi *et al.*, 2003), so most genes probably acquired their function by gradual evolution from a gene doing a related job. The role of gene duplication in this process is easy to see: If a gene is performing an essential role in the cell, it is only when a duplicate copy exists to maintain this role that the gene is free to mutate away from the original function and evolve a new one. Of course, most mutations will reduce the gene's usefulness and many mutations will be silencing. A number of authors have envisaged this process as a "race" between a gene copy acquiring a new function and being silenced and eventually deleted from the genome (Walsh, 1995). Both empirical (Nadeau and Sankoff, 1997) and theoretical (Walsh, 1995) studies support the importance of gene duplication in the evolution of new gene functions.

Detecting paralogy may also be important in studying the pattern (and so the process) of gene duplication (Cotton, 2003; Page and Cotton, 2002). Empirical interest in the pattern of gene duplication has largely focused on testing the importance of polyploidy or genome duplication in evolution, looking at both phylogenetic and map-based data (Skrabanek and Wolfe, 1998), although a few papers have looked more widely at patterns of gene duplication (Lynch and Conery, 2000; Semple and Wolfe, 1999). To highlight the research interest in gene duplication, at least two major journals have published "thematic issues" focusing on evolution by gene duplication within a year (see introductions by Long [2003] and Meyer and Van de Peer [2003]).

Finally, recognizing paralogy is also important in molecular clock dating. If the estimated divergence date of two species is based on paralogous genes, then the event being dated is actually a gene duplication, rather than the speciation event, and the date estimate will be too old (Fig. 2). This could be a significant overestimate, depending on the rates of gene duplication and gene loss, and may be important in explaining at least some of the well-known discrepancies between molecular clock-based date estimates and dates estimated from the fossil record (Benton and Ayala, 2003), although other problems with both clock-based dates and the ways in which the fossil record has been used have also been described (Rodriguez-Trelles et al., 2002; Shaul and Graur, 2002).

Similarity (Homology) is not Orthology

The oldest way of identifying related genes is to identify genes that have similar sequences. Although systematists appear to be aware of the dangers of using sequence similarity alone to select genes for phylogenetic analysis (Bapteste et al., 2002), similarity has been used as a selection criterion in molecular clock-dating studies (Kumar and Hedges, 1998) and to identify genes that have similar functions (Eisen, 1998; Zmasek and Eddy, 2002). This similarity is most often detected by using BLAST or FASTA searches of sequence databases. This is particularly common in functional

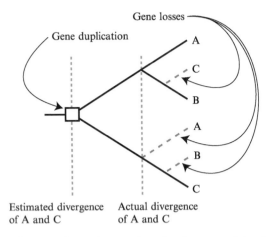

FIG. 2. How paralogy can alter estimates of species divergence dates. Gene duplications and subsequent gene loss will affect molecular estimates of divergence dates if the date of the gene duplication event rather than the actual speciation event is estimated. This will lead to overestimates of divergence dates.

annotation of genes and genomes, where sequence similarity methods are the standard approach. The use of sequence similarity in systematics seems set to become more important as large-scale phylogenetic analyses become possible by combining data from a range of sequencing projects (Rokas *et al.*, 2003).

Against this background, it is important to point out that similarity is not the same as orthology, in the sense that orthologs of a particular gene may not be the most similar genes in a database. One major reason for this is the well-known fact that sequence similarity may not accurately reflect phylogenetic relationships, whether because of failure to correct for multiple substitutions at particular sites or because rates of evolution are unequal (Felsenstein, 2004, p.175) (Fig. 3). A different problem is that the database may not contain orthologs of the sequence, because of either being incomplete or because gene loss has led to the disappearance of these orthologs from extant genomes. This first problem is avoided if we build accurate phylogenies for the sequences involved.

Detecting Paralogy on Phylogenies

As is obvious from the definition discussed earlier, gene duplication events are key to understanding paralogy and orthology. A gene duplication event is one in which a piece of DNA is physically duplicated, forming

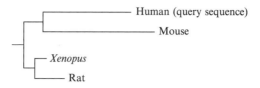

Fig. 3. Sequence similarity can be misleading. Unequal rates of evolution have led to the human query sequence being most similar to the *Xenopus* sequence rather than its ortholog (the mouse sequence). Methods based on sequence similarity alone will suggest that the human sequence is related to the subfamily containing the rat and *Xenopus* gene rather than to the mouse gene (modified, with permission, from Zmasek and Eddy [2002]).

a second copy of the genetic material. This can occur at a range of scales, from a few bases to the entire genome, representing a range of mechanisms from unequal crossing over and slippage during DNA replication to chromosomal non-disjunction and the production of unreduced gametes (Li and Graur, 1991, p.137). For our purposes, gene duplication must have occurred on a sufficiently large scale to have affected an entire locus that could be used for phylogenetic inference. Just as phylogenetic methods make assumptions about the process of nucleotide substitution, we would expect methods for detecting gene duplications to make some assumptions about the process of gene duplication, a point I will return to later.

The internal nodes on a phylogenetic tree represent divergence events, which, in molecular systematics, are usually presumed to be speciation. After a speciation event, the two lineages can no longer interbreed and are free to evolve independently and meet separate evolutionary fates, with the accumulation of mutations leading to them becoming gradually more and more distinct from one another. Gene duplications represent a similar event; after a gene duplication event, the two copies of a gene are free to accumulate independent mutations and diverge (at least in the absence of gene conversion). Gene duplication and speciation are similar splitting events and, in fact, cannot always be distinguished by simple inspection of a molecular phylogeny. If we accept that phylogenetic methods are needed to correctly identify paralogy and orthology, then the problem of detecting paralogy becomes that of identifying which internal nodes of a tree represent gene duplications and which represent speciation events.

Sometimes this can be easy. If two similar genes are present in the same genome, they must be paralogs or at least partial paralogs. Identifying paralogy on larger gene family phylogenies is similarly straightforward if there has been no loss of genes. Multiple copies are descended from a gene duplication, and the most parsimonious placement of the duplication is the least common ancestor (LCA) (the ancestor of both sequences that is furthest away from the root, also called *most recent common ancestor* [MRCA]) of the duplicate copies. This placement will imply the smallest number of subsequent gene losses or deletions. When we have a phylogeny for the gene family, these nodes can then be discerned simply by inspection, and a number of studies have done exactly this (e.g., Katoh and Miyata, 2002). We can see how this is done by looking at Fig. 4D. The

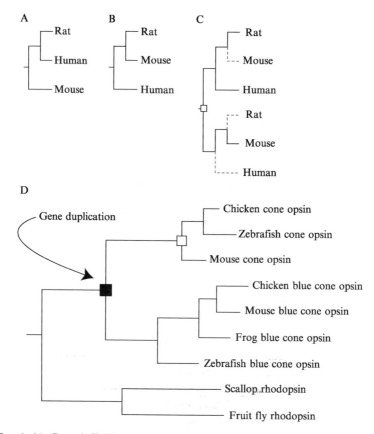

Fig. 4. (A, B, and C) The simplest possible case of gene duplication and gene loss obscuring the pattern of orthology and paralogy. Given a gene tree (A), it might seem that rat and human are more closely related to each other than either is to mouse. The correct species tree is shown (B). The incongruence between (A) and (B) can be explained by postulating a single gene duplication and three gene losses, shown on the reconciled tree (C). (D) A slightly more complex example. One gene duplication is implied by the presence of multiple sequences from mouse, chicken, and zebrafish (shaded box). A second gene duplication (open box) is implied only by the fact that the phylogeny for the top clade of cone opsins does not match the correct species phylogeny. The gene losses are not shown.

marked duplication event is the LCA of the two chicken opsins (and of the two mouse opsins and the two zebrafish opsins). For large, complex gene families (such as in Fig. 5), it is preferable to have some kind of computerized method for doing this, and LCAs can be found in linear time

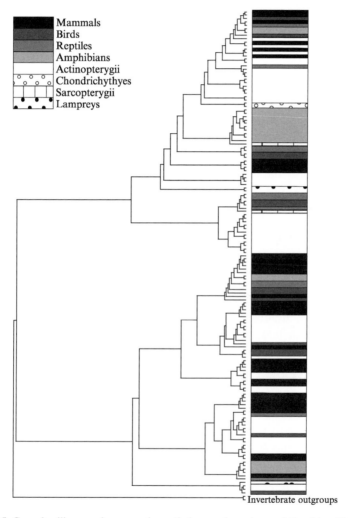

FIG. 5. Gene families can show complex orthology and paralogy relationships. This is the phylogeny for selected vertebrate opsin genes from Hovergen family HBG031788 (Duret *et al.*, 1994), color-coded to reflect the taxonomy of the species included. It is clear that there are a number of clades of related opsin genes and that the phylogeny within these clades does not always reflect the organism phylogeny. The pattern of orthology in this gene family is obscured by both gene loss and failure to sample; even some genomes that are fully sequenced lack certain orthologs.

(Harel and Tarjan, 1984), which means that the worst-case speed of finding the LCA scales as a linear function of the size of the input tree.

Importance of Knowing the Species Tree

The previous discussion ignores one crucial complicating factor: In the presence of gene loss or the absence of some gene copies from a phylogeny, recognizing the pattern of gene duplication (and so of orthology and paralogy) can be much more difficult. To understand this, a conceptually easy case to imagine is if one set of a duplicated pair of genes is lost, there will be no descendants to suggest that the gene duplication occurred at all. Of course, in this case, the survivors are all orthologous and the duplication will have had no effect on the phylogeny of the gene involved. This will not hold in other cases. If one of the two copies from a gene duplication is lost in each descendant lineage, there will be no multiple copies to suggest that a gene duplication occurred, but the two genes will be paralogous.

The most obvious symptom of these kinds of duplication is incongruence between a species tree and the gene tree, and it is this incongruence that allows gene duplication to be correctly inferred in the presence of gene loss, a realization that dates back at least to a seminal paper by Goodman *et al.* (1979). In the absence of any molecular events that introduce differences, we would expect the correct phylogeny for a set of gene sequences to exactly match the phylogeny for the species the genes have been sampled from. By fitting the observed gene tree into the known phylogeny for the species the genes have been sampled from, it is possible to infer evolutionary events, such as gene duplication and gene loss, which have introduced the differences between the two phylogenies.

Figure 4A–C shows this situation. Examining the tree in Fig. 4A by eye does not reveal any evidence of multiple gene lineages produced by a gene duplication event. Most biologists, however, will recognize that mice and rats are more closely related to each other than either is to humans, and the tree in Fig. 4A thus looks wrong. If we assume that the tree is, in fact, a correct estimate of the phylogeny for the gene, we can explain the difference between this tree and what we know to be the correct phylogeny for the three taxa in Fig. 4B as being due to a single gene duplication, followed by three gene losses (Fig. 4C). Even in more complex cases (Figs. 5 and 6), knowing the correct phylogeny for the species involved allows us to infer a scenario of duplications and losses that can explain incongruence between a gene and species tree. This idea of fitting a gene tree into a species tree has become known as tree reconciliation.

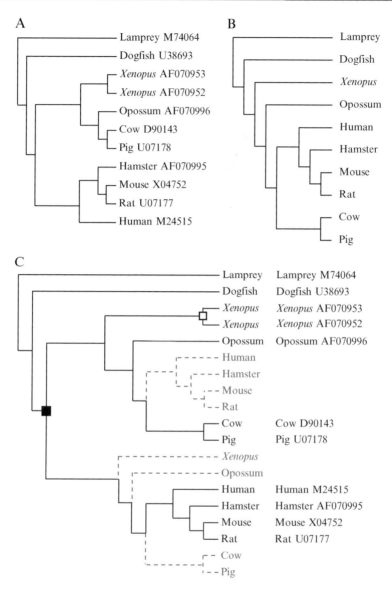

FIG. 6. (A) A molecular phylogeny for some vertebrate lactate dehydrogenase (LDH) genes. GenBank accession numbers are shown for each sequence. (B) The species tree for the organisms included on the tree. (C) A reconciled tree from GeneTree showing the evolutionary history of these genes. The reconciled tree identifies two gene duplications: one within *Xenopus* that is implied by the multiple copies in that species (open box) and another implied by the incongruence between the gene tree and species tree (shaded box). This second duplication correctly separates genes encoding the LDH-A muscle-specific and LDH-C testis-specific isozymes of the LDH enzyme.

Reconciled Trees: A Parsimony Method

Parsimony Mapping and Co-phylogeny

The problem of understanding the difference between two associated trees is a general one and has led to the idea of tree mapping or co-phylogeny. The associated trees can be any two trees that one would expect to be identical in the absence of some specific evolutionary event. If, for example, a parasitic organism has always speciated in response to host speciation, then they will have identical phylogenies, and host–parasite systems are probably the best-known example of associated phylogenies. Such systems can be studied by creating a map between the two trees to elucidate what events could have occurred to introduce any observed differences. For hosts and parasites, these events are things like host switching, in which a parasite population becomes established on a different host species, independent speciation of the parasite without a host speciation, and parasite extinction. In the gene tree–species tree system, the relevant events are lateral gene transfer (LGT), gene duplication, and gene loss. These events have similar phylogenetic effects to the host–parasite events; LGT is equivalent to host switching, gene duplication to independent parasite speciation, and gene loss to parasite extinction (Page and Charleston, 1997). The other commonly discussed system is the biogeographical system of an organism phylogeny and a hierarchy relating the areas the organisms inhabit (Page and Charleston, 1998).

The original concept of Goodman *et al.*, that of producing a map between two associated trees in order to explain differences between them, has since been formalized by Page (1994), who presented the first algorithm for reconciling two trees. The algorithm is very simple, involving constructing a map between each node in the gene tree and each node in the species tree. The map is constructed traveling down the tree from leaves to the root. First, each leaf in the gene tree is mapped onto the corresponding leaf in the species tree. Any nonleaf node N in the gene tree is mapped onto the LCA of the species tree nodes onto which the descendants of N are mapped. When this map is completed, a gene duplication event is inferred wherever a gene tree node is mapped onto the same node as its immediate descendant. The number of gene losses can then be computed by another pass through the gene tree. In fact, a number of ways of speeding up this algorithm have been suggested, leading to two linear-time algorithms for reconciling two trees (Eulenstein, 1997; Zhang, 1997) and a simpler algorithm that has inferior worst-case running time but is claimed to be faster on most biological data (Zmasek and Eddy, 2001b). The Eulenstein (1997) and Zmasek and Eddy (2001b) algorithms are implemented in GeneTree and RIO, respectively (discussed later in this chapter). In addition to counting gene duplications and gene losses, this map can be used to produce a reconciled tree that represents the evolution of the gene within the species phylogeny (Fig. 6C), allowing ready identification of paralogs and orthologs across the gene tree.

The Gene Tree Is not Known without Error

Gene trees inferred by phylogenetic methods from amino acid or nucleotide sequence data are estimates of the true tree. They are unlikely to be estimates without error, whether from sampling error caused by the finite length of sequences used or because of the well-known biases in some phylogenetic methods (Felsenstein, 2004). The reconciled tree methods discussed earlier explicitly assume that the gene tree is known without error, as any incongruence between the gene tree and the species tree is explained in terms of gene duplication and gene loss. Clearly, if some of this incongruence is due to error in the gene tree, some of the implied duplications (and losses) will also be in error. A robust method for detecting paralogy will need to take some account of gene tree error.

A number of ways of dealing with this error have been proposed. One possibility is that an alternative gene tree, less parsimonious, less likely, or less probable than the optimal tree, is, in fact, the correct tree. A number of authors have suggested using the fit between the gene tree and species tree as a criterion for choosing between alternative gene tree topologies. Goodman et al. (1979) assigned each hemoglobin gene tree a score based on both the length of the tree in terms of nucleotide substitutions and the number of gene duplications and losses it implied on a species tree. They thus preferred less parsimonious hemoglobin trees that matched the expected species tree more closely. Fitch (1979) criticized this approach as requiring an arbitrary choice between the "cost" of a substitution event versus a duplication/loss event, although assigning these costs could be explored experimentally in specific circumstances (Ronquist, 2003).

One way to avoid this dilemma is to use some kind of statistical confidence interval around each gene tree, to contain all the gene trees that cannot be rejected by the sequence data (Martin, 2000). This could be a set of credible trees in the Bayesian sense or all the trees inferred from bootstrapping the original sequence data, which would form a (rather conservative) pseudo–confidence interval for each gene tree estimate (Page, 1996; Sanderson, 1989). This kind of bootstrapping procedure has been suggested a number of times (Page and Cotton, 2000; Ronquist, 2003) and has been implemented in the programs GeneTree (Page, 1998), OrthoStrapper (Storm and Sonnhammer, 2002), and RIO (Zmasek and Eddy, 2002) (discussed later in this chapter). An obvious alternative to this bootstrapping approach is to use some kind of likelihood function that incorporates a model of sequence evolution and a model of gene duplication and loss. This has also been suggested previously (Page and Cotton, 2000), initially in the related context of allele coalescence (Maddison, 1997), and has led to the development of statistical methods. Finally, at least one other possibility has been explored—using local rearrangements (nearest-neighbor interchanges) (Waterman and Smith, 1978) around poorly supported nodes to make the gene tree better fit the species tree (Chen *et al.*, 2000; Page, 2000).

Detecting Paralogy (and Inferring a Species Tree) with GeneTree

A popular implementation of reconciled tree methods, and perhaps the easiest to use, is GeneTree (Page, 1998). Given a species tree and one or more gene trees, GeneTree will find the reconciled tree that represents the evolution of each gene tree, counting gene duplications and gene losses. It can also graphically show a reconciled tree for each gene family, allowing the user to see which genes are paralogs and orthologs (Fig. 6). GeneTree is a C++ program with a full graphical user interface (GUI), available for Mac OS and Microsoft Windows from http://taxonomy.zoology.gla.ac.uk/rod/GeneTree/GeneTree.html. A cross-platform command line version is in development.

As the astute reader has probably noticed, detecting paralogy is of particular concern for molecular systematists. Understanding paralogy depends on knowing something about the species tree, so studies intended to elucidate the species tree for a little-known group will have no means of understanding paralogy in the molecular markers used. This is a potentially vicious circle: to get an estimate of a species phylogeny, we need to use orthologous sequences, but to accurately determine orthology, we need to know the species phylogeny accurately! The most common method of breaking this circle is simply to use markers that are considered to be free of paralogy, but other tactics may be available and may even be preferable.

GeneTree implements one approach: it can find the species tree that requires the minimum number of gene duplications to fit it onto the gene trees given. If gene duplications are thought to be sufficiently rare, the species tree minimizing the number of gene duplications (or the total number of gene duplications and gene losses) could be preferred as the best estimate of the species tree. Where gene trees are available from multiple independently evolving gene families, this approach may be particularly powerful and has been advocated as a general approach to molecular systematics (Slowinski and Page, 1999). Note that other approaches to inferring species phylogenies in the presence of paralogy have been proposed (Simmons *et al.*, 2000), which may or may not be preferable to reconciled tree-based methods (Cotton and Page, 2003; Simmons and Freudenstein, 2002). In fact, this idea of searching for a tree that minimizes some cost, or distance, from a set of source trees, is one characterization of supertree methods (Thorley and Wilkinson, 2003), and the use of reconciled trees to infer a species tree (which has become known as *gene tree parsimony*) (Slowinski and Page, 1999) can be usefully compared with other supertree methods (Cotton and Page, 2004).

Identifying Orthologous Genes Using RIO

Although GeneTree is aimed at molecular systematists, RIO is aimed at molecular biologists wanting to identify the functions of newly sequenced genes and, appropriately, takes a rather different approach. RIO (Zmasek and Eddy, 2002) is a suite of C and Java programs connected by a perl pipeline, specifically designed for the inference of orthology and paralogy from a set of sequence data. These programs together automate the entire process of ortholog and paralog identification.

RIO begins by identifying similar sequences in the Pfam protein family database and aligning these sequences using a hidden Markov model approach, using the HMMER package. This alignment is then bootstrap resampled, and a phylogenetic tree constructed by neighbor joining on ML distances inferred under an empirical amino acid substitution matrix. Each of the bootstrap trees are then rooted to give a minimum number of duplications and then compared with a single species tree based on a number of large, published phylogenies to infer gene duplications and losses. These inferences can then be converted into percentage probabilities for orthology and paralogy between the query sequence and each related sequence identified in Pfam (see later discussion and Zmasek and Eddy [2002] for details of each step and references).

Zmasek and Eddy (2002) have recognized that if gene duplications are responsible for much of the origin of new gene functions, simple paralogy versus orthology may not be the only distinction of importance in functional annotation, leading them to introduce some new terminology. They define *superorthologs* as genes where not only is the LCA of two genes a speciation event rather than a gene duplication, but all the nodes on the shortest path connecting the two genes represent speciation events (this path connects the LCA to the two leaves). If gene duplications can lead to the evolution of new function, then superorthologs (which have undergone no gene duplication since their divergence) are most likely to share a common function. Zmasek and Eddy also introduce *ultra-paralogs*, which are genes for which the smallest subtree containing both genes contains only nodes that represent gene duplications. Such subtrees, which will contain sequences from a single species, represent lineage-specific expansion of a particular gene family. Lineage-specific duplication has been reported in a number of cases (e.g., in the lineage separating humans from the great apes [Nahon, 2003]) and seems to represent the selected growth of a functionally important gene family. Despite the large number of gene duplications, these genes share closely related functions because the newly formed gene copies appear to have partitioned the original function of the parental gene, rather than evolving completely new functions (so they are evolving by subfunctionalization rather than neofunctionalization) (Force *et al.*, 1999). Finally, Zmasek and Eddy also introduced the term *subtree neighbors* to define gene copies that are present on the same clade of a certain size, presumably because more closely related genes may sometimes share the same function, whether they are paralogs or orthologs.

Perhaps the greatest strength of RIO is that it automates the entire analysis, performing a number of steps that the user of a program like GeneTree is required to do manually. RIO is designed explicitly for the molecular biologist interested in identifying the orthologs and paralogs of a particular query sequence. Because of its relative ease of use, an already available species tree and its attempt to further dissect paralogy and orthology to make functional annotation more accurate, RIO is likely to be the first choice for this particular application. RIO is also available as a web service (from http://www.rio.wustl.edu), so users will not even need to install a local copy of the software (although RIO is available to download from http://www.genetics.wustl.edu/eddy/forester/). GeneTree may be of more interest to some other users, both because of its ability to infer an optimal species tree and because the reconciled tree allows the user to interpret paralogy and orthology across an entire gene family tree, rather than with respect to a particular sequence.

Other Implementations

Reconciled trees have also been implemented in a few other software packages, which I mention here for completeness. TreeMap (http://taxonomy.zoology.gla.ac.uk/~mac/treemap/index.html) implements reconciled tree methods in the context of host–parasite evolution and can deal with LGT (host switching) and gene duplication and gene loss. DupLoss is a JAVA applet that implements an efficient fixed-parameter tractable algorithm for finding duplication and loss histories (Hallett and Lagergren, 2000) (available from http://www.sable.mcgill.ca/~dbelan2/duploss/applet/duploss.html). The ATV tree editor (Zmasek and Eddy, 2001a) also includes a simple algorithm for locating gene duplications and producing reconciled trees. Finally, OrthoParaMap (Cannon and Young, 2003) integrates both phylogenetic and genetic map data in an attempt to identify duplicated genes and divide them into regional and tandem duplications. This is interesting additional information, but the authors themselves admit that "GeneTree and RIO generally appear to do a better job of identifying probable gene duplications and speciations" than OrthoParaMap. In the example they give, OrthoParaMap worryingly appears to miss duplications that must have occurred, given the presence of multiple descendent copies in the same genome (Cannon and Young, 2003) (Fig. 4).

A Statistical Approach

All the parsimony methods for tree reconciliation share a number of problems. One problem is that a single parsimony mapping assumes that both the gene tree and the species tree are known without error, the problem that the bootstrapped analyses of RIO, OrthoStrapper, and GeneTree are designed to ameliorate. A closely related problem is that parsimony mapping is deterministic, so the reconciled tree algorithms produce only a single mapping and a single inference of the evolutionary history of a gene family. Again, using a bootstrap profile of input trees avoids this problem, by providing a bootstrap-based confidence interval of possible reconstructions. A third problem with parsimony methods is that they make assumptions about the processes of gene duplication and gene loss; they implicitly assume that gene duplication and gene loss are both rare events, necessary conditions if minimizing these events is to give a realistic reconstruction of evolutionary history. This mirrors a similar assumption about nucleotide substitution made by parsimony methods for molecular sequences (Felsenstein, 1978). Although it seems likely that gene duplications and gene losses probably are rather rare, this assumption of parsimony leads to well-known undesirable properties of parsimony methods in phylogenetic reconstruction, not least statistical inconsistency in the phenomenon known as *long-branch attraction* (Felsenstein, 2004, pp. 113–122). It seems likely that similar problems will beset parsimony-based tree reconciliation.

A solution to these problems is to use a probabilistic model of gene duplication and gene loss. This is conceptually rather simple if we assume that both of these processes occur at a constant rate over time and independently of each other, a possibly dubious simplifying assumption that makes the models mathematically tractable. Such constant-rate Markov process models have been quite widely used to study the process of gene duplication and gene loss, as well as the related processes of speciation and extinction (Kubo and Isawa, 1995; Lynch and Conery, 2000; Nee *et al.*, 1992). Under these models, it is relatively straightforward to calculate the probability for any pattern of gene duplications and gene losses in a single lineage. More complexity ensues when the gene lineage is evolving within a tree, as is needed when dealing with a gene family evolving inside a species phylogeny, but these calculations are certainly feasible (Arvestad *et al.*, 2003). If we can calculate the probability of a particular reconstruction, given a gene tree and a species tree, this opens the possibility of both ML and Bayesian methods for reconciliation.

At present, only one implementation of Bayesian tree reconciliation has been reported (Arvestad *et al.*, 2003), and the software for performing this reconciliation is not yet freely available. However, probabilistic methods will clearly be the preferred method for detecting paralogy. The work of Arvestad *et al.* takes trees produced by a Bayesian phylogenetic method (such as MrBayes; Huelsenbeck and Ronquist, 2001) and uses a Markov chain Monte Carlo (MCMC) method to estimate the posterior probability that each node represents a gene duplication (or conversely, that it represents a speciation) under the constant-rate model of duplication and loss. Arvestad *et al.* present a case in which this Bayesian method clearly gives a more sensible result than the bootstrap approach as implemented in OrthoStrapper or RIO. This highlights the important difference between parsimony and probabilistic approach to reconciliation, even when uncertainty in the tree is incorporated in the parsimony framework.

The difference between parsimony-based and probabilistic paralogy detection is easily explained (Fig. 7). A parsimony-based method will assume the minimum possible number of gene duplications (as it assumes that gene duplication and gene loss are both rare) and so the minimum amount of paralogy. In the example in Fig. 7, parsimony methods will report no paralogy in gene tree ((A1,B1),C1), as it matches species tree ((A,B),C) exactly (Fig. 7A). In fact, there is a non-zero probability that gene duplications have occurred along any particular edge of non-zero length (such as the edge *e* in the figure), followed by subsequent gene loss in the descendant lineages leaving only a single copy, in such a way that has not affected the phylogeny of the gene (Fig. 7B). Similarly, a non-zero probability is attached to any number of possible duplications along edge *e*, followed by an appropriate number of later gene deletions. This translates into a non-zero probability that genes A_1 and B_1 are in fact paralogs. Note that bootstrapping will not deal with this problem adequately; the bootstrap probability that A_1 and B_1 are orthologous could still be 100% if the gene tree robustly supports their sister–group relationship.

This Bayesian framework could be extended in a number of ways. Given that the species tree is rarely known without error, it should be

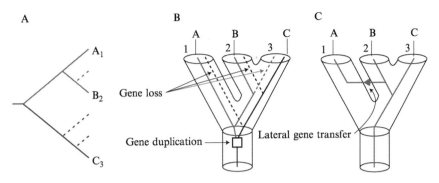

FIG. 7. Lateral gene transfer (LGT) (C) and gene duplication and subsequent loss (B) can have the same phylogenetic effect, introducing incongruence between the gene tree and species tree as shown on the reconciled tree (A). Genes 1, 2, and 3 are evolving inside species A, B, and C respectively.

possible to use MCMC to integrate across a probability distribution of species trees rather than a single estimate; such a distribution could, for example, come from analysis of some other gene or combination of genes that is thought to be more reliable than the gene family under investigation. Another possibility is a Bayesian method for inference of a species tree from a set of gene families in the presence of gene duplication and gene loss. This would be a Bayesian analogue of gene tree parsimony. Such an approach is certainly technically feasible, although it would require assuming that different gene families are statistically independent estimates of the species phylogeny, which may not be the case for linked genes,

Concluding Remarks and Future Prospects

Lateral Gene Transfer

The reconciled tree methods discussed here are designed to deal correctly with gene duplication and gene loss, but they do not distinguish another form of homology, where genes have undergone LGT. This form of nonorthology has become known as *xenology* (Fitch, 2000; Gray and Fitch, 1983). LGT is certainly common among prokaryotes. Bacterial genomes are increasingly seen as dynamic mosaics of genes (Martin, 1999), with LGT considered to have "had an extraordinary effect on bacterial genomes" (Ochman, 2001). Although LGT is of great research interest in its own right, it is directly relevant to studying paralogy, as understanding the pattern of LGT is crucial in understanding the pattern of gene duplication and gene loss. The differences between a gene tree and a species tree

A   B

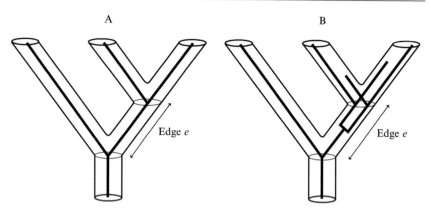

FIG. 8. A difference between parsimony and likelihood tree mapping. The two diagrams show a gene tree evolving within a species tree, where the species tree and gene tree match. In the parsimony case (A), no gene duplications will be inferred, whereas the likelihood method (B) takes into account cases in which gene duplications have occurred, followed by gene losses that reproduce the correct tree. The likelihood method should integrate across any number of duplications, from zero to infinity, along edge *e,* in calculating the probability that the two descendant lineages are orthologous.

introduced by LGT can be identical to those introduced by gene duplication and gene loss (Fig. 8). In many taxa, inferring the pattern of gene duplication and gene loss will thus depend on distinguishing these events from LGT.

The idea of reconciled trees has since been generalized to include potential events such as LGT and the equivalent host switching in the host–parasite setting (Page, 2003). Dealing correctly with this kind of event can become rather complex (Charleston, 1998) and makes the problem of correctly weighting different kinds of event even more difficult (see Ronquist [2003] for the most complete available discussion of this problem). At least two parsimony-based algorithms have been proposed to deal with host switching or LGT in a co-phylogenetic context, but there are problems with each. The Jungles algorithm (Charleston, 1998) is computationally intensive and thus too slow and memory hungry for many realistic problems, while the algorithm implemented in TreeFitter (Ronquist, 2003) does not seem to provide explicit reconstructions of the history of the lineage, making it useless for detecting paralogy. A Bayesian method that correctly deals with one particular model of host switching has been proposed (Huelsenbeck *et al.*, 1997), but this model assumes that only a single lineage is present in a species at any time, making it inapplicable in the context of gene family evolution within a species phylogeny. A more promising algorithmic approach has been described (Hallett and Lagergren, 2001).

Just as methods for detecting duplication and loss need to take into account the confounding effect of LGT, so methods for studying LGT need to take gene duplication and gene loss into account. Existing methods for detecting LGT are widely seen as unsatisfactory (Eisen, 1998; Sicheritz-Pontén and Andersson, 2001) and the increasing amount of genome sequence data is particularly rich for microbes, where LGT is likely to be important. Developing co-phylogenetic methods and software, and particularly statistical methods, that deal with LGT, gene duplication and gene loss together is clearly an important avenue of research for the future.

Independence of Gene Duplications

One concern is that all of the methods described here assume that gene duplications are independent, both within and between gene families, but this is by no means sure to be the case. As gene duplication events can affect an any size piece of DNA from a few bases to the entire genome, a single event can introduce duplications on a number of gene families simultaneously and can introduce multiple duplications on a particular family. At the extreme, in a whole-genome duplication, all the extant members of the family will be duplicated. Such whole-genome duplications or polyploidization events may be rather common; they are certainly very widespread in flowering plants (Otto and Whitton, 2000) and have been recorded in many other lineages (Skrabanek and Wolfe, 1998), including vertebrates (Furlong and Holland, 2002; Page and Cotton, 2002; Taylor *et al.*, 2001). Methods that can deal with large-scale gene duplications may be more reliable in inferring paralogy and could help study these complex patterns of gene duplications. Reconstructing the pattern of large-scale gene duplications from phylogenetic data alone is computationally complex (Guigó *et al.*, 1996; Page and Cotton, 2002). It seems likely that methods integrating phylogenetic information with genetic map data, such as has been attempted with OrthoParaMap, will be needed to infer some events. Statistical models that relax the assumption of independence of duplications should also be possible to formulate, and inference under these models should be possible using MCMC. These complications might be modeled adequately by allowing the rate of gene duplication to vary somehow over the tree, but other complications might serve to make the probability model of duplication and loss more realistic. There is, for example, reason to believe that rates of duplication and loss may not be independent, as duplicate genes may be more likely to die earlier in their life than later (Force *et al.*, 1999; Walsh, 1995).

Conclusion

I have discussed two main reasons for wanting to detect paralogy: for molecular systematists to ensure they are correctly sampling the species tree, and for molecular biologists to improve the assignment of function. Paralogy detection might also be important for some of the growing number of molecular evolution studies that are being carried out at the scale of whole genomes (Wolfe and Li, 2003) and so involve loci beyond the few well-known gene families. Of course, recognizing paralogy is an essential part of understanding the pattern of gene duplication from phylogenetic data (Page and Cotton, 2002). If we are to make wider use of the enormous amount of phylogenetic information contained in nuclear gene families, methods for dealing with paralogy in molecular systematics will become widely needed. A species tree could be estimated from a large sample of loci without assuming orthology of all gene copies by using methods that explicitly deal with paralogy (Page and Cotton, 2000; Slowinski and Page, 1999), by using the potentially riskier strategy of hoping that the weight of evidence will overwhelm any error from paralogous sequences (Brower *et al.*, 1996), or by using a method that is somewhere intermediate (Cotton and Page, 2003; Simmons *et al.*, 2000). Which of these will be most popular and most successful remains unclear. In any event, better understanding of the processes of gene duplication, and particularly of gene loss, and in particular better quantitative data, together with statistical approaches to studying these processes, seems likely to have a considerable impact on methods for detecting paralogy.

Acknowledgments

Thanks to Trevor Cotton, Claire Pickthall, and Mark Wilkinson for constructive comments on the manuscript. I also thank Elizabeth Zimmer and Eric Roalson for the invitation to write this chapter. The author is supported by BBSRC grant no. 40/G18385.

References

Arvestad, L., Berglund, A.-C., Lagergren, J., and Sennblad, B. (2003). Bayesian gene/species tree reconciliation and orthology analysis using MCMC. *Bioinformatics* **19,** 7–15.
Bapteste, E., Brinkmann, H., Lee, J. A., Moore, D. V., Sensen, C. W., Gordon, P., Durufle, L., Gaasterland, T., Lopez, P., Muller, M., and Philippe, H. (2002). The analysis of 100 genes supports the grouping of three highly divergent amoebae: *Dictyostelium, Entamoeba,* and *Mastigamoeba. Proc. Natl. Acad. Sci. USA* **99,** 1414–1419.
Benton, M. J., and Ayala, F. J. (2003). Dating the tree of life. *Science* **300,** 1698–1700.
Brower, A. V. Z., DeSalle, R., and Vogler, A. (1996). Gene trees, species trees and systematics: A cladistic perspective. *Annu. Rev. Ecol. Syst.* **27,** 423–450.

Cannon, S. B., and Young, N. D. (2003). OrthoParaMap: Distinguishing orthologs from paralogs by integrating comparative genome data and gene phylogenies. *BMC Bioinformatics* **4**, 35.

Charleston, M. A. (1998). Jungles: A new solution to the host/parasite phylogeny reconciliation problem. *Math. Biosci.* **149**, 191–223.

Chen, K., Durand, D., and Farach-Colton, M. (2000). NOTUNG: A program for dating gene duplications and optimizing gene family trees. *J. Comput. Biol.* **7**, 429–447.

Cotton, J. A. (2003). "Vertebrate phylogenomics and gene family evolution." Ph.D. Thesis, University of Glasgow, Glasgow, U.K. Available at: http://taxonomy.zoology.gla.ac.uk/ ~jcotton/thesis.htm.

Cotton, J. A., and Page, R. D. M. (2003). Gene tree parsimony vs. uninode coding for phylogenetic reconstruction. *Mol. Phylog. Evol.* **29**, 298–308.

Cotton, J. A., and Page, R. D. M. Tangled trees from molecular markers: Reconciling conflict between phylogenies to build molecular supertrees. *In* "Phylogenetic Supertrees: Combining Information to Reveal the Tree of Life" (O. R. P. Bininda-Emonds, ed.) pp. 107–125. Kluwer Academic, Dordrecht, The Netherlands.

Cruickshank, R. H. (2002). Molecular markers for the phylogenetics of mites and ticks. *Syst. Appl. Acarol.* **7**, 3–14.

Duret, L., Mouchiroud, D., and Gouy, M. (1994). HOVERGEN: A database of homologous vertebrate genes. *Nucleic Acids Res.* **22**, 2360–2365.

Eisen, J. A. (1998). Phylogenomics: Improving functional predictions for uncharacterized genes by evolutionary analysis. *Genome Res.* **8**, 163–167.

Eulenstein, O. (1997). "A Linear Time Algorithm for Tree Mapping." St Augustine, Germany.

Felsenstein, J. (1978). A likelihood approach to character weighting and what it tells us about parsimony and compatibility. *Biol. J. Linn. Soc.* **16**, 183–196.

Felsenstein, J. (2004). "Inferring Phylogenies." Sinauer, Sunderland, MA.

Fitch, W. M. (1970). Distinguishing homologous from analogous proteins. *Syst. Zool.* **19**, 99–113.

Fitch, W. M. (1979). Cautionary remarks on using gene expression events in parsimony procedures. *Syst. Zool.* **28**, 375–379.

Fitch, W. M. (2000). Homology: A personal view on some of the problems. *Trends Genet.* **16**, 227–231.

Fitch, W. M., and Margoliash, E. (1967). Construction of phylogenetic trees. *Science* **155**, 279–284.

Force, A., Lynch, M., Pickett, F. B., Amores, A., Yan, Y. L., and Postlethwait, J. (1999). Preservation of duplicate genes by complementary, degenerate mutations. *Genetics* **151**, 1531–1545.

Furlong, R. F., and Holland, P. W. H. (2002). Were vertebrates octoploid? *Philos. Trans. R. Soc. Lond. Series B* **357**, 531–544.

Goodman, M., Czelusniak, J., William-Moore, G., Romero-Herrera, A. E., and Matsuda, G. (1979). Fitting the gene lineage into its species lineage: A parsimony strategy illustrated by cladograms constructed from globin sequences. *Syst. Zool.* **28**, 132–168.

Gray, G. S., and Fitch, W. M. (1983). Evolution of antibiotic resistance genes: The DNA sequence of a kanamycin resistance gene from *Staphylococcus aureus*. *Mol. Biol. Evol.* **1**, 57–66.

Guigó, R., Muchnik, I., and Smith, T. F. (1996). Reconstruction of ancient molecular phylogeny. *Mol. Phylog. Evol.* **6**, 189–213.

Hallett, M. T., and Lagergren, J. (2000). New algorithms for the duplication-loss model. *In* "RECOMB '00, the Fourth Annual International Conference on Computational

Molecular Biology" (R. Shamir, S. Miyano, S. Istrail, P. Pevzner, and M. Waterman, eds.), pp. 138–146. Association for Computing Machinery, New York.

Hallett, M. T., and Lagergren, J. (2001). Efficient algorithms for lateral gene transfer problems. *In* "RECOMB '01, Proceedings of the Fifth Annual International Conference on Computational Molecular Biology" (T. Lengauer, ed.), pp. 149–156. Association for Computing Machinery, New York.

Harel, D., and Tarjan, R. E. (1984). Fast algorithms for finding nearest common ancestors. *SIAM J. Comput.* **13**, 338–355.

Hayashi, Y., Sakata, H., Makino, Y., Urabe, I., and Yomo, T. (2003). Can an arbitrary sequence evolve towards acquiring a biological function? *J. Mol. Evol.* **56**, 162–168.

Henikoff, S., Greene, E. A., Petrokovski, S., Bork, P., Attwood, T. K., and Hood, L. (1997). Gene families: The taxonomy of protein paralogs and chimaeras. *Science* **278**, 609–614.

Huelsenbeck, J. P., Rannala, B., and Yang, Z. H. (1997). Statistical tests of host–parasite cospeciation. *Evolution* **51**, 410–419.

Huelsenbeck, J. P., and Ronquist, F. (2001). MrBayes: Bayesian inference of phylogenetic trees. *Bioinformatics* **17**, 754–755.

Katoh, K., and Miyata, T. (2002). Cyclostome hemoglobins are possibly paralogous to gnathostome hemoglobins. *J. Mol. Evol.* **55**, 246–249.

Kubo, T., and Isawa, Y. (1995). Inferring the rates of branching and extinction from molecular phylogenies. *Evolution* **49**, 694–704.

Kumar, S., and Hedges, S. B. (1998). A molecular timescale for vertebrate evolution. *Nature* **392**, 917–920.

Kunin, V., Cases, I., Enright, A. J., Lorenzo, V. D., and Ouzounis, C. A. (2003). Myriads of protein families, and still counting. *Genome Biol.* **4**, 401.

Li, W.-H., and Graur, D. (1991). "Fundamentals of Molecular Evolution." Sinauer, Sunderland, MA.

Long, M. (2003). Preface. *Genetica* **118**, 97.

Long, M., and Thornton, K. (2001). Gene duplication and evolution. *Science* **293**, 1551.

Lynch, M., and Conery, J. S. (2000). The evolutionary fate and consequences of duplicate genes. *Science* **290**, 1151–1155.

Maddison, W. P. (1997). Gene trees in species trees. *Syst. Biol.* **46**, 523–536.

Martin, A. P. (2000). Choosing among alternative trees of multi-gene families. *Mol. Phylog. Evol.* **16**, 430–439.

Martin, A. P., and Burg, T. M. (2002). Perils of paralogy: using HSP70 genes for inferring organismal phylogenies. *Syst. Biol.* **51**, 570–587.

Martin, W. (1999). Mosaic bacterial chromosomes: A challenge en route to a tree of genomes. *BioEssays* **21**, 99–104.

Meyer, A., and Van de Peer, Y. (2003). 'Natural selection merely modified while redundancy created'—Susumu Ohno's idea of the evolutionary importance of gene and genome duplications. *J. Struct. Funct. Genomics* **3**, vii–ix.

Nadeau, J. H., and Sankoff, D. (1997). Comparable rates of gene loss and functional divergence after genome duplications early in vertebrate evolution. *Genetics* **147**, 1259–1266.

Nahon, J.-L. (2003). Birth of "human-specific" genes during primate evolution. *Genetica* **118**, 193–208.

Nee, S., Mooers, A. Ø., and Harvey, P. H. (1992). Tempo and modes of evolution revealed from molecular phylogenies. *Proc. Natl. Acad. Sci. USA* **89**, 8322–8366.

Ochman, H. (2001). Lateral gene transfer and the nature of bacterial innovation. *Curr. Opin. Genet. Dev.* **11**, 616–619.

Ohno, S. (1970). "Evolution by Gene Duplication." Springer-Verlag, Berlin.

Otto, S. P., and Whitton, J. (2000). Polyploid incidence and evolution. *Annu. Rev. Genet.* **34,** 401–437.

Page, R. D. M. (1994). Maps between trees and cladistic analysis of historical associations among genes, organisms and areas. *Syst. Biol.* **43,** 58–77.

Page, R. D. M. (1996). On consensus, confidence, and "total evidence." *Cladistics* **12,** 83–92.

Page, R. D. M. (1998). GeneTree: Comparing gene and species phylogenies using reconciled trees. *Bioinformatics* **14,** 819–820.

Page, R. D. M. (2000). Extracting species trees from complex gene trees: Reconciled trees and vertebrate phylogeny. *Mol. Phylog. Evol.* **14,** 89–106.

Page, R. D. M. (2003). Introduction. *In* "Tangled Trees: Phylogeny, Cospeciation and Coevolution" (R. D. M. Page, ed.), pp. 1–21. University of Chicago Press.

Page, R. D. M., and Charleston, M. A. (1997). Reconciled trees and incongruent gene and species trees. *In* "Mathematical Hierarchies in Biology" (B. Mirkin, F. R. McMorris, F. S. Roberts, and A. Rzhetsky, eds.), pp. 57–71. American Mathematical Society, Providence, RI.

Page, R. D. M., and Charleston, M. A. (1998). Trees within trees: Phylogeny and historical associations. *Trends Ecol. Evol.* **13,** 356–359.

Page, R. D. M., and Cotton, J. A. (2000). GeneTree: A tool for exploring gene family evolution. *In* "Comparative Genomics: Empirical and Analytical Approaches to Gene Order Dynamics, Map Alignment and the Evolution of Gene Families" (D. Sankoff and J. H. Nadeau, eds.), pp. 525–536. Kluwer Academic Publishers, Dordrecht.

Page, R. D. M., and Cotton, J. A. (2002). Vertebrate phylogenomics: Reconciled trees and gene duplications. *In* "Proceedings of the Pacific Symposium on Biocomputing 2002" (R. B. Altman, A. K. Dunker, L. Hunter, K. Lauderdale, and T. E. Klein, eds.), pp. 536–547. World Scientific Publishing, Singapore.

Rodriguez-Trelles, F., Tarrio, R., and Ayala, F. J. (2002). A methodological bias toward overestimation of molecular evolutionary time scales. *Proc. Natl. Acad. Sci. USA* **99,** 8112–8115.

Rokas, A., Williams, B. L., King, N., and Carroll, S. B. (2003). Genome-scale approaches to resolving incongruence in molecular phylogenies. *Nature* **42,** 798–804.

Ronquist, F. (2003). Parsimony analysis of coevolving species associations. *In* "Tangled Trees: Phylogeny, Cospeciation and Coevolution" (R. D. M. Page, ed.), pp. 22–64. University of Chicago Press, Chicago.

Sanderson, M. J. (1989). Confidence limits on phylogenies: the bootstrap revisited. *Cladistics* **5,** 113–129.

Sanderson, M. J., and Shaffer, H. B. (2002). Troubleshooting molecular phylogenetic analyses. *Annu. Rev. Ecol. Syst.* **33,** 49–72.

Semple, C., and Wolfe, K. H. (1999). Gene duplication and gene conversion in the *C. elegans* genome. *J. Mol. Evol.* **48,** 555–564.

Shaul, S., and Graur, D. (2002). Playing chicken (*Gallus gallus*): Methodological inconsistencies of molecular divergence date estimates due to secondary calibration points. *Gene* **300,** 59–61.

Sicheritz-Pontén, T., and Andersson, S. G. E. (2001). A phylogenomic approach to microbial evolution. *Nucleic Acids Res.* **29,** 545–552.

Simmons, M. P., Bailey, C. D., and Nixon, K. C. (2000). Phylogeny reconstruction using duplicate genes. *Mol. Biol. Evol.* **17,** 469–473.

Simmons, M. P., and Freudenstein, J. V. (2002). Uninode coding vs gene tree parsimony for phylogenetic reconstruction using duplicate genes. *Mol. Phylog. Evol.* **23,** 481–498.

Skrabanek, L., and Wolfe, K. H. (1998). Eukaryotic genome duplication—where's the evidence? *Curr. Opin. Genet. Dev.* **8,** 694–700.

Slowinski, J. B., and Page, R. D. M. (1999). How should phylogenies be inferred from sequence data? *Syst. Biol.* **48,** 814–825.

Storm, C. E. V., and Sonnhammer, E. L. L. (2002). Automated ortholog inference from phylogenetic trees and calculation of orthology reliability. *Bioinformatics* **18,** 92–99.

Taylor, J. S., Van de Peer, Y., Braasch, I., and Meyer, A. (2001). Comparative genomics provides evidence for an ancient genome duplication event in fish. *Philos. Trans. R. Soc. Lond.* **356,** 1661–1679.

Thorley, J. L., and Wilkinson, M. (2003). A view of supertree methods. *In* "Bioconsensus" (M. F. Janowitz, F. J. Lapoine, F. R. McMorris, and F. S. Roberts, eds.), pp. 185–194. American Mathematical Society, Providence, RI.

Walsh, J. B. (1995). How often do duplicated genes evolve new functions? *Genetics* **139,** 421–428.

Waterman, M. S., and Smith, T. F. (1978). On the similarity of dendrograms. *J. Theor. Biol.* **73,** 789–800.

Wolfe, K. H., and Li, W.-H. (2003). Molecular evolution meets the genomics revolution. *Nat. Genet.* **33**(suppl), 255–265.

Zhang, L. (1997). On a Mirchkin-Muchnik-Smith conjecture for comparing molecular phylogenies. *J. Comput. Biol.* **4,** 177–187.

Zmasek, C. M., and Eddy, S. R. (2001a). ATV: Display and manipulation of annotated phylogenetic trees. *Bioinformatics* **17,** 383–384.

Zmasek, C. M., and Eddy, S. R. (2001b). A simple algorithm to infer gene duplication and speciation events on a gene tree. *Bioinformatics* **17,** 821–828.

Zmasek, C. M., and Eddy, S. R. (2002). RIO: Analyzing proteomes by automated phylogenomics using resampled inference of orthologs. *BMC Bioinformatics* **3,** 14.

[37] Analytical Methods for Studying the Evolution of Paralogs Using Duplicate Gene Datasets

By Sarah Mathews

Abstract

Gene duplication is widely viewed as an important source of raw material for functional innovation in proteins because at least some duplicate copies will evolve new or slightly modified functions. The study of the molecular processes by which functional innovation occurs interests both evolutionary biologists and protein chemists, and the development of methods to investigate these processes has led to a productive meeting of disciplines and an availability of complementary approaches for exploring datasets. This has resulted in insights into past events, prediction of current function, and prediction of future change. The methods fall broadly into two categories: those that rely on detection of shifts in selective constraints and those that rely on detection of correlations between molecular changes and functional shifts. Strengths and limitations of the methods are evaluated here in the context of the question being addressed, the input required, and the specific metric that is evaluated in each test.

Introduction

Phylogenetic trees of duplicate genes provide a rich source of data for studies of evolution at both molecular and phenotypic levels, yielding insights into the processes by which genes and the characters they encode have evolved. Immediately following a gene duplication, two identical copies, referred to as *paralogs* (Fitch, 1970), exist in a single genome. Subsequently, various fates for the paralogs are possible. At one extreme, high sequence identity across both coding and regulatory regions may be maintained, and gene function may be redundant. At the opposite extreme, the coding region of one copy may degenerate and it may become a pseudogene (Haldane, 1933; Lynch and Connery, 2000; Walsh, 1995), although this does not necessarily indicate that the sequence retains no function (Hirotsune *et al.*, 2003). Alternatively, if the ancestral gene encoded multiple functions, mutations may result in the loss of a subset of the functions, or they may lead to a subdivision of the ancestral functions (Force *et al.*, 1999; Hughes, 1994; Lynch and Force, 2000). Finally, mutations may lead to the origin of a novel function (Ohno, 1970). In each of these cases, different selective pressures are expected to predominate. Several approaches have been developed for using multiple species alignments to characterize the selective processes by which single genes evolve and to generate experimentally testable hypotheses about changes at the sequence level that might be associated with functional or phenotypic divergence. These are readily applied to alignments of multiple paralogs for the study of how genes evolve after duplication. Additional approaches have been developed specifically for use with datasets of duplicate genes. Some of these rely on detecting shifts in selective constraints while others do not. Although it also is possible to characterize variable selective pressures acting on paralogs in a single genome (Martínez-Castilla and Alvarez-Buyalla, 2003) or acting on a large number of genes, related and unrelated, across a whole genome (Plotkin *et al.*, 2004), this chapter focuses on examples that use alignments of multiple paralogs sampled from multiple species.

A direct assessment of selective pressure at the molecular level is based on the ratio of nonsynonymous (replacement) substitutions per nonsynonymous site (d_N) to synonymous (silent) substitutions per synonymous site (d_S). In many cases, d_S exceeds d_N, indicating the influence of negative or purifying selection. However, if selective pressures are relaxed and the sequence is evolving neutrally, the ratio of d_N to d_S $(d_N/d_S = \omega)$ will be elevated, approaching but not exceeding one. If a sequence is evolving under positive or diversifying selection, d_N may be greater than d_S, suggesting that nonsynonymous substitutions are being fixed with a higher probability than synonymous substitutions. Thus, finding that d_N significantly exceeds d_S has been considered strong evidence of positive selection (Hill and Hastie, 1987; Hughes and Nei, 1988, 1989).

Tests based on d_N/d_S provide insight into variable selection among paralogs and among sites and allow the temporal patterns of shifts in selective pressures on lineages and/or sites to be determined. Characterization of these patterns may provide data that are generally descriptive of molecular evolution or that test explicit hypotheses about the fates of paralogs. For example, if paralogs encode highly redundant proteins of equal functional importance, there should be evidence of purifying selection acting on both copies to maintain high sequence identity. If a duplicate is only partially redundant, or is a pseudogene, selective constraints on one copy should be relaxed. If paralogs subdivide ancestral functions (Force et al., 1999; Hughes, 1994; Lynch and Force, 2000), purifying selection on both copies may predominate. This is because subfunctionalization may often occur by changes in regulatory regions external to the coding sequences, leading, for example, to spatial or temporal changes in patterns of gene expression (Force et al., 1999). In cases in which changes in coding sequences determine the mechanism of subfunctionalization (Henikoff et al., 1997; Lynch and Force, 2000), selective pressures may change at a small number of codons. If a duplicate evolves a novel function, positive selection may play an important role, but here, too, a small number of sites may be involved.

Although tests based on d_N/d_S may in general reveal the processes by which genes diverge, their utility diminishes as the age of the paralogs increases. This is because mutations at synonymous sites generally occur at higher rates than mutations at nonsynonymous sites. Thus, synonymous sites tend to become saturated with mutations more rapidly than nonsynonymous sites, and d_N/d_S will be a less useful indicator of elevated replacement substitutions. Tests based on d_N/d_S also may not identify all sites at which constraints have changed. One reason is that multiple species alignments retain a limited amount of information about microevolutionary processes; it is difficult to detect episodic changes in selective pressure at a small number of sites. If a few advantageous changes are rapidly fixed, an elevation in d_N may be difficult to detect in interspecific comparisons. Conversely, if a gene or gene region is under ongoing pressure to diversify, nonsynonymous change may occur more often through time and be more detectable in interspecific comparisons. Finally, it is important to remember that these tests rely on the assumption that synonymous changes are neutral, an assumption that is violated in some cases.

An alternative class of approach rests on the assumption that a shift in rates across sites indicates a change in selective constraints, that is, that the mutability of individual sites may change if protein function changes. A fundamental assumption about protein evolution is that at least some sites will be selectively constrained in order to conserve the identity of functionally important residues, whereas constraints at less critical sites will be relaxed. Thus, across a coding sequence, substitution rates will vary. Variation in rates across sites is commonly modeled using the gamma distribution, with the shape parameter, α, describing the degree of rate variation. The model assumes that rates at individual sites remain the same over time, which is reasonable if functional constraints at sites do not change. However, during the divergence of paralogs, constraints at individual sites might change if the function of one or both copies is changing. In these cases, covarion or covariotide models (Fitch and Markowitz, 1970; Shoemaker and Fitch, 1989) may be more useful. Specifically, these models account for the observation that there is a fraction of sites (codon or nucleotide) that is invariable and that although the fraction of sites that is variable may remain relatively constant through time, the identity of these sites may change. For example, if a variable site becomes fixed, a previously invariable site may become variable. Thus, detection of such shifts may provide insights into functional divergence among paralogs. Moreover, these shifts can be detected based on amino acid replacements alone, so these methods are well suited to the study of more ancient paralogs.

A third class of approach is based on tracing amino acid changes on trees and/or protein structures and does not rely on the detection of shifts in selective constraints. One of these, evolutionary tracing, is motivated by the goal of locating functional surfaces and understanding the role of constituent residues (Bennner, 1989; Landgraf et al., 1999, 2001; Lichtarge et al., 1996; Madabushi et al., 2004). It starts with a structural model of a protein or protein domain and seeks to identify amino acid replacements that are associated with functional differences between copies. An alternative approach is explicitly tree based and can be implemented in the absence of detailed structural knowledge. In these studies, a tree inferred from and well supported by whole sequences is the framework for a rigorous analysis of character evolution that provides insight into the correlations between functional shifts and specific amino acid changes. A subset of the replacements detected in character mapping is considered for further analyses to test functional hypotheses. In both approaches, it is difficult to discriminate between replacements that are paralog specific because of their role in functional divergence or because of chance alone. Despite this potential drawback, the predictive power of the approach has been demonstrated (Madabashi et al., 2004). One way to minimize this problem is to focus on mutations that have multiple independent origins on a tree and that are associated with the same functional shift (Chang *et al.*, 1995). An advantage of both methods of tracing is that they identify character states that are fixed in each copy but that differ between them. In an interesting variation of the approach that uses whole sequences, the entire sequence at a tree node of interest is reconstructed and synthesized and the properties of this putatively ancient protein sequence are examined (Adey *et al.*, 1994; Bennner, 2002; Chandrasekharan *et al.*, 1996; Chang, 2003; Chang *et al.*, 2002; Jermann *et al.*, 1995; Shi and Yokoyama, 2003; Stewart, 1995; Sun *et al.*, 2002; Ugalde *et al.*, 2004; Zhang and Rosenberg, 2002).

Methods Based on d_N/d_S

Methods Based on d_N/d_S to Detect Variable Selection among Lineages

Given a tree of duplicate genes that are diverging from a common ancestor under similar evolutionary constraints, we would expect the values of d_N/d_S to be similar on all branches in the tree. Specifically, we expect d_N/d_S to be similar within and between gene clades or clusters. We can test this by using pairwise comparisons of d_N/d_S (Hughes and Nei, 1988, 1989) and by using relative rate tests that separately test for elevated rates of d_N and d_S (Muse and Gaut, 1994; Muse and Weir, 1992). These approaches are easily implemented, allow quick detection of altered constraints, and can identify a group of sequences in which rates are elevated. However, they are not designed to identify the branch in the gene tree in which rates shifted. Thus, branch tests based on d_N/d_S have been proposed. Two of these (Messier and Stewart, 1997; Zhang *et al.*, 1997) rely on the reconstruction of ancestral sequences at internal nodes in the tree, counting of nonsynonymous (N) and synonymous (S) sites, estimation of changes in d_N and d_S along each branch, and a statistical test of the significance in the difference between d_N and d_S. A third test takes a likelihood approach using codon-based substitution models that allow for different d_N/d_S values among branches and significance is tested in likelihood ratio tests (Yang, 1998).

Messier and Stewart (1997) studied the evolution of primate lysozymes to determine whether changes in enzyme function might be associated with nonneutral evolution. Their branch test consisted of three steps: (1) Maximum likelihood (ML) and maximum parsimony (MP) were used to reconstruct ancestral DNA sequences at internal nodes. (2) Pairwise comparisons between sequences at nodes were used to estimate d_N/d_S by the method of Li (1993) for each branch. (3) A *t* test was used to determine the significance of the differences between d_N and d_S. They found that MP and ML inferred identical ancestral sequences and that d_N was significantly greater than d_S on two branches of the lysozyme tree, suggesting that positive selection had occurred in the represented lineages.

Zhang et al. (1997) examined the same dataset to address the concern that use of the t test might be inappropriate because it assumes that d_N and d_S are normally distributed. They simulated data in order to test this and suggested that numbers of both synonymous and nonsynonymous substitution must exceed about 10 in order to avoid a high type I error rate using the t test, a criterion not met by the lysozyme data. As an alternative, they used Fisher's exact test and found they could not reject neutrality on either of the branches on which Messier and Stewart (1997) had inferred positive selection. However, in analyses of other datasets using this approach, episodic selection was inferred after duplication of primate ribonuclease genes (Zhang et al., 1998) and after duplication of a triosephosphate isomerase (TPI) gene in teleosts (Merritt and Quattro, 2001). Except for the significance tests, the d_N/d_S-based branch test of Zhang et al. (1997, 1998) generally is similar to that of Messier and Stewart (1997). They differ in that Zhang et al. (1997, 1998) used Bayesian inference to reconstruct ancestral sequences, and the method of Ina (1995) or a modification of the method of Nei and Gojobori (1986) (Zhang et al., 1998) to estimate N and S.

Yang (1998) developed a likelihood-based approach that does not rely on reconstruction of ancestral sequences. The test, implemented in PAML (Yang, 1997), consists of three basic steps. (1) A series of likelihood analyses using a codon substitution model (Goldman and Yang, 1994) is used to estimate d_N/d_S under varying assumptions of heterogeneity in d_N/d_S values among branches in an unrooted gene tree. The simplest model, the one-ratio model, assumes the same ratio for all branches. A second model may assume that a branch of interest has a d_N/d_S value that differs from the background ratio (the ratio on the remaining branches). (2) Likelihood ratio tests are used to compare the fit of nested models to the data. For example, in the aforementioned case, a likelihood ratio test examines the null hypothesis that the background ratio is the same as the ratio on the branch of interest. (3) If the null hypothesis is rejected, and $d_N/d_S > 1$ along the branch of interest, comparison of the two-ratio models with and without the constraints that d_N/d_S on the branch of interest is 1 or less examines the possibility of positive selection in the lineage represented by this branch.

Yang (1998) also analyzed the lysozyme data, finding clear evidence for variable selection among lineages but evidence of positive selection on only one of the two branches on which Messier and Stewart (1997) had inferred positive selection. The ML approach addressed potential concerns about the approaches of Messier and Stewart (1997) and Zhang *et al.* (1997). First, the reconstruction of ancestral sequences may be biased or erroneous in some cases (Collins *et al.*, 1994; Yang *et al.*, 1995; Zhang and Nei, 1997). Second, the methods used to estimate substitution rates did not account for biases in codon usage and the transition/transversion rate, and this could compromise the estimation of d_N and d_S (Ina, 1995; Li *et al.*, 1985; Yang and Nielsen, 2000). Third, when the test is applied to many or all branches of the tree, significance values must be adjusted to account for the performance of multiple tests on the same data. Although it is possible to incorporate more sophisticated substitution models to estimate d_N and d_S and to correct significance values for multiple tests, the reconstruction of ancestral sequences remains problematic and becomes more difficult as divergence among sequences increases. When multiple states are inferred for a site at a node, the power of d_N/d_S-based branch tests decreases (Zhang *et al.*, 1998).

Branch tests are useful because they may reveal *when* selective pressures shifted. However, they are very conservative tests for selection because they assume that all amino acid sites are under the same selective pressure, averaging ω over all sites. Thus, positive selection will be detected only if it has influenced a large number of sites.

Methods Based on d_N/d_S to Detect Variable Selection among Sites

If amino acid identity is critical to the function of the encoded protein, evolution of the corresponding codon will be constrained by negative selection. Conversely, if a residue is important in recognition processes, it may be under ongoing pressure to diversify, accomplished by positive selection at that site. Alternatively, changes at specific residues may accompany the origin of a novel advantageous function, some of which may be fixed by positive selection. However, nucleotides at many positions will be evolving relatively free of constraints because their identity is less critical to protein function. Thus, characterizing the patterns of site-specific selection provides (1) insight into the relative roles that different selective forces play in the origin and maintenance of differences among paralogs that are known to involve functionally important sites, (2) evidence of functionally important sites where this information is lacking, and (3) evidence from which selective agents can be postulated.

Given an alignment of paralogs from multiple species, several approaches are available for detecting variable selection among sites and these are broadly divisible into those that are alignment based and those that are tree based. Alignment-based tests usually limit the number of codons over which d_N/d_S is averaged in order to increase the sensitivity of the test. When the identity of active sites or functionally important regions is known, it is possible to average d_N/d_S over the codons found in these regions (Hill and Hastie, 1987; Hughes and Nei, 1988, 1989). When such information is lacking, a sliding window analysis may be used, averaging d_N/d_S over 20 codons simultaneously (Endo *et al.*, 1996). These are conservative tests in that elevated ω values will be detected only when selectively favored mutations are clustered within a region (Endo *et al.*, 1996).

Tree-based tests average d_N/d_S of a single site over the entire evolutionary time period that separates genes. Fitch *et al.* (1997) and Bush *et al.* (1999) used a parsimony-based approach to investigate the evolution of the H3 hemagglutinin gene of human influenza virus A. Their test (1) used parsimony to infer a tree and ancestral states, (2) counted the number of replacements per codon position, and (3) inferred positive selection when the distribution of replacements among codons departed from random. They also separately estimated the distribution of replacements among codons for trunk (the strains that persist from year to year), twig, and tip branches, finding evidence that replacements were unevenly distributed among these classes of branch, with trunk branches having the fewest.

Nielsen and Yang (1998) developed codon models for the estimation of d_S and d_N that allow for selection to vary among sites according to different statistical distributions. The test, implemented in PAML (Yang, 1997), consists of three basic steps. (1) A series of likelihood analyses using different codon models is used to estimate d_N/d_S under various assumptions about heterogeneity of d_N/d_S among sites. For example, the neutral model assumes that sites are either evolving neutrally or under negative selection while the selection model allows for a category of sites with $d_N/d_S > 1$. (2) The two models are compared in a likelihood ratio test to determine whether there is evidence of a class of positively selected sites. (3) If there is evidence of positive selection, a Bayesian approach is used to identify which sites fall into this class. Using this approach in a study of the HIV-1 envelope gene, Nielsen and Yang (1998) found evidence for positive selection at a small number of sites in the V3 and neighboring regions. They also found that the results varied considerably among strains from different years, with only one site, which was outside the V3 region, under selection in all years. In a subsequent study, Yang *et al.* (2000) developed additional codon models for the detection of variable selection among sites.

Suzuki and colleagues (Suzuki and Gojobori, 1999; Suzuki *et al.*, 2001) also developed a parsimony-based approach for detecting positive selection at individual amino acid sites. Their test, implemented in ADAPT-SITE (Suzuki *et al.*, 2001), (1) used parsimony to infer ancestral codons at each internal node of a neighbor-joining tree based on synonymous substitutions, (2) counted the total number of synonymous and nonsynonymous changes using the method of Nei and Gojobori (1986), and (3) inferred negative selection when d_S significantly exceeded d_N and positive selection when the opposite was found. Because the test is applied to multiple sites, a correction for multiple tests on the same data should be performed. Suzuki and Gojobori (1999) also analyzed the HIV-1 envelope gene, inferring positive selection at four sites in the V3 region.

Performance of the Sites Tests. Suzuki and Gojobori (1999) found that the efficiency of their sites test increased with the increase of the (1) selective force, (2) number of sequences, and (3) branch lengths. However, with the increase of sequence number and branch lengths, the number of codons that can be statistically tested becomes small because there are many sites at which there are more than 10,000 possible combinations of ancestral codons across the nodes in the tree (Suzuki and Gojobori, 1999). Thus, where the test should perform the best, its utility is limited. Suzuki and Gojobori (1999) recommended that total branch length in the tree should be increased while keeping individual branch lengths relatively short. Conversely, they also noted that too many closely related sequences could confound phylogenetic inference and possibly compromise estimation of N and S. Because the test requires reconstruction of ancestral states at each node of the tree, inference of sites under selection will be affected by tree topology.

The approach developed by Nielsen and Yang (1998) for identifying positively selected sites consists of two basic steps of increasing difficulty. First, likelihood ratio tests are used to detect variable selective constraints among sites, and second, an empirical Bayes approach is used to identify positively selected sites. Yang et al. (2000) found that both the LRTs for detecting selection and the identification of selected sites were robust to changes in tree topology, and their accuracy and power were examined in additional studies (Anisimova et al., 2001, 2002). For detecting variable selection, these studies revealed that (1) the likelihood ratio tests used to detect positive selection were very conservative, (2) the probability of detecting selection was higher for longer sequences, (3) power increases with sequence divergence up to a maximal value, but decreases if sequences are more divergent than this value, (4) power increased with sequence number, and (5) power increased with selective force (Anisimova et al., 2001). For identification of sites under selection, they found that (1) the Bayes approach is unreliable when sequences are few and very similar, (2) increasing the number of sequences improved both accuracy and power, even when sequences were similar, and (3) robustness of the results should be explored using different models, and for some models, running the analysis more than once with different starting values of d_N/d_S

is necessary (Anisimova *et al.*, 2002). The poor performance of Bayes prediction of sites in small datasets of similar sequences may in part explain the controversial results from analyses of the *Sig1* gene from *Thalassiosira weissflogii* (Sorhannus, 2003; Suzuki and Nei, 2004). Recombination at levels more than 10% affects the accuracy of the test (Anisimova *et al.*, 2003) and this should be kept in mind when exploring population data.

The most thorough comparison of the accuracy and power of the MP-based methods (Suzuki and Gojobori, 1999) and the ML-based methods (Nielsen and Yang, 1998; Yang *et al.*, 2000) for detecting adaptive evolution and identifying positively selected sites was conducted by Wong *et al.* (2004). Several important considerations emerge from this study. First, as previously noted (Sorhannus, 2003; Yang *et al.*, 2000), failure to find optimal starting parameter estimates for the ML analyses can lead to erroneous results, based on suboptimal peaks in parameter space. This is particularly important to consider when implementing the more realistic or complex models of codon evolution. Second, neither method could reliably distinguish between sites evolving neutrally ($\omega = 1.0$) and sites under weak positive selection ($\omega = 1.5$). Third, when properly implemented, the ML approach is a powerful test for positive selection *and* accurately identifies positively selected sites. Fourth, the MP approach is very conservative and may fail to detect selection even when sites are under strong selection ($\omega = 5.0$) unless the dataset is very large. In analyses of the simulated datasets, none of the sites under strong positive selection were identified using ADAPTSITE (Wong *et al.*, 2004).

It is important to remember that characterizing selective forces at individual sites is a demanding task. The ML approach is a powerful method for detecting the presence of variable selective pressure among sites, but it may sometimes fail to properly identify positively selected sites. It always should be implemented after an exploration of different initial parameter values in order to identify those that are optimal, and results from tests using different models should be compared. It is the most flexible method for exploring datasets over a range of sizes and degrees of sequence divergence. The MP approach has low power for detecting the presence of variable selection and for identifying selected sites but should be useful for exploring large datasets with sites under strong positive selection.

Methods Based on d_N/d_S *to Detect Variable Selection among Sites along Specific Lineages*

The tests based on d_N/d_S for variable selection among lineages and for variable selection among sites are both conservative tests. In the former case, d_N/d_S is averaged over all sites in a sequence, whereas in the latter case, d_N/d_S at each site is averaged over all branches in a tree. For this reason, it is difficult for the tests to detect episodic selection involving just a few sites (Suzuki and Gojobori, 1999; Yang *et al.*, 2000). To address this concern, Yang and Nielsen (2002) developed models that allow the d_N/d_S ratio to vary both among sites and among lineages. The branch-site models assume that each amino acid site falls into one of four classes of d_N/d_S ratio. Sites that are highly conserved across all branches comprise a class having a small d_N/d_S ratio (ω_0), whereas sites that are neutral or weakly constrained across all branches comprise a class having a ratio near or below 1 (ω_1). Sites that are highly conserved on background branches (ω_0) but that have a d_N/d_S ratio that may be greater than 1 (ω_2) on the branch being tested for selection, the foreground branch, comprise a third class. Sites that are neutral or weakly constrained (ω_1) on the background branches but that may have a ratio greater than 1 (ω_2) on the foreground branch comprise a fourth class. Model A assumes that $\omega_0 = 0$ and that $\omega_1 = 1$, whereas model B allows ω_0 and ω_1 to vary. Both models estimate ω_2 from the data. LRTs compare models A and B with site-specific models M1 and M3, respectively. Models M1 and M3 are identical to models A and B, respectively, except that they assume that ω_0 and ω_1 are the same across all branches of the phylogeny. If $\omega_2 > 1$, and if either model A or B fits the data significantly better than model M1 or model M3, then these LRTs constitute a test of positive selection in a subset of the amino acids on the foreground branch. The specific sites that change along the foreground branch are identified using an empirical Bayes approach, which estimates the posterior probabilities that a site falls into a particular site class (Nielsen and Yang, 1998).

Performance of the Branch-Sites Tests. Zhang (2004) analyzed simulated data sets to examine the accuracy of the branch-sites test (Yang and Nielsen, 2002), concluding that the method generally gives misleading results. However, the analyses of the simulated datasets was not well enough described to determine whether the test was implemented after identification of optimal starting values of parameter estimates. This makes the results difficult to interpret, but they are consistent with the observations about the performance of the sites tests (Wong *et al.*, 2004). For example, Wong *et al.* (2004) found that the test falsely identified positively selected sites when selection is very relaxed, a condition under which both the MP and ML sites tests are known to perform poorly. This finding that the method is sensitive to violation in assumptions about the number of ω classes also is consistent with the observation that estimating the distribution of ω across sites is difficult (Yang *et al.*, 2000). Thus, as with the sites tests, it is important to keep this computational difficulty in mind, as well as the difficulty of using the Bayes calculation of posterior probabilities to identify selected sites (Yang, 2002). For these reasons, it is important to interpret the results with caution and to examine the results in light of multiple lines of evidence.

Methods Based on Using Shifts in Rates across Sites to Detect Changes in Selective Constraints

As noted earlier, the utility of methods based on detecting changes in d_N/d_S decreases with the increasing age of paralogs, and they may be less effective in cases in which assumptions about nucleotide evolution at synonymous sites are violated. In these cases, detection of shifts in rates across sites may be particularly useful. Both tree-based and alignment-based approaches have been developed for detecting sites or regions at which constraints between paralogs have been altered.

Gu (1999, 2001) assessed changes in the positions of invariable sites by estimating the coefficient of rate correlation of sites between paralogs. If constraints are not altered after gene duplication, the coefficient of rate correlation (r_λ) equals 1, whereas altered constraints will reduce the correlation. The probability that the substitution rate at a site is statistically independent between two gene subtrees is assessed by estimating θ ($= 1 - r_\lambda$). The test focuses on sites that are constant in one copy but variable in the other (referred to as *type I functional divergence*). The number of changes at each site is estimated by (1) inferring ancestral sequences and (2) using ML to infer branch lengths and the α shape parameter for the whole tree and each gene subtree. A likelihood ratio test compares the null hypothesis that $\theta = 0$ to the alternative that $\theta > 1$. If the test statistic is significant, a Bayesian approach is used to determine the probability that a site falls into the type I class. A variation of this approach uses MP to infer the number of changes at each site and to generate variability profiles for the estimation of θ (Naylor and Gerstein, 2000).

An alternative approach tests the assumption that the α shape parameter of the gamma distribution is stationary across a tree of duplicate genes (Gaucher *et al.*, 2001). A change in α between gene subtrees and the whole tree is expected when the positions of variable sites changes (Gu, 1999). The test (1) uses ML to estimate α and the rate of replacement per site assuming a best-fit model of protein evolution, (2) uses parametric bootstrapping to calculate the standard deviation of α for the whole tree and each gene subtree, (3) evaluates kurtosis to determine if rate differences per site follow a normal distribution, and (4) chooses sites in the tails of the distribution of rate differences (e.g., >2 SD faster or slower in one gene subtree compared to another) for evaluation in light of structural data in order to identify candidates for functional studies.

A third approach implements a method developed by Tang and Lewontin (1999) for locating sequence regions in which there are real differences in the density of variable sites. This method tests the null hypothesis that events are uniformly distributed along a sequence, where an event is a variable site. If the null hypothesis is rejected, the endpoints of the regions with higher or lower variability than expected are estimated. This region is omitted and the sequences are again tested for heterogeneity. These steps are repeated until no heterogeneity is found. Dermatzakis and Clark (2001) modified this approach in a study of genes that duplicated before the divergence of mice and humans. In pairwise comparisons of paralogs, they found that the regions of differential selection corresponded with domains likely to determine functional specificity. Marín et al. (2001) also used an approach to identify regions of differential constraint between paralogs, but they restricted their analysis to conserved regions of known function.

Conceptually, these methods fall into one of two categories, and this determines to some extent their relative merits and limitations. In the tests developed by Gu (1999, 2001) and Gaucher et al. (2001), shifts in rates across sites are inferred by estimating rates at individual sites and comparing them between gene subtrees. Thus, they may readily identify sites that are variable in one copy and conserved in another, but they will fail to detect sites at which change has led to the constant-but-different condition because rates may not shift at these sites. Some of these sites, however, may have implications for functional divergence. In the heterogeneity tests (Dermatzakis and Clark, 2001; Marín et al., 2001; Tang and Lewontin, 1999), shifts in rates across sites are inferred from shifts in patterns of heterogeneity across the sequence, but the rates themselves are not measured; they simply assess whether a cluster of sites that did not vary becomes variable or vice versa. Although these heterogeneity tests have the advantage that they will not exclude constant-but-different sites, they may fail to detect single changes that occur outside of a run of changes, some of which may have been critical in functional divergence.

Methods Based on Mapping Character Changes on Trees and/or Structures

The third class of approach considered here is conceptually similar to comparative methods of the category that seek to infer correlated character evolution. Central to these methods is the question of whether change in one character effects change in another character. With regard to paralog evolution, one question of interest is whether specific amino acid changes might have led to changes in protein function. Given a phylogenetic tree and data characterizing the distribution of both amino acid states and protein phenotypes, an effort is made to (1) identify potentially critical amino acid sites based on where on the tree they accept mutations and (2) test the effects of altering the identity of residues at these sites in mutational analyses to test their functional importance.

Evolutionary tracing aims explicitly to identify residues of potential functional importance that can be tested in mutational analyses (Lichtarge *et al.*, 1996). It relies on an X-ray or nuclear magnetic resonance structure to define the local structure to be investigated and on a gene tree to identify clusters of similar and/or related sequences. Starting with a three-dimensional model of a functional region, the approach comprises six steps. (1) Sequences encoding the domain of interest are gathered and aligned. (2) A tree is inferred to identify groups of sequences that are closely related and that might encode a similar function. (3) An evolutionary trace, or consensus sequence, is inferred from each group of sequences thus identified. At every position in this trace, there is either a completely conserved residue or a gap, the latter at any position that is variable. (4) The evolutionary traces from the different groups are compared to infer a single trace that has completely conserved residues, residues that are cluster specific (encoded as *X*s), or gaps. (5) This trace is mapped onto the structural model to determine the positions of these residues. (6) Mutational analyses are used to test the functional hypotheses inferred from the identity of the residues at specific positions. In steps 3 and 4, amino acid substitution matrices may be used to evaluate the degree to which amino acid identity is conserved (Landgraf *et al.*, 1999). The approach elegantly demonstrates the utility of taking into account paralog-specific changes at the sequence level in order to understand functional distinctions among paralogs. An important consideration is that the choice of cluster or clade size strongly influences inference of the trace, and the inference of the tree strongly influences the composition of the clusters or clades. For many large families of proteins or domains, a robust tree is difficult to infer. Thus, it may be important to inspect traces from increasingly smaller partitions of the tree to formulate reasonable functional hypotheses (Lichtarge *et al.*, 1996) or from trees of different topologies. Inspecting traces from increasingly smaller partitions of the tree also is useful for defining the limits of structural epitopes (Lichtarge *et al.*, 1996). For purely functional investigations, one possible limitation of this approach is that there may be regions or residues that are similar as a result of convergent or parallel evolution.

In these cases, functionally similar sequences may not occur together in a phylogenetic tree. To address this, Landgraf et al. (2001) developed a modification of evolutionary tracing that relies on a similarity score rather than a phylogenetic tree to predict functionally similar structures. This modified method has also proven useful in detecting clusters of residues that show similarity relationships that differ from other regions in the same proteins (Landgraf et al., 2001).

Although structural databases are expanding and more data from which to infer structural domains are available, insights into protein structure still range from highly resolved three-dimensional structures to little more than rough estimates of where functional domains occur along a sequence. An alternative approach is explicitly tree based and can be implemented in the absence of detailed structural knowledge. In these studies, the tree inferred from sequence data is the framework for a rigorous analysis of character evolution that provides insight into the correlations between functional shifts and specific amino acid changes. The goal of detecting paralog-specific amino acid residues is accomplished in three basic steps. (1) Using complete or nearly complete sequences, a phylogenetic analysis is conducted to infer a tree from the data. (2) Amino acid substitutions at individual amino acid sites are traced on the tree. (3) The traces of individual sites are evaluated to detect paralog-specific changes that might be associated with functional shifts. The success of this approach depends on the accurate reconstruction of ancestral sequences at nodes within the tree and thus requires sequence data that are phylogenetically informative. It also depends on knowing the distribution of functional states. Additionally, the method does not, as evolutionary tracing does not, discriminate between amino acid replacements that are paralog specific by chance alone and those that have a role in functional divergence. The probability that chance explains paralog-specific changes is decreased in cases in which convergent replacements are associated with convergent functional shifts (Chang et al., 1995). In other cases, the evaluation of amino acid replacements may be considered in light of detailed structural and functional data. Thus, although these data are not a prerequisite of the method, they greatly enhance the ability to formulate reasonable hypotheses for further testing in mutational analyses. In an interesting variation of this approach, the ancestral sequences inferred at particular nodes are synthesized in the lab to test hypotheses about the function of ancient proteins (Adey et al., 1994; Bennner, 2002; Chandrasekharan et al., 1996; Chang, 2003; Chang et al., 2002; Jermann et al., 1995; Shi and Yokoyama, 2003; Stewart, 1995; Sun et al., 2002; Ugalde et al., 2004; Zhang and Rosenberg,

2002). Character mapping is highly complementary to methods that rely on detection of altered selective constraints, either at the nucleotide or at the amino acid level because it is particularly good at identifying character states that differ between paralogs, but that are conserved across species, where fixation may have occurred rapidly enough to leave no signal in interspecific datasets.

Concluding Remarks

Table I summarizes the most widely used methods for analyses of duplicate gene data sets to study patterns of paralog evolution, organized according to the questions they are designed to address. It does not include approaches for analyzing single-genome sequences (Plotkin *et al.*, 2004), which are increasingly available and present new analytical challenges. It is clear from Table I that there is no single best approach. The appropriateness of any one method will depend on several features of the data themselves. These include the number and length of sequences available, the degree of divergence among sequences, the degree of divergence across the tree, the strength and frequency of selection, the distribution of ω across sites, and levels of recombination. Because the methods are complementary in their ability to detect changes at different classes of sites, they should be used in combination for a most thorough examination of the data, depending on the question of interest. For example, tests for variable selective pressure, based on either d_N/d_S or a shift in rates across sites, detect changes at sites that are variable in one paralog but conserved in another. Conversely, character tracing does not rely on detection of altered constraints and will detect changes not only at sites that are variable in one copy and conserved in another, but also at sites that differ between paralogs but are fixed in each. Certain caveats also are important to consider in light of the question of interest. For example, changes in selective pressure at individual sites may occur without affecting the overall function of a protein. It is of interest to characterize these patterns, but their implications for functional divergence should be interpreted with caution. Conversely, functional change can occur without an obvious change in constraints as some advantageous mutations may drift to fixation. Together, the use of these methods has provided insights into past events and has demonstrated their utility to predict current function and future change. As users of the methods, it is not only important to keep the aforementioned considerations in mind but to provide feedback for their continued improvement.

TABLE I

METHODS FOR STUDYING THE EVOLUTION OF PARALOGS IN DUPLICATE GENE DATASETS

A. Tests for variable selective pressures among branches or clades

Primary citations	Test	Input	Software[a]
Tests based on d_N/d_S			
Messier and Stewart, 1997; Zhang et al., 1997	Estimate d_N/d_S for each branch to detect cases in which $d_N > d_s$	Alignment and tree	
Yang, 1998	Tests hypothesis that d_N/d_s is the same on all branches of the tree; LRT detects branches under positive selection	Alignment and tree	PAML[b]
Tests for covarion or covariotide structure			
Gu, 1999, 2001	Tests the assumption that site-specific rates are correlated between paralogs	Alignment and tree	DIVERGE[c]
Gaucher et al., 2001	Tests the assumption that alpha is stationary across the tree of paralogs	Alignment and tree	

B. Tests for shifts in substitution rates across sites during the divergence of paralogs

Primary citations	Test	Input	Software
Tests based on d_N/d_S			
Fitch et al., 1997	Tests d_N/d_s of each site against the average value for sequence	Alignment and tree	
Suzuki and Gojobori, 1999	Tests d_N/d_s of each site against the neutral expectation of one	Alignment and tree	ADAPTSITE[d]
Nielsen and Yang, 1998; Yang et al., 2000	Tests for variable d_N/d_s among sites; Bayes prediction of sites in the class with $d_N/d_s > 1$	Alignment and tree	PAML
Yang and Nielsen, 2002	Tests for variable d_N/d_s among sites along a specific branch; Bayes prediction of sites in the class with $d_N/d_s > 1$	Alignment and tree	PAML

Tests for covarion or covariotide structure

	Test	Input	Software
Gu, 1999, 2001	Tests the assumption that site-specific rates are correlated between paralogs; Bayes prediction of sites at which rates have shifted	Alignment and tree	DIVERGE
Gaucher et al., 2001	Tests the assumption that alpha is stationary across the tree of paralogs; evaluates kurtosis to identify sites that are evolving slowly or rapidly	Alignment and tree	
Heterogeneity tests Dermatzakis and Clark, 2001; Tang and Lewontin, 1999	Tests the hypothesis that variable sites are uniformly distributed along a sequence	Alignment	

C. Tracing characters to find paralog-specific amino acid residues

Primary citations	Test	Input	Software
Evolutionary tracing Landgraf et al., 1999; 2001; Lichtarge et al., 1996	Identifies positions of conserved and paralog-specific residues on a protein structure	3D or NMR structural model, alignment, ± tree	
Character state mapping Chang et al., 1995[e]	Reconstructs ancestral states to trace character changes on a tree	Alignment and tree	PAML, MacClade[f]

[a] Additional software or scripts may be available from authors.
[b] Yang, 1997 and http://abacus.gene.ucl.ac.uk/software/paml.html.
[c] Gu and Vander Velden, 2002 and http://xgu.zool.iastate.edu/software.html.
[d] Suzuki et al., 2001 and see also http://www.hyphy.org/.
[e] A particularly useful example of the use of character tracing to study the evolution of paralogs.
[f] PAML is useful for ML reconstruction of character states; MacClade (Maddison and Maddison, 2000) is useful for MP reconstruction of character states and tracing character changes on trees.

References

Adey, N. B., Tollefsbol, T. O., Sparks, A. B., Edgell, M. H., and Hutchison, C. A., III (1994). Molecular resurrection of an extinct ancestral promoter for mouse L1. *Proc. Natl. Acad. Sci. USA* **91,** 1569–1573.

Anisimova, M., Bielawski, J. P., and Yang, Z. H. (2001). Accuracy and power of the likelihood ratio test in detecting adaptive molecular evolution. *Mol. Biol. Evol.* **18,** 1585–1592.

Anisimova, M., Bielawski, J. P., and Yang, Z. H. (2002). Accuracy and power of Bayes prediction of amino acid sites under positive selection. *Mol. Biol. Evol.* **19,** 950–958.

Anisimova, M., Nielsen, R., and Yang, Z. H. (2003). Effect of recombination on the accuracy of the likelihood method for detecting positive selection at amino acid sites. *Genetics* **164,** 1229–1236.

Bennner, S. A. (1989). Patterns of divergence in homologous proteins as indicators of tertiary and quaternary structure. *Adv. Enzymol. Regul.* **28,** 219–236.

Bennner, S. A. (2002). The past as key to the present: Resurrection of ancient proteins from eosinphils. *Proc. Natl. Acad. Sci. USA* **99,** 4760–4761.

Bush, R. M., Fitch, W. M., Bender, C. A., and Cox, N. J. (1999). Positive selection on the H3 hemagglutinin gene of human influenza virus A. *Mol. Biol. Evol.* **16,** 1457–1465.

Chandrasekharan, U. M., Sanker, S., Glynias, M. J., Karnik, S. S., and Husain, A. (1996). Angiotensin II-forming activity in a reconstructed ancestral chymase. *Science* **271,** 502–505.

Chang, B. S. W. (2003). Ancestral gene reconstruction and synthesis of ancient rhodopsins in the laboratory. *Integr. Compar. Biol.* **43,** 500–507.

Chang, B. S. W., Crandall, K. A., Carulli, J. P., and Hartl, D. L. (1995). Opsin phylogeny and evolution: A model for blue shifts in wavelength regulation. *Mol. Phyl. Evol.* **4,** 31–43.

Chang, B. S. W., Jönsson, K., Kazmi, M. A., Donoghue, M. J., and Sakmar, T. P. (2002). Recreating a functional ancestral archosaur visual pigment. *Mol. Biol. Evol.* **19,** 1483–1489.

Collins, T. M., Wimberger, P. M., and Naylor, G. J. P. (1994). Compositional bias, character-state bias, and character-state reconstruction using parsimony. *Syst. Biol.* **43,** 482–496.

Dermatzakis, E. T., and Clark, A. G. (2001). Differential selection after duplication in mammalian developmental genes. *Mol. Biol. Evol.* **18,** 557–562.

Endo, T., Ikeo, K., and Gojobori, T. (1996). Large-scale search for genes on which positive selection may operate. *Mol. Biol. Evol.* **13,** 685–690.

Fitch, W. M. (1970). Distinguishing homologous from analogous proteins. *Syst. Zool.* **19,** 99–113.

Fitch, W. M., and Markowitz, E. (1970). An improved method for determining codon variability in a gene and its application to the rate of fixation of mutations in evolution. *Biochem. Genet.* **4,** 579–593.

Fitch, W. M., Bush, R. M., Bender, C. A., and Cox, N. J. (1997). Long term trends in the evolution of H(3) HA1 human influenza type A. *Proc. Natl. Acad. Sci. USA* **94,** 7712–7718.

Force, A., Lynch, M., Pickett, F. B., Amores, A., Yan, Y. L., and Postlethwait, J. (1999). Preservation of duplicate genes by complementary, degenerative mutations. *Genetics* **151,** 1531–1545.

Gaucher, E. A., Miyamoto, M. M., and Bennner, S. A. (2001). Function-structure analysis of proteins using covarion-based evolutionary approaches: Elongation factors. *Proc. Natl. Acad. Sci. USA* **98,** 548–552.

Goldman, N., and Yang, Z. (1994). A codon-based model of nucleotide substitution for protein-coding DNA sequences. *Mol. Biol. Evol.* **11,** 725–736.

Gu, X. (1999). Statistical method for testing functional divergence after gene duplication. *Mol. Biol. Evol.* **16,** 1664–1674.

Gu, X. (2001). Maximum-likelihood approach for gene family evolution under functional divergence. *Mol. Biol. Evol.* **18**, 453–464.

Gu, X., and Vander Velden, K. (2002). DIVERGE: Phylogeny-based analysis for functional-structural divergence of a protein family. *Bioinformatics* **18**, 500–501.

Haldane, J. B. S. (1933). The part played by recurrent mutation in evolution. *Am. Nat.* **67**, 5–9.

Henikoff, S., Greene, E. A., Pietrokovski, S., Bork, P., Attwood, T. K., and Hood, L. (1997). Gene families: The taxonomy of protein paralogs and chimeras. *Science* **278**, 609–614.

Hill, R. E., and Hastie, N. D. (1987). Accelerated evolution in the reactive centre regions of serine protease inhibitors. *Nature* **326**, 96–99.

Hirotsune, S., Yoshida, N., Chen, A., Garrett, L., Sugiyama, F., Takahashi, S., Yagami, K., Wynshaw-Boris, A., and Yoshiki, A. (2003). An expressed pseudogene regulates the messenger-RNA stability of its homologous coding gene. *Nature* **423**, 91–96.

Hughes, A. L. (1994). The evolution of functionally novel proteins after gene duplication. *Proc. R. Soc. Lond. B* **256**, 119–124.

Hughes, A. L., and Nei, M. (1988). Pattern of nucleotide substitution at major histocompatibility complex class I loci reveals overdominant selection. *Nature* **335**, 167–170.

Hughes, A. L., and Nei, M. (1989). Nucleotide substitution at major histocompatibility complex class II loci: Evidence for overdominant selection. *Proc. Natl. Acad. Sci. USA* **86**, 958–962.

Ina, Y. (1995). New methods for estimating the numbers of synonymous and nonsynonymous substitutions. *J. Mol. Evol.* **40**, 190–226.

Jermann, T. M., Opitz, J. G., Stackhouse, J., and Benner, S. A. (1995). Reconstructing the evolutionary history of the artiodactyls ribonuclease superfamily. *Nature* **374**, 57–59.

Landgraf, R., Fischer, D., and Eisenberg, D. (1999). Analysis of Heregulin symmetry by weighted evolutionary tracing. *Prot. Eng.* **12**, 943–951.

Landgraf, R., Xenarios, I., and Eisenberg, D. (2001). Three-dimensional cluster analysis identifies interfaces and functional residue clusters in proteins. *J. Mol. Biol.* **307**, 1487–1502.

Li, W.-H. (1993). Unbiased estimation of the rates of synonymous and nonsynonymous substitutions. *J. Mol. Evol.* **36**, 96–99.

Li, W.-H., Wu, C.-I., and Luo, C.-C. (1985). A new method for estimating synonymous and nonsynonymous rates of nucleotide substitutions considering the relative likelihood of nucleotide and codon changes. *Mol. Biol. Evol.* **2**, 150–174.

Lichtarge, O., Bourne, H. R., and Cohen, F. E. (1996). An evolutionary trace method defines binding surfaces common to protein families. *J. Mol. Biol.* **257**, 342–358.

Lynch, M., and Conery, J. S. (2000). The evolutionary fate and consequences of duplicate genes. *Science* **290**, 1151–1155.

Lynch, M., and Force, A. (2000). The probability of duplicate gene preservation by subfunctionalization. *Genetics* **154**, 459–473.

Maddison, D. R., and Maddison, W. P. (2000). "MacClade 4. Analysis of phylogeny and character evolution." Sinauer Associates, Sunderland, MA.

Madubashi, S., Gross, A. K., Philippi, A., Meng, E. C., Wensel, T. G., and Lichtarge, O. (2004). Evolutionary trace of G protein-coupled receptors reveals clusters of residues that determine global and class-specific functions. *J. Biol. Chem.* **279**, 8126–8132.

Marín, I., Fares, M. A., González-Candelas, F., Barrio, E., and Moya, A. (2001). Detecting functional constraints of paralogous genes. *Mol. Biol. Evol.* **52**, 17–28.

Martínez-Castilla, L., and Alvarez-Buyalla, E. R. (2003). Adaptive evolution in the *Arabidopsis* MADS-box gene family inferred from its complete resolved phylogeny. *Proc. Natl. Acad. Sci. USA* **100**, 13407–13412.

Merritt, T. J. S., and Quattro, J. M. (2001). Evidence for a period of directional selection following gene duplication in a neurally expressed locus of triosephosphate isomerase. *Genetics* **159**, 689–697.

Messier, W., and Stewart, C.-B. (1997). Episodic adaptive evolution of primate lysozymes. *Nature* **385**, 151–154.

Muse, S. V., and Gaut, B. S. (1994). A likelihood approach for comparing synonymous and nonsynonymous nucleotide substitution rates, with application to the chloroplast genome. *Mol. Biol. Evol.* **11**, 715–724.

Muse, S. V., and Weir, B. S. (1992). Testing for equality of evolutionary rates. *Genetics* **132**, 269–276.

Naylor, G. J. P., and Gerstein, M. (2000). Measuring shifts in function and evolutionary opportunity using variability profiles: A case study of the globins. *J. Mol. Evol.* **51**, 223–233.

Nei, M., and Gojobori, T. (1986). Simple methods for estimating the numbers of synonymous and nonsynonymous nucleotide substitutions. *Mol. Biol. Evol.* **3**, 418–426.

Nielsen, R., and Yang, Z. (1998). Likelihood models for detecting positively selected amino acid sites and applications to the HIV-1 envelope gene. *Genetics* **148**, 929–936.

Ohno, S. (1970). "Evolution by Gene Duplication." Springer-Verlag, Berlin.

Plotkin, J. B., Dushoff, J., and Fraser, H. B. (2004). Detecting selection using a single genome sequence of *M. tuberculosis* and *P. falciparum*. *Nature* **428**, 942–945.

Shoemaker, J. S., and Fitch, W. M. (1989). Evidence from nuclear sequences that invariable sites should be considered when sequence divergence is calculated. *Mol. Biol. Evol.* **6**, 270–289.

Shi, Y., and Yokoyama, S. (2003). Molecular analysis of the evolutionary significance of ultraviolet vision in vertebrates. *Proc. Natl. Acad. Sci. USA* **100**, 8308–8313.

Sorhannus, U. (2003). The effect of positive selection on a sexual reproductive gene in *Thalassiosira weissflogii* (Bacilliariophyta): Results obtained from maximum-likelihood and parsimony-based methods. *Mol. Biol. Evol.* **20**, 1326–1328.

Stewart, C.-B. (1995). Active ancestral molecules. *Nature* **374**, 12–13.

Sun, H. M., Merugu, X., Gu, X., Kang, Y. Y., Dickinson, D. P., Callaerts, P., and Li, W. H. (2002). Identification of essential amino acid changes in paired domain evolution using a novel combination of evolutionary analysis and *in vitro* and *in vivo* studies. *Mol. Biol. Evol.* **19**, 1490–1500.

Suzuki, Y., and Gojobori, T. (1999). A method for detecting positive selection at single amino acid sites. *Mol. Biol. Evol.* **16**, 1315–1328.

Suzuki, Y., and Nei, M. (2004). False-positive selection identified in by ML-based methods: Examples from the *Sig1* gene of the diatom *Thalassiosira weissflogii* and the *tax* gene of a human T-cell lymphotrophic virus. *Mol. Biol. Evol.* **21**, 914–921.

Suzuki, Y., Gojobori, T., and Nei, M. (2001). ADAPTSITE: Detecting natural selection at single amino acid sites. *Bioinformatics* **17**, 660–661.

Tang, H., and Lewontin, R. C. (1999). Locating regions of differential variability in DNA and protein sequences. *Genetics* **153**, 485–495.

Ugalde, J. A., Chang, B. S. W., and Matz, M. V. (2004). Evolution of coral pigments recreated. *Science* **305**, 1433.

Walsh, J. B. (1995). How often do duplicated genes evolve new functions. *Genetics* **139**, 421–428.

Wong, W. S. W., Yang, Z., Goldman, N., Nielsen, R. (2004). Accuracy and power of statistical methods for detecting adaptive evolution in protein coding sequences and for identifying positively selected sites. *Genetics* **168**, 1041–1051.

Yang, Z. (1997). PAML: A program package for phylogenetic analysis by maximum likelihood. *Comput. Appl. Biosci.* **13**, 555–556.

Yang, Z. (1998). Likelihood ratio tests for detecting positive selection and application to primate lysozyme evolution. *Mol. Biol. Evol.* **15,** 568–573.

Yang, Z. (2002). Inference of selection from multiple species alignments. *Curr. Opin. Genet. Dev.* **12,** 688–694.

Yang, Z., Kumar, S., and Nei, M. (1995). A new method of inference of ancestral nucleotide sequences with variable rates over sites: Approximate methods. *Genetics* **141,** 1641–1650.

Yang, Z., and Nielsen, R. (2000). Estimating synonymous and nonsynonymous substitution rates under realistic evolutionary models. *Mol. Biol. Evol.* **17,** 32–43.

Yang, Z., and Nielsen, R. (2002). Codon-substitution models for detecting molecular adaptation at individual sites along specific lineages. *Mol. Biol. Evol.* **19,** 908–917.

Yang, Z. R., Nielsen, R., Goldman, N., and Pedersen, A.-M. K. (2000). Codon-substitution models for heterogenous selection pressure at amino acid sites. *Genetics* **155,** 431–449.

Zhang, J. (2004). Frequent false detection of positive selection by the likelihood method with branch-sites models. *Mol. Biol. Evol.* **21,** 1332–1339.

Zhang, J., and Nei, M. (1997). Accuracies of ancestral amino acid sequences inferred by the parsimony, likelihood, and distance methods. *J. Mol. Evol.* **44,** S139–S146.

Zhang, J., Kumar, S., and Nei, M. (1997). Small-sample test of episodic adaptive evolution: A case study of primate lysozymes. *Mol. Biol. Evol.* **14,** 1335–1338.

Zhang, J., Rosenberg, H. F., and Nei, M. (1998). Positive Darwinian selection after gene duplication in primate ribonuclease genes. *Proc. Natl. Acad. Sci. USA* **95,** 3708–3713.

Zhang, J., and Rosenberg, H. F. (2002). Complementary advantageous substitutions in the evolution of an antiviral RNase of higher primates. *Proc. Natl. Acad. Sci. USA* **99,** 5486–5491.

[38] Supertree Construction in the Genomic Age

By OLAF R. P. BININDA-EMONDS

Abstract

Supertree construction is the process whereby overlapping phylogenetic trees, and not character data, are combined to yield a larger, more comprehensive phylogeny. In this chapter, I review the logic and methodology behind supertree construction and argue that it holds a necessary place in phylogenetic inference. Much of the justification for supertrees is admittedly practical. As I show with an empirical example, most large groups have insufficient sequence data to build complete phylogenies for them. By being able to indirectly combine diverse forms of phylogenetic information, supertrees are the best method for constructing complete phylogenies of groups with hundreds of species. However, supertree construction can also be justified on theoretical grounds. As whole genomic data are obtained for increasing numbers of species, the theoretical and practical advantages of supertrees together will ensure that the method will

0076-6879/05 $35.00

play a necessary analytical role as part of a divide-and-conquer strategy to reconstructing the Tree of Life.

Introduction

Since the beginning of the molecular revolution in the 1960s, a progressive array of molecular data has been used to elucidate the phylogenetic relationships of the species in the world around us. These data types include amino acid and DNA sequences, immunology and serology, DNA–DNA hybridization, isozymes, chromosomal banding patterns and rearrangements, SINEs and LINEs, gene order data, gene composition data, and linkage map data. Many of these data types are described elsewhere in both this volume and its predecessor (Zimmer et al., 1993).

In the age of genomics, the prospect of whole genomic data for phylogenetic inference is an exciting possibility. In fact, as of March 2005, whole genomic data already exist for 179 diverse microbial species in TIGR's Comprehensive Microbial Resource (Peterson et al., 2001), with the sequencing of several additional microbial genomes due for completion in 2005. The situation is not as advanced for eukaryotic organisms, however, where the larger genome sizes mean that sequencing efforts are concentrated in a few model species. Thus, for the moment, building large, comprehensive phylogenies for many large clades involves the combination of existing phylogenetic data. Traditionally, the data that are combined are character data, which has come to be known as the *total evidence approach* (Kluge, 1989) or the *supermatrix approach* (Sanderson et al., 1998) to phylogenetic inference.

In this chapter, I review a different approach for building comprehensive phylogenies in which the data that are combined are overlapping phylogenetic trees rather than the primary character data underlying those trees. This supertree approach has been used increasingly to construct (virtually) complete phylogenetic trees of clades with several hundred species (Davies et al., 2004; Jones et al., 2002; Pisani et al., 2002; Salamin et al., 2002), which in many cases, represent the only complete phylogenies for the groups in question (at the taxonomic level of the study). I first briefly introduce the concept of supertrees before describing the desirable features of supertree construction that will prove necessary in our efforts to reconstruct the Tree of Life, even in an age of whole genomic data.

What Are Supertrees?

The idea underlying supertree construction is that a more comprehensive phylogeny can be constructed by combining two source trees that overlap only partially in their taxon sets. In this way, statements of relationship can be made between two species that do not appear on the same tree (Sanderson *et al.*, 1998) (Fig. 1). Because trees are combined in this approach, supertree construction has many obvious parallels with the more familiar field of consensus trees. However, a distinction is often made between the supertree and consensus settings (Bininda-Emonds *et al.*, 2002), with the latter being a special case of the former where the source trees have identical taxon sets. Thus, although supertree methods will work in the consensus setting (i.e., to combine trees with identical taxon sets), the same is not true in reverse.

The principle of combining overlapping trees to yield a more comprehensive tree is probably as old as systematics itself, where trees were informally pasted together historically. As it is recognized currently, however, the field of supertree construction is only about a dozen years old, stemming from the independent description of the supertree method matrix representation with parsimony (MRP) by Baum (1992) and Ragan (1992). Although the first formal supertree method, and the term *supertree*, is attributable to Gordon (1986), it was the development of MRP with its numerous desirable properties that spurred the growth of supertrees. Among these properties is the ability to combine all possible statements of phylogenetic relatedness as long as they could be represented as a treelike structure, the use of the familiar and well-understood parsimony as an optimization criterion, and the ability to produce well-resolved trees. The potential for MRP (and supertree construction in general) to yield complete phylogenies of large clades was quickly realized by Purvis (1995), who used MRP to produce the first complete phylogeny for all 203 extant species of primates that was based on a rigorous objective methodology. The primate supertree has since gone on to be cited numerous times and used as a framework for understanding the biology of the entire order in an evolutionary perspective and at an unprecedented taxonomic scale.

Today, many supertree methods exist (Bininda-Emonds, 2004), all with slightly different properties. The basis for many of these methods is matrix representation (Fig. 1), whereby the topology of the source trees is coded into a matrix. This matrix is then optimized using any of a number of criteria (e.g., parsimony, compatibility, likelihood, least-squares, and Bayesian methods) to yield the supertree. Although the one-to-one correspondence between any single tree and its matrix representation is well founded in both graph and network theory, no such relationship exists between the joint set of matrix representations and the supertree (Baum and Ragan, 1993). Instead, the supertree must be viewed as the tree with the best fit to the set of source trees according to the given optimization criterion. However, each column (matrix element) in the combined matrix

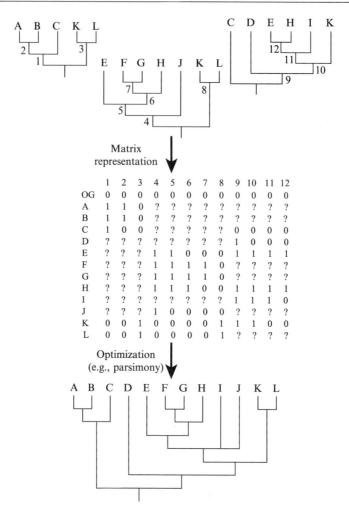

FIG. 1. An example of matrix representation with parsimony (MRP) supertree construction. In the first step, the informative nodes (numbered) of all three source trees are coded such that taxa that are descended from that node are scored as 1; those that are not but are present on the tree are scored as 0; and those that are not present on the tree are scored as ? A hypothetical all-zero outgroup (OG) is added to the matrix to root the trees; it is later pruned from the supertree. In the second step, the combined matrix representations are optimized (here, using parsimony) to yield the supertree. Note that the supertree allows statements of relationship between taxa that do not co-occur on the any single source tree (e.g., taxon I with taxa F and G).

representations does maintain a one-to-one correspondence with a particular node in one source tree. This allows for the differential weighting of matrix elements according to the differential support of the nodes among the set of source trees. For instance, if the bootstrap frequencies are known for the nodes of a source tree, the matrix element representing each node can be weighted in proportion to the bootstrap frequency of that node. Such weighting has been shown to improve the fit of the matrix representation to the source tree (Ronquist, 1996) and to improve the performance of MRP in simulation (Bininda-Emonds and Sanderson, 2001).

As with consensus techniques, the choice of which supertree method to use is partly dependent on the question being asked. Strict (Gordon, 1986; Steel, 1992) and semistrict (Goloboff and Pol, 2002; Lanyon, 1993) supertree methods present the relationships that are common to or uncontradicted among, respectively, the set of source trees. As such, they provide a conservative summary of the information common to a set of source trees. Similarly, MinCutSupertree (Page, 2002; Semple and Steel, 2000) preserves nestings and, like Adams consensus, can be used to detect common statements of relationship among a set of source trees (e.g., A and B are more closely related than either is to C, where A, B, and C need not be each other's closest relatives). Gene tree parsimony (Slowinski and Page, 1999) yields a supertree that explains incongruence among the (molecular) source trees in terms of biological phenomena such as gene losses or duplications. Methods such as the average consensus (Lapointe and Cucumel, 1997) or RankedTree (Bryant et al., 2004) directly use branch-length information from the source trees, which is less easy to accommodate using matrix representation methods (although possible through weighting of the matrix elements; see earlier discussion). Finally, supertree methods derived from the Build algorithm of Aho et al. (1981) (e.g., strict, [modified] MinCutSupertree, RankedTree, AncestralBuild, and Semi-LabeledBuild) run in polynomial time according to the number of taxa. Thus, they are much faster than the remaining supertrees methods (which are NP-complete and have no efficient solution, thereby requiring the use of less-desirable heuristics) and might be particularly well suited for very large supertree problems.

Why Use Supertrees? Supertrees vs. Supermatrices

Practical Considerations

Supertrees are often viewed as being an alternative to conventional, character-based phylogenetic analysis (Gatesy et al., 2004), with critics suggesting that supertrees have been justified largely on the basis of utility and expediency (Gatesy and Springer, 2004; Gatesy et al., 2002). Much of the interest in supertrees does indeed derive from practical considerations. In particular, it is simply not possible to construct a complete phylogeny for most (large) groups of organisms because of a lack of data that can be analyzed using a single optimization criterion (i.e., compatible data). In contrast, combining trees as a supertree allows data of all forms to be combined indirectly (e.g., DNA hybridization, morphological, DNA or amino acid sequences, and immunological distances), thus potentially using the full phylogenetic dataset that exists. Of the complete supertrees that exist for many large clades of hundreds of species (Bininda-Emonds, 2004), probably none could be constructed using a supermatrix approach, although large character-based phylogenies do exist (Källersjö et al., 1998).

The molecular revolution, however, has done much to close this gap by yielding compatible data in great quantities for many species. Even so, data collection remains patchy and incomplete for many groups (Sanderson et al., 2003). Consider, in particular, the Carnivora, a well-sampled order in a well-sampled class (Mammalia). When I completed the supertree for all 271 extant carnivore species (Bininda-Emonds et al., 1999) in January 1996, there was no possibility of producing a molecular phylogeny on the same scale: GenBank (http://www.ncbi.nlm.nih.gov/Genbank/index.html) contained only 677 sequences for 48 species (Fig. 2). In the meantime, however, molecular sampling for the order has increased tremendously, so there were 1,984,623 sequences for 197 species as of March 21, 2004. A molecular phylogeny for the order seems within reach.

However, these raw numbers are somewhat deceptive. Of the nearly two million carnivore sequences, 99.6% and 0.2% are for the domestic dog and domestic cat, respectively (Table I), two carnivore species with active genome projects. Although this still leaves an average of 3900 sequences for each of the remaining 195 species, or 20 sequences per species on average, many species have been sampled repeatedly for the same gene. For example, 191 of the 219 sequences for *Martes americana* are for cytochrome *b*. Thus, many species are represented by very few genes and sequences (Fig. 3), and a complete molecular phylogeny for the Carnivora based on a wide variety of sequence data might be a more distant possibility than it might at first glance seem. The situation for less charismatic groups that have attracted less attention will naturally be even worse (Bininda-Emonds et al., 2002).

Thus, the supertrees of today are providing complete phylogenetic hypotheses for many groups that could not otherwise be achieved. These phylogenies have proven valuable for understanding the biology of the groups in question, with their large size and completeness giving unprecedented statistical power and scope to studies of descriptive systematics,

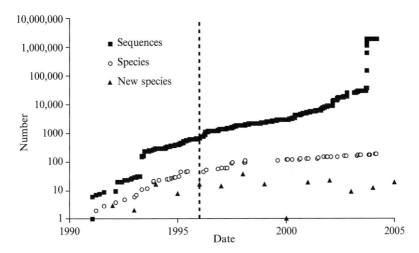

FIG. 2. The number of Carnivora sequences in GenBank and the number of extant species and new species per year represented by those sequences as a function of time. The sequence data are complete to March 12, 2004. The dashed line indicates the completion of the carnivore supertree in January 1996.

TABLE I
THE 10 MOST SEQUENCED CARNIVORA SPECIES IN GENBANK AS OF MARCH 12, 2004

Species	Common name	No. of sequences
Canis lupus	Gray wolf (includes domestic dog)	1,976,358
Felis silvestris	Wild cat (includes domestic cat)	4365
Leopardus pardalis	Ocelot	295
Ursus arctos	Brown bear	253
Martes americana	American marten	219
Panthera onca	Jaguar	173
Ursus americanus	American black bear	168
Mustela vison	American mink	160
Leopardus wiedii	Margay	151
Meles meles	Eurasian badger	141

Note: In total, 1,984,623 sequences were available for 197 of the 271 extant Carnivora species. Taxonomy follows Wozencraft (1993).

evolutionary models, cladogenesis and species richness, evolutionary patterns and comparative biology, and biodiversity and conservation (Gittleman *et al.*, 2004). However, given that more data are becoming available daily, is there a future for supertrees?

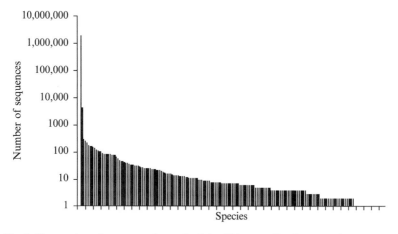

Fig. 3. The number of sequences for each of the 197 extant Carnivora species represented in GenBank. Species are presented in decreasing order according to the number of sequences. The sequence data are complete to March 12, 2004.

Theoretical Considerations

What is less appreciated is that there are also good theoretical arguments for using supertrees (Bininda-Emonds *et al.*, 2002) and ones that will give an increasingly important role to supertree construction in the age of complete genomic information. In so doing, however, the role of supertree construction will change from its current form, from being used mostly to combine phylogenetic estimates derived from the literature, to being an important analytical technique.

The primary hindrance to reconstructing the Tree of Life is its sheer size. It has long been appreciated that the number of possible phylogenetic trees increases superexponentially with the number of taxa being examined (Felsenstein, 1978). Thus, the larger the phylogenetic problem, the greater the number of algorithmic shortcuts that must be taken to derive a solution in a reasonable amount of time, and the greater the probability that the globally optimal solution will not be found. Furthermore, the breadth of the Tree of Life makes deriving a globally informative dataset problematic. For example, a major stumbling block in deriving a morphological phylogeny of the metazoan phyla has been the difficulty in identifying homologous features among the phyla. Molecular data in the form of a few highly conserved genes such as 18S ribosomal DNA (rDNA) have helped to address this problem. Even so, suitable genes at this level (or beyond) are comparatively rare. More importantly, Sanderson *et al.* (1998) have suggested that aligning such genes at such levels (i.e., to identify homologous features) will prove difficult, a situation parallel to that for morphological data.

Instead, any attempt to reconstruct large portions of the Tree of Life will require the use of supertree construction as part of a divide-and-conquer strategy to phylogenetic reconstruction. The principle underlying the divide-and-conquer approach is to break a large phylogenetic problem into numerous smaller subproblems, each of which is solved using conventional analyses. The results of the subproblems are then recombined (as a supertree) to derive the global answer. The reduced sizes of the subproblems make them computationally easier to solve and possibly more accurate because they are both smaller (fewer species) and of reduced breadth, allowing more data to be used (Roshan *et al.*, 2004).

The value of supertrees as part of a divide-and-conquer strategy has been demonstrated nicely by Daubin *et al.* (2001, 2002). In the latter study in particular, Daubin *et al.* (2002) derived a phylogenetic estimate of 45 bacterial species using whole genomic data. Instead of analyzing all 730 orthologous genes that they were able to identify simultaneously, Daubin *et al.* (2002) analyzed each separately and so were able to analyze each according to the most appropriate model of evolution for it. [Although mixed-model analyses are possible in a supermatrix setting, especially in a Bayesian framework, they are more intense computationally. Therefore, many supermatrix studies use a parsimony criterion (Gatesy *et al.*, 2002), which cannot account for models of molecular evolution as fully.] Each gene tree could also be pared to only those species for which data were present, thereby avoiding the adverse effects of including a large amount of missing data in the analysis (Wilkinson, 1995). The final tree was obtained by forming an MRP supertree of the individual gene trees and is regarded as one of the most robust estimates for the phylogenetic relationships of the species it contains.

Although Daubin *et al.* (2002) partitioned their supermatrix into orthologous genes, other possible strategies for the divide step are possible. Two especially promising approaches include disk-covering methods (Huson *et al.*, 1999a,b) and bicliques (Burleigh *et al.*, 2004; Sanderson *et al.*, 2003) [for more detail, see Bininda-Emonds (2004)]. Bicliques in particular can identify portions of a supermatrix that are data rich in terms of both species and characters, thereby again avoiding the inclusion of large amounts of missing data.

As part of a divide-and-conquer strategy, supertree construction shows two desirable properties. The first is that simulation studies have shown that several supertree methods show good accuracy at reconstructing a known model tree under such circumstances (Bininda-Emonds and Sanderson, 2001; Chen *et al.*, 2003; Levasseur and Lapointe, 2003;

Piaggio-Talice *et al.*, 2004). In most cases, this accuracy was usually as good as that achieved by a simultaneous analysis of the combined character data. However, when the differential support within individual source trees was accounted for using weighting, MRP and the average consensus, at least, slightly outperformed the analogous supermatrix analysis (Bininda-Emonds and Sanderson, 2001; Levasseur and Lapointe, 2003). Thus, the inherent loss of information in combining trees as opposed to the primary character data does not appear to be detrimental in practice. Instead, the use of supertrees potentially allows for the inclusion of more information than a supermatrix approach in the form of the use of appropriate models of evolution for each partition (see earlier discussion; Bininda-Emonds *et al.*, 2003).

The second advantage of a supertree-based divide-and-conquer search strategy of genomic data is the promise of decreased analysis time compared to a traditional supermatrix approach. The suitability of such a strategy for parallel processing is immediately clear, with each subproblem forming an independent analysis. As mentioned earlier, the smaller subproblems are also computationally easier to solve. Furthermore, as mentioned earlier, many supertree methods achieve results in polynomial time and are, therefore, much faster than character-based optimization criteria such as maximum likelihood, maximum parsimony, or neighbor joining. Even Bayesian supertrees display a speed advantage over comparable Bayesian analyses of molecular sequence data because of the special properties of Bayesian supertrees that allow a more efficient sampling strategy (Ronquist *et al.*, 2004).

Thus, a divide-and-conquer strategy promises to show gains in both accuracy and speed compared to a conventional phylogenetic analysis. Evidence in support of this was provided by Roshan *et al.* (2004), who showed the postulated performance gains in the analysis of some, but not all, large molecular datasets that they examined. This is clearly an area of great promise and one that needs to be researched in more detail.

Summary: Future of Supertree Construction

Instead of being viewed as an alternative to the supermatrix approach (Gatesy *et al.*, 2004), supertree construction should be viewed as being a complementary approach, both now and in the future. The basis for this is the realization that the two approaches analyze different datasets. The supermatrix approach uses the primary character data, whereas supertree construction analyzes phylogenetic hypotheses in the form of trees. These hypotheses not only derive from the primary character data, but also incorporate the many auxiliary assumptions made in the analysis of them (Bininda-Emonds *et al.*, 2003). Thus, the supermatrix and supertree approaches form important components of a global congruence framework (Lapointe *et al.*, 1999), whereby well-supported relationships are those common to both sets of analyses. In contrast, conflicting sets of relationships indicate the need to identify possible sources of the conflict, be they inadequate analyses, insufficient data, or true conflict. The true complementarity of the supertree and supermatrix approaches, however, will be seen in the future, when their respective strengths contribute to a divide-and-conquer search strategy that probably represents our best opportunity to reconstruct larger portions of the Tree of Life.

References

Aho, A. V., Sagiv, Y., Szymanski, T. G., and Ullman, J. D. (1981). Inferring a tree from lowest common ancestors with an application to the optimization of relational expressions. *SIAM J. Comput.* **10**, 405–421.

Baum, B. R. (1992). Combining trees as a way of combining data sets for phylogenetic inference, and the desirability of combining gene trees. *Taxon.* **41**, 3–10.

Baum, B. R., and Ragan, M. A. (1993). Reply to A.G. Rodrigo's "A comment on Baum's method for combining phylogenetic trees." *Taxon.* **42**, 637–640.

Bininda-Emonds, O. R. P., Gittleman, J. L., and Purvis, A. (1999). Building large trees by combining phylogenetic information: A complete phylogeny of the extant Carnivora (Mammalia). *Biol. Rev.* **74**, 143–175.

Bininda-Emonds, O. R. P., and Sanderson, M. J. (2001). Assessment of the accuracy of matrix representation with parsimony supertree construction. *Syst. Biol.* **50**, 565–579.

Bininda-Emonds, O. R. P., Gittleman, J. L., and Steel, M. A. (2002). The (super)tree of life: Procedures, problems, and prospects. *Annu. Rev. Ecol. Syst.* **33**, 265–289.

Bininda-Emonds, O. R. P., Jones, K. E., Price, S. A., Grenyer, R., Cardillo, M., Habib, M., Purvis, A., and Gittleman, J. L. (2003). Supertrees are a necessary not-so-evil: A comment on Gatesy *et al. Syst. Biol.* **52**, 724–729.

Bininda-Emonds, O. R. P. (2004). The evolution of supertrees. *Trends Ecol. Evol.* **6**, 315–322.

Bryant, D., Semple, C., and Steel, M. (2004). *In* "Phylogenetic Supertrees: Combining Information to Reveal the Tree of Life" (O. R. P. Bininda-Emonds, ed.), Vol. 3, pp. 129–150. Kluwer Academic, Dordrecht, The Netherlands.

Burleigh, J. G., Eulenstein, O., Fernández-Baca, D., and Sanderson, M. J. (2004). Supertree methods for ancestral divergence dates and other applications. MRF Supertrees. *In* "Phylogenetic Supertrees: Combining Information to Reveal the Tree of Life" (O. R. P. Bininda-Emonds, ed.), Vol. 3, pp. 65–85. Kluwer Academic, Dordrecht, The Netherlands.

Chen, D., Diao, L., Eulenstein, O., Fernández-Baca, D., and Sanderson, M. J. (2003). Flipping: A supertree construction method. *In* "Bioconsensus" (M. F. Janowitz, F.-J. Lapointe, F. R. McMorris, B. Mirkin, and F. S. Roberts, eds.), Vol. 61, pp. 135–160. American Mathematical Society, Providence, RI.

Daubin, V., Gouy, M., and Perrière, G. (2001). Bacterial molecular phylogeny using supertree approach. *Genome Inform.* **12**, 155–164.

Daubin, V., Gouy, M., and Perrière, G. (2002). A phylogenomic approach to bacterial phylogeny: Evidence of a core of genes sharing a common history. *Genome Res.* **12**, 1080–1090.

Davies, T. J., Barraclough, T. G., Chase, M. W., Soltis, P. S., Soltis, D. E., and Savolainen, V. (2004). Darwin's abominable mystery: Insights from a supertree of angiosperms. *Proc. Natl. Acad. Sci. USA* **101**, 1904–1909.

Felsenstein, J. (1978). The number of evolutionary trees. *Syst. Zool.* **27,** 27–33.

Gatesy, J., Matthee, C., DeSalle, R., and Hayashi, C. (2002). Resolution of a supertree/supermatrix paradox. *Syst. Biol.* **51,** 652–664.

Gatesy, J., and Springer, M. S. (2004). A critique of matrix representation with parsimony supertrees. *In* "Phylogenetic Supertrees: Combining Information to Reveal the Tree of Life" (O. R. P. Bininda-Emonds, ed.), Vol. 3, pp. 369–388. Kluwer Academic, Dordrecht, The Netherlands.

Gatesy, J., Baker, R. H., and Hayashi, C. (2004). Inconsistencies in arguments for the supertree approach: Supermatrices versus supertrees of Crocodylia. *Syst. Biol.* **53,** 342–355.

Gittleman, J. L., Jones, K. E., and Price, S. A. (2004). Supertrees: Using complete phylogenies in comparative biology. *In* "Phylogenetic Supertrees: Combining Information to Reveal the Tree of Life" (O. R. P. Bininda-Emonds, ed.), Vol. 3, pp. 439–460. Kluwer Academic, Dordrecht, the Netherlands.

Goloboff, P. A., and Pol, D. (2002). Semi-strict supertrees. *Cladistics* **18,** 514–525.

Gordon, A. D. (1986). Consensus supertrees: The synthesis of rooted trees containing overlapping sets of labeled leaves. *J. Classif.* **3,** 31–39.

Huson, D. H., Nettles, S. M., and Warnow, T. J. (1999a). Disk-covering, a fast-converging method for phylogenetic tree reconstruction. *J. Comput. Biol.* **6,** 369–386.

Huson, D. H., Vawter, L., and Warnow, T. J. (1999b). Solving large scale phylogenetic problems using DCM2. *In* "Proceedings of the Seventh International Conference on Intelligent Systems for Molecular Biology" (T. Lengauer, R. Schneider, P. Bork, D. Brutlag, J. Glasgow, H.-W. Mewes, and R. Zimmer, eds.), Vol. 7, pp. 118–129. AAAI Press, Menlo Park, CA.

Jones, K. E., Purvis, A., MacLarnon, A., Bininda-Emonds, O. R. P., and Simmons, N. B. (2002). A phylogenetic supertree of the bats (Mammalia: Chiroptera). *Biol. Rev.* **77,** 223–259.

Källersjö, M., Farris, J. S., Chase, M. L., Bremer, B., Fay, M. F., Humphries, C. J., Petersen, G., Seberg, O., and Bremer, K. (1998). Simultaneous parsimony jackknife analysis of 2538 *rbc*L DNA sequences reveals support for major clades of green plants, land plants, seed plants, and flowering plants. *Pl. Syst. Evol.* **213,** 259–287.

Kluge, A. G. (1989). A concern for evidence and a phylogenetic hypothesis of relationships among *Epicrates* (Boidae, Serpentes). *Syst. Zool.* **38,** 7–25.

Lanyon, S. M. (1993). Phylogenetic frameworks: Towards a firmer foundation for the comparative approach. *Biol. J. Linn. Soc.* **49,** 45–61.

Lapointe, F.-J., and Cucumel, G. (1997). The average consensus procedure: Combination of weighted trees containing identical or overlapping sets of taxa. *Syst. Biol.* **46,** 306–312.

Lapointe, F.-J., Kirsch, J. A. W., and Hutcheon, J. M. (1999). Total evidence, consensus, and bat phylogeny: A distance based approach. *Mol. Phylogenet. Evol.* **11,** 55–66.

Levasseur, C., and Lapointe, F.-J. (2003). Increasing phylogenetic accuracy with global congruence. *In* "Bioconsensus" (M. F. Janowitz, F.-J. Lapointe, F. R. McMorris, B. Mirkin, and F. S. Roberts, eds.), Vol. 61, pp. 221–230. American Mathematical Society, Providence, RI.

Page, R. D. M. (2002). Modified minifut supertrees. *In* "Algorithms in Bioinformatics, Second International Workshop, WABI, 2002, Rome, Italy, September 17–21, 2002, Proceedings" (R. Guigó and D. Gusfield, eds.), Vol. 2452, pp. 537–552. Springer, Berlin.

Peterson, J. D., Umayam, L. A., Dickinson, T., Hickey, E. K., and White, O. (2001). The comprehensive microbial resource. *Nucleic Acids Res.* **29,** 123–125.

Piaggio-Talice, R., Burleigh, J. G., and Eulenstein, O. (2004). Quartet supertrees. *In* "Phylogenetic Supertrees: Combining Information to Reveal the Tree of Life" (O. R. P. Bininda-Emonds, ed.), Vol. 3, pp. 173–191. Kluwer Academic, Dordrecht, the Netherlands.

Pisani, D., Yates, A. M., Langer, M. C., and Benton, M. J. (2002). A genus-level supertree of the Dinosauria. *Proc. R. Soc. Lond. B* **269,** 915–921.

Purvis, A. (1995). A composite estimate of primate phylogeny. *Philos. Trans. R. Soc. Lond. B* **348,** 405–421.

Ragan, M. A. (1992). Phylogenetic inference based on matrix representation of trees. *Mol. Phylogenet. Evol.* **1,** 53–58.

Ronquist, F. (1996). Matrix representation of trees, redundancy, and weighting. *Syst. Biol.* **45,** 247–253.

Ronquist, F., Huelsenbeck, J. P., and Britton, T. (2004). Bayesian supertrees. *In* "Phylogenetic Supertrees: Combining Information to Reveal the Tree of Life" (O. R. P. Bininda-Emonds, ed.), Vol. 3, pp. 193–224. Kluwer Academic, Dordrecht, the Netherlands.

Roshan, U., Moret, B. M. E., Williams, T. L., and Warnow, T. (2004). Performance of supertree mathods on various data set decompositions. *In* "Phylogenetic Supertrees: Combining Information to Reveal the Tree of Life" (O. R. P. Bininda-Emonds, ed.), Vol. 3, pp. 301–328. Kluwer Academic, Dordrecht, the Netherlands.

Salamin, N., Hodkinson, T. R., and Savolainen, V. (2002). Building supertrees: An empirical assessment using the grass family (Poaceae). *Syst. Biol.* **51,** 136–150.

Sanderson, M. J., Purvis, A., and Henze, C. (1998). Phylogenetic supertrees: Assembling the trees of life. *Trends Ecol. Evol.* **13,** 105–109.

Sanderson, M. J., Driskell, A. C., Ree, R. H., Eulenstein, O., and Langley, S. (2003). Obtaining maximal concatenated phylogenetic data sets from large sequence databases. *Mol. Biol. Evol.* **20,** 1036–1042.

Semple, C., and Steel, M. (2000). A supertree method for rooted trees. *Discrete Appl. Math.* **105,** 147–158.

Slowinski, J. B., and Page, R. D. M. (1999). How should species phylogenies be inferred from sequence data? *Syst. Biol.* **48,** 814–825.

Steel, M. (1992). The complexity of reconstructing trees from qualitative characters and subtrees. *J. Classif.* **9,** 91–116.

Wilkinson, M. (1995). Coping with abundant missing entries in phylogenetic inference using parsimony. *Syst. Biol.* **44,** 501–514.

Zimmer, E. A., White, T. J., Cann, R. L. and Wilson, A. C. (eds.) (1993). "Molecular Evolution: Producing the Biochemical Data." *Methods Enzymol.* **224,** 3–725.

[39] Maximum-Likelihood Methods for Phylogeny Estimation

By JACK SULLIVAN

Abstract

Maximum-likelihood (ML) estimation of phylogenies has reached a rather high level of sophistication because of algorithmic advances, improvements in models of sequence evolution, and improvements in statistical approaches and application of cluster computing. Here, I provide a brief basic background in application of the general principle of ML estimation to phylogenetics and provide an example of selecting among a nested set of ML models using a dynamic approach to hierarchical likelihood-ratio tests. I focus attention on PAUP* because it provides unique ease of switching among alternative optimality criteria (e.g., minimum evolution, parsimony, and ML). Further, examples of parametric bootstrap tests are provided that demonstrate statistical tests of phylogenetic hypotheses and model adequacy, in an absolute rather than relative sense. The increasing availability of clustered, parallelized computation makes use of such parametric approaches feasible.

Application of ML as an Optimality Criterion in Phylogeny Estimation

Maximum-likelihood (ML) estimation is a standard and useful statistical procedure that has become widely applied to phylogenetic analysis. Although this application of ML presents some unique issues, the general idea is the same in phylogeny as in any other application. One calculates the likelihood of an observed dataset given a particular hypothesis and some assumed probabilistic model.

$$L = \text{Prob}(\text{data}|\text{hypothesis}) \tag{1}$$

We evaluate several hypotheses and select the one that maximizes the probability of generating the observed data. When applied to phylogeny estimation, the hypotheses that are examined represent alternative phylogenies and the data are the set of aligned sequences. The likelihood of a tree (τ) is

$$L(\tau) = \text{Prob}(D|\tau) \tag{2}$$

simply the probability of the data (the set of aligned sequences), given the tree (and some assumed model of character evolution). Just as the length of a tree can be calculated as its optimality score in parsimony analyses, the likelihood of a tree can be used as its optimality score in ML estimation. We make the assumption that characters are independent (just as in parsimony) so that we may treat likelihoods for each site separately:

$$L(\tau) = \prod_{i=1}^{s} \text{Prob}(D^i|\tau) = \prod_{i=1}^{s} L^i(\tau) \tag{3}$$

where s is the number of sites (characters) and $\text{Prob}(D^i \mid \tau)$ is the probability of site i (character i), given tree τ. This value is the single-site likelihood, and just as the parsimony score for a tree across an entire dataset is the sum of the character lengths, the likelihood of a tree across an entire dataset is the product of the single-site likelihoods. The single-site likelihood is, therefore, analogous to the length of a most parsimonious character reconstruction in MP estimation.

The calculation of single-site likelihoods is accomplished as follows. Let us assume the following rooted, four-taxon tree:

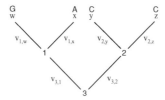

This example is somewhat modified from that provided by Swofford *et al.* (1996), and in this tree, taxa w, x, y, and z have nucleotides G, A, C, and C, respectively, at the first position in the alignment. The branches, which are labeled $v_{x,y}$, and their lengths (in units of expected number of substitutions per site—a function of rate of evolution times the temporal duration of branch) are parameters that need to be estimated. So to calculate the single-site likelihood for this character, we must sum the probabilities for all possible character-state reconstructions. Because there are $n - 1 = 3$ internal nodes (for a rooted tree) and four possible character states at each node, there are $4^{n-1} = 4^3 = 64$ possible reconstructions. So

$$L^i(\tau) = \sum_{r}^{4n-1} \mathrm{Prob}(R_r^i|\tau)$$

or

$$L^i(\tau) = \mathrm{Prob}\left(\begin{smallmatrix}G & A\ C & C\\ & A & \diagdown & A\\ & & A\end{smallmatrix}\right) + \mathrm{Prob}\left(\begin{smallmatrix}G & A\ C & C\\ & A & \diagdown & A\\ & & C\end{smallmatrix}\right)$$

$$+ \mathrm{Prob}\left(\begin{smallmatrix}G & A\ C & C\\ & A & \diagdown & A\\ & & G\end{smallmatrix}\right) \bullet \bullet \bullet + \mathrm{Prob}\left(\begin{smallmatrix}G & A\ C & C\\ & T & \diagdown & T\\ & & T\end{smallmatrix}\right)$$

Of course many reconstructions are extremely unlikely (such as the last one shown, with T at all internal nodes), and they will contribute very little to the single-site likelihood; nevertheless, we consider them as possibilities. So now the issue is how one calculates the probabilities of a particular reconstruction. Let us assume that in reconstruction r, m is the state at the root node 3, k is the state at node 1, and l is the state at node 2. We know from our data that nodes w, x, y, and z have states C, A, C, and C, respectively. So the probability of reconstruction r at site i is

$$P(R_r^i|\tau) = \pi_m \times P_{m,k}(v_{3,1}) \times P_{k,G}(v_{1,w}) \times P_{k,A}(v_{1,x}) \times P_{m,l}(v_{3,2})$$
$$\times P_{l,C}(v_{2,y}) \times P_{l,C}(v_{2,z}), \tag{5}$$

where π_m is the frequency of the nucleotide A (which provides an estimate of the probability of observing state m at the root node). $P_{i,j}$ is the probability of substitution between states i and j, which is derived from the model of sequence evolution that we assume (see below). This can be calculated for all 4^{n-1} reconstructions, and these are then summed across reconstructions to calculate the single-site likelihoods. However, Felsenstein (1973) developed a more efficient way to calculate the same value, which uses the structure of the tree, so that the single-site likelihood for character i is as follows:

$$L^i(\tau) = \sum_m \pi_m \times \left(\sum_k P_{m,k}(v_{3,1}) P_{k,G}(v_{1,w}) P_{k,A}(v_{1,x}) \right)$$
$$\times \left(\sum_l P_{m,l}(v_{3,2}) P_{l,C}(v_{2,y}) P_{l,C}(v_{2,z}) \right) \qquad (6)$$

and each summation is across all four nucleotides. The improvements in efficiency achieved here are attributable to the fact that we can calculate the contributions of various subtrees (indicated by the structure of the parentheses) just once and use the subtree values as they are needed. In almost all applications of ML estimation, rather than dealing with the product of extremely small numbers (the single-site likelihoods), their natural logarithms are usually taken and summed. The overall log likelihood of a tree is, therefore,

$$lnL(\tau) = \sum_{i=1}^{s} lnL^i(\tau) \qquad (7)$$

An additional efficiency is achievable by realizing that if more than one site has the same distribution of character states (i.e., has the same site pattern), we only have to calculate $L^i(\tau)$ once. So, if we consider the following dataset:

```
1          A  G  T  A  C  A  .  .  .  .  .  .  .  .  .  .  .  .  .  .  .  .
2          A  G  T  A  .  .  .  .  .  .  .  .  .  .  .  .  .  .  .  .  .  .
3          A  G  T  A  .  .  .  .  .  .  .  .  .  .  .  .  .  .  .  .  .  .
.          .  .  .  .  .  .  .  .  .  .  .  .  .  .  .  .  .  .  .  .  .  .
.          .  .  .  .  .  .  .  .  .  .  .  .  .  .  .  .  .  .  .  .  .  .
n          A  G  T  A  .  .  .  .  .  .  .  .  .  .  .  .  .  .  .  .  .  .
Pattern    1  2  3  1  .  .  .  .  .  .  .  .  .  .  .  .  .  .  .  . (a)
```

The first and fourth sites have the same site pattern, and with n sequences, there are 4^n possible site patterns (because there are four possible nucleotides). Rather than recalculating the single-site likelihood for sites with the same pattern, the frequency of each site pattern is tallied. The overall likelihood score for a particular tree therefore is as follows:

$$lnL(\tau) = \sum_{a=1}^{4^n} f_a(lnL^a(\tau))$$ (8)

Here, f_a is the frequency of the ath site pattern and $lnL^a(\tau)$ is the single-site log likelihood of tree τ for all sites with pattern a. This value represents a measure of fit between the tree and the data (assuming a model of sequence evolution), and $lnL(\tau)$ is calculated for every tree that is examined during a tree search.

Generating ML Estimates of Phylogeny

Justification for Iterative Approach

In principle, searching trees under the likelihood criterion is no different than doing so under parsimony. However, one qualification is that the optimality score for a given tree under likelihood $lnL^i(\tau)$ is computationally more difficult than the corresponding value (tree length) under parsimony. Furthermore, the $P_{i,j}$ values used in calculating $lnL^i(\tau)$ represent instantaneous rates of substitution from nucleotide i to nucleotide j; these are specified by the model of sequence evolution, and a model must be chosen that makes explicit assumptions. One difficulty is that the optimum values of these parameters are conflated both with each other (i.e., nonindependent) and with topology. The ideal solution is to simultaneously optimize all parameters on every tree that one examines during a tree search, an approach that is not feasible for most empirical studies. Fortunately, this problem can be circumvented by adopting an iterative approach (Sullivan et al., 1996; Sullivan and Swofford, 1997; Swofford et al., 1996), in which one uses a rapid approximate method to find a reasonable initial tree. This initial tree is used both to evaluate alternative models (Frati et al., 1997; Sullivan et al., 1997) and to derive initial estimates of model parameters (such as $P_{i,j}$ parameters). In the next step, the model parameters are held constant and alternative trees are evaluated (usually using some heuristic search). This process is repeated until the same tree (or set of trees) is found in successive iterations. Sullivan et al., 2005 have demonstrated that the iterative method is a useful approximation to the ideal analytical approach.

Starting the Iteration: Selecting a Model

As is true for many statistical methods, there are a number of approaches to model selection, and theory is continually being developed and tested by statisticians. With respect to ML models for phylogeny estimation, the most commonly employed approach is likelihood-ratio tests (LRTs) and there are a number of ways to implement LRTs. Model selection via LRTs can be accomplished in an automated fashion, with programs such as Modeltest (Posada and Crandall, 1998), or it can be conducted in a more interactive fashion (sometimes called *dynamic model selection*). Regardless of how one chooses to proceed, LRTs require that the models being examined form a nested family of models, whereby every model in the

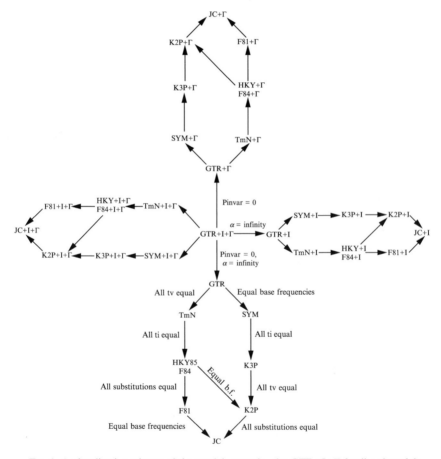

FIG. 1. A visualization of part of the model space for the GTR+I+Γ family of models.

collection is a special case of some parameter-rich general model. In most phylogenetic analyses, attention has focused on the GTR+I+Γ family of models. One visualization of that family is shown in Fig. 1.

Here, the most general and parameter-rich model, the GTR+I+Γ has 10 free parameters: three free base frequencies, 5 free relative instantaneous transformation rates (r_{AC}, r_{AG}, r_{AT}, r_{CG}, and r_{CT}; r_{GT} is arbitrarily set to 1), a proportion of the sites that are invariable (p_{inv}), and the gamma distribution shape parameter (α, which describes rate variation across the potentially variable sites). I focus on describing the steps one takes in using PAUP* (Swofford, 1998) to implement the iterative search strategy, because no other package allows one to switch optimality criteria in the same run as easily as PAUP* (Swofford and Sullivan, 2003). The initial step in implementing the iterative searches is to generate an initial tree. Usually, this is accomplished with a very rapid method such as neighbor joining (NJ), typically applied to a distance matrix generated with LogDet distances. Below, I go through a dynamic, top-down approach to model selection (i.e., starting with the most general and parameter-rich model, GTR+I+Γ) for a dataset containing 22 cytochrome *b* sequences from sigmodontine rodents (Rinehart *et al.*, unpublished). Once the dataset is loaded, the commands are as follows:

```
dset dist = logdet;
nj;
lset nst = 6 rmat = est basefreq = est pinv = est rates = gamma
    shape = est; [This sets the likelihood model to GTR + I+Γ]
lscore;
```

This generates the following output:

```
Tree                         1
_ _ _ _ _ _ _ _ _ _ _ _ _ _ _
-ln L                   6449.72254
Base frequencies:
A                        0.360254
C                        0.316726
G                        0.099847
T                        0.223173
Rate matrix R:
AC                        1.29169
AG                        5.43624
AT                        3.14761
CG                        0.13416
CT                       35.49652
GT                        1.00000
P_inv                     0.446391
Shape                     0.596201
```

Just by looking at these parameter estimates, $r_{AC} \sim r_{GT}$, we should be able to equate r_{AC} with r_{GT} using the rclass command. One would type the following command:

```
lscore / rclass = (a b c d e a);
```

And the output is

```
Tree                          1
_ _ _ _ _ _ _ _ _ _ _ _   _ _
-ln L              6450.04916
Base frequencies:
A                     0.379897
C                     0.324175
G                     0.066497
T                     0.229431
Rate matrix  R:
AC                     1.00000
AG                    10.62273
AT                     1.95152
CG                     0.44845
CT                    28.84545
GT                     1.00000
P_inv                 0.456866
Shape                 0.517319
```

So by eliminating one parameter, we have decreased the likelihood score by 0.326 lnL units. The LRT statistic is $\delta = 2(lnL_{general} - lnL_{restricted})$, so $\delta = 0.652$ and, making assumptions regarding asymptotic properties, we can use the χ^2 distribution with d.f. equal to the difference in number of parameters between the two models (one in this case). Thus, $p = .419$ and we accept the null hypothesis that there is no significant difference in fit between the two models; that is, we accept the simpler model. Now let us see if we can simplify further. The next most similar relative rate parameter is perhaps the r_{CG}, so we can restrict the matrix further.

```
lscore / rclass = (a b c a d a);
```

And the output is

```
Tree                          1
_ _ _ _ _ _ _ _ _ _ _ _   _ _
-ln L              6450.32915
Base frequencies:
A                     0.383239
C                     0.324820
G                     0.062259
T                     0.229682
```

```
Rate  matrix  R:
AC              1.00000
AG             14.46313
AT              2.18638
CG              1.00000
CT             32.44806
GT              1.00000
P_inv           0.459870
Shape           0.512830
```

Again, we see just a slight deterioration in the likelihood score of 0.28 lnL units, so $\delta = 0.560$ and $p = .454$, so again we can accept the simpler model. Let us continue by setting r_{AT} equal as well:

```
lscore / rclass = ( a b a a c a);
```

And the output is

```
Tree                      1
- - - - - - - - - - - - -  - -
-ln  L          6452.67603
Base  frequencies:
A               0.383349
C               0.323744
G               0.059190
T               0.233717
Rate  matrix  R:
AC              1.00000
AG             12.37828
AT              1.00000
CG              1.00000
CT             23.66113
GT              1.00000
P_inv           0.459595
Shape           0.499518
```

Now we have a deterioration in likelihood score of 2.38 *lnL* units and a $\delta = 4.772$. This corresponds to a p value of .028926, and we can reject the simpler model. It is unlikely that any further simplifications of the R matrix would be acceptable, so let us try to simplify the rate heterogeneity among sites by looking at $\alpha = $ infinity (this is equivalent to an equal rates model):

```
lscore / rates = equal;
```

Again, the output is

```
Tree                    1
- - - - - - - - - - - -  - -
-ln L          6673.98739
Base frequencies:
A                  0.305462
C                  0.335057
G                  0.097600
T                  0.261880
Rate matrix R:
AC                  1.00000
AG                  4.12118
AT                  1.95195
CG                  1.00000
CT                  6.94868
GT                  1.00000
P_inv              0.525604
```

This represents a huge deterioration of the likelihood score of about 223.7, for a single parameter, so clearly we cannot use an invariable-sites model alone (and in this case, it really does not matter which null distribution we use: The mixed χ^2 [following Goldman and Whelan, 2000] or the χ^2 with 1 d.f.). The base frequencies are so wildly different that it is pointless to even try restricting basefreq = equal. So now we are at a point at which we have three free base frequencies, three free relative-rate parameters (two transitions and two transversions, one of which is set to one), and two rate heterogeneity parameters. This actually is not a named model, and incidentally, it is not a model that any automated model selection method, such as Modeltest (Posada and Crandall, 1998), would choose. This is not to say that the model chosen here is correct, and an alternative that may be chosen by an alternative implementation of an LRT (or some other criterion) is incorrect. All models are approximations of the true underlying process, and the model selected above is a slightly different approximation of the unknown true model.

Nevertheless, one disadvantage of using automated model selection programs is that one is restricting the outcome of model selection to those cases that are hard-coded into the programs. Another advantage of the dynamic approach shown above is that the act of simplifying models manually generates a much better understanding of one's data (and indeed of the relationships among alternative models) than can be achieved by relying on an automated model-selection approach. Note that there are model-selection approaches other than LRTs, such as Akaike Information Content (AIC), Bayesian Information Criterion (BIC), and a new method based on decision theory (DT-ModSel; Minin *et al.*, 2003). The ability of different model-selection approaches to select an adequate approximation to the unknown true generating process is a question that is being addressed by a number of groups. At this point, there is evidence that at least one of these, DT-ModSel (Minin *et al.*, 2003), will outperform the LRT (as commonly implemented by ModelTest) with respect to accuracy of branch length estimated under the selected models (Minin *et al.*, 2003).

Searching Tree Space Using the Chosen Model

Now that we have selected a model, we can begin the process of searching tree space. The following set of commands implements an ML heuristic search, under the fully defined model we have chosen, with 10 replicate searches, each started with a starting tree generated from random addition sequence and using TBR branch swapping.

```
dset dist = logdet;
nj;
lset nst = 6 rclass = (abcada) rmat = est basefreq = est pinv = est
  rates = gamma shape = est;
lscore;
lset rmat = prev basefreq = prev pinv = prev shape = prev;
set criterion = like;
hs addseq = random nrep = 10;
```

Although most systematists who employ this approach stop there, the process really should be iterated. One would use the ML tree just generated to reoptimize model parameters (unless the NJ tree from LogDet distances is identical to the ML tree, which it almost never is), and then conduct another ML search with the refined parameter estimates. This iteration should continue until the same tree or set of trees is generated in successive iterations.

One potential pitfall that can occur when there is strong rate heterogeneity among sites is that the default number of rate categories (ncat = 4) may overly discretize the gamma distribution (which is a continuous function). This default was selected following Yang (1993), but it is definitely worth assessing the influence of varying the number of rate categories. The indication of such a problem is an estimate of the shape parameter α that is diverging toward zero; the PAUP* output is

Shape < 0.001

If this occurs, more than four rate categories are required, and usually eight (ncat = 8) are sufficient. The `ncat` setting is an option under the `lset` command. This problem is restricted to models in which all sites are assumed to be potentially variable (i.e., pinv = 0).

Assessing Phylogenetic Uncertainty

Again, just as is true for parsimony, non-parametric bootstrap analysis (Felsenstein, 1985) can be used to assess nodal support in ML analyses. Two issues need to be confronted in conducting ML bootstrap analyses. The first of these is that, ideally, one would reevaluate the relative fit of alternative models of sequence evolution and reoptimize model parameters for each pseudo-replicate. This is almost never done because of the computational limits involved. Instead, ML bootstraps are almost always conducted with the model fully defined and fixed to that which was selected in the analyses of the real data. The effect that such an incorrectly defined model for each pseudo-replicate dataset will have on bootstrap values has not been directly studied. However, phylogenetic theory (Waddell, 1995) predicts that bootstrap values for nodes that are poorly supported (i.e., are defined by a short internal branch or a long one defined only by change at high-rate sites) will be underestimated, whereas bootstrap values for well-supported nodes should be relatively unaffected.

The second issue is also related to computational time. In any application of a bootstrap analysis, one would ideally analyze the pseudo-replicate datasets identically to how the original dataset was analyzed. Thus, one would ideally conduct multiple random addition heuristic searches with TBR branch swapping on each pseudo-replicate. Such an approach is particularly problematic for datasets where divergence is relatively low because of the chance of constructing a pseudo-replicate that has little or no phylogenetic information. In such cases, the bootstrap analysis will become bogged down by swapping interminably on a particularly information-poor replicate. The most extreme approximation that can be taken is to omit branch swapping altogether and simply use the stepwise-addition tree as the estimate for each pseudo-replicate. In PAUP*, this is accomplished as follows (assuming the optimality criterion has been set to likelihood and the model has been fully defined previously):

```
bootstrap nrep = 1000 search = faststep;
```

A much less extreme approach that still achieves great time savings is to conduct full TBR branch swapping but to only hold a single optimal tree in memory. This can be accomplished with the following:

```
set maxtrees = 1 increase = no;
bootstrap nrep = 1000 search = hs keepall = yes;
```

DeBry and Olmstead (2000) and Mort *et al.* (2000) have demonstrated that this approach will provide bootstrap values that are not significantly different than those that would be attained by full heuristic searches for large datasets under the parsimony criterion. The same should hold true for ML bootstraps.

Hypothesis Testing

Finding Trees Constrained to Fit Hypotheses

Perhaps the greatest advance in systematic biology over the last 10 years is the development of explicitly statistical approaches to phylogenetic hypothesis testing. Many hypotheses in evolutionary biology make specific predictions about phylogenetic relationships, and these predicted relationships form the basis of phylogenetic hypothesis testing. The idea is that the ML (or MP) tree for a particular dataset may contradict the relationships predicted by some hypothesis one wants to test. By using topological constraints, one may assess how much worse than the optimal tree is the best tree consistent with the predictions of the hypothesis. In order to do this with PAUP, one needs to define a constraint tree in a tree file, which in this example is called "constraint.nex." It is a simple file that contains only a trees block.

```
# nexus
begin trees;
    utree constraint1 = (1 – 5,(6,7,8)); [Taxa 8 – 22 will be
        unresolved in the constraint tree]
end;
```

There are a few important points to note here. First, the constraint tree will not be fully resolved. Ideally, it should be resolved to the minimum extent possible while still fitting the predictions of the hypothesis being tested. The above constraint tree would be used to test some evolutionary hypothesis that predicts that taxa 6, 7, and 8 exclusively share a common ancestor. Second, not all taxa in the data matrix need to be specified in the constraint tree. So, if there were more than eight sequences in the test dataset, but the hypothesis under examination does not address them, they should be left unresolved in the constraint tree and need not even be included in the constraint tree. To test this hypothesis (that taxa 6, 7, and 8 exclusively share a common ancestor), we first need to find the unconstrained ML tree (as above). Let us assume that we have chosen the HKY+I+Γ model and found that the ML tree has a score of −6456.14360, and that the clade (6, 7, 8) is not present on the ML tree. The ML tree must be saved to a file:

```
savetree file = ML.tre;
```

We now need to run a constrained search to find the best tree that contains clade (6, 7, 8).

```
loadconst file = constraint.nex;
showconst; [make sure the constraint tree is correct]
hs enforce = yes;
lsc nst = 2 trat = est ba = est ra = g sha = est pinv = est;
   [we should reoptimize parameters on this tree to find
   the best possible fit between data and hypothesis]
lsc trat = prev ba = prev sha = prev pinv = prev;
savetree file = hypothesis.tre;
```

The showconst command allows one to view the constraint and ensure that the constraint tree was written correctly; it is a good idea to check this before running long searches. In our example, the best tree constrained to contain the clade (6, 7, 8) has a likelihood score of −6481.94451, a deterioration of 25.80091 *lnL* units. It is critical to save the ML branch lengths if one is interested in employing the parametric bootstrap for significance testing.

The two trees are shown in Fig. 2, with the tree on the left being the ML tree and the tree on the right being the best tree constrained for taxa 6, 7, and 8 constrained to form a group.

Note that, because there are no characters supporting that clade (6, 7, 8) in the dataset, the group is united by an internal branch length of zero. This is sometimes the case in constrained trees.

Evaluating the Test Statistic

In this example, the value of the test statistic is, therefore, 25.80. For several years, the only approach available to assess the significance of the test statistic, and therefore test the hypothesis that predicts the presence of clade (6, 7, 8), was through the use of the Kishino-Hasegawa test (K-H test) (Kishino and Hasegawa, 1989). Assuming that there are no trees in the display buffer (i.e., that the best constrained tree was saved to the file "hypothesis.tre" and the ML tree was saved to the file "ml.tre") and we have selected the HKY+I+Γ model of evolution, this is accomplished in PAUP as follows:

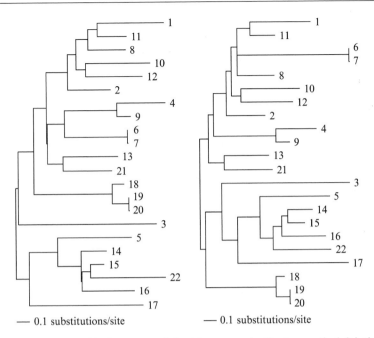

Fig. 2. ML trees used in the parametric bootstrap example. The tree on the left is the ML tree and the tree on the right is the best tree constrained for monophyly of taxa 6, 7, and 8. The best constrained tree is used as the true tree in the simulation.

```
gettree file = ml.tre;
gettree file = hypothesis.tre mode = 7; [mode = 7 retains any
   trees currently in buffer]
lset nst = 2 ba = est trat = est ra = g sha = est pinv = est;
lscore/khtest = normal;
```

The output is as follows

```
Kishino-Hasegawa test:
   KH test using normal approximation, one-tailed test
```

			KH-test
Tree	−ln L	Diff −ln L	P
1	6456.14360	(best)	
2	6481.94451	25.80091	0.000*

*P < 0.05

This has been the most commonly used of the K-H tests, with a normal distribution of single-site lnL differences assumed for the null distribution.

This is a somewhat stringent assumption that can be relaxed by using the reestimated log-likelihood (RELL) bootstrap procedure:

```
lscore/khtest = rell nrep = 1000;
```

with the output as follows:

```
Kishino-Hasegawa test:
   KH test using RELL bootstrap, one-tailed test
   Number of bootstrap replicates = 1000
```

| | | | KH-test |
Tree	−ln L	Diff −ln L	P
1	6456.14360	(best)	
2	6481.94451	25.80091	0.015*

* P < 0.05

Use of the RELL bootstrap to assess significance of the test statistic is preferable because it eliminates assumptions about the actual shape of the null distribution. If the null hypothesis (the support for the two trees is not significantly different) is false, the distribution of single-site *lnL*'s will be skewed (i.e., not normal). Note that this is a one-tailed test. This seems appropriate because we are using the test incorrectly; one tree is the estimated ML tree, whereas one is known *a priori* to be suboptimal (Goldman *et al.*, 2000; Shimodaira and Kishino, 1999). For the test to be used appropriately, the two trees being compared must be selected *a priori* and a two-tailed test would be more appropriate:

```
lscore/khtest = rell nrep = 1000 tailkh = 2;
```

With the output of

```
Kishino-Hasegawa test:
   KH test using RELL bootstrap, two-tailed test
   Number of bootstrap replicates = 1000
```

| | | | KH-test |
Tree	−ln L	Diff −ln L	P
1	6456.14360	(best)	
2	6481.94451	25.80091	0.018**

P < 0.05

Shimodaira and Hasegawa (S-H test) have attempted to correct this bias by including a set of trees into consideration and centering the null distribution. However, the collection of trees to consider still must be erected *a priori,* and with just two trees that were selected *a posteriori* considered, the S-H test reverts to the H-K test.

Therefore, the most appropriate way to assess the significance of the test statistic ($lnL_{MLTree} - lnL_{ConstrainedTree}$) is through use of the parametric bootstrap (Goldman *et al.*, 2000; Hillis *et al.*, 1996). This entails generating the null distribution by simulation, with the best constrained tree used as the model (true) tree for simulation. The idea is that we have the test statistic that measures how much more poorly the data fit the hypothesis than they do the ML tree. We now want to derive a probability (given the hypothesis is true) of observing a test statistic at least as large as that observed in the real data that is simply due to stochasticity. This provides an assessment (conditional on the model; see below) of whether phylogenetic uncertainty can plausibly explain the difference between the ML tree and the relationships predicted by the hypothesis. To accomplish a parametric bootstrap test, one needs a program that can simulate sequence evolution, given a tree with branch lengths and a fully specified model of sequence evolution. For the example above, we can use the tree saved to simulate sequences using the program Seq-Gen (Rambault and Grassley, 1997), which is available for several platforms. Because Seq-Gen does not model gaps, missing data, or ambiguities, the test statistic, branch lengths, and model parameters must be recalculated after excluding any characters that contain any such issues. Assuming the tree is loaded into PAUP's tree buffer, one accomplishes this as follows:

```
exclude gapped missambig;
lsc all/nst = 2 ba = est trat = est ra = g sha = est pinv = est;
savetree file = NoGapHypoth.tre brlens = y;
```

For Mac systems, the easiest thing to do is to convert the tree file that we just saved into one that Seq-Gen can read. This is done by deleting everything from the tree file except the tree (the file should be just a single line that specifies the tree in Newic format with branch lengths); here, I will rename it "NoGapHypothIn.tre." In the dialogue box that appears when starting the program, one hits the "file" button on the bottom left and then navigates to the appropriate file (NoGapHypothIn.tre). It is best to save the output to a file (by clicking the "file" button on the bottom right). In the field labeled "Argument," one specifies the model of sequence evolution (in our case HKY+I+Γ), the sequence length (here, 720 bp), the number of replicate datasets to generate, and any formatting information. This includes type of output file (i.e., PAUP vs. Phylip) and any set of commands that one wants to use in analyzing the simulated sequences. This last task is accomplished by reference to a text file that contains a PAUP block (e.g., PaupBlock.txt). The program allows a number of options, including use of mixed models for multilocus (or otherwise partitioned) data and the ability to use different true trees for different partitions. For the purpose of simulating null distributions to assess the significance of our test statistic, we want to simulate a number of replicate datasets on the best constrained tree (with its ML branch estimates of branch lengths and model parameters). We then must find the difference between the ML tree and the best constrained tree for each replicate. Thus, the argument line for this example would look like this:

```
-mHKY -f0.379386, 0.329111, 0.053277, 0.238226
   -t9.094694 -a0.465274 -i0.461462 -1720 -n100 -on
   -xPaupBlock.txt,
```

where the file PaupBlock.txt contains the following:

```
begin paup;
    set monitor=no autoclose=y;
    dset dist=logdet;
    nj; [get an approximate tree for parameter
      estimation]
    lsc/nst=2 ba=est trat=est ra=g sha=est pinv=est;
    lset ba=prev trat=prev sha=prev pinv=prev;
    set crit=like;
    hs;
    lsc 1/ba=est trat=est sha=est pinv=est
        scorefile=mltree.score append; [reoptimize
          parameters
        on the ML tree to find the best possible fit]
    lset ba=prev trat=prev sha=prev pinv=prev;
    loadconstr file=constraint.tre;
    hs enforce=y; [find the best tree constrained to
        fit the hypothesis]
    lsc 1/ba=est trat=est sha=est pinv=est
        scorefile=hypotree.score append; [reoptimize
        parameters on best constrained tree to find best
        possible fit to hypothesis]
  end;
```

Seq-Gen will generate a single file with 100 simulated datasets (i.e., 100 data blocks) under the hypothesis being examined (i.e., using the best tree constrained to fit the hypothesis of interest). After each of the data blocks, the output file will contain the above PAUP block, which will find the best unconstrained tree (and write the ML score to a file named `mltree.score`) and the best tree constrained to fit the hypothesis (and write the ML score the `hypotree.score` file). These two score files can be merged into a single database file (e.g., using Microsoft Excel) and the

distribution of differences across the 100 (or more replicates) is the null
distribution with which to assess the significance of the test statistic
(25.80091 lnL units). If one is using a Unix version of Seq-Gen, the
simulation is conducted with a single command:

```
seq-gen -mHKY -f0.379386,0.329111,0.053277,0.238226
 -t9.094694 -a0.465274 -i0.461462 -1722 -n100
 <NoGapHypothIn.tre>OutFile.dat -on -xPaupBlock.txt
```

These runs can take substantial CPU time. One way to reduce this is to
confine the tests to the parsimony searches. One then uses the difference in
tree length of constrained versus unconstrained trees as the test statistic,
still uses ML to estimate branch lengths and model parameters of the best
constrained tree as the model tree/parameters, and then searches each
replicated for the best constrained and unconstrained MP trees. A preferable alternative is to use emerging technology in parallel clusters (i.e.,
a Beowulf cluster). For example, at the University of Idaho, the Bioinformatics computing core facility has three clusters. The largest of these, a
modest cluster of 108 2.8-GHz processors can run a 500-replicate full ML
parametric bootstrap analysis for a moderate dataset of 66 taxa and 1.5 Kb
in just 2 days. Given that it takes much more time than that to generate the
real data (and the growing affordability/availability of Beowulf clusters), it
makes little sense to take shortcuts in data analysis.

Parametric Bootstrap Test of Absolute Goodness of Fit

One caveat that must be given in the use of parametric bootstraps is
their reliance on the chosen model of evolution. In relying on the chosen
model to simulate the null distribution, one makes the assumption that the
model is adequate (Felsenstein, 2003). In the example given above, despite
that we have selected the HKY+I+Γ model objectively based on its fit/
performance relative to others examined, we still have no indication about
its absolute goodness of fit. Goldman (1993) introduced an absolute goodness-of-fit test that is based on simulation. Here, the null hypothesis is a
perfect fit between model and data. As we can see from Eqn. 8, the ML
score that a dataset can have occurs when the model predicts the data
exactly, that is, when the probability of observing each site pattern is equal
to the frequency of each site pattern in the dataset. Thus, the maximum
possible likelihood score can be calculated as

$$\ln L_{max} = \sum_{a=1}^{4^n} f_a(\ln f_a) \qquad (9)$$

This is the unconstrained likelihood and is sometimes called the
multinomial likelihood. One can think of it as the likelihood score under

a scenario in which each site is allowed its own model and tree. This value is calculated by PAUP every time one invokes the lscore or hs commands. The difference between the ML score under the model and the unconstrained likelihood measures the deterioration in fit associated with forcing all the data to a single model (in this case HKY+I+Γ) and tree. Thus, we have a test statistic:

$$\delta = \ln L_{max} - \ln L(\text{Data}|\tau_{HKY+I+\Gamma}) \tag{10}$$

We can now use Seq-Gen to simulate sequences under the model and find the distribution of the difference between the $\ln L_{max}$ and the ML score under the model. Here, we know that the fit between model and data is perfect, because the model was used to generate the data, and therefore, any deviation between $\ln L_{max}$ and the ML score is attributable to stochasticity. For the real data, the ML score under HKY+I+Γ is -6374.32274, whereas the $\ln L_{max}$ is -2938.93723; the test statistic is $\delta = 3435.38551$. To use Seq-Gen to simulate the expectation of this statistic under the null hypothesis of a perfect fit, we use the ML tree under the model and the ML estimates of model parameters. Furthermore, we embed the following text (using $-x$filename in the argument field) into after each replicate data block:

```
begin paup;
    log file=ABGoF.log append;
    dset dist=logdet;
    nj;
    lsc/nst=2 ba=est trat=est ra=g sha=est pinv=est;
    lset ba=prev trat=prev sha=prev pinv=prev;
    set crit=like;
    hs;
    lsc 1/ba=est trat=est sha=est pinv=est
        scorefile=ABGoFML.score append;
end;
```

The log file can then be searched to extract lines containing the string "$-$lnL (unconstrained) =" (the "Copy Lines Containing" tool in text editors can be used for this), which will occur three times in the log file for each replicate. The file ABGoFML.score will contain the ML scores under the HKY+I+Γ model for each replicate, as well as the distribution of differences forms the null distribution. Here, any deviation between the ML score under HKY+I+Γ and the unconstrained (multinomial) model is due simply to stochasticity, because HKY+I+Γ was used to generate the data (i.e., the data fit it perfectly).

In our example, the observed difference of $\delta = 3435.38551$ is well within the distribution simulated under the null hypothesis of a perfect fit between the model and the data (Fig. 3).

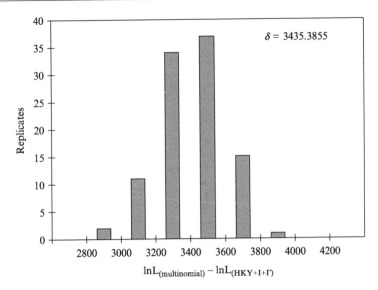

Fig. 3. Results of the Goldman test for absolute goodness of fit. The test statistic falls near the center of the distribution simulated under the null hypothesis of a perfect fit between the data and the HKY+I+Γ model.

The HKY+I+Γ, therefore, seems to be a statistically adequate description of the processes that generated this dataset; the fact that the model assumes independence among sites and ignores codon structure should not lead to biases in application of the model to statistical hypothesis testing.

Concluding Remarks

Advances in model complexity (Yang, 1994; Yang *et al.*, 1994), algorithmic efficiency, and cluster computing have made ML estimation of phylogeny applicable to increasingly large datasets. This is certainly true for phylogeny estimation under a Bayesian framework (Leaché and Reeder, 2002). It is also true under the traditional frequentist framework, in which point estimates of parameters of interest are sought (e.g., optimal topologies) in conjunction with an analysis of the uncertainty associated with the point estimate. Given the amount of time and grant money that are invested in generating sequence data, it makes little sense to analyze data in a less than rigorous fashion.

Acknowledgments

This research is part of the University of Idaho Initiative in Bioinformatics and Evolutionary Studies (IBEST). Funding was provided by NSF EPSCoR EPS-0080935 (to IBEST), NSF Systematic Biology Panel DEB-9974124 (to JS), and NIH NCRR grant NIH

NCRR 1P20RR016448-01 (to IBEST). The following provided much appreciated guidance, advice, editorial comments and/or suggestions with regard to content: Dave Althoff, Ken Berger, Bryan Carstens, Jeremiah Degenhardt, Sarah Hird, Barley Hyde, Eric Roalson, Kari Segraves, Angie Stevenson, Karina Villa, and Liz Zimmer.

References

DeBry, R. W., and Olmstead, R. G. (2000). A simulation study of reduced tree-search effort in bootstrap resampling analysis. *Syst. Biol.* **49,** 171–179.

Felsenstein, J. (1973). Maximum likelihood and minimum-steps methods for estimating evolutionary trees from discrete characters. *Syst. Zool.* **22,** 240–249.

Felsenstein, J. (1985). Confidence limits on phylogeny: An approach using the bootstrap. *Evolution* **39,** 783–791.

Felsenstein, J. (2003). "Inferring Phylogenies." Sinauer, Sunderland, MA.

Frati, F., Simon, C., Sullivan, J., and Swofford, D. L. (1997). Evolution of the mitochondrial COII gene in Collembola. *J. Mol. Evol.* **44,** 145–158.

Goldman, N. (1993). Statistical tests of models of DNA substitution. *J. Mol. Evol.* **36,** 182–198.

Goldman, N., and Whelan, S. (2000). Statistical tests of gamma-distributed rate heterogeneity in model of sequence evolution in phylogenetics. *Mol. Biol. Evol.* **17,** 975–978.

Goldman, N., Andersen, J. P., and Rodrigo, A. G. (2000). Likelihood-based tests of topologies in phylogenetics. *Syst. Biol.* **49,** 652–670.

Hillis, D. M., Mable, B. K., and Moritz, C. (1996). Applications of molecular phylogenetics: The state of the field and a look to the future. *In* "Molecular Systematics," (D. M. Hillis, C. Moritz, and B. K. Mable, eds.), 2nd Ed., pp. 515–543. Sinauer, Sunderland, MA.

Leaché, A. D., and Reeder, T. W. (2002). Molecular Systematics of the Eastern Fence Lizard (Sceloporus undulatus): A Comparison of Parsimony, Likelihood, and Bayesian Approaches. *Syst. Biol.* **51,** 44–68.

Kishino, H., and Hasegawa, M. (1989). Evaluation of the maximum likelihood estimate of the evolutionary tree topologies from DNA sequence data, and the branching order of Hominoidea. *J. Mol. Evol.* **29,** 170–179.

Minin, V., Abdo, Z., Joyce, P., and Sullivan, J. (2003). Performance-based selection of likelihood models for phylogeny estimation. *Syst. Biol.* **52,** 674–683.

Mort, M. E., Soltis, P. S., Soltis, D. E., and Marby, M. L. (2000). A comparison of three methods for estimating internal support on phylogenetic trees. *Syst. Biol.* **49,** 160–170.

Posada, D., and Crandall, K. A. (1998). Modeltest: Testing the model of DNA substitution. *Bioinformatics* **14,** 817–818.

Rambaut, A., and Grassly, N. C. (1997). Seq-Gen: An application for the Monte Carlo simulation of DNA sequence evolution along phylogenetic trees. *Comput. Applied Biosci.* **13,** 235–238.

Shimodaira, H., and Hasegawa, M. (1999). Multiple comparisons of log-likelihoods with application to phylogenetic inference. *Mol. Biol. Evol.* **16,** 1114–1116.

Sullivan, J., and Swofford, D. L. (1997). Are guinea pigs rodents? The importance of adequate models in molecular phylogenetics. *J. Mamm. Evol.* **4,** 77–86.

Sullivan, J., Swofford, D. L. (in preparation). Starting point dependence and the successive approximations approach to maximum likelihood estimation of phylogeny from DNA. *In* "Statistical Methods in Molecular Evolution" (R. Nielson, ed.),

Sullivan, J., Abdo, Z., Joyce, P., and Swofford, D. L. (2005). Comparing successive approximations and simultaneous optimization approaches to maximum-likelihood estimation of phylogeny from DNA sequences. *Mol. Biol. Evol.*, In Press.

Sullivan, J., Holsinger, K. E., and Simon, C. (1996). The effect of topology on estimates of among-site rate variation. *J. Mol. Evol.* **42,** 308–312.

Sullivan, J., Markert, J. A., and Kilpatrick, C. W. (1997). Phylogeography and molecular systematics of the *Peromyscus aztecus* group (Rodentia: Muridae) inferred using parsimony and likelihood. *Syst. Biol.* **46,** 426–440.

Swofford, D. L. (1998). "PAUP*. Phylogenetic Analysis Using Parsimony (*and Other Methods)," Version 4.0b10a. Sinauer Associates, Sunderland, MA.

Swofford, D. L., and Sullivan, J. (2003). Phylogenetic inference using parsimony and maximum likelihood using PAUP*. *In* "The Phylogenetic Handbook" (M. Salemi and A. M. Vandamme, eds.). Cambridge University Press, Cambridge, UK.

Swofford, D. L., Olsen, G. J., Waddell, P. J., and Hillis, D. M. (1996). Phylogenetic inference. *In* "Molecular Systematics," (D. M. Hillis, C. Moritz, and B. K. Mable, eds.), 2nd Ed., pp. 407–514. Sinauer, Sunderland, MA.

Waddell, P. (1995). "Statistical methods of phylogenetic analysis, including Hadamard conjugations, LogDet transforms, and maximum likelihood." Ph.D. Thesis, Massey University, Palmerston North, New Zealande.

Yang, Z. (1993). Maximum likelihood estimation of phylogeny from DNA sequences when substitution rates differ over sites. *Mol. Biol. Evol.* **10,** 1396–1401.

Yang, Z. (1994). Estimating the pattern of nucleotide substitution. *J. Mol. Evol.* **39,** 105–111.

Yang, Z., Goldman, N., and Friday, A. (1994). Comparison of models for nucleotide used in maximum-likelihood phylogenetic estimation. *Mol. Biol. Evol.* **11,** 316–324.

[40] Context Dependence and Coevolution Among Amino Acid Residues in Proteins

By Zhengyuan O. Wang and David D. Pollock

Abstract

As complete genomes accumulate and the generation of genomic biodiversity proceeds at an accelerating pace, the need to understand the interaction between sequence evolution and protein structure and function rises in prominence. The pattern and pace of substitutions in proteins can provide important clues to functional importance, functional divergence, and adaptive response. Coevolution between amino acid residues and the context dependence of the evolutionary process are often ignored, however, because of their complexity, but they are critical for the accurate interpretation of reconstructed evolutionary events. Because residues interact with one another, and because the effect of substitutions can depend on the structural and physiological environment in which they occur, an accurate science of evolutionary functional genomics and a complete understanding of selection in proteins require a better understanding of how context dependence affects protein evolution. Here, we present new evidence from vertebrate cytochrome oxidase sequences that pairwise coevolutionary

METHODS IN ENZYMOLOGY, VOL. 395

interactions between protein residues are highly dependent on tertiary and secondary structure. We also discuss theoretical predictions that impinge on our expectations of how protein residues may interact over long distances because of their shared need to maintain protein stability.

Introduction

Perhaps the most important and well-known use of evolutionary inference in protein biochemistry is the relationship between functional importance and evolutionary conservation. Beginning graduate students studying a novel protein learn that to knock out function, the best places to mutate the protein are the most conserved sites. This relationship is sometimes viewed as almost a tautology, so conserved sites are believed to be functionally important by definition, but surveys of many proteins have revealed that residue conservation can be well predicted based on a combination of distance from active sites and distance from the hydrophobic core (Dean and Golding, 2000). An important development based on this relationship has been that changes in residue conservation can be viewed (again, sometimes tautologically) as strong predictors of changes in the function of those residues. In a somewhat counterintuitive twist, accelerated evolution can also be used as a predictor of functional importance, because the selective forces underlying accelerated evolution (whether long-term diversifying evolution or short-term adaptive bursts) are unlikely to operate on functionally neutral residues.

Although a simple interpretation of the relationship between divergence rates and functional importance has been highly successful (particularly the relationship between absolute conservation and functional importance), it ignores the potential for interaction among residues and the likelihood that functional importance may change over the normal course of evolution. Most evolutionary analyses rely on the assumption that the probabilities of substitution at each site are independent of substitutions at other sites, although protein structure and function result from interactions among amino acids, and this assumption cannot be true in principle. Although hydrophobic effects may be largely additive, hydrogen bonds, charge interactions, and van der Waals interactions among residues are all highly dependent on the size and physicochemical nature of interacting amino acid residues. Such interdependence of physical interactions seems bound to lead to interdependence, or coevolution, in the evolutionary process, and coevolution has indeed been detected on numerous occasions (Atchley et al., 2000; Chelvanayagam et al., 1997; Fukami-Kobayashi et al., 2002; Gobel et al., 1994; Govindarajan et al., 2003; Korber et al., 1993; Lapedes et al., 1997; Neher 1994; Pazos et al., 1997; Pollock and Taylor, 1997; Pollock et al., 1999; Pritchard et al., 2001; Shindyalov et al., 1994;

Taylor and Hatrick, 1994; Tuff and Darlu, 2000; Valencia and Pazos, 2002; Wollenberg and Atchley, 2000). Interdependence should also lead to changes in rates at individual sites during the normal course of evolution, and such rate changes have been found to occur regularly in the absence of functional change (Gribaldo *et al.*, 2003; Lopez *et al.*, 2002; Philippe *et al.*, 2003), sending a loud warning to those who would define functional divergence as synonymous with rate change.

Despite regular detection of coevolution, results have not been obviously consistent as to the conditions and manner in which coevolution apparently occurs. The strongest pairwise signal comes from residues stacked in alpha helices (Pollock *et al.*, 1999), but the strength of pairwise coevolution between more distant residues appears to vary (Pollock, 2002; Pollock *et al.*, 1999), and interaction between protein subunits has had tantalizing but limited success (Fukami-Kobayashi *et al.*, 2002; Pazos *et al.*, 1997; Pazos and Valencia, 2001; Valencia and Pazos, 2002). One reason for the difficulty in consistently detecting coevolution has been that the majority of methods employed ignore phylogenetic relationships, which adds considerable noise and reduces the power of the methods (Pollock, 2002; Pollock and Taylor, 1997). Nevertheless, results from methods that incorporate phylogeny into development of a statistic (Chelvanayagam *et al.*, 1997; Pollock *et al.*, 1999; Shindyalov *et al.*, 1994) indicate that other factors are also at play. These may include the number of sequences analyzed, the depth of the evolutionary relationship between the sequences, the structural or functional context of residues in the sequences analyzed, adaptive bursts, or rate accelerations, and the potentially variable and dispersed nature of coevolutionary interactions between residues.

Using a phylogeny-based method (Pollock *et al.*, 1999), we have analyzed the coevolution of cytochrome *c* oxidase (Wikström, 2004) subunit I (COI) from a large sample of 231 vertebrates, all of which have had their mitochondrial genomes completely sequenced. The large number of genes available from these species allowed us to obtain phylogenetic trees that were only slightly dependent on substitutions in the gene of interest. As the central functional component of the CO complex, a large portion of COI consists of transmembrane helices, heme-binding regions, electron channels, and proton tunnels, as well as some intermembrane and matrix regions, providing many different structural and functional contexts. We are undertaking a detailed serial investigation of all the mitochondrially encoded members of the oxidative phosphorylation complex, and COI was chosen as the first subject partly because of its functional importance and generally conserved evolutionary rate, which indicates that much of the protein will have been in a similar evolutionary context throughout the vertebrate phylogenetic tree. There has been evidence of adaptive evolution in cytochrome oxidase in primates (Goldberg *et al.*, 2003; Wu *et al.*,

2000). We also present some results from COII from the same taxa for comparison. Before analysis, we clustered amino acids at each site according to volume, polarity, and hydrophobicity, and we analyzed the sites with slow substitution rates in greater detail, again to focus on sites for which the structural and functional context might not have evolved much during the range of evolutionary time we are considering. There was some dependency on the physicochemical vector used for clustering, but our main interests here are the stronger correlation of coevolutionary signal with physical distance in the transmembrane domain than within or between other domains, and the tendency for coevolved sites to colocalize with functionally critical regions. We, thus, present only the results for the polarity vector. The weak physical relation of coevolved sites in some protein regions is discussed in terms of theory on protein stability.

Methods

Choice of Sequences

Two critical factors that influence choice of sequence datasets for context-dependent evolutionary analysis are the number of sequences and their distribution, that is, the relationships among them. For there to be coevolution, there must be evolution, and it is, therefore, pointless to include identical or nearly identical sequences, but beyond that it is useful to include sequences that are closely related so that not too many changes (perhaps only a handful) have occurred along most branches. This allows the pinpointing of most replacement changes along the tree, avoids excess random co-occurrence of change along branches, and allows the presumption that the overall context has not changed too much over the course of evolution being examined. If the context changes dramatically and repeatedly, it is to be expected that coevolutionary relationships between sites will also change, and therefore, the signal will be overwhelmed by noise and difficult to detect. For alignment-only methods that ignore phylogenetic relationships, sequences should be as distant as possible to reduce the influence of phylogenetic relationships and to be consistent with the assumption of the methods that all sequences are independent examples of the protein; distant sequences are incompatible with the goal of a relatively constant contextual environment, however, and issues of alignment accuracy can also become a problem for these methods.

Here, vertebrate protein-coding sequences from complete mitochondrial genomes were obtained from GenBank and underwent automated alignment using ClustalW (Thompson *et al.*, 1994) in our EGenBio database. After removing sites involved in multiple insertions and deletions, a phylogenetic tree was constructed from distances calculated using Phylip's ProtDist module (Felsenstein, 1989) and using the neighbor-joining (NJ) heuristic (Saitou and Nei, 1987). Branch lengths were modified using Phylip's ProML and PAM matrices. This tree was trimmed to remove as many long branches or obviously incorrect relationships as possible, ultimately resulting in a dataset of 231 species. The accuracy of the tree topology used and whether to consider a distribution of tree topologies (such as derived from a Bayesian posterior probability distribution or from a bootstrap analysis) are important issues, but they are not central to our discussion of context-dependent change, and we do not consider them here.

Availability of Structure

The availability of three-dimensional structural information for proteins under study is essential for interpreting the relationship of coevolutionary interactions and how they are affected by structure and function. Obviously, we sometimes would like to use coevolutionary analyses to predict structural features and interactions, but to study the question of how structural context affects coevolution one or more high-resolution crystal structures are essential, and it is preferable that at least one should be within the phylogenetic tree under consideration. Homology modeling to predict local structure can be performed if only distantly related structures are available, but this reduces the precision of structural inferences. Here, we visualized coevolved residues on the structure of cytochrome oxidase (including all three mitochondrial-encoded subunits) from bovine heart (1OCR) at 2.35 angstroms resolution (Tsukihara *et al.*, 1996;, 2003). The relationship between coevolution and structure was evaluated by calculating the distance between the Cα atoms (Cα distance) and by the location of the pairs, that is whether they were in the transmembrane domain (TM), or one of the surface domains (S), on either the intermembrane (IM) or matrix (M) side, or between the transmembrane and surface domains (Across, A). We also considered whether pairs were part of secondary structure elements (e.g., alpha helices or beta sheets), but the transmembrane regions are almost entirely alpha helix in nature. Cα distances were clustered into bins of 4.0 angstrom width for comparison of total domain distributions with the distributions of proposed "coevolving" sites, and comparisons were carried out with the standard G test.

Analytical Approach and Statistical Considerations

The choice of an analytical approach will undoubtedly affect the outcome of coevolutionary analysis, but because there is so little information about how coevolution occurs in real proteins, the choice is debatable and not obvious. An approach pioneered by Shindyalov *et al.* (1994) is to evaluate coincident changes along branches. This ignores which amino acids are replaced, although this can be evaluated on an *ad hoc* basis (Chelvanayagam *et al.*, 1997). This method may be strongly affected by inaccuracies in topological inference and by bias in ancestral reconstruction (Krishnan *et al.*, 2004), although such problems can be accounted for, in theory. In principle, this method should do well for detecting coevolution that is nearly simultaneous and if coevolution occurs randomly with respect to physicochemical parameters and amino acids states. Residue-based approaches, in contrast, have potentially greater power to detect coevolution if there is some consistency with regards to amino acids, for example if charge matters, or if there is an energetic need to maintain the volume occupied by hydrophobic side-chain groups in a particular region of the protein structure. The main difficulty with residue-based approaches is that there can be a large number of parameters. Information-theory approaches (Atchley *et al.*, 2000; Korber *et al.*, 1993; Lapedes *et al.*, 1997; Wollenberg and Atchley, 2000), for example, consider whether there are significant associations between states, but for pairs of sites with only 5 of the 20 amino acids each, there are still (at least) 25 parameters to be estimated. Although problems with these methods are often confounded by the absence of phylogenetics in developing the statistic, post-analysis simulations reveal that over-parameterization is a serious hindrance to obtaining reliable results (Wollenberg and Atchley, 2000).

In our "LnLCorr" methodology (Pollock *et al.*, 1999), we avoid problems of over-parameterization by clustering the amino acids into groups or physicochemical "states." The logic behind this is that there may be a primary axis of coevolution with respect to physicochemical properties, and the method will be most powerful if this is so and if the axis is correctly identified. Because the method compares the likelihood ratio of a coevolutionary model of evolution for pairs of sites with that of an independent model, it does not need to estimate ancestral states, and the fewer number of parameters means that it is fairly robust. The power of the method is dependent on the choice of methodology for clustering amino acids, and it is, therefore, generally best to choose at least a few different methods for comparison. Here, for simplicity, we present only the results from clustering according to a vector of polarity. The rate of evolution also matters, both because the rate can affect the ability of a method to detect coevolving sites, and because the same factors that affect the rate of evolution at a site may also affect the likelihood that the site will coevolve with other sites. Here, again for simplicity, we present only the results of coevolutionary analysis among the slowest evolving (most "conserved") half of the sites, which in this analysis had a greater relationship with distance in the three-dimensional structure. Although we have extended our models to allow more than two groups, here we consider only the two-group model delineated by Pollock *et al.* (1999).

A prime reason coevolutionary analysis is difficult and results are hard to interpret is the large number of comparisons, which increase with the square of the number of sites considered. With thousands of comparisons made, this leads to a large multiple-comparisons problem when evaluating significance. One approach is to consider only sites that are still significant after correcting for the number of comparisons (e.g., a Bonferroni correction), but such approaches sacrifice a great deal of power; it cannot be expected that many coevolving sites will be paired strongly enough to lead to extreme levels of significance, and at such extreme levels of significance the lack of data, overparameterization relative to the amount of data, inaccuracies of the model and even small inadequacies of the methodology may overwhelm the results. The approach we take instead is to find the sites with coevolution statistics that are greater than prespecified "significance" levels (i.e., .05, .01, and .002), and consider both whether the number of such sites is greater than expectation and whether the distribution of such sites in the crystal structure is perturbed relative to the distribution of all sites in the same category. By taking such an approach, we can also evaluate the posterior probability that these sites have coevolved or alternatively that they have not coevolved (i.e., the expected number divided by the observed number). Significance levels for values of the likelihood ratio (or any other statistic) need to be determined by parametric bootstrapping, because the chi-square distribution cannot be assumed for coevolution analyses (Pollock *et al.*, 1999). Here, we simulated 6000 pairs for each data comparison, with values sampled randomly from the maximum likelihood estimators.

Results

The strong dependence of coevolutionary results on structural and functional context was demonstrated by the differences among within-domain analyses, across-domain analyses, and the two different subunits. All comparisons showed significantly greater numbers of residue pairs than expected at all significance levels (Table I). COII had the largest excesses, whereas within the surface domain (S) and between domains (A) in COI had the smallest. The relationship between coevolutionary predictions and structural distance also varied greatly among comparisons (Fig. 1). The transmembrane regions showed the clearest relationship between coevolution and distance, with a large excess of closely paired sites in the coevolutionary fractions. The clearest difference between the coevolved

TABLE I
EXPECTED AND OBSERVED COEVOLVING PERCENTAGES AND TOTAL NUMBER OF
PAIRS ANALYZED

Expected	Location[a]	COI all sites		COI conserved		COII conserved	
		%	No.[b]	%	No.[b]	%	No.[b]
5%	S	11.1	1071	12.7	355	17.7	5671
	TM	15.1	5778	16.5	2701		
	A	11.0	7794	13.3	3031		
1%	S	3.7		5.4		7.1	
	TM	6.5		6.7			
	A	3.7		3.9			
0.2%	S	1.3		2.5		4.0	
	TM	2.8		3.9			
	A	1.9		1.9			

[a] COI comparisons were within the surface (S), within the transmembrane (TM), or between the surface and transmembrane domains (A), whereas comparisons in COII were within the entire protein.
[b] The total numbers of sites for each comparison are shown only once, in the top row.

fraction and the total distribution of transmembrane sites was seen for the .2% significance level cutoff (Fig. 1); for higher significance levels, the differences between the distributions are smaller, though still highly significant, and the number of excess close sites is larger than at the .2% level. This indicates that many sites that coevolve due to physical proximity occur within the 5–1% and 1%–0.2% ranges, but that the physically close sites make up a smaller proportion of the sites (this is to be expected, if for no other reason that the expected number of background sites is increasing fivefold between adjacent categories).

Within the surface domain, there are many fewer coevolving sites, but strikingly it appears that coevolution occurs between sites that are close and between those that are distant, but not between those that are moderately close (Fig. 1). This is consistent with earlier results for surface residues of myoglobin (Pollock et al., 1999) and is probably due to maintenance of charge interactions and the charge distribution across the protein. The distance distribution of the closest pairs is different than for the TM analysis, and all four of the more distant coevolving pairs are within the M domain, rather than the IM domain. The coevolving sites from the across-domain comparison (A) show the smallest effect of physical distance (Fig. 1), and most of the excess close sites appear to occur as interactions at the boundary of the transmembrane and surface domain, at the end of the transmembrane helices (unlike many soluble proteins, the domain

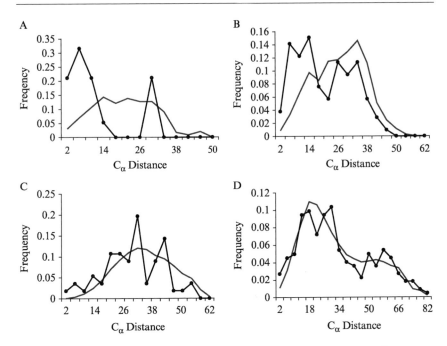

FIG. 1. Structural distance distributions for coevolving residues. Distance (Cα) frequency distributions are shown for all residue pairs (solid lines) and for hypothetical coevolving pairs (circles) within the transmembrane domain of COI (A), within the surface domains of COI (B), across domains of COI (C), and for all pairs in COII (D). The hypothetical coevolving residue pairs shown are for the 0.2% significance level, except for the surface domains that had many fewer sites and for which results at the 1% level are shown.

definitions in COI and many other transmembrane proteins are such that the amino acid chain goes in and out of the different domains repeatedly). As with the TM comparison, for both the S and A comparisons, the larger significance values produce distance distributions more similar to the overall distribution. The distributions of hypothetically coevolved sites in COII (Fig. 1), in contrast to the COI analyses, were not significantly different from the overall distribution of sites.

Concluding Remarks

Pairwise coevolution in vertebrate COI is closely related to distance in the three-dimensional structure, and the correlation with distance is strongest among sites located in the functionally critical transmembrane domain than it is within the two surface domains or across domains. The strongly coevolving pairs were often at the end of helices, echoing the results of Pollock *et al.*

(1999) for vertebrate myoglobin. Interestingly, coevolution appears to be strongest in the functionally critical regions of COI, whereas in COII, which is further from the active site of the CO complex, predicted coevolutionary pairs of sites had no obvious relationship to structural distance.

Although there are clear trends in the relationship of hypothetically coevolving pairs with structural distance, there is also clearly an excess number of coevolving sites that have nothing to do with physical distance. Within COII there was no apparent relationship with distance, even though 7% of the sites were beyond the 1% significance level and 4% were beyond the .2% level (thus, if these predictions are correct, 86% and 95% of the site pairs at these levels have truly coevolved). Possible failures in the topological reconstruction or the model used cannot explain this discrepancy by themselves, because the topology and model are common between the analyses. One possible explanation is that there are adaptive bursts or other forms of variation in the replacement rate along specific branches. Such bursts may tend to be distributed around the protein and would be correlated in evolution only because of a common causal agent. Indeed, in one of the lineages we removed, that of snakes, there were many apparently coevolved pairs that were otherwise conserved throughout vertebrate evolution, and the coevolutionary signal may have been due to an adaptive burst along this lineage. Other explanations may have to do with functionality and exposure to the environment. COI is at the functional core of the CO complex, whereas COII, like COIII and the 10 nuclear-encoded CO complex subunits, is on the periphery, surrounding COI. This may mean that COII has interactions with outside factors that COI is shielded from, and the effect of these outside factors would then be distributed along the elongated COII protein. It may also be that the important functional role of COI, and particularly of the transmembrane helices, leads to tighter pairwise interactions.

Finally, it is worth considering coevolutionary results in an energetic framework. Folding and protein stability may generally be viewed as a global protein variable (Williams *et al.*, 2001; Williams *et al.*, in review; Xu *et al.*, review), and it is easy to conceive that a slightly destabilizing replacement in one part of the protein may be compensated by a replacement leading to greater stability in a distant part of the protein. Certainly mutation studies have long shown that compensatory mutations can occur over long distances in a protein (Brasseura *et al.*, 2001). COII may be selected mostly to bind COI and the other adjacent subunits in the CO complex, so only the overall binding coefficient matters. If this is the case, future developments in coevolutionary analysis should probably be aimed to distinguish which patterns of coevolution are associated with structural distance and which are not, in order to build models that are not only powerful for detecting any kind of coevolution, but also capable of discriminating between different kinds of coevolution, some of which may be of greater interest for a particular goal.

Acknowledgments

This work was supported by grants from the National Institutes of Health (GM065612-01 and GM065580-01), the National Science Foundation (EPS-0346411), and the State of Louisiana Board of Regents (Support Fund, Research Competitiveness Subprogram LEQSF (2001-04)-RD-A-08 and the Millennium Research Program's Biological Computation and Visualization Center), and Governor's Biotechnology Initiative.

References

Atchley, W. R., Wollenberg, K. R., Fitch, W. M., Terhalle, W., and Dress, A. W. (2000). Correlation among amino acid sites in bHLH protein domains: An information theoretic analysis. *Mol. Biol. Evol.* **17,** 164–178.
Brasseura, G., Ragob, J.-P. D., Slonimskic, P. P., and Lemesle-Meuniera, D. (2001). Analysis of suppressor mutation reveals long distance interactions in the bc1 complex of *Saccharomyces cerevisiae. Biochim. Biophys. Acta Bioenergetics* **1506,** 89–102.
Chelvanayagam, G., Eggenschwiler, A., Knecht, L., Connet, G. H., and Benner, S. A. (1997). An analysis of simultaneous variation in protein structures. *Protein Eng.* **10,** 307–316.
Dean, A. M., and Golding, G. B. (2000). Enzyme evolution explained (sort of). *Pac. Symp. Biocomput.* 6–17.
Felsenstein, J. (1989). Phylogeny inference package. *Cladistics.* **5,** 164–166.
Fukami-Kobayashi, K., Schreiber, D. R., and Benner, S. A. (2002). Detecting compensatory covariation signals in protein evolution using reconstructed ancestral sequences. *J. Mol. Biol.* **319,** 729–743.
Gobel, U., Sander, C., Schneider, R., and Valencia, A. (1994). Correlated mutations and residue contacts in proteins. *Proteins.* **18,** 309–317.
Goldberg, A., Wildman, D. E., Schmidt, T. R., Huttemann, M., Goodman, M., Weiss, M. L., and Grossman, L. I. (2003). Adaptive evolution of cytochrome c oxidase subunit VIII in anthropoid primates. *PNAS.* **100,** 5873–5878.
Govindarajan, S., Ness, J. E., Kim, S., Mundorff, E. C., Minshull, J., and Gustafsson, C. (2003). Systematic variation of amino acid substitutions for stringent assessment of pairwise covariation. *J. Mol. Biol.* **328** .
Gribaldo, S., Casane, D., Lopez, P., and Philippe, H. (2003). Functional divergence prediction from evolutionary analysis: A case study of vertebrate hemoglobin. *Mol. Biol. Evol.* **20,** 1754–1759.
Korber, B. T., Farber, R. M., Wolpert, D. H., and Lapedes, A. S. (1993). Covariation of mutations in the V3 loop of human immunodeficiency virus type 1 envelope protein: An information theoretic analysis. *Proc. Natl. Acad. Sci. USA* **90,** 7176–7180.
Krishnan, N. M., Seligmann, H., Stewart, C.-B., de Koning, A. P. J., and Pollock, D. D. (2004). Ancestral sequence reconstruction in primate mitochondrial DNA: Compositional bias and effect on functional inference. *Mol. Biol. Evol* **21**(10), 1871–1883.
Lapedes, A. S., Giraud, B. G., Liu, L. C., and Stormo, G. D. (1997). Correlated Mutations In Protein Sequences: Phylogenetic and Structural Effects. Correlated Mutations in Protein. Proceedings of the AMS/SIAM Conference: Statistics in Molecular Biology. Seattle, WA.
Lopez, P., Casane, D., and Philippe, H. (2002). Heterotachy, an important process of protein evolution. *Mol. Biol. Evol.* **19,** 1–7.

Neher, E. (1994). How frequent are correlated changes in families of protein sequences? *Proc. Natl. Acad. Sci. USA* **91,** 98–102.

Pazos, F., Helmer-Citterich, M., Ausiello, G., and Valencia, A. (1997). Correlated mutations contain information about protein–protein interaction. *J. Mol. Biol.* **271,** 511–523.

Pazos, F., and Valencia, A. (2001). Similarity of phylogenetic trees as indicator of protein–protein interaction. *Protein Eng.* **14,** 609–614.

Philippe, H., Casane, D., Gribaldo, S., Lopez, P., and Meunier, J. (2003). Heterotachy and functional shift in protein evolution. *IUBMB Life* **55,** 257–265.

Pollock, D. D. (2002). Genomic diversity, phylogenetics and coevolution in proteins. *Applied Bioinformatics* **1,** 25–36.

Pollock, D. D., and Taylor, W. R. (1997). Effectiveness of correlation analysis in identifying protein residues undergoing correlated evolution. *Protein Eng.* **10,** 647–657.

Pollock, D. D., Taylor, W. R., and Goldman, N. (1999). Coevolving protein residues: Maximum likelihood identification and relationship to structure. *J. Mol. Biol* **287,** 187–198.

Pritchard, L., Bladon, P., Mitchell, J. M. O., and Dufton, M. J. (2001). Evaluation of a novel method for the identification of coevolving protein residues. *Protein Eng.* **14**(8), 549–555.

Saitou, N., and Nei, M. (1987). The neighbor-joining method: A new method for reconstructing phylogenetic trees. *Mol. Biol. Evol.* **4,** 406–425.

Shindyalov, I., Kolchanov, N., and Sander, C. (1994). Can three-dimensional contacts in protein structures be predicted by analysis of correlated mutations? *Protein Eng.* **7,** 349–358.

Taylor, W., and Hatrick, K. (1994). Compensating changes in protein multiple sequence alignments. *Protein Eng.* **7,** 341–348.

Thompson, J., Higgins, D., and Gibson, T. (1994). CLUSTAL W: Improving the sensitivity of progressive multiple sequence alignment through sequence weighting, position-specific gap penalties and weight matrix choice. *Nucleic Acids Res.* **22,** 4673–4680.

Tsukihara, T., Aoyama, H., Yamashita, E., Tomizaki, T., Yamaguchi, H., Shinzawa-Itoh, K., Nakashima, R., Yaono, R., and Yoshikawa, S. (1996). The whole structure of the 13-subunit oxidized cytochrome *c* oxidase at 2.8 A. *Science* **272,** 1136–1144.

Tsukihara, T., Shimokata, K., Katayama, Y., Shimada, H., Muramoto, K., Aoyama, H., Mochizuki, M., Shinzawa-Itoh, K., Yamashita, E., Yao, M., Ishimura, Y., and Yoshikawa, S. (2003). The low-spin heme of cytochrome c oxidase as the driving element of the proton-pumping process. *Proc. Natl. Acad. Sci. USA* **100,** 15304–15309.

Tuff, P., and Darlu, P. (2000). Exploring a phylogenetic approach for the detection of correlated substitutions in proteins. *Mol. Biol. Evol.* **17,** 1753–1759.

Valencia, A., and Pazos, F. (2002). Computational methods for the prediction of protein interactions. *Curr. Opin. Struct. Biol.* **12,** 368–373.

Wikström, M. (2004). Cytochrome c oxidase: 25 years of the elusive proton pump. *Biochim. Biophys. Acta Bioenergetics* **1655,** 241.

Williams, P. D., Pollock, D. D., and Goldstein, R. A. (2001). Evolution of functionality in lattice proteins. *J. Mol. Graph Model* **19,** 150–156.

Williams, P. D., Pollock, D. D., and Goldstein, R. A. (in revision). Why are proteins marginally stable? II: Functionality strikes back. Proteins: Structure, function, and genetics.

Wollenberg, K. R., and Atchley, W. R. (2000). Separation of phylogenetic and functional association in biological sequences by using the parametric bootstrap. *Proc. Natl. Acad. Sci. USA* **97,** 3288–3291.

Wu, W., Schmidt, T. R., Goodman, M., and Grossman, L. I. (2000). Molecular evolution of cytochrome c oxidase subunit I in primates: Is there coevolution between mitochondrial and nuclear genomes? *Mol. Phylogenet. Evol.* **17,** 294–304.

Xu, Y. O. Hall, R. W., Goldstein, R. A., and Pollock, D. D. (in review). Divergence, recombination, and retention of functionality during protein evolution. Human Genomics.

Author Index

A

Aach, J., 591
Aarts, H. J., 145, 146, 147
Abajian, C., 120, 363, 364
Abatzopoulos, T. J., 165
Abbo, S., 10, 14
Abdo, Z., 767
Abe, S., 223
Abee, T., 39
Abel, Y., 335
Abernethy, K. A., 77
Ackerman, C. M., 419, 434
Acra, A., 88
Adachi, J., 378, 655, 656
Adamczyk, J., 264, 270
Adamowicz, S. J., 475, 477
Adams, B. A., 76
Adams, J. R., 77
Adams, K. L., 312, 378, 573, 580, 592
Adams, M. D., 105, 106, 400, 551
Adams, R. M., 147
Ades, P. K., 280
Adey, N. B., 653, 728, 738
Adoutte, A., 521
Aert, R., 145, 146, 152, 153
Affourtit, J. P., 659
Afshari, C., 599, 605, 614
Agatsuma, T., 311, 330, 334, 340
Aguilar, F., 263
Ahearn, M., 678
Ahlberg, P. E., 541
Ahman, I., 146
Aho, A. V., 749
Ahren, D., 609
Ahrne, S., 268
Ainsworth, C. C., 419
Aitken, E. A. B., 135
Aitken, N., 117, 118, 123
Ajmone-Marsan, P., 146, 147, 150, 151, 153, 154
Akam, M., 521
Akesson, S., 148

Akhter, N., 119
Akkaya, M. S., 147
Akkermans, A. D. L., 39
Albarghouthi, M. N., 248
Albert, V. A., 750
Alberts, S. C., 77
Albertson, D. G., 124
Albertson, R. C., 146, 147, 167
Albrecht, G., 581
Albritton, W. L., 494, 504
Alcivar-Warren, A., 82
Aldwell, F. E. B., 60
Ale, M., 581
Alexander, J., 116, 119, 124
Alexandre, H., 605
Alibardi, L., 637, 638
Alizadeh, A. A., 593
Allard, M. W., 312
Allendorf, F. W., 146, 148
Alm, R. A., 551
Alonso-Blanco, C., 445
Alpers, D. L., 76
Alstrup, V., 52
Altmann, J., 77
Altschul, S. F., 363, 402
Aluru, S., 405
Alvarez-Buyalla, E. R., 725
Amann, R. I., 260, 261, 263, 303, 304
Amato, G., 112
Ambrose, M. J., 182
Amigues, Y., 202
Amores, A., 714, 719
Amos, W., 75, 77, 233
Amsterdam, A., 531
Ananjeva, N. B., 312, 336, 337, 338, 339, 340
Ancrenaz, M., 76
Anders, A. D., 147
Andersen, J. P., 772, 773
Andersen, R. A., 302
Anderson, G. J., 134, 138
Anderson, K. L., 39, 40, 45
Anderson, M. J., 375

Anderson, R., 287, 288, 289
Anderson, R. A., 302
Anderson, S., 88
Anderson, S. L., 146, 153, 154
Andersson, M., 598
Andersson, S. G. E., 719
Andre, C., 135
Andreeva, T., 521
Angel, M. V., 300
Angerbjörn, A., 76
Anh, V., 377, 378
Anisimova, M., 732, 733
Ansanay-Galeote, V., 605
Ansorge, W., 591
Anthony, A., 402
Antoine, N., 335
Antonarakis, S. E., 118
Antonescu, V., 405, 411
Antonovics, J., 280
Aoki, K., 552
Aota, S., 349
Aoyama, H., 783
Apajalahti, J. H. A., 40, 42
Aprelikova, O., 599
Apweiler, R., 408
Arabidopsis Genome Initiative, 583
Archak, S., 146
Archibald, J., 138, 142
Archidiacono, N., 124
Arctander, P., 312
Ardlie, K., 268
Arellano, A. R., 324
Arias, M. C., 385
Armbrust, E. V., 301
Armour, J. A. L., 203
Armstrong, J. L., 59, 62, 63, 64
Armstrong, M. R., 311
Arnason, J. T., 147
Arnaud, J.-F., 281
Arndt, A., 339
Arnheim, N., 424
Arnold, G. M., 146
Arnold, W., 77, 81
Arpin, C., 478, 481
Arvestad, L., 716
Asada, K., 642
Asano, K., 552
Ascenzi, R. A., 404
Ashburner, M., 468
Ashley, M. V., 76, 223

Ast, J. C., 328, 337
Atanassov, I., 420, 422, 423, 425
Atchley, W. R., 780, 784
Atlas, R. M., 42, 62, 63
Attardi, G., 331
Atteson, K., 675
Attwood, T. K., 702, 726
Audiot, P., 280
Auger, D. L., 443, 570
Ausiello, G., 780, 781
Austin, A. D., 340
Austin, C. C., 337
Austin, J. J., 88
Ausubel, F. M., 271, 315, 319, 328, 427, 555
Avery, A., 427
Avgustin, G., 39, 42, 43
Avila, P. C., 124
Avise, J. C., 235
Avner, P., 554
Ayad, W. G., 182, 189, 190, 191, 192, 198
Ayala, F. J., 703
Ayele, K., 119

B

Baarbacar, N., 419, 420, 422, 425, 437
Baba, S., 239
Baba, T., 552
Bachmann, K., 203, 294
Bachtrog, D., 419
Bacon, P. J., 77
Bader, D. A., 335, 684, 685, 689, 691
Baeza, C. M., 134, 138
Bafna, V., 684
Bagley, M. J., 146, 153, 154
Bahrami, A. K., 598
Bailey, C. D., 713, 720
Bailey, J. C., 302
Bailey, J. P., 445
Bailey, L. G., 9
Bailey, M. J., 42
Baillie, B. K., 301
Baillie, L. W., 123
Baker, B. S., 465
Baker, C. S., 118
Baker, C. V., 529
Baker, I., 123
Baker, K., 226
Baker, R. H., 105, 112, 113, 114,
    466, 749, 754

Bakken, L. R., 63
Bakker, F. T., 267
Balavoine, G., 521
Balbiani, E. G., 462
Balcomb, K. C. III, 77, 81
Baldauf, S. L., 116
Baldi, P., 693
Bale, A., 99
Ball, C. A., 591
Ball, S., 107, 109
Ballard, S. G., 425
Balmain, A., 116, 119
Bancroft, D. R., 223
Baner, J., 324
Bankier, A. T., 88
Banks, S. C., 76, 80
Bapteste, E., 116, 703
Barbacar, N., 420, 421
Barbrook, A. C., 378
Barbujani, G., 89
Barch, M. J., 468
Barcia, M., 88
Bargelloni, L., 203, 223, 224
Baril, C. P., 193, 194, 195, 196
Barkay, T., 61, 62, 63
Barker, G. L. A., 301
Barker, J. H., 146
Barker, S. C., 340
Barkman, T. J., 378
Barlow, D. P., 554
Barnes, G. L., 637, 638, 648
Barnes, I., 93, 100
Barnes, W. M., 99, 321
Barnwell, P., 5, 11
Baron, M. H., 541
Barone, E. D., 638, 640
Barraclough, T. G., 746
Barrell, B. G., 88
Barrett, B. A., 193, 194, 196
Barrett, J. A., 223
Barrio, E., 736
Barron, A. E., 248
Barry, S. C., 233
Barsova, E. V., 662, 666
Barthlott, W., 286, 288
Bartley, J., 311, 330, 334, 340
Bartolomei, M. S., 554
Barton, N. H., 443
Bartos, J., 471
Bartoszewski, G., 375

Barwegen, M. W., 147
Basehore, L. S., 599
Baskaran, N., 99
Bass, H. W., 444, 445
Batley, J., 301
Battisti, A., 164, 165, 166
Baulard, A., 497
Baum, B. M., 147
Baum, B. R., 9, 747
Baur, 511
Baxevanis, A. D., 105
Bayes, M. K., 77, 224
Beagley, C. T., 311, 312, 330
Beanan, M. J., 123
Beard, C. E., 39, 40, 42, 43, 45
Beardmore, J. A., 165
Beardsley, P. M., 170
Bearman, G., 538
Beaton, M. J., 311, 312, 330, 473, 476
Beauwens, T., 147, 148, 163, 193, 194, 196
Bebb, C. E., 424, 436
Bechem, C. W., 571, 573
Beck, A., 123
Becker, M., 401, 495
Beckstrom-Sternberg, S. M., 119
Beebe, S., 288, 292
Beggs, M. L., 38, 42
Beheregaray, L. B., 76
Behrens, S., 303
Beisel, K. W., 529
Beitia, F., 165, 166
Bej, A. K., 42, 62, 63
Bekessy, S. A., 280
Belda-Baille, C. A., 301
Bell, J., 289
Bellemain, E., 154
Bellman, S. A., 76
Ben-Amor, K., 39
Bender, C. A., 731, 740
Bender, M., 682
Bender, R. A., 609
Bending, G. D., 61, 63
Benfey, P. N., 106, 123
Benjamin, B., 119
Benjamini, Y., 584, 592
Benkel, B. F., 148
Benner, S. A., 653, 727, 728, 735,
        736, 738, 740, 741, 780, 781, 784
Bennett, J., 165, 166, 167
Bennett, L., 599, 605, 614

Bennett, M. D., 443, 444, 478, 479, 480, 481, 483
Bennetzen, J. L., 134
Bensaid, A., 202
Bensch, S., 147, 148
Bentivenga, S. P., 60
Benton, J. L., 123
Benton, M. J., 703, 746
Bentur, J. S., 165, 166, 167
Berche, P., 497
Berg, T. N., 494, 498, 505, 506, 507
Berger, B., 402
Berghage, R., 61, 67
Berglund, A.-C., 716
Bergman, C. M., 119, 120
Berk, A. J., 555, 559
Berland, K., 522
Bernacchi, D., 385
Bernardello, G., 134, 138
Bernasconi-Quadroni, F., 378
Bernatchez, L., 147, 164
Bernatzky, R., 385
Bernillon, D., 61, 63, 64
Berno, A., 268
Bernstein, C., 165, 168
Berry, K. J., 123
Berthelet, M., 62
Berthier, J., 202
Bertman, J., 312
Bertolla, F., 495, 507, 514
Bertoni, G., 146, 147, 151, 153, 154
Bertorelle, G., 89
Bertranpetit, J., 89
Bertucci, L. A., 385
Beszteri, B., 21, 301
Betancourt, J. L., 81, 82
Bettini, P., 145, 146, 152, 153
Beuchle, D., 531
Beukeboom, L. W., 165, 168
Bevan, R., 119
Bever, J. D., 60
Bhandol, P., 443
Bhargava, A. K., 99
Bhat, K. V., 182, 189, 190, 191, 192, 198
Bidigare, R. R., 302
Biegala, I. C., 261
Bielawski, J. P., 656, 732, 733
Bieler, R., 312, 337
Biémont, C., 478, 481
Binarova, P., 435

Binder, B. J., 260
Bininda-Emonds, O. R. P., 746, 747, 749, 750, 751, 752, 753, 754, 755
Binladen, J., 94
Birch, D. E., 546
Birchler, J. A., 443, 570
Bird, A. R., 39, 45, 151
Birkett, C. L., 402
Birney, E., 118
Birren, B., 385, 386
Birstein, V. J., 123
Bischler, H., 419
Black, M. A., 323, 573, 580, 581, 583, 584, 585, 589, 592, 593, 610
Blackman, R. L., 467
Bladon, P., 780
Blair, D., 311, 330, 334, 340
Blake, J. A., 106
Blakesley, R. W., 119
Blanchard, A. N., 5, 11
Blanchard, A. P., 583
Blanchette, M., 119, 335, 691
Blanco, L., 324
Bland, H., 93, 94
Blanz, P., 55, 56
Blattner, F., 263, 427, 583
Blaut, M., 39, 40
Blaxter, M., 402
Bleaker, M., 427
Blears, M. J., 145, 146, 182, 497
Bleeker, M., 16, 145, 146, 151, 152, 161, 162, 180, 182, 570
Blinn, D. W., 165, 168
Blitchington, R. B., 39, 43
Bloch, W., 432
Blok, V. C., 311
Blondin, D., 605
Blouet, F., 193, 194, 195, 196
Blouin, M. S., 235
Bock, R., 369
Bode, A., 261
Bodeau, J., 451
Boesch, C., 77, 83, 223
Boffelli, D., 107, 116
Bogden, R., 82
Boguski, M. S., 401
Boitani, L., 77
Boivin, A., 471
Bolker, J. A., 541
Bollback, J. P., 654, 655

Bolund, L., 55
Bonaldo, M. F., 401
Bones, A. M., 494, 498, 505, 506, 507
Bonfante, P., 66, 68
Bonierbale, M., 286, 287, 385
Bonin, A., 154
Bonnemaison, D., 39, 45
Bonnet, J., 39, 45
Bontrop, R. E., 122, 123, 609
Bookjans, G., 354, 357
Boom, R., 80, 81
Boonham, N., 123
Boore, J. L., 311, 312, 321, 322, 323, 330, 332, 333, 336, 337, 338, 339, 340, 365, 366, 367, 368, 376, 378, 379, 678, 680
Boorstein, W. R., 554
Bork, P., 112, 702, 726
Borneman, J., 65
Borrone, J. W., 240
Borry, J.-P., 311
Borucki, M. K., 264
Botas, J., 119, 120
Botstein, D., 263, 265, 593, 599, 605
Bouchon, D., 311
Bouffard, G. G., 119
Bougri, O., 404
Bourne, H. R., 727, 737, 741
Bourque, G., 335, 376, 691
Boushey, H. A., 124
Boutin-Ganache, I., 220
Boutte, C., 193, 194, 196
Bouvet, J., 74, 76, 77, 83
Bouzat, J., 223
Bowers, J. E., 385, 445
Bowman, C. M., 372
Bowtell, D., 272
Boyd, A., 444
Boyle, T. J. B., 196
Bozdech, Z., 608, 610
Bozzi, R., 147
Braasch, I., 719
Bradeen, J. M., 147
Bradley, B. J., 77
Bradshaw, H. D., 385
Brady, A., 124
Braid, M. D., 63
Braman, J., 599
Bramel, P., 286, 288, 294
Brand, L. E., 300
Brand, M., 531

Brand, T. B., 94
Brandon, R. C., 105
Brannan, C. I., 554
Brasseura, G., 788
Bratu, D. P., 123
Braun, T. A., 402
Bray-Ward, P., 324, 359
Brazma, A., 591
Breck, C. K. D., 494, 498, 502, 505, 510
Bredemeijer, G., 145, 146, 152, 153
Bremer, K., 105
Brendel, V., 402, 405, 408, 409, 412, 414
Brennecke, A., 420, 433
Brenner, E. D., 106, 123
Brenner, J., 303, 304
Brenner, S., 581
Brent, R., 271, 315, 319, 328
Brettschneider, R., 145, 146, 152, 153
Brettske, I., 261
Bridge, D., 311, 312, 338
Bridge, P. D., 52
Bridgen, J., 638, 639
Bridgham, J., 581
Briles, W. E., 148
Brinkac, L. M., 123
Brinkley, C. P., 119
Brinkman, F. S., 408
Brinkmann, H., 703
Brito, B., 495, 514
Britschgi, T. B., 260, 304
Britton, T., 754
Brochier, C., 116
Brochmann, C., 154
Bronken Eidesen, P., 154
Bronner-Fraser, M., 529
Brookfield, J. F. Y., 223, 233
Brooks, S. Y., 119
Brossard, N., 312
Brouat, C., 280
Broun, P., 385
Brousseau, R., 608
Brower, A. V. Z., 720
Brown, E. D., 551
Brown, E. L., 263
Brown, J. K. M., 182
Brown, J. R., 116
Brown, P., 605
Brown, P. O., 263, 265, 266, 583, 593, 599, 605, 610, 611
Brown, S. K., 4, 12

Brown, W. M., 311, 312, 323, 330, 336, 337, 338, 340, 678, 680
Brownlee, C., 59
Bruce, K. D., 39, 61
Bruce-Johnson, W. A., 495, 509, 513, 515
Bruen, T. C., 119
Bruford, M. W., 76, 77, 224
Bruno, W. J., 675, 689
Bruns, M. A., 59, 62, 63, 64
Bruns, T. D., 50, 59, 60, 62, 63, 66, 67
Bruns, U., 77, 81
Bryant, D., 687, 749
Bryant, J. A., 5, 11
Bryant, J. E., 60, 62, 64
Bucci, G., 281
Bucher, P., 118, 408
Buchner, A., 261
Buck, K. R., 304
Buckley, T. R., 655
Buckton, S. T., 282, 289
Budiman, M. A., 453
Budowle, B., 312
Buell, C. R., 402, 451
Buiatti, M., 145, 146, 152, 153, 182
Buijsman, P., 494, 498, 502, 505, 510
Bult, C. J., 105, 106
Bunce, M., 89, 94
Bunnell, M., 88
Bunstead, N., 148
Buntjer, J. B., 146, 147
Burcham, T., 581
Burchard, J., 583
Burg, T. M., 702
Burgan, M. F., 653
Burger, G., 311, 312
Burghoff, R. L., 61, 62
Burgman, M. A., 280
Burke, D. T., 385
Burke, J. J., 62
Burke, T., 76, 77, 148, 222
Burleigh, J. G., 753, 754
Burow, G. B., 445
Burow, M. D., 385
Burt, D. W., 148
Burton, J., 39, 40, 42, 43, 45
Buscot, F., 68
Bush, R. M., 731, 740
Bushel, P., 599, 605, 614
Buss, L. W., 311, 312, 338
Buzard, G. S., 428

Buzek, J., 423, 424, 425, 435
Byers, B., 571, 573, 581
Byrne, M. C., 263

C

Caballero, P., 288, 289, 292
Cabezas, J. A., 165, 166
Cabrita, J., 39, 45
Cadd, G. G., 76, 77, 81, 82
Cai, W.-W., 116, 119
Cairney, J. W. G., 134, 619
Caithness, N., 288, 289
Califano, J., 124
Call, D. R., 264, 268, 269
Callaerts, P., 653, 728, 738
Calsyn, E., 193, 194, 195, 196
Calvert, R. J., 428
Camacho, F. J., 142
Camarra, J.-J., 76, 77
Cameron, A., 123
Camm, J. D., 286
Campbell, D., 147, 164
Campbell, L. L., 260, 303, 304
Campbell, N., 123
Campbell, N. J., 311, 330, 334, 340
Cangelosi, G., 304
Canman, J. C., 540
Cann, R. L., 746
Cannas, R., 480, 481
Cannon, S. B., 715
Cano, R. J., 88
Canter, D., 268
Cantor, C. R., 655
Cantrell, R. G., 182, 193, 194, 196
Cao, D., 147, 148
Cao, W.-W., 38
Cao, Y., 655, 656
Capella, A. N., 400
Capelli, C., 94
Caperta, A., 555
Caprara, A., 691
Cara, F. A., 400
Caramelli, D., 89
Carapelli, A., 338
Carbone, L., 124
Cardillo, M., 754
Cariaga, K., 119
Carle, G. F., 385
Carlson, C. A., 494, 504

Carmel, G., 551
Carninci, P., 120, 121
Caron, D. A., 261
Carpenter, P. J., 77
Carr, J., 193, 194, 196
Carroll, S. B., 106, 111, 112, 113,
    114, 115, 116, 521, 702, 704
Carson, H. L., 466
Carss, D. N., 77
Carter, N. P., 424, 436
Cartinhour, S. W., 581
Carulli, J. P., 728, 738, 741
Caruso, A., 551
Carver, W. E., 637, 638
Casagranda, F., 638, 639
Casane, D., 781
Cases, I., 702
Casoli, A., 89
Cassandro, M., 146, 147, 151, 153, 154
Cassidy, M. B., 497
Castaglione, S., 145, 146, 152, 153
Castellanos, C., 147, 154
Castillo-Davis, C. I., 610
Castle, J., 411
Cau, A., 480, 481
Causton, H. C., 266, 591
Cavalieri, D., 607
Cavalier-Smith, T., 311, 312, 330
Cavet, G., 583
Cedergren, R., 312, 335
Celniker, S. E., 119, 120
Cembella, A., 303
Cenis, J. L., 165, 166
Centurion-Lara, A., 93, 94, 95, 98
Cerniglia, C. R., 38, 42
Cerrina, F., 263, 583
Cervera, M. T., 147, 151, 154, 165, 166
Chaabane, R., 193, 196
Chada, K., 571
Chadee, D. D., 164
Chakraborty, R., 312
Chakravarti, A., 119, 124
Chamberlin, H. M., 311, 340
Chambers, J. R., 38, 39
Chambers, K. E., 77, 83, 223
Chambers, P. A., 39
Chambers, S. M., 134
Chan, Y. L., 152
Chandel, G., 165, 166, 167
Chandler, D. P., 268, 269

Chandrasekharan, U. M., 653, 728, 738
Chang, B. S., 653, 657, 659
Chang, B. S. W., 653, 728, 738, 741
Chang, C., 385
Chang, H. S., 609
Chang, S., 29, 30, 619
Chang, S.-B., 453
Chang, Z., 349
Chanin, P. R. F., 77
Chapman, J., 324
Charleston, M. A., 678, 710, 718
Charlesworth, B., 419
Charlesworth, D., 419, 420, 421, 422, 423, 424,
    426, 427, 429
Charlton, J. W., 419, 434
Chase, M. W., 110, 182, 378, 746
Chatterjee, S. N., 134
Chavez, F. P., 304
Chee, M. S., 263, 268
Cheeseman, C. L., 77
Chelius, M. K., 65
Chelvanayagam, G., 780, 781, 784
Chen, A., 725
Chen, D., 754
Chen, K., 712
Chen, L., 123
Chen, M., 572, 573, 581, 583, 585, 592
Chen, S., 38, 39
Chen, Y.-Q., 22
Chen, Z. D., 378
Chen, Z. J., 443, 555, 557, 570, 571, 572, 573,
    580, 581, 583, 584, 585, 589, 592, 593, 610
Chenchik, A., 120, 121
Chenery, G., 378
Cheng, F. S., 4, 12
Cheng, H. H., 146, 148, 151
Cheng, R., 571
Cheng, S., 321
Cheng, Z. K., 444, 445, 451
Cheung, F., 405, 411
Cheung, V. G., 263, 436
Chi, M., 116, 119, 124
Chiapelli, B., 401
Chiarelli, B., 89
Childs, G., 263
Childs, K. L., 445, 581
Chin, K., 124
Chisholm, S. W., 260, 263
Chissoe, S., 401
Chiu, W.-L., 372

Cho, J., 404, 405, 411
Choi, S., 386
Chou, H.-H., 402
Choudhuri, J. V., 373
Chowdhary, B. P., 223
Choy, F. Y. M., 659
Chretiennot-Dinet, M.-J., 302
Christopher, D., 351
Chu, K. H., 377, 378
Chuah, A., 146
Chumley, T. W., 376, 378
Chunqongse, J., 349
Church, D. M., 409
Churchill, G. A., 584, 585, 599, 614
Chyi, Y.-S., 134
Cicogna, M., 146, 147, 154
Ciofi, C., 224
Ciuperescu, C., 420
Claassen, V. P., 63
Clague, 171
Claridge, D. E., 77, 81
Clark, A. G., 164, 736, 741
Clark, E., 511
Clark, F., 411
Clarke, J. D., 537
Clary, D. O., 311, 331, 340
Claustre, H., 302, 303
Clayton, D. A., 312, 337
Clayton, R. A., 105, 551
Clegg, M. T., 88
Clement, B. G., 39
Clerc, S., 61, 63, 64
Cleveland, W. S., 592
Clifford, S. L., 77
Coble, M. D., 312
Cockburn, A., 233
Cockerham, C. C., 165
Cocks, K., 288
Coffey, E., 583
Cohan, F. M., 494, 502, 505
Cohen, F. E., 608, 610, 727, 737, 741
Cohen, Y., 124
Cole, J. R., 261
Collard, J. M., 497
Collazo, A., 529
Collier, C. T., 38
Collins, F. S., 213, 642
Collins, M. D., 39, 40
Collins, T. M., 312, 336, 337, 338,
    678, 680, 730

Colot, V., 443, 570, 583, 585, 592
Coltman, D. W., 224
Coluccia, E., 480, 481
Comai, L., 443, 570, 571, 572, 573, 580, 581,
    583, 584, 585, 589, 592, 593, 610
Conant, D., 678
Concato, S., 164, 165, 166
Conery, J. S., 703, 716, 725
Connell, C. R., 432
Connet, G. H., 780, 781, 784
Constable, J. L., 76, 223
Constantino, S., 165, 166, 167
Conway, J. M., 182, 193, 194, 196
Cook, C. E., 521
Cooper, A., 89, 93, 94, 100, 334
Cooper, M. S., 532
Coote, T., 224
Copeland, N. G., 117, 118, 123
Copertino, D. W., 351
Coppens d'Eeckenbrugge, G., 288, 292
Corcoran, K., 581
Cordum, H. S., 419
Corkett, C. J., 473
Cornelsen, S., 349, 351, 367, 375, 377
Cornillet-Lefebvre, P., 465
Coruzzi, G. M., 106, 123
Coryell, V. H., 146, 148
Coscoy, L., 124
Cosner, M. E., 335, 351, 372,
    375, 376, 377, 678
Cotton, J. A., 703, 711, 712, 713, 719, 720
Cotton, M. D., 105
Cottrill, M., 39, 40, 42, 43, 45
Coulibaly, S., 193, 196
Coulson, A. R., 88
Courtois, S., 61, 63, 64
Cowe, E., 658
Cox, N. J., 731, 740
Cox, P. T., 226
Coxon, K. E., 77
Craddock, C., 224
Craig, E. A., 554
Crainey, J., 29, 30
Crandall, K. A., 302, 663, 728, 738,
    741, 762, 766
Crane, C. F., 444
Crawford, A. M., 233
Crawford, D. J., 134, 138, 142
Crawford, R. L., 62
Creel, S., 77

Crepaldi, P., 146, 147, 154
Crespo, A., 52
Cresswell, N., 62, 63
Crews, D., 653
Criel, G. R. L., 165
Crittenden, L. B., 148
Crocker, L. A., 638, 639, 640, 641, 642
Cronn, R., 573, 580, 592
Crooijmans, R. P., 148
Crotty, D., 541
Crow, J. F., 462
Crozier, R. H., 467
Crubezy, E., 89
Cruickshank, R. H., 702
Cruz, M., 259, 287
Csuti, B., 286
Cubero, O. F., 52
Cucumel, G., 755
Cui, C., 351, 372, 375, 377
Cui, L., 376, 378, 552, 687, 691, 692, 693
Cui, X., 424
Cullen, D. W., 60, 61, 62, 63
Culley, T. M., 134, 138, 142
Culpepper, P., 408
Cumming, G., 288, 289
Cunningham, C. W., 311, 312, 338, 654
Curtis, D. E., 268
Cutler, D. J., 119
Cvitkovitch, D. G., 495, 505, 508
Cywinska, A., 107, 109
Czelusniak, J., 708, 711
Czepulkowski, B. H., 468
Cziko, A.-M., 614

D

Daehler, L. L., 336, 338, 678, 680
Dahl, 496
Dai, G., 106, 123
Dai, H., 583
Dalal, S. S., 571
Dalén, L., 76
Dalin, E., 324
Dallai, R., 338
Dallas, J. F., 77, 81
Dalrymple, 171
Daly, A., 145, 146, 152, 153
Daly, D. S., 268, 269
Daly, M. J., 268
D'Amico, L. A., 532

Dangolla, A., 76
Daniels, L. M., 63
Danilchik, M. V., 312
Danilenko, N. G., 356
Danley, P. D., 146, 147, 167
Darlu, P., 780
Dart, R. T., 60
Da Silva, A. M., 408
da Silva, F. R., 400
Datwyler, S. L., 142
Daubin, V., 753
Daugherty, S. C., 123
Davenport, B., 76
David, R. W., 583
Davidson, L., 528, 540
Davies, T. J., 746
Davis, C., 583
Davis, J. I., 375
Davis, R. W., 583, 609
Davison, J., 571, 581
Davydenko, O. G., 356
Day, M. J., 493, 495, 496, 508, 515
Day, W., 676
Dayhoff, M. O., 656
Dean, A. M., 780
Dean, F. B., 324, 359
Dearborn, D. C., 147
Debaud, J.-C., 65
DeBoy, R. T., 123
de Bruijn, M. H., 88
de Bruijn, S. M., 571, 573
DeBry, R. W., 769
De Domenico, E., 260
De Domenico, M., 260
Deetz, K., 123
Degen, B., 292
De Giorgi, C., 90
de Grandis, S. A., 145, 146, 182
de Haas, J., 145, 146, 147
Deiana, A. M., 480, 481
de Jong, H., 453, 456
de Jong, J. H., 425, 426, 445, 451
de Jong, P. J., 119, 120, 398
de Jong, W. W., 106, 117, 118, 123
deJonge, B. L., 551
de Koning, A. P. J., 784
De la Cruz, J., 286
de la Cruz, M., 4, 10
Delage, E., 312
Delahay, R. J., 77

deLange, C., 39, 40, 42, 43, 45
de Leon, A. P., 148
Delichère, C., 420, 421, 422,
    423, 425, 437
DelMonte, T. A., 385
DeLong, E. F., 263
de Loose, M., 193, 194, 195, 196
De los Rios, A., 55, 56
Delpero, M., 66, 68
Demanéche, S., 494, 511, 512
den Bieman, M. G., 146, 147, 148, 151
Denis, M., 407
Dennett, M. R., 261
Deno, H., 349
de Oliveira, A. C., 134
dePamphilis, C. W., 376, 378, 687,
    691, 692, 693
dePamphilis, W., 351, 372, 375, 377
DePriest, P. T., 49, 50, 52
Dequin, S., 605
de Rieks, J., 193, 194, 195, 196
DeRisi, J. L., 124, 263,
    266, 605, 608, 610
De Riva, A., 106, 122
Dermatzakis, E. T., 736, 741
Dermitzakis, E. T., 118
de Rosa, R., 521
de Ruijter, C., 147
DeSalle, R., 88, 105, 106, 111, 112,
    116, 117, 118, 123, 311, 312, 338,
    465, 720, 750, 753
Deschepper, C. F., 220
Desjardins, P., 337
Desmarais, C., 366
Despres, L., 170
Detter, J. C., 324
Deutsch, G., 55, 56
Deutsch, S., 118
Devereux, R., 260
Devos, K. M., 385
de Vos, W. M., 39, 40
de Vries, J., 502, 505
De Waal, E. C., 16
DeWaard, J., 107, 109
De Wachter, R., 260
Dewey, M. J., 203
Dey, D. K., 142, 143
Dhat, A. K., 82
Diachenko, L., 120, 121
Diao, L., 754

Dias, J. M., 282, 289
Dias, Z., 684
Dib, C., 202
Dickie, I. A., 64, 65, 67
Dickinson, D. P., 653, 728, 738
Dickinson, M. E., 538, 541
Dickinson, T., 746
Dietmaier, W., 424, 437
Dietrich, N. L., 119, 401
Díez, B., 260
DiFiore, A., 203, 214
Dijkshoorn, L., 145, 146, 147
Dimcheff, D. E., 328, 338, 339, 340
Ding, B., 76
DiPierro, D., 268
Disotell, T. R., 77
Di Stilio, V. S., 427
Doco-Fenzy, M., 465
Dodds, K. G., 233
Dodgson, J. B., 146, 148, 151
Dodson, R. J., 123, 551
Doebley, J., 385
Doerfler, W., 39, 45
Doerge, R. W., 443, 570, 573, 580,
    581, 583, 584, 585, 589, 592, 593, 610
Dogson, J. B., 146
Doi, R. H., 260
Doi, T., 659
Doig, P. C., 551
Dolezel, J., 420, 422,
    423, 435, 471, 480
Dollinger, G., 544
Don, R. H., 226
Donahoe, P. K., 116, 119
Dong, H., 263
Dong, J., 581
Dong, Q., 402, 409
Donnison, I. S., 422, 423, 424, 437
Donoghue, M. J., 378, 653, 657, 728, 738
Doolittle, W. F., 116
Doré, J., 39, 40
Dorit, R. L., 622
Dortch, Q., 303
Douady, C. J., 106, 117, 118, 123
Double, M. C., 233
Doucette, G. J., 303
Dougherty, B. A., 105, 551
Douhan, G. W., 134
Doukakis, P., 123
Downie, S. R., 349, 351, 372, 678

Downs, B., 286
Dowson, C. G., 494, 502, 505
Dowton, M., 340
Doxiadis, G. M., 122, 123, 609
Doyle, J. J., 22, 26, 135, 362, 375
Doyle, J. L., 22, 26, 362
Draper, D., 288, 293
Drayer, X., 385, 445
Dress, A. W., 780, 784
Drew, R. A., 193, 194, 196
Driessen, G., 165, 168
Driscoll, M., 324, 359
Driskell, A. C., 750, 753
Drouin, J., 88
Drumm, M., 213, 642
Drummond, A., 89
Du, J., 324, 359
Du, Y. F., 324, 359
Du, Z., 150
Dubendiek, S. L., 324
Dubnau, D., 260, 492, 493
Dubnick, M., 400
Dubois-Paganon, C., 76, 77
Dubreuil, P., 193, 194, 195, 196
DuBridge, R. B., 581
DuBuque, T., 401
Duchesne, P., 147, 164
Duddridge, J. A., 59
Dudley, J. W., 193, 194, 196
Dufton, M. J., 780
Duhon, M., 268
Duim, B., 145, 146, 147
du Jardin, P., 193, 194, 196
Dunham, R. A., 147, 148
Dunn, S., 639, 640
Dupanloup, I., 89
Duque, M. C., 193
Durand, D., 712
Durando, C. M., 465
Durban, J. W., 77, 81
Durbin, M., 88
Durbin, R., 332
Duret, L., 707
Durkin, A. S., 123
Durufle, L., 703
Dushoff, J., 725, 739
Dutrillaux, B., 468
Duyk, G., 203
Dyer, A. T., 147
Dyer, T., 372

E

Eannetta, N. T., 400, 411
Earnest-DeYoung, J., 687
Eddy, S. R., 332, 703, 705, 710, 711, 713, 715
Edgell, M. H., 653, 728, 738
Edvardsen, B., 300, 303, 304
Edwards, K. J., 145, 146, 152, 153, 179, 301
Edwards, S. G., 63, 66
Edwards, T. C., Jr., 288
Effosse, A., 65
Efron, B., 196
Egan, S., 497
Eggenschwiler, A., 780, 781, 784
Eggert, J. A., 77
Eggert, L. S., 77
Egholm, M., 324, 359
Eichler, E. E., 124, 461, 472
Eide, D., 605
Eisen, J. A., 106, 107, 116, 123, 703, 719
Eisen, M. B., 593, 599, 610, 611
Eisenberg, D., 727, 737, 738, 741
Eizirik, E., 106, 112, 117, 118, 123
Elbrächter, M., 303
Eldrege, N., 259
Eletr, S., 581
Elkin, 327
Ellegren, H., 223, 224
Ellen, R. P., 495, 505, 508
Ellengren, H., 202
Eller, G., 260, 261
Ellis, L. B., 408
Ellis, T. H. N., 182
Ellisens, W. J., 378
El-Mabrouk, N., 335, 682, 685, 693
Elnitski, L., 119, 372, 373
Elsik, C. G., 385
Elwood, H. J., 20, 25, 271
Enard, W., 122, 123, 609
Endl, E., 424, 437
Endo, T., 731
England, P. R., 163
Ennos, R. A., 280, 281
Enright, A. J., 702
Eperon, I. C., 88
Erlandsson, R., 609
Erlich, H., 104
Ernst, J., 264, 270
Ernsting, G., 165, 167

Errampalli, D., 497
Erschadi, S., 193, 194, 196
Escaravage, N., 147
Escudero, A., 281
Escuyer, V., 497
Esposito, D., 124
Esselman, E. J., 134, 138, 142
Essich, E., 502
Estaban, J. A., 324
Etges, W. J., 465
Eulenstein, O., 710, 750, 753, 754
Evans, J. D., 472
Everaert, I., 193, 194, 195, 196
Evershed, R. P., 93, 94
Ewan, M., 581
Ewing, B., 120, 363, 430
Ewing, E. P., 280, 282
Excaravage, N., 74, 77, 83
Excoffier, L., 143, 165, 197
Eybert, M. C., 147

F

Fabbri, E., 77
Fabian, C., 531
Fahrenkrug, S. C., 408
Fahselt, D., 52
Fain, S. R., 76, 77, 81, 82
Falciani, F., 404
Fallon, S., 268
Faloona, F., 104
Fan, J. B., 268
Fan, W. H., 351
Fan, Y., 467
Fang, L. H., 324, 359
Fang, Z., 336, 337, 338, 339, 340
Fanizza, G., 193, 196
Farach-Colton, M., 712
Faraut, T., 407
Farber, R. M., 780, 784
Farbos, I., 420, 421
Fares, M. A., 736
Farinas, I., 529
Farrell, L. E., 76
Farris, J. S., 105, 750
Farris, R. J., 261
Faruqi, A. F., 324, 359
Fassbender, A., 493
Fatehi, J., 52
Faught, M., 39

Favello, A., 401
Fawley, M. W., 269
Fay, M. F., 182
Fayer, R., 39
Fearon, K., 581
Fedlmann, K. A., 385
Feinberg, A. P., 180
Feldmaier-Fuchs, G., 311, 330
Feldman, L. J., 609
Felgel, V., 118
Felsenstein, J., 105, 196, 378, 655,
    656, 674, 676, 704, 711, 716, 752,
    760, 768, 775, 782
Feng, J., 147
Feng, L., 268, 659
Fensome, R. A., 260
Ferguson, C., 20
Ferguson, J. W. H., 77, 82
Ferguson, M. E., 286, 288, 294
Fernandes, M. L., 76, 77
Fernández-Baca, D., 753
Fernando, K., 165, 166, 167
Fernando, P., 76
Fero, M., 599
Ferreira, P. C., 400
Ferretti, L., 659
Ferretti, V., 680
Ferrier, S., 288, 289
Ferriol, M., 193, 194, 196
Feuk, L., 116, 119
Feulgen, R., 477
Field, K. G., 260, 521
Fielding, A., 289
Fields, C., 105, 106
Fields, R. L., 76, 223
Fieller, E. C., 606
Figurski, D. H., 111, 112, 116
Filatov, D. A., 419, 420, 421,
    422, 423, 425, 429
Fine, D. H., 112, 116
Fine, L. D., 105
Fink, G. R., 583
Fire, A., 324
Firestone, M. K., 511
Fischer, D., 727, 737, 741
Fischer, R. U., 203
Fisher, D., 203
Fisher, D. L., 95, 98
Fisher, M. E., 546, 547, 551, 552
Fitch, M. M. M., 193, 194, 196

Fitch, W. M., 110, 676, 701, 711, 718, 725, 727, 731, 740, 780, 784
Fitter, A. H., 63, 66
Fitzgerald, L. M., 551
FitzHugh, W., 105
Flagstad, Ø., 77
Fleischer, R., 76, 78, 203, 214
Fleischmann, R. D., 105, 551
Flesher, B., 39, 42, 43
Flint, H. J., 39, 42, 43, 495, 496, 509, 513, 515
Flood, S. A. J., 123
Flook, P., 340
Florea, L., 372, 409
Fockler, C., 321, 544
Fodor, S. P., 583
Follettie, M. T., 263
Folmer, O., 323
Force, A., 714, 719, 725, 726
Ford, M. J., 659
Ford, T., 89
Ford-Lloyd, B. V., 178
Forster, R. J., 38, 39, 40, 42, 43, 45
Forster, W., 261
Forterre, P., 116
Fortey, R. A., 88
Fortin, J. A., 59, 63
Foster, B., 525, 526, 528
Foster, P. G., 106, 122
Fourcade, H. M., 311, 312, 330, 337, 338, 339, 340
Fourcade, M., 376, 378
Fouts, D. E., 123
Fowler, J. C. S., 203
Fowles, N. L., 59, 63
Fox, P. N., 193, 196
Foy, M., 581
Fradkov, A. F., 662, 666
Franca, S. C., 400
Francisco-Ortega, J., 134, 138, 142
Franco, M. d. R., 76
Franklin, J., 287, 288
Fransz, P. F., 424, 425, 445, 451, 456
Frantz, A. C., 77
Frantzen, M. A. J., 77, 82
Fraser, C. M., 105, 106, 107, 123
Fraser, H. B., 725, 739
Fraser, S. E., 529, 538, 541
Frati, F., 338, 761
French, J. O., 637, 638, 642, 647, 648, 649
Freudenstein, J. V., 713

Frewen, B. E., 385
Frey, J. E., 124
Freyssinet, G., 496, 516
Friday, A., 656, 777
Friedlander, A. M., 123
Friend, S. H., 583
Frijters, A., 16, 145, 146, 151, 152, 161, 162, 180, 182, 427, 570
Frisch, D. A., 386, 387, 390, 450
Frischer, M. E., 502
Fritchman, J. L., 105
Fritsch, E. F., 34, 75, 173, 204, 206, 213, 214, 421, 438, 571, 573, 575, 579, 581
Fritsch, R. M., 294, 443
Fritzsch, B., 529
Frohman, M. A., 642
Frommer, M., 151
Frost, B. W., 473, 476
Frostegård, Å., 61, 63, 64, 495, 514
Fruth, B., 77
Fry, J. C., 60, 62, 63, 64, 493, 495, 496, 508, 515
Fryxell, G. A., 303
Fuchs, B. M., 303
Fuchs, J., 423
Fujimori, M., 193, 196
Fujimoto, J., 38
Fujisawa, M., 420, 433
Fujiwara, M., 655
Fukamachi, S., 148
Fukami-Kobashi, K., 780, 781
Fukui, K., 444
Fukushima, N., 124
Fukuzawa, H., 349, 420, 433
Fulton, T. M., 385, 400, 411
Funk, S. M., 76, 224
Furlong, R. F., 719
Furnier, G. R., 180
Furutaniseiki, M., 531
Furuya, K., 552
Furuya, M., 420, 421

G

Gaasterland, T., 591, 703
Gage, J. D., 300
Gaikwad, A. B., 146
Gaither, L. A., 605
Galbraith, D. W., 479, 480
Gale, M. D., 385

Galen, C., 281
Galitski, T., 583
Gallacher, S., 20
Galleguillos, R., 164, 165, 167
Gallo, M. V., 263
Galtier, N., 655
Galwey, N., 288, 292
Gan, L. H., 193, 194, 196
Gan, Y., 193, 194, 196
Ganal, M. W., 385
Ganem, D., 124
Gao, W., 571, 581
Garbett, C. A., 203
Garcia, C., 288, 293
Garcia-Gonzales, R., 495, 507
Garcia Sáez, A., 301
Gardes, M., 50, 59, 63, 66, 67
Gardiner, D. M., 148, 151
Gardiner-Garden, M., 151
Garg, K., 401
Gargas, A., 49, 50, 52
Garnier, J. N., 77
Garrett, K., 286, 287
Garrett, L., 725
Garrity, G. M., 261
Garvin, D., 375
Gasch, A. P., 605
Gascuel, O., 675, 676
Gaskins, H. R., 38, 39, 40
Gaspari, Z., 269
Gatesy, J., 88, 105, 112, 113, 114,
    749, 750, 753, 754
Gaucher, E. A., 653, 735, 736, 740, 741
Gaurino, L., 287
Gaut, B. S., 378, 656, 693, 728
Gautam, D., 146
Gavin, A. J., 402
Ge, D., 682
Gebhard, F., 496, 516
Gelfand, D. H., 104, 546, 547, 551, 552
Gelfand, R., 331
Gellissen, G., 340
Gentleman, R., 592
Gentzbittle, L., 407
George, D., 581
George, M., Jr., 312
George, R. A., 119, 120
Georges, M., 202
Georgiou, G., 659
Gepts, P., 193, 196

Gerber, S., 261
Gerloff, U., 77
Gernand, D., 444
Gerstein, M., 735
Getts, R. C., 589
Ghandour, G., 268
Ghiorse, W. C., 60, 62, 63, 64
Ghiselin, M. T., 521
Gianinazzi-Pearson, V., 68
Giannasi, D. E., 88
Giannasi, N., 147
Giavalisco, P., 122, 123, 609
Gibbs, R., 116, 119
Gibson, G., 269, 599, 605, 614, 615
Gibson, R., 551
Gibson, T. J., 641, 782
Giegerich, R., 373
Gielly, L., 170
Giesler, T. L., 324
Gifford, J. A., 99
Gilbert, M. T., 93, 94, 100
Gilichinsky, D. A., 94
Gill, B. S., 444, 452, 453
Gill, S. R., 123, 551
Gillespie, J. M., 638, 639
Gillespie, S. E., 124, 465
Gilliam, T. C., 116, 119
Gilmartin, P. M., 420, 421, 423, 424
Ginevan, M. E., 280, 282
Gingeras, T. R., 583
Ginhart, A. W., 261
Giovannoni, J. J., 385
Giovannoni, S. J., 260, 521
Giraud, B. G., 780, 784
Giribet, G., 105
Gish, W., 401, 411
Gissi, C., 312
Git, Y., 404
Gittleman, J. L., 747, 750, 751, 752, 754, 755
Giuliano, L., 260
Giusti, W., 123
Giver, L. J., 22
Gladkov, A., 288
Glas, A. S., 294
Glazebrook, J., 623, 634
Glenisson, P., 591
Glenn, T. C., 203, 637, 638, 647, 648, 649
Glodek, A., 105, 551
Glover, A. L., 496, 515
Glover, L. A., 495, 509, 513, 515

Glynias, M. J., 653, 728, 738
Glynn, M. W., 99
Gnehm, C. L., 105
Gnirke, A., 119, 120
Gobel, U., 780
Göbel, U. B., 62
Gobert, S., 203
Gocayne, J. D., 105, 400
Godwin, I. D., 134, 135
Goeer, S. L.-D., 302
Goeghagen, S. M., 105
Goggins, M., 124
Gojobori, T., 729, 731, 732, 733, 734, 740, 741
Goksøyr, J., 42, 62, 63, 503
Golda, G., 581
Goldberg, A., 781
Goldberg, M. E., 529
Golding, B., 111
Golding, G. B., 312, 780
Goldman, N., 655, 656, 657, 662, 729, 731,
    732, 733, 734, 740, 766, 772, 773, 775, 777,
    780, 781, 784, 785, 786, 787
Goldstein, D. B., 202, 223
Goldstein, R. A., 655, 788
Golenberg, E. M., 88
Gollotte, A., 68
Goloboff, P. A., 676, 749
Golub, T. R., 266
Gomes, C., 288, 293
Gomez, A. V., 301
Gomez, P. I., 268
Gomez-Laverde, M., 287, 288, 289
Gomez-Uchida, D., 164, 165, 167
GÆmez-Zurita, J., 106, 122
Gong, J., 38, 39, 40, 42, 43, 45
Gonzalez, D. O., 193
Gonzalez, M. A., 268
González-Candelas, F., 736
Goodall, J., 76, 82, 223
Goodman, H. M., 444, 445
Goodman, M., 312, 708, 711, 781
Goodman, S. J., 224
Goodnight, K. F., 235
Goossens, B., 74, 76, 77, 83
Gordon, A. D., 747, 749
Gordon, D., 120, 363, 364, 366
Gordon, P., 703
Goremykin, V., 359, 376, 378
Gorni, C., 146, 147, 154
Gosden, J. R., 468

Götherström, A., 76
Goto, S., 552
Gourbiere, F., 494, 511, 512
Gouy, M., 655, 707, 753
Govindarajan, S., 780
Grabowski, G., 323
Graham, C. D., 659
Graham, I., 123
Graham, J. B., 494, 504
Graham, M. W., 213, 642
Graham, S. W., 378
Grainger, R. M., 529
Grams, K. L., 385
Granato, M., 531
Grandillo, S., 385
Granite, S., 119
Grant, M. C., 280, 282
Grant, S. R., 421, 422, 423,
    424, 426, 435, 437
Gras, N., 45
Grassly, N. C., 773
Graupner, S., 494, 512, 513
Graur, D., 703, 705
Gravendeel, B., 170
Gray, G. S., 718
Gray, J. W., 124
Gray, M. W., 311, 312
Graybeal, A., 608
Grayzel, J., 589
Green, E. D., 119, 372, 373
Green, J. C., 301
Green, M. R., 266
Green, P., 119, 120, 328, 329, 363, 364, 366,
    401, 430
Green, R. D., 263, 583
Greenberg, D. L., 147
Greene, E. A., 702, 726
Greenwood, A. D., 76, 94
Greer, C. W., 62
Gregg, K., 638, 639, 640, 641, 642, 648
Gregorius, H.-R., 292
Gregory, T. R., 471, 472, 473,
    475, 476, 478, 483
Greilhuber, J., 443, 471, 480
Grenier, J. K., 521
Grenyer, R., 754
Gribaldo, S., 781
Griffin, S., 74, 76, 77, 83
Griffiths, R. I., 42, 147
Grimaldi, D., 88

Grime, J. P., 443
Grishin, N. V., 112
Griswold, K. E., 659
Groben, R., 16, 21, 260, 261, 301, 303, 304
Groenen, M. A., 148
Gross, A. K., 727
Gross, M. R., 235
Gross, O., 261
Grossman, L. I., 312, 781
Groth, D., 123
Grube, M., 49, 52, 55, 56
Grum, M., 288, 292
Grumann, S., 261
Gu, X., 653, 728, 735, 736, 738, 740, 741
Guan, X., 119
Guarino, L., 285, 286, 288, 289, 292, 293, 294
Guccione, L., 659
Guerra, C. E., 659
Guicking, D., 134
Guidon, S., 676
Guidot, A., 65
Guigó, R., 719
Guild, B. C., 551
Guillard, R. R. L., 300
Guillaume, G., 497
Guillemaut, P., 3
Guillou, L., 260, 261, 302, 303, 304
Guisan, A., 288
Gullans, S. R., 589
Gundersen-Rindal, D., 472
Gupta, J., 119
Gupta, M., 134
Gupta, V., 193, 194, 196
Gurskaya, N. G., 662
Guschin, D. Y., 264
Gustafsson, C., 780
Guttman, D. A., 420, 422, 426, 427
Gwinn, M., 123
Gyapay, G., 202
Gyllensten, U., 312

H

Haake, A. R., 637, 638, 648
Haberer, G., 193, 194, 196
Habib, M., 754
Habicht, C., 146, 148
Hachtel, W., 372
Hacia, J. C., 123
Hacia, J. G., 608

Hackler, L., 599
Haddad, N., 497
Haeseler, A., 100
Hafellner, J., 49
Haffter, P., 531
Haft, D. H., 123
Haghighi, P., 119
Hagler, A. N., 494, 498, 505, 506, 507
Hagstrom, R., 676
Haig, S. M., 134
Hakki, E. E., 147
Halbleib, D., 613
Haldane, J. B. S., 725
Hall, B. G., 641
Hall, I. R., 60
Hall, R. W., 788
Hallett, M. T., 715, 719
Hallick, R. B., 351
Halling, C., 659
Halloran, N. D., 150
Halpern, A. L., 675, 689
Hamadeh, H., 599, 605, 614
Hamidi, A., 635
Hamilton, M. B., 203, 214
Hamilton, R., 286
Hammerschmidt, M., 531
Hammond, R. L., 224
Hampton, J. N., 351, 356, 372
Han, Y., 124
Hanamura, N., 148
Hanawalt, P. C., 106
Hance, I. R., 123
Hancock, R. E., 408
Hanekamp, E., 147
Hanelt, P., 443
Hanna, M. C., 105
Hanna, P. C., 123
Hannenhalli, S., 684, 689
Hanotte, O., 76, 77, 147
Hansen, A. J., 93, 94, 100, 349, 351,
    367, 375, 377
Hansen, B. D., 76, 80
Hansen, N. F., 119
Hanson, R. E., 444
Harbison, C. T., 266
Hardie, D. C., 475, 476, 478
Hardison, R. C., 372, 373
Hardy, O. J., 164
Harel, D., 708
Hariharan, J., 659

Harland, R. M., 529
Harper, I., 525, 538, 539
Harper, J. L., 259
Harris, C., 116, 119
Harris, H., 351, 372, 375, 377
Harris, P., 304
Harris, S. A., 187, 188, 189, 190, 191
Harris-Warrick, 494
Hart, A. L., 404
Hart, M., 60
Hartl, D. L., 164, 599, 605, 607, 610,
    614, 728, 738, 741
Hartman, T., 684
Hartmann, A., 424, 437
Hartzell, G., 409
Harvey, P. H., 716
Hasebe, M., 678
Hasegawa, K., 148
Hasegawa, M., 349, 351, 367, 375, 377, 378,
    655, 656, 770, 772
Hass, B. J., 374
Hastie, N. D., 726, 731
Hastie, T., 288
Hatrick, K., 780
Hatta, K., 537
Hattori, M., 552
Haught, G., 73
Hauser, L., 164, 165, 167
Hausner, M., 495, 508, 512
Hausser, J., 288
Haussler, C., 119
Hauswaldt, J. S., 203
Havey, M. J., 375
Havlak, P., 116, 119
Hawkins, J. R., 180
Hawkins, T. L., 150, 324
Hawksworth, D. L., 259
Hawley, R. S., 468
Hay, J. M., 337
Hayashi, C., 112, 749, 750, 753, 754
Hayashi, H., 552
Hayashi, K., 428
Hayashi, Y., 702
Hayashida, N., 349
Hayashizaki, Y., 120, 121
Haydock, P., 304
Hayes, H., 468
Hayes, P. K., 301
He, C., 123
He, S., 682

He, T., 281
He, Y. D., 583
Healy, J., 116, 119, 124
Heard, E., 554
Heard, S., 674
Hebert, P. D. N., 107, 109, 473, 475, 476, 478
Hector, A., 259
Hedblom, E., 105
Hedges, S. B., 88, 521, 703
Hedley, A., 402
Hedrick, P. W., 280, 282
Heikkinen, P., 40, 42
Heim, R., 661
Heinmoller, E., 424, 437
Heinrichs, J., 68
Heisenberg, C. P., 531
Heissig, F., 122, 123, 609
Helbig, A. J., 147
Helfenbein, K. G., 311, 312, 330, 338
Helfer, V., 288
Helgason, E., 123
Heller, R. A., 583
Hellingwerf, K. J., 494, 498, 502, 505, 510
Hellwig, F. H., 359, 376, 378
Helmer-Citterich, M., 780, 781
Hemerly, A. S., 400
Henderson, D., 88, 467
Hendrickx, L., 495, 508, 512
Hengartner, C. J., 266
Henikoff, S., 702, 726
Hennig, S., 123
Henningsen, K. W., 354, 357
Henning Sommer, J., 288
Henrion, B., 59, 62, 63, 66
Henry, C. A., 532
Henze, C., 746, 747, 753
Herbergs, J., 148
Herbon, L. A., 351
Heritage, J., 39
Hermann, S., 261
Hernandez, H., 4, 10
Herniou, E. A., 311, 330, 334, 340
Hernould, M., 419, 420, 422, 425, 437
Herrick, J. B., 63
Herrmann, R. G., 372
Herwig, R., 123
Heslop-Harrison, J. S., 444, 445, 458
Hessel, D. C., 288
Hetzel, J. S., 202
Heuer, H., 497

Hewitt, G., 224, 468
Heywood, J. S., 281
Hibino, T., 619
Hickey, E. K., 746
Hicks, J., 116, 119
Hickson, R. E., 334
Hietala, J., 480, 481
Higasa, K., 247
Higashiyama, T., 423, 424, 436
Higazi, T., 311, 340
Higbee, J. A., 82
Higgins, D. G., 641, 658, 782
Higgins, J. A., 39
Higuchi, R., 78, 104, 321, 544, 545,
    546, 547, 551, 552
Hijmans, R. J., 285, 286, 287, 293
Hill, J., 537
Hill, K. K., 268
Hill, M., 182
Hill, R. E., 726, 731
Hillel, J., 148
Hiller, L., 120
Hillier, L. D., 401, 430
Hillis, D. M., 105, 654, 674, 684, 759, 761, 773
Hinchliffe, S. J., 123
Hindkjaer, J., 55
Hinds, J., 123
Hingamp, P., 591
Hinnisdaels, S., 420, 421
Hinton, S. M., 61, 62
Hiorns, W. D., 39
Hipkins, V. D., 372
Hiramatsu, K., 552
Hirons, W. D., 61
Hirotsune, S., 725
Hirsch, P. R., 60, 61, 62, 63
Hirsch-Ernst, K. I., 359, 376, 378
Hirschhorn, J. N., 268
Hirzel, A., 288
Hizume, M., 423, 424
Hlen, M., 268
Hnath, J., 659
Ho, S.-L., 119
Ho, V. V., 608
Hobman, J. L., 39, 61
Hobza, R., 420, 421, 422, 426
Hochberg, Y., 584, 592
Hocking-Murray, D., 598
Hodgkin, T., 182, 189, 190, 191, 192, 198
Hodkinson, T. R., 746

Hoeh, W., 323
Hoepffner, N., 300
Hoffman, M. H., 294
Hoffmann, R. J., 311, 330, 337, 340
Hofle, M. G., 260
Hofreiter, M., 81, 82, 93, 94, 99, 100
Hofstadter, F., 424, 437
Hogers, R., 16, 145, 146, 151, 152, 161, 162,
    180, 182, 427, 570
Hohmann, F., 77
Holben, W. E., 40, 42
Holder, N., 537
Holland, J. L., 39, 40, 45
Holland, M. J., 546, 547, 551, 552
Holland, M. M., 95, 98
Holland, P. W. H., 678, 719
Hollocher, H., 385
Holmes, M. H., 402
Holmes-Davis, R., 571, 573, 581
Holsinger, K. E., 142, 143, 761
Holstege, F. C., 266, 591
Holt, I., 120, 405
Holton, T. A., 213, 642
Holtzapple, E. K., 123
Holzgang, O., 124
Holzmann, M., 19
Honda, D., 302
Hong, Y., 146
Hongyo, T., 428
Hood, L., 104, 702, 726
Hopkins, N., 531
Hoppitt, W., 89
Hoque, M. O., 124
Horn, G. T., 104
Hornes, M., 16, 145, 146, 151, 152, 161, 162,
    180, 182, 427, 570
Hornung, S., 372
Horton, H., 263
Horton, T. R., 60
Hoskins, R. A., 119, 120
Hosono, S., 324, 359
Hosoyama, H., 552
Hoss, M., 82, 94, 95
Houben, A., 423, 424, 425, 435, 444
Houston, C. S., 76, 77, 81, 82
Howard, J., 304
Howe, C. J., 372, 378
Hradecna, Z., 386
Hsie, L., 268
Hu, H., 682

Huaman, Z., 286, 287
Huan, X., 120, 122
Huang, B., 165, 166, 167
Huang, H., 531
Huang, M. C., 119
Huang, X., 324, 405
Hubbell, E., 268
Huber, S., 77, 81
Hubert, R., 424
Hudson, T. J., 268
Hudspeth, A. J., 529
Huelsenbeck, J. P., 105, 653, 654,
    655, 656, 718, 754
Hugall, A. F., 105, 112
Hughes, A. L., 726, 728, 731
Hughes, J. M., 227
Hughes, T. R., 583
Hulbert, S. H., 452, 453
Humair, P. F., 311, 330, 334, 340
Hummert, C., 303
Humphries, C. J., 287
Hunkapiller, T., 104
Hunter, D. W. F., 63, 65
Hunter, W., 619
Hupfer, H., 372
Hurt, A., 76
Husain, A., 653, 728, 738
Huso, M., 286
Huson, D. H., 684, 689, 691, 753
Huson, S., 689, 691
Hutcheon, J. M., 749
Hutchison, C. A. III, 653, 728, 738
Huttemann, M., 781
Huynen, M. A., 112
Huynh, K. D., 554
Hwang, L., 598
Hyman, G., 294
Hyten, D., 148

I

Iafrate, A. J., 116, 119
Ibisch, P. L., 286, 288
Idol, J. R., 119
Igarashi, R. T., 260
Iglesias-Rodriguez, M. D., 301
Igloi, G. L., 351
Ihaka, R., 592
Ikeo, K., 731
Ina, Y., 729, 730

Inazuka, M., 240, 242
Inglis, A. S., 638, 639
Ingman, M., 312
Ingraham, J. L., 494, 504
Ingram, D. S., 300, 301
Innis, M., 546
Inokuchi, H., 349
Inoue, J. G., 328
Inouye, I., 302
Installé, P., 420
Inthavong, S., 165, 166, 167
Intrieri, G., 659
Iriondo, J. M., 281, 293
Irish, V. F., 622
Irwin, D. E., 148
Isaac, P. G., 300, 301
Isaaks, J. A., 165, 167
Isawa, Y., 716
Iseli, C., 408
Isherwood, K. E., 123
Ishibashi, Y., 223
Ishikawa, Y., 637, 638, 648
Ishimura, Y., 783
Islam-Faridi, M. N., 444, 445
Isono, E., 420, 422
Istock, C. A., 494, 504
Italia, M. J., 116
Ito, T., 351, 552
Itoh, M., 120, 121
Ivanova, N. V., 52
Iverson, B. L., 659
Ives, C., 551
Iwagami, M., 311, 330, 334, 340
Iwahana, H., 428
Iwama, N., 552
Iwatsuki, K., 678
Iyer, V. R., 263, 266, 605

J

Jaccard, P., 196
Jackson, D., 445, 458, 623, 634
Jackson, M., 39, 45
Jackson, P. J., 268
Jackson, S. A., 444, 445
Jacob, H. J., 146, 148
Jacobs, M., 420
Jacqmin, S., 193, 194, 196
Jacquot, E., 68
Jaeger, C. H. III, 511

Jaenicke, V., 100
Jakobsen, I., 59
Jakobsen, I. B., 676
Jakobsen, K. S., 77
Jamart, A., 76
Jamieson, A., 233, 234
Janies, D. A., 105
Janousek, B., 423, 424, 425, 435
Jansen, C. L., 80, 81
Jansen, J. L., 65
Jansen, R. K., 332, 333, 335, 349, 351, 352,
    365, 366, 367, 368, 372, 375, 376, 377, 378,
    379, 678, 680, 689
Jarne, P., 223
Jarvis, A., 285, 286, 288, 289, 292, 293, 294
Jastrow, J. D., 60, 63
Jauch, K.-W., 424, 436
Jay, E., 659
Jeanmougin, F., 641
Jeannin, P., 61, 63, 64
Jedrzkiewicz, B., 659
Jefferys, A. J., 203
Jeffrey, S. S., 593
Jegalian, K., 423
Jenkins, G. J. S., 60, 62, 63, 64
Jenkins, M. C., 39
Jenkins, N. A., 117, 118, 123
Jenkins, T., 420, 421
Jennings, E. G., 266
Jensen, K., 124
Jensen, R. J., 134, 138
Jermann, T. M., 653, 728, 738
Jett, J. M., 324
Jiang, C.-X., 385
Jiang, H. M., 573, 580, 581, 583, 584, 585, 589,
    592, 593, 610
Jiang, J., 444, 445, 451, 452, 453
Jiang, L., 123
Jiang, Q., 551
Jiang, T., 116, 119
Jiang, Y. J., 531
Jianqiangt, L., 134, 138, 142
Jin, W., 616
Jin, Y. M., 281
Joachimiak, M. P., 608, 610
Jobb, G., 261
Joe, L., 77
Joerger, R. D., 39, 42, 43
Joger, U., 134
Johannsson, M. L., 268

Johansen, B., 635
Johansson, T., 609
John, K. S., 677
John, U., 21, 260, 261, 301, 303
Johnson, D. A., 147, 581
Johnson, J. M., 411
Johnson, M., 581
Johnson, S., 138, 142
Johnson, W. J., 106, 112, 117, 118, 123
Johnston, D. A., 311, 330, 334, 340
Johnston, J. S., 478, 479, 480, 481
Johnston, M., 583
Jones, A. G., 235
Jones, A. R., 583
Jones, C. J., 145, 146, 152, 153
Jones, D. T., 656, 662
Jones, E. A., 541
Jones, K., 529
Jones, K. E., 746, 751, 754, 755
Jones, P., 286, 288, 292, 294
Jones, R., 105
Jones, S. J., 408
Jong, P. M., 203
Jongeneel, C. V., 118, 408
Jönsson, K., 653, 728, 738
Joos, S., 467
Jordan, E., 678
Jordan, S., 163
Jorde, P. E., 147
Josefsson, C., 583, 585, 592
Jost, R., 261
Jowett, T., 537
Joyce, P., 767
Jukes, T. H., 655
Juni, E., 493, 498
Jurka, J., 269

K

Kachlany, S. C., 112, 116
Kadane, J. B., 376, 694
Kaessman, H., 312
Kaestle, F. A., 94
Kagochi, T., 572, 573, 581
Kaiser, D., 106
Kaiser, R. J., 104
Kaldorf, M., 68
Källersjö, M., 105, 750
Kallioniemi, A., 466
Kalyanaraman, A., 405

Kami, H. G., 337, 339, 340
Kamisugi, Y., 420, 423, 424
Kamogashira, T., 349
Kamradt, D. A., 73
Kan, Z., 411
Kanamori, A., 148
Kanamori, M., 552
Kanazawa, H., 428
Kandpal, G., 203
Kandpal, R. P., 99, 203
Kane, D. A., 531
Kane, J. F., 658
Kanehisa, M., 552
Kaneko, A., 376
Kang, J. J., 546, 547, 551, 552
Kang, T. J., 269
Kang, Y. Y., 653, 728, 738
Kanin, E., 266
Kapala, J., 39
Kaplan, H., 684
Kapoor, S., 351
Karaca, M., 599
Karagouni, A., 497
Karaman, M. W., 608
Karamycheva, S., 120, 405, 411
Karihaloo, J. L., 146
Karl, S., 224
Karlins, E., 119
Karlovsky, P., 165, 166, 167
Karnik, S. S., 653, 659, 728, 738
Karolchik, D., 119
Karp, A., 145, 146, 152, 153, 182, 189, 190,
    191, 192, 198, 300, 301
Karsi, A., 148
Karyala, S., 613
Katari, M. S., 106, 123
Katayama, Y., 783
Kathir, P., 268
Katiyar, S. K., 165, 166, 167
Kato, A., 349
Kato, I., 642
Katoh, K., 706
Katsar, C. S., 385
Katz, E. D., 430
Katzir, N., 375
Kaufman, J. C., 659
Kawachi, M., 302
Kawano, S., 420, 421, 422, 423,
    424, 427, 436, 437
Kawazu, T., 619

Kay, E., 494, 495, 507, 511, 512, 514
Kazmi, M. A., 659, 728, 738
Kearney, M., 314
Keddie, E. M., 311, 340
Kee, L., 193, 194, 196
Keim, L. G., 61, 62
Keim, P., 146, 147, 148, 165, 168, 182, 268
Kejnovsky, E., 420, 422, 423, 426, 435
Keller, R., 528, 540
Kelley, J. M., 105, 400
Kelly, B. T., 77
Kelly, M., 20
Kelmanson, I. V., 662, 665
Kelsh, R. N., 531
Kemp, 90
Kemp, D. J., 638, 639
Kendall, K. C., 76, 82
Kent, W. J., 119, 409, 411
Kephart, S. R., 134, 138, 140, 142
Kerby, S., 39
Kerlavage, A. R., 105, 400, 551
Kerlavage, R., 106
Kerner, T., 424, 437
Kerr, M. K., 584, 585, 599, 614
Kerrigan, J., 134
Kershaw, M., 286
Kesseli, R., 182, 426, 427, 433
Ketchum, K. A., 551
Ketola, M., 480, 481
Keyser-Tracqui, C., 89
Khaitovich, P., 122, 123, 609
Khalak, H. G., 551
Khodursky, A. B., 609
Khorana, H. G., 659
Khouri, H. M., 123
Kidwell, K. K., 193, 194, 196
Kiester, A. R., 286
Kiew, R., 193, 194, 196
Kihara, T., 619
Kijas, J. M. H., 203
Kil, S. H., 529
Kilburn, D., 268
Kilpatrick, C. W., 761
Kim, I. F., 591
Kim, J., 123, 659, 674, 675
Kim, J. M., 351
Kim, J. S., 445
Kim, M. S., 193, 194, 196, 419, 434, 635
Kim, M. Y., 351
Kim, S., 123, 780

Kim, U.-J., 385, 386
Kimmel, C. B., 534
Kimura, H., 148
Kimura, M., 655
King, B. L., 551
King, N., 106, 111, 112, 113, 114, 115, 116, 702, 704
King, R., 462
King, T. J., 202
Kingsley, M. T., 268, 269
Kingsmore, S. F., 324, 359
Kingston, R. E., 271, 315, 319, 328
Kinoshita, M., 148
Kinzler, K. W., 581
Kirch, S., 678
Kirchner, J., 581
Kirk, B. W., 239
Kirkness, E. F., 105, 118, 551
Kirsch, J. A. W., 749
Kirsch, J. F., 653
Kirshnaswamy, H., 659
Kishino, H., 656, 770
Kitts, C. L., 39, 63
Kjøller, R., 64, 68
Klein, P. E., 445, 581
Klein, R. R., 445, 581
Klenk, H. P., 551
Klironomos, J. N., 60
Klose, J., 122, 123, 609
Kluge, A. G., 105, 746
Knaebel, D. B., 62
Knapp, L. W., 637, 638, 640, 647, 648, 649
Knauer, F., 77
Knecht, L., 780, 781, 784
Knorr, C., 146, 148, 151
Knox, M. R., 182
Knudsen, K. L., 146, 148
Knutsen, T., 468
Kobayashi, I., 552
Kobayashi, M., 263
Kobayashi, S., 583
Koch, J., 55
Kochan, K. J., 148
Kocher, T. D., 146, 147, 167, 332, 337
Koehler, G. M., 76, 77, 81, 82
Koehler, T. M., 123
Koenig, G., 637, 638, 648
Koepfli, K.-P., 77
Koh, S. S., 266
Kohchi, T., 349

Kohn, M. H., 73, 77, 82
Koide, R. T., 59, 60, 61, 64, 65, 67
Koike, S., 38
Kolasa, J., 473, 475
Kolchanov, N., 780, 781, 783
Kolliker, R., 193, 196
Kolman, C. J., 93, 94, 95, 98
Kolodner, R., 354, 357
Kolonay, J. F., 123
Kolsto, A. B., 123
Kølvraa, S., 55
Komatsu, T., 193, 196
Kondo, M., 148
Kondo, S., 148
Kondow, A., 311, 330
Kong, P., 242
Konieczny, A., 427, 555
Konig, A., 261
Konnerth, A., 540
Kono, K., 420, 433
Konrad, K., 583
Kool, E. T., 324
Koonin, E. V., 112, 123
Koono, K., 292
Koop, B. F., 104
Korban, S. S., 193, 194, 196
Korbel, J. O., 112
Korber, B. T., 780, 784
Koretke, K. K., 116
Korfanta, N. M., 203
Korpelainen, H., 480, 481
Kortschak, R. D., 123
Kosalski, S. P., 385
Koshi, J. M., 655
Koski, J., 111
Kostrzynska, M., 497
Kothari, S., 405
Kotseruba, V., 444
Koumbaris, G. L., 444, 445
Koutnikova, H., 423, 424, 425, 435
Kovarík, A., 444
Kowalschuk, G. A., 59, 62, 63
Koyama, H., 619
Kozyreva, O., 420, 421, 422
Krainitzki, H., 88, 93, 94
Krakauer, D. C., 312
Kramer, E. M., 622
Kramer, F. R., 123
Kranz, D. M., 659
Krause, D. O., 61, 63

Krauss, S. L., 164, 233
Kreader, C. A., 62
Kremer, A., 292
Kremer, L., 497
Kren, V., 146, 148
Kresovich, S., 182, 189, 190, 191, 192, 198
Kretschmer, E., 146, 148
Krings, M., 88, 93, 94
Krishnan, N. M., 784
Kristiansen, J., 300
Krogenecklenfort, E., 497
Kroken, S., 50
Kronmiller, B., 119, 120
Kruglyak, L., 268
Krumlauf, R., 521
Kruuk, H., 77
Kruuk, L. E. B., 235
Krylov, D. M., 123
Ku, H.-M., 385
Kubelik, A. R., 180, 427
Kubo, T., 716
Kucaba, T., 402
Kuch, M., 81, 82, 99, 100
Kucherlapati, R., 263
Kuehl, J. V., 312, 337, 338, 339, 340, 376, 378
Kugita, M., 376
Kuhara, S., 552
Kuhn, D. N., 240
Kuiper, M., 16, 145, 146, 147, 148, 151, 152,
        153, 154, 161, 162, 180, 182, 427, 570
Kukita, Y., 240
Kulesa, P. M., 541
Kumar, S., 654, 655, 676, 703, 728,
        729, 730, 740
Kumazawa, Y., 334, 337, 338
Kunin, V., 702
Kunisawa, T., 335
Kuo, F. C., 603
Kurabayashi, A., 338
Kuriowa, T., 115, 422, 423, 424, 435, 436
Kuroda, H., 351, 552
Kuroda, M., 552
Kuroda-Kawaguchi, T., 419
Kuroiwa, T., 420, 421, 423, 424, 427, 436, 437
Kuroki, Y., 423, 424
Kurtz, S., 373, 374
Kurtz, T. W., 146, 148
Kusakabe, T., 148
Kustu, S., 609
Kusuda, J., 349

Kusukawa, N., 642
Kuzoff, R. K., 635

L

Labas, Y. A., 662, 666
Labuda, D., 134
Lacailee, V., 235
Lacey, M., 659
Lachenmeier, E., 583
Laflamme, D., 351
Lagergren, J., 716, 719
Lagoda, J. L., 223
Lahn, B. T., 423
Lai, T., 261
Lake, J. A., 112
Laker, M. T., 268
Lakshmi, B., 116, 119
Lal, A. A., 39
Lalueza-Fox, C., 89
Lamaj, F., 193, 196
Lamberti, F., 90
Lamont, S., 148
Lan, T.-H., 385
Lande, R., 282
Landegren, U., 239, 324
Lander, E. S., 266, 268, 583
Landgraf, R., 727, 737, 738, 741
Lane, D. J., 521
Lane, J., 659
Lanfranco, L., 66, 68
Lang, B. F., 311, 312, 335
Lange, M., 22, 260, 261, 300, 302, 303, 304
Langer, M. C., 746
Langley, S., 750, 753
Lankhorst, A. E., 146, 147, 148, 151
Lanni, F., 540
Lansford, R., 538
Lanyon, S. M., 749
Lapchin, L., 165, 168
Lapedes, A. S., 780, 784
Lapitan, N. L., 385
Lapointe, F.-J., 749, 754, 755
Laporte, V., 429
Lardon, A., 420, 421
Larget, B., 376, 655, 694
Largiader, C. R., 148
Lari, M., 89
Laric, P. L., 119
Larkum, A. W. D., 378

Larsen, A., 300
Larsen, L. A., 240
Larson, A., 312, 336, 337, 338, 339, 340
Larsson, S., 146
Lascoux, M., 169
Lasken, R. S., 324, 359
Lathrop, M., 202
Latimer, D. R., 659
Latorre, C., 81, 82
Latour, S., 76
Lau, P. C. Y., 495, 505, 508
Laughlin, T. F., 117, 118, 123
Laukkanen, M., 268
Laviolette, J. P., 268
Lavrov, D. V., 312, 330, 337, 338
Lawrence, C. E., 119
Lawrence, C. J., 402
Le, C. T., 603
Le, T. H., 311, 330, 334, 340
Leach, F., 495, 507
Leatham, T., 300
Lebel-Hardenack, S., 420, 421, 422
Lebepe-Mazur, S., 39, 40, 45
Leblois, R., 280
LeBot, N., 303
Lebrun, M., 496, 516
Lederberg, 494
Ledl, F., 82
Leduc, G., 335
Lee, C., 116, 119
Lee, D., 405
Lee, H., 145, 146, 182, 497
Lee, H. L., 609
Lee, H. S., 443, 570, 571, 572, 573, 580, 581,
    583, 584, 585, 589, 592, 593, 610
Lee, J. A., 703
Lee, J. H., 378, 495, 505, 508, 529
Lee, J. J., 572, 573, 581, 583, 585, 592
Lee, J. T., 554
Lee, L. G., 432
Lee, M.-L. T., 603
Lee, M. S. Y., 105, 112
Lee, N., 264, 270, 551
Lee, S. B., 49, 453
Lee, T. I., 266
Lee, W.-J., 337
Lee, Y., 405, 411
Leebens-Mack, J. H., 376, 378
Lee-Lin, S.-Q., 119
Lefebvre, J.-F., 682, 693

Lefebvre, P. A., 268
Lefkowitz, S. M., 583
Le Gall, F., 61, 63, 64
Legaspi, R., 119
Legendre, P., 196
Lehner, A., 264, 270
Lehrach, H., 123
Leister, D., 349, 351, 367, 375, 377
Leitch, A. R., 444, 445, 458
Leitch, I. J., 443, 444, 445, 458, 478, 481, 483
Lemaire, M., 147, 148, 163, 196
Lemesle-Meuniera, D., 788
Lemieux, C., 312, 372, 375, 376
Lengerova, M., 426
Lenke, M., 261
Lennon, G., 401
Lenstra, J. A., 145, 146, 147
Leonard, J. A., 94
Leonard, K. J., 147
Lepetit, D., 478, 481
Lepingle, A., 202
Le Quere, A., 609
Lerat, E., 687
Lessios, H. A., 233
Le Tacon, F., 59, 62, 63, 66
Letowski, J., 608
Leung, K., 60, 62, 63, 497
Levasseur, C., 754
Levin, D. A., 443
Lewin, H. A., 223
Lewis, D., 76
Lewis, K. D., 107, 116
Lewis, M. S., 555, 567
Lewis, P. O., 142, 143, 655
Lewontin, R. C., 736, 741
Li, G., 689
Li, J., 116, 119
Li, M., 39, 40, 42, 43, 45, 372, 373, 678
Li, P., 123, 147, 148
Li, T., 420, 424
Li, W.-H., 163, 165, 196, 349, 653, 674, 705,
    720, 728, 730, 738
Li, W. K. W., 300
Li, Y., 419
Li, Y.-H., 495, 505, 508
Lian, J., 552
Liang, F., 120, 405
Liang, P., 571, 576
Liang, S., 39, 40
Liao, J. C., 605, 659

Libertini, A., 480, 481
Lichtarge, O., 727, 737, 741
Lichtenstein, C. P., 444
Lightfoot, D. A., 148
Liharska, T. B., 445
Lilburn, T. G., 261
Lilly, W., 351, 372, 375, 377
Lim, E. L., 261
Lim, K. Y., 444
Lim, M. J., 119
Lin, C. H., 412
Lin, J., 603
Lin, S., 531
Lin, Y. R., 385, 445
Linacero, R., 145, 146, 152, 153
Lindblad-Toh, K., 268
Linder, C. R., 385, 678
Lindner, K. R., 146, 148
Lindow, S. E., 511, 514
Lindroos, K., 268
Lindström, K., 60, 62, 64
Ling, L. S., 551
Linhart, Y. B., 280, 282
Link, W., 193, 194, 196
Lins, T., 349, 351, 367, 375, 377
Linsley, P. S., 583
Lipman, D. J., 120, 122, 363, 402
Lippincott-Schwartz, J., 661
Lippman, Z., 583, 585, 592
Lipshutz, R., 268, 583
Lisitsyn, N., 124, 421, 434
Liss, T., 261
Listewnik, M. L., 116, 119
Liston, A., 135
Littlejohn, T. G., 312
Littlewood, D. T., 311, 330, 334, 340
Liu, B., 570
Liu, D., 324
Liu, J., 385
Liu, L. C., 780, 784
Liu, L.-I., 105
Liu, S. C., 385
Liu, T., 685, 692
Liu, Y. G., 390
Liu, Z., 123, 147, 148, 180, 419, 433
Livak, K. J., 123, 180, 427
Lizardi, P. M., 324, 659
Lloyd, D. H., 581
Lloyd-Jones, G., 63, 65
Locht, C., 497

Locke, D. P., 124
Lockhart, D. J., 263, 583
Lockhart, P. J., 378
Loftus, B., 551
Loge, F. J., 264
Loh, J. P., 193, 194, 196
Lohtander, K., 50
Lombard, V., 193, 194, 195, 196
Long, M., 702, 703
Lopez, L., 288, 292
Lopez, P., 703, 781
López-Garcia, P., 260
Lorenz, M. G., 61, 494, 498, 504, 506, 512, 513, 514
Lorenzo, V. D., 702
Lottaz, C., 408
Lotz, S., 235
Lou, Y., 324
Loughry, W. J., 235
Louie, L., 39, 40, 45
Louie, M., 39, 40, 45
Loveless, M. D., 292
Lowe, T. M., 401
Loy, A., 264, 270
Lozano, J. J., 378
Lu, A. T., 583
Lu, B. R., 281
Lu, J., 182
Lu, R., 165, 168
Lucas, S. M., 324
Lucchini, V., 77
Lucito, R., 116, 119, 124
Lucretti, S., 423, 435, 480
Lucsa, W., 635
Ludes, B., 89
Ludwig, T., 261
Ludwig, W., 261, 303
Luikart, G., 73, 74, 163, 233
Lukehart, S. A., 93, 94, 95, 98
Lukens, L., 573, 580, 581, 583, 584, 585, 589, 592, 593, 610
Lukyanov, K. A., 661, 662, 666
Lukyanov, S., 120, 121
Lukyanov, S. A., 661, 662, 666
Lumsden, A., 534
Lunde, C. F., 402
Lundeberg, J., 609
Lundin, P., 116, 119
Luo, C.-C., 730
Luo, S., 581

Lupas, A., 116
Lussmann, R., 261
Lute, J. R., 61, 62
Lutz, A. W., 142
Lutz, R., 323
Lwin, K. S., 338, 340
Lyle, R., 118, 554
Lynch, M., 164, 189, 703, 714, 716, 719, 725, 726
Lynnerup, N., 93, 100
Lyons, L. A., 117, 118, 123
Lyons, T. J., 605
Lyons-Weiler, J., 378
Lysak, M. A., 480

M

Ma, B., 678
Ma, H., 419, 434
Ma, J., 227
Mable, B. K., 105, 773
Mabuza-Dlamini, P., 293
Macas, J., 423, 426
Mace, T. R., 134
Macey, J. R., 312, 334, 336, 337, 338, 339, 340
MacFarlane, J. L., 311, 312, 330, 340
Machado, A. A., 408
Machray, G. C., 135
Mackie, R. I., 38, 39, 40
MacLarnon, A., 746
MacLean, I., 224
Madagan, K., 123
Madan, A., 120, 122, 405
Madden, T. L., 363, 402
Maddison, D. R., 654, 741
Maddison, W. P., 654, 678, 693, 702, 712, 741
Maddox, P. S., 540
Madlung, A., 443, 570, 571, 572, 573, 580, 581, 583, 584, 585, 589, 592, 593, 610
Madpu, R., 123
Madsen, E. L., 60, 62, 63, 64
Madsen, O., 106, 117, 118, 123
Madubashi, S., 727
Maduro, Q. L., 119
Maduro, V. V. B., 119
Maerz, M., 351
Maestri, E., 145, 146, 152, 153
Mahamoud, Y., 123
Mahmood, N. A., 659
Mahoney, J. B., 39

Maidak, B. L., 261
Maier, R. M., 372
Maier, U. G., 351
Maitra, A., 124
Majewski, J., 494, 502, 505
Mäki, B. R. E., 40, 42
Makino, Y., 702
Maklad, A., 529
Malcevschi, A., 145, 146, 152, 153
Malcolm, B. A., 653
Maldonado, J. E., 76
Malhotra, A., 147
Malibari, A., 59
Maliga, P., 351
Mallegni, F., 89
Mambo, E., 124
Manceau, V., 74, 77, 83
Mancino, V., 385, 386
Manel, S., 282, 289
Månér, S., 116, 119
Maniatis, T., 34, 75, 173, 204, 206, 213, 214, 421, 438, 571, 573, 575, 579, 581
Mann, C. C., 259
Mansfield, H. R., 39, 42, 43
Manshardt, R. M., 193, 194, 196
Mansouri, H., 495, 507
Mao, J.-H., 116, 119
Mao, J. X., 196, 581
Mao, M., 583
Marby, M. L., 769
Marcade, I., 311
Marchuk, D., 213, 642
Marec, F., 465
Marechal-Drouard, L., 3
Marescalchi, O., 478, 481
Margoliash, E., 701
Margulies, E. H., 119
Marie, D., 303
Marin, I., 466
Marín, I., 736
Marins, M., 400
Markert, J. A., 146, 147, 167, 761
Markowitz, E., 727
Markowitz, V., 591
Marmaro, J., 123
Marmeisse, R., 65
Marmiroli, N., 145, 146, 152, 153
Marmur, J., 260
Marras, S. A. E., 123
Marron, M., 687

Marshall, D. F., 179
Marshall, F., 77
Marshall, K. A., 372
Marshall, T. C., 235
Martienssen, R. A., 106, 123, 443, 570, 583, 585, 592
Martin, A., 323
Martin, A. P., 702, 711
Martin, F., 59, 62, 63, 66
Martin, G. B., 385
Martin, K. J., 421
Martin, P. S., 93, 94
Martin, W., 349, 351, 367, 375, 377, 378, 718
Martínez-Castilla, L., 725
Martinez-Zapater, J. M., 147, 151, 154
Martínez-Zapater, M. M., 165, 166
Marton, M. J., 583
Marucco, F., 77
Maruyama, A., 552
Maruyama, K., 120, 121
Marx, P. A., 77
Maschinski, J., 182
Masiello, C. A., 119
Maskeri, B., 119
Massa, H., 116, 119
Massana, R., 260
Massimi, A., 263
Masson, L., 608
Masta, S. E., 330
Mastrian, S. D., 119
Masuelli, R. W., 571, 581
Masutti, L., 164, 165, 166
Matese, J. C., 591
Mather, P. B., 227
Mathews, S., 378
Matlock, H., 76
Matsubayashi, T., 349
Matsuda, G., 708, 711
Matsuda, H., 676
Matsuki, T., 38
Matsumaru, H., 552
Matsumoto, K., 38
Matsumoto, T., 376
Matsunaga, S., 420, 421, 422, 423, 424, 427, 435, 436
Matsuoka, T., 148
Matsuoka, Y., 376
Matsuzaki, M., 115
Mattessich, M. J., 659
Matte-Tailliez, O., 116

Matthee, C., 112, 750, 753
Matthes, M., 145, 146, 152, 153
Matthews, B. W., 653
Mattick, J. S., 202, 226
Matyásek, R., 444
Matz, M. V., 661, 662, 665, 666, 728, 738
Maul, J., 351, 372, 375, 377
Maxson, L. R., 521
Maxted, N., 285, 293
May, B., 146, 153, 154
May, M., 261
May, P. E., 202
Mayer, K. F., 106, 123, 405
Mays, B., 637, 638, 648
Mazur, 45
McAuliffe, J., 107, 116
McClelland, M., 571
McCloskey, J. C., 119
McCombie, R., 124
McCombie, W. R., 106, 123, 400
McConnell, K. J., 659
McCormick, C. A., 534
McCormick, S., 635
McCracken, V. J., 38, 39, 40
McCurdy, S., 581
McCusker, J. H., 609
McDonald, L. A., 105
McDonough, C. M., 235
McDowell, J. C., 119
McElfresh, K., 268
McEwan, J. C., 233
McIntosh, C., 76, 78
McKay, J. K., 385
McKelvey, K., 77
McKenney, K., 105, 551
McKnight, T. D., 444
McLaren, I. A., 473, 476
McLellan, B. N., 76
McLysaght, A., 693
McManus, D. P., 311, 330, 334, 340
McMichael, M., 165, 167
McMillan, W. O., 224, 323
McNeal, J. R., 378
McNicol, J. W., 135
McOrist, A. L., 39, 45
McSweeney, C. S., 61, 63
Mechanda, S. M., 147
Medina, M., 312, 321, 322
Medlin, L. K., 16, 20, 21, 22, 25, 260, 261, 264, 267, 270, 271, 300, 301, 302, 303, 304

Medvedovic, M., 613
Mègraud, F., 39, 45
Mei, M., 571, 581
Meidanis, J., 684
Meier, H., 261, 264, 270
Meiklejohn, C. D., 610
Meister, A., 444, 480
Meksem, K., 148
Melnick, D. J., 76
Melville, C., 496, 515
Melville, C. M., 495, 509, 513
Melville, J., 338, 340
Mendel-Hartvig, M., 324
Mendelson, B., 534
Mendelson, T. C., 165, 169, 170, 171, 172
Mendonca-Hagler, L. C., 62, 63
Meng, B. Y., 349, 351
Meng, E. C., 727
Menozzi, P., 281
Menz, M. A., 445
Merberg, D., 551
Mercer, D. K., 495, 496, 509, 513, 515
Merkel, C., 331
Merkel, J. S., 659
Merrick, J. M., 105
Merril, C. R., 400
Merritt, T. J. S., 729
Merugu, S., 653
Merugu, X., 728, 738
Message, B., 495, 507
Messier, W., 728, 729, 730, 740
Metcalfe, W. K., 534
Metfies, K., 264, 270
Metscher, B. D., 541
Metzger, D. A., 78
Meunier, J., 781
Mewes, H.-W., 405
Meyer, A., 224, 338, 703, 719
Meyer, M. R., 583
Meyers, A., 662, 666
Michaelson, C., 659
Michelmore, R., 182, 426, 433
Michener, C. D., 196
Michimoto, T., 423, 424, 436
Mickett, K., 147
Micsik, T., 599
Mielke, P. W., 196
Milanesi, E., 146, 147, 154
Milia, A., 480, 481
Millasseau, P., 202

Miller, D. J., 123
Miller, D. N., 60, 62, 64
Miller, M., 119
Miller, M. P., 165, 168
Miller, P. E., 261, 303, 304
Miller, R. E., 655
Miller, R. M., 60, 63
Miller, W., 363, 372, 373, 402, 409, 511
Millie, D., 349
Milligan, B. G., 164, 189, 351, 356, 372
Mills, D. M., 551
Mills, L. S., 77
Mills, S. D., 551
Minak-Bernero, V., 61, 62
Mindell, D., 338, 339, 340
Mindell, D. P., 328, 337
Ming, R., 193, 194, 196, 385, 419, 434
Minin, V., 767
Minov, R. I., 38, 39, 40
Minshull, J., 780
Minx, P. J., 419
Mir, K. U., 123, 263
Mira, A., 312
Mirsky, A. E., 471, 473
Mirzabekov, A. D., 264
Mishima, M., 444
Misumi, O., 115
Mitani, H., 148
Mitchell, J. M. O., 780
Mitchell, L., 95, 98
Mittmann, M., 263, 268
Miya, M., 328
Miyagi, T., 351
Miyamoto, 38
Miyamoto, M. M., 735, 736, 740, 741
Miyata, T., 656, 706
Mizrahi, Y., 10, 14
Mobarry, B. K., 264
Moberg, K. D., 521
Mochizuki, M., 783
Mock, K. E., 147
Mock, T., 266
Mogg, R., 179
Mohamed, I., 613
Mohandas, T. P., 134
Moir, D. T., 551
Molin, G., 268
Møller, A. P., 202, 224
Møller, A. R., 223
Molloy, P. L., 638, 639, 640, 641, 642

Monger, F., 419, 420, 421, 422, 423, 425, 429, 437
Monje, V., 301
Monrozier, L. J., 61
Monterio, L., 39, 45
Montgomery, L., 39, 42, 43
Mooers, A., 674
Mooers, A. Ø., 716
Moon, K., 581
Moon-van der Staay, S. Y., 260, 261, 303
Moore, D. D., 271, 315, 319, 328
Moore, D. V., 703
Moore, G., 378
Moore, P. H., 193, 194, 196, 419, 434
Moore, R. C., 420, 421, 422, 426
Moore, S. S., 202
Moqadam, F., 120, 121
Morais, R., 337
Morais-Cecílio, L., 555
Moré, M. I., 63
Moree, B., 540
Moreira, D., 116, 260
Morell, P., 260
Moreno, R. F., 400
Moret, B. M. E., 335, 375, 376, 677, 678, 684, 685, 687, 689, 691, 692, 693, 694, 753, 754
Morgante, M., 135
Morin, P. A., 76, 83, 117, 118, 123, 223
Morishige, D. T., 581
Morissette, J., 202
Morita, Y., 148
Moritz, C., 105, 773
Morizot, D. C., 148
Morley, M., 263
Morrell, P. L., 445
Morris, C. P., 638, 639, 640, 641, 642
Morris, M. S., 268
Morrow, D. J., 402
Mort, M. E., 134, 138, 142, 769
Morton, B. R., 349, 378
Morton, C., 147
Morton, J. B., 60
Moseyko, N., 609
Mosig, G., 351
Moss, H., 293
Mottram, G., 288, 289, 292
Mouchiroud, D., 707
Mount, D. W., 105

Mouras, A., 420, 421, 422, 425, 437
Mowat, G., 76
Moya, A., 736
Moyer, C. L., 260
Muchmore, E., 122, 123, 609
Muchnik, I., 719
Mueller, U. G., 145, 146, 147, 162, 182
Muili, J., 411
Mukherjee, A. B., 464
Mukouda, M., 292
Mulcahy, D., 427
Muller, C., 407
Muller, M., 703
Müller, R. T., 286
Mullet, J. E., 351, 445, 581
Mullins, M. C., 531
Mullins, T. D., 134
Mullis, K., 104
Mullis, K. B., 104
Mundorff, E. C., 780
Mungall, C. J., 119, 120
Murakami, H., 552
Murakami, T., 77
Muralhidar, M., 268
Muramatsu, M., 120, 121
Muramoto, K., 783
Murphy, K. M., 77
Murphy, M. A., 76, 82
Murphy, W. J., 106, 112, 117, 118, 123
Murray, H. G., 63
Muse, S. V., 378, 656, 728
Muyzer, G., 16
Myers, N., 259
Myers, R. M., 106
Myslabodski, D., 10, 14

N

Nadeau, J. H., 385, 680, 682, 703
Nadershahi, A., 408
Nagai, Y., 552
Nagaoka, S., 120, 121
Nagley, P., 312
Nagpal, S., 608
Nagy, Z. T., 134
Nagylaki, T., 282
Nahon, J.-L., 714
Naimi, T., 552
Nakano, A., 540

Nakao, S., 420, 423, 424, 427, 436
Nakashima, K., 351
Nakashima, R., 783
Nakayama, S., 420, 433
Nakhleh, L., 677, 678
Nakimura, M., 124
Nalin, R., 495, 507, 514
Nardi, F., 338
Nardi, L., 480
Nardon, C., 478, 481
Naruse, K., 148
Nassal, M., 659
Navarro, A., 443
Navidi, W., 424
Navin, N., 116, 119
Naylor, G. J. P., 730, 735
Neale, D. B., 372
Nedved, O., 475, 477
Nee, S., 716
Need, E., 653
Neff, B. D., 235
Negrini, R., 146, 147, 154
Negrutiu, I., 419, 420, 422, 423, 425, 429, 437
Neher, E., 780
Nei, M., 163, 165, 196, 233, 654, 655, 675, 676, 678, 689, 726, 728, 729, 730, 731, 732, 733, 740, 741, 782
Nelson, A., 286
Nelson, C., 263, 583
Nelson, J. R., 324
Nelson, J. S., 541
Nelson, K. E., 123, 551
Nelson, S. F., 436
Nelson, W. C., 123
Nelson, W. S., 235
Nerenberg, M., 268
Nesme, X., 42, 61, 63, 64, 495, 514
Ness, J. E., 780
Nettles, S., 689, 691, 753
Neumann, R., 203
Neves, N., 555
Nevo, E., 193, 194, 196, 280, 282
Nevoigt, E., 493
Newbold, C. J., 39, 42, 43
Newton, A. C., 280
Nguyen, D. T., 105

Nguyen, K. C., 124
Nichols, R. A., 123
Nickerson, D. A., 401
Nickisch-Rosenegk, M., 311, 338, 340
Nicod, J. C., 148
Nielsen, K. M., 494, 498, 503, 505, 506, 507, 510, 511, 512
Nielsen, R., 656, 662, 730, 731, 732, 733, 734, 740
Nieminen, A. L., 312
Nienhuis, J., 65
Nierlich, D. P., 88
Nieselt-Struwe, K., 122, 123, 609
Nigrutiu, I., 419, 421
Nijman, I. J., 146, 147
Nikaido, M., 312
Nikolakopoulou, D., 497
Nilsen, T. W., 589
Nilsson, M., 324
Nishida, M., 328, 334, 337, 338
Nishigaki, R., 148
Nishikawa, K., 311, 330
Nishiyama, R., 420, 433
Nishiyama, Y., 120, 121
Nissila, E., 193, 194, 196
Nixon, K. C., 713, 720
Noble, L. R., 77, 81
Noethig, E. V., 300, 302
Nonaka, M., 148
Nonhoff, B., 261
Nonnis Marzano, F., 150, 154
Noonan, B., 551
Noor, M. A. F., 385
Nordenskjold, M., 424, 436
Norell, M. D., 112
Norton, L., 124
Not, F., 261
Novoradovskaya, N., 599
Nowicki, C., 286, 288
Nozaki, H., 115
Nuez, F., 193, 194, 196
Nugaliyadde, L., 165, 166, 167
Nugent, J. M., 351
Nunez, J. M., 662, 666
Nurminen, P. H., 40, 42
Nusbaum, C., 268
Nussleinvolhard, C., 531
Nyakaana, S., 312
Nyberg, D., 99

O

Oakley, T. H., 654
Obara, M., 427
Obermayer, R., 480
Obert, J. A., 581
Obokata, J., 349
O'Brien, S. J., 106, 112, 117, 118, 123
Ochieng, J. W., 147
Ochman, H., 718
Odenthal, J., 531
O'Donnell, G., 42
Oefner, P. J., 609
Ogasawara, N., 552
Ogden, R., 147
Ogihara, Y., 376
O'Grady, P. M., 105, 106, 112, 113,
    114, 117, 118, 123, 466
Ogram, A., 61, 62, 63
Oguchi, A., 552
O'Guin, W. M., 638, 647, 648, 649
Ogundiwin, E., 193, 194, 195, 196
Oh, M.-K., 605
O'Hanlon, P. C., 147
Ohba, K., 135
Ohlebusch, E., 373
Ohme, M., 312, 349
Ohmido, N., 444
Ohno, S., 702, 725
Ohri, D., 443
Ohsumi, T., 120, 121
Ohta, N., 115
Ohta, T., 552
Ohto, C., 349
Ohyama, K., 349, 420, 433
Ojala, D., 331
Okada, N., 311, 312, 330
Okada, S., 420, 433
Okamoto, T., 263
Okazaki, Y., 120, 121
O'Kelly, C. J., 302, 312
Okimoto, R., 146, 311,
    312, 330, 340
Oksanen, I., 50
Okstad, O. A., 123
Okura, V. K., 400
Olde, B., 400
O'Leary, B., 402
Oliver, M. J., 62
Olmstead, R. G., 170, 349, 378, 769

Olsen, G. J., 521, 654, 674, 676,
    684, 759, 761
Olsen, J. L., 267
Olsen, R. A., 63
Olshen, A., 124
Olson, B. N., 39, 40, 62, 63
Olson, G. J., 261
Olson, M. V., 385
Olson, R. J., 260
Omland, K. E., 654
Oorschot, R. A. H., 39, 45
Opitz, J. G., 653, 728, 738
Oprian, D. D., 659
Orban, L., 20
Orita, M., 240, 428
Ormerod, S. J., 282, 289
Ormond, R. F., 300
Ornolfsdottir, E., 260, 303, 304
Orr, B., 659
Orr, K., 147
Ortega, J., 76
Ortiz, A. R., 378
Osborn, A. M., 39, 61
Osborn, T. C., 443, 570, 573,
    580, 581, 583, 584, 585, 589,
    592, 593, 610
Osborne, P., 288, 289
Oshima, K., 552
Oshima, T., 330
Osoegawa, K., 119, 398
Ostell, J. M., 409
Ostrander, E. A., 203
O'Sullivan, K. M., 65
Otis, C., 372, 375, 376
Otsen, M., 145, 146, 147, 148
Otto, K. G., 385
Otto, S. P., 152, 719
Otto, T. D., 465
Ouellette, B. F. F., 105
Ouyang, S., 402
Ouzounis, C. A., 702
Ovcharenko, D., 107, 116
Ovcharenko, I., 107, 116
Overbeek, R., 676
Ovilo, C., 147, 154
Owen, J. L., 134
Owsley, D. W., 93, 94, 95, 98
Oyaizu, H., 38
Oyston, P. C., 123
Ozeki, H., 349

P

Paabo, S., 76, 77, 82, 88, 93, 94, 95, 99, 100, 122, 123, 311, 312, 330, 609
Pace, B., 260
Pace, N. R., 22, 263, 521
Pachter, L., 107, 116
Pacleb, J., 119, 120
Padulosi, S., 293
Paetkau, D., 76, 223
Page, R. D. M., 678, 702, 703, 710, 711, 712, 718, 719, 720, 749
Paget, E., 42, 61, 496, 516
Paige, K. N., 223
Painter, T. S., 462
Pallas, M., 581
Palleroni, N. J., 260
Palm, S., 147
Palmen, R., 494, 498, 502, 505, 510
Palmer, J., 678, 680
Palmer, J. D., 312, 349, 351, 354, 355, 356, 372, 375, 378
Palumbi, S. R., 118, 323, 641
Palus, J. A., 65
Pamilo, P., 678
Pan, W. H., 435, 603
Pandya, Y., 39, 42, 43
Panopoulou, G., 123
Papa, R., 150, 154, 193, 196
Papenfuss, T. J., 312, 336, 337, 338, 339, 340
Paquin, B., 335
Paquola, A. C., 408
Paran, I., 426, 433
Pardee, A. B., 421, 571, 576
Parichy, D. M., 148, 151
Park, S., 119, 120
Parker, C. T. J., 261
Parker, J. S., 182
Parker, M., 76
Parkes, R. J., 60, 62, 63, 64
Parkinson, C. L., 378
Parkinson, H., 591
Parkinson, J., 402
Parman, C. E., 659
Parsons, K., 77, 81
Parsons, M., 235
Parsons, M. L., 303
Parsons, T. J., 312
Parsons, Y. M., 165, 169, 174

Partensky, F., 261, 300, 303
Parvizi, B., 405, 411
Paskind, M. P., 302
Pasquet, R. S., 193, 196
Passador-Gurgel, G., 615
Pastinen, T., 268
Patarnello, T., 164, 165, 166, 203, 223, 224
Paterson, A. H., 385, 386, 387, 390, 419, 434, 444, 445, 450
Patterson, G. H., 661
Patterson, N., 116, 119
Patton, J. G., 412
Paul, E. A., 59, 62, 63
Paul, J. H., 267, 502
Paul, S., 182
Paules, R. S., 599, 605, 614
Paull, R. E., 193, 194, 196
Paulsen, I. T., 123
Pawlowski, J., 19
Paxinos, E., 76, 78
Pazos, F., 781
Peacock, B., 311, 330, 334, 340
Peakall, R., 147
Pearce, J. M., 76, 223, 288, 289
Pearl, H. M., 419, 434
Pearson, N. M., 423
Pearson, R., 119
Pearson, W. R., 120, 122
Pedersen, A.-M. K., 656, 662, 731, 732, 733, 734, 740
Pedrosa, G. L., 400
Pedros-Alio, C., 260
Pe'er, I., 691
Peeters, A. J. M., 445
Peleman, J., 16, 145, 146, 151, 152, 161, 162, 165, 427, 570
Pelenchar, P. M., 106, 123
Pelser, P. B., 170
Pemberton, J. M., 223, 224, 235
Penny, D., 334, 349, 351, 367, 375, 377, 378
Pental, D., 193, 194, 196
Pepin, L., 202
Perbal, M.-C., 635
Perbost, C., 659
Percifield, R., 573, 580, 592
Pergamenschikov, A., 593
Pergams, O. R., 99
Perica, C., 495, 507
Perkins, N., 268
Perna, N. T., 332

Pernthaler, A., 263
Pernthaler, J., 263
Perou, C. M., 593, 599
Perrière, G., 753
Perrin, N., 288
Pertea, G., 120, 405, 411
Pesich, R., 599
Pesole, G., 312
Peter, B. J., 609
Peters, J. M., 228
Peterson, A., 287, 288, 289
Peterson, D. G., 386, 387, 390, 445, 450
Peterson, J. D., 123, 746
Peterson, L., 116, 119
Peterson, S. N., 123, 551
Pethiyagoda, R., 338, 340
Petitpierre, 463
Petrokovski, S., 702
Petry, K. G., 39, 45
Petrykowska, H., 372
Pettersson, B., 268
Pevzner, P. A., 376, 684, 689, 691, 693
Pfeiffer, B. D., 119, 120
Pfenninger, M., 281
Pfunder, M., 124
Pham, T., 635
Philippe, H., 116, 703, 781
Philippi, A., 727
Phillips, A., 105, 106, 108, 110
Phillips, C. A., 105
Phillips, C. R., 312
Phillips, M. S., 311
Piaggio-Talice, R., 754
Picard, C., 42
Pickerill, P., 494, 502, 505
Pickett, F. B., 714, 719, 725, 726
Pickup, R. W., 517
Pico, B., 193, 194, 196
Pieniazek, N. J., 88
Piertney, S. B., 77
Pietrokovski, S., 726
Piggott, M. P., 76, 80
Pikaard, C. S., 554, 555, 557, 567, 570
Pike, B. L., 608
Pilgrim, K. L., 77
Pillay, M., 193, 194, 195, 196
Pincus, E. L., 203, 214
Pinkel, D., 124
Pinter, R., 682
Piper, K. R., 659

Pippola, S., 60, 62, 64
Pires, A. E., 76, 77
Pires, J. C., 443, 444, 570, 573, 580,
    581, 583, 584, 585, 589, 592, 593, 610
Pirrung, M. C., 583
Pisani, D., 746
Pitch, U., 54
Pittman, R. N., 286, 288, 294
Planet, P. J., 111, 112, 116
Plant, A., 619
Plante, I., 312
Plaut White, O., 123
Plewniak, F., 105, 641
Plotkin, J. B., 725, 739
Plummer, M. L., 259
Poch, O., 105
Poinar, G. O., Jr., 88
Poinar, H. N., 81, 82, 88, 93, 94, 99, 100
Pol, D., 749
Polacco, M. L., 402
Polasky, S., 286
Pollack, J. R., 593
Pollock, D. D., 780, 781, 784, 785,
    786, 787, 788
Polymeropoulos, M. H., 400
Pombert, J. F., 372, 375
Pompanon, F., 154
Ponder, B. A., 424, 436
Ponsonnet, C., 42
Pontius, J. U., 405
Pont-Kingdon, G. A., 311, 312, 330
Poot, G. A., 134
Pop, M., 123
Pope, L. C., 77
Popescu, P., 467
Porebski, S., 9
Porteous, L. A., 62, 63, 64
Porter, R. D., 502
Portnoy, M. E., 119
Posada, D., 302, 663, 762, 766
Posch, T., 263
Possnert, G., 93, 94
Postlethwait, J., 714, 719, 725, 726
Pot, J., 16, 145, 146, 151, 152,
    161, 162, 180, 182, 427, 570
Potter, S. M., 526, 539
Potylicki, T.-T. F., 637, 638, 647, 648, 649
Poustka, A. J., 123
Powell, B. C., 638, 639, 640
Powell, C. L., 303

Powell, S. W., 659
Powell, W., 135, 182
Powers, S., 124
Pradhan, A. K., 193, 194, 196
Prasad, A. B., 119
Prasher, D. C., 661
Pravenec, M., 146, 148
Prensky, W., 589
Prentice, M. B., 123
Presland, R. B., 638, 639,
    640, 641, 642
Pressey, R. L., 286
Presting, G. G., 445, 451
Price, H. J., 444, 445, 478, 479,
    480, 481, 483
Price, S. A., 751, 754
Primmer, C. R., 202, 223, 224
Prince, J. P., 385
Prince, K. L., 203
Pringle, A., 60
Pritchard, L., 780
Proctor, M., 76
Prodöhl, P. A., 235
Program, N. C. S., 372, 373
Prosser, J. I., 59, 62, 63
Proudnikov, D., 264
Prowell, D. P., 165, 167
Pruett-Jones, S., 235
Pryde, S. E., 39
Pulliam, B., 608, 610
Purvis, A., 259, 746, 747, 750,
    751, 753, 754
Puryear, J., 29, 30, 619
Pusey, A. E., 76, 223
Puskás, L. G., 599

Q

Qamaruz-Zaman, F., 182
Qi, J., 377, 378
Qi, X., 689
Qi, Y., 116, 119
Qian, X. Y., 445
Qiu, T., 165, 168
Qiu, Y. L., 378
Quackenbush, J., 120, 405, 551,
    584, 591, 592
Quattro, J. M., 165, 197, 729
Queller, D. C., 202, 228, 235
Questiau, S., 74, 77, 83, 147

R

Racey, P. A., 77
Rademaker, J. L., 145, 146, 147
Radtkey, R., 268
Radune, D., 123
Rafalski, A., 134, 180
Rafalski, J. A., 135, 427
Raff, E. C., 521
Raff, R. A., 521
Ragan, M. A., 747
Ragiba, M., 419, 434
Ragob, J.-P. D., 788
Raha, D., 659
Raimond, R., 311
Raitio, M., 268
Rajapakse, C., 76
Rakin, A., 123
Rakuya, U., 311, 330
Ralls, K., 76, 78
Ralph, D., 571
Rambaut, A., 773
Rambold, I., 77
Ramirez, F., 4, 10
Ramisse, V., 61, 63, 64
Randi, E., 77
Randolph, J. B., 589
Rannala, B., 656, 718
Ransom, D. G., 148
Ransome, N., 39
Ranz, J. M., 610
Rao, E. V., 146
Rao, P. H., 116, 119
Raposo, M., 220
Rapp, J., 351
Rasch, E. M., 476, 477
Rasner, C. A., 77
Rasplus, J., 280
Rassmann, K., 77
Rastegar-Pouyani, N., 337, 338, 339, 340
Ratnasingham, S., 107, 109
Raubeson, L. A., 335, 351, 352,
    375, 376, 377, 378, 678, 689
Raudsepp, T., 223
Ravid, R., 122, 123, 609
Rawlings, T., 312, 337
Ray, I. M., 182, 193, 194, 196
Ray, J. L., 505, 506
Rayburn, A. L., 479, 480
Raymond, M., 220

Read, D. J., 58, 59, 61, 62
Read, J. L., 583
Read, T. D., 123
Rebelo, A. G., 293
Redlinger, T., 39, 40
Redoutet, W., 170
Ree, R. H., 750, 753
Reed, D. J., 59
Reed, J. Z., 75, 77
Reeder, R. H., 554, 555
Reedy, D. J., 146, 148
Rege, J. E., 147
Rehman, F., 427
Reich, D. E., 268
Reichardt, L. F., 529
Reichel, B., 261
Reijans, M., 16, 145, 146, 151, 152,
    161, 162, 427, 570
Reiland, J., 385
Reineke, A., 165, 166, 167
Reiner, A., 116, 119, 124
Reinhardt, R., 123
Reis, E. M., 408
Reischmann, K. P., 385
Reiseberg, L. H., 385
Relstone, J., 123
Ren, B., 266
Ren, J. C., 240, 248
Ren, S. X., 445
Renker, C., 68
Rensing, S. A., 351
Rentz, M., 180
Renz, D., 39, 45
Repka, J., 235
Rest, J. S., 337
Retamal, M., 164, 165, 167
Reutter, B. A., 288
Reyes, A., 312
Reymond, A., 118
Reynolds, S. H., 571, 573, 581
Reysenbach, A. L., 22
Rhodes, O. E., Jr., 147
Rhodes, W., 637, 638, 648
Ricci, S., 77
Ricciardi, L., 193, 196
Richards, 100
Richards, D. R., 609
Richardson, A. J., 39
Richardson, D., 551
Richardson, P. M., 324

Richly, E., 349, 351, 367, 375, 377
Richter, L., 261
Rico, C., 224
Rico, I., 224
Riddle, A. E., 77
Riedmuller, S. B., 404
Riemer, C., 372, 373
Rieseberg, L. H., 443
Rifkin, S. A., 123
Rigaud, T., 311
Riha, K., 423, 424, 425, 435
Rijans, M., 180, 182
Rikkinen, J., 50
Riley, M., 269
Riley, R. M., 615
Rincón-Limas, D. E., 119, 120
Rine, J., 203, 598
Rios, L., 288
Rioux, J., 268
Rioux, P., 312
Ris, H., 471, 473
Ritchie, D. A., 39, 61
Rittmann, B. E., 264
Rivas, R. D., 378
Rivera, M. C., 112
Rivera, M. N., 116, 119
Roberts, C. A., 402
Roberts, D. B., 467
Robertson, L. H., 38, 42
Robertson, M., 288, 289
Robertson, S. E., 420, 424
Robinson, A., 591
Robinson, E., 268
Robinson, G. E., 614
Robinson, J. P., 187, 188, 189, 190, 191
Robinson, N. A., 76, 80
Robinson, N. L., 402
Rochelle, P. A., 60, 62, 63, 64
Rodgers, L., 124
Rodrigo, A. G., 772, 773
Rodriguez-Trelles, F., 703
Rodriguez-Valera, F., 260
Roe, B. A., 88
Røed, K., 77
Roessli, D., 143
Roger, A. J., 116, 311, 312, 330
Rogers, G. E., 638, 639, 640, 641, 642, 648
Rogers, J. D., 134
Rogozin, I. B., 112, 123
Rohland, N., 81, 82

Rohlf, F. J., 196
Rohlin, L., 605
Rojas, E., 287
Rokas, A., 106, 111, 112, 113, 114, 115, 116, 678, 702, 704
Rokhsar, D., 324
Roldan-Ruiz, I., 147, 148, 163, 196
Rollinson, D., 311, 330, 334, 340
Roman, J., 76
Romano, S., 323
Romanowski, G., 61
Romero-Herrera, A. E., 708, 711
Romero-Severson, J., 134, 164
Ronning, C. M., 404
Ronquist, F., 653, 655, 711, 718, 749, 754
Rooney, A. P., 312
Roper, T. J., 77
Roques, P., 77
Rose, A. M., 408
Rosenberg, H. F., 653, 728, 729, 730, 738
Rosenberg, L. A., 312, 321, 322
Rosendahl, S., 64, 68, 147
Roshan, U., 677, 753, 754
Ross, A. J., 88
Rossello-Graell, A., 288, 293
Rossenbeck, H., 477
Rossier, C., 118
Rostan, S., 124
Rotella, J., 77
Roth, M. E., 659
Roth, R., 581
Rottmann, W. H., 372
Roubik, D. W., 292
Roux, K. H., 227
Rowell, H., 340
Rowley, I., 227
Roxburgh, C. M., 638, 639
Roy, L. M., 402
Rozen, S., 268, 547, 548
Ruben, E., 148
Rubin, E. M., 107, 116
Rubin, G. M., 119, 120, 409
Rudbeck, L., 93, 100
Rudd, S. A., 106, 123, 400, 405
Ruggiero, L. F., 77
Ruiz, E., 134, 138
Ruiz-Garcia, L., 151
Ruiz-Medrano, R., 635
Rujan, T., 349, 351, 367, 375, 377
Rumjanek, N. G., 65

Runko, S. J., 106, 123
Ruschoff, J., 424, 437
Russel, D. W., 620, 627
Russell, D. R., 558, 559, 569
Russell, D. W., 43, 204, 214, 315, 318, 319, 328, 641, 642, 643
Russell, E., 227
Ryan, K., 324
Ryder, O. A., 76, 106, 112, 117, 118, 123
Rygiewicz, P. T., 59, 63
Ryman, N., 147
Rynearson, T. A., 301

S

Saano, A., 60, 62, 64
Sabour, P. M., 38, 39
Saccheri, I. J., 224
Saccone, C., 312
Saedler, H., 422, 635
Sagiv, Y., 749
Sagot, M., 684
Sahara, K., 464, 465
Sahr, K., 286
Saiki, R. K., 104
Saint, R., 123
Saint-Louis, D., 312
Saitoh, T., 223
Saitou, N., 675, 689, 782
Sajantila, A., 268
Sakai, A., 420, 421, 423, 424, 427, 436
Sakai, F., 419, 423, 424
Sakaida, M., 420, 433
Sakata, H., 702
Saki, R., 104
Sakmar, T. P., 653, 659, 728, 738
Sala, F., 145, 146, 152, 153
Salamin, N., 746
Salas, M., 324
Saldanha, A. J., 583
Salesses, G., 420
Salimans, M. M. M., 80, 81
Salmon, E. D., 540
Salomon, M., 147
Salonga, T., 39
Salvadori, S., 480, 481
Salvato, P., 164, 165, 166
Salvatore, B. A., 637, 638, 647, 648, 649
Salyers, A. A., 515

Salzberg, S. L., 120, 123, 374
Sambrook, J., 34, 43, 75, 173, 204, 206, 213,
    214, 272, 315, 318, 319, 328, 421, 438, 558,
    559, 569, 571, 573, 575, 579, 581, 620, 627,
    641, 642, 643
Samuel, G., 123
Sandbrink, J. M., 355
Sander, C., 780, 781, 783
Sanderson, M. J., 702, 711, 746, 747,
    749, 750, 753, 754
Sang, T., 444
Sanger, F., 88
Sanker, S., 653, 728, 738
Sankoff, D., 312, 335, 385, 461, 472,
    680, 682, 691, 693, 703
Sano, S., 349
Sano, T., 349
Santini, M. P., 39, 45
Santos-Guerra, A., 134, 138, 142
Sapolsky, R., 268
Sargeant, L. L., 202
Sarkans, U., 591
Sarkar, I. N., 111
Särkilathi, L. K., 40, 42
Sartor, M., 613
Sasaki, N., 120, 121
SAS Institute, 196
Saudek, D. M., 105
Saulino, A., 213, 642
Saunders, G. W., 302
Sauvajot, R. M., 73
Savelkoul, P. H., 145, 146, 147
Savolainen, V., 110, 378, 746
Sawyer, N. A., 134
Sawyer, R. H., 637, 638, 640, 647, 648, 649
Saxman, P. R., 261
Sayler, G. S., 61, 62, 63
Scali, V., 478, 481
Scally, M., 106, 117, 118, 123
Scamuffa, N., 118
Schable, N. A., 203
Schach, U., 531
Schafer, A. J., 180
Schaffer, A. A., 363, 402
Schaffer, P. J., 659
Scharf, S. J., 104
Scheetz, T. E., 402
Schelter, J. M., 583
Schemske, D. W., 385
Schena, M., 583

Scherer, S. W., 116, 119
Schertz, K., 134
Schertz, K. F., 385
Schierwater, B., 111, 311, 312, 338
Schlegel, R., 435
Schleicher, E., 82
Schleiermacher, C., 373, 374
Schleifer, K. H., 261, 264, 270
Schlötterer, C., 202, 223
Schlueter, S. D., 405, 409, 412, 414, 581
Schmid, R., 402
Schmidt, T. M., 261, 312
Schmidt, T. R., 781
Schmitt, K., 424
Schmitz, J., 312
Schmitz, R. W., 88, 93, 94
Schmuths, H., 294
Schneider, C., 120, 121
Schneider, K., 239
Schneider, M. V., 165, 168
Schneider, R., 780
Schneider, S., 143
Scholin, C. A., 261, 303, 304
Scholz, F., 292
Schön, C., 193, 194, 196
Schöniger, M., 193, 194, 196
Schouls, L., 145, 146, 147
Schouten, J., 152
Schreiber, D. R., 780, 781
Schreiber, E. A., 147
Schreuder, M., 286, 287
Schribner, K. T., 76, 223
Schrier, P. H., 88
Schröder, W., 77
Schroeder, L. J., 148
Schubert, I., 54
Schubert, J. I., 423
Schuler, G. D., 405
Schulte, J. A. II, 312, 337, 338, 339, 340
Schultz, P. A., 60
Schulze-Kremer, S., 591
Schüssler, A., 59
Schutze, K., 423, 424, 437
Schwanekamp, J., 613
Schwartz, M. K., 77
Schwartz, S., 119, 372, 373
Schwarz, C., 117, 118, 123
Schwarz, M. S., 119
Schwarz, S., 118
Schwarzacher, T., 444, 445, 458

Schwarzott, D., 59
Schwarz-Sommer, Z., 635
Schweitzer, M. H., 88
Schwieger, F., 16, 22
Scott, J., 105
Scott, K. P., 39, 42, 43, 495, 496, 509, 513, 515
Scouras, A., 338
Scutt, C. P., 420, 421, 423, 424
Sears, B. B., 372
Sebat, J., 116, 119, 124
Seberg, O., 182
Seeb, J. E., 146, 148
Segraves, R., 124
Segura, S., 288, 292
Seibold, I., 147
Seidman, J. G., 271, 315, 319, 328
Seigfried, T. E., 402
Sekar, R., 263
Sekioka, T., 193, 194, 196
Sekiya, T., 428
Seligmann, H., 784
Selvin, S., 233
Semenova, T. N., 662, 666
Semerikov, V. L., 169
Semple, C., 703, 749
Sennbald, B., 716
Sennedot, F., 280
Sensen, C. W., 703
Sergia, C., 288, 293
Sergovia-Lerma, A., 182, 193, 194, 196
Serre, D., 99, 100
Serres, M. H., 269
Seruga, M., 495, 507
Setlow, J. K., 494, 504
Severson, D. W., 164
Sévigny, J.-M., 473, 476
Sexton, J. P., 302
Sgaramella, V., 659
Shaffer, H. B., 702
Shafizadeh, E., 531
Shagin, D. A., 662, 666
Shah, N., 124, 268
Shalon, D., 583
Shames, R. B., 637, 638, 648
Shamir, R., 684, 691
Shammakov, S. M., 338, 339, 340
Shamshad, S., 147
Shannon, K. W., 583
Shao, R., 340
Shapiro, B., 94

Sharan, R., 684
Sharma, S. K., 182
Sharp, P. A., 555, 559
Sharp, P. M., 349, 658
Shaul, S., 703
Shaw, K. L., 165, 169, 170, 171, 172, 174
Shchepinov, M., 263
Shear, J. B., 539
Shein, A. K., 338, 340
Shen, J., 148
Shen, N., 268
Sherlock, G., 591
Sherry, R. A., 281
Sherwin, W. B., 76
Shi, Y., 728, 738
Shiba, T., 552
Shibata, F., 423, 424
Shields, D. C., 658
Shiki, Y., 349
Shima, A., 148
Shimada, A., 148
Shimada, H., 349, 783
Shimodaira, H., 772
Shimokata, K., 783
Shindyalov, I., 780, 781, 783
Shinozaki, K., 349
Shinzawa-Itoh, K., 783
Shirai, H., 349
Shirley, R., 105
Shizuya, H., 385, 386
Shoemaker, D. D., 583
Shoemaker, J. S., 727
Shore, J. S., 281
Shorthouse, D. P., 473, 478
Shotton, D. M., 540
Shu, C. L., 119
Shumway, D. L., 60
Sialer, M. F., 90
Siao, C. J., 268
Sicheritz-Pontén, T., 719
Sidow, A., 88
Sidransky, D., 124
Siebert, P. D., 120, 121
Siegismund, H., 312
Siepel, A. C., 119, 691, 692
Sikorski, J., 494, 512, 513
Sil, A., 598
Silflow, C. D., 268
Silk, J. B., 77, 82

Silva, M. C., 63, 555
Silva, O. M., 134, 138
Silvestre, V., 301
Simmons, M., 123, 147
Simmons, M. P., 713, 720
Simmons, N. B., 746
Simon, B., 165, 166
Simon, C., 334, 761
Simon, D., 694
Simon, D. L., 376
Simon, M., 385, 386
Simon, N., 260, 261, 267, 303, 304
Simon, P. W., 147
Simonet, P., 42, 61, 63, 64, 494,
    495, 496, 507, 511, 512, 514, 516
Simor, A. E., 39, 40, 45
Simpson, C. E., 286, 288, 294
Simpson, C. L., 351, 375
Simpson, J. M., 38, 39, 40
Simpson, S. P., 233
Sinclair, G., 659
Sinervo, B., 385
Singh, R. J., 445, 446, 458
Singh, R. S., 597
Singh-Gasson, S., 263, 583
Sinha, H., 609
Sinha, N., 635
Sinigalliano, C. D., 504
Siroky, J., 420, 422, 423, 424, 425, 435
Sison, M., 301
Sistonen, P., 268
Sive, H. L., 529
Siwek, M., 148
Skaletsky, H., 419, 547, 548
Skiena, S., 682
Sklar, J., 603
Sklar, P., 268
Skrabanek, L., 703, 719
Skroch, P. W., 65
Slaney, L., 494, 504
Slate, J., 223, 224, 235
Sledzik, P. S., 95, 98
Slepak, T., 385, 386
Sleutels, F., 554
Sloane, M. A., 76
Slonimskic, P. P., 788
Slowinski, J. B., 702, 713, 720, 749
Small, K. V., 105
Smalla, K., 62, 63, 494, 496, 497,
    503, 505, 512, 516

Smiley, C. J., 88
Smirnoff, N., 5, 11
Smit, A. F., 119, 372, 373
Smith, 90
Smith, A. B., 88
Smith, A. J., 88
Smith, C. W., 571, 581
Smith, D., 77
Smith, D. A., 76
Smith, D. G., 94
Smith, D. R., 551
Smith, F. A., 59
Smith, H. O., 105
Smith, I., 260
Smith, J. A., 271, 315, 319, 328
Smith, J. B., 483
Smith, J. J., 148, 151
Smith, K. L., 77, 684
Smith, L. W., 135
Smith, M. C., 76
Smith, M. J., 338, 339
Smith, M. T., 134
Smith, P., 123
Smith, R. F., 408
Smith, S., 117, 118, 123
Smith, S. E., 58, 59, 61
Smith, T. F., 712, 719
Smith, W. J., 61, 63
Smouse, P. E., 165, 197, 233
Sneath, S. K., 165
Snel, B., 112
So, B. G., 378
Sobolik, K., 81
Socci, N. D., 675, 689
Soderstrom, B., 609
Sogin, M. L., 20, 25, 271
Sokal, R. R., 165, 196
Sol, C. J. A., 80, 81
Solas, D., 583
Solís, C. R., 228
Soller, M., 148
Soltis, D. E., 378, 444, 769
Soltis, P. S., 378, 444, 746, 769
Sone, T., 420, 433
Song, W., 324
Song, W. M., 359
Song, X., 39
Soniat, T. M., 303
Sonnhammer, E. L. L., 711
Soreng, R. J., 375

Sorenson, M. D., 114, 328, 338, 339, 340
Sorgeloos, P., 165
Sorhannus, U., 733
Sorkin, B., 123
Sosnowski, R., 268
Soucek, P., 423, 435
Soumi, L., 61, 67
Soupene, E., 609
Souquière, S., 77
Southern, E. M., 123, 263, 319, 328
Souty-Grosset, C., 311
Sparks, A. B., 653, 728, 738
Sparrow, A. H., 483
Spaulding, W. G., 81, 82, 93, 94
Speed, T., 599
Speicher, M. R., 425
Spellman, P., 591
Spencer, J., 268
Spicer, G. S., 334
Spiegelman, J. I., 609
Spinsanti, G., 338
Spong, G., 77
Spooner, D., 293
Spriggs, T., 105
Springer, M. S., 106, 117, 118, 123, 750
Spruell, P., 146, 148
Spurbeck, J. L., 468
Srivastava, A., 193, 194, 196
St. John, K., 677
Stabler, R. A., 123
Stace, C. A., 444, 445
Stackebrandt, E., 62
Stackhouse, J., 653, 728, 738
Stacy, J. E., 77
Staden, R., 88
Stahl, D. A., 39, 42, 43, 260, 264, 303
Stahl, U., 493
Stalker, H. D., 466
Stalker, H. T., 286, 288, 294
Stam, W. T., 267
Stamatakis, A., 261
Stanhope, M. J., 106, 116, 117, 118, 123
Stankiewicz, B. A., 93, 94
Stanton, D., 336, 338, 678, 680
Stanton, M. L., 281
Stantripop, S., 119
Stapleton, M., 119, 120
Stears, R. L., 589
Steel, M., 749
Steel, M. A., 676, 747, 749, 750, 752

Steffan, R. J., 42, 62, 63
Stegalkina, S. S., 404
Stehlik, I., 427
Steiger, D. L., 193, 194, 196
Stein, D., 678
Stein, L., 268
Steiniger, G. E., 464
Steinmetz, L. M., 609
Stekel, D. J., 404
Stelly, D. M., 444, 445, 571, 581
Stem, C., 82
Stephaniants, S. B., 583
Stephen, J. R., 59, 62, 63
Steppan, S., 81, 82
Steppi, S., 261
Stern, B., 351, 372, 375, 377
Stern, D. B., 351, 375
Sterner, K. N., 77
Stevens, C. M., 60
Stevens, S. E. J., 502
Stevens, Y., 571, 573, 581
Stevenson, B. J., 118
Stevenson, D. W., 106, 123
Stewart, C.-B., 659, 728, 729, 730, 738, 740, 784
Stewart, C. S., 39
Stewart, G. J., 493, 494, 495, 496, 502, 504, 508, 515
Stewart, J., 591
Stice, L., 323
Stickel, S., 20, 25, 271
Stiles, J. I., 419, 434
Stoebe, B., 349, 351, 367, 375, 377
Stoeckert, C., 591
Stoeckle, M., 107
Stoffel, S., 104
Stoffella, A., 77
Stoike, L. L., 372
Stone, A., 88, 93, 94
Stoneking, M., 88, 93, 94
Storm, C. E. V., 711
Stormo, G. D., 780, 784
Stoughton, R., 583
Stoye, J., 373
Strassmann, J. E., 202, 228
Straub, T. M., 268, 269
Straus, D., 421
Strauss, S. H., 135, 372
Streelman, J. T., 224
Strehlow, R., 261

Streiff, R., 292
Strike, P., 39, 61
Strobeck, C., 76, 223
Strucki, S., 465
Struhl, K., 271, 315, 319, 328
Strunk, O., 261
Stryer, L., 583
Stuart, B., 314
Stuckmann, N., 261
Stuessy, T. F., 134, 138
Stummann, B. M., 354, 357
Styles, C. A., 583
Suárez-Seone, S., 288, 289
Sudarsanam, P., 605
Sugano, S., 120, 121
Sugita, M., 349, 351
Sugiura, M., 348, 349, 351
Sugiyama, F., 725
Sulawati, T., 147
Sullivan, J., 334, 761, 763, 767
Sullivan, W., 467
Sultana, R., 405, 411
Summers, K., 233
Summers, T. J., 119
Sun, H. M., 653, 728, 738
Sun, J., 677
Sun, M., 169
Sun, Z. Y., 324, 359
Sunnucks, P., 76, 231
Sunquist, M. E., 76
Sussman, M. R., 263, 583
Sutton, G. G., 105, 106, 551
Suzuki, N., 148
Suzuki, T., 263
Suzuki, Y., 619, 731, 732, 733, 734, 740, 741
Swaitek, M., 372
Swamy, K. R., 146
Swann, E., 20
Swanson, W. J., 656
Sweder, K. S., 106
Swenson, K., 687
Swidan, F., 682
Swofford, D. L., 114, 302,
    336, 654, 674, 676, 684,
    759, 761, 763
Syed, N. H., 571, 581
Sykes, T., 77
Syvanen, A. C., 268, 431
Szybalski, W., 386
Szymanski, T. G., 749

T

Taberlet, P., 73, 74, 76, 77, 83,
    147, 154, 163, 170, 233
Tabita, F. R., 267
Tachiiri, Y., 385, 386
Tago-Nakazawa, M., 134, 138
Takada, T., 38
Takahama, M., 420, 433
Takahashi, H., 148
Takahashi, M., 292
Takahashi, S., 725
Takaiwa, F., 349
Takami, 164, 165, 169
Takano, H., 420, 421
Takenaka, M., 420, 433
Takeuchi, F., 552
Takeuchi, M., 349
Takeya, Y., 376
Tallmon, D., 163
Tamhane, A. C., 584, 592
Tamura, K., 676
Tan, M., 678
Tan, Y., 165, 166, 167
Tanaka, M., 349
Tanaka, R., 38
Tang, H., 736, 741
Tang, J. J., 335, 376, 684, 685,
    687, 689, 691, 692, 693, 694
Tanksley, S. D., 385, 400, 411
Tannier, E., 684
Tannock, G. W., 38
Tarabykin, V., 120, 121
Targan, R. E., 708
Tarjan, R., 684
Tarrio, R., 703
Tarvin, K. A., 235
Tas, E., 60, 62, 64
Tate, M. L., 233
Tautz, D., 77, 180, 202
Tavare, L., 662
Taylor, A. C., 76, 80
Taylor, B., 680, 682
Taylor, D., 202
Taylor, D. E., 551
Taylor, E., 224
Taylor, J. S., 719
Taylor, J. W., 20, 49,
    50, 59, 63, 147
Taylor, R., 591

Taylor, W. R., 656, 662, 781, 784, 785, 786, 787
Teather, R. M., 39, 40, 42, 43, 45
Tebbe, C. C., 16, 22, 25, 61, 62
Teeling, E., 106, 117, 118, 123
Teetes, G. L., 479, 480
Tegelström, H., 231
Tehler, A., 52
Telenius, H., 424, 436
Telfer, P. T., 77
Tel-Zur, N., 10, 14
Tenkouano, A., 193, 194, 195, 196
Tepfer, D., 495, 507
Terado, T., 148
Terhalle, W., 780, 784
Terkelsen, C., 55
Tesler, G., 689, 693
Tettelin, H., 123
Tevelev, A., 659
Tewari, K. K., 354, 357
Thamm, S., 123
Thanaraj, T. A., 411
Thaxton, P. M., 571, 581
Theimer, T. C., 147
Theodorides, K., 106, 122
Theriault, T. P., 583
Thin, T., 338, 340
Thindwa, H. P., 479, 480
Thomas, D., 268
Thomas, D. C., 324
Thomas, J. W., 119
Thomas, K. M., 165
Thomas, M. R., 203
Thomas, P. J., 119
Thomas, R. H., 88
Thomason, B., 123
Thompson, E. A., 235
Thompson, J. D., 105, 641, 782
Thompson, P. M., 75, 77, 81
Thompson, W. F., 63
Thomson, J., 678
Thomson, J. M., 653
Thorgaard, G. H., 146, 148
Thorley, J. L., 713
Thornton, J. M., 656, 662
Thornton, J. W., 111, 653
Thorpe, R. S., 147
Tian, L., 572, 573, 580, 581, 583, 584, 585, 589, 592, 593, 610
Tibbetts, E. A., 337

Tibshirani, R. J., 196
Ticknor, L. O., 268
Tiedje, J. M., 62, 63, 64, 261
Tille, S., 538
Tillier, E., 682, 693
Timmerman, M. M., 608
Timmermans, M., 635
Tingey, S. V., 135, 180, 427
Tiongson, E. E., 119
Titball, R. W., 123
Töbe, K., 20
Tohdoh, N., 349
Tohme, J., 193, 288, 292
Tollefsbol, T. O., 653, 728, 738
Tollit, D. J., 75, 77
Tolstoshev, C. M., 401
Tomb, J.-F., 105, 551
Tomiuk, J., 294
Tomizaki, T., 783
Tomkins, J. P., 386, 387, 390, 450
Tomlinson, C. R., 613
Toon, A., 227
Topaloglou, T., 268
Topp, E., 39, 40, 42, 43, 45
Torazawa, K., 349
Torres, M. E., 281
Torres-Ruiz, R. A., 193, 194, 196
Torsvik, V. L., 61, 503
Toth, G., 269
Touchman, J. W., 119
Tourasse, N., 123
Townsend, J. P., 599, 603, 605, 607, 614
Townsend, T., 338, 340
Tran, J. T., 119
Trask, B., 116, 119
Traut, W., 466
Travis, S. E., 182
Trevors, J. T., 60, 62, 63, 145, 146, 182, 497
Triantaphyllidis, G. V., 165
Triboush, S. O., 356
Triplett, E. W., 65
Trivedi, N., 402
Triwitayakorn, K., 148
Troge, J., 116, 119, 124
Troggio, M., 150, 154
Trout, J. M., 39
True, H. L., 266
Trust, T. J., 551
Tsai, J., 411
Tsai, Y. L., 39, 40, 62, 63

Tseng, G. C., 605
Tsien, R. Y., 661
Tsinoremas, N. F., 411
Tsudzuki, J., 351
Tsukihara, T., 783
Tsukomoto, K., 328
Tsumura, Y., 135
Tsunewaki, K., 376
Tsurgeon, C., 119
Tu, E., 268
Tuff, P., 780
Tummino, P. J., 551
Tun, H., 338, 340
Tunlid, A., 609
Tunnacliffe, A., 424, 436
Turley, M., 60, 62, 63, 64
Turmel, M., 312, 372, 375, 376
Tuross, N., 93, 94, 95, 98
Turpeinen, T., 193, 194, 196
Tutin, C. E. G., 76
Tuttle, E. M., 235
Twigg, R. W., 106, 123
Tyagi, A. P., 571, 573, 581
Tyagi, S., 123
Tyler, B. M., 63

U

Ubi, B. E., 193, 196
Uchida, H., 420, 421
Uchiyama, I., 552
Ucla, C., 118
Ude, G., 193, 194, 195, 196
Ueland, P. M., 240, 248
Uemori, T., 642
Ueshima, R., 311, 330, 338
Ugalde, J. A., 662, 666, 728, 738
Uhlen, M., 609
Uhrès, E., 76, 77
Uittrlinden, A. G., 16
Ulanch, P. E., 581
Ullman, J. D., 749
Ulloa, O., 300
Umayam, L. A., 746
Umesono, K., 349
Underbink, A. G., 483
Unnasch, T. R., 311, 340
Urabe, I., 702
Uria-Nickelsen, M., 551
Usary, J., 599

Usha, A. P., 233
Usuka, J., 409
Utterbach, T. R., 404
Utterback, T. R., 105

V

Vahjen, D., 25
Vahjen, W., 61, 62
Vaiman, D., 202
Valencia, A., 780, 781
Valentin, 270
Valentini, A., 146, 147, 151, 153, 154
Valerio, A., 659
Valls, J. F. M., 286, 288, 294
Vanaken, S. E., 404
Van Alphen, J. J. M., 165, 168
van Bockstaele, E., 193, 194, 195, 196
Vanbrederode, J., 355
van Custem, P., 193, 194, 196
van de Lee, T., 16, 145, 146, 151, 152, 161,
    162, 180, 182, 427, 570
Vandenberg, N., 39, 45
Van de Peer, Y., 703, 719
van der Heijden, M. G. A., 60
van der Hoeven, R., 400, 411, 571, 573
van der Meijden, R., 170
van der Noordaa, J., 80, 81
Van der Poel, J. J., 148
van der Staay, G. W. M., 303
Vander Velden, K., 741
van der Wurff, A. W. G., 152, 165, 167
Van de Wiel, C., 145, 146, 152, 153
van Dillen Werthien, P. M. E., 80, 81
Vaneeden, F. J. M., 531
van Elsas, J. D., 60, 62, 63, 64,
    494, 497, 498, 503, 505, 506,
    507, 510, 511, 512
Vane-Wright, R. I., 287
Van Haeringen, W. A., 147, 148, 151
Vanhala, T., 193, 194, 196
Vanham, R., 355
van Hummelen, P., 599
Vanjani, R. G., 311, 312, 330, 338
van Kammen, A., 451, 456
Van Lith, H. A., 147, 148, 151
van Loo, J., 39, 40
van Overbeek, L. S., 497
van Straalen, N. M., 152, 165, 167
van Tuinen, D., 68

Van Weerelt, M. D. M., 494, 498, 505, 506, 507
van Wennekes, E. J., 456
Van Zutphen, B. F., 146, 148
Van Zutphen, L. F., 147, 148, 151
Vargesson, N., 537
Varki, A., 122, 123, 609
Vaudor, A., 196
Vaughn, M., 583, 585, 592
Vaulot, D., 260, 261, 300, 302, 303
Vawter, L., 753
Vaysseix, G., 202
Vazquez, A., 145, 146, 152, 153
Vecchiotti-Antaldi, G., 146, 147, 151, 153, 154
Vekemans, X., 147, 148, 163, 193, 194, 196
Vekris, A., 39, 45
Velculescu, V. E., 581
Vellekoop, P., 355
Velten, J. P., 62
Vendrely, C., 471
Vendrely, R., 471
Venkatraman, E., 124
Venter, J. C., 105, 106, 551
Verjovski-Almeida, S. M., 408
Verkaar, E. L., 147
Verma, A., 334
Vermaas, E., 581
Vernesi, C., 89
Vestal, J. R., 60
Veuskens, J., 419, 420, 422, 424, 437
Vicente, M. C. D., 385
Vidal, C., 76
Vidal, R., 39, 45
Vidya, T. N. C., 76
Viegas, W., 555
Vieira, C., 478, 481
Vigilant, L., 77, 83, 223
Vignal, A., 148, 202
Vilbig, A., 261
Villet, M., 288, 289
Vilo, J., 591
Vincentz, M., 400
Vindum, J. V., 338, 340
Vingron, M., 591
Vision, T., 385
Visser, R. G., 571, 573
Vitte, C., 418, 428
Vogel, P., 288
Vogel, T. M., 495, 507, 514

Vogelsang, E., 531
Vogelstein, B., 180, 581
Vogler, A. P., 106, 122, 720
Voglmayr, H., 471
Vogt, J. L., 119
Voinov, A. V., 268
Volfvovsky, N., 374
Volker, C., 116
Vollmer, D., 224
Vomstein, J., 372
Vos, P., 16, 145, 146, 151, 152, 161, 162, 180, 182, 426, 570
Vosman, B., 145, 146, 152, 153, 494, 498, 502, 505, 510
Voss, S. R., 148, 151
Vovis, G. F., 551
Vrana, J., 423, 434
Vreugdenhil, D., 571, 573
Vrijenhoek, R., 323
Vyskot, B., 420, 421, 422, 423, 424, 425, 434, 435
Vysotskaia, V. S., 268

W

Wachira, F. N., 182
Wackernagel, W., 61, 492, 494, 498, 502, 504, 505, 506, 509, 512, 513, 514
Wada, H., 148
Wada, T., 619
Waddell, P. J., 337, 378, 654, 674, 684, 759, 761, 768
Waerik, B., 267
Wages, J. M., 429
Waggoner, A. S., 589
Wagner, L., 405
Wagner, M., 264, 270
Wagner, R. P., 462
Wahleithner, J. A., 311
Wainwright, B. J., 226
Waits, L. P., 73, 74, 76, 77, 82, 83, 233
Wakasugi, T., 349, 351
Walbot, V., 402
Walker, C., 59
Walker, I. D., 638, 639
Walker, M. A., 116, 119
Walker, P., 288
Walker, W. L., 583
Wallace, D. C., 312

Wallinger, S., 423, 436
Walsh, J. B., 703, 719, 725
Walsh, K., 123
Walsh, P. S., 78
Walter, M., 684
Walton, M., 164
Wan, C. C. Y., 98
Wan, I., 408
Wan, K., 119, 120
Wang, A. M., 119, 120
Wang, C., 263
Wang, D., 124
Wang, D. G., 268
Wang, J. J., 573, 580, 581, 583, 584,
    585, 589, 592, 593, 610
Wang, J. L., 572, 573, 580, 581,
    583, 584, 585, 589, 592, 593, 610
Wang, L.-S., 335, 375, 376, 678,
    682, 684, 685, 689, 691, 694
Wang, M.-L., 324, 444, 445
Wang, R., 659
Wang, R.-F., 38, 42
Wang, S. X., 619
Wang, X., 609, 689
Ward, D. C., 324, 424, 452, 453
Ward, J. H., 196
Warnecke, F., 263
Warnow, A. T., 694
Warnow, T., 335, 375, 376, 674,
    675, 677, 678, 682, 684, 685,
    689, 691, 694, 753, 754
Washington, L. D., 637, 638
Wasmuth, J., 402
Wasser, S. K., 76, 77, 81, 82
Watanabe, K., 38, 311, 330
Watanabe, Y., 330
Waterbury, J. B., 300
Waterman, M. S., 712
Waters, C. W., 541
Watson, B., 571, 572, 573, 580, 581, 583, 584,
    585, 589, 592, 593, 610
Watson, R., 544, 545
Watson, R. M., 546, 547, 551, 552
Watson, S. W., 300
Waught, R., 182
Wayne, R. K., 73, 77, 82, 94
Webb, W. W., 539
Weber, J. L., 202
Weber, R. J., 119
Webster, M., 202

Webster, M. S., 223, 235
Wedel, J. F., 385
Wee, J. L., 349
Weeden, N. F., 4, 12
Weedn, V. W., 95, 98
Weetman, D., 164, 165, 167
Weghorst, C. M., 427
Wei, Y.-H., 312
Weidman, J. F., 105, 123
Weigel, D., 623, 634
Weightman, A. J., 60, 62, 63, 64
Weir, A. F., 5, 11
Weir, B. S., 165, 195, 728
Weirauch, M., 372, 373
Weiss, M. L., 781
Weissenbach, J., 202
Weissman, S. M., 99, 203
Welling, G. W., 39, 40
Wellington, E. M. W., 497
Welsh, J., 571
Wendel, J. F., 570, 573, 580, 592
Wendisch, V. F., 609
Wendl, M., 120, 429
Wenk-Siefert, I., 116
Wensel, T. G., 727
West, A., 583
West, J. A., 124
Westerbergh, A., 385
Westerfield, M., 532, 537
Westergaard, M., 418, 419
Westneat, D. F., 223, 235
Westover, A., 120, 121
Westram, R., 261
Wetherby, K. D., 119
Weyand, N. J., 88
Whalen, B., 312
Wheatcroft, R., 38, 39
Wheelan, S. J., 409
Wheeler, P. A., 146, 148
Wheeler, W., 88, 105
Whelan, S., 656, 766
Whitbread, L. A., 638, 639, 640, 642, 648
White, B. A., 38, 39, 40
White, D. C., 60
White, J., 405, 411
White, J. A., 404
White, K. P., 123, 615
White, M. J. D., 465
White, M. M., 473
White, O., 105, 106, 551, 746

White, T. J., 59, 63, 746
Whiteley, A. S., 42
Whitfield, C. W., 614
Whitfield, M. L., 599
Whiting, M. F., 106
Whitmore, G. A., 603
Whittaker, L. A., 638, 639
Whittier, C. A., 82
Whittier, R. F., 390
Whitton, J., 152, 719
Whyte, L. G., 62
Wickham, G. S., 22, 263
Wickings, E. J., 76, 77
Widder, E. A., 662, 666
Wiese, B. A., 408
Wiggins, L. S., 119
Wigler, M., 116, 119, 124, 421, 434
Wikström, M., 781
Wilcox, A. W., 95, 98
Wilcox, M. H., 39
Wild, J., 386
Wildman, D. E., 781
Wilkins, T. A., 98
Wilkinson, M., 713, 753
Will, J. A. K., 60, 62, 63, 64
Willerslev, E., 89, 93, 94, 100
William-Moore, G., 708, 711
Williams, B. L., 106, 111, 112, 113,
    114, 115, 116, 702, 704
Williams, C. F., 593
Williams, D., 284, 286, 288, 289, 292, 294
Williams, H., 289
Williams, H. G., 493, 495, 496, 508, 515
Williams, J. G. K., 180, 426
Williams, J. L., 224, 233
Williams, K., 288, 289, 292, 294
Williams, P. D., 788
Williams, P. H., 286, 287
Williams, R. M., 539
Williams, S. R., 581
Williams, T., 677
Williams, T. L., 753, 754
Williamson, V. M., 451
Willis, M. E., 419, 424
Wilson, A. C., 88, 99, 312, 653, 746
Wilson, G. J., 77
Wilson, I. G., 61
Wilson, K. H., 39, 43
Wilson, K. P., 653
Wilson, M. R., 312

Wilson, R. K., 150
Wilton, S. D., 639, 640
Wimberger, P. M., 730
Winchester, E., 268
Windus, J. L., 134, 138, 142
Winfield, M. O., 145, 146, 152, 153
Wing, R. A., 386, 445, 450, 451, 453
Wing, R. W., 386, 387, 390
Wink, M., 134
Winston, F., 605
Winter, K., 571, 573, 581
Wintzingerode, F., 62
Winzeler, E. A., 263
Withford, M. F., 39, 40, 42, 43, 45
Witsenboer, H., 182
Wiuf, C. C., 94
Woelfle, M. A., 351
Woese, C. R., 260, 265, 302
Wolf, Y. I., 112, 123
Wolfe, A. D., 134, 135, 138, 140, 142, 378
Wolfe, K. H., 349, 658, 703, 719, 720
Wolfenbarger, L. L., 145, 146, 147, 162, 182
Wolfinger, E. D., 599, 605, 614
Wolfinger, R. D., 599, 605, 614, 615
Wolfl, S., 359, 376, 378
Wolinski, H., 55, 56
Wollenberg, K. R., 780, 784
Wolpert, D. H., 780, 784
Wolstenholme, D. R., 311, 312,
    330, 331, 340
Wolters, A., 62, 63
Womack, J. E., 117, 118, 123
Wong, W. H., 605
Wong, W. K., 599
Wong, W. S. W., 733, 734
Wood, A. M., 300
Wood, J., 39, 42, 43
Woodruff, D. S., 76, 77
Woods, J. G., 76
Woodward, S. R., 88
Worley, K. C., 408
Worthy, T. H., 89
Wozencraft, 751
Wray, C., 88
Wren, B. W., 123
Wright, D. A., 148
Wright, F., 39, 658
Wright, R. J., 385
Wright, S., 281
Wu, A., 400

Wu, C.-I., 385, 730
Wu, L., 280
Wu, M., 123
Wu, W., 312, 781
Wu, X., 324
Wu, X. H., 359
Wu, Y., 419
Wuertz, S., 495, 508, 512
Wunschel, S. C., 268, 269
Wyman, S. K., 332, 333, 335,
    365, 366, 367, 368, 376, 379,
    678, 684, 685, 689, 691
Wyngaard, G. A., 476
Wynshaw-Boris, A., 725
Wyrick, J. J., 266

X

Xenarios, I., 727, 738, 741
Xiang, C., 165, 168
Xiang, Q.-Y., 134, 138, 140, 142
Xiao, H., 400
Xiao, L., 39
Xie, H., 165, 168
Xin, Z., 62
Xing, L., 408, 409
Xoconostle-Cazares, B., 635
Xu, B., 64
Xu, C., 539
Xu, M., 193, 194, 196
Xu, S. Q., 324
Xu, Y. O., 788

Y

Yabuzaki, J., 552
Yadhukumar, 261
Yagami, K., 725
Yahara, T., 444
Yakimov, M. M., 260
Yamada, K., 349
Yamaguchi, H., 783
Yamaguchishinozaki, K., 349
Yamamoto, K., 552
Yamamoto, N., 263
Yamamoto, Y., 376
Yamaoka, S., 419, 433
Yamashita, A., 552
Yamashita, E., 783
Yamato, K. T., 419, 433

Yamato, M., 66
Yamazaki, Y., 376
Yampolsky, C., 418
Yampolsky, H., 418
Yan, G., 164
Yan, M., 335, 684, 689, 691
Yan, Y. L., 714, 719, 725, 726
Yang, R.-C., 196
Yang, T. J., 453
Yang, W., 134
Yang, Y. H., 599
Yang, Z., 654, 655, 656, 728, 729, 730, 731,
    732, 733, 734, 735, 740, 741, 767, 777
Yang, Z. H., 656, 662, 718, 732, 733
Yano, T.-A., 655, 656
Yanushevich, Y. G., 662, 666
Yao, M., 783
Yaono, R., 783
Yates, A. M., 746
Ye, D., 419
Ye, K. Q., 116, 119, 124
Ye, Z.-H., 196
Yeh, F. C., 196
Yen, A., 170
Yi, S., 418
Ying, B., 659
Ying, S.-Y., 120, 121
Yokobori, S.-I., 311, 330
Yokoyama, S., 728, 738
Yomo, T., 702
York, E. C., 73
Yoshida, M. C., 223
Yoshida, N., 725
Yoshikawa, S., 783
Yoshiki, A., 725
Yoshinaga, K., 376
Yoshino, C., 552
Yost, H. J., 312
Young, A. C., 119
Young, I. G., 88
Young, J., 116, 119
Young, J. P. W., 61, 62, 63, 66
Young, N. D., 715
Young, P., 268
Young, R. A., 266
Young, W. P., 146, 148
Yu, H., 38, 39, 40, 42, 43, 45
Yu, Q., 418, 419, 433
Yu, X. B., 105
Yu, Z. G., 377, 378

Yue, G. H., 20
Yue, Y., 263, 583
Yuri, T., 328
Yuste, R., 540
Yuzawa, H., 552

Z

Zabeau, M., 16, 145, 146, 151,
    152, 161, 162, 180, 182, 571, 573
Zabel, P., 424, 425, 445, 451, 456
Zachgo, S., 635
Zahler, A. M., 411
Zaita, N., 349
Zane, L., 164, 165, 166, 203, 223, 224
Zanis, M., 378
Zardoya, R., 224, 338
Zarlenga, D. S., 311, 330, 334, 340
Zasoski, R. J., 63
Zawadzki, P., 494, 502, 505
Zdobnov, E. M., 408
Zebitz, C. P. W., 165, 166, 167
Zee, F., 193, 194, 196
Zee, F. T., 418, 419, 433
Zeid, M., 193, 194, 196
Zeigle, J., 77
Zeigler, R., 116, 117
Zetterberg, A., 116, 119
Zhang, D., 193, 194, 195,
    196, 286, 287, 444
Zhang, H.-B., 453
Zhang, H. Q., 169
Zhang, J., 165, 168, 402, 728,
    729, 730, 734, 738, 740
Zhang, J. H., 363
Zhang, J. Z., 653
Zhang, L., 423, 581, 678, 710
Zhang, L.-H., 119
Zhang, Y., 165, 166, 167
Zhang, Y. H., 427
Zhang, Y.-P., 76, 106, 112, 117, 118, 123

Zhang, Z., 363, 402, 409
Zhao, B., 68, 119
Zhao, X., 429
Zhao, X.-P., 444
Zheng, J., 119
Zhivotovsky, L. A., 164
Zhong, T., 39, 42, 43
Zhong, X., 456
Zhong, X. B., 451
Zhou, J., 62, 63, 64
Zhou, L., 551
Zhu, B., 119, 689
Zhu, J., 608, 610
Zhu, T., 609
Zhu, W., 405, 409, 411, 414
Zhu, X. Y., 39, 42, 43
Zhu, Y., 202, 312
Zhu, Z., 324
Zietkiewicz, E., 134
Zillman, M. A., 324
Zilversmit, M., 106, 112,
    117, 118, 123
Ziman, M., 583
Zimmer, D. P., 609
Zimmer, E. A., 378, 746
Zimmermann, N., 288
Zipfel, W., 539
Zischler, H., 312
Zmasek, C. M., 703, 705, 710,
    711, 713, 714, 715
Zoetendal, E. G., 38, 39
Zola, H., 647
Zolla, B., 659
Zöllner, S., 122, 123, 609
Zon, L. I., 148
Zong, Q., 324, 359
Zuccotti, M., 478, 481
Zurawski, G., 88
Zvara, Á., 599
Zwick, M. S., 444
Zylberberg, M., 124

Subject Index

A

AFLP, *see* Amplified fragment
    length polymorphism
Amplified fragment length polymorphism
    applications and advantages, 146, 174,
        182, 187–188
    arthropod diversity analysis
      bands
        number selection, 170–171
        sequencing for homology testing,
          172–174
      heterozygosity estimation within
        populations, 164–165
      interspecific genetic differentiation
        estimation, 167–168
      intraspecific genetic differentiation
        estimation, 165–167
      phylogeny reconstruction, 169–170
      prospects, 174–175
      testing evolutionary and ecological
        hypothesis, 168–169
    comparison with other markers, 181
    complementary DNA amplified
        fragment length polymorphism
        analysis, *see* Complementary
        DNA amplified fragment length
        polymorphism display
    data analysis
      binary matrix generation, 190
      character estimates, 191–192
      frequency estimates, 191
      similarity estimates, 190–191
    data collection, 185–186
    fingerprinting, 301
    genomic DNA analysis
      DNA extraction and digestion, 148
      enzyme and primer combination
        optimization, 150–152
      gel electrophoresis and analysis, 150
      genome size and complexity effects, 152
      marker selection and scoring, 153
      oligonucleotide adapter ligation, 149

      preselective amplification, 149
      selective amplification, 149–150
    genotyping errors, 153–154
    history of analysis, 145
    limitations, 147–148, 163–164, 188–190
    minimum requirements, 186–187
    plant diversity surveys
      data generation, 193–195
      data management, 195
      genetic diversity level assessment,
        195–196
      genetic structure
        description, 196
        significance testing, 197
      prospects, 197–198
    principles of analysis
      adapter ligation, 183
      DNA restriction, 183
      overview, 145–146, 162–163,
        182–183, 186
      preselective amplification, 184
      selective amplification, 184–185
      visualization, 185
    vertebrate genome analysis
      birds and mammals, 154–155
      non-mammals, 155, 157
      troubleshooting, 156–157
    Y-linked genes in plants, 427
Ancestral protein reconstruction
    artificial gene design, 658–659
    gene synthesis incorporating degenerate
      sites, 659–661
    green fluorescent protein-like protein
      synthesis, 661–666
    phylogenetic analysis for gene inference,
      654–657
    rationale, 652–654, 667
    sequence selection for reconstruction,
      657–658
Ancient DNA isolation
    bones and teeth
      decalcification, 91
      decontamination, 90–91

Ancient DNA isolation *(cont.)*
  dialysis, 91
  modifications, 93, 95
  organic extractions, 91–92
  proteinase K digestion, 91
  purification of DNA, 95–96
  sample preparation, 90
  specimen selection, 92–93
  comparison of techniques, 94
  contamination
    authentication of studies, 100–101
    controls, 88–89
    prevention, 100–101
  examples, 88
  human specimens, 88–89
  polymerase chain reaction
    inhibitor testing, 97
    optimization, 98–99
  postmortem damage to DNA, 99–100
  scat molecular coproscopy, 81–82
  soft tissue samples
    lysis, 96
    organic extraction, 96–97
    polymerase chain reaction inhibitor
      extraction, 97
    sample preparation, 96
Aqueous field sample nucleic acid isolation
  analysis and quality considerations, 20, 22
  biodiversity studies, 15–16
  concentration determination, 32
  contaminants, 23, 25
  disposables, 20
  DNA isolation
    extraction
      cetylammonium bromide
        extraction, 26
      kits, 26–27
      phenol extraction, 25–26
      precipitation, 27–28
      RNA contamination removal, 28–29
  gel electrophoresis for quality
    assessment, 33–35
  general considerations, 17
  hybridization experiments, 36
  lysis, 22–23
  polymerase chain reaction, 35–36
  RNA isolation
    cetylammonium bromide extraction, 30
    DNA contamination removal, 31–32
    poly(A) RNA, 31

  RNeasy kit, 31
    TRIzol reagent, 29–30
  sample treatment for degradation
    prevention, 18–20
  sampling, 17–18
  storage, 35
Arthropods
  amplified fragment length polymorphism
    analysis of diversity
    bands
      number selection, 170–171
      sequencing for homology testing,
        172–174
    heterozygosity estimation within
      populations, 164–165
    interspecific genetic differentiation
      estimation, 167–168
    intraspecific genetic differentiation
      estimation, 165–167
    phylogeny reconstruction, 169–170
    prospects, 174–175
    testing evolutionary and ecological
      hypothesis, 168–169
  chromosomes
    fluorescence *in situ* hybridization, 464
    polytene versus nonpolytene
      chromosomes, 461–462
    preparation for morphology studies
      metaphase chromosome
        preparation, 470
      overview, 467
      resources, 468
      salivary gland polytene chromosome
        preparation, 469
    prospects for study, 485
    squashes, 463
    staining, 464
  cytogenetic and genome databases,
    483–484
  genome size
    arthropod morphology, development,
      and evolution significance, 473–474
    databases, 483–484
    definition, 467
    Feulgen densitometry
      DNA compaction
        considerations, 478
      integrated optical density, 477
      principles, 477–478
      slide preparation, 475–477

flow cytometry
  data acquisition and analysis, 483
  nuclear suspension preparation,
    478–480
  nuclear preparations for
    estimation, 474–475
  relevance for sequence and structure,
    471–473
  units, 471
phylogenetic analysis
  comparative genomic hybridization,
    466–467
  metaphase chromosome spreads,
    465–466
  polytene chromosomes, 466

B

BAC, *see* Bacterial artificial chromosome
Bacterial artificial chromosome
  chloroplast genome cloning, 361–363
  library construction for phylogenetic
    studies
    arraying and storage, 396–397
    DNA insert preparation
      isolation from gel, 394–395
      nuclei isolation, 390–391
      partial digestion, 391–392
      quantification, 395
      size fractionation, 393–394
    equipment, 388
    isolation of inserted DNA, 397–398
    ligation of vector and insert,
      395–396
    overview, 384–386
    plant material, 387
    prospects, 398
    replication, 397
    restriction enzymes, 387
    size of library, 387
    vectors
      choice, 386
      electroporation into *Escherichia coli*,
        389–390
      preparation, 388
      quality assessment, 389
Biodiversity
  definition, 258–259
  global species abundance, 259
  microbe discovery, 259–260

C

CAE-SSCP, *see* Capillary array
  electrophoresis-single-strand
  conformational polymorphism analysis
Capillary array electrophoresis-single-strand
  conformational polymorphism analysis
  germplasm analysis
    allele calling, 254, 257
    data analysis, 249–250, 252, 254
    electrophoresis
      array length, 246
      buffer and polymer selection, 248
      calibration of instrument, 247
      electroinjection parameters, 249
      instrumentation, 247
      run temperature, 246–247
    molecular weight standards, 244–246
    polymerase chain reaction, 243
    primer design and testing
      candidate genes from other plants, 241
      expressed sequence tag sequences,
        241–242
      fluorescently labeled primer testing,
        242–243
      unlabeled primer testing, 242
    prospects, 257
    sample preparation, 243–244
  overview, 240–241
Cetylammonium bromide
  aqueous field samples
    DNA extraction, 26
    RNA extraction, 30
  secondary metabolite removal in plant
    DNA isolation, 4
CGH, *see* Comparative genomic
  hybridization
Chloroplast genomics
  database and software resources, 369–372
  gene annotation using DOGMA, 367–369
  gene content and order for phylogeny
    reconstruction, 375–376
  genome structure, 349, 351–352
  historical perspective, 348–349
  molecular evolutionary analysis, 376–378
  prospects, 378–379
  sequencing
    amplification
      long polymerase chain reaction, 359
      overview, 358–359

Chloroplast genomics *(cont.)*
    rolling circle amplification, 359–360
    approaches, 353–354
    assembly, 363, 365
    cloning, 361–363
    DNA isolation
      caveats, 354, 357
      DNase I treatment, 357
      fluorescence-activated cell sorting of
        chloroplasts, 358
      high-salt buffers, 357–358
      overview, 354
      sequence availability, 350
      sucrose step-gradient centrifugation,
        354–357
    finishing, 365–367
  whole genome comparisons and repeat
    analysis, 372–375
Comparative gene expression, *see* DNA
  microarray; *In situ* hybridization
  Northern blot; Reverse transcriptase-
  polymerase chain reaction
Comparative genomic hybridization,
  arthropods, 465–466
Complementary DNA amplified fragment
  length polymorphism display
  advantages, 571
  gene expression change analysis in
    polyploids
    adapter ligation, 576–577
    caveats, 580–581
    first-strand complementary DNA
      synthesis, 575
    messenger RNA isolation, 574–575
    overview, 573
    polyacrylamide gel electrophoresis,
      579–580
    polymerase chain reaction
      preselective amplification,
        577–578
      reamplification of differentially
        expressed fragments, 580
      selective amplification, 578
    primer labeling, 578–579
    restriction digestion, 576
    second-strand complementary DNA
      synthesis, 575–576
    total RNA isolation, 573–574
  principles, 185, 570–572
  verification, 573

Confocal laser scanning microscopy
  deconvolution software, 540
  developmental study prospects, 541
  fluorescence labeling of
    specimens, 525
  hindbrain neuron antibody staining
    and retrograde labeling in
    zebrafish embryo
  image collection and analysis, 538
  labeling, 537
  mounting, 537
  optimization, 537–538
  overview, 534, 537
  images
    acquisition, 527–528
    analysis, 528
    optimization, 526–527
  instrumentation, 523–524
  β-keratin immunofluorescence
    microscopy, 649–650
  mechanosensory hair cell imaging in
    tadpole inner ear
  image collection and analysis, 531
  labeling, 529
  mounting, 529
  optimization, 529, 531
  mounting specimens, 525–526
  multicolor imaging, 538
  principles, 522–524
  real-time microscopy, 539–540
  time-lapse imaging of otic placode
    development in zebrafish
  image collection and analysis,
    533–534
  labeling, 532
  mounting, 532–533
  optimization, 533
  overview, 531–532
  two-photon microscopy, 539
CTAB, *see* Cetylammonium bromide
C-value, *see* Genome size
Cytochrome *c* oxidase, subunit context
  dependence and coevolution analysis,
  781, 785–788

D

DNA isolation
  ancient DNA, *see* Ancient DNA isolation
  chloroplasts, *see* Chloroplast genomics

environmental aqueous samples, *see*
Aqueous field sample nucleic
acid isolation
fungal DNA isolation, *see* Lichen
DNA isolation; Mycorrhizal fungus
DNA isolation
gut flora, *see* Gut flora DNA isolation
hair, *see* Hair DNA isolation
mitochondrial DNA, *see* Mitochondrial
genome
plants, *see* Plant DNA isolation
scat, *see* Fecal DNA isolation
DNA microarray
comparative gene regulation analysis
within and among species
cyanine labeling of complementary
DNA, 612–613
data analysis, 614–615
experimental design
comparative hybridizations, 598–599
comparisons within and among
species, 608, 609
disturbances in messenger RNA
harvesting, 609
graph theoretical depictions,
599–603
replication, 603, 605
statistical significance, 605, 607
fluorescence scanning, 613–614
hybridization and washing, 613
microarray construction, 609–611
model systems, 597–598
normalization of data, 614
probe types, 598
RNA extraction and reverse
transcription, 611–612
microbe ribosomal RNA analysis
gene expression profiling, 265–267
genotyping, 267–269
hybridization, 271–273
image analysis, 274
nucleic acid labeling, 270–271
overview, 263, 269–270
prospects, 274–275
signal detection, 273
species diversity analysis, 264–265
target nucleic acid preparation
gene expression profiling, 270
species identification, 270
microchips, 263

polyploid gene expression change analysis
advantages, 581, 583
array fabrication, 585–587
caveats, 593
data normalization and analysis, 592
experimental design, 582–585
hybridization, 590–591
oligonucleotide design and
distribution, 585
platforms, 583
primer annealing, 588
probe labeling with direct dye
incorporation, 588
probe purification, 589
reverse transcription, 588–589
RNA
degradation, 589
labeling, 589–590
scanning parameters and data
analysis, 591
slide posttreatment, 587–588
Y-linked gene expression in plants, 421
DOGMA, *see* Dual Organellar GenoMe
Annotator
Dual Organellar GenoMe Annotator,
chloroplast gene annotation, 367–369
DupLoss, species tree–gene tree
reconciliation, 715

E

Ecological Niche Factor Analysis, predictive
species distribution modeling, 288, 291
ENFA, *see* Ecological Niche
Factor Analysis
EST, *see* Expressed sequence tag
Expressed sequence tag
complementary DNA library preparation,
120–121
phylogenomic analysis principles, 120
plant comparative analysis
annotation, 407–408
applications
alternative splicing, 411–413
gene annotation, 413–415
gene discovery, 410–411
gene families conserved across
species, 411
clustering and Unigene assembly,
404–405, 407

Expressed sequence tag *(cont.)*
    contaminant sequence filtering, 402–403
    databases, 400–402
    Internet resources, 404
    library information, 403–404
    prospects, 415
    software, 408–409
    processing
      annotation, 122–123
      sequence assembly, 122
      trace file processing, 122

F

FACS, *see* Fluorescence-activated cell
    sorting
Fecal DNA isolation
    contamination prevention and analysis,
      82–83
    extraction
      Chelex-100 resin, 78
      comparison of techniques, 76–77
      guanidine thiocyanate/silica, 80–81
      kits
        overview, 76–78, 80
        Qiagen extraction kits, 79–80
      molecular coproscopy for ancient
        species, 81–82
      phenol–chloroform, 75
    noninvasive sampling advantages, 73–74
    polymerase chain reaction
      mitochondrial DNA, 74
      optimization, 83
      quantitative polymerase
        chain reaction, 83
Feulgen densitometry, *see* Genome size
FISH, *see* Fluorescence *in situ* hybridization
Flow cytometry, *see* Fluorescence-activated
    cell sorting
Fluorescence-activated cell sorting
    arthropod genome sizing
      data acquisition and analysis, 483
      nuclear suspension preparation, 478–480
    chloroplasts for DNA isolation, 358
    Y chromosome in plants, 434–435
Fluorescence *in situ* hybridization
    arthropods chromosomes
      chromosome preparation
        metaphase chromosome
          preparation, 470

        resources, 468
        salivary gland polytene chromosome
          preparation, 469
      overview, 467
    historical perspective, 444
    phytoplankton fluorescent ribosomal RNA
      probes
      applications, 299–301
      fixation and hybridization conditions,
        304, 306–308
      materials, 306–307
      probe design, 303
    plant chromosome comparative studies
      bacterial artificial chromosome libraries,
        444–445
      chromosome preparation
        overview, 445–446
        pretreatment of chromosomes,
          447–448
        root tip harvesting and
          treatment, 447
        root tip squashing, 448–449
      DNA blocking, 452–453
      fiber fluorescence *in situ* hybridization
        denaturation/hybridization
          conditions, 455
        DNA fiber extension, 450–451
        multilayer antibody detection,
          456–458
        nuclei preparation, 450
        resolution, 449–450
      meiotic fluorescence *in situ* hybridization
        denaturation/hybridization
          conditions, 454–455
        meiocyte preparation, 451
        single-layer antibody detection, 456
      overview, 443–445
      probe
        preparation and labeling, 451–452
        solution, 453–454
      prospects, 458
      washing for antibody detection, 455–456
      ribosomal RNA gene probes, 261
      Y chromosome gene isolation from plants,
        425–426, 438–439
    variations, 445
Fosmid, chloroplast genome cloning, 361–363
Fungal DNA isolation, *see* Lichen DNA
    isolation; Mycorrhizal fungus
    DNA isolation

G

Gene annotation
  chloroplast genes using DOGMA, 367–369
  mitochondrial genome, 329–332
  plant expressed sequence tag comparative
    analysis, 413–415
Gene chip, *see* DNA microarray
Generalized Nadeau-Taylor model,
  phylogenetic analysis, 682
Genetic algorithm, predictive species
  distribution modeling, 289, 291
GeneTree, parology detection and species
  tree inference, 712–713
Genome sequences, availability,
  105–106, 124–127
Genome size
  arthropods
    databases, 483–484
    Feulgen densitometry
      DNA compaction
        considerations, 478
      integrated optical density, 477
      principles, 477–478
      slide preparation, 475–477
    flow cytometry
      data acquisition and analysis, 483
      nuclear suspension preparation,
        478–480
    morphology, development, and
      evolution significance, 473–474
    nuclear preparations for
      estimation, 474–475
  definition, 467
  relevance for sequence and structure,
    471–473
  units, 471
Genotyping, definition, 178
Geographic information system
  data checking, 286
  definition, 285
  diversity analysis tools, 286–287
  genetic diversity models, 289, 292
  genetic structure variation, geographic
    effects, 280–282
  georeferenced data
    biological data, 282–283
    non-biological data, 283–284
  overview of spatial variation in biological
    data, 279–280

  plant genetic resource workers
    examples of use, 292–294
    prospects for use, 295
    spatial questions, 284–285
  predictive species distribution modeling,
    287–289
  software, 285
Germplasm collection
  definition, 177–178
  molecular characterization, overview,
    179–180
  single-strand conformational
    polymorphism analysis, *see* Capillary
    array electrophoresis-single-strand
    conformational polymorphism
    analysis
GFP, *see* Green fluorescent protein
GIS, *see* Geographic information system
Green fluorescent protein, ancestral protein
  reconstruction, *see* Ancestral
  protein reconstruction
Gut flora DNA isolation
  extraction, 42, 44
  flow chart, 40–41
  kits, 39, 45
  lysis, 40, 42, 44
  microorganism identification, 38–39
  precipitation, 43, 45
  sample preparation, 40, 43–44
  scat, *see* Fecal DNA isolation
  yield, 43, 45

H

Hair DNA isolation
  extraction, 75–76
  invasive versus noninvasive sampling,
    73–74
HGT, *see* Horizontal gene transfer
Horizontal gene transfer
  assays
    basic *in vitro* transformation assays
      Acinetobacter sp. BD413, 498, 500–504
      applications, 505
      cell-contact transformation, 504–505
      competent cell preparation, 498, 500
      filter transformation, 500–502
      liquid cultures, 502–503
      lysate or culture supernatant
        transformation, 503–504

Horizontal gene transfer *(cont.)*
  optimization parameters, 498–499
  principles, 497–498
  modified natural systems
    animal feeding experiments, 515
    artificial inoculation in field
      experiments, 515–516
    plant greenhouse and growth chamber
      experiments, 513–515
  natural system components for
    transformations
    Acinetobacter sp. BD413 in sterile soil
      microcosms, 505–507, 510–511
    applications, 510–513
    biofilms in simulated aqueous
      environments, 508–509,
      512–513
    human host tissues and fluids,
      509–510, 513
    plant tissue as donor DNA source,
      507–508, 511
    overview, 493–496
    prospects, 517–518
    selection markers, 493, 497
    in situ systems, 516–517
  biological relevance of transformation
    processes, 492–493
  natural genetic transformation, 492

I

In situ hybridization, plant
  comparative gene expression studies
  imaging, 633–634
  overview, 623
  posthybridization, 631–633
  RNA probe preparation, 625–627
  sectioning and prehybridization, 628–631
  solutions, 627–628
  tissue embedding and fixation, 623–625
  troubleshooting, 634–635
Inter-simple sequence repeat
  amplification, 134–135
  characteristics, 134
  data analysis
    qualitative analysis, 142
    quantitative analysis, 142–143
  plant DNA
    data gathering and scoring, 140–141
    isolation

Elu-Quik DNA cleanup, 137
    precipitation and resuspension,
      136–137
    sample preparation, 135
    polymerase chain reaction
      amplification conditions, 137–139
      optimization, 139–140
ISH, *see In situ* hybridization
ISSR, *see* Inter-simple sequence repeat

K

β-Keratins
  avian subfamilies, 636, 638
  immunostaining of histological sections,
    649–650
  subfamily-specific antibody generation
    amino acid sequence alignment, 644–646
    comparative genomics for peptide
      antigen generation
      automated DNA sequence, 643–644
      cloning of family member genes,
        642–643
      DNA sequence alignment, 640–641
      initial characterization of multigene
        family, 638–640
      polymerase chain reaction primers,
        641–642
    immunization of rabbits, 647
    overview, 637–638
    peptide antigen synthesis, 646–647
    Western blot analysis, 647–649

L

Laser scanning confocal microscopy, *see*
  Confocal laser scanning microscopy
Lichen DNA isolation
  alcohol precipitations, 53
  cetylammonium bromide precipitation, 52
  contamination with other genomes, 48–49
  gel electrophoresis for quality
    assessment, 54
  kits, 56
  polymerase chain reaction
    considerations, 54–55
    simultaneous isolation and
      amplification, 55–56
  sample preparation and symbiont removal,
    49–51

starting material quality consideration, 51–52
Logistic regression, predictive species distribution modeling, 288, 290

M

Maximum likelihood
  ancestral gene inference, 655
  bootstrap test of absolute goodness of fit, 775–777
  generation of phylogeny estimates
    GTR+I+Γ models, 763–767
    iterative approach justification, 761
    likelihood ratio tests, 762, 764, 766–767
    phylogenetic uncertainty assessment, 768–769
    searching of tree space, 767–768
  hypothesis testing, 769–770
  optimality criterion in phylogeny estimation, 758–761
  overview of phylogeny estimation, 676
  test statistic evaluation, 770–775
Microsatellites
  comparison with other markers, 181
  DNA microarray analysis, 268–269
  fingerprinting, 180, 301
  hybridization capture
    colony storage, 215–217
    DNA extraction, 203–204
    DNA selection for marker development, 204–205
    Dynabead enrichment, 209–211
    ligation of recovered DNA into plasmids, 213–214
    linker ligation to DNA fragments, 207–209
    overview, 203, 220
    polymerase chain reaction
      colony screening, 215–217
      product preparation for sequencing, 217
      recovery of enriched DNA, 212–213
    restriction enzyme digestion, 205–206
    sequencing
      automated sequencing reactions, 218–219
      precipitation of sequencing reactions, 219–220
    transfection, 214–215

parentage and kinship analysis in animals
  advantages of microsatellite analysis, 223
  fluorescently-labeled amplification product running on sequencer, 228, 230
  gel scoring, 230–231
  genotype data analysis, 231, 233–235
  limitations of microsatellite analysis, 223
  loci identification, 224
  overview, 223–224
  polymerase chain reaction
    optimization, 224–227
    product visualization and genotype determination, 227–228
  uracil-DNA selection, 202
Minisatellites, *see* Variable number of tandem repeats
Mitochondrial genome
  comparative sequencing
    advantages, 312
    applications, 312
    assembly, 328–329
    cloning, 319–320
    gene annotation, 329–332
    gene-order comparisons, 334–340
    high-throughput sequencing, 313
    mapping of DNA, 320–321
    polymerase chain reaction, 321–324
    primer walk-through of long fragments, 327–328
    rolling circle amplification, 324–326
    shotgun sequencing, 326–327
    software analysis, 332, 334
    template preparation
      DNA isolation, 314–318
      overview, 313–314
  evolution, 312
  genes, 311–312
  internet resources, 314
  sequence availability, 313
ML, *see* Maximum likelihood
Multigene families, *see* β-Keratins
Mycorrhizal fungus DNA isolation
  chlamydospore extraction, 68
  general considerations, 60–61
  identification of fungi, 60
  importance of study, 59–60

Mycorrhizal fungus DNA isolation *(cont.)*
  kits, 64–65, 69
  lysis, 63–64
  overview of approaches, 62–63
  polymerase chain reaction, inhibitor
      coextraction, 61–62
  root sample extraction, 66–68
  soil sample extraction, 65–66
  sporocarp tissue extraction, 68
  symbiosis, 58–59

N

Natural genetic transformation, *see*
      Horizontal gene transfer
Northern blot, plant comparative gene
      expression studies
  guidelines, 620–621
  RNA isolation, 618–620

O

Orthology
  definition, 701
  homology versus orthology, 703–704
  parology versus orthology, 701
  RIO identification of orthologous genes,
      713–715
OrthoParaMap, species tree–gene tree
      reconciliation, 715

P

Paralogy
  definition, 700–701, 725
  detection on phylogenies, 704–708
  duplicate gene data set analysis,
      comparison of techniques, 739–741
  gene duplication in evolution, 702–705
  independence of gene duplications, 719
  lateral gene transfer, 718–719
  least common ancestor, 705, 707–708
  mapping amino acid changes on trees
      and/or protein structures,
      727–728, 736–739
  molecular clock dating, 703
  most common recent ancestor, 706
  nonsynonymous site substitution/
      synonymous site substitution ratio
      testing

  branch tests, 728–730
  branch-sites tests and performance,
      733–735
  gene divergence, 726, 728
  selection pressure at molecular level,
      725–726
  variable selection testing among sites
      and performance, 730–733
  orthology comparison, 701
  problem of paralogy, 701–702
  prospects for study, 720
  shifts in rates across sites to detect changes
      in selective constraints, 727, 735–736
  species tree–gene tree reconciliation
    Bayesian tree reconciliation, 716–717
    bootstrapping by algorithms, 715–716
    DupLoss, 715
    gene tree error correction, 711–712
    GeneTree parology detection and
        species tree inference, 712–713
    OrthoParaMap, 715
    overview, 708
    parsimony-based versus probabilistic
        parology detection, 717–718
    parsimony mapping and co-phylogeny,
        710–711
    RIO identification of orthologous genes,
        713–715
    TreeMap, 715
Parsimony mapping, *see* Paralogy
PCA, *see* Predictive components analysis
PCR, *see* Polymerase chain reaction
Phylogenetic analysis
  ancestral genes, *see* Ancestral protein
      reconstruction
  arthropods, *see* Arthopods
  gene duplication, *see* Paralogy
  gene order and whole-genome evolution
    evolutionary events, 680–682
    generalized Nadeau-Taylor model, 682
    overview, 678, 680
  genomic distances
    breakpoint distance, 683
    distance corrections, 684–685
    distances between multichromosomal
        genomes, 689
    distances between two chromosomes
        with unequal gene content, 685, 687
    expected true evolutionary distance, 683
    inversion distance, 684

true evolutionary distance, 683

maximum likelihood estimation of phylogeny, *see* Maximum likelihood

orthologous genes, *see* Orthology

parsimony mapping, *see* Paralogy

protein context dependence and coevolution

  approaches, 783–785

  cytochrome *c* oxidase subunits, 781, 785–788

  overview, 780–781

  prospects, 788–789

  sequence selection, 782–783

  statistical considerations, 785

  structure availability, 783

sequence analysis

  distance-based algorithms

    assumptions, 675

    maximum likelihood, 676

    maximum parsimony, 676

    neighbor-joining algorithm, 675

    limitations, 676–678

    model trees and stochastic models of evolution, 673–674

    performance issues, 676–677

  supertree construction, *see* Supertree

Phylogenomics

  congruence analysis

    character congruence versus taxonomic congruence, 112

    comparative questions, 113, 115

    hidden support, 112–113

    yeast phylogenomics, 114

  distance-based algorithms, 689–690

  expressed sequence tag

    complementary DNA library preparation, 120–121

    principles of phylogenomic analysis, 120

    processing

      annotation, 122–123

      sequence assembly, 122

      trace file processing, 122

  likelihood-based algorithms, 693–694

  orthology

    determination, 110–111

    problems, 111

  overview, 106–107

  parsimony-based algorithms, 690–693

  primer design for shallow genomics, 117–118

prospects for research, 694

sequence alignment tools, 105

single nucleotide polymorphism studies, 123–124

space

  definition and overview, 107–108

  DNA barcoding, 107, 109

  mitogenomics, 109–110

  transforming molecular systematics, 110

supertree construction, *see* Supertree

targeted genome sequencing, 119–120

whole genomes

  comparison, 115–116

  gene mining, 116

  primer mining, 116–118

Phytoplankton

DNA fingerprinting approaches, 301

fluorescence *in situ* hybridization of fluorescent ribosomal RNA probes

  applications, 299–301

  fixation and hybridization conditions, 304, 306–308

  materials, 306–307

  probe design, 303

  sequence data for phylogenetic analysis, 302

Plant diversity, *see* Amplified fragment length polymorphism; Capillary array electrophoresis-single-strand conformational polymorphism analysis; Chloroplast genomics; Expressed sequence tag; Fluorescence *in situ* hybridization; Geographic information system

Plant DNA isolation

cactus DNA extraction, 10–11

Fragaria DNA extraction, 9–10

inter-simple sequence repeat analysis samples

  Elu-Quik DNA cleanup, 137

  precipitation and resuspension, 136–137

  sample preparation, 135

large-scale preparation from mucilaginous tissues, 6–7

micropreparation, 7–9

polyphenol removal, 4

polysaccharide removal, 4

RNA removal

  lithium chloride precipitation, 13

  overview, 5

Plant DNA isolation *(cont.)*
   ribonuclease A digestion, 14
   secondary metabolite interference, 3–4
   Sedum DNA extraction, 11
   Sephacryl spin column, 12–13
   starch digestion, 7
   tissue selection and preparation, 4–5
   woody material DNA extraction, 12
   yield, 5
Polymerase chain reaction
   ancient DNA
     inhibitor testing, 97
     optimization, 98–99
   aqueous field sample DNA, 35–36
   chloroplast DNA, 359
   fecal DNA
     mitochondrial DNA, 74
     optimization, 83
     quantitative polymerase chain
       reaction, 83
   inter-simple sequence repeats
     overview, 134–135
     plant DNA
       amplification conditions, 137–139
       optimization, 139–140
   lichen DNA
     considerations, 54–55
     simultaneous isolation and
       amplification, 55–56
   microsatellites
     hybridization capture
       colony screening, 215–217
       product preparation for sequencing,
         217
       recovery of enriched DNA, 212–213
     parentage and kinship analysis
       in animals
       optimization, 224–227
       product visualization and genotype
         determination, 227–228
   mitochondrial DNA, 321–324
   mycorrhizal fungus DNA, 61–62
   transcript profiling, *see* Reverse
     transcriptase–polymerase chain
     reaction
   Y chromosome gene isolation and analysis
     from plants
     allele-specific oligonucleotide
       primed polymerase chain reaction,
       430–433

     degenerate oligonucleotide-primed
       polymerase chain reaction,
       436–437
     improved primer-extension-
       premplification polymerase chain
       reaction, 436
Polyploid gene expression change analysis,
   see Complementary DNA amplified
   fragment length polymorphism display;
   DNA microarray
Predictive components analysis, predictive
   species distribution modeling, 288, 290
Primer extension assay
   allele-specific transcript detection in
     hybrids and allopolyploids
     gel analysis, 559
     incubation conditions, 558–559
     prime labeling, 558
     principles, 555–556
     solutions, 557–558
   orthologous allele transcripts, 554–555
PrimerSelect program
   applications, 552
   pathogenic organism isolate analysis,
     551–552
   performance of primer pairs, 549–551
   primer pair design for kinetic reverse
     transcriptase–polymerase chain
     reaction, 547–549

R

Random amplified polymorphic DNA
   comparison with other markers, 181
   fingerprinting, 180, 301
   Y-linked genes in plants, 427
RAPD, *see* Random amplified
   polymorphic DNA
RCA, *see* Rolling circle amplification
RDA, *see* Representational difference
   analysis
Representational difference analysis
   phylogenomic analysis, 124
   Y-linked gene analysis in plants,
     422–423, 434–435
Restriction fragment length polymorphism
   comparison with other markers, 181
   fingerprinting, 180
Reverse transcriptase–polymerase
   chain reaction

fractional polymerase chain reaction cycle
in kinetic reactions, 545
plant comparative gene expression studies
controls, 622
first-strand synthesis, 622
guidelines, 622–623
RNA isolation, 618–621
primer dimers
control of formation, 545–546
cross-priming reaction control,
546–547
kinetics of formation, 544–545
normalized performance in
amplification reactions, 547
PrimerSelect program
applications, 552
pathogenic organism isolate analysis,
551–552
performance of primer pairs,
549–551
primer design, 547–549
Reverse transcriptase–polymerase
chain reaction, 546–547
Reverse transcription–cleaved amplified
polymorphic sequence assay
allele-specific transcript detection in
hybrids and allopolyploids
polymerase chain reaction and gel
analysis, 568
principles, 566–567
reagents, 567
orthologous allele transcripts, 554–555
RFLP, see Restriction fragment length
polymorphism
Ribosomal RNA
DNA microarrays for microbe analysis
gene expression
profiling, 265–267
genotyping, 267–269
hybridization, 271–273
image analysis, 274
nucleic acid labeling, 270–271
overview, 263, 269–270
prospects, 274–275
signal detection, 273
species diversity analysis, 264–265
target nucleic acid preparation
gene expression profiling, 270
species identification, 270
gene probes

design, 261
fluorescence in situ hybridization, 261
flow cytometry, 261
phytoplankton fluorescence in situ
hybridization of fluorescent probes
applications, 299–301
fixation and hybridization conditions,
304, 306–308
materials, 306–307
probe design, 303
taxonomic classification of microbes, 260
RIO, identification of orthologous genes,
713–715
RNA isolation
aqueous field samples
cetylammonium bromide extraction, 30
DNA contamination removal, 31–32
poly(A) RNA, 31
RNeasy kit, 31
TRIzol reagent, 29–30
complementary DNA amplified
fragment length polymorphism
display, 573–574
plant reproductive tissues
guidelines, 620
kits, 619
tissue collection and storage, 618–619
ribonucleases, prevention of
contamination, 618
Rolling circle amplification
chloroplast DNA, 359–360
mitochondrial DNA, 324–326
rRNA, see Ribosomal RNA
RT-CAPS, see Reverse transcription–cleaved
amplified polymorphic sequence assay
RT-PCR, see Reverse
transcriptase–polymerase chain reaction

S

S1 nuclease protection assay
allele-specific transcript detection in
hybrids and allopolyploids
advantages, 568–569
applications, 562–563
principles, 559–561
probe design and generation,
561–562, 564–566
solutions, 564
orthologous allele transcripts, 554

Simple sequence repeats,
  see Microsatellites
Single nucleotide polymorphism
  comparison with other markers, 181
  discovery, 239–240
  DNA microarray analysis, 268
  fingerprinting, 180
  phylogenomic analysis, 123–124
  plant allele frequency, 239
  single-strand conformational
      polymorphism, *see* Capillary array
      electrophoresis-single-strand
      conformational polymorphism
      analysis
  Y-linked gene genotyping in plants,
      430–433
Single-strand conformational polymorphism,
  see also Capillary array
      electrophoresis-single-strand
      conformational polymorphism
      analysis
  polymorphism discovery, 240
  separation techniques, 240
  Y-linked genes in plants, 427–428
SNP, *see* Single nucleotide polymorphism
SSCP, *see* Single-strand conformational
  polymorphism
Supertree
  applications, 749–751
  construction
    algorithms, 749
    matrix representation principles,
        747, 749
    matrix representation with parsimony,
        747–749
    prospects, 754–755
  definition, 746–747
  theory, 752–754

T

Transformation, *see* Horizontal gene
  transfer
TreeMap, species tree–gene tree
  reconciliation, 715

V

Variable number of tandem repeats,
  fingerprinting, 301

VNTRs, *see* Variable number of tandem
  repeats

W

Western blot, β-keratin antibody analysis,
  647–649

X

Xenology, definition, 718

Y

Y chromosome, plants
  abundance, 418
  gene isolation
    amplified fragment length
        polymorphism, 427
    chromosome isolation techniques
      chromosome flow sorting, 434–435
      degenerate oligonucleotide-primed
          polymerase chain reaction,
          436–437
      improved primer-extension-
          preamplification polymerase
          chain reaction, 436–437
      laser microdissection, 436
      library screening with Y-specific
          probes, 424, 436–437
      micromanipulator
          microdissection, 436
      overview, 423–425
    difficulty, 420
    fluorescence *in situ* hybridization,
        425–426, 438–439
    genomics approaches, 433–434
    random amplified
        polymorphic DNA, 427
    representational difference analysis,
        422–423, 434–435
    restriction cleavage-based
        polymorphisms, 428–429
    segregation analysis, 426–434
    sequence analysis, 429
    single nucleotide
        polymorphism genotyping, 430–433
    single-strand conformational
        polymorphism, 428
    subtraction analysis, 421–423
  genetic degeneration, 419

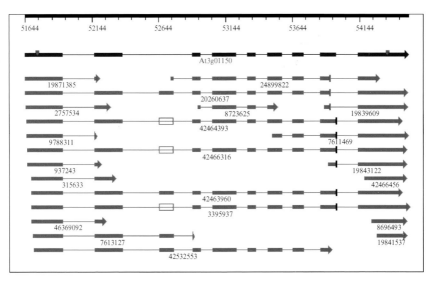

DONG *ET AL.*, CHAPTER 22, FIG. 3. Visualization of expressed sequence tag (EST)/ complementary DNA (cDNA) alignments and alternative splicing. Arabidopsis ESTs and cDNAs were aligned against the genome sequence using GeneSeqer. The black scale on top of the picture indicates the chromosome three coordinates of the alignments. Thick horizontal bars represent exons, and thin lines represent introns. Red indicates EST/cDNAs alignments. Blue indicates GenBank annotation. The green and red triangles above the blue bars indicate the translation start and stop positions, respectively. For alternative splicing, the green box on the red lines indicates an exon-skipping case. For the alternative donor site in the last intron sites, a black vertical bar indicates the most prevalent splice site, and a green vertical bar represents the alternative splice site.

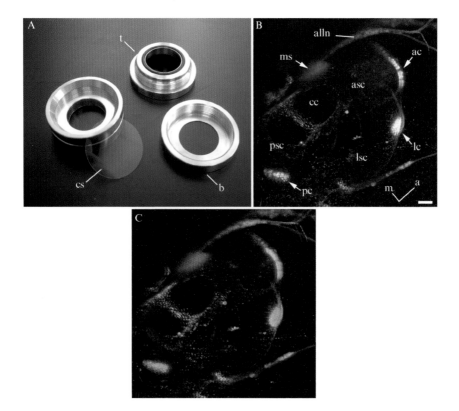

COLLAZO *ET AL.*, CHAPTER 27, FIG. 2. *In vivo* confocal microscope images of fluorescently labeled sensory organs in the inner ear of the frog *Xenopus laevis* and the chamber used to hold the specimen. (A) Stainless-steel chamber used to view living, labeled tadpole. On the left side is the intact chamber with a 25-mm coverslip (cs) resting on it for scale. Chamber unscrews into top (t) and bottom (b) pieces with the coverslip placed in the bottom. Top piece's black gasket provides a seal when it is flipped and screwed into bottom. (B) Maximum projection of confocal z-stack of the dorsal sensory organs of *Xenopus* inner ear. Arrows point to the sensory organs visible. Three most dorsal are the anterior crista (ac), lateral crista (lc), and posterior crista (pc) at the base of the three semicircular canals: anterior (asc), lateral (lsc), and posterior (psc). A fourth sensory organ, macula sacculus (ms), is barely visible because it is too ventral for the working distance of our objective. Anterior and posterior semicircular canals meet in a structure called the *common crus* (cc). The anterior lateral line nerve (alln) outside of the inner ear also labeled with 4-Di-2-ASP that leaked out of the injected ear. Dorsal view with the anterior (a) and medial (m) axes indicated by lines. Scale bar is 50 μm for this and the following image. (C) Three-dimensional view of the same ear shown in (B) is visible with red-green three-dimensional glasses (green over left eye). The positions of the semicircular canals and the relative heights of the cristae are particularly clear in this view with the appropriate glasses.

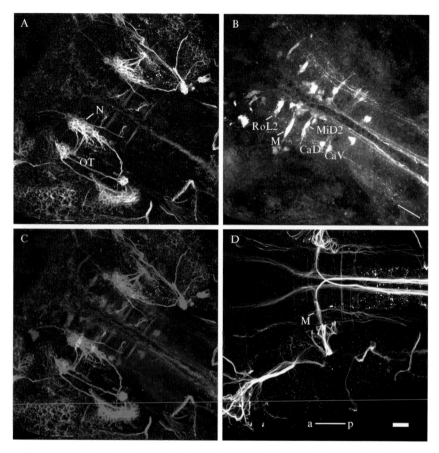

COLLAZO *ET AL.*, CHAPTER 27, FIG. 4. Antibody staining and retrograde labeling of neurons in the hindbrain of zebrafish embryos 3 days post-fertilization (dpf). Ventral view of inner ear and hindbrain. Images in panels A to C are of the same embryo, taken with two different wavelengths as indicated in the text. The anterior/posterior axis for panels A, B, and C is indicated in B. (A) Immunostaining with Zn-8 antibody and Alexa 488 secondary antibody reveals a yet to be identified nucleus (N) just medial to the developing saccule of the zebrafish inner ear. The central part of the otic vesicle (OT) is indicated. (B) Retrograde labeling with lysinated rhodamine dextran of the reticulospinal neurons. Five of the neurons are identified. Four of the names (RoL2, MiD2, CaD, CaV) incorporate their hindbrain position: rostral (Ro), middle (Mi), and caudal (Ca). The cell body of the Mauthner (M) neuron resides in rhombomere 4, in the middle of the hindbrain. (C) A merged pseudo-colored image in which the immunostaining of Zn-8 is green and the retrograde labeling of the reticulospinal neurons is red. This allows for determination of hindbrain position of our Zn-8 antibody labeling. (D) Immunostaining with 3A10 antibody (recognizes neurofilaments). The cell body of the Mauthner (M) neuron is indicated. 3A10 is not a zebrafish-specific antibody, but it labels specific neurons in the central nervous system. Note the similarity to the retrograde labeling in (B) Scale bar, 25 μm (same for all panels).

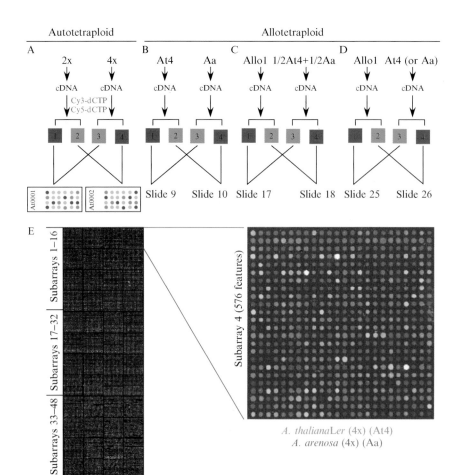

WANG *ET AL.*, CHAPTER 30, FIG. 2. Microarray experimental design and analysis for detecting gene expression changes in autotetraploids and allotetraploids. (A) A dye-swap experiment using messenger RNA (mRNA) from a diploid and an isogenic autotetraploid to study changes of gene expression in response to ploidy increase. Four labeling reactions using Cy3- and Cy5-dCTP are prepared and mixed reciprocally (e.g., 1 + 2 and 3 + 4) to hybridize two slides in one dye-swap experiment. Four dye-swap experiments (or eight slides) are used in each comparison. (B–D) To detect genes that are differentially expressed in an allotetraploid, four sets of dye-swap experiments are performed for each comparison between the two parents (B), between the allotetraploid and two progenitors (C), or between the allotetraploid and each progenitor (D). Results from the experiments in (B) and (C) determine the overall changes of gene expression in the allotetraploid relative to the progenitors but may underestimate the number of homoeologous genes that are silenced (see text). (E) A hybridization image shows 27,648 features in the *Arabidopsis* spotted 70-mer oligogene microarray, as previously described (Chen *et al.*, 2004). The slide is hybridized to complementary DNA (cDNA) prepared from *Arabidopsis thaliana* and *Arabidopsis arenosa*. The red, green, and yellow-white features in the subarray indicate low, high, and equal levels of gene expression, respectively, in *A. thaliana*. The data obtained from four sets of dye-swap experiments in each comparison are analyzed using a linear model (Lee *et al.*, 2004) to detect differentially expressed genes at a significance level ($\alpha = 0.05$, see text).

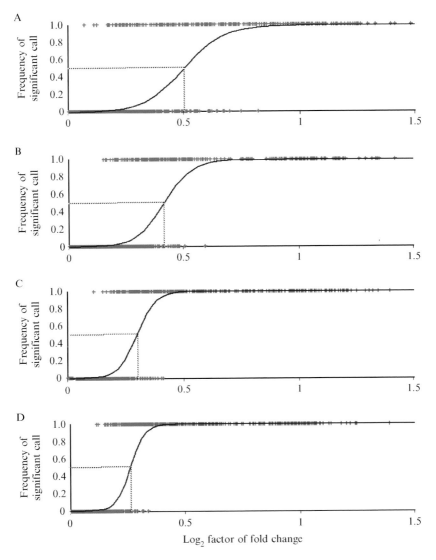

JEFFREY *ET AL.*, CHAPTER 31, FIG. 4. (*continued*)

Jeffrey ET AL., Chapter 31, Fig. 4. Logistic regressions of the frequency of affirmative significance call over $\log_2$ factor of difference in gene expression. The logistic model plotted is that $\log_e p/(1 - p) = mx + b$, where x is the $\log_2$ factor of difference in gene expression. Cross symbols represent estimated expression levels from simulated data using a bayesian analysis of gene expression level with additive small error terms. Each cross is placed on the abscissa at the estimated expression level, either at the top of the plot (significant, S) or at the bottom (not significant, NS). Ratio data were simulated using the probability distribution of (Fieller, 1932) assuming a constant coefficient of variation across samples. Logistic regressions are on the factors of difference estimated from simulated data comparing two samples to each other with different numbers of replicates. (A) An example with four replicates. The model has a highly significant fit ($\chi^2 = 1593.1$, $P < .0001$). The estimated intercept for the log odds, b, of a significant call versus no significant call is -5.4 (significant, $P < .0001$), and the estimated slope with $\log_2$ factor of difference in gene expression, m, is 11.5 (significant, $P < .0001$). The factor of gene expression at which 50% of estimated differences were identified as significant (GEL$_{50}$) was 1.4-fold. (B) An example with six replicates. The model has a highly significant fit ($\chi^2 = 1186.8$, $P < .0001$). The estimated intercept for the log odds, b, of a significant call versus no significant call is -6.1 (significant, $P < .0001$), and the estimated slope with $\log_2$ factor of difference in gene expression, m, is 16.5 (significant, $P < .0001$). The factor of gene expression at which 50% of estimated differences were identified as significant (GEL$_{50}$) was 1.29-fold. (C) An example with eight replicates. The model has a highly significant fit ($\chi^2 = 1186.8$, $P < .0001$). The estimated intercept for the log odds, b, of a significant call versus no significant call is -7.3 (significant, $P < .0001$), and the estimated slope with $\log_2$ factor of difference in gene expression, m, is 26.7 (significant, $P < .0001$). The factor of gene expression at which 50% of estimated differences were identified as significant (GEL$_{50}$) was 1.21-fold. (D) An example with 10 replicates. The model has a highly significant fit ($\chi^2 = 1307.2$, $P < .0001$). The estimated intercept for the log odds, b, of a significant call versus no significant call is -8.6 (significant, $P < .0001$), and the estimated slope with $\log_2$ factor of difference in gene expression, m, is 35.0 (significant, $P < .0001$). The factor of gene expression at which 50% of estimated differences were identified as significant (GEL$_{50}$) was 1.18-fold.

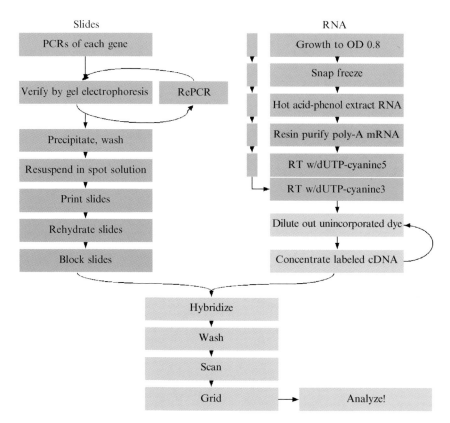

JEFFREY *ET AL.*, CHAPTER 31, FIG. 5. Schematic diagram of the steps employed to create spotted microarrays from polymerase chain reaction products and to perform a comparative microarray hybridization.

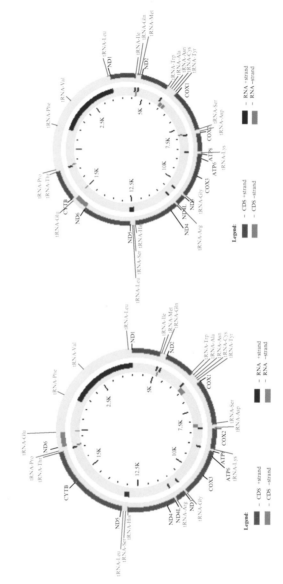

MORET AND WARNOW, CHAPTER 35, FIG. 1. The chloroplast chromosome of *Guillardia* (from NCBI).